Glencoe

ADVANCED
Mathematical
Concepts

*Precalculus
with Applications*

**Glencoe
McGraw-Hill**

New York, New York   Columbus, Ohio   Woodland Hills, California   Peoria, Illinois

Lee Yunker (1941–1994)

This edition of Advanced Mathematical Concepts is dedicated to the memory of Lee E. Yunker. For 31 years, he educated students and teachers alike. He was personally committed to life-long learning and strove to help all teachers best serve the needs of their students. His many contributions to mathematics education continue in this and the 15 other mathematics books he authored.

Cover Photograph

The angle of the photographer's camera when shooting the escalator in the Science Center in Detroit, Michigan, resulted in an interesting elliptical shape.

Glencoe/McGraw-Hill

*A Division of The **McGraw·Hill** Companies*

Send all inquiries to:
GLENCOE/McGRAW-HILL
936 Eastwind Drive
Westerville, Ohio 43081-3329

ISBN: 0-02-834135-X (Student Edition)
ISBN: 0-02-834136-8 (Teacher's Wraparound Edition)

1 2 3 4 5 6 7 8 9 10 071/043 05 04 03 02 01 00 99 98

Berchie W. Gordon-Holliday teaches mathematics at Northwest High School in Cincinnati, Ohio. She also teaches at Cincinnati area junior colleges. Dr. Gordon-Holliday has taught mathematics at every level from junior high school to college. She received her B.S. degree in mathematics from Emory University in Atlanta, Georgia, and her M.A.T. in education from Northwestern University, in Evanston, Illinois. She has done further study at the University of Illinois and the University of Cincinnati, where she received her doctorate in curriculum and instruction. Dr. Gordon-Holliday has developed and conducted numerous in-service workshops in mathematics and computer applications and has traveled nationally to make presentations on graphing calculators to various groups. She has also served as a consultant for IBM and is a co-author of *Merrill Algebra 1* and *Merrill Algebra 2 with Trigonometry*.

Lee E. Yunker was a teacher and chairperson of the Mathematics Department at West Chicago Community High School, West Chicago, Illinois. Mr. Yunker obtained his B.S. degree from Elmhurst College in Elmhurst, Illinois, and his M.Ed. in mathematics from the University of Illinois. Mr. Yunker frequently spoke or conducted workshops on a variety of topics. He participated in the first U.S./Korea Seminar on a Comparative Analysis of Mathematics Education in the United States and Korea at The Seoul National University. Mr. Yunker was a past member of the boards of directors of both the National Council of Teachers of Mathematics and the National Council of Supervisors of Mathematics. Mr. Yunker was a 1985 State Presidential Award Winner for Excellence in Mathematics Teaching and was a coauthor of *Merrill Geometry* and *Fractals for the Classroom: Strategic Activities, Volumes One and Two*, co-published by the NCTM and Springer-Verlag.

F. Joe Crosswhite is professor emeritus of mathematics education at The Ohio State University. He obtained his B.S. degree in mathematics education and M.Ed. degree in secondary administration from Missouri University and his Ph.D. degree in mathematics education from The Ohio State University. Dr. Crosswhite served as president of the National Council of Teachers of Mathematics from 1984–1986. During his tenure, the Commission on Standards for School Mathematics was formed, on which he served. Dr. Crosswhite has also served on the Mathematical Sciences Education Board of the National Research Council and as the chairperson of the Conference Board of Mathematical Sciences. Dr. Crosswhite has published widely in professional journals and is a coauthor of *Merrill Pre-Calculus Mathematics*.

Glen D. Vannatta was a mathematics teacher, department chairperson, and supervisor of mathematics for the Indianapolis Public Schools in Indianapolis, Indiana. He received his B.S., M.S., and Ed.D. degrees in mathematics education from Indiana University. Dr. Vannatta has authored or co-authored over 30 mathematics textbooks from junior high school to the freshman college level. He has provided leadership in mathematics contests and has supervised extensive achievement testing programs. In addition, he has written numerous articles for mathematics publications, including the *Mathematics Teacher*. Dr. Vannatta has also served as president or vice-president of several local, state, and national mathematics organizations.

CONSULTANTS

William Collins
Mathematics Teacher and
    Department Chairperson
James Lick High School
San Jose, California

Dr. Christian Hirsch
Professor of Mathematics and
    Mathematics Education
Western Michigan University
Kalamazoo, Michigan

Valarie Elswick
Mathematics and Computer
    Science Teacher
Roy C. Ketcham High School
Wappingers Falls, New York

Dalia Trevino
Secondary Instructional Facilitator
United High School
Laredo, Texas

TECHNOLOGY WRITERS/CONSULTANTS

Graphing Calculator Explorations and Exercises
Dr. Leo Edwards
Vice Chancellor for Academic Affairs
Fayetteville State University
Fayetteville, North Carolina

Technology Pages
Dr. Jeff Gordon
Director of Instructional Technology
University of Cincinnati
Cincinnati, Ohio

Graphing Calculator Programs
Babs Merket
Mathematics Teacher
Waukesha South High School
Waukesha, Wisconsin

Technology Pages
Dwayne Hickman
Developmental Mathematics Specialist
Columbus State Community College
Columbus, Ohio

Graphing Calculator Lessons
Jill Baumer-Piña
Former Program Assistant
Department of Mathematics
The Ohio State University
Columbus, Ohio

REVIEWERS

Stephanie A. Allison
Mathematics Department
    Chairperson
C. D. Hylton High School
Woodbridge, Virginia

Margaret G. Couvillon
Mathematics Teacher
Bellaire High School
Houston, Texas

Richard F. Dube
Mathematics Supervisor
Taunton High School
Taunton, Massachusetts

Barbara A. Baker
Mathematics Teacher
Thomas A. Edison High School
Alexandria, Virginia

Bill Diedrich
Mathematics Teacher and
    Department Chairperson
Minnetonka High School
Minnetonka, Minnesota

Ron England
Mathematics Department
    Chairperson
Rangeview High School
Aurora, Colorado

Russell Chappell
Mathematics Department
    Chairperson
Granite City Senior High School
Granite City, Illinois

Reid L. Dillon
Mathematics Teacher
Brighton High School
Salt Lake City, Utah

Bill Fisher
Mathematics Teacher and
    Department Chairperson
Emmerich Manual High School
Indianapolis, Indiana

John Frankino
Mathematics Teacher
Capital High School
Helena, Montana

Michael Mizyed Hattar
Mathematics Teacher
Rancho Cucamonga High School
Rancho Cucamonga, California

Judith C. Kemler
Mathematics Teacher
J. Frank Dobie High School
Houston, Texas

Lloyd C. Merick, Jr.
Mathematics Teacher and
    Curriculum Coordinator
Lima Public Schools
Lima, Ohio

Dr. Donna J. Middaugh
Mathematics Curriculum
    Specialist
Canton City Schools
Canton, Ohio

Felix R. Persi
Mathematics Teacher and
    Department Chairperson
Ambridge Area High School
Ambridge, Pennsylvania

Judy Rice
Mathematics Teacher
Alief Hastings High School-
    South
Alief, Texas

Richard C. Shipley
Mathematics Teacher (Retired)
Romulus Central School
Romulus, New York

Bobby G. Tyus
Mathematics Teacher
Mead High School
Spokane, Washington

Allan R. Weinheimer
Mathematics Department
    Chairperson (Retired)
North Central High School
Indianapolis, Indiana

Nanci White
Mathematics Department
    Chairperson
University City High School
San Diego, California

Gerald N. Gambino
Mathematics Supervisor
West Chester Area School District
West Chester, Pennsylvania

Ron Johnson
Mathematics Teacher
Traverse City Senior High School
Traverse City, Michigan

Gerald Martau
Deputy Superintendent
Lakewood City Schools
Lakewood, Ohio

Irene Metviner
High School Peer Intervenor
New York City Board of
    Education
New York, New York

Rod Montanye
Mathematics Department
    Chairperson
Craig High School
Janesville, Wisconsin

Elizabeth Ann Przybysz
Mathematics Teacher
Dr. Phillips High School/
Valencia Community College
Orlando, Florida

LaMar Rogers
Mathematics Teacher
Crawford High School
San Diego, California

Marilyn Stor
Mathematics Teacher
Arvada High School
Arvada, Colorado

Larry E. Wadel
Mathematics Teacher
Williamsport High School
Williamsport, Maryland

Kenneth Welsh
Mathematics Supervisor
Williamsport Area School District
Williamsport, Pennsylvania

Wanda M. White
Secondary Mathematics
    Coordinator
Fulton County Board of Education
Atlanta, Georgia

Ella Brooks Gauthier
Mathematics Teacher and
    Department Chairperson
South Houston High School
South Houston, Texas

Wanda Jones
Mathematics Teacher
Cypress-Fairbanks High School
Houston, Texas

Lynn E. Mendro
Mathematics Teacher
Dundee-Crown High School
Carpentersville, Illinois

Frank Meystrik
Mathematics Department
    Chairperson
Paul D. Schreiber High School
Port Washington, New York

Marianne M. Nicklow
Mathematics Teacher
William Penn Senior High School
York, Pennsylvania

John W. Reed
Mathematics Department
    Chairperson
Lincoln High School
Gahanna, Ohio

Wayne R. Selover
Mathematics Teacher
Charter Oak High School
Covina, California

Leonard F. Thomas
Mathematics Department
    Chairperson
Milby High School
Houston, Texas

Wayne Ward
Mathematics Teacher
Pine View High School
Saint George, Utah

Roy D. Wendling
Mathematics Teacher
Dundee-Crown High School
Carpentersville, Illinois

TABLE OF CONTENTS

TECHNOLOGY

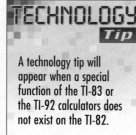

A technology tip will appear when a special function of the TI-83 or the TI-92 calculators does not exist on the TI-82.

What would your life be like without technology? Imagine the absence of computer games, grocery store scanners, digital watches, microwaves, VCRs, or CDs. It's hard to imagine! The technology features in this text will expose you to very powerful applications of mathematics you'll be able to use for a lifetime.

Graphing Calculator Technology

There are 43 pages of **Graphing Calculator Lessons** in this text covering 24 topics. These lessons allow you to use the Texas Instruments TI-82 as a tool for learning and doing mathematics. For example, see page 369 to find out how to use a graphing calculator to determine whether an equation is a trigonometric identity. Also see page 844 to learn how a graphing calculator can be used to draw a scatter plot and a line that best represents the relationship of the points in the scatter plot.

Graphing calculator technology is also integrated in many lessons through **Graphing Calculator Explorations** and **Graphing Calculator and Programming** exercises. For example, in Exercise 38 on page 26, you use a graphing calculator program to plot points in a relation.

Computer Technology

The **Technology** pages in this text let you use technology to explore patterns, make conjectures, and discover mathematics. You will learn to use programs written in BASIC as well as computer software and spreadsheets.

Mathematics Exploration Toolkit (MET) was developed for IBM by Wicat Systems, Inc. *Matrix* was developed by The North Carolina School of Science and Mathematics. *The Function Analyzer* and *Data Insights* were developed by Sunburst Communications, Inc.

Chapter 10 Conics 522

Chapter 11 Exponential and Logarithmic Functions 596

Unit 4 Discrete Mathematics ... 652

Chapter 12 Sequences and Series 654

MAKING A DIFFERENCE WITH MATHEMATICS

When are you ever going to use this stuff? It may be sooner than you think. At the beginning of each chapter, you'll meet real college students and learn about the careers each of them is pursuing. In the **Decision Making** and **Case Study Follow-Up** features, you'll find situations that will enable you to connect mathematics to your real-life experiences as a consumer and a citizen.

A Message to Students About Graphing Calculators

What is it?

What does it do?

How is it going to help me learn math?

These are just a few of the questions many students ask themselves when they first see a graphing calculator. Some students may think, "Oh, no! Do we **have** to use one?", while others may think, "All right! We get to use these neat calculators!" There are as many thoughts and feelings about graphing calculators as there are students, but one thing is for sure: a graphing calculator *can* help you learn mathematics.

So what is a graphing calculator? Very simply, it is a calculator that draws graphs. This means that it will do all of the things that a "regular" calculator will do, *plus* it will draw graphs of simple or very complex equations. In precalculus, this capability is nice to have because the graphs of some complex equations take a lot of time to sketch by hand. Some are even considered impossible to draw by hand. This is where a graphing calculator can be very useful.

But a graphing calculator can do more than just calculate and draw graphs. You can program it, work with matrices, and perform statistical computations, just to name a few things. If you need to generate random numbers, you can do that on the graphing calculator. If you need to find the absolute value of numbers, you can do that, too. You can use it to help you with algebra and geometry as well. It's really a very powerful tool—so powerful that it is often called a pocket computer. But don't let that intimidate you. A graphing calculator can save you time and make doing mathematics easier.

If you are already familiar with graphing calculators, you may remember some of the difficulty you had at first. They can seem a bit confusing and overwhelming, but you can overcome those feelings by watching others who are familiar with these calculators and know how to use them, or by trying to use them yourself and reading the instruction manual. In some ways, using a graphing calculator is like playing a video game. With a video game, you may not be familiar with the controls or understand the game in the beginning, but the more you play it, the more you understand. This is also true of using a graphing calculator. It can feel awkward at first, but it's fun once you get the hang of it.

As you have probably noticed, graphing calculators have some keys that other calculators do not. The up and down arrow keys ($\blacktriangle$ and $\blacktriangledown$) can be used to move from one graph to another when you graph more than one equation at a time and to move up or down when editing or entering a calculation. The left and right arrow keys ($\blacktriangleleft$ and $\blacktriangleright$) can be used to move (or trace) along a graph and to move left or right when editing or entering a calculation. The 2nd and ALPHA keys allow you to access whatever is typed in blue (2nd) or gray (ALPHA) above the keys on the keyboard. For example, pressing the 2nd key followed by the x^2 key will result in the square root symbol "$\sqrt{}$" appearing on the screen. This is because "$\sqrt{}$" is in blue above the x^2 key on the keyboard. Pressing ALPHA followed by x^2 will result in the letter "I" appearing on the screen. This is because "I" is in gray above the x^2 key on the keyboard. In this text, however, the keystrokes with the blue or gray characters will be written in the keystroke boxes, not the specific keys that you need to press. For example, to find "$\sqrt{3}$", the text will give a keying sequence of 2nd $\sqrt{}$ 3 instead of 2nd x^2 3.

There are some keystrokes that can save you time when you are using a graphing calculator. A few of them are listed on the next page.

- [2nd] [ON] turns the calculator off.
- [2nd] [ENTRY] copies the previous calculation so you can edit and use it again.
- Pressing [ON] while the calculator is graphing stops the calculator from completing the graph.
- [2nd] [A-LOCK] locks the [ALPHA] key, which is like pressing "shift lock" or "caps locks" on a typewriter or computer. The result is that all caps will be typed and you do not have to hold the shift key down. (This is good for programming.)
- [ZOOM] 6 redraws a graph in the standard viewing window of [-10, 10] by [-10, 10].
- [2nd] [QUIT] will return you to the home (or text) screen.

Now that we have discussed what a graphing calculator is and some of the things that it can do, we still have the question, "How can it help me learn math?" This may not be an easy question to answer because every student learns things in his or her own way. But one way that a graphing calculator can help everyone learn is by giving you a pictorial representation (that is, a picture) of the equations. The ease of changing the graphs and making new ones can also help you understand what changes in the equations mean. In this way, a graphing calculator can also help you make connections in mathematics that before were difficult to see, and even more difficult to demonstrate.

The advantages of a graphing calculator can be seen if you try to put things in perspective. For example, would you rather use a washing machine or wash clothes by hand? Would you rather cook in an oven or over a fireplace? Would you rather cut wood with a hand saw or a power saw? These may seem like silly questions, but the same type of question can be asked about graphing calculators: Would you rather draw graphs by hand or use a graphing calculator? You may say, "Draw them by hand" now, but once you learn how to use a graphing calculator, it can make learning math easier and more fun. So many things that used to take a lot of time and effort can be done so much more quickly. A graphing calculator can help you gain understanding without some of the frustrations that you face when you draw graphs with paper and a pencil.

While a graphing calculator cannot do everything, it can make some things easier. It would be nice to be able to look ahead and see what we will be doing 10, 20, or even 30 years from now, but none of us can do that. So, to prepare for whatever lies ahead, you should try to learn as much as you can. This future will definitely involve technology, and using a graphing calculator is a good start toward becoming familiar with technology. Who knows? Maybe one day you will be designing the next satellite, building the next skyscraper, or helping students learn mathematics with the aid of a graphing calculator!

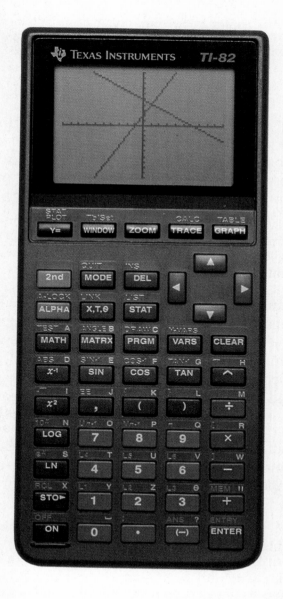

UNIT 1

RELATIONS, FUNCTIONS, AND GRAPHS

The word *relation* is from the Latin root word *relatus*. From this word we get *relate*, which is what you do when you share a story with another person; *related*, as in being a member of someone's family; *relationship*, or an attachment between two people; and *relative*, as in being kin to another person. No matter what form of the word is used, you can see that there is a pairing between things that are related.

In mathematics, the word *relation* also refers to a pairing. A relation is most often represented by a set of ordered pairs. With these ordered pairs, we can examine the concept of function, which is a special kind of relation, and we can graph relations and functions to analyze trends and relationships.

Throughout this text, you will see that many real-world phenomena can be modeled by relations and functions. A mathematical model is a simplified version of a complex problem. Each of the lessons in this unit will contain a real-world problem and its mathematical model. As you work with these models, we hope that you will see the value of using mathematics to solve problems of all types and that you will become more comfortable with using relations and functions to solve these problems.

CHAPTER 1

LINEAR RELATIONS AND FUNCTIONS

HISTORICAL SNAPSHOT

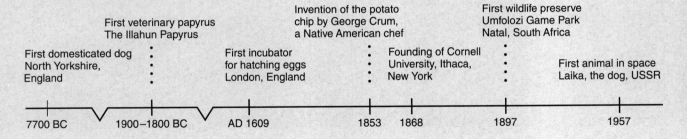

First veterinary papyrus
The Illahun Papyrus

Invention of the potato
chip by George Crum,
a Native American chef

First wildlife preserve
Umfolozi Game Park
Natal, South Africa

First domesticated dog
North Yorkshire,
England

First incubator
for hatching eggs
London, England

Founding of Cornell
University, Ithaca,
New York

First animal in space
Laika, the dog, USSR

| 7700 BC | 1900–1800 BC | AD 1609 | 1853 | 1868 | 1897 | 1957 |

CHAPTER OBJECTIVES

In this chapter, you will:
- Define relations, functions, composites, and inverses.
- Write equations of lines.
- Find the distance between two points, the slope of the line through two points, and the midpoint of a line segment.

CAREER GOAL: Veterinary Medicine

Perhaps it was because she was inspired by James Herriot's touching book *All Creatures Great and Small*. Or maybe she was inspired by growing up with her best friend, her dog Duchess. But whatever the reason, Julie Ferguson has wanted to be a veterinarian since she was in the fifth grade.

Julie studied algebra, geometry, trigonometry, and calculus in high school. She also studied biology, which was one of her favorite subjects. After working in a veterinary clinic for two years while in high school, Julie says "I saw how rewarding the job was. It was a way to use my biology experience, so I decided that veterinary medicine was the career for me."

Julie is currently working as a research assistant at Cornell University, which helps her learn practical applications of her classwork. For example, she often operates a photospectrometer in her job. "In animal science, we learn to generate a standard curve for use in comparing unknown quantities of various substances. For example, suppose we have a fluid with an unknown amount of nitrogen. We would first use fluids with known amounts of nitrogen and read the concentration of nitrogen on a photospectrometer. This equipment measures the absorption of light versus the concentration of nitrogen. We would plot the concentrations of nitrogen onto a graph. This becomes our standard curve." It is then possible to use the graph and photospectrometer to find the concentration of nitrogen in the unknown quantity.

After graduating with a degree in animal science, Julie expects to spend four more years in veterinary school at Cornell. "As a veterinarian, I hope that I will be in the community's eye as a health professional and as a community volunteer. I hope to provide comfort to pet owners wherever I can."

CONNECTION

For up-to-date information on veterinary medicine, visit:
www.glencoe.com/sec/math/amc/mathnet

1-1 Relations and Functions

Objectives

After studying this lesson, you should be able to:
- determine whether a given relation is a function, and
- identify the domain and range of any relation or function.

Application

FYI...

The first artificial satellite to be placed in orbit was Sputnik 1, launched on October 4, 1957, by the former Soviet Union.

A satellite is a natural or artificial object that revolves around a celestial body. The moon is an example of a natural satellite. Artificial satellites are used for scientific research, communication, weather forecasting, and military reconnaissance. The time that it takes an artifical satellite to complete a revolution around Earth depends on its speed and altitude. For a satellite orbiting at 100 miles above Earth, the following equation shows the relationship between speed (s) and time (t).

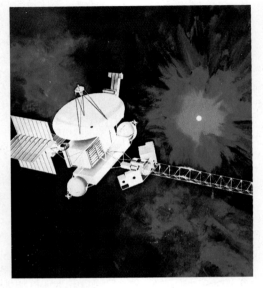

$$t = \frac{1540}{s}$$

The table below contains values for s, in thousands of miles per hour, and corresponding values for t, in minutes, given the equation above.

s	15	20	25	30
t	102.7	77	61.6	51.3

We can write the values in the table as a set of ordered pairs. A pairing of elements of one set with elements of a second set is called a **relation**. The first element of an ordered pair is the *abscissa*. The second element is the *ordinate*.

The following set, from the example above, represents a relation.

$$\{(15, 102.7), (20, 77), (25, 61.6), (30, 51.3)\}$$

The set of abscissas $\{15, 20, 25, 30\}$ of the ordered pairs is called the **domain** of the relation. The set of ordinates $\{102.7, 77, 61.6, 51.3\}$ is called the **range**.

Definition of Relation, Domain, and Range	**A relation is a set of ordered pairs. The domain is the set of all abscissas of the ordered pairs. The range is the set of all ordinates of the ordered pairs.**

A relation can be presented as a set of ordered pairs, a table of values, a graph, or by a rule in words or symbols that determines pairs of values.

Example 1

> **If x is a positive integer less than 6, state the relation representing the equation $y = 5 + x$ by listing a set of ordered pairs. Also, state the domain and range of the relation.**
>
> The relation is $\{(1, 6), (2, 7), (3, 8), (4, 9), (5, 10)\}$.
> The domain is $\{1, 2, 3, 4, 5\}$.
> The range is $\{6, 7, 8, 9, 10\}$.

The relation in Example 1 is a special type of relation called a **function**.

Definition of Function	**A function is a relation in which each element of the domain is paired with exactly one element in the range.**

All functions are relations, but not all relations are functions.

Example 2

> **The table below shows the number of students enrolled in U.S. high schools from 1960 through 2000.**
>
> | 1960 | 1970 | 1980 | 1990 | 2000 (projected) |
> |---|---|---|---|---|
> | 13,000,000 | 19,700,000 | 18,000,000 | 16,700,000 | 19,700,000 |
>
> **Does this relation represent a function? Why or why not?**
>
> This table represents the following set of ordered pairs.
>
> $(1960, 13, 000, 000), (1970, 19, 700, 000), (1980, 18, 000, 000),$
> $(1990, 16, 700, 000), (2000, 19, 700, 000)$
>
> For each member of the domain, there is only one corresponding element in the range. Thus, this relation is a function.

A function can also be defined as a set of ordered pairs in which no two pairs have the same first element. This definition can be applied when a relation is represented by a graph. If every vertical line drawn on the graph of a relation passes through no more than one point of the graph, then the relation is a function. This is called the **vertical line test.** Thus, the graph at the right does not show a function, but it does show a relation.

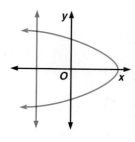

Example 3 | **Determine if the graph of each relation is that of a function.**

a.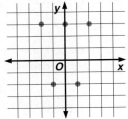

b.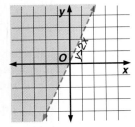

a. Each element of the domain is paired with exactly one element of the range. Therefore, this graph does represent a function.

b. A vertical line would pass through infinitely many points on the graph. Therefore, this graph does not represent a function.

x is called the independent variable, and y is called the dependent variable.

A function is commonly denoted by f. In *function notation*, the symbol $f(x)$ is read "f of x" and should be interpreted as the value of the function f at x. The expression $y = f(x)$ indicates that for each element in the domain that replaces x, the function assigns one and only one replacement for y. The ordered pairs of the function are written in the form (x, y) or $(x, f(x))$.

Example 4 | **Find $f(3)$ if $f(x) = 4x^2 - 2x + 5$.**

$f(3) = 4(3)^2 - 2(3) + 5$ $\quad x = 3$
$\quad\;\; = 35$

Example 5 | **Find $f(4)$ if $f(x) = |-3x + 8|$.**

$f(4) = |-3(4) + 8|$ $\quad$ *The vertical bars denote absolute value.*
$\quad\;\; = |-4|$
$\quad\;\; = 4$

Often an equation for a function is given, but the domain is not specified. In cases such as this, the domain consists of all the real numbers for which the corresponding values in the range are also real numbers.

Example 6 | **Name all values of x that are not in the domain of f.**

a. $f(x) = \dfrac{3x}{x^2 - 5}$

a. Any value that makes the denominator equal zero must be excluded from the domain of f, since division by zero is undefined. To determine the excluded values, let $x^2 - 5 = 0$ and solve.

$x^2 - 5 = 0$
$\quad\; x^2 = 5$
$\quad\;\; x = \pm\sqrt{5}$

Therefore, the domain includes all real numbers except $+\sqrt{5}$ and $-\sqrt{5}$.

b. $f(x) = \sqrt{x + 1}$

b. Any value that makes the radicand negative must be excluded from the domain of f, since the square root of a negative number is not a real number. Let $x + 1 < 0$ and solve for the excluded values.

$$x + 1 < 0$$
$$x < -1$$

Therefore, the domain excludes all real numbers less than –1.

You can use a graphing calculator to help determine the excluded values in a function by examining the graph of the function.

EXPLORATION: Graphing Calculator

Determine the excluded values in the domain of
$$f(x) = \frac{x}{x^2 - x - 12} \text{ with a graphing calculator.}$$

1. Establish the viewing window by using the settings below.

XMIN: -10   XMAX: 10   XSCL: 1
YMIN: -5   YMAX: 5   YSCL: 1

2. Press the Y= key. Enter the function.
3. Press GRAPH.

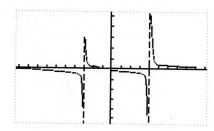

4. What do you notice about the vertical lines on the graph?

CHECKING FOR UNDERSTANDING

Communicating Mathematics

Read and study the lesson to answer each question.

1. Compare and contrast the meanings of the words *relation* and *function*.

2. Describe what the vertical line test does.

3. Draw the graph of a function on a set of coordinate axes.

4. Explain why all functions are relations, but not all relations are functions.

Guided
Practice

State whether each graph represents a function. Write _yes_ or _no_.

5.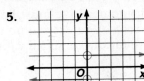

6.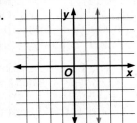

7.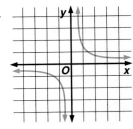

Given $f(x) = 7 - x^2$, find each value.

8. $f(0)$ **9.** $f(4)$ **10.** $f(-3)$ **11.** $f(11)$

12. $f\left(\dfrac{1}{2}\right)$ **13.** $f(3.7)$ **14.** $f(2a)$ **15.** $f(6 + n)$

16. What value(s) of x are not in the domain of $f(x) = \dfrac{x^3 + 5x}{4x}$?

17. What value(s) of x are not in the domain of $f(x) = \dfrac{1}{\sqrt{x - 4}}$?

EXERCISES

Practice

State the domain and range of each relation. Then state whether the relation is a function. Write _yes_ or _no_.

18. $\{(-3, 0), (4, -2), (2, -6)\}$ **19.** $\{(1, 2), (2, 4), (-3, -6), (0, 0)\}$

20. $\{(0, 3), (5, 3), (6, 3), (2, 3)\}$ **21.** $\{(-2, 9), (-2, 8), (-2, 7)\}$

22. $\{(5, 5), (6, 6)\}$ **23.** $\{(1, 5), (2, 6), (3, 7), (4, 8)\}$

24. $\{(4, -2), (4, 2), (9, -3), (9, 3)\}$ **25.** $\{(8, -3), (7, 3), (6, -3)\}$

Given that x is an integer, state the relation representing each of the following by listing a set of ordered pairs. Then state whether the relation is a function. Write _yes_ or _no_.

26. $y = 3x - 3$ and $0 < x < 6$ **27.** $y = 2.5x$ and $4 \leq x \leq 8$

28. $y = 5$ and $4 \leq x \leq 9$ **29.** $y^2 = x - 2$ and $x = 11$

30. $|2y| = x$ and $x = 4$ **31.** $y = |x| - 1.5$ and $-2 \leq x < 4$

The symbol $[x]$ means the greatest integer not greater than x. If $f(x) = [x] + 4$, find each value.

32. $f(-4)$ **33.** $f(2.5)$ **34.** $f(-6.3)$ **35.** $f(\sqrt{2})$

36. $f(-\sqrt{3})$ **37.** $f(\pi)$ **38.** $f(-4 + t)$ **39.** $f(q + 1)$

Given $f(x) = |x^2 - 13|$, find each value.

40. $f(0)$ **41.** $f(-4)$ **42.** $f(-\sqrt{13})$ **43.** $f(2)$

44. $f(4.8)$ **45.** $f\left(1\dfrac{1}{2}\right)$ **46.** $f(n + 4)$ **47.** $f(5m)$

Name all values of x that are not in the domain of the given function.

48. $f(x) = \dfrac{3}{x - 1}$ **49.** $f(x) = \dfrac{3 - x}{5 + x}$ **50.** $f(x) = \dfrac{x^2 - 18}{32 - x^2}$

51. $f(x) = \dfrac{15}{|2x| - 9}$ **52.** $f(x) = \sqrt{x^2 - 9}$ **53.** $f(x) = \dfrac{x + 2}{\sqrt{x^2 - 7}}$

Graphing Calculator

Use a graphing calculator to determine whether each equation is a relation or a function.

54. $y = 3x^2 + 1$ **55.** $y = (x - 5)^{\frac{1}{2}}$ **56.** $x^2 + y^2 = 25$ **57.** $\dfrac{x^2}{9} + \dfrac{y^2}{16} = 1$

Critical Thinking

58. Find $P(6)$ if $P(x)$ is a function for which $P(1) = 1$, $P(2) = 2$, $P(3) = 3$, and $P(x + 1) = \dfrac{P(x - 2)P(x - 1) + 1}{P(x)}$ for $x > 3$.

Applications and Problem Solving

59. Recycling A recycling center accepts aluminum cans for recycling. A weekly record from the center is shown at the right. Can you use the information in the second and third columns of the chart to illustrate a function? Can you use this information to illustrate a relation that is not a function? Explain.

Day	Number of Cans Recycled	Pounds of Aluminum Produced
M	75,000	3000
T	80,000	3150
W	70,000	2850
Th	75,000	3100
F	85,000	3400

60. Finance The formula for the simple interest earned on an investment is $I = prt$, where I is the interest earned, p is the principal, r is the interest rate, and t is the time in years. Assume that $5000 is invested at an annual interest rate of 8% and that the interest is added to the principal at the end of each year.
   a. Find the amount of interest that will be earned each year for five years.
   b. State the domain and range of the relation.
   c. Is this relation a function? Why or why not?

CASE STUDY FOLLOW-UP

Refer to Case Study 2: *Trashing the Planet*, on pages 959–961.

Let $x =$ the number of non-federal hazardous waste sites in a given state, and let $y =$ the number of federal hazardous waste sites in that state. The following ordered pairs (x, y) represent the numbers of sites in the six states with a total of at least 11, but not greater than 15, hazardous waste sites.

Connecticut (14,1)   Tennessee (12,2)   Georgia (11,2)
Alabama (10,2)   Rhode Island (9,2)   Utah (7,4)

1. Is the set of ordered pairs a relation? Is it a function?

2. Let $f(x) = -240x + 482,640$, where x is the year and $f(x)$ is the number of active trash landfills in the United States that year. Evaluate the function for the years 1986, 1993, and 2006.
   a. For which years does the function yield the estimates given in the case study?
   b. Is the function that specifies the number of landfills in the United States as a function of the year a linear function?

3. Research Write a one-page paper on the Superfund, the federal program to clean up hazardous waste sites.

1-2 Composition and Inverses of Functions

Objectives

After studying this lesson, you should be able to:
- perform operations with functions,
- find composite functions, and
- find and recognize inverse functions.

Application

Joyce Wallace is buying a pair of jeans for $39.99. The jeans are on sale at a 20% discount and the sales tax in Joyce's area is 7%. How much will Joyce have to pay for the jeans? *This problem will be solved in Example 3.*

To solve the problem above, two functions, one involving a discount and the other involving sales tax, can be used. If you have two functions f and g, you can form new functions by adding, subtracting, multiplying, or dividing.

Operations With Functions

Sum:	$(f + g)(x) = f(x) + g(x)$
Difference:	$(f - g)(x) = f(x) - g(x)$
Product:	$(f \cdot g)(x) = f(x) \cdot g(x)$
Quotient:	$\left(\dfrac{f}{g}\right)(x) = \dfrac{f(x)}{g(x)}, g(x) \neq 0$

The domain of each new function consists of those values of x common to the domains of f and g. The domain of the quotient function is further restricted by excluding any values that make the denominator, $g(x)$, zero.

Example 1

Given $f(x) = x + 3$ and $g(x) = \dfrac{2x}{x - 5}$, find each function below. Name all values of x that are not in the domain of the new function.

a. $(f + g)(x)$ **b.** $(f - g)(x)$

a. $(f + g)(x) = f(x) + g(x)$

$\qquad = x + 3 + \dfrac{2x}{x - 5}$

$\qquad = \dfrac{(x + 3)(x - 5) + 2x}{x - 5}$

$\qquad = \dfrac{x^2 - 15}{x - 5}, \ x \neq 5$

b. $(f - g)(x) = f(x) - g(x)$

$\qquad = x + 3 - \dfrac{2x}{x - 5}$

$\qquad = \dfrac{(x + 3)(x - 5) - 2x}{x - 5}$

$\qquad = \dfrac{x^2 - 4x - 15}{x - 5}, \ x \neq 5$

c. $(f \cdot g)(x)$ **d.** $\left(\dfrac{f}{g}\right)(x)$

c. $(f \cdot g)(x) = f(x) \cdot g(x)$ **d.** $\left(\dfrac{f}{g}\right)(x) = \dfrac{f(x)}{g(x)}$

$$= (x + 3) \cdot \dfrac{2x}{x - 5}$$

$$= \dfrac{2x^2 + 6x}{x - 5}, x \neq 5$$

$$= (x + 3) \div \dfrac{2x}{x - 5}$$

$$= \dfrac{x^2 - 2x - 15}{2x}, x \neq 0, 5$$

Another way of combining two functions is through **composition** of functions. A function g maps the elements in set R to those in set S. Another function f maps the elements in set S to those in set T. Thus, the range of g is the same as the domain of f. A diagram is shown below.

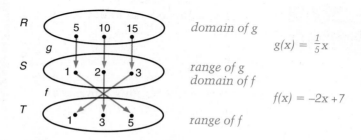

By combining f and g in this way, a function from R to T is defined, since each element of R is associated with exactly one element of T. For example, $g(15) = 3$ and $f(3) = 1$. This new function that maps R onto T is called the **composite** of f and g. It is denoted by $f \circ g$, as shown below.

$f \circ g$ is read "f composition g," or "f of g."

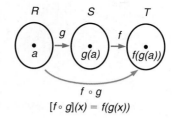

$$[f \circ g](x) = f(g(x))$$

Composition of Functions

Given functions f and g, the composite function $f \circ g$ can be described by the following equation.

$$[f \circ g](x) = f(g(x))$$

The domain of $f \circ g$ includes all of the elements x in the domain of g for which $g(x)$ is in the domain of f.

Example 2

If $f(x) = x + 1$ and $g(x) = \dfrac{1}{x-1}$, find $[f \circ g](x)$.

$$[f \circ g](x) = f(g(x))$$

$$= f\left(\frac{1}{x-1}\right) \qquad \text{\textit{Replace} } g(x) \text{ \textit{with} } \frac{1}{x-1}.$$

$$= \frac{1}{x-1} + 1 \qquad \text{\textit{Substitute} } \frac{1}{x-1} \text{ \textit{for} } x \text{ \textit{in} } x+1.$$

$$= \frac{1}{x-1} + \frac{x-1}{x-1} \qquad \text{\textit{The least common denominator is} } x-1.$$

$$= \frac{x}{x-1}$$

You can use composition of functions to solve the application problem presented at the beginning of the lesson.

Example 3

Refer to the application at the beginning of the lesson. How much will Joyce have to pay for the jeans?

Let $x =$ the original price of the jeans, or $39.99.

Let $T(x) = 1.07x$. This represents the cost of an item with a 7% sales tax rate.

Let $S(x) = 0.80x$. This represents the cost of an item with a 20% discount.

The cost of Joyce's jeans is $[T \circ S](x)$.

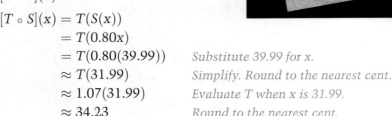

$$[T \circ S](x) = T(S(x))$$

$$= T(0.80x)$$

$$= T(0.80(39.99)) \qquad \text{\textit{Substitute 39.99 for x.}}$$

$$\approx T(31.99) \qquad \text{\textit{Simplify. Round to the nearest cent.}}$$

$$\approx 1.07(31.99) \qquad \text{\textit{Evaluate T when x is 31.99.}}$$

$$\approx 34.23 \qquad \text{\textit{Round to the nearest cent.}}$$

Joyce will have to pay $34.23 for the jeans.

In the examples above, the range of g was the same as the domain of f.

The composite function $f \circ g$ exists only when the range of g is a subset of the domain of f. Otherwise, the composite is undefined. For example, $[f \circ g](x)$ does not exist for the following functions.

$$g(x) = -\sqrt{x}$$
$$f(x) = \sqrt{x} + 1$$

$$[f \circ g](x) = f(g(x)) = \sqrt{(-\sqrt{x})} + 1 \qquad \text{\textit{Does } [g \circ f](x) \text{ \textit{exist?}}}$$

Iteration is a special type of composition, the composition of a function to itself. Chapter 13 is devoted to this topic.

Now consider two functions $f(x) = 3x - 2$ and $g(x) = \dfrac{x+2}{3}$. You can find functions $[f \circ g](x)$ and $[g \circ f](x)$ as follows.

$$[f \circ g](x) = f\left(\frac{x+2}{3}\right) \qquad\qquad [g \circ f](x) = g(3x - 2)$$

$$= 3\left(\frac{x+2}{3}\right) - 2 \qquad\qquad = \frac{(3x-2)+2}{3}$$

$$= x \qquad\qquad\qquad\qquad\qquad = x$$

Since $[f \circ g](x) = [g \circ f](x) = x$ for all values of x, f and g are called **inverse functions**.

Inverse Functions	**Two functions f and g are inverse functions if and only if $[f \circ g](x) = [g \circ f](x) = x$.**

The notation f^{-1} is read "f inverse," or "the inverse of f." The –1 is not an exponent.

The inverse function of f, if it exists, is denoted by f^{-1}. If f and g are inverse functions, then $f = g^{-1}$ and $g = f^{-1}$.

The inverse of a function can be shown as a mapping.

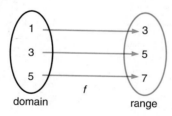
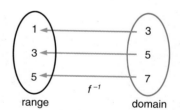

If $f = \{(1, 3), (3, 5), (5, 7)\}$, then $f^{-1} = \{(3, 1), (5, 3), (7, 5)\}$.

Notice that the inverse is found by reversing the order of the coordinates of each ordered pair for the function. The mapping shown above is one-to-one. The mapping shown at the right is not one-to-one. The inverse of f is a function only if f represents a one-to-one mapping.

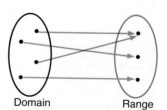

Property of Inverse Functions	**Suppose f and f^{-1} are inverse functions. Then, $f(x) = y$ if and only if $f^{-1}(y) = x$.**

Example 4

Given $f(x) = 4x - 9$, find $f^{-1}(x)$, and show that f and f^{-1} are inverse functions.

$$y = 4x - 9 \qquad f(x) = y$$
$$x = 4y - 9 \qquad \text{Exchange x and y.}$$
$$x + 9 = 4y \qquad \text{Solve for y.}$$
$$\frac{x + 9}{4} = y$$
$$\frac{x + 9}{4} = f^{-1}(x) \qquad \text{Replace y with } f^{-1}(x).$$

Then, show that $[f \circ f^{-1}](x) = [f^{-1} \circ f](x) = x$.

$$[f \circ f^{-1}](x) = f\left(\frac{x+9}{4}\right) \qquad\qquad [f^{-1} \circ f](x) = f^{-1}(4x - 9)$$
$$= 4\left(\frac{x+9}{4}\right) - 9 \qquad\qquad\qquad = \frac{(4x - 9) + 9}{4}$$
$$= x \qquad\qquad\qquad\qquad\qquad = x$$

Since $[f \circ f^{-1}](x) = [f^{-1} \circ f](x) = x$, f and f^{-1} are inverse functions.

Not all functions have inverses that are functions. *Note that the notation f^{-1} is used whether or not the inverse is a function.*

Example 5

Given $f(x) = x^2 - 1$, find $f^{-1}(x)$.

$$y = x^2 - 1 \qquad f(x) = y$$
$$x = y^2 - 1 \qquad \text{Exchange x and y.}$$
$$x + 1 = y^2 \qquad \text{Solve for y.}$$
$$\pm\sqrt{x + 1} = y$$
$$\pm\sqrt{x + 1} = f^{-1}(x) \qquad \text{Replace y with } f^{-1}(x)$$

The equation $f^{-1}(x) = \pm\sqrt{x + 1}$ does not define a function. Therefore, $f(x) = x^2 - 1$ does not have an inverse that is a function.

CHECKING FOR UNDERSTANDING

Communicating Mathematics

Read and study the lesson to answer each question.

1. Is $[f \circ g](x)$ always equal to $[g \circ f](x)$ for two functions $f(x)$ and $g(x)$? Explain your answer.

2. Why does the equation for the inverse in Example 5 not represent a function?

3. **Graph** $f(x) = 4x - 9$ and $f^{-1}(x) = \dfrac{x + 9}{4}$ on the same set of axes, along with $y = x$. What do you notice about the graphs of $f(x)$ and $f^{-1}(x)$?

Guided Practice

Given $f(x) = \dfrac{x}{x+1}$ and $g(x) = x^2 - 1$, find each function below.

4. $(f+g)(x)$ **5.** $(f-g)(x)$ **6.** $(f \cdot g)(x)$ **7.** $\left(\dfrac{f}{g}\right)(x)$

Find $[f \circ g](x)$ and $[g \circ f](x)$.

8. $f(x) = x + 3$
    $g(x) = 2x + 5$

9. $f(x) = x^2 - 9$
    $g(x) = x + 4$

Determine if the given functions are inverses of each other. Write *yes* or *no*. Show your work.

10. $f(x) = 3x + 1$
    $g(x) = \dfrac{x-1}{3}$

11. $f(x) = \dfrac{1}{2}x - 5$
    $g(x) = 2x + 5$

EXERCISES

Practice

Given $f(x) = \dfrac{3}{x-7}$ and $g(x) = x^2 + 5x$, find each function below.

12. $(f+g)(x)$ **13.** $(f-g)(x)$ **14.** $(f \cdot g)(x)$ **15.** $\left(\dfrac{f}{g}\right)(x)$

Find $[f \circ g](x)$ and $[g \circ f](x)$.

16. $f(x) = \dfrac{1}{2}x - 7$
    $g(x) = x + 6$

17. $f(x) = 3x^2$
    $g(x) = x - 4$

18. $f(x) = x^3$
    $g(x) = x + 1$

19. $f(x) = 5x^2$
    $g(x) = x^2 - 1$

20. $f(x) = x^3 + x^2 + 1$
    $g(x) = 2x$

21. $f(x) = x^2 + 5x + 6$
    $g(x) = x + 1$

Determine if the given functions are inverses of each other. Write *yes* or *no*. Show your work.

22. $f(x) = 5x - 6$
    $g(x) = \dfrac{x+6}{5}$

23. $f(x) = x + 5$
    $g(x) = x - 5$

24. $f(x) = \dfrac{x-1}{2}$
    $g(x) = 2x + 1$

25. $f(x) = -x$
    $g(x) = x$

26. $f(x) = -3x + 7$
    $g(x) = 3x - 7$

27. $f(x) = \dfrac{1}{3}x + 2$
    $g(x) = 3(x - 2)$

Find the inverse of each function. Then state whether the inverse is a function.

28. $f(x) = 4x + 4$ **29.** $f(x) = x^3$ **30.** $f(x) = x^2 - 9$

31. Find two functions, $f(x)$ and $g(x)$, such that $[f \circ g](x) = [g \circ f](x)$.

32. Given $f(x) = 2x + 1$, find each value.
　　a. $f(f(2))$　　　　　**b.** $f(f(f(-1)))$　　　　　**c.** $f(f(f(3)))$

Graphing Calculator

Given $f(x) = x^{\frac{1}{2}}$ and $g(x) = x - 4$, use a graphing calculator to graph each function. Then determine the domain and range.

33. $f \circ g$　　　　　　　　　　　　　　　**34.** $g \circ f$

Critical Thinking

35. If $[f \circ g](x) = \dfrac{x^4 + x^2}{1 + x^2}$ and $g(x) = 1 - x^2$, find $f\left(\dfrac{1}{2}\right)$.

Applications and Problem Solving

36. Sales On the same shopping trip described on page 12, Joyce Wallace spotted a sweater that would go perfectly with her new jeans. The original price of the sweater, $59.99, is marked 25% off. Use composition of functions to determine how much Joyce would have to pay for the sweater, including sales tax.

37. Sales Jason and Heather Paulsen bought a vase with a $50 gift certificate that they had received as a present. The vase was marked 33% off, and the sales tax in their area was 5.5%. If they paid $45.95 for the vase, use composition of functions to determine the original price of the vase.

Mixed Review

38. Does the graph at the right represent a function? Write *yes* or *no*. Then explain your answer. **(Lesson 1-1)**

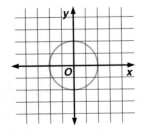

39. Given that x is an integer, state the relation representing $y = 11 - x$ and $-3 \leq x \leq 0$ by listing a set of ordered pairs. Then state whether this relation is a function. **(Lesson 1-1)**

40. Given $f(x) = 4 + 6x - x^3$, find $f(14)$. **(Lesson 1-1)**

41. Name all values of x that are not in the domain of $f(x) = \dfrac{x}{x^2 - 6}$. **(Lesson 1-1)**

42. College Entrance Exam Choose the best answer. $\dfrac{9^5 - 9^4}{8} =$

　　(A) $\dfrac{1}{8}$　　　　**(B)** $\dfrac{9}{8}$　　　　**(C)** $\dfrac{9^3}{8}$　　　　**(D)** 9^4　　　　**(E)** $\dfrac{9^9}{8}$

1-3A Graphing Calculators: Graphing Linear Equations and Inequalities

The graphing calculator is a powerful tool for use in mathematics and other areas of life. On a graphing calculator, the **viewing window** for a graph is that portion of the coordinate grid displayed on the **graphics screen** of the calculator. Sometimes the viewing window is written as [left, right] by [bottom, top] or [Xmin, Xmax] by [Ymin, Ymax]. This means that a viewing window of [–5, 7] by [–10, 10] denotes the domain values of $-5 \le x \le 7$ and the range values of $-10 \le y \le 10$. A viewing window of [–10, 10] by [–10, 10] is called the **standard viewing window** and is a good viewing window to start with to solve a problem. This standard viewing window can be easily obtained by pressing ZOOM 6.

Any viewing window can be set by pressing the WINDOW key. The **window screen** will appear and display the current settings for your viewing window as well as the **Xscl** and **Yscl**, which indicate the x-scale and y-scale. These labels set the frequency of the tick marks along each of the axes. For example, Xscl = 1 means that there will be a tick mark for every one unit along the x-axis.

Example 1

Graph $y = -2x + 33$.

Use the standard viewing window.

Enter: Y= (–) 2 X,T,θ + 33

GRAPH

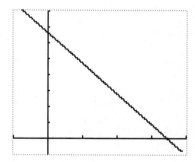

None of the graph is shown when you graph this equation in the standard viewing window. Pressing the WINDOW key and setting the viewing window to [–5, 20] by [–5, 40] with scale factors of 5 for each scale will allow you to see where the graph intersects both the x- and the y-axes.

When the equation in Example 1 is graphed in the viewing window [–5, 20] by [–5, 40], it is a **complete graph**. A complete graph is a graph that shows all of the important characteristics of the graph on the graphics screen. For a linear equation, these are the x- and y-intercepts.

Linear inequalities can be graphed on the TI-82 by using the "Shade(" command and entering a function for a lower boundary and a function for the upper boundary of the inequality. The calculator first graphs both functions and then shades above the first function entered and below the second function entered.

Before graphing an inequality, reset the window values to the standard viewing window and clear any functions in the Y = list. Do this by pressing $\boxed{\text{Y=}}$ and then using the arrow keys and the $\boxed{\text{CLEAR}}$ key to select and clear all functions. Next, return to the text screen by pressing $\boxed{\text{2nd}}$ and $\boxed{\text{QUIT}}$.

When graphing a linear inequality, you can use the Ymin range value as the lower boundary if the inequality asks for "$y \leq$," since the points that satisfy the inequality are below the graph of the related equation. Use the Ymax range value as the upper boundary if the inequality asks for "$y \geq$," since the points that satisfy the inequality are above the graph of the related equation.

Before graphing any inequality, first clear the graphics screen by pressing $\boxed{\text{2nd}}$ $\boxed{\text{DRAW}}$ 1 $\boxed{\text{ENTER}}$.

Example 2

Graph $y \leq 3x + 4$ in the standard viewing window.

The inequality asks for points "less than or equal to", so we will use Ymin or –10 as the lower boundary and the expression for the related equation, $3x + 4$, as the upper boundary. Enter a comma between each boundary.

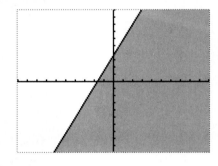

Enter: $\boxed{\text{2nd}}$ $\boxed{\text{DRAW}}$ 7 $\boxed{\text{(–)}}$ 10 $\boxed{,}$
3 $\boxed{\text{X,T,}\theta}$ $\boxed{+}$ 4 $\boxed{)}$ $\boxed{\text{ENTER}}$

Since both the x- and the y-intercept of the line $y = 3x + 4$ are within the standard viewing window, this is a complete graph of the inequality.

You can use values other than Ymin or Ymax as the boundaries of an inequality. Any value below Ymin for "less than or equal to" and any value above Ymax for "greater than or equal to" will allow the calculator to graph the inequality in the viewing window that is set.

EXERCISES

Use a graphing calculator to graph each equation. State the range values that you used to view a complete graph for each equation.

1. $y = 4x - 4$ **2.** $y = -5x + 7$ **3.** $y = 0.02x$

4. $y = 150x + 5$ **5.** $y = -23x - 91$ **6.** $y = 0.8x - 43$

Graph each linear inequality on a TI-82 and sketch the graph on a piece of paper.

7. $y \geq 2x + 5$ **8.** $y \leq -3x + 6$ **9.** $y \leq 0.3x - 0.47$

10. $y \geq 50x - 4$ **11.** $y \leq 0.45x$ **12.** $y \geq 24$

1-3 Linear Functions and Inequalities

Objectives

After studying this lesson, you should be able to:
- find zeros of linear functions, and
- graph linear equations and inequalities.

Application

Most VCRs allow you to record programs in three modes. They are the two-hour standard-play mode (SP), the four-hour long-play mode (LP), and the six-hour super-long-play mode (SLP). The two-hour mode, which allows you to record two hours of programs on the videotape, produces the highest-quality recording; the six-hour mode, which allows you to record six hours of programs, produces the lowest-quality recording.

FYI...

The first practical videotape recorder in the world was introduced in April of 1956 by the California Ampex Corporation, under the name Ampex VR 1000.

Toshiki wants to fill a videotape with a movie that is three hours long. He wants the best quality possible. Since the two-hour mode would not be long enough to record the whole movie, and the four-hour mode would not fill the tape, Toshiki decides to begin in the four-hour mode and switch to the two-hour mode sometime during the program. Let's call the time that he switches the "switching time."

Let x = the switching time in hours. Since the movie is recorded for x hours in the four-hour mode, the fraction of the videotape filled in this mode is $\frac{x}{4}$. That means that $3 - x$ hours of the movie are to be recorded in the two-hour mode. Thus, the fraction of the videotape filled in the two-hour mode is $\frac{3-x}{2}$. Since the entire videotape will be filled with the movie, the sum of the two fractions must be 1. The following equation represents the situation.

$$\frac{x}{4} + \frac{3-x}{2} = 1$$
$$x + 2(3-x) = 4 \qquad \textit{Multiply each side by 4.}$$
$$x = 2$$

The switching time is 2 hours. Thus, Toshiki should record the first two hours in the four-hour mode and the last hour in the two-hour mode.

The equations on the previous page are **linear equations**. A linear equation has the form $Ax + By + C = 0$, where A and B are not both 0. Its graph is always a straight line. The graph of the equation $x = 2$ is shown at the right. *Does this equation represent a function?*

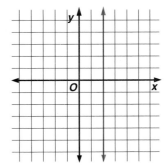

Each solution of a linear equation is an ordered pair. Each ordered pair corresponds to a point in the coordinate plane. Since two points determine a line, only two points are needed to graph a linear equation.

Example 1

Graph $3x - 2y = 6$.

Solve for y. Then find two ordered pairs that are solutions of the equation.

$$-2y = -3x + 6$$
$$y = \frac{3}{2}x - 3$$

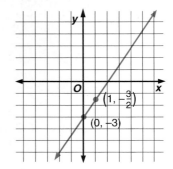

x	y	(x, y)
0	-3	(0, -3)
1	$-\frac{3}{2}$	$\left(1, -\frac{3}{2}\right)$

Graph the ordered pairs and connect them with a line.
Does this equation represent a function?

Not all linear equations represent functions, as shown by the example of $x = 2$ above. A **linear function** is defined as follows.

Linear Function

A linear function is defined by $f(x) = mx + b$, where m and b are real numbers.

Values of x for which $f(x) = 0$ are called **zeros** of the function f. For a linear function, the zeros can be found by solving the equation $mx + b = 0$. If $m \neq 0$, then $-\dfrac{b}{m}$ is the only zero of the function. The graph of the function crosses, or intercepts, the x-axis at the point with coordinates $\left(-\dfrac{b}{m}, 0\right)$. Thus, $-\dfrac{b}{m}$ is called the **x-intercept**.

If $m = 0$, then $f(x) = b$. The graph is a horizontal line. This function is called a **constant function**. A constant function either has *no* zeros ($b \neq 0$), or every value of x is a zero ($b = 0$).

Example 2 | **Find the zero of each function. Then graph the function.**

a. $f(x) = 3x - 1$

a. To find the zeros of $f(x)$, set $f(x)$ equal to 0 and solve for x.

$$3x - 1 = 0$$
$$x = \frac{1}{3}$$

$\frac{1}{3}$ is a zero of the function. So, the coordinates of one point on the graph are $\left(\frac{1}{3}, 0\right)$. Find the coordinates of a second point. When $x = 1$, $f(x) = 3(1) - 1$, or 2. Thus, the coordinates of a second point are $(1, 2)$.

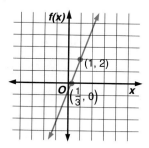

b. $f(x) = 3$

b. Since $m = 0$ and $b = 3$, this function has no zeros. The graph of this function is a horizontal line 3 units above the x-axis.

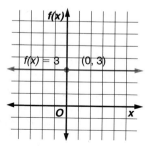

The graph of a **linear inequality** is not a linear function. However, you can use the graphs of linear functions to help graph linear inequalities. The graph of $y = -\frac{1}{2}x + 2$ is a line that separates the coordinate plane into two regions, as shown at the right. The line described by $y = -\frac{1}{2}x + 2$ is called the **boundary** of each region. If the boundary is part of a graph, it is drawn as a solid line. If the boundary is not part of a graph, it is drawn as a dashed line. The graph of $y > -\frac{1}{2}x + 2$ is the region above the line. The graph of $y < -\frac{1}{2}x + 2$ is the region below the line.

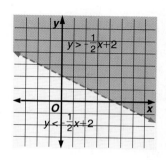

Example 3

M A T H
JOURNAL

Write a paragraph to explain to another person how you graph an inequality.

Graph $8 - y \geq 3x$.

$$8 - y \geq 3x$$
$$-y \geq 3x - 8$$
$$y \leq -3x + 8$$

Remember to reverse the direction of the inequality symbol when you multiply by a negative number.

Notice that the boundary line is included since the inequality symbol is less than or equal to.

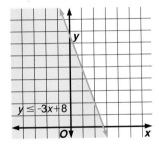

Example 4

APPLICATION

Business

Mike Welz can spend up to $35 per day plus $0.30 per mile when renting a car to use on company business. The total cost of the daily rental (C) is a function of the total miles driven (m).

a. Write a linear inequality that expresses the acceptable daily rental car cost.

b. Graph the inequality.

a. A linear equation that expresses the acceptable daily cost of a rental car is $C = 35 + 0.30m$. Since Mr. Welz's company will only pay up to that amount, the inequality would be $C \leq 35 + 0.30m$.

b. The graph of the inequality is shown at the right. Since costs cannot be negative, the graph is only in the first quadrant.

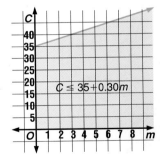

Relations such as $2 < x + y \leq 5$ can also be graphed. The graph of this relation is the intersection of the graph of $2 < x + y$ and the graph of $x + y \leq 5$. Notice that the boundary $x + y = 5$ is part of the graph, but the boundary $2 = x + y$ is not part of the graph.

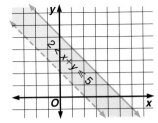

CHECKING FOR UNDERSTANDING

Communicating
Mathematics

Read and study the lesson to answer each question.

1. **Describe** the process you would use to graph $-5 \leq x - 5y \leq 6$.

2. **Describe** how Mike Welz (in Example 4) could determine if he would be completely reimbursed for his rental car travel, if his rental car bill was $72.45 on Thursday, and he drove 186 miles.

Guided Practice

Name the zero of each function whose graph is shown.

3.
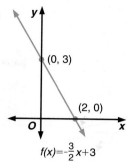
$f(x) = -\frac{3}{2}x + 3$

4.
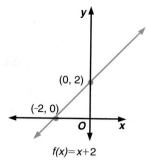
$f(x) = x + 2$

5.
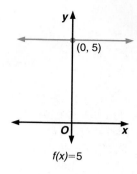
$f(x) = 5$

Of $(0, 0)$, $(3, 2)$, $(-4, 2)$, **or** $(-2, 4)$, **name the ordered pair(s) that satisfy each inequality.**

6. $3x - 4y \geq -5$

7. $y \neq x - 5$

8. $x > 4y + 3$

Write an inequality that describes each graph.

9.
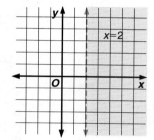

10.
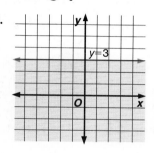

11.
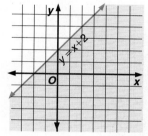

Graph each equation or inequality.

12. $3x = 2y$

13. $2 < 2x + y < 8$

EXERCISES

Practice

Write an inequality that describes each graph.

14.
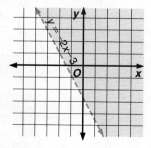

15.
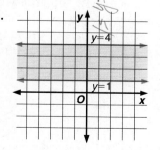

16.
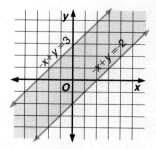

Find the zero of each function.

17. $f(x) = 0.5x + 6$

18. $f(x) = 14x$

19. $f(x) = 9x + 5$

20. $f(x) = 5x - 8$

21. $f(x) = 19$

22. $f(x) = 3x + 1$

Graph each equation or inequality.

23. $y = 3x - 2$

24. $y - 7 = x$

25. $x + 2y = 5$

26. $x = -\dfrac{1}{2}y$

27. $4y = 2 + 3x$

28. $y = |3x - 2|$

29. $x < 5$

30. $y > -x$

31. $x - y < 5$

32. $2y \leq x - 5$

33. $-y < 2x + 1$

34. $y \geq \dfrac{2}{5}x + \dfrac{19}{5}$

35. $-2 \leq x + 2y \leq 4$

36. $y > |x|$

37. $|x + 3| < y - 1$

Programming

38. The following program will plot points in a relation.

```
Prgm 1: PLOTPTS
:FnOff                          Deactivates all current functions.
:ClrDraw                        Clears the graphing screen.
:Lbl 1                          Sets the marker at position 1.
:Disp "X="                      Displays the message "X=".
:Input X                        Accepts an X value.
:Disp "Y="                      Displays the message "Y=".
:Input Y                        Accepts a Y value.
:Disp "PRESS 0 TO QUIT OR 1     Displays memory.
 TO PLOT MORE POINTS"
:Input A                        Accepts an A value.
:If A = 0                       Tests whether A is 0.
:End                            Ends program if A = 0.
:PT-ON(X,Y)                     Plots the point (X, Y).
:Pause                          Waits until the user presses the ENTER key.
:Goto 1                         Goes back to the line Lbl "1" and continues.
```

Find three ordered pairs that are solutions of each equation and graph them.

a. $2x + 5y = 8$

b. $y = 4x - 9$

Critical Thinking

39. A linear function may have zero, one, or infinitely many zeros. Illustrate all three of these situations.

Applications and Problem Solving

40. Economics The average daily demand for the sale of soybean futures may be represented by a function known as a *demand function*. For example, in September of 1990, the average number of bushels demanded daily was 103.7 million. The function $D(p) = \dfrac{560}{p}$ represents the demand for soybeans that particular month, where p is the price in dollars per bushel ($p > 0$).

a. Find $D(6.15)$.

b. Interpret the meaning of $D(6.15)$.

41. Economics The equation $y = 0.82x + 24$, where $x \geq 0$, expresses a relationship between a nation's disposable income, x (in billions of dollars), and personal consumption expenditures, y (in billions of dollars). Economists call this type of equation a *consumption function*.

a. Graph the consumption function.

b. Name the y-intercept.

c. Explain the significance of the y-intercept.

Mixed Review

42. Given that x is an integer, state the relation representing $y = x^2$ and $-4 \leq x \leq -2$ by listing a set of ordered pairs. Then state whether this relation is a function. **(Lesson 1-1)**

43. Given $f(x) = 4 + 6x - x^3$, find $f(2 + a)$. **(Lesson 1-1)**

44. Given $f(x) = 2x$ and $g(x) = x^2 - 4$, find $(f + g)(x)$ and $(f - g)(x)$. **(Lesson 1-2)**

45. Find $[f \circ g](x)$ if $f(x) = \frac{2}{5}x$ and $g(x) = 40x - 10$. **(Lesson 1-2)**

46. Determine if $f(x) = 3x + 5$ and $g(x) = \frac{x + 5}{3}$ are inverses of each other. Write *yes* or *no*. **(Lesson 1-2)**

47. College Entrance Exam Choose the best answer.

If $\dfrac{2x - 3}{x} = \dfrac{3 - x}{2}$, which of the following could be a value for x?

(A) –3 **(B)** –1 **(C)** 37 **(D)** 5 **(E)** 15

MID-CHAPTER REVIEW

1. Consider the relation $\{(2, 3), (5, 2), (3, 0), (6, 1), (5, -1)\}$. **(Lesson 1-1)**

a. State the domain and range of the relation.

b. Is the relation a function? Why or why not?

Given $f(x) = 4 + 6x - x^3$, find each value. **(Lesson 1-1)**

2. $f(0)$ **3.** $f(-1)$ **4.** $f\left(\frac{1}{2}\right)$ **5.** $f(3k)$

6. Name all values of x that are not in the domain of $f(x) = \dfrac{x + 1}{x^2 - 5}$. **(Lesson 1-1)**

7. If $f(x) = x^2$ and $g(x) = \dfrac{1}{x}$, find $[f \circ g](x)$ and $[g \circ f](x)$. **(Lesson 1-2)**

8. Find the inverse of $f(x) = \dfrac{7 - x}{4}$. **(Lesson 1-2)**

Graph each equation or inequality. **(Lesson 1-3)**

9. $2x - 4y = 8$ **10.** $-2x + 7 \geq y$

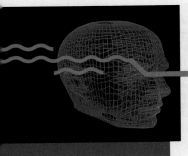

Technology

Graphing Linear Relations

You can use IBM's *Mathematics Exploration Toolkit (MET)* to determine if two functions are inverses.

Example

Determine if $f(x) = \dfrac{x-5}{4}$ and $g(x) = 4x + 5$ are inverses.

First, enter the expression on the right side of the equation. Then, define the function. The steps for $f(x)$ are shown. Follow the same procedure for $g(x)$.

Enter	Result	
$(x-5)/4$	$\dfrac{x-5}{4}$	
def f	$f(x) = \dfrac{x-5}{4}$	*Defines the function f(x).*
sto p		*Stores f(x) in p.*

Then, find $[f \circ g](x)$ and $[g \circ f](x)$.

Enter	Result	Enter	Result
$f(g(x))$	$f(g(x))$	$g(f(x))$	$g(f(x))$
simp	$\dfrac{f(x)-5}{4}$	simp	$4g(x) + 5$
simp	$\dfrac{4x+5-5}{4}$	simp	$4\left(\dfrac{x-5}{4}\right) + 5$
simp	x	simp	x

Therefore, since $[f \circ g](x) = [g \circ f](x) = x$, the functions are inverses. Graph the two equations on the same screen. The steps for $f(x)$ are shown.

Enter	Result	
p	$f(x) = \dfrac{x-5}{4}$	*Recalls p to the screen.*
subs y $f(x)$	$y = \dfrac{x-5}{4}$	*Replaces f(x) with y.*
gra		*Graphs p.*

EXERCISES

Use *MET* to determine if the given functions are inverses of each other. Write *yes* or *no*.

1. $f(x) = 3x + 7$
$g(x) = \dfrac{x-7}{3}$

2. $f(x) = x + 4$
$g(x) = x - 4$

3. $f(x) = -5x + 2$
$g(x) = 5x - 2$

4. Graph each pair of functions and $y = x$ on the same screen. What is true about the intersection of the graphs of $f(x)$ and $g(x)$?

1-4 Distance and Slope

Objectives

After studying this lesson, you should be able to:
- find the distance between two points,
- find the slope of a line through two points, and
- prove geometric theorems involving slope, distance, and midpoints analytically.

Application

The concept of distance is one we use daily in so many different ways that we almost take it for granted. We use distance to estimate the desired number of car lengths between us and the car in front of us, to decide whether a basket made in a basketball game is worth two points or three, to determine the margin of victory in the Indianapolis 500, or to estimate our arrival time when we travel.

In mathematics, the concept of distance is extremely useful and very important. The distance between two points on a number line can be found by using absolute value. Let A and B be two points on a line with coordinates a and b, respectively. The distance between A and B is $|a - b|$ or $|b - a|$.

Recall that RT denotes the length of $\overline{RT}$, the segment with endpoints R and T.

The distance between points in the coordinate plane can also be found. Consider points $R(-3, 2)$ and $S(4, 12)$. To find RS, first choose a point T such that $\overline{RT}$ is parallel to the x-axis and $\overline{ST}$ is parallel to the y-axis. T has coordinates $(4, 2)$. Since S has the same abscissa as T, ST is equal to the absolute value of the difference in the ordinates of S and T, $|12 - 2|$. Similarly, RT is equal to the absolute value of the difference in the abscissas of R and T, $|4 - (-3)|$.

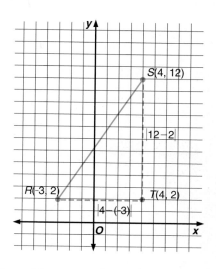

Since $\triangle RST$ is a right triangle, RS can be found by using the Pythagorean theorem, which is illustrated on the next page.

$$(RS)^2 = (RT)^2 + (ST)^2 \quad \textit{Pythagorean theorem}$$

$$\begin{aligned}
RS &= \sqrt{(RT)^2 + (ST)^2} \\
&= \sqrt{|4 - (-3)|^2 + |12 - 2|^2} \\
&= \sqrt{7^2 + 10^2} \\
&= \sqrt{149} \text{ or about } 12.2 \quad \overline{RS} \textit{ is about 12.2 units long.}
\end{aligned}$$

Assume (x_1, y_1) and (x_2, y_2) represent the coordinates of any two points in the plane. The figure at the right illustrates how the formula for finding the distance between (x_1, y_1) and (x_2, y_2) is derived.

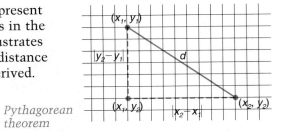

$$d = \sqrt{|x_2 - x_1|^2 + |y_2 - y_1|^2} \quad \textit{Pythagorean theorem}$$

$$= \sqrt{(x_2 - x_1)^2 + (y_2 - y_1)^2}$$

Why does $|x_2 - x_1|^2 = (x_2 - x_1)^2$?

Distance Formula for Two Points

The distance, *d* units, between two points with coordinates (x_1, y_1) and (x_2, y_2) is given by the following formula.
$$d = \sqrt{(x_2 - x_1)^2 + (y_2 - y_1)^2}$$

Example 1

Find the distance between (4, –5) and (–2, 3).

$$\begin{aligned}
d &= \sqrt{(x_2 - x_1)^2 + (y_2 - y_1)^2} \\
&= \sqrt{(-2 - 4)^2 + (3 - (-5))^2} \quad \textit{Let } (x_1, y_1) = (4, -5) \textit{ and } (x_2, y_2) = (-2, 3). \\
&= \sqrt{(-6)^2 + 8^2} \\
&= \sqrt{100} \text{ or } 10
\end{aligned}$$

The distance is 10 units.

Slope is often defined as $\dfrac{rise}{run}$.

Any two distinct points determine a line. The **slope** of the line is the ratio of the change in the ordinates of the coordinates of the points to the corresponding change in the abscissas. The slope of a line is constant.

Definition of Slope

The slope, *m*, of the line through (x_1, y_1) and (x_2, y_2) is given by the following equation, if $x_2 \neq x_1$.

$$m = \frac{y_2 - y_1}{x_2 - x_1}$$

Example 2

A relationship between a nation's disposable income, x (in billions of dollars), and personal consumption expenditures, y (in billions of dollars), is shown in the table at the right. Find the rate of change of personal consumption expenditures with respect to disposable income.

x	y
56	50
76	67.2

The rate of change is the slope of the line containing points at $(56, 50)$ and $(76, 67.2)$. Find the slope of this line.

$$m = \frac{y_2 - y_1}{x_2 - x_1}$$
$$= \frac{67.2 - 50}{76 - 56} \qquad \text{Let } x_1 = 56, y_1 = 50, x_2 = 76, \text{ and } y_2 = 67.2.$$
$$= \frac{17.2}{20} \text{ or } 0.86$$

Economists call this result the *marginal propensity to consume (MPC)*. An MPC of 0.86 means that for each \$1 increase in disposable income, consumption increases \$0.86. In other words, 86% of each additional dollar earned is spent and 14% is saved.

Example 3

Graph the line through $(4, 5)$ and $(4, -3)$. Then, find the slope of the line.

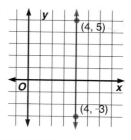

$$m = \frac{-3 - 5}{4 - 4}$$
$$= -\frac{8}{0} \qquad \text{Division by zero is undefined.}$$

Because the abscissas of the two points are the same, the slope is undefined. Any line parallel to the y-axis has undefined slope and is of the form $x = c$.

Many theorems from plane geometry can be more easily proven by analytic methods. That is, they can be proven by placing the figure in a coordinate system and using algebra to express and draw conclusions about the geometric relationships. The study of coordinate geometry from an algebraic perspective is called **analytic geometry**.

The formulas for distance, slope, and midpoints are frequently used in analytic geometry. The distance formula can be used to find the coordinates of the midpoint of a line segment. In the figure at the right, the midpoint of $\overline{P_1P_2}$ is P_m.

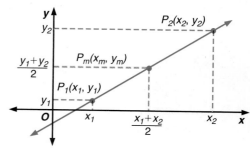

Midpoint of a *Line Segment*	If the coordinates of P_1 and P_2 are (x_1, y_1) and (x_2, y_2), respectively, then the midpoint of $\overline{P_1P_2}$ has coordinates $\left(\dfrac{x_1 + x_2}{2}, \dfrac{y_1 + y_2}{2}\right)$.

Example 4

Find the midpoint of the segment that has endpoints at (5, 8) and (2, 6).

The midpoint is at $\left(\dfrac{5 + 2}{2}, \dfrac{8 + 6}{2}\right)$ or $\left(\dfrac{7}{2}, 7\right)$.

How can you show that the midpoint is equidistant from the endpoints?

When using analytic methods to prove theorems from geometry, the position of the figure in the coordinate plane can be arbitrarily selected as long as size and shape are preserved. This means that the figure may be translated, rotated, or reflected from its original position. For polygons, a vertex is usually located at the origin and one side is made to coincide with the x-axis, as shown below.

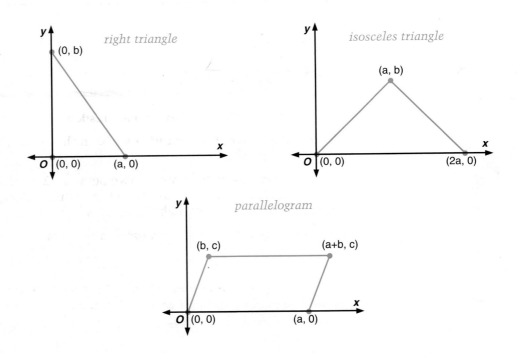

Example 5

Prove that the medians to the congruent sides of an isosceles triangle are congruent.

In $\triangle ABC$, let the vertices be $A(0,0)$ and $B(a, 0)$. Since $\overline{AC}$ and $\overline{BC}$ are congruent sides, let the third vertex be $C\left(\dfrac{a}{2}, b\right)$. Let D be the midpoint of $\overline{AC}$ and let E be the midpoint of $\overline{BC}$. First, find the coordinates of D and E by using the midpoint formula.

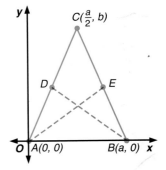

The coordinates of D are: $\left(\dfrac{\frac{a}{2}+0}{2}, \dfrac{b+0}{2}\right)$ or $\left(\dfrac{a}{4}, \dfrac{b}{2}\right)$.

The coordinates of E are: $\left(\dfrac{\frac{a}{2}+a}{2}, \dfrac{b+0}{2}\right)$ or $\left(\dfrac{3a}{4}, \dfrac{b}{2}\right)$.

Then find the measure of each median using the distance formula.

$$AE = \sqrt{\left(\dfrac{3a}{4} - 0\right)^2 + \left(\dfrac{b}{2}\right)^2} = \dfrac{1}{2}\sqrt{\dfrac{9a^2}{4} + b^2}$$

$$BD = \sqrt{\left(a - \dfrac{a}{4}\right)^2 + \left(0 - \dfrac{b}{2}\right)^2} = \dfrac{1}{2}\sqrt{\dfrac{9a^2}{4} + b^2}$$

Since $AE = BD$, the medians to the congruent sides of an isosceles triangle are congruent.

CHECKING FOR UNDERSTANDING

Communicating Mathematics

Read and study the lesson to answer each question.

1. **Describe** three other ways, not mentioned in this lesson, that you use distance in your everyday life.

2. If you were given the coordinates of two points, $(-2, 3)$ and $(7, -9)$, describe how you would decide which point is (x_1, y_1) and which is (x_2, y_2) when finding the slope of the line passing through them.

3. **Research** the meaning of the term *disposable income*.

Guided Practice

Find the distance between the points with the given coordinates. Then, find the slope of the line passing through each pair of points.

4. $(4, 1), (7, 1)$ 5. $(5, 1), (5, 11)$ 6. $(1, 3), (-1, -3)$

7. $(0, 0), (-4, -3)$ 8. $(-1, 1), (4, 13)$ 9. $(-2, 2), (0, 4)$

Find the perimeter of the triangle having vertices with the given coordinates.

10. $(2, 3), (14, 3), (14, 8)$ 11. $(2, 2), (5, 2), (2, 6)$

12. $(1, -1), (1, 3), (-2, -1)$ 13. $(3, 3), (3, -9), (-2, 3)$

14. The vertices of a rectangle are at $(-3, 1), (-1, 3), (3, -1), (1, -3)$. Find the area of the rectangle.

15. The vertices of a triangle are at $(5, 0), (-3, 2)$, and $(-1, -4)$. Find the coordinates of the midpoints of the sides.

EXERCISES

Find the distance between the points with the given coordinates. Then, find the slope of the line passing through each pair of points.

16. $(5, -3), (-1, -6)$

17. $(6, 0), (0, 6)$

18. $(5, 7), (0, 0)$

19. $(1, -5), (-7, 11)$

20. $(3, a), (8, a)$

21. $(b, 6 + a), (b, a + 3)$

22. $(r, s), (r + 2, s - 1)$

23. $(n, 4n), (n + 1, n)$

Determine whether the figure with vertices with the given coordinates is a parallelogram.

24. $(3, 4), (6, 2), (8, 7), (5, 9)$

25. $(4, 11), (8, 14), (4, 19), (0, 15)$

26. $(-2, 1), (-1, 5), (-5, 6), (-6, 2)$

27. $(2, -3), (-2, 3), (-3, -2), (3, 2)$

Collinear points lie on the same line. Find the value of *k* for which points with each set of coordinates is collinear. *Remember, the slope of each line is constant.*

28. $(4, 0), (k, 3), (4, -3)$

29. $(2, -5), (-4, -11), (k, 1)$

30. $(7, -2), (0, 5), (3, k)$

31. $(15, 1), (-3, -8), (3, k)$

32. Consider rectangle $ABCD$.

    a. Prove that $\overline{AC} \cong \overline{BD}$.

    b. Prove that $\overline{AE} \cong \overline{EC}$ and $\overline{BE} \cong \overline{ED}$.

    c. What can you conclude about the diagonals of a rectangle?

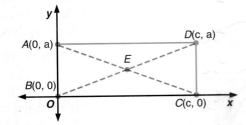

Prove using analytic methods.

33. The line segment joining the midpoints of two sides of a triangle is equal in length to one-half the third side.

34. The median of a trapezoid is equal in length to one-half the sum of the lengths of the parallel sides.

35. Prove analytically that the segments joining midpoints of successive sides of an isosceles trapezoid form a rhombus.

36. Economics In the table at the right, a nation's disposable income is represented by x and its personal consumption expenditures are represented by y. All figures represent billions of dollars.

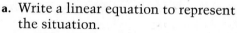

x	y
48	44
68	60

a. Calculate the marginal propensity to consume.

b. Suppose the average disposable income per family increases $1805 in a year. Based on the MPC, by how much would consumption increase?

37. Economics In Example 2, a rate of change called the marginal propensity to consume (MPC) was presented. Another rate of change is the *marginal propensity to save* (MPS): MPS $= 1 -$ MPC.

a. Calculate the MPS for the data in Exercise 36.

b. If disposable income were to increase as in Exercise 36b, how many additional dollars would the average family save?

38. Consumer Products Citydog Screen Printers of Staten Island, New York, makes special-order T-shirts. Recently, Citydog received two orders for a special T-shirt designed for a mathematics symposium. The first order was for 40 T-shirts at a total cost of $295, and the second order was for an additional 80 T-shirts at a total cost of $565. Each order included a standard shipping and handling charge.

a. Write a linear equation to represent the situation.

b. What is the cost per T-shirt?

c. What is the standard shipping and handling charge?

39. Given $f(x) = 5 - x^2$, find $f(3)$. **(Lesson 1-1)**

40. If $f(x) = -2x + 11$ and $g(x) = x - 6$, find $[f \circ g](x)$ and $[g \circ f](x)$. **(Lesson 1-2)**

41. If $f(x) = x^3$ and $g(x) = x^2 - 3x + 7$, find $(f \cdot g)(x)$ and $\left(\dfrac{f}{g}\right)(x)$. **(Lesson 1-2)**

42. Find the zero of $f(x) = 5x - 3$. **(Lesson 1-3)**

43. Graph $y = \dfrac{1}{2}x - 1\dfrac{1}{2}$. **(Lesson 1-3)**

44. College Entrance Exam Choose the best answer.
In the figure at the right, the area of square $OXYZ$ is 2. What is the area of the circle?

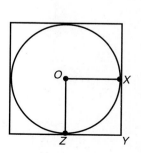

(A) $\dfrac{\pi}{4}$ **(B)** $\pi\sqrt{2}$ **(C)** 2π **(D)** 4π **(E)** 8π

1-5 Forms of Linear Equations

Objectives

After studying this lesson, you should be able to:
- write linear equations using slope-intercept form, and
- write linear equations using point-slope form.

Application

The average American pays about $2000 a year for health insurance.

From 1984 to 1987, medical expenses in the United States rose in a linear pattern from 10.3% of the gross national product in 1984 to 10.9% of the gross national product in 1987. Assuming that the level of growth continues at that rate, write a linear equation to describe the percent of the gross national product devoted to medical expenses, *y*, in year *x*. *This problem will be solved in Example 4.*

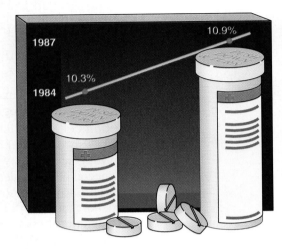

Sometimes the slope and **y-intercept** of a line are easy to determine just by looking at the equation of the line. For example, the graph of $y = 3.5x - 9$ has slope 3.5 and y-intercept –9. The equation $y = 3.5x - 9$ is said to be in **slope-intercept form**.

Slope-Intercept Form	**The slope-intercept form of an equation of a line is $y = mx + b$. The slope is m and the y-intercept is b.**

Example 1

> **Write the equation $4x - 3y + 7 = 0$ in slope-intercept form. Then, name the slope and y-intercept.**
>
> $4x - 3y + 7 = 0$
> $-3y = -4x - 7$
> $y = \frac{4}{3}x + \frac{7}{3}$
>
> The slope is $\frac{4}{3}$ and the y-intercept is $\frac{7}{3}$.

If one point and the slope of a line are known, the slope-intercept form can be used to find the equation of the line.

Example 2

Write the slope-intercept form of the equation of the line through $(3, 7)$ that has a slope of 2.

Substitute the slope and coordinates of the point in the general slope-intercept form of a linear equation. Then, solve for b.

$$y = mx + b$$
$$7 = 2(3) + b$$
$$1 = b$$

The slope-intercept form of the equation of the line is $y = 2x + 1$.

The slope formula can also be used to find the equation of a line when a point and the slope are known.

Example 3

Find the equation of the line through $(-2, 4)$ that has a slope of -1. Then, write the equation in slope-intercept form.

Suppose a second point on the line is at (x, y). Substitute the values into the slope formula.

$$m = \frac{y_2 - y_1}{x_2 - x_1}$$
$$-1 = \frac{y - 4}{x - (-2)}$$
$$-1(x + 2) = y - 4$$
$$-x - 2 = y - 4$$
$$-x + 2 = y$$

The slope-intercept form of the equation is $y = -x + 2$.

The form of a linear equation derived from the slope formula is called the **point-slope form** of the equation. If $\frac{y_2 - y_1}{x_2 - x_1} = m$, then $y - y_1 = m(x - x_1)$.

Point-Slope Form

If the point with coordinates (x_1, y_1) lies on a line having slope m, the point-slope form of the equation of the line can be written as follows.
$$y - y_1 = m(x - x_1)$$

If the coordinates of two points on a line are known, the slope can be found. Then the equation of the line can be written using slope-intercept form or point-slope form.

Example 4

In 1984, medical expenses were 10.3% of the gross national product (GNP), and in 1987, medical expenses were 10.9% of the gross national product. Assuming that the level of growth continues at that rate, write a linear equation in slope-intercept form to describe the percent of the gross national product devoted to medical expenses, y, in year x.

Since y represents the percent of the GNP devoted to medical expenses and x represents the year, let $x_1 = 1984$, $y_1 = 10.3$, $x_2 = 1987$, and $y_2 = 10.9$. Then find the slope of the line through the points $(1984, 10.3)$ and $(1987, 10.9)$.

$$m = \frac{y_2 - y_1}{x_2 - x_1}$$
$$= \frac{10.9 - 10.3}{1987 - 1984}$$
$$= \frac{0.6}{3} \quad \text{or} \quad 0.2$$

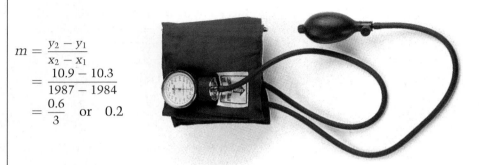

inter NET
CONNECTION

For the latest U.S. economic statistics, visit: **www.glencoe.com/sec/ math/amc/mathnet**

Substitute the slope and coordinates of a point in the general slope-intercept form. Then, solve for b.

$$y = mx + b$$
$$10.9 = 0.2(1987) + b \qquad x = 1987,\ y = 10.9,\ \text{and}\ m = 0.2$$
$$10.9 = 397.4 + b$$
$$-386.5 = b$$

The slope-intercept form of the equation of the line is $y = 0.2x - 386.5$.

Example 5

Find the point-slope form of the equation of the line through $(3, 7)$ and $(4, -1)$. Then, write the equation in slope-intercept form.

First, find the slope.
$$m = \frac{-1 - 7}{4 - 3} \quad \text{or} \quad -8$$

Then, substitute values into the general point-slope form of a linear equation.
$$y - y_1 = m(x - x_1)$$
$$y - 7 = -8(x - 3) \qquad \textit{The coordinates of either point can be used for } (x_1, y_1)$$
$$y = -8x + 31$$

The slope-intercept form of the equation is $y = -8x + 31$.

Example 6

The Fluff Detergent Company manufactures detergent with a fixed overhead cost of $5000. In addition, each box of Fluff Detergent costs $0.40 to produce. Write an equation that represents the total cost, $C(x)$, of producing x boxes of Fluff. Then find the fixed cost and the variable cost per box of Fluff.

The equation is $C(x) = 0.40x + 5000$. This equation is called a *cost function*.

The slope of the function is called the *variable cost*. In this example, $m = 0.40$. The cost of producing 0 boxes is called the *fixed cost*.

$$C(x) = 0.40x + 5000$$
$$C(0) = 0.40(0) + 5000$$
$$= 5000$$

Therefore, the fixed cost is $5000, and the variable cost is $0.40.

CHECKING FOR UNDERSTANDING

Communicating Mathematics

Read and study the lesson to answer each question.

1. **Research** the meaning of the term *gross national product*.

2. If the equation $S = 5.5H - 10$ represents the weekly salary (S) earned by an employee at a fast-food restaurant for H hours worked, explain what 5.5 and –10 represent.

3. **Describe** how you could quickly graph the line $3x + 5y = 30$.

4. **Explain** when you would use the slope-intercept form to find the equation of a line and when you would use the point-slope form to find the equation of a line.

5. **Research** the meaning of the term *overhead costs*.

Guided Practice

Write each equation in slope-intercept form. Then name the slope and y-intercept.

6. $3x - 2y = 7$

7. $8x = 2y - 1$

8. $4x + 3y = 0$

Write the slope-intercept form of the equation of the line through the point with the given coordinates and having the given slope.

9. $(3, 2), 4$

10. $(5, 7), 0$

11. $(-3, -4), -6$

Write the point-slope form of the equation of the line through the points with the given coordinates. Then write the equation in slope-intercept form.

12. $(6, 6), (-6, -6)$ **13.** $(-2, 0), (1, -3)$ **14.** $(4, 2), (7, 2)$

EXERCISES

Practice

Write the slope-intercept form of the equation of the line through the point with the given coordinates and having the given slope.

15. $(-6, 2), 8$ **16.** $(-5, -12), -5$

17. $(3, 5), -3$ **18.** $(-10, 4), \dfrac{3}{4}$

19. $(-7, 3), -\dfrac{1}{4}$ **20.** $(9, 11), \dfrac{2}{3}$

Write the point-slope form of the equation of the line through the points with the given coordinates. Then write the equation in slope-intercept form.

21. $(-1, 4), (-1, 7)$ **22.** $(3, -5), (2, -1)$

23. $(5, 2), (7, 9)$ **24.** $(2, 5), (7, 8)$

25. $(3, 1), (-2, 4)$ **26.** $(-7, -1), (4, -2)$

Graphing Calculator

Use a graphing calculator to graph each function, without erasing, for $m = 1$, $m = -3$, and $m = 5$. Explain how the graphs are alike or different.

27. $f(x) = mx - 1$ **28.** $f(x) = mx - 3$

29. $f(x) = mx + 2$ **30.** $f(x) = mx + 4$

Critical Thinking

31. Write equations for the sides of the triangle that has vertices $A(2, -7)$, $B(5, 1)$, and $C(-3, 2)$.

32. Write equations for the sides of the square that has vertices $P(1, 4)$, $Q(4, -1)$, $R(-1, -4)$, and $S(-4, 1)$.

Applications and Problem Solving

33. Manufacturing It costs Consolidated Cereals Corporation $1050 to produce 100 boxes of corn flakes and $1250 to produce 500 boxes of corn flakes.

    a. Find the cost function.

    b. Determine the fixed cost and the variable cost per box.

    c. Sketch the graph of the cost function.

34. Anatomy The tibia is a bone in the human that connects the knee and the ankle. A man with a tibia 38.5 centimeters long should be about 173 centimeters tall, and a man with a tibia 44.125 centimeters long should be about 188 centimeters tall.

    a. Write a linear equation relating height, y, to the length of the tibia, x.

    b. About how tall should a man be whose tibia is 40 cm long?

35. Anatomy The radius is a bone in the human that connects the elbow and the wrist. A woman with a radius 22 centimeters long should be about 160 centimeters tall, and a woman with a radius 26 centimeters long should be about 174 centimeters tall.

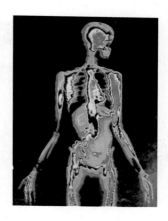

  a. Write a linear equation relating height, y, to the length of the radius, x.

  b. About how long should the radius be of a woman who is 157.5 centimeters tall?

Mixed Review

36. Does the graph at the right represent a function? **(Lesson 1-1)**

37. If $f(x) = x$ and $g(x) = 4x^2 - 7$, find $f \circ g(x)$ and $g \circ f(x)$. **(Lesson 1-2)**

38. Graph $y \leq -\frac{1}{3}x + 2$. **(Lesson 1-3)**

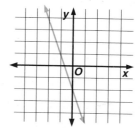

39. Demographics The population of the Dallas-Fort Worth metropolitan area was 2,352,000 in 1970. The population grew to 3,885,000 in 1990. If x represents the year and y represents the population, find the rate of increase for the growth of the population. **(Lesson 1-4)**

40. College Entrance Exam Compare quantities A and B below.
Write A if the quantity in Column A is greater.
Write B if the quantity in Column B is greater.
Write C if the two quantities are equal.
Write D if there is not enough information to determine the relationship.

  (A) 5×6^{12} **(B)** 6×5^{12}

1-6 Parallel and Perpendicular Lines

After studying this lesson, you should be able to:
- write equations of parallel and perpendicular lines, and
- prove geometric theorems involving parallel and perpendicular lines analytically.

Application

The contrails of an airplane are the streaks of condensed water vapor that follow the airplane when the airplane flies at high altitudes. The contrails of the jet aircraft pictured at the right represent a family of lines. Families of lines always have a common feature. In this case, the contrails all represent lines that have the same slope.

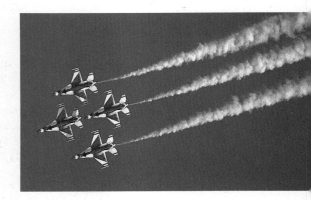

Lines that are parallel or perpendicular can be studied by examining their equations and slopes. Lines that are **parallel** have the same slope. That is, if m_1 is the slope of ℓ_1 and m_2 is the slope of ℓ_2, then ℓ_1 and ℓ_2 are parallel if and only if $m_1 = m_2$. Vertical lines are parallel even though the slope of a vertical line is undefined.

Parallel Lines	**Two nonvertical lines are parallel if and only if their slopes are equal. Any two vertical lines are always parallel.**

A convenient way to write a linear equation is in **standard form.** The standard form of a linear equation will be used to develop relationships among slope and parallel and perpendicular lines.

Standard Form of a Linear Equation	**The standard form of a linear equation is $Ax + By + C = 0$, where A, B, and C are real numbers and A and B are not both zero.**

What is the slope of the line if B is 0?

The slope of a line can be obtained directly from the standard form of the linear equation if B is not 0. By solving $Ax + By + C = 0$ $(B \neq 0)$ for y, the result is $y = -\dfrac{A}{B}x - \dfrac{C}{B}$. Therefore, the slope is $-\dfrac{A}{B}$. For example, the slope of the line $2x - 4y + 7 = 0$ is $-\dfrac{2}{-4}$ or $\dfrac{1}{2}$. *The expression "the line $2x - 4y + 7 = 0$" means "the line that is the graph of $2x - 4y + 7 = 0$."*

Since parallel lines have the same slope, the equation of a line parallel to a given line and through a given point can be written using point-slope form.

Example 1

> Write the standard form of the equation of the line that passes through $(2, -3)$ and is parallel to the line $4x - y + 3 = 0$.
>
> Since $-\dfrac{A}{B} = -\left(\dfrac{4}{-1}\right)$, the slope is 4.
>
> $\quad y - y_1 = m(x - x_1)$ *Use the point-slope form.*
> $\quad\quad y + 3 = 4(x - 2)$ *Replace y_1 with –3, m with 4, and x_1 with 2.*
> $\quad\quad y + 3 = 4x - 8$
> $4x - y - 11 = 0$
>
> The equation is $4x - y - 11 = 0$.

Example 2

APPLICATION

Sales

> **During the month of January, Fransworth Computer Center sold 24 Macintosh IIci computers and 40 Personal LaserWriter NT printers. The total sales on these two items for the month of January was $213,600. In February, they sold 30 Macintosh IIci computers and 50 Personal LaserWriter NT printers. Assuming the prices stayed consistent during the months of January and February, is it possible that their February sales could have totaled $283,000 on these two items?**
>
> The sales function for the month of January is $24x + 40y = 213,600$ or $y = -\dfrac{3}{5}x + 5340$. The sales function for the month of February is $30x + 50y = 283,000$ or $y = -\dfrac{3}{5}x + 5660$. Since these two equations have the same slope, their graphs are parallel lines. As a result, the ratio of total sales should be proportional to the ratio of computer sales or printer sales from month to month. However, since $\dfrac{64}{80} \neq \dfrac{213,600}{283,000}$, the February sales could not have been $283,000.

Quadrilateral $ABCD$ at the right is a rhombus. A rhombus has perpendicular diagonals. The slope of diagonal $\overline{AC}$ is $\dfrac{4}{8}$ or $\dfrac{1}{2}$. The slope of diagonal $\overline{BD}$ is $\dfrac{4}{-2}$ or –2. The two slopes, $\dfrac{1}{2}$ and –2, are negative reciprocals of each other. That is, the product of the slopes is –1. This is true for the slopes of any two nonvertical **perpendicular** lines.

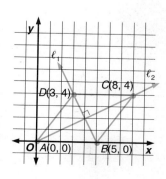

Thus, when ℓ_1 is perpendicular to ℓ_2, the following is true.

$$m_1 m_2 = -1 \text{ or } m_1 = -\frac{1}{m_2}$$

Perpendicular Lines	**Two nonvertical lines are perpendicular if and only if their slopes are negative reciprocals.**

The equation of a line perpendicular to a given line and through a given point can be written using the point-slope form.

Example 3

Write the standard form of the equation of the line that passes through $(3, -5)$ and is perpendicular to the line $2x - 3y + 6 = 0$.

Since $m = -\dfrac{A}{B}$, the line $2x - 3y + 6 = 0$ has slope $\dfrac{2}{3}$. Therefore, the slope of a line that is perpendicular to the line $2x - 3y + 6 = 0$ would be $-\dfrac{3}{2}$.

$y - y_1 = m(x - x_1)$ *Use the point-slope form.*

$y + 5 = -\dfrac{3}{2}(x - 3)$ *Replace y_1 with –5, m with $-\dfrac{3}{2}$, and x_1 with 3.*

$2y + 10 = -3x + 9$

$3x + 2y + 1 = 0$

The equation is $3x + 2y + 1 = 0$.

You can use the definitions of parallel and perpendicular lines to prove theorems by using analytic methods.

Example 4

Prove that the line segment joining the midpoints of two sides of a triangle
a. is parallel to the third side and
b. has a length that is one-half the length of the third side.

a. In $\triangle ABC$, let the vertices be $A(0, 0)$, $B(a, 0)$, and $C(b, c)$. Then, $M\left(\dfrac{b}{2}, \dfrac{c}{2}\right)$ and $N\left(\dfrac{a+b}{2}, \dfrac{c}{2}\right)$ are the midpoints of $\overline{AC}$ and $\overline{BC}$, respectively. The slope of $\overline{AB}$ is $\dfrac{0}{a}$ or 0. The slope of $\overline{MN}$ is as follows.

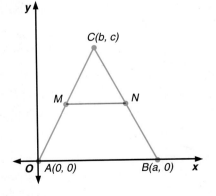

$$\frac{\dfrac{c}{2} - \dfrac{c}{2}}{\dfrac{a+b}{2} - \dfrac{b}{2}} = \frac{0}{\dfrac{a}{2}} \text{ or } 0$$

Since the slopes are the same, $\overline{AB}$ is parallel to $\overline{MN}$.

b. To find the lengths of the horizontal line segments, use the distance formula.

$$MN = \sqrt{\left(\frac{b}{2} - \frac{a+b}{2}\right)^2 + \left(\frac{c}{2} - \frac{c}{2}\right)^2} \qquad AB = \sqrt{(a-0)^2 + (0-0)^2}$$

$$= \sqrt{\left(\frac{b-a-b}{2}\right)^2 + (0)^2} \qquad\qquad = \sqrt{a^2 + 0}$$

$$= \sqrt{\left(-\frac{a}{2}\right)^2} \qquad\qquad\qquad = a$$

$$= \frac{a}{2} \text{ or } \frac{1}{2}a$$

Therefore, the line segment joining the midpoints of two sides of a triangle is parallel to the third side of the triangle, and its length is one-half the length of the third side.

 Portfolio

A portfolio is representative samples of your work, collected over a period of time. Begin your portfolio by selecting an item that shows something new you learned in this chapter.

CHECKING FOR UNDERSTANDING

Communicating Mathematics

Read and study the lesson to answer each question.

1. **Explain** why the definition of parallel lines begins with the wording "Two nonvertical lines...."

2. What should have been the sales for the month of February in Example 2? Describe how you reached your solution.

3. **Discuss** whether it is possible for two perpendicular lines to have slopes that are either both positive or both negative. Explain your answer.

4. **Describe** the slope of a line that is perpendicular to a line that has an undefined slope. Verify that your description of this line is correct.

Guided Practice

Find the slope of each line whose equation is given. Then, determine whether the lines are *parallel*, *perpendicular*, or *neither*.

5. $y = 3x - 2$
   $y = -3x + 2$

6. $y = 6x - 2$
   $y = 6x + 7$

7. $y = x - 9$
   $x + y + 9 = 0$

8. $y = 2x + 4$
   $x + 2y + 10 = 0$

9. $y + 4x - 2 = 0$
   $y + 4x + 1 = 0$

10. $y = 8x - 1$
    $7x - y - 1 = 0$

Write the standard forms of the equations of the lines that are parallel and perpendicular to $y = 2x - 3$ and pass through the point with the given coordinates.

11. $(4, 2)$

12. $(-3, 6)$

13. $(8, -2)$

14. $(-5, -7)$

15. Show that quadrilateral $WXYZ$ is a square if its vertices are $W(-1, 3)$, $X(3, 6)$, $Y(6, 2)$, and $Z(2, -1)$.

16. Prove analytically that the diagonals of a square are perpendicular.

EXERCISES

Practice

Write the standard form of the equation of the line that is parallel to the given line and passes through the given point.

17. $y = 3x - 5; (0, 6)$

18. $y = 2x + 6; (-1, -2)$

19. $y = 6x + 7; (0, -3)$

20. $y = -4x - 3; (5, -7)$

21. $2x + 3y - 5 = 0; (2, 4)$

22. $2x - 7y = 3; (8, 0)$

Write the standard form of the equation of the line that is perpendicular to the given line and passes through the given point.

23. $y = -2x + 5; (0, -3)$

24. $y = 4x - 2; (3, 4)$

25. $5y - 4x = 10; (-15, 8)$

26. $3y + 2x = 3; (-9, -6)$

27. $6x - 4y + 8 = 0; (2, 12)$

28. $3x - y = 8; (-1, 5)$

29. For what value of k is the graph of $kx - 7y + 10 = 0$ parallel to the graph of $8x - 14y + 3 = 0$? For what value of k are the graphs perpendicular?

30. For what value of k is the graph of $2x - ky + 5 = 0$ parallel to the graph of $3x + 7y + 15 = 0$? For what value of k are the graphs perpendicular?

31. Show that triangle ABC is a right triangle if its vertices are $A(-1, 2)$, $B(4, -3)$, and $C(-2, -1)$.

32. Show that quadrilateral $PQRS$ is a rhombus if its vertices are $P(3, 1)$, $Q(8, 1)$, $R(12, 4)$, and $S(7, 4)$.

Critical Thinking

33. Prove analytically that the line joining the midpoints of the nonparallel sides of a trapezoid is parallel to the bases of the trapezoid.

Applications and Problem Solving

34. Sales Lisa owns and operates a pizza and calzone concession at a county fair each summer. On two consecutive days in July, Lisa reported the following sales: on the first day, she sold 520 slices of pizza and 280 calzones; on the second day, she sold 390 slices of pizza and 210 calzones. Write a convincing argument to show whether her sales on the second day could have been $1057.50 if her sales on the first day were $1410. (Assume that the prices remained the same over the two days.)

35. Sports Marquita starts riding her bicycle at 10 miles per hour. At the same time, Heather starts riding her bicycle in the same direction from a point 5 miles north of Marquita. Heather also rides at a rate of 10 miles per hour. If the girls continue to ride at the same speed, will Marquita ever catch up with Heather? Why or why not?

Mixed Review

36. If $f(x) = 3x^2 - 4$ and $g(x) = 5x + 1$, find $f \circ g(4)$ and $g \circ f(4)$. **(Lesson 1-2)**

37. Graph $-6 \leq 3x - y \leq 12$. **(Lesson 1-3)**

38. Find the distance between $(4, -3)$ and $(-2, 6)$. **(Lesson 1-4)**

39. Manufacturing It costs ABC Corporation $3000 to produce 20 color televisions and $5000 to produce 60 of the same color televisions. **(Lesson 1-5)**
  a. Find the cost function.
  b. Determine the fixed cost and the variable cost per unit.
  c. Sketch the graph of the cost function.

40. College Entrance Exam If $2x + y = 12$ and $x + 2y = -6$, then find the value of $2x + 2y$.

DECISION MAKING

Recycling

Americans throw away more than 600 million pounds of paper and 100 million pounds of plastic every day. Much of this material could be recycled. However, some materials are easier and less expensive than others to recycle. How can you best use your resources to start a recycling campaign? Should you decide on one or two materials and try to get others to recycle these? Or should you concentrate your efforts on education by publicizing the importance of recycling?

1. Find out which kinds of materials can be recycled most efficiently. Are there companies or organizations in your city that will buy recycled materials from you? Are there companies that will recycle materials without paying for them? How would you organize a recycling campaign if you were in charge of it?

2. Describe steps you could take to raise student or public awareness of the need to recycle.

3. **Project** Work with your group to complete one of the projects listed at the right based on the information you gathered in Exercises 1–2.

PROJECTS

- *Conduct an interview.*
- *Write a book report.*
- *Write a proposal.*
- *Write an article for the school paper.*
- *Write a report.*
- *Make a display.*
- *Make a graph or chart.*
- *Plan an activity.*
- *Design a checklist.*

VOCABULARY

Upon completing this chapter, you should be familiar with the following terms:

analytic geometry **31**	**42** parallel
boundary **23**	**43** perpendicular
composite **13**	**37** point-slope form
composition **13**	**6** range
constant function **23**	**6** relation
domain **6**	**30** slope
function **7**	**36** slope-intercept form
inverse functions **15**	**42** standard form
iteration **15**	**7** vertical line test
linear equation **22**	**23** x-intercept
linear function **22**	**30** y-intercept
linear inequality **23**	**22** zero of a function

SKILLS AND CONCEPTS

OBJECTIVES AND EXAMPLES

Upon completing this chapter, you should be able to:

■ determine whether a given relation is a function **(Lesson 1-1)**

If x is a positive integer between 1 and 4, determine whether $y^2 = x - 1$ is a function.

The relation is $\{(2, 1), (2, -1), (3, \sqrt{3}),$ $(3, -\sqrt{3})\}$. Since 3 is paired with $\sqrt{3}$ and $-\sqrt{3}$, the relation is not a function.

■ identify the domain and range of any relation or function **(Lesson 1-1)**

Name all values of x that are not in the domain of $f(x) = \dfrac{7x - 1}{x^3 + 1}$.

$x^3 + 1 = 0$

$x = -1$ -1 is not in the domain of f.

REVIEW EXERCISES

Use these exercises to review and prepare for the chapter test.

Given that x is an integer, state the relation representing each of the following by listing a set of ordered pairs. Then state whether the relation is a function. Write *yes* or *no*.

1. $y = 5x - 7$ and $0 \le x \le 3$

2. $y = 3x^3$ and $-2 < x < 3$

3. $|y| = x - 4$ and $5 \le x < 7$

4. $y = |4 + x|$ and $-8 \le x < -2$

Name all values of x that are not in the domain of the given function.

5. $f(x) = \dfrac{5}{x}$ 6. $f(x) = \dfrac{x^2}{x^2 - 3}$

7. $f(x) = \dfrac{2(x^2 - 9)}{x + 3}$ 8. $f(x) = \dfrac{2x^3 - 5}{2|x| - 5}$

■ find the composition of functions
(**Lesson 1-2**)

Find $[f \circ g](x)$ if $f(x) = -5x + 4$ and
$g(x) = 2x - 3$.

$$[f \circ g](x) = f(2x - 3)$$
$$= -5(2x - 3) + 4$$
$$= -10x + 19$$

Find $[f \circ g](x)$ and $[g \circ f](x)$.

9. $f(x) = 3x - 5$
$\quad g(x) = x + 2$

10. $f(x) = -3x^2$
$\quad g(x) = 2x^3$

11. $f(x) = x^2 + 2x + 3$
$\quad g(x) = x + 1$

12. $f(x) = \dfrac{1}{2x}$
$\quad g(x) = \dfrac{1}{2}x - 1$

■ find and recognize inverse functions
(**Lesson 1-2**)

Given $f(x) = \dfrac{1}{2}x^3 + 5$, find $f^{-1}(x)$.

$$x = \dfrac{1}{2}y^3 + 5$$
$$2x - 10 = y^3$$
$$\sqrt[3]{2x - 10} = y \quad \rightarrow \quad f^{-1}(x) = \sqrt[3]{2x - 10}$$

**Find the inverse of each function. Then
state whether the inverse is a function.**

13. $f(x) = x^2 - 12$

14. $f(x) = 6 - 3x$

15. $f(x) = 8x^3$

16. $f(x) = 3x^3 - 2$

■ find the zeros of linear functions
(**Lesson 1-3**)

Find the zero of $f(x) = 4x + 3$.

$$4x + 3 = 0$$
$$x = -\dfrac{3}{4} \quad \text{The zero is } -\dfrac{3}{4}.$$

**Find the zero of each function. Then graph
the function.**

17. $f(x) = 3x - 8$

18. $f(x) = 19$

19. $f(x) = 0.25x - 5$

20. $f(x) = 7x + 1$

■ find the distance between two points
(**Lesson 1-4**)

Find the distance between $(3, 8)$ and
$(-5, 11)$.

$$d = \sqrt{(x_2 - x_1)^2 + (y_2 - y_1)^2}$$
$$= \sqrt{(-5 - 3)^2 + (11 - 8)^2}$$
$$= \sqrt{73}$$

**Find the distance between the points with
the given coordinates.**

21. $(0, 0), (5, 12)$

22. $(1, -6), (-3, -4)$

23. $(a, b), (a + 3, b + 4)$

24. $(2k, 4k), (3k, 6k)$

■ write linear equations using slope-
intercept form (**Lesson 1-5**)

Write the slope-intercept form of the line
through $(2, -4)$ that has a slope of $\dfrac{1}{2}$.

$$m = \dfrac{y_2 - y_1}{x_2 - x_1}$$
$$\dfrac{1}{2} = \dfrac{y - (-4)}{x - 2}$$
$$x - 2 = 2y + 8$$
$$y = \dfrac{1}{2}x - 5$$

**Write the slope-intercept form of the
equation of the line through the point
with the given coordinates and having the
given slope.**

25. $(5, 5), 2$

26. $(-2, 3), -1$

27. $(0, 0), \dfrac{3}{5}$

28. $(1, 4), -\dfrac{4}{3}$

■ write equations of lines using point-slope form **(Lesson 1-5)**

Write the slope-intercept form of the line through the points $(5, -2)$ and $(3, -8)$.

$$m = \frac{-8 - (-2)}{3 - 5}$$
$$= 3$$

$$y - y_1 = m(x - x_1)$$
$$y - (-2) = 3(x - 5)$$
$$y = 3x - 17$$

Find the point-slope form of the equation of the line through the points with the given coordinates. Then write the equation in slope-intercept form.

29. $(3, 7), (6, 10)$

30. $(-1, 0), (5, 9)$

31. $(4, 4), (2, -3)$

32. $(11, -6), (10, -9)$

■ write the equations of parallel and perpendicular lines **(Lesson 1-6)**

Write the standard form of the equation of the line that passes through $(-1, 3)$ and is perpendicular to the line $y = \frac{1}{5}x + 2$.

$$y - y_1 = m(x - x_1)$$
$$y - 3 = -5(x - (-1))$$
$$y = -5x - 2 \quad \text{or} \quad 5x + y + 2 = 0$$

Write the standard form of the equation of the line described below.

33. parallel to $y = 4x - 7$, passes through $(-2, 3)$

34. parallel to $-2x + 3y - 5 = 0$, passes through $(2, 4)$

35. perpendicular to $6x = 2y - 4$, y-intercept 4

36. perpendicular to $5y = 4x + 2$, x-intercept 5

APPLICATIONS AND PROBLEM SOLVING

37. Aviation A jet plane starts from rest on a runway. It accelerates uniformly at a rate of 20 m/s^2. The equation for computing the distance traveled is $d = \frac{1}{2}at^2$. **(Lesson 1-1)**

  a. Find the distance traveled at the end of each second for 5 seconds.

  b. Is this relation a function? Why or why not?

38. Prove that the diagonals of a parallelogram bisect each other. **(Lesson 1-4)**

39. Manufacturing The Yummy Candy Company can produce 100 candy bars at a cost of $150 and 600 candy bars at a cost of $200. Find the cost function and determine the fixed cost and the variable cost per candy bar. **(Lesson 1-5)**

Given that *x* **is an integer, state the relation representing each of the following by listing a set of ordered pairs. Then state whether the relation is a function. Write** *yes* **or** *no.*

1. $y = 3x - 1$ and $-3 < x \leq 2$

2. $|y| = 2x + 5$ and $0 \leq x \leq 4$

Given $f(x) = x - 3x^2$, **find each value.**

3. $f(0)$

4. $f(4)$

5. $f(-6)$

6. $f(7.1)$

Find $[f \circ g](x)$ **and** $[g \circ f](x)$.

7. $f(x) = \sqrt{x}$
$g(x) = 2x^2 - 5$

8. $f(x) = 2x^2$
$g(x) = 5x + 6$

9. $f(x) = -x - 7$
$g(x) = -3x$

Find the inverse of each function. Then state whether the inverse is a function.

10. $f(x) = \dfrac{x^2 + 5}{3}$

11. $f(x) = \dfrac{1}{4}x + 1$

12. $f(x) = -2(x^3 - 1)$

Graph each equation or inequality.

13. $y = 3x - 6$

14. $2x = y + 1$

15. $y + 4x \leq 12$

16. $y > 2x - 2$

Find the distance between the points with the given coordinates. Then, find the slope of the line passing through each pair of points.

17. $(-1, 2)$, $(3, 1)$

18. $(5, 11)$, $(12, 12)$

19. $(3k, k + 1)$, $(2k, k - 1)$

20. Write the slope-intercept form of the equation of the line through $(-1, 3)$ that has a slope of $\dfrac{5}{3}$.

21. Write the slope-intercept form of the equation of the line through $(0, 4)$ and $(8, -2)$.

22. Write the standard form of the equation of the line parallel to the line $3x + 2y - 3 = 0$ and passing through $(3, 4)$.

23. Write the standard form of the equation of the line perpendicular to the line $-x + 5y = -3$ and passing through the origin.

24. Prove that the midpoint of the hypotenuse of a right triangle is equidistant from all of the vertices.

25. Physics The equation of the illuminance, E, of a small light source is $E = \dfrac{P}{4\pi d^2}$, where P is the luminous flux (lm) of the source and d is its distance from the surface. Suppose the luminous flux of a desk lamp is 750 lm.

a. What is the illuminance (in lm/m^2) on a desktop if the lamp is 2.5 m above it?

b. Name the values of d, if any, that are not in the domain of the given function.

Bonus
The function f is defined by $f(x) = \dfrac{kx}{4x + 2}$, where k is a constant and $x \neq -\dfrac{1}{2}$. What is the value of k when $f(f(x)) = x$ for all real numbers except $-\dfrac{1}{2}$?

CHAPTER 2

SYSTEMS OF EQUATIONS AND INEQUALITIES

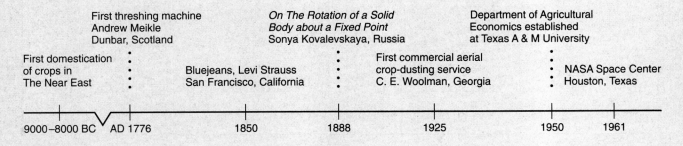

First threshing machine
Andrew Meikle
Dunbar, Scotland

*On The Rotation of a Solid
Body about a Fixed Point*
Sonya Kovalevskaya, Russia

Department of Agricultural
Economics established
at Texas A & M University

First domestication
of crops in
The Near East

Bluejeans, Levi Strauss
San Francisco, California

First commercial aerial
crop-dusting service
C. E. Woolman, Georgia

NASA Space Center
Houston, Texas

9000–8000 BC   AD 1776   1850   1888   1925   1950   1961

CHAPTER OBJECTIVES

In this chapter, you will:
- Solve systems of equations and inequalities.
- Define matrices.
- Add, subtract, and multiply matrices.
- Use linear programming procedures to solve problems.

CAREER GOAL: Agricultural Management

Ask Juan Porras what his career goal is, and you'll be presented with a flurry of options. He says that he is interested in getting into the management of a large agri-business corporation. Or, he may go to law school. Or, he may pursue a research position with the United States Dairy Association (USDA). Or, he may go into farm and ranch management. And oh, by the way, he is also interested in real estate.

Where did all of this ambition come from? Juan says it began with the family business. His father owns several farms and ranches. In his home county, Texas's Starr County, agriculture is big business. "I have always felt that agriculture is a basic profession," Juan says. "That is, no matter what else happens, we will always need agriculture."

In the summer of 1991, Juan did an internship with the Starr County Agricultural Extension agent. He says, "My internship gave me lots of hands-on experiences. Being in the classroom is entirely different from actually being out on the farm, applying what you learn in textbooks." One of the things that Juan has learned is how to use linear programming to maximize return. "For example, assume you know your available constraints in terms of land, labor, and capital. You then identify your potential agricultural activities, such as raising wheat and corn, or having a cow-calf operation, or growing and selling cotton. Linear programming helps you to determine the combination of your potential agricultural activities that will maximize your profit and minimize your costs."

In the future, Juan "would really like to help to expand agriculture in South Texas. The people here are quiet, good, hardworking individuals. If we could attract more agricultural industries to South Texas, there would be many more jobs, and the financial stability of the area would improve greatly."

For up-to-date information on agricultural economics, visit:
www.glencoe.com/sec/math/amc/mathnet

2-1A Graphing Calculators: Graphing Systems of Linear Equations

You can use a graphing calculator to graph and solve systems of linear equations since several equations can be graphed on the screen at the same time. If the graphs of the two equations intersect, then the system has a solution and the solution is the coordinates of the point of intersection of the two lines. These coordinates, (x, y), can be determined to a great degree of accuracy on the graphing calculator by using the TRACE and ZOOM functions.

Graph the system of equations $y = 3.4x + 2.1$ and $y = -5.1x + 8.3$ in the standard viewing window.

Enter: [Y=] 3.4 [X,T,θ] [+]

2.1 [ENTER] [(−)] 5.1 [X,T,θ] [+]

8.3 [GRAPH]

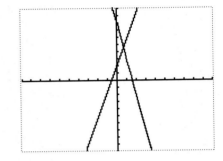

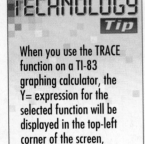

TECHNOLOGY
Tip

When you use the TRACE function on a TI-83 graphing calculator, the Y= expression for the selected function will be displayed in the top-left corner of the screen, unless the **ExprOff** format is selected.

Now use the trace function to determine the coordinates of the intersection point. Press [TRACE]. Then use the left and right arrow keys to move along the function to the intersection point. You can move from one function to another by pressing the up and down arrow keys.

A special "zoom" feature is very useful for determining the coordinates of a solution or point to a great degree of accuracy. With the blinking dot or cursor on the point of intersection, zoom in by pressing [ZOOM] 2 and [ENTER]. This magnifies a portion of the previous graphics screen and creates a new smaller screen with the coordinates of the cursor at the center. Checking your range now will help to illustrate what is happening.

You can continue zooming in until the coordinates of the point of intersection are found to the degree of accuracy that you desire or need. If you would like to zoom in faster, you can set the zoom factors of the calculator higher and zoom in on the point again. Setting the factors to 10 will allow you to zoom in fairly quickly.

Enter: [ZOOM] [▶] 4   10 [ENTER] 10 [ZOOM] 2 [ENTER]

A special function on the CALC menu will identify the point of intersection of two functions to the nearest hundred-thousandth. With the intersection in the viewing window, press [2nd] [CALC] 5.

The current graph and a prompt saying "First curve?" appears. Use the up or down arrow keys to select the first function and press ENTER. Use the same keys to answer the "Second curve?" prompt and press ENTER. Move the cursor to the point of intersection at the "Guess?" prompt to help the TI-82 find the correct intersection more quickly and press ENTER. The cursor automatically moves to the intersection, and the coordinates of that point are displayed at the bottom of the screen.

This process determines that the x- and y-coordinates of the point of intersection of the system $y = 3.4x + 2.1$ and $y = -5.1x + 8.3$ are $(0.73, 4.58)$ to the nearest hundredth.

Example

Graph the system of equations below and find the solution to the nearest hundredth.
$$y = 3.27x + 2.87$$
$$y = -2.55x - 4.58$$

Enter: Y= 3.27 X,T,θ + 2.87 ENTER (−) 2.55 X,T,θ − 4.58 GRAPH

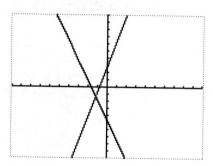

Enter: 2nd CALC 5 ENTER ENTER

Move the cursor to the point of intersection, then press ENTER.

The solution is $(-1.28, -1.32)$ to the nearest hundredth.

EXERCISES

Use a graphing calculator to solve the following systems of equations by graphing. Determine the coordinates of the solution to the nearest hundredth.

1. $y = 6x + 3$
$y = -3x - 4$

2. $y = 4x + 0.342$
$y = 0.5x - 2.905$

3. $y = 7.65x + 3.85$
$y = 4.957x + 1.492$

4. $y = 2.34x - 2$
$y = 5.39x - 14$

5. $4.9x + 0.3y = 1.6$
$-6.2x + 3.2y = -4.4$

6. $8.33x + 3.492y = 3.8$
$4.8x - 1.074y = 2.3$

2-1 Solving Systems of Equations

Objectives

After studying this lesson, you should be able to:
- solve systems of equations graphically, and
- solve systems of equations algebraically.

Application

Heather Washington, a recent mechanical engineering graduate, plans to start her own business manufacturing custom touring bicycle frames. She is planning to call her business Tour de Force, Inc. After careful research, she has decided that in order to be successful, she will need $12,500 for overhead (fixed costs like labor, rent, utilities, and so on), an additional $200 per frame for materials (variable cost), and in order to be competitive, a selling price of $700 per bicycle frame. How many frames must Heather sell in order to break even?

Let's write two equations to represent the constraints in the problem. The equation that represents the cost to produce the frames, called the *cost equation*, is $y = 200x + 12,500$. The equation that represents the sales price of the frames, called the *revenue equation*, is $y = 700x$.

FYI...

In 1990, there were 55.3 million bikers in the United States.

The intersection of the graphs of the cost and revenue equations is called the *break-even point*. The graph at the right shows the break-even point for Tour de Force, Inc. Thus, if Heather sells 25 bicycle frames, she will break even.

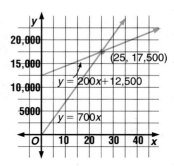

As you can see, graphing more than one linear equation on the same coordinate system has practical applications. If the graphs intersect, the ordered pair for the point of intersection is the solution to the **system of equations**.

Example 1

Solve the system of equations by graphing.
$$y = 5x - 2$$
$$y = -2x + 5$$

Since the two lines have different slopes, the graphs of the equations are intersecting lines.

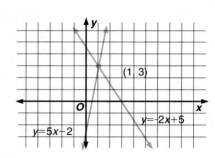

The solution of the system is (1, 3).

A **consistent** system of equations has at least one solution. If there is exactly one solution, the system is **independent**. If there are infinitely many solutions, the system is **dependent**.

Consider the equations $2y + 3x = 6$ and $4y + 6x = 12$. The equations written in slope-intercept form are $y = -\frac{3}{2}x + 3$ and $y = -\frac{3}{2}x + 3$. The graphs of these equations are the same line. Since any ordered pair on the graph satisfies both equations, there are infinitely many solutions to this system of equations.

Some systems of equations have no ordered pairs as the solution. Consider the equations $y = -3x - 2$ and $y = -3x + 3$. The graphs of these two equations are two lines with slopes of –3 and different y-intercepts. These lines are parallel and do not intersect. Therefore, there is no solution to this system of equations. This is called an **inconsistent** system of equations.

The chart below gives a summary of the possibilities for the graphs of two linear equations in two variables.

Graphs of Equations	Slopes of Lines	Type of System	Number of Solutions
lines intersect	different slopes	consistent and independent	one
lines coincide	same slope, same intercepts	consistent and dependent	infinitely many
lines parallel	same slope, different intercepts	inconsistent	none

Systems of linear equations may also be solved algebraically. Two ways of solving systems algebraically are the **elimination method** and the **substitution method**. In some situations, one method may be easier to use than the other.

Example 2

Use the elimination method to solve the system of equations.

$3x - 4y = 360$
$5x + 2y = 340$

To solve this system, first multiply the second equation by 2. Then add the two equations.

$$3x - 4y = 360$$
$$\underline{10x + 4y = 680}$$
$$13x \quad\;\; = 1040$$
$$x = 80$$

Now substitute 80 for x in either of the original equations.

$$5x + 2y = 340$$
$$5(80) + 2y = 340$$
$$y = -30$$

The solution is $(80, -30)$.

Example 3

Use the substitution method to solve the system of equations.
$$x = 7y + 3$$
$$2x - y = -7$$

The first equation is stated in terms of x, so substitute $7y + 3$ for x in the second equation.

$$2x - y = -7$$
$$2(7y + 3) - y = -7$$
$$13y = -13$$
$$y = -1$$

Now solve for x by substituting -1 for y in either of the original equations.

$$x = 7y + 3$$
$$x = 7(-1) + 3$$
$$x = -4$$

The solution is $(-4, -1)$.

You can also use elimination and substitution to solve real-world problems.

Example 4

APPLICATION

Sales

HomeMade Toys manufactures solid pine trucks and cars and usually sells four times as many trucks as cars. The net profit from each truck is $6 and from each car, $5. If the company wants a total profit of $29,000, how many trucks and cars should they sell?

Let t represent the number of trucks sold.
Let c represent the number of cars sold.
Because they usually sell four times as many trucks as cars, $t = 4c$.
The total profit can be represented by the equation $6t + 5c = 29{,}000$.

The first equation is stated in terms of t, so substitute $4c$ for t in the second equation.

$$6t + 5c = 29{,}000$$
$$6(4c) + 5c = 29{,}000$$
$$29c = 29{,}000$$
$$c = 1000$$

Now solve for w by substituting 1000 for c in either of the original equations.

$$t = 4c$$
$$= 4(1000)$$
$$= 4000$$

HomeMade Toys must sell 4000 trucks and 1000 cars to make a total profit of $29,000.

You can use a graphing calculator to show solutions of systems of equations.

EXPLORATION: Graphing Calculator

Suppose the cost function for producing a widget is $y = 3x + \$80$, and the revenue function is $y = 8x$.

1. Complete the table below to find when cost is equal to revenue.

Number of Widgets	Cost	Revenue
4	$3(4) + 80 = 92$	$8(4) = 32$
8	$3(8) + 80 = 104$	$8(8) = 64$
12	?	?
16	?	?
20	?	?

M A T H
JOURNAL

Describe the three different possibilities that may occur when graphing a system of two linear equations. What types of solutions occur with each possibility?

2. Graph the two functions.
   Make sure you use appropriate range values.

3. What is the relationship between the graph and the results in the table?

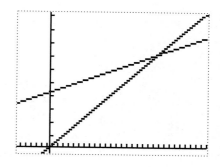

CHECKING FOR UNDERSTANDING

Communicating Mathematics

Read and study the lesson to answer each question.

1. **Describe** the circumstances under which it would be easier to solve a system of equations by using the substitution method rather than the elimination method.

2. What is the name for the point of intersection of the graph of a revenue equation and the graph of a cost equation?

3. When is a system of equations inconsistent?

4. **Describe**, as you would to a friend, how to solve a system of equations by using the elimination method.

Guided Practice

A possible solution is given for each system of equations. Is it the correct solution? Write *yes* or *no*.

5. $\left(\frac{1}{2}, 6\right)$
   $4x + y = 8$
   $6x - 2y = -9$

6. $\left(\frac{1}{3}, \frac{2}{3}\right)$
   $2x + 3y = 3$
   $12x - 15y = -4$

7. $(2, 5)$
   $-3x + 10y = 5$
   $2x + 7y = 24$

8. $(4, 3)$
   $\frac{1}{2}x + 5y = 17$
   $3x + 2y = 18$

9. $(-6, -4)$
   $\frac{1}{3}x - \frac{3}{2}y = -4$
   $5x - 4y = 14$

10. $(4, 7)$
    $1.5x + 2y = 20$
    $2.5x - 5y = -25$

11. Solve the system of equations by graphing.
$$3x - y - 1 = 0$$
$$x + y - 3 = 0$$

12. Solve the system of equations algebraically.
$$5x - y = 16$$
$$2x + 3y = 3$$

EXERCISES

Practice

State whether each system is *consistent and independent*, *consistent and dependent*, or *inconsistent*.

13. $5x - 7y = 70$
    $-10x + 14y = 120$

14. $7x - 8y = -11$
    $-35x + 40y = 55$

15. $3x - 8y = 10$
    $16x - 32y = 75$

Solve each system of equations by graphing.

16. $x = 0$
    $y = 1$

17. $x = 0$
    $4x + 5y = 20$

18. $x + 4y = 12$
    $3x - 2y = -6$

19. $3x - 2y = -6$
    $x + y = -2$

20. $x + y = -2$
    $3x - y = 10$

21. $3x - y = 10$
    $x + 4y = 12$

Solve each system of equations algebraically.

22. $x + 2y = 1$
    $3x - 5y = -8$

23. $2x - y = -7$
    $x = 2y - 8$

24. $3x + 5y = 7$
    $2x - 6y = 11$

25. $3x + 2y = 40$
    $x - 7y = -2$

26. $2x + 3y = 8$
    $x - y = 2$

27. $x + y = 6$
    $x = y + 4.5$

Programming

28. The following program will solve a system of two linear equations written in the form $Ax + By = C$ and $Dx + Ey = F$.

```
Prgm 2: SLVSYSTM
:Disp "FIRST X-COEFF="
:Input A
:Disp "FIRST Y-COEFF="
:Input B
:Disp "FIRST CONSTANT="
:Input C
:Disp "SECOND X-COEFF="
:Input D
:Disp "SECOND Y-COEFF="
:Input E
:Disp "SECOND CONSTANT="
:Input F
:If AE-BD = 0
:Goto 1
```

Displays the quoted message.
Accepts a value for A.

Accepts a value for B.

Accepts a value for C.

Accepts a value for D.

Accepts a value for E.

Accepts a value for F.
If the determinant of coefficients is 0,
program will go to line "Lbl 1."

```
:(CE-BF)/(AE-BD)→ X          Value is calculated and stored for X.
:Disp "X="
:Disp X                      Displays the value of X.
:(AF-CD)/(AE-BD)→ Y          Value is calculated and stored for Y.
:Disp "Y="
:Disp Y
:Goto 2                      Program will go to line "Lbl 2."
:Lbl 1
:Disp "INCONSISTENT OR DEPENDENT"
:Lbl 2
```

Use the program above to solve each system of equations.

a. $x - y = 1$
$ 3x - y = 3$

b. $x + y = 6$
$ 3x + 3y = 3$

c. $a - b = 0$
$ 3a + 2b = -15$

Critical Thinking

29. Find the coordinates of the vertices of the triangle with sides determined by the graphs of the following equations: $4x + 3y + 1 = 0, 4x - 3y - 17 = 0$, and $4x - 9y + 13 = 0$.

Applications and Problem Solving

30. Sales The Cotton Club, a small manufacturer of sweatshirts, has fixed costs of $2000 and a variable cost of $5 per sweatshirt. They sell their sweatshirts for $15 apiece.

a. Find their break-even point.

b. What is the least number of sweatshirts they must sell to make a profit?

31. Agriculture Mrs. Griffin wants to plant soybeans and corn on 100 acres of land. Soybeans require 6 hours of labor per acre, and corn requires 8 hours of labor per acre. If Mrs. Griffin has 660 hours available, how many acres of each crop should she plant?

32. Real Estate AMC Homes, Inc. is planning to build three- and four-bedroom homes in a housing development called Chestnut Hills. Public demand indicates a need for three times as many three-bedroom homes as four-bedroom homes. The net profit from each three-bedroom home is $6000 and from each four-bedroom home, $7000. If AMC Homes must net a total profit of $2,500,000 from this development, how many of each type of home should they build?

Mixed Review

33. State the domain and range of the relation $\{(16, -4), (16, 4)\}$. Is this relation a function? **(Lesson 1-1)**

34. Of $(0, 0), (3, 2), (-4, 2)$, or $(-2, 4)$, which satisfy $x + y \geq 3$? **(Lesson 1-3)**

35. Write the slope-intercept form of the equation of the line through $(1, 4)$ and $(5, 7)$. **(Lesson 1-5)**

36. Show that the points with coordinates $(-1, 3), (3, 6), (6, 2)$, and $(2, -1)$ are the vertices of a square. **(Lesson 1-6)**

37. College Entrance Exam Choose the best answer.
How old was a person exactly 1 year ago if exactly x years ago the person was y years old?

(A) $y - 1$ **(B)** $y - x - 1$ **(C)** $x - y - 1$ **(D)** $y + x - 1$ **(E)** $x - 1$

2-2A Graphing Calculators: Matrices

A graphing calculator can help you with many algebraic and mathematical tasks. One of these is performing operations with matrices. The TI-82 can find determinants and inverses of matrices, as well as perform operations with matrices.

The $\boxed{\text{MATRX}}$ key accesses the matrix operations menus. There are three: the NAMES menu, the MATH menu, and the EDIT menu. The first to appear is the NAMES menu, which lists the matrix locations available. The MATH menu lists the matrix functions available. The EDIT menu allows you to define matrices. You can access the EDIT menu by using the left or right arrow key ($\boxed{\blacktriangleleft}$ or $\boxed{\blacktriangleright}$). This menu displays the dimensions of the matrices. A matrix with a dimension of 2×3 indicates a matrix with 2 rows and 3 columns. The TI-82 will accommodate a maximum of 99 rows and 30 columns or 30 rows and 99 columns in a matrix.

To enter a matrix into your calculator, choose the EDIT menu and select matrix [A]. Then enter the dimensions and elements of the matrix.

Example 1

Enter matrix $A = \begin{bmatrix} 1 & 3 \\ 2 & -2 \end{bmatrix}$ on a TI-82 graphing calculator.

Enter: $\boxed{\text{MATRX}}$ $\boxed{\blacktriangleleft}$ $\boxed{\text{ENTER}}$ 2 $\boxed{\text{ENTER}}$ 2 $\boxed{\text{ENTER}}$
1 $\boxed{\text{ENTER}}$ 3 $\boxed{\text{ENTER}}$ 2 $\boxed{\text{ENTER}}$ $\boxed{(-)}$ 2

You can display the matrix by pressing $\boxed{\text{2nd}}$ $\boxed{\text{QUIT}}$ to return to the text screen, then $\boxed{\text{MATRX}}$ $\boxed{\text{ENTER}}$ $\boxed{\text{ENTER}}$ to display the matrix.

The TI-82 graphing calculator can be used to find the determinant and the inverse of a matrix.

Example 2

Find the determinant and the inverse of matrix A.

First, find the determinant.

Enter: $\boxed{\text{MATRX}}$ $\boxed{\blacktriangleright}$ 1 $\boxed{\text{MATRX}}$ 1 $\boxed{\text{ENTER}}$ *-8*

Then, find the inverse.

Enter: $\boxed{\text{MATRX}}$ 1 $\boxed{x^{-1}}$ $\boxed{\text{ENTER}}$ $\begin{bmatrix} 0.25 & 0.375 \\ 0.25 & -0.125 \end{bmatrix}$

The determinant of matrix A is –8 and the inverse is $\begin{bmatrix} 0.25 & 0.375 \\ 0.25 & -0.125 \end{bmatrix}$.

You can also use a graphing calculator to perform operations on matrices.

Example 3

Enter matrix $B = \begin{bmatrix} 2 & 3 & 6 \\ 4 & -8 & 5 \end{bmatrix}$. **Then find** $\frac{1}{2}B$, AB, A^2, **and** $AB + B$.

First enter matrix B.

Enter: $\boxed{\text{MATRX}}$ $\boxed{\blacktriangleleft}$ 2 2 $\boxed{\text{ENTER}}$ 3 $\boxed{\text{ENTER}}$ 2 $\boxed{\text{ENTER}}$
       3 $\boxed{\text{ENTER}}$ 6 $\boxed{\text{ENTER}}$ 4 $\boxed{\text{ENTER}}$ $\boxed{(-)}$ 8 $\boxed{\text{ENTER}}$ 5
       $\boxed{\text{2nd}}$ $\boxed{\text{QUIT}}$

Find $\frac{1}{2}B$.

Enter: .5 $\boxed{\text{MATRX}}$ 2 $\boxed{\text{ENTER}}$ $\begin{bmatrix} 1 & 1.5 & 3 \\ 2 & -4 & 2.5 \end{bmatrix}$

Then find AB.

Enter: $\boxed{\text{MATRX}}$ 1 $\boxed{\text{MATRX}}$ 2 $\boxed{\text{ENTER}}$ $\begin{bmatrix} 14 & -21 & 21 \\ -4 & 22 & 2 \end{bmatrix}$

Next find A^2.

Enter: $\boxed{\text{MATRX}}$ 1 $\boxed{x^2}$ $\boxed{\text{ENTER}}$ $\begin{bmatrix} 7 & -3 \\ -2 & 10 \end{bmatrix}$

Now find $AB + B$.

Enter: $\boxed{\text{MATRX}}$ 1 $\boxed{\text{MATRX}}$ 2 $\boxed{+}$ $\boxed{\text{MATRX}}$ 2 $\boxed{\text{ENTER}}$ $\begin{bmatrix} 16 & -18 & 27 \\ 0 & 14 & 7 \end{bmatrix}$

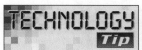

The TI-83 and TI-92 graphing calculators can create a matrix from elements stored in a list.

Therefore, $\frac{1}{2}B = \begin{bmatrix} 1 & 1.5 & 3 \\ 2 & -4 & 2.5 \end{bmatrix}$, $AB = \begin{bmatrix} 14 & -21 & 21 \\ -4 & 22 & 2 \end{bmatrix}$, $A^2 = \begin{bmatrix} 7 & -3 \\ -2 & 10 \end{bmatrix}$,

and $AB + B = \begin{bmatrix} 16 & -18 & 27 \\ 0 & 14 & 7 \end{bmatrix}$.

EXERCISES

Enter the matrices into a graphing calculator in the appropriate locations. Then find each of the following.

$$A = \begin{bmatrix} 5 & 2 & 4 \\ 1 & -4 & -3 \\ 0 & 6 & 3 \end{bmatrix} \qquad B = \begin{bmatrix} 2 & 8 \\ 7 & -5 \\ 2 & 1 \end{bmatrix} \qquad C = \begin{bmatrix} 0 & 3 & 1 \\ 7 & -5 & 6 \end{bmatrix}$$

1. $-2A$

2. determinant of A

3. A^{-1}

4. $3B$

5. AB

6. BC

7. determinant of BC

8. CB

9. determinant of CB

10. CA

11. $C + CA$

12. $BC - A$

13. $(CB)^2$

14. $(CB)^{-1}$

15. $A + BC$

16. ABC

17. $B + AB$

18. $4BC$

2-2 Introduction to Matrices

Objective

After studying this lesson, you should be able to:
- add, subtract, and multiply matrices.

Application

Edwin Yunker, the father of one of the authors of this textbook, is a grain farmer in Mokena, Illinois. Mr. Yunker sold most of his fall harvest for 1992 on the grain futures market for delivery in the summer of 1993. He scheduled his delivery of corn, soybeans, and oats (in bushels) on a wall chart as shown below.

$$\begin{array}{c} \\ \text{June} \\ \text{July} \\ \text{August} \end{array} \begin{array}{ccc} \text{corn} & \text{soybeans} & \text{oats} \\ \left[\begin{array}{ccc} 15,000 & 2000 & 500 \\ 13,500 & 6000 & 1000 \\ 14,000 & 5500 & 1500 \end{array}\right] \end{array}$$

This display is called a **matrix**. A matrix is any rectangular array of terms called **elements**. The elements of a matrix are arranged in rows and columns and are usually enclosed by brackets. A matrix with m rows and n columns is an $\boldsymbol{m \times n}$ **matrix** (read "m by n"). The **dimensions** of the matrix are m and n.

$$\begin{bmatrix} -2 & 3 \\ 11 & -4 \end{bmatrix}$$

2 × 2 matrix

$$\begin{bmatrix} 7 & \frac{1}{2} & 4 & 12 \\ -3 & 6 & 15 & -2 \end{bmatrix}$$

2 × 4 matrix

$$\begin{bmatrix} 2 & -8 \\ 0 & 5 \\ -3 & 6 \end{bmatrix}$$

3 × 2 matrix

Certain matrices have special names. A matrix that has only one row is called a **row matrix**, and a matrix that has only one column is called a **column matrix**. A **square matrix** has the same number of rows as columns. Sometimes square matrices are called matrices of $\boldsymbol{n}$**th order**, where n is the number of rows and columns. The elements of an $m \times n$ matrix can be represented using double subscript notation, as shown below.

$$\begin{bmatrix} a_{11} & a_{12} & a_{13} & \dots & a_{1n} \\ a_{21} & a_{22} & a_{23} & \dots & a_{2n} \\ \dots & \dots & \dots & \dots & \dots \\ a_{m1} & a_{m2} & a_{m3} & \dots & a_{mn} \end{bmatrix}$$

a_{ij} *is the element in the ith row and the jth column.*

Just as with numbers or algebraic expressions, matrices are equal under certain conditions.

Definition of Equal Matrices	Two matrices are equal if and only if they have the same dimensions and are identical, element by element.

Example 1

Find the values of x and y for which the following equation is true.

$$\begin{bmatrix} y - 3 \\ y \end{bmatrix} = \begin{bmatrix} x \\ 2x \end{bmatrix}$$

Since corresponding elements must be equal, the following equations are true.

$$y - 3 = x$$
$$y = 2x$$

Solve the system of equations.

$2x - 3 = x$ *Substitute 2x for y in the first equation.*

$\quad\quad x = 3$ *Solve for x.*

$y = 2(3) \text{ or } 6$ *Substitute 3 for x in the second equation to find y.*

The matrices are equal if $x = 3$ and $y = 6$.

Matrices with the same dimensions can be added. The ijth element of the sum of matrices A and B is $a_{ij} + b_{ij}$.

Definition of Addition of Matrices	The sum of two $m \times n$ matrices is an $m \times n$ matrix in which the elements are the sum of the corresponding elements of the given matrices.

Example 2

Find $A + B$ if $A = \begin{bmatrix} 3 & 5 & -7 \\ -1 & 0 & 4 \end{bmatrix}$ and $B = \begin{bmatrix} -2 & 8 & 6 \\ 5 & -9 & 10 \end{bmatrix}$.

$$A + B = \begin{bmatrix} 3 + (-2) & 5 + 8 & -7 + 6 \\ -1 + 5 & 0 + (-9) & 4 + 10 \end{bmatrix} = \begin{bmatrix} 1 & 13 & -1 \\ 4 & -9 & 14 \end{bmatrix}$$

Matrices, like real numbers, have identity elements. For every matrix A, another matrix can be found such that their sum is A. For example, if $A = \begin{bmatrix} a_{11} & a_{12} \\ a_{21} & a_{22} \end{bmatrix}$, then $\begin{bmatrix} a_{11} & a_{12} \\ a_{21} & a_{22} \end{bmatrix} + \begin{bmatrix} 0 & 0 \\ 0 & 0 \end{bmatrix} = \begin{bmatrix} a_{11} & a_{12} \\ a_{21} & a_{22} \end{bmatrix}$. The matrix $\begin{bmatrix} 0 & 0 \\ 0 & 0 \end{bmatrix}$ is called a **zero matrix**. Thus, the identity matrix for addition for any $m \times n$ matrix is the $m \times n$ zero matrix.

Matrices also have additive inverses. If $A = \begin{bmatrix} a_{11} & a_{12} \\ a_{21} & a_{22} \end{bmatrix}$, the matrix that

must be added to A in order to obtain a zero matrix is $\begin{bmatrix} -a_{11} & -a_{12} \\ -a_{21} & -a_{22} \end{bmatrix}$ or $-A$.

Therefore, $-A$ is the additive inverse of A. The additive inverse is used when subtracting matrices.

Definition of Subtraction of Matrices	The difference $A - B$ of two $m \times n$ matrices is equal to the sum $A + (-B)$, where $-B$ represents the additive inverse of B.

Example 3

Find $A - B$ if $A = \begin{bmatrix} 3 & 6 \\ -2 & 4 \end{bmatrix}$ and $B = \begin{bmatrix} 1 & 5 \\ -7 & 8 \end{bmatrix}$.

$A - B = A + (-B)$

$$= \begin{bmatrix} 3 & 6 \\ -2 & 4 \end{bmatrix} + \begin{bmatrix} -1 & -5 \\ 7 & -8 \end{bmatrix}$$

$$= \begin{bmatrix} 3 + (-1) & 6 + (-5) \\ -2 + 7 & 4 + (-8) \end{bmatrix}$$

$$= \begin{bmatrix} 2 & 1 \\ 5 & -4 \end{bmatrix}$$

A matrix can be multiplied by a number called a **scalar.** The product of a scalar k and a matrix A is defined as follows.

Scalar Product	The product of a scalar k and an $m \times n$ matrix A is an $m \times n$ matrix denoted by kA. Each element of kA is equal to k times the corresponding element of A.

Example 4

Multiply $\begin{bmatrix} 5 & 3 & -1 \\ -2 & 8 & -9 \end{bmatrix}$ by 4.

$$4 \begin{bmatrix} 5 & 3 & -1 \\ -2 & 8 & -9 \end{bmatrix} = \begin{bmatrix} 4(5) & 4(3) & 4(-1) \\ 4(-2) & 4(8) & 4(-9) \end{bmatrix}$$

$$= \begin{bmatrix} 20 & 12 & -4 \\ -8 & 32 & -36 \end{bmatrix}$$

A matrix can also be multiplied by another matrix, provided that the first matrix has the same number of columns as the second matrix has rows. The product of the two matrices is found by multiplying rows and columns.

Suppose $A = \begin{bmatrix} a_1 & b_1 \\ a_2 & b_2 \end{bmatrix}$ and $X = \begin{bmatrix} x_1 & y_1 \\ x_2 & y_2 \end{bmatrix}$. Each element of matrix AX is the product of one row of matrix A and one column of matrix X.

$$AX = \begin{bmatrix} a_1 & b_1 \\ a_2 & b_2 \end{bmatrix}\begin{bmatrix} x_1 & y_1 \\ x_2 & y_2 \end{bmatrix} = \begin{bmatrix} a_1x_1 + b_1x_2 & a_1y_1 + b_1y_2 \\ a_2x_1 + b_2x_2 & a_2y_1 + b_2y_2 \end{bmatrix}$$

In general, the product of two matrices is defined as follows.

Product of Two Matrices	**The product of an $m \times n$ matrix A and an $n \times r$ matrix B is an $m \times r$ matrix AB. The ijth element of AB is the sum of the products of the corresponding elements in the ith row of A and the jth column of B.**

Example 5

Find AB if $A = \begin{bmatrix} 4 & 3 \\ 5 & 2 \end{bmatrix}$ and $B = \begin{bmatrix} 8 & 0 \\ 9 & -6 \end{bmatrix}$.

$$AB = \begin{bmatrix} 4(8) + 3(9) & 4(0) + 3(-6) \\ 5(8) + 2(9) & 5(0) + 2(-6) \end{bmatrix} = \begin{bmatrix} 59 & -18 \\ 58 & -12 \end{bmatrix}$$

Example 6

APPLICATION

Agriculture

Refer to the application at the beginning of the lesson. If Mr. Yunker is paid for his delivery of grain at the end of the month, how much will he receive each month if the prices per bushel are as follows: corn, \$2.85, soybeans, \$6.10, and oats, \$1.85?

Multiply the delivery matrix by a column matrix that contains the grain prices.

$$\begin{matrix} \text{June} \\ \text{July} \\ \text{August} \end{matrix} \begin{bmatrix} 15{,}000 & 2000 & 500 \\ 13{,}500 & 6000 & 1000 \\ 14{,}000 & 5500 & 1500 \end{bmatrix} \times \begin{bmatrix} 2.85 \\ 6.10 \\ 1.85 \end{bmatrix}$$

$$= \begin{bmatrix} 15{,}000(2.85) + 2000(6.10) + 500(1.85) \\ 13{,}500(2.85) + 6000(6.10) + 1000(1.85) \\ 14{,}000(2.85) + 5500(6.10) + 1500(1.85) \end{bmatrix}$$

$$= \begin{bmatrix} 55{,}875 \\ 76{,}925 \\ 76{,}225 \end{bmatrix}$$

inter NET CONNECTION

For the latest grain prices, visit:
www.glencoe.com/sec/math/amc/mathnet

Notice that the product has as many rows as the first matrix and as many columns as the second matrix. The resulting 3×1 matrix shows that Mr. Yunker will receive \$55,875 for grain delivered in June, \$76,925 for grain delivered in July, and \$76,225 for grain delivered in August.

CHECKING FOR UNDERSTANDING

Communicating Mathematics

Read and study the lesson to answer each question.

1. **Define** the dimensions of a matrix.

2. **Write** the element in the fifth column and the fourth row of matrix B using double subscript notation.

3. Give an example of a matrix of 5th order.

4. **Write** two 3×4 matrices that are additive inverses of each other. Explain how you know that they are additive inverses.

Guided Practice

Use matrices A, B, and C to find each of the following.

$$A = \begin{bmatrix} 3 & 0 \\ -1 & 5 \end{bmatrix} \qquad B = \begin{bmatrix} 5 & 1 \\ -3 & 4 \end{bmatrix} \qquad C = \begin{bmatrix} 3 \\ -1 \end{bmatrix}$$

5. $A + B$
6. $A - B$
7. $B - A$

8. $5C$
9. $-3B$
10. AC

11. AB
12. ABC
13. BB

Find the values of x and y for which each matrix equation is true.

14. $\begin{bmatrix} y \\ x \end{bmatrix} = \begin{bmatrix} 2x - 6 \\ 2y \end{bmatrix}$

15. $\begin{bmatrix} 5x - 3y \\ 3y \end{bmatrix} = \begin{bmatrix} 8 \\ 27 \end{bmatrix}$

EXERCISES

Practice

Use matrices A, B, and C to find each sum or difference.

$$A = \begin{bmatrix} 1 & 5 & 7 \\ 5 & 2 & -6 \\ 3 & 0 & -2 \end{bmatrix} \qquad B = \begin{bmatrix} -3 & 6 & -9 \\ 4 & -3 & 0 \\ 8 & -2 & 3 \end{bmatrix} \qquad C = \begin{bmatrix} 6 & 9 & -4 \\ -11 & 13 & -8 \\ 20 & 4 & -2 \end{bmatrix}$$

16. $A + B$
17. $A + C$
18. $B + C$

19. $(A + B) + C$
20. $B + (-A)$
21. $C - B$

22. $B - C$
23. $C - A$
24. $B - A$

Find the values of x and y for which each matrix equation is true.

25. $\begin{bmatrix} x & 2y \end{bmatrix} = \begin{bmatrix} y + 5 & x - 3 \end{bmatrix}$

26. $\begin{bmatrix} 5 & 4x \end{bmatrix} = \begin{bmatrix} 2x & 5y \end{bmatrix}$

27. $\begin{bmatrix} 2x \\ 0 \\ 16 \end{bmatrix} = \begin{bmatrix} 8 - y \\ y \\ 4x \end{bmatrix}$

28. $\begin{bmatrix} y \\ 8x \end{bmatrix} = \begin{bmatrix} 15 + x \\ 2y \end{bmatrix}$

Use matrices D, E, and F to find each product.

$$D = \begin{bmatrix} 7 & 0 \\ 5 & 3 \end{bmatrix} \qquad E = \begin{bmatrix} 2 & 4 \\ 8 & -4 \\ -2 & 6 \end{bmatrix} \qquad F = \begin{bmatrix} 3 & -3 & 6 \\ 5 & 4 & -2 \end{bmatrix}$$

29. $3D$

30. $4E$

31. $2F$

32. $-5D$

33. ED

34. EF

35. FE

36. DF

37. DD

38. $(FE)D$

39. $E(DF)$

40. $-4EF$

41. Find $2A - 3B$ if $A = \begin{bmatrix} 1 & -7 \\ 3 & 2 \end{bmatrix}$ and $B = \begin{bmatrix} -4 & 5 \\ 1 & -1 \end{bmatrix}$.

Prove or disprove each statement.

42. Addition of 2×2 matrices is commutative.

43. Addition of 2×2 matrices is associative.

44. Multiplication of two second-order matrices is commutative.

45. Multiplication of three second-order matrices is associative.

Critical Thinking

46. All integers n, such that $n \geq 0$, except 1, are arranged in a five-column matrix as shown at the right. In which column will the number 1992 be found?

$$\begin{bmatrix} 0 & 2 & 3 & 4 & 5 \\ 9 & 8 & 7 & 6 & 0 \\ 0 & 10 & 11 & 12 & 13 \\ 17 & 16 & 15 & 14 & 0 \\ 0 & 18 & 19 & 20 & 21 \\ \ldots & \ldots & \ldots & \ldots & 0 \end{bmatrix}$$

Applications and Problem Solving

47. **Entertainment** On the opening weekends for *The Flintstones*, *Forrest Gump*, and *Batman Returns*, the Foxfield Theater reported the following ticket sales.

	Adults	Children
The Flintstones	1021	523
Forrest Gump	2547	785
Batman Returns	3652	2456

If the ticket prices were $6 for each adult and $4 for each child, what were the weekend sales for each movie?

48. **Geometry** Scalar multiplication can be used to find the coordinates of the vertices of a geometric figure after the figure has been reduced or enlarged. This type of transformation is called a **dilation**. Find the vertices of $\triangle DEF$, if the perimeter of $\triangle DEF$ is three times the perimeter of $\triangle ABC$, and the vertices of $\triangle ABC$ are $A(2, 5)$, $B(-4, 5)$, and $C(3, -7)$. *Hint: Put the coordinates in a 3×2 matrix.*

49. Find $[f \circ g](x)$ and $[g \circ f](x)$ if $f(x) = x^2 + 3x + 2$ and $g(x) = x - 1$. **(Lesson 1-2)**

50. Graph $y < -2x + 8$. **(Lesson 1-3)**

51. Find the distance between points at $(2t, t)$ and $(5t, 5t)$. Then, find the slope of the line passing between these points. **(Lesson 1-4)**

52. Write the slope-intercept form of the equation of the line through $(1, 5)$ that has a slope of -2. **(Lesson 1-5)**

53. Solve the system of equations. **(Lesson 2-1)**
$$3x + 4y = 375$$
$$5x + 2y = 345$$

54. College Entrance Exam Choose the best answer.
If $xy = 1$, then x is the reciprocal of y. Which of the following equals the arithmetic mean of x and y?

(A) $\dfrac{y^2 + 1}{2y}$ **(B)** $\dfrac{y + 1}{2y}$ **(C)** $\dfrac{y^2 + 2}{2y}$ **(D)** $\dfrac{y^2 + 1}{y}$ **(E)** $\dfrac{x^2 + 1}{y}$

DECISION MAKING

Choosing a College Major

Which college is best for you? There are more than 3600 colleges and junior colleges in the United States, so you have a wide range of options open to you. Would you like a small, intimate college in the east, or a large, bustling university in the midwest? Are there schools with special programs in the field you wish to pursue? How much financial aid is available to you? There are a number of things to consider.

1. Describe a college you think you would like to attend. Include such factors as size, location, proximity to your home, academic credentials, and any other factors you think are important.

2. Go to the library and find a book with descriptions of colleges. Make a list of schools that fit the description that you wrote for Exercise 1.

3. Project Work with your group to complete one of the projects listed at the right, based on the information you gathered in Exercises 1–2 above.

PROJECTS

- *Conduct an interview.*
- *Write a book report.*
- *Write a proposal.*
- *Write an article for the school paper.*
- *Write a report.*
- *Make a display.*
- *Make a graph or chart.*
- *Plan an activity.*
- *Design a checklist.*

2-3 Determinants and Multiplicative Inverses of Matrices

Objectives

After studying this lesson, you should be able to:
- evaluate determinants,
- find inverses of matrices, and
- solve systems of equations by using inverses of matrices.

Application

To make 20 kilograms of aluminum alloy with 70% aluminum, Kim Troy, a metallurgist, wants to use two metals with 55% and 80% aluminum content. How much of each metal should she use? *This problem will be solved in Example 5.*

The term determinant is often used to mean the value of the determinant.

Each square matrix has a **determinant**. The determinant of $\begin{bmatrix} 8 & 7 \\ 4 & 5 \end{bmatrix}$ is denoted by $\begin{vmatrix} 8 & 7 \\ 4 & 5 \end{vmatrix}$ or $\det \begin{bmatrix} 8 & 7 \\ 4 & 5 \end{bmatrix}$. The value of a second-order determinant is a number and is defined as follows.

Second-Order Determinant	The value of $\det \begin{bmatrix} a_1 & b_1 \\ a_2 & b_2 \end{bmatrix}$ is $\begin{vmatrix} a_1 & b_1 \\ a_2 & b_2 \end{vmatrix} = a_1 b_2 - a_2 b_1.$

A matrix that has a nonzero determinant is called a nonsingular matrix.

Example 1

Find the value of $\begin{vmatrix} 7 & 9 \\ 3 & 6 \end{vmatrix}$.

$$\begin{vmatrix} 7 & 9 \\ 3 & 6 \end{vmatrix} = 7(6) - 3(9)$$
$$= 42 - 27$$
$$= 15$$

The **minor** of an element of an nth-order determinant is a determinant of $(n-1)$th order. This minor can be found by deleting the row and column containing the element.

$$\begin{vmatrix} a_1 & b_1 & c_1 \\ a_2 & b_2 & c_2 \\ a_3 & b_3 & c_3 \end{vmatrix} \qquad \textit{The minor of } a_1 \textit{ is } \begin{vmatrix} b_2 & c_2 \\ b_3 & c_3 \end{vmatrix}.$$

To evaluate a determinant of the nth order, expand the determinant by minors, using the elements in one of the rows. Find the product of each term in the row and its respective minor, and then add. The signs of the terms alternate. If the first row is used, the first term will be positive, as shown on the next page.

Third-Order Determinant	$$\begin{vmatrix} a_1 & b_1 & c_1 \\ a_2 & b_2 & c_2 \\ a_3 & b_3 & c_3 \end{vmatrix} = a_1 \begin{vmatrix} b_2 & c_2 \\ b_3 & c_3 \end{vmatrix} - b_1 \begin{vmatrix} a_2 & c_2 \\ a_3 & c_3 \end{vmatrix} + c_1 \begin{vmatrix} a_2 & b_2 \\ a_3 & b_3 \end{vmatrix}$$

Example 2   **Find the value of** $\begin{vmatrix} 8 & 9 & 3 \\ 3 & 5 & 7 \\ -1 & 2 & 4 \end{vmatrix}$.

$$\begin{vmatrix} 8 & 9 & 3 \\ 3 & 5 & 7 \\ -1 & 2 & 4 \end{vmatrix} = 8 \begin{vmatrix} 5 & 7 \\ 2 & 4 \end{vmatrix} - 9 \begin{vmatrix} 3 & 7 \\ -1 & 4 \end{vmatrix} + 3 \begin{vmatrix} 3 & 5 \\ -1 & 2 \end{vmatrix}$$
$$= 8(6) - 9(19) + 3(11)$$
$$= -90$$

The identity matrix for multiplication for any square matrix A is the matrix, I, such that $IA = A$ and $AI = A$. A second-order matrix can be represented by

For any $m \times m$ matrix, the identity matrix, I, must be $m \times m$.

$\begin{bmatrix} a_1 & b_1 \\ a_2 & b_2 \end{bmatrix}$. Since $\begin{bmatrix} a_1 & b_1 \\ a_2 & b_2 \end{bmatrix} \cdot \begin{bmatrix} 1 & 0 \\ 0 & 1 \end{bmatrix} = \begin{bmatrix} 1 & 0 \\ 0 & 1 \end{bmatrix} \cdot \begin{bmatrix} a_1 & b_1 \\ a_2 & b_2 \end{bmatrix} = \begin{bmatrix} a_1 & b_1 \\ a_2 & b_2 \end{bmatrix}$, the

matrix $\begin{bmatrix} 1 & 0 \\ 0 & 1 \end{bmatrix}$ is the identity matrix for multiplication for any second-order matrix.

Identity Matrix under Multiplication	**The identity matrix of nth order, I_n, is the square matrix whose elements in the main diagonal, from upper left to lower right, are 1s, while all other elements are 0s.**

Multiplicative inverses exist for some matrices. Suppose A is equal to $\begin{bmatrix} a_1 & b_1 \\ a_2 & b_2 \end{bmatrix}$, a nonzero matrix of second order. The inverse matrix A^{-1} can be designated as $\begin{bmatrix} x_1 & y_1 \\ x_2 & y_2 \end{bmatrix}$. The product of a matrix A and its inverse A^{-1} must equal the identity matrix, I, for multiplication.

$$\begin{bmatrix} a_1 & b_1 \\ a_2 & b_2 \end{bmatrix} \begin{bmatrix} x_1 & y_1 \\ x_2 & y_2 \end{bmatrix} = \begin{bmatrix} 1 & 0 \\ 0 & 1 \end{bmatrix}$$
$$\begin{bmatrix} a_1x_1 + b_1x_2 & a_1y_1 + b_1y_2 \\ a_2x_1 + b_2x_2 & a_2y_1 + b_2y_2 \end{bmatrix} = \begin{bmatrix} 1 & 0 \\ 0 & 1 \end{bmatrix}$$

From the previous matrix equation, two systems of linear equations can be written as follows.

$$a_1x_1 + b_1x_2 = 1 \qquad\qquad a_1y_1 + b_1y_2 = 0$$
$$a_2x_1 + b_2x_2 = 0 \qquad\qquad a_2y_1 + b_2y_2 = 1$$

By solving each system of equations, values for x_1, x_2, y_1, and y_2 can be obtained.

$$x_1 = \frac{b_2}{a_1 b_2 - a_2 b_1} \qquad\qquad y_1 = \frac{-b_1}{a_1 b_2 - a_2 b_1}$$

$$x_2 = \frac{-a_2}{a_1 b_2 - a_2 b_1} \qquad\qquad y_2 = \frac{a_1}{a_1 b_2 - a_2 b_1}$$

The denominator $a_1 b_2 - a_2 b_1$ is equal to the determinant of A. If the determinant of A is not equal to 0, the inverse exists and can be defined as follows.

| Inverse of a Second-Order Matrix | If $A = \begin{bmatrix} a_1 & b_1 \\ a_2 & b_2 \end{bmatrix}$ and $\begin{vmatrix} a_1 & b_1 \\ a_2 & b_2 \end{vmatrix} \neq 0$, then $A^{-1} = \frac{1}{\begin{vmatrix} a_1 & b_1 \\ a_2 & b_2 \end{vmatrix}} \begin{bmatrix} b_2 & -b_1 \\ -a_2 & a_1 \end{bmatrix}$. |

$A \cdot A^{-1} = A^{-1} \cdot A = I$, where I is the identity matrix.

Example 3

Find the multiplicative inverse of the matrix $\begin{bmatrix} 3 & -1 \\ 4 & 2 \end{bmatrix}$.

First, find the determinant.

$$\begin{vmatrix} 3 & -1 \\ 4 & 2 \end{vmatrix} = 3(2) - 4(-1) = 10$$

The inverse is $\frac{1}{10} \begin{bmatrix} 2 & 1 \\ -4 & 3 \end{bmatrix}$ or $\begin{bmatrix} \frac{1}{5} & \frac{1}{10} \\ -\frac{2}{5} & \frac{3}{10} \end{bmatrix}$. *Check to see if $A \cdot A^{-1} = A^{-1} \cdot A = I$*

You can solve systems of equations by using matrix equations.

Example 4

Solve the system of equations by using matrix equations.
$5x + 4y = -3$
$3x - 5y = -24$

Write the system as a matrix equation.

$$\begin{bmatrix} 5 & 4 \\ 3 & -5 \end{bmatrix} \cdot \begin{bmatrix} x \\ y \end{bmatrix} = \begin{bmatrix} -3 \\ -24 \end{bmatrix}$$

To solve the matrix equation, first find the inverse of the coefficient matrix.

$$\frac{1}{\begin{vmatrix} 5 & 4 \\ 3 & -5 \end{vmatrix}} \begin{bmatrix} -5 & -4 \\ -3 & 5 \end{bmatrix} = -\frac{1}{37} \begin{bmatrix} -5 & -4 \\ -3 & 5 \end{bmatrix} \qquad \begin{vmatrix} 5 & 4 \\ 3 & -5 \end{vmatrix} = 5(-5) - 3(4) \text{ or } -37$$

(continued on the next page)

Now multiply each side of the matrix equation by the inverse and solve.

$$-\frac{1}{37}\begin{bmatrix} -5 & -4 \\ -3 & 5 \end{bmatrix} \cdot \begin{bmatrix} 5 & 4 \\ 3 & -5 \end{bmatrix} \cdot \begin{bmatrix} x \\ y \end{bmatrix} = -\frac{1}{37}\begin{bmatrix} -5 & -4 \\ -3 & 5 \end{bmatrix} \cdot \begin{bmatrix} -3 \\ -24 \end{bmatrix}$$

$$\begin{bmatrix} x \\ y \end{bmatrix} = \begin{bmatrix} -3 \\ 3 \end{bmatrix}$$

The solution is $(-3, 3)$.

Example 5

APPLICATION

Metallurgy

Refer to the application at the beginning of the lesson. How much of the metals with 55% aluminum and 80% aluminum content should Kim Troy use to make 20 kilograms of a 70% aluminum alloy?

Let x represent the amount of metal with 55% aluminum content, and let y represent the amount of metal with 80% aluminum content. Then $x + y = 20$, since 20 kg are needed.

Write an equation in standard form that represents the proportions of each metal needed.

$$55\%x + 80\%y = 70\%(x + y)$$
$$0.55x + 0.8y = 0.7(x + y)$$
$$0.55x + 0.8y = 0.7x + 0.7y$$
$$0.15x - 0.1y = 0$$
$$15x - 10y = 0$$

Write the system as a matrix equation and solve.

$$\begin{aligned} x + y &= 20 \\ 15x - 10y &= 0 \end{aligned} \longrightarrow \begin{bmatrix} 1 & 1 \\ 15 & -10 \end{bmatrix} \cdot \begin{bmatrix} x \\ y \end{bmatrix} = \begin{bmatrix} 20 \\ 0 \end{bmatrix}$$

$$-\frac{1}{25}\begin{bmatrix} -10 & -1 \\ -15 & 1 \end{bmatrix} \cdot \begin{bmatrix} 1 & 1 \\ 15 & -10 \end{bmatrix} \cdot \begin{bmatrix} x \\ y \end{bmatrix} = -\frac{1}{25}\begin{bmatrix} -10 & -1 \\ -15 & 1 \end{bmatrix} \cdot \begin{bmatrix} 20 \\ 0 \end{bmatrix}$$

$$\begin{bmatrix} x \\ y \end{bmatrix} = \begin{bmatrix} 8 \\ 12 \end{bmatrix}$$

The solution is $(8, 12)$. So, Kim should use 8 kg of metal with 55% aluminum content and 12 kg of metal with 80% aluminum content. *Check the solution by substituting these values in the original problem.*

CHECKING FOR UNDERSTANDING

Communicating
Mathematics

Read and study the lesson to answer each question.

1. Do all matrices have a determinant? Explain.
2. When is a matrix considered to be nonsingular?
3. **Describe** the identity matrix under multiplication for any nth order matrix.
4. **Explain** why the inverse matrix under multiplication exists only when the determinant is nonzero.

Guided
Practice

Find the value of each determinant.

5. $\begin{vmatrix} 3 & -5 \\ 7 & 9 \end{vmatrix}$

6. $\begin{vmatrix} 10 & 50 \\ -5 & 25 \end{vmatrix}$

7. $\begin{vmatrix} 7 & 1 & 6 \\ 3 & -1 & 4 \\ -2 & 3 & 0 \end{vmatrix}$

8. $\begin{vmatrix} 2 & 4 & 6 \\ 1 & 2 & 3 \\ 3 & -1 & 4 \end{vmatrix}$

Find the multiplicative inverse of each matrix, if it exists.

9. $\begin{bmatrix} -2 & 3 \\ -4 & 1 \end{bmatrix}$

10. $\begin{bmatrix} 8 & 4 \\ -4 & -2 \end{bmatrix}$

Solve each matrix equation.

11. $\begin{bmatrix} 4 & 8 \\ 2 & -3 \end{bmatrix} \cdot \begin{bmatrix} x \\ y \end{bmatrix} = \begin{bmatrix} 7 \\ 0 \end{bmatrix}$

12. $\begin{bmatrix} 5 & 4 \\ 3 & -5 \end{bmatrix} \cdot \begin{bmatrix} x \\ y \end{bmatrix} = \begin{bmatrix} -3 \\ -24 \end{bmatrix}$

EXERCISES

Practice

Find the value of each determinant.

13. $\begin{vmatrix} 7 & 16 \\ 3 & 8 \end{vmatrix}$

14. $\begin{vmatrix} -4 & 8 \\ 0 & 2 \end{vmatrix}$

15. $\begin{vmatrix} 16 & 17 \\ 15 & 16 \end{vmatrix}$

16. $\begin{vmatrix} 6 & 7 & 4 \\ -2 & -4 & 3 \\ 1 & 1 & 1 \end{vmatrix}$

17. $\begin{vmatrix} 3 & 0 & 2 \\ 0 & -1 & 5 \\ 6 & 7 & 0 \end{vmatrix}$

18. $\begin{vmatrix} 4 & 2 & -3 \\ 5 & 1 & 0 \\ -2 & 1 & 11 \end{vmatrix}$

Find the multiplicative inverse of each matrix, if it exists.

19. $\begin{bmatrix} -1 & -2 \\ 0 & -1 \end{bmatrix}$

20. $\begin{bmatrix} -2 & 2 \\ 1 & -1 \end{bmatrix}$

21. $\begin{bmatrix} 1 & 1 \\ 2 & -3 \end{bmatrix}$

22. $\begin{bmatrix} 2 & -5 \\ 6 & 1 \end{bmatrix}$

23. $\begin{bmatrix} 3 & 1 \\ -4 & 1 \end{bmatrix}$

24. $\begin{bmatrix} 1 & 2 \\ 2 & 1 \end{bmatrix}$

Solve each system by using matrix equations.

25. $5x + y = 26$
    $2x - 3y = 41$

26. $5x + y = 1$
    $9x + 3y = 1$

Solve each matrix equation.

27. $\begin{bmatrix} 1 & 2 & 2 \\ 2 & -1 & 1 \\ 3 & -2 & 3 \end{bmatrix} \cdot \begin{bmatrix} x \\ y \\ z \end{bmatrix} = \begin{bmatrix} 0 \\ -1 \\ -4 \end{bmatrix}$, if the inverse is $-\dfrac{1}{9}\begin{bmatrix} -1 & -10 & 4 \\ -3 & -3 & 3 \\ -1 & 8 & -5 \end{bmatrix}$.

28. $\begin{bmatrix} 3 & 1 & 1 \\ -6 & 5 & 3 \\ 9 & -2 & -1 \end{bmatrix} \cdot \begin{bmatrix} x \\ y \\ z \end{bmatrix} = \begin{bmatrix} -1 \\ -9 \\ 5 \end{bmatrix}$, if the inverse is $-\dfrac{1}{9}\begin{bmatrix} 1 & -1 & -2 \\ 21 & -12 & -15 \\ -33 & 15 & 21 \end{bmatrix}$.

Find the value of each determinant by using expansion by minors.

29. $\begin{vmatrix} 1 & 2 & 3 & 1 \\ 4 & 3 & -1 & 0 \\ 2 & -5 & 4 & 4 \\ 1 & -2 & 0 & 2 \end{vmatrix}$

30. $\begin{vmatrix} 7 & 0 & 9 & 5 \\ 8 & 2 & -1 & 2 \\ -5 & 3 & 7 & 9 \\ 0 & -1 & -4 & -6 \end{vmatrix}$

31. $\begin{vmatrix} 3 & 0 & 0 & 4 & 0 \\ 6 & -3 & 2 & 0 & 7 \\ 0 & 4 & 3 & 0 & 5 \\ 0 & 2 & 1 & 3 & -4 \\ 6 & 0 & -2 & -3 & 0 \end{vmatrix}$

Graphing Calculator

32. Use the MATRIX feature on a graphing calculator to solve the system below.

$3x - 3y + 6z = 30$
$x - 3y + 10z = 50$
$-x + 3y - 5z = 40$

Critical Thinking

Prove each statement.

33. $AI = IA = A$ for second-order matrices.

34. $AA^{-1} = A^{-1}A = I$ for second-order matrices.

Applications and Problem Solving

35. Metallurgy George Johnson is a chemist who is preparing an acid solution to be used as a cleaner for machine parts. The machine shop needs 200 mL of the solution at a 48% concentration. George has only 60% and 40% solutions. How much of each solution should George combine to make 200 mL of the 48% solution?

36. Landscaping Two dump trucks have capacities of 10 tons and 12 tons. They make a total of 20 round trips to haul 226 tons of topsoil to be used to landscape the new skyscraper. How many round trips did each truck make?

37. Find $[f \circ g](4)$ and $[g \circ f](4)$ if $f(x) = x^2 - 1$ and $g(x) = x + 1$. **(Lesson 1-2)**

38. Write the slope-intercept form of the equation of the line that passes through $(1, 6)$ and has a slope of 2. **(Lesson 1-4)**

39. Solve the system of equations below. **(Lesson 2-1)**
$$x = 5y - 2$$
$$3x + y = 10$$

40. Find two square matrices A and B for which $(A + B)^2$ does not equal $A^2 + 2AB + B^2$. **(Lesson 2-2)**

41. **College Entrance Exam** Compare quantities A and B below.
Write A if the quantity in Column A is greater.
Write B if the quantity in Column B is greater.
Write C if the two quantities are equal.
Write D if there is not enough information to determine the relationship.

(A) $3^2 + 4^2$ **(B)** 7^2

MID-CHAPTER REVIEW

Solve each system of equations by graphing. **(Lesson 2-1)**

1. $3x - y - 1 = 0$
$x + y - 3 = 0$

2. $3x - 2y = 5$
$x + 7y = -6$

Solve each system of equations algebraically. **(Lesson 2-1)**

3. $5x - y - 16 = 0$
$2x + 3y - 3 = 0$

4. $-2x + 4y = 3$
$5x - 7y = 11$

Let $A = \begin{bmatrix} 3 & 0 \\ -1 & 5 \end{bmatrix}$ **and let** $B = \begin{bmatrix} 5 & 1 \\ -3 & 4 \end{bmatrix}$. **Find each of the following.** **(Lesson 2-2)**

5. $A + B$ **6.** $A - B$ **7.** AB

8. Find the value of $\begin{vmatrix} 5 & 3 & 2 \\ 6 & 1 & 8 \\ 4 & 2 & 2 \end{vmatrix}$. **(Lesson 2-3)**

Solve each matrix equation. **(Lesson 2-3)**

9. $\begin{bmatrix} -2 & -4 \\ 2 & 3 \end{bmatrix} \cdot \begin{bmatrix} x \\ y \end{bmatrix} = \begin{bmatrix} 6 \\ 5 \end{bmatrix}$

10. $\begin{bmatrix} -1 & 14 \\ 6 & 12 \end{bmatrix} \cdot \begin{bmatrix} x \\ y \end{bmatrix} = \begin{bmatrix} 4 \\ 0 \end{bmatrix}$

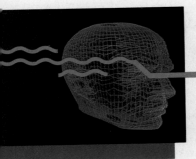
Technology

Finding Determinants of Matrices

Matrix software accompanies Introduction to College Mathematics, developed by the North Carolina School of Science and Mathematics.

The calculations used to evaluate determinants can be very tedious, if done using paper and a pencil. Fortunately, there are computer programs that can calculate determinants for you. One such program is *Matrix*, for IBM-compatible computers. To evaluate the determinant of a 5 × 5 matrix using *Matrix*, use the following steps.

- Select **F1. Define/Edit Matrix** from the main menu.
- Select **1. Define a Matrix (Static)**.
- When the computer prompts you to enter the number of rows and columns of the matrix, type in **5, 5** and press **[Enter]**.
- You enter a matrix by entering its rows at the prompts.

  ROW1: **1, 4, 3, 2, 3**   ROW4: **0, 5, 2, –1, 1**

  ROW2: **–4, 2, –9, 0, 1**   ROW5: **0, 0, 4, 2, 5**

  ROW3: **1, 1, 4, 3, 4**
- When requested, name the matrix **A**.
- Select **F3. Matrix Calculator Menu** from the main menu.
- Select **1. Calculator.**
- At the prompt, type in **DET(A)** and press **[Enter]** to find the determinant of matrix A.
- The computer will respond with **Number result = –110.**

The menu system used by *Matrix* is as follows.

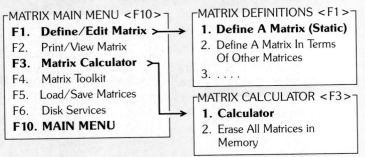

```
┌MATRIX MAIN MENU <F10>┐    ┌MATRIX DEFINITIONS <F1>┐
  F1.   Define/Edit Matrix >─┐  1. Define A Matrix (Static)
  F2.   Print/View Matrix    │  2. Define A Matrix In Terms
  F3.   Matrix Calculator  > │     Of Other Matrices
  F4.   Matrix Toolkit       │  3. . . . .
  F5.   Load/Save Matrices   │  └───────────────────────┘
  F6.   Disk Services        │  ┌MATRIX CALCULATOR <F3>┐
  F10.  MAIN MENU            └> 1. Calculator
                                2. Erase All Matrices in
                                   Memory
```

EXERCISES

Use *Matrix* to evaluate the determinant for each matrix.

1. $\begin{bmatrix} 3 & 5 \\ 6 & -3 \end{bmatrix}$

2. $\begin{bmatrix} 1 & 6 & -2 \\ 8 & -2 & 0 \\ 5 & 9 & -1 \end{bmatrix}$

3. $\begin{bmatrix} 1 & 4 & 8 & 1 \\ 0 & 1 & -3 & 0 \\ -4 & 2 & 8 & 1 \\ 6 & -5 & 8 & 9 \end{bmatrix}$

2-4

Solving Systems of Equations by Using Matrices

Objective

After studying this lesson, you should be able to:
■ solve systems of equations by using matrices.

Application

The *K'iu-ch'ang*, or *Arithmetic in Nine Sections*, is the greatest of the ancient Chinese books on mathematics. A mathematician named Ch'ang Ts'ang is thought to have collected and edited these writings of the ancients around 213 B.C. Chapter 8 was called *Fang cheng*, or *Method of Tables*, and dealt with the solution of systems of equations with two or three unknowns using matrices. This work indicates that Chinese mathematicians were pioneers in establishing the early science of mathematics.

The coefficients and constants of the system of equations shown below can be written in the form of a 3 × 4 **augmented matrix**. An augmented matrix is an array of the coefficients and constants of a system of equations.

System of Equations

$$\left.\begin{array}{c} x - 2y + z = 7 \\ 3x + y - z = 2 \\ 2x + 3y + 2z = 7 \end{array}\right\} \rightarrow \left.\begin{array}{c} 1x - 2y + 1z = 7 \\ 3x + 1y - 1z = 2 \\ 2x + 3y + 2z = 7 \end{array}\right\} \rightarrow$$

Augmented Matrix

$$\left[\begin{array}{ccc:c} 1 & -2 & 1 & 7 \\ 3 & 1 & -1 & 2 \\ 2 & 3 & 2 & 7 \end{array}\right]$$

This system can be solved by transforming the rows of the matrix, since each row represents an equation. Each change in the matrix represents a corresponding change in the system of equations. Any operation that results in an equivalent system of equations is permitted for the matrix. The objective is to obtain the identity matrix on the left side. That is, we want to obtain

$$\left[\begin{array}{ccc:c} 1 & 0 & 0 & x \\ 0 & 1 & 0 & y \\ 0 & 0 & 1 & z \end{array}\right]$$ for which the solution is (x, y, z). In general, any of the

following row operations can be used to transform an augmented matrix.

Row Operations on Matrices

■ **Interchange any two rows.**
■ **Replace any row with a nonzero multiple of that row.**
■ **Replace any row with the sum of that row and a multiple of another row.**

There is no single correct order in which to perform row operations to arrive at the solution. That is, the order in which you transform rows in a matrix may be different from the way your classmate does it, but you may both be correct.

Example 1

Solve the system of equations by using row operations.

$$x - 2y + z = 7$$
$$3x + y - z = 2$$
$$2x + 3y + 2z = 7$$

First, write the augmented matrix.

$$\begin{bmatrix} 1 & -2 & 1 & \vdots & 7 \\ 3 & 1 & -1 & \vdots & 2 \\ 2 & 3 & 2 & \vdots & 7 \end{bmatrix}$$

Multiply row one by –3 and add the result to row two.

$$\begin{bmatrix} 1 & -2 & 1 & \vdots & 7 \\ 0 & 7 & -4 & \vdots & -19 \\ 2 & 3 & 2 & \vdots & 7 \end{bmatrix}$$

Multiply row one by –2 and add the result to row three.

$$\begin{bmatrix} 1 & -2 & 1 & \vdots & 7 \\ 0 & 7 & -4 & \vdots & -19 \\ 0 & 7 & 0 & \vdots & -7 \end{bmatrix}$$

Multiply row two by –1 and add the result to row three.

$$\begin{bmatrix} 1 & -2 & 1 & \vdots & 7 \\ 0 & 7 & -4 & \vdots & -19 \\ 0 & 0 & 4 & \vdots & 12 \end{bmatrix}$$

Add row three to row two.

$$\begin{bmatrix} 1 & -2 & 1 & \vdots & 7 \\ 0 & 7 & 0 & \vdots & -7 \\ 0 & 0 & 4 & \vdots & 12 \end{bmatrix}$$

Multiply row two by $\frac{1}{7}$.

$$\begin{bmatrix} 1 & -2 & 1 & \vdots & 7 \\ 0 & 1 & 0 & \vdots & -1 \\ 0 & 0 & 4 & \vdots & 12 \end{bmatrix}$$

Multiply row three by $\frac{1}{4}$.

$$\begin{bmatrix} 1 & -2 & 1 & \vdots & 7 \\ 0 & 1 & 0 & \vdots & -1 \\ 0 & 0 & 1 & \vdots & 3 \end{bmatrix}$$

Multiply row two by 2 and add the result to row one.

$$\begin{bmatrix} 1 & 0 & 1 & \vdots & 5 \\ 0 & 1 & 0 & \vdots & -1 \\ 0 & 0 & 1 & \vdots & 3 \end{bmatrix}$$

Multiply row three by –1 and add the result to row one.

$$\begin{bmatrix} 1 & 0 & 0 & \vdots & 2 \\ 0 & 1 & 0 & \vdots & -1 \\ 0 & 0 & 1 & \vdots & 3 \end{bmatrix}$$

The solution is $(2, -1, 3)$.   *Notice that the left side is the identity matrix. Check the solution by substituting in the original equations.*

Augmented matrices can be used to solve a variety of problems.

Example 2

Find the equation of a parabola that contains the points at $(1, 9)$, $(4, 6)$, and $(6, 14)$.

The general form of the equation of a parabola is $y = ax^2 + bx + c$. Since each of the ordered pairs satisfies the general equation, the following can be generated.

$a(1)^2 + b(1) + c = 9$ *Use (1, 9).*
$a(4)^2 + b(4) + c = 6$ *Use (4, 6).*
$a(6)^2 + b(6) + c = 14$ *Use (6, 14).*

Simplify.

$a + b + c = 9$
$16a + 4b + c = 6$
$36a + 6b + c = 14$

Use an augmented matrix and row operations to solve.

$$\begin{bmatrix} 1 & 1 & 1 & \vdots & 9 \\ 16 & 4 & 1 & \vdots & 6 \\ 36 & 6 & 1 & \vdots & 14 \end{bmatrix} \rightarrow \begin{bmatrix} 1 & 0 & 0 & \vdots & 1 \\ 0 & 1 & 0 & \vdots & -6 \\ 0 & 0 & 1 & \vdots & 14 \end{bmatrix}$$

The solution is $(1, -6, 14)$, so $a = 1$, $b = -6$, and $c = 14$. Therefore, the equation of the parabola is $y = x^2 - 6x + 14$.   *Check this equation by substituting the original three points.*

CHECKING FOR UNDERSTANDING

Communicating Mathematics

Read and study the lesson to answer each question.

1. **State** the objective desired when you use augmented matrices to solve systems of equations.

2. Could the system of equations in Example 2 have been solved by using another method? If so, show how it could have been done.

Guided Practice

State the row operations you would use to get a zero in the second column of row one. Then state the row operations you would use to get a zero in the first column of row two.

3. $\begin{bmatrix} 3 & 5 & 7 \\ 6 & -1 & -8 \end{bmatrix}$

4. $\begin{bmatrix} 4 & -7 & -2 \\ 1 & 2 & 7 \end{bmatrix}$

5. $\begin{bmatrix} 5 & -3 & 0 \\ 3 & 6 & -6 \end{bmatrix}$

Solve each system of equations by using augmented matrices.

6. $5x - 2y = 5$
   $x + y = 8$

7. $2x + y - 2z - 7 = 0$
   $x - 2y - 5z + 1 = 0$
   $4x + y + z + 1 = 0$

EXERCISES

Practice

Solve each system of equations by using augmented matrices.

8. $3x + 5y = 7$
   $6x - y = -8$

9. $4x - 7y = -2$
   $x + 2y = 7$

10. $3x + 3y = -9$
    $-2x + y = -4$

11. $5x = 3y - 50$
    $2y = 1 - 3x$

12. $x - y + z = 3$
    $2y - z = 1$
    $2y - x + 1 = 0$

13. $x + y + z = -2$
    $2x - 3y + z = -11$
    $-x + 2y - z = 8$

14. $2x + 6y + 8z = 5$
    $-2x + 9y - 12z = -1$
    $4x + 6y - 4z = 3$

15. $x + y + z - 6 = 0$
    $2x - 3y + 4z - 3 = 0$
    $4x - 8y + 4z - 12 = 0$

16. $x + 2y = 5$
    $3x + 4z = 2$
    $2y + 3w = -2$
    $3z - 2w = 1$

17. $w + x + y + z = 0$
    $2w + x - y - z = 1$
    $-w - x + y + z = 0$
    $2x + y = 0$

Find the equation of a parabola that contains the points given.

18. $(1, 4)$, $(5, 40)$, and $(3, 14)$

19. $(2, 3)$, $(3, 6)$, and $(-2, 31)$

20. What does it mean when an augmented matrix has a row containing all zeros?

21. What does it mean when an augmented matrix has a row containing all zeros, except for the last element?

22. Restaurants At Morgan's Fine Cuisine, meals are served *a la carte.* That is, each item on the menu is priced separately. Jackie and Ted Parris went to Morgan's to celebrate their anniversary. Jackie ordered prime rib, 2 side dishes, and a roll. Ted ordered prime rib, 3 side dishes, and 2 rolls. Jackie's meal came to $36 and Ted's meal came to $44. If the prime rib is three times as expensive as a side dish, what is the cost of each item?

23. Consumer Purchasing Janet is spending her clothes allowance on school clothes. If she buys 3 blouses, 2 skirts, and 4 pairs of jeans, she will spend $292. If she buys 4 blouses, 1 skirt, and 3 pairs of jeans, she will spend $252. If jeans cost $4 more than skirts, what is the price of each item?

24. Does the graph at the right represent a function? **(Lesson 1-1)**

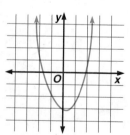

25. Solve the system of equations below. **(Lesson 2-1)**
$$x + 2y = 11$$
$$3x - 5y = 11$$

26. Find $A + B$ if $A = \begin{bmatrix} 3 & 8 \\ -2 & 4 \end{bmatrix}$ and $B = \begin{bmatrix} 1 & 5 \\ -2 & 8 \end{bmatrix}$. **(Lesson 2-2)**

27. Find the value of $\begin{vmatrix} 2 & 3 & 4 \\ 5 & 6 & 7 \\ 8 & 9 & 10 \end{vmatrix}$. **(Lesson 2-3)**

28. College Entrance Exam
What is the sum of four integers whose average is 15?

2-5A Graphing Calculators: Graphing Systems of Linear Inequalities

Graphing a system of linear inequalities is similar to graphing one linear inequality, which is explained in Lesson 1-3A. The "Shade(" command is still used to graph a system of inequalities, but the two equations of the system are used as the upper and lower boundaries instead of entering a minimum or a maximum value.

Once again, prepare to graph by resetting the range values to the standard viewing window and clearing any functions from the Y = list. You will also need to clear the graphics screen before graphing an inequality.

Since the TI-82 graphs functions and shades above the first function entered and below the second function entered, the first step in solving a system is deciding which function to enter first as the lower boundary and which to enter second as the upper boundary.

A "greater than or equal to" symbol indicates that values on and above the line of the related equation will satisfy the inequality. Similarly, a "less than or equal to" symbol indicates that values on and below the line of the related equation satisfy the inequality. Therefore, the related equation of the inequality with the sign "$y \geq$" will be entered first, and the related equation of the inequality with the sign "$y \leq$" will be entered second.

Example 1

Graph the system of inequalities below.
$$y \leq 2x - 3$$
$$y \geq -0.5x - 4$$

Values below the line $y = 2x - 3$ will satisfy the inequality $y \leq 2x - 3$.

Values above the line $y = -0.5x - 4$ will satisfy the inequality $y \geq -0.5x - 4$.

So, we will enter the equation $y = -0.5x - 4$ first and the equation $y = 2x - 3$ second.

Enter: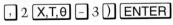

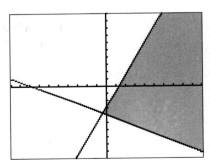

The points in the shaded area satisfy both $y \leq 2x - 3$ and $y \geq -0.5x - 4$.

You may have to solve some inequalities for *y* first to find a solution to the system graphically. Also, remember to clear the graphics screen by pressing 2ND DRAW 1 before graphing another inequality.

Example 2

Graph the system of inequalities below.
$$5x - y \geq 2$$
$$6x + 2y \geq 4$$

Solving both inequalities for *y*, we obtain $y \leq 5x - 2$ for the first inequality and $y \geq -3x + 2$ for the second inequality.

Values above the line $y = -3x + 2$ will satisfy the inequality $6x + 2y \geq 4$.

Values below the line $y = 5x - 2$ will satisfy the inequality $5x - y \geq 2$.

So, we will enter the equation $y = -3x + 2$ first and the equation $y = 5x - 2$ second.

Enter: 2ND DRAW 7
(−) 3 X,T,θ + 2 , 5 X,T,θ − 2) ENTER

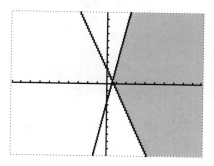

TECHNOLOGY *Tip*

You can specify one of four shading patterns when using a TI-83 graphing calculator. The TI-92 also has four shading patterns. They are automatically selected on a rotating basis if you shade above or below more than one function.

The points in the shaded area satisfy both $5x - y \geq 2$ and $6x + 2y \geq 4$.

EXERCISES

Graph each system of inequalities and sketch the graph on a piece of paper. Solve the inequalities for *y* first if necessary.

1. $y \leq x$
$y \geq -4$

2. $y \leq -3x$
$y \geq 4x$

3. $y \geq 2x - 3$
$y \leq -x + 6$

4. $y \geq -2x + 0.5$
$y \leq 0.5x - 5$

5. $x + y \leq -1$
$-x + y \geq 5$

6. $y \geq -2$
$2x + y \leq 8$

7. $2x - 2y \geq 9$
$-3x - y \leq -4$

8. $3x + 3y \leq 6$
$2x - y \leq 5$

2-5 Solving Systems of Inequalities

Objectives

After studying this lesson, you should be able to:
- graph systems of inequalities, and
- find the maximum or minimum value of a function defined for a polygonal convex set.

Application

Belan Chu, a recent graphic arts graduate, has been doing some research on starting her own custom greeting card business. She will have initial start-up costs of $1500. Belan figures that it will cost 45¢ to produce each card. In order for Belan to remain competitive with the large greeting card manufacturers, she must sell her cards for no more than $1.70 per card, and her income must exceed her costs. Belan needs to determine if and when she might make a profit.

This situation can be modeled by a **system of linear inequalities**. To solve a system of inequalities, you must find the ordered pairs that satisfy both inequalities. One way to do this is to graph both inequalities on the same coordinate plane. The intersection of the two graphs contains points with the ordered pairs in the solution set. If the graphs do not intersect, then the system has no solution.

 FYI...

In ancient Rome, young girls dropped love messages into a large urn in the center of the city. Men of the city drew messages and dated the girls whose messages they had drawn. Hence, the valentine card was born.

In order to solve the application described above, write a system of inequalities to represent the situation. Let x represent the number of greeting cards sold. Let y represent the income in dollars.

$x \geq 0$
$y \geq 0$
$y \geq 0.45x + 1500$ *Income must exceed costs.*

$y \leq 1.70x$ *Cards will be sold for no more than $1.70 each.*

Graph the system of inequalities. The ordered pair for any point in the intersection of the two graphs is a solution to the system. Belan's break-even point is where the lines intersect. She will begin making a profit after she sells 1200 cards. Notice that as the number of cards sold increases beyond 1200, her profit region expands, allowing her an opportunity for greater profit.

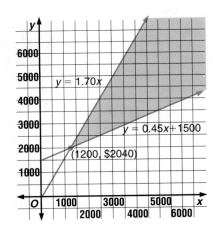

Not every system of inequalities has a solution.

Example 1

Solve the system of inequalities by graphing.
$$y > x + 5$$
$$y < x - 2$$

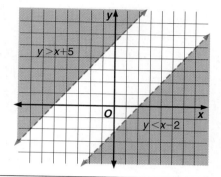

The graphs have no points in common. Therefore, no ordered pairs satisfy both inequalities.

Often the solution set of the system of linear inequalities is a **polygonal convex set**. Such a set consists of all points on or inside a convex polygon.

Example 2

Solve the system of inequalities by graphing. Then name the coordinates of the vertices of the polygonal convex set.
$$x \geq 0$$
$$y \geq 0$$
$$x + y \leq 5$$

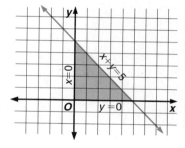

Points in the shaded region satisfy all three inequalities. The coordinates of the vertices of the set are $(0,0)$, $(5,0)$, and $(0,5)$.

f(x, y) represents a function of two variables, x and y.

Sometimes it is necessary to find the maximum or minimum value that a function has for the points of a polygonal convex set. Consider the function $f(x, y) = 5x - 3y$, with the following inequalities forming a polygonal convex set.

$$y \geq 0 \qquad -x + y \leq 2 \qquad 0 \leq x \leq 5 \qquad x + y \leq 6$$

By graphing the inequalities and finding the intersection of the graphs, you can determine a polygonal convex set of points for which the function can be evaluated. The region shown at the right is the polygonal convex set determined by the inequalities listed above. Since the polygonal convex set has infinitely many points, it would be impossible to evaluate the function for all of them. However, according to the **vertex theorem**, a function such as $f(x, y) = 5x - 3y$ need only be evaluated for the coordinates of the vertices of the polygonal boundary in order to find the maximum and minimum values.

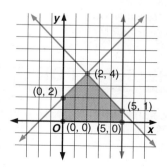

Vertex Theorem	The maximum or minimum value of $f(x, y) = ax + by + c$ on a polygonal convex set occurs at a vertex of the polygonal boundary.

The value of $f(x, y) = 5x - 3y$ at each vertex can be found as follows.

$f(x, y) = 5x - 3y$ $f(2, 4) = 5(2) - 3(4) = -2$

$f(0, 0) = 5(0) - 3(0) = 0$ $f(5, 1) = 5(5) - 3(1) = 22$

$f(0, 2) = 5(0) - 3(2) = -6$ $f(5, 0) = 5(5) - 3(0) = 25$

Therefore, the maximum value of $f(x, y)$ is 25, and the minimum is –6.
The maximum occurs at $(5, 0)$, and the minimum occurs at $(0, 2)$.

Example 3

Find the maximum and minimum values of $f(x, y) = x + 2y + 1$ for the polygonal convex set determined by the following inequalities.

$x \geq 0$ $y \geq 0$ $2x + y \leq 4$ $x + y \leq 3$

First, graph the inequalities and find the coordinates of the vertices of the resulting polygon.

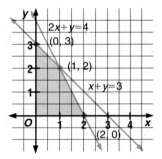

The coordinates of the vertices are $(0, 0)$, $(2, 0)$, $(1, 2)$, and $(0, 3)$.

Then, evaluate the function $f(x, y) = x + 2y + 1$ at each vertex.

$f(0, 0) = 0 + 2(0) + 1 = 1$ $f(1, 2) = 1 + 2(2) + 1 = 6$

$f(2, 0) = 2 + 2(0) + 1 = 3$ $f(0, 3) = 0 + 2(3) + 1 = 7$

Thus, the maximum value of the function is 7, and the minimum value is 1.

CHECKING FOR UNDERSTANDING

Communicating Mathematics

Read and study the lesson to answer each question.

1. Is it possible to have more than one ordered pair that produces a maximum or minimum for a given function? Explain your answer.

2. Can a function within a polygonal convex set have no maximum? Explain your answer.

Given $f(x, y) = 3x + 2y + 1$, find each value.

3. $f(3, 2)$ **4.** $f\left(\frac{1}{3}, \frac{1}{2}\right)$ **5.** $f(-1.5, 4)$

Find the maximum and minimum values of each function defined for the polygonal convex set having vertices at $(0, 0)$, $(4, 0)$, $(3, 5)$, and $(0, 5)$.

6. $f(x, y) = x + y$ **7.** $f(x, y) = 8x + y$ **8.** $f(x, y) = 4y - 3x$

Solve each system of inequalities by graphing and name the coordinates of the vertices of each polygonal convex set. Then, find the maximum and minimum values for the given function on that set.

9. $x + 4y \leq 12$
$3x - 2y \geq -6$
$x + y \geq -2$
$3x - y \leq 10$
$f(x, y) = x - y + 2$

10. $x \leq 3$
$y \leq 5$
$x + y \geq 1$
$x \geq 0$
$y \geq 0$
$f(x, y) = 2x + 8y + 10$

EXERCISES

Solve each system of inequalities by graphing and name the coordinates of the vertices of each polygonal convex set. Then, find the maximum and minimum values for the given function on that set.

11. $x \geq 0$
$y \geq 0$
$2x + y \leq 4$
$f(x, y) = y - x$

12. $x \leq 0$
$y + 3 \geq 0$
$y \leq x$
$f(x, y) = x + 2y$

13. $x \geq 0$
$y \geq 1$
$x + y \leq 4$
$f(x, y) = 4x + 2y + 7$

14. $x + y \leq 5$
$y - x \leq 5$
$y \geq -10$
$f(x, y) = \frac{1}{2}x - \frac{1}{3}y$

15. $y \geq 0$
$0 \leq x \leq 5$
$-x + y \leq 2$
$x + y \leq 6$
$f(x, y) = 3x - 5y$

16. $x \geq 1$
$y \geq 2$
$y \leq 8$
$x + y \geq 5$
$2x + y \leq 14$
$f(x, y) = y - 2x + 5$

17. Write a system of inequalities that determines the polygonal convex set with vertices at $(0, 0)$, $(5, 1)$, $(1, 6)$, and $(6, 4)$.

Solve each system of inequalities. Use the ZOOM feature on a graphing calculator to check your solution. Assume that $x, y \geq 0$.

18. $x - y \leq 4$
$2x + y \leq 14$

19. $2x + 2y \geq 40$
$xy \geq 100$

20. Graph the following system to form a polygonal convex set. Determine which lines intersect and solve pairs of equations to determine the coordinates of each vertex. Then, find the maximum and minimum values of $f(x, y) = 5x + 6y$ for each region.

$$0 \leq 2y \leq 17$$
$$y \geq 2x - 13$$
$$y \leq 3x + 1$$
$$3y \geq \text{-}2x + 11$$
$$y \geq 7 - 2x$$
$$y \leq 16 - x$$

21. Education Bob Chase intends to major in medicine at the Harvard Medical School. He has been told that he must have two outstanding admissions scores in order to be admitted to a pre-med program. He needs an ACT score of at least 30 and an SAT score of at least 1200. Write two inequalities to represent this situation, and graph the inequalities to show the solution.

22. Business Henry Jackson, a recent college graduate, plans to start his own business manufacturing bicycle tires. Henry knows that his start-up costs are going to be $3000 and that each tire will cost him $2 to manufacture. In order to remain competitive, Henry cannot charge more than $5 per tire. Draw a graph to show when Henry will make a profit.

23. State whether the system below is *consistent and independent, consistent and dependent,* or *inconsistent.* **(Lesson 2-1)**

$$4x - 2y = 7$$
$$\text{-}12x + 6y = \text{-}21$$

24. Find the values of x and y for which $\begin{bmatrix} 4x + y \\ x \end{bmatrix} = \begin{bmatrix} 6 \\ 2y - 12 \end{bmatrix}$ is true. **(Lesson 2-2)**

25. Find the determinant for $\begin{bmatrix} 1 & 3 \\ 2 & 5 \end{bmatrix}$. Does an inverse exist for this matrix? **(Lesson 2-3)**

26. Solve the system below by using augmented matrices. **(Lesson 2-4)**

$$4x + 2y + 3z = 6$$
$$2x + 7y = 3z$$
$$\text{-}3x - 9y + 13 = \text{-}2z$$

27. College Entrance Exam Choose the best answer.
$\triangle ABC$ and $\triangle ABD$ are right triangles that share side $\overline{AB}$. $\triangle ABC$ has area x and $\triangle ABD$ has area y. If $\overline{AD}$ is longer than $\overline{AC}$ and $\overline{BD}$ is longer than $\overline{BC}$, which of the following cannot be true?

(A) $y > x$ **(B)** $y < x$ **(C)** $y \geq x$ **(D)** $y \neq x$ **(E)** $y \leq x$

2-6 Linear Programming

Objectives

After studying this lesson, you should be able to:
- use linear programming procedures to solve problems, and
- recognize situations where exactly one solution to a linear programming problem may not exist.

Application

Farmers often have a difficult time in late winter trying to determine exactly what crops to plant in the spring. There are so many factors to be considered: new government regulations on maximum crop production, seed costs, labor costs, fertilizer expenses, fluctuating prices on the grain market, and weather. Joe Washington has a choice of planting a combination of two different crops on 20 acres of land. For crop A, seed costs $120 per acre, and for crop B, seed costs $200 per acre.

Government restrictions limit acreage of crop A to 15 acres but do not limit crop B. Crop A will take 15 hours of labor per acre at a cost of $5.60 per hour, and crop B will require 10 hours of labor per acre at $5.00 per hour. If the expected income from crop A is $600 per acre and from crop B is $520 per acre, how should the 20 acres be apportioned between the two crops in order to maximize profit?

Mr. Washington can use a procedure called **linear programming** to help him maximize his profit. Many practical problems can be solved by using this method. The nature of these problems is that certain **constraints** exist or are placed upon the variables, and some function of these variables must be maximized or minimized. The constraints are often written as a system of linear inequalities.

The following procedure can be used to solve linear programming problems.

Linear Programming Procedure

1. **Define variables.**
2. **Write the constraints as a system of inequalities.**
3. **Graph the system and find the coordinates of the vertices of the polygon formed.**
4. **Write an expression to be maximized or minimized.**
5. **Substitute values from the coordinates of the vertices into the expression.**
6. **Select the greatest or least result.**

Let's use this procedure to solve the application presented at the beginning of the lesson.

Define variables.

Let x = the number of acres of crop A.
Let y = the number of acres of crop B.

Write inequalities.

$x \geq 0, \ y \geq 0$ *Acreage cannot be less than 0.*
$x \leq 15$ *No more than 15 acres of crop A are permitted.*
$x + y \leq 20$ *No more than 20 acres can be planted in all.*

Graph the system.

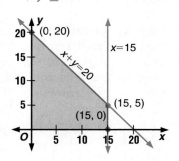

The constraints x ≥ 0 and y ≥ 0 tell you to consider only those points that are in Quadrant I.

The vertices are at $(0, 0)$, $(15, 0)$, $(15, 5)$, and $(0, 20)$.

Write an expression.

Profit equals income less costs. The profit from crop A equals $600x - 120x - 15(5.60)x$, or $396x$. The profit from crop B equals $520y - 200y - 10(5.00)y$, or $270y$. Thus, the profit function is $P(x, y) = 396x + 270y$.

Substitute values.

$P(0,0) = 396(0) + 270(0) = 0$
$P(15,0) = 396(15) + 270(0) = 5940$
$P(15,5) = 396(15) + 270(5) = 7290$
$P(0,20) = 396(0) + 270(20) = 5400$

Answer the problem.

The maximum occurs at $(15, 5)$. Thus, Mr. Washington should plant 15 acres of crop A and 5 acres of crop B to obtain the maximum profit of $7290.

In certain circumstances, the use of linear programming is not helpful. Consider the graph at the right, based on the following constraints.

$x \geq 0$
$y \geq 0$
$y \geq 6$
$4x + 3y \leq 12$

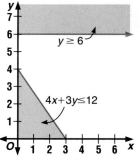

The constraints do not define a region with any points in common in Quadrant I. When the constraints of a linear programming problem cannot be satisfied simultaneously, then **infeasibility** is said to occur. This may mean that the constraints have been formulated incorrectly, certain requirements need to be changed, or that additional resources are required before the problem can be solved.

The solution of a linear programming problem is **unbounded** if the region defined by the constraints is infinitely large. Consider the graph at the right. While this graph would be helpful in a problem where the objective is to minimize cost, it is not possible to determine how to maximize revenue.

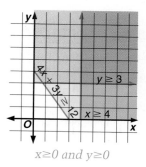

$x \geq 0$ and $y \geq 0$

It is also possible for a linear programming problem to have two or more optimal solutions. When this occurs, the problem is said to have **alternate optimal solutions**. This usually occurs when the function to be maximized or minimized is parallel to one side of the polygonal convex set.

Example

APPLICATION

Manufacturing

The AC Telephone Company manufactures two styles of cordless telephones, deluxe and standard. Each deluxe telephone nets the company $9 in profit, and each standard telephone nets $6. Machines A and B are used to make both styles of telephones. Each deluxe telephone requires three hours of machine A time and one hour of machine B time. Each standard telephone requires two hours of machine A time and two hours of machine B time. An employee has an idea that frees twelve hours of machine A time and eight hours of machine B time. Determine the mix of telephones that can be made during the free time that most effectively generates profit for the company within the given constraints.

Define variables.

Let $d =$ the number of deluxe telephones.
Let $s =$ the number of standard telephones.

Write inequalities.

$d \geq 0$
$s \geq 0$
$3d + 2s \leq 12$ *machine A*
$d + 2s \leq 8$ *machine B*

Graph the system.

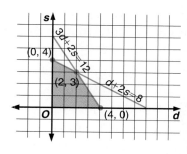

Write an expression.

Since the profit on each deluxe telephone is $9 and the profit on each standard telephone is $6, the profit function is $P(d, s) = 9d + 6s$.

(continued on the next page)

Substitute values.

$$P(0,0) = 9(0) + 6(0) = 0$$
$$P(0,4) = 9(0) + 6(4) = 24$$
$$P(2,3) = 9(2) + 6(3) = 36$$
$$P(4,0) = 9(4) + 6(0) = 36$$

Answer the problem.

This problem has alternate optimal solutions. The company will make the same profit if they make and sell 2 deluxe telephones and 3 standard telephones as it will from making and selling 4 deluxe telephones and no standard telephones.

CHECKING FOR UNDERSTANDING

Communicating Mathematics

Read and study the lesson to answer each question.

1. **Define** linear programming in your own words.

2. **Explain** why the vertex theorem is useful when solving linear programming problems.

3. **Explain**, in your own words, under what conditions a linear programming problem has alternate optimal solutions.

Guided Practice

The Champion Lumber Company converts logs into lumber or plywood. In a given week, the total production cannot exceed 800 units, of which 200 units of lumber and 300 units of plywood are required by regular customers. The profit on a unit of lumber is $20, and the profit on a unit of plywood is $30. Let x represent the units of lumber and let y represent the units of plywood.

4. Write an inequality to represent the total production.

5. Write an inequality to represent the production of lumber.

6. Write an inequality to represent the production of plywood.

7. Write an equation to represent the total profit.

8. Graph the polygonal convex set determined by the inequalities.

9. Name the vertices of the polygon.

10. Find the number of units each of lumber and plywood that should be produced to maximize profit.

11. What is the maximum profit?

Graph each system of inequalities. Then, state whether the situation is *infeasible*, has *alternate optimal solutions*, or is *unbounded*. In each system, assume that $x \geq 0$ and $y \geq 0$.

12. $3y \geq 25$
    $x + y \leq 30$
    $f(x, y) = 3x + 3y$

13. $2x + y \geq 15$
    $3x + y \leq 12$
    $f(x, y) = 2x + 2y$

14. $2x + 2y \geq 10$
    $2x + y \geq 8$
    $2x + 3y \geq 30$
    $f(x, y) = x + y$

EXERCISES

Solve each problem, if possible. If not possible, state whether the problem is *infeasible*, has *alternate optimal solutions*, or is *unbounded*.

15. **Manufacturing**   The Cruiser Bicycle Company makes two styles of bicycles: the Traveler, which sells for $200, and the Tourister, which sells for $600. Each bicycle has the same frame and tires, but the assembly and painting time required for the Traveler is only one hour, while it takes three hours for the Tourister. There are 300 frames and 360 hours of labor available for production. How many of each model should be produced to maximize revenue?

16. **Manufacturing**   A manufacturer makes widgets and gadgets. At least 500 widgets and 700 gadgets are needed to meet minimum daily demands. The machinery can produce no more than 1200 widgets and 1400 gadgets per day. The combined number of widgets and gadgets that the packaging department can handle is 2300 per day.

    a. If the company sells widgets for 40 cents each and gadgets for 50 cents each, how many of each item should be produced for maximum daily income? What is the maximum daily income?

    b. Suppose the cost of producing a widget is 7 cents and the cost of producing a gadget is 18 cents. How many widgets and gadgets should be produced for maximum daily profit? What is the maximum daily profit?

17. **Nutrition**   A diet is to include at least 140 mg of Vitamin A and at least 145 mg of Vitamin B. These requirements can be obtained from two types of food. Type X contains 10 mg of Vitamin A and 20 mg of Vitamin B per pound. Type Y contains 30 mg of Vitamin A and 15 mg of Vitamin B per pound. If type X food costs $12 per pound and type Y food cost $8 per pound, how many pounds of each type of food should be purchased to satisfy the requirements at the minimum cost?

18. **Manufacturing**   The SwingWell Company produces two types of golf clubs: the Driver, which sells for $30, and the Master, which sells for $40. SwingWell has more orders for the upcoming month than it is capable of producing. Using the production schedule below, what is the maximum revenue that SwingWell should anticipate for the coming month?

Process	Driver	Master	Time Available
Cutting	2 min	2 min	$166\frac{2}{3}$ h
Assembly	1 min	3 min	150 h
Finishing	2 min	3 min	200 h

19. **Contracting** The BJ Electrical Company needs to hire master electricians and apprentices for a one-week project. Master electricians receive a salary of $750 per week and apprentices receive $350 per week. As part of its contract, the company has agreed to hire at least 30 workers. The local Building Safety Council recommends that each master electrician spend three hours for inspection time during the project. This project should require 25 hours of inspection time. How many of each type of worker should be hired to accomplish the project and still meet the contract safety requirements?

20. **Manufacturing** A company makes two models of light fixtures, A and B, each of which must be assembled and packed. The time required to assemble model A is 12 minutes, and model B takes 18 minutes. It takes 2 minutes to package model A and 1 minute to package model B. Each week there are available 240 hours of assembly time and 20 hours for packing.

    a. If model A sells for $1.50 and model B sells for $1.70, how many of each model should be made to obtain the maximum weekly income? What is the maximum weekly income?

    b. Suppose the cost to produce model A is 75¢ and the cost to produce model B is 85¢. How many model A light fixtures and model B light fixtures should be produced for maximum weekly profit? What is the maximum weekly profit?

21. **Energy** A chemical company uses two types of fuel for heating and processing. At least 3800 gallons of fuel are used each day. The burning of each gallon of #1 crude leaves a residue of 0.02 pounds of ash and 0.06 pounds of soot. Each gallon of #2 crude leaves a residue of 0.05 pounds of ash and 0.01 pounds of soot. The factory needs at least 120 pounds of ash and at least 136 pounds of soot each day. If #1 crude costs $1.50 per gallon and #2 crude costs $1.10 per gallon, then how many gallons of each type should be purchased in order to minimize costs?

22. **Manufacturing** The File-Away Company sells letter-size file cabinets for $138 and legal-size file cabinets for $176.50. The company has 120,000 square feet of sheet metal in stock. It takes 36 square feet to make the letter-size file cabinet and 54 square feet to make the legal-size file cabinet. The board of directors has allocated the following labor costs.

Cost	Letter Size	Legal Size	Total Available
Direct Labor	$42	$52	$375,000
Indirect Labor	$18	$21	$194,500

Determine the optimum number of each type of cabinet the company should make to maximize their revenues.

23. **Marketing** Yummy Ice Cream conducted a survey and found that people liked their black walnut flavor three times more than their tutti-frutti flavor. One distributor wants to order at least 20,000 gallons of the tutti-frutti flavor. The company has all of the ingredients to produce both flavors, but it has only 45,000 gallon-size containers available. If each gallon of ice cream sells for $2.95, how many gallons of each type flavor should the company produce?

24. The system below forms a polygonal convex set.

$$x \le 0$$
$$y \le 10$$
$$y \ge -x, \text{ when } -6 \le x \le 0$$
$$2x + 3y \ge 6 \text{ when } -12 \le x \le -6$$

What is the area of the closed figure?

25. Multiply $\begin{bmatrix} 8 & -7 \\ -4 & 0 \end{bmatrix}$ by $\frac{3}{4}$. **(Lesson 2-2)**

26. Find the multiplicative inverse of $\begin{bmatrix} 2 & 1 \\ -3 & 2 \end{bmatrix}$. **(Lesson 2-3)**

27. State the row operations you would use to get a zero in the second column of row one of $\begin{bmatrix} 3 & 3 & -9 \\ -2 & 1 & -4 \end{bmatrix}$. Then state the row operations you would use to get a zero in the first column of row two. **(Lesson 2-4)**

28. Find the maximum and minimum values of $f(x, y) = x + 2y$ if it is defined for the polygonal convex set having vertices at $(0, 0)$, $(4, 0)$, $(3, 5)$, and $(0, 5)$. **(Lesson 2-5)**

29. College Entrance Exam Choose the best answer. $\sqrt{\dfrac{\sqrt{25}}{5}} =$

(A) 1 **(B)** $\sqrt{2}$ **(C)** 2 **(D)** 5 **(E)** $5\sqrt{2}$

CASE STUDY FOLLOW-UP

Refer to Case Study 1: *Buying a Home*, on pages 956–958.

Manuel Tijera plans to rent one house and purchase another. He estimates that mortgage interest deductions on his federal income tax will save him 25% on his mortgage payments. He wants to spend no more than $800 per month on rent and no more than $2000 per month on rent and mortgage payments, including tax savings, combined.

1. Use linear programming to find the maximum total monthly amount he might have to pay, factoring in his tax savings.

2. If he chooses this option, what is the maximum amount he can borrow on a 15-year 9% FRM?

3. Research Find out about local median rental costs and new home prices. Write a 1-page paper analyzing the advantages and disadvantages of renting versus buying a home in your city.

VOCABULARY

Upon completing this chapter, you should be
familiar with the following terms:

alternate optimal solutions	**93**	**91**	linear programming
augmented matrix	**79**	**64**	matrix
column matrix	**64**	**71**	minor
consistent	**57**	**64**	nth order
constraint	**91**	**87**	polygonal convex set
dependent	**57**	**64**	row matrix
determinant	**71**	**66**	scalar
dimensions	**64**	**64**	square matrix
element	**64**	**57**	substitution method
elimination method	**57**	**56**	system of equations
inconsistent	**57**	**86**	system of linear inequalities
independent	**57**	**93**	unbounded
infeasibility	**92**	**87**	vertex theorem
$m \times n$ matrix	**64**	**65**	zero matrix

SKILLS AND CONCEPTS

OBJECTIVES AND EXAMPLES

Upon completing this chapter, you should
be able to:

- solve systems of equations algebraically
  (Lesson 2-1)

Solve the system of equations below.
$y = -x + 2$
$3x + 4y = 2$

$$3x + 4y = 2 \qquad y = -x + 2$$
$$3x + 4(-x + 2) = 2 \qquad = -6 + 2$$
$$x = 6 \qquad = -4 \qquad (6, -4)$$

- add, subtract, and multiply matrices
  (Lesson 2-2)

$$-2\begin{bmatrix} 1 & -4 \\ 0 & 3 \end{bmatrix} = \begin{bmatrix} -2(1) & -2(-4) \\ -2(0) & -2(3) \end{bmatrix}$$
$$= \begin{bmatrix} -2 & 8 \\ 0 & -6 \end{bmatrix}$$

REVIEW EXERCISES

Use these exercises to review and prepare for
the chapter test.

**Solve each system of equations
algebraically.**

1. $y = -2x$
   $x + y = -2$

2. $x - y = 5$
   $x = 6y$

3. $5y - 2x = 0$
   $3y + x = -1$

4. $y - 6x = 1$
   $-15x + 2y = -4$

**Use matrices A, B, and C to find each sum,
difference, or product.**

$$A = \begin{bmatrix} 7 & 8 \\ 0 & -4 \end{bmatrix} \quad B = \begin{bmatrix} -3 & -5 \\ 2 & -2 \end{bmatrix} \quad C = \begin{bmatrix} 2 \\ -5 \end{bmatrix}$$

5. $A + B$ **6.** $B - A$ **7.** $3B$

8. $-4C$ **9.** AB **10.** BC

OBJECTIVES AND EXAMPLES	REVIEW EXERCISES

■ evaluate determinants **(Lesson 2-3)**

Find the value of $\begin{vmatrix} -2 & 5 \\ 6 & -7 \end{vmatrix}$.

$$\begin{vmatrix} -2 & 5 \\ 6 & -7 \end{vmatrix} = (-2)(-7) - 6(5) = -16$$

Find the value of each determinant.

11. $\begin{vmatrix} 7 & -4 \\ 5 & -3 \end{vmatrix}$ 12. $\begin{vmatrix} 8 & -4 \\ -6 & 3 \end{vmatrix}$

13. $\begin{vmatrix} 5 & 0 & 4 \\ 7 & 3 & -1 \\ 2 & -2 & 6 \end{vmatrix}$ 14. $\begin{vmatrix} 3 & -1 & 4 \\ 5 & -2 & 6 \\ 7 & 3 & -4 \end{vmatrix}$

■ solve systems of equations by using inverses of matrices. **(Lesson 2-3)**

Solve $\begin{bmatrix} 3 & -5 \\ -2 & 2 \end{bmatrix} \cdot \begin{bmatrix} x \\ y \end{bmatrix} = \begin{bmatrix} 1 \\ -2 \end{bmatrix}$.

$$-\frac{1}{4}\begin{bmatrix} 2 & 5 \\ 2 & 3 \end{bmatrix} \cdot \begin{bmatrix} 3 & -5 \\ -2 & 2 \end{bmatrix} \cdot \begin{bmatrix} x \\ y \end{bmatrix} = -\frac{1}{4}\begin{bmatrix} 2 & 5 \\ 2 & 3 \end{bmatrix} \cdot \begin{bmatrix} 1 \\ -2 \end{bmatrix}$$

$$\begin{bmatrix} x \\ y \end{bmatrix} = \begin{bmatrix} 2 \\ 1 \end{bmatrix}$$

Solve each matrix equation.

15. $\begin{bmatrix} 2 & 5 \\ -1 & -3 \end{bmatrix} \cdot \begin{bmatrix} x \\ y \end{bmatrix} = \begin{bmatrix} 1 \\ 2 \end{bmatrix}$

16. $\begin{bmatrix} 3 & 2 \\ -6 & 4 \end{bmatrix} \cdot \begin{bmatrix} x \\ y \end{bmatrix} = \begin{bmatrix} -3 \\ 6 \end{bmatrix}$

17. $\begin{bmatrix} -3 & 5 \\ -2 & 4 \end{bmatrix} \cdot \begin{bmatrix} x \\ y \end{bmatrix} = \begin{bmatrix} 1 \\ -2 \end{bmatrix}$

■ solve systems of equations by using augmented matrices **(Lesson 2-4)**

The solution represented by the reduced

augmented matrix $\begin{bmatrix} 1 & 0 & 0 & -4 \\ 0 & 1 & 0 & \frac{1}{3} \\ 0 & 0 & 1 & 6 \end{bmatrix}$ is

$(-4, \frac{1}{3}, 6)$.

Solve each system of equations by using augmented matrices.

18. $x - 2y - 3z = 2$
    $x - 4y + 3z = 14$
    $-3x + 5y + 4z = 0$

19. $2x + 3y - 4z = 5$
    $x + y + 2z = 3$
    $-x + 2y - 6z = 4$

■ find the maximum or minimum value of a function defined for a polygonal convex set **(Lesson 2-5)**

Find the maximum and minimum values of $f(x, y) = 4y + x - 3$ for the polygonal convex set graphed at right.

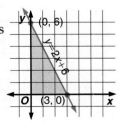

$f(0, 0) = 4(0) + 1(0) - 3 = -3$ *minimum*

$f(3, 0) = 4(0) + 1(3) - 3 = 0$

$f(0, 6) = 4(6) + 1(0) - 3 = 21$ *maximum*

Solve each system of inequalities by graphing and name the coordinates of the vertices of each polygonal convex set. Then, find the maximum and minimum values for the function $f(x, y) = 3y + 2x - 4$.

20. $x \geq 1$
    $y \geq -2$
    $y \leq 6 - x$
    $y + 2x \leq 10$

21. $x \geq 0$
    $y \geq 4$
    $2y \leq 18 - x$
    $x \leq 6$
    $y \leq 11 - x$

■ recognize situations where exactly one solution to a linear programming problem may not exist **(Lesson 2-6)**

The system of inequalities graphed at the right represents an *unbounded* situation because the constraints cannot be satisfied simultaneously.

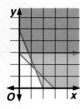

Graph each system of inequalities. Then, state whether the situation is *infeasible*, has *alternate optimal solutions*, or is *unbounded*. In each system, assume that $x \geq 0$ and $y \geq 0$.

22. $x + 3y \leq 9$
$6x + 3y \leq 18$
$f(x, y) = x + y$

23. $2x \geq 14$
$x + 3y \leq 6$
$f(x, y) = 2x + y$

APPLICATIONS AND PROBLEM SOLVING

24. Sports In a three-team track meet, the following numbers of first-, second-, and third-place finishes were recorded.

School	First Place	Second Place	Third Place
Boardman	2	5	5
Girard	8	2	3
Niles	6	4	1

Use matrices to find the final scores for each school if 5 points are awarded for first place, 3 for second place, and 1 for third place. **(Lesson 2-2)**

25. Geometry The perimeter of a triangle is 83 inches. The longest side is three times the length of the shortest side and 17 inches more than one-half the sum of the other two sides. Find the length of each side. **(Lesson 2-4)**

26. Manufacturing A toy manufacturer produces two types of model spaceships, the Voyager and the Explorer. Each of the toys requires the same three operations: plastic molding, machining, and bench assembly. Each Voyager requires 5 minutes for molding, 3 minutes for machining, and 5 minutes for assembly. Each Explorer requires 6 minutes for molding, 2 minutes for machining, and 18 minutes for assembly. The manufacturer can afford a daily schedule of not more than 4 hours for molding, 2 hours for machining, and 9 hours for assembly. **(Lesson 2-6)**

  a. If the profit is $2.40 on each Voyager and $5.00 on each Explorer, how many of each toy should be produced for maximum profit?

  b. What is the maximum daily profit?

Solve each system of equations.

1. $x - 4 = y$
   $y = 2x - 8$

2. $x + 5 = y$
   $3x = y - 1$

3. $y - 3x = 8$
   $x + y = 4$

4. $5 - 6x = y$
   $2x + 7y = 5$

Use matrices $A, B, C,$ and D to find each sum, difference, or product.

$$A = \begin{bmatrix} 5 & 4 \\ -1 & -2 \end{bmatrix} \qquad B = \begin{bmatrix} -1 & -2 \\ 5 & 4 \end{bmatrix} \qquad C = \begin{bmatrix} -2 & 4 & 6 \\ 5 & -7 & -1 \end{bmatrix} \qquad D = \begin{bmatrix} 1 & -2 \\ 0 & 4 \\ -3 & 4 \end{bmatrix}$$

5. $5D$

6. $2A + B$

7. $2B - A$

8. CD

9. $AB + CD$

Find the value of each determinant.

10. $\begin{vmatrix} -3 & 5 \\ 1 & -4 \end{vmatrix}$

11. $\begin{vmatrix} 8 & 5 \\ -3 & -2 \end{vmatrix}$

12. $\begin{vmatrix} 2 & 1 & -1 \\ 6 & 4 & -3 \\ 0 & 2 & -2 \end{vmatrix}$

13. $\begin{vmatrix} 3 & 1 & 2 \\ -2 & 0 & 4 \\ 3 & 5 & 2 \end{vmatrix}$

14. Solve $\begin{bmatrix} -1 & 4 \\ 7 & 0 \end{bmatrix} \cdot \begin{bmatrix} x \\ y \end{bmatrix} = \begin{bmatrix} -1 \\ -14 \end{bmatrix}$.

Solve each system of equations using augmented matrices.

15. $x + 2y + z = 3$
    $2x - 3y + 2z = -1$
    $x - 3y + 2z = 1$

16. $-3x + y + z = 2$
    $5x + 2y - 4z = 21$
    $x - 3y - 7z = -10$

Solve each system of inequalities by graphing and name the coordinates of the vertices of each polygonal convex set. Then, find the maximum and minimum values for the function $f(x, y) = 5y + 3x$.

17. $x \geq 0$
    $y \geq 0$
    $y \leq x + 3$
    $2x + y \leq 8$

18. $y \geq 0$
    $x \geq 1$
    $x + y \leq 6$
    $y + 3x \leq 12$

19. $y \leq 5$
    $x \leq 3$
    $y + 2x \geq 0$
    $y \geq x - 2$

20. **Carpentry**   Emilio has a small carpentry shop in his basement to make bookcases. He makes two sizes, large and small. His profit on a large bookcase is $80, and his profit on a small bookcase is $50. It takes Emilio 6 hours to make a large bookcase and 2 hours to make a small one. He can spend only 24 hours each week on his carpentry work. He must make at least two of each size each week. What is the maximum weekly profit?

Bonus

Name the polygonal boundary formed by the set of all points with coordinates (x, y) that satisfy all of the constraints below when n is a positive number.

$n \leq 2y \leq 4n$ $x + n \geq y$

$2n \leq x + y \leq 5n$ $y + n \geq x$

$n \leq 2x \leq 4n$

CHAPTER 3

THE NATURE OF GRAPHS

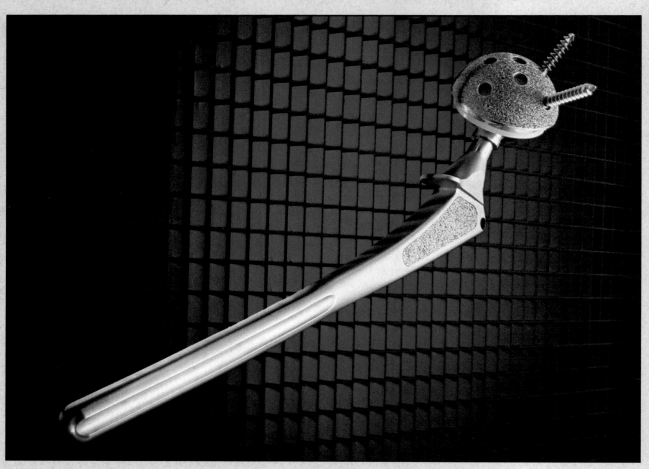

HISTORICAL SNAPSHOT

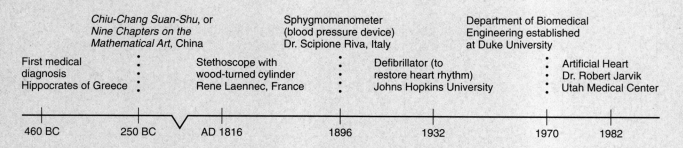

	Chiu-Chang Suan-Shu, or *Nine Chapters on the* *Mathematical Art*, China		Sphygmomanometer (blood pressure device) Dr. Scipione Riva, Italy		Department of Biomedical Engineering established at Duke University	
First medical diagnosis Hippocrates of Greece		Stethoscope with wood-turned cylinder Rene Laennec, France		Defibrillator (to restore heart rhythm) Johns Hopkins University		Artificial Heart Dr. Robert Jarvik Utah Medical Center
460 BC	250 BC	AD 1816	1896	1932	1970	1982

CHAPTER OBJECTIVES

In this chapter, you will:
- Graph relations and functions.
- Analyze families of graphs.
- Study about symmetry, continuity, and transformations of graphs.
- Find asymptotes, intercepts, and critical points of graphs.

CAREER GOAL: Biomedical Research

Around the world, more than 3 million people live with artificial body parts. Sheila Chuang wants to be a part of this emerging technology. As a biomedical researcher, Sheila wants "to help people and to have an impact on the health system." She is also considering going to medical school because "I dream of developing and implanting artificial organs."

Sheila wanted to choose a field that would combine her love for mathematics with her interests in chemistry and biology. She found that "biomedical engineering provided the challenge of engineering, yet it was the one area of engineering that seemed to have the most direct contact with human beings."

Biomedical researchers use mathematics in many ways. For example, Sheila says, "After angioplasty to open up a clogged artery, a permanent metal *stent* is implanted to hold the artery open. Duke is currently working on the development of a bioabsorbable stent that will degrade over time and be eliminated by the body. However, we want to be able to put drugs into the stent for delivery into the body. We use graphs to measure the rate at which a given drug will be delivered by the stent."

One summer, Sheila worked in biomechanics under a fellowship from the University of Minnesota Biomedical Engineering Center. She worked with orthopedic surgeons on advances in prostheses for knees and hips, like the one shown in the photo. She says, "I really saw the close relationship between biomedical engineering and medicine."

Most biomedical experts agree that sometime in the near future, an artificial part will be able to take the place of almost any part of the human anatomy, except the brain. Sheila says, "Biomedical engineering is a new and growing field, and it is really neat to be a part of it!"

interNET CONNECTION

For up-to-date information on biomedical engineering, visit:
www.glencoe.com/sec/math/amc/mathnet

3-1A Graphing Calculators: Graphing Polynomial Functions

You can use a graphing calculator to graph polynomial functions and approximate the real zeros of a function. When using a graphing calculator to approximate zeros, it is important to view a complete graph of the function before zooming in on a certain point. Zeros may be overlooked if they are not in the original viewing window. A complete graph of a polynomial function shows all of the important characteristics of the graph, such as all the *x*- and *y*-intercepts, all relative minimum and maximum points, and the end behavior of the graph.

Example 1

Graph $f(x) = 2x^3 + 5x^2 - 8x + 4$ so that a complete graph appears. Then approximate each of the real zeros to the nearest hundredth.

First, graph the function in the standard viewing window.

Enter: [Y=] 2 [X,T,θ] [^] 3 [+] 5 [X,T,θ] [x²] [−] 8 [X,T,θ] [+] 4 [GRAPH]

This viewing window does not accommodate a complete graph because the relative maximum cannot be seen. Change the viewing window to [–5, 4] by [–5, 30] with a scale factor of 1 for the *x*-axis and a scale factor of 5 for the *y*-axis. Regraph by pressing [GRAPH].

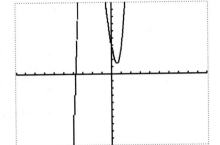

This viewing window accommodates the complete graph. We can tell by the graph that there is exactly one real zero. Tracing and zooming in reveals that the real zero is approximately –3.72.

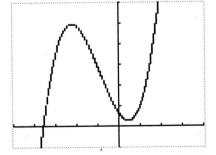

The function in Example 1 is a third-degree function because the highest power of *x* in the equation is 3. A third-degree function has three zeros. For a second-degree function, the highest power of *x* in the equation is 2, and a second-degree function has two zeros, and so on. When using your graphing calculator to approximate real zeros, it is helpful to know that a function of degree *n* has *n* zeros. The zeros may not all be real, however. Complex zeros, or imaginary zeros, occur in conjugate pairs, but they do not lie on the *x*-axis in the coordinate plane.

You can always determine the number of real zeros of a function by finding the number of times the graph crosses the x-axis. Functions of odd degree must have at least one real zero because they must cross the x-axis at least once. Looking at various graphs of linear and cubic equations can illustrate this point. Functions of even degree, however, may not have any real roots and may not cross the x-axis at all. The graph of a parabola, such as $y = x^2 + 4$, is one such function. Graphs of polynomial functions of degree 2 or higher can cross the x-axis more than once.

Example 2

Graph $p(x) = x^4 - 3x^3 - 7x^2 - x + 2$ so that a complete graph is shown. Then approximate each of the real zeros to the nearest hundredth.

First, try the standard viewing window.

Enter: [Y=] [X,T,θ] [^] 4 [−] 3 [X,T,θ] [^] 3
[−] 7 [X,T,θ] [x²] [−] [X,T,θ] [+]
2 [GRAPH]

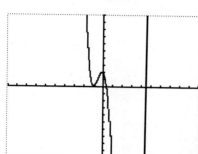

A complete graph is not shown in the standard viewing window. Try changing the range to [-4, 6] by [-70, 10] with a scale factor of 1 for the x-axis and a scale factor of 10 for the y-axis.

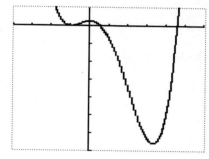

This shows a complete graph. Now trace and zoom in on the real zeros. They are $-1, 4.56$, and 0.44 to the nearest hundredth.

EXERCISES

Graph each function so that a complete graph is shown. Then approximate each of the real zeros to the nearest hundredth and sketch the graph on a piece of paper.

1. $f(x) = x^5 + 2x^2 - 4$

2. $h(x) = x^4 - 3x^3 + x - 4$

3. $p(x) = 3x^3 + 4x^2 - 8$

4. $m(x) = 2x^4 - 3x^2 + 9$

5. $g(x) = x^3 - 2x^2 - 3x + 6$

6. $f(x) = x^3 - 7x^2 + 6x - 1$

7. $r(x) = x^4 + 5x^3 - 3x^2 - 7x - 2$

8. $g(x) = 4x^4 + 7x + 2$

9. $q(x) = x^5 + 2x^4 - 11x^3 + 4x^2 - 3x + 4$

10. $p(x) = -x^5 - 3x^4 + 9x^3 + 18x^2 - 7x - 13$

3-1 Symmetry

Objectives

After studying this lesson, you should be able to:
- identify symmetrical graphs,
- use symmetry to complete a graph, and
- identify an odd function and an even function.

Application

One of the smallest butterflies is the Western Pygmy Blue butterfly of North America. Its wingspan is only 1 centimeter. The largest butterfly, the Queen Alexandra Birdwing butterfly of Papua, New Guinea, has a wingspan of 28 centimeters. All butterflies have wings that are symmetrical.

Butterflies cannot fly if their body temperature is less than 86° F. They warm themselves in the sun or flutter their wings while perched to generate enough body heat to fly when the air temperature is below 86°.

A knowledge of symmetry can often help you sketch and analyze graphs. A graph may have **point symmetry**, **line symmetry**, neither, or both. If a graph has point or line symmetry, you can plot points to sketch part of the graph, then use symmetry to sketch the rest.

Point Symmetry	**Two distinct points *P* and *P′* are symmetric with respect to a point, *M*, if and only if *M* is the midpoint of $\overline{PP'}$. Point *M* is symmetric with respect to itself.**

When the definition of symmetry is extended to a set of points, it must be true for each point in the set. A figure with point symmetry can be turned about a center point and, in less than a full turn, the image coincides with the original figure. Each of the figures below has point symmetry with respect to the point labeled in the diagram.

Symmetry with respect to a given point M can also be expressed as symmetry about point M.

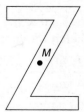

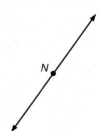

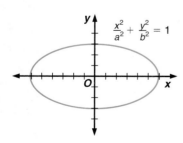

The graph of a function is a set of points. The graph of a function may or may not have point symmetry.

Determine if the graph of $f(x) = x - 3$ has point symmetry about the x-intercept 3.

1. Graph the function.
2. Draw a vertical line, v, through the x-intercept parallel to the y-axis. Fold the graph paper along the x-axis. Then fold it along line v.
3. Hold the paper up to the light. Notice that the part of the graph above the x-axis now matches up with the part of the graph below the x-axis. This shows that the graph of $f(x)$ has point symmetry about the x-intercept.

One of the most common points of symmetry is the origin. Observe that the graph of $f(x) = \dfrac{1}{x}$, shown below, has symmetry with respect to the origin. Use this table to compare the range of $f(x)$ when the domain is positive with the range of $f(x)$ when the domain is negative.

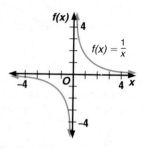

$f(x), x > 0$	$f(x), x < 0$
$f(1) = 1$	$f(-1) = -1$
$f(2) = \dfrac{1}{2}$	$f(-2) = -\dfrac{1}{2}$
$f(3) = \dfrac{1}{3}$	$f(-3) = -\dfrac{1}{3}$
$f(4) = \dfrac{1}{4}$	$f(-4) = -\dfrac{1}{4}$
$f(n) = \dfrac{1}{n}, n \neq 0$	$f(-n) = -\dfrac{1}{n}, n \neq 0$

The values in the table above suggest that $f(-x) = -f(x)$, when a function is symmetric with respect to the origin.

Symmetry with Respect to the Origin	**The graph of a relation S is symmetric with respect to the origin if and only if $(-a, -b) \in S$ whenever $(a, b) \in S$. A function $f(x)$ has a graph that is symmetric with respect to the origin if and only if $f(-x) = -f(x)$.**

$\in$ *means belongs to.*

You can verify symmetry with respect to the origin algebraically, as shown in Example 1.

Example 1

Determine whether the graph of $f(x) = x^5$ is symmetric with respect to the origin.

The graph of $f(x) = x^5$ appears to be symmetric with respect to the origin. Verify this algebraically.

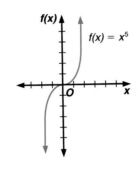

$f(x) = x^5$
$f(-x) = (-x)^5$ *Replace x with –x.*
$f(-x) = -(x^5)$
$f(-x) = -f(x)$ *Replace x^5 with f(x).*

Since $f(-x) = -f(x)$, the graph of $f(x) = x^5$ is symmetric with respect to the origin.

Another type of symmetry is line symmetry.

Line Symmetry	**Two distinct points P and P' are symmetric with respect to a line ℓ if and only if ℓ is the perpendicular bisector of $\overline{PP'}$. A point P is symmetric to itself with respect to line ℓ if and only if P is on ℓ.**

Each of the graphs below has line symmetry. The equation of each line of symmetry is given. Graphs that have line symmetry can be folded along the line of symmetry so that the two halves match exactly. Some graphs, such as the graph of an ellipse, have more than one line of symmetry.

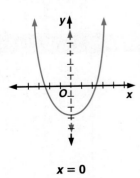

$x = 0$

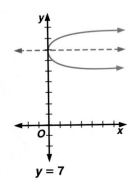

$y = 7$

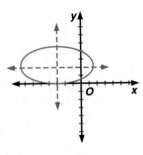

$x = -3$ and $y = 2$

Some common lines of symmetry are the *x*-axis, the *y*-axis, the line $y = x$, and the line $y = -x$. The following table shows how the coordinates of symmetric points are related for each of these common lines of symmetry.

Symmetry with Respect to	Definition	Sample Graph
x-axis	$(a, -b) \in S$ if and only if $(a, b) \in S$. *Example:* $(2, \sqrt{6})$, and $(2, -\sqrt{6})$ are on the graph.	 $(2, \sqrt{6})$ $(2, -\sqrt{6})$ $x = y^2 - 4$
y-axis	$(-a, b) \in S$ if and only if $(a, b) \in S$. *Example:* $(2, 8)$ and $(-2, 8)$ are on the graph.	 $(-2, 8)$ $(2, 8)$ $y = -x^2 + 12$
the line $y = x$	$(b, a) \in S$ if and only if $(a, b) \in S$. *Example:* $(2, 3)$ and $(3, 2)$ are on the graph.	 $(2, 3)$ $y = x$ $(3, 2)$ $xy = 6$
the line $y = -x$	$(-b, -a) \in S$ if and only if $(a, b) \in S$. *Example:* $(4, 5)$ and $(-5, -4)$ are on the graph.	 $(4, 5)$ $y = -x$ $(-5, -4)$

Example 2

Determine whether the graph of
$f(x) = x^4 + 3x^2 + 8$ is symmetric with
respect to the x-axis, the y-axis, neither,
or both.

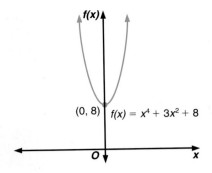

The graph appears to be symmetric with
respect to the y-axis, but not the x-axis.

Test for symmetry with respect to
the x-axis. For the graph of $f(x)$ to be
symmetric with respect to the x-axis,
point (a, b) is on the graph if and only if
point $(a, -b)$ is on the graph.

Suppose $a = 2$. Then $b = 2^4 + 3(2)^2 + 8$ or 36, and $-b = -36$. The point $(2, 36)$
is on the graph. However, $(2, -36)$ cannot be on the graph since b is never
negative. While a single point cannot prove that symmetry exists, it can
show that symmetry does not exist. Thus, there is no symmetry with
respect to the x-axis.

Test for symmetry with respect to the y-axis. For the graph of $f(x)$ to be
symmetric with respect to the y-axis, point (a, b) is on the graph if and only
if point $(-a, b)$ is on the graph.

$$\text{For } a: \quad b = a^4 + 3a^2 + 8$$
$$\text{For } -a: \quad b = (-a)^4 + 3(-a)^2 + 8$$
$$= a^4 + 3a^2 + 8$$

So, in general, point $(-a, b)$ is on the graph if and only if point (a, b) is on the
graph. Thus, the graph is symmetric with respect to the y-axis.

The line $y = x$ includes the origin and forms a 45° angle with both the x- and
y-axes. It has a slope of 1. The line $y = -x$ also includes the origin and forms a
45° angle with both the axes, but its slope is −1.

Example 3

Determine whether the graph of the function
$xy = 6$ is symmetric with respect to the line
$y = x$, the line $y = -x$, neither, or both.

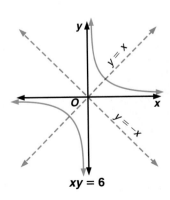

If the graph is symmetric with respect to the
line $y = x$, (a, b) and (b, a) must both be on the
graph. The point (a, b) is on the graph if $ab = 6$.
The point (b, a) is on the graph if $ba = 6$. Since
these statements are equivalent, the graph is
symmetric with respect to the line $y = x$.

If the graph is symmetric with respect to the line $y = -x$, (a, b) and $(-b, -a)$ must both be on the graph. The point (a, b) is on the graph if $ab = 6$. The point $(-b, -a)$ is on the graph if $(-b)(-a) = 6$. But $(-b)(-a) = ba$ and $ba = ab$. So both of these statements are true, and the graph is symmetric with respect to the line $y = -x$.

Symmetry can be used to sketch the graph of a function. If you have part of the graph and know what types of symmetry the function has, the rest of the graph can be sketched without graphing more ordered pairs.

Example 4

The path of a moon orbiting a planet can be modeled by the graph of an ellipse with the equation $\dfrac{x^2}{25} + \dfrac{y^2}{16} = 1$. The graph is symmetric with respect to both the x- and y-axes. If the moon travels through $(5, 0)$, $(0, -4)$, and $(-3, -3.2)$ on the model, complete a sketch of the path of the moon.

Graph the given ordered pairs.

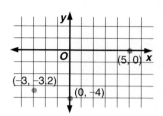

Since the graph is symmetric with respect to the x-axis, every point on one side of the x-axis has a corresponding point on the other side of the axis.

$$(0, -4) \quad \rightarrow \quad (0, 4)$$
$$(-3, -3.2) \quad \rightarrow \quad (-3, 3.2)$$

Graph these ordered pairs.

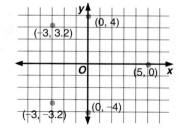

Since the graph is symmetric with respect to the y-axis, every point on either side of the y-axis has a corresponding point on the other side of the axis.

$$(-3, -3.2) \quad \rightarrow \quad (3, -3.2)$$
$$(-3, 3.2) \quad \rightarrow \quad (3, 3.2)$$
$$(5, 0) \quad \rightarrow \quad (-5, 0)$$

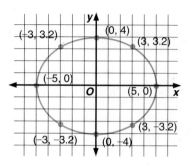

Graph these ordered pairs.
Connect all the points to form the ellipse.

You will learn more about polynomial functions in Lesson 4.1.

Functions whose graphs are symmetric with respect to the *y*-axis are **even functions**. If a function of the form $f(x) = a_0x^n + a_1x^{n-1} + \ldots + a_{n-2}x^2 + a_{n-1}x + a_n$, called a polynomial function, has exponents that are all even, then the function is an even function. Functions whose graphs are symmetric with respect to the origin are **odd functions**. Odd polynomial functions have exponents that are all odd. Some functions are neither even nor odd.

even functions
$f(-x) = f(x)$

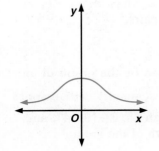 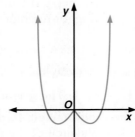

odd functions
$f(-x) = -f(x)$

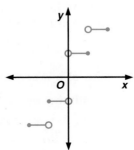 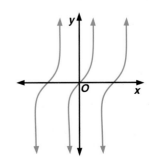

You can check whether a function is odd, even, or neither by using the TRACE feature on a graphing calculator.

EXPLORATION: Graphing Calculator

Determine whether $f(x) = x^8 - 3x^4 + 2x^2 + 2$ is odd, even, or neither.

1. Press the [Y=] key and enter the equation. Press [GRAPH].

2. The graph appears to be symmetric about the *y*-axis. Check this by determining whether $f(-x) = f(x)$. Use the TRACE function.

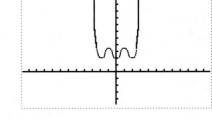

    X= .94736842   Y=2.0273182
    X=- .94736842   Y=2.0273182

    These are sample comparisons.

3. It appears that $f(-x) = f(x)$. Verify this algebraically.

$f(x) = x^8 - 3x^4 + 2x^2 + 2$
$f(-x) = (-x)^8 - 3(-x)^4 + 2(-x)^2 + 2$ *Replace x with –x.*
$f(-x) = x^8 - 3x^4 + 2x^2 + 2$ *Simplify.*
$f(-x) = f(x)$ *Replace $x^8 - 3x^4 + 2x^2 + 2$ with $f(x)$.*

The graph is symmetric about the *y*-axis and all the exponents of the polynomial are even. So, $f(x) = x^8 - 3x^4 + 2x^2 + 2$ is an even function.

CHECKING FOR UNDERSTANDING

Communicating Mathematics

Read and study the lesson to answer each question.

1. **Tell** what point is one of the most common points of symmetry.

2. **Sketch** a graph that has more than one line of symmetry.

3. **Complete** each sentence.
   a. If a graph of a function $f(x)$ is symmetric with respect to the origin, then $f(-x) = $ __?__ .
   b. If a graph of a function $f(x)$ is symmetric with respect to the y-axis, then $f(-x) = $ __?__ .
   c. The line $y = x$ forms a __?__ angle with both the x- and y-axes.

4. **Explain** why $f(x)$ cannot be negative in Example 2.

Guided Practice

State whether each figure has point symmetry, line symmetry, neither, or both.

5.

6.

7.

8.

9.

10.

Determine the line(s) of symmetry for each graph.

11.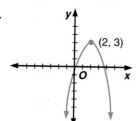

12.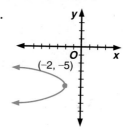

13.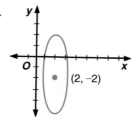

Graph each point. Then graph and record the coordinates of the point P' for each of the following conditions: P and P' are symmetric with respect to (a) the x-axis, (b) the y-axis, (c) the line $y = x$, and (d) the line $y = -x$.

14. $P(3, 4)$

15. $P(-2, -5)$

16. $P(4, -8)$

EXERCISES

Practice   **Find the coordinates of P' if P and P' are symmetric with respect to point M.**

17. $P(-2, -2), M(1, 1)$ **18.** $P(-3, 2), M(0, 0)$ **19.** $P(-3, 7), M(-3, 1)$

The graphs below are portions of complete graphs. Sketch a complete graph for each of the following symmetries: with respect to (a) the x-axis, (b) the y-axis, (c) the line $y = x$, and (d) the line $y = -x$.

20. 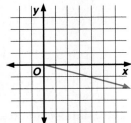   **21.** 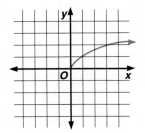   **22.**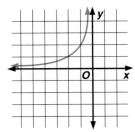

23. 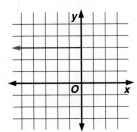   **24.** 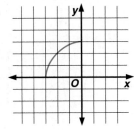   **25.**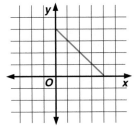

Determine whether each function is an even function, an odd function, or neither.

26. $y = x^5 - 4x$ **27.** $y = 6x^3 - 3x + 5$ **28.** $y = x^2 - 64$

29. $y = 6 - 6x^2 + x^8$ **30.** $y = 5x^2 + 6x - 9$ **31.** $y = -3x^9 + 4x^5$

Determine whether the graph of each equation is symmetric with respect to the origin, the x-axis, the y-axis, the line $y = x$, or the line $y = -x$.

32. $y = -6x$ **33.** $y = 6x^4 - 3x^2 + 1$ **34.** $x = y^2 + 4$

35. $y = 2x^9 - 7x^2 + 4$ **36.** $x^2 + y^2 = 9$ **37.** $y^2 = \dfrac{4x^2}{9} - 4$

Graphing Calculator   **Use a graphing calculator to graph each function and determine any lines of symmetry.**

38. $y = -(x - 2)^{-2}$ **39.** $y = \dfrac{4x^2}{1 + x^2}$

40. Graph $y = \dfrac{1}{4}x^2 - x + 1$ and $y = x^2 - 7$ on the same screen. Use the ZOOM and TRACE functions to determine their points of intersection.

Critical Thinking   **41.** Plot points to graph the parabola whose equation is $y = 3x^2 - 4x - 4$ and the circle whose equation is $(x - 2)^2 + (y - 4)^2 = 16$. One of the points of intersection of these two graphs is at $(2, 0)$. Do you expect another point of intersection to be at $(-2, 0)$? Why or why not?

42. The following program can be used to check whether a function is even, odd, or neither. The function to be checked must be in Y_1. You may get an error message if –A is not in the domain of the function. If this occurs, the function is neither even nor odd.

```
Prgm 3: ODDREVEN
:ClrHome
:Disp "X-COOR OF POINT ON GRAPH"
:Disp "(NOT 1 OR 0)"
:Input A
:A → X               Stores the value of A in X.
:Y₁ → B              Calculates Y₁, stores in B.
:-A → X              Stores the opposite of A in X.
:Y₁ → C              Calculates Y₁, stores in C.
:If B = C            If f(A) = f(–A),
:Disp "EVEN FUNCTION"    calculator displays the message.
:If B = C            If f(A) = f(–A),
:Goto 1              program goes to line "Lbl 1."
:If C = -B           If f(–A) = –f(A),
:Disp "ODD FUNCTION"    calculator displays the message.
:If C = -B           If f(–A) = –f(A),
:Goto 1              program goes to line "Lbl 1."
:Disp "NOT ODD OR EVEN"
:Lbl 1
```

Determine whether each function is an even function, an odd function, or neither.

a. $y = x + 1$ **b.** $y = 3x^9 + 4x^7 - x^3 + 2x$

Applications and Problem Solving

43. Physics The light pattern from a fog light at the airport can be modeled by the equation $\frac{x^2}{4} - \frac{y^2}{16} = 1$. One of the points on the graph is at $(3, 4.5)$, and one of the x-intercepts is –2. Find the coordinates of three additional points on the graph and the other x-intercept.

44. Gardening A garden containing flowers called impatiens is planted in a region shaped like a parabola. Since impatiens grow better in shade, the gardener is designing a circular arbor for a climbing vine that will be placed above the garden. Four posts, placed on the circle, support the arbor. The gardener models this plan with the graphs of a circle whose equation is $x^2 + y^2 = 25$ and a parabola whose equation is $y = x^2 - 6$. The two poles are positioned at $(3.2, 3.9)$ and $(1.1, -4.9)$. Find the coordinates of the position of the other two poles.

Mixed Review

45. Find $[f \circ g](4)$ and $[g \circ f](4)$ if $f(x) = x^2 - 4x + 5$ and $g(x) = x - 2$. **(Lesson 1-2)**

46. Find the coordinates of the midpoint of $\overline{FG}$ given its endpoints $F(-3, -2)$ and $G(8, 4)$. **(Lesson 1-4)**

47. The slope of $\overleftrightarrow{AB}$ is 0.6. The slope of $\overleftrightarrow{CD}$ is $\frac{3}{5}$. State whether the lines are parallel, perpendicular, or neither. Explain. **(Lesson 1-6)**

48. Find the multiplicative inverse of $\begin{bmatrix} 8 & -3 \\ 4 & -5 \end{bmatrix}$. **(Lesson 2-3)**

49. Use matrices to solve the system of equations. **(Lesson 2-4)**
$$8m - 3n - 4p = 6$$
$$4m + 9n - 2p = -4$$
$$6m + 12n + 5p = -1$$

50. **Retail**   Amanda Haines, a sales associate at a paint store, plans to mix as many gallons as possible of colors A and B. She has exactly 32 units of blue dye and 54 units of red dye. Each gallon of color A requires 4 units of blue dye and 1 unit of red dye. Each gallon of color B requires 1 unit of blue dye and 6 units of red dye. Use linear programming to answer the following questions. **(Lesson 2-6)**

   a. Let a be the number of gallons of color A and let b be the number of gallons of color B. Write the inequalities that describe this situation.

   b. Find the maximum number of gallons possible.

51. **College Entrance Exam**   Student A is 15 years old. Student B is one-third older. How many years ago was student B twice as old as student A?

DECISION MAKING

The Ozone Layer

Earth's ozone layer blocks ultraviolet rays that can cause skin cancer and other serious environmental hazards. Many scientists believe that some commonly-used chemicals, especially chlorofluorocarbons (CFCs), are destroying the ozone layer. What are your responsibilities as a citizen in the face of this evidence? How can we weigh the benefits that would accrue from banning CFCs against the financial losses that companies which produce CFCs will suffer?

1. Research the evidence for ozone-layer destruction by CFCs and other chemicals.

2. If you were a legislator, what steps would you take to resolve the conflict between the need for environmental safety and the need to protect businesses and jobs?

3. List three things you can do to address the problem.

4. **Project**   Work with your group to complete one of the projects listed at the right, based on the information you gathered in Exercises 1–3.

PROJECTS

- *Conduct an interview.*
- *Write a book report.*
- *Write a proposal.*
- *Write an article for the school paper.*
- *Write a report.*
- *Make a display.*
- *Make a graph or chart.*
- *Plan an activity.*
- *Design a checklist.*

3-2 Families of Graphs

Objectives

After studying this lesson, you should be able to:
- identify the graphs of simple polynomial functions, absolute value functions, and step functions, and
- sketch the graphs of these functions.

Application

Max Dunn of Protem, Missouri, has a dog breeding business. One of his beagles, Belle, recently had a litter of puppies. Mr. Dunn's children noticed that all the puppies looked like Belle and the father, King. However, they could tell the puppies apart because each has distinctively different characteristics from its parents and from the other puppies.

As a member of a family, the puppies have physical characteristics similar to those of their parents as well as their siblings. However, each puppy also has physical characteristics that makes it look different from the other members of the family. Graphs can also be related in a similar manner.

A **parent graph** is an anchor graph from which other graphs in the family are derived. Some different types of parent graphs are shown below. Notice that the coefficient of x in each parent graph is 1.

$y = x^2$ $y = x^3$ $y = x^4$ $y = x^5$

polynomial functions

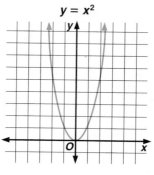 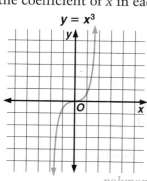 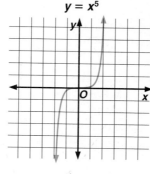 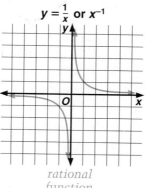

$y = \sqrt{x}$ $y = |x|$ $y = [x]$ $y = \frac{1}{x}$ or x^{-1}

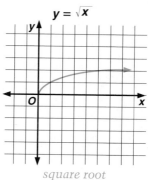 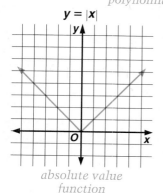 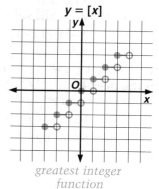

square root function   *absolute value function*   *greatest integer function*   *rational function*

Changes to the equation of the parent graph affect the appearance of the graph. The changed graph still resembles the parent graph, but it may appear in a different location on the coordinate plane.

One way to relocate the graph is by using a **reflection**. A reflection *flips* a figure over a line called the *axis of symmetry*. Every point in the original figure has a corresponding point in the reflected image on the opposite side of the axis of symmetry. The exception is any point that lies on the axis of symmetry. Then the point is its own reflected image.

The axis of symmetry is also called the line of symmetry.

Example 1

Graph $f(x) = x^2$ and $g(x) = -x^2$. Describe how the graphs of $g(x)$ and $f(x)$ are related.

$f(x) = x^2$	
x	$f(x)$
-2	4
-1	1
0	0
1	1
2	4

$g(x) = -x^2$	
x	$g(x)$
-2	-4
-1	-1
0	0
1	-1
2	-4

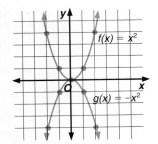

The graph of $g(x)$ is a reflection of the graph of $f(x)$ over the x-axis, and vice versa. The symmetry of these two graphs is shown algebraically by $g(x) = -f(x)$, or $f(x) = -g(x)$. *Notice that when the function has a negative coefficient, the graph turns downward.*

The reflection in Example 1 is one type of **linear transformation**. A linear transformation relocates the graph on the coordinate plane, but does not change its shape or size. Another example of a linear transformation is a **translation**. A translation *slides* the graph vertically and/or horizontally, but does not change the shape.

Example 2

Use the parent graph $y = x^3$ given on the previous page to sketch the graphs of each function. Describe how each graph is related to the parent graph.

You can use a graphing calculator to graph the parent function and the related function at the same time.

Notice that the point at $(0, 0)$ is the point of symmetry for the parent graph.

a. $y = x^3 + 3$

a. When a number is added to the parent function, the result is a translation up or down, the y-axis. Since 3 is added to the original function or $y = f(x) + 3$, the graph of the parent function slides 3 units up. The point of symmetry is now at $(0, 3)$.

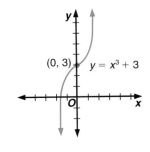

b. $y = (x + 3)^3$

b. When a positive number is added to x in the parent function, the result is a translation to the left along the x-axis. Since 3 is added to x in the original function or $y = f(x + 3)$, the graph of the parent function slides 3 units left. The point of symmetry is now at $(-3, 0)$.

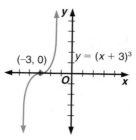

c. $y = (x + 3)^2 - 4$

c. In $y = (x + 3)^3 - 4$, two translations occur. The 3 added to x indicates a slide 3 units to the left. The 4 subtracted indicates a slide 4 units down from the graph of the function $y = (x + 3)^3$. The point of symmetry is now at $(-3, -4)$.

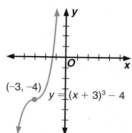

The **greatest integer function**, $f(x) = [x]$, is a type of step function. The symbol $[x]$ means the greatest integer not greater than x. For example, $[5.3]$ is 5 and $[-2.9]$ is -3. The graphs of step functions are often used to model real-world problems.

Example 3

APPLICATION

Business

InfoSystem is a database company that organizes payroll files for other companies. They charge for their services by the minute. If they work on your files for less than 1 minute, you pay nothing. If they work for at least 1 minute but less than 2 minutes, the charge is $1. If they work for at least 2 minutes but less than 3 minutes, the charge is $2, and so on. If you require them to service your files more than once in a calendar week, a $25 charge is added. Graph the function that would determine the cost for 12.8 minutes of service occurring the second time during the week.

The function $f(x) = [x]$ describes the charges for servicing accounts once a week.

x	f(x)
0.75	0
1	1
1.5	1
1.9	1
2	2
2.9	2
3.5	3
4	4

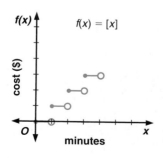

(continued on the next page)

A charge of $25 added to the function translates the function 25 units up. The function $f(x) = [x] + 25$ describes these charges.

The cost of their services for 12.8 minutes the second time during the week would be $37 since $f(12.8) = [12.8] + 25$ or 37.

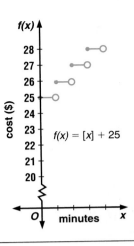

A geometric transformation of a linear graph changes the slope of the line.

Another way in which a parent graph can be modified is by a **geometric transformation**. With a geometric transformation, a nonlinear graph is *stretched* or *shrunk*. This results from multiplying the function by a number. *Geometric transformations are sometimes called* <u>dilations</u>.

Example 4

Graph each function. Then describe how each graph is related to the parent graph.
a. $f(x) = |x|$

a. $f(x) = |x|$ is the absolute value function.

x	f(x)
-2	2
-1	1
0	0
1	1
2	2

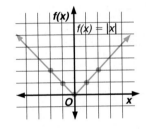

This is the parent function. Notice that the vertex is at $(0, 0)$.

b. $g(x) = 2|x|$

b.

x	g(x)
-2	4
-1	2
0	0
1	2
2	4

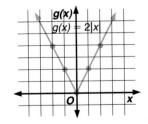

The graph has a V-shape like the parent graph. However, the graph is narrower. Multiplying a function by a number greater than 1 stretches the graph vertically. The vertex of the graph is still at $(0, 0)$.

c. $h(x) = -0.5|x| - 3$

c. Multiplying by a negative number reflects the graph over the *x*-axis. Multiplying by 0.5 widens the graph. Subtracting 3 translates the graph 3 units down. The vertex of the graph is now at $(0, -3)$.

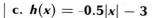

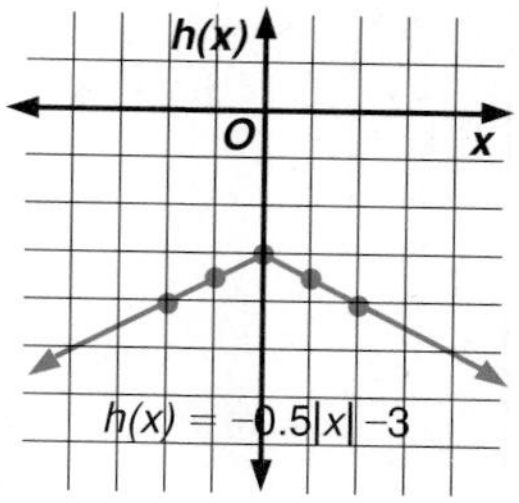

x	$-0.5|x|$	$h(x)$
-2	-1	-4
-1	-0.5	-3.5
0	0	-3
1	-0.5	-3.5
2	-1	-4

Notice that multiplying by a positive number less than 1 shrinks the graph vertically.

TECHNOLOGY *Tip*

If you are using a graphing calculator to graph a family of functions, the TI-83 has five different graph styles from which to choose, and the TI-92 has six different graph styles.

   The following chart summarizes the relationships in families of graphs. The parent graph may differ, but the transformations of the graphs have the same effect. Remember that more than one transformation may affect a parent graph.

Change to the Parent Function $y = f(x)$, $c \geq 0$	Change to Parent Graph	Examples
Reflections $y = -f(x)$ $y = f(-x)$	Is reflected over the *x*-axis. Is reflected over the *y*-axis.	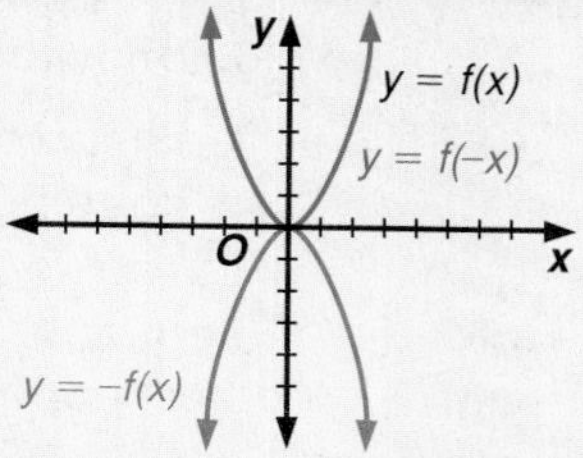
Translations $y = f(x) + c$ $y = f(x) - c$	Translates the graph *c* units up. Translates the graph *c* units down.	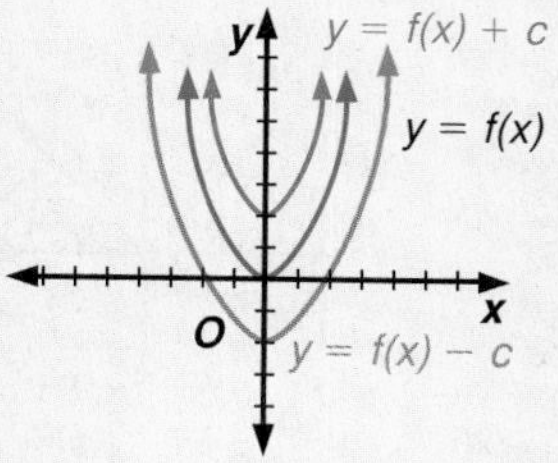
$y = f(x + c)$ $y = f(x - c)$	Translates the graph *c* units left. Translates the graph *c* units right.	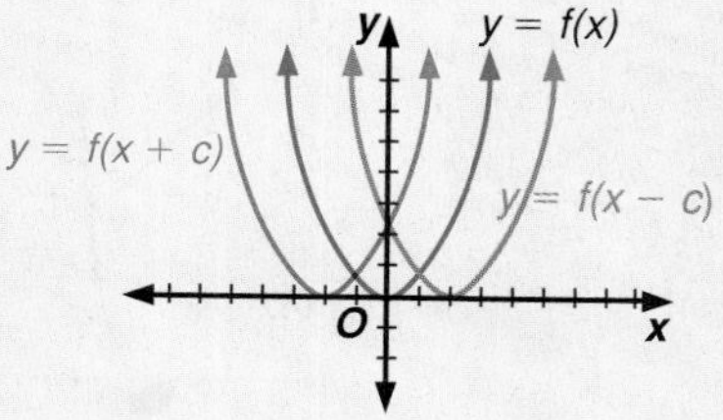

Dilations		
$y = c \cdot f(x), c > 1$ $y = c \cdot f(x), 0 < c < 1$	Stretches the graph vertically. Shrinks the graph vertically.	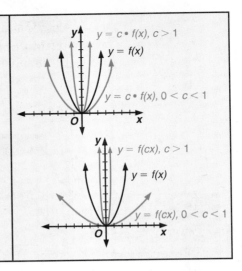
$y = f(cx), c > 1$ $y = f(cx), 0 < c < 1$	Shrinks the graph horizontally. Stretches the graph horizontally.	

CHECKING FOR UNDERSTANDING

Communicating Mathematics

Read and study the lesson to answer each question.

1. **Explain** the similarities and differences among reflections, translations, and dilations.

2. **Describe** the changes in the parent graph $f(x) = x^4$ for the function $f(x) = 3x^4$.

3. **Show** how the point at $(4, 5)$ on the graph of $f(x)$ translates for $f(x - 2) + 7$.

4. **Draw** one of the parent graphs on page 117 and then draw a related graph that is stretched horizontally. Write the equation for the related graph.

Guided Practice

The graph on the left in each set of graphs is the parent graph. Describe how the other graphs in the set are related to the parent graph.

5. $f(x) = |x|$ **a.** $f(x) = 0.5|x|$ **b.** $f(x) = |x| + 2$ **c.** $f(x) = |3x|$

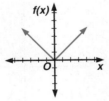 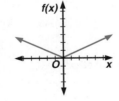 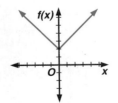 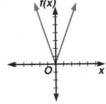

6. $f(x) = x^2$ **a.** $f(x) = -x^2 + 3$ **b.** $f(x) = (x + 1)^2$ **c.** $f(x) = (x - 3)^2$

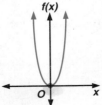 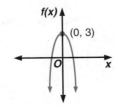 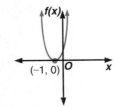 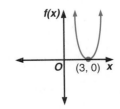

Express each function graphed at the right in terms of $f(x)$.

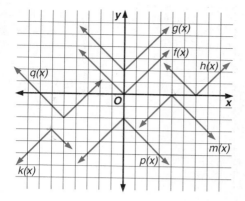

7. $g(x)$

8. $h(x)$

9. $m(x)$

10. $p(x)$

11. $q(x)$

12. $k(x)$

EXERCISES

Practice

The graph of $f(x)$ is shown at the right. Sketch a graph of each function based on the graph of $f(x)$.

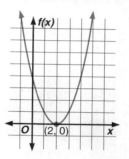

13. $g(x) = f(x + 3)$

14. $h(x) = f(x) - 4$

15. $j(x) = -f(x)$

16. $k(x) = 3f(x)$

17. $s(x) = -0.5f(x)$

18. $m(x) = 2f(x) + 3$

19. $n(x) = f(x - 5) + 4$

20. $p(x) = 3f(x + 2) - 7$

For each parent graph, describe the transformation(s) that have taken place in the related graph of each function.

21. $f(x) = x^2$
   a. $y = 3x^2$
   b. $y = -1.4x^2$
   c. $y = 4(x - 2)^2$
   d. $y = \frac{1}{2}(x + 1)^2$
   e. $y = 2x^2 - 5$

22. $f(x) = [x]$
   a. $y = [x] - 5$
   b. $y = [x + 7]$
   c. $y = 3[x]$
   d. $y = 2[x] + 4$
   e. $y = -[x]$

23. $f(x) = |x|$
   a. $y = |x| + 6$
   b. $y = |x - 4|$
   c. $y = 0.4|x|$
   d. $y = 2|x + 3| - 8$
   e. $y = -|x - 3| + 6$

Sketch the graph of each function.

24. $f(x) = 2(x - 3)^4 - 7$

25. $h(x) = -4|x - 5| - 1$

26. $p(x) = 0.33(x + 2)^3 + 2$

27. $q(x) = -(x + 2)^2 + 5$

Graphing Calculator

Use a graphing calculator to graph each set of functions on the same screen. Name the x-intercepts of the graphs.

28. $y = x^2$
   $y = (2x + 5)^2$
   $y = (3x - 1)^2$

29. $y = x^3$
   $y = (2x + 1)^3$
   $y = (5x - 2)^3$

30. $y = |x|$
   $y = |4x - 2|$
   $y = |5x + 1|$

31. Study the coordinates of the x-intercepts you found in the related graphs in Exercises 28–30. Make a conjecture about the x-intercept of $y = (ax + b)^n$ if $y = x^n$ is the parent function.

Critical Thinking

32. Study the parent graphs at the beginning of this lesson.
  a. Select a parent graph or a modification of the parent graph that meets each of the following specifications.
  (1) positive at its leftmost points and positive at its rightmost points
  (2) negative at its leftmost points and positive at its rightmost points
  (3) negative at its leftmost points and negative at its rightmost points
  (4) positive at its leftmost points and negative at its rightmost points
  b. Sketch the related graph for each parent graph that is translated 3 units right and 5 units down.
  c. Write an equation for each related graph.

Applications and Problem Solving

33. Service Charges Gatsby's Automotive Shop charges $35 per hour or any fraction of an hour for labor when servicing cars.
  a. Graph the function that would determine the cost of 5.25 hours of labor.
  b. Graph the function that would show a $15 discount if you leave your car overnight.
  c. What would be the cost of 3.45 hours labor on a car left overnight?

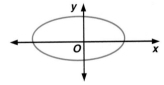

34. Geometry The length of a rectangular prism is 3 inches greater than the width. The height is 2 inches less than the width.
  a. Write the volume of the prism as the function $V(w)$, where w represents the width.
  b. Sketch the graph of $V(w)$.
  c. Sketch the graph of the function describing the volume of a prism whose width is 2 inches greater than that of the prism described by $V(w)$.
  d. In reality, which values of w would not be considered in the domain of $V(w)$? Why?

Mixed Review

35. Determine whether the graph at the right is the graph of a function. **(Lesson 1-1)**

36. Write the equation $15y - x = 1$ in slope-intercept form. **(Lesson 1-5)**

37. Find the determinant of $\begin{bmatrix} 5 & 9 \\ 7 & -3 \end{bmatrix}$. Then state whether an inverse exists for the matrix. **(Lesson 2-3)**

38. Determine whether the graph of $6x^4 - 3x^2 + 1$ is symmetric with respect to the x-axis, the y-axis, the line $y = x$, the line $y = -x$, or the origin. **(Lesson 3-1)**

39. College Entrance Exam Choose the best answer.
  The product of 75^3 and 75^7 is
  (A) 75^5 **(B)** 75^{10} **(C)** 150^{10} **(D)** 5625^{10} **(E)** 75^{21}

3-3A Graphing Calculators: Graphing Radical Functions

You can use a graphing calculator to graph various types of radical functions quickly and easily. To graph a radical function on the TI-82 graphing calculator you will need to rewrite some radical equations to enter them into the calculator. If you have an equation of the form $y = \sqrt[n]{x}$, enter it into the calculator as $y = x^{\frac{1}{n}}$. For example, the equation $y = \sqrt[4]{x^3}$ can be graphed by entering the equivalent equation of $y = x^{\frac{3}{4}}$ using the $\boxed{\wedge}$ key and parentheses around the $\frac{3}{4}$. The parentheses are very important because without them, the calculator will graph $y = x^3 \div 4$.

You can also use the $\sqrt[x]{}$ command found in the MATH menu. The equation $y = \sqrt[4]{x^2}$ can be graphed by entering $\boxed{Y=}$ 4 $\boxed{MATH}$ 5 $\boxed{X,T,\theta}$ $\boxed{x^2}$ $\boxed{GRAPH}$.

Example

Graph $y = \sqrt{x - 3} + 1$ in the viewing window $[-1, 6]$ by $[-1, 6]$, with a scale factor of 1 for both axes. Determine the domain and range.

Enter: $\boxed{Y=}$ $\boxed{2nd}$ $\boxed{\sqrt{}}$ $\boxed{(}$ $\boxed{X,T,\theta}$ $\boxed{-}$ 3 $\boxed{)}$ $\boxed{+}$ 1 $\boxed{GRAPH}$

Tracing and zooming in on the initial point of the graph shows that there are only values for $x \geq 3$ and $y \geq 1$. So the domain is $x \geq 3$, and the range is $y \geq 1$.

It may be difficult to find the domain and range on a calculator using TRACE and ZOOM. Try using the ZOOM "box" command to keep the initial point in the box.

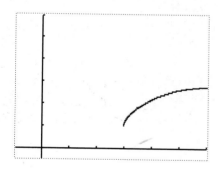

EXERCISES

Graph each function so that a complete graph is shown. Then state the domain and range of each function and sketch the graph on a piece of paper.

1. $y = \sqrt{x - 1} + 5$

2. $y = \sqrt{x + 2} - 6$

3. $y = \sqrt[3]{x - 2}$

4. $y = \sqrt{6x + 5}$

5. $y = \sqrt[4]{x} + 2$

6. $y = \sqrt[5]{3x + 4} - 4$

3-3 Inverse Functions and Relations

Objectives

After studying this lesson, you should be able to:
- determine the inverse of a relation or function, and
- graph a function and its inverse.

Application

The economic recession of the early 1990s caused many businesses to declare bankruptcy and, in some cases, collapse totally. In an attempt to avoid bankruptcy, Burdette's Clothing Store had a storewide sale. Every item was marked 20% off at the register. In order to satisfy the bankruptcy lawyers, they had to calculate what their income would have been if that merchandise had been sold at retail value.

To find the reduced price of each sale, the registers' computer had been programmed with the formula $S = x(1 - R)$, where S is the sale price, x is the original price and R is the rate of discount. Since $R = 0.20$, the formula that results is $S = 0.80x$.

In order to find the original price if you are given the reduced price, you would have to work backward, reversing the procedure. Instead of multiplying the original cost by 0.80 to find the sale price, you divide the sale price by 0.80 to arrive at the original price. These two procedures are examples of inverse operations.

Relations also have inverses, and these inverses are themselves relations. Since functions are relations, all functions have inverses, but the inverse may not be a function.

Definition of *Inverse Relations*	**Two relations are inverse relations if and only if one relation contains the element (b, a), whenever the other relation contains the element (a, b).**

The inverse of $f(x)$ is written $f^{-1}(x)$.

Example 1

You can review more about inverses in Lesson 1-2.

Graph $f(x) = x^3$ and its inverse.

To graph the inverse of a function, the first step is to interchange the x- and y-coordinates of the ordered pairs of the original function. Then graph $f(x)$ and $f^{-1}(x)$.

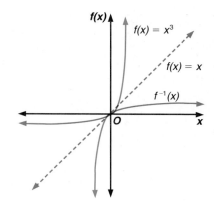

$f(x) = x^3$		$f^{-1}(x)$	
x	y	x	y
-3	-27	-27	-3
-2	-8	-8	-2
-1	-1	-1	-1
0	0	0	0
1	1	1	1
2	8	8	2
3	27	27	3

Notice that $f^{-1}(x)$ is a function.

The graph of $f^{-1}(x)$ is the reflection of $f(x)$ over the line $f(x) = x$.

You can find the inverse of a function algebraically. Review the property of inverse functions in Lesson 1-2 and then study the following example.

Example 2

Find the inverse of $f(x) = x^2 + 2$. Then graph $f(x)$ and its inverse.

Let $f(x) = y$.	$y = x^2 + 2$
Interchange x and y.	$x = y^2 + 2$
Solve for y.	$x - 2 = y^2$
	$\pm\sqrt{x - 2} = y$

Replace y with $f^{-1}(x)$. $f^{-1}(x) = \pm\sqrt{x - 2}$

You can use a graphing calculator to graph $f(x) = x^2 + 2$ and $f^{-1}(x) = \pm\sqrt{x - 2}$.

Press the $\boxed{Y =}$ key. Enter the three equations: $y = x^2 + 2$, $y = \sqrt{x - 2}$, and $y = -\sqrt{x - 2}$.

Press $\boxed{\text{GRAPH}}$.

Sketch the images shown on your screen.

Remember to use the vertical line test to verify that a graph represents a function.

Notice that $f^{-1}(x)$ is *not* a function.

You have learned to use the vertical line test to determine if a graph is the graph of a function. You can use the **horizontal line test** to determine if the graph of the inverse of a function is also a function. If any horizontal line drawn on the graph of a function passes through no more than one point of the graph, then the function has an inverse that is also a function.

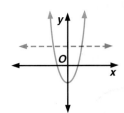

The inverse is not a function.

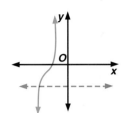

As a general rule for polynomial functions,
- *if n is even, the inverse of $f(x) = x^n$ is not a function;*
- *if n is odd, the inverse of $f(x) = x^n$ is a function.*

The inverse is a function.

The transformations you learned in Lesson 3-2 can also be applied to graphing inverses. Recall that the inverse of $y = x^5$ is the radical function $y = \sqrt[5]{x}$.

Example 3

Sketch $y = -\sqrt[5]{x}$.

First sketch the parent graph of $y = x^5$.

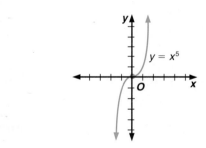

Reflect the graph over the line $y = x$.

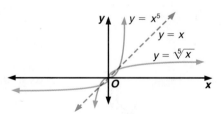

A coefficient of –1 indicates that the graph should be reflected over the x-axis.

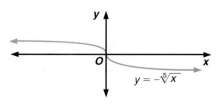

Example 4

Sketch the graph of $y = \pm 2\sqrt{x + 3} - 6$.

The parent graph is the graph of $y = x^2$.

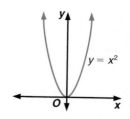

Reflect the parent graph over the line $y = x$ to obtain its inverse whose equation is $y = \pm\sqrt{x}$.

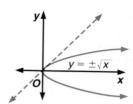

Sketch the graph of $y = \pm 2\sqrt{x}$, which is wider than the reflection of the parent graph.

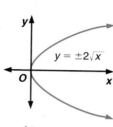

The graph of $y = \pm 2\sqrt{x + 3}$ is the graph of $y = \pm 2\sqrt{x}$, translated 3 units left. Subtracting 6 translates the graph 6 units down.

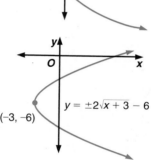

CHECKING FOR UNDERSTANDING

Communicating Mathematics

Read and study the lesson to answer each question.

1. **Describe** the relationship between the coordinates of the ordered pairs of a relation and its inverse.

2. **Demonstrate** how transformations are used to graph $y = \sqrt[3]{x - 2} + 3$.

3. **Find a counterexample** to this statement: The inverse of a function is also a function.

4. **Show** how you know whether the inverse of a function is also a function without graphing the inverse.

Guided Practice

Given point P of the function $f(x)$, state the corresponding point P' in the inverse of the function.

5. $P(-4, 5)$ 6. $P(-3, -2)$ 7. $P(-2, 8)$ 8. $P(3t, 8u)$

Determine if the inverse of each relation graphed below is a function. Write *yes* or *no*.

9.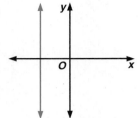

10.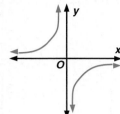

11.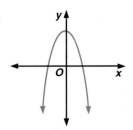

12. Find the inverse of $y = x^2 - 7$. Graph the function and its inverse.

EXERCISES

Practice

Find the inverse of each function.

13. $y = 2x + 3$

14. $y = 0.5x - 8$

15. $y = 4 - 3x^2$

16. $y = x^2 + 4$

17. $y = (x - 2)^3$

18. $y = (x + 3)^4 - 5$

Find the inverse of each function. Sketch the function and its inverse. Is the inverse a function? Write *yes* or *no*.

19. $y = x^2 + 2$

20. $y = (x + 3)^3$

21. $y = (x - 4)^4$

22. $y = (x + 2)^2 - 5$

23. $y = (x - 3)^3 + 6$

24. $y = x^5 + 5$

For each parent graph, describe the transformation(s) that have taken place in the related graph of each function.

25. $y = x^2$

   a. $y = \pm\sqrt{x}$

   b. $y = \pm\sqrt{x - 3}$

   c. $y = \pm\sqrt{x} + 6$

   d. $y = \pm 3\sqrt{x}$

   e. $y = \pm 0.2\sqrt{x} + 4$

   f. $y = \pm\sqrt{x - 5} + 3$

26. $y = x^3$

   a. $y = \sqrt[3]{x}$

   b. $y = \sqrt[3]{x} + 6$

   c. $y = \sqrt[3]{x} - 4$

   d. $y = 0.4\sqrt[3]{x}$

   e. $y = \sqrt[3]{x + 3} - 8$

   f. $y = -\sqrt[3]{x - 3} + 6$

Graphing Calculator

Graph the two relations on the same screen. Determine if the two relations are inverses. Write *yes* or *no*. If the answer is no, write the equation that is the inverse of the first equation.

27. $y = (x + 1)^2$
   $y = \pm\sqrt{x - 1}$

28. $y = x^2 - 2$
   $y = \pm\sqrt{x + 2}$

29. $y = x^3 - 5$
   $y = \sqrt[3]{x} + 5$

30. Find a function that is its own inverse. Can you find more than one?

31. **Academics** In Beth's physics class, the lab work score counts the same as a test score when the grades are averaged for the quarter. No other scores are used to find the average grade. She scored 75%, 80% and 72% on her tests. What lab work score must she receive to average 80% for the quarter?

32. **Fire Fighting** The Fairfield Fire Department must purchase a pump that is powerful enough to propel water 80 feet into the air. One pump is advertised to project water with a velocity of 75 feet/second. The velocity of the water can be described by the equation $v = \sqrt{2gh}$, where v is the velocity of the water, g is the acceleration due to gravity (32 ft/s^2), and h is the maximum height of water flow.
  a. Write the simplified equation after replacing g with 32.
  b. Graph the relation.
  c. Use the graph to determine the maximum height of the water possible from this pump.
  d. Does this pump meet the department's needs?
  e. Another pump is being considered that propels water with a velocity of 70 feet/second. Will this pump meet the department's needs? Explain.

33. Graph $0 < x - y < 2$. **(Lesson 1-3)**

34. Solve the system of equations by graphing. **(Lesson 2-1)**

$$3x - 8y = 4$$
$$6x - 42 = 16y$$

35. State whether the figure at the right has point symmetry, line symmetry, neither, or both. **(Lesson 3-1)**

36. Describe the transformation(s) that have taken place between the parent graph of $y = x^2$ and the graph of $y = 0.5(x + 1)^2$. **(Lesson 3-2)**

37. **College Entrance Exam** Choose the best answer.
  If $d = m - \dfrac{50}{m}$, and m is a positive number that increases in value, then d

  (A) increases in value.
  (B) decreases in value.
  (C) remains unchanged.
  (D) increases, then decreases.
  (E) decreases, then increases.

3-4A Graphing Calculators: Graphing Rational Functions

A rational function is an equation in which one polynomial is divided by another, or in mathematical terms, $f(x) = \dfrac{p(x)}{q(x)}$, where $q(x) \neq 0$. Rational functions have some features that other polynomial functions do not have, and the graphing calculator is a good tool to use to explore these graphs.

Some graphs of rational functions are **discontinuous.** Discontinuous means that if you had to draw the graph with a piece of paper and a pencil, you would have to pick up your pencil at least once in order to completely draw the graph. These breaks in continuity can appear as asymptotes or as point discontinuities. A point discontinuity is like a hole in a graph. A break in continuity may not be visible in the first viewing window. You may have to zoom in to see it.

Example 1

Graph $y = \dfrac{x^2 - 1}{x + 1}$. Use the viewing window $[-4.4, 5]$ by $[-5, 5]$ with scale factors of 1 for both axes.

Enter: [Y=] [(] [X,T,θ] [x²] [−] 1 [)] [÷] [(] [X,T,θ] [+] 1 [)] [GRAPH]

Note: To see the break in continuity, your calculator must be in dot mode.

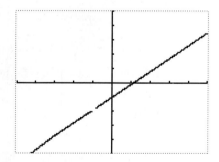

The graph looks like a line with a break in continuity at $x = -1$. Remember that the denominator cannot be zero, so $x + 1 \neq 0$, or $x \neq -1$. The rest of the graph looks like the line $y = x - 1$ because the function $y = \dfrac{x^2 - 1}{x + 1}$ simplifies to $y = x - 1$.

Rational functions can also have asymptotes at values of x for which discontinuities occur. An asymptote can either be vertical or horizontal. Vertical asymptotes can be found algebraically by finding where the denominator of the function equals zero. Horizontal asymptotes can be found algebraically by solving the equation for x and looking where y is undefined in the denominator of the function. Both asymptotes can also be found graphically by using the TRACE and ZOOM functions.

Example 2

Graph $y = \dfrac{2x + 3}{x - 1}$. Use the viewing window [−7, 7] by [−5, 10] with scale factors of 1 on both axes. Then find the vertical and horizontal asymptotes.

Enter: Y= ((2 X,T,θ + 3) ÷ ((X,T,θ − 1) GRAPH

First, find the vertical asymptote by tracing along the graph and observing the x-values near the discontinuity. As |y| increases, x approaches 1. By pressing ZOOM 8 ENTER TRACE , you can trace to a point where x = 1 and the calculator gives no y-value. This shows that the vertical asymptote is at x = 1.

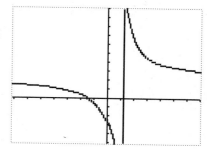

The horizontal asymptotes can be found by looking at the end behavior of the function. You can trace along the function to find that as |x| increases, y approaches 2. This can also be seen by graphing the function in a viewing window with large values of x, such as [1000, 1001] by [−5, 10], and tracing along the function. The horizontal asymptote is at y = 2.

Sometimes it may appear as though the graphing calculator has drawn a vertical asymptote on the screen, but this is not the case. If it appears as though a vertical asymptote has been drawn, notice that the calculator does not plot the line through the exact point where the vertical asymptote occurs. In Example 2, the calculator appeared to draw the asymptote, but not at x = 1.

EXERCISES

Graph each function on a graphing calculator so that a complete graph is shown. Then sketch the graph on a piece of paper. Determine the vertical and horizontal asymptotes and the point discontinuities, if any.

1. $y = \dfrac{1}{x^2}$

2. $y = \dfrac{5}{x^3}$

3. $y = \dfrac{x^2 + 6}{2x + 3}$

4. $y = \dfrac{x + 3}{x^3 - 2x^2 - 8x}$

5. $y = \dfrac{4x}{x^2 + 9}$

6. $y = \dfrac{x^2 - 4}{x - 2}$

7. $y = \dfrac{x - 5}{3x}$

8. $y = \dfrac{x^2 + 2x + 5}{2x^2}$

9. $y = \dfrac{3x^2 + 1}{0.5x^2 + x}$

10. $y = \dfrac{x^3 - 1}{x^3 + 1}$

3-4 Rational Functions and Asymptotes

Objectives

After studying this lesson, you should be able to:
- determine horizontal, vertical, and slant asymptotes, and
- graph rational functions.

Application

Music stars on tour often use specialized buses to travel from one concert city to the next. On her first solo tour in 1992, Wynona Judd traveled in her bus from Live Oak, Florida, where she did a concert on October 31, to Atlanta for a concert at the Fox Theater the next day. The trip is 297 miles long and lasted approximately 5.5 hours. Her bus driver keeps a record of their average travel time in order to better plan future tours.

The average speed for this part of the trip can be determined by the formula $r = \dfrac{d}{t}$. In this case, $d = 297$. You could use your graphing calculator to graph $r = \dfrac{297}{t}$. Then find the point at which the t-coordinate is 5.5 and read the r-coordinate. In this case, the point is at (5.5, 54). The average speed of the bus for this part of the tour was 54 miles per hour.

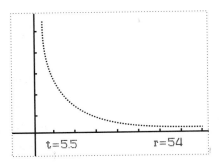

The function $r = \dfrac{297}{t}$ is an example of a **rational function**. A rational function is the quotient of two polynomials. It has the form $f(x) = \dfrac{g(x)}{h(x)}$, where $h(x) \neq 0$. The parent rational function is $f(x) = \dfrac{1}{x}$. The graph of $f(x) = \dfrac{1}{x}$ consists of two parts, one in Quadrant I and one in Quadrant III. Neither part of the graph has a point on the y-axis, where $x = 0$. The parent function $f(x) = \dfrac{1}{x}$ is undefined at $x = 0$, since division by zero is undefined. Also neither part of the graph has a point on the x-axis, where $y = 0$, since $\dfrac{1}{x} \neq 0$.

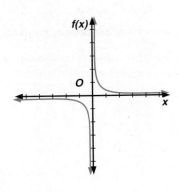

For a rational function $f(x) = \dfrac{g(x)}{h(x)}$, where $h(x) \neq 0$, the branches of the graph of a rational function approach lines called **asymptotes**. If the function is not defined when $x = a$, then the line with the equation $x = a$ is a **vertical asymptote**.

*"x approaches ∞" is
usually written x → ∞.*

Notice that as the value of x approaches zero in the first quadrant of the graph, the value of $f(x)$ increases without bound toward positive infinity (∞). In the third quadrant, as x approaches zero, the value of $f(x)$ decreases without bound toward negative infinity ($-\infty$).

Vertical Asymptote	The line $x = a$ is a vertical asymptote for a function $f(x)$ if $f(x) \to \infty$ or $f(x) \to -\infty$ as $x \to a$ from either the left or the right.

Also notice that as the value of x increases and approaches positive infinity in the first quadrant, the value of $f(x)$ approaches zero. The same pattern can be observed in the third quadrant. As the value of x decreases and approaches negative infinity, the value of $f(x)$ approaches zero. Since there is no value of x such that $y = 0$, the line $y = 0$, which is the x-axis, is a **horizontal asymptote**.

Horizontal Asymptote	The line $y = b$ is a horizontal asymptote for a function $f(x)$ if $f(x) \to b$ as $x \to \infty$ or as $x \to -\infty$.

Example 1

Determine the asymptotes for the graph of $f(x) = \dfrac{x}{x + 2}$.

$f(-2)$ is not defined. So there is a vertical asymptote at $x = -2$.

*You can use the TRACE
function on your
graphing calculator
to approximate the
horizontal asymptote.*

To find the horizontal asymptote, let $f(x) = y$ and solve for x in terms of y. Then find where the function is undefined for values of y.

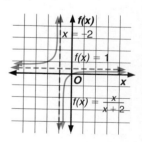

$$y = \frac{x}{x + 2}$$
$$y(x + 2) = x$$
$$xy + 2y = x$$
$$2y = x - xy$$
$$2y = x(1 - y)$$
$$\frac{2y}{1 - y} = x$$

The rational expression $\dfrac{2y}{1 - y}$ is undefined for $y = 1$. The horizontal asymptote is the line $y = 1$, or $f(x) = 1$.

Example 2

Use the parent graph $f(x) = \dfrac{1}{x}$ to graph each equation. Describe the transformation(s) that have taken place.

a. $g(x) = \dfrac{1}{x+1}$

a. To graph $g(x) = \dfrac{1}{x+1}$, translate the parent graph 1 unit to the left. When the parent graph is translated, the asymptote it approaches is also translated in the same direction. The new vertical asymptote is $x = -1$. Since the graph did not move up or down, horizontal asymptote is still $y = 0$.

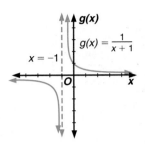

b. $j(x) = \dfrac{3}{x}$

b. The graph of $j(x) = \dfrac{3}{x}$ does not come as close to the origin as the parent graph, but no translation has taken place. The vertical asymptote is still $x = 0$, and the horizontal asymptote is $y = 0$.

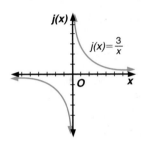

c. $k(x) = \dfrac{-2}{x}$

c. The graph of $k(x) = \dfrac{-2}{x}$ involves two transformations. The graph is not as close to the origin as the parent graph, and because it is multiplied by –1, it is reflected over the x-axis. The asymptotes remain the same.

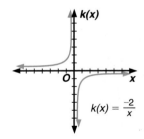

d. $m(x) = \dfrac{1}{x-4} + 3$

d. The graph of $m(x) = \dfrac{1}{x-4} + 3$ is the parent graph translated 4 units to the right and 3 units up. The new vertical asymptote is $x = 4$, and the new horizontal asymptote is $y = 3$.

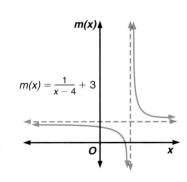

Example 3

As you get farther from Earth's surface, gravity has less effect and you actually weigh less at higher altitudes. A person's weight, $W(h)$, at a specific height h above Earth's surface can be found by the formula $W(h) = \dfrac{r}{h + r}(w)$, where r is Earth's radius and w is the person's weight at sea level. Earth's radius is about 6400 kilometers. Graph this function for a person weighing 50 kilograms at sea level, and explain the meaning of the asymptotes.

Substitute all known values into the formula.

$$W(h) = \left(\frac{6400}{h + 6400}\right) 50$$

$$= \frac{320,000}{h + 6400}$$

The vertical asymptote of the function is $h = -6400$. However, for this problem, we will only consider values for which $h \geq 0$.

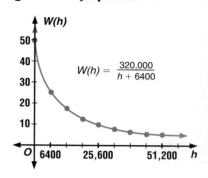

The horizontal asymptote of the function is $W(h) = 0$, because the weight of a person cannot be zero.

A third type of asymptote is the **slant asymptote**. Slant asymptotes occur when the degree of the numerator is exactly one greater than that of the denominator. For example, in $f(x) = \dfrac{x^2}{x - 6}$, the exponents are 2 and 1 and $2 - 1 = 1$. Thus, this function has a slant asymptote, as shown in the graph at the right.

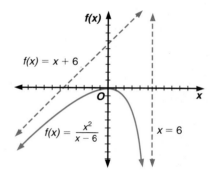

When the degrees are the same or the denominator has the greater degree, the function has vertical and horizontal asymptotes.

Slant Asymptote	The oblique line ℓ is a slant asymptote for a function $f(x)$ if the graph of $f(x)$ approaches ℓ as $x \to \infty$ or as $x \to -\infty$.

An oblique line is a line that is neither vertical nor horizontal.

Example 4

Determine the slant asymptote for $f(x) = \dfrac{x^2 - 2x + 1}{x}$.

First use division to rewrite the rational expression as a quotient. *Remember to use the same procedure as with whole numbers. That is, $7 \div 3 = 2\ R\ 1$ or $2\frac{1}{3}$.*

Therefore, $f(x) = x - 2 + \dfrac{1}{x}$.

$$\begin{array}{r} x - 2 \\ x\overline{)x^2 - 2x + 1} \\ \underline{x^2} \\ -2x + 1 \\ \underline{-2x} \\ 1 \end{array}$$

(continued on the next page)

As the value of x increases positively and negatively, the value of $\frac{1}{x}$ approaches zero. Therefore, the value of $x - 2 + \frac{1}{x}$ approaches $x - 2$. So, the line $g(x) = x - 2$ is the slant asymptote for $f(x)$.

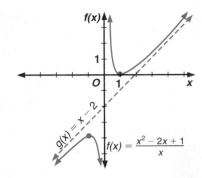

The line $x = 0$ is a vertical asymptote for $f(x)$, because $f(x)$ increases or decreases without bound close to $x = 0$.

Whenever the denominator and numerator of a rational function contain a common factor, a **hole** may appear in the graph of the funtion. For example, in $f(x) = \frac{x^2 - 9}{x + 3}$, the numerator and denominator both have a factor of $x + 3$. The function is undefined if $x = -3$. If numerator and denominator are both divided by the common factor, the function $f(x) = x - 3$ results. The graph of $f(x) = \frac{x^2 - 9}{x + 3}$ looks like the graph of $f(x) = x - 3$, but it has a hole at $x = -3$.

Example 5

Graph $y = \dfrac{x^2 + 3x - 10}{x - 2}$.

$$y = \frac{x^2 + 3x - 10}{x - 2}$$
$$= \frac{(x + 5)(x - 2)}{x - 2}$$
$$= x + 5, x \neq 2$$

The graph is the graph of $y = x + 5$ with a hole at $(2, 7)$.

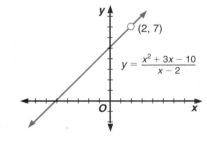

Example 6

Graph $y = \dfrac{(x - 4)(x - 2)}{x(x - 3)^2(x - 4)}$.

$$y = \frac{(x - 4)(x - 2)}{x(x - 3)^2(x - 4)}$$
$$= \frac{x - 2}{x(x - 3)^2}, x \neq 4$$

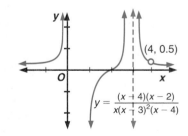

Since $x - 4$ is a common factor, there is a hole at $x = 4$. Because y increases or decreases without bound close to $x = 0$ and $x = 3$, there are vertical asymptotes at $x = 0$ and $x = 3$. There is also a horizontal asymptote at $y = 0$ for $x > 3$ and $x < 0$.

The graph is the graph of $y = \dfrac{x - 2}{x(x - 3)^2}$, with a hole at $(4, 0.5)$.

CHECKING FOR UNDERSTANDING

Read and study the lesson to answer each question.

1. **Describe** how a graph behaves when an asymptote is present.

2. **Explain** under what conditions a rational function has a slant asymptote and how you find its equation.

3. **Graph** $f(x) = \dfrac{1}{x}$ after it has been translated 2 units right and 6 units down.
   a. What are its asymptotes?
   b. Write the equation of the translated graph.

4. **Explain** why only values for $h \geq 0$ were considered in Example 3.

Guided
Practice

Given the graphs of transformations of $f(x) = \dfrac{1}{x}$, write the equation of each graph.

5.

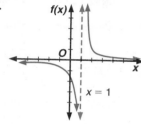

6.

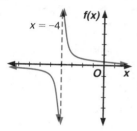

7.

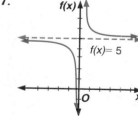

8.

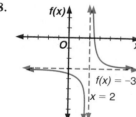

9.

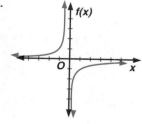

10.

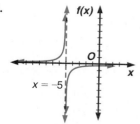

Determine the asymptotes for each function.

11. $y = \dfrac{-2}{(x+1)^2}$

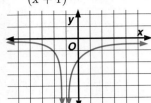

12. $y = \dfrac{x}{x-2}$

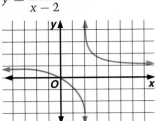

Determine the values of x for which a hole appears in the graph of each function.

13. $y = \dfrac{(x-3)(x+5)}{(x+5)(x-1)}$

14. $y = \dfrac{x^2 + 3x - 28}{x^3 - 11x^2 + 28x}$

EXERCISES

Practice

Determine any horizontal, vertical, or slant asymptotes in the graph of each function.

15. $y = \dfrac{x}{x^2 + x - 6}$

16. $y = \dfrac{2}{x^5 - 3x^3}$

17. $y = \dfrac{3x^3 - 2x + 1}{2x^2}$

18. $y = \dfrac{x^2 + 4x + 4}{x^2 - 11x + 30}$

For each parent graph, describe the transformation(s) that have taken place in the related graph of each function. Then sketch each graph.

19. $y = \dfrac{1}{x}$

    a. $y = \dfrac{1}{x} - 1$

    b. $y = \dfrac{1}{x+5}$

    c. $y = \dfrac{1}{4x}$

    d. $y = \dfrac{1}{0.2x} + 8$

    e. $y = \dfrac{1}{x-3} + 7$

20. $y = \dfrac{1}{x^2}$

    a. $y = \dfrac{1}{x^2} + 3$

    b. $y = \dfrac{1}{(x-9)^2}$

    c. $y = \dfrac{1}{0.4x^2}$

    d. $y = \dfrac{1}{(x+5)^2} - 1$

    e. $y = \dfrac{-1}{(x-2)^2} + 4$

Create a function of the form $y = f(x)$ that satisfies each set of conditions.

21. vertical asymptote at $x = 4$, hole at $x = 0$

22. vertical asymptotes at $x = -5$ and $x = 1$, hole at $x = -1$

23. holes at $x = 3$ and $x = -7$, resembles $y = x$

Graph each rational function.

24. $y = \dfrac{3}{x + 2}$

25. $y = \dfrac{x - 5}{x + 1}$

26. $y = \dfrac{-4}{x - 1}$

27. $y = \dfrac{(x + 2)(x - 2)}{x - 2}$

28. $y = \dfrac{x}{x - 5}$

29. $y = \dfrac{-2}{(x - 3)^2}$

30. $y = \dfrac{x^2 - x}{x}$

31. $y = \dfrac{-5}{(x - 3)(x + 1)}$

32. $y = \dfrac{x^2}{x(x - 1)}$

33. $y = \dfrac{3}{(x - 4)^2}$

34. $y = \dfrac{x^2 + 3x}{x}$

35. $y = \dfrac{x^2 + 3x - 4}{x}$

36. $y = \dfrac{x}{1 - x^2}$

37. $y = \dfrac{-x}{x^2 - 4}$

38. $y = \dfrac{x - 1}{x^2 - 9}$

Critical Thinking

39. Graph $y = \dfrac{3x^2(x + 1)}{2(x - 2)^2(x - 1)}$.

Applications and Problem Solving

40. Amusements Louis was baby-sitting his neighbor's two children and decided to take them to the park. The children wanted to play on the seesaw. When they got on the seesaw, the older child was too heavy for the younger one. Louis knew that if the older child sat closer to the center of the seesaw, it would work better. The equation relating the two weights and the distances from the center of the seesaw is $w_1 d_1 = w_2 d_2$, where w_1 is the weight of one child and d_1 is the distance of that child from the center of the seesaw and w_2 and d_2 are the weight and distance for the other child.

a. Suppose the children are balanced on the seesaw. One child weighs 30 pounds and sits 6 feet from the center of the seesaw. Write a rational function to find the weight of the other child.

b. Graph the function in part a.

c. If the older child weighs 45 pounds, how far must she sit from the center of the seesaw?

d. Make a general conjecture about the relationship of a person's weight to the distance from the center of the seesaw in order to make it work properly.

41. **Music**   The frequency of a sound wave is called its pitch. The pitch p of a musical tone and its wavelength w are related by the equation $p = \dfrac{v}{w}$, where v is the velocity of sound through air. Suppose a sound wave has a velocity of 1056 feet/second.

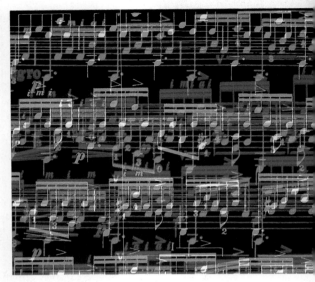

a. Graph the equation $p = \dfrac{v}{w}$. What lines are close to the maximum values for the pitch and the wavelength?

b. What happens to the pitch of the tone as the wavelength decreases?

c. If the wavelength is doubled, what happens to the pitch of the tone?

42. **Physics**   The average speed v of a certain particle in meters per second is given by the equation $v = \dfrac{2t^2 + 7t + 5}{t + 3}$, where t is the time in seconds.

a. For what values of t would an asymptote exist?

b. What is the value of v if $t = 2$?

c. Use a graphing calculator to graph this function. Describe the appearance of the graph.

Mixed Review

43. Write an example of a relation that is not a function. Tell why.   **(Lesson 1-1)**

44. Find the length of the sides of a triangle whose vertices are $A(-1, 3)$, $B(-1, -3)$, and $C(3, 0)$.   **(Lesson 1-4)**

45. Find the slope of a line perpendicular to a line whose equation is $3x - 4y = 0$.   **(Lesson 1-6)**

46. Find the maximum and minimum values of $f(x, y) = y - x$ defined for the polygonal convex set having vertices at $(0, 0)$, $(4, 0)$, $(3, 5)$, and $(0, 5)$.   **(Lesson 2-5)**

47. Find the inverse of $x^2 - 9 = y$.   **(Lesson 3-3)**

48. **College Entrance Exam**   Compare quantities A and B below.

Write A if quantity A is greater.

Write B if quantity B is greater.

Write C if the two quantities are equal.

Write D if there is not enough information to determine the relationship.

(A) the volume of a cube with side $x, x > 1$

(B) the volume of a rectangular solid with sides measuring $x, x + 1$, and $x - 1, x > 1$

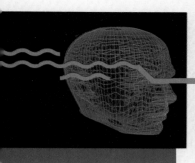

Technology

Graphing Rational Functions

The graphs of rational functions are intriguing. However, much time can be spent sketching a single graph by hand. Graphing calculators and graphing software graph rational functions, but not always correctly. Many graphing utilities fail to properly represent the asymptotes and holes that occur when you graph rational functions. *The Function Analyzer*, by Sunburst Communications, correctly graphs rational functions.

To generate the graph of the rational function $f(x) = \dfrac{x^3 + 4x^2 - 2x + 3}{x^3 - 4x^2 - x - 1}$ using *The Function Analyzer*, use the following steps.

- Select **Function** from the menu and then press <u>**N**</u> to use a new function.
- Type in the function: $(\mathbf{X^\wedge 3 + 4X^\wedge 2 - 2X + 3})/(\mathbf{X^\wedge 3 - 4X^\wedge 2 - X - 1})$. Then press **[Enter]**. *The caret symbol (^) does not show on the screen when typed.*

After the graph is drawn, find the equation of the vertical asymptote between $x = 4$ and $x = 5$.

- Press **[Esc]** twice to return to the main menu.
- Select **Scale** and then select **Enter New Coordinate Values**.
- To set the scale, type **100[↓] - 100 [←] 4 [→] 5** and press **[Enter]**. The graph will be redrawn using the new scale.
- Then set the scale to **500[↓] - 500 [←] 4.2 [→] 4.4** and press **[Enter]**.
- Finally, set the scale to **2000 [↓] -2000 [←] 4.25 [→] 4.35** and press **[Enter]**.
- Press **[Esc]** twice to return to the main menu and then select **Values**.
- Press **[←]** eight times to show the coordinates of the points that are near the asymptote.

We can conclude that the asymptote is somewhere between $x = 4.286$ and $x = 4.288$. *The equation of the asymptote is actually $x = 4.28762526163747$.*

EXERCISES

Use *The Function Analyzer* to graph each rational function. Select the Clear Screen option before beginning each exercise. Then find the equation(s) of the asymptote(s) for each function.

1. $f(x) = \dfrac{(x - 3)(x + 5)}{(x + 1)(x + 3)}$

2. $g(x) = \dfrac{x^2 + x - 6}{x^2 - x - 2}$

3. $h(x) = \dfrac{3.12x^3 - 4.5x + 8.92}{-1.02x^4 + 11.41x^2 - 9.15x + 7.11}$

3-5 Graphs of Inequalities

Objective

After studying this lesson, you should be able to:
- graph polynomial, absolute value, and radical inequalities.

Application

Call it Magic Mitts®, Scatch®, Katch-a-Roo®, or Super Grip Ball®. It's basically a Velcro™ ball with Velcro™ paddles and lots of imagination. (The photo at the left is a close-up of Velcro™ magnified 18 times.) Mark Paliafito was 26 when he founded Paliafito America Inc. in 1990. His company distributes Super Grip Ball®. In July of 1991, they had sold 650,000 of their product and had orders for one million more.

Recreation Limited sells sports equipment. According to their accountants, their profits $f(x)$ from the sale of Super GripBall® can be modeled by the relation $f(x) < -0.003x^2 + 9x - 1500$, where x is the number of sets of Super Grip Ball® sold in a week and $0 \leq x < 1500$.

To graph this inequality, first graph the function $f(x) = -0.003x^2 + 9x - 1500$. Since the inequality does not contain the equality, graph the function as a broken graph. Next, draw a dotted line at $x = 1500$ to show that the boundary we are considering is the graph of the function up to $x = 1500$.

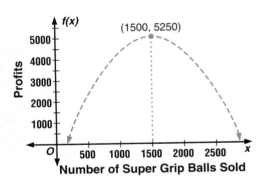

The inequality contains infinitely many solutions. These solutions are indicated in the graph as a shaded region. To find which region should be shaded, test a point that is not on the boundary.

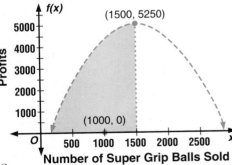

Test (1000,0):   *This point is inside the boundary.*

Is $0 < -0.003(1000)^2 + 9(1000) - 1500$?

$0 < 4500$ *true*

Since $(1000, 0)$ is a point that satisfies the inequality and $0 \leq 1000 < 1500$, the region that contains this point should be shaded.

Notice that the graph is a member of the $y = x^2$ family of graphs. You can use the skills that you learned with families of graphs to graph other inequalities.

Example 1

Graph $y \leq \sqrt[3]{x+2} - 4$.

First graph the inverse of the parent function $y = x^3$.
Then translate the inverse 2 units left and 4 units down.
Since the boundary is included in the inequality, the graph is drawn as a solid line.

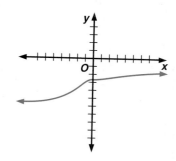

Test a point not on the boundary.

Test $(0, 0)$:

$y \leq \sqrt[3]{x+2} - 4$

$\overset{?}{0 \leq} \sqrt[3]{0+2} - 4$ *Replace (x, y) with $(0, 0)$.*

$\overset{?}{0 \leq} \sqrt[3]{2} - 4$ *$\sqrt[3]{2} \approx 1.26$*

$0 \not\leq -2.74$

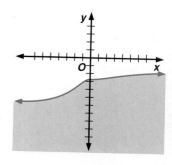

The origin is often used as a test point because $(0, 0)$ is easily tested in any inequality.

Since $(0, 0)$ does not satisfy the inequality, the region containing $(0, 0)$ should *not* be shaded. So, shade the region below the graph.
Check your solution by testing a point in the region you shaded.

Some inequalities involve absolute value functions. The same process used in Example 1 is used for determining which region should be included in the solution.

Example 2

Graph $y > |x - 3| - 5$.

Graph the parent function $y = |x|$ and translate it 3 units right and 5 units down. The boundary is not included in the inequality, so it is dashed.

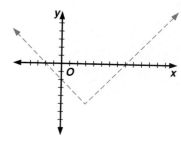

Test a point not on the boundary to determine which region should be included in the graph.

Test $(0, 0)$:

$y > |x - 3| - 5$

$\overset{?}{0 >} |0 - 3| - 5$ *Let $(x, y) = (0, 0)$.*

$0 > -2$ *true*

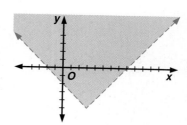

Since $(0, 0)$ satisfies the inequality, the region containing the origin is included in the graph and is shaded.

When solving absolute value inequalities algebraically, the definition of absolute value is used to determine the solution set. That is, if $a < 0$, then $|a| = -a$, and, if $a > 0$, then $|a| = a$.

Example 3

Solve $|x - 4| < 0.1$.

There are two cases that must be solved. In one case, $x - 4$ is negative, and in the other, it is positive.

An inequality in the form $|x| < n$, where $n \leq 0$, has no solution.

If $a < 0$, then $|a| = -a$.
$-(x - 4) < 0.1$
$-x + 4 < 0.1$
$-x < -3.9$
$x > 3.9$

If $a > 0$, then $|a| = a$.
$x - 4 < 0.1$
$x < 4.1$

The solution set is $\{x \mid 3.9 < x < 4.1\}$ $\{x \mid 3.9 < x < 4.1\}$ *is read "the set of all numbers x such that x is between 3.9 and 4.1."*

Example 4

The AccuData Company makes 3.5-inch computer disks. So the disks can fit into the drives of various computers, the dimensions of the plastic case about the magnetic disk can vary by no more than 0.11 inch.
a. Write an inequality to represent the width of the plastic case.
b. What are the largest and smallest lengths of the case allowable?
c. Sketch a graph to represent the range of values allowed for the case.

a. If x represents the width of the new disk, then $|x - 3.5|$ equals the acceptable variance from the 3.5-inch norm. This variance can be no more than 0.11 inches. So, $|x - 3.5| \leq 0.11$.

b. *If $a < 0$, then $|a| = -a$.*
$-(x - 3.5) \leq 0.11$
$-x + 3.5 \leq 0.11$
$-x \leq -3.39$
$x \geq 3.39$

If $a > 0$, then $|a| = a$.
$x - 3.5 \leq 0.11$
$x \leq 3.61$

The solution is $\{x \mid 3.39 \leq x \leq 3.61\}$. The length of the disk can be as great as 3.61 inches and as small as 3.39 inches.

c. The function $|x - 3.5| < y$ describes all variances less than y. The values of the function below the line $y = 0.11$ describe the acceptable variances.

Describe something new you learned in this lesson. Be sure to give examples.

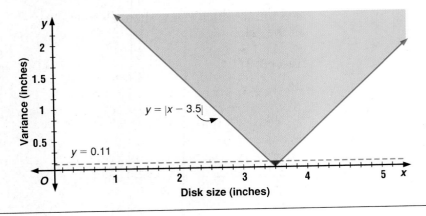

CHECKING FOR UNDERSTANDING

Communicating Mathematics

Read and study the lesson to answer each question.

1. **Describe** how you determine which region of the graph of an inequality should be shaded.

2. **Write** an absolute value inequality that has no solution.

Guided Practice

Determine if each ordered pair is a solution for the given inequality. Write _yes_ or _no_.

3. $y > x^4 - 5x^2 + 2, (2, 3)$

4. $y < -0.2x^2 + 9x - 7, (10, 20)$

5. $y < |x - 2| + 7, (3, 8)$

6. $y > -\sqrt{x + 11} + 1, (-2, -4)$

Graph each inequality.

7. $y < |x + 4|$

8. $y > \sqrt[3]{x} + 3$

9. $x^3 - 5x^2 + 6x - 9 > y$

EXERCISES

Practice

Graph each inequality.

10. $y > 2x + 3$

11. $y < -4x + 9$

12. $y > \sqrt{x + 2}$

13. $y > x^2 - 5$

14. $|x - 5| < 0.05 + y$

15. $|x + 2| > 5 + y$

16. $y < |x + 9|$

17. $y > |x + 3| - 2$

18. $y > |x| + 2$

19. $y > (x - 4)^2$

20. $y < |3x - 4| + 5$

21. $y < \sqrt[3]{x + 8} - 8$

Solve each inequality.

22. $|x + 2| > 3$

23. $|2x + 7| < 0$

24. $|3x + 12| > 42$

25. $|x| \leq x$

26. $|x| > x$

27. $|x + 2| - x \geq 0$

Graphing Calculator

Use a TI-82 calculator to graph the boundaries of each inequality. Then use the SHADE feature to complete the graph.

28. $y > 3x + 2$

29. $y < x^2 + 2x - 3$

30. $y < x^3 + 5x^2 - 18x - 72$

31. $y < |x| + 2$

Critical Thinking

32. Solve $|x + 1| + |x - 1| \leq 2$.

33. Solve the system of inequalities by graphing.
$$y > 2(x - 2)^2 - 5$$
$$y < -(x - 2)^2$$

Applications and Problem Solving

34. **Woodworking** Marcia Gardlik makes customized picture frames. A customer came into her shop and requested that she make a square frame that used wood 1.5 inches wide and was sized so that the area of the family crest was at most two-thirds the total area of the picture and frame. Write an inequality to describe the possible lengths of one side of the frame.

35. Elections The 1992 presidential campaign had three prominent candidates. In April, a television news program poll stated that Ross Perot was favored by 35% of the American voters, with a margin of error of 3.3%. Write an inequality to describe the range of the percents of American voters who favored Ross Perot at that time.

Mixed Review

36. Find $[f \circ g](x)$ if $f(x) = x - 0.2x$ and $g(x) = x - 0.3x$. **(Lesson 1-2)**

37. Name the slope and y-intercept of the graph of $x - 2y - 4 = 0$. **(Lesson 1-5)**

38. Find $3A + 2B$ if $A = \begin{bmatrix} 4 & -2 \\ 5 & 7 \end{bmatrix}$ and $B = \begin{bmatrix} -3 & 5 \\ -4 & 3 \end{bmatrix}$. **(Lesson 2-2)**

39. Solve $\begin{bmatrix} -1 & 2 \\ -4 & 3 \end{bmatrix} \cdot \begin{bmatrix} x \\ y \end{bmatrix} = \begin{bmatrix} 0 \\ 15 \end{bmatrix}$. **(Lesson 2-3)**

40. Graph $y = \dfrac{4x}{x-1}$. **(Lesson 3-4)**

41. College Entrance Exam Choose the best answer. Nine playing cards from the same deck are placed to form a large rectangle whose area is 180 in^2. There is no space between the cards and no overlap. What is the perimeter of this rectangle?

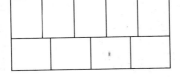

(A) 29 in.   **(B)** 58 in.   **(C)** 64 in.

(D) 116 in.   **(E)** 210 in.

MID-CHAPTER REVIEW

1. Determine whether the graph of $y = |x| + 1$ is symmetric with respect to the x-axis, the y-axis, the line $y = x$, the line $y = -x$, or the origin. In your answer, include all symmetries that apply. **(Lesson 3-1)**

2. For each parent graph, describe the transformations that have taken place in the related graph of each function. **(Lesson 3-2)**
   $f(x) = x^3$
   a. $y = x^3 + 3$
   b. $y = (x + 2)^3$
   c. $y = 2x^3 - 4$

3. Sketch the graph of $y = |2x - 1| + 2$ reflected over the line $y = x$. **(Lesson 3-2)**

4. Find the inverse of $y = x^2 + 6$. Then sketch the function and its inverse. **(Lesson 3-3)**

5. Determine any holes in the graph of $y = \dfrac{x^2 - x}{x^2 - 1}$. Then graph the function. **(Lesson 3-4)**

6. Graph $y \le |x + 2| - 5$. **(Lesson 3-5)**

3-6 Tangent to a Curve

Objectives

After studying this lesson, you should be able to:
- find the derivative of a function, and
- find the slope and the equation of a line tangent to the graph of a function at a given point.

Application

The Department of Transportation in each state is responsible for marking hills with restrictions on the weight of vehicles, such as trucks or buses, that are allowed on the hill. One of the factors affecting this limit is the steepness or slope of the hill at given points. The steepness of the hill can be determined from the slope of the line tangent to the hill at those points.

A line tangent to a curve at a point on the curve is the line that passes through that point and has a slope that is the same as the slope of the curve at that point. In the figure at the right, two lines tangent to the graph of $f(x) = x^2$ are shown. One line is tangent at $(-1, 1)$, and the other line is tangent at $(2, 4)$. Lines that are tangent to a curve are also called **tangents** to a curve.

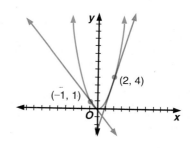

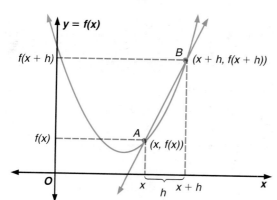

Suppose you graph a function $y = f(x)$. Suppose points A and B are two points on the graph near each other. The slope of line AB can be found by using the coordinates of A and B. Let the coordinates of A and B be $(x, f(x))$ and $(x + h, f(x + h))$, respectively. The slope is

$$\frac{f(x + h) - f(x)}{(x + h) - x} \text{ or } \frac{f(x + h) - f(x)}{h}.$$

Line AB is called a **secant line** because it intersects the curve more than once.

Suppose A remains fixed while you move B along the curve toward A. As B approaches A, the values of h become smaller and smaller, approaching zero. When $h = 0$, B coincides with A. Then the secant line becomes the line tangent to the curve at point A.

In the expression for the slope $\overleftrightarrow{AB}$, notice that $\frac{f(x + h) - f(x)}{h}$ is undefined when $h = 0$. However, the secant is a tangent when $h = 0$. So, the definition of the slope of the tangent line must be defined in a special way.

Let's consider the slope of the line tangent to the graph of $f(x) = 3x^2 - 4x + 1$ at $(2, 5)$. The slope of the secant line through two general points $A(x, f(x))$ and $B(x + h, f(x + h))$ is equal to $\dfrac{f(x + h) - f(x)}{h}$, where $h \neq 0$. Find the slope of the secant line to the graph of $f(x) = 3x^2 - 4x + 1$. That is, find $f(x + h)$.

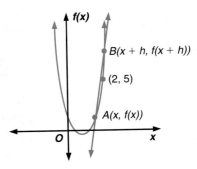

$$f(x + h) = 3(x + h)^2 - 4(x + h) + 1$$
$$= 3x^2 + 6xh + 3h^2 - 4x - 4h + 1$$

Now, find the slope of $\overleftrightarrow{AB}$.

$$
\begin{aligned}
m &= \frac{f(x + h) - f(x)}{h} \qquad \textit{Definition of slope} \\
&= \frac{(3x^2 + 6xh + 3h^2 - 4x - 4h + 1) - (3x^2 - 4x + 1)}{h} \\
&= \frac{6xh + 3h^2 - 4h}{h} \\
&= 6x + 3h - 4
\end{aligned}
$$

A general expression for the slope of the secant line for $f(x) = 3x^2 - 4x + 1$ is $6x + 3h - 4$. The slope of the secant line through specific points can be found by substituting values for x and h.

x	h	slope of the secant line
2	2	$6(2) + 3(2) - 4 = 14$
2	1	$6(2) + 3(1) - 4 = 11$
2	0.5	$6(2) + 3(0.5) - 4 = 9.5$
2	0.25	$6(2) + 3(0.25) - 4 = 8.75$
2	0.1	$6(2) + 3(0.1) - 4 = 8.3$
2	0.01	$6(2) + 3(0.01) - 4 = 7.97$

As h approaches zero, the middle term of $6x + 3h - 4$ also approaches zero, which means that $6x + 3h - 4$ approaches $6x - 4$. As h approaches zero, the secant line approaches the tangent; so, as h approaches zero, the slope of the secant line approaches the slope of the tangent. Therefore, the slope of the line tangent to the graph of $f(x) = 3x^2 - 4x + 1$ at $(x, f(x))$ is $6x - 4$.

The slope of the tangent at $(2, 5)$ can be found by substituting 2 for x in the expression $6x - 4$. The slope of the tangent at $(2, 5)$ is $6(2) - 4$, or 8.

The symbol $f'(x)$ is often used to denote the slope of the tangent line. It is defined as follows.

$f'(x)$ is read "f prime of x."

$$f'(x) = \lim_{h \to 0} \frac{f(x + h) - f(x)}{h} \qquad \textit{lim is read "the limit as h approaches zero"}$$

The symbol $\displaystyle\lim_{h \to 0} \frac{f(x + h) - f(x)}{h}$ means the limiting value of the slope as the value of h approaches zero. The function $f'(x)$ is called the **derivative** of $f(x)$.

As you saw, finding the derivative of a function can be time consuming and complex. There are rules that will make the process easier.

Rules for Derivatives

Constant Rule:	**The derivative of a constant function is zero. If $f(x) = c$, then $f'(x) = 0$.**
Power Rule:	**If $f(x) = x^n$, where n is a rational number, then $f'(x) = n \cdot x^{n-1}$.**
Constant Multiple of a Power Rule:	**If $f(x) = cx^n$, where c and n are rational numbers, $f'(x) = n \cdot cx^{n-1}$.**
Sum Rule:	**If $f(x) = g(x) + h(x)$, then $f'(x) = g'(x) + h'(x)$.**

Example 1

Find the derivative of each function.

a. $f(x) = 6$ b. $f(x) = \dfrac{1}{x}$ c. $f(x) = 6x^3 - 4x^2 + 2x - 9$

a. $f(x) = 6$
$f'(x) = 0$ *Use the constant rule.*

b. $f(x) = \dfrac{1}{x}$
$f(x) = x^{-1}$
$f'(x) = -1x^{-2}$ *Use the power rule.*

c. $f(x) = 6x^3 - 4x^2 + 2x - 9$
Use the power rule and constant rule for each term. Then use the sum rule.
$f(x) = 3 \cdot 6x^{3-1} - 2 \cdot 4x^{2-1} + 1 \cdot 2x^{1-1} - 0$
$f'(x) = 18x^2 - 8x + 2$

Since the derivative is also the expression for the slope of the line tangent to the graph of a function, you can use the derivative to find the slope of the tangent at a given point on the graph.

Example 2

Find the slope of the line tangent to the graph of $f(x) = x^4 - 2x^3 + 5x^2 - 8$ at the point (-2, 44).

Use the rules for derivatives to find $f'(x)$.
$f'(x) = 4x^{4-1} - 2(3)x^{3-1} + 5(2)x^{2-1} - 0$
$\quad\ = 4x^3 - 6x^2 + 10x$

Now evaluate $f'(x)$ for $x = -2$.

$f'(-2) = 4(-2)^3 - 6(-2)^2 + 10(-2)$
$\qquad\ = -32 - 24 - 20 \text{ or } -76$

The slope of the line tangent to the graph of $f(x) = x^4 - 2x^3 + 5x^2 - 8$ at the point (-2, 44) is -76.

The derivative is also useful in many real-life applications.

Example 3

APPLICATION

Amusement

Cedar Point Amusement Park in Sandusky, Ohio, holds the Guinness World Record for the most roller coasters at a single site, with 12. The slope of one of the hills on the *Mean Streak*, a wooden roller coaster, can be modeled by the graph of the function $f(x) = \dfrac{-x^2}{2} + 5x + 50$. Find the slope of the line tangent to this curve at the point where $x = 1$.

$$f(x) = \frac{-x^2}{2} + 5x + 50$$
$$f'(x) = 2(-\frac{1}{2})x^{2-1} + 1(5)x^{1-1} + 0$$
$$f'(x) = -x + 5$$
$$f'(1) = -(1) + 5$$
$$\quad\ = 4 \qquad \text{The slope of the tangent line at the point where } x = 1 \text{ is 4.}$$

If you know the coordinates of one point on a curve and the slope of the line tangent to a curve at that point, you can find the equation of the tangent line. This process uses the point-slope form of the equation.

Example 4

Find the equation of the line tangent to the graph of $y = 3x^3 - 2x^2 + 4x - 2$ at the point $(-2, -42)$. Write the equation in slope-intercept form. Then graph the function and the tangent line.

First, use the derivative to find the slope of the tangent line. Then, evaluate the derivative for $x = -2$.

$$f(x) = 3x^3 - 2x^2 + 4x - 2$$
$$f'(x) = 9x^2 - 4x + 4$$
$$f'(-2) = 9(-2)^2 - 4(-2) + 4$$
$$f'(-2) = 48 \qquad \textit{The slope of the tangent at } (-2, -42) \textit{ is 48.}$$

Use the slope and the point $(-2, -42)$ to write the equation of the tangent line in slope-intercept form.

You can review the point-slope form of an equation in Lesson 1-5.

$$y - y_1 = m(x - x_1)$$
$$y - (-42) = 48(x - (-2)) \qquad (x, y) = (-2, -42), m = 48$$
$$y = 48x + 54$$

You can use your graphing calculator to graph the function and the tangent line.
You can use the ZOOM and TRACE features to verify that the line is tangent to the graph at $(-2, -42)$.

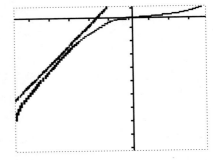

CHECKING FOR UNDERSTANDING

Communicating Mathematics

Read and study the lesson to answer each question.

1. **Explain** the difference between a secant line and a tangent line.

2. **Describe** how to find the derivative of a function.

3. **Illustrate** at what point the line tangent to the graph of a function representing a highway hill would have a slope of zero.

4. **Write** the symbol for the derivative of $f(x)$.

Guided Practice

Find the derivative of each function.

5. $f(x) = x^2 + 5$

6. $f(x) = 2x^2 - 11$

7. $\frac{1}{2}x^4$

8. $f(x) = -2x^2 + 3x + 1$

9. $f(x) = 0.6x^4 - 0.4x^2 - 0.8$

10. $f(x) = 2x^6 + 8x^3 + 4x - 120$

Find the slope of the line tangent to the graph of each function at the given point.

11. $y = x^2,\ (3, 9)$

12. $y = x^2 + 4,\ (0, 4)$

13. $y = 2x^2,\ (0, 0)$

14. Find the equation of the line tangent to the graph of $y = x^2$ at $(2, 4)$. Write the equation in slope-intercept form.

EXERCISES

Practice

Find the derivative of each function.

15. $f(x) = x^3 - 2x^2 + 4x$

16. $f(x) = 3x^2 - 8x + 5$

17. $f(x) = 0.3x^3 - 4x^2 + 5$

18. $f(x) = 4x^{-2} + 2x^{-3} + 4$

19. $f(x) = -4x^5 + 7x^3 + x$

20. $f(x) = \frac{3}{x^4} + x^2 - \frac{6}{13}$

Find the slope of the line tangent to the graph of each function at the given point.

21. $y = 2x^2,\ (-2, 8)$

22. $y = \frac{1}{2}x^2 + 12,\ (-3, 16\frac{1}{2})$

23. $y = -2x^2 + 3x + 10,\ (1, 11)$

24. $y = 0.5x^2 - 0.4x - 0.5,\ (1, -0.4)$

25. $y = x^2 + \frac{1}{6}x + 1,\ \left(\frac{1}{2}, \frac{4}{3}\right)$

26. $y = \frac{x^3 - 4}{6},\ \left(2, \frac{2}{3}\right)$

27. Find the slope of the line tangent to the graph of $y = 2x^2 - 3x - 4$ at the point where $x = -2$.

28. Find the slope of the line tangent to the graph of $y = \frac{1}{2}x^2 + \frac{1}{4}x + \frac{1}{8}$ at the point where $x = \frac{1}{2}$.

Find the equation of the line tangent to the graph of each function at the given point. Write the equation in slope-intercept form. Graph the function and the tangent.

29. $y = x^2 - 3,\ (3, 6)$

30. $y = x^2 - 3x + 2,\ (1, 0)$

31. $y = -x^2 - x + 2,\ (0.5, 1.25)$

32. $y = x^2 - 5x + 6,\ (2.5, -0.25)$

33. $y = 0.5x^2 + x - 1,\ (-4, 3)$

34. $y = -3x^2 + x + 5,\ \left(\frac{1}{6}, 5\frac{1}{12}\right)$

Find the coordinates of the point(s) at which the line tangent to the graph of the function $f(x)$ has the designated slope.

35. $f(x) = x^2 + 4x + 4, m = -2$

36. $f(x) = 3x^3 - 36x, m = 0$

37. $f(x) = \dfrac{-4}{x}, m = 1$

38. $f(x) = 4x^4 - 8x^2 - 5, m = 0$

Graphing Calculator

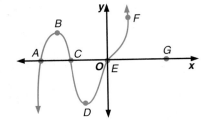

Use a graphing calculator to graph the derivative of each function. Use the TRACE function to find the values of x for which the slope of the tangent line (Y) is positive, negative, and zero.

39. $y = x^2 + 2x$

40. $y = 12 - x^2$

41. $y = x^3 - 3x^2 - 7$

Critical Thinking

42. Write a convincing argument to prove that the derivative of a constant is zero.

43. The graph of $g(x)$ is shown at the right. At which of the labeled points on the graph does $g'(x)$ appear to equal 0? Explain your reasoning.

Applications and Problem Solving

44. Physics Instantaneous velocity is the instantaneous rate of change in position of a particle with respect to time, t seconds. If $f(t)$ is a function describing the position of the particle, then $f'(t)$ is the instantaneous velocity of that particle. Find the instantaneous velocity in meters per second of a particle whose function is $f(t) = 2t^2 - 8t + 3$, when $t = 5$.

45. Entertainment Blimpy is a clown who entertains at children's birthday parties. One of his tricks is magically blowing up spherical balloons with the tip of his finger. Actually, Blimpy has a hose from an air tank connected to his finger tip.

   a. Find the instantaneous rate of change in the volume of a balloon Blimpy is inflating with respect to the radius when the radius is 2 inches. *The formula for volume is $V = \dfrac{4}{3}\pi r^3$.*

   b. What does this rate of change mean?

Mixed Review

46. Graph $y + 6 \geq 4$. **(Lesson 1-3)**

47. Find the solution to the system of equations. **(Lesson 2-1)**
$x + y = 1$
$3x + 5y = 7$

48. Use an augmented matrix to solve the system of equations. **(Lesson 2-4)**
$2x + y + z = 0$
$3x - 2y - 3z = -21$
$4x + 5y + 3z = -2$

49. Find $f(-7, 4)$ if $f(x, y) = 5x - 2y + 1$. **(Lesson 2-5)**

50. Graph $y = (x + 2)^2 - 3$ and its inverse. **(Lesson 3-3)**

51. Graph $y \geq x^2 - 25$. **(Lesson 3-5)**

52. College Entrance Exam Choose the best answer.
A rectangular block of metal weighs 3 ounces. How many pounds will a similar block of the same metal weigh if the edges are twice as long?

   (A) 0.375   **(B)** 0.75   **(C)** 1.5   **(D)** 3   **(E)** 24

3-7A   Graphing Calculators: Locating Critical Points of Polynomial Functions

A graphing calculator can help you locate critical points of a polynomial function. A critical point is a point where the graph of the function changes direction or where the graph changes concavity. A critical point can be a relative minimum, a relative maximum, or a point of inflection.

Using the zoom function is a good way to locate critical points of a function. You can use the ZOOM key if you are using a TI-82 graphing calculator. You can also reset your range values manually or zoom in by a factor on both graphing calculators.

Example

Approximate the critical points on the graph of $y = x^4 - 3x^3 + 2x + 4$ to the nearest hundredth.

Enter: Y= X,T,θ ^ 4 − 3 X,T,θ ^ 3 + 2 X,T,θ + 4 GRAPH

The viewing window of $[-5, 5]$ by $[-5, 10]$ contains a complete graph of the function. Looking at the graph, we can see that there are three critical points: one relative maximum and two relative minima.

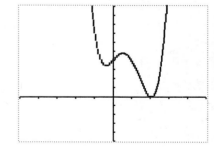

Using TRACE and ZOOM, we find that the relative maximum occurs at $(0.54, 4.69)$, to the nearest hundredth.

The same process can be used to find the two relative minima. They are at $(-0.43, 3.41)$ and $(2.14, -0.15)$, to the nearest hundredth.

EXERCISES

Graph each function so that a complete graph is shown. Then approximate each of the critical points to the nearest hundredth and sketch the graph on a piece of paper.

1. $y = x^2 + 2$

2. $y = x^2 + 3x - 6$

3. $y = (x - 4)^3$

4. $y = x^3 - 4x + 1$

5. $y = x^3 - 2x^2 - 6x - 3$

6. $y = 2x^4 + x^3 - x^2 + 4$

7. $y = -x^4 + 3x^2 - 4$

8. $y = x^5 + 4x^3 + 2$

3-7 Graphs and Critical Points of Polynomial Functions

Objective

After studying this lesson, you should be able to:
- find the critical points of the graph of a polynomial function and determine if each is a minimum, maximum, or point of inflection.

Application

On May 5, 1961, astronaut Alan B. Shepard, Jr. was launched from Cape Canaveral, Florida, on the first manned space flight from the United States. The Redstone booster rocket launched him into flight at an initial velocity of 6274 feet per second. Then the Project Mercury spacecraft fell to Earth off the coast of Florida. How long into the flight did the spacecraft, *Freedom 7*, reach its maximum height? *This problem will be solved in Example 1.*

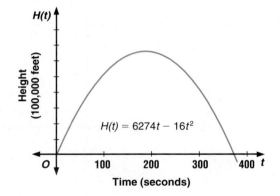

As an object is propelled upward, gravity tends to pull it back to Earth. This relationship can be expressed by the formula $H(t) = v_0 t - 0.5gt^2$, where $H(t)$ is the distance above the starting point, v_0 is the initial velocity, t is the time elapsed, and g is the acceleration due to gravity. Earth has a constant acceleration of 32 feet/second^2 (9.8 meters/second^2).

If 6274 ft/s is substituted for v_0 and 32 ft/s^2 for g, the formula becomes $H(t) = 6274t - 16t^2$. The graph of this function is shown at the left.

Remember that this is a graph of the function, not the actual path of the object being projected into the air.

Look at the graph of $H(t)$. The highest point on the graph is the point at which the slope of the tangent at that point is zero. We can use the derivative to determine the t-coordinate or time of that point.

$$H(t) = 6274t - 16t^2$$
$$H'(t) = 6274 - 32t \qquad \textit{Find the derivative.}$$
$$0 = 6274 - 32t \qquad \textit{Replace the slope with 0.}$$
$$-6274 = -32t \qquad \textit{Solve for t.}$$
$$196 \approx t$$

The spacecraft reached its maximum height at 196 seconds after liftoff.

Example 1

Refer to the application at the beginning of the lesson. Find the maximum height of the Freedom 7 spacecraft.

We found that the maximum height occurred at 196 seconds. To find the maximum height, find $H(196)$.

$$H(t) = 6274t - 16t^2$$
$$H(196) = 6274(196) - 16(196)^2 \quad \text{$t = 196$}$$
$$H(196) = 615{,}048$$

The maximum height is 615,048 feet which is about 116.5 miles.

The coordinates of the maximum point on the graph of H(t) are (196, 615,048).

A curve may possess three types of points called **critical points.** They are called critical points because at these points the nature of the graph changes. A critical point may be a **maximum**, a **minimum**, or a **point of inflection**. For a maximum, the curve changes from an increasing curve to a decreasing curve. For a minimum, the curve changes from a decreasing curve to an increasing curve. A point of inflection is a point where the graph changes its curvature as described below.

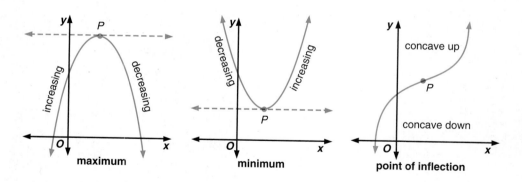

maximum   minimum   point of inflection

The plural of maximum is maxima. The plural of minimum is minima.

Some graphs may have more than one critical point. In the graph at the right, points A, B, and C are critical points. Point A is called a **relative maximum**. The y-coordinate of A is not the greatest value of the function, but it does represent the greatest value in an interval. Point B is a **relative minimum.** Likewise, its y-coordinate is not the least value of the function, but is a least value for an interval. Point C is a point of inflection.

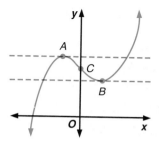

When $f'(x) = 0$, the line tangent to the curve at $(x, f(x))$ is parallel to the x-axis. To find the relative maximum or relative minimum of a function, you need to find the points on the graph where the slope of the tangent is 0. To do this, find the derivative, which is the function for the slope of the tangent. Then set it equal to zero and solve.

To find the point of inflection, you need to find the derivative of the derivative and set it equal to zero and solve. This is called the second derivative and is represented by $f''(x)$.

Example 2

Determine the critical points for the graph of $f(x) = 3x^3 - 9x + 5$.

$f(x) = 3x^3 - 9x + 5$

$f'(x) = 9x^2 - 9$ *Find the derivative.*

$0 = 9x^2 - 9$ *Replace $f'(x)$ with 0.*

$9 = 9x^2$ *Solve for x.*

$\pm 1 = x$

$f'(x) = 9x^2 - 9$

$f''(x) = 18x$ *Find the second derivative.*

$0 = 18x$ *Replace $f''(x)$ with 0.*

$0 = x$ *Solve for x.*

The graph of the function has three critical points, one when $x = 1$, one when $x = -1$, and one when $x = 0$. To find the y-coordinates of each point, find $f(1), f(-1)$, and $f(0)$.

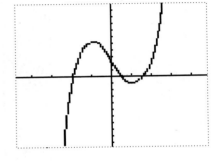

$f(1) = 3(1)^3 - 9(1) + 5$

$f(1) = -1$

$f(-1) = 3(-1)^3 - 9(-1) + 5$

$f(-1) = 11$

$f(0) = 3(0)^3 - 9(0) + 5$

$ = 5$

The critical points are at $(1, -1)$, at $(-1, 11)$, and at $(0, 5)$.   *Use your graphing calculator to verify the critical points.*

Once you have found the critical points of the graph of a function, you can determine if each point is a minimum, maximum, or point of inflection. You can determine this by testing points to either side of the critical point.

For $f(x)$ with $(a, f(a))$ as a critical point and h is a small value greater than zero:		
$f(a - h)$	$f(a + h)$	type of critical point
$f(a - h) < f(a)$	$f(a + h) < f(a)$	maximum
$f(a - h) > f(a)$	$f(a + h) > f(a)$	minimum
$f(a - h) < f(a)$	$f(a + h) > f(a)$	point of inflection
$f(a - h) > f(a)$	$f(a + h) < f(a)$	point of inflection

Example 3

The point at $(1, -4)$ is a critical point of the graph of $f(x) = x^3 - 3x - 2$. Determine whether this point represents a maximum, a minimum, or a point of inflection.

Evaluate the function for a point on either side of $(1, -4)$.

$f(1) = -4$

Let $h = 0.01$. Now find $f(1 + 0.01)$ and $f(1 - 0.01)$.

$f(1.01) = -3.9997$ $f(1.01) > f(1)$
$f(0.99) = -3.9997$ $f(0.99) > f(1)$

The point at $(1, -4)$ is a minimum of the graph.

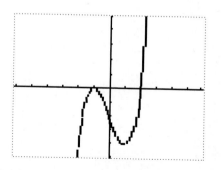

Example 4

Find the critical points of the graph of $f(x) = 3x^3 - 18x^2 - 4$. Then determine whether each point represents a maximum, a minimum, or a point of inflection.

$f(x) = 3x^3 - 18x^2 - 4$
$f'(x) = 9x^2 - 36x$
$\quad 0 = 9x^2 - 36x$ *Replace $f'(x)$ with 0.*
$\quad 0 = x^2 - 4x$ *Solve for x.*
$\quad 0 = x(x - 4)$ *Factor.*
$x = 0$ or $x - 4 = 0$
$\qquad\qquad x = 4$

$f'(x) = 9x^2 - 36x$
$f''(x) = 18x - 36$ *Find the second derivative.*
$\quad 0 = 18x - 36$ *Replace $f''(x)$ with 0.*
$\quad 2 = x$ *Solve for x.*

The critical points occur when $x = 0$, $x = 4$, and $x = 2$. Find $f(0), f(4)$, and $f(2)$ to determine the y-coordinates of the points.

$f(0) = 3(0)^3 - 18(0)^2 - 4$
$f(0) = -4$
$f(4) = 3(4)^3 - 18(4)^2 - 4$
$f(4) = -100$
$f(2) = 3(2)^3 - 18(2)^2 - 4$
$f(2) = -52$

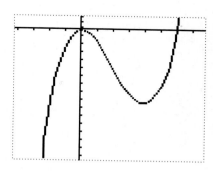

The critical points are at $(0, -4)$, at $(4, -100)$, and at $(2, -52)$. Now test points on each side of the critical points to determine their nature.

$f(0) = -4$
$f(0.01) = -4.0018$ $f(0.01) < f(0)$
$f(-0.01) = -4.0018$ $f(-0.01) < f(0)$
$(0, -4)$ is a relative maximum.

$f(4) = -100$
$f(4.01) = -99.9982$ $f(4.01) > f(4)$
$f(3.99) = -99.9982$ $f(3.99) > f(4)$
$(4, -100)$ is a relative minimum.

(continued on the next page)

$f(2) = -52$

$f(2.01) = -52.3600$ *f(2.01) < f(2)*

$f(1.99) = -51.6400$ *f(1.99) > f(2)*

$(2, -52)$ is a point of inflection.

The graphing calculator screen shown above verifies these results.

You can use critical points from the graphs of a function to solve real-life problems involving maximum and minimum.

Example 5

The Ripe and Juicy Fruit Company is developing plans for an open fruit bin that is 6 feet high and has a volume of 294 cubic feet. What are the dimensions of a rectangular base that will minimize the amount of material used (and thus, the cost of the material) to manufacture the open bin?

Let ℓ = the length of the base of the bin.
Find the value of ℓ terms of its width (w).

$V = \ell w h$

$294 = \ell w(6)$

$\dfrac{49}{w} = \ell$

Only positive values for ℓ and w are shown in the graph of $\ell = \dfrac{49}{w}$ at the right. Why?

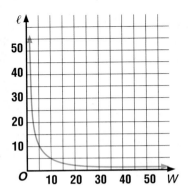

The amount of material used to manufacture the bin is the surface area of the base plus the four sides.

surface area =	*area of base*	+	*area of front/back*	+	*area of 2 sides*
$S(w) =$	ℓw	+	$2\ell h$	+	$2wh$
$=$	$\left(\dfrac{49}{w}\right)w$	+	$2\left(\dfrac{49}{w}\right)(6)$	+	$2w(6)$

$S(w) = 12w + 588w^{-1} + 49$

Now, find $S'(w)$ to determine maxima and/or minima.

$S'(w) = 12 - 588w^{-2}$

$0 = 12 - 588w^{-2}$ *Replace S'(w) with 0.*

$\dfrac{588}{w^2} = 12$ *Solve for w.*

$588 = 12w^2$

$\pm 7 = w$ *Disregard the negative value.*

If the width is 7 feet, the surface area is 217 square feet.

Now determine what type of critical point $(7, 217)$ is.

$f(7) = 217$

$f(6.99) = 217.0002$ *f(6.99) > f(7)*

$f(7.01) = 217.0002$ *f(7.01) > f(7)*

$(7, 217)$ are the coordinates of a minimum point.

If the width is 7 feet, the length is $\frac{49}{7}$ or 7 feet. The dimensions of the bin should be 7 feet by 7 feet by 6 feet.

Other important points on a graph are the *x*-intercepts and *y*-intercepts. Remember that a *y*-intercept of a relation is the *y*-coordinate of the point where $x = 0$. The *x*-intercept is the *x*-coordinate of the point where $y = 0$. It is possible for a graph to have no *x*- or *y*-intercept. Functions have at most one *y*-intercept. The number of *x*-intercepts may vary depending on the function.

In Chapter 1, you learned that the zero of a linear function is the point where $f(x) = 0$. The same is true for other types of functions. The zeros of a function are also the *x*-intercepts. The *y*-intercept can be determined by finding $f(0)$.

Example 6

Find the *x*-intercept(s) and *y*-intercept for $f(x) = x^3 + 6x^2 + 5x$. Then use a graphing calculator to verify your results.

To find the *x*-intercept(s), find the values of *x* for which $f(x) = 0$.

$$f(x) = x^3 + 6x^2 + 5x$$
$$0 = x^3 + 6x^2 + 5x \qquad \text{\textit{Replace f(x) with 0.}}$$
$$0 = x(x + 5)(x + 1) \qquad \text{\textit{Factor.}}$$

$x = 0$ or $x + 5 = 0$ or $x + 1 = 0$
$$x = -5 \qquad\qquad x = -1$$

The three *x*-intercepts are 0, –5, and –1.

To find the *y*-intercept, find $f(0)$.

$$f(x) = x^3 + 6x^2 + 5x$$
$$f(0) = (0)^3 + 6(0)^2 + 5(0)$$
$$f(0) = 0$$

The *y*-intercept is 0.

Use your graphing calculator to graph the function. Use the TRACE function to check the accuracy of the intercepts you found algebraically.

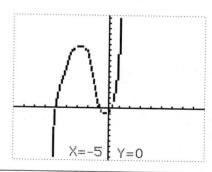

X=-5   Y=0

CHECKING FOR UNDERSTANDING

Read and study the lesson to answer each question.

1. **Describe** the three types of critical points.

2. **Explain** how you determine each type of critical point.

3. **Sketch** an example of a function with two relative maxima and one relative minimum.

Find the critical points for each function. Then determine whether each point is a *minimum*, a *maximum*, or a *point of inflection*.

4. $f(x) = x^2 + 4x - 12$

5. $f(x) = -2x^2 - 6x + 5$

6. $f(x) = x^2 - x - 6$

7. $f(x) = 12x^3$

Find the *x*- and *y*-intercepts of the graph of each function.

8. $f(x) = 3x^2 - 8x + 5$

9. $f(x) = x^2 - x$

EXERCISES

Find the critical points for each function. Then determine whether each point is a *minimum*, a *maximum*, or a *point of inflection*.

10. $f(x) = x^2 - 8x + 10$

11. $g(t) = t^2 + 2t - 15$

12. $D(r) = -r^2 - 2r + 8$

13. $f(x) = 3x^2 - 4x + 1$

14. $S(w) = w^3 - w^2 + 3$

15. $f(x) = 2x^3 - x^2 + 1$

16. $D(r) = r^4 - 8r^2 + 16$

17. $V(w) = w^5 - 28$

Find the *x*- and *y*-intercepts of the graph of each function.

18. $f(x) = x^2 + 12x + 32$

19. $f(x) = 4x^2 + 16x + 15$

20. $f(x) = (x + 3)(x - 3)(x + 1)$

21. $f(x) = x^3 - 12x^2 + 35x$

22. $f(x) = (x + 7)^5$

23. $f(x) = x^4 - 13x^2 + 36$

Graph each function.

24. $f(x) = x^4 - 2x^2 - 8$

25. $D(t) = t^3 + t$

26. $S(w) = w^3 - 7w - 6$

Use a graphing calculator to graph each function. Use the TRACE and BOX functions to determine the *x*-intercept(s), *y*-intercept, relative minima, and relative maxima.

27. $y = x^4 - 26x^2 + 25$

28. $y = 6x^3 + x^2 - 5x - 2$

29. $y = x^3 - 4x^2 - 25x + 28$

30. $y = x^4 + 4x^3 + 3x^2 - 4x - 4$

31. Find the derivative of $f(x) = 2\sqrt{x} + 3\sqrt[3]{x^2} - 4\sqrt[4]{x}$.

32. Which families of graphs have points of inflection but no maximum or minimum points?

33. Horticulture *Ameriflora* was an international flower exhibit held in Columbus, Ohio, in 1992. A rectangular region of the exhibit was fenced off to landscape for the fall season. The area of the region was 125,000 square feet. A type of fence costing $20 per foot was used along the back and front of the region because these were high-traffic areas. A less expensive fence costing $10 per foot was used for the other sides. What were the dimensions of the region that minimized the cost of the fence?

34. **Agriculture** Marta Yoshiki raises cotton. If she harvests her crop now, the yield will average 120 pounds of cotton per acre and will sell for $0.48 per pound. However, she knows that if she waits, her yield will increase by about 10 pounds per week, but the price will decrease by $0.03 per pound per week.
a. How many weeks should she wait in order to maximize her profit?
b. What is the maximum profit?

Mixed Review

35. Show that $P(4, 2)$ is the midpoint of $\overline{AB}$ that has endpoints $A(9, 3)$ and $B(-1, 1)$. **(Lesson 1-4)**

36. Explain how you perform scalar multiplication on a 2×2 matrix. Give an example. **(Lesson 2-2)**

37. Find the determinant for $\begin{bmatrix} -15 & 5 \\ -9 & 3 \end{bmatrix}$. Tell whether an inverse exists for the matrix. **(Lesson 2-3)**

38. Describe the transformation(s) of the parent graph of $f(x) = x^2$ that are required to graph $(x + 4)^2 - 8 = y$. **(Lesson 3-2)**

39. Find the equation of the line tangent to the graph of $y = 2x^2 - 3x$ at $(-1, 5)$. **(Lesson 3-6)**

40. **College Entrance Exam** Choose the best answer.
The area of a right triangle is 12 in^2. The ratio of the lengths of its legs is 2:3. Find the length of the hypotenuse.
(A) $\sqrt{13}$ in. **(B)** 26 in. **(C)** $2\sqrt{13}$ in. **(D)** 52 in. **(E)** $4\sqrt{13}$ in.

CASE STUDY FOLLOW-UP

Refer to Case Study 4: *The U.S. Economy* on pages 966–969.

Spider Industries, a U.S.-based company, test-marketed an electronic cobweb destroyer in Japan. The function $R = 160x - \frac{1}{5}x^2$ specifies the company's revenues (R) in terms of the number of destroyers sold.

1. Find the revenue generated by sales of the following number of destroyers.
   a. 20   b. 100   c. 500

2. How many destroyers must the company sell to maximize the revenue? At what selling price must they sell destroyers to accomplish this?

3. The company sold all of its destroyers, maximizing its revenue. How many times would they have to repeat their accomplishment to make up the United States' 1990 trade imbalance with Japan?

4. **Research** Find out about the U.S. trade imbalance with Japan. What is its cause? What are U.S. companies doing to overcome it? Summarize your findings in a 1-page paper.

3-8 Continuity and End Behavior

Objectives

After studying this lesson, you should be able to:
- determine continuity or discontinuity of functions, and
- identify the end behavior of graphs.

Application

Jeffry Stuart developed an interactive videodisc that is used to assist diabetic teenagers understand their disease better. Each videodisc sells for $600. Mr. Stuart receives an 8% royalty for every disc sold up to 200 discs. For the 200th disc sold, he will receive a one-time payment of $100. For every disc sold after the 200th, he will receive a 16% royalty.

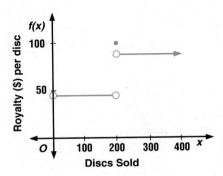

The graph at the left demonstrates the relationship between the number of discs sold and the amount of money Jeffry Stuart receives for each disk.

If $0 < x < 200$, $f(x) = 0.08(\$600)$ or $48.

If $x = 200$, $f(x) = \$100$.

If $x > 200$, $f(x) = 0.16(\$600)$ or $96.

All polynomial functions are continuous.

This graph represents a type of step function. Step functions have breaks in the function. That is, you cannot trace the graph of the function without lifting your pencil. A function is said to be **continuous** at point (x, y) if it is defined at that point and passes through that point without a break. If there is a break in the graph through that point, the function is said to be **discontinuous** at that point. In the function shown above, $f(x)$ is discontinuous at the point where $x = 200$.

Each of the functions graphed below are discontinuous at some point. Each illustrates a different type of discontinuity.

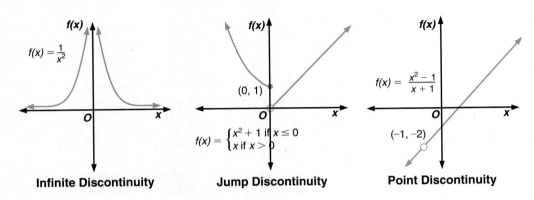

| Infinite Discontinuity | Jump Discontinuity | Point Discontinuity |

Asymptotes in a graph indicate discontinuity.

The application above illustrates a function with jump discontinuity.

A function may have discontinuity but be continuous over an interval. A function is continuous on an interval if it is continuous at every point in that interval without exception. In the graph of $f(x) = \dfrac{1}{x^2}$ on the previous page, the function is continuous for $x > 0$ and $x < 0$, but is discontinuous for $x = 0$.

The function $f(x) = \begin{cases} x^2 + 1 \text{ if } x \leq 0 \\ x \text{ if } x > 0 \end{cases}$, shown on the previous page, has a segmented definition. That is, it includes different expressions for different parts of its domain. While this function is discontinuous, other segmented functions are continuous. The graph of $f(x) = \begin{cases} x \text{ if } x \geq 0 \\ x^2 \text{ if } x < 0 \end{cases}$, shown at the right, is an example of a segmented function that is continuous.

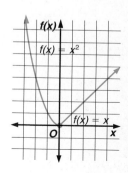

Continuity on an Interval

A function $f(x)$ is continuous on an interval if it is continuous for each value of x in that interval.

There are functions that are impossible to graph in the real number system. Some of these functions are said to be *everywhere discontinuous*. An example of such a function is $f(x)$, where $\begin{cases} f(x) = 1 \text{ when } x \text{ is rational} \\ f(x) = 2 \text{ when } x \text{ is irrational} \end{cases}$.

Example 1

Determine whether each function is continuous.

a. $y = 3x^2 + 7$

a. Graph the function.
The function is defined for every point in the domain.

The function is continuous.

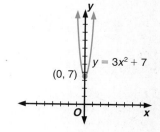

b. $f(x) = \dfrac{|2x|}{x}$

b. When $x > 0$, $f(x) = 2$.
When $x < 0$, $f(x) = -2$.
This function is undefined at $x = 0$ in both intervals.

This function has jump discontinuity at $x = 0$.

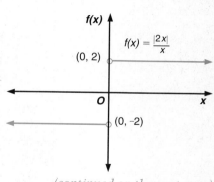

(continued on the next page)

c. $y = \dfrac{1}{2x^2}$

c. The graph of this function has an asymptote
at $x = 0$.
For $x > 0$, as $x \to 0, y \to \infty$.
For $x < 0$, as $x \to 0, y \to \infty$.

This function has infinite discontinuity at
$x = 0$.

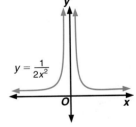

d. $f(x) = \dfrac{x^2 - 9}{x - 3}$

d. This function is undefined at $x = 3$.
Since $\dfrac{x^2 - 9}{x - 3} = x + 3$, the graph of
$f(x)$ resembles $g(x) = x + 3$, but has a
point discontinuity at $x = 3$.

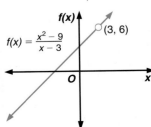

The **end behavior** of the graph of a function refers to the behavior of $f(x)$ as
$|x|$ becomes very large. Notice in the graph of $f(x) = \dfrac{x^2 - 9}{x - 3}$ shown above, as
$x \to \infty, f(x) \to \infty$ and as $x \to -\infty, f(x) \to -\infty$.

Example 2

**Describe the end behavior of the graphs of $f(x) = 4x^3$ and
$g(x) = 4x^3 - 3x^2 + 2x - 5$.**

Use your calculator to observe what happens to $f(x)$ and $g(x)$ as $|x|$ becomes
increasingly great in each function.

*Values for $f(x)$ and
$g(x)$ may vary for
different calculators.
However, the pattern
remains the same.*

x	$f(x)$
$-100{,}000$	-4×10^{15}
$-10{,}000$	-4×10^{12}
-1000	$-4{,}000{,}000{,}000$
-100	$-4{,}000{,}000$
-10	-4000
-1	-4
0	0
1	4
10	4000
100	$4{,}000{,}000$
1000	$4{,}000{,}000{,}000$
$10{,}000$	4×10^{12}
$100{,}000$	4×10^{15}

x	$g(x)$
$-100{,}000$	-4.00003×10^{15}
$-10{,}000$	-4.0003×10^{12}
-1000	$-4{,}003{,}002{,}005$
-100	$-4{,}030{,}205$
-10	-4325
-1	-14
0	-5
1	-2
10	3715
100	$3{,}970{,}195$
1000	$3{,}997{,}001{,}995$
$10{,}000$	3.9997×10^{12}
$100{,}000$	3.99997×10^{15}

Notice that $f(x)$ and $g(x)$ have relatively small absolute values in the
$-10 < x < 10$ interval. However, as the x values become increasingly greater
(or less), the values of the two functions approach the same two extremes.
That is, as $x \to \infty, f(x) \to \infty$ and $g(x) \to \infty$. As $x \to -\infty, f(x) \to -\infty$ and
$g(x) \to -\infty$.

Notice that the terms are written in the order of descending exponents.

As you can see from Example 2, calculations of $f(x)$ and $g(x)$ for $|x|$ are tedious and time-consuming. The behavior of $g(x)$ was the same as its related function $f(x)$. Suppose $P(x) = a_0x^n + a_1x^{n-1} + a_2x^{n-2} + \ldots + a_{n-1}x + a_n$ with $n \geq 1$. The end behavior of $P(x)$ will be the same as the end behavior of $f(x) = a_0x^n$. The values of a_0 and n can indicate the end behavior of the graph of any polynomial function.

End Behavior of Graphs of Polynomial Functions ($n \geq 1$)

a_0	n	$x \rightarrow$	$P(x) \rightarrow$
positive	even	∞	∞
positive	even	$-\infty$	∞
positive	odd	∞	∞
positive	odd	$-\infty$	$-\infty$

a_0	n	$x \rightarrow$	$P(x) \rightarrow$
negative	even	∞	$-\infty$
negative	even	$-\infty$	$-\infty$
negative	odd	∞	$-\infty$
negative	odd	$-\infty$	∞

EXPLORATION: Graphing Calculator

Determine the end behavior of $f(x) = x^3 + 5x^2 - 4$ with a graphing calculator.

To determine the end behavior of a function using a graphing calculator, view the graph of $f(x)$ using successively larger viewing windows. Establish the viewing windows by setting and resetting the range values.

1. XMIN: -10 XMAX: 10 XSCL: 1
YMIN: -10 YMAX: 10 YSCL: 1

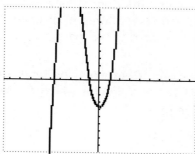

2. XMIN: -10 XMAX: 10 XSCL: 1
YMIN: -100 YMAX: 100 YSCL: 10

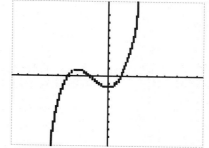

Notice that the larger the viewing window becomes, the more the function resembles the graph of the related function $f(x) = x^3$.

3. XMIN: -20 XMAX: 20 XSCL: 2
YMIN: -400 YMAX: 400 YSCL: 40

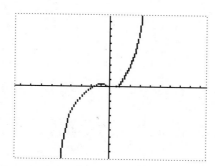

4. XMIN: -100 XMAX: 100
YMIN: -10000 YMAX: 10000
XSCL: 10 YSCL: 1000

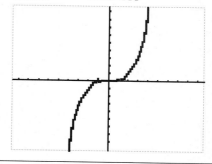

In Lesson 3-7, you learned how to determine whether a critical point was a maximum, a minimum, or a point of inflection by testing points on each side of the critical point. Functions can also be described as increasing or decreasing functions.

Increasing and Decreasing Functions	A function $f(x)$ is increasing if and only if $f(x_1) < f(x_2)$ whenever $x_1 < x_2$. A function $f(x)$ is decreasing if and only if $f(x_1) > f(x_2)$ whenever $x_1 < x_2$.

The function must be defined for the values x_1 and x_2.

Observe the two graphs below. The graph of $f(x) = x^3$ illustrates an increasing function. That is, as you trace the function from left to right, $f(x)$ increases. The graph of $f(x) = (-x)^3$ illustrates a decreasing function. As you trace the function from left to right, $f(x)$ decreases.

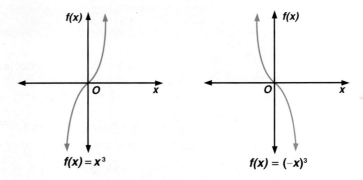

The graphs of many functions illustrate characteristics of both increasing and decreasing functions for different intervals of the function. These intervals often occur between the relative maximum(s) and relative minimum(s).

Example 3

Graph each function. Determine the interval(s) for which the function is increasing and the interval(s) for which the function is decreasing.

a. $f(x) = -x^2 + 12$

a. The graph has a relative maximum at $x = 0$.
For $x < 0$, the function is increasing.
For $x > 0$, the function is decreasing.

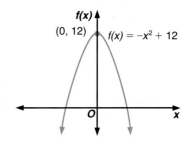

b. $g(x) = \dfrac{1}{x}$

b. The graph has asymptotes at $x = 0$ and $y = 0$.
For $x < 0$, the function is decreasing.
For $x > 0$, the function is decreasing.

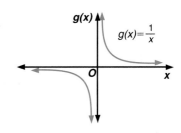

c. $h(x) = 2x^3 + 8x^2 + 8x + 5$

c. $h'(x) = 6x^2 + 16x + 8$

$$0 = 6x^2 + 16x + 8$$
$$0 = 3x^2 + 8x + 4$$
$$0 = (3x + 2)(x + 2)$$

The graph has a relative maximum at $x = -2$ and a relative minimum at $x = -\dfrac{2}{3}$.

For $x < -2$, the function is increasing.

For $-2 < x < -\dfrac{2}{3}$, the function is decreasing.

For $x > -\dfrac{2}{3}$, the function is increasing.

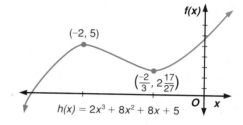

$h(x) = 2x^3 + 8x^2 + 8x + 5$

CHECKING FOR UNDERSTANDING

Communicating Mathematics

Read and study the lesson to answer each question.

1. **Describe** three kinds of discontinuity.

2. **Explain** when a function is decreasing and when it is increasing.

3. **Demonstrate** how to tell if the graph of a function is continuous.

4. **Compare** the end behavior of $g(x) = x^5 + 4x^4 + 3x^2 + 8$ with the end behavior of the parent function $f(x) = x^5$. Make a conjecture about the end behavior of any function $g(x)$ and its parent function $f(x)$.

Guided Practice

Determine whether each graph has *infinite discontinuity*, *jump discontinuity*, or *point discontinuity*, or is *continuous*. Then describe the end behavior of each graph.

5.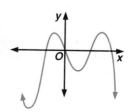

6.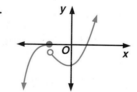

7.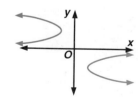

8.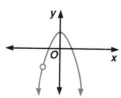

9.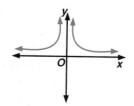

10.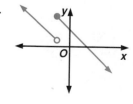

Determine the intervals for which each function is *increasing* or *decreasing*.

11. $y = 2x + 5$ **12.** $y = \dfrac{1}{x}$ **13.** $y = x^2$

EXERCISES

Practice

Determine whether each graph has *infinite discontinuity, jump discontinuity,* or *point discontinuity,* or is *continuous.* Then graph each function.

14. $y = \dfrac{12}{5x^2}$ **15.** $y = \dfrac{x^2 - 4}{x + 2}$

16. $y = \dfrac{|x|^3}{x}$ **17.** $y = 8 - 2x - x^2$

18. $y = \begin{cases} 3x - 6, & \text{if } x > 0 \\ 3x^2 - 6, & \text{if } x < 0 \end{cases}$ **19.** $y = \begin{cases} x, & \text{if } x < 0 \\ 0, & \text{if } x = 0 \\ x, & \text{if } x > 0 \end{cases}$

20. $y = \begin{cases} -x + 5, & \text{if } x < 1 \\ x^2 - 2x + 5, & \text{if } x > 1 \end{cases}$ **21.** $y = \dfrac{6}{x^2} - 4$

22. $y = x^3 - x^2 - 14x + 16$ **23.** $y = \dfrac{4}{x^2} + x$

Without graphing, describe the end behavior of each function.

24. $y = 6x^5 + 7x^3 + 1$ **25.** $y = x^2 - 25$ **26.** $y = 24 - 18x - 45x^3$

Complete each segmented function so that the function is continuous.

27. $y = \begin{cases} \dfrac{x^2 - 11x + 28}{x - 4} \\ \underline{\ ?\ }, & \text{if } x = \underline{\ ?\ } \end{cases}$ **28.** $y = \begin{cases} \dfrac{x^2 - 11}{x + \sqrt{11}} \\ \underline{\ ?\ }, & \text{if } x = \underline{\ ?\ } \end{cases}$

29. $y = \begin{cases} x^3, & \text{if } x < 4 \\ 68 - x, & \text{if } x > 4 \\ \underline{\ ?\ }, & \text{if } x = \underline{\ ?\ } \end{cases}$ **30.** $y = \begin{cases} \dfrac{x^3 + 27}{x + 3} \\ \underline{\ ?\ }, & \text{if } x = \underline{\ ?\ } \end{cases}$

**Graphing
Calculator**

Use a graphing calculator to graph each function.

a. Determine whether the graph is continuous.

b. State the intervals for which the function is increasing or decreasing.

31. $y = x^2 - 5x + 4$ **32.** $y = \dfrac{1}{x + 1}$ **33.** $y = |x + 9|$

**Critical
Thinking**

34. Find the value for p and q so that the function is continuous if

$$f(x) = \begin{cases} x - 2, & \text{if } x \geq 2 \\ \sqrt{p - x^2}, & \text{if } -2 < x < 2 \\ q - x, & \text{if } x \leq -2 \end{cases}$$

**Applications
and Problem
Solving**

35. Manufacturing The Twin Forks packing plant is developing a new type of can that will contain peaches. The can is made of a special alloy that reduces the number of preservatives needed to keep the fruit fresh. If the can has to have a volume of 300 cm^3, determine the radius, height, and minimum surface area needed to manufacture the most cost effective can.

36. Government The table below appeared in the Federal Income Tax instructions booklet for 1991.

Schedule Y-1 Use if your filing status is Married filing jointly or Qualifying widow(er)

If the amount on Form 1040, line 37, is: Over	But not over	Enter on Form 1040, line 38	of the amount over
$0	$34,000	...... 15%	$0
34,000	82,150	$5,100.00 + 28%	34,000
82,150	......	18,582.00 + 31%	82,150

a. Write a function that relates the three different income levels.
b. Graph the function.
c. Determine if the graph is continuous. Explain.
d. Determine the amount of tax individuals owe if the amount on line 37 of Form 1040 is:

   (1) $73,000   **(2)** $32,050   **(3)** $22,174   **(4)** $87,234

Mixed Review

37. Find $[f \circ g](4)$ if $f(x) = 5x + 9$ and $g(x) = 0.5x - 1$. **(Lesson 1-2)**

38. Manufacturing Picto Inc. makes console and wide-screen televisions. The equipment in the factory allows for making at most 450 console televisions and 200 wide-screen televisions in one month. The console television costs $600 per unit to produce and is sold at $125 profit. The wide-screen television costs $900 per unit to produce and is sold at $200 profit. During the month of November, the company can spend $3,600,000 to produce these televisions. **(Lesson 2-6)**

a. To maximize profit, how many of each type should they make?
b. What is the maximum profit possible with these amounts?

39. Find the inverse of $f(x) = (x - 9)^2$. **(Lesson 3-3)**

40. Graph $y \geq \sqrt[3]{x + 4} - 2$. **(Lesson 3-5)**

41. Find the critical points for the graph of $y = x^3 - x^2 + 3$. Determine whether each point is a relative maximum, a relative minimum, or a point of inflection. **(Lesson 3-7)**

42. College Entrance Exam Choose the best answer.

A college graduate goes to work for $x per week. After several months the company falls on hard times and gives all the employees a 10% pay cut. A few months later, business picks up and the company gives all the employees a 10% raise. What is the college graduate's new salary?

  (A) $0.90x   **(B)** $0.99x   **(C)** $x   **(D)** $1.01x   **(E)** $1.11x

VOCABULARY

Upon completing this chapter, you should be
familiar with the following terms:

134 asymptote
164 continuous
157 critical points
150 derivative
164 discontinuous
166 end behavior
112 even function
120 geometric
    transformation
119 greatest integer
    function

138 hole
135 horizontal asymptote
128 horizontal line test
126 inverse relations
108 line symmetry
118 linear transformation
157 maximum
157 minimum
112 odd function
117 parent graph
157 point of inflection

106 point symmetry
134 rational function
118 reflection
157 relative maximum
157 relative minimum
149 secant line
137 slant asymptote
149 tangent
118 translation
134 vertical asymptote

SKILLS AND CONCEPTS

OBJECTIVES AND EXAMPLES

Upon completing this chapter, you should
be able to:

■ identify an odd function and an even
function   **(Lesson 3-1)**

Determine whether $f(x) = x^3 + x$ is odd,
even, or neither.

$f(-x) = -x^3 - x$
$-f(x) = -x^3 - x$

Since $f(-x) = -f(x), f(x) = x^3 + x$ is odd.

■ identify the graphs of simple polynomial
functions, absolute value functions, and
step functions.   **(Lesson 3-2)**

The graph of
$g(x) = (x - 2)^2$
is the translation
of the graph of
$f(x) = x^2$ two
units to the right.

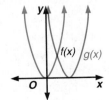

REVIEW EXERCISES

Use these exercises to review and prepare for
the chapter test.

**Determine whether the graph of each
relation is symmetric with respect to the
x-axis, the y-axis, the line $y = x$, the line
$y = -x$, or the origin. Then, state whether
each relation is *even*, *odd*, or *neither*.**

1. $xy = 8$ **2.** $x = y^2 - 3$
3. $y = -2x^3 + x^2 - 3$ **4.** $y = 8 - 4x^2$

**Use the graph of $f(x) = x^4$ to sketch a
graph for each function. Then, describe
the transformation(s) of $f(x)$ that have taken
place.**

5. $y = -f(x)$

6. $y = f(x) - 3$

7. $y = f(x + 5)$

8. $y = 6(f(-x))$

- determine the inverse of a function or relation **(Lesson 3-3)**

  Suppose $f(x) = 4(x-3)^2$. Find f^{-1}.

  $x = 4(y-3)^2$ *Interchange x and y.*

  $\dfrac{x}{4} = (y-3)^2$ *Solve for y.*

  $\pm\dfrac{\sqrt{x}}{2} + 3 = y$ $\rightarrow$ $f^{-1}(x) = \pm\dfrac{\sqrt{x}}{2} + 3$

Find the inverse of each function. Sketch the function and its inverse. Is the inverse a function? Write *yes* or *no*.

9. $y = 6 - 2x$

10. $y = \dfrac{1}{4}(x-3)^2$

11. $y = (x+1)^3 - 2$

- determine horizontal, vertical, and slant asymptotes **(Lesson 3-4)**

  Determine the asymptotes for the graph of $f(x) = \dfrac{x-1}{x}$.

  vertical asymptote: $x = 0$

  Solve for x in terms of y: $x = \dfrac{1}{1-y}$

  horizontal asymptote: $y = 1$

Determine any horizontal, vertical, or slant asymptotes or holes for each function. Then, graph each function.

12. $y = \dfrac{(x+2)(x+1)}{(3x-1)(x+2)}$

13. $y = \dfrac{x}{(x+3)(x-4)}$

14. $y = \dfrac{x^2-9}{x-3}$

- graph polynomial, absolute value, and radical inequalities **(Lesson 3-5)**

  Graph $y < x^2 + 1$.

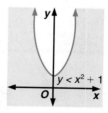

Graph each inequality.

15. $y < \sqrt[3]{x+2}$

16. $y \geq |x| - 3$

17. $y > (x-3)^2$

18. $y \leq x^2 - 4x - 5$

- find the slope and the equation of a line tangent to the graph of a function at a given point **(Lesson 3-6)**

  Find the slope of the tangent to the graph of $y = x^2 + 3x - 7$ at the point $(5, 33)$.

  $f'(x) = 2x + 3$ *Find the derivative.*

  $f'(5) = 2(5) + 3$ or 13

  The slope is 13.

Find the equation of the line tangent to the graph of each function at the given point. Write the equation in slope-intercept form.

19. $y = x^2 + 5x - 2,\ (1, 4)$

20. $y = -x^2 + 6x + 11,\ (-2, -5)$

21. $y = 3x^2 + 4x - 2,\ (1, 5)$

22. $y = 2x^2 - 9x + 5,\ (2, -5)$

■ find the critical points of a graph and determine if each is a minimum, maximum, or point of inflection **(Lesson 3-7)**

Determine whether the point at $(-1, 4)$ on the graph of $f(x) = 2x^3 + x^2 - 4x + 1$ is a minimum or a maximum.

$f(-1) = 4$
$f(-1.01) = 3.9995 \quad \rightarrow \quad f(-1.01) < f(-1)$
$f(-0.99) = 3.9995 \quad \rightarrow \quad f(-0.99) < f(-1)$

The point at $(-1, 4)$ is a maximum.

Find the critical points of each function. Then, determine whether each point represents a *maximum*, a *minimum*, or a *point of inflection*.

23. $f(x) = 4 + x - x^2$

24. $f(x) = x^3 - 6x^2 + 9x$

25. $f(x) = x^3 + 3x^2 - 4$

26. $f(x) = 2x^3 - 5$

■ determine continuity or discontinuity of functions **(Lesson 3-8)**

Determine whether $y = \dfrac{x(x-3)}{x-3}$ is continuous.

The graph resembles the graph of $y = x$, but has a hole at $x = 3$. It has point discontinuity.

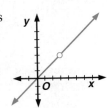

Determine whether each graph has *infinite discontinuity*, *jump discontinuity*, *point discontinuity*, or is *continuous*. Then graph each function.

27. $y = \dfrac{-x}{(x-2)^2}$

28. $y = \begin{cases} \dfrac{|x|}{x} & \text{if } x \neq 0 \\ 1 & \text{if } x = 0 \end{cases}$

29. $y = \begin{cases} x + 1 & \text{if } x < 0 \\ 1 - x & \text{if } x \geq 0 \end{cases}$

APPLICATIONS AND PROBLEM SOLVING

30. Sports One of the most spectacular long jumps ever performed was by Bob Beaman in the 1968 Olympics. His jump of 29 feet 2.5 inches surpassed the world record at that time by over 2 feet! The function $h(t) = 4.6t - 4.9t^2$ describes the height of Beaman's jump (in meters) with respect to time (in seconds). **(Lesson 3-7)**

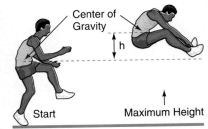

a. Draw a graph of this function.

b. What was the maximum height of his jump?

31. Physics The position of a particle with respect to time can be described by the function $f(t) = 3t^2 + 5t - 8$, where $f(t)$ is in meters and t is in seconds. **(Lesson 3-7)**

a. Find the instantaneous velocity, $f'(t)$, of the particle when $t = 4.29$m/s.

b. The acceleration of the particle can be found by finding the second derivative of the function, $f''(t)$. Find the acceleration of the particle.

Use the graph of $f(x) = x^2 + 1$ to sketch a graph for each function. Describe the transformations of $f(x)$ that have taken place.

1. $y = f(x - 5)$
2. $y = -f(x) + 2$
3. $y = \frac{1}{2}(f(x + 3))$

Find the inverse of each function. Sketch the function and its inverse. Is the inverse a function? Write *yes* or *no*.

4. $y = (x - 4)^2$
5. $y = 8x^3 + 1$
6. $y = (x + 3)^4 - 7$

Find the asymptotes, if they exist, and determine whether each relation is *even*, *odd*, or *neither*.

7. $y = \dfrac{x + 5}{(x - 3)(x + 1)}$
8. $y = \dfrac{x}{x^2 - 4}$
9. $y = x^7 - 5x^3 - 8$

Graph each inequality.

10. $y > |x - 4|$
11. $y \leq 2x^2 + 3$

Find the equation of the line tangent to the graph of each function at the given point. Write the equation in slope-intercept form.

12. $y = x^2 - 4x + 1, (1, -2)$
13. $y = 3x^2 - 2x + 1, (2, 9)$

Find the critical points of each function. Then, determine whether each point represents a *maximum*, a *minimum*, or a *point of inflection*.

14. $y = x^2 - 8x + 4$
15. $y = -x^3 - 3x^2 + 3$

Determine whether each relation has *infinite discontinuity, jump discontinuity, point discontinuity*, or is *continuous*. Then graph each function.

16. $y = \dfrac{-x^2(x - 2)}{x - 2}$
17. $y = \begin{cases} x + 1, \text{ if } x > 0 \\ 3, \text{ if } x \leq 0 \end{cases}$

18. Find the x- and y- intercepts of $y = 3x^3 - 2x^2 - 5x$. Then, graph the function.

19. Without graphing, describe the end behavior of the graph of $y = x^4 - 2x^2 - 1$.

20. Auto Safety Airbags are becoming a standard safety feature in many cars today. They protect people from experiencing the full impact of a collision. Without airbags, a person or object would move forward at the velocity that the car is traveling, v_i, and then after impact, be repelled backward at a velocity of $v_f = \dfrac{m_1 - m_2}{m_1 + m_2} \cdot v_i$, where m_1 and m_2 are the masses of the car and the person, respectively.

  a. Graph the function if $v_i = 5$ m/s and $m_1 = 7{,}000$ kg.

  b. Find the value of v_f when the value of m_2 is 50 kg.

Bonus Determine the constants b and c so that the function

$$f(x) = \begin{cases} x + 1 \text{ if } 1 < x < 3 \\ x^2 + bx + c \text{ if } |x - 2| \geq 1 \end{cases} \text{ is continuous.}$$

CHAPTER 4

POLYNOMIAL AND RATIONAL FUNCTIONS

HISTORICAL SNAPSHOT

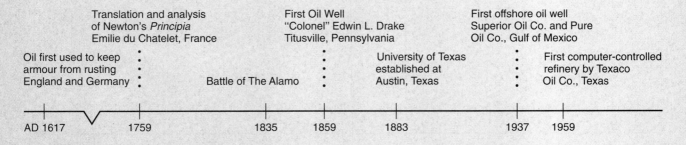

	Translation and analysis of Newton's *Principia* Emilie du Chatelet, France		First Oil Well "Colonel" Edwin L. Drake Titusville, Pennsylvania			First offshore oil well Superior Oil Co. and Pure Oil Co., Gulf of Mexico
Oil first used to keep armour from rusting England and Germany		Battle of The Alamo		University of Texas established at Austin, Texas		First computer-controlled refinery by Texaco Oil Co., Texas
AD 1617	1759	1835	1859	1883	1937	1959

CHAPTER OBJECTIVES

In this chapter, you will:
- Solve polynomial equations.
- Solve inequalities.
- Solve rational equations.

CAREER GOAL: Petroleum Engineering

Imagine doing something you love while living in a faraway, exotic land. Deborah Hempel would like to live and work in Russia for five to ten years. "I really want to work internationally," she says. "The oil company for which a petroleum engineer works must prove that each petroleum engineer it sends to work in that country is better than any of that country's own citizens. Therefore, five years of experience is necessary."

Deborah says that "the petroleum industry is really dynamic and ever-changing." However, she adds, "I do dislike the job uncertainty. For example, in 1986, the petroleum industry hit rock bottom. Then during 1990 and 1991, it was on the upswing. It seems to have stabilized, but where will it go from here?"

Deborah uses equations in her daily work. For example, "Each oil reservoir is made of sand and shale. Two characteristics of the reservoir are porosity and permeability. Porosity refers to the size of the pores in the shale. The greater the porosity, the more oil it can store. Permeability refers to how the pores are interconnected. Unless the pores are interconnected well, you can't get the oil out.

"Sometimes it's necessary to drill a second well next to a potential oil well. Then, water is pumped down the second well to create enough pressure to drive the oil up the first well. Petroleum engineers use an equation to calculate when the water will push the oil through the first well and up to the surface. The equation includes the variables of porosity, permeability, viscosity, or stickiness of the oil, the change in pressure between the two wells, and the distance between the two wells."

Deborah wants to be "a petroleum engineer who doesn't have a lax attitude about the environment," she says. "These days, petroleum engineers are learning more about how they can drill for valuable supplies of oil and, at the same time, protect the environment."

inter NET CONNECTION

For up-to-date information on petroleum engineering, visit:
www.glencoe.com/sec/math/amc/mathnet

4-1 Polynomial Functions

Objectives
After studying this lesson, you should be able to:
- determine roots of polynomial equations, and
- apply the fundamental theorem of algebra.

Application

Instead of investing in the stock market or other financial institutions, many people invest in collectibles, such as antiques. Lesley Yoshito is an investor in vintage automobiles. Three years ago, she purchased a 1953 Chevrolet Corvette roadster for $99,000. Two years ago, she purchased a 1929 Pierce-Arrow Model 125 for $55,000. A year ago, she purchased a 1909 Cadillac Model Thirty for $65,000. If the cars appreciate at a rate of 15% per year, determine the current value of her total investment.

Appreciation is the increase in value of an item over a period of time. The formula for compound interest can be used to find the value of Ms. Yoshito's cars after appreciation. The formula is $A = P(1 + R)^t$, where P is the original amount of money invested, R is the rate of return (written as a decimal), and t is the time invested (in years). The value of her total investment in the cars is the sum of the current value of her investments.

The rate of appreciation in this case is 15%. So $R = 0.15$. Let $x = 1 + R$ or 1.15. Let $T(x)$ represent the total current value of the three cars. Then find $T(x)$.

Total	=	Corvette	+	Pierce-Arrow	+	Cadillac	
$T(x)$	=	$99{,}000(x)^3$	+	$55{,}000(x)^2$	+	$65{,}000(x)^1$	
$T(1.15)$	=	$99{,}000(1.15)^3$	+	$55{,}000(1.15)^2$	+	$65{,}000(1.15)$	$x = 1.15$
$T(1.15)$	=	$298{,}054.13$					

The present value of Ms. Yoshito's investment is $298,054.13.

The function above, $T(x)$, contains a **polynomial in one variable**.

Definition of Polynomial in One Variable	**A polynomial in one variable, x, is an expression of the form $a_0x^n + a_1x^{n-1} + \ldots + a_{n-2}x^2 + a_{n-1}x + a_n$. The coefficients $a_0, a_1, a_2, \ldots, a_n$ represent complex numbers (real or imaginary), a_0 is not zero, and n represents a nonnegative integer.**

Complex numbers are defined on the next page.

The terms of a polynomial are usually written in order of decreasing degree.

The **degree** of a polynomial in one variable is the greatest exponent of its variable. The coefficient of the variable with the greatest exponent is called the *leading coefficient*. The degree of the polynomial developed above, $99{,}000x^3 + 55{,}000x^2 + 65{,}000x$, is 3. The leading coefficient is 99,000.

Example 1

> **Determine if each expression is a polynomial in one variable. If the expression is a polynomial in one variable, state the degree.**
>
> **a.** $c^4 + 2a^2 + 4c$ **b.** $w^4 - w^2 + 5w^5 - 6w^3$ **c.** $3x^2 + \dfrac{3}{x} + 8$
>
> **a.** $c^4 + 2a^2 + 4c$ is not a polynomial in one variable, because it has two variables.
>
> **b.** $w^4 - w^2 + 5w^5 - 6w^3$ is a polynomial in one variable. It has a degree of 5.
>
> **c.** $3x^2 + \dfrac{3}{x} + 8$ is not a polynomial in one variable, because $\dfrac{3}{x}$ is equivalent to $3x^{-1}$. A polynomial cannot have a term in which the variable has a negative exponent.

If a function f is a polynomial in one variable, then f is a **polynomial function**. If a domain value for the function is known, then a unique range value can be determined.

A root of P(x) = 0 is a zero, where f(x) = P(x).

If $P(x)$ represents a polynomial, then $P(x) = 0$ is called a **polynomial equation**. A **root**, or solution, of the equation is a value of x for which the value of the polynomial $P(x)$ is 0. Thus, 5 is a root of the equation $x^2 - 6x + 5 = 0$, since $5^2 - 6(5) + 5 = 0$; 5 is also called a **zero** of the function $f(x) = x^2 - 6x + 5$. On the graph of a function, the zeros are the x-intercepts.

Example 2

> **Determine whether 3 is a root of $x^3 - 2x^2 - 5x + 6 = 0$.**
>
> Evaluate the function $P(x) = x^3 - 2x^2 - 5x + 6$ for $P(3)$.
>
> $$P(3) = 3^3 - 2(3)^2 - 5(3) + 6$$
> $$= 27 - 18 - 15 + 6$$
> $$= 0$$
>
> Since $P(3) = 0$, 3 is a root of $x^3 - 2x^2 - 5x + 6 = 0$.

The root of a polynomial function may be an **imaginary number**, such as $2i$. By definition, the imaginary unit, i, equals $\sqrt{-1}$. The imaginary numbers combined with the real numbers make up the set of **complex numbers**. A complex number is any number in the form $a + bi$, where a and b are real numbers. If $b = 0$, then the complex number is a real number. If $a = 0$ and $b \neq 0$, then the complex number is called a *pure imaginary number*.

One of the most important theorems in mathematics is the **fundamental theorem of algebra**.

Fundamental Theorem of Algebra	**Every polynomial equation with degree greater than zero has at least one root in the set of complex numbers.**

The degree of a polynomial indicates how many roots are possible. This is stated in a corollary to the fundamental theorem of algebra.

Corollary to the Fundamental Theorem of Algebra

Every polynomial $P(x)$ of degree n ($n > 0$) can be written as the product of a constant k ($k \neq 0$) and n linear factors.

$$P(x) = k(x - r_1)(x - r_2)(x - r_3) \ldots (x - r_n)$$

Thus, a polynomial equation of degree n has exactly n complex roots, namely $r_1, r_2, r_3, \ldots, r_n$.

The general shapes of the graphs for polynomial functions with positive leading coefficients and degree greater than 0 are shown below. These graphs also show the maximum number of times the graph of each type of polynomial may cross the x-axis.

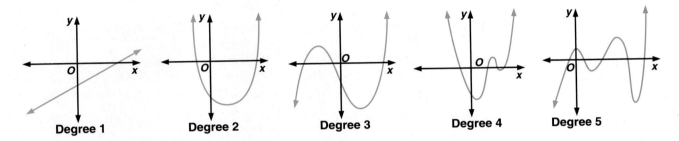

Degree 1   Degree 2   Degree 3   Degree 4   Degree 5

The graph of a polynomial function with odd degree *must* cross the x-axis at least once. The graph of a function with even degree may or may not cross the x-axis. If it does, it will be an even number of times. Each x-intercept represents a real root of the corresponding polynomial equation.

Example 3

State the number of complex roots of the equation $x^3 + 2x^2 - 8x = 0$. Then find the roots and graph the related polynomial function.

The polynomial has degree 3, so there are 3 complex roots. In this case, you can factor the equation to find these roots.

$$x^3 + 2x^2 - 8x = 0$$
$$x(x^2 + 2x - 8) = 0$$
$$x(x + 4)(x - 2) = 0$$

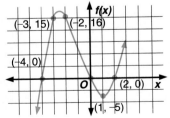

The graph of $f(x) = x^3 + 2x^2 - 8x$ crosses the x-axis at −4, 0, and 2.

To find each root, set each factor equal to zero.

$$x = 0 \qquad x + 4 = 0 \qquad x - 2 = 0$$
$$x = -4 \qquad x = 2$$

The roots are 0, −4, and 2.

Example 4

Write the polynomial equation of least degree with roots –3, 2*i*, and –2*i*.

The linear factors for the polynomial are $x - (-3)$, $x - 2i$, and $x - (-2i)$.
Now find the product of these factors.

$$0 = (x+3)(x-2i)(x+2i)$$
$$0 = (x+3)(x^2 - 4i^2) \qquad i^2 = -1$$
$$0 = (x+3)(x^2 + 4) \qquad -4i^2 = -4(-1)$$
$$0 = x^3 + 3x^2 + 4x + 12$$

The simplest polynomial equation with roots –3, 2*i*, and –2*i* is
$0 = x^3 + 3x^2 + 4x + 12$.

Example 5

APPLICATION

Construction

The Santa Fe Recreation Department has a 50 foot by 70 foot area for construction of a new public swimming pool. The pool will be surrounded by a concrete sidewalk of constant width. Because of water restrictions, the pool can have a maximum area of 2400 square feet. What should be the width of the sidewalk that surrounds the pool?

Let x = the width of the sidewalk.
The length of the pool would be $70 - 2x$ feet.
The width of the pool would be $50 - 2x$ feet.

$$A = \ell w$$
$$2400 = (70 - 2x)(50 - 2x)$$
$$2400 = 3500 - 240x + 4x^2$$
$$0 = 1100 - 240x + 4x^2$$
$$0 = 275 - 60x + x^2$$
$$0 = (55 - x)(5 - x)$$

$$55 - x = 0 \quad \text{or} \quad 5 - x = 0$$
$$x = 55 \qquad\qquad x = 5$$

Use $x = 5$, since 55 is an unreasonable solution. The sidewalk around the pool should be 5 feet wide.

[Diagram: a 50 ft by 70 ft area with a pool inside, sidewalk width labeled x on the left and bottom.]

CHECKING FOR UNDERSTANDING

Communicating Mathematics

Read and study the lesson to answer each question.

1. **Explain** the relationship between a polynomial equation and its corresponding function.

2. **Define** a complex number and tell under what conditions it will be a real number.

3. **Sketch** the general graph of a sixth degree function.

4. **Explain** why the solution of 55 is unreasonable in Example 5.

Guided Practice

Determine if each expression is a polynomial in one variable. If the expression is a polynomial in one variable, state the degree.

5. $a^3 + 2a + \sqrt{3}$

6. $\dfrac{1}{x} = \dfrac{1}{2x}$

7. $2m^3 + 5m^7 + 9$

Determine whether each number is a root of $x^4 - 4x^3 - x^2 + 4x = 0$.

8. 2   **9.** 0   **10.** –1   **11.** –2   **12.** 4

13. Write the polynomial equation of least degree whose roots are 1, –1, 2, and –2.

14. Solve $(u + 2)(u^2 - 4) = 0$ and graph the related polynomial function.

EXERCISES

Practice

State the number of complex roots of each equation. Then find the roots and graph the related function.

15. $x - 2 = 0$ **16.** $x^2 - 144 = 0$

17. $r^2 - 14r + 49 = 0$ **18.** $x^2 + 25 = 0$

19. $12x^2 + 8x - 15 = 0$ **20.** $18x^2 + 3x - 1 = 0$

21. $6c^3 - 3c^2 - 45c = 0$ **22.** $n^3 - 9n = 0$

Write the polynomial equation of least degree for each set of roots given.

23. $-3, 2$ **24.** $-2, -0.5, 4$ **25.** $-1, -1, 4, 4, 4$

26. $-5i, 5i, i, -i$ **27.** $1, -1, 1 + i, 1 - i$ **28.** $2, 2 \pm 3i, -1 \pm i$

Solve each equation and graph the related function.

29. $x^3 - 6x^2 + 10x - 8 = 0$ **30.** $x^4 - 10x^2 + 9 = 0$

31. $x^4 + x^2 - 2 = 0$ **32.** $4m^4 + 17m^2 + 4 = 0$

33. Sketch a fourth degree equation for each situation.
    a. no x-intercept   **b.** one x-intercept
    c. two x-intercepts   **d.** three x-intercepts
    e. four x-intercepts   **f.** five x-intercepts

Graphing Calculator

Graph the related function for each equation. Use the TRACE and BOX functions to find the roots of each equation to the nearest hundredth. State the range values for the window that shows both the x-intercepts and the y-intercept.

34. $x^2 + 4x + 4 = 0$ **35.** $x^3 - 4x^2 - 320x = 0$ **36.** $x^4 - 26x^2 + 25 = 0$

37. Graph the function $f(x) = x^5 - 15x^3 - 10x^2 + 60x + 72$.
    a. What is the maximum number of x-intercepts possible for this function?
    b. How many x-intercepts are there and what are they?
    c. Why are there fewer x-intercepts than the maximum number? *Hint: The factored form of the polynomial is $(x + 2)^3(x - 3)^2$.*

Critical Thinking

38. If B and C are the real roots of $x^2 + Bx + C = 0$, where $B \neq 0$ and $C \neq 0$, find the product of B and C.

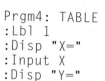

39. The following program will compute the value of a function that is stored in Y_1, given a value for X. To end the program, press 2nd QUIT.

```
Prgm4: TABLE
:Lbl 1
:Disp "X="
:Input X
:Disp "Y="
:Disp Y₁        Displays the value of Y₁ using the input value for X.
:Goto 1
```

Given $f(x) = x^4 + 3x^3 - 2x^2 + x - 4$, use the program shown above to find each value.

a. $f(2)$ **b.** $f(-3)$ **c.** $f(0.5)$

Applications and Problem Solving

40. Aeronautics At liftoff, the space shuttle *Discovery* has a constant acceleration, a, of 16.4 ft/s². The initial velocity, v_0, due to the rotation of Earth is 1341 ft/s. Use the function $d(t) = v_0 t + \frac{1}{2}at^2$ to determine the distance from Earth for each time interval, t, after liftoff.

a. 30 seconds

b. 1 minute

c. 2 minutes

d. If the time the space shuttle is in flight doubles, does the distance double? Explain.

ⓘ inter NET CONNECTION

For information on the latest space shuttle mission, visit:
www.glencoe.com/ sec/math/amc/ mathnet

41. Finance The Lancaster Art Commission has invested in works by American painters. Three years ago, they bought a painting for $30,000. Two years ago, another painting was bought for $55,000. Last year, a third painting was purchased for $75,000.

a. If the appreciation of the paintings is approximately 13% per year, write a function that represents the commission's total investment and calculate the current value.

b. Suppose each year the commission deposited the money in a savings account rather than purchasing works of art. If the rate of interest was 6.5%, what would be the value of the investment now? How does this differ from the art investment?

42. Entertainment The scenery for a new children's show has a playhouse with painted window panes. A special gloss paint covers the area of the painted windows to make them look like glass. If the gloss only covers 315 square inches and the window must be 6 inches taller than it is wide, how large should the scenery painters make the window?

43. Find $[g \circ h](x)$ if $g(x) = x - 1$ and $h(x) = x^2$. **(Lesson 1-2)**

44. Write the slope-intercept form of the equation of the line that passes through $(0, 7)$ and $(5, 2)$. **(Lesson 1-5)**

45. Solve the system of equations algebraically. **(Lesson 2-1)**
$$x - 2y = -1$$
$$2x + 3y = -16$$

46. Find the values of x, y, and z for which $\begin{bmatrix} x^2 & 7 & 9 \\ 5 & 12 & 6 \end{bmatrix} = \begin{bmatrix} 25 & 7 & y \\ 5 & 2z & 6 \end{bmatrix}$ is true.

(Lesson 2-2)

47. Graph $y = \dfrac{2}{x^2 - x - 2}$ **(Lesson 3-4)**

48. Determine whether the graph of $y = \dfrac{x^2 - 1}{x + 1}$ has infinite discontinuity, jump discontinuity, or point discontinuity, or is continuous. Then graph the function. **(Lesson 3-8)**

49. College Entrance Exam   Compare quantities A and B below.
Write A if quantity A is greater.
Write B if quantity B is greater.
Write C if the two quantities are equal.
Write D if there is not enough information to determine the relationship.

(A) the slope of $2x + 3y = 7$ **(B)** the slope of $3x - 2y = 7$

CASE STUDY FOLLOW-UP

Refer to Case Study 3: *The Legal System*, on pages 962–965.

Alcohol may be involved in as many as half of all fatal traffic accidents. Police routinely test drivers suspected of drinking to find their blood-alcohol levels, the portion of their blood that is alcohol.

1. Assuming that the overall male/female arrest proportion holds for all crimes, how many men were arrested for drunken driving in 1990?

2. The polynomial function $A(x) = -0.0015x^2 + 0.1058x$ approximates a person's blood-alcohol level x hours after drinking 8 ounces of 100-proof whiskey.
   a. Find the blood-alcohol level after 5 hours.
   b. In five states—California, Maine, Oregon, Utah, and Vermont—the legal blood-alcohol level is 0.08. In all other states, it is 0.1. Calculate the blood-alcohol level in fifths of an hour beginning with $c = 0.2$ to find the first time after drinking 8 ounces of 100-proof whiskey that a person could be arrested for drunken driving in these five states.

3. **Research**   Collect statistics on drunken driving and accidents in your state and write a 1-page paper summarizing your findings. In your paper, give your ideas about what can be done to reduce the amount of drinking and driving.

4-2A Graphing Calculators: Graphing Quadratic Functions

A graphing calculator can graph many different types of equations and can be used to analyze many different types of functions. In Chapter 3, we examined polynomial, radical, and rational functions. In this lesson, we will graph quadratic functions.

A quadratic function is a function of the form $y = ax^2 + bx + c$. This means that any equation in one variable, where the highest power is 2 is a quadratic equation. The graph of a quadratic equation is called a **parabola**.

Example 1

Graph $y = -16x^2 + 11$.

Enter: [Y=] [(−)] 16 [X,T,θ] [x²] [+] 11 [GRAPH]

The standard viewing window does not show a complete graph of the equation. Try the viewing window [−3, 3] by [−5, 15] with a scale factor of 1 for the x- and y-axes.

This view shows a complete graph of the function because you can see the x- and y-intercepts, the maximum point, and the end behavior of the graph.

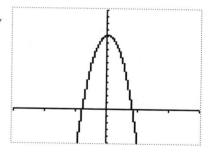

Notice from the graph that the shape of the graph of a quadratic equation is a parabola. Any quadratic equation will have this same general shape, although it may be thinner, fatter, shorter, or taller. Also, a parabola can open up, down, to the left, or to the right. However, graphs of quadratic equations that are functions open up or down.

Quadratic equations can also be solved. That is, you can let y equal 0 and solve the equation for x. The equation is of the form $0 = ax^2 + bx + c$, and you can use a graphing calculator to find the solutions if they are real. The **solutions** are the points where the graph crosses the x-axis because these are the points where $y = 0$.

There are three possible outcomes when solving a quadratic equation. The equation will have either two real solutions, one real solution, or no real solutions. Each of these outcomes is illustrated on the next page.

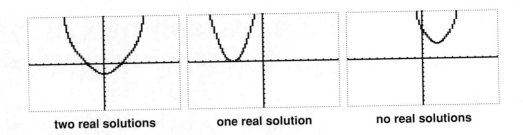

two real solutions   one real solution   no real solutions

Example 2

Solve $0 = 2x^2 - 5x + 1$ by using the graphing calculator. Round solutions to the nearest hundredth.

Graph the equation in the standard viewing window.

Enter: [Y=] 2 [X,T,θ] [x^2] [−] 5 [X,T,θ] [+] 1 [GRAPH]

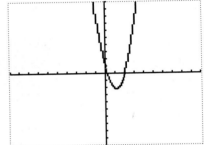

The standard viewing window shows a complete graph of the equation. Now zoom in on each of the x-intercepts to find the solutions to the equation. The solutions of this equation are 0.22 and 2.28 to the nearest hundredth.

It should be noted that solving quadratic equations graphically produces only approximate solutions. While approximations are adequate for most applications, if an exact solution is needed, the equation should be solved by factoring or by using the quadratic formula.

EXERCISES

Use a graphing calculator to show a complete graph of each function. Then sketch the graph on a piece of paper.

1. $y = x^2 + 2x + 9$

2. $y = -3x^2 - 0.5x + 8$

3. $y = 1.1x^2 + 6.3x - 2.4$

4. $y = 12x^2 + 44x + 192$

5. $y = 0.9x^2 - 0.21x - 0.643$

6. $y = -\frac{1}{4}x^2 + 2x - 4$

Find the solutions of each quadratic equation to the nearest hundredth by using a graphing calculator.

7. $\frac{1}{2}x^2 + 3x - 5 = 0$

8. $14x^2 + 3.5x - 12.1 = 0$

9. $x^2 + 36.15x + 209.31 = 0$

10. $22x^2 + 35x - 19 = 0$

11. $0.3x^2 - 0.225x - 0.068 = 0$

12. $-5x^2 + 3.27x + 19.11 = 0$

Quadratic Equations and Inequalities

Objectives

After studying this lesson, you should be able to:
- solve quadratic equations,
- use the discriminant to describe the roots of quadratic equations, and
- graph quadratic equations and inequalities.

Application

The Thespian Club at Colerain Senior High School needs to purchase an additional wireless microphone for the character Albert in the play *Bye Bye Birdie*. Each cast member will contribute an equal share for the cost of the microphone, which is $300. If 5 more students had joined the original cast, the cost per cast member would have been $10 less. How many cast members are there in this production of *Bye Bye Birdie*?

Let x = the number of cast members, and let y = the contribution of each student.

$$(x + 5)(y - 10) = 300$$
$$(x + 5)\left(\frac{300}{x} - 10\right) = 300 \qquad y = \frac{300}{x}$$
$$(x + 5)(300 - 10x) = 300x$$
$$300x - 10x^2 + 1500 - 50x = 300x$$
$$-10x^2 - 50x + 1500 = 0$$

FYI...

The musical *Bye Bye Birdie* opened on Broadway on April 14, 1960, at the Martin Beck Theater. The show completed 607 performances.

One way to determine the roots, or solutions, of this quadratic equation would be to examine the graph of the related function, $y = -10x^2 - 50x + 1500$. The x-intercepts of the graph are the zeros of the function, which are also the solutions of the quadratic equation. So, the solutions are $x = 10$ and $x = -15$. Since the number of cast members cannot be negative, there must be 10 members in the cast of *Bye Bye Birdie*.

The graph of a quadratic function is called a parabola. The vertex of the parabola is its critical point.

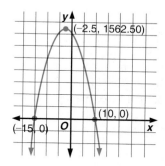

You have solved quadratic equations by factoring and by graphing. Another method for solving a quadratic equation is by **completing the square**. Completing the square is a helpful method to use when the quadratic is not easily factorable. It can be used to solve any quadratic equation. Remember that, for any number b, the square of the binomial $x + b$ has the form $x^2 + 2bx + b^2$. When completing the square, you know the first term and middle term and need to supply the last term. This term equals the square of half the coefficient of the middle term. The leading coefficient must be 1.

Example 1

APPLICATION

Entertainment

Refer to the application at the beginning of the lesson. Solve the problem by completing the square.

$$0 = -10x^2 - 50x + 1500$$
$$0 = x^2 + 5x - 150 \qquad \text{Divide each side by } -10.$$
$$150 = x^2 + 5x \qquad \text{Add 150 to each side.}$$
$$150 + 6.25 = x^2 + 5x + 6.25 \qquad \text{Add } \left(\frac{5}{2}\right)^2 \text{ or 6.25 to each side.}$$
$$156.25 = (x + 2.5)^2 \qquad \text{Factor the perfect square trinomial.}$$
$$\pm 12.5 = x + 2.5 \qquad \text{Take the square root of each side.}$$

$$12.5 = x + 2.5 \qquad \text{or} \qquad -12.5 = x + 2.5$$
$$10 = x \qquad\qquad -15 = x$$

The solutions for the equation are $x = 10$ and $x = -15$. Since the number of cast members cannot be negative, the number is 10. *This agrees with the previous solution.*

Completing the square can be used to develop a general formula for solving any quadratic equation of the form $ax^2 + bx + c = 0$. This formula is called the **quadratic formula**.

Quadratic Formula	**The roots of a quadratic equation of the form $ax^2 + bx + c = 0$ with $a \neq 0$ are given by the following formula.** $$x = \frac{-b \pm \sqrt{b^2 - 4ac}}{2a}$$

You will be asked to derive this formula in Exercise 42.

The quadratic formula can be used to find the roots of any quadratic equation.

Example 2

Solve $4x^2 - 8x + 3 = 0$ using the quadratic formula. Then graph the related function.

$$x = \frac{-b \pm \sqrt{b^2 - 4ac}}{2a}$$
$$= \frac{-(-8) \pm \sqrt{(-8)^2 - 4(4)(3)}}{2(4)} \qquad a = 4,\ b = -8,\ c = 3$$
$$= \frac{8 \pm \sqrt{16}}{8}$$
$$= \frac{8 \pm 4}{8}$$

$$x = \frac{8 + 4}{8} \qquad \text{or} \qquad x = \frac{8 - 4}{8}$$
$$x = \frac{12}{8} \text{ or } \frac{3}{2} \qquad\qquad x = \frac{4}{8} \text{ or } \frac{1}{2}$$

To graph the related function, you need to know the x-intercepts, y-intercept, and the vertex. You already know the x-intercepts: $\frac{3}{2}$ and $\frac{1}{2}$.

To find the y-intercept, find $f(0)$ for $f(x) = 4x^2 - 8x + 3$.
$f(0) = 4(0)^2 - 8(0) + 3$ or 3

To find the vertex, find the derivative.

$f(x) = 4x^2 - 8x + 3$
$f'(x) = 8x - 8$ *Find the derivative.*
$\quad 0 = 8x - 8$ *Replace $f'(x)$ with 0.*
$\quad 1 = x$ *Solve for x.*
$f(1) = 4(1)^2 - 8(1) + 3$
$\quad\ = -1$

The vertex is at $(1, -1)$.

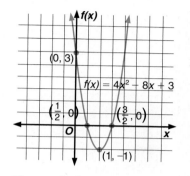

Graph the ordered pairs corresponding to the intercepts and the vertex.

These are the points at $(\frac{3}{2}, 0)$, $(\frac{1}{2}, 0)$, $(0, 3)$, $(1, -1)$.

The graph shows there is an x-intercept at $\frac{3}{2}$ and one at $\frac{1}{2}$.

In the quadratic formula, the expression under the radical, $b^2 - 4ac$, is called the **discriminant**. The discriminant tells us the nature of the roots of a quadratic equation.

Nature of the Roots of a Quadratic Equation	Discriminant	Nature of the Roots
	$b^2 - 4ac > 0$	**two distinct real roots**
	$b^2 - 4ac = 0$	**exactly one real root**
	$b^2 - 4ac < 0$	**no real roots (two distinct imaginary roots)**

In Example 2, the discriminant has a value of 16. Therefore, there are two distinct real roots, namely $\frac{3}{2}$ and $\frac{1}{2}$.

Example 3

Determine the discriminant of $x^2 - 6x + 13 = 0$. Use the quadratic formula to find the roots. Then graph the related function.

The value of the discriminant, $b^2 - 4ac$, is $(-6)^2 - 4(1)(13)$ or -16. Since the value of the discriminant is less than 0, there are no real roots. Verify this by using the quadratic formula.

$x = \dfrac{-b \pm \sqrt{b^2 - 4ac}}{2a}$

$\ \ = \dfrac{-(-6) \pm \sqrt{-16}}{2}$ $b^2 - 4ac = -16$

$\ \ = \dfrac{6 \pm 4i}{2}$

$\ \ = 3 \pm 2i$ Both roots are imaginary.

(continued on the next page)

Find the y-intercept.

$f(0) = (0)^2 - 6(0) + 13$ or 13

Find the vertex.

$$f'(x) = 2x - 6$$
$$0 = 2x - 6$$
$$3 = x$$

$$f(3) = 3^2 - 6(3) + 13$$
$$= 4$$

The vertex is at $(3, 4)$.

Use the y-intercept and vertex to graph the function.

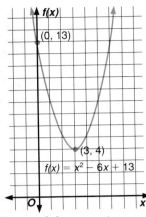

The graph has no x-intercepts, verifying there are no real roots.

The roots of the equation in Example 3 are the complex numbers $3 + 2i$ and $3 - 2i$. Complex numbers in the form $a + bi$ and $a - bi$ are called **conjugates**. Imaginary roots of polynomial equations always occur in conjugate pairs.

Complex Conjugates Theorem

Suppose a and b are real numbers with $b \neq 0$. If $a + bi$ is a root of a polynomial equation with real coefficients, then $a - bi$ is also a root of the equation.

Example 4

APPLICATION

Stationery

Determine if a rectangular piece of wrapping paper with a perimeter of 36 inches can have an area of 81 square inches. State the dimensions of the paper if it does exist.

Use the perimeter and area formulas to produce a quadratic equation.

$$P = 2\ell + 2w \qquad \textit{Perimeter formula}$$
$$36 = 2\ell + 2w \qquad \textit{P = 36}$$
$$36 - 2w = 2\ell \qquad \textit{Subtract 2w from each side.}$$
$$18 - w = \ell \qquad \textit{Divide each side by 2.}$$

Now substitute the value of ℓ into the area formula.

$$A = \ell w \qquad \textit{Area formula}$$
$$81 = (18 - w)w \qquad \textit{ℓ = 18 - w and A = 81}$$
$$0 = -81 + 18w - w^2 \qquad \textit{Subtract 81 from each side and multiply.}$$
$$0 = w^2 - 18w + 81 \qquad \textit{Multiply each side by -1.}$$

Now use the discriminant to determine how many roots there are.

$$b^2 - 4ac = (-18)^2 - 4(1)(81) \text{ or } 0$$

Because the discriminant is 0, there will be one distinct real root.

The quadratic is a perfect square trinomial. You can solve this equation by factoring.

$$w^2 - 18w + 81 = 0$$
$$(w - 9)^2 = 0 \quad \text{\textit{Factor the trinomial.}}$$
$$w = 9 \quad \text{\textit{The width of the rectangle is 9 inches.}}$$

Since the width is 9 inches and $81 = \ell w$, the length is also 9 inches. The dimensions of the paper are 9 inches by 9 inches.

In Chapter 3, you learned to graph different types of inequalities. You can use the methods of graphing quadratic equations in this lesson to graph **quadratic inequalities**.

Example 5

Graph $y > x^2 + 8x - 20$.

First graph the related function, $f(x) = x^2 + 8x - 20$, by finding the x- and y-intercepts and vertex.

Find the x-intercepts.

$$x = \frac{-(8) \pm \sqrt{8^2 - 4(1)(-20)}}{2(1)}$$
$$x = -10 \text{ or } x = 2$$

Find the y-intercept.

$$f(0) = (0)^2 + 8(0) - 20 \text{ or } -20$$

Find the vertex.

$$f'(x) = 2x + 8$$
$$0 = 2x + 8$$
$$-4 = x$$

$$f(-4) = (-4)^2 + 8(-4) - 20$$
$$= -36$$

The vertex is at $(-4, -36)$.

Graph the parabola. Because the inequality does not include the equation $y = x^2 + 8x - 20$, the parabola is dashed.

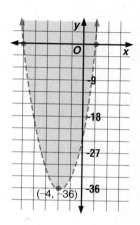

Determine which region is shaded by testing a point inside the parabola.

Test $(0, 0)$: $y > x^2 + 8x - 20$
$$0 > 0^2 + 8(0) - 20$$
$$0 > -20$$

The inequality is true, so $(0, 0)$ is a solution. Shade the region that contains the point at $(0, 0)$.

Test a point outside the parabola to check your solution.

CHECKING FOR UNDERSTANDING

Communicating Mathematics

Read and study the lesson to answer each question.

1. **Discuss** three different methods for determining the zeros of a quadratic equation and the ways in which the solutions might differ.

2. **Write** the expression for the discriminant of a quadratic equation $ax^2 + bx + c = 0$.

3. **Explain** how the discriminant can be used to describe the nature of the roots of a quadratic equation.

Guided Practice

Find the value of c that makes each trinomial a perfect square.

4. $x^2 + 4x + c$

5. $p^2 - p + c$

6. $y^2 + \frac{3}{2}y + c$

Solve each equation by completing the square. Then graph the related function.

7. $x^2 - 5x + 9 = 0$

8. $z^2 - 10z + 10 = -11$

Find the discriminant of each equation and describe the nature of the roots of the equation.

9. $4x^2 + 6x + 25 = 0$

10. $x^2 + 12x + 36 = 0$

11. Name the conjugate of $5 - i\sqrt{2}$.

Solve each equation by using the quadratic formula.

12. $t^2 - 3t - 28 = 0$

13. $3x^2 - 5x + 9 = 0$

14. Graph $10 - 3x - x^2 \geq y$.

EXERCISES

Practice

Solve each equation by completing the square. Then graph the related function.

15. $x^2 - 3x - 88 = 0$

16. $x^2 - \frac{3}{4}x + \frac{1}{8} = 0$

17. $2x^2 + 11x - 21 = 0$

18. $x^2 - 3x - 7 = 0$

19. $z^2 - 2z = 24$

20. $3x^2 - 12x = -4$

Find the discriminant of each equation and describe the nature of the roots of the equation. Then solve each equation by using the quadratic formula and graph the related function.

21. $4x^2 + 19x - 5 = 0$

22. $3 - 7m - 6m^2 = 0$

23. $2w^2 + 3w + 3 = 0$

24. $2k^2 + 5k - 9 = 0$

25. $6b^2 - 39b + 45 = 0$

26. $4x^2 - 2x + 9 = 0$

Graph each quadratic inequality.

27. $y < 3 + 7x - 6x^2$

28. $y < 3x^2 + 7x + 4$

29. $y \geq 2x^2 - 9x - 5$

30. $y > 4x^2 - 8x + 3$

31. $y \leq -x^2 - 7x + 10$

32. $y > -4x^2 - 3x - 6$

Determine the critical point(s) of the graph of each function to the nearest tenth. State if the point is a relative maximum, a relative minimum, or a point of inflection.

33. $f(x) = x^3 - 2x^2 - 35x$

34. $f(x) = 15x^3 - 16x^2 - x + 2$

35. $f(x) = x^3 - 9x^2 + 23x - 15$

36. $f(x) = x^3 + 6x^2 + 8x$

37. Consider the functions $f(x) = x^2 - 3x - 54$ and $g(x) = 54 + 3x - x^2$.
   a. Determine the zeros of each function.
   b. Sketch a graph of each function.
   c. Write a paragraph in which you compare and contrast the two graphs.

Graphing Calculator

Use the TRACE and BOX commands to graph each function and determine the x-intercepts. State the Xmin, Xmax, Ymin, and Ymax of the window that shows the x-intercepts and y-intercept.

38. $x^2 - 6x + 9 = y$

39. $x^2 - 3x - 700 = y$

40. $x^4 - 50x^2 + 49 = y$

Critical Thinking

41. Find the difference of the roots of the quadratic equation below.
$$(3 + 2\sqrt{2})x^2 + (1 + \sqrt{2})x = 2$$

42. Derive the quadratic formula by completing the square.

Applications and Problem Solving

43. **Gardening**   The length of Mary Adam's rectangular flower garden is 6 feet more than its width. A walkway 3 feet wide surrounds the outside of the garden. The total area of the walkway itself is 288 square feet. Find the dimensions of the garden.

44. **Physics**   The distance, $d(t)$, fallen by a free-falling body can be represented by the formula $d(t) = v_0 t - \frac{1}{2}gt^2$, where v_0 is the initial velocity and g represents the acceleration due to gravity. The acceleration due to gravity is 9.8 m/s^2. If a rock is thrown downward with an initial velocity of 4 m/s from the edge of the North Rim of the Grand Canyon, which is 1750 meters deep, determine how long it will take the rock to reach the bottom of the Grand Canyon.

45. **Baseball**   When an object is projected upward, its height in feet above the ground, $d(t)$, is given as a function of time, t in seconds, by the formula $d(t) = v_0 t - \frac{1}{2}gt^2$. The acceleration due to gravity g is 32 ft/s^2. If Jason Kendall of the Pittsburgh Pirates hits a ball straight upward with an intitial velocity, v_0, of 80 ft/s, determine the time that the ball is in the air.

Mixed Review

46. Graph $2y - 5x \leq 8$.   **(Lesson 1-3)**

47. Write the standard form of an equation of the line that passes through $(4, 2)$ and is parallel to the line whose equation is $y = 2x - 4$.   **(Lesson 1-6)**

48. Solve the system of equations by using augmented matrices.   **(Lesson 2-4)**
$$6r + s = 9$$
$$3r + 2s = 0$$

49. Determine whether the graph of $y = |x| + 1$ is symmetric with respect to the x-axis, the y-axis, the line $y = x$, the line $y = -x$, or the origin. Include all symmetries in your answer. **(Lesson 3-1)**

50. Find the equation of the tangent to the graph of $y = -3x^2 + 5$ at $(-2, -7)$. **(Lesson 3-6)**

51. Write the simplest polynomial equation with roots $-5, -6,$ and 10. **(Lesson 4-1)**

52. **College Entrance Exam** Compare quantities A and B below.
Write A if quantity A is greater.
Write B if quantity B is greater.
Write C if the quantities are equal.
Write D if there is not enough information to determine the relationship.

Given: $p > 0, q < 0$

(A) $p + q$ **(B)** $p - q$

DECISION MAKING

Choosing a Vacation Spot

Where is the best place in the United States for you to spend a two-week vacation? What do you want to accomplish on a vacation? Should fun and relaxation be your only goals, or should you try to learn something about America's history, culture, or nature at the same time? Planning for a vacation includes thinking not only about destinations but about food, lodging, methods of transportation, and, of course, how you are going to pay for it all!

1. Without naming a specific place, describe your ideal vacation. Tell what kinds of activities you most want to participate in during your trip and what goals you want to accomplish.

2. Research possible places that will meet your vacation needs. Talk to friends, consult books in the library, contact state tourism departments, or talk to a travel agent. Make a list of three potential vacation destinations.

3. For each destination, outline a two-week agenda and prepare a budget specifying anticipated costs for the entire trip.

4. **Project** Work with your group to complete one of the projects listed at the right, based on the information you gathered in Exercises 1–3.

PROJECTS

- *Conduct an interview.*
- *Write a book report.*
- *Write a proposal.*
- *Write an article for the school paper.*
- *Write a report.*
- *Make a display.*
- *Make a graph or chart.*
- *Plan an activity.*
- *Design a checklist.*

4-3 The Remainder and Factor Theorems

Objective

After studying this lesson, you should be able to:
- find the factors of polynomials using the remainder and factor theorems.

Application

On July 26, 1992, Miguel Indurain of Spain won his second consecutive Tour de France competition in cycling. He averaged a record-setting speed of almost 40 km/h or about 11 m/s. The average recreational cyclist pedals at a rate of about 4 m/s. Suppose you were cycling along at the speed of 4 m/s and headed down a hill, accelerating at a rate of 0.4 m/s². How far down the hill would you be after 5 seconds?

The formula that represents this relationship is $D(t) = v_i t + \frac{1}{2}at^2$, where $D(t)$ is the distance from the starting point, v_i is the initial velocity, t is time, and a is the acceleration. By substituting the initial velocity and the acceleration from this problem into the formula, we get $D(t) = 4t + \frac{1}{2}(0.4)t^2$, or $D(t) = 4t + 0.2t^2$.

To find the distance after 5 seconds, evaluate this function for $t = 5$.

$$D(5) = 4(5) + 0.2(5)^2 \text{ or } 25$$

After 5 seconds, you would be 25 meters down the hill.

Suppose you divide the polynomial $4t + 0.2t^2$ by $t - 5$.

$$
\begin{array}{r}
0.2t + 5 \\
t - 5 \overline{\smash{)}\ 0.2t^2 + 4t + 0} \\
\underline{0.2t^2 - 1t} \\
5t + 0 \\
\underline{5t - 25} \\
25
\end{array}
$$

From arithmetic, you may remember that the dividend equals the product of the divisor and the quotient plus the remainder. For example, $32 \div 5 = 6 \text{ R}2$, so $32 = 5(6) + 2$. This relationship can be applied to polynomials.

$$D(t) = (t - 5)(0.2t + 5) + 25$$

Now let $t = 5$. What do you find?

$$D(5) = (5 - 5)[0.2(5) + 5] + 25 \text{ or } 25$$

Notice that the value of D(5) is the same as the remainder when the polynomial is divided by $t - 5$. This example illustrates the **remainder theorem**.

The Remainder Theorem	If a polynomial $P(x)$ is divided by $x - r$, the remainder is a constant, $P(r)$, and

$$P(x) = (x - r) \cdot Q(x) + P(r)$$

where $Q(x)$ is a polynomial with degree one less than the degree of $P(x)$.

The remainder theorem provides another way to find the value of the polynomial function $P(x)$ for a given value r. The value will be the remainder when $P(x)$ is divided by $x - r$.

Example 1

Let $P(x) = x^3 + 3x^2 - 2x - 8$. Show that the value of $P(-2)$ is the remainder when $P(x)$ is divided by $x + 2$.

Use long division to find the remainder.

$$
\begin{array}{r}
x^2 + x - 4 \\
x + 2 \overline{\smash{)}\, x^3 + 3x^2 - 2x - 8} \\
\underline{x^3 + 2x^2} \\
x^2 - 2x \\
\underline{x^2 + 2x} \\
-4x - 8 \\
\underline{-4x - 8} \\
0
\end{array}
$$

Evaluate $P(x)$ for $x = -2$.

$$P(x) = x^3 + 3x^2 - 2x - 8$$
$$P(-2) = (-2)^3 + 3(-2)^2 - 2(-2) - 8$$
$$P(-2) = -8 + 12 + 4 - 8$$
$$P(-2) = 0$$

The remainder after division is 0, and the value of $P(-2)$ is also 0.

Long division can be very time consuming. **Synthetic division** is a shortcut for dividing a polynomial by a binomial of the form $x - r$. Let's divide the polynomial $x^3 + 3x^2 - 2x - 8$ by $x + 2$ using synthetic division. For $x + 2$, the value of r is -2.

Step 1 Arrange the terms of the polynomial in descending powers of x. Then, write the coefficients as shown.

$$x^3 + 3x^2 - 2x - 8$$
$$1 \quad 3 \quad -2 \quad -8$$

Step 2 Write the constant r of the divisor $x - r$. In this case, write -2.

$$-2 \underline{|} \; 1 \quad 3 \quad -2 \quad -8$$

Step 3 Bring down the first coefficient.

$$
\begin{array}{r}
-2 \underline{|} \; 1 \quad 3 \quad -2 \quad -8 \\
1
\end{array}
$$

Step 4 Multiply the first coefficient by r. Then write the product under the second coefficient. Add.

$$
\begin{array}{r}
-2 \underline{|} \; 1 \quad 3 \quad -2 \quad -8 \\
\underline{\quad -2} \\
1 \quad 1
\end{array}
$$

Step 5 Multiply the sum by r. Then write the product under the next coefficient. Add.

$$
\begin{array}{r}
-2 \underline{|} \; 1 \quad 3 \quad -2 \quad -8 \\
\underline{\quad -2 \; -2} \\
1 \quad 1 \quad -4
\end{array}
$$

Step 6 Repeat Step 5 until all coefficients in the dividend have been used.

$$\begin{array}{r} -2\,\underline{|}\ \ 1 \quad 3 \quad -2 \quad -8 \\ -2 \quad -2 \quad \ \ 8 \\ \hline 1 \quad 1 \quad -4 \,\underline{|}\ \ 0 \end{array}$$

The final sum represents the remainder, which in this case, is zero. The other numbers are the coefficients of the polynomial that is the quotient.

Step 7 Remember that the quotient has a degree one less than the dividend. Write the quotient as $x^2 + x - 4$.

So, $x^3 + 3x^2 - 2x - 8 = (x + 2)(x^2 + x - 4)$. This result agrees with the long division in Example 1.

Example 2

Use synthetic division to divide $m^5 - 3m^2 - 20$ by $m - 2$.

$$\begin{array}{r} 2\,\underline{|}\ \ 1 \quad 0 \quad 0 \quad -3 \quad 0 \quad -20 \\ 2 \quad 4 \quad 8 \quad 10 \quad 20 \\ \hline 1 \quad 2 \quad 4 \quad 5 \quad 10 \,\underline{|}\ \ 0 \end{array}$$

Notice there are no m^4, m^3, or m terms. Zeros are placed in their positions as place holders.

The quotient is $m^4 + 2m^3 + 4m^2 + 5m + 10$.

If the remainder is zero when $P(x)$ is divided by $x - r$, then $x - r$ is a factor of $P(x)$. A corollary to the remainder theorem, called the **factor theorem**, states this more formally.

The Factor Theorem	**The binomial $x - r$ is a factor of the polynomial $P(x)$ if and only if $P(r) = 0$.**

Example 3

Let $P(x) = x^3 - 4x^2 - 7x + 10$. Determine if $x - 5$ is a factor of $P(x)$.

Let's use synthetic division. In this case, $r = 5$.

$$\begin{array}{r} 5\,\underline{|}\ \ 1 \quad -4 \quad -7 \quad 10 \\ 5 \quad 5 \quad -10 \\ \hline 1 \quad 1 \quad -2 \,\underline{|}\ \ 0 \end{array}$$

Since the remainder is 0, $x - 5$ is a factor of $x^3 - 4x^2 - 7x + 10$.

Check: $P(5) = 5^3 - 4(5^2) - 7(5) + 10$
$= 125 - 100 - 35 + 10$ or 0

TECHNOLOGY *Tip*

You can use the TI-92 graphing calculator to divide a polynomial by a monomial. If the remainder is not zero, you can use the **propFrac** (proper fraction) function to display the result as the sum of the remainder and the quotient.

When a polynomial is divided by one of its binomial factors, the quotient is called a **depressed polynomial**. A depressed polynomial has a degree less than the original polynomial. In Example 3, $x^2 + x - 2$ is the depressed polynomial. This polynomial may also be factorable, which would give you other zeros of the function. In this case, $x^2 + x - 2$ can be factored as $(x + 2)(x - 1)$. So, the other zeros of the function are –2 and 1.

You can also find factors of a polynomial such as $x^3 - 4x^2 - 7x + 10$ by using a shortened form of synthetic division to test several values of r.

r	1	-4	-7	10
1	1	-3	-10	0
2	1	-2	-11	-12
3	1	-1	-10	-20
4	1	0	-7	-18
5	1	1	-2	0
-1	1	-5	-2	12
-2	1	-6	5	0

In the table at the left, the first column contains various values for r. The next three columns show the coefficients of the depressed polynomial in the last line of the synthetic division and the fifth column shows the remainder. Any value of r that results in a remainder of zero indicates a factor of the polynomial. The factors of the original polynomial are $x - 1$, $x - 5$, and $x + 2$.

Look at the pattern of values in the last column. Notice that when $r = 2, 3$, and 4, the values of $f(x)$ decrease and then increase. This indicates that there is an x-coordinate of a critical point between 2 and 4.

You can use a graphing calculator to investigate polynomials and their factors.

EXPLORATION: Graphing Calculator

1. Graph each polynomial on the same screen. What do you notice about these graphs?

   a. $P(x) = (x - 1)(x + 2)(x + 3)$
   b. $P(x) = (x^2 + x - 2)(x + 3)$
   c. $P(x) = x^3 + 4x^2 + x - 6$

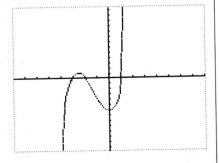

2. Graph $P(x) = \dfrac{(x^3 + 4x^2 + x - 6)}{(x - 1)}$.

   a. Use the TRACE function to find $P(-2)$ and $P(-3)$.
   b. Graph $y = (x + 2)(x + 3)$.
   c. How does this compare with the graph of
   $$P(x) = \frac{(x^3 + 4x^2 + x - 6)}{(x - 1)}?$$
   d. What does this indicate?

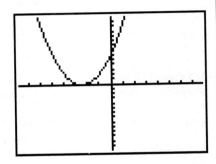

Example 4

APPLICATION

Cycling

Refer to the application at the beginning of the lesson. If the hill on which you were riding is 55 meters long, how long will it take you to reach the bottom of the hill?

Use the formula for the distance and replace $D(t)$ with 55.

$$D(t) = 4t + 0.2t^2$$

$55 = 4t + 0.2t^2$ *Replace D(t) with 55.*

$0 = 0.2t^2 + 4t - 55$ *Subtract 55 from each side.*

$0 = t^2 + 20t - 275$ *Multiply each side by 5 so the coefficient of t^2 is 1.*

You could use synthetic division to factor this polynomial. However, since 275 is such a large number, it may be easier to use the quadratic formula to find the values of t.

$$t = \frac{-20 \pm \sqrt{(20)^2 - 4(1)(-275)}}{2(1)}$$

$$t = \frac{-20 \pm \sqrt{1500}}{2}$$ $\sqrt{1500} \approx 38.73$

$$t \approx 9.4 \text{ or } -29.4$$

The value of –29.4 is not a reasonable answer. So you will reach the bottom of the hill in about 9.4 seconds.

CHECKING FOR UNDERSTANDING

Communicating Mathematics

Read and study the lesson to answer each question.

1. **Write** the expression that completes this sentence: When a polynomial is divided by the factor $(x - r)$, the remainder is the same as **?** .

2. **Define** a depressed polynomial.

3. **Explain** what is meant by synthetic division.

Guided Practice

Find $P(3)$ for each polynomial. State whether $x - 3$ is a factor of the polynomial. Write *yes* or *no*.

4. $P(x) = x^2 - 6x + 0$

5. $P(x) = x^2 + 2x - 15$

6. $P(x) = x^2 - 5x + 6$

7. $P(x) = x^4 + x^2 - 2$

8. Let $f(x) = x^7 + x^9 + x^{12} - 2x^2$.
   a. State the degree of $f(x)$.
   b. State the number of complex zeros that $f(x)$ has.
   c. State the degree of the depressed polynomial that would result from dividing $f(x)$ by $x - a$.
   d. Find one factor of $f(x)$.

9. Use synthetic division to find all the factors of $x^3 + 7x^2 - x - 7$ if one of the factors is $x + 1$.

EXERCISES

Practice

Divide using synthetic division.

10. $(x^2 - x - 56) \div (x + 7)$

11. $(x^2 - x + 4) \div (x - 2)$

12. $(x^3 + x^2 - 17x + 15) \div (x + 5)$

13. $(x^4 + x^3 - 1) \div (x - 2)$

14. $(x^3 - 9x^2 + 27x - 28) \div (x - 3)$

15. $(2x^3 - 2x - 3) \div (x - 1)$

Find the remainder for each division. Is the divisor a factor of the polynomial?

16. $(x^3 - 30x) \div (x + 5)$

17. $(x^4 - 6x^2 + 8) \div (x - \sqrt{2})$

18. $(5x^2 - 2x + 6) \div \left(x - \dfrac{2}{5}\right)$

19. $(x^5 + 32) \div (x + 2)$

Use the remainder theorem to find the remainder for each division. State whether the binomial is a factor of the polynomial.

20. $(x^2 + 1) \div (x + 1)$

21. $(x^2 + 5x - 2) \div (x + 5)$

22. $(2x^2 - x + 3) \div (x - 3)$

23. $(2x^3 - 3x + x) \div (x - 1)$

24. $(x^4 + x^2 + 2) \div (x - 3)$

25. $(2x^4 - x^3 + 1) \div (x + 3)$

Find the value of k so that each remainder is zero.

26. $(x^3 + 8x^2 + kx + 4) \div (x + 2)$

27. $(x^3 + kx^2 + 4x + 1) \div (x + 1)$

28. Determine how many times 1 is a root of $x^3 - 3x + 2 = 0$.

29. Determine how many times 2 is a root of $x^6 - 9x^4 + 24x^2 - 16 = 0$.

30. Determine how many times –1 is a root of $x^3 + 2x^2 - x - 2 = 0$. Then find the other roots.

31. Find a, b, and c for $P(x) = ax^2 + bx + c$ if $P(3 + 4i) = 0$ and $P(3 - 4i) = 0$.

32. Physics As an object moves farther from Earth, its weight decreases due to the diminished effect of Earth's gravity. The force of gravity decreases with the square of the distance from the center of Earth. The distance, R, from the center of Earth to the surface is about 3960 miles. The relationship between the weight of a body on earth, w_E, and the weight of a body a certain distance, r, from Earth, w_S, is $(R + r)^2 = \dfrac{R^2 \cdot w_E}{w_S}$. Determine the distance the spacecraft is from Earth if an astronaut weighs 200 pounds on Earth and 175 pounds in space.

33. Manufacturing A soft drink can has a height 4 inches greater than the radius of its lid. Determine the length of the radius of the lid and the height of the can if the volume of the can is about 15.71 cubic inches. *Use $V = \pi r^2 h$.*

34. If $f(x) = x^3$ and $g(x) = 3x$, find $g[f(-2)]$. **(Lesson 1-2)**

35. Find the critical points of the graph of $f(x) = x^5 - 32$. Determine whether each represents a maximum, a minimum, or a point of inflection. **(Lesson 2-5)**

36. Graph $y = \dfrac{x - 3}{x - 2}$. **(Lesson 3-4)**

37. Solve $z^2 + 4z = 96$ by completing the square. **(Lesson 4-2)**

38. College Entrance Exam Choose the best answer.
The distance from City A to City B is 150 miles. From City A to City C is 90 miles. Therefore, it is necessarily true that

(A) the distance from B to C is 60 miles.

(B) six times the distance from A to B equals 10 times the distance from A to C.

(C) the distance from B to C is 240 miles.

(D) the distance from A to B exceeds by 30 miles twice the distance from A to C.

(E) three times the distance from A to C exceeds by 30 miles twice the distance from A to B.

4-4 The Rational Root Theorem

Objectives

After studying this lesson, you should be able to:

- identify all possible rational roots of a polynomial equation by using the rational root theorem, and
- determine the number of positive and negative real zeros a polynomial function has.

Application

Universal Studios Hollywood has built a new ride called *Backdraft*, patterned after the movie by the same name. The special effects in the ride include scenes where fire fighters must battle blazes that reach temperatures of 2000°F. Most fire departments do not battle blazes of this intensity.

The traditional fire truck is capable of pumping 750–1500 gallons of water per minute. The truck draws water from a fire hydrant or other water source and boosts the water pressure before pumping it through smaller diameter water hoses called booster hoses. As water fills the hose from the fire hydrant to the truck, it becomes a long cylinder. Suppose the volume of the hose is 11,775 cubic inches and the length of the hose is 150 inches less than 300 times the radius of the hose. What are the radius and length of the hose? *This problem will be solved in Example 1.*

> *FYI...*
>
> The first successful steam fire engine was built in 1852 by Moses Latta of Cincinnati, Ohio.

Let's write an equation in terms of r, the radius. The length of the hose is the height of the cylinder. So, $h = 300r - 150$.

$$V = \pi r^2 h \qquad \text{\textit{Volume of a cylinder}}$$
$$11,775 = 3.14r^2(300r - 150) \qquad \text{\textit{Let } } \pi = 3.14, r = r, V = 11,775 \text{ \textit{and}}$$
$$\text{\textit{h} } = 300r - 150.$$
$$11,775 = 942r^3 - 471r^2 \qquad \text{\textit{Multiply.}}$$
$$0 = 942r^3 - 471r^2 - 11,775 \qquad \text{\textit{Subtract 11,775 from each side.}}$$
$$0 = 2r^3 - r^2 - 25 \qquad \text{\textit{Divide each side by 471.}}$$

We know there are three complex roots for this polynomial equation. You could use synthetic division to test the possible roots. However, you may have to test hundreds of values before finding a rational root, if it exists. The **rational root theorem** provides a means for dramatically lowering the number of rational values that you might test to find rational roots of a polynomial equation with integral coefficients.

Rational Root Theorem

Let $a_0x^n + a_1x^{n-1} + \cdots + a_{n-1}x + a_n = 0$ represent a polynomial equation of degree n with integral coefficients. If a rational number $\dfrac{p}{q}$, where p and q have no common factors, is a root of the equation, then p is a factor of a_n and q is a factor of a_0.

For the polynomial equation $0 = 2r^3 - r^2 - 25$ in the fire hose problem, if $\frac{p}{q}$ is a rational root, then p is a factor of 25 and q is a factor of 2.

possible values for p: $\pm 1, \pm 5, \pm 25$

possible values for q: $\pm 1, \pm 2$

possible rational roots, $\frac{p}{q}$: $\pm 1, \pm \frac{1}{2}, \pm 5, \pm \frac{5}{2}, \pm 25, \pm \frac{25}{2}$

Example 1

APPLICATION

Firefighting

Refer to the application at the beginning of the lesson. Find the radius and the length of the fire hose.

Use the shortened form of synthetic division to determine which of the possible roots listed above, if any, are roots of the equation $0 = 2r^3 - r^2 - 25$.

root	2	-1	0	-25
1	2	1	1	-24
5	2	9	45	200
25	2	49	1225	30,600
0.5	2	0	0	-25
2.5	2	4	10	0

To simplify computation, write the fractions as decimals.

← 2.5 is a root of the equation.

You need not continue the process of synthetic division to find roots, since the depressed polynomial is a quadratic. Use the discriminant to determine the nature of the other two roots.

Remember the roots of a polynomial equation are the zeros of the polynomial function.

The discriminant of the depressed polynomial $2r^2 + 4r + 10$ is $(4)^2 - 4(2)(10)$ or -64. A negative discriminant tells us that the two other roots are imaginary. The graph of the function $V(r) = 2r^3 - r^2 - 25$ is shown at the right. There is one real zero at the r-intercept, 2.5. The other two zeros of the function are imaginary, so the graph intersects the r-axis once.

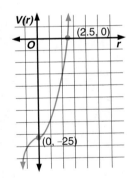

Since the only real root of the equation is 2.5, the radius of the fire hose is 2.5 inches, and the length of the hose is $300(2.5) - 150$ or 600 inches.

A corollary to the rational root theorem, called the **integral root theorem**, states that if the leading coefficient (a_0) has a value of 1, then any rational roots must be factors of a_n, $a_n \neq 0$.

Integral Root Theorem	Let $x^n + a_1 x^{n-1} + \cdots + a_{n-1}x + a_n = 0$ represent a polynomial equation that has leading coefficient of 1, integral coefficients, and $a_n \neq 0$. Any rational roots of this equation must be integral factors of a_n.

Example 2

Find the roots of $x^3 + 6x^2 + 10x + 3 = 0$.

There are three complex roots. According to the integral root theorem, the possible rational roots of the equation are factors of 3. The possibilities are ± 3 and ± 1.

r	1	6	10	3
3	1	9	37	114
-3	1	3	1	0

← There is one root at $x = -3$.

The depressed polynomial is $x^2 + 3x + 1$. Use the quadratic formula to find the other two roots.

$$x = \frac{-b \pm \sqrt{b^2 - 4ac}}{2a}$$

$$= \frac{-3 \pm \sqrt{3^2 - 4(1)(1)}}{2(1)} \qquad a = 1, b = 3, c = 1$$

$$= \frac{-3 \pm \sqrt{5}}{2}$$

The three roots of the equation are -3, $\dfrac{-3 + \sqrt{5}}{2}$, and $\dfrac{-3 - \sqrt{5}}{2}$.

Descartes' rule of signs can be used to determine the possible numbers of positive real zeros a polynomial has. It is named after the French mathematician René Descartes, who first proved the theorem in 1637. *In Descartes' rule of signs, when we speak of a zero of the polynomial, we mean a zero of the corresponding polynomial function.*

Descartes'
Rule of Signs

Suppose $P(x)$ is a polynomial whose terms are arranged in descending powers of the variable. Then the number of positive real zeros of $P(x)$ is the same as the number of changes in sign of the coefficients of the terms, or is less than this by an even number. The number of negative real zeros of $P(x)$ is the same as the number of changes in sign of the coefficients of the terms of $P(-x)$, or is less than this number by an even number.

Ignore zero coefficients when using this rule.

Example 3

State the number of possible complex zeros, the number of positive real zeros, and the number of possible negative real zeros for $h(x) = x^4 - 2x^3 + 7x^2 + 4x - 15$.

Because the degree of $h(x)$ is 4, $h(x)$ has four complex zeros.
To determine the number of possible positive real zeros, count the sign changes for the coefficients.

$h(x) = x^4 - 2x^3 + 7x^2 + 4x - 15$

1   -2   7   4   -15
   yes   yes   no   yes

(continued on the next page)

There are 3 changes. So, there are either 3 positive real zeros or 1 positive real zero.

To determine the number of possible negative real zeros, find $h(-x)$ and count the number of sign changes.

$$h(-x) = (-x)^4 - 2(-x)^3 + 7(-x)^2 + 4(-x) - 15$$

$$x^4 \;+\; 2x^3 \;+\; 7x^2 \;-\; 4x \;-\; 15$$

$$1 \quad\quad 2 \quad\quad 7 \quad\quad -4 \quad\quad -15$$

$$\text{no} \qquad \text{no} \qquad \text{yes} \qquad \text{no}$$

There is 1 change. So, there is 1 negative real zero.
The polynomial $h(x)$ has either 3 or 1 positive real zeros and exactly 1 negative real zero.

Example 4

Find the zeros of $M(x) = x^4 + 4x^3 + 3x^2 - 4x - 4$. Then graph the function.

There are 4 complex zeros. Using Descartes' rule of signs, we find that there is 1 positive real zero and 3 or 1 negative real zeros. The integral root theorem tells us the possible rational zeros are $\pm 1, \pm 2$, and ± 4.

Let's use synthetic division to test our possibilities.

r	1	4	3	-4	-4
1	1	5	8	4	0

The remainder is 0. So, 1 is a zero and $x - 1$ is one factor of the polynomial.

The depressed polynomial is $x^3 + 5x^2 + 8x + 4$. Use this polynomial and synthetic division to find other factors of the polynomial.

r	1	5	8	4
-1	1	4	4	0

The remainder is 0. So, -1 is a zero and $x + 1$ is another factor of the polynomial.

The depressed polynomial from this division is $x^2 + 4x + 4$, which can be factored as $(x + 2)(x + 2)$. This means that both zeros of $x^2 + 4x + 4$ are at $x = -2$.

$$M(x) = (x - 1)(x + 1)(x + 2)^2$$

The zeros of $M(x)$ are $1, -1,$ and -2.

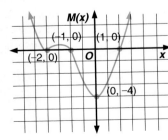

The graph of $M(x)$ is shown at the right. Notice that because there is a double zero at $x = -2$, the graph is tangent to the x-axis at that point and does not cross it. Thus, Descartes' rule of signs holds.

Example 5

A manufacturing company produces rectangular polystyrene packing fillers for an electronics firm that produces computer parts. The filler has a length (in cm) that is the square of the width and a height three times the width. In developing a new filler for another product, the width is increased by 1 cm, the length is increased by 2 cm, and the height is increased by 3 cm. The volume of the new filler is 162 cm². Find the dimensions of the original filler.

Let's define the measurements of each filler.

	original filler	*new filler*
width	x	$x + 1$
length	x^2	$x^2 + 2$
height	$3x$	$3x + 3$

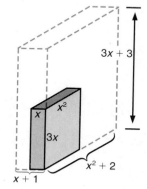

Volume of new filler $= (x + 1)(x^2 + 2)(3x + 3)$
$$162 = 3x^4 + 6x^3 + 9x^2 + 12x + 6$$
$$0 = x^4 + 2x^3 + 3x^2 + 4x - 52$$

There are 4 complex roots. There is one possible positive real root and 3 or 1 possible negative real roots. The possible rational roots are $\pm 1, \pm 2, \pm 4, \pm 13, \pm 26,$ and ± 52.

Use synthetic division to test the possibilities.

r	1	2	3	4	-52
1	1	3	6	10	-42
2	1	4	11	26	0

2 is a root. Since there is only one possible positive root and negative roots are not reasonable for this problem, we can end our search for roots. The original filler is 2 cm by 4 cm by 6 cm.

CHECKING FOR UNDERSTANDING

Communicating Mathematics

Read and study the lesson to answer each question.

1. **Discuss** why you may not need to use synthetic division to find more zeros of a polynomial function once you have found one zero of the function.

2. **Sketch** a graph of a polynomial function with zeros at –3 and 0 and a double zero at 5.

3. **Show** how the integral root theorem is derived from the rational root theorem.

4. **Demonstrate** how Descartes' rule of signs works.

Guided Practice

List all possible rational zeros of each function. Then determine the rational zeros.

5. $f(x) = x^3 + 2x^2 - 5x - 6$

6. $f(x) = 2x^3 + 3x^2 - 8x + 3$

State the number of possible positive real zeros and the number of possible negative real zeros of each function. Then find the rational zeros.

7. $f(x) = x^3 + 2x^2 - 5x - 6$

8. $f(x) = 2x^3 + 3x^2 - 8x + 3$

9. $f(x) = 6x^3 - 11x^2 - 24x + 9$

10. $f(x) = 4x^3 + 5x^2 + 2x - 6$

EXERCISES

Practice

List all possible rational zeros of each function. Then determine the rational zeros.

11. $f(x) = x^4 + 5x^3 + 5x^2 - 5x - 6$

12. $f(x) = x^3 - 4x^2 + x + 2$

13. $f(x) = x^3 - 5x^2 - 4x + 20$

14. $f(x) = x^3 - 2x^2 + x + 18$

15. $f(x) = x^4 - 5x^3 + 9x^2 - 7x + 2$

16. $f(x) = 2x^4 - x^3 - 6x + 3$

Find the number of possible positive real zeros and the number of possible negative real zeros. Determine all of the rational zeros.

17. $f(x) = x^3 - 2x^2 - 8x$

18. $f(x) = x^3 + 7x^2 + 7x - 15$

19. $f(x) = 8x^3 - 6x^2 - 23x + 6$

20. $f(x) = 2x^3 - 5x^2 - 28x + 15$

21. $f(x) = 6x^3 + 19x^2 + 2x - 3$

22. $f(x) = x^3 - 7x - 6$

23. $f(x) = x^4 - 5x^2 + 4$

24. $f(x) = x^4 + 2x^3 - 9x^2 - 2x + 8$

Graphing Calculator

Use the rational root theorem and Descartes' rule of signs to determine the nature of the possible rational roots for each equation. Use this information and the graphs of the related functions to approximate all the rational roots to the nearest tenth.

25. $24x^3 - 26x^2 + 9x - 1 = 0$

26. $25x^3 + 105x^2 - 96x + 20 = 0$

27. $6x^3 - 7x^2 - 29x - 12 = 0$

28. $x^4 - 10x^3 + 35x^2 - 50x + 24 = 0$

Critical Thinking

29. Write a third-degree polynomial function for each restriction.
  a. no positive real roots
  b. no negative real roots

Applications and Problem Solving

30. Driving The formula for calculating the distance traveled for a given time is $D(t) = v_i t + 0.5at^2$, where v_i is the initial velocity and a is the acceleration. A car is moving at 12 m/s and coasts up a hill with an acceleration of -1.6 m/s^2.
  a. Explain what negative accelaration is.
  b. For what values of t does $D(t) = 43$?
  c. Explain why there are two times for the same distance.

31. Cardiology Doctors can measure cardiac output in potential heart attack patients by monitoring the concentration of dye after a known amount is injected in a vein near the heart. In a normal heart, the concentration of the dye is given by the function $g(x) = -0.006x^4 + 0.140x^3 - 0.053x^2 + 1.79x$, where x is the time in seconds.
  a. Graph $g(x)$.
  b. Find all the zeros of this function.

32. **Business** Crystal Clear Water Company delivers spring water in water coolers to companies in the Cincinnati area. They also deliver cone-shaped drinking cups, whose volume is approximately 84.82 cm³. The height of each cup is 6 cm more than the radius. Determine the dimensions of the cone-shaped cup if $V = \frac{1}{3}\pi r^2 h$.

Mixed Review

33. Explain three types of discontinuity in graphs. Sketch an example of each. **(Lesson 3-8)**

34. Find the slope of the tangent to $y = x^2 + 4x - 5$ at $(1, 0)$. **(Lesson 3-6)**

35. Find the value of the discriminant for $3m^2 + 5m + 10 = 0$. Fully describe the nature of its roots. **(Lesson 4-2)**

36. Use synthetic division to determine if $x + 2$ is a factor of $x^3 + 6x^2 + 12x + 12$. Explain your answer. **(Lesson 4-3)**

37. **College Entrance Exam** Refer to the figure at the right. Choose the best answer. What percent of the area of rectangle $PQRS$ is shaded?

    (A) 20%   **(B)** 25%   **(C)** 30%

    (D) $33\frac{1}{3}$%   **(E)** 40%

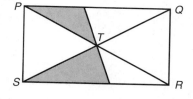

MID-CHAPTER REVIEW

1. Find the roots of $x^3 + 2x^2 - 80x = 0$. **(Lesson 4-1)**

2. Solve $p^2 + 6p + 3 = 0$ by completing the square. **(Lesson 4-2)**

3. Find the discriminant of $7m^2 - 13m + 2 = 0$. Fully describe the nature of its roots. Then solve the equation. **(Lesson 4-2)**

4. Use the remainder theorem to find the remainder for $(x^3 - 4x^2 + 2x - 6) \div (x - 4)$. State whether the binomial is a factor of the polynomial. **(Lesson 4-3)**

5. List all the possible rational zeros of $n(x) = 2x^3 - 13x^2 - 17x + 12$. Then determine the rational zeros. **(Lesson 4-4)**

4-5A Graphing Calculators: Locating Zeros of Polynomial Functions

A graphing calculator can be used to find the coordinates of a point to a great degree of accuracy. This feature is a big advantage when you are trying to locate the zeros of a polynomial function. With some knowledge of polynomials and the ZOOM and TRACE functions on a graphing calculator, you can locate the zeros of polynomial functions without using paper and a pencil.

A **zero** of a function is the x-coordinate of the point where the graph crosses the x-axis. It is also called a root, or a solution, of the equation. Since $y = 0$ on the x-axis, the x-coordinate is the solution.

Example

Approximate the zeros of the function $y = 2x^3 + x^2 - 5x - 3$ to the nearest hundredth.

Enter: [Y=] 2 [X,T,θ] [^] 3 [+] [X,T,θ] [^] 2 [-] 5 [X,T,θ] [-] 3 [GRAPH]

The standard viewing window contains a complete graph of the function. Looking at the graph, we can see that the graph crosses the x-axis three times.

Using TRACE and ZOOM, we find that the zeros are approximately at -1.50, -0.62, and 1.62.

After you have zoomed in on one zero and need to find another, it helps to regraph the function in the original viewing window. Then zoom in and trace again to locate the next zero.

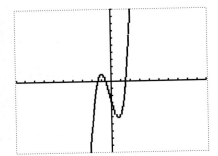

The TI-82 can also calculate the root of a function. Press [2nd] [CALC] 2. Then use the arrow and [ENTER] keys to select the lower bound, the upper bound, and a guess for the x-intercept. The cursor will be on the solution and the coordinates will be displayed.

EXERCISES

Use a graphing calculator to approximate the real zeros of each equation to the nearest hundredth.

1. $y = x^3 - 4x$

2. $y = -x^3 + 4x^2 - 6x + 8$

3. $y = 4x^4 - 6x^2 - 2x + 1$

4. $y = 2x^5 + 3x^4 - 12x + 4$

5. $y = -2x^4 + 5$

6. $y = 3x^3 - x^2 + 6x - 1$

4-5 Locating the Zeros of a Function

Objectives

After studying this lesson, you should be able to:
- approximate the real zeros of a polynomial function, and
- graph polynomial functions.

Application

The earliest jigsaw puzzles were made as "dissected maps" by John Spilsbury in London about 1762. Today there are hundreds of companies making jigsaw puzzles of varying difficulty and size.

The Fully-Interlocking Puzzle Company has been producing 1000-piece jigsaw puzzles. These puzzles have sold so well that they would like to market 1500-piece puzzles. Before manufacturing these puzzles, the company needs to determine the size of the box in which the puzzles will be marketed. The 1000-piece puzzle is packaged in a box 25 cm by 30 cm by 5 cm. The marketing department would like to increase each of these dimensions by the same amount to create the box for the 1500-piece puzzle. The volume of the new box must be 1.5 times the volume of the original box. What should be the dimensions of the box for the 1500-piece puzzle?

Use the information from the problem to write a polynomial function that describes the volume of the 1500-piece puzzle box.

Dimension	Original Box	New Box
width	25 cm	$25 + x$ cm
length	30 cm	$30 + x$ cm
height	5 cm	$5 + x$ cm
volume	3750 cm^3	1.5(3750) or 5625 cm^3

We can substitute the dimensions of the new box into the formula $V = \ell w h$ to find a polynomial equation to solve this problem.

$$V = \ell w h$$
$$5625 = (30 + x)(25 + x)(5 + x)$$
$$5625 = x^3 + 60x^2 + 1025x + 3750$$
$$0 = x^3 + 60x^2 + 1025x - 1875$$

Let $V(x) = x^3 + 60x^2 + 1025x - 1875$. There are three complex zeros, with one possible positive zero and two possible negative zeros. The possible rational zeros are $\pm1, \pm3, \pm5, \pm15, \pm25, \pm75, \pm125, \pm375, \pm625$, and ±1875. The company used a spreadsheet to evaluate these values and found that none of these values is a zero of the function. This means that the zeros are not rational values. We can use another method to help you approximate the zeros of this function. This method is called **the location principle**.

The Location Principle	**Suppose $y = f(x)$ represents a polynomial function. If a and b are two numbers with $f(a)$ negative and $f(b)$ positive, the function has at least one real zero between a and b.**

If $f(a) > 0$ and $f(b) < 0$, then the function also has at least one real zero between a and b.

This principle is illustrated by the graph at the right. The graph of $y = f(x)$ is a continuous curve. At $x = a$, $f(a)$ is negative. At $x = b$, $f(b)$ is positive. As we watch the curve go from negative to positive, we see that it must cross the x-axis. Thus, a zero exists somewhere between a and b.

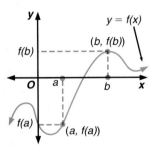

Example 1

APPLICATION

Packaging

Find the dimensions of the box for the 1500-piece puzzle.

When you test the possible rational roots for an equation, you can often narrow down the range of values in which a zero exists. The value of $V(1)$ is –789. The value of $V(3)$ is 1767. A zero must exist somewhere between $V(1)$ and $V(3)$. Let's try other values in that range.

x	V(x)
1	–789
1.5	–199.125
2	423
2.5	1078.125
3	1767

The sign change indicates there is a zero between 2 and 1.5.

You can use the location principle again to fine tune your search for the zero between 2 and 1.5.

x	V(x)
1.5	–199.125
1.6	–77.304
1.7	45.813

The sign change indicates there is a zero between 1.6 and 1.7.

You could again use the location principle to find the zero to the nearest hundredth. However, finding the zero to the nearest tenth is appropriate for the problem. Since $V(1.7)$ is closer to zero than $V(1.6)$, we will use 1.7.

To find the dimensions of the puzzle box, add 1.7 cm to each measurement. The box for the 1500-piece puzzle will be 26.7 cm by 31.7 cm by 6.7 cm.

Example 2

Determine between which consecutive integers the real zeros of
$f(x) = x^3 + 2x^2 - 3x - 5$ **are located.**

By Descartes' rule of signs, the function must have 1 positive real zero and 2 or 0 negative real zeros. Use substitution or synthetic division to evaluate the function for successive integral values of x. Since the greatest possible integral zero is 5 and the least possible integral zero is –5, we will test values in that range until we locate the three possible zeros.

You could also use your graphing calculator to find the range of values in which to find a zero.

r	1	2	-3	-5
-5	1	-3	12	-65
-4	1	-2	5	-25
-3	1	-1	0	-5
-2	1	0	-3	1
-1	1	1	-4	-1
0	1	2	-3	-5
1	1	3	0	-5
2	1	4	5	5

Changes in sign indicate that the three zeros are located between –3 and –2, between –2 and –1, and between 1 and 2. According to Descartes' rule of signs, this result is reasonable.

Example 3

Approximate to the nearest tenth the real zeros of $f(x) = x^4 - 3x^3 - 2x^2 + 3x - 5$**. Then sketch the graph of the function, given that the relative maximum is at** $(0.4, -4.3)$**, and the relative minima are at** $(-0.7, -6.8)$ **and** $(2.5, -17.8)$**.**

By Descartes' rule of signs, the function must have 3 or 1 positive real zeros and 1 negative real zero. The possible rational zeros are ± 1 and ± 5. Use substitution or synthetic division to evaluate the function for several values of x between 5 and –5 until you locate the zeros.

r	1	-3	-2	3	-5
5	1	2	8	43	210
4	1	1	2	11	39
3	1	0	-2	-3	-14
2	1	-1	-4	-5	-15
1	1	-2	-4	-1	-6
0	1	-3	-2	3	-5

r	1	-3	-2	3	-5
-1	1	-4	2	1	-6
-2	1	-5	8	-13	21
-3	1	-6	16	-45	130
-4	1	-7	26	-101	399
-5	1	-8	38	-187	930

By the location principle, there is a negative zero between –1 and –2 and a positive zero between 3 and 4. Evaluate values between these pairs of integers to the nearest tenth.

x	f(x)
-1.1	-5.3
-1.2	-4.2
-1.3	-2.8
-1.4	-1.0
-1.5	1.2

x	f(x)
3.1	-11.9
3.2	-9.3
3.3	-6.1
3.4	-2.2
3.5	2.4

The real zeros are approximately –1.4 and 3.4.

(continued on the next page)

Now sketch the graph using the maximum and minima given and the zeros. You can also use the values you found in evaluating the function to give you other points to sketch the graph.

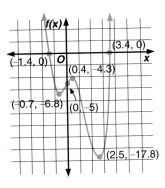

In Example 3, the function $f(x) = x^4 - 3x^3 - 2x^2 + 3x - 5$ had 3 or 1 positive real zeros. Instead of testing greater values of x to determine if additional positive zeros exist, the **upper bound theorem** can be used to determine an **upper bound** of the zeros of a function. An upper bound is the integer greater than or equal to the greatest real zero.

Upper Bound Theorem	**Suppose c is a positive integer and $P(x)$ is divided by $x - c$. If the resulting quotient and remainder have no change in sign, then $P(x)$ has no real zeros greater than c. Thus, c is an upper bound of the zeros of $P(x)$.**

Zero coefficients are ignored when counting sign changes.

The synthetic division in Example 3 indicates that 4 and 5 are upper bounds of the zeros of $f(x) = x^4 - 3x^3 - 2x^2 + 3x - 5$ because there are no change of signs in the quotient and remainder. In this case, the least positive integral upper bound of the zeros is 4.

A **lower bound** of the zeros of $P(x)$ can be found by determining an upper bound for the zeros of $P(-x)$. Therefore, if c is an upper bound of the zeros of $P(-x)$, then $-c$ is a lower bound of the zeros of $P(x)$.

Example 4

Find a lower bound of the zeros of $f(x) = x^4 - 3x^3 - 2x^2 + 3x - 5$.

$f(-x) = x^4 + 3x^3 - 2x^2 - 3x - 5$

r	1	3	–2	–3	–5
1	1	4	2	–1	–6
2	1	5	8	13	21

The last row has no changes in sign, so 2 is an upper bound of $f(-x)$.

Since 2 is an upper bound of $f(-x)$, –2 is a lower bound of $f(x)$.

CHECKING FOR UNDERSTANDING

Communicating Mathematics

Read and study the lesson to answer each question.

1. **Explain** the location principle and why it works.

2. **Tell** why negative zeros of the function $V(x)$ in Example 1 were not determined.

3. **Explain** how to use the location principle to find the real zeros of a function to the nearest hundredth.

4. **Describe** how you find an upper bound and a lower bound for the zeros of a polynomial function.

Guided Practice

Use the synthetic division for each function to determine between which two consecutive x-values the zeros occur. State the location of the real zeros. State the least integral upper bound.

5. $f(x) = x^3 - 3x^2 - 2x + 4$

r	1	-3	-2	4
-2	1	-5	8	-12
-1	1	-4	2	2
0	1	-3	-2	4
1	1	-2	-4	0
2	1	-1	-4	-4
3	1	0	-2	-2
4	1	1	2	12
5	1	2	8	44

6. $f(x) = x^4 - 8x^2 + 10$

r	1	0	-8	0	10
-4	1	-4	8	-32	138
-3	1	-3	1	-3	19
-2	1	-2	-4	8	-6
-1	1	-1	-7	7	3
0	1	0	-8	0	10
1	1	1	-7	-7	3
2	1	2	-4	-8	-6
3	1	3	1	3	19

7. Use substitution or synthetic division to find the zeros of $f(x) = x^2 + 3x + 1$ to the nearest tenth.

8. Find the least integral upper bound and the greatest integral lower bound of the zeros of $f(x) = x^2 - x - 1$.

EXERCISES

Practice

Determine between which consecutive integers the real zeros of each function are located.

9. $f(x) = x^2 - 4x - 2$

10. $f(x) = 2x^2 - 5x + 1$

11. $f(x) = x^3 - 2$

12. $f(x) = x^3 - 3x + 1$

13. $f(x) = x^4 - 2x^3 + x - 2$

14. $f(x) = 2x^4 + x^2 - 3x + 3$

Approximate the real zeros of each function to the nearest tenth.

15. $f(x) = x^2 + 3x + 2$

16. $f(x) = 2x^3 - 4x^2 - 3$

17. $f(x) = x^3 - 4x + 6$

18. $f(x) = 3x^4 + x^2 - 1$

19. $f(x) = 2x^4 - x^3 + x - 2$

20. $f(x) = -x^3 + x^2 - x + 1$

Use the upper bound theorem to find the least integral upper bound and the greatest integral lower bound of the zeros of each function.

21. $f(x) = x^3 + 3x^2 - 5x - 10$

22. $f(x) = x^4 - 8x + 2$

23. $f(x) = 3x^3 - 2x^2 + 5x - 1$

24. $f(x) = x^5 + 5x^4 - 3x^3 + 20x^2 - 15$

For each function, determine the number and type of possible complex zeros. Use the location principle to determine the real zeros to the nearest tenth. Determine the relative maxima and relative minima. Then sketch the graph.

25. $f(x) = 2x^3 + 9x^2 - 12x - 40$

26. $f(x) = x^4 - 3x^2 - 9$

Critical Thinking

27. Suppose the graph of $g(x)$ has two relative maxima and one relative minimum.

   a. What is the minimum degree of $g(x)$?

   b. Draw a graph to support your answer.

Applications and Problem Solving

28. Ecology In the early 1900s, the deer population of the Kaibab Plateau in Arizona experienced a rapid increase because hunters had reduced the number of natural predators. The food supply was not great enough to support the increased population, eventually causing the population to decline. The deer population for the period 1905–1930 can be approximated by the function $f(x) = -0.125x^5 + 3.125x^4 + 4000$, where x is the number of years from 1905.

   a. Graph this function.

   b. Describe the pattern of the deer population in terms of the concepts you have learned in this lesson.

29. Pets Rick Glatfelter built a kennel for his dog by attaching two sides of fencing to the existing corner of the yard to create a rectangle that is 4 meters wide by 6 meters long. As his dog has grown, he wishes to double the kennel's area by increasing each side of the rectangle by the same amount. By how much should the dimensions be increased?

30. Astronomy In 1609, Johannes Kepler discovered two of his three laws that describe the behavior of every planet and satellite in our solar system. His third law, discovered in 1618, states that the ratio of the squares of the periods of any two moons revolving about a planet is equal to the ratio of the cubes of their average distances from the planet. This can be written as $\left(\dfrac{T_a}{T_b}\right)^2 = \left(\dfrac{r_a}{r_b}\right)^3$, where T_a and T_b are the periods of the moons and r_a and r_b are their average distances from the planet. In 1610, Galileo discovered four moons of Jupiter. The first moon, Io, has a period of 1.8 days. Another moon, Ganymede, shown at the left, has a period of 7.3 days and is 6.5 units farther from Jupiter than Io is from Jupiter. Use Kepler's third law to determine the distance of Io and Ganymede from Jupiter.

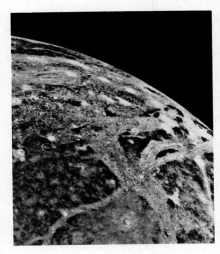

Mixed Review

31. Find the slope of the line whose equation is $3y + 8x = 12$. **(Lesson 1-4)**

32. Graph the parent function and the translated function $f(x) = |x + 2| - 4$. **(Lesson 3-2)**

33. Graph $x^2 + 8x - 20 > 0$. **(Lesson 4-1)**

34. Determine the zeros and critical points of $f(x) = x^3 - 4x^2 - 25x + 28$. Then graph the function. **(Lesson 4-4)**

35. College Entrance Exam If 15 cans of food are needed for seven men for two days, how many are needed for four men for seven days?

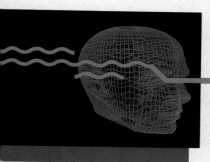

Technology

Locating Zeros and Critical Points

```
100 PRINT
110 PRINT "F(X) = Ax^4+Bx^3+Cx^2+Dx+E"
120 PRINT "ENTER A, B, C, D, E";
130 INPUT A, B, C, D, E
140 PRINT "ENTER STARTING X POINT";
150 INPUT MINIMUM
160 PRINT "ENTER STOPPING X POINT";
170 INPUT MAXIMUM
180 PRINT "ENTER INCREMENT SIZE";
190 INPUT INC
200 PRINT "FINDING ZEROS..."
210 SIGN = 0
220 FOR X = MINIMUM TO MAXIMUM STEP INC
230 F = E+X*(D+X*(C+X*(B+X*A)))
240 OLDSIGN = SIGN
250 IF F = 0 THEN SIGN = 0
260 IF F < 0 THEN SIGN = -1
270 IF F > 0 THEN SIGN = 1
280 IF F = 0 THEN PRINT "ZERO AT X="; X:GOTO 310
290 IF SIGN * OLDSIGN <> -1 THEN GOTO 310
300 PRINT "ZERO BETWEEN X="; X - INC:" AND X=";X
310 NEXT X
320 PRINT "NOW FINDING CRITICAL POINTS..."
330 SIGN = 0
340 FOR X = MINIMUM TO MAXIMUM STEP INC
350 F = D+X*(2*C+X*(3*B+X*4*A))
360 OLDSIGN = SIGN
370 IF F = 0 THEN SIGN = 0
380 IF F < 0 THEN SIGN = -1
390 IF F > 0 THEN SIGN = 1
400 IF F = 0 THEN PRINT "CRITICAL POINT AT X=";X:
    GOTO 430
410 IF SIGN * OLDSIGN <> -1 THEN GOTO 430
420 PRINT "CRITICAL POINT BETWEEN X="; X -INC:"
    AND X=";X
430 NEXT X
440 PRINT "DONE!"
450 END
```

Engineers and scientists use computer programs to find zeros and critical points of large polynomial functions. The programs that they use are powerful, accurate, and fast. A printout of one of these programs might exceed one hundred pages!

The BASIC program shown at the left can help you find zeros and critical points of polynomial functions of degree four or less. This program will either find the zeros or critical points, or it will tell you where possible zeros or critical points may be found. Then you can refine your search to find the exact values.

To execute the program, type **RUN** and then press **[Enter]**. The sample shown below uses $f(x) = x^3 - x^2 - 17x + 20$.

```
RUN

F(X) = Ax^4 + Bx^3 + Cx^2 + Dx + E
ENTER A, B, C, D, E? 0,1,-1,-17,20
ENTER STARTING X POINT? -5
ENTER STOPPING X POINT? 5
ENTER INCREMENT SIZE? .1
FINDING ZEROS...
ZERO BETWEEN X=-4.20000001 AND X=-4.1
ZERO BETWEEN X=1.09999999 AND X=1.19999999
ZERO BETWEEN X=3.99999999 AND X=4.09999999
NOW FINDING CRITICAL POINTS...
CRITICAL POINT BETWEEN X=-2.10000001
AND X=-2.00000001
CRITICAL POINT BETWEEN X=2.69999999
AND X=2.79999999
DONE!
```

EXERCISES

Use the BASIC program to find the zeros and critical points for each polynomial.

1. $f(x) = x^3 - x^2 - 4x + 4$

2. $f(x) = x^4 + x^3 + x^2 - 4x - 20$

3. $f(x) = 3x^3 + 7x - 8$

4-6 Rational Equations and Partial Fractions

Objectives

After studying this lesson, you should be able to:
- find the least common denominator of rational expressions,
- solve rational equations and inequalities, and
- decompose a fraction into partial fractions.

Application

At Wet and Wild Water Park, the Raging River is one of the most popular rides. The ride involves a wide stream of water in which your raft meanders downstream for 3 kilometers in the same amount of time that it takes the park workers to return the rafts directly upstream 2 kilometers to the beginning of the ride. If the raft travels at a rate of 1.5 km/h in still water, what is the rate of the current of the water in the ride?

> **FYI...**
>
> In 1963 and 1964, 70-year-old William Willis of the United States sailed 10,850 miles from Peru to Australia on his raft, the *Age Unlimited* The trip took him 204 days.

Each way on the ride involves the same amount of time. So, solve the formula $d = rt$ for t. Thus, $t = \dfrac{d}{r}$. The current works with the raft going downstream and against the raft going upstream, but the rate, c, of the current is constant. So for the rate downstream, add the rate of the current, and for the rate upstream, subtract the rate of the current.

	d	r	t
downstream	3	$1.5 + c$	$\dfrac{3}{1.5 + c}$
upstream	2	$1.5 - c$	$\dfrac{2}{1.5 - c}$

$$t_d = t_u \qquad t_d = \text{time downstream, } t_u = \text{time upstream}$$

$$\frac{3}{1.5 + c} = \frac{2}{1.5 - c}$$

$$(1.5 + c)(1.5 - c)\left(\frac{3}{1.5 + c}\right) = \left(\frac{2}{1.5 - c}\right)(1.5 + c)(1.5 - c) \qquad \text{\textit{Multiply by the common denominator.}}$$

$$3(1.5 - c) = 2(1.5 + c) \qquad \text{\textit{Simplify each side.}}$$

$$4.5 - 3c = 3 + 2c$$

$$1.5 = 5c$$

$$0.3 = c \qquad \text{The rate of the current is 0.3 km/h.}$$

An equation, like the one above, that consists of one or more rational expressions is called a **rational equation**. One way to solve a rational equation is to multiply each side of the equation by the least common denominator (LCD). The LCD is the least common multiple of the denominators.

You should always check your solutions by substituting them into the original equation. Remember that the denominator in a fraction can never be equal to zero. Any possible solution that results in a zero in the denominator must be excluded from your list of solutions.

Example 1

Solve $\dfrac{t+4}{t} + \dfrac{3}{t-4} = \dfrac{-16}{t^2 - 4t}$.

$$\dfrac{t+4}{t} + \dfrac{3}{t-4} = \dfrac{-16}{t^2 - 4t} \qquad \textit{The LCD is } t(t-4).$$

$$t(t-4)\left(\dfrac{t+4}{t}\right) + t(t-4)\left(\dfrac{3}{t-4}\right) = t(t-4)\left(\dfrac{-16}{t^2 - 4t}\right)$$

$$(t-4)(t+4) + t(3) = -16$$

$$t^2 - 16 + 3t = -16$$

$$t^2 + 3t = 0$$

$$t(t+3) = 0$$

$$t = 0 \text{ or } t = -3$$

When you check your solutions, you find that t cannot equal 0, so $t = 0$ cannot be a solution. Therefore, the only solution is –3.

Example 2

The Northern Hills swimming pool can be filled by a vent in 10 hours. It can be emptied by a drain pipe in 20 hours. The pool manager mistakenly leaves the drain pipe open while trying to fill the pool before Memorial Day. How long will it take to fill the pool?

Let h represent the number of hours it will take to fill the pool.

The fact that it takes 10 hours to fill the pool means that in 1 hour, $\dfrac{1}{10}$ of the pool is filled; 20 hours to drain the pool means that in 1 hour, $\dfrac{1}{20}$ of the pool is drained. A full pool can be represented by 1. The equation that represents this situation is $\dfrac{1}{10}h - \dfrac{1}{20}h = 1$ or $\dfrac{h}{10} - \dfrac{h}{20} = 1$.

$$\dfrac{h}{10} - \dfrac{h}{20} = 1$$

$$20\left(\dfrac{h}{10}\right) - 20\left(\dfrac{h}{20}\right) = 20(1) \qquad \textit{The LCD is 20.}$$

$$2h - h = 20$$

$$h = 20$$

It will take 20 hours to fill the pool.

MATH JOURNAL

Explain why it is always important to examine your solutions when working through a word problem.

When adding or subtracting fractions with unlike denominators, you must first find a common denominator. Suppose you have a rational expression and you want to know what fractions were added or subtracted and resulted in that expression. This process is called decomposing a fraction into **partial fractions**. You can use a rational equation to do this.

Example 3

Decompose $\dfrac{x-11}{x^2-2x-3}$ into partial fractions.

First factor the denominator.

$$\frac{x-11}{x^2-2x-3}=\frac{x-11}{(x-3)(x+1)}$$

Express the factored form as the sum of two fractions using A and B as numerators and the factors as denominators.

$$\frac{x-11}{(x-3)(x+1)}=\frac{A}{x-3}+\frac{B}{x+1}$$

Eliminate the denominators by multiplying each side by the LCD.

$$x-11=A(x+1)+B(x-3)$$

The excluded values for this rational function are –1 and 3. By substituting each of these values into the equation, you can find the values of A and B.

Let x = -1.

$$x-11=A(x+1)+B(x-3)$$
$$-1-11=A(-1+1)+B(-1-3)$$
$$-12=-4B$$
$$3=B$$

Let x = 3.

$$3-11=A(3+1)+B(3-3)$$
$$-8=4A$$
$$-2=A$$

Now substitute the values for A and B into the original fractions.

$$\frac{A}{x-3}+\frac{B}{x+1}=\frac{-2}{x-3}+\frac{3}{x+1}$$

So, $\dfrac{x-11}{x^2-2x-3}=\dfrac{-2}{x-3}+\dfrac{3}{x+1}$.

The processes used in solving rational equations can also be used in determining the solutions of rational inequalities.

Example 4

Solve $\dfrac{(x-3)(x-4)}{(x-5)(x-6)^2}<0$.

Let a number line represent all possible values for x. Plot all values of x that cause $f(x)=\dfrac{(x-3)(x-4)}{(x-5)(x-6)^2}$ to equal 0. Those zeros are 3 and 4. Then plot 5 and 6 as circles since they must be excluded from the list of possible solutions.

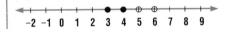

Draw vertical lines to separate the number line as shown below.

Select a sample value within each region and test it in the rational expression to see if it yields a positive or negative result. It is not necessary to find the exact value of the expression.

Test $f(0)$: $\quad \dfrac{(0-3)(0-4)}{(0-5)(0-6)^2} \quad \rightarrow \quad \dfrac{(-)(-)}{(-)(-)(-)} \quad \rightarrow \quad -$

So in the region $x < 3$, $f(x) < 0$. Thus, $x < 3$ is a solution.

Test $f(3.5)$: $\quad \dfrac{(3.5-3)(3.5-4)}{(3.5-5)(3.5-6)^2} \quad \rightarrow \quad \dfrac{(+)(-)}{(-)(-)(-)} \quad \rightarrow \quad +$

In the region $3 < x < 4$, $f(x) > 0$. This is not a solution.

Continue to test a value in each region of the number line to determine all regions in which $f(x) < 0$.

The solution is $x < 3$ or $4 < x < 5$.

Example 5

Solve $\dfrac{1}{4a} + \dfrac{5}{8a} > \dfrac{1}{2}$.

The inequality can be written as $\dfrac{1}{4a} + \dfrac{5}{8a} - \dfrac{1}{2} > 0$. The related function is $f(a) = \dfrac{1}{4a} + \dfrac{5}{8a} - \dfrac{1}{2}$. Find the zeros of this function.

$$\frac{1}{4a} + \frac{5}{8a} - \frac{1}{2} = 0$$

$$8a\left(\frac{1}{4a}\right) + 8a\left(\frac{5}{8a}\right) - 8a\left(\frac{1}{2}\right) = 8a(0) \qquad \textit{The LCD is 8a.}$$

$$2 + 5 - 4a = 0$$

$$a = \frac{7}{4}$$

The value $a = 0$ is a point of discontinuity. Graph the zero and the point of discontinuity on a number line. Draw vertical lines to separate the number line into regions.

Now test a sample value in each region to determine if the values in the region satisfy the inequality.

(continued on the next page)

$$\text{Test } a = -1: \quad \frac{1}{4(-1)} + \frac{5}{8(-1)} \overset{?}{>} \frac{1}{2}$$

$$-\frac{1}{4} - \frac{5}{8} \overset{?}{>} \frac{1}{2}$$

$$-\frac{7}{8} > \frac{1}{2} \quad \textit{false}$$

$$a < 0 \text{ is not a solution.}$$

$$\text{Test } a = 1: \quad \frac{1}{4(1)} + \frac{5}{8(1)} \overset{?}{>} \frac{1}{2}$$

$$\frac{1}{4} + \frac{5}{8} \overset{?}{>} \frac{1}{2}$$

$$\frac{7}{8} > \frac{1}{2} \quad \textit{true}$$

$$0 < a < \frac{7}{4} \text{ is a solution.}$$

$$\text{Test } a = 2: \quad \frac{1}{4(2)} + \frac{5}{8(2)} \overset{?}{>} \frac{1}{2}$$

$$\frac{1}{8} + \frac{5}{16} \overset{?}{>} \frac{1}{2}$$

$$\frac{7}{16} > \frac{1}{2} \quad \textit{false}$$

$$a \geq \frac{7}{4} \text{ is not a solution.}$$

The solution is $0 < a < \frac{7}{4}$.

CHECKING FOR UNDERSTANDING

Communicating Mathematics

Read and study the lesson to answer each question.

1. **Demonstrate** how to find the LCD of two or more fractions.

2. **Explain** why all solutions of a rational equation must be checked.

3. **Describe** what is meant by decomposing into partial fractions.

4. **Compare** the two methods used to solve rational inequalities presented in Examples 4 and 5.

Guided Practice

Find the LCD for the rational expressions in each equation. Solve each equation.

5. $\dfrac{1}{m} = \dfrac{m-34}{2m^2}$

6. $\dfrac{9}{b+5} = \dfrac{3}{b-3}$

7. $\dfrac{7a}{3(a+1)} - \dfrac{5}{4(a-1)} = \dfrac{3a}{2(a+1)}$

Solve each equation. Check your solution.

8. $\dfrac{3}{x} + \dfrac{7}{x} = 8$

9. $1 + \dfrac{5}{a-1} = \dfrac{7}{6}$

10. $\dfrac{3y+2}{4} = \dfrac{9}{4} - \dfrac{3-2y}{6}$

11. Decompose $\dfrac{-x+5}{(x-1)(x+1)}$ into partial fractions.

12. Solve $\dfrac{2}{w}+3>\dfrac{29}{w}$.

EXERCISES

Practice

Solve each equation. Check your solution.

13. $b-\dfrac{5}{b}=4$

14. $\dfrac{6}{p+3}+\dfrac{p}{p-3}=1$

15. $\dfrac{2}{x+2}+\dfrac{3}{x}=\dfrac{-x}{x+2}$

16. $\dfrac{12}{t}+t-8=0$

17. $1=\dfrac{1}{1-y}+\dfrac{y}{y-1}$

18. $\dfrac{1}{3m}+\dfrac{6m-9}{3m}=\dfrac{3m-3}{4m}$

19. $1+\dfrac{n+6}{n+1}=\dfrac{4}{n-2}$

20. $\dfrac{2q}{2q+3}-\dfrac{2q}{2q-3}=1$

Solve each inequality.

21. $5+\dfrac{1}{x}>\dfrac{16}{x}$

22. $\dfrac{2y+1}{5}-\dfrac{2+7y}{15}<\dfrac{2}{3}$

23. $1+\dfrac{5}{a-1}<\dfrac{7}{6}$

24. $\dfrac{2a-5}{6}-\dfrac{a-5}{4}<\dfrac{3}{4}$

25. $x^3-11x^2+18x\geq 0$

26. $x^5+x^4-16x-16<0$

Decompose each expression into partial fractions.

27. $\dfrac{3p-1}{p^2-1}$

28. $\dfrac{-4y}{3y^2-4y+1}$

29. $\dfrac{2m+1}{m^2+m}$

30. Solve $\dfrac{1}{2a-2}=\dfrac{a}{a^2-1}+\dfrac{2}{a+1}$.

31. Solve $\dfrac{x^2-16}{x^2-4x-5}>0$.

32. Tell how you could use the graph of $f(x)=8x^3-22x^2-5x+12$ to solve $8x^3-22x^2-5x+12\geq 0$.

Critical
Thinking

33. Solve $\dfrac{7}{y+1}>7$.

34. A dish of mixed nuts contains cashews and peanuts. Two more ounces of peanuts are added to the dish, making the new mixture 20% cashews. Sarah likes cashews, so she adds 2 ounces of them to the dish. The mixture in the dish is now $33\dfrac{1}{3}$% cashews. What percent of the original mixture of nuts was cashews?

35. **Number Theory** Four times the multiplicative inverse of a number is added to the number. The result is $10\frac{2}{5}$. What is the number?

36. **Candle Making** A beeswax candle is made by repeatedly dipping the wick into the molten wax. Paraffin candles are usually made by pouring the molten wax into forms with wick shafts, and the wick is added as the last step. One type of beeswax candle burns up in 6 hours. Another paraffin candle of the same length burns up in 9 hours. If both candles are lit at the same time, how much time will pass before one is twice as long as the other?

37. **Neighborhoods** Rosea and Tai live on the same street in the same neighborhood in West Branch, Iowa, the birthplace of President Herbert Hoover. When Rosea jogs to Tai's house, it takes 10 minutes. When Tai rides his bike to Rosea's house, it takes 6 minutes. Suppose they both leave their houses at the same time headed for the other person's house. In how many minutes will they meet?

38. Describe the difference between a relation and a function. How do you test a graph to determine if it is the graph of a function? **(Lesson 1-1)**

39. **Education** You may answer up to 30 questions on your final exam in history class. It consists of multiple-choice and essay questions. Two 48-minute class periods have been set aside for taking the test. It will take you 1 minute to answer each multiple-choice question and 12 minutes for each essay question. Correct answers on multiple-choice questions earn 5 points and correct essay answers earn 20 points. If you are confident that you will answer all of the questions you attempt correctly, how many of each type of question should you answer to receive the highest score? **(Lesson 2-6)**

40. Graph the inverse of $f(x) = x^2 - 16$. **(Lesson 3-3)**

41. List all possible rational roots for $6x^3 + 6x^2 - 15x - 2 = 0$. **(Lesson 4-4)**

42. Approximate the real zeros of $h(x) = 3x^3 - 16x^2 + 12x + 6$ to the nearest tenth. **(Lesson 4-5)**

43. **College Entrance Exam** Choose the best answer.
An automobile travels m miles in h hours. At this rate, how far will it travel in x hours?

(A) $\dfrac{m}{x}$ (B) $\dfrac{m}{xh}$ (C) $\dfrac{m}{h}$ (D) $\dfrac{mh}{x}$ (E) $\dfrac{mx}{h}$

4-7A Graphing Calculators: Solving Radical Equations and Inequalities

You can use a graphing calculator to graph and solve many types of equations and inequalities. In Lesson 3-3A, we graphed radical functions on the graphing calculator. In this lesson, we will solve radical equations and inequalities by graphing.

Radical equations can be solved on the graphing calculator by using two different methods, but both methods will produce the same solution. One method is to look at each side of the equation and graph both of them as a separate function. The solution is where the two equations intersect on the graph. The other method is to first solve the equation so that it equals zero and then graph it. In this case, the solution is where the graph crosses the x-axis.

Example 1

Solve $\sqrt{3x + 7} - \sqrt{4x - 3} = 1$.

Try the separate function method. Use the viewing window $[-2, 8]$ by $[-3, 6]$ with a scale factor of 1 for both axes.

Enter: Y= 2nd √ ((3 X,T,θ + 7)) − 2nd √
((4 X,T,θ − 3)) ENTER 1 GRAPH

Trace to the intersection point of the two graphs and zoom in. This method produces a solution of 3.

The TI-82 can also find the intersection of two functions. Press 2nd CALC 5. Use the arrow and ENTER keys to select each function. The cursor will be on the intersection and the coordinates will be displayed.

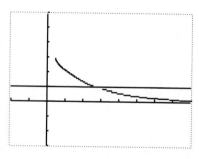

Now try to solve the equation by setting it equal to zero and graphing the equation. The equation is then $\sqrt{3x + 7} - \sqrt{4x - 3} - 1 = 0$.

Enter: Y= 2nd √ ((3 X,T,θ + 7)) − 2nd √
((4 X,T,θ − 3)) − 1 GRAPH

Tracing to the zero of the function and zooming in produces a solution of 3, which is identical to the solution from the first method.

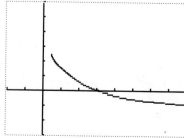

The equation in Example 1 has a solution, namely 3. However, some radical equations do not have a solution in the set of real numbers. This happens when there are no *x*-values that will satisfy the equation. You can tell if an equation has no real solution on a graph if the graph does not cross the *x*-axis.

The TI-82 graphing calculator can also be used to solve radical inequalities by shading the area on the graph that makes the inequality true. Using each side of the inequality as a separate function, use the "Shade(" command and enter the "lesser" expression first as the lower bound and the "greater" expression second as the upper bound.

Example 2

Solve $\sqrt{5x - 3} \le 4$. Round your solution to the nearest hundredth.

First, clear any functions listed in the $\boxed{Y=}$ list and clear the graphics screen by pressing $\boxed{\text{2nd}}$ $\boxed{\text{DRAW}}$ 1 $\boxed{\text{ENTER}}$. Then, enter the inequality. For example, $\sqrt{5x - 3}$ is the lesser expression and 4 is the greater expression.

Enter: $\boxed{\text{2nd}}$ $\boxed{\text{DRAW}}$ 7 $\boxed{\text{2nd}}$ $\boxed{\sqrt{\ }}$ $\boxed{(}$ 5 $\boxed{\text{X,T,}\theta}$ $\boxed{-}$ 3 $\boxed{)}$ $\boxed{,}$ 4 $\boxed{)}$ $\boxed{\text{ENTER}}$

A good graph of the inequality can be seen in the viewing window of [-2, 8] by [-3, 6] with scale factors of 1 for both axes.

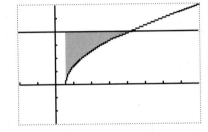

You can use the arrow keys and zoom in to approximate a solution to the inequality. After moving the cursor to the edge of the shade with an arrow key, zoom in. Then press $\boxed{\text{2nd}}$ $\boxed{\text{QUIT}}$ and $\boxed{\text{ENTER}}$ to redraw the graph.

The solution is $0.63 \le x \le 3.79$.

EXERCISES

Graph and solve each equation to the nearest hundredth. Then sketch the graph on a piece of paper.

1. $\sqrt{10x - 1} = 7$

2. $\sqrt{6x - 2} - 1 = 3$

3. $\sqrt{5x - 4} - \sqrt{16x - 7} = -2$

4. $\sqrt{\frac{1}{2}x + 2} + \sqrt{x} = 4$

5. $\sqrt{3x} - \sqrt{\frac{1}{3}x} = 1$

6. $\sqrt{x + 5} + \sqrt{2x + 7} = 2\sqrt{3}$

7. $\sqrt[3]{4x + 13} - 6 = -9$

8. $\sqrt{9 - 3x} + \sqrt{2 - 5x} = -5$

Use a TI-81 graphing calculator to graph each inequality. Then sketch the graph on a piece of paper.

9. $\sqrt{x + 1} \le 3$

10. $\sqrt{2x - 5} \le 6$

11. $\sqrt{4x - 7} \ge 5$

12. $\sqrt[3]{x + 3} \ge -2$

4-7 Radical Equations and Inequalities

Objective

After studying this lesson, you should be able to:
- solve radical equations and inequalities.

Application

The Media Arts class at a local university created a science fiction movie to enter in the state film contest. In their film, Lucky Winslow falls asleep on her space ship and awakens to find that she has landed on a planet. However, she doesn't know whether she has landed on Venus or Mars. The Media Arts class sought the help of the physics classes to add authenticity to their film. They asked how Lucky could figure out where she had landed.

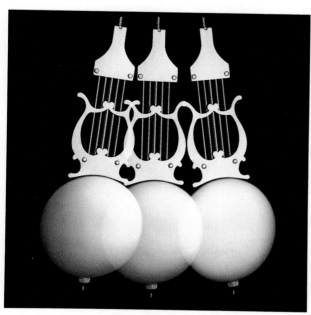

The physics class told them that if Lucky could make a pendulum, she could use a formula to determine the acceleration of gravity of the planet and then use her charts to discover where she was. Suppose Lucky makes a 1-meter pendulum. When she allows the pendulum to oscillate, the period of the oscillation is 3.27 seconds. The local acceleration due to gravity can be determined by the formula,

$T = 2\pi\sqrt{\dfrac{\ell}{g}}$, where T is the period of the oscillation

of the pendulum, ℓ is the length of the pendulum, and g is acceleration due to gravity. Suppose the acceleration due to gravity on the surface of Mars is 3.7 m/s^2 and on the surface of Venus is 8.9 m/s^2. On which planet did Lucky land?

$$T = 2\pi\sqrt{\dfrac{\ell}{g}}$$

$$3.27 = 2\pi\sqrt{\dfrac{1}{g}} \qquad \textit{T = 3.27, \ell = 1}$$

$$\dfrac{3.27}{2\pi} = \sqrt{\dfrac{1}{g}} \qquad \textit{Divide each side by } 2\pi.$$

$$\left(\dfrac{3.27}{2\pi}\right)^2 = \left(\sqrt{\dfrac{1}{g}}\right)^2 \qquad \textit{Square each side.}$$

$$\left(\dfrac{3.27}{2\pi}\right)^2 = \dfrac{1}{g}$$

$$g = \dfrac{1}{\left(\dfrac{3.27}{2\pi}\right)^2}$$

$$g = 3.692021585 \qquad \textit{Use your calculator.}$$

Since this number rounds to 3.7, Lucky must be on Mars.

Equations, such as $T = 2\pi\sqrt{\dfrac{\ell}{g}}$, in which the variables occur in radical expressions are known as **radical equations**. To solve radical equations, you must isolate the radical on one side of the equation and raise each side of the equation to the proper power to eliminate the radical expression.

The process of raising each side of an equation to a power sometimes produces extraneous solutions. These are solutions that do not satisfy the original equation. Therefore, it is important to check all possible solutions in the original equation to determine if any of them should be eliminated from the solution set.

Example 1

Solve $5 + \sqrt{x - 4} = 2$.

$$5 + \sqrt{x - 4} = 2$$
$$\sqrt{x - 4} = -3 \qquad \textit{Isolate the radical.}$$
$$x - 4 = 9 \qquad \textit{Square each side.}$$
$$x = 13$$

Now check the possible solution.   **Check:** $5 + \sqrt{x - 4} = 2$
$$5 + \sqrt{13 - 4} \stackrel{?}{=} 2 \qquad x = 13$$
$$5 + \sqrt{9} \stackrel{?}{=} 2$$
$$5 + 3 \neq 2$$

The solution does not check. The equation has no real solutions.

Example 2

Solve $3 = \sqrt[3]{x + 4} + 12$.

$$3 = \sqrt[3]{x + 4} + 12$$
$$-9 = \sqrt[3]{x + 4} \qquad \textit{Isolate the cube root.}$$
$$(-9)^3 = \left(\sqrt[3]{x + 4}\right)^3 \qquad \textit{Cube each side.}$$
$$-729 = x + 4$$
$$-733 = x \qquad \textit{-733 is a possible solution.}$$

Check : $3 = \sqrt[3]{x + 4} + 12$
$$3 \stackrel{?}{=} \sqrt[3]{-733 + 4} + 12 \qquad x = -733$$
$$3 \stackrel{?}{=} \sqrt[3]{-729} + 12$$
$$3 \stackrel{?}{=} -9 + 12$$
$$3 = 3$$

The solution checks. The solution of the equation is –733.

If there is more than one radical in an equation, you may need to repeat the process for solving radical equations until all radicals have been eliminated.

Example 3

Solve $\sqrt{3x + 4} - \sqrt{2x - 7} = 3$.

$$\sqrt{3x + 4} - \sqrt{2x - 7} = 3$$

$$\sqrt{3x + 4} = 3 + \sqrt{2x - 7} \qquad \textit{Isolate one of the radicals.}$$

$$3x + 4 = 9 + 6\sqrt{2x - 7} + 2x - 7 \qquad \textit{Square each side.}$$

$$x + 2 = 6\sqrt{2x - 7} \qquad \textit{Simplify.}$$

$$x^2 + 4x + 4 = 36(2x - 7) \qquad \textit{Square each side.}$$

$$x^2 - 68x + 256 = 0 \qquad \textit{Simplify.}$$

$$x = \frac{-(-68) \pm \sqrt{(-68^2 - 4(1)(256)}}{2(1)} \qquad \textit{Use the quadratic formula.}$$

$$x = 4 \text{ or } 64$$

Check both possible solutions.

Check $x = 4$:

$$\sqrt{3x + 4} - \sqrt{2x - 7} = 3$$

$$\sqrt{3(4) + 4} - \sqrt{2(4) - 7} \stackrel{?}{=} 3$$

$$\sqrt{16} - \sqrt{1} \stackrel{?}{=} 3$$

$$4 - 1 \stackrel{?}{=} 3$$

$$3 = 3$$

Check $x = 64$:

$$\sqrt{3x + 4} - \sqrt{2x - 7} = 3$$

$$\sqrt{3(64) + 4} - \sqrt{2(64) - 7} \stackrel{?}{=} 3$$

$$\sqrt{196} - \sqrt{121} \stackrel{?}{=} 3$$

$$14 - 11 \stackrel{?}{=} 3$$

$$3 = 3$$

Both solutions check. The solutions are 4 and 64.

To determine the solution set of radical inequalities, use the same procedures that you used to solve radical equations.

Example 4

Solve $\sqrt{5x + 4} \leq 8$.

$$\sqrt{5x + 4} \leq 8$$

$$5x + 4 \leq 64 \qquad \textit{Square each side.}$$

$$5x \leq 60$$

$$x \leq 12$$

In order for $\sqrt{5x + 4}$ to be a real number, $5x + 4$ must be greater than or equal to zero.

$$5x + 4 \geq 0$$

$$5x \geq -4$$

$$x \geq -0.8$$

So, the solution is $-0.8 \leq x \leq 12$.

You can check the solution by substituting values of x from each interval into the original inequality.

Example 5

APPLICATION

History

Argentine cowboys called gauchos rode the Argentine plains in the 18th and 19th centuries. One of the tools used by the gauchos was the bola. Bolas were made of three iron balls or stones attached to strings that met at the other end. The gauchos would use these bolas to entangle their prey's feet. If a gaucho swung his bola over his head forming a circle, the relationship between the period of the revolution T, the mass, m, of the ball, the length of the string, r, and the force, F (in newtons), exerted on the string could be expressed as $T = 2\pi \sqrt{\dfrac{mr}{F}}$. Suppose one of the balls of the bola has a mass of 0.2 kg and the string is 0.8 meters long. If you wanted the bola to make at least one complete revolution every second, what is the minimum force exerted on the string?

$$T \geq 2\pi \sqrt{\dfrac{mr}{F}}$$

$$1 \geq 2\pi \sqrt{\dfrac{(0.2)(0.8)}{F}} \qquad \begin{array}{l} T = 1, \\ m = 0.2, \\ r = 0.8 \end{array}$$

$$\dfrac{1}{2\pi} \geq \sqrt{\dfrac{0.16}{F}}$$

$$\dfrac{1}{4\pi^2} \geq \dfrac{0.16}{F} \qquad \textit{Square each side.}$$

$$F \geq (0.16)(4\pi^2)$$

$$F \geq 6.316546817$$

The force must be at least 6.32 newtons.

CHECKING FOR UNDERSTANDING

Communicating Mathematics

Read and study the lesson to answer each question.

1. Refer to the application at the beginning of the lesson.
   a. **Determine** what the period (in oscillations/second) of a 1-meter pendulum is on Earth if the acceleration due to gravity at Earth's surface is 9.8 m/s^2.
   b. **Calculate** the period of the pendulum if Lucky had landed on Venus.

2. **Demonstrate** how you would solve $\sqrt{x + 5} = 2$.

3. **Explain** the difference between solving an equation with one radical and solving an equation with more than one radical.

4. What should be the last step in solving any radical equation or inequality?

Guided Practice

Solve each equation. Check your solutions.

5. $\sqrt{x + 8} - 5 = 0$

6. $\sqrt[3]{y - 7} = 4$

7. $\sqrt[4]{3x} - 2 = 0$

EXERCISES

Practice

Solve each equation. Check your solution.

8. $\sqrt{8n-5}-1=2$

9. $\sqrt{1-4t}=2$

10. $\sqrt[4]{7v-2}+12=7$

11. $\sqrt[3]{6u-5}+2=\text{-}3$

12. $\sqrt{6x-4}=\sqrt{2x+10}$

13. $\sqrt{9u-4}=\sqrt{7u-20}$

14. $\sqrt{k+9}-\sqrt{k}=\sqrt{3}$

15. $\sqrt{x+10}+\sqrt{x-6}=8$

16. $\sqrt{x+2}-7=\sqrt{x+9}$

17. $\sqrt{4x^2-3x+2}-2x-5=0$

18. $\sqrt{x+4}+\sqrt{x-3}=7$

19. $\sqrt{x-9}-\sqrt{x+7}=2$

Solve each inequality. Check your solution.

20. $\sqrt{x+4}\le 6$

21. $\sqrt{2x-7}\ge 5$

22. $\sqrt[3]{3x-8}\ge 1$

23. $\sqrt[4]{5x-9}\le 2$

Solve each equation. Check your solution.

24. $\sqrt{3x+10}=\sqrt{x+11}-1$

25. $\sqrt{2x+1}+\sqrt{2x+6}=5$

Critical Thinking

26. Physics Refer to the application at the beginning of the lesson.
   a. On Earth, if a pendulum has a length of 0.5 meters, what is its period? If the length of the pendulum is doubled, does the period double? Explain.
   b. Suppose you doubled the period you found in part a. What would be the length of the new pendulum and how does it relate to the original length?
   c. Suppose you triple the period you found in part a. What would be the length of the new pendulum and how does it relate to the original length?
   d. Write a conjecture about the relationship of the original pendulum length and the new pendulum length for n times the period T.

Applications and Problem Solving

27. Physics Acceleration due to gravity on Earth is 9.8 m/s^2. If a pendulum has an oscillation period on Earth of 1.6 seconds, determine the length of the pendulum.

28. Demolition A wrecking ball suspended from a crane is a type of pendulum. Determine the force on a 14-meter chain that is attached to a 60-kilogram ball, if the ball moves at a rate of 0.5 revolutions/second.

Mixed Review

29. Write $y=\text{-}7x+2$ in standard form. **(Lesson 1-5)**

30. Find the determinant of $\begin{bmatrix} 5 & \text{-}4 \\ 8 & 2 \end{bmatrix}$. **(Lesson 2-3)**

31. Find the critical points of the graph of $y=x^3-3x+5$. Determine if each point is a maximum, a minimum, or a point of inflection. **(Lesson 3-7)**

32. Solve $\dfrac{x-4}{x-2}=\dfrac{x-2}{x+2}+\dfrac{1}{x-2}$. **(Lesson 4-6)**

33. College Entrance Exam Choose the best answer.
John is now three times Pat's age. Four years from now John will be x years old. In terms of x, how old is Pat now?

(A) $\dfrac{x+4}{3}$ **(B)** $3x$ **(C)** $x+4$ **(D)** $x-4$ **(E)** $\dfrac{x-4}{3}$

VOCABULARY

Upon completing this chapter, you should be familiar with the following terms:

187 completing the square
179 complex number
190 conjugate
203 Decartes' rule of signs
178 degree
197 depressed polynomial
189 discriminant
197 factor theorem
179 fundamental theorem
    of algebra

179 imaginary number
202 integral root theorem
210 location principle
212 lower bound
217 partial fraction
179 polynomial equation
179 polynomial function
178 polynomial in one
    variable
188 quadratic formula

191 quadratic inequality
226 radical equation
216 rational equation
201 rational root theorem
196 remainder theorem
179 root
196 synthetic division
212 upper bound
212 upper bound theorem
179 zero

SKILLS AND CONCEPTS

OBJECTIVES AND EXAMPLES

Upon completing this chapter, you should be able to:

■ determine roots of polynomial equations **(Lesson 4-1)**

Solve $x^2 - 3x = 0$.

$$x^2 - 3x = 0$$
$$x(x - 3) = 0$$
$$x = 0 \text{ or } x - 3 = 0$$
$$x = 3$$

The roots are 0 and 3.

■ solve quadratic equations **(Lesson 4-2)**

Solve $3x^2 - 2x - 5 = 0$ using the quadratic formula.

$$x = \frac{-b \pm \sqrt{b^2 - 4ac}}{2a}$$
$$= \frac{-(-2) \pm \sqrt{(-2)^2 - 4(3)(-5)}}{2(3)}$$
$$= \frac{2 \pm \sqrt{64}}{6} \text{ or } \frac{2 \pm 8}{6}$$
$$x = \frac{10}{6} \text{ or } \frac{5}{3} \qquad x = \frac{-6}{6} \text{ or } -1$$

REVIEW EXERCISES

Use these exercises to review and prepare for the chapter test.

Solve each equation and sketch a graph of the related function.

1. $a + 4 = 0$
2. $0 = t^2 + 6t + 9$
3. $6y^2 + y - 2 = 0$
4. $x^3 + 2x^2 - 3x = 0$

Find the discriminant of each equation and describe the nature of the roots of the equation. Then solve each equation using the quadratic formula.

5. $2x^2 - 7x - 4 = 0$
6. $3m^2 - 10m + 5 = 0$
7. $4a^2 + a + 4 = 0$
8. $-2y^2 + 3y + 8 = 0$

■ find the factors of polynomials using the remainder and factor theorems **(Lesson 4-3)**

Use the Remainder Theorem to determine if $x + 3$ is a factor of $P(x)$ if $P(x) = x^3 + 2x^2 - 5x - 9$.

$$P(-3) = (-3)^3 + 2(-3)^2 - 5(-3) - 9$$
$$= -3$$

The remainder is not 0, so $x + 3$ is not a factor of $P(x)$.

Find the remainder for each division. Is the divisor a factor of the polynomial?

9. $(x^3 - x^2 - 10x - 8) \div (x + 2)$

10. $(2x^3 - 5x^2 + 7x + 1) \div (x - 5)$

11. $(4x^3 - 7x + 1) \div (x + \frac{1}{2})$

12. $(x^4 - 10x^2 + 9) \div (x - 3)$

■ determine the number of positive and negative real zeros a polynomial function has **(Lesson 4-4)**

$f(x) = 3x^4 - 9x^3 + 4x - 6$

Since $f(x)$ has 3 sign changes, there are 3 or 1 positive real zeros.

$f(-x) = 3x^4 + 9x^3 - 4x - 6$

Since $f(-x)$ has 1 sign change, there is 1 negative real zero.

State the number of complex zeros for each function. Then find the number of possible positive real zeros and the number of possible negative real zeros. Determine all of the rational zeros.

13. $f(x) = x^3 - x^2 - 34x - 56$

14. $f(x) = 2x^3 - 11x^2 + 12x + 9$

15. $f(x) = x^4 - 13x^2 + 36$

16. $f(x) = x^4 + x^3 - 9x^2 - 17x - 8$

■ approximate the real zeros of a polynomial function **(Lesson 4-5)**

Determine between which successive integers the real zeros of $f(x) = x^3 + 4x^2 + x - 2$ are located.

r	1	4	1	-2	
-4	1	0	1	-6	] *a zero*
-3	1	1	-2	4	
-2	1	2	-3	4	
-1	1	3	-2	0	← *a zero*
0	1	4	1	-2	] *a zero*
1	1	5	6	4	

One zero is -1. Another is located between -4 and -3. The other is between 0 and 1.

Determine between which successive integers the real zeros of each function are located.

17. $g(x) = x^2 - 3x - 3$

18. $f(x) = x^3 - x^2 + 1$

19. $g(x) = 4x^3 + x^2 - 11x + 3$

20. $f(x) = x^4 - 9x^3 + 25x^2 - 24x + 6$

■ solve rational equations and inequalities
(Lesson 4-6)

Solve $\dfrac{1}{9} + \dfrac{1}{2a} = \dfrac{1}{a^2}$

$\dfrac{1}{9} + \dfrac{1}{2a} = \dfrac{1}{a^2}$

$(18a^2)\left(\dfrac{1}{9} + \dfrac{1}{2a}\right) = \left(\dfrac{1}{a^2}\right)(18a^2)$

$2a^2 + 9a = 18$

$2a^2 + 9a - 18 = 0$

$(2a - 3)(a + 6) = 0$

$a = \dfrac{3}{2}$ or $a = -6$

Solve each equation or inequality. Check your solution.

21. $n + 5 = \dfrac{6}{n}$

22. $\dfrac{5}{6} - \dfrac{2m}{2m + 3} = \dfrac{19}{6}$

23. $3x^2 - x - 4 \geq 0$

24. $\dfrac{2}{x + 1} - 1 < \dfrac{1}{6}$

■ solve radical equations and inequalities
(Lesson 4-7)

Solve $9 + \sqrt{x - 1} = 1$.

$9 + \sqrt{x - 1} = 1$

$\sqrt{x - 1} = -8$

$x - 1 = 64$

$x = 65$

Solve each equation or inequality. Check your solution.

25. $5 - \sqrt{x + 2} = 0$

26. $\sqrt[3]{4a - 1} - 3 = 0$

27. $\sqrt{x + 8} - \sqrt{x + 35} = -3$

28. $\sqrt{x - 5} - 7 < 0$

APPLICATIONS AND PROBLEM SOLVING

29. Sports Shalonda hit a home run that traveled in a path whose height is described by the function $h(x) = -0.003x^2 + x + 4$, where x represents the number of feet the ball has traveled from the plate and $h(x)$ represents the height of the ball. If the scoreboard is 30 feet high and 410 feet from the plate, show that Shalonda's home run ball did not hit the scoreboard. **(Lesson 4-2)**

30. Art James wants to paint a watercolor for his living room. He has 6 feet of framing material to frame the finished painting. What should the dimensions of his canvas be for the painting to be of maximum area? **(Lesson 4-5)**

Solve each equation or inequality. Check your solution.

1. $n^2 - 5n + 4 = 0$

2. $6z^3 - 7z^2 - 3z = 0$

3. $2a^2 - 5a + 4 = 0$

4. $\dfrac{1}{80} + \dfrac{1}{a} = \dfrac{1}{10}$

5. $3y^2 + 4y - 15 \leq 0$

6. $\dfrac{5}{x+2} > \dfrac{5}{x} + \dfrac{2}{3x}$

7. $\sqrt{y - 2} - 3 = 0$

8. $\sqrt{2x + 2} = \sqrt{3x - 5}$

9. $\sqrt{11 - 10m} > 9$

10. Graph $y \leq -3x^2 + 11x + 4$.

Divide using synthetic division.

11. $(2x^3 - 3x^2 + 3x - 4) \div (x - 2)$

12. $(x^4 - 5x^3 - 13x^2 + 53x + 60) \div (x + 1)$

Use the remainder theorem to find the remainder of each division. State whether the binomial is a factor of the polynomial.

13. $(x^3 + 8x^2 + 2x - 11) \div (x + 2)$

14. $(4x^4 - 2x^2 + x - 3) \div (x - 1)$

State the number of complex zeros for each function. Then find the number of possible positive real zeros and the number of possible negative zeros. Determine all of the rational zeros.

15. $f(x) = 6x^3 + 11x^2 - 3x - 2$

16. $p(x) = x^4 + x^3 - 9x^2 - 17x - 8$

17. $h(x) = x^4 - 3x^3 - 53x^2 - 9x$

18. $g(x) = 8x^3 - 36x^2 + 22x + 21$

Decompose each expression into partial fractions.

19. $\dfrac{5z - 11}{2z^2 + z - 6}$

20. $\dfrac{7x^2 + 18x - 1}{(x^2 - 1)(x + 2)}$

Approximate the real zeros of each function to the nearest tenth.

21. $g(x) = x^2 - 3x - 3$

22. $f(x) = x^3 - x + 1$

23. Find the greatest integral lower bound of the zeros of $f(x) = x^3 + 3x^2 - 5x - 10$.

24. Manufacturing   The volume of a fudge tin must be 120 cubic centimeters. The tin is 7 centimeters longer than it is wide and six times longer than it is tall. Find the dimensions of the tin.

25. Travel   A car travels 300 km in the same time that a freight train travels 200 km. The speed of the car is 20 km/h more than the speed of the train. Find the speed of the freight train.

Bonus

Determine the values of k such that $x^3 - 4x^2 - 2x + k = 0$ has a root between –1 and –2.

Given that x is an integer, state the relation representing each of the following by listing a set of ordered pairs. Then, state whether the relation is a function. Write *yes* or *no*. (Lesson 1-1)

1. $y = 3x + 1$ and $-1 \le x \le 3$
2. $y = |2 - x|$ and $-2 < x \le 2$

Find $[f \circ g](x)$ and $[g \circ f](x)$.   (Lesson 1-2)

3. $f(x) = 2x - 1$
   $g(x) = x + 3$
4. $f(x) = 4x^2$
   $g(x) = -2x^3$
5. $f(x) = x^2 - 25$
   $g(x) = 2x - 4$

Find the zero of each function. (Lesson 1-3)

6. $f(x) = 4x - 10$
7. $f(x) = 15x$
8. $f(x) = 0.75x + 3$

Graph each equation or inequality. (Lesson 1-3)

9. $4 - 3x = y$
10. $x + 3y < 12$
11. $y \ge -\frac{2}{3}x + 5$

Find the distance between the points with the given coordinates. Then, write the slope-intercept form of the equation of the line through those points. (Lessons 1-4, 1-5)

12. $(2, 5), (-1, 3)$
13. $(-3, 2), (0, -5)$
14. $(4, -4), (6, -10)$

Write the standard form of the equation of each line described below.   (Lesson 1-6)

15. parallel to $y = 3x - 1$,
    passes through $(-1, 4)$
16. perpendicular to $2x - 3y = 6$,
    x-intercept 2

Solve each system of equations. (Lesson 2-1)

17. $y = -4x$
    $x - y = 5$
18. $x + y = 12$
    $2x - y = -4$
19. $3x - 2y = 10$
    $4x + 3y = 2$

Use matrices A, B, C, and D to find each sum, difference, or product. (Lessons 2-2, 2-3)

$$A = \begin{bmatrix} 6 & 2 \\ 3 & -3 \end{bmatrix} \quad B = \begin{bmatrix} -4 & 6 \\ 5 & 7 \end{bmatrix} \quad C = \begin{bmatrix} 3 & 2 & -1 \\ -5 & -8 & 1 \end{bmatrix}$$

$$D = \begin{bmatrix} 2 & 0 \\ 6 & -3 \\ -5 & -1 \end{bmatrix} \quad E = \begin{bmatrix} -3 & 1 & 5 \\ -1 & -4 & -2 \\ 3 & 2 & -1 \end{bmatrix}$$

20. $A + B$
21. $2A - B$
22. CD
23. $AB + CD$
24. Evaluate the determinant of matrix A.
25. Evaluate the determinant of matrix E.
26. Find the inverse of matrix B.

Solve each system of equations by using augmented matrices.   (Lesson 2-4)

27. $x - 3y + z = 0$
    $x + 3y - z = 2$
    $-2x + 6y + 2z = 1$
28. $2x + 2y - 2z = 3$
    $3x - y + z = 2$
    $x - 3y - z = 5$

Solve each system of inequalities by graphing and name the coordinates of the vertices of each polygonal convex set. Then, find the maximum and minimum values for the function $f(x, y) = 2y - 2x - 3$. (Lesson 2-5)

29. $x \ge 0$
    $y \ge 0$
    $2y + x \le 1$
30. $x \ge 2$
    $y \ge -3$
    $y \le 5 - x$
    $y + 2x \le 8$

Determine whether each function is an even function, an odd function, or neither. (Lesson 3-1)

31. $y = -3x^3$

32. $y = 2x^4 - 5$

33. $y = x^3 + 3x^2 - 6x - 8$

Use the graph of $f(x) = x^3$ to sketch a graph for each function. Then, describe the transformations that have taken place in the related graphs. (Lesson 3-2)

34. $y = -f(x)$

35. $y = f(x - 2)$

Find the inverse of each function. Sketch the function and its inverse. Is the inverse a function? Write *yes* or *no*. (Lesson 3-3)

36. $y = \frac{1}{2}x - 5$

37. $y = (x - 1)^3 + 2$

Determine any horizontal, vertical, or slant asymptotes or holes in the graph of each function. Then, graph each function. (Lesson 3-4)

38. $y = \frac{x}{(2x + 1)(x + 2)}$

39. $y = \frac{x^2 - 9}{x + 3}$

Graph each inequality. (Lesson 3-5)

40. $y \leq |x + 3|$

41. $y > \sqrt[3]{x + 4}$

Find the equation of the line tangent to the graph of each function at the given point. Write the equation in slope-intercept form. (Lesson 3-6)

42. $y = x^2 + 3x - 7, (1, -3)$

43. $y = -2x^2 + 5x + 1, (-1, -6)$

Find the critical points for each function. Then, determine whether each point is a *maximum*, a *minimum*, or a *point of inflection*. (Lesson 3-7)

44. $f(x) = 2 + x + x^2$

45. $f(x) = x^3 - 3x + 4$

Determine whether each graph has *infinite discontinuity, jump discontinuity, point discontinuity*, or is *continuous*. Then, graph each function. (Lesson 3-8)

46. $y = \frac{x^2 - 1}{x + 1}$

47. $y = \begin{cases} x - 1, & \text{if } x < 0 \\ x - 3, & \text{if } x \geq 0 \end{cases}$

Solve each equation or inequality. (Lessons 4-1, 4-2, 4-6, 4-7)

48. $x^2 - 8x + 16 = 0$

49. $4x^2 - 4x - 10 = 0$

50. $\frac{x + 2}{4} + \frac{x - 3}{4} = 6$

51. $2 + \frac{1}{x - 1} > \frac{1}{2}$

52. $9 + \sqrt{x - 1} = 1$

53. $\sqrt{x + 8} - \sqrt{x + 35} \leq -3$

Use the remainder theorem to find the remainder for each division. (Lesson 4-3)

54. $(x^2 - x + 4) \div (x - 6)$

55. $(2x^3 - 3x + 1) \div (x - 2)$

Find the number of possible positive real zeros and the number of possible negative real zeros. Determine all of the rational zeros. (Lesson 4-4)

56. $f(x) = 3x^2 + x - 2$

57. $f(x) = x^4 + x^3 - 2x^2 + 3x - 1$

Approximate the real zeros of each function to the nearest tenth. (Lesson 4-5)

58. $f(x) = x^2 - 2x - 5$

59. $f(x) = x^3 + 4x^2 + x - 2$

UNIT 2

TRIGONOMETRY

The simplest definition of trigonometry would go something like this: Trigonometry is the study of angles and triangles. The word *trigonometry* literally means "triangle measurement." However, the study of trigonometry is much more complex and comprehensive than that. In this unit alone, you will find trigonometric functions of angles, solve triangles, find the area of triangles, verify trigonometric identities, solve trigonometric equations, graph the trigonometric functions, and use vectors to solve parametric equations and to model motion.

Trigonometry has applications in construction, geography, physics, acoustics, medicine, meteorology, and navigation, among other fields. As a result, you will very likely use trigonometry in your chosen field, no matter what it may be. Such was the case in ancient times. Navigators, surveyors, and astronomers have used triangles to measure distances for thousands of years. Ancient Egyptian papyri and Babylonian clay tablets, dating from around 1600 B.C., also show evidence of practical problem solving by using triangle measurement.

As you can see, trigonometry has many applications to many fields of study. As you study this unit, it is our hope that you will become more familiar with the opportunities that await you when you master this powerful area of mathematics.

CHAPTER 5

THE TRIGONOMETRIC FUNCTIONS

HISTORICAL SNAPSHOT

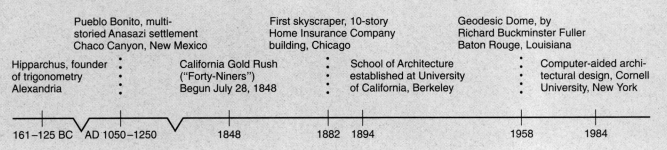

Hipparchus, founder
of trigonometry
Alexandria

Pueblo Bonito, multi-
storied Anasazi settlement
Chaco Canyon, New Mexico

California Gold Rush
("Forty-Niners")
Begun July 28, 1848

First skyscraper, 10-story
Home Insurance Company
building, Chicago

School of Architecture
established at University
of California, Berkeley

Geodesic Dome, by
Richard Buckminster Fuller
Baton Rouge, Louisiana

Computer-aided archi-
tectural design, Cornell
University, New York

161–125 BC   AD 1050–1250   1848   1882 1894   1958   1984

CHAPTER OBJECTIVES

In this chapter, you will:
- Measure angles in radians and degrees.
- Evaluate the trigonometric functions.
- Solve triangles and find the area of triangles.

CAREER GOAL: Construction Engineering Management

Marlene Watson is pursuing a master's degree in construction engineering and management. She would like to become registered as both an architect and a civil engineer. "I eventually intend to become a construction engineering manager for an architectural/engineering firm. This will allow me to be involved in all phases of building construction, including the initial design phase, the final design, budgeting, scheduling, and final construction.

Marlene was inspired to choose this area of study because she has always been interested in traditional Native American architecture. "Most people think Indians only lived in teepees and huts. But Native American architecture was structurally designed for adaption to each group's respective environment. For example, the Miwok Indians of California designed their pithouse to provide protection during earthquakes. The beams rested in Y-shaped supports with enough overlap to allow the roof to move back and forth during an earthquake."

Marlene took courses in algebra, geometry, and trigonometry in high school. She says that "architects use trigonometry a great deal in building design. For example, it is used to measure sunlight and shade at different times of each day of the year in a given geographical location. We use a sun-angle calculator to measure the azimuth angle and bearing angle for a building. Bearing angles indicate where a building will cast a shadow at a given time of the day. Applying trigonometric functions to the bearing angle determines the length of the building's shadow."

Marlene is the only American Indian to have graduated from her high school and gone on to Berkeley. She says, "I want to help change that by serving as a role model for American Indian students. I want them to know that mathematics and science have always been a significant part of Native American cultures."

*inter*NET CONNECTION

For up-to-date information on civil engineering, visit:
www.glencoe.com/sec/math/amc/mathnet

SPOTLIGHT ON...

Marlene Watson
Age: 30
Class: Graduate
School: University of California
   Berkeley, California
Majors: Architecture and Civil
   Engineering

"Mathematics and science have always been a significant part of Native American cultures."

FOR MORE INFORMATION

If you would like more information about architecture or civil engineering, contact:

American Institute of
   Architects
1735 New York Avenue NW
Washington, DC 20006

American Society of Civil
   Engineers (ASCE)
345 E. 47th Street
New York, NY 10017

5-1 Angles and Their Measure

Objectives

After studying this lesson, you should be able to:
- change from radian to degree measure, and vice versa,
- find angles that are coterminal with a given angle, and
- find the reference angle for a given angle.

Application

In order to locate every point on Earth, cartographers use a grid that contains circles through the poles, called longitude lines, and circles parallel to the equator, called latitude lines. Point P is located by traveling north from the equator through a central angle of $a°$ to a circle of latitude and then west along that circle through an angle of $b°$. If you consult an atlas, you will see, for example, that Houston, Texas, is located at north latitude $29°45'26''$ and west longitude $95°21'37''$.

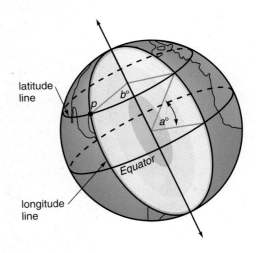

An angle may be generated by the rotation of two rays that share a fixed endpoint. Let one ray remain fixed to form the *initial side* of the angle, and let the second ray rotate to form the *terminal side*. If the rotation is in a counterclockwise direction, the angle formed is a *positive angle*. If the rotation is clockwise, it is a *negative angle*. An angle with its vertex at the origin and its initial side along the positive x-axis is said to be in **standard position**. If the terminal side of an angle in standard position coincides with one of the axes, the angle is called a **quadrantal angle**.

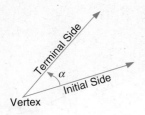

In the figures below, all of the angles are in standard position. The measure of angle A is positive, angle B has a negative measure, and angle C is a quadrantal angle with a positive measure.

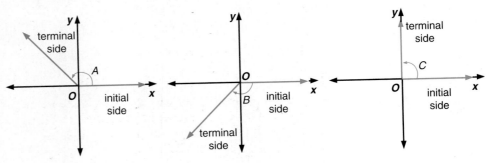

The two most common units used to measure angles are **degrees** and **radians**. You may be more familiar with degree measure. An angle has a measure of one degree (written 1°) if it results from $\frac{1}{360}$ of a complete revolution in the positive direction. Each degree is comprised of 60 **minutes** (written 60′) and each minute is comprised of 60 **seconds** (written 60″). The latitude of Houston, Texas, shown in the application at the beginning of the lesson, would be read "29 degrees, 45 minutes, 26 seconds."

Example 1

Change 29°45′26″ to a decimal number of degrees to the nearest thousandth.

$$29°45′26″ = 29° + 45'\left(\frac{1°}{60'}\right) + 26''\left(\frac{1°}{3600''}\right)$$
$$= 29.757°$$

Some scientific calculators have keys that allow you to change automatically from degrees, minutes, and seconds to decimal values of degrees, and vice versa. Check your calculator's instruction booklet to see if your calculator does this.

Another unit of angle measure is the radian. The definition of a radian is based on the concept of a **unit circle**. A unit circle is a circle of radius 1 whose center is at the origin of a rectangular coordinate system. The unit circle is symmetric with respect to the x-axis, the y-axis, and the origin. A point $P(x, y)$ is on the unit circle if and only if its distance from the origin is 1. Thus, for each point $P(x, y)$ on the unit circle, the distance from the origin is represented by the following equation.

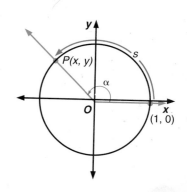

$$\sqrt{(x - 0)^2 + (y - 0)^2} = 1$$

If each side of this equation is squared, the result is an equation of the unit circle.

$$x^2 + y^2 = 1$$

Consider an angle α in standard position, shown at the right. Let $P(x, y)$ be the point of intersection of its terminal side with the unit circle. The radian measure of an angle in standard position is defined as the length of the corresponding arc on the unit circle. Thus, the measure of angle α is s radians.

There is an important relationship between radian and degree measure. Since an angle of one complete revolution can be represented either by 360° or by 2π radians, $360° = 2\pi$ radians. Thus, $180° = \pi$ radians, and $90° = \frac{\pi}{2}$ radians. The following formulas relate degree and radian measures.

Degree/Radian Conversion Formulas	**1 radian** $= \dfrac{180}{\pi}$ **degrees or about 57.3°** **1 degree** $= \dfrac{\pi}{180}$ **radians or about 0.017 radians**

Angles expressed in radians are often written in terms of π. The term *radians* is also usually omitted when writing angle measures. However, the degree symbol is always used in this book to express the measure of angles in degrees.

Examples

2 Change 30° to radian measure in terms of π.

$$30° = 30° \times \frac{\pi}{180°}$$

$$= \frac{\pi}{6}$$

3 Change $\dfrac{3\pi}{4}$ radians to degree measure.

$$\frac{3\pi}{4} = \frac{3\pi}{4} \times \frac{180°}{\pi}$$

$$= 135°$$

Angles whose measures are multiples of 30° and 45° are commonly used in trigonometry. These angle measures correspond to radian measures of $\frac{\pi}{6}$ and $\frac{\pi}{4}$, respectively. The diagrams below can help you to make these conversions mentally.

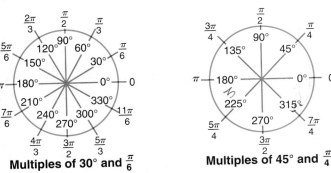

Multiples of 30° and $\dfrac{\pi}{6}$ **Multiples of 45° and $\dfrac{\pi}{4}$**

Two angles in standard position are called **coterminal angles** if they have the same terminal side. Since angles differing in radian measure by multiples of 2π are equivalent, and angles differing in degree measure by multiples of 360° are equivalent, every angle has infinitely many coterminal angles.

| Coterminal Angles | If α is the degree measure of an angle, then all angles of the form $\alpha + 360k°$, where k is an integer, are coterminal with α. If β is the radian measure of an angle, then all angles of the form $\beta + 2k\pi$, where k is an integer, are coterminal with β. |

k is generally assumed to be nonzero.

Example 4

Find one positive angle and one negative angle that are coterminal with an angle having measure $\dfrac{11\pi}{4}$.

A positive angle is $\dfrac{11\pi}{4} - 2\pi$ or $\dfrac{3\pi}{4}$.

A negative angle is $\dfrac{11\pi}{4} - 4\pi$ or $-\dfrac{5\pi}{4}$.

Example 5

Identify all angles that are coterminal with a 60° angle.

All angles having a measure $60 + 360k°$, where k is an integer, are coterminal with a 60° angle.

If α is a nonquadrantal angle in standard position, its **reference angle** is defined as the acute angle formed by the terminal side of the given angle and the x-axis. You can use the figures and the rule below to find the reference angle for any angle α that $0 < \alpha < 2\pi$. If the measure of α is greater than 2π or less than 0, it can be associated with a coterminal angle of positive measure between 0 and 2π.

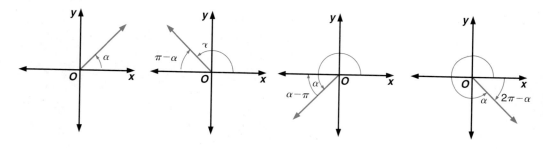

| Reference Angle Rule | For any angle α, $0 < \alpha < 2\pi$, its reference angle α' is defined by
a. α, when the terminal side is in Quadrant I,
b. $\pi - \alpha$, when the terminal side is in Quadrant II,
c. $\alpha - \pi$, when the terminal side is in Quadrant III, and
d. $2\pi - \alpha$, when the terminal side is in Quadrant IV. |

Example 6 | **Find the measure of the reference angle for each angle.**

a. $\dfrac{5\pi}{4}$

b. $-\dfrac{13\pi}{3}$

a. $\alpha = \dfrac{5\pi}{4}$, so its terminal side is in Quadrant III.

$$\alpha' = \alpha - \pi \qquad \text{Reference Rule c}$$

$$= \dfrac{5\pi}{4} - \pi$$

$$= \dfrac{\pi}{4}$$

b. $-\dfrac{13\pi}{3}$ is coterminal with $\dfrac{5\pi}{3}$ in Quadrant IV.

$$\alpha' = 2\pi - \alpha \qquad \text{Reference Rule d}$$

$$= 2\pi - \dfrac{5\pi}{3}$$

$$= \dfrac{\pi}{3}$$

c. $510°$

c. $510°$ is coterminal with $150°$ in Quadrant II.

$$\alpha' = \pi - \alpha \qquad \text{Reference Rule b}$$

$$= 180° - 150 \qquad \pi = 180°$$

$$= 30°$$

CHECKING FOR UNDERSTANDING

Communicating Mathematics

Read and study the lesson to answer each question.

1. **Define** the terms *positive angle* and *negative angle* in your own words.

2. **Explain** how to change the measure of an angle from degrees to radians and from radians to degrees.

3. **Sketch** a $30°$ angle in standard position and a positive angle greater than $360°$ that is coterminal with it.

4. **Write** an expression for the measures of all angles that are coterminal with an angle whose measure is θ.

Guided Practice

If each angle has the given measure and is in standard position, determine the quadrant in which its terminal side lies.

5. $\dfrac{15\pi}{4}$

6. $210°$

7. $-220°$

8. $\dfrac{11\pi}{6}$

9. $-\dfrac{4\pi}{3}$

10. $750°$

11. $\dfrac{14\pi}{3}$

12. $-475°$

Change each degree measure to radian measure in terms of π.

13. $18°$

14. $240°$

15. $1°$

16. $-45°$

Change each radian measure to degree measure.

17. π **18.** $\dfrac{3\pi}{2}$ **19.** $-\dfrac{7\pi}{6}$ **20.** $\dfrac{5\pi}{4}$

State whether each pair of angles is coterminal. Write *yes* or *no*.

21. $\dfrac{\pi}{4},\ \dfrac{9\pi}{4}$ **22.** $-30°,\ 390°$ **23.** $120°,\ \dfrac{14\pi}{3}$

EXERCISES

Practice If each angle has the given measure and is in standard position, determine the quadrant in which its terminal side lies.

24. $-\dfrac{8\pi}{3}$ **25.** $-167°$ **26.** $\dfrac{7\pi}{8}$ **27.** $227°$

28. $\dfrac{13\pi}{3}$ **29.** $-730°$ **30.** $-\dfrac{3\pi}{5}$ **31.** $-\dfrac{11\pi}{5}$

Change each degree measure to radian measure in terms of π.

32. $200°$ **33.** $-150°$ **34.** $75°$ **35.** $105°$

36. $570°$ **37.** $-450°$ **38.** $405°$ **39.** $-1250°$

Change each radian measure to degree measure.

40. $\dfrac{\pi}{3}$ **41.** -3.5 **42.** $\dfrac{4\pi}{3}$ **43.** $-\dfrac{\pi}{2}$

44. 1.75 **45.** $-\dfrac{7\pi}{12}$ **46.** 17.46 **47.** $\dfrac{17\pi}{6}$

Find one positive angle and one negative angle that are coterminal with each angle.

48. $-60°$ **49.** $\dfrac{5\pi}{12}$ **50.** $\dfrac{11\pi}{6}$ **51.** $-310°$

Find the reference angle for each angle with the given measure.

52. $-30°$ **53.** $\dfrac{12\pi}{5}$ **54.** $130°$ **55.** $-210°$

56. $\dfrac{9\pi}{4}$ **57.** $-420°$ **58.** $\dfrac{23\pi}{6}$ **59.** $-\dfrac{5\pi}{3}$

Change each degree measure to radian measure to the nearest thousandth.

60. $55°22'$ **61.** $-110°50'28''$ **62.** $250°49'15''$

63. Geography   Earth rotates on its axis every 24 hours. Through how many degrees does a point on the equator travel in one hour? in one week?

64. Navigation   Navigators often use angles measured clockwise from due north. A ship sets sail from port with a heading of 35° east of north. After sailing for many hours, the ship's captain changes the heading to 154°15′ east of north. At what angle did the ship turn?

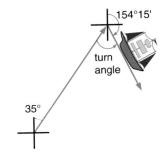

65. Mechanics   A lathe is a machine in which wood or metal is rotated about a horizontal axis and shaped by a tool that is fixed. A point on the rim of a lathe pulley turns at the rate of 2.22×10^4 degrees per second.

  a. How many revolutions per minute is this?

  b. How many radians per minute is this? Express your answer in terms of π.

66. Find $f(3)$ if $f(x) = 4 + 6x - x^3$.   **(Lesson 1-1)**

67. Find the slope of the line through $(-3, 2)$ and $(5, 7)$.   **(Lesson 1-4)**

68. Write the standard form of the equation of the line through $(3, -2)$ that is parallel to the line $3x - y + 7 = 0$.   **(Lesson 1-6)**

69. Solve the system of equations $3x + y = 6$ and $4x + y = 7$.   **(Lesson 2-1)**

70. Find matrix X in the equation $\begin{bmatrix} 1 & 1 \\ 1 & 1 \end{bmatrix} \begin{bmatrix} 3 & 5 \\ -3 & -5 \end{bmatrix} = X$.   **(Lesson 2-3)**

71. Manufacturing   A manufacturer can show a profit of $6 on a bicycle and a profit of $4 on a tricycle. Department A requires 3 hours to manufacture the parts for a bicycle and 4 hours to manufacture parts for a tricycle. Department B takes 5 hours to assemble a bicycle and 2 hours to assemble a tricycle. How many bicycles and tricycles should be produced to maximize the profit if the total time available in department A is 450 hours and in department B is 400 hours? **(Lesson 2-6)**

72. Create a function of the form $y = f(x)$ that has a vertical asymptote at $x = 5$ and a hole at $x = 0$.   **(Lesson 3-4)**

73. Is 6 a root of $x^3 - 5x^2 - 3x - 18 = 0$? Write *yes* or *no*.   **(Lesson 4-1)**

74. Solve $\dfrac{3}{x} + \dfrac{5}{x} = 10$.   **(Lesson 4-6)**

75. Solve $5 - \sqrt{b + 2} = 0$.   **(Lesson 4-7)**

76. College Entrance Exam   If a walker can cover 2 miles in 25 minutes, what is the rate in miles per hour?

5-2 Central Angles and Arcs

Objectives

After studying this lesson, you should be able to:
- find the length of an arc, given the measure of the central angle,
- find linear and angular velocities, and
- find the area of a sector.

Application

Civil engineers must deal with angle measures and distances frequently when they design roadways. The map below shows a stretch of Route 3 outside of Boston, Massachusetts. Two portions of the roadway are arcs of circles.

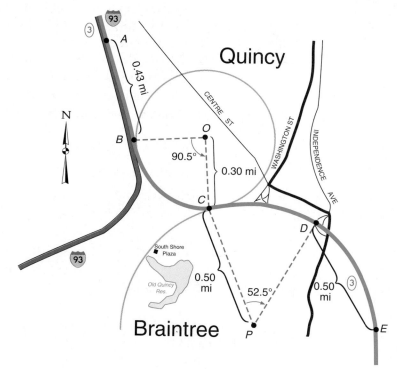

How many miles is it along the roadway from point *A* to point *E*? To solve this problem, you need to be familiar with central angles. *This problem will be solved in Example 2.*

A **central angle** of a circle is an angle whose vertex lies at the center of the circle. If two central angles in different circles are congruent, the ratio of the lengths of their intercepted arcs is equal to the ratio of the measures of their radii. For example, given $\odot O$ and $\odot Q$, if $\angle O \cong \angle Q$, then $\dfrac{m\widehat{AB}}{m\widehat{CD}} = \dfrac{OA}{QC}$.

θ is the Greek letter theta.

We say that an arc <u>subtends</u> its central angle.

Let O be the center of two concentric circles. Let r be the measure of the radius of the larger circle, and let the smaller circle be a unit circle. A central angle of θ radians is drawn in the two circles that intercept $\overset{\frown}{RT}$ on the unit circle and $\overset{\frown}{SW}$ on the other circle. Suppose $\overset{\frown}{SW}$ is s units long. $\overset{\frown}{RT}$ is θ units long since it is an arc of a unit circle intercepted by a central angle of θ radians. Thus, we can write the following proportion.

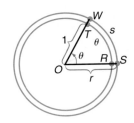

$$\frac{s}{\theta} = \frac{r}{1} \text{ or } s = r\theta$$

Length of an Arc	The length of any circular arc, s, is equal to the product of the measure of the radius of the circle, r, and the radian measure of the central angle, θ, that it subtends. $$s = r\theta$$

Example 1

Find the length of an arc that subtends a central angle of 42° in a circle of radius 8 cm.

$42° = 42 \times \dfrac{\pi}{180}$ *Find the radian measure of the central angle.*

$\quad = \dfrac{7\pi}{30}$

$\quad s = r\theta$ *Find the length of the arc.*

$\quad = (8)\left(\dfrac{7\pi}{30}\right)$

$\quad \approx 5.86$

The length of the arc is approximately 5.86 centimeters.

Example 2

Refer to the application at the beginning of the lesson. How far is it along the roadway from point A to point E?

To find the length of $\overset{\frown}{BC}$, change 90.5° to radians   90.5° ≈ 1.58 radians
Thus, the length of $\overset{\frown}{BC}$ is about 0.30×1.58 or 0.47 miles $r = 0.30$

To find the length of $\overset{\frown}{CD}$, change 52.5° to radians.   52.5° ≈ 0.92 radians
Thus, the length of $\overset{\frown}{CD}$ is about 0.50×0.92, or 0.46 miles. $r = 0.50$

Find the sum of the distances. $0.43 + 0.50 + 0.47 + 0.46 = 1.86$

Therefore, the distance from point A to point E is about 1.86 miles.

The arc length formula can be used to find the relationship between the linear and angular velocities of an object moving in a circular path. If the object moves with constant **linear velocity** (v) for a period of time (t), the distance (s) it travels is given by the formula $s = vt$. Thus, the linear velocity is $v = \frac{s}{t}$.

As the object moves along the circular path, the radius, r, forms a central angle of measure θ. The change in the central angle with respect to time, $\frac{\theta}{t}$, is the **angular velocity** of the object. Since the length of the arc is $s = r\theta$, the following is true.

$$s = r\theta$$

$$\frac{s}{t} = \frac{r\theta}{t} \qquad \textit{Divide each side by t.}$$

$$v = r\frac{\theta}{t} \qquad \textit{Remember that } v = \frac{s}{t}.$$

Linear and Angular Velocity	If an object moves along a circle of radius r units, then its linear velocity, v, is given by $$v = r\frac{\theta}{t},$$ where $\frac{\theta}{t}$ represents the angular velocity in radians per unit of time.

Example 3

A pulley of radius 12 cm turns at 7 revolutions per second. What is the linear velocity of the belt driving the pulley in meters per second?

$$v = r\frac{\theta}{t}$$

$$= 12\frac{(7 \cdot 2\pi)}{1} \qquad \textit{One revolution is } 2\pi, \textit{ so 7 revolutions is } 7 \cdot 2\pi \textit{ radians.}$$

$$= 168\pi$$

$$\approx 527.788$$

$$\approx 5.278 \qquad \textit{Divide by 100 to change cm to m.}$$

The linear velocity of the belt is approximately 5.28 m/s.

Example 4

A trucker drives 55 miles per hour. His truck's tires have a diameter of 26 inches. What is the angular velocity of the wheels in revolutions per second?

First, convert 55 miles per hour to inches per second.

$$\frac{55 \text{ miles}}{1 \text{ hour}} \cdot \frac{5280 \text{ feet}}{1 \text{ mile}} \cdot \frac{12 \text{ inches}}{1 \text{ foot}} \cdot \frac{1 \text{ hour}}{3600 \text{ seconds}} = \frac{968 \text{ inches}}{1 \text{ second}}$$

(continued on the next page)

Then use the formula for linear velocity to find θ.

$$v = r\frac{\theta}{t}$$

$$968 = 13\frac{\theta}{1} \qquad \textit{Since the diameter is 26 inches, the radius is 13 inches.}$$

$$\frac{968}{13} = \theta$$

Since $\theta = \dfrac{968}{13}$ radians, the angular velocity is $\dfrac{968}{13} \div 2\pi$, or about 12 revolutions per second.

A **sector** of a circle is a region bounded by a central angle and the intercepted arc. For example, figure $ORTS$ is a sector of $\odot O$. The ratio of the area of a sector to the area of a circle is equal to the ratio of its arc length to the circumference. Let A represent the area of the sector. Then, $\dfrac{A}{\pi r^2} = \dfrac{\text{length of } \widehat{RTS}}{2\pi r}$. But the length of $\widehat{RTS}$ is $r \cdot \theta$. Thus, the following formula is true.

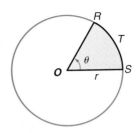

Area of a Circular Sector

If θ is the measure of the central angle expressed in radians and r is the measure of the radius of the circle, then the area of the sector, A, is as follows.

$$A = \frac{1}{2}r^2\theta$$

Example 5

A sector has arc length of 16 cm and a central angle measuring 0.95 radians. Find the radius of the circle and the area of the sector.

First, find the radius of the circle.

$$s = r\theta$$
$$16 = r \cdot 0.95$$
$$r = 16.84 \qquad \text{The radius is about 16.84 cm long.}$$

Then find the area of the sector.

$$A = \frac{1}{2}r^2\theta$$

$$= \frac{1}{2}(16.84)^2(0.95)$$

$$= 134.70 \qquad \text{The area is about 134.70 cm}^2.$$

CHECKING FOR UNDERSTANDING

Communicating Mathematics

Read and study the lesson to answer each question.

1. **Describe** how the length of an arc is a proportional part of the circumference of a circle and the area of a sector is a proportional part of the area of the circle.

2. **Compare and contrast** linear and angular velocity.

3. **Describe** how to change an angular velocity from revolutions per minute to degrees per second.

4. **Explain** why an ice skater at the outside of a chain of skaters must skate faster than a skater near the inside of the chain in order to keep all of the skaters in line.

5. **Show** that when the radius of a circle is doubled, the length of the arc subtended and the linear velocity of a point on the circle are also doubled.

Guided Practice

Find the arc length in terms of π for each central angle shown.

6.
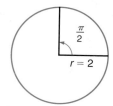

$\frac{\pi}{2}$

$r = 2$

7.
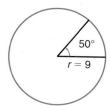

$50°$

$r = 9$

8.
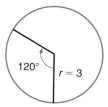

$120°$

$r = 3$

Complete.

9. $60 \text{ mph} = \blacksquare \text{ ft/s}$

10. $8 \text{ rpm} = \blacksquare \text{ rad/s}$

11. $10 \text{ ft/min} = \blacksquare \text{ in./s}$

12. $15 \text{ cm/s} = \blacksquare \text{ m/min}$

Find the area of each sector, given its central angle, θ, and the radius of the circle. Round answers to the nearest tenth.

13. $\theta = \frac{\pi}{8}, r = 7 \text{ m}$

14. $\theta = 48°, r = 22 \text{ in.}$

15. Find the area of the sector swept by the spoke of a wheel of radius 15 cm as the wheel rotates through an angle of $270°$. Round your answer to the nearest tenth.

EXERCISES

Practice

Given the radian measure of a central angle, find the measure of its intercepted arc in terms of π in a circle of radius 10 cm.

16. $\frac{\pi}{4}$

17. $\frac{2\pi}{3}$

18. $\frac{5\pi}{6}$

19. $\frac{2\pi}{5}$

Given the measurement of a central angle, find the measure of its intercepted arc in terms of π in a circle of diameter 30 in.

20. $30°$

21. $5°$

22. $77°$

23. $57°18'$

Given the measure of an arc, find the degree measure to the nearest tenth of the central angle it subtends in a circle of radius 8 cm.

24. 5   **25.** 14   **26.** 24   **27.** 12.5

Find the area of each sector to the nearest tenth, given its central angle, θ, and the radius of the circle.

28. $\theta = \dfrac{5\pi}{12}, r = 10$ ft   **29.** $\theta = 54°, r = 6$ in.

30. $\theta = \dfrac{2\pi}{3}, r = 1.36$ m   **31.** $\theta = 82°, r = 7.3$ km

32. $\theta = 45°, r = 9.75$ mm   **33.** $\theta = 12°, r = 14$ yd

For Exercises 34-38, round answers to the nearest tenth.

34. An arc is 6.5 cm long and it subtends a central angle of 45°. Find the radius of the circle.

35. An arc is 70.7 m long and it subtends a central angle of $\dfrac{2\pi}{7}$. Find the diameter of the circle.

36. A sector has arc length 6 ft and central angle of 1.2 radians. Find the radius and area of the circle.

37. A sector has area of 15 in² and central angle of 0.2 radians. Find the radius of the circle and arc length of the sector.

38. A sector has a central angle of 20° and arc length of 3.5 mm. Find the radius and area of the circle.

Critical Thinking

39. The figure at the right shows two gears that mesh. Make a conjecture about the angular velocity of the gear with radius r if the angular velocity of the gear with radius R is a.

Applications and Problem Solving

40. Astronomy Earth revolves around the sun in an elliptical orbit. However, for this problem, you can assume that the orbit is circular with a radius of about 9.3×10^7 miles.

    a. At what speed in feet per second does Earth travel around the sun?

    b. An astronomer estimates the angle subtended by the width of the sun as 0.5°. If this is the case, what is the approximate diameter of the sun?

41. Mechanics A wheel has a radius of 3.5 feet. As it turns, a cable connected to a box winds onto the wheel. To the nearest foot, how far does the box move if the wheel turns 130° in the counterclockwise direction?

42. **Geography**   The position of a point on Earth is often given in terms of the latitude and longitude. Jackson, Mississippi, and St. Louis, Missouri, lie along the same longitude line. The latitude of Jackson, Mississippi, is 32°17′ N, and the latitude of St. Louis, Missouri, is 38°37′ N. The radius of Earth is about 4000 miles.

    a. About how far apart are the two cities?

    b. If Earth makes one revolution every 24 hours, what is the linear velocity of a point near the equator?

43. **Interior Design**   A carpenter is building a window frame that consists of a semi-circular region, divided as shown at the right. Find the area of panes *A*, *B*, and *C* to the nearest hundredth of a square foot.

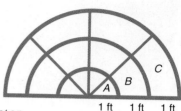

1 ft   1 ft   1 ft

44. **Sports**   Jason's bicycle wheel is 26 inches in diameter.

    a. To the nearest revolution, how many times will the wheel turn if it is ridden for 1 mile?

    b. Suppose the wheel turns at a constant rate of 2.5 revolutions per second. To the nearest tenth, what is the linear speed of a point on the tire in feet per second? in miles per hour?

45. **Civil Engineering**   The figure below shows a stretch of roadway where the curves are arcs of circles.

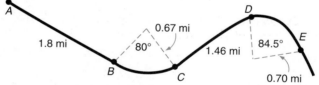

How many miles is it along the road from point *A* to point *E*?

46. **Physics**   A pendulum hangs on a 2.5-meter rod. Every two seconds, the pendulum swings 5° left and then 5° right of center. How many meters does the pendulum swing in one hour?

47. **Manufacturing**   The figure at the right is a side view of three rollers that are tangent to one another.

    a. If roller *A* turns counterclockwise, in which directions do rollers *B* and *C* turn?

    b. If roller *A* turns at 120 revolutions per minute, at what angular velocities do rollers *B* and *C* turn?

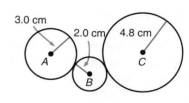

Mixed Review

48. If $f(x) = \dfrac{1}{x-1}$ and $g(x) = x + 1$, find $[f \circ g](x)$.   **(Lesson 1-2)**

49. Find the value of the determinant for $\begin{bmatrix} 4 & 3 \\ 8 & 6 \end{bmatrix}$. Then state whether an inverse exists for the matrix. Write *yes* or *no*.   **(Lesson 2-3)**

50. Find the slope of the line tangent to the graph of $f(x) = 2x^2 - 3x + 3$ at the point $(-2, 17)$.   **(Lesson 3-6)**

51. Solve $6x^2 + 7x + 2 = 0$ using the quadratic formula.
(Lesson 4-2)

52. Landscaping A landscaper works on a job for 10 days and is then joined by a helper. Together they finish the job in 6 more days. The helper could have done the job alone in 30 days. How long would it have taken the landscaper to do the job alone?
(Lesson 4-6)

53. Solve $\sqrt[3]{3y - 1} - 2 = 0$. **(Lesson 4-7)**

54. Change $60°$ to radian measure in terms of π. **(Lesson 5-1)**

55. College Entrance Exam Choose the best answer.
The volume of a cylinder can be found by using the formula $V = \pi r^2 h$. How many cubic inches of sand can be placed into a cylindrical can that has a radius of 4 inches and a height of 1.5 feet?

(A) 288π **(B)** 144π **(C)** 96π **(D)** 24π **(E)** 6π

DECISION MAKING

Education and Literacy

No resource is more important to a democratic nation than an educated citizenry. Yet more than one fourth of American high school students drop out before graduation, and 20% of Americans are functionally illiterate—that is, they are unable to deal with everyday life because of inadequate education. What can you do to help alleviate this problem? How can you turn your talents in a positive direction by working to raise the educational level of your community?

1. Describe the personal skills you have that you might be able to apply to the problem of illiteracy in your community. Do you enjoy working with children? senior citizens? Could you best use your talents by working directly with people seeking to learn to read or to improve their math skills, or should you channel your efforts toward educating the public or politicians about the critical need for educational reform?

2. Find out what organizations and agencies in your community are concerned with this problem. Contact several of them to find out what they do and whether they can use the help of volunteers.

3. **Project** Work with your group to complete one of the projects listed at the right based on the information you gathered in Exercises 1–2.

PROJECTS

- *Conduct an interview.*
- *Write a book report.*
- *Write a proposal.*
- *Write an article for the school paper.*
- *Write a report.*
- *Make a display.*
- *Make a graph or chart.*
- *Plan an activity.*
- *Design a checklist.*

5-3 Circular Functions

Objective After studying this lesson, you should be able to:
- find the values of the six trigonometric functions of an angle in standard position given a point on its terminal side.

Application At the scene of many traffic accidents, police officers often use a trundle wheel to determine the length of skid marks. When the wheel is turned through one complete revolution, there is a one-to-one correspondence between the points on the rim of the wheel and the real numbers representing the length of the skid marks on the ground. As the wheel completes more revolutions, there is a many-to-one correspondence between the points on the rim of the wheel and the real numbers.

In a similar manner, two important **trigonometric functions**, the sine and cosine functions, can be defined in terms of the unit circle.

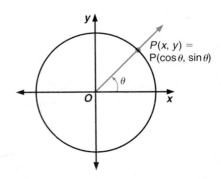

Consider an angle θ standard position. The terminal side of the angle intersects the unit circle at a unique point, $P(x, y)$. The y-coordinate of this point is called **sine θ**. The abbreviation for sine is *sin*. The x-coordinate of this point is called **cosine θ**. The abbreviation for cosine is *cos*.

Definition of Sine and Cosine If the terminal side of an angle θ in standard position intersects the unit circle at $P(x, y)$, then $\cos \theta = x$ and $\sin \theta = y$.

Since there is exactly one point $P(x, y)$ for any angle θ, the relations $\cos \theta = x$ and $\sin \theta = y$ are functions of θ. Because they are both defined using a unit circle, they are often called **circular functions**.

Example 1

Find each value.
a. sin 90°

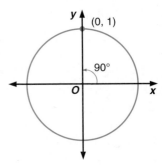

a. The terminal side of a 90° angle in standard position is the positive y-axis, which intersects the unit circle at (0, 1). The y-coordinate of this ordered pair is sin 90°.

Therefore, sin 90° = 1.

b. cos π

b. The terminal side of an angle in standard position measuring π radians is the negative x-axis, which intersects the unit circle at $(-1, 0)$. The x-coordinate of this ordered pair is cos π.

Therefore, cos $\pi = -1$.

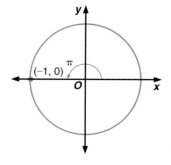

EXPLORATION: Graphing Calculator

1. Use the range values below to set up a viewing window.

   TMIN = 0   XMIN = -2.4   YMIN = -1.6
   TMAX = 360   XMAX = 2.35   YMAX = 1.55
   TSTEP = 15   XSCL = 0.5   YSCL = 0.5

2. Define the unit circle with the definition $X_{1T} = \cos T$ and $Y_{1T} = \sin T$.

3. Activate the TRACE function to move around the circle.
   a. What does T represent?
   b. What does the x-value represent?
   c. What does the y-value represent?

4. Determine the trigonometric functions of the angles whose terminal sides lie at $0°$, $90°$, $180°$, and $360°$.

The sine and cosine functions of an angle in standard position may be defined in terms of the ordered pair for *any* point on its terminal side and the distance between that point and the origin.

Suppose $P(x, y)$ and $P'(x', y')$ are two points on the terminal side of an angle with measure θ, where P' is on the unit circle. Let $OP = r$. By the Pythagorean theorem, $r = \sqrt{x^2 + y^2}$. Since P' is on the unit circle, $OP' = 1$. Triangles $OP'Q'$ and OPQ are similar. Thus, the lengths of corresponding sides are proportional.

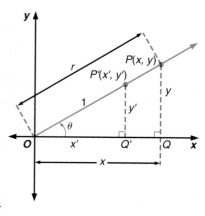

$$\frac{x'}{1} = \frac{x}{r} \text{ and } \frac{y'}{1} = \frac{y}{r}$$

Therefore, $\cos \theta = x'$ or $\dfrac{x}{r}$ and $\sin \theta = y'$ or $\dfrac{y}{r}$.

The ratios $\frac{x}{r}$ and $\frac{y}{r}$ do not depend on the choice of P. They depend only on the measure of θ and thus, are the basic trigonometric functions of θ.

Sine and Cosine Functions of an Angle in Standard Position	For any angle in standard position with measure θ, a point $P(x, y)$ on its terminal side, and $r = \sqrt{x^2 + y^2}$, the sine and cosine functions of θ are as follows. $$\sin \theta = \frac{y}{r} \qquad \cos \theta = \frac{x}{r}$$

Example 2

Find the values of the sine and cosine functions of an angle in standard position with measure θ if the point with coordinates (3, 4) lies on its terminal side.

You know that $x = 3$ and $y = 4$. You need to find r.

$$r = \sqrt{x^2 + y^2}$$
$$= \sqrt{3^2 + 4^2}$$
$$= 5$$

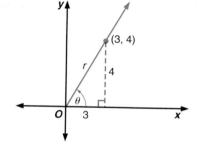

Then write the sine and cosine ratios.

$$\sin \theta = \frac{y}{r} \text{ or } \frac{4}{5} \qquad \cos \theta = \frac{x}{r} \text{ or } \frac{3}{5}$$

Example 3

Find $\sin \theta$ when $\cos \theta = \dfrac{5}{13}$ and the terminal side of θ is in the first quadrant.

Since $\cos \theta = \frac{x}{r} = \frac{5}{13}$ and r is always positive, $r = 13$ and $x = 5$.
Now, find y.

$$r = \sqrt{x^2 + y^2}$$
$$13 = \sqrt{5^2 + y^2}$$
$$169 = 25 + y^2$$
$$144 = y^2$$
$$\pm 12 = y$$

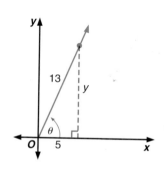

Since θ is in the first quadrant, y must be positive.

Thus, $\sin \theta = \frac{y}{r}$ or $\frac{12}{13}$.

In addition to the ratios $\frac{y}{r}$ and $\frac{x}{r}$ that are used to define the sine and cosine functions, four other ratios may be formed using x, y, and r. These are the **tangent**, **cotangent**, **secant**, and **cosecant** functions, which are abbreviated *tan*, *cot*, *sec*, and *csc*, respectively. These ratios depend only on the measure of θ and thus, provide additional trigonometric functions.

Trigonometric Functions of an Angle in Standard Position	For any angle in standard position with measure θ, a point $P(x, y)$ on its terminal side, and $r = \sqrt{x^2 + y^2}$, the trigonometric functions of θ are as follows.

$$\sin \theta = \frac{y}{r} \qquad \cos \theta = \frac{x}{r} \qquad \tan \theta = \frac{y}{x}$$

$$\csc \theta = \frac{r}{y} \qquad \sec \theta = \frac{r}{x} \qquad \cot \theta = \frac{x}{y}$$

We can write the tangent, cosecant, secant, and cotangent functions in terms of the sine, cosine, and tangent.

$$\tan \theta = \frac{\sin \theta}{\cos \theta} \qquad \csc \theta = \frac{1}{\sin \theta} \qquad \sec \theta = \frac{1}{\cos \theta} \qquad \cot \theta = \frac{1}{\tan \theta}$$

Thus, $\sin \theta$ and $\csc \theta$ are reciprocals, as are $\sec \theta$ and $\cos \theta$, and $\tan \theta$ and $\cot \theta$.

Example 4

The terminal side of an angle θ in standard position contains the point with coordinates $(8, -15)$. Find $\tan \theta$, $\cot \theta$, $\sec \theta$, and $\csc \theta$.

You know that $x = 8$ and $y = -15$. You need to find r.

$r = \sqrt{x^2 + y^2}$

$ = \sqrt{8^2 + (-15)^2}$

$ = \sqrt{289}$ or 17

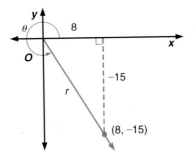

Now, write the ratios.

$$\tan \theta = \frac{y}{x} \text{ or } \frac{-15}{8} \qquad \cot \theta = \frac{x}{y} \text{ or } \frac{8}{-15}$$

$$\sec \theta = \frac{r}{x} \text{ or } \frac{17}{8} \qquad \csc \theta = \frac{r}{y} \text{ or } \frac{17}{-15}$$

If you know the value of one of the trigonometric functions and the quadrant in which the terminal side of θ lies, you can find the values of the remaining five functions.

Example 5 | **If $\csc \theta = -2$ and θ lies in Quadrant III, find $\sin \theta$, $\cos \theta$, $\tan \theta$, $\cot \theta$, and $\sec \theta$.**

Since $\csc \theta$ and $\sin \theta$ are reciprocals, $\sin \theta = -\frac{1}{2}$.

To find the other function values, you must find the coordinates of a point on the terminal side of θ. Since $\sin \theta = -\frac{1}{2}$ and r is always positive, let $r = 2$ and let $y = -1$. Find x.

$$r^2 = x^2 + y^2$$
$$2^2 = x^2 + (-1)^2$$
$$3 = x^2$$
$$\pm\sqrt{3} = x$$

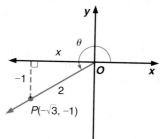

Since the terminal side of θ lies in Quadrant III, $x = -\sqrt{3}$.

Now, write the ratios.

$$\cos \theta = \frac{-\sqrt{3}}{2} \text{ or } -\frac{\sqrt{3}}{2}$$

$$\tan \theta = \frac{-1}{-\sqrt{3}} \text{ or } \frac{\sqrt{3}}{3}$$

$$\sec \theta = \frac{2}{-\sqrt{3}} \text{ or } -\frac{2\sqrt{3}}{3}$$

$$\cot \theta = \frac{-\sqrt{3}}{-1} \text{ or } \sqrt{3}$$

As illustrated in Examples 4 and 5, the values of the trigonometric functions may be either positive, negative, or 0. Since r is always positive, the signs of the functions are determined by the signs of x and y.

CHECKING FOR UNDERSTANDING

Communicating Mathematics

Read and study the lesson to answer each question.

1. **Explain** why the $\sin a°$ and $\sin(a° + 360k°)$, where k is an integer, are equal. Use an example as part of your explanation.

2. **State** the quadrant or quadrants in which $\sin \theta$ and $\cos \theta$ are both positive.

3. **Show** that as the measure of θ increases from $0°$ to $90°$, the value of $\cos \theta$ decreases from 1 to 0.

4. **Tell** which is greater, $\sin 12°$ or $\sin 13°$.

5. **Complete** the chart below that indicates the sign of the trigonometric functions in each quadrant.

Function	Quadrant			
	I	II	III	IV
$\sin \alpha$ or $\csc \alpha$	$+$			$-$
$\cos \alpha$ or $\sec \alpha$	$+$		$-$	
$\tan \alpha$ or $\cot \alpha$		$-$	$+$	

Guided Practice

Find the values of the six trigonometric functions of an angle in standard position if a point with the given coordinates lies on its terminal side.

6. $(3, -4)$ 7. $(3, 3)$ 8. $(-4, 0)$

Suppose $\sin \theta = \dfrac{1}{2}$ and the terminal side is in the second quadrant. Find each value.

9. $\cos \theta$ 10. $\tan \theta$ 11. $\sec \theta$

Suppose θ is an angle in standard position with the given conditions. State the quadrant or quadrants in which the terminal side of θ lie.

12. $\sin \theta > 0$ 13. $\tan \theta < 0$

14. $\sin \theta > 0$, $\cos \theta < 0$ 15. $\tan \theta > 0$, $\cos \theta < 0$

Find two values of x that make each statement true.

16. $\cos x = 1$ 17. $\sin x = 1$

18. $\cos x = 0$ 19. $\sin x = 0$

20. $\cos x = -1$ 21. $\sin x = -1$

EXERCISES

Practice

Find the values of the six trigonometric functions of an angle in standard position if a point with the given coordinates lies on its terminal side.

22. $(5, 12)$ 23. $(15, 8)$ 24. $(3, 4)$ 25. $(1, -8)$

26. $(-3, 0)$ 27. $(-\sqrt{2}, \sqrt{2})$ 28. $(5, -3)$ 29. $(0, 2)$

Suppose θ is an angle in standard position whose terminal side lies in the given quadrant. For each function, find the values of the remaining five trigonometric functions of θ.

30. $\sin \theta = -\dfrac{4}{5}$; Quadrant IV   31. $\cos \theta = -\dfrac{1}{2}$; Quadrant II

32. $\tan \theta = 2$; Quadrant I   33. $\sec \theta = \sqrt{3}$; Quadrant IV

Tell whether the value of each trigonometric function is *positive, negative, zero,* or *undefined*.

34. $\sin 2\pi$

35. $\cos \dfrac{\pi}{4}$

36. $\tan 315°$

37. $\cos \dfrac{5\pi}{4}$

38. $\sin \dfrac{11\pi}{4}$

39. $\tan 90°$

40. $\cos 450°$

41. $\sin (-45°)$

Graphing Calculator

42–45. Graph each angle described in Exercises 34–37 on a unit circle, using the method described in the Exploration. Use the TRACE function to verify whether the value of each trigonometric function is *positive, negative, zero,* or *undefined*.

Critical Thinking

46. Explain why the statement $\sin \theta = 2$ is not possible.

47. Show that $\sin 90k° = 0$, if k is any even integer.

48. Show that $\cos 90k° = 0$, if k is any odd integer.

Applications and Problem Solving

49. Surveying Two surveyors at points R and S sight a marker at point T. Write an equation that relates $\tan 35°$ with the distance across Hidden Lake (d) and the distance between the surveyors.

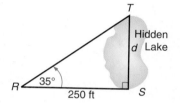

50. Construction A ramp is designed to help people in wheelchairs move more easily from one level to another. If a ramp 16 feet long forms an angle of $12°$ with the level ground, write an equation that relates $\sin 12°$ with the length of the ramp and the vertical rise, v.

Mixed Review

51. Name all values of x that are not in the domain of f for $f(x) = \dfrac{3x}{x^2 - 5}$.
(Lesson 1-1)

52. Find the additive inverse of $\begin{bmatrix} 6 & 5 \\ 8 & 4 \end{bmatrix}$.
(Lesson 2-2)

53. Solve $3x^2 - 7x - 20 = 0$ using the quadratic formula. **(Lesson 4-2)**

54. Find the critical points of the graph of $y = x^4 - 8x^2 + 16$. Then determine whether each point represents a maximum, a minimum, or a point of inflection. **(Lesson 3-7)**

55. Use synthetic division to divide $(x^3 + 2x + 3)$ by $(x - 2)$. **(Lesson 4-3)**

56. Change $\dfrac{5\pi}{6}$ to degree measure. **(Lesson 5-1)**

57. Mechanics A pulley of radius 10 cm turns at 5 revolutions per second. Find the linear velocity of the belt driving the pulley in meters per second. **(Lesson 5-2)**

58. College Entrance Exam Compare quantities A and B below.
Write A if quantity A is greater.
Write B if quantity B is greater.
Write C if the two quantities are equal.
Write D if there is not enough information to determine the relationship.

(A) the maximum number of boxes of pens costing $2.98 each that can be bought for $15, with no tax

(B) the maximum number of boxes of pens costing $3.25 each that can be bought for $20, with no tax

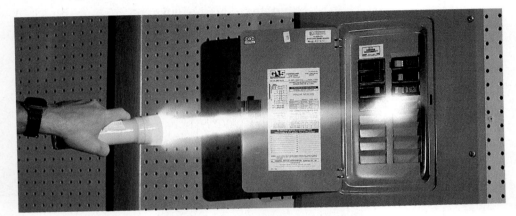

MID-CHAPTER REVIEW

1. Change $-\dfrac{3\pi}{4}$ to degree measure. **(Lesson 5-1)**

2. Change $405°$ to radian measure in terms of π. **(Lesson 5-1)**

3. Find one positive angle and one negative angle that are coterminal with an angle measuring $225°$. **(Lesson 5-1)**

4. Find the reference angle for $\dfrac{5\pi}{8}$. **(Lesson 5-1)**

5. Find the length of an arc to the nearest tenth that intercepts a central angle of $32°$ in a circle of radius 11 cm. **(Lesson 5-2)**

6. A sector has a central angle of 0.8 radians and a radius of 7.5 cm. Find the area of the sector to the nearest tenth. **(Lesson 5-2)**

7. A flashlight has a range of 100 feet and casts its beam through an angle of $100°$. To the nearest 100 square feet, what area is illuminated by the flashlight? **(Lesson 5-2)**

8. Find the values of the six trigonometric functions of an angle α in standard position if the point with coordinates (4, 3) lies on its terminal side. **(Lesson 5-3)**

5-4 Trigonometric Functions of Special Angles

Objectives

After studying this lesson, you should be able to:

- find exact values for the six trigonometric functions of special angles, and
- find decimal approximations for the values of the six trigonometric functions of any angle.

Application

You see an object because light is reflected from the object into your eyes. However, light travels faster in air than it does in water. This can be a bit confusing if you are about to step on a submerged rock while crossing a stream. The rock is not exactly where it appears to be. The displacement of the light ray depends on the angle at which the light strikes the surface of the water from below, A, the depth of the rock, t, and the angle at which the light leaves the surface of the water, B. The measure of displacement, x, is given by the formula below.

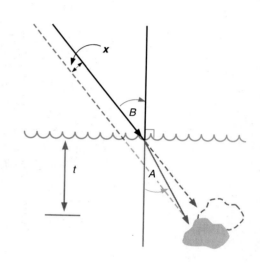

$$x = t\left(\frac{\sin(B - A)}{\cos A}\right)$$

Find the displacement if t measures 10 centimeters, the measure of angle A is 30°, and the measure of angle B is 42°. *This problem will be solved in Example 4.*

Before you find the values of the trigonometric functions in the formula above, it is helpful to study the values of trigonometric functions of special angles. One such group of angles are quadrantal angles. It is easy to find the trigonometric functions for quadrantal angles since their terminal sides lie along an axis. Using a unit circle, let (x, y) be the coordinates of the point of intersection of the circle with the terminal side of the angle. For example, $\cos 180° = -1$ and $\sin 180° = 0$.

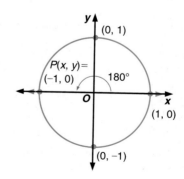

Example 1

Find the values of the six trigonometric functions for an angle in standard position that measures 90°.

$\sin 90° = 1$ $\qquad\qquad$ $\cos 90° = 0$

$\tan 90°$ is undefined because division by zero is undefined.

$\cot 90° = 0$ $\qquad\qquad$ $\csc 90° = 1$

$\sec 90°$ is undefined because division by zero is undefined.

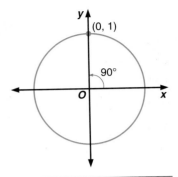

The domain of the sine and cosine functions is the set of real numbers, since $\sin\theta$ and $\cos\theta$ are defined for any angle θ. The range of the sine and cosine functions is the set of real numbers between –1 and 1 inclusive, since $(\sin\theta, \cos\theta)$ are the coordinates of points on the unit circle. Since division by zero is undefined, there are several angle measures that are excluded from the domain of the tangent, cotangent, secant, and cosecant functions. The table below summarizes the values of the trigonometric functions of common quadrantal angles. The dashes represent undefined values.

Angle	sin	cos	tan	csc	sec	cot
$0°$	0	1	0	—	1	—
$90°$ or $\dfrac{\pi}{2}$	1	0	—	1	—	0
$180°$ or π	0	-1	0	—	-1	—
$270°$ or $\dfrac{3\pi}{2}$	-1	0	—	-1	—	0
$360°$ or 2π	0	1	0	—	1	—

You can find the trigonometric functions of other special angles by using relationships from geometry. Recall that in a $30°$–$60°$ right triangle, the lengths of the sides are in the ratio $1:\sqrt{3}:2$. In a $45°$–$45°$ right triangle, the lengths of the sides are in the ratio $1:1:\sqrt{2}$.

Example 2

Find sin 60°, cos 60°, and tan 60°.

Sketch a $30°$–$60°$ right triangle so that the $60°$ angle is in standard position.

Choose $P(x, y)$ on the terminal side of the angle so that $r = 2$. It follows that $x = 1$ and $y = \sqrt{3}$.

$\sin 60° = \dfrac{\sqrt{3}}{2}$

$\cos 60° = \dfrac{1}{2}$

$\tan 60° = \sqrt{3}$

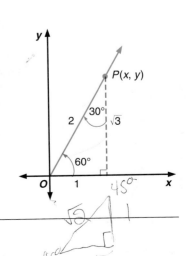

The values found in Example 2 are also those for angles of radian measure $\frac{\pi}{3}$ since $60° = \frac{\pi}{3}$. The chart below summarizes the sine and cosine values for selected angles from 0 to π. *It is a good idea to memorize these values since you will use them frequently.*

FYI...

Al-Battani (A.D. 850-929), an Islamic mathematician, calculated the first trigonometry tables used by modern surveyors.

θ (in radians)	0	$\frac{\pi}{6}$	$\frac{\pi}{4}$	$\frac{\pi}{3}$	$\frac{\pi}{2}$	$\frac{2\pi}{3}$	$\frac{3\pi}{4}$	$\frac{5\pi}{6}$	π
θ (in degrees)	0	30°	45°	60°	90°	120°	135°	150°	180°
$\cos \theta$	1	$\frac{\sqrt{3}}{2}$	$\frac{\sqrt{2}}{2}$	$\frac{1}{2}$	0	$-\frac{1}{2}$	$-\frac{\sqrt{2}}{2}$	$-\frac{\sqrt{3}}{2}$	-1
$\sin \theta$	0	$\frac{1}{2}$	$\frac{\sqrt{2}}{2}$	$\frac{\sqrt{3}}{2}$	1	$\frac{\sqrt{3}}{2}$	$\frac{\sqrt{2}}{2}$	$\frac{1}{2}$	0

You can also use the reference angle for certain angles to find the value of trigonometric functions. Recall that the reference angle is the acute angle formed by the terminal side of a given angle and the x-axis.

Example 3

You may wish to review the reference angle rules in Lesson 5–1.

Find each value.

a. $\cos \dfrac{5\pi}{6}$

a. $\alpha = \dfrac{5\pi}{6}$, so its terminal side is in Quadrant II.

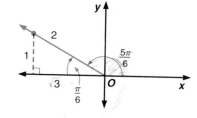

$$\alpha' = \pi - \alpha \qquad \textit{Reference rule b}$$

$$= \pi - \frac{5\pi}{6}$$

$$= \frac{\pi}{6} \qquad \frac{\pi}{6} = 30°$$

$$\cos \frac{5\pi}{6} = -\cos \frac{\pi}{6} \qquad \textit{cos } \alpha < 0 \textit{ in Quadrant II}$$

$$= -\frac{\sqrt{3}}{2}$$

b. $\tan \left(-\dfrac{11\pi}{4} \right)$

b. $-\dfrac{11\pi}{4}$ is coterminal with $\dfrac{5\pi}{4}$ in Quadrant III.

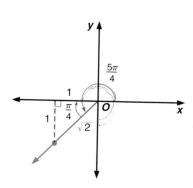

$$\alpha' = \alpha - \pi \qquad \textit{Reference rule c}$$

$$= \frac{5\pi}{4} - \pi$$

$$= \frac{\pi}{4} \qquad \frac{\pi}{4} = 45°$$

$$\tan\left(-\frac{11\pi}{4}\right) = \tan\frac{\pi}{4} \qquad \textit{tan } \alpha > 0 \textit{ in Quadrant III}$$
$$= 1$$

c. $\csc \dfrac{29\pi}{3}$

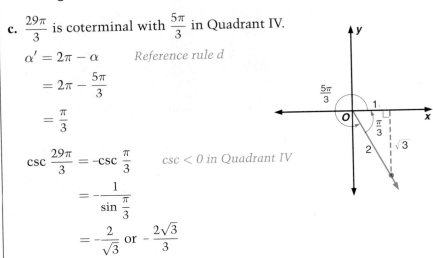

c. $\dfrac{29\pi}{3}$ is coterminal with $\dfrac{5\pi}{3}$ in Quadrant IV.

$$\alpha' = 2\pi - \alpha \qquad \textit{Reference rule d}$$
$$= 2\pi - \frac{5\pi}{3}$$
$$= \frac{\pi}{3}$$
$$\csc\frac{29\pi}{3} = -\csc\frac{\pi}{3} \qquad \textit{csc} < 0 \textit{ in Quadrant IV}$$
$$= -\frac{1}{\sin\dfrac{\pi}{3}}$$
$$= -\frac{2}{\sqrt{3}} \text{ or } -\frac{2\sqrt{3}}{3}$$

The values of the trigonometric functions of *any* angle can be approximated using a scientific calculator. Most approximate values are given to four decimal places. Always check to see whether the calculator is in radian or degree mode.

Example 4

Refer to the application at the beginning of the lesson. Find the displacement if *t* measures 10 cm, the measure of angle *A* is 30°, and the measure of angle *B* is 42°.

$$x = t\left(\frac{\sin (B - A)}{\cos A}\right)$$
$$= 10\left(\frac{\sin (42° - 30°)}{\cos 30°}\right)$$
$$= 10\left(\frac{\sin 12°}{\cos 30°}\right) \qquad \textit{Use your calculator. Make sure it is in degree mode.}$$

$$10 \boxed{\times} \boxed{(} 12 \boxed{\text{SIN}} \boxed{\div} 30 \boxed{\text{COS}} \boxed{)} \boxed{=} \; 2.4007574$$

The rock submerged 10 centimeters under water is actually about 2.4 centimeters from where it appears.

CHECKING FOR UNDERSTANDING

Communicating Mathematics

Read and study the lesson to answer each question.

1. **Explain** why some of the trigonometric functions for quadrantal angles are undefined. Use an example as part of your explanation.

2. **Explain** how you would evaluate sec 79° using a calculator.

3. **Show** how you can use the figure at the right to find the value of the trigonometric functions for a 45° angle.

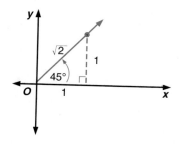

4. **Name** three angle measures that are excluded from the domain of the tangent function.

Guided Practice

Find each exact value. Do not use a calculator.

5. tan 60°

6. cos 450°

7. $\sin\left(-\dfrac{3\pi}{4}\right)$

8. $\csc \dfrac{\pi}{2}$

9. sec 30°

10. $\cot \dfrac{8\pi}{3}$

Use a calculator to approximate each value to four decimal places.

11. cos 42°

12. csc 36.7°

13. $\tan \dfrac{7\pi}{6}$

14. Find the values of the six trigonometric functions of an angle in standard position with measure $\dfrac{3\pi}{2}$.

EXERCISES

Practice

Find each exact value. Do not use a calculator.

15. csc 90°

16. cos 60°

17. $\sin \dfrac{\pi}{3}$

18. $\tan \dfrac{9\pi}{4}$

19. $\sec \dfrac{7\pi}{3}$

20. cot 45°

21. sec 270°

22. $\cos \dfrac{5\pi}{6}$

23. $\sin \dfrac{7\pi}{6}$

24. $\csc\left(-\dfrac{7\pi}{2}\right)$

25. tan 3π

26. $\cot \dfrac{19\pi}{3}$

Use a calculator to approximate each value to four decimal places.

27. $\cot\left(-\dfrac{4\pi}{9}\right)$

28. sin 710°

29. sec (−112°)

30. sin 7

31. cot 11.55π

32. csc 34.78°

33. tan 115°40′

34. cos 72°30′30″

35. tan (16.4 + π)

36. The terminal side of an angle θ in standard position coincides with the line $y = 5x$ and lies in Quadrant III. Find the six trigonometric functions of θ.

37. **Physics** The application at the beginning of the lesson suggests that as light passes from air to water, the light ray is bent toward the normal, the line perpendicular to the boundary between the two substances. The relation between the angle of incidence θ_1 and angle of refraction θ_2 is given by Snell's Law, $\dfrac{\sin \theta_1}{\sin \theta_2} = k$, where k is a constant called the index of refraction. Suppose a ray of light passes from air to glass. The measure of the angle of incidence is $52°$ and the measure of the angle of refraction is $31.3°$. Find the index of refraction for glass.

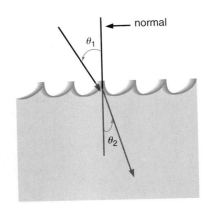

38. **Optics** Refer to the application at the beginning of the lesson. Suppose a fisherman is trying to net a fish under water. Find the measure of displacement if t measures 10 centimeters, the measure of angle A is $41°$, and the measure of angle B is $60°$.

39. Write the slope-intercept form of the equation of the line that passes through $(2, 5)$ and $(6, 3)$. **(Lesson 1-5)**

40. Find the slope of the line tangent to the graph of $y = 2x^2$ at $(2, 8)$. **(Lesson 3-6)**

41. Solve $\dfrac{1}{9} + \dfrac{1}{2a} = \dfrac{1}{a}$. **(Lesson 4-6)**

42. Change $35°20'55''$ to a decimal number of degrees to the nearest thousandth. **(Lesson 5-1)**

43. **Agriculture** A center-pivot irrigation system with a 75-meter radial arm completes one revolution every 6 hours. Find the linear velocity of a nozzle at the end of the arm. **(Lesson 5-2)**

44. Find $\sin \theta$ when $\tan \theta = -\dfrac{3}{4}$ and the terminal side of θ is in Quadrant IV. **(Lesson 5-3)**

45. **College Entrance Exam** Choose the best answer.
    x, y, and z are different positive integers. $\dfrac{x}{y}$ and $\dfrac{y}{z}$ are also positive integers. Which of the following cannot be a positive integer?

    (A) $\dfrac{x}{z}$ **(B)** $(x)(y)$ **(C)** $\dfrac{z}{x}$ **(D)** $(x + y)z$ **(E)** $(x - z)y$

5-5 Right Triangles

Objective

After studying this lesson, you should be able to:
- solve right triangles.

Application

The longest truck-mounted ladder used by the Dallas Fire Department is 108 feet long and consists of four hydraulic sections. Gerald Travis, aerial expert for the department, indicates that the optimum operating angle of this ladder is 60°. Outriggers, with an 18-foot span between each, are used to stabilize the ladder truck and permit operating angles greater than 60°, allowing the ladder truck to be closer to buildings in the downtown streets of Dallas. Assuming the ladder is mounted 8 feet off the ground, how far from an 84-foot burning building should the base of the ladder be placed to achieve the optimum operating angle of 60°? How far should the ladder be extended to reach the roof? *This problem will be solved in Example 4.*

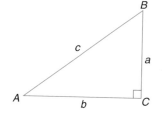

Right triangles can be used to define trigonometric functions. Let A, B, and C designate the vertices of a right triangle and the angles at those vertices. The measures of the sides opposite angles A, B, and C are designated by a, b, and c, respectively. All right triangles having acute angles congruent to angles A and B are similar. Thus, the ratios of corresponding sides are equal. These ratios are determined by the measures of the acute angles. Therefore, any two congruent angles of different right triangles will have the same ratios associated with them.

Trigonometric Functions in a Right Triangle

For an acute angle A in right triangle ABC, the trigonometric functions are as follows.

$$\sin A = \frac{\text{side opposite}}{\text{hypotenuse}} = \frac{a}{c} \qquad \csc A = \frac{\text{hypotenuse}}{\text{side opposite}} = \frac{c}{a}$$

$$\cos A = \frac{\text{side adjacent}}{\text{hypotenuse}} = \frac{b}{c} \qquad \sec A = \frac{\text{hypotenuse}}{\text{side adjacent}} = \frac{c}{b}$$

$$\tan A = \frac{\text{side opposite}}{\text{side adjacent}} = \frac{a}{b} \qquad \cot A = \frac{\text{side adjacent}}{\text{side opposite}} = \frac{b}{a}$$

SOH-CAH-TOA is a mnemonic device commonly used for remembering the first three equations.

$$\text{Sin } \theta = \frac{\text{Opposite}}{\text{Hypotenuse}} \qquad \text{Cos } \theta = \frac{\text{Adjacent}}{\text{Hypotenuse}} \qquad \text{Tan } \theta = \frac{\text{Opposite}}{\text{Adjacent}}$$

Example 1

A right triangle has sides whose lengths are 5 cm, 12 cm, and 13 cm. Find the values of the six trigonometric functions of α.

$$\sin \alpha = \frac{\text{side opposite}}{\text{hypotenuse}} = \frac{5}{13}$$

$$\cos \alpha = \frac{\text{side adjacent}}{\text{hypotenuse}} = \frac{12}{13}$$

$$\tan \alpha = \frac{\text{side opposite}}{\text{side adjacent}} = \frac{5}{12}$$

$$\csc \alpha = \frac{\text{hypotenuse}}{\text{side opposite}} = \frac{13}{5}$$

$$\sec \alpha = \frac{\text{hypotenuse}}{\text{side adjacent}} = \frac{13}{12}$$

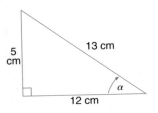

$$\cot \alpha = \frac{\text{side adjacent}}{\text{side opposite}} = \frac{12}{5}$$

You can use trigonometric functions to solve right triangles. To **solve a triangle** means to find all of the measures of its sides and angles. Usually, two measures, such as a side and an angle, are given. Then you can find the remaining measures.

Example 2

Solve right triangle *ABC*. Round angle measures to the nearest degree and side measures to the nearest tenth.

$$49° + B = 90° \qquad \textit{Angles A and B are complementary.}$$

$$B = 41°$$

$$\sin 49° = \frac{7}{c} \qquad\qquad \tan 49° = \frac{7}{b}$$

$$0.7547 \approx \frac{7}{c} \qquad\qquad 1.1504 \approx \frac{7}{b}$$

$$c \approx 9.3 \qquad\qquad\qquad b \approx 6.1$$

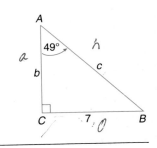

Therefore, $B = 41°$, $c = 9.3$, and $b = 6.1$.

Example 3

In $\triangle RST$, find the measure of $\angle R$ to the nearest degree.

$$\sin R = \frac{\text{side opposite}}{\text{hypotenuse}}$$

$$= \frac{8}{14} \qquad \textit{Use your calculator.}$$

8 ÷ 14 = 2nd SIN 34.849905

$$R \approx 35°$$

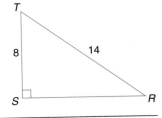

270   CHAPTER 5   THE TRIGONOMETRIC FUNCTIONS

There are many useful applications of trigonometry involving right triangles.

Example 4

Refer to the application at the beginning of the lesson. Assume that the ladder is mounted 8 feet off the ground.
a. How far from an 84-foot burning building should the base of the ladder be placed to achieve the optimum operating angle of 60°?
b. How far should the ladder be extended to reach the roof?

a. Begin by drawing a diagram. In the diagram at the right, d feet represents the distance between the base of the truck-mounted ladder and the base of the wall of the building, and ℓ feet represents the length of the ladder when it is extended.

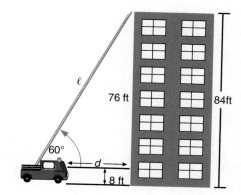

Use the trigonometric ratios to find d.

$$\tan 60° = \frac{76}{d}$$

$$1.7321 \approx \frac{76}{d}$$

$$d \approx 43.9$$

So, the base of the ladder should be placed 43.9 feet from the wall.

b. Then use the trigonometric ratios to find ℓ.

$$\sin 60° = \frac{76}{\ell}$$

$$0.8660 \approx \frac{76}{\ell}$$

$$\ell \approx 87.8 \qquad \text{The hydraulic ladder should be extended 87.8 feet.}$$

You can use right triangle trigonometry to solve problems involving geometric figures. The *apothem* of a regular polygon is the measure of a line segment from the center of the polygon to the midpoint of one of its sides.

Example 5

A regular hexagon is inscribed in a circle with diameter 7.52 cm. Find the apothem. Round the answer to the nearest hundredth.

First, draw a diagram.

If the diameter of the circle is 7.52 cm, the radius is 7.52 ÷ 2 or 3.76 cm.

The measure of the central angle is 360° ÷ 12 or 30°.

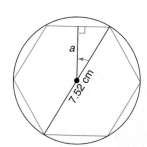

(continued on the next page)

Write an equation using trigonometric ratios.

$$\cos 30° = \frac{a}{3.76} \qquad \cos = \frac{side\ adjacent}{hypotenuse}$$

$$3.76 \cos 30° = a$$
$$3.2562 \approx a$$

Thus, the apothem of the hexagon is about 3.26 cm.

There are many other applications that require trigonometric solutions. For example, surveyors use special instruments to find the measures of angles of elevation and angles of depression. An **angle of elevation** is the angle between a horizontal line and the line of sight from an observer to an object at a higher level. An **angle of depression** is the angle between a horizontal line and the line of sight from the observer to an object at a lower level.

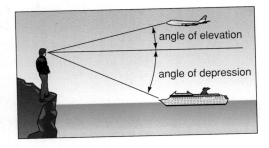
angle of elevation
angle of depression

Example 6

APPLICATION

Engineering

A flagpole 40 feet high stands on top of the Wentworth Building. From a point P in front of Bailey's Drugstore, the angle of elevation of the top of the pole is 54°54′, and the angle of elevation of the bottom of the pole is 47°30′. To the nearest foot, how high is the building?

Draw a diagram to model the situation.
Let x = the height of the building.
Let a = the distance from P to the foot of the building.

$$\tan 47°30' = \frac{x}{a} \qquad \tan \alpha = \frac{side\ opposite}{side\ adjacent}$$

$$\tan 54°54' = \frac{40 + x}{a}$$

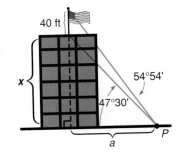

40 ft

x

54°54′

47°30′

a

P

Solve each equation for a. Then, the following is true.

$$\frac{40 + x}{\tan 54°54'} = \frac{x}{\tan 47°30'}$$

$$\tan 47°30'(40 + x) = x(\tan 54°54')$$
$$40\tan 47°30' = x(\tan 54°54' - \tan 47°30')$$
$$43.6523 = 0.3315x$$
$$x = 131.68$$

Use your calculator to change degrees and minutes to a decimal.

So, the building is about 132 feet tall.

MATH JOURNAL

Make up a problem that uses right triangles and solve.

CHECKING FOR UNDERSTANDING

Communicating Mathematics

Read and study the lesson to answer each question.

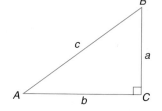

1. **Write** an equation that relates $\sin A$ and $\cos B$. Describe the equation in words. Use the drawing at the right.

2. **Sketch** a right triangle in which $\sin A = \dfrac{3}{7}$ and $\tan B = \dfrac{2\sqrt{10}}{3}$.

3. **Describe** one way to use trigonometry and right triangles to find the height of your school building.

Guided Practice

Write an equation that would enable you to solve for the indicated measures. Do *not* solve.

4. If $A = 20°$ and $c = 35$, find a.

5. If $b = 13$ and $A = 76°$, find a.

6. If $a = 6$ and $c = 12$, find B.

7. If $a = 21.2$ and $b = 9$, find A.

8. If $B = 16°$ and $c = 13$, find a.

9. If $A = 49°13'$ and $a = 10$, find b.

10. If $c = 16$ and $a = 7$, find b.

11. If $a = 7$ and $b = 12$, find A.

12. If $a = 5$ and $b = 6$, find c.

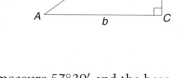

13. **Geometry** The base angles of an isosceles triangle measure $57°30'$ and the base is 7.5 cm long. Find the lengths of the congruent sides and the altitude to the base to the nearest tenth.

14. A tower 250 meters high casts a shadow 176 meters long. Find the angle of elevation of the sun to the nearest minute.

EXERCISES

Practice

Solve each triangle described, given the triangle below. Round angle measures to the nearest minute and side measures to the nearest tenth.

15. $A = 41°$, $b = 7.44$

16. $B = 42°10'$, $a = 9$

17. $b = 22$, $A = 22°22'$

18. $a = 21$, $c = 30$

19. $A = 45°$, $c = 7\sqrt{2}$

20. $a = 31.2$, $c = 42.4$

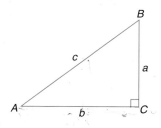

21. $A = 37°15'$, $b = 11$

22. $a = 11$, $b = 21$

23. $A = 55°55'$, $c = 16$

24. $B = 78°8'$, $a = 41$

25. **Geometry**   A regular pentagon has an apothem of 7.43 centimeters. Find the length of a side of the pentagon and the length of the radius of the circumscribed circle to the nearest hundredth.

26. **Geometry**   A rectangle is 17.5 centimeters by 26.2 centimeters. Find the angle formed by the longer side and a diagonal to the nearest minute.

27. **Geometry**   A 7.4-centimeter chord subtends a central angle of 41° in a circle. Find the length of the radius of the circle to the nearest hundredth.

Critical Thinking

28. Derive two formulas for the length of the altitude of the triangle shown at the right, given that a, b, and θ are known. Justify each of the steps you take in your reasoning.

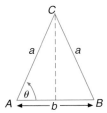

Applications and Problem Solving

29. **Home Maintenance**   Mrs. James is using a 6-meter ladder to clean the windows on her second floor. Her ladder stands on level ground and rests against the side of her house at a point 4 meters from the ground. How far from the side of her house is the foot of the ladder? Round your answer to the nearest hundredth.

30. **Aeronautics**   A hot air balloon rises at the rate of 70 feet per minute. An observer 420 feet from the place of ascent watches the balloon rise.

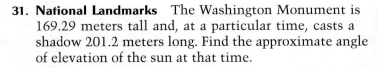

   a. Write an expression for the altitude of the balloon in terms of time, t minutes, and the angle of elevation, θ.
   b. What is the altitude of the balloon after 3.5 minutes, 22 minutes, and 1 hour?
   c. What are the angles of elevation to the nearest minute for each amount of time?

31. **National Landmarks**   The Washington Monument is 169.29 meters tall and, at a particular time, casts a shadow 201.2 meters long. Find the approximate angle of elevation of the sun at that time.

32. **Geography**   To find the height of the Charles Mound in Illinois, Jim located two points, C and D, on a plain in line with the mound. The angle of depression from the mound to C is 49°42′, and the angle of depression to D is 26°27′. Jim reads in an encyclopedia that the mound is 1235 feet high . Find the distance from C to D to the nearest foot.

33. **Cartography**   Find the bearing of a road that runs directly from A to B, with B being 3 miles north and 1.7 miles east of A. (The bearing of B from A is the positive angle with vertex at A measured clockwise from north to B.)

34. Aeronautics A satellite S is in a geosynchronous orbit; that is, it stays over the same point T on Earth as Earth rotates on its axis. From the satellite, a spherical cap of Earth is visible. The circle bounding this cap is called the horizon circle. The line of sight from the satellite is tangent to a point Q on the surface of Earth. If the radius of Earth, CQ, is 3963 miles and the satellite is 23,300 miles from the surface of Earth at Q, what is the radius of the horizon circle, PQ, in miles?

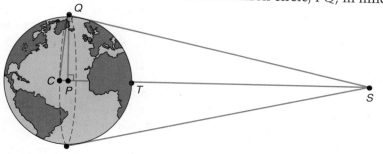

Mixed Review

35. Determine if -2 is a root of $x^3 - 3x^2 - 2x + 4 = 0$. **(Lesson 4-1)**

36. Find the measure of the reference angle for an angle of $210°$. **(Lesson 5-1)**

37. Suppose θ is an angle in standard position and $\tan \theta > 0$. State the quadrants in which the terminal side of θ can lie. **(Lesson 5-3)**

38. Find $\sin \frac{\pi}{4}$, $\cos \frac{\pi}{4}$, and $\tan \frac{\pi}{4}$. **(Lesson 5-4)**

39. College Entrance Exam Choose the best answer.
In the figure at the right, four semicircles are drawn on the four sides of a rectangle. What is the total area of the shaded regions?

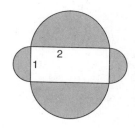

(A) 5π **(B)** $\dfrac{5\pi}{2}$ **(C)** $\dfrac{5\pi}{4}$ **(D)** $\dfrac{5\pi}{8}$ **(E)** $\dfrac{5\pi}{16}$

CASE STUDY FOLLOW-UP

Refer to Case Study 1: *Buying a Home*, on pages 956–958.

The angle of elevation of a roof is called the pitch.

1. The pitch of the roof shown is $28°$. The width of the house is 48 feet. Find h, the maximum height of the attic.

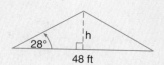

2. The area of a one-story square house with a pitched roof is the same as the mean area of a one-family dwelling built in 1990. The pitch of the roof is $20°$. Find the maximum height of the attic.

3. Research Adobe homes in New Mexico have flat roofs. Farm houses in Maine have roofs with steep pitches. Research the relationship between local weather and roof design. Then write a 1-page paper describing your findings.

5-6 The Law of Sines

Objectives

After studying this lesson, you should be able to:
- determine whether a triangle has zero, one, or two solutions, and
- solve triangles by using the law of sines.

Application

The blades of many power lawn mowers are rotated by a two-stroke engine with a piston sliding back and forth in the engine cylinder. As the piston moves back and forth, the connecting rod rotates the circular crankshaft. In order to ensure smooth operation of your lawn mower, design engineers synchronize the linear motion of the piston with the circular motion of the crankshaft. Suppose that the crankshaft rotates at 1200 revolutions per minute and that P is at the vertical position when it begins to rotate. How many centimeters is the piston from the rim of the crankshaft after 0.01 second? *This problem will be solved in Example 2.*

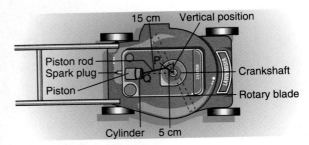

The law of sines can be used to solve triangles that are not right triangles. Consider $\triangle ABC$ inscribed in $\odot O$ with diameter DB. Let $2r$ be the measure of the diameter. Draw $\overline{AD}$. Then $\angle D \cong \angle C$ since they intercept the same arc. So, $\sin C = \sin D$. $\angle BAD$ is inscribed in a semicircle, so it is a right angle. So, $\sin D = \dfrac{c}{2r}$. Thus, since $\sin D = \sin C$, it follows that

$$\sin C = \frac{c}{2r} \text{ or } \frac{c}{\sin C} = 2r.$$

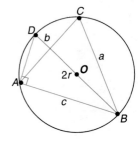

Similarly, by drawing diameters through A and C, $\dfrac{b}{\sin B} = 2r$ and $\dfrac{a}{\sin A} = 2r$. Thus, the following is true.

$$\frac{a}{\sin A} = \frac{b}{\sin B} = \frac{c}{\sin C}$$

These equations state that the ratio of any side of a triangle to the sine of the angle opposite that side is a constant for a given triangle. These equations are collectively called the law of sines.

Law of Sines

Let $\triangle ABC$ be any triangle with a, b, and c representing the measures of the sides opposite the angles with measurements A, B, and C respectively. Then, the following is true.

$$\frac{a}{\sin A} = \frac{b}{\sin B} = \frac{c}{\sin C}$$

For example, in a $30° - 60°$ right triangle with sides measuring 1, $\sqrt{3}$, and 2, the ratios are as follows.

$$\frac{1}{\sin 30°} = \frac{\sqrt{3}}{\sin 60°} = \frac{2}{\sin 90°}$$

Example 1

Solve $\triangle ABC$ if $A = 29°10'$, $B = 62°20'$, and $c = 11.5$. Round angle measures to the nearest minute and side measures to the nearest tenth.

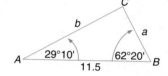

First, find the measure of angle C.
$C = 180° - (29°10' + 62°20')$ or $88°30'$

Use the law of sines to find a and b.

$$\frac{a}{\sin A} = \frac{c}{\sin C} \qquad\qquad \frac{b}{\sin B} = \frac{c}{\sin C}$$

$$\frac{a}{\sin 29°10'} = \frac{11.5}{\sin 88°30'} \qquad\qquad \frac{b}{\sin 62°20'} = \frac{11.5}{\sin 88°30'}$$

$$a = \frac{11.5 \sin 29°10'}{\sin 88°30'} \qquad\qquad b = \frac{11.5 \sin 62°20'}{\sin 88°30'}$$

$$\approx \frac{11.5(0.4874)}{0.9997} \qquad\qquad \approx \frac{11.5(0.8857)}{0.9997}$$

$$\approx 5.6 \qquad\qquad\qquad \approx 10.2$$

Therefore, $C = 88°30'$, $a \approx 5.6$, and $b \approx 10.2$.

Example 2

APPLICATION

Mechanics

Refer to the application at the beginning of the lesson. How many centimeters is the piston from the rim of the crankshaft after 0.01 second? Round your answer to the nearest tenth.

The crankshaft rotates at 1200 revolutions per minute or 20 revolutions per second. So, point P turns through $20(360°)$ or $7200°$ every second or $72°$ every 0.01 second. Thus, the measure of $\angle O$ is $90° - 72°$ or $28°$.

$$\frac{PQ}{\sin O} = \frac{OP}{\sin Q}$$

$$\frac{15}{\sin 28°} = \frac{5}{\sin Q}$$

Apply the law of sines to $\angle Q$.

$$\sin Q = \frac{5 \sin 28°}{15}$$

$$\sin Q \approx 0.1565$$

$$Q \approx 9°$$

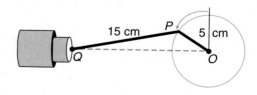

(continued on the next page)

So after 0.01 second, $\angle P = 180° - (28° + 9°)$ or $143°$. Apply the law of sines again.

$$\frac{QO}{\sin 143°} = \frac{15}{\sin 28°}$$

$$QO = \frac{15 \sin 143°}{\sin 28°}$$

$$\approx \frac{15(0.6018)}{0.4695}$$

$$\approx 19.2$$

Therefore, the piston is about $19.2 - 5$ or 14.2 centimeters from the rim of the crankshaft after 0.01 second.

When the measures of two sides of a triangle and the measure of the angle opposite one of them are given, there may not always be one solution. However, one of the following will be true.

1. No triangle exists.
2. Exactly one triangle exists.
3. Two triangles exist.

In other words, there may be no solution, one solution, or two solutions. Suppose you know the measures of a, b, and A. Consider the following cases.

Case 1: $m \angle A < 90°$

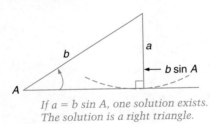

If $a = b \sin A$, one solution exists. The solution is a right triangle.

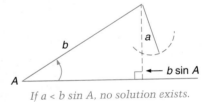

If $a < b \sin A$, no solution exists.

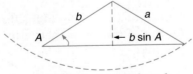

If $a > b \sin A$ and $a \geq b$, one solution exists.

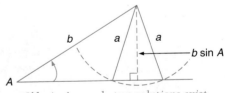

If $b \sin A < a < b$, two solutions exist.

Case 2: $m \angle A \geq 90°$

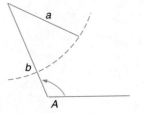

If $a \leq b$, no solution exists.

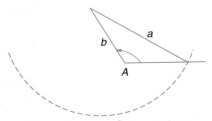

If $a > b$, one solution exists.

Example 3

Solve $\triangle ABC$ if $A = 63°10'$, $b = 18$, and $a = 17$. Round angle measures to the nearest minute and side measures to the nearest tenth.

To determine the number of solutions, first compare a with $b \sin A$.

$$b \sin A = 18 \sin 63°10'$$
$$= 18(0.8923)$$
$$= 16.1$$

Since $63°10' < 90°$ and $b \sin A < a < b$ ($16.1 < 17 < 18$), two solutions exist. Use the law of sines to find B.

$$\frac{17}{\sin 63°10'} = \frac{18}{\sin B}$$

$$\sin B = \frac{18 \sin 63°10'}{17}$$

$$\sin B \approx \frac{18(0.8923)}{17}$$

$$\sin B \approx 0.9448 \qquad \textit{Use a calculator to find B.}$$
$$B \approx 70°52' \qquad \textit{Round to the nearest minute.}$$

Since we know that there are two solutions, another measurement for B is about $180° - 70°52'$ or $109°8'$. Now, solve each triangle.

Solution 1

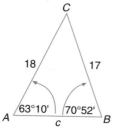

$$C \approx 180° - (63°10' + 70°52')$$
$$\approx 45°58'$$

$$\frac{17}{\sin 63°10'} \approx \frac{c}{\sin 45°58'}$$

$$c \approx \frac{17 \sin 45°58'}{\sin 63°10'}$$

$$\approx \frac{17(0.7189)}{0.8923}$$

$$\approx 13.7$$

One solution is $B \approx 70°52'$, $C \approx 45°58'$, and $c \approx 13.7$.

Solution 2

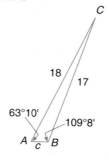

$$C \approx 180° - (63°10' + 109°8')$$
$$\approx 7°42'$$

$$\frac{17}{\sin 63°10'} \approx \frac{c}{\sin 7°42'}$$

$$c \approx \frac{17 \sin 7°42'}{\sin 63°10'}$$

$$\approx \frac{17(0.1340)}{0.8923}$$

$$\approx 2.6$$

Another solution is $B \approx 109°8'$, $C \approx 7°42'$, and $c \approx 2.6$.

Example 4

Solve $\triangle ABC$ if $A = 43°, b = 20$, and $a = 11$. Round angle measures to the nearest minute and side measures to the nearest tenth.

$$b \sin A = 20 \sin 43°$$
$$\approx 20(0.6820)$$
$$\approx 13.64$$

Since $43° < 90°$ and $11 < 13.64$, no solution exists.

CHECKING FOR UNDERSTANDING

Communicating Mathematics

Read and study the lesson to answer each question.

1. **Describe** one set of conditions for which the law of sines can be used to solve a triangle.

2. **Explain** how you know that a triangle has no solution.

3. **Sketch** $\triangle ABC$ if $m\angle ABC = 40°$, BC measures 4 inches, and $m\angle BCA = 60°$.

Guided Practice

Write an equation that would enable you to solve for the indicated measures. Do *not* solve.

4. If $A = 40°, B = 60°$, and $a = 20$, find b.

5. If $b = 2.8, A = 53°$, and $B = 61°$, find a.

6. If $b = 10, a = 14$, and $A = 50°$, find B.

7. If $b = 16, c = 12$, and $B = 42°$, find C.

Determine the number of possible solutions. If a solution exists, solve the triangle. Round angle measures to the nearest minute and side measures to the nearest tenth.

8. $A = 140°, b = 10, a = 3$

9. $C = 17°, a = 10, c = 11$

10. $A = 30°, a = 4, b = 8$

11. $B = 160°, a = 10, A = 41°$

12. $A = 60°, b = 2, a = \sqrt{3}$

13. $A = 38°, b = 10, a = 8$

EXERCISES

Practice

Determine the number of possible solutions. If a solution exists, solve the triangle. Round angle measures to the nearest minute and side measures to the nearest tenth.

14. $a = 8, A = 49°, B = 57°$

15. $a = 6, b = 8, A = 150°$

16. $a = 26, b = 29, A = 58°$

17. $A = 40°, B = 60°, c = 20$

18. $B = 70°, C = 58°, a = 84$

19. $a = 12, b = 14, A = 90°$

20. $A = 25°, a = 125, b = 150$

21. $A = 76°, a = 5, b = 20$

22. $A = 37°20', B = 51°30', c = 125$

23. $b = 40, a = 32, A = 125°20'$

24. $A = 107°13', a = 17.2, c = 12.2$

Use the law of sines to show that each statement is true.

25. $\dfrac{a - c}{c} = \dfrac{\sin A - \sin C}{\sin C}$

26. $\dfrac{b + c}{b - c} = \dfrac{\sin B + \sin C}{\sin B - \sin C}$

27. $\dfrac{a}{b} = \dfrac{\sin A}{\sin B}$

28. $\dfrac{b}{a + b} = \dfrac{\sin B}{\sin A + \sin B}$

Critical
Thinking

29. Suppose the angle measures of $\triangle ABC$ are equal to the angle measures of $\triangle XYZ$. Use the law of sines to show that the triangles are similar, but not necessarily congruent.

Applications
and Problem
Solving

30. Navigation A ship's captain plans to sail to a port that is 450 miles away and 12° east of north. The captain sails the ship due north, turns the ship, and sails 316 miles.

a. Through what angle should the captain turn the ship to arrive at port?

b. How many hours will it take to arrive at the turning point if the captain chooses a speed of 23 miles per hour?

c. Instead of the plan above, the captain decides to sail 200 miles north, turn through an angle of 20° east of north, and then sail along a straight course. Will the ship reach the port by following this plan?

31. Landscaping A corner of McCormick Park occupies a triangular area that faces two streets that meet at an angle measuring 85°. The sides of the area facing the streets are each 60 feet in length. The park's landscaper wants to plant begonias around the edges of the triangular area. Find the perimeter of the triangular area.

Mixed
Review

32. Solve $\dfrac{8}{x - 3} = \dfrac{x + 5}{x - 3}$. **(Lesson 4-6)**

33. Identify all angles that are coterminal with a 45° angle. **(Lesson 5-1)**

34. Find $\sin \dfrac{13\pi}{6}$. **(Lesson 5-4)**

35. When the angle of elevation of the sun is 27°, the shadow of a tree is 25 meters long. How tall is the tree? Round to the nearest tenth. **(Lesson 5-5)**

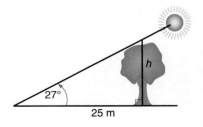

36. College Entrance Exam Choose the best answer.

A person is hired for a job that pays $500 per month and receives a 10% raise in each following month. In the fourth month, how much will that person earn?

(A) 550   **(B)** 600.50   **(C)** 650.50   **(D)** 665.50   **(E)** 700

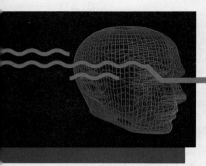

Technology

Solving Triangles

▶ BASIC
Spreadsheets
Software

```
10   INPUT AA, AB, SA
15   IF AA=0 THEN 180
20   LET AC=180 - AA - AB
30   IF AC < = 0 THEN 130
40   LET SB = SA*SIN (AB*3.14159/180)/
     SIN(AA*3.14159/180)
50   LET SC = SA*SIN (AC*3.14159/180)/
     SIN(AA*3.14159/180)
60   PRINT "ANGLE A = ";AA
70   PRINT "ANGLE B = ";AB
80   PRINT "ANGLE C = ";AC
90   PRINT "SIDE A = ";SA
100  PRINT "SIDE B = ";INT (SB + .5)
110  PRINT "SIDE C = ";INT (SC + .5)
120  GOTO 10
130  PRINT "ANGLE A = ";AA
140  PRINT "ANGLE B = ";AB
150  PRINT "SIDE A = ";SA
160  PRINT "NO TRIANGLE EXISTS."
170  GOTO 10
180  END
```

The BASIC program shown at the left solves triangles by using the law of sines, if solutions exist. In the program, AA represents the degree measure of angle A and SA represents the measure of side a. To begin using the program, enter the degree measures of two angles and the measure of a side opposite one of the angles. To exit the program, enter **0,0,0** at the prompt and then press **[Enter]**.

For a triangle in which the measure of angle A is 35°, the measure of angle B is 45°, and the measure of side A is 6 units, the display is as shown at the right.

```
RUN
ANGLE A = 35
ANGLE B = 45
ANGLE C = 100
SIDE A = 6
SIDE B = 8
SIDE C = 17
```

EXERCISES

Use the BASIC program to solve each triangle.

1. $A = 43°, B = 58°, a = 17$
2. $A = 40°, B = 60°, a = 20$
3. $A = 53°, B = 61°, a = 2.8$

The Law of Cosines

Objective

After studying this lesson, you should be able to:
- solve triangles by using the law of cosines.

Application

Suzanne Chu made the diagram at the right based on the measurements listed on the deed to her property. She wants to lay a gravel pathway as shown by the dashed line in the diagram. The pathway will lead from two opposite corners to a circular fountain in the center. The circular region is to have a radius of 10 feet. How long will the pathway be, not including the fountain area? *This problem will be solved in Example 1.*

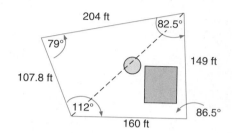

If either two sides and the included angle or three sides of a triangle are given, the law of sines cannot be used to solve the triangle. Another formula is needed. Consider $\triangle ABC$ with a height of h units and sides measuring a units, b units, and c units. Suppose $\overline{DC}$ is x units long. Then $\overline{BD}$ is $(a - x)$ units long.

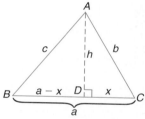

The Pythagorean theorem and the definition of the cosine function can be used to show how $\angle C, a, b,$ and c are related.

$c^2 = (a - x)^2 + h^2$ *Apply the Pythagorean theorem to $\triangle ADB$.*

$c^2 = a^2 - 2ax + x^2 + h^2$ *Expand $(a - x)^2$.*

$c^2 = a^2 - 2ax + b^2$ *$b^2 = x^2 + h^2$ in $\triangle ADC$.*

$c^2 = a^2 - 2a(b \cos C) + b^2$ *$\cos C = \dfrac{x}{b}$, so $x = b \cos C$.*

$c^2 = a^2 + b^2 - 2ab \cos C$

By drawing altitudes from B and C, you can derive similar formulas for a^2 and b^2. All three formulas, which make up the law of cosines, can be summarized as follows.

Law of Cosines

Let $\triangle ABC$ be any triangle with $a, b,$ and c representing the measures of sides opposite angles with measurements $A, B,$ and C, respectively. Then, the following are true.

$$a^2 = b^2 + c^2 - 2bc \cos A$$
$$b^2 = a^2 + c^2 - 2ac \cos B$$
$$c^2 = a^2 + b^2 - 2ab \cos C$$

You can use the law of cosines to solve the application at the beginning of the lesson.

Example 1

APPLICATION

Landscaping

How many feet long is the pathway on Suzanne Chu's property, not including the fountain area, which has a radius of 10 feet?

The proposed gravel pathway is the third side of a triangle whose other two sides measure 107.8 feet and 204 feet with an included angle of 79°. Use the law of cosines to find its length, ℓ, in feet.

$\ell^2 = (204)^2 + (107.8)^2 - 2(204)(107.8)\cos 79°$

$\ell^2 \approx 41{,}616 + 11{,}620.84 - 43{,}982.4(0.1908)$

$\ell^2 \approx 44{,}844.998$

$\ell \approx 211.8$

Since the fountain will have a diameter of 20 feet, the length of the gravel pathway, exclusive of the fountain area, is about $212 - 20$ or 192 feet.

Many times, you will have to use both the law of cosines and the law of sines to solve triangles.

Example 2

Solve $\triangle ABC$ if $A = 52°10'$, $b = 6$, and $c = 8$. Round angle measures to the nearest minute and side measures to the nearest tenth.

Use the law of cosines to find a.

$a^2 = b^2 + c^2 - 2bc \cos A$

$a^2 = 6^2 + 8^2 - 2(6)(8) \cos 52°10'$

$a^2 \approx 36 + 64 - 96(0.6134)$

$a^2 \approx 41.11$

$a \approx 6.4$

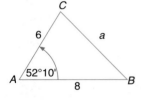

Use the law of sines to find B.

$$\frac{6.4}{\sin 52°10'} = \frac{6}{\sin B}$$

$$\sin B = \frac{6 \sin 52°10'}{6.4}$$

$$\sin B \approx \frac{6(0.7898)}{6.4}$$

$$\sin B \approx 0.7404$$

$$B \approx 47°46' \qquad \textit{Round to the nearest minute.}$$

$C = 180° - (52°10' + 47°46')$ or $80°4'$

Therefore, $a = 6.4$, $B = 47°46'$, and $C = 80°4'$.

Example 3

Solve △ABC if *a* = 21, *b* = 16.7, and *c* = 10.3. Round angle measures to the nearest minute.

Use the law of cosines.

$$a^2 = b^2 + c^2 - 2bc \cos A$$
$$(21)^2 = (16.7)^2 + (10.3)^2 - 2(16.7)(10.3) \cos A$$
$$441 = 278.89 + 106.09 - 344.02 \cos A$$
$$\cos A = \frac{441 - 278.89 - 106.09}{-344.02}$$
$$\cos A \approx -0.1628$$
$$A \approx 99°22' \quad \textit{Round to the nearest minute.}$$

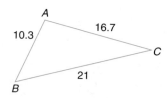

Use the law of sines.

$$\frac{21}{\sin 99°22'} = \frac{16.7}{\sin B}$$
$$\sin B = \frac{16.7 \sin 99°22'}{21}$$
$$\sin B \approx \frac{16.7(0.9867)}{21}$$
$$\sin B \approx 0.7847$$
$$B \approx 51°42' \quad \textit{Round to the nearest minute.}$$

$$C \approx 180° - (99°22' + 51°42')$$
$$\approx 28°56'$$

Therefore, *A* ≈ 99°22', *B* ≈ 51°42' and *C* ≈ 28°56'.

CHECKING FOR UNDERSTANDING

Communicating Mathematics

Read and study the lesson to answer each question.

1. **State** the law of cosines in your own words.

2. **Explain** how the Pythagorean theorem is a special case of the law of cosines.

3. **Explain** how to find *C* in the triangle below. Do not solve.

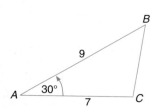

4. **a. Sketch** a triangle in which *AC* is 4 inches, *AB* is 6 inches, and *C* = 53°.
   b. Are there any other triangles that satisfy these conditions?

The measures of three parts of a triangle are given. Determine whether the law of sines or the law of cosines should be used to solve the triangle. Then solve each triangle. Round angle measures to the nearest minute and side measures to the nearest tenth.

5. $a = 14, b = 15, c = 16$

6. $C = 35°, a = 11, b = 10.5$

7. $a = 10, A = 40°, c = 8$

8. $A = 40°, b = 6, c = 7$

9. $c = 21, a = 14, B = 60°$

10. $A = 40°, C = 70°, c = 14$

11. $b = 17, B = 45°28', a = 12$

12. $A = 28°50', b = 4, c = 2.9$

13. **Aviation**   Natalie is flying from Dallas to Little Rock, a distance of 319 miles. She starts her flight 15° off course and flies on this course for 75 miles. How far is she from Little Rock?

EXERCISES

Solve each triangle. Round angle measures to the nearest minute and side measures to the nearest tenth.

14. $a = 4, b = 5, c = 7$

15. $b = 7, c = 10, A = 51°$

16. $A = 52°40', b = 540, c = 490$

17. $a = 5, b = 6, c = 7$

18. $A = 61°25', b = 191, c = 205$

19. $a = 3, b = 7, c = 5$

20. $b = 13, a = 21.5, C = 39°20'$

21. $a = 11.4, b = 13.7, c = 12.2$

22. $A = 40°, B = 59°, c = 14$

23. $a = 9, c = 5, B = 120°$

24. The sides of a triangle measure 6.8 cm, 8.4 cm, and 4.9 cm. Find the measure of the smallest angle to the nearest minute.

25. A parallelogram has sides of 55 cm and 71 cm. Find the length of each diagonal to the nearest tenth if the largest angle measures 106°.

26. **Carpentry**   Two carpenters are trying to carry a board down a corridor and around a corner in the Lawson Building. The figure at the right is an overhead view of the corridor. What is the length of the longest board that can be carried parallel to the floor through the corridor, to the nearest tenth of a foot? *Hint: You may want to make a scale drawing and experiment with thin strips of paper.*

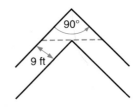

27. Navigation   Two ships leave San Francisco at the same time. One travels 40° west of north at a speed of 20 knots. The other travels 10° west of south at a speed of 15 knots. How far apart are they after 11 hours? *1 knot = 1 nautical mile per hour.*

28. Surveying   A triangular plot of land was measured as shown in the figure at the right. Trigonometry was used to calculate *BC*. However, it was discovered that the measure of angle *A* was really 36.95°. By how many feet was the calculation of *BC* in error?

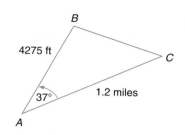

29. Find the slope of the line through $(-3, 2)$ and $(5, 7)$.   **(Lesson 1-4)**

30. Find the value of $\begin{vmatrix} 7 & 9 \\ 3 & 6 \end{vmatrix}$.   **(Lesson 2-3)**

31. Find the reference angle for an angle that measures 150°.   **(Lesson 5-1)**

32. Find $\sin \theta$ when the point with coordinates $(-7, 24)$ lies on the terminal side of the angle in standard position.   **(Lesson 5-3)**

33. Surveying   Two surveyors 560 yards apart sight a boundary marker C on the other side of a canyon at angles of 27° and 38°. Their measurements will be used to plan a bridge that spans the canyon. How long will the bridge be, to the nearest tenth of a yard? **(Lesson 5-6)**

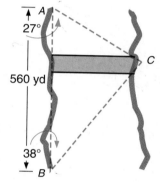

34. College Entrance Exam   Sue's average score for three bowling games was 162. In the second game, Sue scored 10 less than in the first game. In the third game, she scored 13 less than in the second game. What was her score in the first game?

5-8 Area of Triangles

Objective

After studying this lesson, you should be able to:
- find the area of triangles.

Application

Surveyors calculate measures of distances and angles so that they can represent boundary lines of parcels of land. The diagram at the right is a plot of John and Reneé Walters' land. What is the area of the region to the nearest square foot? *This problem will be solved in Example 5.*

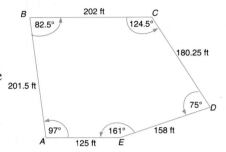

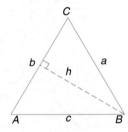

The area of any triangle can be expressed in terms of two sides of the triangle and the measure of the included angle. Suppose you know the measures of $\overline{AC}$ and $\overline{AB}$ and the measure of the included angle A in $\triangle ABC$, shown at the left. Let K represent the measure of the area of $\triangle ABC$ and let h represent the measure of the altitude from B. Then, $K = \frac{1}{2}bh$.

But, $\sin A = \frac{h}{c}$ or $h = c \sin A$. If you substitute c $\sin A$ for h, the result is the following formula.

$$K = \frac{1}{2}bc \sin A$$

If you drew altitudes from A and C, you could also develop the formulas $K = \frac{1}{2}ab \sin C$ and $K = \frac{1}{2}ac \sin B$.

Example 1

Find the area of $\triangle ABC$ if $a = 7.5$, $b = 9$, and $C = 100°$. Round your answer to the nearest tenth.

$K = \frac{1}{2}ab \sin C$

$= \frac{1}{2}(7.5)(9) \sin 100°$

$\approx (33.75)(0.9848)$

≈ 33.237

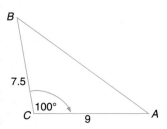

The area is about 33.2 square units.

You can also find the area of a triangle if you know the measures of one side of the triangle and two angles. By the law of sines, $\frac{b}{\sin B} = \frac{c}{\sin C}$ or $b = \frac{c \sin B}{\sin C}$. If you substitute $\frac{c \sin B}{\sin C}$ for b in $K = \frac{1}{2}bc \sin A$, the result is the following formula.

$$K = \frac{1}{2}c^2 \frac{\sin A \sin B}{\sin C}$$

Similarly, $K = \frac{1}{2}a^2 \frac{\sin B \sin C}{\sin A}$ and $K = \frac{1}{2}b^2 \frac{\sin A \sin C}{\sin B}$.

Example 2

Find the area of $\triangle ABC$ if $a = 18.6$, $A = 19°20'$, and $B = 63°50'$. Round your answer to the nearest tenth.

First, find the measure of angle C.
$$C = 180° - (19°20' + 63°50')$$
$$= 96°50'$$

Then, find the area of the triangle.
$$K = \frac{1}{2}a^2 \frac{\sin B \sin C}{\sin A}$$
$$= \frac{1}{2}(18.6)^2 \frac{\sin 63°50' \sin 96°50'}{\sin 19°20'}$$
$$\approx \frac{1}{2}(345.96)\frac{(0.8975)(0.9929)}{(0.3311)}$$
$$\approx 465.6$$

The area is about 465.6 square units.

If you know the measures of three sides of a triangle, you can find the area by using the law of cosines and the formula $K = \frac{1}{2}bc \sin A$.

Example 3

Find the area of $\triangle ABC$ if $a = \sqrt{2}$, $b = 2$, and $c = 3$. Round your answer to the nearest tenth.

First, solve for A by using the law of cosines.
$$a^2 = b^2 + c^2 - 2bc \cos A$$
$$(\sqrt{2})^2 = 2^2 + 3^2 - 2(2)(3) \cos A$$
$$\cos A = \frac{(\sqrt{2})^2 - 2^2 - 3^2}{-2(2)(3)}$$
$$\cos A \approx 0.9167$$
$$A \approx 23°33' \qquad \textit{Round to the nearest minute.}$$

(continued on the next page)

Then, find the area.

$$K = \frac{1}{2}bc \sin A$$

$$\approx \frac{1}{2}(2)(3) \sin 23°33'$$

$$\approx 3(0.3995)$$

$$\approx 1.199$$

The area is about 1.2 square units.

If you know the measures of three sides of any triangle, you can also use Hero's formula to find the area of a triangle.

Hero's Formula

If the measures of the sides of a triangle are a, b, and c, then the area, K, of the triangle is found as follows.

$$K = \sqrt{s(s-a)(s-b)(s-c)}, \text{ where } s = \frac{1}{2}(a+b+c)$$

s is called the semiperimeter.

Example 4

Use Hero's formula to find the area of $\triangle ABC$ if $a = 20$, $b = 30$, and $c = 40$. Round your answer to the nearest tenth.

$$s = \frac{1}{2}(20 + 30 + 40)$$

$$= 45$$

$$K = \sqrt{s(s-a)(s-b)(s-c)}$$

$$= \sqrt{45(45-20)(45-30)(45-40)}$$

$$= \sqrt{84,375}$$

$$\approx 290.5$$

The area is about 290.5 square units.

To solve application problems, you may have to use more than one formula for the area of a triangle.

Example 5

APPLICATION

Land Measure

Refer to the application at the beginning of the lesson. What is the area of the Walters' land to the nearest square foot?

Separate the region as shown at the right. The area of region $ABCDE$ is the sum of the areas of $\triangle ABC$, $\triangle ACE$, and $\triangle ECD$. Let a be the side with measure 202 ft, let b be the side with measure 158 ft, let c be the side with measure 201.5 ft, and let e be the side with measure 180.25 ft.

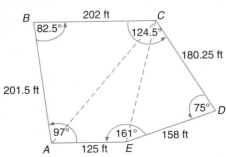

Area of $\triangle ABC = \frac{1}{2}ac \sin B$

$$= \frac{1}{2}(202)(201.5) \sin 82.5°$$

$$\approx (20,351.5)(0.9914)$$

$$\approx 20,176.5$$

Area of $\triangle ECD = \frac{1}{2}eb \sin D$

$$= \frac{1}{2}(180.25)(158) \sin 75°$$

$$\approx (14,239.75)(0.9659)$$

$$\approx 13,754.2$$

To find the area of $\triangle ACE$, first use the law of cosines to find AC and AE.

$AC = \sqrt{(202)^2 + (201.5)^2 - 2(202)(201.5) \cos 82.5°}$
    ≈ 266.0

$CE = \sqrt{(180.25)^2 + (158)^2 - 2(180.25)(158) \cos 75°}$
    ≈ 206.7

Then apply Hero's formula. $\qquad s \approx \frac{1}{2}(125 + 266 + 206.7)$ or 298.85

$K = \sqrt{s(s-a)(s-b)(s-c)}$
  $\approx \sqrt{298.85(298.85 - 125)(298.85 - 266)(298.85 - 206.7)}$
  $\approx 12,540.9$

Area $ABCDE$	=	area of $\triangle ABC$	+	area of $\triangle ECD$	+	area of $\triangle ACE$
	$\approx$	20,176.5	+	13,754.2	+	12,540.9
	$\approx$	46,471.6				

So, the area of region $ABCDE$ is about 46,472 square feet, which is a little more than 1 acre.

A **segment** of a circle is the region bounded by an arc and its chord. If the arc is a minor arc, then the area of the segment can be found by subtracting the area of $\triangle ODE$ from the area of sector $ODFE$, where F is a point on the arc. Let S represent the area of the segment.

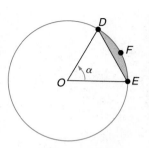

$S = ($area of sector $ODFE) - ($ area of $\triangle ODE)$

$\quad = \frac{1}{2}r^2\alpha - \frac{1}{2}de \sin O$

$\quad = \frac{1}{2}r^2\alpha - \frac{1}{2}r \cdot r \sin \alpha \qquad d = r, e = r,$ and $C = \alpha$

$\quad = \frac{1}{2}r^2(\alpha - \sin \alpha)$

<table>
<tr><td>

*Area of a
Circular Segment*

</td><td>

If α is the measure of the central angle expressed in radians and the radius of the circle measures r units, then the area of the segment, S, is as follows.

$$S = \frac{1}{2}r^2(\alpha - \sin \alpha)$$

</td></tr>
</table>

Example 6

A sector has a central angle of $150°$ in a circle with a radius of 11.5 inches. Find the area of the circular segment to the nearest tenth.

First, convert $150°$ to radian measure.

$150° = 150 \times \dfrac{\pi}{180}$

$\qquad = \dfrac{5\pi}{6}$ or 2.6180 *Round to four decimal places.*

Then, find the area of the segment.

$S = \dfrac{1}{2}r^2(\alpha - \sin \alpha)$

$\quad \approx \dfrac{1}{2}(11.5)^2(2.6180 - \sin 2.6180)$

$\quad \approx \dfrac{1}{2}(132.25)(2.6180 - 0.5000)$

$\quad \approx 140.1$

The area of the circular segment is 140.1 in^2.

Portfolio

Select an item that shows something new you learned in this chapter and place it in your portfolio.

CHECKING FOR UNDERSTANDING

*Communicating
Mathematics*

Read and study the lesson to answer each question.

1. **Explain** how to use triangle area formulas to find the area of the parallelogram $ABCD$ shown below. Do *not* solve.

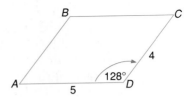

2. **Describe** how to find the length of the third side of a triangle if its area is known and the length of two sides are known.

3. **Sketch** the plot of land on which your school is located. List the measurements you would take to find its area. What is the fewest number of measurements needed?

Write an equation that can be used to find the area of each triangle described below. Then find the area of each triangle to the nearest tenth.

    4. $a = 3, b = 4, C = 120°$ **5.** $c = 20, A = 45°, B = 30°$

    6. $a = 4, b = 6, c = 8$ **7.** $A = 43°, b = 16, c = 12$

    8. $a = 6, B = 52°, c = 4$ **9.** $b = 12, B = 135°, C = 30°$

Find the area of each circular segment to the nearest tenth, given its central angle θ, and the measure of the radius of the circle.

    10. $\theta = 81°, r = 16$ **11.** $\theta = \dfrac{5\pi}{6}, r = 15$

EXERCISES

Find the area of each triangle to the nearest tenth.

    12. $c = 3.2, A = 16°, B = 31°45'$

    13. $a = 2, b = 7, c = 8$

    14. $A = 60°, a = 2, B = 75°$

    15. $a = 174, b = 138, c = 188$

    16. $a = 8, B = 60°, C = 75°$

    17. $a = 11, B = 50°6', c = 5$

    18. $a = 17, b = 13, c = 19$

    19. $b = 146.2, c = 209.3, A = 62°12'$

    20. $a = 19.42, c = 19.42, B = 31°16'$

Find the area of each circular segment to the nearest tenth, given its central angle, θ, and the measure of the radius of the circle.

    21. $\theta = \dfrac{3\pi}{4}, r = 24$ **22.** $\theta = 120°, r = 8$

    23. $\theta = 85°, r = 2.1$ **24.** $\theta = \dfrac{5\pi}{8}, r = 6$

    25. $\theta = 26°, r = 42$ **26.** $\theta = \dfrac{\pi}{5}, r = 16.25$

For Exercises 27—30, round all answers to the nearest tenth.

    27. The adjacent sides of a parallelogram measure 8 cm and 12 cm, and one angle measures 60°. Find the area of the parallelogram.

    28. A rhombus has sides of 5 cm each and one diagonal is 6 cm long. Find the area of the rhombus.

    29. A regular pentagon is inscribed in a circle whose radius measures 7 cm. Find the area of the pentagon.

    30. A regular octagon is inscribed in a circle with a radius of 5 cm. Find the area of the octagon.

31. The program below finds the area of a triangle given the lengths of all three sides of the triangle.

```
Prgm 5:HEROS
:ClrHome                           Clears the text screen.
:Disp "SIDE 1="                    Displays the quoted message.
:Input A                           Accepts a value for A.
:Disp "SIDE 2="
:Input B
:Disp "SIDE 3="
:Input C
:If A+B≤C                          Checks for the sum of 2 sides less than or equal
:Goto 1                            to the third side; goes to line Lbl 1.
:If A+C≤B
:Goto 1
:If B+C≤A
:Goto 1
:(A+B+C)/2→S                       Calculates the semiperimeter, stores it in S.
:√(S(S−A)(S−B)(S−C))→K             Calculates the area, stores it in K.
:Disp "AREA="
:Disp K                            Displays the value stored in K.
:Goto 2
:Lbl 1                             Sets a marker at position 1.
:Disp "NO TRIANGLE"
:Lbl 2
```

Use the program to find the area of each triangle to the nearest tenth. First, find the lengths of the three sides.

a. $a = 17.7$, $b = 21$, $C = 78°10'$

b. $a = 10$, $A = 75°20'$, $B = 49°40'$

32. Derive a formula for the area of $\triangle ABC$ if you know the measures of $\angle B$, $\angle C$ and side AC.

33. Model Building The diagram below shows the dimensions for a sail on a wooden model ship.

a. Find the area of the sail to the nearest square inch.

b. If the scale of the length of the real ship to the model ship is 1 foot : $\frac{1}{2}$ inch, find the area of the real sail in square feet.

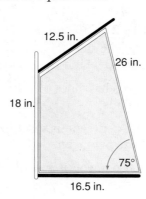

34. Exterior Design The diagram below is of the covering for an exterior doorway to a home. The owner wants to cover it with copper sheeting. If the center piece is an isosceles trapezoid, find the area to be covered to the nearest square foot.

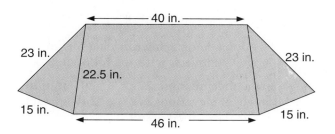

35. Manufacturing The diagram at the right shows a storage bin with a hopper at the bottom. The storage bin is shaped like a rectangular prism.

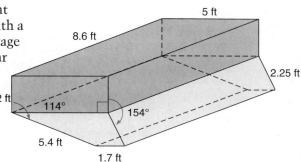

  a. Find the area of a vertical cross section to the nearest square foot.

  b. Find the total volume of the bin and hopper.

Mixed Review

36. Write the standard form of the equation of the line through $(3, -2)$ that is parallel to the line $3x - y + 7 = 0$. **(Lesson 1-6)**

37. Determine whether the graph of $f(x) = x^4 + 3x^2 + 2$ is symmetric with respect to the x-axis, the y-axis, neither, or both. **(Lesson 3-1)**

38. Use the remainder theorem to find the remainder when $x^3 + 8x + 1$ is divided by $x - 2$. **(Lesson 4-3)**

39. Mechanics A circular saw 18.4 cm in diameter rotates at 2400 revolutions per second. What is the linear velocity at which a saw tooth strikes the cutting surface in centimeters per second? **(Lesson 5-2)**

40. Solve $\triangle ABC$ if $A = 36°$, $b = 13$, and $c = 6$. Round angle measures to the nearest minute and side measures to the nearest tenth. **(Lesson 5-7)**

41. College Entrance Exam Compare quantities A and B below.
Write A if quantity A is greater.
Write B if quantity B is greater.
Write C if the two quantities are equal.
Write D if there is not enough information to determine the relationship.

$x - y = 4$
$x + y = 3$

(A) x **(B)** y

VOCABULARY

Upon completing this chapter, you should be
familiar with the following terms:

angle of depression	**272**	**241**	minute
angle of elevation	**272**	**240**	quadrantal angle
angular velocity	**249**	**241**	radian
central angle	**247**	**243**	reference angle
circular functions	**255**	**241**	second
coterminal angle	**242**	**250**	sector
degree	**241**	**291**	segment
Hero's formula	**290**	**240**	standard position
law of cosines	**283**	**255**	trigonometric functions
law of sines	**277**	**241**	unit circle
linear velocity	**249**		

SKILLS AND CONCEPTS

OBJECTIVES AND EXAMPLES	REVIEW EXERCISES

Upon completing this chapter, you should
be able to:

Use these exercises to review and prepare for
the chapter test.

■ change from radian to degree measure, and
vice versa   (**Lesson 5-1**)

Change $-\dfrac{5\pi}{3}$ radians to degree measure.

$$-\frac{5\pi}{3} = -\frac{5\pi}{3} \times \frac{180°}{\pi} = -300°$$

**Change each radian measure to degree
measure.**

1. $\dfrac{\pi}{3}$ **2.** $-\dfrac{5\pi}{12}$ **3.** $\dfrac{4\pi}{3}$

**Change each degree measure to radian
measure in terms of π.**

4. $150°$ **5.** $-315°$ **6.** $270°$

■ find the reference angle for a given angle
(**Lesson 5-1**)

Find the measure of the reference angle
for $\dfrac{8\pi}{3}$.

$\dfrac{8\pi}{3}$ is coterminal with $\dfrac{2\pi}{3}$ in Quadrant II.

$\alpha' = \pi - \dfrac{2\pi}{3}$

$\quad = \dfrac{\pi}{3}$ *Reference rule b*

**Find the reference angle for each angle with
the given measure.**

7. $\dfrac{7\pi}{4}$ **8.** $405°$

9. $-\dfrac{4\pi}{3}$ **10.** $-60°$

11. $\dfrac{11\pi}{6}$ **12.** $870°$

■ find the length of an arc, given the measure of the central angle (Lesson 5-2)

Find the length of the arc intercepted by a central angle of $\frac{2\pi}{3}$ in a circle of radius 10 in.

$s = r\theta$

$\quad = 10\left(\frac{2\pi}{3}\right)$

$\quad \approx 20.9$ \qquad The length is about 20.9 in.

Given the measure of a central angle, find the measure of its intercepted arc in terms of π in a circle of diameter 30 cm.

13. $\frac{3\pi}{4}$ \qquad\qquad **14.** $\frac{\pi}{5}$

15. $75°$

16. $150°$

■ find the values of the six trigonometric functions of an angle (Lesson 5-3)

Find $\tan \theta$ if θ is in standard position and the point with coordinates $(5, 8)$ lies on the terminal side of the angle in standard position.

$\tan \theta = \frac{y}{x}$

$\qquad = \frac{8}{5}$

Find the value of the given trigonometric function of an angle in standard position if a point with the given coordinates lies on its terminal side.

17. $\sin \theta$; $(3, 3)$

18. $\cos \theta$; $(-5, 12)$

19. $\cot \theta$; $(8, -2)$

20. $\sec \theta$; $(-2, 0)$

■ find the values of the six trigonometric functions of special angles (Lesson 5-4)

Find $\sin \frac{4\pi}{3}$.

$\sin \frac{4\pi}{3} = -\sin \frac{\pi}{3}$

$\qquad\quad = -\frac{1}{2}$

Find each exact value. Do not use a calculator.

21. $\cos \frac{3\pi}{4}$ \qquad\qquad **22.** $\tan \frac{7\pi}{3}$

23. $\csc 120°$ \qquad\qquad **24.** $\cot 315°$

25. $\sin \frac{18\pi}{4}$ \qquad\qquad **26.** $\sec -\frac{5\pi}{6}$

■ solve right triangles (Lesson 5-5)

Solve $\triangle ABC$ if $C = 90°, c = 10$, and $A = 74°$.

$\frac{a}{10} = \sin 74°$ \qquad\qquad $\frac{b}{10} = \cos 74°$

$a \approx 9.6$ \qquad\qquad\quad $b \approx 2.8$

$B = 90° - 74° = 16°$

Solve each triangle ABC described below. Angle C is the right angle. Round angle measures to the nearest minute and side measures to the nearest tenth.

27. $A = 63°, a = 9.7$

28. $a = 2, b = 7$

29. $B = 83°, b = \sqrt{31}$

30. $a = 44, B = 44°44'$

OBJECTIVES AND EXAMPLES	REVIEW EXERCISES

■ solve triangles by using the law of sines
(Lesson 5-6)

In $\triangle ABC$, find a if $A = 51°$, $C = 32°$, and $c = 18$.

$$\frac{a}{\sin A} = \frac{c}{\sin C}$$

$$\frac{a}{\sin 51°} = \frac{18}{\sin 32°}$$

$$a = 26.4$$

Determine the number of possible solutions. If a solution exists, solve the triangle. Round angle measures to the nearest minute and side measures to the nearest tenth.

31. $A = 38°42'$, $a = 172$, $c = 203$

32. $a = 12$, $b = 19$, $A = 57°$

33. $A = 29°$, $a = 12$, $b = 15$

34. $A = 45°$, $a = 83$, $b = 79$

■ solve triangles by using the law of cosines
(Lesson 5-7)

In $\triangle ABC$, find a if $A = 63°$, $b = 20$, and $c = 14$.

$$a^2 = b^2 + c^2 - 2bc \cos A$$

$$a^2 = 20^2 + 14^2 - 2(20)(14) \cos 63°$$

$$a^2 \approx 341.77$$

$$a \approx 18.5$$

Solve each triangle. Round angle measures to the nearest minute and side measures to the nearest tenth.

35. $A = 51°$, $b = 40$, $c = 45$

36. $B = 19°$, $a = 51$, $c = 61$

37. $a = 11$, $b = 13$, $c = 20$

38. $B = 24°$, $a = 42$, $c = 6.5$

■ find the area of triangles   **(Lesson 5-8)**

Find the area of $\triangle ABC$ if $a = 6$, $b = 4$, and $C = 54°$.

$$K = \frac{1}{2}ab \sin C$$

$$= \frac{1}{2}(6)(4) \sin 54°$$

$$\approx 9.7$$

Find the area of each triangle to the nearest tenth. All answers are in square units.

39. $A = 20°$, $a = 19$, $C = 64°$

40. $a = 5$, $b = 7$, $c = 9$

41. $a = 11.7$, $b = 13.5$, $C = 81°20'$

42. $A = 42°$, $B = 65°$, $a = 63$

APPLICATIONS AND PROBLEM SOLVING

43. **Entertainment**   Shannon rides the Ferris wheel at the Franklin County Fair. She makes 12 revolutions during a 3-minute ride. If the diameter of the Ferris wheel is 30 feet, what is Shannon's linear velocity in miles per hour? Round your answer to the nearest tenth.   **(Lesson 5-2)**

44. **Landscaping**   Rusty wants to plant a circular flower bed around a tree in his front yard. He wants it to look like the drawing at the right.   **(Lesson 5-8)**

   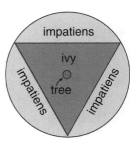

   a. If the radius of the entire flower bed measures 4 feet, and each chord is the same length, what is the area of each circular segment where he plans to plant impatiens? Round your answer to the nearest tenth.

   b. A flat of impatiens holds 24 plants. If he can plant 3 plants per square foot, how many flats should he buy?

Change each degree measure to radian measure in terms of π. Change each radian measure to degree measure.

1. $135°$

2. $-\dfrac{\pi}{5}$

3. $480°$

4. $\dfrac{7\pi}{12}$

Use the circle at the right to answer the following questions to the nearest tenth.

5. What is the length of the intercepted arc?

6. What is the area of the sector?

7. What is the area of the circular segment?

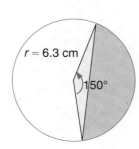

$r = 6.3$ cm

$150°$

Find each exact value. Do not use a calculator.

8. $\sin\dfrac{5\pi}{6}$

9. $\cot 210°$

10. $\sec 135°$

11. $\cos -\dfrac{\pi}{6}$

12. $\tan 225°$

Solve each right triangle. Angle C is the right angle. Round angle measures to the nearest minute and side measures to the nearest tenth.

13. $b = 42, A = 77°$

14. $c = 13, a = 12$

15. $c = 14, B = 32°$

Determine the number of possible solutions. If a solution exists, solve the triangle. Round angle measures to the nearest minute and side measures to the nearest tenth.

16. $a = 64, c = 90, C = 98°$

17. $a = 9, b = 20, A = 31°$

18. $a = 13, b = 7, c = 15$

19. $a = 20, c = 24, B = 47°$

Find the area of each triangle to the nearest tenth.

20. $A = 70°11', B = 43°55', b = 16.7$

21. $b = 11.5, c = 14, A = 20°$

22. Astronomy The linear velocity of Earth's moon is about 2300 mph. If the average distance from the center of Earth to the center of the moon is 2.4×10^5 miles, how long does it take the moon to make one revolution about Earth? Assume the orbit is circular.

23. Recreation At ground level, the measurement of the angle of elevation of a kite is $70°$. It is held by a string 65 meters long. How far is the kite above the ground?

24. Navigation A ship at sea is 70 miles from one radio transmitter and 130 miles from another. The measurement of the angle between signals is $130°$. How far apart are the transmitters?

25. Geometry An isosceles triangle has a base of 22 centimeters and a vertex angle measuring $36°$. Find its perimeter.

Bonus In $\triangle ABC$, where a, b, and c are measures of the sides, $\dfrac{a+b+c}{2+\sqrt{3}} = \dfrac{ab}{a+b-c}$. Find the measure of $\angle C$.

CHAPTER 6

GRAPHS AND INVERSES OF THE TRIGONOMETRIC FUNCTIONS

HISTORICAL SNAPSHOT

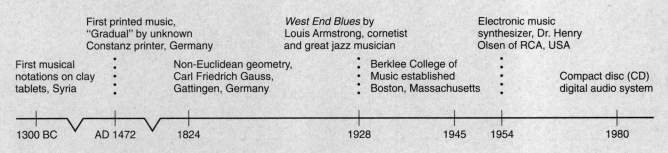

First musical notations on clay tablets, Syria

First printed music, "Gradual" by unknown Constanz printer, Germany

Non-Euclidean geometry, Carl Friedrich Gauss, Gattingen, Germany

West End Blues by Louis Armstrong, cornetist and great jazz musician

Berklee College of Music established Boston, Massachusetts

Electronic music synthesizer, Dr. Henry Olsen of RCA, USA

Compact disc (CD) digital audio system

1300 BC   AD 1472   1824   1928   1945   1954   1980

CHAPTER OBJECTIVES

In this chapter, you will:

- Graph trigonometric functions and compound functions.
- Determine the amplitude, period, and phase shift for a graph.
- Evaluate and graph inverse trigonometric functions.
- Solve problems involving simple harmonic motion.

CAREER GOAL: Independent Music Production

Did you ever take music lessons? Eric Essix began playing the guitar when he was ten years old. Then ten years ago, he began "dabbling in studio work," and today, he has two record albums to his credit. Although Eric has been a performer for over half of his life, he says, "I like the challenge of taking sounds that you hear in your head and reproducing them on tape. I like the challenge of taking an abstract idea and turning it into something that other people can hear and enjoy." Eric's goal is to "produce contemporary jazz music as an independent producer. As such, I will have opportunities to produce jazz music for advertising companies, recording companies, television, and films."

You may not think that musicians use mathematics in their work, but Eric disagrees. "Music production is so computer-based that it is impossible to produce music in the '90s without the use of mathematics," Eric says. "We frequently use mathematics to derive a particular sound or effect. For instance, we can place music in a sonic environment that is totally different from the one in which it was recorded. You may listen to a CD that sounds like the artist was in a large concert hall. However, he may have been performing in a small recording studio.

"If you shout into the Grand Canyon, you can count the number of times that your echo repeats. That is called the *delay time.* Delay times are different for sounds produced in a large concert hall versus sound produced in a small recording studio. However, we can mathematically change the number and frequency of delay times of a sound recorded in a small studio to make the sound simulate that produced in a large concert hall."

As to why he loves music, Eric says, "Music is probably the most powerful force in the universe. Nothing else affects people and covers the entire spectrum of emotions as music does. I want to be a part of that."

For up-to-date information on music production, visit:
www.glencoe.com/sec/math/amc/mathnet

Graphs of the Trigonometric Functions

Objective

After studying this lesson, you should be able to:
- use the graphs of the trigonometric functions.

Application

We live in an electronic age. Electricity powers our lights, heats our homes, and even runs our cars. However, unlike light or sound, under normal circumstances we cannot sense this form of energy directly. We must use tools, like an oscilloscope, to measure the behavior of electric currents.

FYI...

An electrocardiograph is a special device that uses an oscilloscope to monitor the activity of the heart. Dutch physician Willem Einthoven was the first to consider using an oscilloscope in this way. He received the 1924 Nobel Prize for Medicine for his research.

The electric power that is used in American homes is delivered in alternating current. An alternating current surges in one direction and then the other in a way that can be represented on the oscilloscope by a sine curve.

You could generate the coordinates of points on the parent graphs of trigonometric functions with a calculator or spreadsheet program. The graphs may be generated by using a graphing calculator or graphing software. The parent graphs of the six basic trigonometric functions are shown below and on the next page.

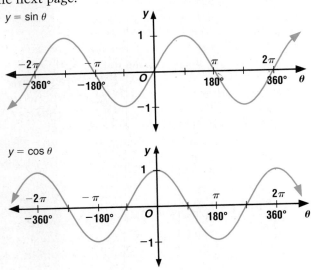

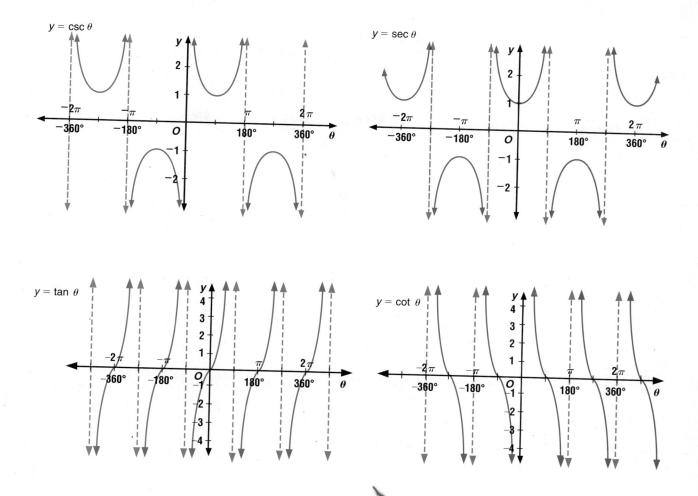

Do you notice a similarity between the sine and cosine curves? The cosine curve is the sine curve translated 90° to the left along the θ-axis. Since the cosecant and secant ratios are the reciprocals of the sine and cosine ratios, the same relationship holds true for their curves. However, this does not hold true for the tangent and cotangent curves. *Why?*

A quadrantal angle is an angle in standard position whose terminal side coincides with one of the coordinate axes. So the measures of the quadrantal angles are ⋯, –270°, –180°, –90°, 0°, 90°, 180°, 270°, ⋯ . Knowing the characteristics of the basic trigonometric graphs can help you quickly determine quadrantal values of the trigonometric functions. The value of the sine function is 0 at –360°, –180°, 0°, 180°, 360°, and at all other integral multiples of 180°. The maximum value of the sine function is 1 for 90° or –270°, and the minimum is –1 for 270° or –90°. The fact that the trigonometric functions are **periodic** allows us to find more function values easily.

Periodic Function and Period	A function is periodic if, for some real number α, $f(x + \alpha) = f(x)$ for each x in the domain of f. The least positive value of α for which $f(x) = f(x + \alpha)$ is the period of the function.

Since the sine function is periodic, its graph repeats itself every 360°. So we can find other zero, maximum, and minimum values easily. For example, since we know that $\sin 90° = 1$, $\sin(90° + 360k°) = 1$ for any integral value of k. The same methods can be used to find special values of the other functions.

Example 1

Use the graph of the cosine function to find the values of θ for which $\cos \theta = 1$.

When $\cos \theta = 1$ and $-360 \leq \theta \leq 360°$, the value of θ is $-360°, 0°$, or $360°$. Since the cosine function has a period of $360°$, the values of θ for which $\cos \theta = 1$ are given by $0° + 360k°$ or simply $360k°$, where k is any integer.

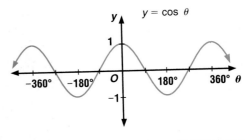

The graph of a trigonometric function may be drawn from the knowledge of its shape and the values of the function at integral multiples of $90°$.

Example 2

Graph the sine curve in the interval $-540° \leq \theta \leq 0°$.

Find the value of sine θ for $\theta = -540°, -450°, -270°, -180°, -90°,$ and $0°$.

Plot the points from these ordered pairs.
$(-540°, 0)$, $(-450°, -1)$, $(-360°, 0)$, $(-270°, 1)$, $(-180°, 0)$, $(-90°, -1)$, $(0°, 0)$

Connect these points with a smooth continuous curve.

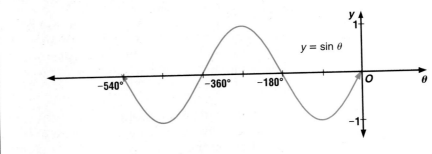

CHECKING FOR UNDERSTANDING

Communicating
Mathematics

Read and study the lesson to answer each question.

1. **Describe** how a cosine curve can be translated to coincide with the graph of a sine curve.

2. The vertical lines in the graphs of the tangent, cotangent, secant, and cosecant functions are asymptotes. What do they indicate about the values of θ where the asymptotes cross the θ-axis?

3. **Explore** the graphs of the trigonometric functions on your graphing calculator. What are the values of θ, for which the graphs of the sine and the cosecant functions are tangent to each other in the interval from $-360°$ to $360°$?

4. **Explain** which transformation(s) changes the graph of the tangent curve to the graph of the cotangent curve.

Guided Practice

Find each value by referring to the graphs of the sine and cosine functions.

5. $\cos 270°$
6. $\sin 90°$
7. $\sin 360°$
8. $\cos 450°$
9. $\cos 90°$
10. $\cos (-90°)$
11. $\sin 180°$
12. $\sin 450°$
13. $\sin (-90°)$

Find the values of θ for which each equation is true.

14. $\cos \theta = 1$
15. $\tan \theta = 1$
16. $\sin \theta = 0$
17. $\cot \theta = 0$
18. Graph the function $y = \cos x$ on the interval $-360°$ to $360°$.

EXERCISES

Practice

Find each value by referring to the graphs of the trigonometric functions.

19. $\sin 810°$
20. $\cos (-540°)$
21. $\tan 45°$
22. $\cot (-45°)$
23. $\csc (-90)°$
24. $\sec 0°$
25. $\tan 720°$
26. $\csc 270°$
27. $\cot 180°$

Find the values of θ for which each equation is true.

28. $\sin \theta = 1$
29. $\cos \theta = -1$
30. $\csc \theta = -1$
31. $\tan \theta = -1$
32. $\sec \theta = 1$
33. $\cot \theta = -1$

State the domain and range for each function.

34. $y = \sin \theta$
35. $y = \cos x$
36. $y = \cot x$
37. $y = \sec \theta$
38. $y = \tan \theta$
39. $y = \csc \theta$

Graph each function on the given interval.

40. $y = \sin x;\ -180° \leq x \leq 180°$
41. $y = \cos x;\ 270° \leq x \leq 630°$
42. $y = \tan x;\ 90° \leq x \leq 450°$
43. $y = \csc x;\ -540° \leq x \leq 0°$
44. $y = \sec x;\ -180° \leq x \leq 360°$
45. $y = \cot x;\ -90° \leq x \leq 360°$

Graphing Calculator

Use a graphing calculator to graph the sine and cosine functions on the same set of axes for $0° \leq x \leq 360°$. Use the graphs to find values of x, if any, for which each of the following is true.

46. $\sin x = -\cos x$
47. $\sin x \leq \cos x$
48. $\sin x \cos x > 1$
49. $\sin x \cos x \leq 0$
50. $\sin x + \cos x = 1$
51. $\sin x - \cos x = 0$

52. Graph the functions $y = \sin x$ and $y = 2 \sin x$ on the same set of axes on a graphing calculator. How do the graphs differ? Make a conjecture as to the shape of the graph of $y = 3 \sin x$. Then graph this function on the same axes. Was your conjecture correct?

53. Electronics The frequency of an alternating current is the number of complete back-and-forth cycles it goes through each second. The frequency is measured in hertz. The formula for the effective current, I_{eff}, of a maximum current of I_{max} volts with a frequency of f hertz is $I_{eff} = I_{max}\sqrt{2} \sin (2\pi ft)$, where t is measured in seconds.

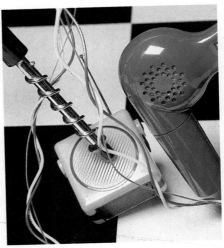

   a. The electric power delivered for household use in the United States is 110 volts and has a frequency of 60 hertz. Write the function that describes the effective current related to time.

   b. Graph the function on a graphing calculator using a range representing $\frac{1}{60}$ of a second. Then sketch the graph.

   c. Use the TRACE function of the calculator to find the times when the effective current is zero.

54. Physics A torsion pendulum is an object suspended by a wire or rod so that its plane is horizontal and it rotates back and forth around the wire without losing energy. Suppose that the pendulum is rotated θ_m radians and released. Then the angular displacement, or the angle of rotation, θ, at time t is $\theta = \theta_m \cos \omega t$, where ω is the angular frequency in radians per second. Suppose the angular frequency of a certain torsion pendulum is π radians per second and its initial rotation is $\frac{\pi}{4}$ radians.

   a. Write the equation for the angular displacement of the pendulum.

   b. What are the first two values of t for which the angular displacement of the pendulum is 0?

55. Write the slope-intercept form of the equation of the line through $(5, 2)$ and $(-1, 4)$. **(Lesson 1-5)**

56. Solve the system of equations below by using augmented matrices. **(Lesson 2-4)**

$$4x + 2y = 10$$
$$y = 6 - x$$

57. Determine whether the graph of $y = \pm\sqrt{x - 49}$ is symmetric with respect to the x-axis, the y-axis, the line $y = x$, the line $y = -x$, or the origin. **(Lesson 3-1)**

58. Is -4 a root of $x^4 + 3x^3 + 10x - 20 = 0$? **(Lesson 4-1)**

59. Solve $\dfrac{4}{x+1} = \dfrac{6}{x-3}$. **(Lesson 4-6)**

60. Change $\dfrac{5\pi}{16}$ radians to degree measure to the nearest minute. **(Lesson 5-1)**

61. Solve $\triangle ABC$ if $A = 40°$, $b = 16$, and $a = 9$. **(Lesson 5-6)**

62. Navigation Two boats carrying marine researchers leave McMurdo port at the same time. One boat travels east at a speed of 12 mph. The other boat travels northeast at 16 mph. If both boats keep on the same course, how far apart will they be after 2 hours? **(Lesson 5-7)**

63. Find the area to the nearest square unit of $\triangle ABC$ if $a = 17$, $b = 13$, and $c = 19$. **(Lesson 5-8)**

64. College Entrance Exam A certain fraction is equivalent to $\dfrac{2}{5}$. If the numerator is decreased by 2 and the denominator is increased by 1, then the new fraction is equivalent to $\dfrac{1}{4}$. What is the numerator of the original fraction?

DECISION MAKING

Drunk Driving

Drunk driving has become epidemic in the United States, particularly among teenagers. About 43% of deaths of 16- to 20-year-olds result from motor vehicle accidents, about half of which are alcohol related. In that age group there were about 3000 alcohol-related deaths in 1991. What can you do to raise awareness of the problem among your fellow students and help avert a tragedy in your community?

1. Describe some of the reasons that teenagers drink and drive. Outline an argument you could use to refute each of these reasons.

2. Research the problem locally. Find local statistics or stories you could use to illustrate the seriousness of the problem. Contact members of the community who have expertise in this area to come to your school to speak about what you can do to help.

3. Project Work with your group to complete one of the projects listed at the right based on the information you gathered in Exercises 1–2.

PROJECTS

- *Conduct an interview.*
- *Write a book report.*
- *Write a proposal.*
- *Write an article for the school paper.*
- *Write a report.*
- *Make a display.*
- *Make a graph or chart.*
- *Plan an activity.*
- *Design a checklist.*

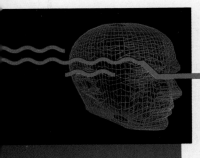

Technology

Ordered Pairs for Graphing Trigonometric Functions

Spreadsheet programs are among the most popular software for businesses. Spreadsheets are used in accounting, for generating charts, and when working with databases. Spreadsheets are also used in mathematics, mainly to examine functions by generating ordered pairs.

```
= = = = = A = = = = = B = = = = = = = C = = = = =
1    START  X =        -2
2    STOP   X =         2
3    STEP      =    (B2-B1)/10
4
5         ^X                        ^Y
6    +B1+0*B3                  @SIN(A6+@PI/3)
7    +B1+1*B3                  @SIN(A7+@PI/3)
8    +B1+2*B3                  @SIN(A8+@PI/3)

16   +B1+10*B3                 @SIN(A16+@PI/3)
```

The spreadsheet shown at the left calculates and displays a list of ordered pairs for the function $y = \sin\left(x + \dfrac{\pi}{3}\right)$. While only eleven ordered pairs are shown on the screen, you can generate others by changing the **Start X** and **Stop X** values. As soon as you change an X value, new ordered pairs are calculated and displayed.

The display at the right shows what the spreadsheet looks like for the domain $-2 \le x \le 2$. Only the first four ordered pairs are shown.

If you want to generate ordered pairs for a trigonometric function that is different from the one used, you can change

```
= = = = = A = = = = = B = = = = C = = =
1    START  X =        -2
2    STOP   X =         2
3    STEP      =       0.4
4
5           X                        Y
6          -2                    -0.81504
7          -1.6                  -0.52507
8          -1.2                  -0.15220
9          -0.8                   0.244687
```

the function when typing in the new information. Note that you will have to change all of the expressions in cells C6 through C16.

EXERCISES

Make a detailed graph of the function $y = \sin\left(x + \dfrac{\pi}{3}\right)$ by using the spreadsheet to generate ordered pairs over the following domains.

1. $-4 \le x \le -2$ **2.** $-2 \le x \le 0$ **3.** $0 \le x \le 3$ **4.** $2 \le x \le 4$

6-2 Amplitude, Period, and Phase Shift

Objectives

After studying this lesson, you should be able to:
- find the amplitude, period, and phase shift for a trigonometric function, and
- write equations of trigonometric functions given the amplitude, period, and phase shift.

Application

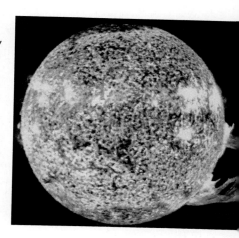

A *sunspot* is a relatively dark area on the surface of the sun. These spots are cooler areas, with temperatures of about 7000°F, compared with temperatures of about 11,000°F on the rest of the sun's surface. Astronomers have long been studying the number of sunspots that occur on the surface of the sun each year. They have found that the number of sunspots counted in a given year varies periodically from about 10 to 110 per year. The most sunspots occurred in the years 1750 and 1948, and 18 complete cycles occurred between these years. If the number of sunspots, y, is graphed with respect to years since 1750, x, the graph is similar to a modified cosine curve.

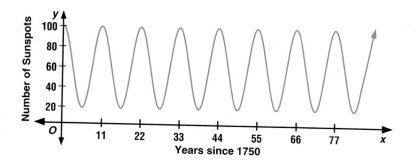

Recall from Chapter 3 that changes to the equation of the parent graph affect the appearance of the graph. One way that the parent graphs of the trigonometric functions can be modified is to multiply by constants. Let's consider an equation of the form $y = A \sin \theta$. We know that the maximum absolute value of $\sin \theta$ is 1. If every value of $\sin \theta$ is multiplied by A, the maximum value of $A \sin \theta$ is $|A|$. Similarly, the maximum value of $A \cos \theta$ is $|A|$. The absolute value of A, or $|A|$, is called the **amplitude** of the functions $y = A \sin \theta$ and $y = A \cos \theta$.

If $A < 0$ in the equation $y = A \sin \theta$ or $y = A \cos \theta$, the curve is the reflection of the graph of the function $y = |A| \sin \theta$ or $y = |A| \cos \theta$ over the θ-axis.

| Amplitude of Sine Cosine Functions | The amplitude of the functions $y = A \sin \theta$ and $y = A \cos \theta$ is the absolute value of A, or $|A|$. |

The amplitude can also be described as the absolute value of one-half the difference of the maximum and minimum function values.

$$\left| \frac{A - (-A)}{2} \right| = |A|$$

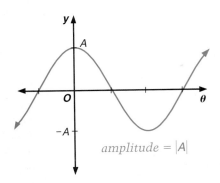

amplitude = |A|

The tangent, cotangent, secant, and cosecant functions do not have amplitudes because their values increase and decrease without bound for values that approach certain multiples of 90°.

Example 1

State the amplitude of the function $y = 3 \cos \theta$. Graph $y = 3 \cos \theta$ and $y = \cos \theta$ on the same set of axes. Compare the graphs.

According to the definition of amplitude, the amplitude of $y = A \cos \theta$ is $|A|$. So the amplitude of $y = 3 \cos \theta$ is $|3|$ or 3.

Make a table of values. Then graph the points and draw a smooth curve.

θ	0°	45°	90°	135°	180°	225°	270°	315°	360°
$\cos \theta$	1	$\frac{\sqrt{2}}{2}$	0	$-\frac{\sqrt{2}}{2}$	-1	$-\frac{\sqrt{2}}{2}$	0	$\frac{\sqrt{2}}{2}$	1
$3 \cos \theta$	3	$\frac{3\sqrt{2}}{2}$	0	$-\frac{3\sqrt{2}}{2}$	-3	$-\frac{3\sqrt{2}}{2}$	0	$\frac{3\sqrt{2}}{2}$	3

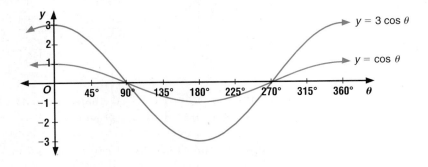

The graphs cross the θ-axis at $\theta = 90°$ and $\theta = 270°$. Also, both functions reach their maximum value at $\theta = 0°$ and $\theta = 360°$ and their minimum value at $\theta = 180°$. But the maximum and minimum values of the function $y = \cos \theta$ are 1 and –1, and the maximum and minimum values of the function $y = 3 \cos \theta$ are 3 and –3.

EXPLORATION: Graphing Calculator

1. Use the range values below to set up a viewing window. Mode should be in radians.

$$-4.7 \leq X \leq 4.8 \qquad -3 \leq Y \leq 3 \qquad XSCL = 1 \qquad YSCL = 1$$

2. Graph the following functions on the same screen.
   a. $\sin x$ **b.** $2 \sin x$ **c.** $3 \sin x$

3. Describe the change in the graph of $f(x) = A \sin x$, where $A > 0$, as A increases.

4. Make a conjecture about the behavior of the graph of $f(x) = A \sin x$, if $A < 0$. Test your conjecture.

The graphs of trigonometric functions can be modified by changing the amplitude of the function. We can modify graphs of trigonometric functions in a second way. Consider an equation of the form $y = \sin k\theta$, where k is any integer. Since the period of the sine function is $360°$, the following identity can be developed.

If θ is expressed in radians, the period of $y = \sin k\theta$ and $y = \cos k\theta$ is $\dfrac{2\pi}{k}$. The period of $y = \tan k\theta$ is $\dfrac{\pi}{k}$.

$$y = \sin k\theta$$
$$= \sin(k\theta + 360°)$$
$$= \sin k \left(\theta + \frac{360°}{k} \right)$$

Therefore, the period of $y = \sin k\theta$ is $\dfrac{360°}{k}$. Similarly, the period of $y = \cos k\theta$ is $\dfrac{360°}{k}$. The period of $y = \tan k\theta$ is $\dfrac{180°}{k}$ since the period of the tangent function is $180°$.

Period of Sine, Cosine, and Tangent Functions	**The period of the functions $y = \sin k\theta$ and $y = \cos k\theta$ is $\dfrac{360°}{k}$. The period of the function $y = \tan k\theta$ is $\dfrac{180°}{k}$.**

Example 2

State the period of the function $y = \sin 4\theta$. Then graph the function and $y = \sin \theta$ on the same set of axes.

The definition of the period of $y = \sin k\theta$ is $\dfrac{360°}{k}$. Therefore, the period of $y = \sin 4\theta$ is $\dfrac{360°}{4}$ or $90°$.

If you are using a graphing calculator, set the range first: Xmin = 0, Xmax = 360, Xscl = 30, Ymin = –1.5, Ymax = 1.5, Yscl = 1, Xres = 1. Mode should be in degrees.

(continued on the next page)

Then enter both $y = \sin x$ and $y = \sin 4x$ and press GRAPH.

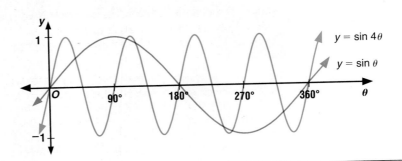

EXPLORATION: Graphing Calculator

1. Use the range values below to set up a viewing window. Mode should be in radians.

$$-4.7 \le X \le 4.8 \qquad -3 \le Y \le 3 \qquad XSCL = 1 \qquad YSCL = 1$$

2. Graph the following functions on the same screen.

   a. $\sin x$ **b.** $\sin \left(x + \dfrac{\pi}{2} \right)$ **c.** $\sin \left(x + \dfrac{\pi}{3} \right)$

3. Describe the behavior of the graph of $f(x) = \sin(x + c)$, where $c > 0$, as c increases.

4. Make a conjecture about what happens to the graph of $f(x) = \sin(x + c)$ if $c < 0$ and continues to decrease. Test your conjecture.

The phase shift is the least value of $|k\theta + c|$ for which $y = 0$.

In addition to changing the amplitude or period, the third way that a trigonometric graph can be modified is by a **phase shift**. Consider an equation of the form $y = A \sin(k\theta + c)$, where $A \ne 0$, $k \ne 0$, and $c \ne 0$. To find a zero of the function, find a value of θ for which $0 = A \sin(k\theta + c)$. Since $\sin 0 = 0$, solving $k\theta + c = 0$ will yield a zero of the function.

$$k\theta + c = 0$$
$$\theta = -\frac{c}{k}$$

Therefore, $y = 0$ when $\theta = -\dfrac{c}{k}$. The value of $-\dfrac{c}{k}$ is the phase shift. When $c > 0$, the graph of $y = A \sin(k\theta + c)$ is the graph of $y = A \sin k\theta$, shifted $\left| \dfrac{c}{k} \right|$ to the left. When $c < 0$, the graph of $y = A \sin(k\theta + c)$ is the graph of $y = A \sin k\theta$ shifted $\left| \dfrac{c}{k} \right|$ to the right.

Phase Shift of All Trigonometric Functions	The phase shift of the function $y = A \sin(k\theta + c)$ is $-\frac{c}{k}$. If $c > 0$, the shift is to the left. If $c < 0$, the shift is to the right. This definition applies to all of the trigonometric functions.

Example 3

State the phase shift of the function $y = \tan(\theta - 45°)$. Then use a graphing calculator to graph the function and $y = \tan\theta$ on the same set of axes.

The phase shift is $-\frac{c}{k} = -\frac{-45°}{1}$ or 45°. Since c is less than 0, the shift is to the right.

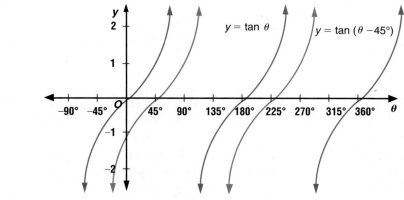

We can write an equation for a trigonometric function if we are given the amplitude, period, and phase shift.

Example 4

Find the possible equations of a cosine function with amplitude 3, period 90°, and phase shift 45°.

The form of the equation will be $y = A \cos(k\theta + c)$. First, find the possible values of A for an amplitude of 3.

$|A| = 3$
$A = 3 \text{ or } -3$

Now find the value of k when the period is 90°.

$\frac{360°}{k} = 90°$
$k = 4$

Then find the value of c for a phase shift of 45°.

$-\frac{c}{k} = 45°$

$-\frac{c}{4} = 45°$
$c = -180°$

The possible equations are $y = 3 \cos(4\theta - 180°)$ or $y = -3 \cos(4\theta - 180°)$.

Many real-world situations have cyclical characteristics that can be described with trigonometric functions. When you are writing an equation to describe a situation, remember the characteristics of the sine and cosine graphs. If you know the function value when $x = 0$ and whether the function is increasing or decreasing, you can choose the appropriate trigonometric function to write an equation for the situation.

Function value at $x = 0$	Sign of A	Function
maximum	positive	cosine
minimum	negative	cosine
0, with values increasing	positive	sine
0, with values decreasing	negative	sine

Example 5

The motion of a weight on a certain kind of spring can be described by a modified trigonometric function. At time 0, Carrie pushes the weight upward 3 inches from its equilibrium point and releases it. She finds that the weight returns to the point three inches above the equilibrium point after 2 seconds.

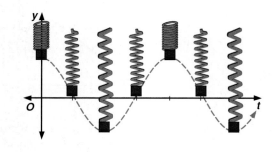

a. **Write a function that represents the position of the spring in reference to its equilibrium point, y, in terms of the time, t, in seconds after it was released.**

a. At time 0, the weight is 3 inches above the equilibrium point and at its maximum displacement from 0. Since the weight will fall, the values will get smaller. The function will be a cosine function with a positive value of A.

The maximum and minimum points are 3 and –3. Thus, the amplitude is $\left| \dfrac{3 - (-3)}{2} \right|$ or 3. Thus, $A = 3$.

The weight makes a complete cycle in 2 seconds. Thus, the period is 2.

$$\dfrac{2\pi}{k} = 2 \qquad \textit{Use radians.}$$
$$k = \pi$$

Now write the equation.

$$y = A \cos kt$$
$$y = 3 \cos \pi t \qquad A = 3, k = \pi$$

MATH JOURNAL

Write how you would explain amplitude, period, and phase shift to another student and write the examples you would use.

b. What will the position of the weight be after 15.5 seconds?

b. Use a calculator to find y when $t = 15.5$ to find the position of the weight after 15.5 seconds. *Make sure your calculator is in radian mode.*

$$y = 3\cos\pi(15.5)$$

15.5 $\boxed{\times}$ $\boxed{\pi}$ $\boxed{=}$ $\boxed{\text{COS}}$ $\boxed{\times}$ 3 $\boxed{=}$ *0*

After 15.5 seconds, the weight will be at the equilibrium point.

CHECKING FOR UNDERSTANDING

Communicating Mathematics

Read and study the lesson to answer each question.

1. **Graph** the functions $y = \cos x$ and $y = \sin(x + 90°)$ on a graphing calculator. What is true of these graphs?

2. **Explain** what happens to the graph of $y = A\sin\theta$ if A is increased.

3. **Write** two sentences to describe phase shift to a classmate.

Guided Practice

State the amplitude, period, and phase shift for each function.

4. $y = 4\sin\theta$

5. $y = 2\sin 5\theta$

6. $y = 2\cos 2\theta$

7. $y = \tan(2x - \pi)$

8. $y = 2\cos 4\theta$

9. $y = 3\cos(\theta - 90°)$

10. $y = \tan 2(\theta - 180°)$

11. $y = 243\sin(15\theta - 40°)$

Write an equation for each function.

12. a sine function with amplitude 3, period 720°, and phase shift 60°

13. a cosine function with amplitude 4, period 4π, and phase shift $\dfrac{\pi}{2}$

14. a tangent function with period 180° and phase shift 25°

EXERCISES

Practice

State the amplitude, period, and phase shift for each function.

15. $y = 2\cos\theta$

16. $y = 10\tan 4\theta$

17. $y = 110\sin 20\theta$

18. $y = 2\sin\theta$

19. $y = -7\sin 6\theta$

20. $y = 4\sin\dfrac{\theta}{2}$

21. $y = \dfrac{1}{4}\cos\dfrac{\theta}{2}$

22. $y = 12\cos 3\left(x - \dfrac{\pi}{2}\right)$

23. $y = -6\cos(180° - \theta)$

24. $y = 10\sin\left(\dfrac{1}{3}\theta - 300°\right)$

Write an equation of the sine function with each amplitude, period, and phase shift.

25. amplitude $= 5$, period $= 360°$, phase shift $= 60°$

26. amplitude $= \dfrac{2}{3}$, period $= \pi$, phase shift $= \dfrac{\pi}{4}$

27. amplitude $= 17$, period $= 45°$, phase shift $= -60°$

28. amplitude $= \dfrac{1}{2}$, period $= \dfrac{3\pi}{2}$, phase shift $= -\dfrac{\pi}{4}$

29. amplitude $= 7$, period $= 225°$, phase shift $= -90°$

Write an equation of the cosine function with each amplitude, period, and phase shift.

30. amplitude $= \dfrac{1}{3}$, period $= 180°$, phase shift $= 0°$

31. amplitude $= 3$, period $= 180°$, phase shift $= 120°$

32. amplitude $= 100$, period $= 630°$, phase shift $= -90°$

33. amplitude $= \dfrac{7}{3}$, period $= 150°$, phase shift $= 270°$

34. amplitude $= 1$, period $= \dfrac{3\pi}{4}$, phase shift $= -\dfrac{\pi}{3}$

Graph each function.

35. $y = \dfrac{1}{2}\cos\theta$

36. $y = 3\sin\theta$

37. $y = 3\sec\theta$

38. $y = 4\sin\dfrac{\theta}{2}$

39. $y = \sin(\theta - 45°)$

40. $y = \cos(\theta + 30°)$

41. $y = -\dfrac{1}{2}\cos\dfrac{3}{4}\theta$

42. $y = -6\sin\left(2x + \dfrac{\pi}{4}\right)$

43. The graph of $y = \tan x$ approaches an asymptote at $x = 90°$. What is the first positive value of x for which each of the following has a vertical asymptote?
  a. $y = \tan(x - 45°)$ **b.** $y = \tan 3x$
  c. $y = \tan(2x - 90°)$ **d.** $y = \tan(kx - c)$

Graphing Calculator

Use a graphing calculator to graph the following functions on the same screen. Then write an equation that relates the sine and cosine functions.

44. $\sin\left(x + \dfrac{\pi}{2}\right)$

45. $\cos x$

Critical Thinking

46. Graph the functions $y = \sin 2x$ and $y = 2\cos x$ on a graphing calculator. Make a conjecture as to what the graph of $y = \sin 2x + 2\cos x$ will look like. Then check your results with the graphing calculator.

47. Zoology In predator-prey situations, the number of animals in each category tends to vary periodically. A certain region has pumas as predators and deer as prey. The number of pumas varies with time according to the function $P = 500 + 200 \sin 0.4(t - 2)$, and the number of deer varies according to the function $D = 1500 + 400 \sin 0.4t$, with time, t, in years.

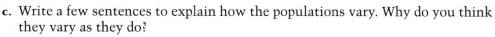

a. How many pumas and deer will there be in the region in 15 years?

b. Graph both of the equations on a graphing calculator. Then sketch the graph on paper.

c. Write a few sentences to explain how the populations vary. Why do you think they vary as they do?

48. Entertainment As you ride a Ferris wheel, the height that you are above the ground varies periodically. Consider the height of the center of the wheel to be the equilibrium point. A particular wheel has a diameter of 38 feet and travels at a rate of 4 revolutions per minute.

a. Write an equation to describe the changes in height, h, of the seat that was filled last before the ride began in terms of time, t, in seconds. *Hint: This seat was at the minimum height at $t = 0$.*

b. Find the height of the seat after 22 seconds.

c. If the seats of the Ferris wheel clear the ground by 3 feet, how long after the ride starts will the seat first be 27 feet above the ground?

49. Hydraulics The branch of physics that involves the behavior of liquids at rest and in motion is called hydraulics. A reservoir is a part of a scientific apparatus in which liquid is held. An important formula in analyzing reservoirs involves the capillary pressure, P. If t is the tension between two fluids, r is the radius of the capillary tube, and θ is the angle between the interface and the capillary wall, then $P = \dfrac{2t}{r} \cos \theta$. A certain reservoir with an interfacial tension between fluids of 75 d/cm has a capillary tube of radius 0.02 cm.

a. Draw a graph of the capillary pressure in the apparatus for $\theta = 0°$ to $90°$.

b. What is the amplitude of this curve?

c. Write a few sentences to explain the meaning of the graph.

50. Graph the inequality $6 < 4x + 2 < 18$. **(Lesson 1-3)**

51. Find the sum of the matrices $\begin{bmatrix} 4 & -3 & 2 \\ 8 & -2 & 0 \\ 9 & 6 & -3 \end{bmatrix}$ and $\begin{bmatrix} -2 & 2 & -2 \\ -5 & 1 & 1 \\ -7 & 2 & -2 \end{bmatrix}$. **(Lesson 2-2)**

52. Find the critical points of the graph of $y = 3x^4 - 8x^3 - 144x^2 + 6$. Then, determine whether each point represents a maximum, a minimum, or a point of inflection.   **(Lesson 3-7)**

53. Approximate the real zero(s) of $f(x) = x^5 + 3x^3 - 4$ to the nearest tenth.
    (Lesson 4-5)

54. **Physics**   A pendulum 20 centimeters long swings $3°30'$ on each side of its vertical position. Find the length of the arc formed by the tip of the pendulum as it swings.   **(Lesson 5-2)**

55. Find the values of the six trigonometric functions of an angle in standard position if the point at $(7, -3)$ lies on its terminal side.   **(Lesson 5-3)**

56. Use a graphing calculator to graph the sine and cosine functions on the same set of axes for $0° \leq x \leq 360°$. Use the graphs to find the values of x, if any, for which $\sin x + \cos x = -1$.   **(Lesson 6-1)**

57. **College Entrance Exam**   Choose the best answer.
    The coordinates of two vertices of an equilateral triangle are $(-6, 0)$ and $(6, 0)$. The coordinates of the third vertex may be

    (A) $(0, 3\sqrt{3})$ **(B)** $(6, 6\sqrt{3})$ **(C)** $(0, 6\sqrt{3})$ **(D)** $(6\sqrt{3}, 0)$ **(E)** $(0, 6)$

CASE STUDY FOLLOW-UP

Refer to Case Study 1: *Buying a Home*, on pages 956–958.

Sound waves are *longitudinal* waves because the particles of a sound wave vibrate in the direction of the motion of the wave. The *frequency* of a wave, f, is the number of vibrations occurring in one second. Frequency is measured in hertz. One hertz (Hz) is one vibration per second. The human ear can hear a range of frequencies of sound waves from a low of about 20 Hz to a high of about 20,000 Hz. The relationship between the frequency, wavelength, λ, and speed, v, of a sound wave is given by the formula $v = f\lambda$.

1. A dog that lives next door to the Arlen residence barks constantly at a frequency of 7000 Hz. The speed of sound is 1085 feet/second. Find the wavelength of the dog's bark.

2. The cost of soundproofing a room, C, depends on the frequency of the sound. One dealer quoted Mrs. Arlen a price, in dollars, of $C = 450 \sin [(\frac{f}{20,000})90]°$
   per room. The Arlen house has the same number of rooms as the largest typical American house. How much will it cost the Arlens to soundproof all of the rooms?

3. **Research**   Find out about the *amplitude* of a sound wave. Research the connection between the frequency and amplitude of a sound wave and the difficulty of soundproofing a room. Write about your findings in a 1-page paper.

6-3A Graphing Calculators: Graphing Trigonometric Functions

The graphing calculator is a good tool for graphing trigonometric functions. With the calculator, you can see how changing or modifying a trigonometric function alters its graph. The functions can be graphed in degrees or radians; in this lesson we will use degrees. Check to see that you are using degrees by pressing the [MODE] key. If not, use the arrow keys to move to "Degree" and press [ENTER].

Example 1

Graph $y = \cos x$.

The TI-82 graphing calculator does not have any built-in functions, so the viewing window must be set. Set the window to $[-360°, 360°]$ by $[-3, 3]$ with a scale factor of $90°$ for the x-axis and 1 for the y-axis. You can also obtain a similar viewing window by pressing [ZOOM] 7; the calculator will set the range values to $[-352.5°, 352.5°]$ by $[-4, 4]$ with scale factors of $90°$ for the x-axis and 1 for the y-axis.

Enter: [Y=] [COS] [X,T,θ] [GRAPH]

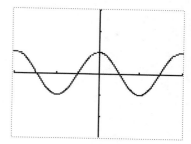

Notice that the amplitude of the graph in Example 1 is 1. The period of the function is $360°$.

Changing a trigonometric equation will also change its graph.

Example 2

Graph $y = 2 \sin \dfrac{1}{2}x$. Use the viewing window of $[-540°, 540°]$ by $[-3, 3]$ with a scale factor of $90°$ for the x-axis and 1 for the y-axis. State the amplitude and period of the function.

Enter: [Y=] 2 [SIN] [(] [X,T,θ] [÷] 2 [)] [GRAPH]

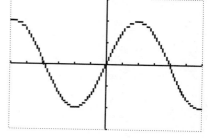

The amplitude of the function is $\dfrac{|-2 - 2|}{2} = \dfrac{4}{2} = 2$. The period of the function is $720°$.

You can also graph the other trigonometric functions on the graphing calculator, namely cot x, sec x, and csc x. These functions are also periodic and changing their equations will also change their graphs.

Example 3

Graph $y = \sec x$ and $y = \sec (x + 90°)$ at the same time. Use the trig viewing window. State the phase shift of the function.

Recall that $\sec x = \dfrac{1}{\cos x}$.

Enter: $\boxed{Y=}$ 1 $\boxed{\div}$ $\boxed{COS}$ $\boxed{X,T,\theta}$ $\boxed{ENTER}$
1 $\boxed{\div}$ $\boxed{COS}$ $\boxed{(}$ $\boxed{X,T,\theta}$ $\boxed{+}$ 90 $\boxed{)}$
$\boxed{ZOOM}$ 7

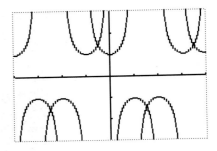

Watching the calculator as it graphs shows that the 90° in the second equation moves the graph of $y = \sec x + 90°$ to the left. Thus, the phase shift of the function is –90°.

The period of this function is 360°, there is no amplitude, and the phase shift is –90°.

EXERCISES

Graph each equation so that a complete graph is shown. Find the amplitude, period, and phase shift for each function. Then sketch the graph on a piece of paper.

1. $y = \sin x$ **2.** $y = \csc x$ **3.** $y = -3 \tan 2x$

4. $y = \cot \dfrac{1}{2}x$ **5.** $y = \sec 3x$ **6.** $y = \tan (3x - 90°)$

7. $y = 3 \sin (x + 45°)$ **8.** $y = -4 \cos (3x + 60°)$ **9.** $y = \cot (-x)$

10. $y = \csc (45° - x)$ **11.** $y = \dfrac{1}{2} \cos (2x - 90°)$ **12.** $y = -\sec (2x + 180°) + 1$

6-3 Graphing Trigonometric Functions

Objective

After studying this lesson, you should be able to:
■ graph various trigonometric functions.

Application

The forces due to the gravity of the sun and moon pull upon Earth and cause tides, which are the rise and fall of the ocean waters. In most locations, there are two low tides and two high tides each day, with a period of about 12 hours and 25 minutes. The beaches of the United States have tides that rise and fall between 1 and 20 feet. In the funnel-shaped Bay of Fundy in Nova Scotia, Canada, the tides rise and fall 43 feet or more.

If the rise and fall of the tides is graphed with respect to time, the graph is similar to a sine or cosine graph. On a June day in Daytona, Florida, the high tides occur at 1:04 A.M. and 1:30 P.M., and the low tides occur at 7:15 A.M. and 7:39 P.M. Kelly anchored a stick in the sand and began graphing the distance from the stick to the water at 9:58 P.M. the day before. She marked the stick with a 0 at the level of the water. At this time, the water was midway between high tide and low tide. Her results are shown on the graph below.

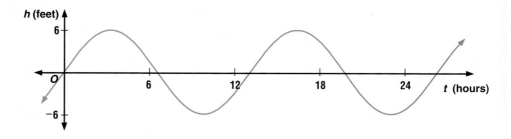

Suppose Kelly had measured the distance from the low-tide position of the water instead of from the equilibrium point. The graph would look like the one below.

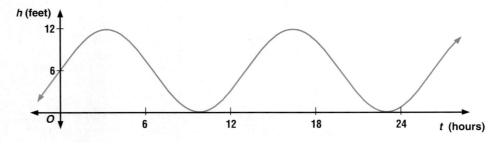

Do you see the similarities between the two graphs? The shapes are the same, but the second graph has been translated vertically. If (t, h) are the coordinates of a point on the first graph, $(t, h + 6)$ are the coordinates of a point on the second graph. So if an equation of the first graph is $h = 6 \sin t$, then an equation of the second graph is $h = 6 \sin t + 6$.

Vertical Displacement	**The graph of $y = A \sin (k\theta + c) + h$ is the graph of $y = A \sin (k\theta + c)$ translated h units vertically. If $h > 0$, the graph is moved up, and if $h < 0$, the graph is moved down. This definition applies to all the trigonometric functions.**

Example 1

Graph $y = \cos \theta$ and $y = \cos \theta - 1$ on the same set of axes. Then explain the similarities and differences between the graphs.

Generate ordered pairs for the two functions.

θ	0°	45°	90°	135°	180°	225°	270°	315°	360°
$\cos \theta$	1	$\dfrac{\sqrt{2}}{2}$	0	$-\dfrac{\sqrt{2}}{2}$	−1	$-\dfrac{\sqrt{2}}{2}$	0	$\dfrac{\sqrt{2}}{2}$	1
$\cos \theta - 1$	0	$\dfrac{\sqrt{2}}{2} - 1$	−1	$-\dfrac{\sqrt{2}}{2} - 1$	−2	$-\dfrac{\sqrt{2}}{2} - 1$	−1	$\dfrac{\sqrt{2}}{2} - 1$	0

Plot the points and draw the curves.

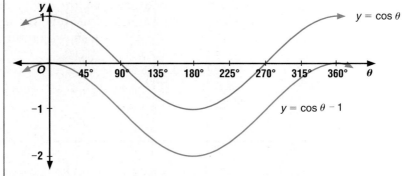

The graph of $y = \cos \theta - 1$ is the graph of $y = \cos \theta$ shifted 1 unit down.

You can graph any trigonometric function by finding its amplitude (if it has an amplitude), its period, and its phase shift and then using your general knowledge of the shape of the trigonometric curves.

Example 2

Graph $y = 4 \sin 2\theta$.

Find the amplitude, period, and phase shift.

amplitude:	*period:*	*phase shift:*
$|A| = 4$	$\dfrac{360°}{k} = \dfrac{360°}{2}$ or $180°$	$-\dfrac{c}{k} = \dfrac{0}{2}$ or 0

The amplitude of 4 tells us that the values of the function vary from −4 to 4. Since the period is 180°, the curve will repeat every 180°. There is no phase shift.

It may be helpful to graph the basic sine curve first, and then graph $y = 4 \sin 2\theta$.

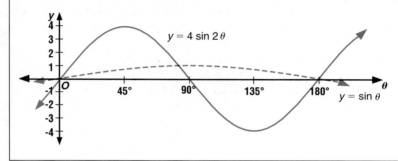

The functions $y = A \cos(k\theta + c)$ and $y = A \tan(k\theta + c)$ can be graphed using the same process.

Example 3

Graph $y = \tan\left(\dfrac{x}{2} - \dfrac{\pi}{6}\right)$.

Find the period.
$$\frac{\pi}{k} = \frac{\pi}{\frac{1}{2}} \text{ or } 2\pi$$

Determine the phase shift.
$$-\frac{c}{k} = -\left(\frac{-\frac{\pi}{6}}{\frac{1}{2}}\right) \text{ or } \frac{\pi}{3}$$

Since $c < 0$, the graph is shifted to the right $\dfrac{\pi}{3}$.

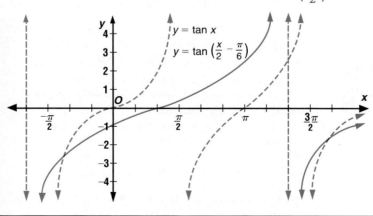

Compound functions may consist of sums or products of trigonometric functions. For example, $y = \sin x \cdot \cos x$ is a compound function that contains a product of trigonometric functions. Compound functions may also include sums or products of trigonometric functions and other functions. For example, $y = \sin x + x$ is a compound function that is the sum of a trigonometric function and a linear function.

You can graph some compound functions by graphing each function involved separately on the same coordinate axes and then adding the ordinates. After you find a few of the critical points in this way, you can sketch the rest of the curve of the compound function.

Physicists call this method the principle of superposition.

Example 4

Graph $y = \sin x + x$.

First graph $y = x$ and $y = \sin x$ on the same axes. Then add the corresponding ordinates of the functions. Finally, sketch the graph.

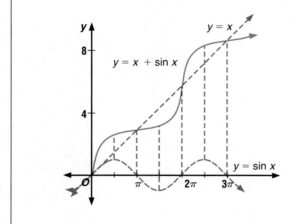

x	$\sin x$	$x + \sin x$
0	0	0
$\dfrac{\pi}{2}$	1	$\dfrac{\pi}{2} + 1$ or 2.57
π	0	π or 3.14
$\dfrac{3\pi}{2}$	-1	$\dfrac{3\pi}{2} - 1$ or 3.71
2π	0	2π or 6.28
$\dfrac{5\pi}{2}$	1	$\dfrac{5\pi}{2} + 1$ or 8.85
3π	0	3π or 9.42

Some real-world applications involve finding the result of two influences acting on a system.

Example 5

A guitar string is fixed at both ends and plucked in the middle. Waves travel down the string toward either end and are reflected. The reflected wave from the right side can be described by the function $y = \sin \pi(x - t)$, and the wave from the right side can be described by $y = \sin 2\pi(x + t)$, where t is time in seconds. Draw a sketch of the wave pattern on the string after 4 seconds.

The equations of the two waves are $y = \sin \pi(x - 4)$ and $y = \sin 2\pi(x + 4)$.

Graph both functions. Then find the sum.

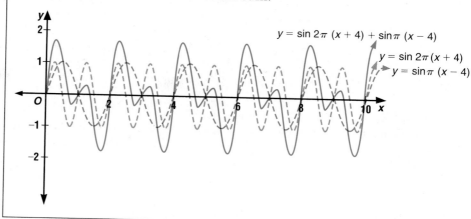

$y = \sin 2\pi (x + 4) + \sin \pi (x - 4)$

$y = \sin 2\pi (x + 4)$

$y = \sin \pi (x - 4)$

Example 6 involves the graph of a compound function that is the product of a linear function and a trigonometric function.

Example 6

Graph $y = x \cos x$.

First graph $y = x$ and $y = \cos x$ on the same set of axes. Then find some key points, such as zeros, maxima, and minima.

Since the maximum or minimum points of $y = \cos x$ are ± 1, the corresponding values of $x \cos x$ at these points are $\pm x$. At points where $\cos x = 0$, $x \cos x = 0$. Plot these points and sketch a smooth curve.

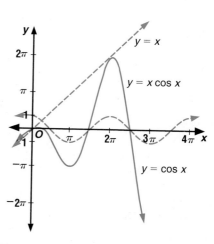

CHECKING FOR UNDERSTANDING

Communicating Mathematics

Read and study the lesson to answer each question.

1. **Analyze** the function $y = A \sin (k\theta + c) + h$. Which variable could you increase or decrease to have each of the following effects on the graph?
   a. stretch the graph vertically
   b. translate the graph downward vertically
   c. shrink the graph horizontally
   d. translate the graph to the left

2. **Explain** why the method of separately graphing two functions that are being added, then adding their ordinates works for graphing compound functions.

3. **Write** a short paragraph that compares the graphs of $y = 3 \sin 2x$ and $y = 3 \sin 2x + 4$.

Graph each function.

4. $y = \frac{1}{2} \cos 2\theta$

5. $y = 6 \sin 4\theta$

6. $y = \tan \left(\frac{x}{2} + \frac{\pi}{2} \right)$

7. $y = \sec 3\theta$

8. $y = -\sin(\theta - 45°)$

9. $y = 5 \cos (2\theta + 180°)$

10. $y = x + \cos x$

11. $y = \sin \frac{x}{3} + \frac{x}{3}$

12. $y = \sin x - \cos x$

EXERCISES

Practice   **Graph each function.**

13. $y = \sin (\theta + 90°)$

14. $y = \sin (\theta - 180°)$

15. $y = -3 \sin (\theta - 45°)$

16. $y = 3 \cos (\theta - 90°)$

17. $y = \frac{1}{2} \cos \left(\frac{\theta}{2} - 180° \right)$

18. $y = -\frac{1}{3} \sin (2\theta + 45°)$

19. $y = \tan (\theta + 90°)$

20. $y = \cot (\theta - 90°)$

21. $\frac{1}{2} y = \sin (3\theta + 180°)$

22. $2y = 10 \sin \left(\frac{\theta}{2} + 90° \right)$

23. $y = \sin x + \sin 2x$

24. $y = \cos x - \sin x$

25. $y = \cos 2x - \cos 3x$

26. $y = 2 \sin x + 3 \cos x$

27. $y = \sin x + \sin \left(x + \frac{\pi}{2} \right)$

28. $y = \frac{1}{2} \sin x - \cos 3x$

29. $y = 2x \sin 2x$

30. $y = \sin^2 x$

**Critical
Thinking**

31. Graph $y = |\sin 2x|$.

**Applications
and Problem
Solving**

32. Acoustics   Oscilloscopes are used to observe the behavior of waves. Pure sound waves are translated into electrical signals and displayed on an oscilloscope. They are represented as sine curves. If each square has a vertical measure of 1 and a horizontal measure of 20°, find the amplitude and period of each graph. Write the equations for the graphs.

a.   **b.**

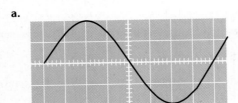

33. Electronics   In electrical circuits, the voltage and current can be described by sine or cosine functions. If the graphs of these functions have the same period, but do not pass through their zero points at the same time, they are said to have a *phase difference*. For example, if the voltage is 0 at 90° and the current is 0 for 180°, they are 90° out of phase. Suppose the voltage across an inductor of a circuit is represented by $y = 2 \cos 2x$ and the current across the component is represented by $y = \cos (2x - 90°)$. What is the phase relationship between the signals?

34. Physics Many natural phenomena, such as the conduction of heat and the weather, can be described using complicated periodic functions. The *sawtooth wave form* is one of the periodic functions that is observed frequently. The French mathematician Joseph

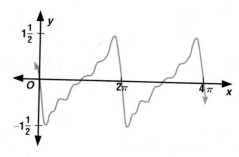

Fourier showed that a complex wave can be represented using an infinite series of simple waves like sine or cosine waves.

  a. The first two terms in the infinite series that generates the sawtooth wave are $-\sin t$ and $-\frac{1}{2}\sin 2t$. Sketch the function $y = -\sin t - \frac{1}{2}\sin 2t$.

  b. Research Write a one-page paper on Fourier's life and discoveries.

Mixed Review

35. Solve the following system of equations algebraically. **(Lesson 2-1)**
$$3x + 5y = 4$$
$$14x - 35y = 21$$

36. Find the value of k so that $(x^3 - 7x^2 - kx + 6) \div (x - 3)$ results in a remainder of zero. **(Lesson 4-3)**

37. Carpentry Carpenters use circular sanders to smooth rough surfaces, such as wood or plaster, before it is finished. The disk of a sander has a radius of 6 inches and is rotating at a speed of 5 revolutions per second. **(Lesson 5-2)**

  a. Find the angular speed of the sander disk in radians per second.

  b. What is the linear velocity of a point on the edge of the sander disk in feet per second?

38. Utilities A utility pole is braced by a cable attached to it at the top and anchored in a concrete block at ground level, a distance of 4 meters from the base of the pole. If the angle between the cable and the ground is $73°$, find the height of the pole and the length of the cable to the nearest tenth of a meter. **(Lesson 5-5)**

39. Write an equation of the cosine function with an amplitude of 4, a period of $180°$, and a phase shift of $20°$. **(Lesson 6-2)**

40. College Entrance Exam Compare quantities A and B below.
Write A if quantity A is greater.
Write B if quantity B is greater.
Write C if the two quantities are equal.
Write D if there is not enough information to determine the relationship.

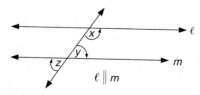

(A) $x + y$ **(B)** $x + z$

6-4 Inverse Trigonometric Functions

Objective

After studying this lesson, you should be able to:

■ evaluate inverse trigonometric functions.

Application

FYI...

→ The Maya were an American Indian people who lived in Central America and southern Mexico from A.D. 250 to 850. They produced exceptional architecture and art sculpture and made notable advancements in math and astronomy.

The Landsat satellites orbit Earth at a height of 920 km. They scan Earth collecting data from reflected sunlight and transmitting it back. The transmissions allow scientists to study the surface features of our planet. The information they have collected has led to the discovery of 112 possible Mayan ruins and 2 confirmed ruins in the Yucatán Peninsula of Mexico.

From a spacecraft in orbit above Earth, like a Landsat satellite, only a portion of the surface of the Earth, called the horizon circle, is visible. Trigonometry can help astronauts determine the size of the horizon circle. Let h represent the height of the spacecraft above Earth, S is the position of the satellite, C is the center of Earth, H is the farthest point on Earth's surface that the satellite can see, r is the radius of Earth, and θ is the angle between $\overline{CH}$ and $\overline{CS}$. Right triangle trigonometry shows us that $\cos \theta = \dfrac{r}{r+h}$. We can use an inverse trigonometric function to find θ, the angular radius of the Landsat satellites. *This problem will be solved in Example 2.*

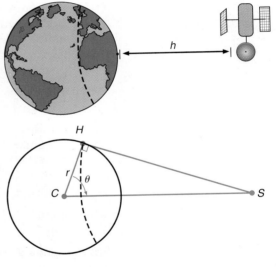

Recall that the inverse of a function may be found by interchanging the coordinates of the ordered pairs of the function. In other words, the domain of the function becomes the range of its inverse, and the range of the function becomes the domain of its inverse. For example, the inverse of $y = 2x + 5$ is $x = 2y + 5$ or $y = \dfrac{x-5}{2}$. Also remember that the inverse of a function *may not* be a function.

The sine function is the set of all ordered pairs $(x, \sin x)$. So the inverse of the sine function, the **arcsine relation**, is the set of all ordered pairs $(\sin x, x)$. The graphs of $y = \sin x$ and $y = \arcsin x$ are shown at the right. The domain of $y = \arcsin x$ is $-1 \leq x \leq 1$ or $|x| \leq 1$, and the range is the set of all real numbers. Notice that the graph of $y = \arcsin x$ fails the vertical line test, so $y = \arcsin x$ is not a function. None of the inverses of the trigonometric functions are functions.

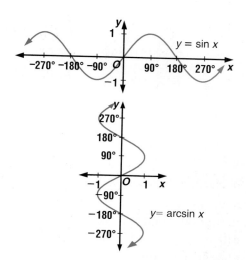

An equation such as $\sin x = 0.3393$ can be written as $x = \arcsin 0.3393$. The latter equation is read "x is an angle whose sine is 0.3393," or "x equals the arcsin of 0.3393." The solution, x, consists of all angles that have 0.3393 as their sine. Infinitely many such angles exist.

Inverses of the Trigonometric Functions	The inverse of **sin x** is **arcsin x** or **sin⁻¹ x**. The inverse of **cos x** is **arccos x** or **cos⁻¹ x**. The inverse of **tan x** is **arctan x** or **tan⁻¹ x**.

The equations in each row of the table below are equivalent. You can use these expressions to rewrite and solve trigonometric equations.

Trigonometric Function	Inverse Trigonometric Function
$y = \sin x$	$x = \sin^{-1} y$ or $x = \arcsin y$
$y = \cos x$	$x = \cos^{-1} y$ or $x = \arccos y$
$y = \tan x$	$x = \tan^{-1} y$ or $x = \arctan y$

Example 1

Find all positive values of x for which $\cos x = \dfrac{\sqrt{3}}{2}$.

If $\cos x = \dfrac{\sqrt{3}}{2}$, then x is an angle or a real number whose cosine is $\dfrac{\sqrt{3}}{2}$.

$x = \arccos \dfrac{\sqrt{3}}{2}$

Therefore, $x = 30°, 330°, 390°, 690°, \cdots$, or $\dfrac{\pi}{6}, \dfrac{11\pi}{6}, \dfrac{13\pi}{6}, \dfrac{23\pi}{6} \cdots$.

Many application problems involve finding the inverse of a trigonometric function.

Example 2

Refer to the application at the beginning of the lesson. Find the angular radius of the Landsat satellites if $0° \leq \theta \leq 90°$. Use 6378 km as the radius of Earth.

We are given that $\cos \theta = \dfrac{r}{r+h}$.

The satellites are 920 km above Earth, so $h = 920$.

$$\cos \theta = \frac{r}{r+h}$$
$$\cos \theta = \frac{6378}{6378 + 920}$$
$$\theta = \arccos \frac{6378}{6378 + 920}$$

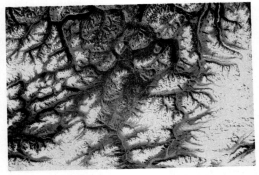

Landsat satellite photograph

Use a calculator to find θ.

6378 ÷ ((6378 + 920)) = 2nd COS *29.080456*

The angular radius of the Landsat satellites is about $29°$.

Trigonometric inverses can be used to evaluate expressions.

Example 3

Evaluate each expression. Assume that all angles are in Quadrant I.

a. sin(arcsin 0.4212)

a. Let $X = \arcsin 0.4212$.
   Then $\sin X = 0.4212$ by the definition of inverse.
   Therefore, by substitution, $\sin (\arcsin 0.4212) = 0.4212$.

b. tan $\left(\sin^{-1} \dfrac{5}{13} \right)$.

b. Let $A = \sin^{-1} \dfrac{5}{13}$. Then $\sin A = \dfrac{5}{13}$.

Since $\sin A > 0$, A must be in
Quadrant I or II, as shown in the
sketch at the right. Assume A is
in Quadrant I.

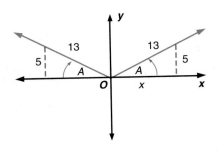

Since $\tan A = \dfrac{\sin A}{\cos A}$, find $\cos A$.

$$x^2 + y^2 = r^2 \qquad \text{\textit{Pythagorean theorem}}$$
$$x^2 + 5^2 = 13^2$$
$$x = 12$$

Therefore, $\cos A = \dfrac{12}{13}$.

Now solve.

$$\tan\left(\sin^{-1}\frac{5}{13}\right) = \tan A$$

$$= \frac{\sin A}{\cos A} \qquad \textit{Definition of tangent ratio}$$

$$= \frac{\frac{5}{13}}{\frac{12}{13}} \qquad \textit{Substitute } \frac{5}{13} \textit{ for sin A and } \frac{12}{13} \textit{ for cos A.}$$

$$= \frac{5}{12} \qquad \textit{Simplify.}$$

Therefore, $\tan\left(\sin^{-1}\frac{5}{13}\right) = \frac{5}{12}$.

CHECKING FOR UNDERSTANDING

Communicating Mathematics

Read and study the lesson to answer each question.

1. **Compare** $y = \sin^{-1} x$, $y = (\sin x)^{-1}$, and $y = \sin(x^{-1})$.

2. **Explain** why $y = \arctan x$ is not a function.

Guided Practice

Write each equation in the form of an inverse relation.

3. $x = \sin\theta$

4. $\cos\alpha = \frac{1}{3}$

5. $\tan y = -3$

6. $\frac{4}{3} = \tan\theta$

7. $y = \cos x$

8. $\sin x = 1$

Find the values of x in the interval $0° \leq x < 360°$ that satisfy each equation.

9. $x = \arcsin 0$

10. $x = \arctan 1$

11. $x = \arcsin\frac{\sqrt{3}}{2}$

Evaluate each expression. Assume that all angles are in Quadrant I.

12. $\cos\left(\arccos\frac{4}{5}\right)$

13. $\sec(\cos^{-1} 1)$

14. $\tan\left(\cos^{-1}\frac{3}{5}\right)$

EXERCISES

Practice

Write each equation in the form of an inverse relation.

15. $n = \sin\theta$

16. $\cos\beta = \frac{1}{3}$

17. $\frac{3}{2} = \tan\delta$

18. $\sin\alpha = 1$

19. $\cos\theta = y$

20. $\tan A = \sqrt{3}$

Find the values of x in the interval $0° \leq x < 360°$ that satisfy each equation.

21. $x = \cos^{-1} 0$

22. $x = \sin^{-1}\frac{1}{\sqrt{2}}$

23. $x = \arctan\frac{\sqrt{3}}{3}$

24. $x = \sec^{-1} 2$

25. $\arctan 0 = x$

26. $\arcsin\frac{1}{2} = x$

27. $\text{arccot } 2.1445 = x$

28. $x = \arcsin(-0.5)$

29. $\cot^{-1} 0 = x$

Evaluate each expression. Assume that all angles are in Quadrant I.

30. $\sin\left(\sin^{-1}\dfrac{1}{2}\right)$

31. $\cot\left(\arctan\dfrac{4}{5}\right)$

32. $\sin\left(\cos^{-1}\dfrac{\sqrt{3}}{2}\right)$

33. $\sec\left(\cos^{-1}\dfrac{1}{2}\right)$

34. $\cos\left(\text{arccot}\dfrac{4}{3}\right)$

35. $\tan\left(\sec^{-1}2\right)$

36. $\sin\left(\arctan\sqrt{3}+\text{arccot}\sqrt{3}\right)$

37. $\sin\left(\tan^{-1}1\right)+\cos\left(\cos^{-1}0.5\right)$

38. $\tan\left(\arcsin\dfrac{\sqrt{2}}{2}\right)-\cot\left(\arccos\dfrac{\sqrt{2}}{2}\right)$

39. $\tan\left(\sin^{-1}\dfrac{\sqrt{3}}{2}-\cos^{-1}\dfrac{\sqrt{3}}{2}\right)$

Verify each equation. Assume that all angles are in Quadrant I.

40. $\sin^{-1}\dfrac{\sqrt{2}}{2}+\cos^{-1}\dfrac{\sqrt{2}}{2}=\tan^{-1}\dfrac{\sqrt{3}}{3}+\tan^{-1}\sqrt{3}$

41. $\arccos\dfrac{\sqrt{3}}{2}+\arcsin\dfrac{\sqrt{3}}{2}=\arctan 1+\text{arccot}\,1$

42. $\arcsin\dfrac{2}{5}+\arccos\dfrac{2}{5}=90°$

43. $\tan^{-1}\dfrac{3}{4}+\tan^{-1}\dfrac{5}{12}=\tan^{-1}\dfrac{56}{33}$

44. $\tan^{-1}1+\cos^{-1}\dfrac{\sqrt{3}}{2}=\sin^{-1}\dfrac{1}{2}+\sec^{-1}\sqrt{2}$

45. $\arcsin\dfrac{3}{5}+\arccos\dfrac{15}{17}=\arctan\dfrac{77}{36}$

Critical Thinking

46. Explain why the equation $[\cos\circ\cos^{-1}](x)=[\cos^{-1}\circ\cos](x)$ is not always true. Could you restrict the domain and range to make the equations true for an interval?

Applications and Problem Solving

47. Navigation   Earth has been charted with vertical and horizontal lines so that points can be named with coordinates. The horizontal lines are called latitude lines. The equator is latitude line 0°. Parallel lines are numbered up to 90° to the north and to the south. The length of any parallel of latitude is equal to the distance around Earth times the cosine of the latitude angle, if we assume a spherical Earth.

a. If the radius of Earth is about 6400 km, which latitude lines are about 5543 km long?

b. What is the length of the 90° parallel? Why?

48. Optics   Light travels like waves in many different planes. If light is polarized, all of the waves are traveling in parallel planes. You may have polarized sunglasses that eliminate glare by polarizing the light. Suppose vertically-polarized light with intensity I_o strikes a polarizing filter with its axis at an angle of θ with the vertical. The intensity of the transmitted light, I_t, and θ are related by the equation $\cos\theta=\sqrt{\dfrac{I_t}{I_o}}$

a. If one-fourth of the polarized light is transmitted through the lens, at what angle is the lens being held?

b. At what angle will none of the polarized light be transmitted through the lens?

49. Civil Engineering Highway curves are usually banked, or tilted inward, so that cars can negotiate the curve more safely. The proper banking angle, θ, for a car making a turn of radius r feet at a velocity v feet per second is given by $\tan \theta = \dfrac{v^2}{gr}$ where g is the acceleration due to gravity. Assume that $g = 32$ ft/s². An engineer is designing a curve with a radius of 1000 feet. If the speed limit on the curve will be 55 mph, at what angle should the curve be banked?

Mixed Review

50. Geometry The endpoints of $\overline{AB}$ are $A(9, 3)$ and $B(-1, 1)$. Use the distance formula to show that $P(4, 2)$ is the midpoint of $\overline{AB}$. **(Lesson 1-4)**

51. Find the value of the determinant $\begin{vmatrix} -2 & 4 & -1 \\ 1 & -1 & 0 \\ -3 & 4 & 5 \end{vmatrix}$. **(Lesson 2-3)**

52. Find the inverse of the function $y = \dfrac{3}{2}x - 2$. **(Lesson 3-3)**

53. Solve $3x^2 - 7x + 2 = 0$ by using the quadratic formula. **(Lesson 4-2)**

54. Find the value of $\sin 360°$. **(Lesson 5-4)**

55. Angle C of $\triangle ABC$ is a right angle. Solve the triangle if $A = 20°$ and $c = 35$. **(Lesson 5-5)**

56. Which value is greater, $\cos 70°$ or $\cos 170°$? **(Lesson 6-1)**

57. Graph $y = 2 \sin x - \dfrac{1}{2} \cos x$. **(Lesson 6-3)**

58. College Entrance Exam Choose the best answer. In parallelogram $ABCD$, the ratio of the shaded area to the unshaded area is
(A) 1:2   **(B)** 1:1
(C) 4:3   **(D)** 2:1
(E) It cannot be determined from the information given.

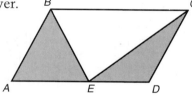

MID-CHAPTER REVIEW

Find the values of θ for which each of the following is true. (Lesson 6-1)

1. $\tan \theta = 0$ **2.** $\sec \theta$ is undefined   **3.** $\csc \theta = 1$

State the amplitude, period, and phase shift for each function. Then graph the functions. (Lesson 6-2 and 6-3)

4. $y = \sin \left(\theta + \dfrac{\pi}{2} \right)$ **5.** $y = \tan (\theta + 60°)$ **6.** $y = 5 \sin 2\theta$

Graph each function. (Lesson 6-3)

7. $y = x - \sin x$ **8.** $y = 2 \sin x - \dfrac{1}{2}x$ **9.** $y = 2x + 2 \sin x$

10. Show that $\cos^{-1} \dfrac{\sqrt{2}}{2} = \tan^{-1} 1$. Assume that all angles are in Quadrant I. **(Lesson 6-4)**

6-5 Principal Values of the Inverse Trigonometric Functions

Objective

After studying this lesson, you should be able to:
- find principal values of inverse trigonometric functions.

Application

The ancient Hindu and Arab peoples had a great interest in astronomy. It was this interest that drove them to master trigonometry. None of the works of the Hindu mathematicians still exist today, but we do know that they made an extensive table of sines using only a few equations. Their work in trigonometry united the fields of algebra, geometry, and arithmetic.

The concept of a function can be applied to the trigonometric ratios and their inverses. Recall that the inverse of a function can be graphed by reflecting the graph of the given function over the line $y = x$. Reflecting the graph of $y = \sin x$ over the line $y = x$ shows that $y = \arcsin x$ does not pass the vertical line test for functions. However, we can limit the domain of the sine function so that its inverse is a function.

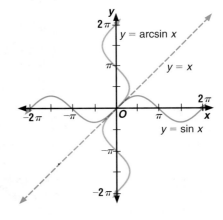

Capital letters are used to distinguish the function with restricted domains from the usual trigonometric functions.

Consider only a part of the domain of the sine function, namely any x such that $-90° \leq x \leq 90°$. The range then contains all of the possible values from -1 to 1. It is possible to define a new function, called Sine, whose inverse is a function.

$$y = \text{Sin } x \text{ if and only if } y = \sin x \text{ and } -90° \leq x \leq 90°$$

The values in the domain of Sine are called **principal values.** Other new functions can be defined as follows.

$$y = \text{Cos } x \text{ if and only if } y = \cos x \text{ and } 0° \leq x \leq 180°$$
$$y = \text{Tan } x \text{ if and only if } y = \tan x \text{ and } -90° < x < 90°$$

You will learn more about the graphs of these functions in Lesson 6-6.

The inverses of the Sine, Cosine, and Tangent functions are called Arcsine, Arccosine, and Arctangent, respectively. They are defined as follows.

Arcsine Function	**Given $y = \text{Sin } x$, the inverse Sine function is defined by the equation** $$y = \text{Sin}^{-1} x \text{ or } y = \text{Arcsin } x.$$
Arccosine Function	**Given $y = \text{Cos } x$, the inverse Cosine function is defined by the equation** $$y = \text{Cos}^{-1} x \text{ or } y = \text{Arccos } x.$$
Arctangent Function	**Given $y = \text{Tan } x$, the inverse Tangent function is defined by the equation** $$y = \text{Tan}^{-1} x \text{ or } y = \text{Arctan } x.$$

Example 1

Find each value.

a. Arccos $\dfrac{1}{2}$

a. Let $\theta = \text{Arccos } \dfrac{1}{2}$.

$\text{Cos } \theta = \dfrac{1}{2}$ *Definition of Arccos function*

$\theta = 60°$ *Why is θ not 300°?*

Therefore, $\text{Arccos } \dfrac{1}{2} = 60°$.

b. $\text{Sin}^{-1} \left(\tan \dfrac{\pi}{4} \right)$

b. Let $x = \tan \dfrac{\pi}{4}$.

$x = 1$

$\text{Sin}^{-1} \left(\tan \dfrac{\pi}{4} \right) = \text{Sin}^{-1} 1$ *Replace $\tan \dfrac{\pi}{4}$ in the original expression with 1.*

$= \dfrac{\pi}{2}$

c. $\sin \left(\text{Sin}^{-1} 1 - \text{Cos}^{-1} \dfrac{1}{2} \right)$

c. Let $\alpha = \text{Sin}^{-1} 1$ and $\beta = \text{Cos}^{-1} \dfrac{1}{2}$.

$\text{Sin } \alpha = 1$ $\text{Cos } \beta = \dfrac{1}{2}$

$\alpha = 90°$ $\beta = 60°$

$\sin \left(\text{Sin}^{-1} 1 - \text{Cos}^{-1} \dfrac{1}{2} \right) = \sin(\alpha - \beta)$

$= \sin(90° - 60°)$ $\alpha = 90°, \beta = 60°$

$= \sin 30°$

$= \dfrac{1}{2}$

Scientists use trigonometry to describe physical properties. So inverse trigonometric functions are often used to solve scientific problems.

Example 2

APPLICATION

Optics

Malus' Law describes the amount of light transmitted through two polarizing lenses. If the axes of two lenses are at an angle of θ degrees, the intensity of the light transmitted through them, I, is determined by the equation $I = I_o \cos^2 \theta$, where I_o is the intensity of light that shines on the lenses. At what angle should the axes be held so that one-half of the transmitted light passes through the lenses?

If one-half of the transmitted light passes through the lenses, then $I = \frac{1}{2}I_o$.

$$I = \frac{1}{2}I_o$$

$$I_o \cos^2 \theta = \frac{1}{2}I_o$$

$$\cos^2 \theta = \frac{1}{2} \qquad \textit{Divide each side by } I_o.$$

$$\cos \theta = \frac{\sqrt{2}}{2} \qquad \textit{Take the square root of each side.}$$

$$\theta = \text{Arccos}\,\frac{\sqrt{2}}{2}$$

$$\theta = 45°$$

The lenses should be held so that their axes are at an angle of $45°$.

Photograph taken without polarizing lens

Photograph taken with polarizing lens

You can use a calculator to find inverse trigonometric functions. The calculator will always give the least, or principal, value of the inverse trigonometric function.

Example 3

Find the radian measure of an angle θ in Quadrant I with a tangent of 1.3284.

Arctan $1.3284 = \theta$ *Make sure your calculator is in radian mode.*

1.3284 [2nd] [TAN] *0.925515*

The radian measure of θ is about 0.926.

A calculator can also be helpful in finding the value of a complicated trigonometric expression.

Example 4

Find $\sin\left(\text{Arcsin}\,\frac{5}{12} - \text{Arccot}\,\frac{5}{3}\right)$.

5 [÷] 12 [=] [2nd] [SIN] [−] [(] 5 [÷] 3 [)] [1/x]
[2nd] [TAN] [=] [SIN] *−0.1104185* *Arccot x = Arctan $\frac{1}{x}$. Why?*

Thus, $\sin\left(\text{Arcsin}\,\frac{5}{12} - \text{Arccot}\,\frac{5}{3}\right) \approx -0.1104$.

CHECKING FOR UNDERSTANDING

Read and study the lesson to answer each question.

1. **Describe** the relationship between the functions $y = \text{Sin } x$ and $y = \text{Arcsin } x$.

2. **Explain** why the domains of the trigonometric functions must be restricted in order to ensure that the inverse is a function.

3. How can you tell if the domain of a trigonometric function is restricted?

4. Use your calculator to solve Example 4 by using degrees. Is the result the same? Explain.

Guided Practice

Find each value.

5. $\text{Sin}^{-1}\left(-\dfrac{\sqrt{3}}{2}\right)$

6. $\text{Arctan } 1$

7. $\text{Sin}^{-1} 0$

8. $\text{Sin}^{-1}\left(-\dfrac{1}{2}\right)$

9. $\text{Tan}^{-1}\left(\dfrac{\sqrt{3}}{3}\right)$

10. $\text{Arcsin } 1$

11. $\text{Arccos } 0$

12. $\cos\left(\text{Cos}^{-1}\dfrac{4}{5}\right)$

13. $\tan\left[\text{Cos}^{-1}\left(-\dfrac{3}{5}\right)\right]$

14. $\cos\left(2\,\text{Tan}^{-1}\sqrt{3}\right)$

15. $\sin\left[\text{Arctan}\left(-\dfrac{3}{4}\right) + \text{Arccot}\left(-\dfrac{4}{3}\right)\right]$

EXERCISES

Practice

Find each value.

16. $\text{Arctan } 1$

17. $\text{Cos}^{-1}\left(-\dfrac{\sqrt{3}}{2}\right)$

18. $\text{Arcsin}\left(-\dfrac{\sqrt{2}}{2}\right)$

19. $\text{Arctan }\dfrac{3}{4}$

20. $\text{Cos}^{-1}\left(-\dfrac{1}{2}\right)$

21. $\text{Arctan }\sqrt{3}$

22. $\sin\left(\text{Sin}^{-1}\dfrac{1}{2}\right)$

23. $\text{Sin}^{-1}\left(\cos\dfrac{\pi}{2}\right)$

24. $\cos\left(\text{Cos}^{-1}\dfrac{1}{2}\right)$

25. $\cos\left(\text{Tan}^{-1}\sqrt{3}\right)$

26. $\tan\left(\text{Sin}^{-1}\dfrac{5}{13}\right)$

27. $\sin\left(\text{Sin}^{-1}\dfrac{\sqrt{3}}{2}\right)$

28. $\sin\left(2\,\text{Cos}^{-1}\dfrac{3}{5}\right)$

29. $\cos\left[\text{Arcsin}\left(-\dfrac{1}{2}\right)\right]$

30. $\sin\left(2\,\text{Sin}^{-1}\dfrac{1}{2}\right)$

31. $\sin\left[\text{Arctan}\left(-\sqrt{3}\right)\right]$

32. $\cos(\text{Tan}^{-1} 1)$

33. $\sin\left(2\,\text{Sin}^{-1}\dfrac{\sqrt{3}}{2}\right)$

34. $\tan\left(\dfrac{1}{2}\,\text{Sin}^{-1}\dfrac{15}{17}\right)$

35. $\sin\left(\dfrac{1}{2}\,\text{Arctan }\dfrac{3}{5}\right)$

36. $\cos\left(\text{Tan}^{-1}\sqrt{3} - \text{Sin}^{-1}\dfrac{1}{2}\right)$

37. $\cos\left(\text{Cos}^{-1} 0 + \text{Sin}^{-1}\dfrac{1}{2}\right)$

38. $\sin\left(\text{Tan}^{-1} 1 - \text{Sin}^{-1} 1\right)$

39. $\cos\left[\text{Cos}^{-1}\left(-\dfrac{1}{2}\right) - \text{Sin}^{-1} 2\right]$

40. $\cos\left[\text{Cos}^{-1}\left(-\dfrac{\sqrt{2}}{2}\right) - \dfrac{\pi}{2}\right]$

41. $\cos\left[\dfrac{4}{3}\pi - \text{Cos}^{-1}\left(-\dfrac{1}{2}\right)\right]$

42. $\sin\left[\dfrac{\pi}{2} - \text{Cos}^{-1}\left(\dfrac{1}{2}\right)\right]$

43. $\tan\left(\text{Cos}^{-1}\dfrac{3}{5} - \text{Sin}^{-1}\dfrac{5}{13}\right)$

44. Express sin (Arcsin u − Arccos v) in terms of u and v.

45. Electronics   The average power P of an electrical circuit with alternating current is determined by the equation $P = VI$ Cos ϕ, where V represents the voltage, I represents the current, and ϕ is the measure of the phase angle. If a certain circuit has a voltage of 120 volts, a current of 0.75 amperes, and produces 7.2 watts of power, what is the measure of the phase angle?

46. Optics   When you look at an object that is under water, refraction bends the light rays so that the object appears to be closer to the surface than it really is. According to Snell's Law, the angle between a perpendicular to the surface of the water and a beam of light entering the water, I, is related to the angle at which the light travels in the water, r, by the equation $\dfrac{\sin I}{\text{speed of light in air}} = \dfrac{\sin r}{\text{speed of light in water}}$. The speed of light in air is approximately 3.00×10^8 meters per second, and the speed of light in water is approximately 2.00×10^8 meters per second.

a. Is there any angle at which you could direct a beam of light at the surface of a pool so that the light is not bent?

b. Place a pencil in a clear glass jar full of water. Look at it from different angles until you find the one where you can see the whole pencil undistorted. How does the angle at which you are looking at the pencil relate to the angle you found for Exercise 46a?

47. Is the relation $\{(4, 0), (3, 0), (5, -2), (4, -3), (0, -13)\}$ a function?   **(Lesson 1-1)**

48. Manufacturing   The Eastern Minnesota Paper Company can convert wood pulp to either newsprint or notebook paper. The mill can produce up to 200 units of paper a day, and regular customers require 10 units of notebook paper and 80 units of newsprint per day. If the profit on a unit of notebook paper is $500 and the profit on a unit of newsprint is $350, how much of each should the plant produce?   **(Lesson 2-6)**

49. Without graphing, describe the end behavior of the function $y = 4x^5 - 2x^2 + 4$.   **(Lesson 3-8)**

50. Find all the rational roots of the equation $3x^4 + 20x^3 - 4x^2 + 20x - 7 = 0$.   **(Lesson 4-4)**

51. Determine the number of possible solutions to $\triangle ABC$ if $A = 152°$, $b = 12$, and $a = 10.2$. If a solution exists, solve the triangle.   **(Lesson 5-6)**

52. Find the area of a triangle whose sides are 7, 9, and 12 inches long.   **(Lesson 5-8)**

53. Find the value of $\sin \left(2 \operatorname{Sin}^{-1} \dfrac{1}{2}\right)$.   **(Lesson 6-4)**

54. College Entrance Exam   Choose the best answer.

Find an expression equivalent to $\left(\dfrac{2x^2}{y}\right)^3$.

(A) $\dfrac{8x^6}{y^3}$ **(B)** $\dfrac{64x^6}{y^3}$ **(C)** $\dfrac{6x^5}{y^3}$ **(D)** $\dfrac{8x^5}{y^3}$ **(E)** $\dfrac{2x^5}{y^4}$

6-6A Graphing Calculators: Graphing Inverses of Trigonometric Functions

You can use a graphing calculator to graph inverses of trigonometric functions.

Example 1

Graph $y = $ Arccos x in the viewing window $[-2, 2]$ by $[-360°, 360°]$ with a scale factor of 1 for the x-axis and a scale factor of $90°$ for the y-axis.

You must manually set the range for the graph on the TI-82 graphing calculator. Then enter the equation.

Enter: $\boxed{Y=}$ $\boxed{2nd}$ $\boxed{COS^{-1}}$ $\boxed{X,T,\theta}$ $\boxed{GRAPH}$

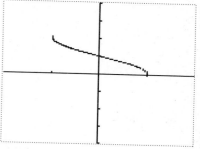

The function is defined on the interval $[-1, 1]$ for x and $[0°, 180°]$ for y.

Example 2

Graph $y = \sin (\text{Tan}^{-1}x)$ in the viewing window $[-5, 5]$ by $[-2, 2]$ with a scale factor of 1 for both axes.

Enter: $\boxed{Y=}$ $\boxed{SIN}$ $\boxed{(}$ $\boxed{2nd}$ $\boxed{TAN^{-1}}$
$\boxed{X,T,\theta}$ $\boxed{)}$ $\boxed{GRAPH}$

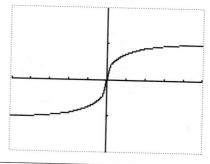

The function is defined on the interval $(-\infty, \infty)$ for x and $(-1, 1)$ for y.

EXERCISES

Graph each equation so that a complete graph is shown. Then sketch the graph on a piece of a paper.

1. $y = $ Arctan x

2. $y = \cos (2\text{Sin}^{-1}x)$

3. $y = \cos (\text{Tan}^{-1}x - \text{Cos}^{-1}x)$

4. $y = \sin (\text{Sin}^{-1}x)$

5. $y = \cos (\text{Arctan } 2x)$

6. $y = \sin (\text{Sin}^{-1}x - \text{Cos}^{-1}\frac{x}{2})$

6-6 Graphing Inverses of Trigonometric Functions

Objectives

After studying this lesson, you should be able to:

- write equations for inverses of trigonometric functions, and
- graph inverses of trigonometric functions.

Application

Mathematicians have been trying to calculate the value of the number π accurately for centuries. In 1992, brothers David (*top*) and Gregory (*bottom*) Chudnovsky of Columbia University extended the calculation of the digits of π to 2,160,000,000 places. They performed the calculations using a supercomputer that they assembled from mail-order parts. Their calculation doubled the number of digits previously calculated. The Chudnovsky brothers are studying the digits in hopes of finding patterns within them, but with no results thus far.

One of the methods for approximating π uses inverse trigonometric functions. Gregory's series, named for Scottish mathematician James Gregory, is $\arctan x = x - \dfrac{x^3}{3} + \dfrac{x^5}{5} - \dfrac{x^7}{7} + \cdots$. In 1706, John Machin used Gregory's series to obtain the formula $\dfrac{\pi}{4} = 4 \arctan\left(\dfrac{1}{5}\right) - \arctan\left(\dfrac{1}{239}\right)$. With it, he determined the value of π to 100 decimal places.

Just as the graphs of the sine and arcsine functions are related, the inverse cosine and tangent functions can be graphed using the graphs of the cosine and tangent functions. The values of Arccos x are shown as the colored portion of the curve at the right. There are an unlimited number of values of the arccos x for a given x, but there is only one value of Arccos x for each x. The range of Arccos x is $0° \leq \text{Arccos } x \leq 180°$.

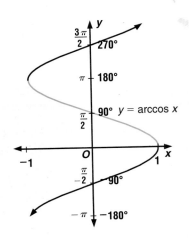

Notice the similarity of the graph of the inverse of the tangent function to the graph of $y = \tan x$ with the axes interchanged. Values of Arctan x are shown as the colored portion of the curve. For each value of x, there is only one value of Arctan x, but infinitely many values for arctan x. The range of Arctan x is $-90° < y < 90°$.

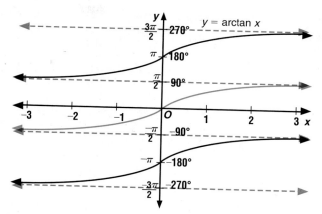

Example 1

You may wish to review finding inverse functions in Lesson 1-2.

Find the inverse of the function $y = \text{Cot } x$. Then graph both functions.

To find the inverse of a trigonometric function, exchange x and y.

$y = \text{Cot } x$

$x = \text{Cot } y$ *Exchange x and y.*

$y = \text{Arccot } x$ or $y = \text{Cot}^{-1} x$

Now graph the functions.

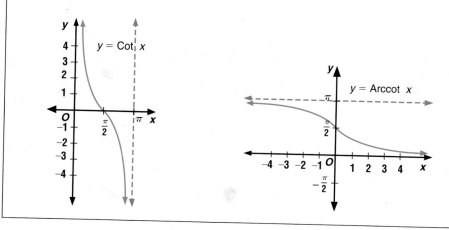

Example 2

Determine if $\text{Cos}^{-1}(\cos x) = x$ for all values of x. If false, give a counterexample.

Try several values of x to see if we can find a counterexample.

When $x = 270°$, $\text{Cos}^{-1}(\cos x) \neq x$. So $\text{Cos}^{-1}(\cos x) = x$ is not true for all values of x.

x	$\cos x$	$\text{Cos}^{-1}(\cos x)$
$0°$	1	$0°$
$180°$	-1	$180°$
$270°$	0	$90°$

Inverse trigonometric functions are used to solve practical problems in many fields.

Example 3

APPLICATION

Medicine

When surgeons join arteries, as when grafting an artery for heart bypass surgery, they must ensure that the friction of the blood passing through the artery at the juncture is minimized. Suppose the angle between the joined arteries measures θ degrees and the radii of the smaller and larger arteries measure r and R millimeters respectively. To minimize the friction, the equation $\text{Cos}\,\theta = \left(\dfrac{r^4}{R^4}\right)$ must hold. Graph the function for $R = 5$.

If $R = 5$, then $\theta = \text{Cos}^{-1}\left(\dfrac{r^4}{5^4}\right)$ or

$\theta = \text{Cos}^{-1}\left(\dfrac{r^4}{625}\right)$.

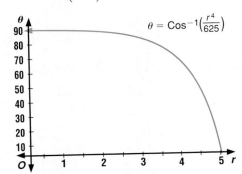

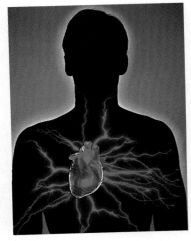

Portfolio

Select an item from this chapter that you feel shows your best work and place it in your portfolio. Explain why you selected it.

CHECKING FOR UNDERSTANDING

Communicating Mathematics

Read and study the lesson to answer each question.

1. **Describe** how to draw the graph of an inverse trigonometric function from the graph of the related trigonometric function.

2. **Write** a few sentences describing how to find the equation of the inverse of a trigonometric function.

3. How are the domain and range of a relation related to the domain and range of its inverse?

Guided Practice

State the domain and range of each relation.

4. $y = \sin x$ 5. $y = \tan x$ 6. $y = \text{Cos}\,x$

7. $y = \arcsin x$ 8. $y = \text{Arccos}\,x$ 9. $y = \text{Arctan}\,x$

10. Write the equation for the inverse of the function $y = \cos x$. Then graph the function and its inverse.

Determine if each of the following is *true* or *false*. If false, give a counterexample.

11. $\tan\left(\text{Tan}^{-1} x\right) = x$ for all x 12. $\text{Cot}^{-1}(\cot x) = x$ for all x

EXERCISES

Practice

State the domain and range of each relation.

13. $y = \cos x$

14. $y = \text{Sin } x$

15. $y = \text{Tan } x$

16. $y = \arccos x$

17. $y = \text{Arcsin } x$

18. $y = \arctan x$

Write the equation for the inverse of each function. Then graph the function and its inverse.

19. $y = \text{Arctan } x$

20. $y = \text{Arcsin } x$

21. $y = \text{Sin } x$

22. $y = \dfrac{\pi}{2} + \text{Arcsin } x$

23. $y = \text{Arctan } 2x$

24. $y = \text{Cos } (x + 90°)$

Determine if each of the following is *true* or *false*. If false, give a counterexample.

25. $\text{Sin}^{-1} x = -\text{Sin}^{-1}(-x),\ -1 \leq x \leq 1$

26. $\text{Arccos } x = \text{Arccos } (-x),\ -1 \leq x \leq 1$

27. $\text{Cos}^{-1}(-x) = -\text{Cos}^{-1} x,\ -1 \leq x \leq 1$

28. $\text{Sin}^{-1} x + \text{Cos}^{-1} x = \dfrac{\pi}{2},\ -1 \leq x \leq 1$

29. $\text{Cos}^{-1} x = \dfrac{1}{\text{Cos } x}$

30. $\text{Tan}^{-1} x = \dfrac{1}{\text{Tan } x}$

Sketch the graph of each equation.

31. $y = \tan (\text{Tan}^{-1} x)$

32. $y = \sin (\text{Tan}^{-1} x)$

Critical
Thinking

33. Graph $y = \arcsin x$ and $y = \arccos x$ on the same coordinate axes. Give four values of y corresponding to points of intersection of the graphs.

Applications
and Problem
Solving

34. Navigation Navigators aboard ships and airplanes use nautical miles to measure distance. A nautical mile is equal to an arc length of one minute of a degree. The actual length varies slightly since Earth is not a perfect sphere. The formula for the length of a nautical mile in feet, ℓ, on the latitude line θ is $\ell = 6077 - 31 \cos 2\theta$.

  a. Solve the formula for θ.

  b. Find the length of a nautical mile on the 40th parallel.

  c. Where is the length of a nautical mile approximately 6060 feet?

35. **Coordinate Geometry**   Suppose ℓ_1 and ℓ_2 are two nonvertical nonperpendicular lines with slopes of m_1 and m_2 respectively. Then the tangent of the angle formed by ℓ_1 and ℓ_2 is given by the formula $\tan\theta = \dfrac{m_2 - m_1}{1 + m_1 m_2}$.

   a. The slope of line a is 2. If lines a and b form an angle of $37°$, what is the slope of line b? *Hint: Let a be m_1.*

   b. The slopes of lines s and t are $-\dfrac{1}{2}$ and $\dfrac{3}{4}$, respectively. What is the measure of the angle from s to t? What is the measure of the angle from t to s?

36. **History**   Gregory's series is $\arctan x = x - \dfrac{x^3}{3} + \dfrac{x^5}{5} - \dfrac{x^7}{7} + \cdots$. In 1706, John Machin used the Gregory's series to obtain the formula $\dfrac{\pi}{4} = 4\arctan\left(\dfrac{1}{5}\right) - \arctan\left(\dfrac{1}{239}\right)$.

   a. Use the first six terms of Gregory's series and your calculator to approximate $\arctan\left(\dfrac{1}{5}\right)$ and $\arctan\left(\dfrac{1}{239}\right)$.

   b. Use Machin's formula and the values you calculated for Exercise 36a to find an approximation for π. How accurate is your approximation?

   c. Research   Write a 1-page paper on approximations of π. Include the methods of approximation other than approximation by Gregory's series and Machin's formula. Find out what method David and Gregory Chudnovsky used with their computer.

Mixed Review

37. Find $[f \circ g](x)$ and $[g \circ f](x)$ if $f(x) = x^3 - 1$ and $g(x) = 3x$.   **(Lesson 1-2)**

38. Write the standard form of the equation of the line that is parallel to the line $2x - 6y = -3$ and passes through the point with coordinates $(9, -4)$. **(Lesson 1-6)**

39. Solve $\sqrt{2y - 3} - \sqrt{2y + 3} = -1$.   **(Lesson 4-7)**

40. Find the value of $\cos\alpha$ if $(4, 3)$ lies on the terminal side of the angle in standard position.   **(Lesson 5-3)**

41. Find $\cos\left(\operatorname{Tan}^{-1}\sqrt{3} - \operatorname{Sin}^{-1}\dfrac{1}{2}\right)$.   **(Lesson 6-5)**

42. **College Entrance Exam**   Compare quantities A and B below.
Write A if quantity A is greater.
Write B if quantity B is greater.
Write C if the quantities are equal.
Write D if there is not enough information to determine the relationship.

   Given: $2 < b < a$

   (A) $\dfrac{120a^5 b^6}{8a^3 b^2}$ **(B)** $\dfrac{120a^4 b^{\frac{13}{2}}}{6a^0 b^{\frac{1}{2}}}$

6-7 Simple Harmonic Motion

Objective
After studying this lesson, you should be able to:
■ solve problems involving simple harmonic motion.

Application
Buoys are floating markers placed in water to aid in navigation. The centers and sides of navigation channels, wrecks, rocks, and underwater cables are a few of the hazards buoys indicate. Boaters can tell what the buoy indicates by the shape, color, and lights on it. Some buoys also have bells or whistles and are made to reflect radar.

A buoy in a lake channel bobs up and down as the waves move past. Suppose the buoy moves a total of 6 feet from its high point to its low point and returns to its high point every 10 seconds. Assuming that at $t = 0$ the buoy is at its high point, write an equation to describe its motion. *This problem will be solved in Example 1.*

A wide variety of problems involve the rhythmic motion of an object. The motion of a buoy is one example. Other examples of rhythmic motion are the vibrations of a guitar string, the pistons of an engine, and a pendulum moving back and forth. When friction and other factors affecting such motion are ignored, this movement is called **simple harmonic motion.**

The angular velocity of an object that is turning is the number of degrees or radians that a point on the edge of the object moves in a unit of time. Suppose a wheel is turning counterclockwise at a constant angular velocity and P is a point on the rim of the wheel. If the wheel is modeled by a circle with its center at the origin of a rectangular coordinate system, the motion of point P can be described in terms of the coordinates of its position. Since the angular velocity is constant, the measure of angle θ as a function of time is

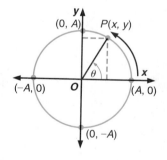

$\theta = kt$, where k is the constant and t represents time. Let A be the measure of the radius of the wheel and substitute kt for θ. Then $\cos \theta$ or $\cos kt = \dfrac{x}{A}$ and $\sin \theta = \sin kt = \dfrac{y}{A}$. Solving for x and y produces the following equations for the horizontal and vertical motion of the point P if its initial position was at $(A, 0)$.

$$x = A \cos kt \qquad\qquad y = A \sin kt$$

These equations assume that $\theta = 0$ at time $t = 0$. If angle θ measures c radians at time $t = 0$, then $\theta = kt + c$ would express θ as a function of time. Thus, the more general equations for the motion of point P are as follows.

$$x = A \cos (kt + c) \qquad \qquad y = A \sin (kt + c)$$

$-\dfrac{c}{k}$ *represents the phase shift in terms of t. The angular phase shift is* $-c$.

You know that the amplitude of each of the functions above is $|A|$, the period is $\dfrac{2\pi}{k}$ (or $\dfrac{360°}{k}$), and the phase shift is $-\dfrac{c}{k}$.

These equations can be used to model problems involving simple harmonic motion. The description of the motion of point P above was with respect to the coordinate system used, rather than the true orientation of the wheel. In an applied problem, the appropriate equation will have to be chosen to conform to the actual physical conditions of the problem. Since sine and cosine curves having the same amplitude and period can be made to coincide by a phase shift, several equations may be used to describe the same simple harmonic motion.

Example 1

APPLICATION

Navigation

Refer to the application problem at the beginning of the lesson.

a. Write two possible equations to describe the motion of the buoy.

a. First, find the amplitude, period, and phase shift of the function.

The amplitude, $|A|$, is $\dfrac{6}{2}$ or 3 feet.

The period is 10 seconds, so $10 = \dfrac{2\pi}{k}$ and $k = \dfrac{\pi}{5}$.

The initial state can be represented by an equation of the form $x = A \cos kt$. Thus, the equation is $x = 3 \cos \dfrac{\pi t}{5}$.

An alternate equation can be written using the sine function. Consider that 2.5 seconds before the cycle begins, or $\dfrac{1}{4}$ of a complete cycle, the buoy is at 0 feet. Thus, the phase shift in terms of time would be –2.5 seconds. Since $-2.5 = -\dfrac{c}{k}$ and $k = \dfrac{\pi}{5}$, $c = \dfrac{\pi}{2}$.

An alternate equation is $y = 3 \sin \left(\dfrac{\pi t}{5} + \dfrac{\pi}{2} \right)$.

Why are the equations equivalent?

b. Draw a graph to show the motion of the buoy in 30 seconds.

b. Graph one of the functions.

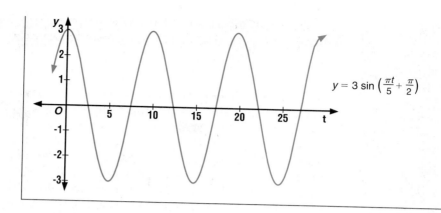

$$y = 3 \sin\left(\frac{\pi t}{5} + \frac{\pi}{2}\right)$$

The concept of **frequency** is one that is useful in some applied problems, particularly those involving sound waves or electricity. Frequency is the number of cycles per unit of time and is equal to the reciprocal of the period. For instance, the frequency of the function in Example 1 is $\frac{1}{10}$ or $\frac{1}{10}$ of a cycle per second.

Cycles per second are called <u>hertz</u>.

Example 2

APPLICATION

Entertainment

FYI...

Will Rogers was an American entertainer in the 1920s, famous for his homespun humor and down-to-earth philosophy. During his lectures, he chewed gum and performed rope tricks while joking about business, government, people, and politics. His best known statement is, "I never met a man I didn't like."

Will Rogers spun a lasso in a vertical circle. The diameter of the loop was 6 feet, and the loop spun 50 times each minute. If the lowest point on the rope was 6 inches above the ground, write an equation to describe the height of this point above the ground after *t* seconds.

The frequency of the lasso was 50 revolutions per minute, or 50 rpm. At this rate, the lasso made one revolution every $\frac{60}{50}$ or 1.2 seconds. So the period is $1.2 = \frac{2\pi}{k}$ and $k = \frac{5\pi}{3}$ radians/second.

We know that a point on the rope was 6 inches above the ground at time $t = 0$. So let $c = 0$ at $t = 0$. Use the center of the rotation as the origin of a coordinate system. Relative to the center of rotation, the initial position was at $(0, -3)$. *The radius is 3 feet.*

$h = A \cos (kt + c)$

$\quad = -3 \cos \left(\frac{5\pi}{3} t + 0\right)$

$\quad = -3 \cos \frac{5\pi}{3} t$

This equation gives the position of the point relative to the center of rotation, which was 3 feet 6 inches or 3.5 feet above the ground. Therefore, the equation for the height of the point relative to the ground is $h = -3 \cos \frac{5\pi}{3} t + 3.5$.

The study of physics offers many examples of harmonic motion. The movement of a pendulum is one example.

EXPLORATION: Physics

Materials: string, small weight like an eraser or washer

Work in groups to complete this exploration. Make a pendulum with a piece of string about 3 feet long and a small weight. Have one group member hold the end of the string or attach it to a door frame so that it can swing freely. Pull the string back and measure the displacement from the equilibrium point. Release the weight and use a stopwatch to time ten full swings of the pendulum. A full swing means that the weight swings over and then back to its starting point. Divide the time for ten swings of the pendulum by 10 to find the time for one full swing. This is the period of the pendulum.

1. Use the distance that you displaced the weight and the time that you measured to write an equation for the position of the weight after t seconds.

2. Use your equation to determine at what time the pendulum will complete two cycles. Then use the stopwatch to measure the time experimentally. Is your equation accurate?

3. Repeat the experiment in Step 1 using a different displacement. How does the equation change?

4. Change the length of the pendulum and repeat the experiment in Step 1 using the same displacement as you did for the first experiment. How does the equation change?

CHECKING FOR UNDERSTANDING

Communicating Mathematics

Read and study the lesson to answer each question.

1. **Describe** what it means for an object to be in simple harmonic motion. Give an example of an object that demonstrates this type of motion.

2. **Explain** the relationship between frequency and period.

3. The frequency of the vibrations of an E string on a guitar is about 330 hertz. How long does it take for the string to make one full vibration?

Guided Practice

Find the amplitude, period, frequency, and phase shift.

4. $s = 7 \cos \left(\dfrac{\pi t}{2} + \dfrac{\pi}{4} \right)$

5. $V = -8 \sin (4t - \pi)$

6. $E = 120 \sin 100\pi t$

7. $y = -4 \sin \left(3t + \dfrac{3\pi}{4} \right)$

Write an equation with phase shift 0 to represent simple harmonic motion under each set of circumstances.

8. initial position –3, amplitude 3, period 2

9. initial position 0, amplitude 5, period 2

10. initial position 0, amplitude 7, period 10

11. initial position 7, amplitude 7, period 10

12. the y-coordinate of a point P on the rim of a wheel 12 feet in diameter that is turning at 10 rpm; let P be $(6, 0)$ at $t = 0$ seconds

13. **Navigation**   Ken first observes a buoy at the bottom of a wave bobbing up and down a total distance of 10 feet. The buoy completes a full cycle every 12 seconds.

EXERCISES

Practice

Find the amplitude, period, frequency, and phase shift.

14. $y = 5 \sin\left(\pi t - \dfrac{\pi}{2}\right)$

15. $h = -15 \cos\left(\dfrac{\pi t}{3}\right) + 21$

16. $y = 10 \sin(2\pi x - \pi)$

17. $E = 150 \cos 80\pi t$

18. $n = -0.1 \sin\left(\dfrac{\pi t}{4} + \dfrac{\pi}{4}\right)$

19. $W = -25 \cos 8t$

Write an equation with phase shift 0 to represent simple harmonic motion under each set of circumstances.

20. initial position –7, amplitude 7, period 4

21. initial position 0, amplitude 0.5, period 1

22. initial position 0, amplitude 22, period 12

23. initial position 10, amplitude 10, period $\dfrac{1}{2}$

Critical Thinking

24. **Physics**   A weight in a spring-mass system is oscillating in simple harmonic motion. The amplitude is doubled. How are the period and phase shift affected?

Applications and Problem Solving

25. **Entertainment**   At one time, Cedar Point Amusement Park in Sandusky, Ohio, had a ride called the Rotor in which riders stood against the walls of a spinning cylinder. As the cylinder spun, the riders were held up by friction and the floor of the ride dropped out. The cylinder of the Rotor had a radius of 3.5 meters and rotated counterclockwise at a rate of 14 rpm. Suppose the center of rotation of the Rotor is at the origin of a rectangular coordinate system.

a. What is the period of the function, in terms of t seconds, that represents the position of the door of the ride's cylinder?

b. If the initial coordinates of the door of the ride's cylinder are $(3.5, 0)$, what are its coordinates after t seconds?

c. Under the conditions of part b, what are the coordinates of the door's position after 4 seconds?

d. Under the conditions of part b, what are the coordinates of the position of the door after 19.25 seconds? Round to the nearest hundredth.

e. If the initial coordinates of the position of the door were $(0, 3.5)$, what would the coordinates be after t seconds?

f. If the initial y-coordinate of the position of the door was 2.2, what was the initial x-coordinate?

26. **Automotive Engineering** In a gasoline engine, the explosion of the fuel exerts a thrust of pressure on the pistons which turn the *crankshaft* and power the engine. One thrust of the pistons turns the crankshaft 360°. As the crankshaft turns, the radius of the throw makes an angle θ with the line from the shaft to the end of the rod. The distance between the end of the rod and the crankshaft center, h, is given by the formula
$h = r\cos\theta + \sqrt{d^2 + r^2(\sin\theta)^2}$,
where r is the measure of the radius of the throw, and d is the measure of the rod. For a certain engine, r is 2.2 inches and d is 5.8 inches.

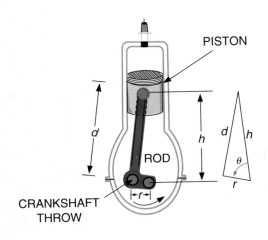

a. Find the distance from the end of the rod to the crankshaft center when $\theta = 30°$.

b. Graph the distance from the end of the rod to the crankshaft center for values of θ from 0° to 360° using a graphing calculator. Then sketch the graph.

27. **Physics** A weight in a spring-mass system exhibits harmonic motion. The system is in equilibrium when the weight is motionless. If the weight is pulled down or pushed up and released, it would tend to oscillate freely if there were no friction. In a certain spring-mass system, the weight is 5 feet from a 10-foot ceiling when it is at rest. The

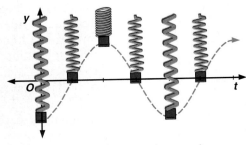

motion of the weight can be described by the equation $y = 3\sin\left(\pi t - \dfrac{\pi}{2}\right)$, where y is the distance from the equilibrium point, and t is measured in seconds.

a. Was the weight pulled down or pushed up before it was released?

b. How far was the weight from the ceiling when it was released?

c. How close will the weight come to the ceiling?

d. When does the weight first pass its equilibrium point?

e. What is the greatest distance that the weight will be from the ceiling?

f. Find the period of the motion.

g. Find the amplitude of the motion.

h. What is the frequency of the motion?

i. How far from the ceiling is the weight after 2.5 seconds?

28. **Zoology**   The number of predators and the number of prey in a predator-prey system tend to vary periodically. In a certain region with hawks as predators and rodents as prey, the rodent population M varied according to the equation $M = 1200 + 300 \sin \dfrac{\pi t}{2}$ and the hawk population varied with the equation $H = 250 + 25 \sin \left(\dfrac{\pi t}{2} - \dfrac{\pi}{4} \right)$, with t measured in years since January 1, 1970.

a. What was the population of rodents on January 1, 1970?

b. What was the population of hawks on January 1, 1970?

c. What are the maximum populations of rodents and hawks? Do these maxima ever occur at the same time?

d. On what date was the first maximum population of rodents achieved?

e. What is the minimum population of hawks? On what date was the minimum population of hawks first achieved?

f. Use a graphing calculator to graph both equations on the same set of axes. Write a paragraph to explain why the populations fluctuate in the way that they do.

Mixed Review

29. Graph the rational function $y = \dfrac{-4}{x^2 - 2}$.   **(Lesson 3-4)**

30. Approximate the real zeros of the function $f(x) = 5x^4 - 8x^2 + 2$ to the nearest tenth.   **(Lesson 4-5)**

31. Are the angles $-15°$ and $345°$ coterminal?   **(Lesson 5-1)**

32. **Geometry**   A regular hexagon is inscribed in a circle with a radius 6.4 centimeters long. Find the apothem; that is, the distance from the center of the circle to the midpoint of a side.   **(Lesson 5-5)**

33. **College Entrance Exam**   The sum of the reciprocals of a and b is 7 and the difference of the reciprocals of a and b is 3. Find $\dfrac{1}{a^2} - \dfrac{1}{b^2}$.

VOCABULARY

Upon completing this chapter, you should be
familiar with the following terms:

amplitude **309**   **312** phase shift
arcsine **329**   **334** principal values
compound function **324**   **345** simple harmonic motion
frequency **347**   **322** vertical displacement
period **311**

SKILLS AND CONCEPTS

OBJECTIVES AND EXAMPLES	REVIEW EXERCISES

Upon completing this chapter, you should
be able to:

Use these exercises to review and prepare for
the chapter test.

■ graph the trigonometric functions
(Lesson 6-1)

**Find each value by referring to the graphs
of the trigonometric functions on pages
310–311.**

Graph $y = \cos x$ on the interval
$-90° \le x \le 450°$.

1. $\tan 180°$ **2.** $\sin 450°$

3. $\sec(-270°)$ **4.** $\cos 540°$

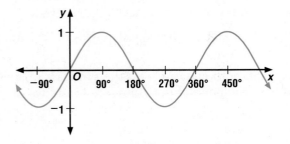

Graph each function on the given interval.

5. $y = \sin x, 90° \le x \le 450°$

6. $y = \tan x, -\pi \le x \le \pi$

7. $y = \cot x, -180° \le x \le 360°$

■ find the amplitude, period, and phase shift
for a trigonometric function   **(Lesson 6-2)**

**State the amplitude (if it exists), period, and
phase shift for each function.**

State the amplitude, period, and phase
shift for $y = -\sqrt{3}\cos(x - \pi)$.

8. $y = 4\cos 2x$

amplitude: $|A| = |-\sqrt{3}| = \sqrt{3}$

9. $y = 15\sin\left(\dfrac{3}{2}x + 90°\right)$

period: $\dfrac{360°}{k} = \dfrac{360°}{1} = 360°$

10. $y = 5\cot\left(\dfrac{x}{2} - 45°\right)$

phase shift: $-\dfrac{c}{k} = -\dfrac{-\pi}{1} = \pi$ or $180°$

11. $y = 2\tan 5x$

OBJECTIVES AND EXAMPLES	REVIEW EXERCISES

■ graph various trigonometric functions
(Lesson 6-3)

Graph $y = \sin\left(4x + \dfrac{\pi}{3}\right)$.

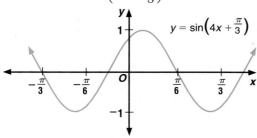

Graph each function.

12. $y = 3\cos\dfrac{\theta}{2}$

13. $y = 2\tan(3\theta + 90°)$

14. $y = -3\sin(\theta - 45°)$

15. $y = 2\sin x - \cos x$

16. $y = x + 2\sin 2x$

■ evaluate inverse trigonometric functions
(Lesson 6-4)

Evaluate $y = \sin\left(\arccos\dfrac{1}{2}\right)$. Assume that the angle is in Quadrant I.

Let $A = \arccos\dfrac{1}{2}$.

$\cos A = \dfrac{1}{2}$

$\quad A = 60°$

$\sin\left(\arccos\dfrac{1}{2}\right) = \sin 60°$

$\qquad\qquad = \dfrac{\sqrt{3}}{2}$

Write each equation in the form of an inverse relation.

17. $y = \sin\alpha$ **18.** $\tan y = x$

19. $n = \cos\theta$ **20.** $\sec\theta = y$

Evaluate each expression. Assume that all angles are in Quadrant I.

21. $\cos\left(\arccos\dfrac{1}{2}\right)$

22. $\tan\left(\operatorname{arccot}\dfrac{4}{5}\right)$

23. $\sin(\tan^{-1}1) + \cos(\sin^{-1}1)$

24. $\tan\left(\arcsin\dfrac{\sqrt{3}}{2} + \arccos\dfrac{\sqrt{3}}{2}\right)$

■ find principal values of inverse trigonometric functions   **(Lesson 6-5)**

Find $\cos(\operatorname{Tan}^{-1}1)$.

Let $\theta = \operatorname{Tan}^{-1}1$.

$\operatorname{Tan}\theta = 1$

$\quad \theta = 45°$

So, $\cos 45° = \dfrac{\sqrt{2}}{2}$.

Find each value.

25. $\cos\left(\operatorname{Sin}^{-1}\dfrac{1}{2}\right)$

26. $\sin\left(3\operatorname{Sin}^{-1}\dfrac{\sqrt{3}}{2}\right)$

27. $\sin 2\left(\operatorname{Arcsin}\dfrac{1}{2}\right)$

28. $\cos\left(\dfrac{\pi}{2} - \operatorname{Cos}^{-1}\dfrac{\sqrt{2}}{2}\right)$

29. $\cos\left(\operatorname{Arctan}\sqrt{3} + \operatorname{Arcsin}\dfrac{1}{2}\right)$

■ graph inverses of trigonometric functions **(Lesson 6-6)**

Graph the inverse of $y = \cos x$.

The inverse is $y = \text{Arccos } x$.

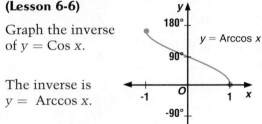

Write the equation for the inverse of each function. Then graph the function and its inverse.

30. $y = \text{Arcsin } x$

31. $y = \text{Csc } x$

32. $y = \cot x$

33. $y = \text{arcsec } x$

■ solve problems involving simple harmonic motion **(Lesson 6-7)**

Two equations that describe simple harmonic motion are:

$x = A \cos (kt + c)$, and

$y = A \sin (kt + c)$.

The point P is on the circumference of a circle of radius 5 inches that is rotating in a counterclockwise direction at the rate of 4 revolutions per second. Let the initial position of P at $t = 0$ be at (5, 0).

34. Write two equations that describe the position of P after t seconds.

35. What is the frequency of the motion?

36. What are the coordinates of P after $\frac{1}{6}$ second?

APPLICATIONS AND PROBLEM SOLVING

37. Entertainment A popular, but dangerous, activity is bungee jumping. A person, whose ankles are attached to a bungee cord, jumps off a ledge and bounces upside-down in mid-air. Suppose a person jumps off the ledge and falls 122 feet before rebounding. Consider the point at which the person rebounds to be $t = 0$. The person then rebounds to a distance of 46 feet from the ledge after 3 seconds. Assume there is no resistance in the bungee cord. **(Lesson 6-2)**

  a. Write a function that represents the person's distance from the equilibrium point as a function of time.

  b. Write a function that represents the person's vertical distance from the ledge as a function of time.

  c. What would be the position of the person after 30 seconds?

38. Physics The strength of a magnetic field is called magnetic induction. An equation for magnetic induction is $B = \dfrac{F}{IL \sin \theta}$, where F is a force on a current I which is moving through a wire of length L at an angle θ to the magnetic field. A wire within a magnetic field is 1 meter long and carries a current of 5.0 A. The force on the wire is 0.2 N and the magnetic induction is 0.04 N/Am. What is the angle of the wire to the magnetic field? **(Lesson 6-5)**

Find the values of x in degrees for which each equation is true.

1. $\tan x = 0$

2. $\sin x = -1$

3. $\cos x = \dfrac{\sqrt{2}}{2}$

Graph each function on the given interval.

4. $y = \sin x; -180° \leq x \leq 180°$

5. $y = \tan x; 0 \leq x \leq 2\pi$

State the amplitude, period, and phase shift for each function.

6. $y = 3 \cos 4\theta$

7. $y = 110 \sin (15\theta - 40°)$

8. $y = 10 \sin (\pi - x)$

Graph each function.

9. $y = 3 \cos \dfrac{\theta}{2}$

10. $y = \tan \left(2x - \dfrac{\pi}{4}\right)$

11. $y = 2 \cos x - x$

Evaluate each expression. Assume that all angles are in Quadrant I.

12. $\sin \left(\arccos \dfrac{\sqrt{3}}{2}\right)$

13. $\tan \left(\cos^{-1} \dfrac{5}{13}\right)$

14. $\cos (\arctan \sqrt{3} + \operatorname{arccot} \sqrt{3})$

Find each value.

15. $\sin \left(\operatorname{Arccos} \dfrac{1}{2}\right)$

16. $\cos \left(\dfrac{1}{2} \operatorname{Tan}^{-1} \dfrac{3}{4}\right)$

17. $\tan \left(\pi + \operatorname{Arcsin} \dfrac{2}{3}\right)$

Write the equation for the inverse of each function. Then graph the function and its inverse.

18. $y = \operatorname{Arccsc} x$

19. $y = \tan x$

A point P is on the circumference of a circle of radius 10 centimeters with its center at the origin. The circle is rotating in a counterclockwise direction at the rate of 2 radians per second. Let P have coordinates $(10, 0)$ at $t = 0$.

20. Write an equation that describes the x-coordinate of P after t seconds.

21. Write an equation that describes the y-coordinate of P after t seconds.

22. What is the amplitude of the motion?

23. What is the frequency of this motion?

24. Oceanography The tides of the ocean can be described by simple harmonic motion. At midnight and at noon, the high tide rises to the 6-foot mark on a retaining wall. At 6 A.M. and at 6 P.M., the low tide falls to the 2-foot mark. At what mark is the water at 2 P.M.?

25. Civil Engineering Ms. Jefferson, a civil engineer, is designing a curve for a new highway. She uses the equation, $\tan \theta = \dfrac{v^2}{gr}$, where the radius r will be 1200 feet, the curve will be designed for a maximum velocity of 65 mph, and the acceleration due to gravity is 32 ft/s². At what angle θ should the curve be banked?

Bonus
Determine whether the relation $y = x^2 - \cos x$ is *even*, *odd*, or *neither*.

CHAPTER 7

TRIGONOMETRIC IDENTITIES AND EQUATIONS

HISTORICAL SNAPSHOT

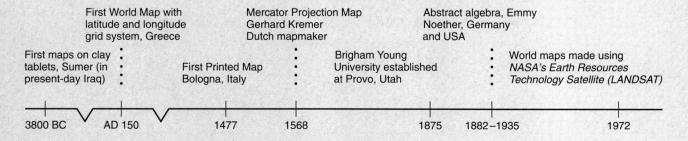

First World Map with
latitude and longitude
grid system, Greece

Mercator Projection Map
Gerhard Kremer
Dutch mapmaker

Abstract algebra, Emmy
Noether, Germany
and USA

First maps on clay
tablets, Sumer (in
present-day Iraq)

First Printed Map
Bologna, Italy

Brigham Young
University established
at Provo, Utah

World maps made using
*NASA's Earth Resources
Technology Satellite (LANDSAT)*

3800 BC   AD 150   1477   1568   1875   1882–1935   1972

CHAPTER OBJECTIVES

In this chapter, you will:
- Use and verify trigonometric identities.
- Solve trigonometric equations.
- Write the normal form of a linear equation.
- Find the distance from a point to a line and the distance between parallel lines.

CAREER GOAL: Cartography

The political face of the world is constantly changing. Think about the demise of the Soviet Union or the turmoil in Eastern Europe, Western Asia, and Africa. When new political boundaries are formed, cartographers are called upon to make the maps that reflect the changes. Brandon Plewe wants to be a cartographer. He says that "cartography never seems to get stale. It is always new and changing with changes in the world."

Brandon has always had an interest in maps, but never thought it would lead to a career. Instead, he enrolled at BYU in electrical engineering and mathematics. After two years, he decided to take a minor in cartography. "I began to realize that cartography offered new and attractive opportunities for me. Cartography uses computers extensively to draw and analyze maps. It has allowed me to pursue a field in which I have always had an interest, plus allows me to utilize my interest and experience in computers."

Today, cartography involves transforming satellite images into map projections. A satellite image is made up of digital data. Each point, or *pixel*, on this digital image must be converted to its appropriate latitude and longitude. In order to identify the location, each pixel's digital data must be plugged into several different formulas. These cartography formulas often include trigonometry.

Brandon eventually wants to own his own business as a mapmaker, even though he knows that the equipment needed to analyze and produce top quality maps is very expensive. "The world is always changing, so cartographers will always have a job. Making maps is a way to always stay on top of current changes in the world."

inter NET CONNECTION

For up-to-date information on cartography, visit:
www.glencoe.com/sec/math/amc/mathnet

7-1 Basic Trigonometric Identities

Objective

After studying this lesson, you should be able to:
- identify and use reciprocal identities, quotient identities, Pythagorean identities, and symmetry identities.

Application

A fly ball soars toward center field as the batter heads toward first base. The ball is an example of a projectile. A formula for the height, h, of a projected object is $h = \dfrac{v_o^2 \tan^2 \theta}{2g \sec^2 \theta}$, where θ is the measure of the angle of elevation of the initial path of the object, v_o is the initial velocity of the object, and g is the acceleration due to gravity. Is there a simpler way of writing this formula? *This problem will be solved in Example 5.*

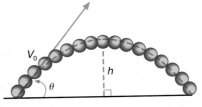

In algebra, variables and constants usually represent real numbers. The values of trigonometric functions are also real numbers. Therefore, the language and operations of algebra also apply to trigonometry. Algebraic expressions involve the operations of addition, subtraction, multiplication, division, and exponentiation. These operations are used to form trigonometric expressions. Each expression below is a trigonometric expression.

$$\cos x - x \qquad \sin^2 a + \cos^2 a \qquad \frac{1 - \sec A}{\tan A}$$

Note $\sin^2 A$ and $(\sin A)^2$ are the same, but $\sin A^2$ is not.

A statement of equality between two expressions that is true for *all* values of the variable(s) for which the expressions are defined is called an **identity**. An identity involving trigonometric expressions is a **trigonometric identity**.

Some trigonometric identities were introduced in the discussion of the trigonometric functions. Some of these are reviewed here because they are important in the development of new identities. For example, the definitions of the trigonometric functions are used to derive the **reciprocal identities**.

The following trigonometric identities hold for all values of A except those for which any function is undefined.

Reciprocal Identities

$$\sin A = \frac{1}{\csc A} \qquad \cos A = \frac{1}{\sec A} \qquad \tan A = \frac{1}{\cot A}$$

$$\csc A = \frac{1}{\sin A} \qquad \sec A = \frac{1}{\cos A} \qquad \cot A = \frac{1}{\tan A}$$

You can use the reciprocal identities to find the values of trigonometric functions.

Example 1

If $\tan A = 0.8$, find $\cot A$.

$$\cot A = \frac{1}{\tan A} \qquad \text{\textit{Reciprocal identity}}$$

$$= \frac{1}{0.8} \text{ or } 1.25 \qquad \text{\textit{Replace tan A with 0.8.}}$$

The initial definition of the tangent function, $\tan A = \dfrac{\sin A}{\cos A}$, and the reciprocal identities lead us to the **quotient identities**.

Quotient Identities	The following trigonometric identities hold for all values of A, except those for which any function is undefined.
	$\dfrac{\sin A}{\cos A} = \tan A \qquad \rightarrow \qquad \sin A = \cos A \tan A$
	$\dfrac{\cos A}{\sin A} = \cot A \qquad \rightarrow \qquad \cos A = \sin A \cot A$

You can use the fact that the graph of $(\cos A, \sin A)$ is a point on the unit circle for all values of A to derive other identities.

$$\sin^2 A + \cos^2 A = 1$$

$$\frac{\sin^2 A}{\cos^2 A} + \frac{\cos^2 A}{\cos^2 A} = \frac{1}{\cos^2 A} \qquad \text{\textit{Divide each side by } } \cos^2 A \text{\textit{, where } } \cos^2 A \neq 0.$$

$$\tan^2 A + 1 = \sec^2 A \qquad \text{\textit{Reciprocal identities}}$$

The identity $1 + \cot^2 A = \csc^2 A$ can be derived when each side of the equation $\sin^2 A + \cos^2 A = 1$ is divided by $\sin^2 A$. These identities are **Pythagorean identities**.

Pythagorean Identities	The following trigonometric identities hold for all values of A, except those for which any function is undefined.
	$\sin^2 A + \cos^2 A = 1 \qquad \tan^2 A + 1 = \sec^2 A \qquad 1 + \cot^2 A = \csc^2 A$

Example 2

If $\tan A = \dfrac{2}{5}$, find $\cos A$.

To find $\cos A$, first find $\sec A$.

$$\tan^2 A + 1 = \sec^2 A \qquad \text{\textit{Pythagorean identity}}$$

$$\left(\frac{2}{5}\right)^2 + 1 = \sec^2 A \qquad \text{\textit{Replace tan A with } } \frac{2}{5}.$$

$$\frac{29}{25} = \sec^2 A$$

$$\pm\frac{\sqrt{29}}{5} = \sec A$$

Then, find $\cos A$.

$$\cos A = \frac{1}{\sec A} \qquad \text{\textit{Quotient identity}}$$

$$= \pm\frac{5}{\sqrt{29}} \text{ or about } \pm 0.93$$

To determine the sign of a function value, you need to know the quadrant in which the angle terminates. The signs of function values in different quadrants are related according to the symmetries of the unit circle. Since we can determine the values of tan A, cot A, sec A, and csc A in terms of sin A and/or cos A with the reciprocal and quotient identities, we only need **symmetry identities** for sin A and cos A.

Symmetry Identities	**The following trigonometric identities hold for any integer k and all values of A.**
Case 1:	$\sin (A + 360k°) = \sin A \qquad \cos (A + 360k°) = \cos A$
Case 2:	$\sin (A + 180°(2k - 1)) = -\sin A \quad \cos (A + 180°(2k - 1)) = -\cos A$
Case 3:	$\sin (360k° - A) = -\sin A \qquad \cos (360k° - A) = \cos A$
Case 4:	$\sin (180°(2k - 1) - A) = \sin A \quad \cos (180°(2k - 1) - A) = -\cos A$

An illustration of each case where A is in Quadrant I and $k = 1$ follows.

If A is a quadrantal angle, then neither angle lies within a quadrant. However, the identities still apply.

Case 1: Multiples of 360° are added to A. Since angles measuring A and $A + 360k°$ are coterminal, there is no sign change in the values of sine or cosine.

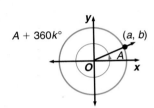

Case 2: Odd multiples of 180° are added to A and the terminal sides lie in diagonally opposite quadrants. In this case, the signs of both functions change.

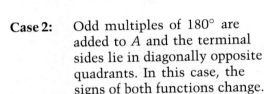

Case 3: A is subtracted from multiples of 360° and the terminal sides lie in vertically adjacent quadrants. So only the sign of the sine function is changed.

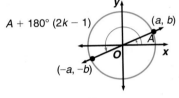

Case 4: When A is subtracted from odd multiples of 180°, the terminal sides lie in horizontally adjacent quadrants. Only the sign of the cosine function is changed.

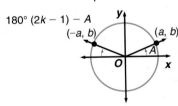

To use the symmetry identities with radian measure, replace 180° with π.

Example 3 | **Express each value as a function of an angle in Quadrant I.**

a. sin 765°

b. cos –135°

a. $765° = 2(360°) + 45°$ *Case 1,*
$\sin 765° = \sin 45°$ *with $k = 2$*

b. $-135° = 45° + (-180°)$ *Case 2,*
$\cos -135° = -\cos 45°$ *with $k = 0$*

c. tan 315°

c. $315° = 360° - 45°$ *Case 3, with k = 1*

$$\tan 315° = \frac{\sin 315°}{\cos 315°}$$

$$= \frac{-\sin 45°}{\cos 45°}$$

$$= -\tan 45°$$

The basic trigonometric identities can be used to simplify trigonometric expressions. Simplifying an expression that contains trigonometric functions means that the expression is written as a numerical value or in terms of a single trigonometric function, if possible.

Example 4

Simplify $\sin^2 x + \sin^2 x \tan^2 x$.

$$\sin^2 x + \sin^2 x \tan^2 x = \sin^2 x(1 + \tan^2 x)$$ *Factor.*

$$= \sin^2 x \sec^2 x$$ *Pythagorean identity*

$$= \sin^2 x \cdot \frac{1}{\cos^2 x}$$ *Reciprocal identity*

$$= \tan^2 x$$ *Quotient identity*

Now we can solve the application problem presented at the beginning of the lesson.

Example 5

APPLICATION

Physics

Simplify the formula $h = \dfrac{v_o^2 \tan^2 \theta}{2g \sec^2 \theta}$.

$$h = \frac{v_o^2 \tan^2 \theta}{2g \sec^2 \theta}$$

$$= \frac{v_o^2 \left(\dfrac{\sin^2 \theta}{\cos^2 \theta}\right)}{2g \left(\dfrac{1}{\cos^2 \theta}\right)}$$ $\tan^2 \theta = \dfrac{\sin^2 \theta}{\cos^2 \theta}$ *and* $\sec^2 \theta = \dfrac{1}{\cos^2 \theta}$

$$= \frac{\dfrac{v_o^2 \sin^2 \theta}{\cos^2 \theta}}{\dfrac{2g}{\cos^2 \theta}}$$ *Simplify.*

$$= \frac{v_o^2 \sin^2 \theta}{2g}$$ A simpler formula is $h = \dfrac{v_o^2 \sin^2 \theta}{2g}$.

CHECKING FOR UNDERSTANDING

Communicating Mathematics

Read and study the lesson to answer each question.

1. **Describe** what it means to simplify a trigonometric expression.

2. Scientific calculators have keys that will find the sine, cosine, or tangent ratio, but no keys to find the secant, cosecant, or cotangent. Explain why these keys are not necessary.

3. Explain why there are some values of θ for which the identity $\cot \theta = \dfrac{\cos \theta}{\sin \theta}$ does not hold and list two specific examples.

Guided
Practice

If $\sin A = \dfrac{4}{5}$, and A is in Quadrant I, find each of the following.

4. $\cos A$ **5.** $\tan A$ **6.** $\csc A$

Solve for values of θ between and $0°$ and $90°$.

7. If $\tan \theta = 3$, find $\cot \theta$.   **8.** If $\sin \theta = \dfrac{5}{13}$, find $\cos \theta$.

9. If $\cos \theta = \dfrac{2}{3}$, find $\tan \theta$.   **10.** If $\csc \theta = 5$, find $\sec \theta$.

Express each value as a function of an angle in Quadrant I.

11. $\sin 400°$ **12.** $\tan 475°$ **13.** $\cos 220°$

Simplify.

14. $\csc^2 \theta - \cot^2 \theta$ **15.** $\dfrac{\sin^2 x + \cos^2 x}{\cos^2 x}$

16. $\cos y \csc y$ **17.** $2 \csc^2 \alpha - \csc^4 \alpha + \cot^4 \alpha$

18. $\dfrac{\tan z}{\sin z}$ **19.** $\tan A \csc A$

EXERCISES

Practice

Solve for values of θ between $0°$ and $90°$.

20. If $\cot \theta = 2$, find $\tan \theta$.   **21.** If $\sin \theta = 0$, find $\csc \theta$.

22. If $\sec \theta = 4.5$, find $\cos \theta$.   **23.** If $\tan \theta = 1$, find $\cot \theta$.

24. If $\sin \theta = \dfrac{1}{2}$, find $\cos \theta$.   **25.** If $\tan \theta = \dfrac{\sqrt{3}}{2}$, find $\sec \theta$.

26. If $\cot \theta = 0.8$, find $\csc \theta$.   **27.** If $\tan \theta = \dfrac{\sqrt{11}}{2}$, find $\sec \theta$.

28. If $\sin \theta = \dfrac{40}{41}$, find $\tan \theta$.   **29.** If $\cos \theta = \dfrac{2}{3}$, find $\csc \theta$.

30. If $\cos \theta = \dfrac{3}{5}$, find $\tan \theta$.   **31.** If $\tan \theta = \dfrac{7}{2}$, find $\sec \theta$.

32. If $\cos \theta = \dfrac{3}{10}$, find $\cot \theta$.   **33.** If $\tan \theta = \dfrac{1}{2}$, find $\sin \theta$.

Express each value as a function of an angle in Quadrant I.

34. $\sin 665°$ **35.** $\cos 562°$ **36.** $\tan -342°$

37. $\sin (-792°)$ **38.** $\csc 850°$ **39.** $\sec (-210°)$

Simplify.

40. $\dfrac{\tan x \csc x}{\sec x}$ **41.** $\dfrac{\cos \theta}{\sec \theta - \tan \theta}$ **42.** $\tan A \cos^2 A$

43. $\sin A \cot A$ **44.** $\dfrac{\tan \beta}{\cot \beta}$ **45.** $\cos x \tan x \csc x$

46. $\dfrac{1}{\sin^2 \theta} - \dfrac{\cos^2 \theta}{\sin^2 \theta}$ **47.** $\dfrac{\cos^2 A}{1 + \sin A}$ **48.** $\dfrac{\csc \beta}{1 + \cot^2 \beta}$

49. $\sin^2 A \cos^2 A + \sin^4 A$

50. $(1 - \sin x)(1 + \sin x)$

51. $\cos^4 \alpha + 2 \cos^2 \alpha \sin^2 \alpha + \sin^4 \alpha$

52. $\sin x + \cos x \tan x$

Write an expression for each of the following in terms of the given function.

53. $\cos \theta$ in terms of $\sin \theta$

54. $\sec \theta$ in terms of $\cot \theta$

55. $\tan \theta$ in terms of $\cos \theta$

56. $\csc \theta$ in terms of $\cos \theta$

Critical Thinking

57. Which of the symmetry identities represents a reflection over the x-axis? over the y-axis? Explain your answer.

Applications and Problem Solving

58. Geometry The circle at the right is a unit circle with its center at the origin. $\overleftrightarrow{AB}$ and $\overleftrightarrow{CD}$ are tangent to the circle. State which segments represent the ratios $\sin \theta$, $\cos \theta$, $\tan \theta$, $\sec \theta$, $\cot \theta$, and $\csc \theta$. Justify your answers.

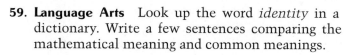

59. Language Arts Look up the word *identity* in a dictionary. Write a few sentences comparing the mathematical meaning and common meanings.

Mixed Review

60. Find the distance between the points at $(2, -5)$ and $(8, -2)$. Then find the slope of the line passing through these points. **(Lesson 1-4)**

61. Is the function $y = 5x^3 - 2x + 5$ odd, even, or neither? **(Lesson 3-1)**

62. Without graphing, describe the end behavior of the graph of $y = 2x^2 + 2$. **(Lesson 3-8)**

63. State the number of possible positive and negative real zeros of the function $f(x) = x^4 + 2x^3 - 6x - 1$. **(Lesson 4-4)**

64. Change $\dfrac{15\pi}{16}$ radians to degree measure to the nearest minute. **(Lesson 5-1)**

65. Geometry Each side of a rhombus is 30 units long. One diagonal makes a 25° angle with a side. What is the length of each diagonal to the nearest tenth of a unit? **(Lesson 5-6)**

66. Navigation A ship at sea is 70 miles from one radio transmitter and 130 miles from another. The angle formed by the rays from the ship to the transmitters measures 130°. How far apart are the transmitters? **(Lesson 5-7)**

67. Find the area of $\triangle ABC$ if $a = 6.2, b = 7.5,$ and $C = 97°$. **(Lesson 5-8)**

68. Write an equation with phase shift 0 to represent simple harmonic motion in which the initial position is -6, the amplitude is 6, and the period is 2. **(Lesson 6-7)**

69. College Entrance Exam Three times the least of three consecutive odd integers is 3 greater than two times the greatest. Find the greatest of the three integers.

7-2 Verifying Trigonometric Identities

Objectives

After studying this lesson, you should be able to:
- use the basic trigonometric identities to verify other identities, and
- find numerical values of trigonometric functions.

Application

The amount of light that a source provides to a surface is called the illumination. The illumination, E, in footcandles on a surface that is R feet from a source of light with intensity I candelas is $E = \dfrac{I \cos \theta}{R^2}$, where θ is the measure of the angle between the direction of the light and a line perpendicular to the surface being illuminated. Is the formula $E = \dfrac{I \cot \theta}{R^2 \csc \theta}$ an equivalent formula? *This problem will be solved in Example 2.*

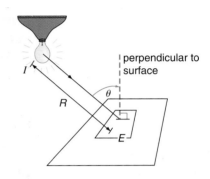

Verifying trigonometric identities involves transforming one side of the equation into the same form as the other side by using the basic trigonometric identities and the properties of algebra. Either side may be transformed into the form of the other side, or both sides may be transformed separately into other forms that are the same.

Suggestions for Verifying Trigonometric Identities

- **Transform the more complicated side of the equation into the form of the simpler side.**
- **Substitute one or more basic trigonometric identities to simplify the expression.**
- **Factor or multiply to simplify the expression.**
- **Multiply both the numerator and denominator by the same trigonometric expression.**

Example 1

Remember that you cannot add or subtract quantities from both sides of an identity like we often do with equations.

Verify that $\tan \beta = \dfrac{\cos \beta}{\sin \beta \cot^2 \beta}$ is an identity.

Transform the more complicated side of the equation, $\dfrac{\cos \beta}{\sin \beta \cot^2 \beta}$, into the form of the simpler side, $\tan \beta$.

$$\tan \beta \stackrel{?}{=} \frac{\cos \beta}{\sin \beta \cot^2 \beta}$$

$$\tan \beta \stackrel{?}{=} \frac{\cos \beta}{\sin \beta} \cdot \frac{1}{\cot^2 \beta} \qquad \textit{Write as a product of two fractions.}$$

$$\tan \beta \stackrel{?}{=} \cot \beta \cdot \frac{1}{\cot^2 \beta} \qquad \frac{\cos \beta}{\sin \beta} = \cot \beta$$

$$\tan \beta \stackrel{?}{=} \frac{1}{\cot \beta} \qquad \textit{Multiply.}$$

$$\tan \beta = \tan \beta \qquad \textit{Reciprocal identity}$$

The transformation of the right side has produced an expression that is the same as the left side. The identity is verified.

Example 2

APPLICATION

Optics

Refer to the application in the beginning of the lesson. Is the formula $E = \dfrac{I \cos \theta}{R^2}$ equivalent to the formula $E = \dfrac{I \cot \theta}{R^2 \csc \theta}$?

$$\frac{I \cos \theta}{R^2} \stackrel{?}{=} \frac{I \cot \theta}{R^2 \csc \theta}$$

$$\frac{I \cos \theta}{R^2} \stackrel{?}{=} \frac{I \dfrac{\cos \theta}{\sin \theta}}{R^2 \dfrac{1}{\sin \theta}} \qquad \cot \theta = \frac{\cos \theta}{\sin \theta}, \csc \theta = \frac{1}{\sin \theta}$$

$$\frac{I \cos \theta}{R^2} \stackrel{?}{=} \frac{I \dfrac{\cos \theta}{\sin \theta}}{R^2 \dfrac{1}{\sin \theta}} \cdot \frac{\sin \theta}{\sin \theta} \qquad \textit{Multiply the numerator and denominator by } \sin \theta.$$

$$\frac{I \cos \theta}{R^2} = \frac{I \cos \theta}{R^2} \qquad \textit{Simplify.}$$

Since the expression $\dfrac{I \cot \theta}{R^2 \csc \theta}$ can be transformed into $\dfrac{I \cos \theta}{R^2}$, the two formulas for E are equivalent.

You can use the techniques that you use to verify trigonometric identities to find numerical values of trigonometric functions.

Example 3

Find a numerical value of one trigonometric function of x if $\sin x \sec x = 1$.

$\sin x \sec x = 1$

$\sin x \cdot \dfrac{1}{\cos x} = 1 \qquad \sec x = \dfrac{1}{\cos x}$

$\dfrac{\sin x}{\cos x} = 1 \qquad \text{Multiply.}$

$\tan x = 1 \qquad \tan x = \dfrac{\sin x}{\cos x}$

If $\sin x \sec x = 1$, then $\tan x = 1$.

EXPLORATION: Graphing Calculator

You can use a graphing calculator to verify that
$$\csc x = \dfrac{1 + \sec x}{\sin x + \tan x} \text{ is an identity.}$$

1. Use the range values below to set up a viewing window. Mode should be in degrees.

$-360 \le X \le 360 \qquad -3 \le Y \le 3 \qquad XSCL = 90 \qquad YSCL = 0.25$

2. Let $Y_1 = (1 + 1/\cos x)/(\sin x + \tan x)$ and let $Y_2 = 1/\sin x$.
3. Graph the two functions on the same screen. What do you notice?

CHECKING FOR UNDERSTANDING

Communicating Mathematics

Read and study the lesson to answer each question.

1. **Explain** why you may not use an operation involving both sides of the equation when you are verifying a trigonometric identity.

2. **Describe** the process you use to verify a trigonometric identity.

Guided Practice

Verify that each of the following is an identity.

3. $\tan^2 x \cos^2 x = 1 - \cos^2 x$

4. $\csc A \sec A = \cot A + \tan A$

5. $\tan \beta \csc \beta = \sec \beta$

6. $\csc \alpha \cos \alpha \tan \alpha = 1$

7. $\sin \theta \sec \theta \cot \theta = 1$

8. $\sec^2 y - \tan^2 y = \tan y \cot y$

Find a numerical value of one trigonometric function of each x.

9. $\sin x = \tan x$

10. $2 \tan x = \cot x$

11. $\sin x = 2 \cos x$

12. $\tan x \cos x = \dfrac{1}{2}$

EXERCISES

Verify that each of the following is an identity.

13. $\dfrac{1}{\sec^2 \theta} + \dfrac{1}{\csc^2 \theta} = 1$

14. $\dfrac{\tan x \cos x}{\sin x} = 1$

15. $\dfrac{\sin A}{\csc A} + \dfrac{\cos A}{\sec A} = 1$

16. $\dfrac{1 + \tan^2 \theta}{\csc^2 \theta} = \tan^2 \theta$

17. $\dfrac{1 + \tan \gamma}{1 + \cot \gamma} = \dfrac{\sin \gamma}{\cos \gamma}$

18. $\dfrac{\sec \alpha}{\sin \alpha} - \dfrac{\sin \alpha}{\cos \alpha} = \cot \alpha$

19. $\cos^2 x + \tan^2 x \cos^2 x = 1$

20. $\tan^2 \theta - \sin^2 \theta = \tan^2 \theta \sin^2 \theta$

21. $1 - \cot^4 x = 2 \csc^2 x - \csc^4 x$

22. $\sin \theta + \cos \theta = \dfrac{1 + \tan \theta}{\sec \theta}$

23. $\dfrac{\sec x - 1}{\sec x + 1} + \dfrac{\cos x - 1}{\cos x + 1} = 0$

24. $\sec^4 \alpha - \sec^2 \alpha = \dfrac{1}{\cot^4 \alpha} + \dfrac{1}{\cot^2 \alpha}$

25. $\dfrac{\cos x}{1 + \sin x} + \dfrac{\cos x}{1 - \sin x} = 2 \sec x$

26. $\dfrac{\sec B}{\cos B} - \dfrac{\tan B}{\cot B} = 1$

27. $1 + \sec^2 x \sin^2 x = \sec^2 x$

28. $\dfrac{1 - 2 \cos^2 \theta}{\sin \theta \cos \theta} = \tan \theta - \cot \theta$

Find a numerical value of one trigonometric function of each x.

29. $2 \sin^2 x = 3 \cos^2 x$

30. $\dfrac{\tan x}{\sin x} = \sqrt{2}$

31. $1 - \sin^2 x = \dfrac{1}{9}$

32. $\dfrac{\sin x \sec x}{\cot x} = \dfrac{9}{16}$

33. $1 + \tan^2 x = \sin^2 x + \dfrac{1}{\sec^2 x}$

34. $\dfrac{\cos x \tan x}{\csc x} = \dfrac{1}{9}$

Graphing Calculator

Use the **SIN, COS,** and **TAN** keys on a graphing calculator, together with the basic identities, to find the six trigonometric function values for each degree measure given below.

35. $420°$

36. $-650°$

Critical Thinking

37. Prove each of the following.

   a. $\sin x + \cos x \geq 1$ if $0° \leq x \leq 90°$

   b. $\tan x + \cot x \geq 2$ if $0° < x < 90°$

Applications and Problem Solving

38. Physics A mass weighing w kilograms is at rest on an adjustable plane. The mass begins to slide down the plane when that angle between the plane and the horizontal reaches a certain value θ, known as the angle of repose. The weight of the mass can be expressed in terms of a force parallel to the plane, F_w, and a force perpendicular to the plane,

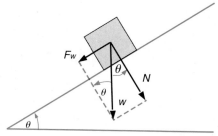

N. Formulas for the magnitudes of F_w and N are $F_w = w \sin \theta$ and $N = w \cos \theta$. Another formula for F_w is F_w is $F_w = \mu_s N$, where μ_s is the coefficient of static friction

   a. Solve $F_w = \mu_s N$ for the coefficient of static friction in terms of θ.

(continued on next page)

b. Peg is standing on the wooden plank between the dock and her boat. She is wearing leather-soled shoes, and the plank makes an angle of 24° with the horizontal. What must be the minimum coefficient of static friction of leather on wood? *Hint: Use the formula you found in Exercise 38a.*

39. Civil Engineering The surface of a section of highway makes an angle of A degrees with the horizontal. An engineer has found that for a particular section, $\dfrac{d^2 - d^2 \cos^2 A}{d^2 - d^2 \sin^2 A} = \dfrac{9}{400}$, where d is the distance along the surface of the road.

a. Make a labeled drawing of the situation.

b. Find a numerical value for one trigonometric function of A.

Mixed Review

40. Find $[f \circ g](x)$ and $[g \circ f](x)$ if $f(x) = \dfrac{2}{3}x - 2$ and $g(x) = x^2 - 6x + 9$. **(Lesson 1-2)**

41. Use the remainder theorem to find the remainder for the quotient $(x^4 - 4x^3 - 2x^2 - 1) \div (x - 5)$. Then, state whether the binomial is a factor of the polynomial. **(Lesson 4-3)**

42. Child Care Melanie King is the manager for the Learning Loft Day Care Center. The center offers all day service for preschool children for $18 per day and after school only service for $6 per day. Fire codes permit only 50 people in the building at one time. State law dictates that a child care worker can be responsible for a maximum of 3 preschool children or 5 school-age children at one time. Ms. King has ten child care workers available to work at the center during the week. How many children of each age group should Ms. King accept to maximize the daily income of the center? **(Lesson 2-6)**

43. A circle has a radius of 12 inches. Find the degree measure of the central angle subtended by an arc 11.5 inches long. **(Lesson 5-2)**

44. Simplify $4 \csc \theta \cos \theta \tan \theta$. **(Lesson 7-1)**

45. College Entrance Exam Choose the best answer.
You have added the same positive quantity to the numerator and denominator of a fraction. The result is

(A) greater than the original fraction,

(B) less than the original fraction,

(C) equal to the original fraction,

(D) one-half the original fraction, or

(E) not determinable with the information given.

7-2B   Graphing Calculators: Verifying Trigonometric Identities

You can use a graphing calculator to determine whether an equation may be a trigonometric identity by graphing both sides of the equation as two separate functions and then comparing the graphs. If the graphs appear to lay on top of each other, then the equation may be a trigonometric identity. If the graphs do not match, then the equation is not a trigonometric identity. Any equation must be verified algebraically to prove that it is an identity.

Example 1

Use a graphing calculator to determine whether $\tan x + \cot x = \sec x \cdot \csc x$ may be an identity.

Graph the equations $y = \tan x + \cot x$ and $y = \sec x \cdot \csc x$ on the same graph to verify the identity. Use the trig viewing window.
Remember that $\cot x = \dfrac{1}{\tan x}$, $\sec x = \dfrac{1}{\cos x}$, and $\csc x = \dfrac{1}{\sin x}$.

Enter: [Y=] [TAN] [X,T,θ] [+] [(] 1 [÷]
[TAN] [X,T,θ] [)] [ENTER]
[(] 1 [÷] [COS] [X,T,θ] [)] [(] 1
[÷] [SIN] [X,T,θ] [)] [ZOOM] 7

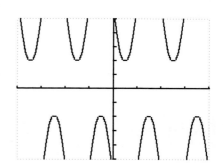

Since the graphs of the two functions coincide, the equation may be an identity. Verifying algebraically shows that the equation is, in fact, an identity.

The use of parentheses is important when you are graphing complicated fractions or equations involving fractions. Without the parentheses, the calculator may graph something other than the equation you would like it to graph. For example, parentheses must be included when entering the trigonometric function $\tan\left(\dfrac{x}{2}\right)$ or the calculator will graph it as $(\tan x) \div 2$.

Example 2

Use a graphing calculator to determine whether $\cot\left(\dfrac{x}{2}\right) = \dfrac{2\cos x}{1 + 2\sin x}$ may be an identity.

Graph the two equations $y = \cot\left(\dfrac{x}{2}\right)$ and $y = \dfrac{2\cos x}{1 + 2\sin x}$ on the same graph.

(continued on the next page)

Use the trig viewing window. Remember to use parentheses around the numerator and denominator of the second equation.

Enter: $\boxed{Y=}$ 1 $\boxed{\div}$ $\boxed{TAN}$ $\boxed{(}$ $\boxed{X,T,\theta}$ $\boxed{\div}$
2 $\boxed{)}$ $\boxed{ENTER}$ $\boxed{(}$ 2 $\boxed{COS}$ $\boxed{X,T,\theta}$
$\boxed{)}$ $\boxed{\div}$ $\boxed{(}$ 1 $\boxed{+}$ 2 $\boxed{SIN}$ $\boxed{X,T,\theta}$ $\boxed{)}$
$\boxed{ZOOM}$ 7

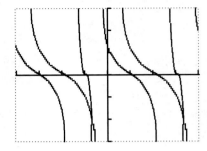

Since the graphs of the functions do not coincide, the equation is not an identity.

It is important to know that when trying to verify an identity with a squared function, you must first enter the function in parentheses and then square it. For example, to verify $\sin^2 x + \cos^2 x = 1$, you must enter it into your calculator as $(\sin x)^2 + (\cos x)^2$.

EXERCISES

Use a graphing calculator to determine whether each equation *may* be an identity. Write *yes* or *no*.

1. $\cot x - \tan x = 2\cot(2x)$

2. $\tan x \cdot \sin x = \sec x + \cos x$

3. $\cos^4 x - \sin^4 x = \cos(2x)$

4. $\dfrac{1 + \cot x}{1 + \tan x} = \dfrac{\sin x}{\cos x}$

5. $\cos(x + 90°) = \sin x$

6. $\cos(x + 540°) = -\cos x$

7. $\cos^2 x + 1 = 2\cos^2 x + \sin^2 x$

8. $\sec x + \cot x = \dfrac{\sin x + \cos x}{\sin x \cdot \cos x}$

9. $2 + \sec^2 x = \tan^2 x$

10. $\cot^2 x \cdot \sec^2 x = \csc^2 x$

11. $\cos x = \dfrac{\sin(2x)}{2\sin x}$

12. $\tan\left(\dfrac{x}{2}\right) = \dfrac{\sin x}{1 - \cos x}$

13. $\dfrac{1}{\cos x} + \dfrac{1}{\sin x} = 1$

14. $\dfrac{\sec^2 x}{1 + \tan^2 x} = \sin^2 x + \cos^2 x$

15. $\sin x(\sin x + \tan x) = 1 + \cos^2 x$

16. $\dfrac{\sin x - \cos x}{\cos x} + 1 = \tan x$

17. $\dfrac{\sin x}{\csc x - 1} + \dfrac{\sin x}{\csc x + 1} = 2\tan^2 x$

18. $\csc x + \cot x = \dfrac{\sin x}{1 + \cos x}$

7-3 Sum and Difference Identities

Objective

After studying this lesson, you should be able to:
- use the sum and difference identities for sine, cosine, and tangent functions.

Application

Terri Cox is an electrical engineer designing a three-phase AC-generator. Three-phase generators produce three currents of electricity at one time. They can generate more power for the amount of materials used and lead to better transmission and use of power than single-phase generators can. The three phases of the generator Ms. Cox is making are expressed as $I \cos \theta$, $I \cos (\theta + 120°)$, and $I \cos (\theta + 240°)$. She must show that each phase is equal to the sum of the other two phases but opposite in sign. To do this, she will show that $I \cos \theta + I \cos (\theta + 120°) + I \cos (\theta + 240°) = 0$. *This problem will be solved in Example 2.*

Remember $\overparen{P_1 P_3}$ is the minor arc from P_1 to P_3.

Consider the unit circle on the right. The points P_1, P_2, and P_3 lie on the circle. The measure of $\overparen{AP_1}$ is s_1. The measure of $\overparen{P_1 P_2}$ and $\overparen{AP_3}$ is s_2. Therefore, $\overparen{AP_2}$ is congruent to $\overparen{P_1 P_3}$ since the measure of each is the sum $s_1 + s_2$. Since congruent arcs in the same circle have congruent chords, $\overline{AP_2} \cong \overline{P_1 P_3}$.

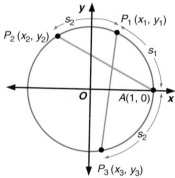

$$AP_2 = P_1 P_3$$
$$\sqrt{(x_2 - 1)^2 + (y_2 - 0)^2} = \sqrt{(x_1 - x_3)^2 + (y_1 - y_3)^2} \quad \textit{Use the distance formula.}$$
$$x_2{}^2 - 2x_2 + 1 + y_2{}^2 = x_1{}^2 - 2x_1 x_3 + x_3{}^2 + y_1{}^2 - 2y_1 y_3 + y_3{}^2$$
$$(x_2{}^2 + y_2{}^2) - 2x_2 + 1 = (x_1{}^2 + y_1{}^2) + (x_3{}^2 + y_3{}^2) - 2x_1 x_3 - 2y_1 y_3$$
$$1 - 2x_2 + 1 = \quad 1 \quad + \quad 1 \quad - 2x_1 x_3 - 2y_1 y_3 \quad \textit{On the unit circle,}$$
$$x_2 = x_1 x_3 + y_1 y_3 \qquad\qquad\qquad\qquad \textit{x}^2 + \textit{y}^2 = 1.$$

FYI...

Generators with more than one phase, called polyphase generators, were developed by Nikola Tesla, a Serbian engineer who immigrated to the United States in 1884. Tesla's invention made it possible to generate current far away from where it was used.

The coordinates of the points on the unit circle allow us to relate s to x and y. Substitute the appropriate functions of s in the equation. Use $x_1 = \cos s_1$, $x_2 = \cos (s_1 + s_2)$, $x_3 = \cos (-s_2)$, $y_1 = \sin s_1$, $y_2 = \sin (s_1 + s_2)$, and $y_3 = \sin(-s_2)$.

$$x_2 = x_1 x_3 + y_1 y_3$$
$$\cos (s_1 + s_2) = \cos s_1 \cos (-s_2) + \sin s_1 \sin (-s_2) \quad \textit{Substitute.}$$
$$\cos (s_1 + s_2) = \cos s_1 \cos s_2 - \sin s_1 \sin s_2 \quad \textit{cos} (-s_2) = \cos s_2; \sin (-s_2) = -\sin s_2$$

To find an identity for $\cos (s_1 - s_2)$, write $\cos (s_1 - s_2)$ as $\cos [s_1 + (-s_2)]$. Then use the identity above and simplify.

$$\cos [s_1 + (-s_2)] = \cos s_1 \cos (-s_2) - \sin s_1 \sin (-s_2)$$
$$\cos (s_1 - s_2) = \cos s_1 \cos s_2 + \sin s_1 \sin s_2$$

Replace s_1 with α and s_2 with β so that the cosine of the sum and difference of the measures of two angles are given by the following identities.

Sum and Difference Identities for the Cosine Function	**If α and β represent the measures of two angles, then the following identities hold for all values of α and β.** $$\cos(\alpha \pm \beta) = \cos\alpha\cos\beta \mp \sin\alpha\sin\beta$$

Note how the addition and subtraction symbols are related in the sum and difference identities.

You can use the sum and difference identities and the values of the trigonometric functions of common angles to find the values of trigonometric functions of other angles. Note that the sum and difference identities may be used if α and β are measured in either degrees or radians. So, α and β may be interpreted as either real numbers or angle measures.

Example 1

Find $\cos 15°$ from values of functions of $30°$ and $45°$.

$$\cos 15° = \cos(45° - 30°)$$
$$= \cos 45° \cos 30° + \sin 45° \sin 30°$$
$$= \frac{\sqrt{2}}{2} \cdot \frac{\sqrt{3}}{2} + \frac{\sqrt{2}}{2} \cdot \frac{1}{2}$$
$$= \frac{\sqrt{6} + \sqrt{2}}{4}$$
$$\cos 15° \approx 0.9659$$

We can use the sum identity for the cosine function to solve the application problem from the beginning of the lesson.

Example 2

APPLICATION

Engineering

Show that $I\cos\theta + I\cos(\theta + 120°) + I\cos(\theta + 240°) = 0$.

$$I\cos\theta + I\cos(\theta + 120°) + I\cos(\theta + 240°) \overset{?}{=} 0$$
$$I\cos\theta + I(\cos\theta\cos 120° - \sin\theta\sin 120°) +$$
$$I(\cos\theta\cos 240° - \sin\theta\sin 240°) \overset{?}{=} 0$$
$$I\cos\theta + I\left(\left(-\frac{1}{2}\right)\cos\theta - \left(\frac{\sqrt{3}}{2}\right)\sin\theta\right) + I\left(\left(-\frac{1}{2}\right)\cos\theta - \left(-\frac{\sqrt{3}}{2}\right)\sin\theta\right) \overset{?}{=} 0$$
$$I\cos\theta - \frac{1}{2}I\cos\theta - \frac{\sqrt{3}}{2}I\sin\theta - \frac{1}{2}I\cos\theta + \frac{\sqrt{3}}{2}I\sin\theta \overset{?}{=} 0$$
$$0 = 0$$

If we replace α with $\frac{\pi}{2}$ and β with s in the identities for $\cos(\alpha \pm \beta)$, the following equations result.

$$\cos\left(\frac{\pi}{2} + s\right) = -\sin s$$
$$\cos\left(\frac{\pi}{2} - s\right) = \sin s$$

Replace s with $\frac{\pi}{2} + s$ in the equation for $\cos\left(\frac{\pi}{2} + s\right)$ and with $\frac{\pi}{2} - s$ in the equation for $\cos\left(\frac{\pi}{2} - s\right)$ to obtain the following equations.

$$\cos s = \sin\left(\frac{\pi}{2} + s\right)$$
$$\cos s = \sin\left(\frac{\pi}{2} - s\right)$$

Replace s with $(\alpha + \beta)$ in the equation for $\cos\left(\frac{\pi}{2} - s\right)$ to derive an identity for the sine of the sum of two real numbers.

$$\cos\left[\frac{\pi}{2} - (\alpha + \beta)\right] = \sin(\alpha + \beta)$$
$$\cos\left[\left(\frac{\pi}{2} - \alpha\right) - \beta\right] = \sin(\alpha + \beta)$$
$$\cos\left(\frac{\pi}{2} - \alpha\right)\cos\beta + \sin\left(\frac{\pi}{2} - \alpha\right)\sin\beta = \sin(\alpha + \beta) \qquad \textit{Use identity for } \cos(\alpha - \beta).$$
$$\sin\alpha\cos\beta + \cos\alpha\sin\beta = \sin(\alpha + \beta) \qquad \textit{Substitute.}$$

Replace β with $(-\beta)$ in the identity for $\sin(\alpha + \beta)$ to derive an identity for the sine of the difference of two real numbers.

$$\sin[\alpha + (-\beta)] = \sin\alpha\cos(-\beta) + \cos\alpha\,\sin(-\beta)$$
$$\sin(\alpha - \beta) = \sin\alpha\cos\beta - \cos\alpha\sin\beta$$

Sum and Difference Identities for the Sine Function	**If α and β represent the measures of two angles, then the following identities hold for all values of α and β.** $$\sin(\alpha \pm \beta) = \sin\alpha\cos\beta \pm \cos\alpha\sin\beta$$

Example 3

Find $\sin 15°$ from values of functions of $30°$ and $45°$.

$$\sin 15° = \sin(45° - 30°)$$
$$\sin 15° = \sin 45°\cos 30° - \cos 45°\sin 30°$$
$$= \frac{\sqrt{2}}{2} \cdot \frac{\sqrt{3}}{2} - \frac{\sqrt{2}}{2} \cdot \frac{1}{2}$$
$$= \frac{\sqrt{6} - \sqrt{2}}{4} \text{ or about } 0.2588$$

We can use the sum and difference identities for the cosine and sine functions to find sum and difference identities for the tangent function.

$$\tan(\alpha + \beta) = \frac{\sin(\alpha + \beta)}{\cos(\alpha + \beta)} \qquad\qquad \tan x = \frac{\sin x}{\cos x}$$

$$\tan(\alpha + \beta) = \frac{\sin\alpha\cos\beta + \cos\alpha\sin\beta}{\cos\alpha\cos\beta - \sin\alpha\sin\beta} \qquad \textit{Divide the numerator and denominator by } \cos\alpha\cos\beta.$$

$$\tan(\alpha + \beta) = \frac{\dfrac{\sin\alpha\cos\beta}{\cos\alpha\cos\beta} + \dfrac{\cos\alpha\sin\beta}{\cos\alpha\cos\beta}}{\dfrac{\cos\alpha\cos\beta}{\cos\alpha\cos\beta} - \dfrac{\sin\alpha\sin\beta}{\cos\alpha\cos\beta}} \qquad \textit{Assume } \cos\alpha \neq 0 \textit{ and } \cos\beta \neq 0.$$

$$\tan(\alpha + \beta) = \frac{\tan\alpha + \tan\beta}{1 - \tan\alpha\tan\beta}$$

Replace β with $-\beta$ to find $\tan(\alpha - \beta)$.

$$\tan(\alpha + (-\beta)) = \frac{\tan\alpha + \tan(-\beta)}{1 - \tan\alpha\tan(-\beta)} \qquad \tan(-\beta) = -\tan\beta$$

$$\tan(\alpha - \beta) = \frac{\tan\alpha - \tan\beta}{1 + \tan\alpha\tan\beta}$$

Sum and Difference Identities for the Tangent Function	**If α and β represent the measures of two angles, then the following identities hold for all values of α and β.** $$\tan(\alpha \pm \beta) = \frac{\tan\alpha \pm \tan\beta}{1 \mp \tan\alpha\tan\beta}$$

Example 4

Find $\tan 105°$ from values of functions of $60°$ and $45°$.

$$\tan 105° = \tan(60° + 45°)$$
$$= \frac{\tan 60° + \tan 45°}{1 - \tan 60° \tan 45°}$$
$$= \frac{\sqrt{3} + 1}{1 - \sqrt{3}\cdot 1} \qquad Multiply\ by\ \frac{1 + \sqrt{3}}{1 + \sqrt{3}}\ to\ simplify.$$
$$= -2 - \sqrt{3}\ or\ about\ -3.7321$$

You can use sum and difference identities to verify other identities.

Example 5

Verify that $\cot x = \tan\left(\dfrac{\pi}{2} - x\right)$ is an identity.

$$\cot x \stackrel{?}{=} \tan\left(\frac{\pi}{2} - x\right)$$

$$\cot x \stackrel{?}{=} \frac{\sin\left(\dfrac{\pi}{2} - x\right)}{\cos\left(\dfrac{\pi}{2} - x\right)}$$

$$\cot x \stackrel{?}{=} \frac{\sin\dfrac{\pi}{2}\cos x - \cos\dfrac{\pi}{2}\sin x}{\cos\dfrac{\pi}{2}\cos x + \sin\dfrac{\pi}{2}\sin x} \qquad Use\ the\ difference\ identities.$$

$$\cot x \stackrel{?}{=} \frac{1\cdot\cos x - 0\cdot\sin x}{0\cdot\cos x + 1\cdot\sin x} \qquad sin\ \frac{\pi}{2} = 1,\ cos\ \frac{\pi}{2} = 0$$

$$\cot x \stackrel{?}{=} \frac{\cos x}{\sin x}$$

$$\cot x = \cot x$$

CHECKING FOR UNDERSTANDING

Communicating Mathematics

Read and study the lesson to answer each question.

1. Is it true that $\cos(A + B) = \cos A + \cos B$? Explain your answer.

2. **Describe** a method for finding the exact value of $\sin 105°$. Then find the exact value.

3. **Compare** the identities for $\cos(\alpha + \beta)$ and $\cos(\alpha - \beta)$. Explain why the signs are different.

Guided Practice

Use the sum and difference identities to find the exact value of each function.

4. Find $\cos 105°$ from values of functions of $60°$ and $45°$.
5. Find $\sin 150°$ from values of functions of $120°$ and $30°$.
6. Find $\tan 75°$ from values of functions of $45°$ and $30°$.
7. Find $\cos 150°$ from values of functions of $180°$ and $30°$.

Find the value of each function for $0° < x < 90°$, $0° < y < 90°$, $\sin x = \dfrac{4}{5}$, and $\cos y = \dfrac{3}{5}$.

8. $\sin (x + y)$
9. $\cos (x - y)$
10. $\tan (x + y)$

Verify that each of the following is an identity.

11. $\sin (180° - \theta) = \sin \theta$
12. $\cos \left(\dfrac{3\pi}{2} + \theta \right) = \sin \theta$
13. $\tan (270° - x) = \cot x$
14. $\cos (360° - \theta) = \cos \theta$

EXERCISES

Practice

Use the sum and difference identities to find the exact value of each function.

15. $\sin 195°$
16. $\cos 255°$
17. $\tan (-105°)$
18. $\sin 75°$
19. $\tan (-195°)$
20. $\cos 195°$
21. $\tan 165°$
22. $\cos 345°$
23. $\sin 285°$

If α and β are the measures of two first quadrant angles, find the exact value of each function.

24. If $\sin \alpha = \dfrac{5}{13}$ and $\cos \beta = \dfrac{4}{5}$, find $\cos (\alpha + \beta)$.
25. If $\tan \alpha = \dfrac{4}{3}$ and $\cot \beta = \dfrac{5}{12}$, find $\sin (\alpha - \beta)$.
26. If $\cos \alpha = \dfrac{5}{13}$ and $\cos \beta = \dfrac{35}{37}$, find $\tan (\alpha + \beta)$.
27. If $\sin \alpha = \dfrac{8}{17}$ and $\tan \beta = \dfrac{7}{24}$, find $\cos (\alpha - \beta)$.
28. If $\csc \alpha = \dfrac{13}{5}$ and $\tan \beta = \dfrac{3}{4}$, find $\tan (\alpha + \beta)$.
29. If $\cos \alpha = \dfrac{15}{17}$ and $\cot \beta = \dfrac{24}{7}$, find $\sin (\alpha - \beta)$.

Verify that each of the following is an identity.

30. $\sin (270° + x) = -\cos x$
31. $\cos (90° + \theta) = -\sin \theta$
32. $\tan (90° + \theta) = -\cot \theta$
33. $\sin \left(\dfrac{\pi}{2} + x \right) = \cos x$
34. $-\cos \theta = \cos (\pi + \theta)$
35. $\tan (\pi - \theta) = -\tan \theta$
36. $-\sin \theta = \cos \left(\dfrac{\pi}{2} + \theta \right)$
37. $\dfrac{\sin (\beta - \alpha)}{\sin \alpha \sin \beta} = \cot \alpha - \cot \beta$
38. $\sin^2 \alpha - \sin^2 \beta = \sin (\alpha + \beta) \sin (\alpha - \beta)$
39. $\cos (30° - x) + \cos (30° + x) = \sqrt{3} \cos x$
40. $\cos (\alpha + \beta) + \cos (\alpha - \beta) = 2 \cos \alpha \cos \beta$

Derive formulas for each of the following.

41. $\cot(\alpha + \beta)$ in terms of $\cot \alpha$ and $\cot \beta$

42. $\sin(\alpha + \beta + \gamma)$ in terms of functions of α, β, and γ

Critical
Thinking

43. Find $\alpha + \beta$ in radians if $\tan \alpha = M$ and $\tan \beta = N$ and $(3M + 3)(2N + 2) = 12$.

Applications
and Problem
Solving

44. Optics The index of refraction for a medium through which light is passing is the ratio of the velocity of light in free space to the velocity of light in the medium. For light passing symmetrically through a glass prism, the index of refraction, n, is given by the equation $n = \dfrac{\sin\left[\frac{1}{2}(\alpha + \beta)\right]}{\sin\left(\frac{\beta}{2}\right)}$, where α is the deviation angle and β is the angle of the apex of the prism. If $\beta = 60°$, show that $n = \sqrt{3}\sin\left(\dfrac{\alpha}{2}\right) + \cos\left(\dfrac{\alpha}{2}\right)$.

45. Geometry The slope of a line, m, is equal to the tangent of the angle that the line makes with the x-axis. Let θ be the measure of the smaller angle formed by the intersection of the lines ℓ and n. Lines ℓ and n make angles measuring α and β, respectively, with the x-axis. The slopes of lines ℓ and n are m_1 and m_2, respectively. Find an expression for $\tan \theta$ in terms of m_1 and m_2.

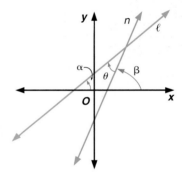

Mixed
Review

46. Solve the system algebraically. **(Lesson 2-1)**
$5x - 2y = 9$
$y = 3x - 1$

47. Find the equation of the line tangent to the graph of $y = \dfrac{1}{2}x^2 - 2x + 1$ at $(4, 1)$. Write the equation in slope-intercept form. **(Lesson 3-6)**

48. Solve the equation $y = 4x^2 - 6x + 11$ by using the quadratic formula. **(Lesson 4-2)**

49. Find the value of $\csc 270°$ without using a calculator. **(Lesson 5-3)**

50. Solve $\triangle ABC$ if $a = 7$, $b = 9$, and $c = 13$. Round angle measures to the nearest minute. **(Lesson 5-7)**

51. Verify that $2 \sec^2 x = \dfrac{1}{1 + \sin x} + \dfrac{1}{1 - \sin x}$ is an identity. **(Lesson 7-2)**

52. College Entrance Exam Compare quantities A and B below.
Write A if quantity A is greater.
Write B if quantity B is greater.
Write C if the quantities are equal.
Write D if there is not enough information to determine the relationship.
(A) $7x + 1$ **(B)** $7x - 1$

Double-Angle and Half-Angle Identities

Objective

After studying this lesson, you should be able to:
- use the double- and half-angle identities for the sine, cosine, and tangent functions.

Application

On October 14, 1947, Charles Yeager flew a Bell X-1 rocket airplane to become the first person to fly faster than the speed of sound. The speed of sound is called mach 1. When an airplane exceeds the speed of sound, about 739 mph, a shock wave in the shape of a cone is formed from the nose of the plane. The shock wave

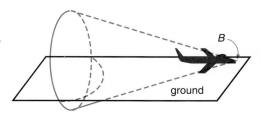

produces a sonic boom as it passes. If B is the measure of the angle at the vertex of the cone, and M is the mach number where $M > 1$, the equation $\frac{1}{M} = \sin\left(\frac{B}{2}\right)$ holds. Write this equation in terms of functions of B. Suppose an aircraft forms a cone whose vertex angle measures 45°. What is its speed in relation to the speed of sound; that is, what is its mach number? *This problem will be solved in Example 3.*

As demonstrated in this application, it is sometimes useful to have identities to find the value of a function of twice an angle or half an angle. We can substitute θ for both α and β in $\sin(\alpha + \beta)$ to find an identity for $\sin 2\theta$.

$$\sin 2\theta = \sin(\theta + \theta) \qquad \textit{Substitute } \theta \textit{ for } \alpha \textit{ and } \beta.$$
$$= \sin\theta\cos\theta + \cos\theta\sin\theta$$
$$= 2\sin\theta\cos\theta$$

The same method can be used to find an identity for $\cos 2\theta$.

$$\cos 2\theta = \cos(\theta + \theta)$$
$$= \cos\theta\cos\theta - \sin\theta\sin\theta$$
$$= \cos^2\theta - \sin^2\theta$$

If we substitute $1 - \cos^2\theta$ for $\sin^2\theta$ or $1 - \sin^2\theta$ for $\cos^2\theta$, we will have two alternate identities for $\cos 2\theta$.

$$\cos 2\theta = 2\cos^2\theta - 1$$
$$\cos 2\theta = 1 - 2\sin^2\theta$$

These identities may be used if θ is measured in degrees or radians. So, θ may represent either a degree measure or real number.

The tangent of a double angle can be found by substituting θ for both α and β in $\tan(\alpha + \beta)$.

$$\tan(\theta + \theta) = \frac{\tan\theta + \tan\theta}{1 - \tan\theta\tan\theta}$$
$$= \frac{2\tan\theta}{1 - \tan^2\theta}$$

Double-Angle Identities

If θ represents the measure of an angle, then the following identities hold for all values of θ.

$$\sin 2\theta = 2\sin\theta\cos\theta$$
$$\cos 2\theta = \cos^2\theta - \sin^2\theta$$
$$= 2\cos^2\theta - 1$$
$$= 1 - 2\sin^2\theta$$
$$\tan 2\theta = \frac{2\tan\theta}{1 - \tan^2\theta}$$

Example 1

If $\cos\theta = \dfrac{\sqrt{5}}{5}$ and θ terminates in the first quadrant, find the exact value of each function.

a. $\cos 2\theta$

a. Since we know $\cos\theta$, we will use the identity $\cos 2\theta = 2\cos^2\theta - 1$.

$$\cos 2\theta = 2\cos^2\theta - 1$$
$$= 2\left(\frac{\sqrt{5}}{5}\right)^2 - 1$$
$$= -\frac{3}{5}$$

b. $\sin 2\theta$

b. To use the identity for $\sin 2\theta$, we must first find $\sin\theta$.

$$\sin^2\theta + \cos^2\theta = 1 \qquad\qquad \sin 2\theta = 2\sin\theta\cos\theta$$
$$\sin^2\theta + \left(\frac{\sqrt{5}}{5}\right)^2 = 1 \qquad\qquad = 2\left(\frac{2\sqrt{5}}{5}\right)\left(\frac{\sqrt{5}}{5}\right)$$
$$\sin^2\theta = \frac{4}{5} \qquad\qquad\qquad = \frac{4}{5}$$
$$\sin\theta = \frac{2\sqrt{5}}{5}$$

c. $\tan 2\theta$

c. We must find $\tan\theta$ to use the identity for $\tan 2\theta$.

$$\tan\theta = \frac{\sin\theta}{\cos\theta} \qquad\qquad \tan 2\theta = \frac{2\tan\theta}{1 - \tan^2\theta}$$
$$= \frac{\frac{2\sqrt{5}}{5}}{\frac{\sqrt{5}}{5}} \qquad\qquad = \frac{2(2)}{1 - (2)^2}$$
$$= 2 \qquad\qquad\qquad = -\frac{4}{3}$$

If we solve the two alternate forms of the identity for $\cos 2\theta$ for $\cos \theta$ and $\sin \theta$ respectively, the following equations result.

$$\cos \theta = \pm\sqrt{\frac{1 + \cos 2\theta}{2}}$$

$$\sin \theta = \pm\sqrt{\frac{1 - \cos 2\theta}{2}}$$

Since θ is a real number, 2θ can be replaced with α and θ can be replaced with $\frac{\alpha}{2}$ to derive the identities for half of any angle measure. The identity for the tangent of half of an angle is found by dividing $\sin \frac{\alpha}{2}$ by $\cos \frac{\alpha}{2}$.

$$\tan \frac{\alpha}{2} = \frac{\pm\sqrt{\dfrac{1 - \cos \alpha}{2}}}{\pm\sqrt{\dfrac{1 + \cos \alpha}{2}}}$$

$$= \pm\sqrt{\frac{1 - \cos \alpha}{1 + \cos \alpha}}$$

Half-Angle Identities

If α represents the measure of an angle, then the following identities hold for all values of α.

$$\sin \frac{\alpha}{2} = \pm\sqrt{\frac{1 - \cos \alpha}{2}} \qquad \cos \frac{\alpha}{2} = \pm\sqrt{\frac{1 + \cos \alpha}{2}}$$

$$\tan \frac{\alpha}{2} = \pm\sqrt{\frac{1 - \cos \alpha}{1 + \cos \alpha}} \quad (\cos \alpha \neq -1)$$

Example 2

Use a half-angle identity to find the exact value of each function.
a. sin 15°

a. $\sin 15° = \sin \dfrac{30°}{2}$

$$= \pm\sqrt{\frac{1 - \cos 30°}{2}}$$

$$= \pm\sqrt{\frac{1 - \dfrac{\sqrt{3}}{2}}{2}}$$

$$= \pm\frac{\sqrt{2 - \sqrt{3}}}{2}$$

Since 15° is a first quadrant angle and sine is positive in the first quadrant, choose the positive value. Thus, $\sin 15° = \dfrac{\sqrt{2 - \sqrt{3}}}{2}$.

(continued on the next page)

b. cos 135°

b. $\cos 135° = \cos \dfrac{270°}{2}$

$\qquad\qquad = \pm\sqrt{\dfrac{1 + \cos 270°}{2}}$

$\qquad\qquad = \pm\sqrt{\dfrac{1 + 0}{2}}$

$\qquad\qquad = \pm\dfrac{\sqrt{2}}{2}$

Since 135° is a second quadrant angle and cosine is negative in the second quadrant, choose the negative value. Thus, $\cos 135° = -\dfrac{\sqrt{2}}{2}$.

We can use the half-angle identity for the sine function to solve the application problem from the beginning of the lesson.

Example 3

APPLICATION

Aeronautics

Write the equation $\dfrac{1}{M} = \sin\left(\dfrac{B}{2}\right)$ in terms of functions of B. What is the speed in relation to sound or the mach number of an aircraft that forms a cone whose vertex angle measures 45°?

$\dfrac{1}{M} = \sin\left(\dfrac{B}{2}\right)$

$\dfrac{1}{M} = \pm\sqrt{\dfrac{1 - \cos B}{2}}$

If the angle of the cone measures 45°, $B = 45$.

$\dfrac{1}{M} = \pm\sqrt{\dfrac{1 - \cos 45°}{2}}$

$\dfrac{1}{M} = \sqrt{\dfrac{1 - \dfrac{\sqrt{2}}{2}}{2}}$ *45° is in Quadrant I, so use the positive value.*

$\dfrac{1}{M} = \dfrac{\sqrt{2 - \sqrt{2}}}{2}$

$M = \dfrac{2}{\sqrt{2 - \sqrt{2}}}$

$M \approx 2.6$

The speed of the aircraft is about mach 2.6. So, the speed of the aircraft is about 2.6 times the speed of sound.

The double- and half-angle identities can also be used to verify other identities.

Example 4

Verify that $\tan\dfrac{x}{2} = \dfrac{1-\cos x}{\sin x}$ is an identity for $0° < x < 90°$.

$$\tan\frac{x}{2} \overset{?}{=} \frac{1-\cos x}{\sin x}$$

$$\sqrt{\frac{1-\cos x}{1+\cos x}} \overset{?}{=} \left(\frac{1-\cos x}{\sin x}\right)\left(\frac{1+\cos x}{1+\cos x}\right)$$

Use the half-angle identity for tangent. Consider only the positive value since x is in the first quadrant.

$$\left(\sqrt{\frac{1-\cos x}{1+\cos x}}\right)\left(\sqrt{\frac{1+\cos x}{1+\cos x}}\right) \overset{?}{=} \frac{1-\cos^2 x}{\sin x\,(1+\cos x)}$$

$$\sqrt{\frac{1-\cos^2 x}{(1+\cos x)^2}} \overset{?}{=} \frac{\sin^2 x}{\sin x\,(1+\cos x)}$$

Simplify.

$$\sqrt{\frac{\sin^2 x}{(1+\cos x)^2}} \overset{?}{=} \frac{\sin x}{(1+\cos x)}$$

$$\frac{\sin x}{1+\cos x} = \frac{\sin x}{1+\cos x}$$

Make a list of the trigonometric identity formulas you have studied thus far. Include an example for each type of identity.

CHECKING FOR UNDERSTANDING

Communicating Mathematics

Read and study the lesson to answer each question.

1. **Compare** the three identities for $\cos 2\theta$. Under what conditions would you choose to use each one?

2. **Explain** how you determine the sign of a result of a double- or half-angle identity.

3. For each angle in standard position, name the quadrant in which the terminal side lies.

   a. x is a second quadrant angle. In which quadrant does $2x$ lie?

   b. $\dfrac{x}{2}$ is a first quadrant angle. In which quadrant does x lie?

   c. $2x$ is a second quadrant angle. In which quadrant does $\dfrac{x}{2}$ lie?

Guided Practice

If $\sin A = \dfrac{3}{5}$ and A is in the first quadrant, find each value.

4. $\cos 2A$ 5. $\tan 2A$ 6. $\sin 2A$

7. $\sin\dfrac{A}{2}$ 8. $\cos\dfrac{A}{2}$ 9. $\tan\dfrac{A}{2}$

Use a half-angle identity to find each value.

10. $\sin 22°30'$ 11. $\cos 22°30'$ 12. $\tan 22°30'$

Verify that each of the following is an identity.

13. $\dfrac{1}{2}\sin 2A = \dfrac{\tan A}{1+\tan^2 A}$ 14. $\tan 2x\tan x + 2 = \dfrac{\tan 2x}{\tan x}$

15. $\sin 2x = 2\cot x\sin^2 x$ 16. $\sin^2\theta = \dfrac{1}{2}(1-\cos 2\theta)$

EXERCISES

If $\tan y = \dfrac{5}{12}$ and y is in the third quadrant, find each value.

17. $\sin 2y$

18. $\tan 2y$

19. $\sin \dfrac{y}{2}$

20. $\cos 2y$

21. $\tan \dfrac{y}{2}$

22. $\cos \dfrac{y}{2}$

Use a half-angle identity to find each value.

23. $\sin 105°$

24. $\cos \dfrac{13\pi}{12}$

25. $\tan 195°$

26. $\cos \dfrac{19\pi}{12}$

27. $\sin \dfrac{7\pi}{8}$

28. $\tan \dfrac{13\pi}{12}$

Verify that each of the following is an identity.

29. $1 + \cos 2A = \dfrac{2}{1 + \tan^2 A}$

30. $\cos^2 2x + 4\sin^2 x \cos^2 x = 1$

31. $\csc A \sec A = 2\csc 2A$

32. $\dfrac{1 - \tan^2 \theta}{1 + \tan^2 \theta} = \cos 2\theta$

33. $\cot X = \dfrac{\sin 2X}{1 - \cos 2X}$

34. $\dfrac{1 + \cos x}{\sin x} = \cot \dfrac{x}{2}$

35. $\sin 2B \left(\cot B + \tan B\right) = 2$

36. $1 - \sin A = \left(\sin \dfrac{A}{2} - \cos \dfrac{A}{2}\right)^2$

37. $\cot \dfrac{\alpha}{2} = \dfrac{\sin \alpha}{1 - \cos \alpha}$

38. $\tan \dfrac{x}{2} = \dfrac{\sin x}{1 + \cos x}$

39. $\dfrac{\cos 2A}{1 + \sin 2A} = \dfrac{\cot A - 1}{\cot A + 1}$

40. $\dfrac{\sin \alpha + \sin 3\alpha}{\cos \alpha + \cos 3\alpha} = \tan 2\alpha$

Derive a formula for each of the following.

41. $\sin 3\alpha$ in terms of $\sin \alpha$

42. $\cos 3\alpha$ in terms of $\cos \alpha$

43. Explain how you would find $\sin A$ if $\sin 4A = \dfrac{2}{3}$ and the terminal side of $4A$ lies in the second quadrant. Then find $\sin A$.

44. Sports Betsy King, one of the champions of women's golf, hit a golf ball with an initial velocity of 100 feet per second. The distance that a golf ball travels is found by the formula $d = \dfrac{v_o^2}{g} \sin 2\theta$, where v_o is the initial velocity, g is the acceleration due to gravity, and θ is the measure of the angle that the initial path of the ball makes with the ground. The acceleration due to gravity is 32 ft/s².

a. Write an expression for the distance the ball travels in terms of θ.

b. Use a calculator to find the distance Ms. King's ball traveled if the angle between the initial path of the ball and the ground measured $60°$.

c. At what angle should the ball leave the head of the golf club if Ms. King wants it to travel the maximum distance? Explain.

45. **Optics** A glass prism has an apex angle of measure α and an angle of deviation of measure β. The index of refraction, n, of the prism is equal to $\dfrac{\sin\left(\dfrac{\alpha+\beta}{2}\right)}{\sin\left(\dfrac{\alpha}{2}\right)}$.

Show that $\sqrt{\dfrac{1-\cos\alpha\cos\beta+\sin\alpha\sin\beta}{1-\cos\alpha}}$ is an equivalent expression for n. Assume that $n > 0$.

Mixed Review

46. State the domain and range of the relation $\{(3, 7), (-1, 4), (0, 0), (3, -6)\}$. Is the relation a function? **(Lesson 1-1)**

47. Write the standard form of the equation of the line that is parallel to the line with equation $3x - y = 10$ and passes through $(0, -2)$. **(Lesson 1-6)**

48. Solve the equation $4x^3 + 3x^2 - x = 0$. **(Lesson 4-1)**

49. Does $A = 120°$, $b = 12$, and $a = 4$ determine one triangle, two triangles, or no triangles? **(Lesson 5-5)**

50. If $\cos\theta = \dfrac{2}{3}$ and $0° \leq \theta \leq 90°$, find $\sin\theta$. **(Lesson 7-1)**

51. Find $\tan 150°$ from values of functions of $180°$ and $30°$. **(Lesson 7-3)**

52. **College Entrance Exam** Choose the best answer.
    x is an integer greater than 1. What is the least x for which $a^2 = b^3 = x$ for some integers a and b?

    (A) 81   **(B)** 64   **(C)** 4   **(D)** 2   **(E)** 9

MID-CHAPTER REVIEW

Solve for values between $0°$ and $90°$. (Lesson 7-1)

1. If $\sin\theta = \dfrac{4}{5}$, find $\cos\theta$.

2. If $\tan\theta = \dfrac{\sqrt{2}}{5}$, find $\cos\theta$.

Simplify. (Lesson 7-1)

3. $\dfrac{1-\sin^2\alpha}{\sin^2\alpha}$

4. $\cos\beta\csc\beta$

Verify that each of the following is an identity. (Lessons 7-2, 7-3, 7-4)

5. $\sin\theta\,(1+\cot^2\theta) = \csc\theta$

6. $\cos^2 A = (1-\sin A)(1+\sin A)$

7. $-\cos\theta = \cos(180°-\theta)$

8. $\dfrac{\cos A + \sin A}{\cos A - \sin A} = \dfrac{1+\sin 2A}{\cos 2A}$

9. $\sin^2\theta = \dfrac{1}{2}(1-\cos 2\theta)$

10. $\sin 2x = 2\cot x \sin^2 x$

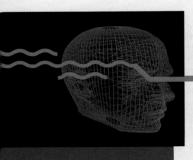

Technology

Double and Half Angles

The BASIC program below finds the values of cos 2x and sin 2x given the value of sin x and the quadrant in which x lies.

```
10 INPUT "ENTER THE VALUE OF SIN X: "; S1
20 INPUT "ENTER THE QUADRANT OF ANGLE X: "; Q
30 C1 = SQR (1-S1^2)
40 IF Q = 1 THEN 70
50 IF Q = 4 THEN 70
60 C1 = (-1)*C1
70 PRINT "SIN 2X = ";2*S1*C1
80 PRINT "COS 2X = ";1-2*S1^2
90 END
```

To run the program, type RUN and press the enter key. Enter each value at the colon. This sample is for sin x = –0.9, where x is in Quadrant IV.

```
RUN
ENTER THE VALUE OF SIN X: -0.9
ENTER THE QUADRANT OF ANGLE X: 4
SIN 2X = -0.784601809
COS 2X = -0.62
```

EXERCISES

Use the BASIC program to find sin 2x and cos 2x for each value of sin x in the quadrant given. Round your answers to the nearest thousandth.

1. $\sin x = 0.5$, Quadrant I

2. $\sin x = 0.9$, Quadrant II

3. $\sin x = -0.65$, Quadrant IV

4. $\sin x = -0.01$, Quadrant III

5. Modify the program to find sin 2x and cos 2x given the value of cos x and the quadrant in which x lies.

Use the BASIC program from Exercise 5 to find sin 2x and cos 2x for each value of cos x in the quadrant given. Round your answers to the nearest thousandth.

6. $\cos x = -0.5$, Quadrant II

7. $\cos x = 0.8$, Quadrant IV

8. Write a BASIC program to find $\sin \frac{x}{2}$ and $\cos \frac{x}{2}$ given the value of sin x and the quadrant in which x lies.

7-5A   Graphing Calculators: Solving Trigonometric Equations

Solving a trigonometric equation means to find all of the values of x that satisfy the equation, or make its value 0. Setting an equation equal to zero and finding the x-intercepts is one method used to solve an equation. This method was discussed in Lesson 4-5A when we located the zeros of polynomial functions. Another method is to graph each side of the equation as a separate function and find where the graphs intersect. This method was discussed in Lesson 2-1A when we graphed a system of equations and found the point of intersection. The solutions will be the same no matter which method you use.

Example 1

Solve tan $x =$ sin $x \cdot$ cos x if $-360° \leq x \leq 360°$.

First, graph each side of the equation and solve by finding the intersection points. Use the trig viewing window.

Enter: [Y=] [TAN] [X,T,θ] [ENTER]
[SIN] [X,T,θ] [COS] [X,T,θ]
[GRAPH]

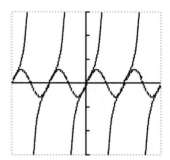

By tracing and zooming in, the solutions are found to be $-360°, -180°, 0°, 180°$, and $360°$.

Subtracting sin $x \cdot$ cos x from each side of the equation produces the equivalent equation tan $x -$ sin $x \cdot$ cos $x = 0$. Graphing this equation will give you the same solutions, except the solutions are now where the graph crosses the x-axis.

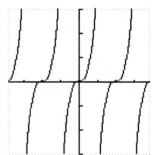

Remember that the solutions will be the same no matter which method you choose, but you need to know what to look for on the graph. If you use the method of graphing two equations, you need to look for the point of intersection because the coordinates of this point satisfy both equations. If you use the method of solving the equation for zero, you need to find the x-intercepts because that is where the equation equals zero.

Example 2

Solve $2 \cos x - 4 = 0$ if $0 \leq x \leq 360°$.

Since the equation already equals zero, graph $y = 2 \cos x - 4$ and find the x-intercepts. Use the viewing window $[-90°, 450°]$ by $[-8, 2]$ with a scale factor of 90° for the x-axis and 1 for the y-axis.

Enter: [Y=] 2 [COS] [X,T,θ] [−] 4 [GRAPH]

The graph does not cross the x-axis, so there are no solutions to the equation.

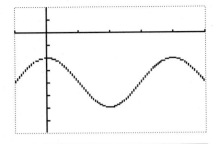

As you have done with other equations, you can use the TRACE and ZOOM features or the root (CALC item 2) feature of the calculator to find solutions to trigonometric equations.

Example 3

Solve $\sin 3x + \cos 2x = 1$ if $0 \leq x \leq 360°$.

Let's use the method of solving the equation for zero and finding the x-intercepts. The equation to be graphed then becomes $\sin 3x + \cos 2x - 1 = 0$. Use the viewing window $[-90°, 450°]$ by $[-5, 5]$ with a scale factor of 90° for the x-axis and 1 for the y-axis.

Enter: [Y=] [SIN] [(] 3 [X,T,θ] [)] [+]
[COS] [(] 2 [X,T,θ] [)] [−] 1
[GRAPH]

Now trace and zoom in on the points of intersection to find solutions: 0°, 40.6°, 139.4°, 180°, and 360°.

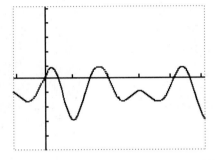

EXERCISES

Graph each equation so that a complete graph is shown. Then find the solutions to the nearest tenth.

1. $\tan x = \cos x, 0° \leq x \leq 360°$

2. $\sin x + \cos x = 2, 0° \leq x \leq 360°$

3. $4 \sin 2x - \sin x = 2, 0° \leq x \leq 360°$

4. $\sin 3x = \sin x, -360° \leq x \leq 360°$

5. $2 \cos^2 x = \sin^2 x, 0° \leq x \leq 360°$

6. $2 \cos x = 2 \tan x, -360° \leq x \leq 360°$

7. $3 \sin^2 x + 2 \sin x - 4 = 0, -360° \leq x \leq 360°$

8. $\sin^2 x - \cos x = 0, -360° \leq x \leq 360°$

9. $\sin x \cdot \cos x = \tan 2x, -360° \leq x \leq 360°$

10. $2 \sin 2x + \cos 2x + 1 = 0, 0° \leq x \leq 360°$

7-5 Solving Trigonometric Equations

Objective

After studying this lesson, you should be able to:
- solve trigonometric equations.

Application

The beautiful sparkle of a diamond is created by *refracted* light. Light travels at different speeds in different mediums. When light rays pass from one medium to another in which they travel at a different velocity, the light is bent, or refracted. According to Snell's Law, $n_1 \sin i = n_2 \sin r$, where n_1 is the index of refraction of the medium the light is exiting, n_2 is the index of refraction of the medium it is entering, i is the angle of incidence, and r is the angle of refraction.

The index of refraction of a diamond is 2.42 and the index of refraction of air is 1.00. If a beam of light strikes a diamond at an angle of 35°, what is the angle of refraction? *This problem will be solved in Example 4.*

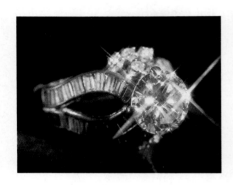

Trigonometric identities, like all identities, are true for *all* values of the variable for which the expression is defined. However, most **trigonometric equations**, like most algebraic equations, are true for *some* but not *all* values of the variable. Trigonometric equations do not have unique solutions. They have infinitely many solutions, differing by the period of the function, 2π or 360° for the sine and cosine functions, and π or 180° for the tangent function.

If the value of a function is restricted to two adjacent quadrants, a trigonometric equation will have unique solutions. These solutions are called **principal values**. For sin x and tan x, the principal values are in Quadrants I or IV. So x is in the interval $-90° \leq x \leq 90°$. For cos x, the principal values are in Quadrants I or II, so x is in the interval $0° \leq x \leq 180°$.

Example 1

Solve $2 \cos^2 x - 5 \cos x + 2 = 0$ for principal values of x. Express solutions in degrees.

$$2 \cos^2 x - 5 \cos x + 2 = 0$$
$$(2 \cos x - 1)(\cos x - 2) = 0 \qquad \textit{Factor.}$$

$$\begin{aligned} 2 \cos x - 1 &= 0 & \text{or} & & \cos x - 2 &= 0 \\ 2 \cos x &= 1 & & & \cos x &= 2 \\ \cos x &= \frac{1}{2} & & & & \textit{There is no solution for } \cos x = 2 \\ x &= 60° & & & & \textit{since } -1 \leq \cos x \leq 1. \end{aligned}$$

The solution is 60°.

When all of the values of x are required, the solution should be represented as $x + 360k°$ for sin x and cos x and $x + 180k°$ for tan x, where k is any integer.

Example 2

Solve $2\tan x \sin x + 2\sin x = \tan x + 1$ for all values of x. Express solutions in degrees.

$$2\tan x \sin x + 2\sin x = \tan x + 1$$

$2\tan x \sin x + 2\sin x - \tan x - 1 = 0$ *Subtract tan x + 1 from each side.*

$(\tan x + 1)(2\sin x - 1) = 0$ *Factor.*

$\tan x + 1 = 0$	$2\sin x - 1 = 0$
$\tan x = -1$	$\sin x = \dfrac{1}{2}$

or

$x = -45° + 180k°$	$x = 30° + 360k°$
	or $x = 150° + 360k°$

The solutions are $-45° + 180k°, 30° + 360k°$, or $150° + 360k°$.

If an equation cannot be solved easily by factoring, try writing the expressions in terms of only one trigonometric function. Remember to use your knowledge of identities.

Example 3

Solve $\sin^2 x + \cos 2x - \cos x = 0$ for principal values of x.

The first step in solving this equation is to express $\cos 2x$ in terms of $\sin x$.

$$\sin^2 x + \cos 2x - \cos x = 0$$
$$\sin^2 x + (1 - 2\sin^2 x) - \cos x = 0 \quad\quad cos\ 2x = 1 - 2\sin^2 x$$
$$1 - \sin^2 x - \cos x = 0$$
$$\cos^2 x - \cos x = 0 \quad\quad cos^2 x = 1 - \sin^2 x$$
$$\cos x(\cos x - 1) = 0 \quad\quad Factor.$$

$\cos x = 0$	or	$\cos x - 1 = 0$
$x = 90°$		$\cos x = 1$
		$x = 0°$

The solutions are $0°$ and $90°$.

Some application problems can be solved by using a calculator to solve a trigonometric equation.

Example 4

Refer to the application at the beginning of the lesson. If a beam of light strikes a diamond at an angle of 35°, what is the angle of refraction?

$$n_1 \sin i = n_2 \sin r$$
$$1.00 \sin 35° = 2.42 \sin r \quad\quad n_1 = 1.00,\ n_2 = 2.42,\ i = 35°$$
$$\sin r = \frac{\sin 35°}{2.42}$$
$$\sin r = 0.2370 \quad\quad Use\ a\ calculator\ to\ solve.$$
$$r = 13.710433$$

The angle of refraction is about $14°$.

It is important to always check your solutions. Some algebraic operations may introduce answers that are *not* solutions to the original equation.

Example 5

Solve $\cos x = 1 + \sin x$ for $0° \leq x < 360°$.

$$\cos x = 1 + \sin x$$
$$\cos^2 x = (1 + \sin x)^2 \qquad \textit{Square each side of the equation.}$$
$$1 - \sin^2 x = 1 + 2\sin x + \sin^2 x \qquad \textit{cos}^2 x = 1 - \sin^2 x$$
$$0 = 2\sin x + 2\sin^2 x$$
$$0 = 2\sin x(1 + \sin x) \qquad \textit{Factor.}$$

$$2\sin x = 0 \qquad\qquad \text{or} \qquad 1 + \sin x = 0$$
$$\sin x = 0 \qquad\qquad\qquad\qquad \sin x = -1$$
$$x = 0° \text{ or } 180° \qquad\qquad\qquad x = 270°$$

Check:

$$\cos x = 1 + \sin x \qquad\qquad \cos x = 1 + \sin x \qquad\qquad \cos x = 1 + \sin x$$
$$\cos 0° \overset{?}{=} 1 + \sin 0° \qquad \cos 180° \overset{?}{=} 1 + \sin 180° \qquad \cos 270° \overset{?}{=} 1 + \sin 270°$$
$$1 = 1 \checkmark \qquad\qquad -1 \overset{?}{=} 1 + 0 \qquad\qquad 0 \overset{?}{=} 1 + (-1)$$
$$-1 \neq 1 \qquad\qquad\qquad 0 = 0 \checkmark$$

Based on the check, $180°$ is not a solution. The solutions are $0°$ and $270°$.

EXPLORATION: Graphing Calculator

You can use a graphing calculator to verify the solution of the equation in Example 5.

1. Use the range values below to set up a viewing window. Mode should be in degrees.

$$0 \leq X \leq 360 \qquad -3 \leq Y \leq 3 \qquad XSCL = 90 \qquad YSCL = 0.25$$

2. Let $Y_1 = \cos x$ and let $Y_2 = 1 + \sin x$.
3. Graph the two functions on the same screen.
4. There are two intersection points for the two graphs. Use the TRACE function to find the coordinates of the points of intersection.

CHECKING FOR UNDERSTANDING

Communicating Mathematics

Read and study the lesson to answer each question.

1. **Explain** how solving a trigonometric equation is different from verifying a trigonometric identity.

2. **State** the number of solutions that each equation has.

   a. $\sin x = 0.5$ **b.** $\cos \theta = -\dfrac{\sqrt{3}}{2}$ if $0° \leq \theta \leq 180°$ **c.** $\sin 2\alpha = -2$

Solve each equation for principal values of x.

3. $2\sin x + 1 = 0$ **4.** $2\cos x - 1 = 0$

5. $\sqrt{2}\sin x - 1 = 0$ **6.** $2\cos x + 1 = 0$

7. $2\cos x - \sqrt{3} = 0$ **8.** $\sin 2x - 1 = 0$

9. $\cos 3x - 0.5 = 0$ **10.** $\tan 2x - \sqrt{3} = 0$

Solve each equation for all values of x.

11. $\cos 2x = \cos x$ **12.** $\sin x = \tan x$

13. $\sin x + \sin x \cos x = 0$ **14.** $\sin x = \cos x$

15. $\cos 2x + \cos x + 1 = 0$ **16.** $\tan^2 x - \sqrt{3}\tan x = 0$

EXERCISES

Solve each equation for $0° \leq x \leq 180°$.

17. $4\sin^2 x - 3 = 0$ **18.** $2\sin^2 x + \sin x = 0$

19. $\sqrt{3}\tan x + 1 = 0$ **20.** $\sqrt{2}\cos x - 1 = 0$

21. $\tan 2x = \cot x$ **22.** $2\cos^2 x = \sin x + 1$

23. $\sin 2x = \cos x$ **24.** $\sin^2 x - 3\sin x + 2 = 0$

25. $\sin x + \cos x = 0$ **26.** $\cos^2 x - \dfrac{7}{2}\cos x - 2 = 0$

27. $3\cos 2x - 5\cos x = 1$ **28.** $\tan^2 x = 3\tan x$

29. $3\tan^2 x + 4\sec x = -4$ **30.** $\sin 2x = \cos 3x$

31. $\sin 2x \sin x + \cos 2x \cos x = 1$ **32.** $3\sin^2 x - \cos^2 x = 0$

33. $\cos 2x + 3\cos x - 1 = 0$ **34.** $4\tan x + \sin 2x = 0$

35. $2\sin x \cos x + 4\sin x = \cos x + 2$

36. $\sqrt{3}\cot x \sin x + 2\cos^2 x = 0$

Solve each equation for all values of x.

37. $2\sin^2 x - 1 = 0$ **38.** $\cos x - 2\cos x \sin x = 0$

39. $\sin^2 x - 2\sin x - 3 = 0$ **40.** $3\cos 2x - 5\cos x = 1$

41. $\cos x \tan x - \sin^2 x = 0$ **42.** $\sin^2 x - \sin x = 0$

43. $4\cos^2 x - 4\cos x + 1 = 0$ **44.** $\cos x = 3\cos x - 2$

45. Solve $\dfrac{\tan x - \sin x}{\tan x + \sin x} = \dfrac{\sec x - 1}{\sec x + 1}$ for all values of x.

46. Physics Ted set up a physics experiment with a weighted spring system. He determined that the motion of the weight can be described by the equation $y = 2\sin\left(\pi t - \dfrac{\pi}{2}\right)$, where y is the distance in inches from the equilibrium point and t is the time in seconds.

    a. What is the position of the weight after 3 seconds?

    b. At what times is the weight at the equilibrium point?

    c. Graph the motion of the weight for the first 5 seconds.

47. **Optics**   Lori has placed a small lamp behind her freshwater aquarium. A beam of light strikes the glass of the aquarium at an angle of 10° and passes through the glass and into the water of the aquarium.

   a. Use Snell's Law to find the angle of refraction. Use 1.5 as the index of refraction for glass and 1.0 as the index of refraction for air.

   b. What is the angle of refraction of the beam as it passes from the glass into the water? Use 1.33 as the index of refraction of water.

   c. At what angle will the beam exit through the front of the aquarium?

48. **Gemology**   Explain how a gemologist might use Snell's Law to determine if a diamond is genuine.

Mixed Review

49. State whether each of the points at $(9, 3)$, $(-1, 2)$, and $(2, -2)$ satisfy the inequality $2x - 4y \leq 7$.   **(Lesson 1-3)**

50. Describe how the graphs of $y = 2x^3$ and $y = 2x^3 + 1$ are related. **(Lesson 3-2)**

51. Find the critical points of the graph of $y = x^3 - 4x^2 + 4x + 6$. Then determine whether each point is a *minimum*, a *maximum*, or a *point of inflection*. **(Lesson 3-7)**

52. Find the values of the six trigonometric functions of an angle θ in standard position whose terminal side contains the point at $(-3, 2)$.   **(Lesson 5-3)**

53. Evaluate $\sec\left(\cos^{-1}\dfrac{2}{5}\right)$ if the angle is in Quadrant I.   **(Lesson 6-4)**

54. If $\csc\theta = 3$ and $0° \leq \theta \leq 90°$, find $\sin\theta$.   **(Lesson 7-1)**

55. Use a half-angle identity to find the value of $\cos 7°30'$.   **(Lesson 7-4)**

56. **College Entrance Exam**   Choose the best answer.
    Let $*x$ be defined as $*x = x^3 - x$. What is the value of $*4 - *(-3)$?

    (A) 84   **(B)** 55   **(C)** –10   **(D)** 22   **(E)** 4

CASE STUDY FOLLOW-UP

Refer to Case Study 4: *The U.S. Economy,* on pages 966–969.

Economic changes often follow cyclical patterns. If there is a reason for a cycle and it is not simply accidental, it can be used to predict future trends.

1. Describe the apparent cycle in annual percentage changes in the CPI of all items, 1976–1990.

2. If the cycle continues as it has, when will the annual percentage in the CPI next reach +2%?

3. If $f(x)$ defines the apparent inflation cycle after 1976, where x equals the year, and $f(1983) = 0$, would you write $f(x)$ using the sine or cosine function?

4. **Research**   Find an example of a widely accepted business cycle. In a one-page paper, give evidence to support and deny the existence of the cycle and tell whether you think the cycle exists.

7-6 Normal Form of a Linear Equation

Objective

After studying this lesson, you should be able to:
- write a linear equation in normal form.

Application

Don Eger operates a grinding wheel in a large factory. When he uses the grinding wheel to sharpen a dull cutting blade, the sparks fly in a line that is tangent to the wheel. If Mr. Eger holds a blade at a 45° angle to a vertical line and the grinding wheel has a diameter of 8 inches, how far must the lamp be placed above the grinding wheel so that it will not be showered with sparks? *This problem will be solved in Example 2.*

Trigonometric functions sometimes appear in equations whose graphs are not like those of the trigonometric functions themselves. For example, the **normal form** of linear equations can be written in terms of trigonometric functions.

A **normal** is a line that is perpendicular to another line, curve, or surface. The normal form of the equation of a line is written in terms of the length of the normal from the line to the origin. This form uses the trigonometric functions of the positive angle the normal makes with the *x*-axis.

Suppose ℓ is a line that does not pass through the origin and p units is the length of the normal from the origin. Let C be the point of intersection of the line ℓ with the normal and let ϕ be the positive angle formed by the *x*-axis and $\overline{OC}$. Draw $\overline{MC}$ perpendicular to the *x*-axis. Since ϕ is in standard position, $\cos \phi = \dfrac{OM}{p}$ or $OM = p \cos \phi$ and $\sin \phi = \dfrac{MC}{p}$ or $MC = p \sin \phi$. So $\dfrac{p \sin \phi}{p \cos \phi}$ or $\dfrac{\sin \phi}{\cos \phi}$ is the slope of $\overline{OC}$.

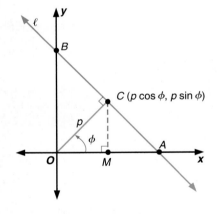

Since ℓ is perpendicular to $\overline{OC}$, its slope is the negative reciprocal of the slope of $\overline{OC}$, or $-\dfrac{\cos \phi}{\sin \phi}$.

Since ℓ contains C, we can use the point-slope form to write the equation of line ℓ.

$$y - y_1 = m(x - x_1)$$

$$y - p \sin \phi = -\frac{\cos \phi}{\sin \phi}(x - p \cos \phi) \qquad m = -\frac{\cos \phi}{\sin \phi}, (x_1, y_1) = (p \cos \phi, p \sin \phi)$$

$$y \sin \phi - p \sin^2 \phi = -x \cos \phi + p\ \cos^2 \phi \qquad \textit{Multiply each side by } \sin \phi.$$

$$x \cos \phi + y \sin \phi = p(\sin^2 \phi + \cos^2 \phi)$$

$$x \cos \phi + y \sin \phi - p = 0 \qquad \textit{sin}^2 \phi + \cos^2 \phi = 1$$

<table>
<tr><td>Normal Form</td><td>The normal form of a linear equation is
$$x \cos \phi + y \sin \phi - p = 0,$$
where p is the length of the normal from the line to the origin and ϕ is the positive angle formed by the positive x-axis and the normal.</td></tr>
</table>

You can write the normal form of a linear equation if you are given the values of ϕ and p.

Example 1

Write the standard form of the equation of a line for which the length of the normal is 3 units and the normal makes an angle of 60° with the positive x-axis.

$$x \cos \phi + y \sin \phi - p = 0 \qquad \textit{Normal form}$$
$$x \cos 60° + y \sin 60° - 3 = 0 \qquad \textit{$\phi = 60°$ and $p = 3$}$$
$$\frac{1}{2}x + \frac{\sqrt{3}}{2}y - 3 = 0$$
$$x + \sqrt{3}y - 6 = 0 \qquad \textit{Multiply each side by 2.}$$

The equation is $x + \sqrt{3}y - 6 = 0$.

We can use the normal form of a linear equation to solve the application problem presented at the beginning of the lesson.

Example 2

Refer to the application at the beginning of the lesson. How far must the lamp be placed above the grinding wheel so that it will not be showered with sparks?

If we consider the center of the cutting wheel to be the origin, the equation that represents the edge of the cutting wheel is $x^2 + y^2 = 16$. The line of sparks will be at a 45° angle to the vertical. So, in the normal form of the equation, $p = 4$ and $\phi = 90° - 45°$ or 45°.

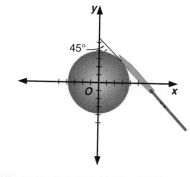

$$x \cos \phi + y \sin \phi - p = 0$$
$$x \cos 45° + y \sin 45° - 4 = 0$$
$$\frac{\sqrt{2}}{2}x + \frac{\sqrt{2}}{2}y - 4 = 0$$
$$x + y - 4\sqrt{2} = 0$$

The line of sparks will cross the y-axis $4\sqrt{2}$ inches above the center of the wheel. Thus, the lamp must be at least $4\sqrt{2} - 4$ or about 1.65 inches above the wheel to avoid being showered with sparks.

We can transform the standard form of a linear equation, $Ax + By + C = 0$, into normal form if the relationship between the coefficients in the two forms is known. The equations will represent the same line if and only if their corresponding coefficients are proportional. That is, if

$$\frac{A}{\cos\phi} = \frac{B}{\sin\phi} = \frac{C}{-p}, \text{ then } \sin\phi = \frac{-Bp}{C}, \text{ and } \cos\phi = \frac{-Ap}{C}.$$

We can divide $\sin\phi = \frac{-Bp}{C}$ by $\cos\phi = \frac{-Ap}{C}$, where $\cos\phi \neq 0$.

$$\frac{\sin\phi}{\cos\phi} = \frac{-\dfrac{Bp}{C}}{-\dfrac{Ap}{C}}$$

$$\tan\phi = \frac{B}{A}$$

Refer to the diagram at the right. Consider an angle ϕ in standard position such that $\tan\phi = \dfrac{B}{A}$. The length of $\overline{OP}$ is $\sqrt{A^2 + B^2}$. Thus,

$$\sin\phi = \frac{B}{\pm\sqrt{A^2 + B^2}} \text{ and } \cos\phi = \frac{A}{\pm\sqrt{A^2 + B^2}}.$$

Since we know that $\dfrac{B}{\sin\phi} = \dfrac{C}{-p}$, we can substitute to get the result $\dfrac{B}{\dfrac{B}{\pm\sqrt{A^2 + B^2}}} = \dfrac{C}{-p}.$

Therefore, for this proof only, $p = \dfrac{C}{\pm\sqrt{A^2 + B^2}}.$

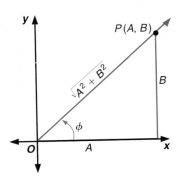

The $\pm$ sign is used since p is a measure and must be positive in the equation $x\cos\phi + y\sin\phi - p = 0$. Therefore, the sign must be chosen as opposite of the sign of C. That is, if C is positive, use $-\sqrt{A^2 + B^2}$, and, if C is negative, use $\sqrt{A^2 + B^2}$.

If $C = 0$, the sign is chosen so that $\sin\phi$ is positive, that is, the same sign as that of B.

Substitute the values for $\sin\phi$, $\cos\phi$, and p into the normal form.

$$\frac{Ax}{\pm\sqrt{A^2 + B^2}} + \frac{By}{\pm\sqrt{A^2 + B^2}} - \frac{C}{\pm\sqrt{A^2 + B^2}} = 0$$

Notice that the standard form is closely related to the normal form.

Changing the Standard Form to Normal Form	**The standard form of a linear equation, $Ax + By + C = 0$, can be changed to normal form by dividing each term by $\pm\sqrt{A^2 + B^2}$. The sign is chosen opposite the sign of C.**

If the equation of a line is in normal form, you can find the length of the normal, p units, directly from the equation. You can find the angle ϕ by using the relation $\tan\phi = \dfrac{B}{A}$. However, you must find the quadrant in which the normal lies to find the correct angle for ϕ. When the equation of a line is in normal form, the coefficient of x is equal to $\cos\phi$ and the coefficient of y is equal to $\sin\phi$. Thus, the correct quadrant can be determined by studying the signs of $\cos\phi$ and $\sin\phi$. For example, if $\sin\phi$ is negative and $\cos\phi$ is positive, the normal lies in the fourth quadrant.

Example 3

> **Write the equation $2x - 5y + 3 = 0$ in normal form. Then, find the length of the normal and the angle it makes with the positive x-axis.**
>
> Since C is positive in $2x - 5y + 3 = 0$, use $-\sqrt{A^2 + B^2}$.
>
> $-\sqrt{A^2 + B^2} = -\sqrt{2^2 + (-5)^2}$ or $-\sqrt{29}$
>
> The normal form is $\dfrac{2x}{-\sqrt{29}} - \dfrac{5y}{-\sqrt{29}} + \dfrac{3}{-\sqrt{29}} = 0$ or $-\dfrac{2x}{\sqrt{29}} + \dfrac{5y}{\sqrt{29}} - \dfrac{3}{\sqrt{29}} = 0$.
>
> Therefore, $\sin \phi = \dfrac{5}{\sqrt{29}}$, $\cos \phi = -\dfrac{2}{\sqrt{29}}$, and $p = \dfrac{3}{\sqrt{29}}$.
>
> $\tan \phi = -\dfrac{5}{2}$ *Why?*
>
> $\quad\quad = -2.5$ *Since $\sin \phi$ is positive and $\cos \phi$ is negative, the*
>
> $\quad\quad \phi \approx 112°$ *terminal side of angle ϕ is in Quadrant II.*
>
> Thus, angle ϕ measures $112°$, and the length of the normal is $\dfrac{3}{\sqrt{29}}$ or about 0.56 units.

CHECKING FOR UNDERSTANDING

Communicating Mathematics

Read and study the lesson to answer each question.

1. **Explain** how you tell in which quadrant the normal to a line lies by examining the x- and y-coefficients of the equation in normal form. Make a table of all of the combinations of signs of x, and y, and the quadrants in which the normal lies.

2. **Describe** how you would write the normal form of the equation of a line for which $p = 7$ and $\phi = 225°$.

Guided Practice

Simplify.

3. $x \cos 45° + y \sin 45° - 11 = 0$ 4. $x \cos 60° + y \sin 60° - 3 = 0$

5. $x \cos 135° + y \sin 135° = 0$ 6. $x \cos 225° + y \sin 225° - 6 = 0$

Write the standard form of the equation of each line given p, the measure of its normal, and ϕ, the angle the normal makes with the positive x-axis.

7. $p = 2, \phi = 30°$ 8. $p = 1, \phi = 135°$ 9. $p = 4, \phi = 270°$

Write each equation in normal form. Then find p, the measure of the normal, and ϕ, the angle the normal makes with the positive x-axis.

10. $5x - y + 3 = 0$ 11. $3x - y = 4$ 12. $5x + y = 7$

EXERCISES

Write the standard form of the equation of each line given p, the measure of its normal, and ϕ, the angle the normal makes with the positive x-axis.

13. $p = 5, \phi = 45°$ **14.** $p = 3, \phi = 60°$ **15.** $p = 25, \phi = 225°$

16. $p = 2, \phi = 150°$ **17.** $p = 8, \phi = 240°$ **18.** $p = 32, \phi = 120°$

Write each equation in normal form. Then find p, the measure of the normal, and ϕ, the angle that the normal makes with the positive x-axis.

19. $y = x + 6$ **20.** $2x - 3y - 1 = 0$ **21.** $x + y - 8 = 0$

22. $3x + 4y - 1 = 0$ **23.** $x - 3y - 2 = 0$ **24.** $6x - 8y - 15 = 0$

25. Write the standard form of the equation of a line if a point on the line nearest to the origin is $(-4, 4)$.

26. The point nearest to the origin on a line is $(3, 3)$. Write the standard form of the equation of the line.

27. Two different lines make an angle of $150°$ with the positive x-axis and are 1 unit from the origin. Write the standard form of the equations of the lines.

28. Write the standard form of the equations of two lines that make an angle of $135°$ with the positive x-axis and are 3 units from the origin.

29. The point on a line nearest to the origin has coordinates $(4, 3)$. The point on a second line nearest to the origin has coordinates $(-3, 1)$. Where do the two lines intersect?

30. Coordinate Geometry The three sides of a triangle are tangent to a unique circle called the *incircle*. On the coordinate plane, the incircle of $\triangle ABC$ has its center at the origin. The lines whose equations are $x + 4y = 6\sqrt{17}$, $2x + \sqrt{5}y = -18$, and $2\sqrt{2}x = y + 18$ contain the sides of $\triangle ABC$. What is the length of the radius of the incircle?

31. History Ancient slingshots were made from straps of leather that cradled a rock until it was released. One would spin the slingshot in a circle and the initial path of the released rock would be a straight line tangent to the circle at the point of release. The rock will travel the greatest distance if it is released when the angle between the normal to the path and the horizontal is $-45°$. The center of the circular path is the origin and the radius of the circle measures 1.25 feet.

  a. Draw a labeled diagram of the situation.

  b. Write the equation of the initial path of the rock in standard form.

Mixed Review

32. Write the slope-intercept form of the equation of the line that passes through $(5, 2)$ and $(-4, 4)$. **(Lesson 1-5)**

33. Are $y = 4x + 4$ and $y = -4x - 4$ inverses of each other? **(Lesson 3-3)**

34. State the number of roots of the equation $4x^3 - 4x^2 + 13x - 6 = 0$. Then solve the equation. **(Lesson 4-1)**

35. Find a numerical value of one trigonometric function of S if $\tan S \cos S = \frac{1}{2}$. **(Lesson 7-2)**

36. Solve the equation $\tan x + \cot x = 2$ for principal values of x. **(Lesson 7-5)**

37. **College Entrance Exam** Choose the best answer.

Divide $\frac{a - b}{a + b}$ by $\frac{b - a}{b + a}$.

(A) 1 **(B)** $\frac{(a - b)^2}{(a + b)^2}$ **(C)** $\frac{1}{a^2 - b^2}$ **(D)** -1 **(E)** 0

DECISION MAKING

Writing a Budget

The United States has become a nation that operates on credit. In 1990, the federal deficit amounted to about $1000 per citizen, while personal debt averaged more than $3200 per person. One way for individuals to keep control of their finances is to write and stick to a budget. A budget allows you to see where you spend money and readjust your financial priorities when necessary.

1. Make a list of your basic personal needs like food, clothing, and so on. Then make a second list of things such as rent and utilities that you would have to pay for if you were living by yourself.

2. Research the costs of the items on your lists. Talk to utility companies, read classified ads to learn of apartment rental costs, visit grocery stores to find the cost of food. Then make a realistic estimate of the monthly cost of each item on your lists.

3. Make a reasonable estimate of the amount of money you will need to earn to live within your budget and also be able to save for the future.

4. **Project** Work with your group to complete one of the projects listed at the right based on the information you gathered in Exercises 1–3.

PROJECTS

- *Conduct an interview.*
- *Write a book report.*
- *Write a proposal.*
- *Write an article for the school paper.*
- *Write a report.*
- *Make a display.*
- *Make a graph or chart.*
- *Plan an activity.*
- *Design a checklist.*

7-7 Distance from a Point to a Line

Objectives

After studying this lesson, you should be able to:
- find the distance from a point to a line,
- find the distance between parallel lines, and
- write the equations of lines that bisect angles formed by intersecting lines.

Application

Your blood pressure is the pressure that your blood exerts against the walls of your arteries. The pressure depends upon the strength and rate of your heart's contractions, the volume of blood in your circulatory system, and the elasticity of your arteries. Doctors report blood pressures as two measurements, the systolic pressure and the diastolic pressure. The systolic pressure is the pressure of the blood when the heart contracts, and the diastolic pressure is the pressure when the heart is between contractions. As a person ages, his or her arteries tend to lose elasticity, and as a result the blood pressure rises.

> **FYI...**
>
> Divide your weight in pounds by 16 to find the approximate number of pints of blood in your body.

In statistics, *prediction equations* are often used to show the relationship between two quantities. A prediction equation relating a person's systolic blood pressure, y, to their age, x, is $4x - 3y + 228 = 0$. If an actual data point is close to the graph of a prediction equation, the equation gives a good approximation for the coordinates of that point. Linda is 19 years old, and her systolic blood pressure is 112. Her father, who is 45 years old, has a systolic blood pressure of 120. For whom is the given prediction equation a better predictor? *This problem will be solved in Example 2.*

The normal form of a linear equation can be used to find the distance from a point to a line. Let $\overleftrightarrow{RS}$ be a line in the coordinate plane, and let $P(x_1, y_1)$ be a point not on $\overleftrightarrow{RS}$. P may lie on the same side of $\overleftrightarrow{RS}$ as the origin does or it may lie on the opposite side. If a line segment joining P to the origin does not intersect $\overleftrightarrow{RS}$, point P is on the same side of the line as the origin. Construct $\overleftrightarrow{TV}$ parallel to $\overleftrightarrow{RS}$ and passing through P. The distance d between the parallel lines is the distance from P to $\overleftrightarrow{RS}$. We will use a negative distance, d, if point P and the origin are on the same side of the line.

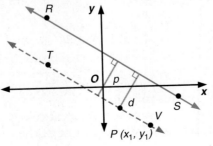

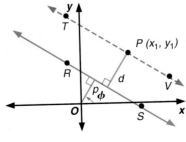

Let $x \cos \phi + y \sin \phi - p = 0$ be the equation of $\overleftrightarrow{RS}$ in normal form. Since $\overleftrightarrow{TV}$ is parallel to $\overleftrightarrow{RS}$, they have the same slope. The equation for $\overleftrightarrow{TV}$ can be written as $x \cos \phi + y \sin \phi - (p + d) = 0$. Solve this equation for d.

$$d = x \cos \phi + y \sin \phi - p$$

Since $P(x_1, y_1)$ is on $\overleftrightarrow{TV}$, its coordinates satisfy this equation.

$$d = x_1 \cos \phi + y_1 \sin \phi - p$$

We can use an equivalent form of this expression to find d when the equation of a line is in standard form.

Distance from a Point to a Line

The following formula can be used to find the distance from a point (x_1, y_1) to a line with equation $Ax + By + C = 0$.

$$d = \frac{Ax_1 + By_1 + C}{\pm \sqrt{A^2 + B^2}}$$

The sign of the radical is chosen opposite the sign of C.

The distance will be positive if the point and the origin are on opposite sides of the line. The distance will be negative if the origin is on the same side of the line as the point. If you are solving an application problem, the absolute value of d will probably be required.

Example 1

Find the distance between $P(3, -1)$ and the line with equation $2x + 5y - 2 = 0$.

$$d = \frac{Ax_1 + By_1 + C}{\pm \sqrt{A^2 + B^2}}$$

$$= \frac{2(3) + 5(-1) + (-2)}{\pm \sqrt{2^2 + 5^2}} \qquad A = 2, B = 5, C = -2, x_1 = 3, y_1 = -1$$

$$= \frac{-1}{\sqrt{29}} \qquad \qquad \text{Since } C \text{ is negative, use } +\sqrt{A^2 + B^2}.$$

$$= -\frac{\sqrt{29}}{29} \text{ or about } -0.19 \qquad |d| = 0.19$$

Therefore, point P is $-\dfrac{\sqrt{29}}{29}$ or about 0.19 units from the graph of $2x + 5y - 2 = 0$ and is on the same side of the line as the origin.

We can use the formula for the distance between a point and a line to solve the application presented at the beginning of the lesson.

Example 2

APPLICATION

Health

Recall that Linda is 19 years old and her father is 45. If Linda's systolic blood pressure is 112 and her father's is 120, for whom is the given prediction equation $4x - 3y + 228 = 0$ a better predictor?

Find the distance between the line and each data point. The prediction equation $4x - 3y + 228 = 0$ is a better predictor for the point that is closer to the line.

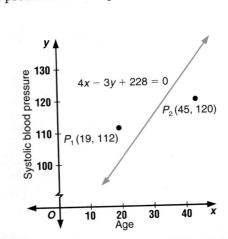

Linda:

$P_1 = (19, 112)$

$$d = \frac{Ax_1 + By_1 + C}{\pm\sqrt{A^2 + B^2}}$$

$$= \frac{4(19) + (-3)(112) + 228}{-\sqrt{4^2 + (-3)^2}}$$

$$= \frac{-32}{-5}$$

$$= 6.4$$

Father:

$P_2 = (45, 120)$

$$d = \frac{Ax_1 + By_1 + C}{\pm\sqrt{A^2 + B^2}}$$

$$= \frac{4(45) + (-3)(120) + 228}{-\sqrt{4^2 + (-3)^2}}$$

$$= \frac{48}{-5}$$

$$= -9.6$$

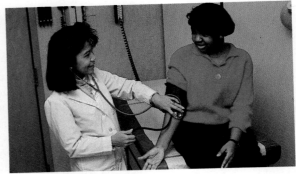

The prediction equation is a better predictor for Linda since her data point, $(19, 112)$ is closer to the graph of the line.

You can use the formula for the distance from a point to a line to find the distance between two parallel lines. To do this, choose a point on one of the lines and use the formula to find the distance from that point to the other line.

Example 3

Find the distance between the lines with equations $3x + 2y = 10$ and $y = -\frac{3}{2}x + 7$.

Since $y = -\frac{3}{2}x + 7$ is in slope-intercept form, we know that $(0, 7)$ are the coordinates of a point on the line, given that 7 is the y-intercept. Use this point to find the distance between the lines.

$$3x + 2y = 10$$
$$3x + 2y - 10 = 0 \qquad \textit{Write the equation in standard form.}$$

$$d = \frac{Ax_1 + By_1 + C}{\pm\sqrt{A^2 + B^2}}$$ *Distance from a point to a line*

$$= \frac{3(0) + 2(7) - 10}{\sqrt{3^2 + 2^2}}$$ *$A = 3$, $B = 2$, $C = -10$, $x_1 = 0$, $y_1 = 7$*

$$= \frac{4}{\sqrt{13}} \text{ or } \frac{4\sqrt{13}}{13}$$

So, the distance between the lines is $\dfrac{4\sqrt{13}}{13}$ or about 1.11 units.

An equation of the bisector of an angle formed by two lines in the coordinate plane can also be found using the formula for the distance between a point and a line. The bisector of an angle is the set of all points in the plane equidistant from the sides of the angle. Using this definition, equations of the bisectors of the angles formed by two lines can be found.

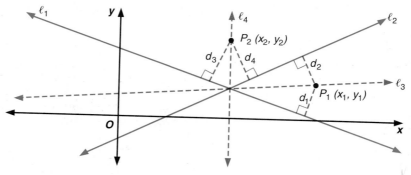

In the figure, ℓ_3 and ℓ_4 are the bisectors of the angles formed by ℓ_1 and ℓ_2. $P_1(x_1, y_1)$ is a point on ℓ_3 and $P_2(x_2, y_2)$ is a point on ℓ_4. Let d_1 be the distance from ℓ_1 to P_1, and let d_2 be the distance from ℓ_2 to P_1.

Notice that P_1 and the origin lie on opposite sides of ℓ_1. So d_1 is positive. Since the origin and P_1 are on opposite sides of ℓ_2, d_2 is also positive. Therefore, for any point $P_1(x_1, y_1)$ on ℓ_3, $d_1 = d_2$. However, d_3 is positive and d_4 is negative. *Why?* Therefore, for any point $P_2(x_2, y_2)$ on ℓ_4, $d_3 = -d_4$.

The origin is in the interior of the angle that is bisected by ℓ_3, but in the exterior of the angle bisected by ℓ_4. So, a good way for you to determine whether to equate distances or to let one distance equal the opposite of the other is to observe the position of the origin.

Relative Position of the Origin	If the origin lies within the angle being bisected or the angle vertical to it, the distances from each line to a point on the bisector have the same sign. If the origin does not lie within the angle being bisected, the distances have opposite signs.

To find the equation of a specific angle bisector, first graph the lines. Then, determine whether to equate the distances or to let one distance equal the opposite of the other.

Example 4

Find an equation of the line that bisects the acute angle formed by the graphs of the equations $3x + y - 9 = 0$ and $2x - 3y + 6 = 0$.

Graph each equation and sketch the bisector. The origin is in the interior of the acute angle. So $d_1 = d_2$.

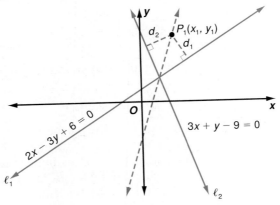

Write expressions for the distances from each line to $P_1(x_1, y_1)$.

$$d_1 = \frac{2x_1 - 3y_1 + 6}{-\sqrt{13}} \qquad \textit{Use the formula } \frac{Ax_1 + By_1 + C}{\pm\sqrt{A^2 + B^2}}.$$

$$d_2 = \frac{3x_1 + y_1 - 9}{\sqrt{10}}$$

Since $d_1 = d_2$, use substitution.

$$\frac{2x_1 - 3y_1 + 6}{-\sqrt{13}} = \frac{3x_1 + y_1 - 9}{\sqrt{10}}$$

If we simplify the expressions and drop the subscripts, we find that an equation of the bisector is

$$(2\sqrt{10} + 3\sqrt{13})x + (-3\sqrt{10} + \sqrt{13})y + 6\sqrt{10} - 9\sqrt{13} = 0.$$

CHECKING FOR UNDERSTANDING

Communicating Mathematics

Read and study the lesson to answer each question.

1. **Analyze** the situation where a point lies on a line. Does the equation for the distance between a point and a line give a valid result under these conditions?

2. In Example 2, the distances between the data points and the graph of the prediction equation had opposite signs. Explain what this means for the predictions for these points.

Guided Practice

Find the distance between the point with the given coordinates and line with the given equation. Round to the nearest tenth.

3. $(4, 2)$, $6x - 8y + 1 = 0$

4. $(-3, 5)$, $12x + 5y - 3 = 0$

5. $(-1, 4)$, $3x - 7y - 1 = 0$

6. $(-5, 0)$, $x - 3y + 11 = 0$

Name the coordinates of one point that satisfy the first equation. Then find the distance from the point to the graph of the second equation.

7. $3x - y + 9 = 0$
   $6x - 2y - 4 = 0$

8. $x + 4 = -2y$
   $2x + 4y - 7 = 0$

9. $\frac{1}{3}x - 4 = y$
   $2x - 6y = 4$

10. $2x - 7y = -9$
    $7y = 2x - 9$

Find an equation of the line that bisects the acute angle formed by the graphs of each pair of equations.

11. $4x - 3y + 12 = 0$
    $5x + 12y - 1 = 0$

12. $9x + 40y = 1$
    $y = -\frac{3}{4}x - 2$

13. $y - 5 = 0$
    $x - y + 2 = 0$

14. $y = x + 3$
    $2x + y - 1 = 0$

EXERCISES

Practice   Solve.

15. Find the distance from the graph of $x - 7y + 4 = 0$ to the point $C(-4, 2)$.

16. Find the distance from the origin to the graph of $3x - y + 1 = 0$.

17. Find the distance from the point $A(2, -2)$ to a line with equation $2x + 3y + 2 = 0$. Explain what the distance means.

18. Find the distance between the point $B(-3, 2)$ and the graph of $6x - 5y = 2$.

19. The lines whose equations are $3x - 4y - 12 = 0$ and $6x - 8y - 48 = 0$ are parallel. Find the distance between these lines.

20. Find the distance between the lines with equations $x + y - 1 = 0$ and $y = -x + 6$.

21. The graphs of $2x - 3y + 1 = 0$ and $3y - 2x = 5$ are parallel lines. Find the distance between the lines.

22. Find the distance between the lines with equations $3x - 5y + 7 = 0$ and $6x - 10y - 2 = 0$.

Find the equation of the line that bisects the acute angle formed by the graphs of each pair of equations.

23. $y = 8 - x$
    $2x - y - 4 = 0$

24. $4x + y + 3 = 0$
    $y + x = -2$

25. $x + 4 = y$
    $x + 4y + 6 = 0$

26. $3x + 2y - 2 = 0$
    $2x + 3y + 2 = 0$

Find the equation of the line that bisects the obtuse angle formed by the graphs of each pair of equations.

27. $y = 5 - x$
    $x = 5$

28. $3x + y - 7 = 0$
    $2x + 5y + 3 = 0$

29. $y = 3 - 6x$
    $x + 3y + 1 = 0$

30. $y = 3x - 1$
    $x + y + 2 = 0$

31. Write the equations of the lines that are 3 units from the line with equation $x - 5y + 10 = 0$.

32. Find the lengths of the three altitudes of a triangle whose vertices are at $(5, 3)$, $(1, -4)$, and $(-4, 1)$.

Programming

33. The following program will calculate the distance from a point to a line.

```
Prgm 7: PTTOLINE
:ClrHome                              Clears the text screen.
:Disp "AX + BY + C = 0"               Displays the quoted message.
:Disp "A ="
:Input A                              Asks the user to input value A.
:Disp "B ="
:Input B
:Disp "C ="
:Input C
:Disp "X-COORDINATE OF POINT ="
:Input X
:Disp "Y-COORDINATE OF POINT ="
:Input Y
:If C > 0                             If C > 0, the denominator is negative.
:-√(A² + B²)→ D                       Calculates denominator and stores it as D.
:If C ≤ 0                             If C ≤ 0, the denominator is positive.
:√(A² + B²)→ D
:AX + BY + C → N                      Calculates numerator and stores it as N.
:N/D → E                              Calculates distance and stores it as E.
:Disp "DISTANCE ="
:Disp E
```

Use the program to find the distance between the point with the given coordinates and line with the given equation. Round to the nearest tenth.

a. $(8, 5)$; $7x - 2y - 9 = 0$

b. $(10, 8)$; $6x + y + 12 = 0$

c. $(-4, 0)$; $y = 9x + 3$

d. $(-12, 0)$; $y = -0.5x - 6$

Critical Thinking

34. Circle P has its center at $(-5, 6)$ and passes through $(-2, 2)$. Show that the line with equation $5x - 12y + 32 = 0$ is tangent to circle P.

Applications and Problem Solving

35. Entomology Jenny is working on an advanced biology project in which she is studying old-fashioned methods used to predict the weather. One of the theories she is investigating is the one that relates the number of times a cricket chirps in a minute to the temperature. She conducted experiments and determined the prediction equation $y = 3x + 28$, where y is the temperature in degrees Fahrenheit and x is the number of times a cricket chirps in a minute. Jenny wished to test her equation with three more data points. They were $(16, 72)$, $(18, 84)$, and $(16, 76)$.

a. Find the distance between each of the data points and the graph of the prediction equation line.

b. For which data point is Jenny's prediction equation a very good predictor?

c. For which data point is the prediction equation the worst predictor?

Mixed Review

36. Consumer Awareness Bill and Liz are going on a vacation in Jamaica. Bill bought 8 rolls of film and 2 bottles of sunscreen for $35.10. The next day, Liz paid $14.30 for three rolls of film and one bottle of sunscreen. If the price for each bottle of sunscreen and each roll of film are the same, what is the price of a roll of film and a bottle of sunscreen? **(Lesson 2-1)**

37. Create a function in the form $y = f(x)$ that has vertical asymptotes at $x = -2$ and $x = 0$ and a hole at $x = 2$. **(Lesson 3-4)**

38. Determine whether the function $y = \dfrac{4}{x^3} + 1$ is *continuous*, has *infinite discontinuity*, *jump discontinuity*, or *point discontinuity*. **(Lesson 3-8)**

39. Simplify the expression $\dfrac{1 - \sin^2 \alpha}{\sin^2 \alpha}$. **(Lesson 7-1)**

40. Change $2x - 5y + 3 = 0$ to normal form. Then find p, the measure of the normal, and ϕ, the angle it makes with the positive x-axis. **(Lesson 7-6)**

41. College Entrance Exam Compare quantities A and B.

Write A if quantity A is greater.
Write B if quantity B is greater.
Write C if the quantities are equal.
Write D if there is not enough information to determine the relationship.

(A) the area of the shaded region

(B) the area of the small circle

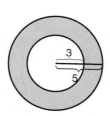

VOCABULARY

Upon completing this chapter, you should be
familiar with the following terms:

double-angle identities **378**
half-angle identities **379**
identity **358**
normal **392**
normal form **392**
principal values **387**
Pythagorean identities **359**

359 quotient identities
358 reciprocal identities
372, 373   sum and difference identities
360 symmetry identities
387 trigonometric equations
358 trigonometric identity

SKILLS AND CONCEPTS

OBJECTIVES AND EXAMPLES

Upon completing this chapter, you should
be able to:

- identify and use identities   **(Lesson 7-1)**

If θ is in Quadrant I and $\cos\ \theta = \frac{1}{3}$, find
$\sin\ \theta$.

$$\sin^2 A + \cos^2 A = 1$$
$$\sin^2 A + \left(\frac{1}{3}\right)^2 = 1$$
$$\sin^2 A + \frac{1}{9} = 1$$
$$\sin^2 A = \frac{8}{9}$$
$$\sin A = \frac{2\sqrt{2}}{3}$$

- use the basic trigonometric identities to
verify other identities   **(Lesson 7-2)**

Verify that $\csc\ x \sec\ x = \cot\ x + \tan\ x$ is
an identity.

$$\csc\ x \sec\ x = \cot\ x + \tan\ x$$
$$\frac{1}{\sin\ x} \cdot \frac{1}{\cos\ x} = \frac{\cos\ x}{\sin\ x} + \frac{\sin\ x}{\cos\ x}$$
$$\frac{1}{\sin\ x \cos\ x} = \frac{\cos^2 x + \sin^2 x}{\sin\ x \cos\ x}$$
$$\frac{1}{\sin\ x \cos\ x} = \frac{1}{\sin\ x \cos\ x}$$

REVIEW EXERCISES

Use these exercises to review and prepare for
the chapter test.

Solve for values of θ between $0°$ and $90°$.

1. If $\sin\ \theta = \frac{1}{2}$, find $\csc\ \theta$.

2. If $\tan\ \theta = 4$, find $\sec\ \theta$.

3. If $\csc\ \theta = \frac{5}{3}$, find $\cos\ \theta$.

4. If $\cos\ \theta = \frac{4}{5}$, find $\tan\ \theta$.

**Verify that each of the following is an
identity.**

5. $\cos^2 x + \tan^2 x \cos^2 x = 1$

6. $\dfrac{1 - \cos\ \theta}{1 + \cos\ \theta} = (\csc\ \theta - \cot\ \theta)^2$

7. $\dfrac{\sec\ \theta + 1}{\tan\ \theta} = \dfrac{\tan\ \theta}{\sec\ \theta - 1}$

8. $\dfrac{\sin^4 x - \cos^4 x}{\sin^2 x} = 1 - \cot^2 x$

■ use the sum and difference identities for sine, cosine, and tangent functions
(Lesson 7-3)

Find $\sin 105°$.

$$\sin 105° = \sin (60 + 45)°$$
$$= \sin 60° \cos 45° + \cos 60° \sin 45°$$
$$= \frac{\sqrt{3}}{2} \cdot \frac{\sqrt{2}}{2} + \frac{1}{2} \cdot \frac{\sqrt{2}}{2}$$
$$= \frac{\sqrt{6} + \sqrt{2}}{4}$$

Use the sum and difference identities to find the exact value of each function.

9. $\cos 240°$

10. $\cos 15°$

11. $\sin (-255°)$

12. $\tan 165°$

Verify that each of the following is an identity.

13. $\cos (90° - \theta) = \sin \theta$

14. $\cos (60° + \theta) + \cos (60° - \theta) = \cos \theta$

■ use the double- and half-angle identities for the sine, cosine, and tangent functions
(Lesson 7-4)

If $\sin x = \frac{3}{4}$ and x is in the first quadrant, find $\cos 2x$.

$$\cos 2x = 1 - 2 \sin^2 x$$
$$= 1 - 2 \left(\frac{3}{4}\right)^2$$
$$= -\frac{1}{8}$$

If $\cos \theta = \frac{3}{5}$ and θ is in the first quadrant, find each of the following.

15. $\sin 2\theta$

16. $\cos 2\theta$

17. $\sin \dfrac{\theta}{2}$

18. $\cos \dfrac{\theta}{2}$

19. $\tan 2\theta$

20. $\tan \dfrac{\theta}{2}$

■ solve trigonometric equations
(Lesson 7-5)

Solve $2 \cos^2 x - 1 = 0$ for all values of x.

$$2 \cos^2 x - 1 = 0$$
$$\cos^2 x = \frac{1}{2}$$
$$\cos x = \pm \frac{\sqrt{2}}{2}$$
$$x = 45° + 90k°$$

Solve each equation for all values of x.

21. $\tan x + 1 = \sec x$

22. $\sin^2 x + \cos 2x - \cos x = 0$

23. $\sin x \tan x - \tan x = 0$

24. $\cos 2x \sin x = 1$

■ write linear equations in normal form
(Lesson 7-6)

Write $3x + 2y - 6 = 0$ in normal form.

Since C is negative, use the positive value of $\sqrt{A^2 + B^2} = \sqrt{3^2 + 2^2} = \sqrt{13}$.

The normal form is $\dfrac{3x}{\sqrt{13}} + \dfrac{2y}{\sqrt{13}} - \dfrac{6}{\sqrt{13}} = 0$.

Write each equation in normal form. Then find p, the measure of the normal, and ϕ, the angle the normal makes with the positive x-axis.

25. $7x + 3y - 8 = 0$

26. $6x = 4y - 5$

27. $9x = -5y + 3$

28. $x - 7y = -5$

■ find the distance from a point to a line
(Lesson 7-7)

Find the distance between $(-1, 3)$ and the graph of $-3x + 4y = -5$.

$$d = \frac{Ax_1 + By_1 + C}{\pm\sqrt{A^2 + B^2}}$$

$$= \frac{-3x_1 + 4y_1 + 5}{-\sqrt{(-3)^2 + 4^2}}$$

$$= \frac{-3(-1) + 4(3) + 5}{-\sqrt{25}}$$

$$= \frac{20}{-5} \text{ or } -4 \text{ units}$$

Find the distance from the point with the given coordinates to the line with the given equation. Round answers to the nearest tenth.

29. $(5, 6); 2x - 3y + 2 = 0$

30. $(-3, -4); 2y = -3x + 6$

31. $(-2, 4); 4y = 3x - 1$

32. $(21, 20); y = \frac{1}{3}x + 6$

■ write the equations of lines that bisect angles formed by intersecting lines
(Lesson 7-7)

Find an equation of the line that bisects the acute angle formed by the lines $-x + 3y - 2 = 0$ and $y = \frac{3}{5}x + 3$.

$$\frac{-x + 3y - 2}{\sqrt{10}} = -\left(\frac{-3x + 5y - 15}{\sqrt{34}}\right)$$

$$-\sqrt{34}x + 3\sqrt{34}y - 2\sqrt{34} = 3\sqrt{10}x - 5\sqrt{10}y + 15\sqrt{10}$$

$$(-\sqrt{34} - 3\sqrt{10})x + (3\sqrt{34} + 5\sqrt{10})y - 2\sqrt{34} - 15\sqrt{10} = 0$$

Find an equation of the line that bisects the acute angle formed by the graphs of each pair of equations.

33. $3x + y - 2 = 0, x + 2y - 3 = 0$

34. $y = \frac{1}{3}x - 6, y = \frac{3}{4}x + 2$

APPLICATIONS AND PROBLEM SOLVING

35. Physics   While studying two different physics books, Jeremy notices that two different formulas are used to find the height of a projectile. One formula is $h = \frac{v_0^2 \sin^2 \theta}{2g}$; the other is $h = \frac{v_0^2 \tan^2 \theta}{2g \sec^2 \theta}$. Are these two formulas equivalent or is there an error in one of the books? Show your work.   **(Lesson 7-2)**

36. Surveying   Talitia Jones is surveying a piece of property. She measures the angle between one side of a rectangular lot and the line from her position to the opposite corner of the lot as 30°. She then measures the angle between that line and the line to a telephone pole on the edge of the property as 45°. If Ms. Jones stands 100 yards from the opposite corner of the property, how far is she from the telephone pole?   **(Lesson 7-3)**

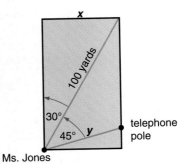

CHAPTER 7   TEST

Solve for values of θ between $0°$ and $90°$.

1. If $\sin \theta = \dfrac{1}{2}$, find $\cos \theta$.

2. If $\csc \theta = \dfrac{5}{3}$, find $\cos \theta$.

3. If $\sec \theta = 3$, find $\tan \theta$.

4. If $\sin \theta = \dfrac{4}{5}$, find $\sec \theta$.

Verify that each of the following is an identity.

5. $\tan \theta(\cot \theta + \tan \theta) = \sec^2 \theta$

6. $\sin^2 A \cot^2 A = (1 - \sin A)(1 + \sin A)$

7. $\dfrac{\sec x}{\sin x} - \dfrac{\sin x}{\cos x} = \cot x$

8. $\dfrac{\cos x}{1 + \sin x} + \dfrac{\cos x}{1 - \sin x} = 2 \sec x$

Use the sum and difference identities to find the exact value of each function.

9. $\sin 255°$

10. $\cos 165°$

11. $\sin(-195°)$

If $\cos x = \dfrac{3}{4}$ and x is in the fourth quadrant, find each value.

12. $\sin 2x$

13. $\cos \dfrac{x}{2}$

14. $\tan 2x$

Solve each equation for $0° \le x \le 180°$.

15. $\sin x - \cos x = 0$

16. $2 \cos^2 x + 3 \sin x - 3 = 0$

17. $\tan^2 x - \sqrt{3} \tan x = 0$

18. $\tan 2x \cot x - 3 = 0$

Write each equation in normal form. Then find p, the measure of its normal, and ϕ the angle the normal makes with the positive x-axis.

19. $-x + y - 3 = 0$

20. $-3x = 6y - 7$

21. $-10x + 5 = -5y$

Find the distance from the point with the given coordinates to the line with the given equation. Round answers to the nearest tenth.

22. $(-5, 8); 2x + y - 6 = 0$

23. $(-6, 8); -3x - 4y = 2$

24. Find the equation of the line that bisects the acute angle formed by the graphs of $7 = 5x + 2y$ and $y = -\dfrac{3}{4}x + 1$.

25. Physics The range of a projected object is the distance that it travels from the point where it is released. In the absence of air resistance, a projectile released at an angle of elevation, θ, with an initial velocity of v_0 has a range of $R = \dfrac{v_0^2}{g} \sin 2\theta$, where g is the acceleration due to gravity. Find the range of a projectile with an initial velocity of 88 feet per second if $\sin \theta = \dfrac{3}{5}$ and $\cos \theta = \dfrac{4}{5}$. The acceleration due to gravity is 32 feet per second squared.

Bonus If $0° < \theta < 90°, x > y > 0$, and $\cot \theta = \dfrac{x^2 - y^2}{2xy}$, find the value of $\sec \theta$.

CHAPTER 8

VECTORS AND PARAMETRIC EQUATIONS

HISTORICAL SNAPSHOT

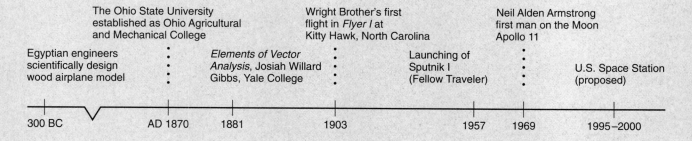

The Ohio State University
established as Ohio Agricultural
and Mechanical College

Wright Brother's first
flight in *Flyer I* at
Kitty Hawk, North Carolina

Neil Alden Armstrong
first man on the Moon
Apollo 11

Egyptian engineers
scientifically design
wood airplane model

*Elements of Vector
Analysis*, Josiah Willard
Gibbs, Yale College

Launching of
Sputnik I
(Fellow Traveler)

U.S. Space Station
(proposed)

| 300 BC | AD 1870 | 1881 | 1903 | 1957 | 1969 | 1995–2000 |

CHAPTER OBJECTIVES

In this chapter, you will:
- Study about vector notation, magnitude, and amplitude.
- Add, subtract, and multiply vectors.
- Find the inner product and the cross product of vectors.
- Solve and use parametric equations to solve problems.

CAREER GOAL: Aerospace Medicine

At the end of this school year, Regina Brooks will be the first African-American woman to graduate with a degree in aeronautical and astronautical engineering from The Ohio State University. After graduation, Regina is seriously considering going to medical school with a concentration in aerospace medicine. "I hope to be a professional who can bring on and advance space technology." Among other things, a career in aerospace medicine will allow Regina to engage in research on the effects of space on astronauts. "I would love to be part of the space station program or even part of an expedition to Mars," she says.

To prepare herself for college, Regina took courses in algebra, analytic geometry, trigonometry, and precalculus in high school. However, "If I had it to do all over again," says Regina, "I would have taken calculus in high school—it would have made my math courses in college much easier!"

She adds that she has found her mathematics courses to be invaluable in her studies. "When studying actual flight in aeronautics, the use of vectors becomes a very important part of air flow dynamics," Regina explains. "As air flows over each airfoil, the way the air is flowing and its velocity are always identified with vectors."

Regina's major in aeronautical and astronautical engineering has provided her the opportunity to spend summers working for NASA at the Goddard Flight Center in Greenbelt, Maryland, where she helped design the instrument carrier for Spartan 201. She says, "I feel that a part of me is in space aboard the Spartan satellite!"

For up-to-date information on aeronautical and astronautical engineering, visit:
www.glencoe.com/sec/math/amc/mathnet

SPOTLIGHT ON...

Regina Brooks
Age: 21
Class: Senior
School: The Ohio State University
Columbus, Ohio
Major: Aeronautical and Astronautical Engineering

"I hope to be a professional who can bring on and advance space technology."

FOR MORE INFORMATION

If you would like more information about aeronautical and astronautical engineering, contact:

American Institute of Aeronautics and Astronautics (AIAA)
370 L'Enfant Promenade SW
Washington, DC 20024

8-1 Geometric Vectors

Objectives

After studying this lesson, you should be able to:
- find equal, opposite, and parallel vectors, and
- add and subtract vectors geometrically.

Application

The release of a hot-air balloon begins the annual United States National Hot-Air Balloon Championships. The Championships have been held in Indianola, Iowa, each year since 1963. One year, a balloon rose at a rate of 2 m/s, and it was blown horizontally by a wind of 3 m/s. These velocities can be represented mathematically by **vectors**. A vector is a quantity, or directed distance, that has both magnitude and direction. A vector is represented geometrically by a directed line segment.

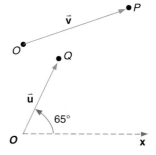

A directed line segment with an initial point at O and terminal point at P is shown at the right. The direction of the arrowhead indicates the direction of the vector. The vector at the right can be designated $\vec{v}$ or $\overrightarrow{OP}$. The length of the directed line segment indicates the vector's **magnitude**. The magnitude of $\vec{v}$ is denoted by $|\vec{v}|$.

If a vector has its initial point at the origin, it is in **standard position**. The **amplitude** of the vector is the directed angle between the positive x-axis and the vector. The amplitude of $\vec{u}$, shown above, is 65°.

Example 1

Determine the magnitude (in centimeters) and the amplitude of $\vec{a}$.

 Using a ruler and a protractor, sketch the vector in standard position and measure the magnitude and amplitude.

The magnitude is 1.7 cm, and the amplitude is 50°.

Two vectors are equal if and only if they have the same amplitude, or direction, and the same magnitude. Five vectors are shown at the right. $\vec{a}$ and $\vec{b}$ are equal since they have the same direction and $|\vec{a}| = |\vec{b}|$. $\vec{c}$ and $\vec{d}$ have the same direction but $|\vec{c}| \neq |\vec{d}|$, so $\vec{c} \neq \vec{d}$. Since $\vec{c}$ and $\vec{e}$ have different directions, they are not equal, even though $|\vec{c}| = |\vec{e}|$.

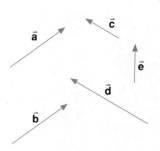

The sum of two or more vectors is called the **resultant** of the vectors. You can find the resultant using either the *parallelogram method* or the *triangle method*. To find the sum of two vectors using the parallelogram method, draw the vectors so that their initial points coincide. Then draw lines to make a complete parallelogram. The diagonal from the initial point to the opposite vertex of the parallelogram is the resultant. *Notice that the parallelogram method cannot be used to find the sum of a vector and itself.*

Example 2

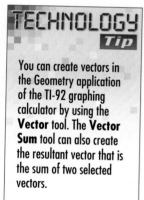

TECHNOLOGY Tip

You can create vectors in the Geometry application of the TI-92 graphing calculator by using the **Vector** tool. The **Vector Sum** tool can also create the resultant vector that is the sum of two selected vectors.

Find the sum of $\vec{p}$ and $\vec{q}$ by using the parallelogram method.

Copy $\vec{p}$. Then copy $\vec{q}$, placing the initial point at the initial point of $\vec{p}$.

Form a parallelogram that has $\vec{p}$ and $\vec{q}$ as two of its sides. Draw broken lines to represent the other two sides.

The resultant is the vector from the vertex of $\vec{p}$ and $\vec{q}$ to the opposite vertex of the parallelogram.

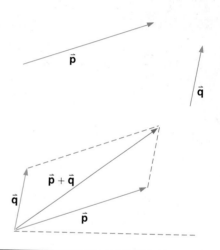

To use the triangle method to find the resultant of two vectors, draw the vectors so that the initial point of the second vector is the terminal point of the first vector. The vector from the initial point of the first vector to the terminal point of the second vector is the resultant.

Example 3

APPLICATION

Sports

At the U.S. National Hot-Air Balloon Championships, a hot-air balloon rose at a rate of 2 m/s, and was blown horizontally by a wind of 3 m/s. Draw a vector representing the resulting velocity of the balloon using the triangle method.

Suppose 1 cm represents 1 m/s. Draw two vectors, $\vec{r}$ and $\vec{w}$, to represent the upward velocity of the balloon and the velocity due to the wind, respectively.

(continued on the next page)

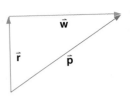

Copy $\vec{r}$. Then copy $\vec{w}$, placing the initial point of $\vec{w}$ on the terminal point of $\vec{r}$.

The resultant velocity of the balloon is the vector $\vec{p}$ from the initial point of $\vec{r}$ to the terminal point of $\vec{w}$.

The triangle method is also called the *tip-to-tail method* of adding vectors. You can use this method to add several vectors. Draw the vectors one after another, placing the initial point of each successive vector at the terminal point of the previous one. Then draw the resultant from the initial point of the first vector to the terminal point of the last one.

Two vectors are **opposites** if they have the same magnitude and opposite directions. The opposite of $\vec{b}$ is denoted by $-\vec{b}$. You can use opposite vectors to subtract vectors. To find $\vec{a} - \vec{b}$, find $\vec{a} + (-\vec{b})$.

A scalar possesses only magnitude. Real numbers are scalars.

The product of a scalar k and a vector $\vec{a}$ is a vector with the same direction as $\vec{a}$, if $k > 0$, and a magnitude of $k|\vec{a}|$. If $k < 0$, the vector has the opposite direction. In the figure at the right, $\vec{d} = 3\vec{c}$.

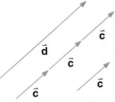

Example 4

Find $2\vec{r} - \vec{s}$.

$2\vec{r} - \vec{s} = 2\vec{r} + (-\vec{s})$.

Two vectors are **parallel** if and only if they have the same or opposite directions.

Equal vectors: $\vec{a}$ and $\vec{g}$

Opposite vectors: $\vec{a}$ and $\vec{d}$,
$\vec{c}$ and $\vec{f}$,
$\vec{d}$ and $\vec{g}$

Parallel vectors: $\vec{a}$ and $\vec{d}$, $\vec{a}$ and $\vec{e}$,
$\vec{a}$ and $\vec{g}$, $\vec{b}$ and $\vec{c}$,
$\vec{b}$ and $\vec{f}$, $\vec{c}$ and $\vec{f}$,
$\vec{d}$ and $\vec{e}$, $\vec{d}$ and $\vec{g}$,
$\vec{e}$ and $\vec{g}$

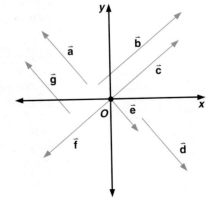

Two or more vectors whose sum is a given vector are called **components** of the given vector. Components can have any direction. Often it is useful to express a vector in terms of two perpendicular components. In the figure at the right, $\vec{y}$ is the vertical component of $\vec{a}$, and $\vec{x}$ is the horizontal component of $\vec{a}$.

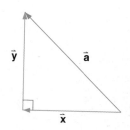

Example 5

$\vec{v}$ **has a magnitude of 2.6 cm and an amplitude of 43°. Find the magnitude of its vertical and horizontal components.**

Draw $\vec{v}$. Then draw a horizontal vector through the initial point and a vertical vector through the terminal point. The vectors will form a right triangle, so we can use the sine and cosine ratios to find the magnitude of the components.

$$\sin 43° = \frac{y}{2.6}$$
$$y \approx 1.8$$

$$\cos 43° = \frac{x}{2.6}$$
$$x \approx 1.9$$

The magnitude of the vertical component is approximately 1.8 cm, and the magnitude of the horizontal component is approximately 1.9 cm.

Vectors are used in physics to represent motion or forces acting upon objects.

Example 6

APPLICATION

Soccer

Two players kick a soccer ball at exactly the same moment. One player's foot exerts a force of 66 N (newtons) north. The other player's foot exerts a force of 88 N east.

a. **What is the magnitude of the resulting force upon the ball?**

b. **What is the direction of the resulting force upon the ball?**

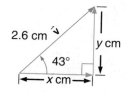

→ *FYI...*

A newton (N) is a unit of force used in physics. A force of one newton will accelerate a one-kilogram mass at a rate of one meter per second squared.

Let $\vec{a}$ represent the force from player 1.
Let $\vec{b}$ represent the force from player 2. Draw the resultant, $\vec{c}$. This represents the force upon the ball. Let B represent the measure of the angle that the resultant makes with $\vec{a}$.

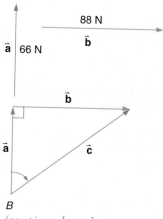

(continued on the next page)

a. We can use the Pythagorean theorem to find the magnitude of the resultant.

$$|\vec{c}| = \sqrt{66^2 + 88^2} \qquad \textit{Pythagorean theorem}$$
$$= \sqrt{4356 + 7744}$$
$$= \sqrt{12,100} \text{ or } 110$$

The magnitude of $\vec{c}$ is 110 N.

b. The direction of the resultant can be found by using the tangent ratio.

$$\tan B = \frac{\vec{b}}{\vec{a}}$$

$$\tan B = \frac{88}{66}$$

$$B \approx 53°$$

The resultant makes an angle of 53° with $\vec{a}$. The direction of the resulting force is 53° east of north.

EXPLORATION: Graphing Calculator

Determine at what initial velocity a ball must be thrown from the ground at an angle of 30° from the horizontal in order to travel a horizontal distance of 235 feet.

1. The calculator should be in PARAMETRIC mode.

2. Use the range values below to set up a viewing window.

   TMIN = 0   XMIN = 0   YMIN = −5
   TMAX = 4   XMAX = 250   YMAX = 35
   TSTEP = 0.05   XSCL = 30   YSCL = 5

3. Use the equations $X_{1T} = ? T \cos 30$ and $Y_{1T} = ? T \sin 30 - 16T^2$. Guess replacement values for ? to solve the problem. Check by using the TRACE function.

CHECKING FOR UNDERSTANDING

Communicating Mathematics

Read and study the lesson to answer each question.

1. **Compare** a line segment and a vector.

2. **Draw** diagrams illustrating the two methods of adding vectors.

3. Is $\overrightarrow{BA}$ the same as $\overrightarrow{AB}$? Explain.

4. **Define** the components of a vector.

Use a metric ruler and a protractor to draw a vector with the given magnitude and amplitude.

5. 1 cm, 50°   **6.** 4 cm, 180°   **7.** 3 cm, 120°

8. 2.5 cm, 270°   **9.** 18 mm, 45°   **10.** 35 mm, 65°

Copy each pair of vectors and draw the resultant vector.

11.   **12.**   **13.**

14. A vector has a magnitude of 2.2 cm and an amplitude of 30°. Find the magnitude of its vertical and horizontal components.

15. A hiker walks 13 km due west, then turns and walks 7 km due north. How far is the hiker from his starting point?

EXERCISES

Use a metric ruler and a protractor to find each sum or difference. Then, find the magnitude and amplitude of each resultant.

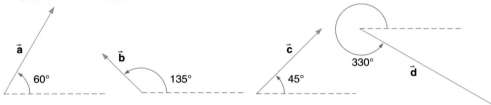

16. $\vec{a} + \vec{c}$ **17.** $\vec{a} + \vec{b}$ **18.** $\vec{d} + \vec{b}$

19. $\vec{a} + \vec{d}$ **20.** $2\vec{b}$ **21.** $3\vec{a}$

22. $3\vec{d} + 2\vec{b}$ **23.** $\vec{c} - \vec{a}$ **24.** $\vec{a} - \vec{d}$

25. $3\vec{c} - 2\vec{a}$ **26.** $\vec{a} + \vec{c} + \vec{d}$

27. $\vec{b} + \vec{d} - \vec{c}$ **28.** $3\vec{a} + \vec{c} - \vec{d}$

Find the magnitude of the vertical and horizontal components of each vector shown for Exercises 16—28.

29. $\vec{a}$ **30.** $\vec{b}$ **31.** $\vec{c}$ **32.** $\vec{d}$

33. The magnitude of $\vec{x}$ is 7.3 m, and the magnitude of $\vec{y}$ is 8.8 m. If $\vec{x}$ and $\vec{y}$ are perpendicular, what is the magnitude of their sum?

34. Is addition of vectors commutative? Justify your answer. *Hint: Find the sum of two vectors, $\vec{r} + \vec{s}$ and $\vec{s} + \vec{r}$, using the triangle method.*

35. In the parallelogram drawn for the parallelogram method, what does the diagonal between the two terminal points of the vectors represent? Explain your answer.

Critical
Thinking

36. Could the sum of two vectors ever be zero? Could the sum of three vectors be zero? Justify your answer and draw a diagram if possible.

Applications
and Problem
Solving

37. Aviation   An airplane is flying due east at a velocity of 100 m/s. The wind is blowing out of the north at 5 m/s.
   a. Draw a labeled diagram of the situation.
   b. What is the airplane's resulting velocity?

38. Physics   Nina is pushing a lawn mower with a force of 95 N along the handle of the mower.
   a. The handle makes a 60° angle with the horizontal. What are the magnitudes of the horizontal and vertical components of the force?
   b. If Nina lowers the handle so that it makes a 30° angle with the horizontal, what are the magnitudes of the horizontal and vertical components of the force?

39. Physics   Shirley, Scott, and Dan are pushing a footlocker across the floor. Shirley is pushing with a force of 185 N at 0°, Scott pushes with a force of 165 N at 30°, and Dan pushes with a force of 195 N at 300°.
   a. Draw a labeled diagram of the situation.
   b. What are the magnitude and direction of the resulting force on the footlocker?

Mixed
Review

40. Write the standard form of the equation of the line that is parallel to the graph of $y = x - 8$ and passes through $(-3, 1)$.   **(Lesson 1-6)**

41. Find the equation of the line tangent to the graph of $y = 3x^2$ at $(2, 12)$.
   (Lesson 3-6)

42. Solve the equation $2x^3 + 5x^2 - 12x = 0$.   **(Lesson 4-1)**

43. Gardening   A sprinkler is set to rotate 65° and spray a distance of 6 feet. What is the area of the ground being watered?   **(Lesson 5-2)**

44. Write an equation with phase shift 0 to represent simple harmonic motion in which the initial position is 0, the amplitude is 3, and the period is 2.   **(Lesson 6-7)**

45. College Entrance Exam   Compare quantities A and B below.
   Write A if quantity A is greater.
   Write B if quantity B is greater.
   Write C if the two quantities are equal.
   Write D if there is not enough information to determine the relationship.
   $z = 6y$, where y is a positive integer, and
   $z = 2x$, where x is a positive integer
   (A) z **(B)** 10

8-2 Algebraic Vectors

Objectives After studying this lesson, you should be able to:
- find ordered pairs that represent vectors, and
- add, subtract, multiply, and find the magnitude of vectors algebraically.

Application

The Lynch family is building a log cabin for weekend getaways. Mr. Lynch and his son James have erected a scaffold to stand on while they build the walls of the cabin. They are able to stand on the scaffold and pull up the logs to erect the walls of the cabin. A rope has been attached to a large log. If James pulls with a force of 400 N at an angle of 65° with the horizontal and Mr. Lynch pulls with a force of 600 N at an angle of 110° with the horizontal, what is the net force on the log? *This problem will be solved in Example 3.*

We could draw vectors to represent the forces on the log and then measure to find the magnitude and amplitude of the resultant. However, drawings can be inaccurate and drawings for more complicated systems could be quite confusing. In cases where we want a more precise answer or where the system is more complicated, we can add the vectors algebraically.

Vectors can be represented algebraically using ordered pairs of real numbers. For example, the ordered pair $(1, 2)$ represents the vector from the origin to the point at $(1, 2)$. You can think of this vector as the resultant of a horizontal vector with a magnitude of 1 unit and a vertical vector with a magnitude of 2 units.

Since vectors with the same magnitude and amplitude are equal, many vectors can be represented by an ordered pair. Each vector on the graph at the right can be represented by the ordered pair $(1, 2)$. The initial point of a vector can be any point in the plane.

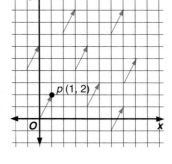

Assume that P_1 and P_2 are any two distinct points in the coordinate plane. Drawing the horizontal and vertical components of $\overrightarrow{P_1P_2}$ yields a right triangle. So the magnitude of the vector can be found by using the Pythagorean theorem. *Assume that $\overrightarrow{P_1P_2}$ is neither horizontal nor vertical.*

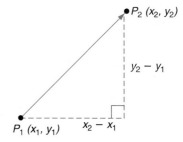

<table>
<tr><td>

**Representation
of a Vector as
an Ordered Pair**

</td><td>

Suppose $P_1(x_1, y_1)$ is the initial point of a vector and $P_2(x_2, y_2)$ is the terminal point. The ordered pair that represents $\overrightarrow{P_1P_2}$ is $(x_2 - x_1, y_2 - y_1)$. Its magnitude is given by $\overrightarrow{P_1P_2} = \sqrt{(x_2 - x_1)^2 + (y_2 - y_1)^2}$.

</td></tr>
</table>

The order of the coordinates representing a vector is important.
P_2P_1 is represented by the ordered pair $(x_1 - x_2, y_1 - y_2)$.

Example 1

Find the ordered pair that represents the vector from $A(5, 4)$ to $B(0, -3)$. Then find the magnitude of $\overrightarrow{AB}$.

$\overrightarrow{AB} = (0 - 5, -3 - 4)$
$\qquad = (-5, -7)$

$|\overrightarrow{AB}| = \sqrt{(0 - 5)^2 + (-3 - 4)^2}$
$\qquad = \sqrt{5^2 + (-7)^2}$
$\qquad = \sqrt{74}$

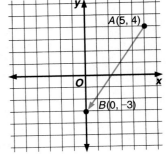

We can add or subtract vectors, and multiply vectors by a scalar when they are represented as ordered pairs. The rules for these operations on vectors are similar to those for matrices. In fact, vectors in a plane can be represented as row matrices of dimension 1×2.

Vector Operations

The following operations are defined for $\vec{a} = (a_1, a_2)$, $\vec{b} = (b_1, b_2)$, and any real number k.

Addition: $\vec{a} + \vec{b} = (a_1, a_2) + (b_1, b_2) = (a_1 + b_1, a_2 + b_2)$

Subtraction: $\vec{a} - \vec{b} = (a_1, a_2) - (b_1, b_2) = (a_1 - b_1, a_2 - b_2)$

Scalar multiplication: $k\vec{a} = k(a_1, a_2) = (ka_1, ka_2)$

Example 2

If $\vec{q} = (3, 9)$ and $\vec{r} = (-1, 6)$, find each of the following.

a. $\vec{q} + \vec{r}$

b. $\vec{q} - \vec{r}$

a. $\vec{q} + \vec{r} = (3, 9) + (-1, 6)$
$\qquad = (2, 15)$

b. $\vec{q} - \vec{r} = (3, 9) - (-1, 6)$
$\qquad = (4, 3)$

c. $5\vec{q}$

c. $5\vec{q} = 5(3, 9)$
$\qquad = (15, 45)$

If we write the vectors described in the application at the beginning of the lesson as ordered pairs, we can find the resultant vector using vector addition.

Example 3

Refer to the application at the beginning of the lesson. If James pulls with a force of 400 N at an angle of 65° with the horizontal and Mr. Lynch pulls with a force of 600 N at an angle of 110° with the horizontal, what is the net force on the log?

Draw a diagram of the situation. Let $\vec{\mathbf{F}}_1$ represent the force James exerts, and let $\vec{\mathbf{F}}_2$ represent the force that Mr. Lynch exerts.

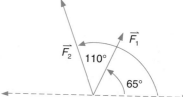

Write each vector as an ordered pair by finding its horizontal and vertical components. Let $\vec{\mathbf{F}}_{1x}$ and $\vec{\mathbf{F}}_{1y}$ represent the x- and y-components of $\vec{\mathbf{F}}_1$, and let $\vec{\mathbf{F}}_{2x}$ and $\vec{\mathbf{F}}_{2y}$ represent the x- and y-components of $\vec{\mathbf{F}}_2$.

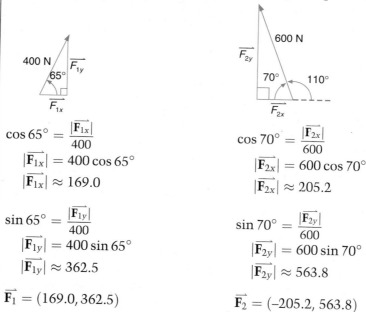

$$\cos 65° = \frac{|\vec{\mathbf{F}}_{1x}|}{400}$$
$$|\vec{\mathbf{F}}_{1x}| = 400 \cos 65°$$
$$|\vec{\mathbf{F}}_{1x}| \approx 169.0$$

$$\sin 65° = \frac{|\vec{\mathbf{F}}_{1y}|}{400}$$
$$|\vec{\mathbf{F}}_{1y}| = 400 \sin 65°$$
$$|\vec{\mathbf{F}}_{1y}| \approx 362.5$$

$$\vec{\mathbf{F}}_1 = (169.0, 362.5)$$

$$\cos 70° = \frac{|\vec{\mathbf{F}}_{2x}|}{600}$$
$$|\vec{\mathbf{F}}_{2x}| = 600 \cos 70°$$
$$|\vec{\mathbf{F}}_{2x}| \approx 205.2$$

$$\sin 70° = \frac{|\vec{\mathbf{F}}_{2y}|}{600}$$
$$|\vec{\mathbf{F}}_{2y}| = 600 \sin 70°$$
$$|\vec{\mathbf{F}}_{2y}| \approx 563.8$$

$$\vec{\mathbf{F}}_2 = (-205.2, 563.8)$$

Find the sum of the vectors.
$$\vec{\mathbf{F}}_1 + \vec{\mathbf{F}}_2 = (169.0, 362.5) + (-205.2, 563.8)$$
$$= (-36.2, 926.3)$$

The net force on the log is the magnitude of the sum.

$$|\vec{\mathbf{F}}_1 + \vec{\mathbf{F}}_2| = \sqrt{(-36.2)^2 + (926.3)^2}$$
$$= \sqrt{859,342.13}$$
$$\approx 927.007$$

The net force on the log is about 927 N.

A vector that is parallel to the x- or y-axis and has a magnitude of one unit is called a **unit vector**. A unit vector in the direction of the positive x-axis is represented by $\vec{i}$, and a unit vector in the direction of the positive y-axis is represented by $\vec{j}$. So $\vec{i} = (1, 0)$ and $\vec{j} = (0, 1)$.

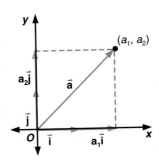

Any vector $\vec{a} = (a_1, a_2)$ can be expressed as $a_1 \vec{i} + a_2 \vec{j}$.

$$
\begin{aligned}
a_1 \vec{i} + a_2 \vec{j} &= a_1(1, 0) + a_2(0, 1) \qquad \text{\textit{$\vec{i} = (1, 0)$ and $\vec{j} = (0, 1)$}}\\
&= (a_1, 0) + (0, a_2) \qquad \text{\textit{Scalar product}}\\
&= (a_1 + 0, 0 + a_2) \qquad \text{\textit{Addition of vectors}}\\
&= (a_1, a_2)
\end{aligned}
$$

Since $(a_1, a_2) = \vec{a}$, $a_1 \vec{i} + a_2 \vec{j} = \vec{a}$. Therefore, any vector that is represented by an ordered pair can also be written as the sum of unit vectors.

Example 4

Write $\vec{CD}$ as the sum of unit vectors for points $C(6, 3)$ and $D(-4, 8)$.

First, write $\vec{CD}$ as an ordered pair.

$$
\begin{aligned}
\vec{CD} &= (-4 - 6, 8 - 3)\\
&= (-10, 5)
\end{aligned}
$$

Then, write $\vec{CD}$ as the sum of unit vectors.

$$\vec{CD} = -10\vec{i} + 5\vec{j}$$

CHECKING FOR UNDERSTANDING

Communicating Mathematics

Read and study the lesson to answer each question.

1. Could the magnitude of a resultant vector ever be less than the magnitude of one of its components? Explain.

2. **Compare and contrast** the vectors $(2, 4)$ and $(4, 2)$.

3. If two vectors have the same magnitude, are the ordered pairs representing them necessarily identical? Explain.

4. If two vectors have the same amplitude, are the ordered pairs representing them necessarily identical? Explain.

Guided Practice

Find the magnitude of each vector and write each vector as the sum of unit vectors.

5. $(4, 3)$ 6. $(6, 7)$ 7. $(-2, -3)$ 8. $(-5, 15)$

Find the ordered pair that represents the vector from *A* to *B*. Then find the magnitude of $\overrightarrow{AB}$.

9. $A(4, 2), B(2, 8)$

10. $A(0, 4), B(3, 1)$

11. $A(-4, 0), B(1, 9)$

12. $A(-5, 7), B(-1, 2)$

Find the sum or difference of the given vectors algebraically.

13. $(4, 5) + (2, 1)$

14. $(-1, 2) + (3, 5)$

15. $(-1, 6) + (-8, -5)$

16. $(2, 4) + (-2, -3)$

17. $\overrightarrow{i} + \overrightarrow{j}$

18. $2\overrightarrow{i} - \overrightarrow{j}$

19. $5\overrightarrow{i} + 7\overrightarrow{j}$

20. $4\overrightarrow{i} - \overrightarrow{j}$

EXERCISES

Practice

Find the ordered pair that represents the vector from *A* to *B*. Then find the magnitude of $\overrightarrow{AB}$.

21. $A(7, 7), B(-2, -2)$

22. $A(-2, 5), B(1, 3)$

23. $A(0, 5), B(-5, 0)$

24. $A(5, 0), B(7, 6)$

25. $A(4, -5), B(5, -4)$

26. $A(-9, 2), B(-4, -3)$

27. $A(7, 6), B(8, 6)$

28. $A(12, -4), B(19, 1)$

Find an ordered pair to represent $\overrightarrow{u}$ in each equation if $\overrightarrow{v} = (4, -3)$ and $\overrightarrow{w} = (-6, 2)$.

29. $\overrightarrow{u} = \overrightarrow{v} + \overrightarrow{w}$

30. $\overrightarrow{u} = \overrightarrow{v} - \overrightarrow{w}$

31. $\overrightarrow{u} = 3\overrightarrow{v}$

32. $\overrightarrow{u} = 4\overrightarrow{w}$

33. $\overrightarrow{u} = \overrightarrow{w} - 2\overrightarrow{v}$

34. $\overrightarrow{u} = \overrightarrow{v} - 3\overrightarrow{w}$

35. $\overrightarrow{u} = 2\overrightarrow{v} + 3\overrightarrow{w}$

36. $\overrightarrow{u} = 4\overrightarrow{w} - 3\overrightarrow{v}$

37. $\overrightarrow{u} = 6\overrightarrow{w} - 2\overrightarrow{v}$

38. Prove that addition of vectors is commutative.

39. Find the vector that is the additive identity for vectors. By definition, this is the vector that does not alter any vector when added to it.

Critical Thinking

40. Suppose the points *A*, *B*, *C*, and *D* are noncollinear, and $\overrightarrow{AB} + \overrightarrow{CD} = 0$.
  a. What is the relationship between $\overrightarrow{AB}$ and $\overrightarrow{CD}$?
  b. What is true of the quadrilateral with vertices *A*, *B*, *C*, and *D*?

Applications and Problem Solving

41. Sports In the 1992 Winter Olympic trials, Hershel Walker pushed the *USA I* bobsled with a force of 110 N at an angle of 42° with the horizontal. Find the horizontal and vertical components of the force.

42. Shipping Two tugboats are towing a ship into Boston Harbor. Although the tugboats are traveling at the same speed, one tugboat is pulling at an angle of 30°, and the second boat is pulling at an angle of 330°. If each tugboat exerts a force of 6 tons, what is the resultant force on the ship?

43. **Recreation** Traci is pulling her sled up the hill with a force of 58 N at an angle of 50° with the horizontal. If Traci were to pull with the same force and lower the angle, what effect would this have on the horizontal and vertical components?

Mixed Review

44. Determine if the graph of $y^2 = 121 - x^2$ is symmetric with respect to the x-axis, the y-axis, the line $y = x$, the line $y = -x$, or the origin.   **(Lesson 3-1)**

45. Use the quadratic formula to solve $5x^2 - 8x + 12 = 0$.   **(Lesson 4-2)**

46. Solve $\dfrac{x+3}{x+2} = 2 - \dfrac{3}{x^2 + 5x + 6}$. Check the solution.   **(Lesson 4-6)**

47. Solve $2\tan x - 4 = 0$ for principal values of x.   **(Lesson 7-5)**

48. **Navigation** A boat heads due west across a lake at 8 m/s. If a current of 5 m/s moves due south, what is the boat's resultant velocity and direction?
(Lesson 8-1)

49. **College Entrance Exam** Choose the best answer.
If $a(b + 1) = C$, then $b =$
  (A) $C - 1$ **(B)** $\dfrac{C-1}{a}$ **(C)** $C - a + 1$ **(D)** $\dfrac{C}{a} - 1$ **(E)** $C - a$

DECISION MAKING

Choosing a Used Car

Because costs are so high, a new car is out of reach for many people. But car values depreciate rapidly after two or three years, making an almost-new car a very economical choice for a prospective buyer. What factors do you need to take into consideration when buying a used car?

1. Describe what you would look for in a used car. Include such factors as the make of the car, its condition, mileage, and any features or accessories that you feel are important.

2. Outline a plan to make sure that you get the best value for your money. Talk with mechanics and car owners about what to look for in a used car. Contact your city or state consumer protection agency to find out your rights as a used-car buyer.

3. Research ways to finance the purchase of a used car. Can you come up with the cash to buy a car outright? Or, will you need to borrow money from your family or a bank?

4. **Project** Work with your group to complete one of the projects listed at the right, based on the information you gathered in Exercises 1–3.

PROJECTS

- *Conduct an interview.*
- *Write a book report.*
- *Write a proposal.*
- *Write an article for the school paper.*
- *Write a report.*
- *Make a display.*
- *Make a graph or chart.*
- *Plan an activity.*
- *Design a checklist.*

8-3 Vectors in Three-Dimensional Space

Objectives

After studying this lesson, you should be able to:
- add and subtract vectors in three-dimensional space, and
- find the magnitude of vectors in three-dimensional space.

Application

The Sears Tower in Chicago rises an imposing 443 meters above street level. It is the tallest building in the world. Especially in the "Windy City," tall buildings are vulnerable to the force of strong winds. Engineers have designed tuned mass dampers for buildings to help them keep from swaying too much, and possibly tumbling over, when the winds blow. Mass dampers are large blocks of cement or lead which are controlled by computerized hydraulic systems. When the winds blow, the computers move the weights in the direction opposite the force of the wind so that the resultant force reduces the motion of the building. The forces of the winds and the dampers can be represented by vectors in space.

FYI...

The frame of the Sears Tower is made up of vertical tubes that provide the rigidity necessary to limit swaying of the building due to the wind.

Vectors in three-dimensional space can be described by coordinates in a way similar to the way we describe vectors in a plane. Imagine three real number lines intersecting at the zero point of each so that each line is perpendicular to the plane determined by the other two. To show this arrangement on paper, a figure is used to convey the feeling of depth. The axes are named the x-axis, the y-axis, and the z-axis.

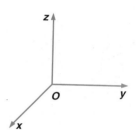

Each point in space corresponds to an ordered triple of real numbers. To locate a point P with coordinates (x_1, y_1, z_1), first find x_1 on the x-axis, y_1 on the y-axis, and z_1 on the z-axis. Then imagine a plane perpendicular to the x-axis at x_1 and planes perpendicular to the y- and z-axes at y_1 and z_1, respectively. The three planes will intersect at the point P.

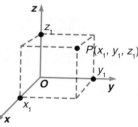

Example 1

Locate the point (3, 5, 4).

Locate 3 on the x-axis, 5 on the y-axis, and 4 on the z-axis.

Now draw broken lines for parallelograms to represent the three planes.

The planes intersect at $(3, 5, 4)$.

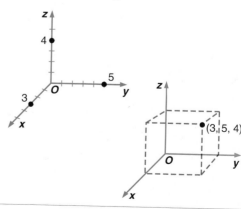

Ordered triples, like ordered pairs, can be used to represent vectors. The geometric interpretation is the same for a vector in space as it is for a vector in a plane. A directed line segment from the origin O to $P(x, y, z)$ is called vector $\overrightarrow{OP}$, corresponding to vector (x, y, z).

An extension of the formula for the distance between two points in a plane allows us to find the distance between two points in space. The distance from the origin to a point (x, y, z) is $\sqrt{x^2 + y^2 + z^2}$. So the magnitude of vector (x, y, z) is $\sqrt{x^2 + y^2 + z^2}$.

Representation of a Vector as an Ordered Triple	**Suppose $P_1(x_1, y_1, z_1)$ is the initial point of a vector in space and $P_2(x_2, y_2, z_2)$ is the terminal point. The ordered triple that represents $\overrightarrow{P_1 P_2}$ is $(x_2 - x_1, y_2 - y_1, z_2 - z_1)$. Its magnitude is given by** $$\lvert \overrightarrow{P_1 P_2} \rvert = \sqrt{(x_2 - x_1)^2 + (y_2 - y_1)^2 + (z_2 - z_1)^2}.$$

Example 2

Find the ordered triple that represents the vector from $A(3, 7, -1)$ to $B(10, -4, 0)$.

$$\overrightarrow{AB} = (10 - 3, -4 - 7, 0 - (-1))$$
$$= (7, -11, 1)$$

Example 3

Refer to the application at the beginning of the lesson. The force of the wind blowing against the Sears Tower at a certain moment can be expressed as the vector (132, 3454, 0), where each measure in the ordered triple represents the force in newtons. What is the magnitude of this force?

magnitude $= \sqrt{132^2 + 3454^2 + 0^2}$
$= \sqrt{11,947,540}$
≈ 3456.5

The magnitude of the force is about 3457 N.

You can perform operations on vectors represented by ordered triples in the same way that you perform operations on vectors represented by ordered pairs.

Example 4 | **Find an ordered triple that represents $3\vec{v} - \vec{u}$, if $\vec{v} = (4, -1, 7)$ and $\vec{u} = (9, 11, -3)$.**

$$3\vec{y} - \vec{u} = 3(4, -1, 7) - (9, 11, -3)$$
$$= (12, -3, 21) - (9, 11, -3)$$
$$= (3, -14, 24)$$

Three unit vectors are used as components of vectors in space. The unit vectors on the x-, y-, and z-axes are $\vec{i}$, $\vec{j}$, and $\vec{k}$, respectively, where $\vec{i} = (1, 0, 0)$, $\vec{j} = (0, 1, 0)$, and $\vec{k} = (0, 0, 1)$. The unit vectors are shown at the right with a vector $\vec{a} = (a_1, a_2, a_3)$. The component vectors of $\vec{a}$ along the three axes are $a_1\vec{i}$, $a_2\vec{j}$, and $a_3\vec{k}$. $\vec{a}$ can be written as the sum of unit vectors; that is, $\vec{a} = a_1\vec{i} + a_2\vec{j} + a_3\vec{k}$.

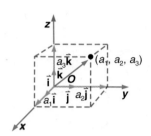

Example 5 | **Write $\overrightarrow{GH}$ as the sum of unit vectors for $G(10, -3, 15)$ and $H(4, 1, -11)$.**

First, express $\overrightarrow{GH}$ as an ordered triple. Then write as the sum of unit vectors $\vec{i}$, $\vec{j}$, and $\vec{k}$.

$$\overrightarrow{GH} = (4 - 10, 1 - (-3), -11 - 15)$$
$$= (-6, 4, -26)$$
$$= -6\vec{i} + 4\vec{j} - 26\vec{k}$$

CHECKING FOR UNDERSTANDING

Communicating Mathematics

Read and study the lesson to answer each question.

1. **Describe** the process you would use to locate $(3, -1, 2)$ in space. Then sketch the point on a coordinate system.

2. **Compare** the formula for the magnitude of a vector in space to the formula for the magnitude of a vector in a plane.

3. Suppose the unit vectors in a four-dimensional system are $\vec{i}$, $\vec{j}$, $\vec{k}$, and $\vec{\ell}$. Name the coordinates of the terminal points of $\vec{i}$, $\vec{j}$, $\vec{k}$, and $\vec{\ell}$.

Guided Practice

Locate points with the given coordinates. Then find the magnitude of a vector from the origin to each point.

4. $(3, 4, 9)$ 5. $(-2, 1, 3)$ 6. $(-1, 0, 4)$

7. $(7, 2, 4)$ 8. $(4, 1, -3)$ 9. $(-1, -1, 5)$

For each pair of points A and B, find an ordered triple that represents $\overrightarrow{AB}$. Then write $\overrightarrow{AB}$ as the sum of unit vectors.

10. $A(2, 5, 4), B(3, 1, 0)$

11. $A(-1, 3, 10), B(3, -5, -4)$

12. $A(-11, 4, -2), B(0, -2, 13)$

13. $A(-3, 0, 0), B(4, 3, -1)$

EXERCISES

Practice

For each pair of points A and B, find an ordered triple that represents $\overrightarrow{AB}$. Then find the magnitude of $\overrightarrow{AB}$.

14. $A(8, 1, 1), B(4, 0, 1)$

15. $A(3, 7, -1), B(5, 7, 2)$

16. $A(-2, 5, 8), B(3, 9, -3)$

17. $A(-2, 4, 7), B(-3, 5, 2)$

18. $A(-2, -4, 1), B(20, 5, 11)$

19. $A(-12, 3, 0), B(3, -21, 4)$

Write each vector as the sum of unit vectors.

20. $(9, 3, -1)$

21. $(3, 0, 1)$

22. $(0, 2, -2)$

23. $(-5, -8, 1)$

24. $(-15, 7, 0)$

25. $(8, -3, 11)$

Find an ordered triple to represent $\vec{u}$ in each equation if $\vec{v} = (1, -3, -8)$ and $\vec{w} = (3, 9, -1)$.

26. $\vec{u} = \vec{v} + \vec{w}$

27. $\vec{u} = \vec{w} - \vec{v}$

28. $\vec{u} = 3\vec{v} + \vec{w}$

29. $\vec{u} = \vec{v} - 2\vec{w}$

30. $\vec{u} = 4\vec{v} - 3\vec{w}$

31. $\vec{u} = 2\vec{w} - 5\vec{v}$

32. Show that $|\overrightarrow{P_1 P_2}| = |\overrightarrow{P_2 P_1}|$.

33. If $\vec{a} = (a_1, a_2, a_3)$, then $-\vec{a}$ is defined to be $(-a_1, -a_2, -a_3)$. Show that $|-\vec{a}| = |\vec{a}|$.

Critical Thinking

Find a unit vector that has the same direction as each given vector.

34. $\vec{a} = \vec{i} + \vec{j} - \vec{k}$

35. $\vec{b} = 4\vec{i} - 3\vec{j} - 12\vec{k}$

Applications and Problem Solving

36. Electrostatics Charles Augustin de Coulomb was a French scientist, inventor, and engineer in the late 1700s. He made contributions to the fields of friction, electricity, and magnetism. He also formulated Coulomb's Law, which states that if the electric force F exerted by a charge q_1 at the point Q_1 on a charge q located at point Q is given by the equation $F = \dfrac{q_1 q}{4\pi \varepsilon_0 r^2} \vec{u}$, where $\vec{u}$ is the unit vector in the direction of $\overrightarrow{Q_1 Q}$, r is the distance between the charges, and $\varepsilon_0 = 8.854 \times 10^{-12}$. If the charge of a proton located at $(0, 2 \times 10^{-11}, 0)$ is 1.6×10^{-19} coulombs, find the electric force exerted on an electron of charge -1.6×10^{-19} located at the origin.
Hint: In this case, $\vec{u} = \vec{j}$.

37. Physics If vectors working on an object are in equilibrium, then their resultant is zero.

a. Find the vector for the force of the damper that results in equilibrium with the force created by the wind in Example 3.

b. Two forces on an object are represented by $(3, -2, 4)$ and $(6, 2, 5)$. Find a third vector that will place the object in equilibrium.

Mixed Review

38. Manufacturing The Simply Sweats Corporation makes high quality sweatpants and sweatshirts. Each garment passes through the cutting and sewing departments of the factory. The cutting and sewing departments have 100 and 180 worker-hours available each week, respectively. The fabric supplier can provide 195 yards of fabric each week. The hours of work and yards of fabric required for each garment are shown in the table below. If the profit from a sweatshirt is $5.50 and the profit from a pair of sweatpants is $4.00, how many of each should the company make for maximum profit? **(Lesson 2-6)**

clothing	cutting	sewing	fabric
shirt	1 h	2.5 h	1.5 yd
pants	1.5 h	2 h	3 yd

39. Determine whether the functions $y = \dfrac{3x}{x+1}$ and $y = \dfrac{x+1}{3x}$ are inverses of one another. **(Lesson 3-3)**

40. Find the critical points of the graph of $f(x) = \dfrac{1}{2}x^4 + 4x^3 - 7x^2 + 1$. Then determine whether each point represents a maximum, a minimum, or a point of inflection. **(Lesson 3-7)**

41. Simplify the expression $1 - 2\sin^2 10°$ without using a calculator. **(Lesson 7-4)**

42. Write the standard form of the equation of the line that has a normal 3 units long and makes an angle of $60°$ with the positive x-axis. **(Lesson 7-6)**

43. Find an ordered pair that represents $\overrightarrow{AB}$ for $A(5, -6)$ and $B(6, -5)$. Then find the magnitude of $\overrightarrow{AB}$. **(Lesson 8-2).**

44. College Entrance Exam Compare quantities A and B below.
Write A if quantity A is greater.
Write B if quantity B is greater.
Write C if the quantities are equal.
Write D if there is not enough information to determine the relationship.
Given: $a > 0$
$\quad\quad\quad b > 0$
$\quad\quad\quad x > 1$

(A) $(x^a)^b$ **(B)** $x^a x^b$

8-4 Perpendicular Vectors

Objectives

After studying this lesson, you should be able to:
- find the inner and cross products of two vectors, and
- determine whether two vectors are perpendicular.

Application

Hold a pencil horizontally between two fingers of your left hand and use your right hand to press up on one end of the pencil. The pencil will rotate about the point where your left hand is holding it. In physics, the torque of a force is the measure of the effectiveness of a force in turning an object about a pivot point. We can use perpendicular vectors to find the torque of a force.

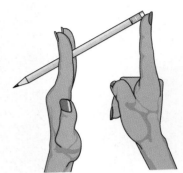

Let $\vec{a}$ and $\vec{b}$ be perpendicular vectors, and let $\overrightarrow{BA}$ be a vector between their terminal points as shown. Then the magnitudes of $\vec{a}$, $\vec{b}$, and $\overrightarrow{BA}$ must satisfy the Pythagorean theorem.

$$|\overrightarrow{BA}|^2 = |\vec{a}|^2 + |\vec{b}|^2$$

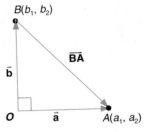

Use the definition of magnitude of a vector to evaluate $|\overrightarrow{BA}|^2$.

$$
\begin{aligned}
|\overrightarrow{BA}|^2 &= \left(\sqrt{(a_1 - b_1)^2 + (a_2 - b_2)^2}\right)^2 \\
&= (a_1 - b_1)^2 + (a_2 - b_2)^2 \\
&= a_1{}^2 - 2a_1b_1 + b_1{}^2 + a_2{}^2 - 2a_2b_2 + b_2{}^2 \\
&= (a_1{}^2 + a_2{}^2) + (b_1{}^2 + b_2{}^2) - 2(a_1b_1 + a_2b_2) \\
&= |\vec{a}|^2 + |\vec{b}|^2 - 2(a_1b_1 + a_2b_2)
\end{aligned}
$$

Compare this equation with the original one.

Therefore, $|\overrightarrow{BA}|^2 = |\vec{a}|^2 + |\vec{b}|^2$ if and only if $a_1b_1 + a_2b_2 = 0$.

The expression $a_1b_1 + a_2b_2$ is used often in the study of vectors. It is called the **inner product** of $\vec{a}$ and $\vec{b}$.

Inner Product of Vectors in a Plane

If $\vec{a}$ and $\vec{b}$ are two vectors, (a_1, a_2) and (b_1, b_2), the inner product of $\vec{a}$ and $\vec{b}$ is defined as $\vec{a} \cdot \vec{b} = a_1b_1 + a_2b_2$.

$\vec{a} \cdot \vec{b}$ is read "$\vec{a}$ dot $\vec{b}$" and is often called the dot product.

Two vectors are perpendicular if and only if their inner product is zero.

Example 1

Find the inner products of $\vec{a}$ and $\vec{b}$ and of $\vec{a}$ and $\vec{c}$ if $\vec{a} = (3, 12)$, $\vec{b} = (8, -2)$, and $\vec{c} = (3, -2)$. Is either pair of these vectors perpendicular?

$$\vec{a} \cdot \vec{b} = 3(8) + 12(-2)$$
$$= 24 - 24$$
$$= 0$$

$\vec{a}$ and $\vec{b}$ are perpendicular.

$$\vec{a} \cdot \vec{c} = 3(3) + 12(-2)$$
$$= 9 - 24$$
$$= -15$$

$\vec{a}$ and $\vec{c}$ are not perpendicular.

We can also find the inner product of two vectors in space. Two vectors in space are perpendicular if and only if their inner product is zero.

Inner Product of Vectors in Space

If $\vec{a} = (a_1, a_2, a_3)$ and $\vec{b} = (b_1, b_2, b_3)$, then
$\vec{a} \cdot \vec{b} = a_1 b_1 + a_2 b_2 + a_3 b_3$.

Example 2

Find the inner product of $\vec{v}$ and $\vec{w}$ if $\vec{v} = (-6, 2, 10)$ and $\vec{w} = (4, 1, 3)$. Are $\vec{v}$ and $\vec{w}$ perpendicular?

$$\vec{v} \cdot \vec{w} = (-6)(4) + 2(1) + 10(3)$$
$$= -24 + 2 + 30$$
$$= 8$$

$\vec{v}$ and $\vec{w}$ are not perpendicular.

Another important product involving vectors in space is the **cross product**. The cross product of two vectors is a vector. This vector does not lie in the plane of the given vectors, but is perpendicular to each of them. Therefore, it is perpendicular to the plane containing the two vectors.

Cross Product of Vectors in Space

If $\vec{a} = (a_1, a_2, a_3)$ and $\vec{b} = (b_1, b_2, b_3)$, then the cross product of $\vec{a}$ and $\vec{b}$ is defined as follows.

$$\vec{a} \times \vec{b} = \begin{vmatrix} a_2 & a_3 \\ b_2 & b_3 \end{vmatrix} \vec{i} - \begin{vmatrix} a_1 & a_3 \\ b_1 & b_3 \end{vmatrix} \vec{j} + \begin{vmatrix} a_1 & a_2 \\ b_1 & b_2 \end{vmatrix} \vec{k}$$

An easy way to remember the coefficients of $\vec{i}$, $\vec{j}$, and $\vec{k}$ is to set up a determinant as shown and expand by minors using the first row. *You may want to review expansion by minors in Lesson 2-3.*

$$\begin{vmatrix} \vec{i} & \vec{j} & \vec{k} \\ a_1 & a_2 & a_3 \\ b_1 & b_2 & b_3 \end{vmatrix}$$

Example 3

Find the cross product of $\vec{a}$ and $\vec{b}$ if $\vec{a} = (5, 2, 3)$ and $\vec{b} = (-2, 5, 0)$. Then verify that the resulting vector is perpendicular to $\vec{a}$ and $\vec{b}$.

$$\vec{a} \times \vec{b} = \begin{vmatrix} \vec{i} & \vec{j} & \vec{k} \\ 5 & 2 & 3 \\ -2 & 5 & 0 \end{vmatrix}$$

$$= \begin{vmatrix} 2 & 3 \\ 5 & 0 \end{vmatrix} \vec{i} - \begin{vmatrix} 5 & 3 \\ -2 & 0 \end{vmatrix} \vec{j} + \begin{vmatrix} 5 & 2 \\ -2 & 5 \end{vmatrix} \vec{k}$$

$$= -15\vec{i} - 6\vec{j} + 29\vec{k} \text{ or } (-15, -6, 29)$$

Find the inner products $(-15, -6, 29) \cdot (5, 2, 3)$ and $(-15, -6, 29) \cdot (-2, 5, 0)$.

$$-15(5) + (-6)(2) + 29(3) = 0 \qquad -15(-2) + (-6)(5) + 29(0) = 0$$

Since the inner products are zero, $(-15, -6, 29)$ is perpendicular to both $(5, 2, 3)$ and $(-2, 5, 0)$.

Example 4

APPLICATION

Mechanics

The moment, $\vec{M}$, of a force, $\vec{F}$, at a point Q about another point P is defined by the equation $\vec{M} = \vec{PQ} \times \vec{F}$. The magnitude of $\vec{M}$ represents the torque of the force in pounds per foot. Sumi places a wrench on a nut and is applying a downward force of 32 pounds to tighten the nut. If the center of the nut is at the origin, the force is applied at the point (0.65, 0, 0.3). Find the torque of the force.

First, find $\vec{PQ}$ and $\vec{F}$.

$\vec{PQ} = (0.65 - 0, 0 - 0, 0.3 - 0)$
or $(0.65, 0, 0.3)$

$\vec{F}$ is in a downward direction, so
$\vec{F} = -32\vec{k}$ or $(0, 0, -32)$.

32 lb | $\vec{F}$

Then, find $\vec{M}$.

$$\vec{M} = \vec{PQ} \times \vec{F}$$

$$= \begin{vmatrix} \vec{i} & \vec{j} & \vec{k} \\ 0.65 & 0 & 0.3 \\ 0 & 0 & -32 \end{vmatrix}$$

$$= \begin{vmatrix} 0 & 0.3 \\ 0 & -32 \end{vmatrix} \vec{i} - \begin{vmatrix} 0.65 & 0.3 \\ 0 & -32 \end{vmatrix} \vec{j} + \begin{vmatrix} 0.65 & 0 \\ 0 & 0 \end{vmatrix} \vec{k}$$

$$= 0\vec{i} + 20.8\vec{j} + 0\vec{k} \text{ or } (0, 20.8, 0)$$

The torque is the magnitude of $\vec{M}$.

$$|\vec{M}| = \sqrt{0^2 + 20.8^2 + 0^2} \text{ or } 20.8$$

The torque of the force is 20.8 lb/ft.

MATH JOURNAL

Write a sentence explaining how you can tell if two vectors are perpendicular.

CHECKING FOR UNDERSTANDING

Read and study the lesson to answer each question.

1. If the inner product of $\vec{w}$ and $\vec{v}$ is nonzero, what can you say about $\vec{w}$ and $\vec{v}$?

2. Could the inner product of a nonzero vector and itself ever be zero? Explain.

3. **Show** that $\vec{i} \times \vec{j} = \vec{k}$.

Guided
Practice

Find each inner product and state whether the vectors are perpendicular. Write *yes* or *no*.

4. $(3, 5) \cdot (4, -2)$

5. $(8, 4) \cdot (2, 4)$

6. $(5, -1) \cdot (2, 3)$

7. $(-6, 3) \cdot (1, 2)$

8. $(4, 2) \cdot (-3, 6)$

9. $(11, 2) \cdot (-3, 16)$

10. $(7, -2, 4) \cdot (3, 8, 1)$

11. $(-2, 4, 8) \cdot (16, 4, 2)$

Find each cross product.

12. $(7, 2, 1) \times (2, 5, 3)$

13. $(-2, -3, 1) \times (2, 3, -4)$

14. $(1, -3, 2) \times (5, 1, -2)$

15. $(-1, 0, 4) \times (5, 2, -1)$

EXERCISES

Practice

Find each inner product and state whether the vectors are perpendicular. Write *yes* or *no*.

16. $(-6, 1) \cdot (-1, 2)$

17. $(2, 0) \cdot (0, 4)$

18. $(2, -2) \cdot (5, -5)$

19. $(2, 5) \cdot (0, 1)$

20. $(5, 2) \cdot (-3, 7)$

21. $(-8, 2) \cdot (4.5, 18)$

22. $(3, -2, 4) \cdot (1, -4, 0)$

23. $(-4, 9, 8) \cdot (3, 2, -2)$

24. $(4, 9, -3) \cdot (-6, 7, 5)$

25. $(-6, 2, 10) \cdot (4, 1, 9)$

Find each cross product. Then verify that the resulting vector is perpendicular to the given vectors.

26. $(1, 3, 2) \times (2, -1, -1)$

27. $(2, 1, 2) \times (1, -1, 3)$

28. $(1, -3, 2) \times (-2, 1, -5)$

29. $(4, 0, -2) \times (-7, 1, 0)$

Find a vector perpendicular to the plane containing the given points.

30. $(1, 2, 3), (-4, 2, -1)$, and $(5, -3, 0)$

31. $(0, 1, 2), (-2, 2, 4)$, and $(-1, -1, -1)$

32. Prove that $\vec{\mathbf{a}} \cdot \vec{\mathbf{b}} = \vec{\mathbf{b}} \cdot \vec{\mathbf{a}}$ for two-dimensional vectors.

33. Prove that $\vec{\mathbf{a}} \cdot \vec{\mathbf{b}} = \vec{\mathbf{b}} \cdot \vec{\mathbf{a}}$ for three-dimensional vectors.

34. Prove that $\vec{\mathbf{a}} \times \vec{\mathbf{a}} = 0$.

35. Use the definition of cross product to prove that $\vec{\mathbf{a}} \times (\vec{\mathbf{b}} + \vec{\mathbf{c}}) = (\vec{\mathbf{a}} \times \vec{\mathbf{b}}) + (\vec{\mathbf{a}} \times \vec{\mathbf{c}})$.

36. Is it true that $\vec{\mathbf{a}} \times \vec{\mathbf{b}} = \vec{\mathbf{b}} \times \vec{\mathbf{a}}$? Explain.

Programming

37. The program below calculates the cross product of two vectors.

```
:Prgm8:XPRODUCT
:ClrHome                        Clears the text screen.
:Disp "1ST VECTOR, A1="         Displays the quoted message.
:Input A                        Accepts a value for A.
:Disp "A2="
:Input B
:Disp "A3="
:Input C
:Disp "2ND VECTOR, B1="
:Input D
:Disp "B2="
:Input E
:Disp "B3="
:Input F
:BF-EC→I                        Calculates i coefficient, stores in I.
:DC-AF→J
:AE-BD→K
:Disp "CROSS PRODUCT"
:Disp "COEFF. OF I="
:Disp I                         Displays the value stored in I.
:Disp "COEFF. OF J="
:Disp J
:Disp "COEFF. OF K="
:Disp K
```

Find each cross product.

a. $(1, 3, -2) \times (1, -1, 2)$ **b.** $(-3, 6, 5) \times (-2, -7, 2)$

Critical Thinking

38. If $\vec{\mathbf{a}} \neq 0$, $\vec{\mathbf{a}} \cdot \vec{\mathbf{b}} = \vec{\mathbf{a}} \cdot \vec{\mathbf{c}}$, and $\vec{\mathbf{a}} \times \vec{\mathbf{b}} = \vec{\mathbf{a}} \times \vec{\mathbf{c}}$, must it be true that $\vec{\mathbf{b}} = \vec{\mathbf{c}}$? Justify your answer.

Applications and Problem Solving

39. Physics The handle of a jack is at an angle of 30° with the horizontal. A downward force of 55 pounds is applied to the jack at a point 1.5 feet from the point where the handle meets the body of the jack.

a. Find the moment of the force.

b. Find the torque of the force.

c. An object tends to rotate around the line that passes through the point P and is parallel to the moment. This line is called the axis of rotation. Find the axis of rotation for the force applied to the jack handle.

40. Geometry Find the perimeter of the triangle with vertices at the points $A(2,-1)$, $B(5,3)$, and $C(-3,11)$. **(Lesson 1-4)**

41. Without graphing, determine the end behavior of the graph of the function $y = 11 - 15x - x^3$. **(Lesson 3-8)**

42. Approximate the real zeros of the function $f(x) = 4x^4 + 5x^3 - x^2 + 1$ to the nearest tenth. **(Lesson 4-5)**

43. Find the value of $\cos\left(\text{Tan}^{-1} \dfrac{\sqrt{33}}{4}\right)$. **(Lesson 6-5)**

44. Find an ordered triple that represents $\overrightarrow{AB}$ for $A(5, 7, -2)$ and $B(0, -11, 8)$. **(Lesson 8-3)**

45. College Entrance Exam Choose the best answer.
The sum of $\sqrt{12} + \sqrt{27}$ is

(A) $\sqrt{39}$ **(B)** $5\sqrt{3}$ **(C)** $13\sqrt{3}$ **(D)** $7\sqrt{3}$ **(E)** $6\sqrt{3}$

MID-CHAPTER REVIEW

Use a metric ruler and a protractor to draw a vector with the given magnitude and amplitude. Then find the magnitude of the horizontal and vertical components of the vector. (Lesson 8-1)

1. 1 cm, 65°   **2.** 3.1 cm, 15°   **3.** 0.8 cm, 120°   **4.** 2.2 cm, 160°

Find the ordered pair or ordered triple that represents the vector from _A_ to _B_. Then find the magnitude of $\overrightarrow{AB}$. (Lessons 8-2 and 8-3)

5. $A(5, 2)$, $B(-3, 3)$ **6.** $A(-4, 3)$, $B(-3, 3)$ **7.** $A(8, -7)$, $B(-2, 5)$

8. $A(3, 1, -2)$, $B(4, 0, -1)$ **9.** $A(6, 10, -11)$, $B(-2, 4, -12)$

Find an ordered pair or ordered triple to represent $\vec{u}$ in each equation if $\vec{v} = (-2, 7)$, $\vec{w} = (7, -10)$, $\vec{s} = (-2, 1, 3)$ and $\vec{t} = (-1, -1, 5)$. (Lessons 8-2 and 8-3)

10. $\vec{u} = \vec{v} + \vec{w}$ **11.** $\vec{u} = 2\vec{v} - 4\vec{w}$ **12.** $\vec{u} = 2\vec{w} - \vec{v}$

13. $\vec{u} = \vec{s} + \vec{t}$ **14.** $\vec{u} = \vec{t} - 2\vec{s}$ **15.** $\vec{u} = 4\vec{t}$

Find each inner product and state whether the vectors are perpendicular. (Lesson 8-4)

16. $(2, -1) \cdot (3, 4)$ **17.** $(-1, 5) \cdot (10, 2)$ **18.** $(9, 2, -4) \cdot (2, -7, 3)$

19. Find a vector perpendicular to the plane determined by the points at $(2, -1, 5)$, $(0, 2, 1)$, and $(8, 2, -1)$. **(Lesson 8-4)**

8-5 Applications with Vectors

Objective

After studying this lesson, you should be able to:
- solve problems using vectors and right triangle trigonometry.

Application

When 1992 Olympic champion Kerrin Lee-Gartner is skiing, the force due to gravity acts vertically. Physicists resolve this force into two perpendicular component vectors. One vector is parallel to the slope of the hill and accelerates Kerrin down the hill, while the other vector is perpendicular to the hill and helps to hold her skis on the surface of the snow.

Vectors can be used to represent any quantity, such as gravity, that has magnitude and direction. Velocity, acceleration, weight, and force are some of the quantities that physicists represent with vectors.

Example 1

APPLICATION

Skiing

Kerrin Lee-Gartner is on a snowy ski slope that makes an angle of 30° with the horizontal.
a. Draw a labeled diagram that represents the forces at work.
b. If she weighs 125 pounds, find the force that propels Kerrin down the slope.
c. Use the formulas $W = mg$ and $F = ma$, where W represents weight in pounds, m represents mass in slugs, $g = 32$ ft/s², F represents force in pounds, and a represents acceleration in ft/s², to find Kerrin's acceleration down the slope.

a. Let $\vec{F_1}$ represent the force parallel to the slope. This is the force that propels Kerrin down the slope. Let $\vec{F_2}$ represent the force perpendicular to the slope. Kerrin's weight, W, is the resultant of $\vec{F_1}$ and $\vec{F_2}$.

b. Use trigonometry to find $|\vec{F_1}|$.

$$\sin 30° = \frac{|\vec{F_1}|}{125}$$

$$|\vec{F_1}| = 125 \sin 30°$$

$$|\vec{F_1}| = 62.5$$

The force propelling Kerrin down the slope is 62.5 pounds.

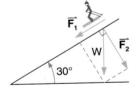

c. To find Kerrin's acceleration, first find her mass.

$$W = mg$$
$$125 = m(32)$$
$$m \approx 3.91$$

Kerrin's mass is 3.91 slugs.

The slug is the gravitational unit of mass.

Then use the formula $F = ma$ to find the acceleration.

$$F = ma$$
$$62.5 = 3.91a$$
$$a \approx 15.98$$

Kerrin's acceleration is 15.98 feet per second squared.

Example 2

APPLICATION

Physics

Charlie and Jean Hughes are attempting to haul a large sign to the roof of their antique store. They are standing at the edge of the roof pulling on ropes attached to the sign. The ropes make a 30° angle with each other. If Jean is pulling with a force of 105 N and Charlie is pulling with a force of 110 N, what is the magnitude and direction of the net force on the sign?

Sketch the given vector quantities and the resultant. $\vec{\mathbf{r}}$ is the resultant force. θ is the angle that the resultant makes with the 105 N force.

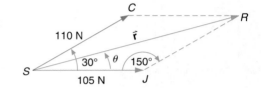

$\angle SJR$ is the supplement of 30°. Since we know two sides and the included angle in $\triangle SJR$, we can use the law of cosines to determine $|\vec{\mathbf{r}}|$.

$$|\vec{\mathbf{r}}|^2 = 105^2 + 110^2 - 2(105)(110)\cos 150°$$
$$|\vec{\mathbf{r}}|^2 = 11,025 + 12,100 - 23,100\cos 150°$$
$$|\vec{\mathbf{r}}| \approx 207.68$$

The magnitude of $\vec{\mathbf{r}}$ is about 207.68 N.

Now we can use the law of sines to find the direction of $\vec{\mathbf{r}}$.

$$\frac{207.68}{\sin 150°} = \frac{110}{\sin \theta}$$
$$\sin \theta = \frac{110 \sin 150°}{207.68}$$
$$\sin \theta \approx 0.2648$$
$$\theta \approx 15°20' \qquad \textit{Round to the nearest ten minutes.}$$

The force on the sign has a magnitude of about 207.68 N and makes an angle of 15°20′ with the 105 N force.

Sometimes there is no motion when several forces are at work on an object. This situation, when the forces balance one another, is called *equilibrium*. If the forces at equilibrium are drawn tip-to-tail, they will form a polygon.

Example 3

A traffic light is supported equally by two cables that make a **160°** angle with each other. If the light weighs **1200 pounds**, what is the tension of each of the cables?

Draw a sketch of the situation. Then draw the vectors tip-to-tail.

Since the triangle is isosceles, the base angles are congruent. We can use the law of sines to find the tension of the cables.

$$\frac{1200}{\sin 20°} = \frac{x}{\sin 80°}$$
$$x = \frac{1200 \sin 80°}{\sin 20°}$$
$$x \approx 3455.26$$

The tension of each cable is about 3455 pounds.

CHECKING FOR UNDERSTANDING

Communicating Mathematics

Read and study the lesson to answer each question.

1. As you sit still in a chair, the force of gravity pushes you against the chair and the chair pushes back. What can you say about these forces?

2. If the angle between the cables in Example 3 is increased, will there be more or less tension on the cables?

Guided Practice

Make a sketch to show the given vectors.

3. a ship traveling at 16 knots at an angle of 25° with the current

4. a force of 35 N acting on an object at an angle of 30° with the level ground

5. a force of 18 N acting on an object while a force of 51 N acts on the same object at an angle of 45° with the first force

Find the magnitude and direction of the resultant vector for each diagram.

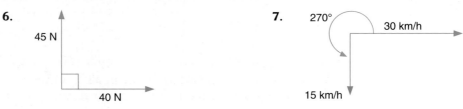

6.

45 N

40 N

7.   270°

30 km/h

15 km/h

8.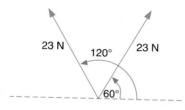

9.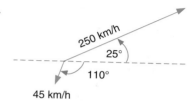

EXERCISES

Practice

Two 10-lb forces act on the same object at the same time. Find the magnitude of their resultant if each of the following is the measurement of the angle between them.

10. $30°$

11. $90°$

12. $120°$

13. $135°$

14. $170°$

15. $180°$

16. A 100 N force and a 50 N force act on the same object. The angle between the forces measures $90°$. Find the magnitude and direction of the resultant force.

17. Rick and Julie are unloading boxes from a truck. Rick places a box at the top of the loading ramp and lets it slide to the ground. If the ramp makes an angle of $40°$ with the ground and the box weighs 25 pounds, find the acceleration of the box. Assume that there is no friction.

18. Denzel pulls a wagon along level ground with a force of 18 N in the handle. If the handle makes an angle of $30°$ with the horizontal, find the horizontal and vertical components of the force.

19. A 33 N force acting at $90°$ and a 44 N force acting at $60°$ act concurrently on a point. What is the magnitude and direction of a third force that produces equilibrium at the point?

20. A force F_1 of 36 N pulls at an angle of $20°$ above due east. Pulling in the opposite direction is a force F_2 of 48 N acting at an angle of $42°$ below due west. Find the magnitude and direction of the resultant force.

21. A block of ice weighing 300 pounds is held on an ice slide by a rope parallel to the slide. If the slide is inclined at an angle of $22°$, find the tension on the rope and the force of the ice perpendicular to the slide.

22. Three forces in a plane act on an object. The forces are 7 N, 11 N, and 15 N. The angle between the 7 N and 11 N forces is $105°$, between the 11 N and 15 N forces is $147°$, and between the 15 N and 7 N forces is $108°$.

  a. Are the vectors in equilibrium?

  b. If not, find the magnitude and the direction of the resultant force.

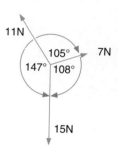

23. Five forces are acting on a point P. They are 60 N at 90°, 40 N at 0°, 80 N at 270°, 40 N at 180°, and 50 N at 60°. What is the magnitude and direction of the vector that would produce equilibrium at point P?

Critical
Thinking

24. Jeff is trying to use a long piece of rope to remove a car from some deep mud. Which method should he use to exert more force on the car? Assume that Jeff will exert the same amount of force using either method. Explain your answer.

   a. Tie one end of the rope to the car and pull.

   b. Tie one end to the car and the other end to a nearby tree. Then push on the rope perpendicular to it at a point about halfway between the car and the tree.

Applications
and Problem
Solving

25. Aviation An airplane flies east for 210 km before turning 70° south and flying for 100 km. Find the distance and direction of the plane from its starting point.

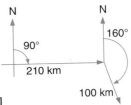

26. Aerospace A descent vehicle landing on the moon is approaching the surface of the moon with a vertical velocity of 35 m/s. The vehicle also has a horizontal velocity of 55 m/s.

   a. What is the speed of the vehicle along its path?

   b. What is the angle of the path with the vertical?

27. Transportation Two riverboat landings are directly across the river from each other. A boat traveling at a speed of 12 mph is attempting to cross directly from one landing to the other. If the current of the river has a speed of 4 mph, at what angle should the skipper head?

28. Gardening Kristin is exerting a force of 40 pounds along the handle of the lawn mower as she pushes it to cut the grass. The mower handle makes an angle of 30° with the ground. How much force is Kristin exerting forward and downward?

29. Sports The engine on a boat is propelling it forward with a force of 125 N. The wind is blowing the boat with a force of 85 N at an angle of 72° from the force of the engine. Find the magnitude and direction of the path of the boat.

30. Aviation An airplane is heading due north at 260 mph. A 16-mph wind blows from the east. Find the ground speed and direction of the plane.

31. Physics Quentin is trying to pull a disabled snowmobile to the lodge. The snowmobile could be pulled with a force of 120 pounds parallel to the ground. If Quentin is pulling at an angle of 25° to the ground, with how much force must he pull?

32. Aviation An airplane is heading due north at 260 mph. A 16-mph wind blows from the northwest at an angle of 110° clockwise from north. Find the ground speed and direction of the plane.

33. Find the value of the determinant $\begin{vmatrix} 2 & 3 & 6 \\ 3 & -1 & -2 \\ 3 & 4 & 2 \end{vmatrix}$ using expansion by minors.
(Lesson 2-3)

34. Graph the rational function $y = \dfrac{6}{(x-1)(x+3)}$. **(Lesson 3-4)**

35. Find all the rational roots of the equation $3x^3 - 4x^2 - 5x + 2 = 0$.
(Lesson 4-4)

36. Use a sum or difference identity to find the exact value of $\tan(-75°)$.
(Lesson 7-3)

37. Find the inner product of vectors $(4, -1, 8)$ and $(-5, 2, 2)$. Are the vectors perpendicular?   **(Lesson 8-4)**

38. **College Entrance Exam**   Choose the best answer.
$\overline{AB}$ is a diameter of circle O and $m\angle BOD = 15°$. If $m\angle EOA = 85°$, find $m\angle ECA$.
(A) $85°$ (B) $50°$ (C) $70°$
(D) $35°$ (E) $45°$

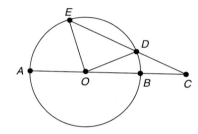

CASE STUDY FOLLOW-UP

Refer to Case Study 2: *Trashing the Planet,* on pages 959–961.

Toxic waste equivalent in weight to the residential and commercial trash thrown out by 10,000 Americans in one day is stored in cylindrical containers in underground storage facilities.

1. Find the weight of the total amount of stored waste. Assume that there are 250 million Americans.

2. During an earthquake, the containers are toppled against one wall of the facility at a 52° angle. Find the magnitude of the horizontal force pressing against the wall.

3. The wall is designed to withstand a horizontal force of 32,000 pounds. Find the container angle at which the wall will collapse.

4. **Research**   Find out about the safety of landfill and hazardous waste sites in your area. Determine if there are any potential dangers to which designers have not, in your opinion, paid adequate attention. Write about your findings in a 1-page paper.

Vectors and Parametric Equations

Objectives

After studying this lesson, you should be able to:
- write vector and parametric equations of lines, and
- graph parametric equations.

Application

Davey Allison passed Morgan Shepherd to win the 1992 Daytona 500. Time-lapse photography has allowed us to freeze the positions of their cars as they sped past the finish line. The relative positions of the cars are dependent upon their velocities at a given moment in time and their starting positions. Vector equations and equations known as **parametric equations** allow us to model their movement.

In Lesson 8-2, we described a vector $\overrightarrow{P_1P_2}$ as the ordered pair $(x_2 - x_1, y_2 - y_1)$ for $P_1(x_1, y_1)$ and $P_2(x_2, y_2)$. The slope of a line through P_1 and P_2 is $m = \dfrac{y_2 - y_1}{x_2 - x_1}$. Since the ordered pair representing a vector is related to slope, we can use vectors to write equations of lines. A vector used to describe the slope of a line is called a **direction vector**.

Example 1

Write a vector equation describing a line passing through $P_1(-2, -3)$ and parallel to the vector $\vec{a} = (5, 4)$.

Let ℓ represent the line through $P_1(-2, -3)$ parallel to $\vec{a}$. For any point $P_2(x, y)$ on ℓ, $\overrightarrow{P_1P_2} = (x - (-2), y - (-3))$. Since $\overrightarrow{P_1P_2}$ is on ℓ and is parallel to $\vec{a}$, $\overrightarrow{P_1P_2} = t\vec{a}$, for some value t. By substitution, we have $(x + 2, y + 3) = t(5, 4)$.

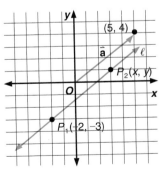

Therefore, the equation $(x + 2, y + 3) = t(5, 4)$ is a vector equation describing all of the points (x, y) on ℓ parallel to $\vec{a}$ through $P_1(-2, -3)$.

Using a process similar to the one in Example 1 for any point $P_1(x_1, y_1)$ and vector $\vec{a}$ leads us to the following definition.

	A line through $P_1(x_1, y_1)$ parallel to the vector $\vec{a} = (a_1, a_2)$ is defined by the set of points $P_2(x, y)$ such that $\overrightarrow{P_1P_2} = t\vec{a}$ for some real number t. Therefore, $(x - x_1, y - y_1) = t(a_1, a_2)$.
Vector Equation of a Line	

The independent variable t in the vector equation of a line is called a **parameter**. We can use vector equations to model physical situations, such as the Daytona 500, where t represents time.

The vector equation $(x - x_1, y - y_1) = t(a_1, a_2)$ can be written as two equations relating the horizontal and vertical components of these two vectors separately.

$$x - x_1 = ta_1 \qquad\qquad y - y_1 = ta_2$$
$$x = x_1 + ta_1 \qquad\qquad y = y_1 + ta_2$$

The resulting equations, $x = x_1 + ta_1$ and $y = y_1 + ta_2$, are known as parametric equations of the line through $P_1(x_1, y_1)$ parallel to $\vec{a} = (a_1, a_2)$.

	A line through $P_1(x_1, y_1)$ that is parallel to the vector $\vec{a} = (a_1, a_2)$ has the following parametric equations, where t is any real number.
Parametric Equations of a Line	$$x = x_1 + ta_1$$ $$y = y_1 + ta_2$$

If we know the coordinates of a point on a line and its direction vector, we can write its parametric equations.

Example 2

Find the parametric equations for a line parallel to $\vec{b} = (-2, 5)$ and passing through $(3, -5)$. Then make a table of values and graph the line.

Use the general form of the parametric equations of a line with $(a_1, a_2) = (-2, 5)$ and $(x_1, y_1) = (3, -5)$.

$$
\begin{aligned}
x &= x_1 + ta_1 & y &= y_1 + ta_2 \\
&= 3 + t(-2) & &= -5 + t(5) \\
&= 3 - 2t & &= -5 + 5t
\end{aligned}
$$

Now make a table of values and graph the line.

t	x	y
-1	5	-10
0	3	-5
1	1	0
2	-1	5

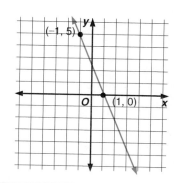

Notice in Example 2 that each value of t establishes a point (x, y) on the graph. You can think of these points as representing positions of an object, and of t as representing time in seconds. Evaluating the parametric equations for a value of t gives us the coordinates of the position of the object after t seconds have passed.

EXPLORATION: Graphing Calculator

We can simulate a NASCAR race, like the Daytona 500, between two cars by using a graphing calculator and parametric equations. The race is a 500-mile race conducted on a 2-mile oval. One car averages 105 mph. A second car averages 120 mph, but is delayed 30 minutes at the start of the race due to an electrical problem on the pace lap. Which of these two cars will finish before the other, assuming they both finish the race?

First, write a set of parametric equations to represent each car's position at time t hours. Let's simulate the race on two parallel race tracks so that we can make visual comparisons. We will use the formula $d = rt$. *Distance, d, is measured in miles, time, t, is in hours, and the average rate, r, is in miles per hour.*

Car 1 : $x = 105t$ *$d = rt$*

              $y = 1$ *y represents the position of car 1 on track 1.*

Car 2 : $x = 120(t - 0.5)$ *Remember that car 2 started $\frac{1}{2}$ hour late.*

              $y = 2$ *Car 2 is on track 2.*

We must put the graphing calculator in the correct mode for graphing parametric equations. To do this on the TI-82, press the MODE button and select the Par and Simul modes.

Now set your viewing window to the following values:
TMIN = 0, TMAX = 5, TSTEP = .01, XMIN = 0, XMAX = 500, XSCL = 10, YMIN = 0, YMAX = 5, and YSCL = 1. Next enter the parametric equations on the Y= menu. Press GRAPH to see "the race."

Notice that the line on top reached the edge of the screen first. That line represented the position of the second car, so car 2 finished first. *You can confirm the conclusions using the TRACE function. The car with the lesser t-value when x = 500 is the winner.*

If we are given the equation of a line in slope-intercept form, we can write parametric equations for that line.

Example 3 | **Write parametric equations for the line $y = 2x - 3$.**

In the equation $y = 2x - 3$, x is the independent variable and y is the dependent variable. In parametric equations, t is the independent variable and x and y are both dependent variables. If we set the independent variables x and t equal, we can write two parametric equations in terms of t.

$$x = t$$
$$y = 2t - 3$$

Parametric equations for the line are $x = t$ and $y = 2t - 3$.

We can also write the equation of a line in slope-intercept or standard form if we are given the parametric equations of the line.

Example 4 | **Write an equation of the line in slope-intercept form whose parametric equations are $x = 3 + 2t$ and $y = -1 + 5t$.**

Solve each parametric equation for t.

$$x = 3 + 2t \qquad\qquad\qquad y = -1 + 5t$$
$$x - 3 = 2t \qquad\qquad\qquad y + 1 = 5t$$
$$\frac{x - 3}{2} = t \qquad\qquad\qquad \frac{y + 1}{5} = t$$

Write an equation involving the expressions for t.

$$\frac{x - 3}{2} = \frac{y + 1}{5}$$
$$5(x - 3) = 2(y + 1)$$
$$5x - 15 = 2y + 2$$
$$y = \frac{5}{2}x - \frac{17}{2}$$

CHECKING FOR UNDERSTANDING

Communicating Mathematics

Read and study the lesson to answer each question.

1. **Explain** how the slope of a line is related to its direction vector.

2. **Confirm** the results of the Exploration by finding the time that each car finished the race. How long after the winning car did the other car finish?

3. **Compare** the paths of the cars in the Graphing Calculator Exploration. When did car 2 pass car 1?

A line that passes through the point with the given coordinates is parallel to
$\vec{a} = (3, 7)$.
a. **Write a vector equation for each line.**
b. **Write parametric equations for each line.**
c. **Use the parametric equations to write the equation of the line in slope-intercept form.**

4. $(-5, 8)$ **5.** $(-1, -5)$ **6.** $(6, 2)$

7. $(5, -9)$ **8.** $(-6, 0)$ **9.** $(11, -4)$

Write parametric equations for each linear equation.

10. $y = 9x - 1$ **11.** $y = 3x + 11$ **12.** $2x + y - 6 = 0$

EXERCISES

Practice

Write a vector equation of the line that passes through point P and is parallel to $\vec{a}$. Then write parametric equations for the line.

13. $P(-4, -11)$, $\vec{a} = (-3, 8)$ **14.** $P(1, 5)$, $\vec{a} = (-7, 2)$
15. $P(-1, 0)$, $\vec{a} = (3, 2)$ **16.** $P(-4, 1)$, $\vec{a} = (-6, 10)$

Write the equation of each line in parametric form.

17. $y = -2x + 3$ **18.** $y = 4x - 2$ **19.** $3x + 2y = 5$

Write an equation in slope-intercept form of the line with the given parametric equations.

20. $x = 3t - 5$ **21.** $x = -t + 6$ **22.** $x = -4t + 3$
       $y = -2t + 7$ $y = t + 2$ $y = 5t - 3$

23. $x = 4t - 11$ **24.** $x = 9t$ **25.** $x = 8$
       $y = t + 3$ $y = 4t + 2$ $y = 2t + 1$

Set up a table of values and then graph each line from its parametric form.

26. $x = 2 + 4t$ **27.** $x = -3 + 5t$ **28.** $x = 1 + t$
       $y = -1 + t$ $y = 2 - 4t$ $y = 1 - t$

Graphing
Calculator

Graph the parametric equations using a graphing calculator. Then sketch the graph.

29. $x = \cos t$ **30.** $x = 4 \cos t$
       $y = \sin t$ $y = 7 \sin t$

Critical
Thinking

31. Find parametric equations for the line through $(-4, -3)$ and $(5, 2)$.

Applications
and Problem
Solving

32. Language Arts Look up the definition of the word "parameter" in a dictionary. Which definition most closely relates to your mathematical understanding of the word? How do the other definitions relate to the mathematical definition?

33. **Transportation** Two semi-trucks are driving loads from Chicago to Denver, a distance of 1125 miles. The first truck leaves at 8:00 A.M. and averages 50 mph. The second truck leaves at 9:00 A.M. Since it has a lighter load, the second truck averages 54 mph. Set up two sets of parametric equations to model this situation and use a graphing calculator to analyze the model.

   a. How long is it until the second truck overtakes the first truck?

   b. How far are they from Chicago when the second truck overtakes the first truck?

   c. If each of the drivers stops for meals for a total of 3 hours, what time will it be in Chicago when each truck reaches Denver?

   d. How much would the driver of the first truck have to increase her speed in order to arrive in Denver first?

Mixed Review

34. Name all the values of x that are not in the domain of the function $f(x) = \dfrac{x}{|3x| - 12}$. **(Lesson 1-1)**

35. Find the product of the matrices $\begin{bmatrix} 4 & -1 & 6 \\ 4 & 0 & 2 \end{bmatrix}$ and $\begin{bmatrix} 0 & 3 \\ 2 & -2 \\ 5 & 1 \end{bmatrix}$. **(Lesson 2-2)**

36. **Surveying** A surveyor finds that the angle of elevation from a certain point to the top of a cliff is $60°$. From a point 45 feet further away, the angle of elevation to the top of the cliff is $52°$. How high is the cliff to the nearest foot? **(Lesson 5-5)**

37. State the amplitude, period, and phase shift for the function $y = 8 \cos (\theta - 30°)$. **(Lesson 7-2)**

38. A 30-pound force is applied to an object at an angle of $60°$ with the horizontal. Find the magnitude of the horizontal and vertical components of the force. **(Lesson 8-5)**

39. **College Entrance Exam** Choose the best answer.
    The trinomial $x^2 + x - 20$ is exactly divisible by

    (A) $x - 4$ **(B)** $x + 4$ **(C)** $x + 6$ **(D)** $x - 10$ **(E)** $x - 5$

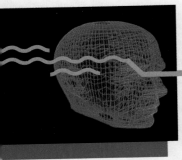

Technology

Using Parametric Equations

BASIC
Spreadsheets
▶ **Software**

You can use IBM's *Mathematics Exploration Toolkit* (*MET*) to solve problems involving parametric equations.

Example

> **The third hole of the Westerville Municipal Golf Course is a 200-yard par three. Ed Gallagher chooses a five iron. His swing launches the ball with an initial velocity of 150 feet per second, and the ball leaves the tee at an initial angle of 25°. How far will the ball travel before it hits the ground?**
>
> The parametric equations for the path of the ball are as follows.
>
> $x = t|150| \cos 25°$
>
> $y = t|150| \sin 25° - 16t^2$
>
> Use MET's CALC commands to rewrite the equation.
>
> | **Enter:** | degrees | *Puts the computer in degree mode.* |
> |---|---|---|
> | | lims 0 600 0 100 | *Sets viewing window to [0, 600] by [0, 100].* |
> | | $x = t150 \cos(25)$ | |
> | | solvefor t | *Solves the expression for t.* |
> | | getright | *Displays the right side of the equation.* |
> | | store t | *Stores the current expressions as t.* |
> | | $y = t150 \sin(25) - 16t\char94 2$ | |
> | | replace | *Replaces t in the expression with stored value of t.* |
> | | simplify | *Simplifies the expression.* |
> | | graph | *Graphs the equation.* |
>
> Pressing ALT and F3 together will allow you to move the cursor around the graphics window and determine the coordinates of different points. Zoom in on the graph where it crosses the x-axis by pressing ALT and F3, then ENTER. Use the arrow keys to box in the area and press ENTER. You can see that the ball will travel about 537 feet.

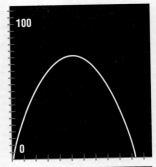

EXERCISES

1. What is the maximum height of Mr. Gallaghers ball?

2. If Mr. Gallagher hits the ball so that the initial velocity is 160 feet per second and the initial angle is 25°, would his ball travel past the pin? That is, will the ball travel more than 200 yards?

8-7 Using Parametric Equations to Model Motion

Objectives

After studying this lesson, you should be able to:
- model the motion of a projectile using parametric equations, and
- solve problems related to the motion of a projectile, its trajectory, and range.

Application

When Jack Nicklaus hits a golf ball, the path of the ball is shaped like a parabola. We can use parametric equations to represent the position of the ball relative to its starting point in terms of the parameter time.

Objects that are launched, like the golf ball, are called projectiles. The path of a projectile is called its trajectory. The horizontal distance that a projectile travels is its range. Physicists describe the motion of a projectile in terms of its position, velocity, and acceleration. These are all vector quantities.

Suppose Nicklaus hits an approach shot with his five-iron. The figure at the right illustrates what happens to the ball after it leaves the head of the club. The magnitude of the initial velocity, $|\vec{v}|$, and direction, θ, of the ball can be described by a vector that can be expressed as the sum of its horizontal and vertical components, $\vec{v}_x$ and $\vec{v}_y$.

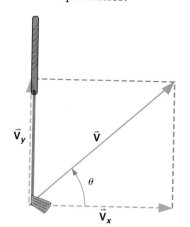

As the ball moves, gravity will act on it in the vertical direction. The horizontal component will be unaffected by gravity. So, discounting wind resistance, the horizontal speed is constant throughout the flight of the ball. The vertical speed of the ball is large and positive at the beginning, decreasing to zero at the top of its trajectory, then increasing in the negative direction as it falls. When the ball returns to the ground, its vertical speed is the same as when it left the club head, but in the opposite direction.

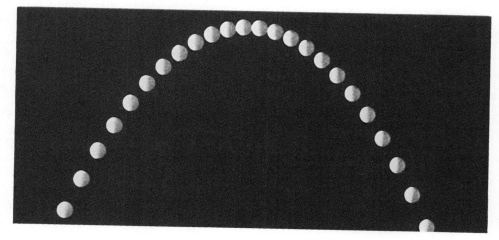

In order to find parametric equations that represent the path of the projectile, we must resolve the initial velocity into its horizontal and vertical components.

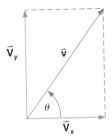

$$\cos \theta = \frac{|\vec{v}_x|}{|\vec{v}|} \qquad \sin \theta = \frac{|\vec{v}_y|}{|\vec{v}|}$$

$$|\vec{v}_x| = |\vec{v}| \cos \theta \qquad |\vec{v}_y| = |\vec{v}| \sin \theta$$

Since it is unaffected by gravity, horizontal speed will be the magnitude of the horizontal component of the initial velocity. Therefore, the horizontal position of a projectile after t seconds is given by the following equation.

horizontal distance = rate · time

$$x = |\vec{v}| \cos \theta \cdot \quad t$$
$$= t |\vec{v}| \cos \theta$$

The vertical distance is the sum of the distance traveled due to the initial velocity and the distance traveled due to gravity. The height of an object affected by gravity is given by the equation $h = \frac{1}{2}gt^2$, where $g = -9.8$ m/s^2 or -32 ft/s^2 according to the units of the problem.

vertical distance = distance due to initial velocity + distance due to gravity

$$y = \qquad (|\vec{v}| \sin \theta)t \qquad + \qquad \frac{1}{2}gt^2$$
$$= t|\vec{v}| \sin \theta + \frac{1}{2}gt^2$$

Therefore, the path of a projectile can be expressed in terms of parametric equations.

Parametric Equations for the Path of a Projectile

If a projectile is launched at an angle of θ with the horizontal, with an initial velocity of magnitude, $|\vec{v}|$, the path of the projectile may be described by the equations

$$x = t|\vec{v}| \cos \theta \text{ and}$$
$$y = t|\vec{v}| \sin \theta + \frac{1}{2}gt^2,$$

where t is time, and g is acceleration due to gravity.

Example 1

APPLICATION

Soccer

Zoe kicked a soccer ball with an initial velocity of 45 ft/s at an angle of 32° to the horizontal. After 0.6 seconds, how far has the ball traveled horizontally and vertically?

First, write the position of the ball as a pair of parametric equations defining the path of the ball for any time, t, in seconds.

$$x = t|\vec{v}| \cos \theta \qquad\qquad y = t|\vec{v}| \sin \theta + \frac{1}{2}gt^2$$

$$= t(45) \cos 32° \qquad\qquad = t(45) \sin 32° + \frac{1}{2}(-32)t^2 \qquad \text{Let } g = -32.$$

$$= 45t \cos 32° \qquad\qquad = 45t \sin 32° - 16t^2$$

Then, find x and y when $t = 0.6$.

$$x = 45(0.6) \cos 32° \qquad y = 45(0.6) \sin 32° - 16(0.6)^2$$
$$\approx 22.9 \qquad\qquad\qquad \approx 8.5$$

After 0.6 seconds, the ball has traveled 22.9 feet horizontally and 8.5 feet vertically.

The parametric equations describe the path of an object that is launched from ground level. If an object is launched from above ground level, as a baseball would be, you must add the vertical height to the expression for y. This accounts for the fact that at time 0, the object will be above the ground.

Example 2

APPLICATION

Archery

Denise Parker was a member of the U.S. Olympic Archery Team in 1988 and in 1992. Denise shoots an arrow with an initial velocity of 65 m/s at an angle of 4.5° with the horizontal at a target 70 meters away. If Denise holds the bow 1.5 meters above the ground when she shoots the arrow, how far above the ground will the arrow be when it hits the target?

First, write parametric equations to model the path of the arrow.

$$x = t|\vec{\mathbf{v}}| \cos \theta \qquad\qquad y = t|\vec{\mathbf{v}}| \sin \theta + \frac{1}{2}gt^2 + h$$
$$= t(65) \cos 4.5° \qquad\qquad = t(65) \sin 4.5° + \frac{1}{2}(-9.8)t^2 + 1.5 \qquad \textit{Let } g = -9.8.$$
$$= 65t \cos 4.5° \qquad\qquad\quad = 65t \sin 4.5° - 4.9t^2 + 1.5$$

Then, find the amount of time that it will take the arrow to travel 70 meters horizontally. This is when it will hit the target.

$$70 = 65t \cos 4.5°$$
$$t = \frac{70}{65 \cos 4.5°}$$
$$t = 1.0802531 \qquad \textit{Use a calculator.}$$

The arrow will hit the target in about 1.08 seconds.

To find the vertical position of the arrow at that time, find y when $t = 1.08$.

$$y = 65t \sin 4.5° - 4.9t^2 + 1.5$$
$$= 65(1.08) \sin 4.5° - 4.9(1.08)^2 + 1.5$$
$$= 1.2924685 \qquad \textit{Use a calculator.}$$

The arrow will be about 1.3 meters above the ground when it hits the target.

Graphing calculators are helpful when we are using parametric equations to study the paths of projectiles.

Example 3

On the sixth hole, Jack Nicklaus selects a five iron. He estimates the distance to the pin to be 200 yards. Nicklaus' swing provides an initial velocity of 150 ft/s to the ball at an angle of 25° above the horizontal. Use a graphing calculator to determine if he will hit the pin. (Assume that he has directed the ball straight at the pin.)

First, write the position of the ball as a pair of parametric equations defining the path of the ball for any time, t, in seconds.

$$x = t|\vec{v}| \cos \theta$$

$$= t(150) \cos 25°$$

$$= 150t \cos 25°$$

$$y = t|\vec{v}| \sin \theta + \frac{1}{2}gt^2$$

$$= t(150) \sin 25° + \frac{1}{2}(-32)t^2 \quad \textit{Use –32 for g.}$$

$$= 150t \sin 25° - 16t^2$$

 Portfolio

Review items in your portfolio. Make a table of contents of the items, noting why each item was chosen. Replace any items that are no longer appropriate.

Then, we can use a graphing calculator to see the path of the ball. Use the range values $TMIN = 0$, $TMAX = 5$, $TSTEP = 0.1$, $XMIN = 0$, $XMAX = 600$, $XSCL = 20$, $YMIN = -20$, $YMAX = 50$, and $YSCL = 10$. *Make sure your calculator is in degree mode.*

Graph the equations and estimate the range of the ball. The point where the ball hits the ground is the point where $y = 0$, so use the TRACE function to find the x-coordinate of that point. The ball traveled about 540 feet in the air, so it hits 60 feet short of the pin.

CHECKING FOR UNDERSTANDING

Communicating Mathematics

Read and study the lesson to answer each question.

1. **Explain** why the vertical spacings between successive positions of the ball are unequal in the photo at the beginning of the lesson.

2. The faces of golf clubs are angled and the angle determines the angle of the initial velocity of the ball. A hockey stick is made with little or no angle to its face. Why?

3. **Analyze** the situation where a projectile is launched at an angle of 90° to the horizontal. How far does the projectile travel horizontally? Can you think of a situation when this happens?

4. What factor affects the vertical velocity of a projectile that does not affect its horizontal velocity?

5. What is the vertical velocity of a golf ball at the highest point in its trajectory? Explain.

6. What is the relationship between the vertical velocities of a projectile at its launch and at its landing?

7. Gretchen Austgen, an outfielder for the West Chicago Wildcats, is 215 feet from home plate after catching a fly ball. The runner tags third and heads for home. Gretchen releases the ball at an initial velocity of 75 ft/s at an angle of 25° with the horizontal. Assume Gretchen releases the ball 5 feet above the ground and aims it directly in line with the plate.

   a. Write two parametric equations that represent the path of the ball.

   b. Use a calculator to graph the path of the ball. Sketch the graph shown on the screen.

   c. How far will the ball travel horizontally before hitting the ground?

   d. What is the maximum height of the trajectory?

   e. The ball will fall short of reaching home plate. Could Gretchen change the initial angle in order for the throw to reach the plate? If so, find the angle she should use.

EXERCISES

8. Find the initial vertical velocity of a stone thrown with an initial velocity of 50 ft/s at an angle of 40° with the horizontal. The vertical velocity is the vertical component of the initial velocity vector.

9. Find the initial horizontal velocity of a discus thrown with an initial velocity of 62 ft/s at an angle of 42° with the horizontal. The horizontal velocity is the horizontal component of the initial velocity vector.

10. An airplane flying at an altitude of 3500 feet is dropping supplies to researchers on an island. The path of the plane is parallel to the ground at the time the supplies are released and the plane is traveling at a speed of 300 mph.

   a. Write the parametric equations that represent the path of the supplies.

   b. Graph the path of the supplies on your graphing calculator and sketch the graph.

   c. How long will it take for the supplies to reach the ground?

   d. How far will the supplies travel horizontally before they land?

A projectile is fired from ground level with an initial velocity of 100 feet per second and at an angle of $\theta°$ with the horizontal.

11. Find the range and the time in the air of the projectile for each value of θ.
   a. 10°   **b.** 30°   **c.** 45°
   d. 60°   **e.** 80°   **f.** 90°

12. At what angle should a projectile be fired in order for it to have maximum range? Explain your answer.

13. Use your answers for Exercises 11a-11f to compare the range and time in the air for angles of α° and $(90 - \alpha)^\circ$. What do you notice?

14. Two stones are launched at the same time from the top of a building 850 feet high. The first stone is simply dropped over the edge, while the second stone is thrown horizontally at 45 ft/s.

   a. Which stone will reach the ground first? Explain your answer.

   b. About how far apart will the stones be when they land?

Critical Thinking

15. If a circle of radius 1 unit is rolled along the x-axis at a rate of 1 unit per second, then the path of a point P on the circle is called a **cycloid**.

   a. Sketch what you think a cycloid must look like. *Hint: Use a coin or some other circular object to simulate the situation.*

   b. The parametric equations of a cycloid are $x = t - \sin t$ and $y = 1 - \cos t$ where t is measured in radians. Use a graphing calculator to graph the cycloid. An appropriate range is $TMIN = 0$, $TMAX = 18.8$, $TSTEP = 0.2$, $XMIN = -6.5$, $XMAX = 25.5$, $XSCL = 2$, $YMIN = -8.4$, $YMAX = 12.9$, and $YSCL = 1$. Was your sketch correct? *Make sure your calculator is in radian mode.*

Applications and Problem Solving

16. **Sports** Kevin Butler, a place kicker for the Chicago Bears, kicks a football in a game with the Dallas Cowboys. The ball leaves the ground with a velocity of 88 ft/s at an angle of 30.0° above the horizontal.

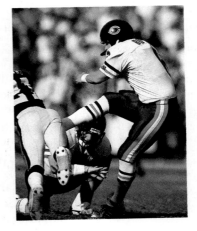

   a. Find the time the ball is in the air.

   b. If the ball is kicked from the Bears' 30 yard line and is aimed straight down the field, where does it land?

   c. Find the maximum height of the ball.

17. **Sports** Juan Gonzalez, outfielder for the Texas Rangers, comes to bat with the bases loaded and the Rangers down 4 to 1 against the Seattle Mariners. The count on Gonzalez is 3 and 2, and the pitcher throws a smoking fastball about waist high (3 feet above the ground). Gonzalez connects and the ball jumps off the bat with an initial velocity of 155 ft/s at an angle of 22° above the horizontal. As Gonzalez rounds the bases, the ball flies straight at the 420-foot marker in center field where the wall is 15 feet tall.

   a. Write the parametric equations that describe the path of the ball.

   b. Find the height of the ball after it has traveled 420 feet horizontally. Do you think it will clear the fence to be a home run or could the center fielder jump to catch the ball?

   c. How far does the ball travel before it hits the ground?

18. **Sports** In football, hang time is the time that the ball takes to land after it is kicked. Punters want to give their teammates as much time as possible to get downfield, so they wish to increase the hang time of their punts.

   a. In order for a punter to increase the hang time, should he increase or decrease the angle the ball makes with the horizontal as it leaves his foot?

   b. Assume the magnitude of the initial velocity of the ball remains constant. How will this affect the range of the ball?

19. **Physics** Erin is performing a physics experiment on projectile motion. She has fired a projectile and measured its range. She hypothesizes that if the magnitude of the initial velocity is doubled and the angle of the velocity remains the same, the projectile will travel twice as far as it did before. Do you agree with her hypothesis? Explain.

20. **Sports** A sport rifle used for target shooting has a muzzle velocity of 1200 ft/s.

   a. What is the maximum range of the rifle if it is fired about 5 feet above the ground?

   b. How long would a bullet be in the air if it was fired for the maximum range?

21. **Entertainment** The Fourth of July fireworks at Schrock Park are fired at an angle of 85° with the horizontal. The technicians firing the shells expect them to explode about 250 feet in the air 4 seconds after they are fired.

   a. Find the initial velocity of a shell fired from ground level.

   b. The technicians will place barriers so that spectators will not be in danger of being showered with debris from the shells. If the barriers will be placed 100 yards from the point directly below the explosion of the shells, how far should the barriers be from the point where the fireworks will be launched?

Mixed Review

22. Find $[f \circ g](x)$ and $[g \circ f](x)$ for the functions $f(x) = x^2 - 4$ and $g(x) = \frac{1}{2}x + 6$. **(Lesson 1-2)**

23. Graph the inequality $y \leq |x + 4|$. **(Lesson 1-3)**

24. **Statistics** The prediction equation $y = -0.13x + 37.8$ gives the fuel economy, y, for a car with a horsepower of x. Is the equation a better predictor for Car 1, which has a horsepower of 135 and averages 19 miles per gallon, or for Car 2, which has a horsepower of 245 and averages 16 miles per gallon? **(Lesson 7-7)**

25. Write an equation in standard form of the line with the parametric equations $x = 4t + 1$ and $y = 5t - 7$. **(Lesson 8-6)**

26. **College Entrance Exam** Bob Perry invests $2400 in the Fidelity Bank at 5% APR. How much additional money must he invest at 8% so that the total annual income from his investments is 6% per year?

VOCABULARY

Upon completing this chapter, you should be familiar with the following terms:

amplitude	**412**	**414**	parallel
components	**415**	**443**	parameter
cross product	**431**	**442**	parametric equation
cycloid	**454**	**413**	resultant
direction vector	**442**	**412**	standard position
inner product	**430**	**422**	unit vector
magnitude	**412**	**412**	vector
opposite	**414**		

SKILLS AND CONCEPTS

OBJECTIVES AND EXAMPLES	REVIEW EXERCISES

Upon completing this chapter, you should be able to:

■ add and subtract vectors geometrically **(Lesson 8-1)**

Find the sum of $\vec{a}$ and $\vec{b}$.

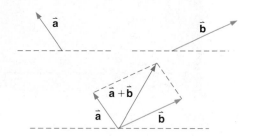

Use these exercises to review and prepare for the chapter test.

Use a metric ruler and a protractor to find each sum or difference. Then, find the magnitude and amplitude of each resultant.

1. $\vec{p} + \vec{q}$ 2. $2\vec{p} + \vec{q}$

3. $3\vec{q} - 2\vec{p}$ 4. $4\vec{p} - \vec{q}$

Find the magnitude of the vertical and horizontal components of each vector above.

5. $\vec{p}$ 6. $\vec{q}$

■ find ordered pairs that represent vectors **(Lesson 8-2)**

Find the ordered pair that represents the vector from $M(3, 1)$ to $N(-7, 4)$. Then find the magnitude of $\overrightarrow{MN}$.

$$\overrightarrow{MN} = (-7 - 3, 4 - 1) \text{ or } (-10, 3)$$
$$|\overrightarrow{MN}| = \sqrt{(-7 - 3)^2 + (4 - 1)^2} \text{ or } \sqrt{109}$$

Find the ordered pair that represents the vector from C to D. Then find the magnitude of $\overrightarrow{CD}$.

7. $C(2, 3), D(7, 15)$

8. $C(-2, 8), D(4, 12)$

9. $C(2, -3), D(0, 9)$

10. $C(-6, 4), D(-5, -4)$

OBJECTIVES AND EXAMPLES

- add, subtract, multiply, and find the magnitude of vectors algebraically **(Lesson 8-2)**

Find the ordered pair that represents $\vec{a} + \vec{b}$, if $\vec{a} = (1, -5)$ and $\vec{b} = (-2, 4)$.

$$\vec{a} + \vec{b} = (1, -5) + (-2, 4)$$
$$= (-1, -1)$$

- add, subtract, and find the magnitude of vectors in 3-dimensional space **(Lesson 8-3)**

Find the ordered triple that represents the vector from $R(-2, 0, 8)$ to $S(5, -4, -1)$. Write it in terms of unit vectors.

$$\vec{RS} = (5 - (-2), -4 - 0, -1 - 8)$$
$$= (7, -4, -9)$$
$$= 7\vec{i} - 4\vec{j} - 9\vec{k}$$

- find the inner and cross product of two vectors **(Lesson 8-4)**

Find the inner product of $\vec{a}$ and $\vec{b}$ if $\vec{a} = (3, -1, 7)$ and $\vec{b} = (0, -2, -4)$.

$$\vec{a} \cdot \vec{b} = 3(0) + (-1)(-2) + 7(-4)$$
$$= -26$$

- solve problems using vectors and right triangle trigonometry **(Lesson 8-5)**

Find the magnitude of $\vec{r}$.

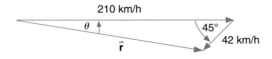
210 km/h
θ
45°
42 km/h
$\vec{r}$

Use the law of cosines.
$$\left|\vec{r}\right|^2 = 210^2 + 42^2 - 2(210)(42)\cos 45°$$
$$\left|\vec{r}\right| \approx 182.7$$

The magnitude is about 182.7 km/h.

REVIEW EXERCISES

Find an ordered pair to represent $\vec{u}$ in each equation if $\vec{v} = (2, -5)$ and $\vec{w} = (3, -1)$.

11. $\vec{u} = \vec{v} + \vec{w}$

12. $\vec{u} = \vec{v} - \vec{w}$

13. $\vec{u} = 3\vec{v} + 2\vec{w}$

14. $\vec{u} = 3\vec{v} - 2\vec{w}$

For each pair of points E and F, find an ordered triple that represents $\vec{EF}$. Then write $\vec{EF}$ as the sum of unit vectors.

15. $E(2, -1, 4); F(6, -2, 1)$

16. $E(9, 8, 5); F(-1, 5, 11)$

17. $E(-4, -3, 0); F(2, -1, 7)$

18. $E(3, 7, -8); F(-4, 0, 5)$

Find each inner or cross product.

19. $(5, -1) \cdot (-2, 6)$

20. $(4, 1, -2) \cdot (3, -4, 4)$

21. $(2, -1, 4) \times (6, -2, 1)$

22. $(5, 2, -1) \times (2, -4, -4)$

Find the magnitude and direction of the resultant vector for each diagram.

23.

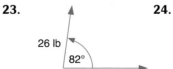
26 lb
82°
32 lb

24.

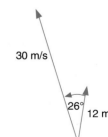
30 m/s
26°
12 m/s

■ write vector and parametric equations of a line **(Lesson 8-6)**

Find parametric equations for the line parallel to $\vec{c} = (-6, 3)$ and passing through $(1, 4)$.

$x = x_1 + ta_1$ $\qquad$ $y = y_1 + ta_2$

$x = 1 + t(-6)$ $\qquad$ $y = 4 + t(3)$

$x = 1 - 6t$ $\qquad$ $y = 4 + 3t$

Write a vector equation of the line that passes through point P and is parallel to $\vec{v}$. Then write parametric equations for the line.

25. $P(3, -5), \vec{v} = (4, 2)$

26. $P(-1, 9), \vec{v} = (-7, -5)$

27. $P(4, 0), \vec{v} = (3, -6)$

28. $P(-2, -7), \vec{v} = (0, 8)$

■ solve problems related to the motion of a projectile, its trajectory, and range **(Lesson 8-7)**

The parametric equations for the path of a projectile are:

$x = t |\vec{v}| \cos \theta$, and

$y = t |\vec{v}| \sin \theta + \dfrac{1}{2} gt^2$.

A soccer ball is kicked with an initial velocity of 30 ft/s at an angle of 28° to the horizontal.

29. Write two parametric equations that represent the path of the ball.

30. If the ball is not touched before its first bounce, how far will the ball travel horizontally before it hits the ground?

APPLICATIONS AND PROBLEM SOLVING

Draw a diagram to help you solve each problem.

31. **Physics** Karen uses a large wrench to change the tire on her car. If she applies a downward force of 50 pounds at an angle of 60° one foot from the center of the lug nut, find the torque of the force. **(Lesson 8-4)**

32. **Physics** Mike and Marie are moving a stove. They are applying forces of 70 N and 90 N at an angle of 30° to each other. Find the resultant force and the angle it makes with the larger force. **(Lesson 8-5)**

33. **Sports** José punts a football with an initial velocity of 38 ft/s at an angle of 40° to the horizontal. If the ball is 2 feet above the ground when it is kicked, how high is it after 0.5 seconds? **(Lesson 8-7)**

Use a metric ruler, a protractor, and
vectors $\vec{a}$ and $\vec{b}$ to solve each problem.

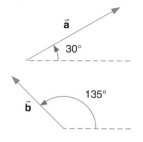

1. Find the magnitude and amplitude of $\vec{a} + \vec{b}$.

2. Find the magnitude and amplitude of $2\vec{a} - 3\vec{b}$.

3. Find the magnitude of the vertical and horizontal
   components of $\vec{a}$.

4. Find the magnitude of the vertical and horizontal
   components of $\vec{b}$.

Find an ordered pair or ordered triple that represents the vector from A to B.

5. $A(3, 6), B(-1, 9)$ 6. $A(-2, 7), B(3, 10)$

7. $A(2, -4, 5), B(9, -3, 7)$ 8. $A(-4, -8, -2), B(-8, -10, 2)$

Let $\vec{r} = (-1, 3, 4)$ and $\vec{s} = (4, 3, -6)$.

9. Find $\vec{r} - \vec{s}$.   10. Find $3\vec{s} - 2\vec{r}$.

11. Find $\vec{r} + 3\vec{s}$.   12. Find $|\vec{r}|$.

13. Find $|\vec{s}|$.   14. Write $\vec{r}$ as the sum of unit vectors.

15. Write $\vec{s}$ as the sum of unit vectors.   16. Find $\vec{r} \cdot \vec{s}$.

17. Find $\vec{r} \times \vec{s}$.   18. Is $\vec{r}$ perpendicular to $\vec{s}$?

Write parametric equations for the line that passes through point P and is parallel to $\vec{a}$.

19. $P(3, 11), \vec{a} = (2, -5)$ 20. $P(-2, 0), \vec{a} = (1, 9)$ 21. $P(12, -8), \vec{a} = (-4, -7)$

22. Write an equation in slope-intercept form of the line with the parametric
    equations $x = -2t + 6$ and $y = 9t - 8$.

23. **Navigation**   A boat that travels at 16 knots in calm water is sailing across a
    current of 3 knots on a river 250 m wide. The boat makes an angle of 35° with
    the current heading into the current.
    a. Find the resultant velocity of the boat.
    b. How far upstream is the boat when it reaches the other shore?

24. **Physics**   A downward force of 110 pounds is applied to the end of a 1.5-foot
    lever. What is the torque of this force about the axle if the angle of the lever to
    the horizontal is 60°?

25. **Gardening**   Tei uses a sprinkler to water his garden. The sprinkler discharges
    water with a velocity of 28 feet per second. If the angle of the water with the
    ground is 35°, how far will the water travel in the horizontal direction?

Bonus
What can you conclude about θ, the angle between two vectors $\vec{u}$ and $\vec{v}$, if $\vec{u} \cdot \vec{v} < 0$?

2 UNIT REVIEW

Change each radian measure to degree measure. (Lesson 5-1)

1. $\dfrac{\pi}{2}$

2. $\dfrac{3\pi}{4}$

3. $\dfrac{7\pi}{2}$

4. $-\dfrac{7\pi}{12}$

Solve. (Lesson 5-2)

5. Find the length of the arc intercepted by a central angle of $60°$ in a circle of radius 6 inches.

Find the value of the given trigonometric function of an angle in standard position if a point with the given coordinates lies on its terminal side. (Lesson 5-3)

6. $\cos\theta; (2,3)$

7. $\tan\theta; (10,2)$

8. $\sin\theta; (-4,1)$

9. $\sec\theta; (1,0)$

Find each exact value. Do not use a calculator. (Lesson 5-4)

10. $\sin\pi$

11. $\cot\dfrac{\pi}{3}$

12. $\sec\dfrac{3\pi}{4}$

13. $\csc -\dfrac{2\pi}{3}$

Solve each triangle ABC described below. Angle C is the right angle. Round angle measures to the nearest minute and side measures to the nearest tenth. (Lesson 5-5)

14. $A = 25°, \ a = 12.1$

15. $a = 3, \ b = 5$

16. $c = 24, \ B = 63°$

Determine the number of possible solutions. If a solution exists, solve the triangle. Round angle measures to the nearest minute and side measures to the nearest tenth. (Lessons 5-6 and 5-7)

17. $A = 46°, \ a = 86, \ c = 200$

18. $a = 19, \ b = 20, \ A = 65°$

19. $A = 73°, \ B = 65°, \ b = 38$

Find the area of each triangle to the nearest tenth. (Lesson 5-8)

20. $a = 5, \ b = 9, \ c = 6$

21. $a = 22, \ A = 63°, \ B = 17°$

State the amplitude (if it exists), period, and phase shift for each function. (Lesson 6-2)

22. $y = 2\cos 3x$

23. $y = -5\tan 5x$

24. $y = 4\cot\left(\dfrac{x}{2} + 90°\right)$

Graph each function. (Lesson 6-3)

25. $y = \dfrac{1}{2}\cos 2x$

26. $y = 3\tan(2x - 90°)$

27. $y = x + 2\sin 3x$

Evaluate each expression. Assume that all angles are in Quadrant I. (Lesson 6-4)

28. $\cos\left(\arccos\dfrac{1}{4}\right)$

29. $\cot\left(\cos^{-1}\dfrac{2}{3}\right)$

30. $\cos(\sin^{-1}0) + \sin(\tan^{-1}0)$

Find each value. (Lesson 6-5)

31. $\cos\left(\text{Arccos }\dfrac{1}{2}\right)$

32. $\sin(\text{Tan}^{-1}1)$

33. $\cos\left(\dfrac{\pi}{2} - \text{Arccot }\dfrac{\sqrt{3}}{3}\right)$

Write the equation for the inverse of each function. Then graph the function and its inverse. (Lesson 6-6)

34. $y = \arccos x$

35. $y = \cot x$

36. $y = \text{Tan } x$

37. $y = \text{Arcsin } 2x$

Solve for values of θ between $0°$ and $90°$. (Lesson 7-1)

38. If $\sec\theta = \dfrac{4}{3}$, find $\cos\theta$.

39. If $\cos\theta = \dfrac{1}{3}$, find $\sin\theta$.

40. If $\sin\theta = \dfrac{1}{2}$, find $\cot\theta$.

Verify that each of the following is an identity. (Lesson 7-2)

41. $\tan x + \tan x \cot^2 x = \sec x \csc x$

42. $\sin(180° - \theta) = \tan \theta \cos \theta$

Use the sum and difference identities to find the exact value of each function. (Lesson 7-3)

43. $\sin 105°$ **44.** $\cos 135°$

45. $\cos 15°$ **46.** $\sin(-210°)$

If $\sin x = \dfrac{2}{5}$ and x is in the first quadrant, find each value. (Lesson 7-4)

47. $\cos 2x$ **48.** $\sin \dfrac{x}{2}$

49. $\tan \dfrac{x}{2}$ **50.** $\sin 2x$

Solve each equation for $0° \leq x \leq 180°$. (Lesson 7-5)

51. $\sin^2 x - \sin x = 0$

52. $\cos 2x = 4 \cos x - 3$

53. $5 \cos x + 1 = 3 \cos 2x$

Write each equation in normal form. Then find p, the measure of the normal and ϕ, the angle that the normal makes with the positive x-axis. (Lesson 7-6)

54. $2x + 3y - 2 = 0$

55. $5x = -2y + 8$

56. $y = 3x - 7$

Find the distance from the point with the given coordinates to the line with the given equation. Round answers to the nearest tenth. (Lesson 7-7)

57. $(2, 5); 2x - 2y + 3 = 0$

58. $(-2, 2); -x + 4y = -6$

59. $(1, -3); 4x - y - 1 = 0$

Use vectors $\vec{a}$ and $\vec{b}$ for Exercises 60–61. (Lesson 8-1)

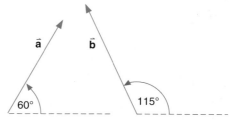

60. Find the magnitude (in cm) and amplitude of $\vec{a} + \vec{b}$.

61. Find the vertical and horizontal components of $\vec{a}$.

Find an ordered pair to represent $\vec{u}$ in each equation if $\vec{v} = (1, -3)$ and $\vec{w} = (2, -2)$. (Lesson 8-2)

62. $\vec{u} = \vec{v} + \vec{w}$ **63.** $\vec{u} = \vec{w} - \vec{v}$

64. $\vec{u} = 3\vec{v} + 2\vec{w}$ **65.** $\vec{u} = -3\vec{v} + \vec{w}$

Find an ordered triple to represent $\vec{u}$ in each equation if $\vec{v} = (3, 1, -1)$ and $\vec{w} = (-5, 2, 3)$. Then write $\vec{u}$ as the sum of unit vectors. (Lesson 8-3)

66. $\vec{u} = 2\vec{v} + \vec{w}$ **67.** $\vec{u} = \vec{v} - 2\vec{w}$

68. $\vec{u} = 3\vec{v} + 3\vec{w}$ **69.** $\vec{u} = 4\vec{v} - 2\vec{w}$

Find each inner product or cross product. (Lesson 8-4)

70. $(4, -2) \cdot (-2, 3)$

71. $(3, -4, 1) \cdot (4, -2, 2)$

72. $(5, -2, 5) \times (-1, 0, -3)$

Write a vector equation of the line that passes through point P and is parallel to $\vec{v}$. Then write parametric equations for the line. (Lesson 8-6)

73. $P(0, 5), \vec{v} = (-1, 5)$

74. $P(4, -3), \vec{v} = (-2, -2)$

UNIT 3

ADVANCED FUNCTIONS AND GRAPHING

You are now ready to apply what you have learned in earlier units to more complex functions. The three chapters in this unit contain very different topics; however, there are some similarities among them. The most striking similarity is that all three chapters require that you have had some experience with graphing.

The chapter on polar coordinates will require you to use your graphing skills on a new set of axes, the polar axes. The graphs that you will construct will not look like the graphs that you have seen thus far; however, you will be able to use the same skills to construct these graphs. In the chapter on conics, you will study the graphs of circles, ellipses, parabolas, and hyperbolas. In this chapter, you will use much of what you have studied about graphing up to this point. Finally, you will examine the exponential and logarithmic functions. These functions are used quite often to model a variety of real-life problems. Examining the graphs of these functions should prove quite illuminating.

As you work through this unit, try to make connections between what you have already studied and what you are currently studying. This will help you to be able to use the skills you have already mastered more effectively in this unit.

CHAPTER 9

POLAR COORDINATES AND COMPLEX NUMBERS

HISTORICAL SNAPSHOT

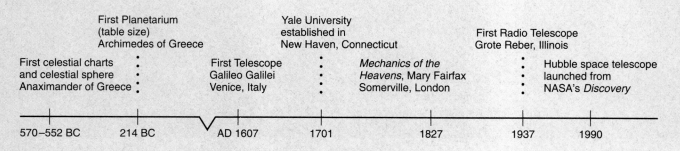

First Planetarium
(table size)
Archimedes of Greece

Yale University
established in
New Haven, Connecticut

First Radio Telescope
Grote Reber, Illinois

First celestial charts
and celestial sphere
Anaximander of Greece

First Telescope
Galileo Galilei
Venice, Italy

*Mechanics of the
Heavens*, Mary Fairfax
Somerville, London

Hubble space telescope
launched from
NASA's *Discovery*

| 570–552 BC | 214 BC | AD 1607 | 1701 | 1827 | 1937 | 1990 |

CHAPTER OBJECTIVES

In this chapter, you will:
- Graph polar equations.
- Convert complex numbers from rectangular to polar form and vice versa.
- Perform basic operations on complex numbers in both rectangular and polar form.

CAREER GOAL: Astronomical Research

Some people have a hard time choosing a career, but Greg Howard has always known what he wanted to do with his life. "I've been interested in astronomy for as long as I can remember," he says. "I got a small telescope when I was eight years old, and a larger one when I was fifteen. I spent many hours late at night seeing what I could find. I thought astronomy was beautiful and interesting. I still find it as intriguing as I ever did."

Greg stays busy year-round, always taking the opportunity for summer internships. "Two summers ago," Greg says, "I collected data about *water masers*, using a radio telescope. Water masers are clouds surrounding stars that give off powerful radio energy. Last summer, my advisor and I catalogued a new type of hydrogen cloud. This summer, I'm doing analyses of clusters of galaxies." Greg says that his summer internships were "incredible. I found them to be a good way to apply what I had learned in class to the real world of research."

Greg uses mathematics to find objects in the sky. "Coordinate systems are very important in astronomy," he says. "Finding coordinates is one of the first things that we learn. The *celestial sphere* is an imaginary globe surrounding Earth. We use coordinate systems to locate objects in the sky in terms of their locations on the celestial sphere. Once you define the coordinates for the celestial sphere, and you apply fundamentals about the speed of Earth's rotation and its motion around the sun, then you can determine which parts of the solar system are overhead at any given time, on any given day."

Greg says, "I like being on the edge of new knowledge. There is plenty to be learned. I hope that I will contribute useful research that will enlarge the current knowledge about the universe in which we live."

interNET CONNECTION

For up-to-date information on astronomy, visit:
www.glencoe.com/sec/math/amc/mathnet

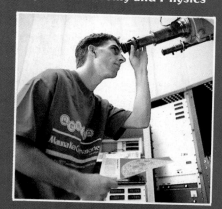

9-1 Polar Coordinates

Objective

After studying this lesson, you should be able to:
- graph polar coordinates and simple polar equations.

Application

In recent years, the portable cellular telephone has gained in popularity, making the office as mobile as a moving vehicle. However, these phones work only if you are in an area serviced by a cellular microwave tower. The Motorola Corporation plans to alleviate this drawback by launching 77 low-orbiting satellites able to relay cellular signals to the entire world. These Iridium satellites, named after the 77 electrons in the iridium atom, are scheduled for launch beginning in 1994. They will circle Earth in seven polar orbits, with each orbit containing 11 satellites.

The map at the right shows an aerial view of the North Pole with latitude and longitude lines. The proposed paths of the seven orbits are shown in red. The latitude lines are concentric circles with the North Pole as their center. The longitude lines intersect the North Pole. These lines are identified by the angles they make with respect to the prime meridian (0° longitude) which goes through Greenwich, England. In mathematics, a coordinate system similar to this one is called a **polar coordinate system**.

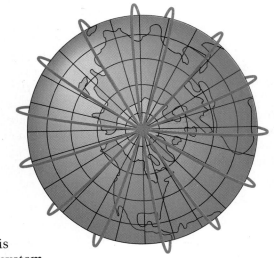

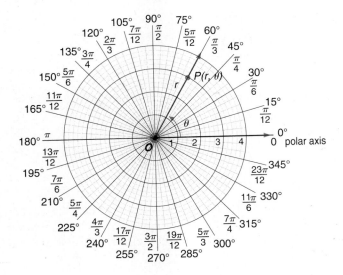

In a polar coordinate system, a fixed point O is called the **pole** or origin. The **polar axis** is usually a horizontal ray directed toward the right from the pole. The location of point P in the polar coordinate system can be identified by polar coordinates in the form (r, θ). If a ray is drawn from the pole through point P, the distance from the pole to point P is $|r|$. The measure of the angle formed by $\overrightarrow{OP}$ and the polar axis is θ. The angle can be measured in degrees or radians. *This grid is sometimes called the polar plane.*

If r is positive, θ is the measure of any angle in standard position that has $\overrightarrow{OP}$ as its terminal side.

If r is negative, θ is the measure of any angle that has the ray opposite $\overrightarrow{OP}$ as its terminal side.

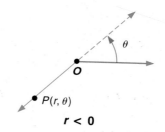

$P(r, \theta)$

θ

O

$r > 0$

θ

O

$P(r, \theta)$

$r < 0$

Example 1

Graph each point.
a. $P(3, 120°)$

a. On a polar plane, sketch the terminal side of a 120° angle in standard position.

Since r is positive, find the point along the terminal side 3 units from the pole.

Notice that point P is on the third circle from the pole.

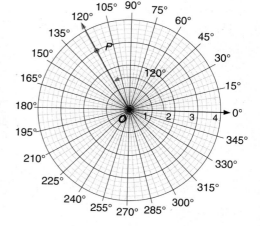

b. $Q\left(-2.5, \dfrac{\pi}{2}\right)$

b. Sketch the terminal side of an angle of $\dfrac{\pi}{2}$ radians in standard position.

Since r is negative, extend the terminal side in the opposite direction. Find the point along this ray 2.5 units from the pole.

Notice that point Q is halfway between the second and third circle from the pole.

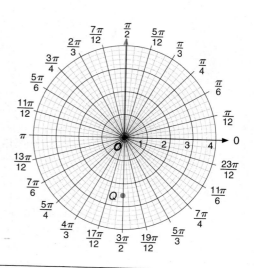

You have seen that r can either be negative or positive. The angle measure θ can also be either negative or positive. If θ is positive, the angle is measured counterclockwise from the polar axis. If θ is negative, the angle is measured clockwise from the polar axis.

Example 2

Notice that the terminal side of $-\frac{\pi}{3}$ is coincident with that of $\frac{5\pi}{3}$. Also, $\left|-\frac{\pi}{3}\right| + \left|\frac{5\pi}{3}\right| = 2\pi$.

Graph the point P that has polar coordinates $\left(-1, -\frac{\pi}{3}\right)$.

Negative angles are measured clockwise. Sketch the terminal side of the angle. Since r is negative, the point $\left(-1, -\frac{\pi}{3}\right)$ is 1 unit from the pole along the ray *opposite* the terminal side of the angle.

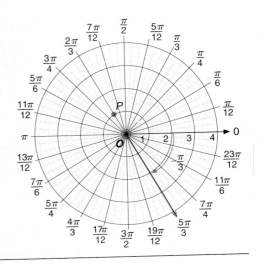

Example 3

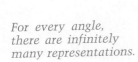

The figure at the right shows a baseball diamond placed on a rectangular coordinate plane. The coordinates of second base are $(90, 90)$ since the bases are 90 feet apart. Find a pair of polar coordinates for second base if the first-base line is the polar axis and home plate is at the pole.

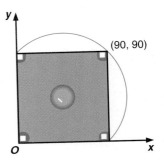

Since the baselines form a square, a ray from home plate through second base forms a $45°$ angle with the first-base line. The polar coordinate for θ is $45°$.

The value of r can be found by using the Pythagorean theorem.

$r = \sqrt{90^2 + 90^2}$

$r = \sqrt{16,200}$

$r \approx 127.28$

A pair of polar coordinates for second base is $(127.28, 45°)$.

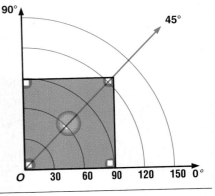

Remember from your study of trigonometry that an angle can be represented in more than one way. For example, a $390°$ angle is the same as a $30°$ angle. Likewise on the polar plane, a point can be represented by more than one pair of polar coordinates. Point P is named in six ways below.

For every angle, there are infinitely many representations.

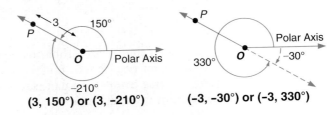

(3, 150°) or (3, −210°)   **(−3, −30°) or (−3, 330°)**

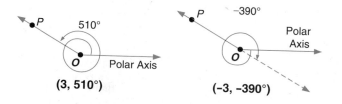

(3, 510°)   **(−3, −390°)**

Multiple Representations of (r, θ)	If P is a point with polar coordinates (r, θ), then P can also be represented by the polar coordinates $(-r, \theta + (2k + 1)\pi)$ or $(r, \theta + 2k\pi)$, where k is any integer.

Example 4

Name four different pairs of polar coordinates that represent point R with the restriction that $-360° \leq \theta \leq 360°$.

Point R can be represented by $(3, 80°)$. Use the formulas for multiple representations of polar coordinates to find three more pairs.

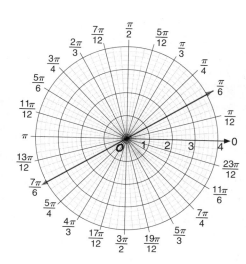

$(-r, \theta + (2k + 1)180°) \rightarrow (-(3), 80° \pm (1)180°)$
$\rightarrow (-3, 260°)$ or $(-3, -100°)$
$(r, \theta + 360k°) \rightarrow (3, 80° + (-1)(360°))$
$\rightarrow (3, -280°)$

An equation that uses polar coordinates is called a **polar equation**. For example, $r = 4 \cos \theta$ is a polar equation. A **polar graph** represents the solution set which is the set of points whose coordinates (r, θ) satisfy a given polar equation.

In the Cartesian coordinate system, you studied special types of equations involving constants, such as $y = 3$, $x = -4$, $y^2 = 4$, and $x^3 = 27$. The solutions to these equations had a unique representation on the coordinate plane. The same is true for solutions of polar equations.

Example 5

Graph each polar equation.

a. $\theta = \dfrac{\pi}{6}$

a. The solution to this equation consists of all points with coordinates of the form $\left(r, \dfrac{\pi}{6}\right)$. That is, r can have any value as long as θ is $\dfrac{\pi}{6}$. The graphs of these points form a line composed of the ray that forms an angle of $\dfrac{\pi}{6}$ with the polar axis and, to represent negative values of r, the ray opposite that ray.

b. $r = 2$

b. The solution to this equation consists of all points with coordinates of the form $(2, \theta)$. That is, θ can have any value as long as r is 2. The graphs of these points form a circle with radius 2.

This circle is also the graph of $r = -2$. Why?

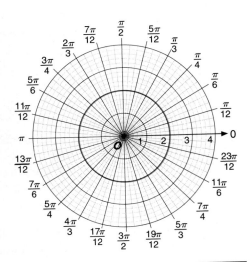

CHECKING FOR UNDERSTANDING

Communicating Mathematics

Read and study the lesson to answer each question.

1. **Determine** which two locations on Earth each of the Iridium satellites will pass over in their orbit.

2. **Explain** why a point in the polar coordinate system cannot be named by a unique ordered pair (r, θ).

3. **Explain** why the graph in Example 5b is also the graph of $r = -2$.

Guided Practice

Graph the point represented by the given polar coordinates. Then name three other pairs of polar coordinates that represent the same point with $-360° \leq \theta \leq 360°$.

4. $(100, 50°)$

5. $(12.8, \frac{\pi}{6})$

6. $(1.2, 70°)$

Graph each polar equation.

7. $r = 3$

8. $\theta = 10°$

9. $\theta = \frac{\pi}{2}$

EXERCISES

Practice

Graph the point that has the given polar coordinates.

10. $(3, 0)$

11. $(2, \frac{\pi}{4})$

12. $(\frac{1}{4}, \frac{2\pi}{3})$

13. $(\frac{1}{2}, \frac{-3\pi}{2})$

14. $(-2, \frac{5\pi}{6})$

15. $(-3, \frac{\pi}{3})$

16. $(0.5, 1)$

17. $(4, -3)$

18. $(6, -45°)$

19. $(8, 585°)$

20. $(-4, 135°)$

21. $(2, -30°)$

Name four different pairs of polar coordinates that represent point A.

22.

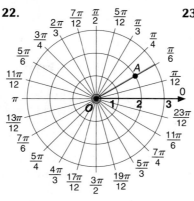

23.

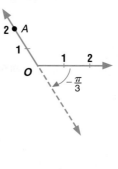

24.

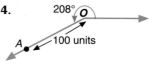

Graph each polar equation.

25. $r = 6$

26. $r = \sqrt{3}$

27. $r = (2)^3$

28. $\theta = \dfrac{\pi}{4}$

29. $\theta = -\pi$

30. $r = 1.27$

31. $\theta = 145°$

32. $\theta = -\dfrac{5\pi}{6}$

33. $\theta = -220°$

34. $r = -4$

35. $r = 1.5$

36. $\theta = 360°$

37. Find an ordered pair of polar coordinates to represent the point whose rectangular coordinates are $(-3, 4)$.

Critical Thinking

38. Three circles are externally tangent to each other. If each circle has a radius of 4 units and each side of the triangle is tangent to two of the circles, find the area of the triangle.

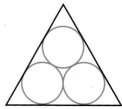

Applications and Problem Solving

39. Baseball   Refer to Example 3.
   a. What are the polar coordinates for first and third bases?
   b. What are the polar coordinates for the pitcher's mound if the pitcher's mound is located 60.5 feet from home plate on a line from home plate to second base?

40. Air Safety   The largest airport in the world is the King Khalid International Airport outside Riyadh, Saudi Arabia. It covers 86 square miles. It also has the largest control tower, which is 243 feet high. The radar at the control tower is at the center of a large polar coordinate system that monitors flights in and out of the airport. The polar axis is due east and r is measured in miles.

   a. If an airplane is on its final approach and is at a point with coordinates $(5, 270°)$, which direction is the airplane heading as it lands?

   b. How far is the airplane from the control tower?

41. Solve $4x^2 - 9x + 5 = 0$ by using the quadratic formula.   **(Lesson 4-2)**

42. Change $\dfrac{3\pi}{8}$ radians to degree measure to the nearest minute.   **(Lesson 5-1)**

43. State which is greater, $\cos 20° + \sin 50°$ or $\cos 80° + \sin 40°$.   **(Lesson 6-1)**

44. For $0° < \theta < 90°$, find $\cot \theta$ if $\tan \theta = \dfrac{\sqrt{2}}{5}$.   **(Lesson 7-1)**

45. Jessie kicked a football with an initial velocity of 60 ft/s at an angle of 60° to the horizontal. After 0.5 seconds, how far has the ball traveled horizontally and vertically? *Use* $x = t|\vec{v}| \cos \theta,\ y = t|\vec{v}| \sin \theta + \dfrac{1}{2}gt^2,\ and\ g = 32\ ft/s^2.$ **(Lesson 8-7)**

46. College Entrance Exam   Choose the best answer.
The number of degrees through which the hour hand of a clock moves in 2 hours 12 minutes is

(A) 66°   **(B)** 72°   **(C)** 126°   **(D)** 732°   **(E)** 792°

DECISION MAKING

The Budget Deficit

Each year the federal government spends several hundred billion dollars more than it takes in, falling further and further into debt. Because all possible solutions to the problem are politically unpopular, Congress and the President have been slow to address the problem. What do you think should be done to get the budget under control? Since belt-tightening will require sacrifice on everyone's part, how would you convince people that such sacrifices are necessary? What can you do to raise public and congressional awareness of the problem and help in the search for a solution?

1. Research causes and effects of the federal deficit. How long has it been a problem? Why have legislators been negligent in coming to grips with the deficit? How does the deficit affect the country?

2. Describe steps that could be taken to reduce the budget deficit. If you were a legislator, what would you do to help get the deficit under control?

3. Describe what you can do to help find a solution to the problem. Would it help to write a letter to your congressional representative? Could you work to raise public consciousness about the dangers of the deficit through a publicity campaign?

4. Project   Work with your group to complete one of the projects listed at the right based on the information you gathered in Exercises 1–3.

PROJECTS

- *Conduct an interview.*
- *Write a book report.*
- *Write a proposal.*
- *Write an article for the school paper.*
- *Write a report.*
- *Make a display.*
- *Make a graph or chart.*
- *Plan an activity.*
- *Design a checklist.*

9-2A Graphing Calculators: Graphing Polar Equations

You can graph polar equations and read polar coordinates on a TI-82 graphing calculator by changing the MODE settings. To begin, press MODE and select polar graphing by using the arrow keys to move to the word "Pol" and pressing ENTER when it is blinking. Also, be sure the calculator is in radian mode. Then press CLEAR to exit the mode screen.

Check the WINDOW. Make sure you have selected an appropriate viewing window.

Example 1

Graph the polar equation $r = 2 \cos \theta$ in the viewing window $[-5, 5]$ by $[-2, 2]$ with a scale factor of 1 for both axes and θmin $= 0$, θmax $= 3.1416$, and θstep $= 0.05$.

Make sure the calculator is in radian mode.

Enter:

The graph appears to be an ellipse, but pressing ZOOM 5 to graph the equation squarely shows that it is actually a circle with radius 2 and center at $(1, 0)$.

You can also read both the rectangular and polar coordinates of a graph by pressing WINDOW ▶ and the arrow keys to highlight "RectGC" to read the rectangular coordinates, and "PolarGC" to read the polar coordinates. Pressing the TRACE key to trace along a graph will tell you which coordinates are currently being read.

Graph the polar equation $r = \dfrac{3}{\sin \theta}$ **in the viewing window** $[-5, 5]$ **by** $[-5, 5]$
with scale factors of 1 for both axes and θmin $= 0$, θmax $= 2\pi$, **and**
θstep $= 0.05$.

Enter: [Y=] 3 [÷] [SIN]
[X,T,θ] [GRAPH]

Graphing the equation shows that
the graph is a line.

There are several different types of curves that can be formed by graphing
polar equations.

Graph the polar equation $r = 8 \cos 4\theta$ **in the standard viewing window**
with scale factors of 1 for both axes and θmin $= 0$, θmax $= 2\pi$, **and**
θstep $= 0.05$.

Enter: [Y=] 8 [COS] 4 [X,T,θ]
[GRAPH]

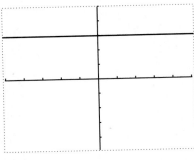

The graph is an eight-petaled flower
with its center at the origin.

You can experiment with θmin, θmax, and θstep values to get an idea as
to how they affect the graph of a polar equation. For instance, in Example 3,
changing θmax to π will only produce half of the rose, and changing θstep to 1
will completely distort the graph because only 7 points will be plotted, namely
$\theta = 0, 1, 2, 3, 4, 5$, and 6.

EXERCISES

**Graph each equation on a TI-82 graphing calculator. Then sketch the graph on a
piece of paper.**

1. $r = 0.50\theta, 0 \le \theta \le 8\pi$

2. $r = 3\sqrt{\cos 2\theta}, 0 \le \theta \le 2\pi$

3. $r = 6 + 6 \sin \theta, 0 \le \theta \le 2\pi$

4. $r = 7 + 4 \cos \theta, 0 \le \theta \le 2\pi$

5. $r = 8 \sin 4\theta, 0 \le \theta \le 2\pi$

6. $r = 4\sqrt{\sin 2\theta}, 0 \le \theta \le 2\pi$

7. $r = 1 + 2 \sin \theta, 0 \le \theta \le 2\pi$

8. $r = 10 + 10 \cos \theta, 0 \le \theta \le 2\pi$

9-2 Graphs of Polar Equations

Objective

After studying this lesson, you should be able to:
■ graph polar equations.

Application

A nautilus is a mollusk native to the southern Pacific and Indian Oceans. Inside its shell are spiraling chambers that appear to be lined with a pearl-like substance. The chambers are connected by a tube that absorbs gases from the chambers allowing the shell to act as a float.

Curves like the one suggested in this shell are called **classical curves** and can be graphed using polar coordinates.

Polar graphs can be classified in families. The graphs below show the parent graphs of $r = \sin\theta$ and $r = \cos\theta$.

$r = \sin\theta$

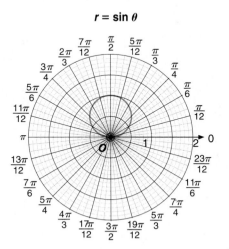

$r = \cos\theta$

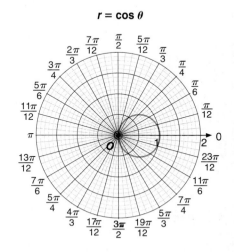

Notice that both of these parent graphs are circles with a diameter of 1 unit. Both pass through the origin. As with the families of graphs you studied in Chapter 3, you can alter the position and shape of the graph by multiplying the function by a number or by adding to it. You can also multiply θ by a number or add a number to it in order to alter the graph. However, the changes in the graphs of polar equations can be quite different from those you studied in Chapter 3.

Example 1

Graph each polar equation.
a. $r = 6 \sin \theta$

a. Make a table of values. Round the values of r to the nearest tenth. Graph the ordered pairs and connect them with a smooth curve.

θ	$\sin \theta$	$6 \sin \theta$	(r, θ)
$0°$	0	0	$(0, 0°)$
$30°$	0.5	3	$(3, 30°)$
$60°$	0.9	5.4	$(5.4, 60°)$
$90°$	1	6	$(6, 90°)$
$120°$	0.9	5.4	$(5.4, 120°)$
$150°$	0.5	3	$(3, 150°)$
$180°$	0	0	$(0, 180°)$
$210°$	-0.5	-3	$(-3, 210°)$
$240°$	-0.9	-5.4	$(-5.4, 240°)$
$270°$	-1	-6	$(-6, 270°)$
$300°$	-0.9	-5.4	$(-5.4, 300°)$
$330°$	-0.5	-3	$(-3, 330°)$
$360°$	0	0	$(0, 360°)$

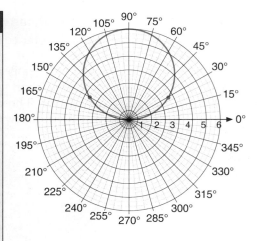

The complete graph is determined by values for θ between $0°$ and $180°$. The graph of $r = 6 \cos \theta$ looks like the parent graph of $r = \sin \theta$, but has a diameter 6 times that of the parent.

b. $r = 2 + 3 \cos \theta$

b.

θ	$2 + 3 \cos \theta$	(r, θ)
0	5	$(5, 0)$
$\frac{\pi}{6}$	4.6	$(4.6, \frac{\pi}{6})$
$\frac{\pi}{3}$	3.5	$(3.5, \frac{\pi}{3})$
$\frac{\pi}{2}$	2	$(2, \frac{\pi}{2})$
$\frac{2\pi}{3}$	0.5	$(0.5, \frac{2\pi}{3})$
$\frac{5\pi}{6}$	-0.6	$(-0.6, \frac{5\pi}{6})$
π	-1	$(-1, \pi)$
$\frac{7\pi}{6}$	-0.6	$(-0.6, \frac{7\pi}{6})$
$\frac{4\pi}{3}$	0.5	$(0.5, \frac{4\pi}{3})$
$\frac{3\pi}{2}$	2	$(2, \frac{3\pi}{2})$
$\frac{5\pi}{3}$	3.5	$(3.5, \frac{5\pi}{3})$
$\frac{11\pi}{6}$	4.6	$(4.6, \frac{11\pi}{6})$
2π	5	$(5, 2\pi)$

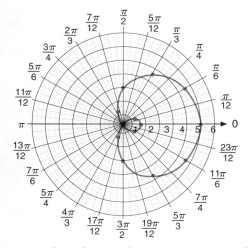

Notice that the graph is symmetric with respect to the polar axis, as is its parent graph. However, the appearance of a small loop inside the outside graph, which is not a circle, does not resemble the parent graph. This graph is called a <u>limaçon</u>.

In Lesson 9-1, you studied two types of polar equations involving constants, for example, $r = 6$ and $\theta = \frac{\pi}{3}$. Another polar equation of this type is $r = n\theta$, where n is a real number and θ is expressed in terms of radians. *In order to determine the exact shape of the graph of a polar equation, you may need to graph many polar coordinates.*

Example 2

Graph $r = 2\theta$.

θ	2θ	(r, θ)
$\frac{\pi}{6}$	$\frac{\pi}{3}$	$\left(1, \frac{\pi}{6}\right)$
$\frac{\pi}{3}$	$\frac{2\pi}{3}$	$\left(2.1, \frac{\pi}{3}\right)$
$\frac{\pi}{2}$	π	$\left(3.1, \frac{\pi}{2}\right)$
$\frac{2\pi}{3}$	$\frac{4\pi}{3}$	$\left(4.2, \frac{2\pi}{3}\right)$
$\frac{5\pi}{6}$	$\frac{5\pi}{3}$	$\left(5.2, \frac{5\pi}{6}\right)$
π	2π	$(6.3, \pi)$
$\frac{7\pi}{6}$	$\frac{7\pi}{3}$	$\left(7.3, \frac{7\pi}{6}\right)$
$\frac{4\pi}{3}$	$\frac{8\pi}{3}$	$\left(8.4, \frac{4\pi}{3}\right)$
$\frac{3\pi}{2}$	3π	$\left(9.4, \frac{3\pi}{2}\right)$
$\frac{5\pi}{3}$	$\frac{10\pi}{3}$	$\left(10.5, \frac{5\pi}{3}\right)$
$\frac{11\pi}{6}$	$\frac{11\pi}{3}$	$\left(11.5, \frac{11\pi}{6}\right)$
2π	4π	$(12.6, 2\pi)$

In this example, θ must be expressed in radians since r is a real number.

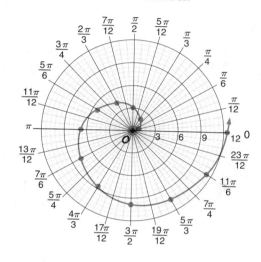

This type of graph is called the spiral of Archimedes, named after the Greek mathematician who discovered it.

The limaçon in Example 1b and the *spiral of Archimedes* in Example 2 are two of the classical curves, which can be formed by graphing polar equations. The equations that distinguish these and other classical curves are listed in the chart below.

Classical Curves		
Curve Name	**Polar Equation**	**General Graph**
rose	$r = a \cos n\theta$ $r = a \sin n\theta$ *n is a positive integer.*	

lemniscate (pronounced lehm NIHS kuht)	$r^2 = a^2 \cos 2\theta$ $r^2 = a^2 \sin 2\theta$	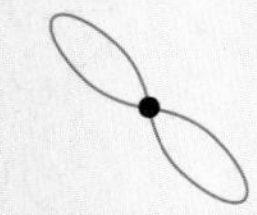
limaçon (pronounced lee muh SOHN)	$r = a + b \cos \theta$ $r = a + b \sin \theta$	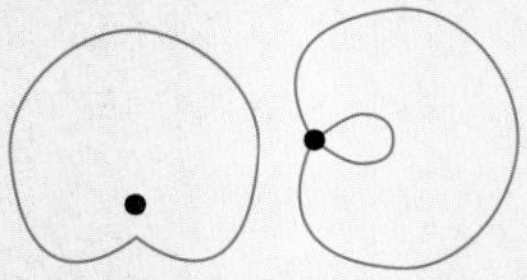
cardioid (pronounced KARD ee oyd)	$r = a + a \cos \theta$ $r = a + a \sin \theta$	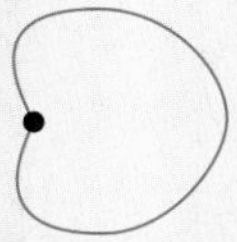
spiral of Archimedes (pronounced ar kih MEED eez)	$r = a\theta$	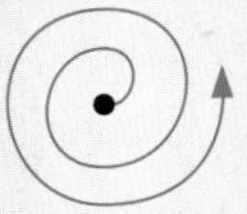

As with the equations you have graphed in a rectangular coordinate system, you can graph more than one polar equation on the same polar coordinate system. However, the points where the graphs intersect do not always represent common solutions to equations, since each point can be represented by infinitely many polar coordinates.

Example 3

Graph the system of polar equations below.

$r = 3 \cos 2\theta$

$r = -3$

Then solve the system and compare the common solutions to the points of intersection of the polar graphs.

Graph each of the polar equations. The graphs intersect at the points with coordinates $(3, 0°), (3, 90°), (3, 180°)$, and $(3, 270°)$.

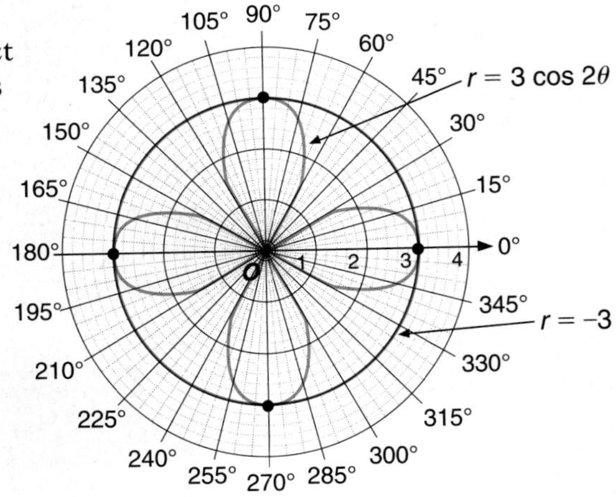

To solve the system of equations, substitute -3 for r in the first equation and solve for θ.

$3 \cos 2\theta = -3$

$\cos 2\theta = -1$

$2\theta = \text{Arccos}(-1)$

$2\theta = 180° \text{ or } -180°$

$\theta = 90° \text{ or } -90°$

Thus, the common solutions are $(-3, 90°)$ and $(-3, -90°)$.

Now compare the common solutions to the coordinates of the points of intersection of the graphs.

The point represented by $(3, 90°)$ is the same point represented by $(-3, -90°)$. The point at $(3, 270°)$ is the same as the point at $(-3, 90°)$. So the other two points of intersection do not represent common solutions to the polar equations.

Polar equations can be written in parametric form and graphed on a graphing calculator. The parametric equations for a polar equation $r = f(\theta)$ are $x = f(t) \cos t$ and $y = f(t) \sin t$. For example, the parametric equations for $r = 3 \cos 2\theta$ are $x = 3 \cos 2t \cos t$ and $y = 3 \cos 2t \sin t$.

EXPLORATION: Graphing Calculator

Graph $r = 2 + 2 \sin \theta$ and $r = 2 \sin 2\theta$ on a graphing calculator and determine the common solutions.

1. Set the mode and the viewing window by using the settings below.

RAD MODE	PARAM MODE	POLAR MODE
TMIN: 0	TMAX: 7	TSCL: 0.05
XMIN: -4	XMAX: 5	XSCL: 1
YMIN: -2	YMAX: 5	YSCL: 1

2. Write the polar equations in parametric form. The parametric equations for $r = 2 + 2 \sin \theta$ are $x = (2 + 2 \sin t) \cos t$ and $y = (2 + 2 \sin t) \sin t$ and the parametric equations for $r = 2 \sin 2\theta$ are $x = 2 \sin 2t \cos t$ and $y = 2 \sin 2t \sin t$. Press the $\boxed{Y=}$ key and enter the equations.

3. Press $\boxed{\text{GRAPH}}$.

4. Press $\boxed{\text{TRACE}}$ and position the cursor over each point of intersection. The up and down arrows will allow us to move back and forth between the two graphs. If the r- and θ-coordinates of the points of intersection are the same for both graphs, then the point is a common solution to the system of equations. What is the common solution to this system of equations?

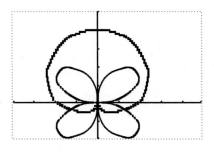

CHECKING FOR UNDERSTANDING

Communicating Mathematics

Read and study the lesson to answer each question.

1. **Determine** which of the classical polar curves best represents the shape of the chambered nautilus.

2. **Explain** why you cannot use a polar coordinate plane scaled in degrees in order to graph $r = a\theta$.

Copy and complete the table for each polar equation. Then graph the equation.

3. $r = 8 \cos \theta$

θ	$\cos \theta$	$8 \cos \theta$	(r, θ)
0°			
30°			
60°			
90°			
120°			
150°			
180°			
210°			
240°			
270°			
300°			
330°			

4. $r = 2 + 2 \sin \theta$

θ	$\sin \theta$	$2 + 2 \sin \theta$	(r, θ)
0°			
30°			
60°			
90°			
120°			
150°			
180°			
210°			
240°			
270°			
300°			
330°			

5. Identify the classical curve that the graph of $r = 2 + 2 \sin \theta$ represents.

EXERCISES

Graph each polar equation. Identify the classical curve it represents.

6. $r = 6\theta$

7. $r = \dfrac{\theta}{2}$

8. $r = 5 + 2 \cos \theta$

9. $r = 5 \sin 5\theta$

10. $r = 3 + 3 \sin \theta$

11. $r = 4 + 6 \cos \theta$

12. $r = 5 \cos 2\theta$

13. $r^2 = 16 \cos 2\theta$

14. $r^2 = 4 \sin 2\theta$

15. $r = 4 + 4 \sin \theta$

16. $r = 6 \cos 3\theta$

17. $r = 3 \cos 3\theta$

Graph each system of equations. Solve the system and compare the common solutions to the points of intersection of the polar graphs. Assume $-2\pi \le \theta \le 2\pi$.

18. $r = \sin \theta$
     $r = 1 - \sin \theta$

19. $r = 2\sqrt{3} \cos \theta$
     $r = 2 \sin \theta$

20. $r = 4 \cos \theta$
     $r = 2 \sin 90°$

Graph each system of equations using a graphing calculator. Then determine the common solutions. Round your answers to the nearest tenth.

21. $r = 3 \sin \theta$
     $r = 3 \cos \theta$

22. $r = 3 \sin \theta$
     $r = 3 \sin 3\theta$

23. $r = 3 \cos \theta - 1$
     $r = 3 \sin 2\theta$

24. If the polar equation of a four-petaled rose can be generalized as $r = a \cos 2\theta$ or $r = a \sin 2\theta$, make a conjecture about the relationship between the value of a and the graph.

25. If the polar equation of a rose can be generalized as $r = a \cos n\theta$ or $r = a \sin n\theta$, make a conjecture about the relationship between the value of n and the graph.

26. What effect does replacing the cosine function by the sine function have on the graph of a polar rose such as $r = a \cos n\theta$?

Applications and Problem Solving

27. Toys In the 1960s, the Kenner Toy Company introduced an art toy, called a Spirograph®, in which a child guided a notched circle inside a notched circular frame to create designs. The type of design created depended on the size of the circle chosen and the size of the frame. One of the designs you could create can also be generated by graphing $r^2 = \sin 10\theta$. Graph the design. Then describe the shape of the graph.

28. History In addition to his other mathematical accomplishments, the French mathematician Pierre de Fermat (1601–1665) is credited with discovering a spiral defined by the equation $r^2 = \theta$. Graph Fermat's spiral.

29. Research Use a dictionary or encyclopedia to investigate the origin of the terms *lemniscate, limaçon,* and *cardioid.* In your own words, explain why you think these names were chosen for the classical curves.

Mixed Review

30. Surveying To find the height of a mountain peak, points A and B were located on a plain in line with the peak and the angle of elevation was measured from each point. The angle at A was $36°40'$, and the angle at B was $21°10'$. The distance from A to B was 570 feet. How high is the peak above the level of the plain? **(Lesson 5-5)**

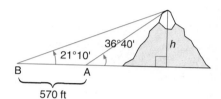

31. Use the sum and difference identities to find the exact value of $\sin 255°$. **(Lesson 7-3)**

32. Find the inner product $(3, -2, 4) \cdot (1, -4, 0)$. Then state whether the vectors are perpendicular. Write *yes* or *no.* **(Lesson 8-4)**

33. Name three other pairs of polar coordinates that represent the point at $(10, -20°)$. Suppose $-360° \le \theta \le 360°$. **(Lesson 9-1)**

34. College Entrance Exam What is the least positive number greater than 2 that has a remainder of 2 when it is divided by 3, 4, or 5?

CASE STUDY FOLLOW-UP

Refer to Case Study 2: *Trashing the Planet,* on pages 959–961.

Use tracing paper to transfer the photographs of the holes in the ozone layer onto polar graph paper.

1. Use polar coordinates to describe changes in the hole from 1989 to 1991. Estimate the increases in the radius and the area of the hole.

2. Describe the shapes of the holes in terms of classical curves.

3. How many more deaths due to skin cancer can be expected in the next 50 years as a result of the depletion of the ozone layer?

4. **Research** Find out about changes in the ozone layer since the photos were taken. In a 1-page paper, describe the changes and current efforts to solve the problem.

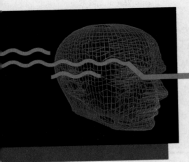

Technology

▶ **BASIC**
Spreadsheets
Software

Graphing Polar Equations

Polar equations are often used to represent complex graphs, such as roses and spirals. Some computer-generated animations use these graphs to create visual effects.

The BASIC program shown at the right will generate twenty-five ordered pairs for the polar equation $r = 2\cos\theta + 3\sin\theta$. You can plot the points and connect them with a smooth curve to graph the equation.

```
120  PRINT "PLOT POINTS, THEN CONNECT"
130  PRINT "THEM IN THE ORDER SHOWN."
140  PRINT "R", "THETA"
150  PRINT "-------", "-------"
160  FOR N=0 TO 24
170  THETA=N/12*3.1415926
180  R=2*COS(THETA)+3*SIN(THETA)
190  PRINT R, N; "PI/12"
200  IF N<>12 THEN GOTO 230
210  PRINT "PRESS [ENTER] FOR MORE...";
220  INPUT A$
230  NEXT N
240  END
```

```
PLOT POINTS, THEN CONNECT
THEM IN THE ORDER SHOWN.
R                      THETA
-------                -------
2                      0 PI/12
2.708309               1 PI/12
3.232051               2 PI/12
3.535534               3 PI/12
3.598076               4 PI/12
3.415416               5 PI/12
3                      6 PI/12
2.38014                7 PI/12
1.598076               8 PI/12
.7071068               9 PI/12
-.2320506             10 PI/12
-1.155394             11 PI/12
-2                    12 PI/12
PRESS [ENTER] FOR MORE...
-2.708309             13 PI/12
-3.232051             14 PI/12
```

To execute the program, type <u>RUN</u> and then press [Enter]. Part of the program output is shown at the left. You can change line 180 of the program to generate ordered pairs for any polar equation of the form $r = f(\theta)$. The graph of $r = 2\cos\theta + 3\sin\theta$ is shown below.

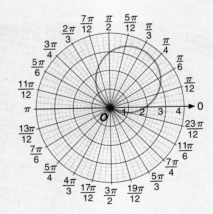

EXERCISES

Use the program to generate ordered pairs for each equation. Then sketch the graph.

1. $r = \cos\theta - 2\sin\theta$

2. $r = 4 - 3\sin\theta$

3. $r = \cos\left(2\theta + \dfrac{\pi}{3}\right)$

4. $r = 0.75\theta$

9-3 Polar and Rectangular Coordinates

Objective

After studying this lesson, you should be able to:
- convert from polar coordinates to rectangular coordinates and vice versa.

Application

In Lesson 9-1, we talked about longitude lines and latitude lines forming a polar coordinate system. However, when was the last time you saw a map of the world done on a polar plane? Instead, you have probably seen the land masses and waterways of the world portrayed on a rectangular system. On maps of this type, the actual shapes of Earth's features are distorted to fit the rectangular system.

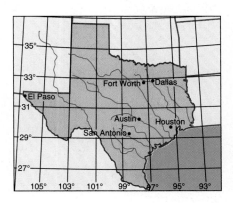

While some real-world phenomena cannot be modeled easily using a rectangular coordinate system, it is possible to write polar coordinates as rectangular coordinates and vice versa.

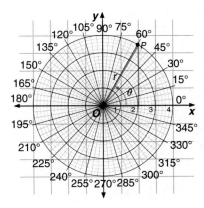

Suppose a rectangular coordinate system is superimposed on a polar coordinate system so that the origins coincide and the x-axis aligns with the polar axis. Let P be any point in the plane. In polar coordinates, P is identified by (r, θ). In rectangular coordinates, P is identified by (x, y). Trigonometric functions can be used to find the rectangular coordinates if you know the polar coordinates. *You will develop this formula in Exercise 37.*

Conversion from Polar Coordinates to Rectangular Coordinates

The rectangular coordinates (x, y) of a point named by the polar coordinates (r, θ) can be found by using the following formulas.

$$x = r \cos \theta$$
$$y = r \sin \theta$$

Example 1 | **Find the rectangular coordinates of each point.**

a. $A\left(-2, \dfrac{3\pi}{4}\right)$

a. For $\left(-2, \dfrac{3\pi}{4}\right)$, $r = -2$ and $\theta = \dfrac{3\pi}{4}$.

$x = r\cos\theta$ $\qquad\qquad$ $y = r\sin\theta$

$\qquad = -2\cos\left(\dfrac{3\pi}{4}\right)$ $\qquad\qquad = -2\sin\left(\dfrac{3\pi}{4}\right)$

$\qquad = -2\left(-\dfrac{\sqrt{2}}{2}\right)$ $\qquad\qquad = -2\left(\dfrac{\sqrt{2}}{2}\right)$

$\qquad = \sqrt{2}$ $\qquad\qquad\qquad = -\sqrt{2}$

The rectangular coordinates of A are $(\sqrt{2}, -\sqrt{2})$ or about $(1.41, -1.41)$.

b. $B(6, -50°)$

b. $x = r\cos\theta$ $\qquad\qquad\qquad$ $y = r\sin\theta$

$\quad = 6\cos(-50°)$ $\qquad\qquad\qquad = 6\sin(-50°)$

$\quad = 6\cos 50°$ $\quad cos\,(-50°) = cos\,50°$ $\qquad = 6(-\sin 50°)$ $\quad sin(-50°) = -sin\,50°$

$\quad \approx 6(0.64279)$ $\qquad\qquad\qquad \approx 6(-0.76604)$

$\quad \approx 3.86$ $\qquad\qquad\qquad\qquad \approx -4.60$

The rectangular coordinates of B are about $(3.86, -4.60)$.

If a point is named by the rectangular coordinates (x, y), you can find the corresponding polar coordinates by using the Pythagorean theorem and the Arctangent function. Since the Arctangent function only determines angles in the first or fourth quadrants, you must add π radians to the value of θ for points with coordinates (x, y) that lie in the second or third quadrants.

When x is zero,
$\theta = \pm\dfrac{\pi}{2}$. *Why?*

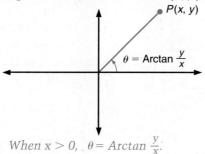

When x > 0, $\theta = Arctan\,\dfrac{y}{x}$.

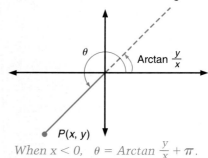

When x < 0, $\theta = Arctan\,\dfrac{y}{x} + \pi$.

Conversion from Rectangular Coordinates to Polar Coordinates

The polar coordinates (r, θ) of a point named by the rectangular coordinates (x, y) can be found by the following formulas.

$$r = \sqrt{x^2 + y^2}$$

$$\theta = \text{Arctan}\,\dfrac{y}{x}, \quad \text{when } x > 0$$

$$\theta = \text{Arctan}\,\dfrac{y}{x} + \pi, \quad \text{when } x < 0$$

In this formula, θ is expressed in radians.

Example 2

You can review the inverse trigonometric functions in Lesson 6-4.

Find the polar coordinates of $C(-3, 5)$.

$$r = \sqrt{x^2 + y^2} \qquad\qquad \theta = \text{Arctan } \frac{y}{x} + \pi \qquad (x < 0)$$

$$= \sqrt{(-3)^2 + 5^2} \qquad\qquad = \text{Arctan } \left(\frac{5}{-3}\right) + \pi$$

$$= \sqrt{34} \qquad\qquad\qquad \approx 2.11$$

$$\approx 5.83$$

The polar coordinates of C are $(5.83, 2.11)$.
Other polar coordinates can also represent this point.

Equations expressed in rectangular form can also be written as polar equations and vice versa.

Example 3

Write the rectangular equation $x^2 + y^2 = 25$ in polar form.

$$x^2 + y^2 = 25$$

$$(r\cos\theta)^2 + (r\sin\theta)^2 = 25 \qquad x = r\cos\theta, y = r\sin\theta$$

$$r^2(\cos^2\theta + \sin^2\theta) = 25$$

$$r^2 = 25 \qquad \cos^2\theta + \sin^2\theta = 1$$

$$r = \pm 5$$

The polar equation is $r = 5$ or $r = -5$.

Polar equations are excellent models for natural phenomena.

Example 4

APPLICATION

Geology

A drumlin is an elliptical, streamlined hill composed of till deposited beneath glacial ice. In 1959, Richard J. Chorley concluded that the shape of a drumlin could best be modeled by the petal of a rose curve. The equation for a drumlin can be expressed as $r = \ell \cos k\theta$ for $\frac{-\pi}{2k} \le \theta \le \frac{\pi}{2k}$, where ℓ is the length of the drumlin and $k > 1$ is a parameter that is the ratio of the length to the width. Chorley's analysis revealed that the area covered by a drumlin was approximated by $A = \frac{\ell^2\pi}{4k}$. Find the area in square meters of a drumlin modeled by the equation $r = 420\cos 3\theta$.

In the equation $r = 420\cos 3\theta$, $\ell = 420$ and $k = 3$. $\quad r = \ell\cos k\theta$

$$A = \frac{\ell^2\pi}{4k}$$

$$= \frac{420^2\pi}{4(3)}$$

$$= 46{,}181.41201$$

The area of this drumlin is about 46,181.4 square meters.

CHECKING FOR UNDERSTANDING

Communicating Mathematics

Read and study the lesson to answer each question.

1. **Describe** any differences that appear in the graphs of the two polar equations found in Example 3.

2. **Explain** how you compensate for an angle whose terminal side lies in the second or third quadrant when using the Arctan function to change from rectangular to polar coordinates.

Guided Practice

Find each value. Express θ in radians to the nearest hundredth.

3. Arctan 1
4. Arctan 4.5
5. Arctan $\sqrt{6}$

Find the polar coordinates of each point with the given rectangular coordinates.

6. $(5, 12)$
7. $(-2, -2)$
8. $(-1, -3)$

Find the rectangular coordinates of each point with the given polar coordinates.

9. $\left(3, \dfrac{\pi}{2}\right)$
10. $\left(-2, \dfrac{\pi}{4}\right)$
11. $(2.5, 2)$

12. Write $x^2 + y^2 = 9$ in polar form.

13. Write $r = 7$ in rectangular form.

EXERCISES

Practice

Find the polar coordinates of each point with the given rectangular coordinates.

14. $(0, 1.5)$
15. $(0, 3)$
16. $\left(\dfrac{\sqrt{3}}{2}, \dfrac{1}{2}\right)$
17. $(2, 0)$
18. $(-0.25, 0)$
19. $(-\sqrt{2}, -\sqrt{2})$

Find the rectangular coordinates of each point with the given polar coordinates.

20. $(\sqrt{2}, 45°)$
21. $(5, 0.93)$
22. $(\sqrt{29}, -1.19)$
23. $(\sqrt{2}, 4.39)$
24. $(\sqrt{13}, -0.59)$
25. $(3.464, 2.09)$

Write each rectangular equation in polar form.

26. $y = -5$
27. $x = 10$
28. $x^2 + y^2 = 7$
29. $2x^2 + 2y^2 = 5y$

Write each polar equation in rectangular form.

30. $r = 12$
31. $\theta = -45°$
32. $r \sin \theta = 4$
33. $r = -2 \sec \theta$

34. Write $x = y$ in polar form.

35. Write $r = \cos \theta + \sin \theta$ in rectangular form.

Critical Thinking

36. Find the equation in rectangular form that has the same graph as the polar equation $r = \dfrac{5}{3\cos\theta + 8\sin\theta}$. Write the equation in standard form.

37. Write a convincing argument to prove that when converting polar coordinates to rectangular coordinates, the formulas $x = r\cos\theta$ and $y = r\sin\theta$ hold true. Include a labeled drawing in your answer.

Applications and Problem Solving

38. **Geology** Refer to Example 4. The area of a drumlin formed near Salamander Glacier in Glacier National Park in Montana is 4380 square yards. It has a length of 132 yards. Find its polar equation.

39. **Geology** If a drumlin is modeled by the equation $r = \ell\cos 5\theta$ and has an area of 7542 square meters, find its length.

40. **Roller Coasters** The thrill of riding a roller coaster down a hill comes from the feeling that the rider and the train are indeed falling together. The seat force is all that riders feel as they sit in the car of the coaster. For example, a seat force of 1.00 means that a person feels 100% of their body weight resting on the seat. A seat force of $0.70W$ means that the person feels only 70% of their body weight, giving the sensation that they are coming out of their seat. The seat force down a slope can be described as $F = W\cos\theta$, where W is the person's weight and θ is the angle of the downslope. The maximum downslope being used to construct roller coasters today is $60°$.

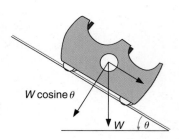

$W\cos\theta$

W θ

a. Find the seat force for a hill with this downslope.

b. Describe what this seat force means and the effect the rider feels.

Mixed Review

41. Create a function of the form $y = f(x)$ whose graph resembles $y = x^3$, but has holes at $x = -2$ and $x = 0$. **(Lesson 3-4)**

42. Find the equation of the line that bisects the acute angle formed by the lines $x - y + 2 = 0$ and $y - 5 = 0$. **(Lesson 7-7)**

43. State whether $\overrightarrow{AB}$ and $\overrightarrow{CD}$ are *equal, opposite, parallel,* or *none* of these for $A(-2, 5)$, $B(3, -1)$, $C(2, 6)$, and $D(7, 0)$. **(Lesson 8-1)**

44. Name three different pairs of polar coordinates that represent point R. Suppose $-360° \leq \theta \leq 360°$. **(Lesson 9-1)**

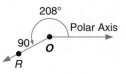

$208°$

Polar Axis

90 O

R

45. Graph $r = 12\cos 2\theta$. **(Lesson 9-2)**

46. **College Entrance Exam** Compare quantities A and B.

Write A if quantity A is greater.

Write B if quantity B is greater.

Write C if the quantities are equal.

Write D if there is not enough information to determine the relationship.

(A) $\dfrac{\dfrac{1}{8} + \dfrac{6}{4}}{\dfrac{3}{16}}$

(B) $\dfrac{\dfrac{2}{12} + \dfrac{1}{3}}{\dfrac{12}{16}}$

Polar Form of a Linear Function

Objectives

After studying this lesson, you should be able to:
- write the polar form of a linear equation, and
- graph the polar form of a linear equation.

Application

In 1691, Jakob Bernouilli introduced the polar coordinate system. However, Isaac Newton was actually the first to think of using polar coordinates. In a paper entitled *Method of Fluxions*, written about 1671, he showed ten types of coordinate systems that could be used to graph equations. One of these systems was the polar coordinate system. The reason that Newton is usually not credited with the first published use of the coordinate system is that his paper was not published until 1736, nine years after his death.

Newton suggested that different coordinate systems were necessary to graph different types of equations. While using different coordinate systems may make graphing easier, it is also convenient to be able to graph equations of different forms on the same coordinate system.

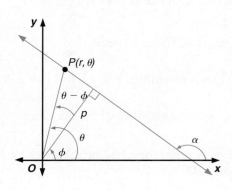

The polar form of the equation for line ℓ is closely related to the normal form, which is $x \cos \phi + y \sin \phi - p = 0$. But we have learned that $x = r \cos \theta$ and $y = r \sin \theta$. The polar form of the equation of line ℓ can be obtained by substituting these values into the normal form.

$$x \cos \phi + y \sin \phi - p = 0$$
$$(r \cos \theta) \cos \phi + (r \sin \theta) \sin \phi - p = 0$$
$$r(\cos \theta \cos \phi + \sin \theta \sin \phi) = p$$
$$r \cos(\theta - \phi) = p$$

$$\cos \theta \cos \phi + \sin \theta \sin \phi = \cos(\theta - \phi)$$

Polar Form of a Linear Equation

The polar form of a linear equation, where p is the length of the normal and ϕ is the positive angle between the positive x-axis and the normal, is
$$p = r \cos(\theta - \phi).$$

You can review the normal form of a linear equation in Lesson 7-6.

In the polar form of a linear equation, θ and r are variables, and p and ϕ are constants. Values for p and ϕ can be obtained from the normal form of the standard equation of a line. Remember to choose the value for ϕ according to the quadrant in which the normal lies.

Example 1

Write each equation in polar form.
a. $2x + 3y - 1 = 0$

a. Write the equation in normal form.

$$\frac{Ax}{\pm\sqrt{A^2 + B^2}} + \frac{By}{\pm\sqrt{A^2 + B^2}} + \frac{C}{\pm\sqrt{A^2 + B^2}} = 0 \qquad \text{\textit{Normal form of the standard form, } } Ax + By + C = 0.$$

We need to find the value of $\pm\sqrt{A^2 + B^2}$.

$$\pm\sqrt{A^2 + B^2} = \pm\sqrt{2^2 + 3^2} \text{ or } \pm\sqrt{13} \qquad \text{\textit{Since C is negative, use }} \sqrt{13}.$$

The normal form is $x\left(\dfrac{2}{\sqrt{13}}\right) + y\left(\dfrac{3}{\sqrt{13}}\right) - \dfrac{1}{\sqrt{13}} = 0$.

The general equation for the normal form is $x\cos\phi + y\sin\phi - p = 0$.

So, $\cos\phi = \dfrac{2}{\sqrt{13}}$, $\sin\phi = \dfrac{3}{\sqrt{13}}$, and $p = \dfrac{1}{\sqrt{13}}$ or $\dfrac{\sqrt{13}}{13}$.

Since $\cos\phi$ and $\sin\phi$ are positive, the normal is in the first quadrant. You can use a calculator to find ϕ.

$$\phi = \arctan\frac{B}{A}$$

$$= \arctan\frac{3}{2}$$

$$= 56.30993247 \text{ or about } 56° \qquad \text{\textit{Round to the nearest degree.}}$$

Substitute values for p and ϕ into the polar form.

$$p = r\cos(\theta - \phi)$$

$$\frac{\sqrt{13}}{13} = r\cos(\theta - 56°)$$

The polar form of $2x + 3y - 1 = 0$ is $\dfrac{\sqrt{13}}{13} = r\cos(\theta - 56°)$.

b. $3x - 4y + 5 = 0$

b. Write the equation in normal form.

$$\frac{3x}{-\sqrt{3^2 + (-4)^2}} + \frac{-4y}{-\sqrt{3^2 + (-4)^2}} + \frac{5}{-\sqrt{3^2 + (-4)^2}} = 0 \qquad \text{\textit{Since C is positive, use }} -\sqrt{A^2 + B^2}.$$

The normal form is $-\dfrac{3}{5}x + \dfrac{4}{5}y - 1 = 0$. So, $\cos\phi = -\dfrac{3}{5}$, $\sin\phi = \dfrac{4}{5}$, and $p = 1$.

Since $\sin\phi$ is positive and $\cos\phi$ is negative, the normal is in the second quadrant.

$$\phi = \arctan\frac{-4}{3}$$

$$\approx -53° \qquad \text{\textit{Round to the nearest degree.}}$$

To find the angle in the second quadrant, add $180°$ to ϕ.
$$-53° + 180° = 127°$$

Substitute the values for p and ϕ into the polar form.
$$p = r\cos(\theta - \phi)$$
$$1 = r\cos(\theta - 127°)$$

The polar form of $3x - 4y + 5 = 0$ is $1 = r\cos(\theta - 127°)$.

The polar form of a linear equation can be written as an equation in rectangular form.

Example 2

Write $1 = r\cos(\theta + 30°)$ in rectangular form.

$$1 = r\cos(\theta + 30°)$$
$$1 = r(\cos\theta\cos 30° - \sin\theta\sin 30°) \qquad \textit{Double-angle identity}$$
$$1 = r\left(\frac{\sqrt{3}}{2}\cos\theta - \frac{1}{2}\sin\theta\right) \qquad \textit{cos 30° = }\frac{\sqrt{3}}{2}\textit{ and sin 30° = }\frac{1}{2}$$
$$1 = \frac{\sqrt{3}}{2}\cdot r\cos\theta - \frac{1}{2}\cdot r\sin\theta \qquad \textit{Multiply.}$$
$$1 = \frac{\sqrt{3}}{2}x - \frac{1}{2}y \qquad \textit{r cos }\theta\textit{ = x and r sin }\theta\textit{ = y}$$
$$2 = \sqrt{3}x - y \qquad \textit{Multiply each side by 2.}$$
$$0 = \sqrt{3}x - y - 2$$

The rectangular form of $1 = r\cos(\theta + 30°)$ is $0 = \sqrt{3}x - y - 2$.

The polar form of a linear equation can be graphed by preparing a table of coordinates and then graphing the ordered pairs on the polar system. *If the graphs of the ordered pairs do not lie in a straight line, then an error has occurred.*

Example 3

Graph the equation $3 = r\cos(\theta + 15°)$.

Write the equation in the form $r = \dfrac{3}{\cos(\theta + 15°)}$. Use your calculator to make a table of values. Graph the ordered pairs on a polar plane.

θ	0°	30°	60°	90°	120°	150°	180°
r	3.1	4.24	11.59	−11.59	−4.24	−3.1	−3.1

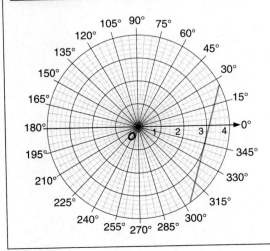

Example 4

Use a graphing calculator to graph $3 = r \cos(\theta - 135°)$.

Write the equation in the form $r = \dfrac{3}{\cos(\theta - 135°)}$.

Convert the polar equation to its equivalent parametric form.

$$x = \frac{3}{\cos(t - 135°)} \cdot \cos t$$

$$y = \frac{3}{\cos(t - 135°)} \cdot \sin t$$

Before plotting these parametric equations, be sure to check your MODE and RANGE settings.

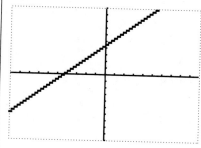

CHECKING FOR UNDERSTANDING

Communicating Mathematics

Read and study the lesson to answer each question.

1. **Explain** how you determine which value of ϕ you use when converting the polar form of an equation to rectangular form.

2. **Explain** why only values of θ from $0°$ through $180°$ were used for calculating values of r in Example 3.

Guided Practice

Write the normal form of each linear equation.

3. $2x + 3y - 5 = 0$

4. $2x - y = -6$

5. Write $x - 3y - 4 = 0$ in polar form. Round ϕ to the nearest degree.

6. Write $2 = r \cos\left(\theta - \dfrac{\pi}{4}\right)$ in rectangular form.

Complete the table for each equation. Then graph the equation.

7. $r \cos(\theta + 10°) = 2$

θ	20°	25°	50°	60°
r				

8. $r \cos(\theta + 20°) = 3$

θ	10°	25°	40°	60°
r				

EXERCISES

Practice **Graph each polar equation.**

9. $2 = r \cos (\theta + 20°)$

10. $5 = r \cos \left(\theta + \dfrac{\pi}{4}\right)$

11. $1.5 = r \cos (\theta - 10°)$

12. $1.6 = r \cos(\theta - 15°)$

13. $r = \dfrac{2.5}{\cos \left(\theta - \dfrac{\pi}{2}\right)}$

14. $r = \dfrac{2.4}{\cos(\theta + 60°)}$

Write each equation in polar form. Round ϕ to the nearest degree.

15. $x = 12$

16. $y - x = 0$

17. $y = 3x + 4$

18. $5 = y + 2x$

Write each equation in rectangular form.

19. $r \cos \left(\theta + \dfrac{\pi}{2}\right) = 0$

20. $r = \dfrac{1}{\cos \theta}$

21. $r \cos \theta - 3 = 0$

22. $r \cos \left(\theta - \dfrac{\pi}{3}\right) - 4 = 0$

Find the polar form of the equation of a line that passes through the pair of points with the given coordinates.

23. $(2, -4)$ and $(1, 3)$

24. $(4, 2)$ and the origin

Find the polar form of the equation of a line for each situation.

25. has a slope of 0.6 and a y-intercept of 4

26. passes through $(3, -2)$ and has a slope of $-\dfrac{1}{2}$

Graphing
Calculator

27. Describe the appropriate mode and range settings for Example 5.

28. If you were to graph the equation $2 = r \cos(\theta + 20°)$ on a graphing calculator, what parametric equations would you use?

29. Graph $2 = r \cos \left(\theta - \dfrac{\pi}{4}\right)$ on a graphing calculator. Sketch the graph on a piece of paper.

Critical
Thinking

30. Describe how you would modify the polar equation in Example 4 to create a new linear polar equation whose graph intersects the given line at an angle of 90°. Sketch the graph of both lines.

Applications
and Problem
Solving

31. Research In 1729, Jacob Hermann proclaimed that polar coordinates were just as efficient as Cartesian coordinates for studying geometric loci. However, Hermann was not a well-known mathematician, and, as a result, few people paid attention to his theories. Twenty years later, a famous Swiss mathematician made the polar coordinate system more popular. Who was that mathematician?

32. Navigation A ship is being tracked by the sonar of a submerged submarine. The sonar screen is a polar system in which the submarine is the origin. The submarine pilot records the path of the ship as $8 = 13r\cos(\theta - 67°)$. Find the linear equation of the path of the ship.

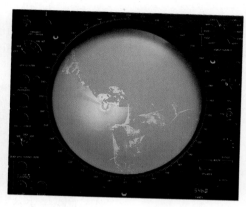

Mixed Review

33. Find $\cos\left(\arcsin\dfrac{\sqrt{3}}{2}\right)$. Assume the angles are in Quadrant I. **(Lesson 6-4)**

34. Engineering A wheel in a motor is turning counterclockwise at 2 radians per second. There is a small hole in the wheel 3 centimeters from its center. Suppose a model of the wheel is drawn on a rectangular coordinate system with the wheel centered at the origin. If the hole has initial coordinates $(3, 0)$, what are its coordinates after t seconds? **(Lesson 8-5)**

35. Graph $r = 6\cos 3\theta$. **(Lesson 9-2)**

36. Write the polar equation $r = 6$ in rectangular form. **(Lesson 9-3)**

37. College Entrance Exam Choose the best answer.
A typist uses a sheet of paper that is 9 inches wide by 12 inches long. He leaves a 1-inch margin on each side and a 1.5-inch margin on top and bottom. What part of the page is used for typing?

(A) $\dfrac{5}{12}$ **(B)** $\dfrac{7}{12}$ **(C)** $\dfrac{5}{9}$ **(D)** $\dfrac{3}{4}$ **(E)** none of these

---

MID-CHAPTER REVIEW

1. Graph point $P(-6, -45°)$ on a polar plane. Name three other pairs of polar coordinates that represent the same point. **(Lesson 9-1)**

2. Graph $r = 3\theta$. **(Lesson 9-2)**

3. Graph $r = 6 + 6\sin\theta$. Identify the classical curve it represents. **(Lesson 9-2)**

4. Find the rectangular coordinates of $D(\sqrt{2}, 135°)$. **(Lesson 9-3)**

5. Find the polar coordinates of $E(1, \sqrt{3})$. **(Lesson 9-3)**

6. Write $x + y - 6 = 0$ in polar form. **(Lesson 9-4)**

9-5 Simplifying Complex Numbers

Objective

After studying this lesson, you should be able to:
- add, subtract, multiply, and divide complex numbers in rectangular form.

Application

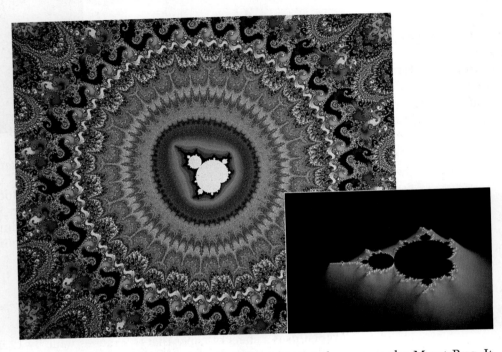

You will learn more about fractals in Chapter 13.

The computer-generated image shown above is known as the M-set Rug. It is a magnified portion of the boundary of a fractal known as the Mandelbrot set. It is formed by repeated evaluation of complex numbers through the function $f(z) = z^2 + c$. Generating images such as this requires multiplication and addition of complex numbers.

Recall that complex numbers are numbers in the form $a + bi$, where a and b are real numbers and i, the imaginary unit, is defined by $i^2 = -1$. Various powers of i are shown below.

$$i^1 = i \qquad i^2 = -1 \qquad i^3 = i^2 \cdot i = -i \qquad i^4 = (i^2)^2 = 1$$
$$i^5 = i^4 \cdot i = i \qquad i^6 = i^4 \cdot i^2 = -1 \qquad i^7 = i^4 \cdot i^3 = -i \qquad i^8 = (i^2)^4 = 1$$

Notice the repeating pattern of the powers of i.

$$i, -1, -i, 1, i, -1, -i, 1$$

In general, the value of i^n, where n is a whole number, can be found by dividing n by 4 and examining the remainder.

To find the value of i^n, let R = the remainder when n is divided by 4.	
if R = 0	$i^n = 1$
if R = 1	$i^n = i$
if R = 2	$i^n = -1$
if R = 3	$i^n = -i$

You can also find the value of any power of i by rewriting the power in terms of a power of a power. For example, $i^{35} = i^{4(8) + 3}$ or $(i^4)^8 \cdot i^3$.

Example 1

Find i^{27}.

Method 1

$27 \div 4 = 6 \text{ R } 3$
If $\text{R} = 3, i^n = -i$.
$i^{27} = -i$

Method 2

$i^{27} = (i^2)^{13} \cdot i$
$\phantom{i^{27}} = (-1)^{13} \cdot i$
$\phantom{i^{27}} = -i$

You have learned that a complex number is composed of a real part a and an imaginary part bi. When $b = 0$, the complex number is a real number. When $b \neq 0$, as in $6 + 5i$, the complex number is an *imaginary number*. When $b \neq 0$ and $a = 0$, as in $-8i$, the imaginary number becomes a *pure imaginary number*. Complex numbers can be added and subtracted by combining their real parts and their imaginary parts.

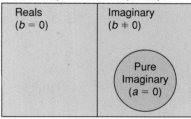
Complex Numbers ($a + bi$)

| Reals ($b = 0$) | Imaginary ($b \neq 0$) |

Pure Imaginary ($a = 0$)

Example 2

Simplify each expression.
a. $(2 + 5i) + (-8 - i)$

a. $(2 + 5i) + (-8 - i) = [2 + (-8)] + [5i + (-i)]$
$ = -6 + 4i$

b. $(9 + 6i) - (i + 5)$

b. $(9 + 6i) - (i + 5) = 9 + 6i - i - 5$
$ = (9 - 5) + (6i - i)$
$ = 4 + 5i$

TECHNOLOGY *Tip*

You can use the TI-83 and TI-92 graphing calculators to enter complex numbers and simplify expressions involving complex numbers.

The product of two or more complex numbers can be found by using the same procedures you use when multiplying binomials.

Example 3

Simplify $(1 + 3i)(2 + 4i)$.

$(1 + 3i)(2 + 4i) = 1(2 + 4i) + 3i(2 + 4i)$
$ = 1(2) + 1(4i) + 3i(2) + 3i(4i)$

(continued on the next page)

$$= 2 + 4i + 6i + 12i^2$$
$$= 2 + 10i + 12(-1) \qquad i^2 = -1$$
$$= -10 + 10i$$

In Chapter 4, you learned that the solution to a quadratic equation sometimes involves an imaginary number. Remember that when this happens, there are always two imaginary solutions. Pairs of imaginary solutions in the form $a + bi$ and $a - bi$ are called **conjugates**. The product of two conjugates is always a real number.

Example 4

Solve $x^2 - 7x + 15 = 0$. Check your solution.

Use the quadratic formula.

$$x = \frac{-(-7) \pm \sqrt{(-7)^2 - 4(1)(15)}}{2(1)}$$

$$= \frac{7 \pm \sqrt{-11}}{2} \text{ or } \frac{7 \pm i\sqrt{11}}{2}$$

Check: $x = \dfrac{7 + i\sqrt{11}}{2}$

$$\left(\frac{7 + i\sqrt{11}}{2}\right)^2 - 7\left(\frac{7 + i\sqrt{11}}{2}\right) + 15 \stackrel{?}{=} 0 \quad x = \frac{7 + i\sqrt{11}}{2}$$

$$\frac{49 + 14i\sqrt{11} + (-11)}{4} - \frac{49 + 7i\sqrt{11}}{2} + 15 \stackrel{?}{=} 0 \quad \textit{Expand.}$$

$$49 + 14i\sqrt{11} - 11 - 98 - 14i\sqrt{11} + 60 \stackrel{?}{=} 0 \quad \textit{Multiply each side by 4.}$$

$$0 = 0 \quad \checkmark$$

$$x = \frac{7 - i\sqrt{11}}{2}$$

$$\left(\frac{7 - i\sqrt{11}}{2}\right)^2 - 7\left(\frac{7 - i\sqrt{11}}{2}\right) + 15 \stackrel{?}{=} 0 \quad x = \frac{7 - i\sqrt{11}}{2}$$

$$\frac{49 - 14i\sqrt{11} + (-11)}{4} - \frac{49 - 7i\sqrt{11}}{2} + 15 \stackrel{?}{=} 0 \quad \textit{Expand.}$$

$$49 - 14i\sqrt{11} - 11 - 98 + 14i\sqrt{11} + 60 \stackrel{?}{=} 0 \quad \textit{Multiply each side by 4.}$$

$$0 = 0 \quad \checkmark$$

Whenever a fraction has a radical expression in the denominator, the denominator is usually rationalized. For example, $\dfrac{1}{3 + \sqrt{2}}$ becomes $\dfrac{1}{3 + \sqrt{2}} \cdot \dfrac{3 - \sqrt{2}}{3 - \sqrt{2}}$ or $\dfrac{3 - \sqrt{2}}{7}$. A similar process is used to rationalize the denominators of fractions that contain imaginary numbers.

Example 5

Simplify $\dfrac{4 - 3i}{2 + i}$.

$\dfrac{4 - 3i}{2 + i} = \left(\dfrac{4 - 3i}{2 + i}\right)\left(\dfrac{2 - i}{2 - i}\right)$ *Multiply the numerator and denominator by the conjugate of the denominator.*

$= \dfrac{8 - 4i - 6i + 3i^2}{4 - i^2}$

$= \dfrac{8 - 10i + 3(-1)}{4 - (-1)}$

$= \dfrac{5 - 10i}{5}$

$= 1 - 2i$

The list below summarizes the operations with imaginary numbers presented in this lesson.

Operations with Complex Numbers

For any complex numbers $a + bi$ and $c + di$, the following is true.

$(a + bi) + (c + di) = (a + c) + (b + d)i$

$(a + bi) - (c + di) = (a - c) + (b - d)i$

$(a + bi)(c + di) = (ac - bd) + (ad + bc)i$

$\dfrac{a + bi}{c + di} = \dfrac{(ac + bd) + (bc - ad)i}{c^2 + d^2}$

Complex numbers are often used in the study of electricity, specifically alternating current (AC) electricity. In a simplified electrical circuit, there are three basic components to be considered:

• the flow of electrical **current**, I,
• the resistance to that flow, Z, called **impedance**, and
• the electromotive force, E, called **voltage**.

Impedance is the result of having resistors, coils, and capacitors in a circuit, and is measured in ohms. Voltage refers to the electrical potential in a circuit and is measured in volts. Current is measured in amperes (amps). The formula $E = I \cdot Z$ illustrates the relationship among these components.

Current, impedance, and voltage are often expressed as complex numbers. Electrical engineers use j instead of i to represent the imaginary unit. For electrical engineers, $j = \sqrt{-1}$ and $j^2 = -1$. When they write a complex number, they use the form $a + bj$.

For the total impedance, $a + bj$, the real part a represents the opposition to current flow due to resistors, and the imaginary part bj represents the opposition due to coils and capacitors. If two circuits are connected so that the current flows through them one after the other, the circuits are said to be connected in series. The total voltage and total impedance of circuits in series are found by adding the voltages and adding the impedances of the two circuits, respectively.

Example 6

Two AC circuits are connected in series, one with an impedance of $3 + 5j$ ohms and the other with an impedance of $2 - 4j$ ohms.
a. Find the total impedance.
b. Then find the current if the total voltage is 110 volts.

a. The total impedance is the sum of the individual impedances.
$(3 + 5j) + (2 - 4j) = 5 + j$
The total impedance is $5 + j$ ohms.

b. Substitute the total impedance and the total voltage into the formula $E = I \cdot Z$.

$$E = I \cdot Z$$
$$110 = I \cdot (5 + j) \qquad \text{\textit{E = 110 volts and Z = 5 + j ohms}}$$
$$\frac{110}{5 + j} = I \qquad \text{\textit{Solve for I.}}$$
$$\left(\frac{110}{5 + j}\right)\left(\frac{5 - j}{5 - j}\right) = I \qquad \text{\textit{Rationalize the denominator.}}$$
$$\frac{550 - 110j}{26} = I$$
$$21.15 - 4.23j = I \qquad \text{The current is } 21.15 - 4.23j \text{ amperes.}$$

CHECKING FOR UNDERSTANDING

Communicating Mathematics

Read and study the lesson to answer each question.

1. **Write** a few sentences to show how operations with imaginary numbers are similar to operations with binomials.

2. **Explain** how the set of real numbers is related to the set of complex numbers.

3. What units are used to measure current, impedance, and voltage?

Guided Practice

Simplify.

4. $i^{15} + i^{28}$

5. $(4 + 5i) + (3 + 2i)$

6. $(8 + 3i) - (1 - 6i)$

7. $(4 + 3i)(2 - i)$

8. $(i - 6)^2$

9. $\dfrac{1 + i}{3 + 2i}$

10. $\dfrac{2}{12 + i}$

11. $\dfrac{7i}{-5i}$

EXERCISES

Practice   **Simplify.**

12. $i^{24} + i^{61}$ **13.** $i^4(7 + 2i)$

14. $(7 - 6i) + (9 + 11i)$ **15.** $(\text{-}5 + 2i\sqrt{7}) - (2 - 7i\sqrt{7})$

16. $(\text{-}3 - 10i) - (\text{-}5 - 4i)$ **17.** $\text{-}6(2 - 8i) + 3(5 + 7i)$

18. $4(7 - i) - 5(2 - 6i)$ **19.** $(3 + 5i)(4 - i)$

20. $(4 - 2i\sqrt{3})(1 + 5i\sqrt{3})$ **21.** $(3 - 4i)^2$

22. $(\sqrt{5} + 2i)^2$ **23.** $(8 - \sqrt{\text{-}11})(8 + \sqrt{\text{-}11})$

24. $(6 - 4i)\,(6 + 4i)$ **25.** $(5 + \sqrt{\text{-}8}) + (\text{-}13 + 4\sqrt{\text{-}2})$

26. $\dfrac{2 - 4i}{1 + 3i}$ **27.** $\dfrac{3 - i}{2 - i}$

Solve each equation. Check the solution.

28. $x^2 - x + 1 = 0$ **29.** $x^2 + 4x + 29 = 0$

Find values for x and y that make each sentence true.

30. $3x - 5yi = 15 - 20i$ **31.** $\sqrt{3}x + 7yi = 6 - 2i$

32. $(x - y) + (2x + y)i = \text{-}3 + 9i$ **33.** $(2x - y) + (x + y)i = \text{-}4 - 5i$

Simplify.

34. $(3 - i)(1 + 2i)(2 + 3i)$ **35.** $(4 + 3i)(2 - 5i)(4 - 3i)$

36. $\dfrac{3}{\sqrt{2} - 5i}$ **37.** $\dfrac{2 + i\sqrt{3}}{12 + i\sqrt{3}}$

38. $\dfrac{(1 - 2i)^2}{(2 - i)^2}$ **39.** $\dfrac{2 + i}{(1 - i)^2}$

Critical Thinking   **40.** Write a paragraph comparing the addition of vectors to addition of complex numbers.

Applications and Problem Solving   **41. Flashlights** In a two-battery flashlight, the positive terminal of the first battery touches the negative terminal of the second. The positive terminal of the second battery touches the center terminal of the light bulb. A metal strip connects the bulb to the switch which is connected to the negative terminal of the first battery. When you turn the flashlight on, the switch completes the circuit, and the bulb lights up. Each of the batteries are 1.5-volt batteries, and the current is $(1 + j\sqrt{3})$ amps.

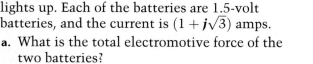

     a. What is the total electromotive force of the two batteries?

     b. What is the total impedance of the circuit in the flashlight?

42. **Electricity** A circuit has a current of $(6 - 8j)$ amps and an impedance of $(14 + 8j)$ ohms. Find the voltage of this circuit.

43. **Electricity** Find the current of a circuit of $(70 + 226j)$ volts with an impedance of $(6 + 8j)$ ohms.

44. **Electricity** What is the impedance of a circuit whose current is $(-6 - 2j)$ amps and whose electromotive force is $(-50 + 100j)$ volts?

Mixed Review

45. Write the equation $2x + 5y - 10 = 0$ in slope-intercept form. Then name the slope and y-intercept. **(Lesson 1-5)**

46. Solve the system below by using matrices. **(Lesson 2-4)**
$a + b + c = 6$
$2a - 3b + 4c = 3$
$4a - 8b + 4c = 12$

47. **Education** The semester test in your English class consists of short answer and essay questions. Each short answer question is worth 5 points and each essay question is worth 15 points. You may choose up to 20 questions of any type to answer. It takes 2 minutes to answer each short answer question and 12 minutes to answer each essay question. **(Lesson 2-6)**

   a. You have one hour to complete the test. Assuming that you answer all of the questions that you attempt correctly, how many of each type should you answer to earn the highest score?

   b. You have two hours to complete the test. Assuming that you answer all of the questions that you attempt correctly, how many of each type should you answer to earn the highest score?

48. Write the equation for the inverse of $y = \cos x$. Then graph the function and its inverse. **(Lesson 6-6)**

49. Verify that $\sin^4 A + \cos^2 A = \cos^4 A + \sin^2 A$ is an identity. **(Lesson 7-2)**

50. Find the sum of the vectors $(3, 5)$ and $(-1, 2)$ algebraically. **(Lesson 8-2)**

51. Write $x^2 + y^2 = 25$ in polar form. **(Lesson 9-3)**

52. Graph $r = \dfrac{1}{\cos(\theta + 15°)}$. **(Lesson 9-4)**

53. **College Entrance Exam** Compare quantities A and B.
   Write A if quantity A is greater.
   Write B if quantity B is greater.
   Write C if the quantities are equal.
   Write D if there is not enough information to determine the relationship.

   Given: $m\angle K = 60°$

   (A) the area of the smaller triangle

   (B) the area of the shaded region

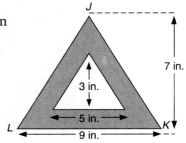

9-6 Polar Form of Complex Numbers

Objective

After studying this lesson, you should be able to:
- change complex numbers from rectangular to polar form and vice versa.

Application

In 1833, at the Royal Irish Academy in Dublin, Ireland, William Rowan Hamilton presented the concept that complex numbers could be expressed in the form $a + bi$. He further put forth the idea that if a were represented on the x-axis and b were represented on the y-axis, each complex number $a + bi$ could be graphed on this rectangular plane. His proposals are still being followed today.

This coordinate system is also called the complex plane.

When complex numbers are expressed in the form $x + yi$, where x is the real part and yi is the imaginary part, the complex number is said to be in **rectangular form**. Sometimes the rectangular form is written as an ordered pair (x, y).

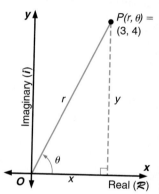

To graph a complex number in rectangular form, the horizontal axis is the real axis and the vertical axis is the imaginary axis. Point P on the graph at the left is the graph of the complex number $3 + 4i$.

The angle θ is called the **amplitude**, and r, the length of the hypotenuse, is called the **modulus**. Two complex numbers are equal if and only if their moduli are equal and their amplitudes differ by integral multiples of 2π radians. This means that θ can be replaced by $\theta + 2k\pi$ for any integer k.

EXPLORATION: Modeling

1. Graph points A and B on the complex plane so that A is the graph of $6 + 7i$ and B is the graph of $-2 + 3i$. Label the origin O.
2. Draw a line through B that is parallel to $\overrightarrow{OA}$.
3. Draw a line through A that is parallel to $\overrightarrow{OB}$.
4. Label the point where the two lines intersect as point C.
5. What are the coordinates of C and what imaginary number does $\overrightarrow{OC}$ represent?
6. How does this number compare with the sum of the complex numbers originally graphed?
7. What conjecture might you make based on this information?
8. Test your conjecture by graphing several other pairs of complex numbers and following the steps given above.

Recall that $x = r \cos \theta$ and $y = r \sin \theta$. As with other rectangular coordinates, complex coordinates can be written in polar form by substituting these values for x and y. This form of a complex number is often called the **polar** or **trigonometric form**.

Polar Form of a Complex Number (Trigonometric Form)	The polar form or trigonometric form of a complex number is $$x + yi = r(\cos \theta + i \sin \theta).$$

$r(\cos \theta + i \sin \theta)$ is often abbreviated as r cis θ.

Values for r and θ can be found by using the same process used when you changed other rectangular coordinates to polar coordinates. For $x + yi$,

$$r = \sqrt{x^2 + y^2} \text{ and } \theta = \text{Arctan } \frac{y}{x} \text{ if } x > 0 \text{ or } \theta = \text{Arctan } \left(\frac{y}{x} + \pi\right) \text{ if } x < 0.$$

The amplitude θ is usually expressed in radian measure, and the angle is in standard position along the polar axis.

Example 1

Graph each complex number. Then express it in polar form.

a. $-2 + i$

a. Graph the complex number on the complex plane.

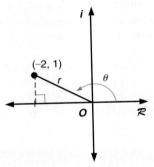

Since the point lies in the second quadrant, the value of θ should be greater than $\frac{\pi}{2}$ and less than π radians.

Find values for r and θ.

$$r = \sqrt{x^2 + y^2}$$
$$= \sqrt{(-2)^2 + 1^2}$$
$$= \sqrt{5}$$

$\theta = \text{Arctan } \dfrac{y}{x} + \pi$ *Notice that $x < 0$.*

$ = \text{Arctan } \dfrac{1}{-2} + \pi$ *Notice that $\frac{\pi}{2} < 2.68 < \pi$.*

$ \approx 2.68$ *Use your calculator.*

Therefore, $-2 + i \approx \sqrt{5} \, (\cos 2.68 + i \sin 2.68)$.

b. $1 + i$

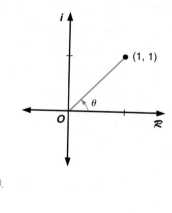

b. Graph $1 + i$
on the complex plane.

$0 < \theta < \dfrac{\pi}{2}$

Find values for r and θ.

$r = \sqrt{1^2 + 1^2}$
$\quad = \sqrt{2}$

$\theta = \text{Arctan } \dfrac{1}{1}$ *Notice that x > 0.*

$\quad = \dfrac{\pi}{4}$

Therefore, $1 + i = \sqrt{2} \left(\cos \dfrac{\pi}{4} + i \sin \dfrac{\pi}{4} \right)$.

If you know the values of r and θ, you can write a complex number in the form $x + yi$.

Example 2

Graph $8 \left(\cos \dfrac{7\pi}{6} + i \sin \dfrac{7\pi}{6} \right)$. Then express it in rectangular form.

Graph the complex number on the complex plane.

The value of θ is $\dfrac{7\pi}{6}$, and the value of r is 8.

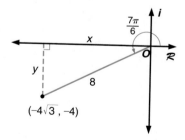

The terminal side is in the third quadrant. So, the values x and y should both be negative.

Find values for x and y.

$\begin{aligned} x &= r \cos \theta \\ &= 8 \cos \dfrac{7\pi}{6} \\ &= -8 \cos \dfrac{\pi}{6} \\ &= -8 \left(\dfrac{\sqrt{3}}{2} \right) \\ &= -4\sqrt{3} \end{aligned}$

$\begin{aligned} y &= r \sin \theta \\ &= 8 \sin \dfrac{7\pi}{6} \\ &= -8 \sin \dfrac{\pi}{6} \\ &= -8 \left(\dfrac{1}{2} \right) \\ &= -4 \end{aligned}$

Therefore, $8 \left(\cos \dfrac{7\pi}{6} + i \sin \dfrac{7\pi}{6} \right) = -4\sqrt{3} - 4i$.

CHECKING FOR UNDERSTANDING

Communicating
Mathematics

Read and study the lesson to answer each question.

1. **Define** the term *modulus* when referring to complex numbers.

2. **Describe** the amplitude of a complex number.

3. **Make a conjecture** about why mathematicians use the form r cis θ for a complex number in polar form.

4. **Describe** the relationship between α and β if r cis α and r cis β both represent the same complex number.

Guided
Practice

Graph each complex number. Then express the number in polar form.

5. $-1 + i$

6. $-2(1 - i)$

7. $0 + 70i$

8. $-6 + 0i$

9. $3 - i\sqrt{3}$

10. $2(1 - i\sqrt{3})$

Graph each complex number. Then express the number in rectangular form.

11. $4(\cos 0 + i \sin 0)$

12. $3(\cos \pi + i \sin \pi)$

13. $3 \cos \dfrac{\pi}{2} + 3i \sin \dfrac{\pi}{2}$

14. $0.5 \left(\cos \dfrac{5\pi}{6} + i \sin \dfrac{5\pi}{6} \right)$

EXERCISES

Practice

Express each complex number in polar form.

15. $-1 - i$

16. $6i$

17. $10i$

18. $-5 - i$

19. $4 + 3i$

20. $2\sqrt{3} - 3i$

21. $1 + i\sqrt{3}$

22. $-4i$

23. -5

Express each complex number in rectangular form.

24. $6 \left(\cos \dfrac{3\pi}{2} + i \sin \dfrac{3\pi}{2} \right)$

25. $24 \left(\cos \dfrac{5\pi}{3} + i \sin \dfrac{5\pi}{3} \right)$

26. $\sqrt{2} \left(\cos \dfrac{5\pi}{4} + i \sin \dfrac{5\pi}{4} \right)$

27. $3(\cos 2 + i \sin 2)$

28. $4(\cos 3 + i \sin 3)$

29. $2 \left(\cos \dfrac{\pi}{6} + i \sin \dfrac{\pi}{6} \right)$

30. $\sqrt{2} \, i \sin \left(\dfrac{-\pi}{2} \right) + \sqrt{2} \cos \left(\dfrac{-\pi}{2} \right)$

31. $5 \left(\cos \dfrac{17\pi}{6} + i \sin \dfrac{17\pi}{6} \right)$

Critical
Thinking

32. Let C represent a complex number. If $C = x + yi$ is a real number, show why its modulus is equal to its absolute value.

33. Refer to the Exploration on page 501. Make a general conjecture about how to find the sum of two complex numbers graphically.

Applications
and Problem
Solving

34. **Algebra** Find the zeros of $f(x) = x^2 - 4x + 8$. Graph them on the complex plane.

35. Electricity   Find the current of an electrical circuit whose voltage is $(70 + 226j)$ volts and impedance is $(6 + 8j)$ ohms. Then write the three complex numbers in this problem in polar form. Round values to the nearest tenth. *The relationship among current (I), voltage (E), and impedance (Z) is $E = I \cdot Z$.*

Mixed Review

36. Hobbies   Sara likes bird watching. She wishes to build a rectangular bird feeder that is 6 inches tall and has a volume of 192 cubic inches. The material for the front and back costs 5¢ per square inch, the materials for the sides cost 10¢ per square inch, and the material for the bottom costs 20¢ per square inch. Find the dimensions of the bird feeder that will be the most economical to build.
(Lesson 3-7)

37. Air Traffic Safety   The traffic pattern for airplanes into San Diego's airport is over the heart of downtown. Therefore, there are restrictions on the heights of new construction. The owner of an office building wishes to erect a microwave tower on top of the building. According to the architect's design, the angle of elevation from a point on the ground to the top of the 40-foot tower is 56°. The angle of elevation from the ground point to the top of the building is 42°. The maximum allowed height of any structure is 100 feet. Will the city allow the building of this tower? Explain.
(Lesson 5-6)

38. Solve $\sin 2x + 2 \sin x = 0$ for $0° \le x \le 360°$.
(Lesson 7-5)

39. Find an ordered triple to represent $\vec{u}$ if $\vec{u} = \vec{v} + \vec{w}$, $\vec{v} = (2, -5, -3)$, and $\vec{w} = (-3, 4, -7)$. Then write $\vec{u}$ in terms of unit vectors.   **(Lesson 8-3)**

40. Graph $3 = r \cos\left(\theta - \dfrac{\pi}{3}\right)$.   **(Lesson 9-4)**

41. Electricity   The reactance of an electrical circuit is found by the formula $X = X_L - X_C$, where X_L represents the reactance from the inductions and X_C represents the reactance from all capacitors. Suppose a circuit has a total reactance of $8i$ ohms. If the reactance of the capacitors is $7i$ ohms, find the reactance of the inductions.   **(Lesson 9-5)**

42. College Entrance Exam   Choose the best answer.
A pulley having a 9-inch diameter is belted to a pulley having a 6-inch diameter, as shown in the figure. If the large pulley runs at 120 rpm (revolutions per minute), how fast does the small pulley run?

(A) 80 rpm
(B) 100 rpm
(C) 160 rpm
(D) 180 rpm
(E) 240 rpm

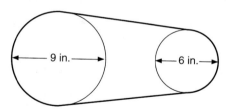

9-7 Products and Quotients of Complex Numbers in Polar Form

Objective

After studying this lesson, you should be able to:
- find the product and quotient of complex numbers in polar form.

Application

The image of the M-set Rug, shown below and in Lesson 9-5, was created by Dr. Richard F. Voss. Dr. Voss has been a Research Staff Member at the IBM Thomas J. Watson Research Laboratory in Yorktown Heights, New York, since 1975. His research has focused on condensed matter physics. The M-set Rug is now shown at a magnification of 10^{23}. At this magnification, the entire Mandelbrot set would be approximately 1,000,000 light years across.

Without the ability to multiply and add complex numbers rapidly on a computer, images like the one above would be impossible to create. When two complex numbers are written in polar form, their product or quotient can easily be computed.

FYI...

Fractal geometry was used to create a landscape that was used in the Death Star sequence of the movie *Star Wars*.

Let $r_1(\cos\theta_1 + i\sin\theta_1)$ and $r_2(\cos\theta_2 + i\sin\theta_2)$ represent two complex numbers. A formula for the product of the two numbers can be obtained by multiplying the two numbers directly and then simplifying the result.

$$r_1(\cos\theta_1 + i\sin\theta_1) \cdot r_2(\cos\theta_2 + i\sin\theta_2)$$

$$= r_1 r_2(\cos\theta_1\cos\theta_2 + i\cos\theta_1\sin\theta_2 + i\sin\theta_1\cos\theta_2 + i^2\sin\theta_1\sin\theta_2)$$

$$= r_1 r_2\left[(\cos\theta_1\cos\theta_2 - \sin\theta_1\sin\theta_2) + i(\sin\theta_1\cos\theta_2 + \cos\theta_1\sin\theta_2)\right]$$

$$= r_1 r_2\left[\cos(\theta_1 + \theta_2) + i\sin(\theta_1 + \theta_2)\right]$$

Product of Complex Numbers in Polar Form	**The product of two complex numbers can be found as follows.** $$r_1(\cos\theta_1 + i\sin\theta_1) \cdot r_2(\cos\theta_2 + i\sin\theta_2) =$$ $$r_1 r_2[\cos(\theta_1 + \theta_2) + i\sin(\theta_1 + \theta_2)]$$

Notice that the modulus $(r_1 r_2)$ of the product of the two complex numbers is the product of their moduli. The amplitude of the product $(\theta_1 + \theta_2)$ is the sum of the amplitudes.

Example 1 | Find the product $8\left(\cos\dfrac{3\pi}{4}+i\sin\dfrac{3\pi}{4}\right)\cdot 2\left(\cos\dfrac{5\pi}{4}+i\sin\dfrac{5\pi}{4}\right)$. Then express the product in rectangular form.

Find the modulus and amplitude of the product.

$$r = r_1 r_2$$
$$= 8(2)$$
$$= 16$$

$$\theta = \theta_1 + \theta_2$$
$$= \frac{3\pi}{4} + \frac{5\pi}{4}$$
$$= 2\pi$$

The product is $16(\cos 2\pi + i\sin 2\pi)$.

Now find the rectangular form of the product.

$$8\left(\cos\frac{3\pi}{4}+i\sin\frac{3\pi}{4}\right)\cdot 2\left(\cos\frac{5\pi}{4}+i\sin\frac{5\pi}{4}\right) = 16(\cos 2\pi + i\sin 2\pi)$$
$$= 16(1 + i\cdot 0)\text{ or }16$$

The rectangular form of this product is 16.

Suppose the quotient of two complex numbers is expressed as a fraction. A formula for this quotient can be found by rationalizing the denominator.

$$\frac{r_1(\cos\theta_1 + i\sin\theta_1)}{r_2(\cos\theta_2 + i\sin\theta_2)}$$

$$= \frac{r_1(\cos\theta_1 + i\sin\theta_1)}{r_2(\cos\theta_2 + i\sin\theta_2)}\cdot\frac{(\cos\theta_2 - i\sin\theta_2)}{(\cos\theta_2 - i\sin\theta_2)}$$

$$= \frac{r_1}{r_2}\cdot\frac{(\cos\theta_1\cos\theta_2 + \sin\theta_1\sin\theta_2) + i(\sin\theta_1\cos\theta_2 - \cos\theta_1\sin\theta_2)}{\cos^2\theta_2 + \sin^2\theta_2}$$

$$= \frac{r_1}{r_2}\left[\cos(\theta_1 - \theta_2) + i\sin(\theta_1 - \theta_2)\right] \qquad \textit{sin}^2\,\theta + \textit{cos}^2\,\theta = 1$$

Quotient of
Complex Numbers
in Polar Form

The quotient of two complex numbers can be found as follows.
$$\frac{r_1(\cos\theta_1 + i\sin\theta_1)}{r_2(\cos\theta_2 + i\sin\theta_2)} = \frac{r_1}{r_2}\left[\cos(\theta_1 - \theta_2) + i\sin(\theta_1 - \theta_2)\right]$$

Notice the modulus $\left(\dfrac{r_1}{r_2}\right)$ of the quotient of two complex numbers is the quotient of their moduli. The amplitude $(\theta_1 - \theta_2)$ of the quotient is the difference of the amplitudes.

Example 2

Find the quotient $2\left(\cos\dfrac{\pi}{6}+i\sin\dfrac{\pi}{6}\right)\div\left(\cos\dfrac{\pi}{3}+i\sin\dfrac{\pi}{3}\right)$. Then express the quotient in rectangular form.

Find the modulus and the amplitude.

$$r=\frac{r_1}{r_2}\qquad\qquad\theta=\theta_1-\theta_2$$
$$=\frac{2}{1}\qquad\qquad=\frac{\pi}{6}-\frac{\pi}{3}$$
$$=2\qquad\qquad=-\frac{\pi}{6}$$

Now find the rectangular form of the quotient, $2(\cos\left(-\dfrac{\pi}{6}\right)+i\sin\left(-\dfrac{\pi}{6}\right))$.

$$2\left(\cos\frac{\pi}{6}+i\sin\frac{\pi}{6}\right)\div\left(\cos\frac{\pi}{3}+i\sin\frac{\pi}{3}\right)=2\left[\cos\left(-\frac{\pi}{6}\right)+i\sin\left(-\frac{\pi}{6}\right)\right]$$
$$=2\left[\frac{\sqrt{3}}{2}+\left(-\frac{1}{2}i\right)\right]\text{ or }\sqrt{3}-i$$

The rectangular form is $\sqrt{3}-i$.

CHECKING FOR UNDERSTANDING

Communicating Mathematics

Read and study the lesson to answer each question.

1. **Rewrite** the polar form of the product in Example 1 using r cis θ notation.
2. **Describe** the process you would use to find $(r\text{ cis }\theta)^2$ and the results obtained.

Guided Practice

Find each product or quotient. Then write the result in rectangular form.

3. $5(\cos\pi+i\sin\pi)\cdot6(\cos2\pi+i\sin2\pi)$

4. $3\left(\cos\dfrac{3\pi}{4}+i\sin\dfrac{3\pi}{4}\right)\cdot8\left(\cos\dfrac{\pi}{2}+i\sin\dfrac{\pi}{2}\right)$

5. $15(\cos2\pi+i\sin2\pi)\div5(\cos\pi+i\sin\pi)$

6. $6\left(\cos\dfrac{\pi}{6}+i\sin\dfrac{\pi}{6}\right)\div2\left(\cos\dfrac{2\pi}{3}+i\sin\dfrac{2\pi}{3}\right)$

EXERCISES

Practice

Find each product or quotient.

7. $(-1-i)(1+i)$

8. $(-4-4i\sqrt{3})\div2i$

9. $(3+3i)\div(-2+2i)$

10. $(-2+2i)(\sqrt{3}+i)$

Find each product or quotient. Then write the result in rectangular form.

11. $3(\cos30°+i\sin30°)\cdot2(\cos60°+i\sin60°)$

12. $2\left(\cos\dfrac{2\pi}{3}+i\sin\dfrac{2\pi}{3}\right)\cdot3\left(\cos\dfrac{7\pi}{6}+i\sin\dfrac{7\pi}{6}\right)$

13. $4\left(\cos\dfrac{3\pi}{4}+i\sin\dfrac{3\pi}{4}\right)\div\sqrt{2}\left(\cos\dfrac{2\pi}{3}+i\sin\dfrac{2\pi}{3}\right)$

14. $6\left(\cos\dfrac{\pi}{6}+i\sin\dfrac{\pi}{6}\right)\cdot 2\left(\cos\dfrac{2\pi}{3}+i\sin\dfrac{2\pi}{3}\right)$

15. $3\left(\cos\dfrac{7\pi}{6}+i\sin\dfrac{7\pi}{6}\right)\cdot 6\left(\cos\dfrac{\pi}{6}+i\sin\dfrac{\pi}{6}\right)$

16. $3\sqrt{2}\left(\cos\dfrac{\pi}{4}+i\sin\dfrac{\pi}{4}\right)\div\sqrt{2}\left(\cos\dfrac{\pi}{6}+i\sin\dfrac{\pi}{6}\right)$

17. $2\left(\cos\dfrac{\pi}{4}+i\sin\dfrac{\pi}{4}\right)\cdot 3\left(\cos\dfrac{\pi}{2}+i\sin\dfrac{\pi}{2}\right)$

18. $12(\cos\pi+i\sin\pi)\div 4\left(\cos\dfrac{\pi}{6}+i\sin\dfrac{\pi}{6}\right)$

19. $9.24(\cos 1.8+i\sin 1.8)\div 3.1(\cos 0.7+i\sin 0.7)$

20. $2(\cos 0.8+i\sin 0.8)\cdot 3.2(\cos 1.5+i\sin 1.5)$

21. $\dfrac{1}{3}\left(\cos\dfrac{7\pi}{8}+i\sin\dfrac{7\pi}{8}\right)\cdot 3\sqrt{3}\left[\cos\left(-\dfrac{\pi}{4}\right)+i\sin\left(-\dfrac{\pi}{4}\right)\right]$

22. $6\sqrt{3}\left(\cos\dfrac{5\pi}{4}+i\sin\dfrac{5\pi}{4}\right)\div\sqrt{3}\left(\cos\dfrac{\pi}{6}+i\sin\dfrac{\pi}{6}\right)$

23. $8\left(\cos\dfrac{3\pi}{2}+i\sin\dfrac{3\pi}{2}\right)\div\dfrac{4}{5}\left(\cos\dfrac{\pi}{2}+i\sin\dfrac{\pi}{2}\right)$

Critical Thinking

24. Suppose $a+bi$ is multiplied by $1+i\sqrt{3}$. By how many degrees must the radius from the origin to (a, b) be rotated to coincide with the radius from the pole to the graph of the product?

Applications and Problem Solving

25. Algebra If s_1 and s_2 are roots of a quadratic equation $ax^2+bx+c=0$, then $s_1+s_2=\dfrac{-b}{a}$ and $s_1\cdot s_2=\dfrac{c}{a}$. Verify that $4\sqrt{2}\left(\cos\dfrac{\pi}{4}+i\sin\dfrac{\pi}{4}\right)$ and $4\sqrt{2}\left[\cos\left(-\dfrac{\pi}{4}\right)+i\sin\left(-\dfrac{\pi}{4}\right)\right]$ are the polar forms of the roots of $x^2-8x+32=0$.

26. Electricity The polar form of the voltage of a circuit is $161(\cos 337°+i\sin 337°)$ volts. The polar form of the impedance is $32(\cos 300°+i\sin 300°)$ ohms. Find the polar form of the current. *Use $E=I\cdot Z$.*

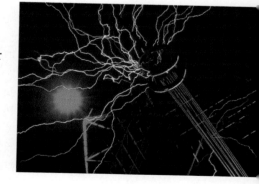

Mixed Review

27. Geometry Two sides of a triangle are 400 feet and 600 feet long, and the included angle measures $46°20'$. Find the perimeter and area of the triangle. **(Lesson 5-8)**

28. Graph $y=\cos 3\theta$ on a rectangular coordinate plane. **(Lesson 6-3)**

29. Simplify $2(4-3i)(7-2i)$. **(Lesson 9-5)**

30. Express $5\left(\cos\dfrac{5\pi}{6}+i\sin\dfrac{5\pi}{6}\right)$ in rectangular form. **(Lesson 9-6)**

31. College Entrance Exam Choose the best answer.
Six quarts of a 20% solution of alcohol in water are mixed with 4 quarts of a 60% solution of alcohol in water. The alcoholic strength of the mixture is

(A) 36%   **(B)** 40%   **(C)** 48%   **(D)** 60%   **(E)** 80%

9-8 Powers and Roots of Complex Numbers

Objective

After studying this lesson, you should be able to:
- find powers and roots of complex numbers in polar form using De Moivre's theorem.

Application

Abraham De Moivre (1667–1754) began his work in mathematics with his study of probability. He published his first work, entitled *Doctrine of Chances*, in 1718. He also worked extensively in number theory and in 1730, presented a theorem for finding the power of a complex number written in polar form. You will learn about this theorem, which bears his name, in this lesson.

You can use the formula for the product of complex numbers to find the square of a complex number.

$$[r(\cos\theta + i\sin\theta)]^2 = [r(\cos\theta + i\sin\theta)] \cdot [r(\cos\theta + i\sin\theta)]$$
$$= r^2[\cos(\theta + \theta) + i\sin(\theta + \theta)]$$
$$= r^2(\cos 2\theta + i\sin 2\theta)$$

Other powers of complex numbers can be found by using the theorem developed by Abraham De Moivre.

De Moivre's Theorem	$[r(\cos\theta + i\sin\theta)]^n = r^n(\cos n\theta + i\sin n\theta)$

Mathematical induction will be formally introduced in Chapter 12.

A method of proof called **mathematical induction** can be used to prove that De Moivre's theorem is valid for any positive integer n. The theorem has been shown to be valid when $n = 1$ and $n = 2$. Assume the theorem is valid for $n = k$. That is, assume that

$$[r(\cos\theta + i\sin\theta)]^k = r^k(\cos k\theta + i\sin k\theta).$$

Multiply each side of the equation by $r(\cos\theta + i\sin\theta)$.

$[r(\cos\theta + i\sin\theta)]^{k+1}$
$$= [r^k(\cos\ k\theta + i\sin k\theta)] \cdot [r(\cos\theta + i\sin\theta)]$$
$$= r^{k+1}(\cos\ k\theta\cos\theta + (\cos k\theta)(i\sin\theta) + i\sin k\theta\cos\theta + i^2\sin k\theta\sin\theta)$$
$$= r^{k+1}[(\cos\ k\theta\cos\theta - \sin k\theta\sin\theta) + i(\sin k\theta\cos\theta + \cos k\theta\sin\theta)]$$
$$= r^{k+1}[\cos\ (k\theta + \theta) + i\sin\ (k\theta + \theta)] \quad \textit{Sum identities for sine and cosine}$$
$$= r^{k+1}[\cos\ (k+1)\theta + i\sin\ (k+1)\theta]$$

Since the right side of the last equation gives the same result as can be obtained directly by substituting $k + 1$ for n, the formula is valid for all positive integral values of n.

Example 1

Find $(1 - i)^8$.

Write $1 - i$ in polar form. Find r and θ.

$$r = \sqrt{1^2 + (-1)^2}$$
$$= \sqrt{2}$$

$$\theta = \text{Arctan}\ \frac{-1}{1} \qquad \theta = Arctan\ \frac{v}{x}\ if\ x > 0$$
$$= -\frac{\pi}{4}\ \text{or}\ \frac{7\pi}{4}$$

$$1 - i = \sqrt{2}\left(\cos\frac{7\pi}{4} + i\sin\frac{7\pi}{4}\right)$$

Use De Moivre's theorem to find the 8th power of the complex number in polar form.

$$(1 - i)^8 = \left[\sqrt{2}\left(\cos\frac{7\pi}{4} + i\sin\frac{7\pi}{4}\right)\right]^8$$
$$= (\sqrt{2})^8\left[\cos 8\left(\frac{7\pi}{4}\right) + i\sin 8\left(\frac{7\pi}{4}\right)\right]$$
$$= 16(\cos 14\pi + i\sin 14\pi)$$

Write the result in rectangular form.

$$16(\cos 14\pi + i\sin 14\pi) = 16[1 + i(0)]$$
$$= 16$$

Thus, $(1 - i)^8 = 16$.

It can be proven that De Moivre's theorem is also valid when n is a rational number. Therefore, the roots of complex numbers can be found by letting $n = \frac{1}{2}, \frac{1}{3}, \frac{1}{4}, \ldots$. In general, the pth principle root of a complex number can be found as follows.

$$(x+yi)^{\frac{1}{p}} = [r(\cos\theta + i\sin\theta)]^{\frac{1}{p}}$$

$$= r^{\frac{1}{p}}\left(\cos\frac{\theta}{p} + i\sin\frac{\theta}{p}\right)$$

Example 2 | **Find $\sqrt{i}$.**

$\sqrt{i} = i^{\frac{1}{2}}$

$= (0 + i)^{\frac{1}{2}}$

$= \left[1\left(\cos\frac{\pi}{2} + i\sin\frac{\pi}{2}\right)\right]^{\frac{1}{2}}$
     $x = 0$ and $y = 1$
     $r = \sqrt{0^2 + 1^2}$ or 1
     $\theta = $ Arctan $\frac{1}{0}$ or $\frac{\pi}{2}$

$= 1^{\frac{1}{2}}\left[\cos\left(\frac{1}{2}\right)\left(\frac{\pi}{2}\right) + i\sin\left(\frac{1}{2}\right)\left(\frac{\pi}{2}\right)\right]$
     De Moivre's theorem

$= 1\left(\cos\frac{\pi}{4} + i\sin\frac{\pi}{4}\right)$

$= \dfrac{\sqrt{2}}{2} + \dfrac{i\sqrt{2}}{2}$

Thus, $\sqrt{i} = \dfrac{\sqrt{2}}{2} + \dfrac{\sqrt{2}}{2}i$. *This is the principal root of i.*

Example 3 | **Find $(1 + i)^{\frac{1}{3}}$. Express the answer in the form $a + bi$ with a and b to the nearest hundredth.**

In this complex number, $x = 1$ and $y = 1$.

$(1 + i)^{\frac{1}{3}} = \left[\sqrt{2}\left(\cos\frac{\pi}{4} + i\sin\frac{\pi}{4}\right)\right]^{\frac{1}{3}}$
     $r = \sqrt{1^2 + 1^2}$ or $\sqrt{2}$
     $\theta = $ Arctan $\frac{1}{1}$ or $\frac{\pi}{4}$

$= 2^{\frac{1}{2}\cdot\frac{1}{3}}\left(\cos\frac{1}{3}\cdot\frac{\pi}{4} + i\sin\frac{1}{3}\cdot\frac{\pi}{4}\right)$
     $\sqrt{2} = 2^{\frac{1}{2}}$

$= \sqrt[6]{2}\left(\cos\frac{\pi}{12} + i\sin\frac{\pi}{12}\right)$

$= 1.084215081 + 0.290514555i$
     Use a calculator.

The cube root of $(1 + i)^{\frac{1}{3}}$ is about $1.08 + 0.29i$.

You can use the TI-83 and the TI-92 graphing calculators to find powers of complex numbers.

It can be proven that any complex number has p distinct pth roots. That is, it has two square roots, three cube roots, four fourth roots, five fifth roots, and so on. Since $\cos\theta = \cos(\theta + 2n\pi)$ and $\sin\theta = \sin(\theta + 2n\pi)$, when $n = 1, 2, 3,...$, a more general formula for finding pth roots of complex numbers can be written in the following way.

The **p** distinct **p**th roots of $x + yi$ can be found by replacing n with 0, 1, 2, . . ., $p - 1$, successively, in the following equation.

$$(x + yi)^{\frac{1}{p}} = (r[\cos(\theta + 2n\pi) + i\sin(\theta + 2n\pi)])^{\frac{1}{p}}$$

$$= r^{\frac{1}{p}}\left(\cos\frac{\theta + 2n\pi}{p} + i\sin\frac{\theta + 2n\pi}{p}\right)$$

Example 4

Find three cube roots of –8.

In rectangular form, $-8 = -8 + 0i$. Write the polar form of $-8 + 0i$.

$-8 + 0i = 8[\cos(\pi + 2n\pi) + i\sin(\pi + 2n\pi)]$ $r = \sqrt{8^2 + 0^2}$ or 8

$\theta = Arctan\,\dfrac{0}{8} + \pi$ or π

Now write an expression for the cube roots of $-8 + 0i$.

$(-8 + 0i)^{\frac{1}{3}} = (8[\cos(\pi + 2n\pi) + i\sin(\pi + 2n\pi)])^{\frac{1}{3}}$

$= 2\left(\cos\dfrac{\pi + 2n\pi}{3} + i\sin\dfrac{\pi + 2n\pi}{3}\right)$

Let $n = 0, 1,$ and 2 successively to find the three cube roots.

Let n = 0. $2\left[\cos\left(\dfrac{\pi + 2(0)\pi}{3}\right) + i\sin\left(\dfrac{\pi + 2(0)\pi}{3}\right)\right] = 2\left(\cos\dfrac{\pi}{3} + i\sin\dfrac{\pi}{3}\right)$

$= 2\left(\dfrac{1}{2} + i\dfrac{\sqrt{3}}{2}\right)$

$= 1 + i\sqrt{3}$

Let n = 1. $2\left[\cos\left(\dfrac{\pi + 2(1)\pi}{3}\right) + i\sin\left(\dfrac{\pi + 2(1)\pi}{3}\right)\right] = 2(\cos\pi + i\sin\pi)$

$= 2(-1) + i(0)$

$= -2$

Let n = 2. $2\left[\cos\left(\dfrac{\pi + 2(2)\pi}{3}\right) + i\sin\left(\dfrac{\pi + 2(2)\pi}{3}\right)\right] = 2\left(\cos\dfrac{5\pi}{3} + i\sin\dfrac{5\pi}{3}\right)$

$= 2\left[\dfrac{1}{2} + i\left(\dfrac{-\sqrt{3}}{2}\right)\right]$

$= 1 - i\sqrt{3}$

The three cube roots of –8 are $-2, 1 + i\sqrt{3}$, and $1 - i\sqrt{3}$.
These roots can be checked by multiplication.

You can also determine the nth roots of a number on a graphing calculator.

EXPLORATION: Graphing Calculators

Find the three cube roots of 1.

1. Select the DEG and PARAM modes.
2. Set the parametric range parameters to: *TMIN: 0, TMAX: 360,* and *TSTEP: 120.*
3. Set the viewing window parameters to: *XMIN: -2, XMAX: 2, XSCL: 1, YMIN: -2, YMAX: 2,* and *YSCL: 1.*
4. Enter the parametric equations $x_{1T} = \cos T$ and $y_{1T} = \sin T$.
5. Graph the equations.
6. The vertices of the figure represent the cube roots of 1. Use the TRACE function to find all three cube roots.

Find the three cube roots of –1.

1. Use the same parametric range parameters as above.
2. Enter $x_{1T} = -\cos T$ and $y_{2T} = -\sin T$. Then TRACE.
3. How does the graph of the cube roots of –1 compare with the graph of the cube roots of 1?

You can solve some polynomial equations by using De Moivre's theorem.

Example 5

Solve $x^4 + 1 = 0$.

$$x^4 + 1 = 0$$
$$x^4 = -1$$
$$x^4 = -1 + 0i$$

Write $-1 + 0i$ in polar form.

$$-1 + 0i = \cos \pi + i \sin \pi \qquad r = \sqrt{1^2 + 0^2} \text{ or } 1, \theta = \text{Arctan } \frac{0}{-1} + \pi \text{ or } \pi$$

Then write an expression for the fourth roots of $-1 + 0i$.

$$(-1 + 0i)^{\frac{1}{4}} = [\cos (\pi + 2n\pi) + i \sin (\pi + 2n\pi)]^{\frac{1}{4}}$$
$$= \cos \frac{\pi + 2n\pi}{4} + i \sin \frac{\pi + 2n\pi}{4}$$

Let $n = 0, 1, 2$, and 3 successively to find the fourth roots, x_1, x_2, x_3, and x_4.

Let n = 0. $\quad x_1 = \cos \dfrac{\pi}{4} + i \sin \dfrac{\pi}{4} = \dfrac{\sqrt{2}}{2} + \dfrac{\sqrt{2}}{2}i$

Let n = 1. $\quad x_2 = \cos \dfrac{3\pi}{4} + i \sin \dfrac{3\pi}{4} = -\dfrac{\sqrt{2}}{2} + \dfrac{\sqrt{2}}{2}i$

Let n = 2. $\quad x_3 = \cos \dfrac{5\pi}{4} + i \sin \dfrac{5\pi}{4} = -\dfrac{\sqrt{2}}{2} - \dfrac{\sqrt{2}}{2}i$

Let n = 3. $\quad x_4 = \cos \dfrac{7\pi}{4} + i \sin \dfrac{7\pi}{4} = \dfrac{\sqrt{2}}{2} - \dfrac{\sqrt{2}}{2}i$

The roots of $x^4 + 1 = 0$ are $\dfrac{\sqrt{2}}{2} \pm \dfrac{\sqrt{2}}{2}i$ and $-\dfrac{\sqrt{2}}{2} \pm \dfrac{\sqrt{2}}{2}i$.

The roots of a complex number are cyclical in nature. That means, when the roots are graphed on the complex plane, the roots are equally spaced around a circle. The graphs below show the cube roots of 1 and the fourth roots of –1.

Compare the graph of the fourth roots of –1 with the results from Example 5.

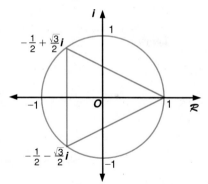

The cube roots of 1 form the vertices of an equilateral triangle.

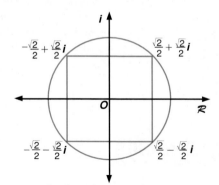

The fourth roots of –1 form the vertices of a square.

If one pth root of a complex number is known, all the pth roots can be graphed on the complex plane.

Example 6

Graph the three cube roots of 27.

The modulus of $27 + 0i$ is 27, and $27^{\frac{1}{3}} = 3$. Graph $r = 3$ on the complex plane.

One of the cube roots is $(3 + 0i)$. Graph this point.

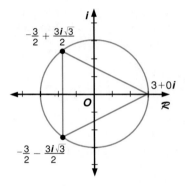

Since the roots are cyclic, they will occur at equal intervals. To find these intervals, divide $360°$ by 3. Each interval is $120°$. Graph $(3, 120°)$ and $(3, 240°)$.

You can solve the equation $x^3 - 27 = 0$ to find the exact roots.

The three roots are $3 + 0i$, $-\dfrac{3}{2} + \dfrac{3i\sqrt{3}}{2}$, and $-\dfrac{3}{2} - \dfrac{3i\sqrt{3}}{2}$.

MATH JOURNAL

Write a paragraph explaining how the roots of complex numbers are related to regular polygons.

CHECKING FOR UNDERSTANDING

Communicating Mathematics

Read and study the lesson to answer each question.

1. **Find** the product $(1 + i)(1 + i)(1 + i)(1 + i)(1 + i)$ by traditional multiplication. Compare the results with the results using De Moivre's theorem on $(1 + i)^5$. Which method do you prefer?

2. **Describe** the geometric figures formed by the graphs of the roots of each of the following.

   a. $x^{\frac{1}{3}}$ **b.** $x^{\frac{1}{4}}$ **c.** $x^{\frac{1}{6}}$

3. **Determine** which degree intervals you would use to graph the 6th roots of 64. What is the principal 6th root?

Find each power. Express the result in rectangular form.

4. $\left(\cos \dfrac{\pi}{6} + i \sin \dfrac{\pi}{6} \right)^4$

5. $\left[3 \left(\cos \dfrac{\pi}{3} + i \sin \dfrac{\pi}{3} \right) \right]^2$

6. $(\sqrt{2} + i\sqrt{2})^5$

7. $\left(-\dfrac{\sqrt{3}}{2} - \dfrac{i}{2} \right)^3$

Find each root. Express the result in rectangular form.

8. $(4i)^{\frac{1}{2}}$

9. $[-1(\cos \pi + i \sin \pi)]^{\frac{1}{3}}$

EXERCISES

Practice

Find each power. Express the result in rectangular form.

10. $(1 - i)^5$

11. $(3 + 4i)^4$

12. $[-3(1 - i)]^3$

13. $(1 + i)^{20}$

14. $(-2 + 2i\sqrt{3})^4$

15. $(12i - 5)^2$

Find each root. Express the result in the form $a + bi$, with a and b to the nearest hundredth.

16. $(0 + 8i)^{\frac{1}{3}}$

17. $[-2(1 + i)]^{\frac{1}{4}}$

18. $\sqrt[10]{-4i}$

19. $(0 + 1i)^{\frac{1}{3}}$

20. $(-1)^{\frac{1}{5}}$

21. $\sqrt[4]{-2 + 2i\sqrt{3}}$

Solve each equation. Then graph the roots on the complex plane.

22. $x^3 + 1 = 0$

23. $x^5 = 1$

24. $x^3 - 8 = 0$

25. $x^6 = -1$

26. $x^4 + 1 = 2$

27. $x^5 - 6 = -7$

**Graphing
Calculator**

Use a graphing calculator and parametric equations to find the roots.

28. $\sqrt[4]{1}$

29. $\sqrt[6]{1}$

30. $\sqrt[5]{1}$

Programming

31. The following program will calculate powers of complex numbers.

```
Prgm 9: CMPLXPWR
:ClrHome                        Clears the text screen
:Disp "ENTER A IN A + BI."      Displays message.
:Input A                        Accepts a value of a.
:Disp "ENTER B IN A + BI."
:Input B
:Disp "ENTER POWER."
:Input P
:R ▶ P(A, B)                    Converts rectangular coordinates to polar.
:R^Pcos Pθ→X                    Calculates real part of power.
:R^Psin Pθ→Y                    Calculates imaginary part of power.
```

```
:Disp "REAL PART = "
:Disp X
:Disp "IMAGINARY PART = "
:Disp Y
```
Use the program to find each power.

a. $(2 + i)^3$ **b.** $(1 - 2i)^4$ **c.** $(0 - 3i)^8$

d. Write a similar program to find the roots of a complex number.

Critical Thinking

32. How many distinct integral divisors exist for 42^3, excluding 1 and 42^3? Explain how you arrived at your solution.

33. Refer to the Graphing Calculator Exploration on page 514.

    a. What Tstep values should be set in order to find the nth roots of 1?

    b. What would you see after graphing the parametric equations associated with this process?

Applications and Problem Solving

34. Science A buzzard will soar for hours circling its prospective meal. Buzzards actually circle in a pattern that can be modeled by successive powers of z^n, where $z = 1.1(\cos \frac{\pi}{3} + i \sin \frac{\pi}{3})$. Graph the course of the buzzard for $n = 8, 7, 6, \cdots, 2, 1, \frac{1}{2}, \frac{1}{3}, \frac{1}{4}, \frac{1}{5}$.

Mixed Review

35. Find the slope of the line tangent to the graph of $y = -2x^2 + 3x + 1$ at $(1, 2)$. **(Lesson 3-6)**

36. Travel Martina went to Acapulco, Mexico, on a vacation with her parents. One of the sights they visited was a cliff-diving exhibition into the waters of the Gulf of Mexico. Martina stood at a lookout site on top of a 200-foot cliff. A team of medical experts were in a boat below in case of an accident. The angle of depression to the boat was $21°$. How far is the boat from the base of the cliff? **(Lesson 5-5)**

37. Food Industry Fishmongers will often place ice over freshly-caught fish that are to be shipped to preserve the freshness without freezing. Suppose a block of ice weighing 300 kilograms is held on an ice slide by a rope parallel to the slide. The slide is inclined at an angle of $22°$. **(Lesson 8-5)**

    a. What is the pull on the rope?

    b. What is the force on the slide?

38. Express $\sqrt{2} \left(\cos \frac{-\pi}{2} + i \sin \frac{-\pi}{2} \right)$ in rectangular form. **(Lesson 9-6)**

39. Find the product $5 \left(\cos \frac{3\pi}{4} + i \sin \frac{3\pi}{4} \right) \cdot 2(\cos \frac{2\pi}{3} + i \sin \frac{2\pi}{3})$. Then express it in rectangular form. **(Lesson 9-7)**

40. College Entrance Exam Choose the best answer.

    $\star x$ is defined such that $\star x = x^2 - 2x$. The value of $\star 2 - \star 1$ is

    (A) -1 **(B)** 0 **(C)** 1 **(D)** 2 **(E)** 4

VOCABULARY

Upon completing this chapter, you should be
familiar with the following terms:

501 amplitude   **466** polar axis   **469** polar graph
475 classical curve   **466** polar coordinate system   **466** pole
496 conjugate   **469** polar equation   **501** rectangular form
501 modulus   **502** polar form   **502** trigonometric form

SKILLS AND CONCEPTS

OBJECTIVES AND EXAMPLES	REVIEW EXERCISES

Upon completing this chapter, you should
be able to:

■ graph polar coordinates and simple polar
equations **(Lesson 9-1)**

Graph point P
that has polar
coordinates
$\left(-2, \dfrac{5\pi}{6}\right)$.

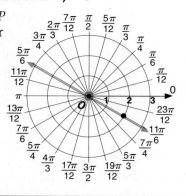

Use these exercises to review and prepare for
the chapter test.

**Graph the point that has the given polar
coordinates. Then, name three other pairs
of polar coordinates for each point.**

  1. $(-3, 50°)$ **2.** $(1.5, -110°)$

  3. $\left(2, \dfrac{\pi}{4}\right)$ **4.** $\left(-3, -\dfrac{\pi}{2}\right)$

Graph each polar equation.

  5. $r = \sqrt{7}$ **6.** $r = -2$

  7. $\theta = -80°$ **8.** $\theta = \dfrac{3\pi}{4}$

■ graph polar equations **(Lesson 9-2)**

Graph $r =$
$3 + 3 \sin \theta$.

This graph
is a cardioid.

**Graph each polar equation. Identify the
classical curve it represents.**

  9. $r = 7 \cos \theta$

  10. $r = 2 + 4 \cos \theta$

  11. $r = 3\theta$

  12. $r = 6 \sin 2\theta$

OBJECTIVES AND EXAMPLES	REVIEW EXERCISES

■ change from polar coordinates to rectangular coordinates and vice versa **(Lesson 9-3)**

Find the rectangular coordinates of $C(-3, 0.8)$.

$x = r \cos \theta \qquad y = r \sin \theta$
$\quad = -3 \cos (0.8) \quad = -3 \sin (0.8)$
$\quad = -2.09 \qquad\quad = -2.15$

The rectangular coordinates of C are $(-2.09, -2.15)$.

Find the rectangular coordinates of each point with the given polar coordinates.

13. $\left(6, \dfrac{\pi}{4}\right)$ **14.** $\left(2, -\dfrac{\pi}{6}\right)$

15. $(-2, 2.3)$ **16.** $(-1, -4.5)$

Find the polar coordinates of each point with the given rectangular coordinates.

17. $(-\sqrt{3}, -3)$ **18.** $(5, 5)$

19. $(3, -2)$ **20.** $(-4, 2)$

■ write the polar form of a linear equation **(Lesson 9-4)**

Write $\sqrt{2}x - \sqrt{2}y - 6 = 0$ in polar form.

The normal form is $\dfrac{\sqrt{2}}{2}x - \dfrac{\sqrt{2}}{2}y - 3 = 0$.

So, $\cos \phi = \dfrac{\sqrt{2}}{2}$, $\sin \phi = -\dfrac{\sqrt{2}}{2}$, and $p = 3$. Thus, $\phi = -45°$ or $315°$.

The polar form is $3 = r \cos (\theta - 315°)$.

Write each equation in polar form. Round θ to the nearest degree.

21. $2x + y = -3$

22. $y = -3x - 4$

Write each equation in rectangular form.

23. $3 = r \cos \left(\theta - \dfrac{\pi}{3}\right)$

24. $4 = r \cos \left(\theta + \dfrac{\pi}{2}\right)$

■ add subtract, multiply, and divide complex numbers in rectangular form. **(Lesson 9-5)**

Simplify $(2 - 4i) - (5 + 2i)$.
$(2 - 4i) - (5 + 2i) = 2 - 4i - 5 - 2i$
$\qquad\qquad\qquad = (2 - 5) + (-4i - 2i)$
$\qquad\qquad\qquad = -3 - 6i$

Simplify.

25. $(2 + 3i) + (4 - 4i)$

26. $(-3 - i) - (2 + 7i)$

27. $i^{10} \cdot i^{25}$ **28.** $i^3(4 - 3i)$

29. $(-i - 7)(i - 7)$ **30.** $\dfrac{4 + i}{5 - 2i}$

31. $\dfrac{5}{\sqrt{2} - 4i}$ **32.** $\dfrac{8 - i}{2 + 3i}$

■ convert complex numbers from rectangular to polar form and vice versa **(Lesson 9-6)**

Express $4 - 3i$ in polar form.

$r = \sqrt{4^2 + (-3)^2} \qquad \theta = \text{Arctan } \dfrac{y}{x} + \pi$

$\quad = 5 \qquad\qquad\qquad = \text{Arctan } \dfrac{-3}{4} + \pi$

$\qquad\qquad\qquad\qquad\quad = 2.50$

The polar form is $(5, 2.5)$.

Express each complex number in polar form.

33. $-2 + 2i\sqrt{3}$ **34.** $6 - 8i$

Express each complex number in rectangular form.

35. $4 \left(\cos \dfrac{5\pi}{6} + i \sin \dfrac{5\pi}{6} \right)$

36. $8 \left(\cos \dfrac{7\pi}{4} + i \sin \dfrac{7\pi}{4} \right)$

■ find the product and quotient of complex numbers in polar form **(Lesson 9-7)**

Find the product
$$3\left(\cos\frac{\pi}{2}+i\sin\frac{\pi}{2}\right)\cdot 2\left(\cos\frac{3\pi}{4}+i\sin\frac{3\pi}{4}\right).$$
Then express in rectangular form.

$$
\begin{aligned}
r &= r_1 r_2 & \theta &= \theta_1 + \theta_2 \\
&= 3(2) & &= \frac{\pi}{2} + \frac{3\pi}{4} \\
&= 6 & &= \frac{5\pi}{4}
\end{aligned}
$$

The product is $6\left(\cos\frac{5\pi}{4}+i\sin\frac{5\pi}{4}\right)$ or $-3\sqrt{2}+3\sqrt{2}i$, in rectangular form.

Find each product or quotient. Then write the result in rectangular form.

37. $2\left(\cos\frac{\pi}{3}+i\sin\frac{\pi}{3}\right)\cdot$
$4\left(\cos\frac{\pi}{3}+i\sin\frac{\pi}{3}\right)$

38. $1.9(\cos 2.1+i\sin 2.1)\cdot$
$3(\cos 0.8+i\sin 0.8)$

39. $8\left(\cos\frac{7\pi}{6}+i\sin\frac{7\pi}{6}\right)\div$
$2\left(\cos\frac{5\pi}{3}+i\sin\frac{5\pi}{3}\right)$

40. $6\left(\cos\frac{\pi}{2}+i\sin\frac{\pi}{2}\right)\div$
$4\left(\cos\frac{\pi}{6}+i\sin\frac{\pi}{6}\right)$

41. $2.2(\cos 1.5+i\sin 1.5)\div$
$4.4(\cos 0.6+i\sin 0.6)$

■ find powers and roots of complex numbers in polar form using De Moivre's theorem **(Lesson 9-8)**

Find $(3-3i)^{\frac{1}{2}}$.

$$(-3+3i)^{\frac{1}{2}}=\left[3\sqrt{2}\left(\cos\frac{3\pi}{4}+i\sin\frac{3\pi}{4}\right)\right]^{\frac{1}{2}}$$

$$=(3\sqrt{2})^{\frac{1}{2}}\left(\cos\frac{3\pi}{8}+i\sin\frac{3\pi}{8}\right)$$

$$=2.06(0.38+0.92i)$$

$$=0.78+1.90i$$

Find each power. Express the result in rectangular form.

42. $(2+2i)^8$ 43. $(\sqrt{3}-i)^7$

44. $(-1+i)^4$ 45. $(-2-2i\sqrt{3})^3$

Find each root. Express the result in the form $a+bi$, with a and b to the nearest hundredth.

46. $\sqrt[4]{i}$ 47. $(\sqrt{3}+i)^{\frac{1}{3}}$

Solve each equation. Then graph the roots.

48. $x^5-32=0$ 49. $x^6-1=0$

APPLICATIONS AND PROBLEM SOLVING

50. **Chemistry** An electron moves about the nucleus of an atom at such a high speed that if it were visible to the eye, it would appear as a cloud. Identify the classical curve represented by the electron cloud at the right. **(Lesson 9-2)**

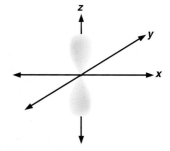

51. **Electricity** Find the current of a circuit of $(50+180j)$ volts with an impedance of $(4+5j)$ ohms. *Use $E = I \cdot Z$.* **(Lesson 9-5)**

Graph the point that has the given polar coordinates. Then, name three other pairs of polar coordinates for each point.

1. $\left(-2, \frac{5\pi}{4}\right)$ 2. $\left(3, -\frac{\pi}{6}\right)$ 3. $(2.5, 140°)$ 4. $(-1.7, 25°)$

Graph each polar equation.

5. $r = -4$ 6. $\theta = \frac{3\pi}{2}$ 7. $r = 8 \sin \theta$

8. $r = 10 \sin 2\theta$ 9. $r = 6 \cos 3\theta$ 10. $r = 6 + \sin \theta$

Find the rectangular coordinates of each point with the given polar coordinates.

11. $\left(3, -\frac{5\pi}{4}\right)$ 12. $(-4, 1.4)$ 13. $\left(2\sqrt{2}, \frac{\pi}{4}\right)$

Write each equation in polar form. Round ϕ to the nearest degree.

14. $5x + 6y = -3$ 15. $2x - 4y = 1$ 16. $y = -\frac{1}{3}x + 2$

Simplify.

17. i^{93} 18. $(2 - 5i) + (-2 + 4i)$ 19. $(-4 + i) - (4 - 2i)$

20. $(3 + 5i)(3 - 2i)$ 21. $(7 + i)^2$ 22. $\frac{6 - 2i}{2 + i}$

Express each complex number in polar form.

23. $-4 + 4i$ 24. -5 25. $6 - 6i\sqrt{3}$

Find each product or quotient. Then write the result in rectangular form.

26. $4\left(\cos \frac{3\pi}{2} + i\sin \frac{3\pi}{2}\right) \cdot 3\left(\cos \frac{\pi}{4} + i\sin \frac{\pi}{4}\right)$

27. $(\sqrt{3} - 3i)(\sqrt{3} + i)$

28. $2\sqrt{3}\left(\cos \frac{2\pi}{3} + i\sin \frac{2\pi}{3}\right) \div \sqrt{3}\left(\cos \frac{\pi}{6} + i\sin \frac{\pi}{6}\right)$

29. Find $(1 - i)^8$.

30. Find $\sqrt[3]{-8i}$.

31. Find all roots of $x^8 - 1 = 0$.

32. **Electricity**   The polar form of the current of a circuit is $8(\cos 307° + j\sin 307°)$. The polar form of the impedance is $20(\cos 115° + j\sin 115°)$. Find the polar form of the voltage of this circuit. *Use $E = I \cdot Z$.*

33. **Number Theory**   Find the zeros of $f(x) = 3x^2 + 6x + 7$. Graph them on the complex plane.

Bonus   Graph $r = 3 \tan \theta$.

CHAPTER 10

CONICS

HISTORICAL SNAPSHOT

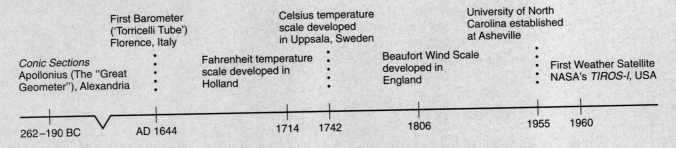

Conic Sections
Apollonius (The "Great Geometer"), Alexandria

First Barometer ('Torricelli Tube') Florence, Italy

Fahrenheit temperature scale developed in Holland

Celsius temperature scale developed in Uppsala, Sweden

Beaufort Wind Scale developed in England

University of North Carolina established at Asheville

First Weather Satellite NASA's *TIROS-I*, USA

262–190 BC   AD 1644   1714 1742   1806   1955   1960

CHAPTER OBJECTIVES

In this chapter, you will:
- Write equations of circles, parabolas, ellipses, and hyperbolas.
- Graph circles, parabolas, ellipses, and hyperbolas.
- Recognize conic sections by their equations.
- Write equations of tangents and normals to conic sections.

CAREER GOAL: Meteorology

James Martin was born with spinal muscular atrophy, a disorder in which there is a bad connection between the nerves and muscles in his legs and arms. He credits his family with being his inspiration. "My mom and dad put me in public school and always encouraged me to pursue my dreams."

James hasn't wasted any time pursuing his dreams. He works part-time as a computer specialist at the National Climatic Center in Asheville, part of the National Oceanic and Atmospheric Administration. At the the University of North Carolina, James majors in computer sciences and atmospheric sciences "so that I can become more familiar with the climate that I work with every day." James hopes that with a bachelor's degree, he will achieve a higher grade in civil service, with supervisory responsibilities. Having the title of meteorologist will also lead to more career opportunities.

Meteorologists use conics and vectors, especially in the study of the cone-shaped hurricane. If you cut horizontally across a hurricane, you would find a circle. A velocity vector drawn tangent to this circle represents the circulation speed of the hurricane. James says that the algebra, geometry, and trigonometry courses he took in high school prepared him well for his college mathematics courses.

From James' perspective, both computer science and atmospheric science offer opportunities for the physically challenged. "I hope that I can encourage other people who are physically challenged to be active in society. I also want to show society that we who are physically challenged don't have to be pushed back into a corner. We have a lot to offer if just given the opportunity."

interNET CONNECTION

For up-to-date information on meteorology, visit:
www.glencoe.com/sec/math/amc/mathnet

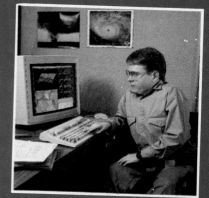

10-1 The Circle

Objectives

After studying this lesson, you should be able to:
- use the standard and general forms of the equation of a circle, and
- graph circles.

Application

The third game of the 1989 World Series between the San Francisco Giants and the Oakland Athletics was about to begin on Tuesday, October 17, when the Loma Prieta earthquake struck, registering 7.1 on the Richter Scale. The epicenter of the quake was in the Santa Cruz Mountains, approximately 56 miles south of San Francisco. The seismic waves traveled outward from the epicenter.

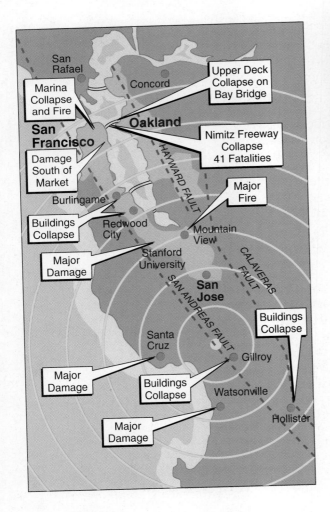

A **locus** is a set of points (and only those points) that satisfy a given set of conditions. A **circle** is the locus of all points in a plane at a given distance from a fixed point on the plane called the **center**. The distance from the center to any point on the circle is called the **radius** of the circle. In the map at the left, each circle has the same center, but a different radius. Seismologists use these circles to determine the strength of the quake as you get farther away from its epicenter.

Each successive circle on the map is 10 miles farther from the epicenter, and each circle is centered at the epicenter. Circles that have the same center, but not necessarily the same radius, are called **concentric circles**. The equations for some of this family of circles are given below.

$x^2 + y^2 = 10^2$ *includes the immediate area around the epicenter*

$x^2 + y^2 = 30^2$ *includes the area out to San Jose*

$x^2 + y^2 = 60^2$ *includes the area out to San Francisco and Oakland*

Materials: graph paper, corrugated cardboard, 10-centimeter piece of string, thumb tack

1. Tape a piece of centimeter graph paper onto a large piece of corrugated cardboard.

2. Draw the *x*-axis and *y*-axis. Plot the point $C(-2, 3)$.

3. Attach one end of the string to the thumb tack. Press the tack into the corrugated cardboard at point *C*.

4. Tie the string around your pencil.

5. Move your pencil around the tack, keeping the string taut.

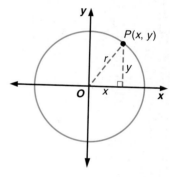

In the Exploration above, the length of string from the thumbtack to the pencil acted as a radius from point *C*. This activity models the set of all points that are at a fixed distance from a given point, in other words, a circle.

The word <u>radius</u> can refer to the distance from the center of a circle to a point on the circle, or to the segment that connects those two points.

In the figure at the right, the center of the circle is at the origin. By drawing a perpendicular from any point $P(x, y)$ on the circle but not on an axis, you form a right triangle. The Pythagorean theorem can be used to write an equation that describes every point on a circle whose center is located at the origin.

$$x^2 + y^2 = r^2$$

This is the equation of the parent graph of all circles.

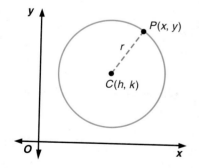

Suppose $C(h, k)$ is the center of a circle, $P(x, y)$ is any point on the circle, and *r* is the radius of the circle. You can use the distance formula to write an equation for the circle.

$$d = \sqrt{(x_2 - x_1)^2 + (y_2 - y_1)^2}$$

$$r = \sqrt{(x - h)^2 + (y - k)^2} \qquad (x_2, y_2) = (x, y), (x_1, y_1) = (h, k), \text{ and } d = r$$

$$r^2 = (x - h)^2 + (y - k)^2$$

This is known as the standard form of the equation of a circle.

Standard Form of the Equation of a Circle	The standard form of the equation of a circle with radius *r* and center at (h, k) is $$(x - h)^2 + (y - k)^2 = r^2.$$

Example 1

Write the equation of a circle that is tangent to the *y*-axis and has its center at $(-5, 6)$. Then graph the equation.

Since the circle is tangent to the *y*-axis, the distance from the center to the *y*-axis is the radius. Since the center is 5 units left of the *y*-axis, the radius is 5.

The equation of the circle can be found by substituting 5 for *r*, -5 for *h*, and 6 for *k*.

The equation is $(x - (-5))^2 + (y - 6)^2 = (5)^2$ or $(x + 5)^2 + (y - 6)^2 = 25$.

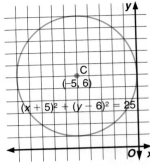

You can graph an equation for a circle by using a graphing calculator.

Example 2

Graph $(x - 4)^2 + (y + 5)^2 = 16$ by using a graphing calculator.

Solve the equation for *y*.

$$(x - 4)^2 + (y + 5)^2 = 16$$
$$(y + 5)^2 = 16 - (x - 4)^2$$
$$y + 5 = \pm\sqrt{16 - (x - 4)^2} \qquad \textit{Take the square root of each side.}$$
$$y = \pm\sqrt{16 - (x - 4)^2} - 5$$

Select a standard range on the calculator. Input the parts of the radical equation as separate equations.

$$Y_1 = \sqrt{16 - (x - 4)^2} - 5$$

$$Y_2 = -\sqrt{16 - (x - 4)^2} - 5$$

Note that the display does not look like a circle. On the TI-81, you can make it look like a circle by pressing ZOOM and 5. On Casio calculators, you must alter your range and scale values to achieve a circular appearance.

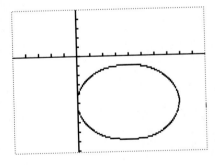

The standard form of the equation of a circle can be expanded to obtain a general form of the equation.

If r = 0, the circle is a point. If r < 0, a real circle does not exist.

$$(x - h)^2 + (y - k)^2 = r^2$$
$$(x^2 - 2hx + h^2) + (y^2 - 2ky + k^2) = r^2$$
$$x^2 + y^2 + (-2h)x + (-2k)y + (h^2 + k^2) = r^2$$

Since h, k, and r are constants, $-2h, -2k, (h^2 + k^2)$, and r^2 are also constants. The equation can be written as $x^2 + y^2 + Dx + Ey + F = 0$. This is called the general form of the equation of a circle.

The general form of the equation of a circle is
$$x^2 + y^2 + Dx + Ey + F = 0,$$
where D, E, and F are constants.

Notice that the coefficients of x^2 and y^2 in the general form must be 1. If those coefficients are not 1, the equation can be transformed by division so that they are 1. Also notice that there is no term containing the product of the variables, xy.

When the equation of a circle is given in general form, it can be rewritten in standard form by completing the square for the terms in x and the terms in y.

Example 3

You can review completing the square in Lesson 4-2.

The equation of a circle is $x^2 + y^2 - 10x + 4y + 17 = 0$. Find the radius and the coordinates of the center. Then graph the equation.

$$x^2 + y^2 - 10x + 4y + 17 = 0$$
$$(x^2 - 10x + \ ?\ ) + (y^2 + 4y + \ ?\ ) = -17 \qquad \textit{Complete the squares.}$$
$$(x^2 - 10x + 25) + (y^2 + 4y + 4) = -17 + 25 + 4$$
$$(x - 5)^2 + (y + 2)^2 = 12$$
$$(x - 5)^2 + (y + 2)^2 = \left(\sqrt{12}\right)^2$$

The center of the circle is located at $(5, -2)$, and the radius is $\sqrt{12}$ or $2\sqrt{3}$ units.

Notice that this is the result of translating the parent graph with radius $\sqrt{12}$ 5 units right and 2 units down.

You know that any two points in the coordinate plane determine a unique line. Any three noncollinear points in the coordinate plane determine a unique circle. An equation of the circle passing through three such points can be found by substituting the coordinates of each point into the general form of the equation. This will produce a system of three equations in three variables $(D, E,$ and $F)$. By solving this system of equations, the general form of the equation of the circle that passes through the three given points can be found.

Example 4

Find the equation of the circle that passes through $(-2, 3)$, $(6, -5)$, and $(0, 7)$. Then identify the center and radius of the circle.

Substitute each ordered pair for (x, y) in $x^2 + y^2 + Dx + Ey + F = 0$.

$$(-2)^2 + (3)^2 + D(-2) + E(3) + F = 0 \qquad (x, y) = (-2, 3)$$
$$(6)^2 + (-5)^2 + D(6) + E(-5) + F = 0 \qquad (x, y) = (6, -5)$$
$$(0)^2 + (7)^2 + D(0) + E(7) + F = 0 \qquad (x, y) = (0, 7)$$

(continued on the next page)

You can also solve this system by using matrices.

Simplify the system of equations.

$$-2D + 3E + F + 13 = 0$$
$$6D - 5E + F + 61 = 0$$
$$7E + F + 49 = 0$$

The solution to the system is $D = -10$, $E = -4$, and $F = -21$.

The equation of the circle is $x^2 + y^2 - 10x - 4y - 21 = 0$ or $(x - 5)^2 + (y - 2)^2 = 50$, after completing the square. Its center is at $(5, 2)$, and its radius is $\sqrt{50}$ units.

Satellites often orbit in circles around Earth. The radius of the orbit determines where the satellite is in relation to a fixed point on Earth. Satellites such as *Syncom, Early Bird, Intelsat,* and *ATS* are placed in geosynchronous orbits about the equator. This means that, since they complete one orbit in 24 hours, each satellite is above the same spot on Earth at all times. When the satellites are "parked" in this orbit, antennas can be pointed at a fixed point on Earth, making it possible to process directional signals very rapidly.

Example 5

APPLICATION

Communications

For the latest satellite statistics, visit:
www.glencoe.com/sec/ math/amc/mathnet

The distance that a satellite in geosynchronous orbit is above the equator can be found by subtracting the radius of Earth (6400 km) from the distance, r, that the satellite is from the center of Earth. The formula for the distance is $r = \sqrt[3]{\dfrac{GM_e t^2}{4\pi^2}}$, where

G = **the universal constant (6.67×10^{-11} newtons m^2/kg^2),**

M_e = **the mass of Earth (5.98×10^{24} kg), and**

t = **period of one orbit (86,400 seconds).**

Find the distance a satellite in geosynchronous orbit is above the equator.

$$r = \sqrt[3]{\frac{GM_e t^2}{4\pi^2}}$$

$$= \sqrt[3]{\frac{(6.67 \times 10^{-11} \text{ newton m}^2/\text{kg}^2)(5.98 \times 10^{24} \text{ kg})(86,400 \text{ s})^2}{4\pi^2}} \quad \begin{array}{l} \textit{1 newton =} \\ \textit{1 kg} \cdot m/s^2 \end{array}$$

$$= \sqrt[3]{7.542143065 \times 10^{22}} \text{ m} \quad \textit{Use a calculator.}$$

$$\approx 42{,}250{,}474.3 \text{ m}$$

$$\approx 42{,}250.5 \text{ km}$$

To find the distance above the equator, subtract the radius of Earth from r.

$$d = r - r_E$$
$$= 42{,}250.5 - 6400$$
$$= 35{,}850.5$$

The satellite is 35,850.5 kilometers above the equator.

EXPLORATION: Graphing Calculator

1. Graph $y_1 = \sqrt{25 - x^2}$ and $y_2 = -\sqrt{25 - x^2}$.

Remember to use ZOOM and 5 on the TI-82 to achieve a circular appearance.

2. For each situation, make a conjecture about the behavior of the graph. Then verify by graphing.
   a. x is replaced with $(x - 2)$.
   b. x is replaced with $(x + 2)$.
   c. y is replaced with $(y - 2)$.
   d. y is replaced with $(y + 2)$.
   e. 25 is replaced with 16.

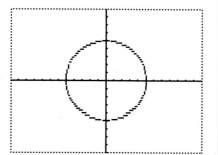

CHECKING FOR UNDERSTANDING

Communicating Mathematics

Read and study the lesson to answer each question.

1. **Write** the equations of a family of five concentric circles whose center is at $(-2, 6)$.

2. **Describe** the coefficients of the x^2 and y^2 terms in the equation of a circle written in standard form.

3. **Explain** the process for rewriting an equation of a circle in standard form as an equation in general form.

4. **Define** the term *geosynchronous* and tell how it relates to concentric circles.

Guided Practice

Write the standard form of the equation of a circle with the given center coordinates and radius. Then graph the equation.

5. $(0, 0), 9$

6. $(0, 0), \sqrt{6}$

7. $(2, -7), 10$

8. $(3, -2), 7$

Find the radius and the coordinates of the center of the circle for each equation. Then graph the equation.

9. $(x - 4)^2 + (y - 3)^2 = 9$

10. $(x + 4)^2 + (y - 7)^2 = 7$

11. $4(x - 3)^2 + 4(y - 2)^2 = 10$

12. $9(x + 5)^2 = 4 - 9(y + 3)^2$

13. Write the standard form of the equation of the circle that passes through the origin, $(2.8, 0)$, and $(5, 2)$.

EXERCISES

Write the standard form of each equation. Then graph the equation.

14. center at $(-3, -5)$, radius 4

15. center at $(7, 4)$, radius $\sqrt{5}$

16. $x^2 = 64 - y^2$

17. $x^2 + y^2 + x = \dfrac{3}{4}$

18. $x^2 + y^2 - 4x + 6y - 12 = 0$

19. $3x^2 + 3y^2 - 27 = 0$

20. $x^2 + y^2 + 8x + 2y - 8 = 0$

21. $16x^2 + 16y^2 + 8x - 32y = 127$

22. $6x^2 - 12x + 6y^2 + 36y = 36$

23. $16x^2 - 48x - 75 + 16y^2 + 8y = 0$

24. $x^2 + y^2 - 4x - 12y + 30 = 0$

25. $x^2 + y^2 + 14x + 24y + 157 = 0$

Write the standard form of the equation of the circle that passes through the points with the given coordinates. Then identify the center and the radius of the circle.

26. $(7, -1), (11, -5), (3, -5)$

27. $(1, 3), (5, 5), (5, 3)$

28. $(5, 3), (-2, 2), (-1, -5)$

29. $(7, -1), (7, 5), (1, -1)$

30. $(-10, -5), (-2, 7), (-9, 0)$

31. $(2, -1), (-3, 0), (1, 4)$

Write the equation of the circle that satisfies each set of conditions.

32. The circle passes through $(7, -1)$ and has its center at $(-2, 4)$.

33. The circle passes through the origin and has its center at $(-3, 4)$.

34. The endpoints of a diameter are at $(-2, -3)$ and at $(4, 5)$.

35. The endpoints of a diameter are at $(-3, 4)$ and at $(2, 1)$.

36. The center of the circle is on the x-axis, its radius is 1, and it passes through $\left(\dfrac{\sqrt{2}}{2}, \dfrac{\sqrt{2}}{2} \right)$.

37. The circle is tangent to the line $2x - y = -3$ and has its center at $(5, 12)$.

Use a graphing calculator to graph each equation.

38. $(x - 4)^2 + (y - 1)^2 = 25$

39. $(x + 3)^2 + (y + 5)^2 = 16$

40. $4x^2 + 4y^2 = 49$

41. $x^2 + 14x + y^2 + 6y + 50 = 0$

42. The following program will determine the radius and the coordinates of the center of a circle from an equation written in general form.

```
Prgm 10: CIRCLEQ
:ClrHome
:Disp "X² + Y² + DX + EY + F = 0"
:Disp "D ="
:Input D
:Disp "E ="
:Input E
:Disp "F ="
:Input F
:-D/2 → H          Calculates x-coordinate of center, stores in H.
:-E/2 → K          Calculates y-coordinate of center, stores in K.
:-F + D²/4 + E²/4 → R   Calculates square of radius value, stores in R.
```

```
:If R < 0                    If the value of R is negative, no circle exists.
:Goto 1
:√R → R                      Calculates radius, stores in R.
:Disp "H ="
:Disp H
:Disp "K ="
:Disp K
:Disp "R ="
:Disp R
:Goto 2
:Lbl 1
:Disp "NO CIRCLE"
:Lbl 2
```

For each equation, use the program shown above to find the radius of the circle and coordinates of its center.

a. $x^2 + y^2 - 6y - 16 = 0$ **b.** $x^2 + y^2 + 2x - 10 = 0$

Critical Thinking

43. Find the radius and the coordinates of the center of a circle defined by the equation $x^2 + y^2 - 2x + 4y + 5 = 0$. Describe the graph of this circle.

44. Write the equation of the family of circles in which $h = k$ and the radius is 8. Let k be any real number. Describe this family of circles.

45. If the equation of a circle is written in standard form and $h^2 + k^2 - r^2 = 0$, what is true of the graph of that equation?

Applications and Problem Solving

46. **Communications** Refer to Example 5. Determine the velocity (v) in meters/second of a communication satellite in a geosynchronous circular orbit if $v = \sqrt{\dfrac{GM}{r}}$, where $r =$ the distance from the center of Earth to the satellite, $G =$ the universal constant, and $M =$ the mass of Earth.

47. **Spacecraft** The formula for determining the velocity of a craft in a parked position above Earth, presented in Exercise 46, can also be used for a craft parked above any other planet or moon. In NASA's history, there were six Apollo spacecraft from 1969–1972 that had to maintain a parked position above the LEM landing craft on the lunar surface. Determine the velocity of the Apollo spacecraft necessary to be parked above the moon's surface at an altitude of 110 kilometers. The mass of the moon is 0.012 times the mass of Earth, and the radius of the moon is 1740 kilometers.

48. Technology Diego has a computer program that will graph all of the trigonometric functions. However, it only accepts angle measures in terms of radians. Find the radian measure (to the nearest thousandth) equivalent to each degree measure he needs to input in the software. **(Lesson 5-1)**

a. 60°   **b.** –105°   **c.** 1000°   **d.** –90°

49. State which value is greater, $\sin 10°$ or $\sin 100°$. **(Lesson 6-1)**

50. Find $\tan \theta$ if $\sec \theta = \dfrac{5}{3}$ and $0° \leq \theta \leq 90°$. **(Lesson 7-1)**

51. Find the coordinates of point D such that $\overrightarrow{AB}$ and $\overrightarrow{CD}$ are equal vectors for points $A(5, 2)$, $B(-3, 3)$, and $C(0, 0)$. **(Lesson 8-1)**

52. Simplify $(\sqrt{2} + i)(4\sqrt{2} + i)$. **(Lesson 9-5)**

53. Find $(2\sqrt{3} + 2i)^{\frac{1}{5}}$. Express the answer in the form $a + bi$ with a and b to the nearest hundredth. **(Lesson 9-8)**

54. College Entrance Exam Compare quantities A and B below.
Write A if quantity A is greater.
Write B if quantity B is greater.
Write C if the two quantities are equal.
Write D if there is not enough information to determine the relationship.

Eight separate cubes, each with an edge of 1 inch, are positioned together to create a large cube.

(A) surface area of one small cube   **(B)** $\dfrac{\text{surface area of the large cube}}{8}$

DECISION MAKING

Choosing a Home Computer

Since the introduction of the personal computer, millions of Americans have purchased computers for home use. Many buyers have been happy with their purchases, but others have discovered that they bought too much or too little, or worse yet, that they didn't need a computer at all.

1. Describe the features you would look for in a home computer. Include such factors as capacity, capabilities, programs you want to run, type of printer, and any other features you feel are important.

2. Research computers and costs. Talk to computer owners and retailers, read articles and ads, and contact computer manufacturers. Then make a list of three computers that would adequately meet your needs.

3. Outline a plan for financing the cost of each computer on your list.

4. **Project** Work with your group to complete one of the projects listed at the right based on the information you gathered in Exercises 1–3.

PROJECTS

- *Conduct an interview.*
- *Write a book report.*
- *Write a proposal.*
- *Write an article for the school paper.*
- *Write a report.*
- *Make a display.*
- *Make a graph or chart.*
- *Plan an activity.*
- *Design a checklist.*

10-2A Graphing Calculators: Locating the Vertex of a Parabola

You can use a graphing calculator to find the coordinates of the vertex of a parabola by using the TRACE and ZOOM functions.

Example

Graph the equation $y = (x + 4)^2 - 2$ and determine the coordinates of the vertex of the parabola.

First, graph the parabola in the standard viewing window.

Enter: $\boxed{Y=}$ $\boxed{(}$ $\boxed{X,T,\theta}$ $\boxed{+}$ 4 $\boxed{)}$ $\boxed{x^2}$ $\boxed{-}$
2 $\boxed{GRAPH}$

Tracing and zooming in shows the coordinates of the vertex to be $(-4, -2)$.

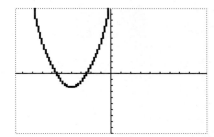

As you zoom in on the vertex of the parabola, the function begins to look more and more like a straight line, and it becomes more and more difficult to find the high or low point. This is one instance when it is better to use the "ZBox" command.

To zoom in using the box, first press $\boxed{ZOOM}$ 1. Then use the arrow keys to move the blinking dot to where you want one of the corners of your box to be. Press $\boxed{ENTER}$. Now use the arrow keys to move the blinking dot to the opposite corner of the box. Press $\boxed{ENTER}$ again. The calculator will redraw the graph with your box being the new viewing window.

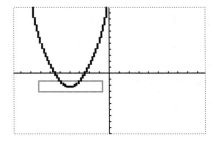

The TI-82 can also calculate the vertex of a parabola. Press $\boxed{2nd}$ $\boxed{CALC}$ 3. Then use the arrow and $\boxed{ENTER}$ keys to select the lower bound, the upper bound, and a guess for the vertex. The cursor will be on the vertex and the coordinates will be displayed correct to 6 digits.

EXERCISES

Graph each equation so that a complete graph is shown. Then find the coordinates of the vertex to the nearest tenth.

1. $y = x^2 + 4x - 6$
2. $y = (x - 3)^2 + 7$
3. $y = 3(x + 1)^2 - 4$

4. $y = 6x^2 + 10x + 12$
5. $y = 9x^2 + 66x - 120$
6. $y = 0.5x^2 - 11x + 289$

10-2 The Parabola

Objectives

After studying this lesson, you should be able to:
- use the standard and general forms of the equation of a parabola, and
- graph parabolas.

Application

FYI...

A communications satellite in geosynchronous orbit above Earth's equator remains at a fixed point above Earth because it moves at the same speed as Earth.

To facilitate worldwide communications, satellites are placed in geosynchronous orbits approximately 22,400 miles above Earth's surface over the equator. The satellites act as space bridges that receive signals from a ground-based parabolic dish antenna and then retransmit them at a lower frequency to a parabolic dish located in another geographic region. In a parabolic reception dish, all of the rays are directed to a feedhorn, located at the focal point of the parabola.

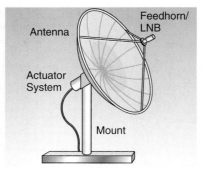

In previous chapters, you have seen that the graphs of quadratic equations like $y = x^2$ are called **parabolas**. A parabola is defined as the locus of all points in a given plane that are the same distance from a given point, called the **focus**, and a given line, called the **directrix**. *Remember that the distance from a point to a line is the length of the segment from the point perpendicular to the line.*

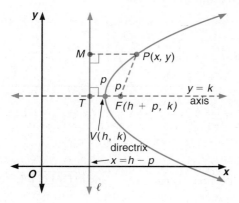

In the figure at the left, F is the focus of the parabola, and ℓ is the directrix. This parabola is symmetric with respect to the line $y = k$, which passes through the focus. This line is called the **axis of symmetry**, or, more simply, the **axis**, of the parabola. The point at which the axis intersects the parabola is called the **vertex**. Suppose the vertex V has coordinates (h, k). Let p be the distance from the focus to the vertex, FV. By the definition of a parabola, the distance from any point on the parabola to the focus must equal the distance from that point to the directrix. So, if $FV = p$, then $VT = p$. The coordinates of F are $(h + p, k)$ and the equation of the directrix is $x = h - p$.

Now suppose $P(x, y)$ is any other point on the parabola. From the definition of parabola, you know that $PF = PM$. Since M lies on the directrix, the coordinates of M are $(h - p, y)$. You can use the distance formula to determine an equation for the parabola.

$$PF = PM$$
$$\sqrt{[x - (h + p)]^2 + (y - k)^2} = \sqrt{[x - (h - p)]^2 + (y - y)^2}$$
$$[x - (h + p)]^2 + (y - k)^2 = [x - (h - p)]^2$$
$$x^2 - 2x(h + p) + (h + p)^2 + (y - k)^2 = x^2 - 2x(h - p) + (h - p)^2$$

This equation can be simplified to obtain the equation
$$(y - k)^2 = 4p(x - h).$$

When p is positive, the parabola opens to the right.
When p is negative, the parabola opens to the left.

Unlike the equation of a circle, the equation of a parabola has only one squared term.

This is the equation of a parabola whose directrix is parallel to the y-axis. The equation of a parabola whose directrix is parallel to the x-axis can be obtained by switching the terms in the parentheses of the previous equation.
$$(x - h)^2 = 4p(y - k)$$

When p is positive, the parabola opens upward.
When p is negative, the parabola opens downward.

Standard Form of the Equation of a Parabola

The standard form of the equation of a parabola with vertex at (h, k) and directrix parallel to the y-axis is
$$(y - k)^2 = 4p(x - h),$$
where p is the distance from the vertex to the focus.

The standard form of the equation of a parabola with vertex at (h, k) and directrix parallel to the x-axis is
$$(x - h)^2 = 4p(y - k),$$
where p is the distance from the vertex to the focus.

Example 1

Find the coordinates of the focus and the vertex and the equations of the directrix and the axis of symmetry for the parabola with equation $x^2 - 12y = 0$. Then graph the equation.

First write the equation in the form $(x - h)^2 = 4p(y - k)$.
$$(x - 0)^2 = 4(3)(y - 0) \qquad 4p = 12, \text{ so } p = 3$$

Since $(h, k) = (0, 0)$, the vertex is at the origin. Since x is squared, the directrix of the parabola is parallel to the x-axis. Since p is positive, the parabola opens upward. The distance from the vertex to the focus is 3 units. So, the coordinates of the focus are $(0, 0 + 3)$ or $(0, 3)$.

The axis of symmetry contains the vertex and the focus. The axis of symmetry is the line $x = 0$, or the y-axis.

Since the directrix is 3 units below the vertex, the equation of the directrix is $y = -3$.

(continued on the next page)

Graph the directrix, the vertex, and the focus. To determine the shape of the parabola, graph several other ordered pairs that satisfy the equation and connect them with a smooth curve.

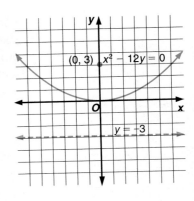

You can also graph a parabola by using a graphing calculator.

Example 2

Use a graphing calculator to graph $(x + 7) = (y - 2)^2$.

Solve for y.

$$x + 7 = (y - 2)^2$$
$$\pm\sqrt{x + 7} = y - 2$$
$$\pm\sqrt{x + 7} + 2 = y$$

Select an appropriate range. As with circles, you need to graph both equations to complete the parabola.

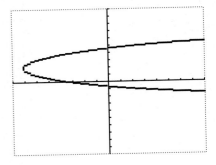

In Example 2, notice that the graph of $(x + 7) = (y - 2)^2$ has the same shape as the graph of $x = y^2$, but it is translated 7 units left and 2 units up. The values of h and k in the standard form of the equation of a parabola will tell you how many units to translate the parent graphs $y^2 = 4px$ and $x^2 = 4py$.

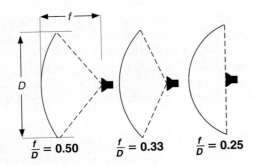

$\frac{f}{D} = 0.50$ $\frac{f}{D} = 0.33$ $\frac{f}{D} = 0.25$

Parabolic satellite dishes are categorized as deep, average, or shallow by using the ratio of the focal length (f) to the diameter (D) of the dish at its widest part. If $\frac{f}{D} < 0.3$, the dish is categorized as deep. If $\frac{f}{D} > 0.45$, the dish is categorized as shallow. If $0.3 \leq \frac{f}{D} \leq 0.45$, the dish is categorized as average. Deeper dishes are preferred because they are less susceptible to interference from environmental noise.

Example 3

APPLICATION

Communications

The Winegard Company produces the QuadStar Perforated Antenna, which is a satellite dish made from aluminum with 2-millimeter perforations. The diameter of the satellite dish is 3 meters, and it has a $\frac{f}{D}$ ratio of 0.278. Write an equation that models the shape of this dish. Assume that the vertex is at the origin and the parabola opens to the right.

Find the value of f, which is also the value of p.

$$\frac{f}{D} = 0.278$$

$$\frac{p}{3} = 0.278 \qquad f = p \text{ and } D = 3$$

$$p = 0.834$$

Since the parabola opens to the right, the directrix is parallel to the y-axis.

$$(y - k)^2 = 4p(x - h)$$

$$(y - 0)^2 = 4(0.834)(x - 0) \qquad (h, k) = (0, 0), \text{ and } p = 0.834$$

$$y^2 = 3.336x$$

The equation that models the shape of the satellite dish is $y^2 = 3.336x$.

You can use the same process you used with circles to rewrite the standard form of the equation of a parabola as the general form of the equation of a parabola.

$$(y - k)^2 = 4p(x - h)$$

$$y^2 - 2ky + k^2 = 4px - 4ph$$

$$y^2 - 4px - 2ky + k^2 + 4ph = 0 \qquad h, k, \text{ and } p \text{ are constants.}$$

$$y^2 + Dx + Ey + F = 0 \qquad D = -4p, E = -2k, F = k^2 + 4ph$$

$$(x - h)^2 = 4p(y - k)$$

$$x^2 - 2hx + h^2 = 4py - 4pk$$

$$x^2 - 2hx - 4py + h^2 + 4pk = 0 \qquad h, k, \text{ and } p \text{ are constants.}$$

$$x^2 + Dx + Ey + F = 0 \qquad D = -2h, E = -4p, F = h^2 + 4pk$$

General Form for the Equation of a Parabola

The general form of the equation of a parabola is
$y^2 + Dx + Ey + F = 0$, when the directrix is parallel to the y-axis, or
$x^2 + Dx + Ey + F = 0$, when the directrix is parallel to the x-axis.

An equation in general form can be rewritten in standard form to determine the coordinates of the vertex, (h, k), and the distance from the vertex to the focus, p.

Example 4

Write the standard form of $y^2 - 4x + 2y + 5 = 0$. Then graph the equation.

Since y is squared, the directrix of the graph of this parabola is parallel to the y-axis.

$$y^2 - 4x + 2y + 5 = 0$$
$$y^2 + 2y = 4x - 5$$
$$y^2 + 2y + 1 = 4x - 5 + 1 \qquad \textit{Complete the square.}$$
$$y^2 + 2y + 1 = 4x - 4 \qquad \textit{Simplify.}$$
$$(y + 1)^2 = 4(x - 1) \qquad \textit{Factor.}$$

The standard form of the equation is $(y + 1)^2 = 4(x - 1)$.

The vertex of the parabola is at $(1, -1)$. The distance from the vertex to the focus is 1.

Sketch the parabola.

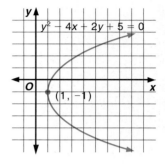

EXPLORATION: Graphing Calculator

1. Graph $y_1 = x^2$.

2. For each situation, make a conjecture about the location of the vertex. Then verify by graphing.
   a. x is replaced with $(x - 2)$.
   b. x is replaced with $(x + 2)$.
   c. y is replaced with $(y - 2)$.
   d. y is replaced with $(y + 2)$.

Parabolas are often used to demonstrate maximum or minimum points in real-world situations.

Example 5

APPLICATION

Entertainment

The amusement park has been charging $12 per person for admission and averaging 1000 patrons per day. The directors of the park are considering a $2 increase in the price for next season. They have calculated that for each $2 increase they propose, they will probably lose 100 patrons per day. What should be the ideal admission fee for the greatest income on an average day?

Let $y =$ the income from admissions.
Let $x =$ the number of $2 price increases.

$$\text{Income} = (\text{number of customers}) \cdot (\text{cost of a ticket})$$
$$y = (1000 - 100x) \cdot (12 + 2x)$$
$$y = 12{,}000 + 2000x - 1200x - 200x^2$$
$$y = -200(x^2 - 4x) + 12{,}000$$
$$y - 12{,}000 = -200(x^2 - 4x)$$
$$y - 12{,}000 + (-800) = -200(x^2 - 4x + 4) \quad \textit{Complete the square.}$$
$$y - 12{,}800 = -200(x - 2)^2$$
$$\frac{-1}{200}(y - 12{,}800) = (x - 2)^2$$

The vertex of the parabola is at (2, 12,800), and because p is negative, it opens downward.

Remember that the vertex is the maximum or minimum point of a parabola. Since the parabola opens downward, the vertex is the maximum.

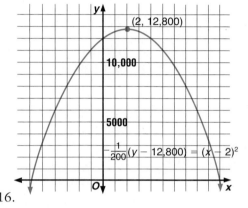

The x-coordinate of the vertex, 2, represents two increases of $2. The y-coordinate, 12,800, represents a maximum income of $12,800.

Maximum income will occur with an admission fee of $12 + 2($2) or $16.

CHECKING FOR UNDERSTANDING

Communicating Mathematics

Read and study the lesson to answer each question.

1. **Describe** how you can determine from the equation of a parabola in what direction the parabola opens.

2. **Describe** how you can determine from the equation of a parabola which axis the directrix is parallel to.

3. **Explain** how you might distinguish an equation for a parabola from an equation for a circle.

Guided Practice

Find the coordinates of the vertex and the value of p for the equation of each parabola. State the direction in which the parabola opens.

4. $16(y - 4) = x^2$
5. $20(x - 9) = (y + 12)^2$
6. $(x - 3)^2 = -8(y + 4)$
7. $2(y + 8) = (x - 6)^2$

Name the coordinates of the vertex and focus, and the equation of the directrix of the parabola defined by each equation. Then graph the equation.

8. $x^2 - 8y = 0$
9. $(x - 2)^2 = 8(y + 1)$
10. $(y + 2)^2 = -16(x - 3)$

Write the standard form of each equation.

11. $y^2 - 6y - 4x = -9$
12. $y^2 + 12y - 16x - 12 = 0$

EXERCISES

Practice **For each equation,**
a. write the standard form,
b. find the coordinates of the focus and vertex, and the equation of the directrix and axis of symmetry, and
c. graph the equation.

13. $y^2 - 4y + 4 = x - 7$ **14.** $-4x + 4 = y^2 + 10y + 25$

15. $x^2 + 8x + 4y + 8 = 0$ **16.** $x^2 - 2x - 12y + 13 = 0$

17. $y^2 + 2x = 0$ **18.** $3x^2 - 19y = 0$

19. $4x^2 - 40y - 24x - 4 = 0$ **20.** $x^2 + 4x + 2y + 10 = 0$

21. $y^2 + 3x = 6y$ **22.** $2x^2 - 16x + 16y + 64 = 0$

Write the equation of the parabola that meets each set of conditions. Then graph the equation.

23. The vertex is at the origin, and the focus is at $(-3, 0)$.

24. The focus is at $(2, 5)$, and the equation of directrix is $x = 4$.

25. The parabola passes through $(5, 2)$, has a vertical axis, and has a maximum at $(4, 3)$.

26. The parabola passes through $(2, -1)$, has its vertex at $(-7, -5)$, and opens to the right.

27. The equation of axis is $x = 2$, the focus is at $(2, -6)$, and $p = -2$.

28. The focus is at $(3, 0)$, the length from the focus to the vertex is 2 units, and the parabola has a minimum.

29. The parabola has a horizontal axis and passes through the origin, $(3, -2)$, and $(-1, 2)$.

30. The parabola has a vertical axis and passes through the origin, $(3, -2)$, and $(-2, 2)$.

31. Consider the standard form of the equation of a parabola in which the vertex is known but the value of p is not known.
   a. As the value of $|p|$ becomes increasingly great, what happens to the shape of the parabola?
   b. As the value of $|p|$ becomes increasingly small, what happens to the shape of the parabola?

Graphing Calculator **Graph each parabola. Then state the coordinates of the vertex and tell whether the graph is the graph of a function. Write *yes* or *no*.**

32. $x^2 - 4x - 4y - 12 = 0$ **33.** $x^2 + 2x + 2y - 5 = 0$

34. $y^2 + 2y - x + 6 = 0$ **35.** $y^2 + 6y - 8x + 65 = 0$

Critical Thinking **The *latus rectum* of a parabola is the line segment through the focus that is perpendicular to the axis and has endpoints on the parabola. The length of the latus rectum is $|4p|$ units, where p is the distance from the vertex to the focus.**

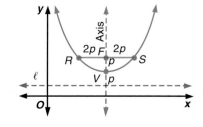

36. Write the equation of a parabola with vertex at $(-2, 1)$, axis $y = 1$, and latus rectum 4 units long.

37. The latus rectum of the parabola with equation $(x - 4)^2 = 8(y - 1)$ coincides with the diameter of a circle. Write the equation of the circle.

38. Communications If a parabolic satellite dish is 10 feet in diameter, determine the range of focal distances for each type of dish.
  a. a shallow dish
  b. an average dish
  c. a deep dish

39. Entertainment The local theater guild puts on plays during the summer at the outdoor amphitheater at a cost of $4 per person. The average attendance at each performance is 500 people. Next summer, they would like to increase the cost of the tickets, but they also want to maximize their profits to keep their program alive. The director estimates that for every $1 increase in admission cost, the attendance at each performance will decrease by 50.
  a. What price should the director propose to maximize their income?
  b. What is the maximum income that might be expected?

40. Baseball Mark McGwire of the St. Louis Cardinals was the 1997 major league home run champion. One Day, Mr. McGwire popped up a baseball at an initial velocity, v_0, of 64 feet/second. Its distance, s, above the ground after t seconds is described by $s = v_0 t - 16t^2$.
  a. Graph the function $s = v_0 t - 16t^2$.
  b. Name the coordinates of the vertex.
  c. What does the value of s at the vertex represent?
  d. How many seconds is the ball in the air?

41. Graph $y = \sin 4\theta$. **(Lesson 6-2)**

42. Verify that $\csc \theta \cos \theta \tan \theta = 1$ is an identity. **(Lesson 7-2)**

43. Find an ordered pair to represent $\vec{u}$ if $\vec{u} = \vec{v} + \vec{w}$, if $\vec{v} = (3, -5)$ and $\vec{w} = (-4, 2)$. **(Lesson 8-2)**

44. Find $\frac{1}{3} \left(\cos \frac{7\pi}{8} + i \sin \frac{7\pi}{8} \right) \cdot 3\sqrt{3} \left[\cos \left(-\frac{\pi}{4} \right) + i \sin \left(-\frac{\pi}{4} \right) \right]$. Then express the product in rectangular form with a and b to the nearest hundredth. **(Lesson 9-7)**

45. Write the standard form of the equation of a circle with a center at $(7, 4)$ and a radius of $\sqrt{2}$ units. **(Lesson 10-1)**

46. College Entrance Exam Choose the best answer.
A swimming pool is 75 feet long and 42 feet wide. If 7.48 gallons equals 1 cubic foot, how many gallons of water are needed to raise the level of the water 4 inches?
  (A) 140   **(B)** 7854   **(C)** 31,500   **(D)** 94,500   **(E)** 727,650

10-3 The Ellipse

Objectives

After studying this lesson, you should be able to:
- use the standard and general forms of the equation of an ellipse, and
- graph ellipses.

Application

Satellites, including spacecraft, travel around planets and their moons in an elliptical orbit with the center of the planet as one of the two foci. The point at which the satellite is closest to a planet is the perihelion, and the point at which it is farthest is the aphelion.

An **ellipse** is the locus of all points in a plane such that the sum of the distances from two given points in the plane, called foci, is constant. In the figure at the right, F_1 and F_2 are the foci of the ellipse. P and Q are any two points on the ellipse. By definition, $PF_1 + PF_2 = QF_1 + QF_2$.

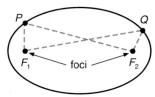

Foci is the plural of focus.

EXPLORATION: Modeling

Materials: 2 pushpins, corrugated cardboard, paper, 10-inch piece of string, pencil

1. Attach a piece of graph paper to the corrugated cardboard. Draw two points on the paper approximately 4 inches apart.

2. Insert one pushpin at each point.

3. Tie the ends of the string together and loop it around the pins.

4. Insert your pencil and pull outward until the string is taut. Move the pencil around the pins, keeping the string taut.

5. Write a paragraph telling why this is a model of an ellipse.

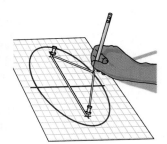

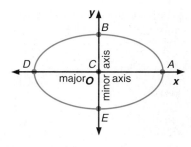

The parent graph of an ellipse is shown at the left. An ellipse has two axes of symmetry, in this case, the x-axis and the y-axis. The point where these axes meet is called the center of the ellipse. The center of the parent graph of an ellipse is the origin.

Notice that the ellipse intersects each axis of symmetry two times. The longer line segment, $\overline{AD}$, which contains the foci, is called the **major axis**. The shorter segment, $\overline{BE}$, is called the **minor axis**. The endpoints of each axis are the vertices of the ellipse.

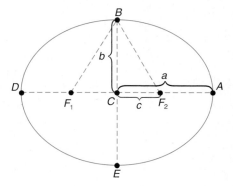

The center separates each axis into two congruent segments. Suppose we let b represent the length of the **semi-minor axis** $\overline{BC}$ and a represent the length of the **semi-major axis** $\overline{CA}$. The foci are located along the major axis, c units from the center. There is a special relationship among the values a, b, and c.

Suppose you draw $\overline{BF_1}$ and $\overline{BF_2}$. The lengths of these segments are equal. Since B and A are two points on the ellipse, you can use the definition of an ellipse to find the length of $\overline{BF_2}$.

$BF_1 + BF_2 = AF_1 + AF_2$	*Definition of ellipse*
$BF_1 + BF_2 = AF_1 + DF_1$	$AF_2 = DF_1$
$BF_1 + BF_2 = AD$	
$BF_1 + BF_2 = 2a$	
$2(BF_2) = 2a$	$BF_1 = BF_2$
$BF_2 = a$	

Since $BF_2 = a$ and $\triangle BCF_2$ is a right triangle, $b^2 + c^2 = a^2$, by the Pythagorean theorem.

Example 1

The graph below models the elliptical path of a space probe around two moons of a planet. The foci of the path are the centers of the moons. Find the coordinates of the foci.

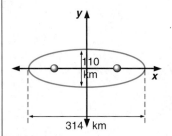

The center of the ellipse is the origin.

$a = \dfrac{1}{2}(314)$ or 157

$b = \dfrac{1}{2}(110)$ or 55

The distance from the origin to the foci is c km.

$$c^2 + b^2 = a^2$$
$$c^2 + (55)^2 = (157)^2$$
$$c \approx 147$$

The foci are at $(147, 0)$ and $(-147, 0)$.

The standard form of the equation of an ellipse can be derived from the definition and the distance formula. Consider the special case when the center is at the origin.

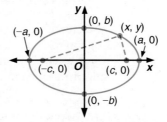

Suppose the foci are at $(c, 0)$ and $(-c, 0)$, and (x, y) are the coordinates of any point on the ellipse. By definition, the sum of the distances from (x, y) to the foci is a constant. Let $2a$ represent this constant. Using the distance formula, $2a = \sqrt{(x + c)^2 + y^2} + \sqrt{(x - c)^2 + y^2}$. Now simplify to find an expression that does not contain radicals.

$$\sqrt{(x + c)^2 + y^2} = 2a - \sqrt{(x - c)^2 + y^2}$$ *Isolate a radical.*

$$(x + c)^2 + y^2 = 4a^2 + (x - c)^2 + y^2 - 4a\sqrt{(x - c)^2 + y^2}$$ *Square each side.*

$$a^2 - xc = a\sqrt{(x - c)^2 + y^2}$$ *Simplify.*

$$a^4 - 2a^2xc + x^2c^2 = a^2[(x - c)^2 + y^2]$$ *Square each side.*

$$x^2(a^2 - c^2) + a^2y^2 = a^2(a^2 - c^2)$$ *Simplify.*

$$\frac{x^2}{a^2} + \frac{y^2}{a^2 - c^2} = 1$$ *Divide each side by $a^2(a^2 - c^2)$.*

$$\frac{x^2}{a^2} + \frac{y^2}{b^2} = 1$$ $b^2 = a^2 - c^2$

This is the equation of an ellipse whose center is the origin and whose foci are on the x-axis. When the foci are on the y-axis, the equation is of the form

$$\frac{y^2}{a^2} + \frac{x^2}{b^2} = 1.$$

The standard form of the equation of an ellipse with a center other than the origin is a translation of the parent graph to a center at (h, k). The standard form of the equation of ellipses and their graphs are given below.

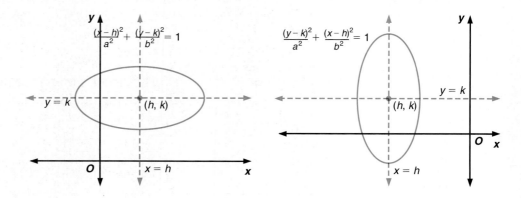

Standard Form of the Equation of an Ellipse

The standard form of the equation of an ellipse with center at (h, k) and major axis of length $2a$ units, where $b^2 = a^2 - c^2$, is as follows.

$$\frac{(x-h)^2}{a^2} + \frac{(y-k)^2}{b^2} = 1, \text{ when the major axis is parallel to the } x\text{-axis,}$$

or

$$\frac{(y-k)^2}{a^2} + \frac{(x-h)^2}{b^2} = 1, \text{ when the major axis is parallel to the } y\text{-axis.}$$

In all ellipses, $a^2 > b^2$. You can use this information to determine the orientation of the major axis from the values given in the equation. If a^2 is the denominator of the x terms, the major axis is parallel to the x-axis. If a^2 is the denominator of the y terms, the major axis is parallel to the y-axis.

Example 2

Graph $\dfrac{(x-4)^2}{121} + \dfrac{(y+5)^2}{64} = 1.$

Determine the values of $a, b, c, h,$ and k.

$a = \sqrt{121}$ or 11 $b = \sqrt{64}$ or 8 $c = \sqrt{a^2 - b^2}$
$\phantom{a = \sqrt{121} \text{ or } 11 \quad b = \sqrt{64} \text{ or } 8 \quad c}= \sqrt{57}$ or about 7.55

$h = 4$ $k = -5$

If the center of the ellipse were the origin, the vertices of the major axis would be at $(11, 0)$ and $(-11, 0)$, the vertices of the minor axis would be at $(0, 8)$ and $(0, -8)$, and the foci would be at $(7.55, 0)$ and $(-7.55, 0)$. However, the center is at $(4, -5)$, which indicates that each of these vertices is translated 4 units to the right and 5 units down.

$$
\begin{aligned}
(0, 0) &\rightarrow (0 + 4, 0 - 5) \text{ or } (4, -5) \\
(11, 0) &\rightarrow (11 + 4, 0 - 5) \text{ or } (15, -5) \\
(-11, 0) &\rightarrow (-11 + 4, 0 - 5) \text{ or } (-7, -5) \\
(0, 8) &\rightarrow (0 + 4, 8 - 5) \text{ or } (4, 3) \\
(0, -8) &\rightarrow (0 + 4, -8 - 5) \text{ or } (4, -13) \\
(7.55, 0) &\rightarrow (7.55 + 4, 0 - 5) \text{ or } (11.55, -5) \\
(-7.55, 0) &\rightarrow (-7.55 + 4, 0 - 5) \text{ or } (-3.55, -5)
\end{aligned}
$$

Graph these translated points. Other points on the ellipse can be found by substituting values for x and y. Complete the ellipse.

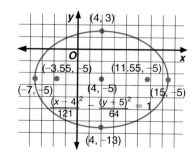

You can also use a graphing calculator to graph an ellipse.

Example 3

Use a graphing calculator to graph $\dfrac{(x+2)^2}{16} + \dfrac{(y-3)^2}{49} = 1.$

First solve the equation for y.

$$\frac{(x+2)^2}{16} + \frac{(y-3)^2}{49} = 1$$

$$49(x+2)^2 + 16(y-3)^2 = 784 \quad \textit{Multiply by the least common denominator, 784.}$$

$$(y-3)^2 = \frac{784 - 49(x+2)^2}{16} \qquad \textit{Isolate the y term.}$$

$$y - 3 = \pm\frac{\sqrt{784 - 49(x+2)^2}}{4} \qquad \textit{Take the square root of each side.}$$

$$y = \pm\frac{\sqrt{784 - 49(x+2)^2}}{4} + 3$$

The last step represents two equations. Graph both equations on the same screen. Our viewing window goes from –15 to 12 on the x-axis and –6 to 12 on the y-axis.

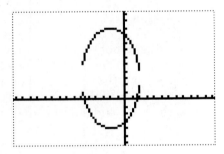

As with circles and parabolas, the standard form of the equation of an ellipse can be expanded to obtain the general form. The result is a second-degree equation of the form $Ax^2 + Cy^2 + Dx + Ey + F = 0$, where $A \neq 0$ and $C \neq 0$, and A and C have the same sign. An equation in general form can be rewritten in standard form to determine the center at (h, k), the length of the semi-major axis (a), and the length of the semi-minor axis (b).

Example 4

Find the coordinates of the center, the foci, and the vertices of the ellipse with the equation $4x^2 + y^2 - 8x + 6y + 9 = 0$. Then graph the equation.

First write the equation in standard form.

$$4x^2 + y^2 - 8x + 6y + 9 = 0$$

$$4(x^2 - 2x + ?) + (y^2 + 6y + ?) = -9 + ? + ?$$

$$4(x^2 - 2x + 1) + (y^2 + 6y + 9) = -9 + 4(1) + 9 \qquad \textit{Complete the square.}$$

$$4(x-1)^2 + (y+3)^2 = 4 \qquad \textit{Factor.}$$

$$\frac{(x-1)^2}{1} + \frac{(y+3)^2}{4} = 1 \qquad \textit{Divide each side by 4.}$$

The center is at $(1, -3)$. Since $a^2 > b^2$, $a^2 = 4$ and $b^2 = 1$. Thus, the major axis is parallel to the y-axis. Since $c^2 = a^2 - b^2$, $c = \sqrt{3}$.

If the center of the ellipse were the origin, the foci would be at $(0, \sqrt{3})$ and $(0, -\sqrt{3})$. The vertices would be at $(1, 0), (-1, 0), (0, 2)$, and $(0, -2)$. However, the center is at $(1, -3)$, so translate each point 1 unit right and 3 units down.

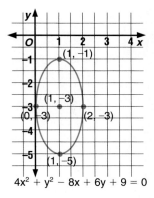

$4x^2 + y^2 - 8x + 6y + 9 = 0$

center: $(1, -3)$

foci: $(1, \sqrt{3} - 3), (1, -\sqrt{3} - 3)$

vertices: $(2, -3), (0, -3),$
$\qquad\qquad (1, -1), (1, -5)$

Sketch the ellipse.

EXPLORATION: Graphing Calculator

1. Graph $y_1 = \left(16 - \dfrac{16}{9}x^2\right)^{\frac{1}{2}}$ and $y_2 = -\left(16 - \dfrac{16}{9}x^2\right)^{\frac{1}{2}}$.

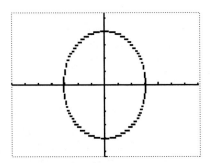

2. For each situation, make a conjecture about the behavior of the graph. Then verify by graphing.
   a. x is replaced with $(x - 2)$.
   b. x is replaced with $(x + 2)$.
   c. y is replaced with $(y - 2)$.
   d. y is replaced with $(y + 2)$.
   e. 9 is switched with 16.

CHECKING FOR UNDERSTANDING

Communicating Mathematics

Read and study the lesson to answer each question.

1. **Explain** how you can determine whether the major axis of an ellipse is parallel to the x-axis or the y-axis.

2. **Explain** how you would graph an ellipse on a graphing calculator.

3. **Describe** the shape of an ellipse in which a equals b.

4. What is the relationship among a, b, and c in reference to an ellipse?

Guided
Practice

For each equation, find the coordinates of the center, foci, and vertices of the ellipse. Then graph the equation.

5. $\dfrac{x^2}{4} + \dfrac{y^2}{9} = 1$

6. $\dfrac{(x-4)^2}{16} + \dfrac{(y+6)^2}{9} = 1$

7. $\dfrac{(x+2)^2}{81} + \dfrac{y^2}{49} = 1$

8. $\dfrac{(x-6)^2}{100} + \dfrac{(y-7)^2}{121} = 1$

9. $\dfrac{x^2}{16} + \dfrac{(y+9)^2}{64} = 1$

10. $\dfrac{(x+10)^2}{225} + \dfrac{(y-9)^2}{64} = 1$

Write the standard equation of each ellipse.

11.

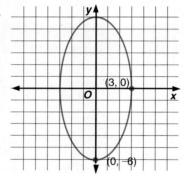

12.

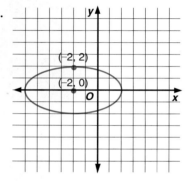

13.

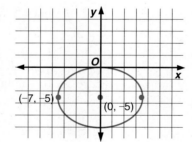

14.

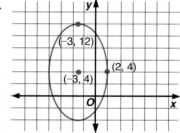

EXERCISES

Practice

For each equation, find the coordinates of the center, foci, and vertices of the ellipse. Then graph the equation.

15. $\dfrac{(x-3)^2}{25} + \dfrac{(y-4)^2}{16} = 1$

16. $\dfrac{(x+2)^2}{4} + \dfrac{(y-1)^2}{25} = 1$

17. $4x^2 + 9y^2 = 36$

18. $9x^2 + 4y^2 - 18x + 16y = 11$

19. $4y^2 - 8y + 9x^2 - 54x + 49 = 0$ **20.** $x^2 - 2x + y^2 - 2y - 6 = 0$

21. $9y^2 + 108y + 4x^2 - 56x = -484$ **22.** $18x^2 + 12y^2 - 144x - 48y = -120$

Write the equation of the ellipse that meets each set of conditions.

23. The center is the origin, $a = 8, b = 6$, and the major axis is parallel to y-axis.

24. The center is at $(-3, -1)$, the length of the horizontal semi-major axis is 7 units, and the length of semi-minor axis is 5 units.

25. The length of the semi-minor axis is $\frac{2}{3}$ the length of the horizontal semi-major axis, the center is the origin, and $a = 6$.

26. The center is the origin, $\frac{1}{2} = \frac{c}{a}$, and the length of the horizontal semi-major axis is 10 units.

27. The foci are at $(-2, 0)$ and $(2, 0)$, and $a = 7$.

28. The semi-major axis has length 4 units, and foci are at $(2, 3)$ and $(2, -3)$.

State whether the graph of each equation is a circle, parabola, or ellipse. Justify your answer.

29. $x^2 + 3y^2 + 2x - 5y = 129$ **30.** $x^2 + y + 4x - 2 = 0$

31. $2x^2 - 5x + y - 19 = 0$ **32.** $y^2 + 2y = x - x^2 + 12$

33. $5x^2 - 2x + 2y^2 - 9y = 22$ **34.** $12x^2 + 24x + 12y^2 + 72y = 124$

35. Write the equation of the ellipse that passes through $(4, 2)$ and has foci at $(1, -1)$ and $(1, 5)$.

36. An ellipse has values a and b such that $b^2 = a^2(1 - 0.7^2)$ and the length of the vertical major axis is 20 units. Write the equation of the ellipse if its center is at $(3, 0)$.

37. Draw an ellipse that is tangent to the x- and y-axes and has its center at $(-3, 7)$. Then write the equation of the ellipse in general form.

Graphing Calculator

Graph each equation. Then use the TRACE function to approximate the coordinates of the vertices to the nearest integer.

38. $4x^2 + 9y^2 - 16x + 18y = 11$ **39.** $x^2 + 4y^2 - 6x + 24y = -41$

40. $16x^2 + 25y^2 + 32x - 150y = 159$ **41.** $4x^2 + y^2 - 8x - 2y = 1$

Critical Thinking

42. One circle is internally tangent to an ellipse with equation $4x^2 + 9y^2 = 36$. Another circle is externally tangent to the same ellipse. Write the equations of the two circles if all three graphs have the same center.

43. Astronomy All of the planets in our solar system have elliptical orbits with the sun as one focal point. If the orbit of Mercury is about 46 million kilometers from the sun at its closest point and 70 million kilometers from the sun at its farthest point, what is the length of the major axis?

44. Astronomy The orbit of Earth is about 9.1×10^7 miles from the sun at its closest point and 9.3×10^7 miles from the sun at its farthest point.
  a. What is the length of the major axis?
  b. What is the distance from the sun to the other focus?

45. Graph $y = \csc(\theta + 60°)$. **(Lesson 6-2)**

46. Find $\cos(A + B)$ if $\cos A = \dfrac{5}{13}$ and $\cos B = \dfrac{35}{37}$ and A and B are first quadrant angles. **(Lesson 7-3)**

47. Aviation An airplane flies at an air speed of 425 mph on a heading due south. It flies against a headwind of 110 mph from a direction 30° east of south. Find the airplane's ground speed and direction. **(Lesson 8-5)**

48. Express $-5 - i$ in polar form. **(Lesson 9-6)**

49. Write the standard form of the equation of the circle that passes through $(0, -9)$, $(7, -2)$, and $(-5, -10)$. **(Lesson 10-1)**

50. Find the coordinates of the vertex and focus and the equation of the directrix of the parabola whose equation is $(x + 5)^2 = 12y$. **(Lesson 10-2)**

51. College Entrance Exam If the base of a rectangle is increased by 30% and the altitude of the rectangle is decreased by 20%, by what percent is the area increased?

CASE STUDY FOLLOW-UP

Refer to Case Study 3: *The Legal System,* on pages 962–965.

An elliptical rotunda on the north side of the building that symbolizes the American legal system is a "whispering gallery." The major axis of the rotunda measures 44.4 feet. The minor axis measures 35.6 feet.

1. Where is the rotunda?

2. Two people standing at the two foci can easily hear each other whispering, even if there is other noise in the room. How far apart are the whisperers standing?

3. Research Find out about the size and shape of the dome of the U.S. Capitol. Can the dome be described in terms of conic sections? Write about your findings in a 1-page paper.

10-4 The Hyperbola

Objectives After studying this lesson, you should be able to:
- use the standard and general forms of the equation of a hyperbola, and
- graph hyperbolas.

Application

Cassegrain antenna satellite dishes are used in extremely hot climates, because the feedhorn is behind the antenna, protected from direct exposure to solar energy. The Cassegrain antenna satellite dish has two different reflectors. A parabolic surface redirects the rays to a second reflective hyperbolic surface, which directs the rays to the feedhorn.

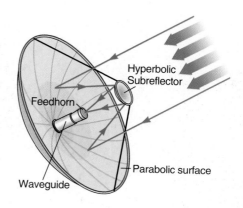

A **hyperbola** is the locus of all points in the plane such that the absolute value of the differences of the distance from two given points in the plane, called **foci**, is constant. That is, if F_1 and F_2 are the foci of a hyperbola and P and Q are any two points on the hyperbola, $|PF_1 - PF_2| = |QF_1 - QF_2|$.

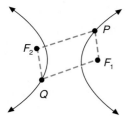

The center of a hyperbola is the midpoint of the segment whose endpoints are the foci. The point on each branch of the hyperbola nearest the center is called a vertex. The **asymptotes** of a hyperbola are lines that the curve approaches as it recedes from the center. As you move farther out along the branches, the distance between points on the hyperbola and the asymptotes approaches zero. The distance from the center to a vertex is a units, and the distance from the center to a focus is c units.

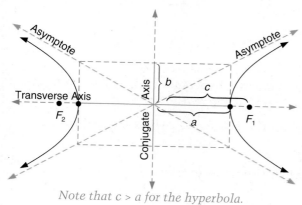

Note that c > a for the hyperbola.

A hyperbola has two axes of symmetry. The line segment that has its endpoints at the vertices is called the **transverse axis** and has a length of $2a$ units. The segment perpendicular to the transverse axis through the center is called the **conjugate axis** and has length $2b$ units. For a hyperbola, the relationship among $a, b,$ and c is represented by $a^2 + b^2 = c^2$. The asymptotes contain the diagonals of the rectangle that is $2a$ units by $2b$ units. The point at which the diagonals meet coincides with the center of the hyperbola.

The standard form of the equation of a hyperbola with the origin as its center can be derived from the definition and the distance formula. Suppose the foci are on the x-axis at $(c, 0)$ and $(-c, 0)$, and (x, y) are the coordinates of any point on the hyperbola. Let $2a$ represent the common difference.

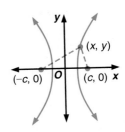

$$\sqrt{(x - c)^2 + y^2} - \sqrt{(x + c)^2 + y^2} = 2a \qquad \textit{Apply the distance formula.}$$

$$\sqrt{(x - c)^2 + y^2} = 2a + \sqrt{(x + c)^2 + y^2} \qquad \textit{Rearrange terms.}$$

$$(x - c)^2 + y^2 = 4a^2 + 4a\sqrt{(x + c)^2 + y^2} + (x + c)^2 + y^2 \qquad \textit{Square each side.}$$

$$-4xc - 4a^2 = 4a\sqrt{(x + c)^2 + y^2} \qquad \textit{Simplify.}$$

$$xc + a^2 = -a\sqrt{(x + c)^2 + y^2}$$

$$x^2c^2 + 2a^2xc + a^4 = a^2x^2 + 2a^2xc + a^2c^2 + a^2y^2 \qquad \textit{Square each side.}$$

$$(c^2 - a^2)x^2 - a^2y^2 = a^2(c^2 - a^2) \qquad \textit{Simplify.}$$

$$\frac{x^2}{a^2} - \frac{y^2}{c^2 - a^2} = 1 \qquad \textit{Divide by } a^2(c^2 - a^2).$$

$$\frac{x^2}{a^2} - \frac{y^2}{b^2} = 1 \qquad \textit{By the Pythagorean theorem, } c^2 - a^2 = b^2.$$

If the foci are on the y-axis, the equation is $\dfrac{y^2}{a^2} - \dfrac{x^2}{b^2} = 1$.

As with the other graphs we have studied in this chapter, the standard form of the equation of a hyperbola with center other than the origin is a translation of the parent graph to a center at (h, k).

Notice that a^2 is always the denominator of the first term.

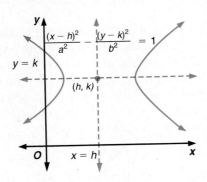

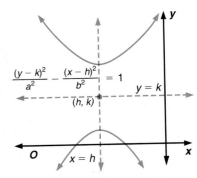

Standard Form of the Equation of a Hyperbola

The standard form of the equation of a hyperbola with center at (h, k) and transverse axis of length $2a$ units, where $b^2 = c^2 - a^2$, is as follows.

$\dfrac{(x - h)^2}{a^2} - \dfrac{(y - k)^2}{b^2} = 1$, when the transverse axis is parallel to the x-axis, or

$\dfrac{(y - k)^2}{a^2} - \dfrac{(x - h)^2}{b^2} = 1$, when the transverse axis is parallel to the y-axis.

Example 1

Find the equation of a hyperbola if the foci are at (2, 5) and (-4, 5) and the transverse axis is 4 units long.

A sketch of the graph is helpful. Let F_1 and F_2 be the foci, let V_1 and V_2 be the vertices, and let C be the center.

Find the center.

The center is the midpoint of F_1F_2.

The center is at $\left(\dfrac{-4+2}{2}, \dfrac{5+5}{2}\right)$ or $(-1, 5)$.

Thus, $h = -1$ and $k = 5$.

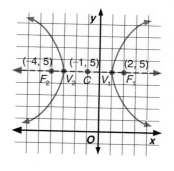

Find c, a^2, and b^2.

$c = CF_1$ $2a = V_1V_2$ $b^2 = c^2 - a^2$
$c = 3$ $2a = 4$ $b^2 = 9 - 4$
             $a = 2$ $b^2 = 5$
             $a^2 = 4$

Use the standard form when the transverse axis is parallel to the x-axis.

$$\frac{(x-h)^2}{a^2} - \frac{(y-k)^2}{b^2} = 1$$
$$\frac{(x+1)^2}{4} - \frac{(y-5)^2}{5} = 1$$

Before graphing a hyperbola, it is often helpful to graph the asymptotes. As noted in the beginning of the lesson, the asymptotes contain the diagonals of the rectangle defined by the transverse and conjugate axes. Suppose the center of a hyperbola is the origin and the transverse axis lies along the x-axis. Study the figure below.

Notice that the x-coordinates of the vertices correspond to the x-coordinates of the endpoints of the transverse axis. The y-coordinates correspond to the y-coordinates of the endpoints of the conjugate axis.

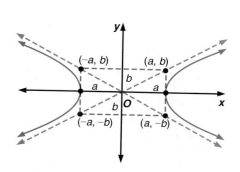

The endpoints of the transverse and conjugate axes are also the midpoints of the sides of the rectangle.

When the transverse axis lies along the x-axis, the slopes of the asymptotes are $\frac{b}{a}$ and $-\frac{b}{a}$. Both of the x-intercepts are 0. The equations of the asymptotes are $y = \frac{b}{a}x$ and $y = -\frac{b}{a}x$. If the transverse axis of a hyperbola is along the y-axis, the slopes of the asymptotes are $\pm\frac{a}{b}$. Thus, the equations of the asymptotes would be $y = -\frac{a}{b}x$ and $y = \frac{a}{b}x$.

The equations of the asymptotes of any hyperbola can be determined by a translation of the graph to a center at (h, k).

	The equations of the asymptotes of a hyperbola are
Equations of the Asymptotes of a Hyperbola	$y - k = \pm\dfrac{b}{a}(x - h)$, for a horizontal transverse axis, or $y - k = \pm\dfrac{a}{b}(x - h)$, for a vertical transverse axis.

Example 2

Find the coordinates of the center, foci, and vertices, and the equations of the asymptotes of the graph of $\dfrac{(y - 3)^2}{25} - \dfrac{(x - 2)^2}{16} = 1$. Then graph the equation.

Since the y terms are in the first expression, the hyperbola has a vertical transverse axis.

The center is at $(2, 3)$, $a = 5$, and $b = 4$.

You can graph a hyperbola on a graphing calculator. First solve for y. Then graph the two resulting equations on the same screen.

The equations of the asymptotes are $y - 3 = \pm\dfrac{5}{4}(x - 2)$.

The vertices are at $(2, 8)$ and $(2, -2)$.

Since $c^2 = a^2 + b^2$, $c = \sqrt{41}$. Thus, the foci are at $(2, 3 + \sqrt{41})$ and $(2, 3 - \sqrt{41})$. $\sqrt{41} \approx 6.4$

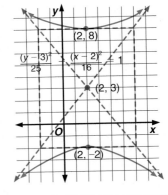

Graph the vertices and the asymptotes. Then sketch the hyperbola.

By expanding the standard form for a hyperbola, you can find the general form, $Ax^2 + Cy^2 + Dx + Ey + F = 0$, where $A \neq 0$, $C \neq 0$, and A and C have different signs. As with the other general forms we have studied, the general form of a hyperbola can be rewritten in standard form. This is helpful since the standard form of the equation of a hyperbola contains important information about the hyperbola and thus, is easier to graph.

Example 3

Find the coordinates of the center, foci, and vertices, and the equations of the asymptotes of the graph of $25y^2 - 9x^2 - 100y - 72x - 269 = 0$. Then graph the equation.

Write the equation in standard form. Use the same process you used with ellipses.

$$25y^2 - 9x^2 - 100y - 72x - 269 = 0$$
$$(25y^2 - 100y + \ ?) - (9x^2 + 72x + \ ?) = 269$$
$$25(y^2 - 4y + 4) - 9(x^2 + 8x + 16) = 269 + 25(4) - 9(16)$$
$$25(y - 2)^2 - 9(x + 4)^2 = 225$$
$$\frac{(y - 2)^2}{9} - \frac{(x + 4)^2}{25} = 1$$

The center is at $(-4, 2)$. Since the y terms are in the first expression, the transverse axis is vertical.

$a = 3$, $b = 5$, and $c = \sqrt{34}$

The foci are at $(-4, 2 + \sqrt{34})$ and $(-4, 2 - \sqrt{34})$.

The vertices are at $(-4, -1)$ and $(-4, 5)$.

The asymptotes are the lines $y - 2 = \pm\dfrac{3}{5}(x + 4)$.

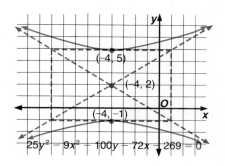

Graph the vertices and asymptotes. Then sketch the hyperbola.

In the standard form of a hyperbola, if $a = b$ the hyperbola is an **equilateral hyperbola**. The asymptotes of an equilateral hyperbola are perpendicular. A special case of the equilateral hyperbola is when the coordinate axes are the asymptotes. The general equation of such a hyperbola is $xy = c$, where c is a nonzero constant. The branches of the equilateral hyperbola lie in the first and third quadrants if c is positive and in the second and fourth quadrants if c is negative.

Example 4

Graph $xy = -25$.

Since c is negative, the hyperbola lies in the second and fourth quadrants.

The transverse axis is along the line $y = -x$.

The coordinates of the vertices must satisfy the equation of the hyperbola and also be points on the transverse axis. Thus, the vertices are at $(5, -5)$ and $(-5, 5)$.

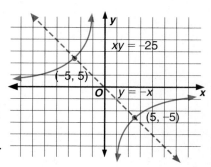

The graphs of functions that vary inversely are hyperbolas with equations of the form $xy = c$. By definition, x varies inversely as y if there is some constant c such that $xy = c$. Boyle's Law in chemistry is an example of a function that varies inversely.

Example 5

Chemistry

Boyle's Law states that the pressure (P) exerted by a gas varies inversely as the volume (V) of a gas if the temperature remains constant. A specific gas is collected in a 242-cm³ container. The pressure is measured and is found to be 87.6 kilopascals. Use the formula $PV = c$ to find c.

$$c = PV$$
$$= (87.6)(242)$$
$$= 21{,}299.2 \qquad \text{The value of } c \text{ is } 21{,}299.2.$$

CHECKING FOR UNDERSTANDING

Communicating Mathematics

Read and study the lesson to answer each question.

1. **Describe** how you can determine whether the transverse axis of a hyperbola is horizontal or vertical.

2. **Give an example** of the equation of an equilateral hyperbola whose axes are the coordinate axes. Tell how its graph differs from other hyperbolas in this lesson.

3. Refer to Example 5.
   a. If the pressure is increased, what happens to the volume?
   b. If the pressure is decreased, what happens to the volume?

Guided Practice

Find the coordinates of the center, the foci, and the vertices and the equations of the asymptotes of the graph of each equation. Then graph the equation.

4. $\dfrac{y^2}{36} - \dfrac{x^2}{49} = 1$

5. $\dfrac{(x+3)^2}{25} - \dfrac{(y-4)^2}{49} = 1$

6. $\dfrac{x^2}{16} - \dfrac{(y-5)^2}{81} = 1$

7. $\dfrac{(y-5)^2}{4} - \dfrac{(x-8)^2}{100} = 1$

8. $\dfrac{y^2}{169} - \dfrac{(x+12)^2}{289} = 1$

9. $\dfrac{(x+11)^2}{144} - \dfrac{(y+7)^2}{36} = 1$

Write the equation of each hyperbola.

10.
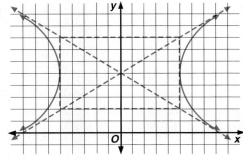

11.
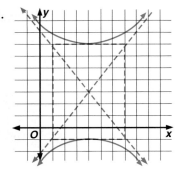

12.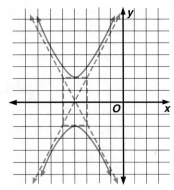

13.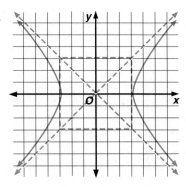

14. Graph $xy = 81$.

EXERCISES

Practice

Find the coordinates of the center, the foci, and the vertices, and equations of the asymptotes of the graph of each equation. Then graph the equation.

15. $\dfrac{x^2}{25} - \dfrac{y^2}{16} = 1$

16. $\dfrac{(y-3)^2}{16} - \dfrac{(x-2)^2}{25} = 1$

17. $81x^2 - 36y^2 = 2916$

18. $(x+6)^2 - 4(y+3)^2 = 36$

19. $9x^2 - 4y^2 - 54x - 40y - 55 = 0$

20. $y^2 - 5x^2 + 20x = 50$

21. $-4y^2 + 9x^2 - 90x - 24y = -153$

22. $49x^2 - 25y^2 + 294x + 200y = 1184$

State whether the graph of each equation is a circle, ellipse, parabola, or hyperbola.

23. $x^2 + 4y^2 + 5x - 2y = 144$

24. $x^2 = y^2 + 8x + 12$

25. $2x^2 + 4x + 2y^2 + 6y = 66$

26. $4y^2 + 6x - 3y = 33$

27. $4y^2 - 3x + 7x^2 - 7y = 18$

28. $y^2 + 9y = 7x + 3x^2 + 52$

Graph each equation.

29. $xy = 9$ **30.** $xy = -16$ **31.** $4xy = -9$ **32.** $9xy = 25$

33. $\dfrac{(x+1)^2}{4} - \dfrac{(y-4)^2}{16} = 1$ **34.** $\dfrac{y^2}{16} - \dfrac{x^2}{9} = 1$ **35.** $\dfrac{(y-2)^2}{4} - \dfrac{(x-3)^2}{9} = 1$

Write an equation of the hyperbola that meets each set of conditions.

36. The center is at $(-1, 4)$, $a = 2$, $b = 3$; it has a horizontal transverse axis.

37. The length of the transverse axis is 8 units; the foci are at $(0, -5)$ and $(0, 5)$.

38. The length of the conjugate axis is 6 units; the vertices are at $(3, 4)$ and $(3, 0)$.

39. The vertices are at $(6, 3)$ and $(0, 3)$, and a focus is at $(8, 3)$.

40. The hyperbola is equilateral and has foci at $(8, 0)$ and $(-8, 0)$.

41. The hyperbola is equilateral and has foci at $(0, 5)$ and $(0, -5)$.

42. The equation of one asymptote is $2x - 9 = 3y$. The hyperbola has its center at $(3, -1)$ and a vertex at $(6, -1)$.

43. The vertex is at $(4, 5)$, the center is at $(4, 2)$, and an equation of one asymptote is $4y + 4 = 3x$.

44. The slopes of the asymptotes are ± 4, and the foci are at $(4, 0)$ and $(-2, 0)$.

45. The foci are at $(1, 5)$ and $(1, -3)$, and the slopes of the asymptotes are ± 2.

Graph each equation. Then state whether the graph is a hyperbola or an ellipse.

46. $16x^2 - y^2 + 96x + 8y = -112$

47. $7x^2 + 3y^2 - 28x - 12y = -19$

48. $y^2 - 4x^2 - 2y - 16x + 1 = 0$

49. $x^2 - 3y^2 + 6y + 6x = 18$

50. Determine the equations of the asymptotes of an equilateral hyperbola whose transverse axis is on the x-axis and center is the origin.

51. Write an equation of the hyperbola that passes through $(2, 0)$ and has asymptotes with equations $x - 2y = 0$ and $x + 2y = 0$.

52. Write an equation of the hyperbola that passes through $(4, 2)$ and has asymptotes with equations $y = 2x$ and $y = -2x + 4$.

53. Chemistry According to Boyle's law, the pressure, P, (in kilopascals) exerted by a gas varies inversely as the volume, V, (in cubic centimeters) of a gas if the temperature remains constant. That is, $PV = c$. Suppose the constant for neon gas is 10,440.

  a. Graph the function $PV = c$, for $c = 10,440$.

  b. Determine the volume of neon if the pressure is 90 kilopascals.

  c. Determine the volume of neon if the pressure is 180 kilopascals.

  d. Study your results for parts b and c. If the pressure is doubled, make a conjecture about the effect on the volume of the gas.

54. Analytical Geometry Two hyperbolas in which the transverse axis of one is the conjugate axis of the other are called *conjugate hyperbolas*. In equations of conjugate hyperbolas, the x^2 and y^2 terms are reversed. For example, $\frac{x^2}{9} - \frac{y^2}{4} = 1$ and $\frac{y^2}{4} - \frac{x^2}{9} = 1$ are equations of conjugate hyperbolas.

  a. Graph $\frac{x^2}{9} - \frac{y^2}{4} = 1$ and $\frac{y^2}{4} - \frac{x^2}{9} = 1$ on the same coordinate plane.

  b. What is true of the asymptotes of conjugate hyperbolas?

  c. Write the equation of the conjugate hyperbola for $\frac{(y - 5)^2}{25} - \frac{(x - 3)^2}{9} = 1$.

  d. Graph the conjugate hyperbolas in part c.

55. Space Travel On March 2, 1972, the *Pioneer 10* space probe was launched from Cape Canaveral. It became the first spacecraft to travel beyond all the planets in our solar system. The rate the spacecraft travels, r, varies inversely as the time spent in traveling, t, if the distance, d, remains constant.

  a. How long did it take *Pioneer 10* to travel from Earth to Pluto, a distance of 4.58×10^9 kilometers, if the average speed was 4.69×10^4 km/h?

  b. In approximately what month and year did it pass by Pluto?

56. Find sin (arctan $\sqrt{3}$). Assume the angle is in Quadrant I.   **(Lesson 6-4)**

57. Verify that $\cot X = (\sin 2X) \div (1 - \cos 2X)$ is an identity.   **(Lesson 7-4)**

58. Use determinants to show how to find the cross product of the vectors $(5, 2, 3)$ and $(-2, 5, 0)$. Do not evaluate.   **(Lesson 8-4)**

59. Write $x = y$ in polar form.   **(Lesson 9-3)**

60. Write the normal form of $x - y + 4 = 0$.   **(Lesson 7-6)**

61. Write the standard form of $y^2 - 4x + 2y + 5 = 0$.   **(Lesson 10-2)**

62. Find the coordinates of the center and vertices of the ellipse whose equation is $4x^2 + 25y^2 + 250y + 525 = 0$.   **(Lesson 10-3)**

63. **College Entrance Exam**   Choose the best answer.
Let $^\star x^\star$ be defined such that $^\star x^\star = x + \dfrac{1}{x}$. Find the value of $^\star 6^\star + {}^\star 4^\star + {}^\star 2^\star$.

(A) 12 (B) $12\dfrac{7}{12}$ (C) $12\dfrac{11}{12}$ (D) $13\dfrac{1}{12}$ (E) $13\dfrac{5}{12}$

MID-CHAPTER REVIEW

1. Write the standard form of the equation of a circle with center at $(2, -7)$ and radius of 9 units.   **(Lesson 10-1)**

2. Graph the circle whose equation is $x^2 + y^2 - 6x + 4y - 3 = 0$.   **(Lesson 10-1)**

3. Find the coordinates of the vertex and the focus, and equation of the directrix of the parabola whose equation is $(y - 3)^2 - 4x = 0$.   **(Lesson 10-2)**

4. Graph the parabola whose equation is $y - (x - 3)^2 = 0$.   **(Lesson 10-2)**

5. Find the coordinates of the center and the foci, the length of the major axis, and the length of the minor axis for the graph of $x^2 + 16y^2 - 6x - 64y + 57 = 0$.   **(Lesson 10-3)**

6. Graph the ellipse described in Exercise 5.   **(Lesson 10-3)**

7. **Space Travel**   A space module is in an elliptical orbit around a planet. The maximum distance from the center of the planet to the module is 300 kilometers, and the minimum distance is 200 kilometers. If a model of this orbit is graphed with the center of the ellipse at the origin, write the equation that represents the path of the space module.   **(Lesson 10-3)**

8. Write $3y^2 + 24y - x^2 - 2x + 41 = 0$ in standard form and identify the shape of its graph.   **(Lesson 10-4)**

9. Graph the equation in Exercise 8.   **(Lesson 10-4)**

10. Graph $xy = -100$.   **(Lesson 10-4)**

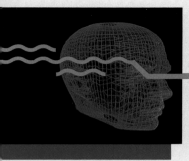

Technology

Asymptotes of a Hyperbola

BASIC
Spreadsheets
▶ **Software**

IBM's *Mathematics Exploration Toolkit (MET)* software allows you to graph hyperbolas written in the form $\dfrac{(x-0)^2}{a^2} - \dfrac{(y-0)^2}{b^2} = c$. You can use MET to explore a family of hyperbolas with centers at $(0, 0)$. To do this, replace c in the equation above by $1, 0.5, 0.1,$ and 0. Observe what happens to the shape of the graph as c gets closer to 0. Use the following steps.

Enter	**Result**
hyper 0 0 4 5	*Displays $\dfrac{(x-0)^2}{4^2} - \dfrac{(y-0)^2}{5^2} = 1$.*
scale 10	*Sets the scale from -10 to 10.*
graph	*Graphs the hyperbola.*
substitute 0.5 1	*Replaces 1 with 0.5.*
graph	
substitute 0.1 0.5	*Replaces 0.5 with 0.1.*
graph	
color 2	
substitute 0 0.1	*Changes color of graph to red.*
graph	*Replaces 0.1 with 0.*

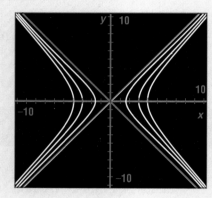

Notice that as the values move from 1 toward 0, the hyperbolas seem to be straightening. Finally, when the constant term is 0, the result is the asymptote lines. Solve to find the equations of the asymptotes.

$$\frac{x^2}{4^2} - \frac{y^2}{5^2} = 0$$

$$\frac{y^2}{5^2} = \frac{x^2}{4^2}$$

$$y^2 = \frac{5^2 x^2}{4^2}$$

$$y = \pm \frac{5x}{4} \qquad \text{Thus, the equations of the asymptotes are } y = \pm \frac{xb}{a}.$$

EXERCISES

Use *MET* to graph each hyperbola for $c = 1, 0.5, 0.1,$ and 0. Then sketch the graph and find the equations for the asymptotes.

1. $\dfrac{(x-0)^2}{7^2} - \dfrac{(y-0)^2}{4^2} = 1$

2. $\dfrac{(x-0)^2}{3^2} - \dfrac{(y-0)^2}{2^2} = 1$

3. $\dfrac{(x-0)^2}{4^2} - \dfrac{(y-0)^2}{7^2} = 1$

4. $\dfrac{(x-0)^2}{6^2} - \dfrac{(y-4)^2}{2^2} = 1$

10-5 Conic Sections

Objectives

After studying this lesson, you should be able to:
- recognize conic sections by their equations, and
- find the eccentricity of conic sections.

Application

The words *ellipse, hyperbola,* and *parabola* were first used by members of the Pythagorean society in ancient Greece around 540 B.C. These terms were used in connection with regions instead of curves as we refer to them today. Menaechmus (350 B.C.) is credited with the first treatment of the terms as curves generated by sections of geometric solids. However, it was Apollonius of Perga (about 225 B.C.), an astronomer of some fame in Greece, who wrote an eight-book essay entitled *Conic Sections*. His work differed from others in that he obtained all of the **conic sections** from one double right cone intersected by a plane. Apollonius stated that all variations in the shape of a conic section could be obtained by varying the slope of the plane intersecting the conical surface, as shown in the figures below.

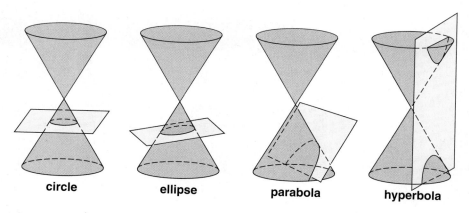

circle   **ellipse**   **parabola**   **hyperbola**

If the plane passes through the vertex of the conical surface, as illustrated below, the intersection is a **degenerate case**. The degenerate cases are a point, a line, and two intersecting lines.

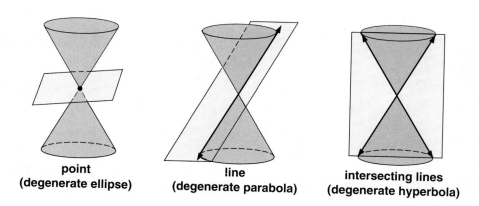

point
(degenerate ellipse)

line
(degenerate parabola)

intersecting lines
(degenerate hyperbola)

A general equation for each of the conic sections we have studied has been determined. All of these equations are forms of the general equation for conic sections.

General Equation for Conic Sections	**The equation of a conic section can be written in the form $Ax^2 + Bxy + Cy^2 + Dx + Ey + F = 0$, where $A, B,$ and C are not all zero.**

The graph of a second-degree equation in two variables always represents a conic or a degenerate case, unless the equation has no graph at all in the real number plane. Most of the conic sections that we have studied have axes that are parallel to the coordinate axes. The general equations of these conics have no xy term; thus, $B = 0$. The hyperbola whose equation is $xy = k$ does not have axes parallel to the coordinate axes. In its equation, $B \neq 0$.

To identify the conic section represented by a given equation, it is helpful to write the equation in standard form. However, you can also identify the conic section by how the equation compares to the general equation. The table below summarizes the standard forms and differences among the general forms.

The circle is actually a special form of the ellipse, $a^2 = b^2 = r^2$.

Conic Section	Standard Form of Equation	Variation of General Form of Conic Equations
circle	$(x - h)^2 + (y - k)^2 = r^2$	$A = C$
parabola	$(y - k)^2 = 4p(x - h)$ or $(x - h)^2 = 4p(y - k)$	Either A or C is zero.
ellipse	$\dfrac{(x - h)^2}{a^2} + \dfrac{(y - k)^2}{b^2} = 1$ or $\dfrac{(y - k)^2}{a^2} + \dfrac{(x - h)^2}{b^2} = 1$	A and C have the same sign, and $A \neq C$.
hyperbola	$\dfrac{(x - h)^2}{a^2} - \dfrac{(y - k)^2}{b^2} = 1$ or $\dfrac{(y - k)^2}{a^2} - \dfrac{(x - h)^2}{b^2} = 1$ or $xy = k$	A and C have opposite signs.

Example 1

Identify the conic section represented by each equation.
a. $5y^2 - 3x^2 + 4x - 3y - 100 = 0$

a. $A = -3$ and $C = 5$. Since $A \neq C$ and they have different signs, the conic is a hyperbola.

b. $2x^2 + 4y^2 - 8x - 8y + 36 = 0$

b. $A = 2$ and $C = 4$. Since A and C have the same sign and are not equal, the conic is an ellipse.

c. $2x^2 + 2y^2 - 9x + 3y - 100 = 0$

c. $A = 2$ and $C = 2$. Since $A = C$, the conic is a circle.

d. $7x^2 + 3x - 8y - 25 = 0$

d. $A = 7$ and $C = 0$. Since $C = 0$, the conic is a parabola.

Another way to define conics is by using eccentricity. A conic section is defined to be the locus of points such that, for any point P in the locus, the ratio of the distance between that point and a fixed point to the distance between that point and a fixed line ℓ is constant. That ratio is the **eccentricity** of the curve, denoted by e.

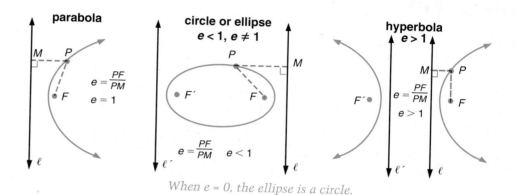

When $e = 0$, the ellipse is a circle.

The eccentricity for ellipses can also be defined as $e = \dfrac{c}{a}$, where c is the distance from the center to the focus and a is the distance from the center to the vertex. Since $0 < c < a$, you can divide by a to show that $0 < e < 1$. If e is closer to zero, the two foci are near the center of the ellipse. In the figure at the right, when $e = \dfrac{1}{5}$, the ellipse looks more like a circle. When e is closer to 1, the ellipse is elongated as in the graph where $e = \dfrac{5}{6}$.

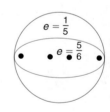

Sometimes, you may need to find the value of b when you know the values of a and e. In an ellipse, $b^2 = a^2 - c^2$ and $\dfrac{c}{a} = e$. By using the two equations, it can be shown that $b^2 = a^2(1 - e^2)$.

$$\frac{c}{a} = e$$

$c = ae$ *Multiply each side by a.*

$c^2 = a^2 e^2$ *Square each side.*

$b^2 = a^2 - c^2$

$b^2 = a^2 - a^2 e^2$ $c^2 = a^2 e^2$

$b^2 = a^2(1 - e^2)$

Example 2

Write the equation of the ellipse with center at $(-2, 3)$, $a = 7$, and $e = 0.5$. The major axis is horizontal.

Find b^2.

$b^2 = a^2(1 - e^2)$
$\quad = 49(1 - 0.25) \qquad a = 7 \text{ and } e = 0.5$
$\quad = 36.75$

Since the major axis is horizontal, use the form $\dfrac{(x - h)^2}{a^2} + \dfrac{(y - k)^2}{b^2} = 1$.

$\dfrac{(x - (-2))^2}{7^2} + \dfrac{(y - 3)^2}{36.75} = 1 \qquad h = -2, k = 3, a = 7, b^2 = 36.75$

The equation of the ellipse is $\dfrac{(x + 2)^2}{49} + \dfrac{(y - 3)^2}{36.75} = 1$.

The orbits of astronomical bodies are often described by their eccentricity.

Example 3

Halley's comet last passed near Earth in April of 1986. Its next appearance is expected in the year 2061. The comet has an elliptical orbit with the sun as one of its foci. Astronomers have determined that the orbit is about 0.5871045 AU (astronomical units) at its closest point (perihelion) to the sun. The eccentricity of the orbit is 0.9672759 and the length of the semi-major axis is 17.94104 AU. *1 AU = the average distance between the sun and Earth, about 9.3×10^7 miles.*

a. **Sketch the orbit of Halley's comet showing the sun in its respective position.**

b. **Find the length of the semi-minor axis of the orbit.**

c. **Find the distance of the comet from the sun at its farthest point (aphelion).**

FYI...

Even though Halley's comet is only visible to Earth about once every 77 years, Earth passes through the debris of its orbit each May and October. Pieces of dust left behind by the comet enter Earth's atmosphere and burn, producing meteor showers visible two hours after sunset.

a. The sketch at the right shows the sun as a focus for the elliptical orbit.

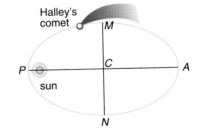

b. b is the length of the semi-minor axis.

$b^2 + a^2(1 - e^2)$
$b = \sqrt{a^2(1 - e^2)}$
$b = \sqrt{(17.94104)^2(1 - 0.9672759^2)}$

Use your calculator.

$b = 4.552124232$ AU

c. *aphelion = length of major axis – perihelion*
$\quad$ SA $=\quad 2(17.94104) \quad\quad – \quad 0.5871045$
$\quad$ SA $=\quad 35.2949755 \quad$ *Use your calculator.*

The aphelion is about 35 AU.

The eccentricity for a hyperbola can also be defined as $e = \frac{c}{a}$. However, in a hyperbola, $0 < a < c$. So, $0 < 1 < e$ or $e > 1$. Since $c^2 = a^2 + b^2$ in hyperbolas, it can be shown that $b^2 = a^2(e^2 - 1)$. *You will derive this formula in Exercise 4.*

Example 4

Write the equation of the hyperbola with center at $(-3, 1)$, a focus at $(2, 1)$, and eccentricity $\frac{5}{4}$.

A general sketch using the points given is often helpful.

Since the center and focus have the same y-coordinate, the transverse axis is horizontal. Use the form

$$\frac{(x - h)^2}{a^2} - \frac{(y - k)^2}{b^2} = 1.$$

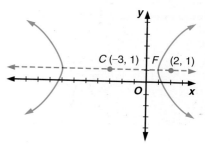

The focus is 5 units to the right of the center, so $c = 5$. Now use the eccentricity to find the value of a and b^2.

$$e = \frac{c}{a}$$

$$\frac{5}{4} = \frac{5}{a} \qquad c = 5 \text{ and } e = \frac{5}{4}$$

$$4 = a$$

$$16 = a^2$$

$$b^2 = a^2(e^2 - 1)$$

$$b^2 = 16\left(\frac{25}{16} - \frac{16}{16}\right)$$

$$b^2 = 9$$

The equation is $\dfrac{(x + 3)^2}{16} - \dfrac{(y - 1)^2}{9} = 1.$

CHECKING FOR UNDERSTANDING

Communicating Mathematics

Read and study the lesson to answer each question.

1. **Explain** in your own words what a conic section is.

2. **Tell** how the general form of the equation of a conic section is modified for each of the following.
   a. ellipse   **b.** hyperbola   **c.** circle   **d.** parabola

3. **Describe** a circle in terms of the equation of an ellipse.

4. **Derive** the equation $b^2 = a^2(e^2 - 1)$ for a hyperbola.

Guided Practice

Identify the conic section represented by each equation.

5. $x^2 - 6x - 4y + 9 = 0$

6. $x^2 + y^2 + 6y - 8x + 24 = 0$

7. $x^2 - 4y^2 + 10x - 16y + 5 = 0$

8. $9y^2 + 27x^2 - 6y - 108x + 82 = 0$

Identify the conic section represented by each equation and write the equation in standard form. Then graph the equation.

9. $x^2 - 8x + 11 = -y^2$

10. $9x^2 + 25y^2 - 54x - 50y - 119 = 0$

11. $4y^2 + 4y + 8x = 15$

12. $x^2 - 3y^2 + 2x - 24y = 41$

13. Write the equation of the ellipse whose center is the origin, $a = 2$, $e = \frac{3}{4}$, and the major axis is parallel to the y-axis.

EXERCISES

Determine the eccentricity of the conic section represented by each equation.

14. $(y - 5)^2 = 4(x - 5)$

15. $4y^2 - 8y + 9x^2 - 54x + 49 = 0$

16. $5x^2 - 9y^2 = 45$

17. $(x + 5)^2 + (y - 1)^2 = 100$

18. $25x^2 + y^2 - 100x + 6y + 84 = 0$

19. $x^2 - y^2 + 2y - 5 = 0$

Identify the conic section represented by each equation. Write the equation in standard form and graph the equation.

20. $y^2 - 8x = -8$

21. $9xy = 4$

22. $2x^2 + 5y^2 = 0$

23. $x^2 - 4y - 28 = 0$

24. $x^2 - 6x + y^2 - 12y + 41 = 0$

25. $x^2 - 4x - y^2 - 5 - 4y = 0$

26. $(x + 1)^2 + 9(y - 6)^2 = 9$

27. $x^2 = y + 8x - 16$

28. $y^2 = 9x^2$

29. $(x - 2)^2 + (y - 3)^2 = 0$

Write the equation of the conic that meets each set of conditions.

30. an ellipse with center at $(3, 1)$, a vertical semi-major axis 6 units long, and $e = \frac{1}{3}$

31. an ellipse with eccentricity 0.25, foci at $(3, 5)$ and $(1, 5)$

32. a hyperbola with foci at $(5, 0)$ and $(-5, 0)$ and eccentricity $\frac{5}{3}$

33. a hyperbola with foci at $(0, 8)$ and $(0, -8)$ and eccentricity $\frac{4}{3}$

34. a hyperbola with eccentricity $\frac{5}{4}$, foci at $(0, 9)$ and $(0, -1)$

35. an ellipse with a vertical major axis 20 units long, center at $(3, 0)$, and $e = \frac{7}{10}$

36. a hyperbola with center at $(3, -1)$, a focus at $(3, -4)$, and $e = 1.5$

37. ellipse passing through $(4, 2)$ with foci at $(1, -1)$ and $(1, 5)$

38. Find the eccentricity of the hyperbola that has foci at $(1, -4)$ and $(1, 6)$ and passes through $(1, 4)$.

39. Find the eccentricity of the hyperbola that passes through $(0, 2)$ and has foci at $(0, 3)$ and $(0, -3)$.

Critical
Thinking

40. Identify the conic section represented by $x^2 - 2xy + 3y^2 = 1$. Discuss how this conic section differs from others presented in this lesson.

Applications
and Problem
Solving

41. Astronomy   A satellite orbiting Earth follows an elliptical path with the center of Earth as one focus. The eccentricity of the orbit is 0.16, and the major axis is 10,440 miles long.

   a. If the mean diameter of Earth is 7920 miles, find the greatest and least distance of the satellite from the surface of Earth.

   b. Assuming that the center of the ellipse is the origin and the foci lie on the x-axis, write the equation of the orbit of the satellite.

42. Construction   The 50-foot tall arch of a bridge over a small stream is shaped like half an ellipse. The span of the bridge is 120 feet.

   a. Suppose a model of this bridge is drawn on a coordinate plane with the origin located on the bridge midway between the ends of the bridge. Draw the model and write the equation of the ellipse representing this model.

   b. Suppose a model of this bridge is drawn on a coordinate plane with one end of the bridge at the origin. Draw the model and write the equation of the ellipse representing this model.

Mixed
Review

43. Engineering   A metallic ring used in a sprinkler system has a diameter of 13.4 cm. Find the length of the metallic cross brace if it subtends a central angle of $26°20'$. **(Lesson 5-6)**

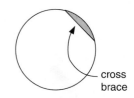

cross brace

44. Find Arccos $\dfrac{\sqrt{3}}{2}$.   **(Lesson 6-5)**

45. Solve $\sin^2 A = \cos A - 1$ for principal values of A.   **(Lesson 7-5)**

46. Graph $r = 5 \cos 2\theta$.   **(Lesson 9-2)**

47. Write the equation of a hyperbola with center at $(3, -1)$, a focus at $(3, -4)$ and eccentricity 1.5.   **(Lesson 10-4)**

48. College Entrance Exam   Choose the best answer.
If one half of the female students in a certain school eat in the cafeteria and one third of the male students eat there, what fractional part of the student body eats in the cafeteria?

   (A) $\dfrac{5}{12}$ **(B)** $\dfrac{2}{5}$ **(C)** $\dfrac{3}{4}$ **(D)** $\dfrac{5}{6}$ **(E)** not enough information given

10-6A Graphing Calculators: Conic Sections

The parabola, circle, ellipse, and hyperbola are all types of conic sections. You can use a graphing calculator to graph them; however, some algebra is often required before you can enter the equations into the graphing calculator. This is because most of the conic sections are relations, but not functions.

Since a graphing calculator will only plot functions, we must first manipulate the equations before entering them into the calculator to be graphed. For example, the equation of the circle, $x^2 + y^2 = 9$, cannot be entered directly since the equation must be in a "$y =$" format. We must solve the equation for y and then enter each equation into the calculator as a function.

Example 1

Graph the circle whose equation is $x^2 + y^2 = 16$.

First, solve the equation for y.

$$x^2 + y^2 = 16$$
$$y^2 = 16 - x^2$$
$$y = \pm\sqrt{16 - x^2}$$

Now enter two equations into the calculator: $y = \sqrt{16 - x^2}$ and $y = -\sqrt{16 - x^2}$. Since both of these equations are functions, they can be graphed by the calculator.

Enter: Y= 2nd √ ((16 − X,T,θ
x²)) ENTER ((−)) 2nd √
((16 − X,T,θ x²))
GRAPH

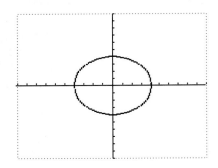

The graph appears to be an ellipse even though it is actually a circle. The viewing window is distorted and must be adjusted so that the scale of the x- and y-axes are equal in length.

You can make the graph square on the TI-82 by pressing ZOOM 5. However, this range does not show a complete graph for this example.

The TI-82 can draw a circle using a special circle function on the DRAW menu. First, clear all drawn images and graphs.

Enter: 2nd DRAW 1 ENTER 2nd Y-VARS 5 2 ENTER

Then select the circle function and enter the coordinates of the center and the length of the radius.

From the previous graph you can see that the center is at (0, 0) and the radius is 4.

Enter: [2nd] [DRAW] 9 0 [,] 0 [,] 4 [)]

[ENTER]

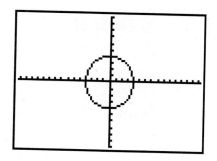

Other types of conic sections that are relations can be graphed on a graphing calculator by solving the equations for y and graphing two separate equations that are functions.

Example 2

Graph the hyperbola $x^2 - y^2 + 8x + 4y + 9 = 0$.

Solve the equation for y by completing the square.

$$y^2 - 4y = x^2 + 8x + 9$$
$$(y^2 - 4y + 4) + 16 = (x^2 + 8x + 16) + 9 + 4 \qquad \textit{Complete the squares.}$$
$$(y - 2)^2 + 16 = (x + 4)^2 + 13$$
$$(y - 2)^2 = (x + 4)^2 - 3$$
$$y - 2 = \pm\sqrt{(x + 4)^2 - 3}$$
$$y = 2 \pm \sqrt{(x + 4)^2 - 3}.$$

Now enter the two equations to graph the relation.

Enter: [Y=] 2 [+] [2nd] [√] [(] [(]
[X,T,θ] [+] 4 [)] [x²] [−] 3 [)]
[ENTER] 2 [−] [2nd] [√] [(]
[(] [X,T,θ] [+] 4 [)] [x²] [−] 3 [)]
[GRAPH]

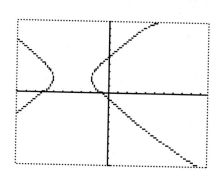

EXERCISES

Solve each equation so that it can be entered into a graphing calculator. Then graph each conic section and sketch the graph on a piece of paper.

1. $x = y^2 + 5y + 22$

2. $x^2 + (y - 3)^2 = 4$

3. $5x^2 + 10y^2 = 150$

4. $12x^2 + 4y^2 - 2x = 44$

5. $4x^2 - y^2 + 24x - 8y + 12 = 0$

6. $4x^2 - 9y^2 - 16x + 144y - 540 = 0$

10-6 Transformations of Conics

After studying this lesson, you should be able to:
- find the equations of conic sections that have been translated or rotated, and
- find the angle of rotation for a given equation and sketch its graph.

Application

A cam is a rotating piece of machinery that gives motion to another piece of machinery. Commonly, a cam is used to convert circular motion into back-and-forth motion. In a sewing machine, cams convert the rotary motion of the motor into the up-and-down motion of the needle. The camshaft in an automobile engine converts the rotary motion of the engine into the up-and-down motion of the pistons.

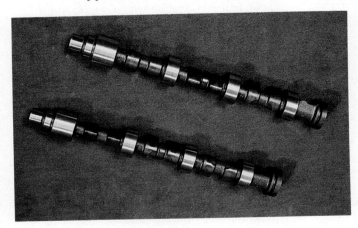

In a punch press, the cam is elliptical with the camshaft passing through one of the foci. The simplified drawing at the right shows that when the elongated part of the cam is up, the punch is up. As the cam rotates, it moves the press stem in a smooth downward motion.

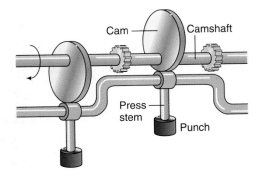

Thus far, we have used a transformation called a translation to show how the parent graph of each of the conic sections is translated to a center other than the origin. For example, the equation of the parabola $y = 4px^2$ became $y - k = 4p(x - h)^2$ for a center at (h, k). You can also translate points of a figure by using a **translation matrix**. The coordinates of the points you want to translate are placed in a matrix with the first row containing the x-coordinates and the second row containing the y-coordinates. The translation matrix has the same number of rows and columns as the matrix containing the coordinates of the points. If the points are to be translated in relation to a point at (h, k), all elements of the first row of the translation matrix equal h and all elements of the second row equal k.

A translation of a set of points with respect to (h, k) is often written $T_{(h, k)}$.

Example 1

Use a translation matrix to find the coordinates of the vertices of the ellipse with equation $\dfrac{x^2}{25} + \dfrac{y^2}{9} = 1$, with respect to $T_{(-8,\ 3)}$.

The original equation is an ellipse whose center is the origin. Its vertices are at $(-5, 0)$, $(5, 0)$, $(0, 3)$, and $(0, -3)$. Write these ordered pairs as a matrix and add the translation matrix for $T_{(-8,\ 3)}$.

$$\begin{bmatrix} -5 & 5 & 0 & 0 \\ 0 & 0 & 3 & -3 \end{bmatrix} + \begin{bmatrix} -8 & -8 & -8 & -8 \\ 3 & 3 & 3 & 3 \end{bmatrix} = \begin{bmatrix} -13 & -13 & -8 & -8 \\ 3 & 3 & 6 & 0 \end{bmatrix}$$

The vertices of the translated ellipse are at $(-13, 3)$, $(-3, 3)$, $(-8, 6)$, and $(-8, 0)$.

The graph shows the ellipse and its translation.

The equation of the translated ellipse is $\dfrac{(x+8)^2}{25} + \dfrac{(y-3)^2}{9} = 1$.

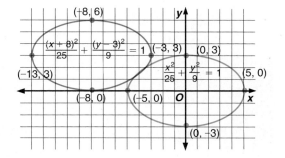

Except for hyperbolas whose equations are of the form $xy = k$, all of the conic sections we have studied thus far have been oriented with their axes parallel to the coordinate axes. In the general form of these conics, $B = 0$. Whenever $B \neq 0$, the axes of the conic section are not parallel to the coordinate axes. That is, the graph is rotated.

The figures below show a hyperbola whose center is the origin and its rotation. Notice that the angle of rotation has the same measure as the angles formed by the positive x-axis and the transverse axis and the positive y-axis and the conjugate axis.

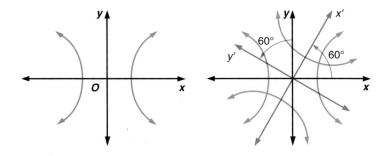

The coordinates of the points of a rotated figure can be found by using a rotation matrix.

Rotation Matrix

A rotation of θ about the origin can be described by the matrix

$$\begin{bmatrix} \cos\theta & -\sin\theta \\ \sin\theta & \cos\theta \end{bmatrix}.$$

You may wish to review matrix multiplication in Lesson 2–2.

A positive value of θ indicates a counterclockwise rotation. A negative value of θ indicates a clockwise rotation.

Suppose $P(x, y)$ is a point on the graph of a conic section and $P'(x', y')$ is its image after a counterclockwise rotation of θ. The values of x' and y' can be found by matrix multiplication.

$$\begin{bmatrix} x' \\ y' \end{bmatrix} = \begin{bmatrix} \cos\theta & -\sin\theta \\ \sin\theta & \cos\theta \end{bmatrix} \cdot \begin{bmatrix} x \\ y \end{bmatrix}$$

The inverse of the rotation matrix represents a rotation of $-\theta$. Multiply each side of the equation by the inverse rotation matrix to solve for x and y.

$$\begin{bmatrix} \cos\theta & \sin\theta \\ -\sin\theta & \cos\theta \end{bmatrix} \cdot \begin{bmatrix} x' \\ y' \end{bmatrix} = \begin{bmatrix} \cos\theta & \sin\theta \\ -\sin\theta & \cos\theta \end{bmatrix} \cdot \begin{bmatrix} \cos\theta & -\sin\theta \\ \sin\theta & \cos\theta \end{bmatrix} \cdot \begin{bmatrix} x \\ y \end{bmatrix}$$

$$\begin{bmatrix} x'\cos\theta + y'\sin\theta \\ -x'\sin\theta + y'\cos\theta \end{bmatrix} = \begin{bmatrix} 1 & 0 \\ 0 & 1 \end{bmatrix} \cdot \begin{bmatrix} x \\ y \end{bmatrix}$$

$$\begin{bmatrix} x'\cos\theta + y'\sin\theta \\ -x'\sin\theta + y'\cos\theta \end{bmatrix} = \begin{bmatrix} x \\ y \end{bmatrix}$$

The result is the following two equations.

$$x = x'\cos\theta + y'\sin\theta$$
$$y = -x'\sin\theta + y'\cos\theta$$

This means that to find the equation of a conic section with respect to a rotation of θ, you can replace x with $x\cos\theta + y\sin\theta$ and replace y with $-x\sin\theta + y\cos\theta$.

Example 2

Find an equation of a 60° rotation about the origin of the graph of $\dfrac{x^2}{25} + \dfrac{y^2}{9} = 1$. Then sketch the graph and its rotation.

The graph of this equation is an ellipse.
Find the expressions to replace x and y.

Replace x with $x\cos 60° + y\sin 60°$ or $\dfrac{1}{2}x + \dfrac{\sqrt{3}}{2}y$.

Replace y with $-x\sin 60° + y\cos 60°$ or $-\dfrac{\sqrt{3}}{2}x + \dfrac{1}{2}y$.

Computation is often easier if the equation is rewritten as an equation with denominators of 1.

$$\frac{x^2}{25} + \frac{y^2}{9} = 1$$

$$9x^2 + 25y^2 = 225 \qquad \textit{Multiply each side by 225.}$$

$$9\left(\frac{1}{2}x + \frac{\sqrt{3}}{2}y\right)^2 + 25\left(-\frac{\sqrt{3}}{2}x + \frac{\sqrt{1}}{2}y\right)^2 = 225 \qquad \textit{Replace x and y.}$$

$$\frac{9}{4}(x^2 + 2\sqrt{3}xy + 3y^2) + \frac{25}{4}(3x^2 - 2\sqrt{3}xy + y^2) = 225 \qquad \textit{Expand the binomial.}$$

$$84x^2 - 32\sqrt{3}xy + 52y^2 = 900 \qquad \textit{Simplify.}$$

$$21x^2 - 8\sqrt{3}xy + 13y^2 = 225 \qquad \textit{Divide each side by 4.}$$

The equation of the ellipse after the 60° rotation is $21x^2 - 8\sqrt{3}xy + 13y^2 = 225$.

The graph below shows the ellipse and its rotation.

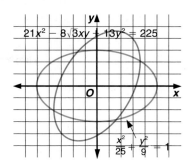

The discriminant of a second-degree equation is defined as $B^2 - 4AC$ for the general equation $Ax^2 + Bxy + Cy^2 + Dx + Ey + F = 0$. The discriminant remains unchanged under any rotation. That is, $B^2 - 4AC = (B')^2 - 4A'C'$. In Lesson 10–5, you learned to identify a conic from its general form. You can also use the discriminant to identify a conic section from its equation.

Identifying Conics By Using the Discriminant	For the general equation $Ax^2 + Bxy + Cy^2 + Dx + Ey + F = 0$: • if $B^2 - 4AC < 0$, the graph is a circle ($A = C$, $B = 0$) or an ellipse ($A \neq C$); • if $B^2 - 4AC > 0$, the graph is a hyperbola; • if $B^2 - 4AC = 0$, the graph is a parabola.

Remember that the graphs can also be degenerate cases.

Example 3

Identify the graph of the equation $4x^2 - 5xy + 16y^2 = 32$.

Find the discriminant.
$B^2 - 4AC = (-5)^2 - 4(4)(16)$ or -231

Since $B^2 - 4AC < 0$ and $A \neq C$, the graph of the equation is an ellipse.

You can also use values from the general form to find the angle of rotation about the origin.

Angle of Rotation About the Origin	For the general equation $Ax^2 + Bxy + Cy^2 + Dx + Ey + F = 0$, the angle of rotation θ about the origin can be found by $$\theta = \frac{\pi}{4}, \text{ if } A = C, \text{ or}$$ $$\tan 2\theta = \frac{B}{A - C}, \text{ if } A \neq C.$$

Example 4

Identify the graph of the equation $16x^2 - 24xy + 9y^2 - 30x - 40y = 0$. Then find θ and use a graphing calculator to draw the graph.

$$B^2 - 4AC = (-24)^2 - 4(16)(9) \qquad \text{\small A = 16, B = -24, and C = 9}$$
$$= 0$$

The graph is a parabola.

$$\tan 2\theta = \frac{B}{A - C} \qquad A \neq C$$
$$= \frac{-24}{16 - 9} \text{ or } \frac{-24}{7}$$

$$\tan 2\theta = \frac{-24}{7} \text{ or } -3.428571429$$

$$2\theta = -74° \qquad \text{\small Round to the nearest degree.}$$
$$\theta = -37°$$

To graph the equation, you must solve for y. Rewrite the equation in quadratic form, $ay^2 + by + c = 0$.

$$\overset{a}{\downarrow} \qquad \overset{b}{\downarrow} \qquad \overset{c}{\downarrow}$$
$$9y^2 + (-24x - 40)y + (16x^2 - 30x) = 0$$

Now use the quadratic formula to solve for y.

$$y = \frac{-(-24x - 40) \pm \sqrt{(-24x - 40)^2 - 4(9)(16x^2 - 30x)}}{2(9)}$$
$$= \frac{24x + 40 \pm \sqrt{(24x + 40)^2 - 576x^2 + 1080x}}{18}$$

Enter each equation and graph.

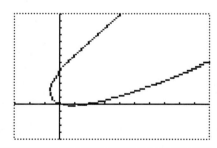

MATH JOURNAL

Write an explanation of how you can use the quadratic formula to identify a conic section.

CHECKING FOR UNDERSTANDING

Communicating Mathematics

Read and study the lesson to answer each question.

1. **Write** the matrix for $T_{(-4, 4)}$ used to translate six points of a hyperbola.

2. **Sketch** an ellipse with a vertical major axis. Then sketch the ellipse rotated $90°$.

3. **Explain** how you can use the quadratic formula to graph the equation $8x^2 + 4xy + 5y^2 = 40$.

Identify the graph of each equation. Write the equation of the translated graph for $T_{(-5, 6)}$ in general form. Then draw the graph.

4. $4x^2 + 5y^2 = 20$

5. $3x^2 - 16 = -3y^2$

6. $3y^2 - 7x^2 = 21$

7. $(x + 3)^2 = 4y$

Suppose the graph of each equation is rotated about the origin for the given angle. Find an equation of the rotated graph.

8. $x^2 - 8y = 0, \theta = 90°$

9. $6x^2 + 5y^2 = 30, \theta = 60°$

10. $49x^2 - 16y^2 = 784, \theta = \dfrac{\pi}{4}$

11. $y^2 + 8x = 0, \theta = \dfrac{\pi}{6}$

12. Identify the graph of the equation $5y^2 + 4xy + 8x^2 - 40 = 0$. Then find θ.

EXERCISES

Identify the graph of each equation. Write the equation of the translated graph in general form. Then draw the graph.

13. $x^2 + y^2 = 6$ for $T_{(2, 3)}$

14. $9x^2 - 25y^2 = 225$ for $T_{(0, -5)}$

15. $y = 2x^2 - 7x + 5$ for $T_{(-3, -4)}$

16. $3x^2 + y^2 = 9$ for $T_{(-1, 2)}$

Suppose the graph of each equation is rotated about the origin for the given angle. Find an equation of the rotated graph.

17. $x^2 - y^2 = 9, \theta = 45°$

18. $x^2 - 5x + y^2 = 3, \theta = \dfrac{\pi}{4}$

19. $xy = -6, \theta = \dfrac{\pi}{4}$

20. $7x^2 - 6\sqrt{3}xy + 13y^2 = 16, \theta = 30°$

Identify the graph of each equation. Then find θ to the nearest degree.

21. $8x^2 + 5xy - 4y^2 + 2 = 0$

22. $9x^2 + 4xy + 6y^2 = 20$

23. $2x^2 + 9xy + 14y^2 - 5 = 0$

24. $x^2 - 2xy + y^2 - 5x - 5y = 0$

25. $2x^2 + 4\sqrt{3}xy + 6y^2 + \sqrt{3}x - y = 0$

26. $2x^2 + 4xy + 2y^2 + 2\sqrt{2}x - 2\sqrt{2}y + 12 = 0$

The graph of each equation is a degenerate case. Identify the graph and then draw it.

27. $3(x - 1)^2 + 4(y + 4)^2 = 0$

28. $xy - y^2 + 2x^2 = 0$

29. $(x - 3)^2 - (x + 4)^2 = -4(y - 1)$

30. $(x + 3)^2 + (y + 3)^2 - 6(x + y) = 18$

Use the quadratic formula to solve each equation for y. Then use a graphing calculator to draw the graph.

31. $8x^2 + 5xy - 4y^2 = -2$

32. $9x^2 + 4xy + 6y^2 - 20 = 0$

33. $2x^2 + 9xy + 14y^2 = 5$

34. $x^2 - 2xy + y^2 - 5x - 5y = 0$

35. $2x^2 + 4\sqrt{3}xy + 6y^2 + \sqrt{3}x = y$

36. $2x^2 + 4xy + 2y^2 + 2\sqrt{2}x - 2\sqrt{2}y = -12$

37. Determine the rotation and/or translation needed to transform the graph of $5x^2 + 4xy + 8y^2 - 36 = 0$ to a graph whose center is the origin and whose axes are on the x- and y-axes.

38. Determine the minimum angle of rotation for the graphs of each equation so that the rotated graph coincides with the original graph. Sketch the graph.
  a. $6x^2 + 8y^2 = 24$
  b. $5y^2 - x^2 = 25$
  c. $12x^2 + 12y^2 = 108$
  d. $y = (x+3)^2$

39. **Manufacturing** A cam in a punch press is shaped like an ellipse with the equation $\dfrac{x^2}{169} + \dfrac{y^2}{25} = 1$. The camshaft goes through the focus on the positive axis.
  a. Graph a model of the cam.
  b. Find an equation that translates the model so that the camshaft is at the origin.
  c. Find the equation of the model in part b when the cam is rotated to an upright position.

40. **Sports** A football is a model of an *ellipsoid*. An ellipsoid is formed when an ellipse centered at the origin is rotated in space around the x- or y-axis. The figure at the right shows an ellipsoid rotated about the y-axis. Suppose the equation of the ellipsoid that models a football is

$\dfrac{y^2}{30.25} + \dfrac{x^2}{11.56} + \dfrac{z^2}{11.56} = 1$. Find the equation of the circle that intersects the x- and z-axes.

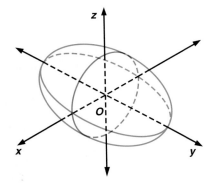

41. Given $A = 43°, b = 20$, and $a = 11$, do these measurements determine one triangle, two triangles, or no triangle? **(Lesson 5-6)**

42. **Physics** Write an equation with phase shift 0 to represent simple harmonic motion for an initial position of 0, amplitude 3, and period 2. **(Lesson 6-7)**

43. Graph $3r = 12$ on a polar plane. **(Lesson 9-1)**

44. Find $(-5 + 12i)^2$. **(Lesson 9-8)**

45. Write the standard form of $x^2 - 8x + y^2 = -11$ and identify the conic section it represents. **(Lesson 10-5)**

46. **College Entrance Exam** Compare quantity A and quantity B below.
Write A if quantity A is greater than quantity B.
Write B if quantity B is greater than quantity A.
Write C if the two quantities are equal.
Write D if there is not enough information to determine the relationship.

  (A) $(\sqrt{x})^2$ **(B)** $\dfrac{1}{x^2}$

10-7A   Graphing Calculators: Solving Quadratic Systems

As we have seen, the graphing calculator is a good tool for solving systems of equations, including those involving quadratic equations. By using the techniques used to graph conic sections in Lesson 10-6A, we can graph and solve systems involving quadratics.

Example 1

Graph the system of equations below and find its solution to two decimal places.

$$4y^2 + 2x = 5$$
$$4x^2 + 6y^2 = 12$$

First, solve each equation for y.

$$4y^2 = 5 - 2x \qquad\qquad 6y^2 = 12 - 4x^2$$

$$y^2 = \frac{1}{4}(5 - 2x) \qquad\qquad y^2 = 2 - \frac{2}{3}x^2$$

$$y = \pm\frac{1}{2}\sqrt{5 - 2x} \qquad\qquad y = \pm\sqrt{2 - \frac{2}{3}x^2}$$

Then, graph the equations using the viewing window $[-4.5, 4.5]$ by $[-3, 3]$. Use a scale factor of 1 for both axes.

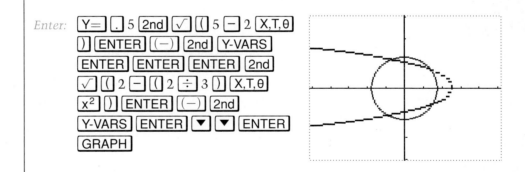

Enter: Y= . 5 2nd √ ((5 − 2 X,T,θ
) ENTER (−) 2nd Y-VARS
ENTER ENTER ENTER 2nd
√ ((2 − ((2 ÷ 3)) X,T,θ
x²)) ENTER (−) 2nd
Y-VARS ENTER ▼ ▼ ENTER
GRAPH

The solutions are $(-0.75, -1.27)$, $(-0.75, 1.27)$, $(1.5, -0.71)$, and $(1.5, 0.71)$, accurate to two decimal places. Notice the symmetry of the points and the symmetry of the graph.

You can find the solutions to any combination of equations in a system involving quadratics using a graphing calculator. The solutions to the equations of two conic sections or the solutions to one conic section and a line can be found.

Example 2 | **Graph the system of equations below and find its solution to two decimal places.**

$$x^2 = 7 - y^2$$
$$y^2 + y - 4x^2 = 10$$

First, solve both equations for y.

$$-y^2 = x^2 - 7 \qquad\qquad y^2 + y = 10 + 4x^2$$
$$y^2 = 7 - x^2 \qquad\qquad y^2 + y + \frac{1}{4} = 10 + 4x^2 + \frac{1}{4}$$
$$y = \pm\sqrt{7 - x^2} \qquad\qquad (y + \frac{1}{2})^2 = \frac{41}{4} + 4x^2$$
$$y + \frac{1}{2} = \pm\sqrt{\frac{41}{4} + 4x^2}$$
$$y = -\frac{1}{2} \pm \sqrt{\frac{41}{4} + 4x^2}$$

Then enter the equations into a graphing calculator. Use the viewing window $[-9.4, 9.4]$ by $[-6.2, 6.2]$ with a scale factor of 1 for all axes.

Enter: [Y=] [2nd] [√] [(] 7 [−] [X,T,θ]
[x²] [)] [ENTER] [(−)] [2nd]
[Y-VARS] [ENTER] [ENTER] [ENTER]
[(−)] .5 [+] [2nd] [√] [(] 10.25
[+] 4 [X,T,θ] [x²] [)] [ENTER]
[(−)] .5 [−] [2nd] [√] [(] 10.25
[+] 4 [X,T,θ] [x²] [)] [GRAPH]

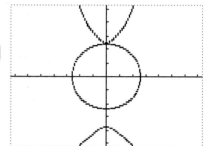

Although it appears that the graphs intersect, using zoom-in shows that there are no solutions to this system because the two graphs do not intersect.

EXERCISES

Use a graphing calculator to graph each system of equations and find the solutions to two decimal places.

1. $x^2 + y^2 = 6$
$2x^2 - 4y^2 = 8$

2. $2x^2 = 3 - y$
$x^2 + \frac{y^2}{9} = 1$

3. $y^2 = 4 - x$
$3x^2 - y^2 = 5$

4. $(x - 3)^2 + (y - 1)^2 = 4$
$(2x + 1)^2 + (4y - 1)^2 = 4$

5. $4x^2 - 9y^2 = -36$
$4x^2 + 9y^2 = 36$

6. $x^2 = 10 - y^2$
$y^2 = 0.2x - 0.6$

10-7 Systems of Second-Degree Equations and Inequalities

Objective

After studying this lesson, you should be able to:
- graph and solve systems of second-degree equations and inequalities.

Application

For hundreds of years, people have reported seeing mysterious lights or objects in the sky. The reports of these unidentified flying objects (UFOs) increased significantly during World War II. The term flying saucer was first used in 1947 after a pilot flying near Mount Ranier, Washington, described seeing a group of strange objects that looked like saucers skipping over water.

Suppose sightings of a UFO are reported by the residents of Waco, Hillsboro, and Groesbeck, Texas. Waco reports that the object seems to be about 50 miles away. Hillsboro reports that the UFO seems to be about 40 miles away, and Groesbeck has the nearest sightings of the UFO estimated at only 13 miles away. The UFO vanishes into thin air shortly after being spotted.

FYI...

There are two major organizations that study UFOs and publish information on them: CUFOS (Center for UFO Studies) in Glenview, Illinois, and MUFON (Mutual UFO Network) in Seguin, Texas.

Students at Baylor University in Waco use a coordinate grid with each square representing 1 square mile to pinpoint the location of the UFO. They plot each of the towns, placing Waco at the origin, Hillsboro at $(0, 30)$, and Groesbeck at $(35, 18)$. Then they use each town as the center for a circle with radius equaling the estimated distance to the UFO. *Where do you think the UFO was?*

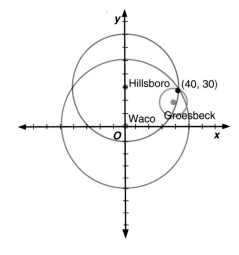

The equation of each circle is a second-degree equation. The three circles represent a system of second-degree equations. The coordinates of the point that satisfies all three equations is the solution to the system. You can solve this system graphically by locating the point where all three circles intersect. In this case, the solution is represented by the point at $(40, 30)$.

With three circles, you saw that the number of possible solutions was one. The number of solutions of a system of second-degree equations is related to the number of times the graphs of the equations intersect.

If the graphs do not intersect, there is no solution.

If the system is composed of a line and a conic, there may be 0, 1, or 2 solutions. *Remember that a line is a degenerate conic.*

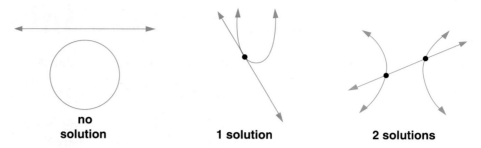

no solution   1 solution   2 solutions

If the system is composed of two conics, there may be 0, 1, 2, 3, or 4 solutions.

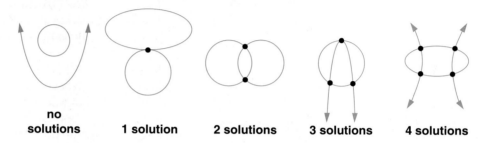

no solutions   1 solution   2 solutions   3 solutions   4 solutions

While you can tell the number of solutions from graphing the equations of a system, the exact solution is not always apparent. To find the exact solution, you must use algebra.

Example 1

Graph the system of equations. Then solve.

$$x^2 + 4y^2 = 36$$
$$x^2 = -y + 3$$

The graph of the first equation is an ellipse. The graph of the second equation is a parabola. When a system is composed of a parabola and an ellipse, there may be 0, 1, 2, 3, or 4 possible solutions. Graph each equation. There appear to be 3 solutions.

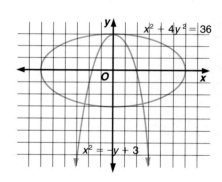

Since both equations contain a single term involving x, x^2, you can solve the system as follows. First, subtract the equations.

$$\begin{aligned} x^2 + 4y^2 \quad\;\; - 36 &= 0 \\ x^2 \qquad + y - \;\; 3 &= 0 \\ \hline 4y^2 - y - 33 &= 0 \end{aligned}$$

Use the quadratic formula to find y. $y = 3$ or $-\dfrac{11}{4}$

Now find the appropriate values of x by substituting these values for y in one of the original equations.

You could also graph the two equations on a graphing calculator and use the ZOOM and TRACE functions to find the intersections of the graphs.

$x^2 = -y + 3$ $x^2 = -y + 3$

$x^2 = -3 + 3$ $y = 3$ $x^2 = -\frac{11}{4} + 3$ $y = -\frac{11}{4}$

$x^2 = 0$ $x^2 = \frac{23}{4}$

$x = 0$ $x = \pm\frac{\sqrt{23}}{2}$

The solutions are $(0, 3)$, $\left(\frac{\sqrt{23}}{2}, -\frac{11}{4}\right)$, and $\left(-\frac{\sqrt{23}}{2}, -\frac{11}{4}\right)$.

Example 2

APPLICATION

Manufacturing

The lid of a crate used to ship peppers is rectangular and must be reinforced by a board placed diagonally across the lid. If the board is 25 inches long and the area of the lid is 300 square inches, determine the dimensions of the lid.

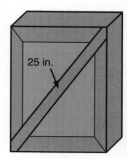

A = 300 in²

From the information in the problem, we can write two equations, each of which is the equation of a conic section.

Use the Pythagorean theorem: $\ell^2 + w^2 = 25^2$ *equation of a circle*
Use the area formula: $\ell w = 300$ *equation of a special hyperbola*

There are 0, 1, 2, 3, or 4 solutions for a system involving a circle and a hyperbola. To solve the equations algebraically, use substitution. You can rewrite the equation of the hyperbola as $\ell = \frac{300}{w}$.

$$\ell^2 + w^2 = 25^2$$
$$\left(\frac{300}{w}\right)^2 + w^2 = 625$$
$$90{,}000 + w^4 = 625w^2$$
$$w^4 - 625w^2 + 90{,}000 = 0$$

You can use the quadratic formula or factoring to find that $w^2 = \pm 225$ or ± 400. Thus, $w = \pm 15$ or ± 20. If $w = 15$, then $\ell = 20$. If $w = 20$, then $\ell = 15$. The lid is 15 by 20 inches.

You can use a graphing calculator to check your solution. Solve each equation for ℓ. Let y represent ℓ and x represent w. Graph the equations. *Remember that the graph of a circle may not appear as a circle on the screen.*

$$y = \pm\sqrt{625 - x^2}$$
$$y = \frac{300}{x}$$

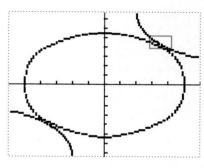

(continued on the next page)

LESSON 10-7 SYSTEMS OF SECOND-DEGREE EQUATIONS AND INEQUALITIES 581

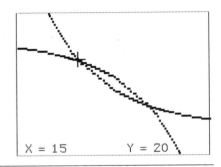

Use the ZOOM feature to enlarge the section of the graph containing the intersections in the first quadrant. Use the TRACE function to find the coordinates of the solutions, $(15, 20)$ and $(20, 15)$.

X = 15   Y = 20

In Unit 1, you learned to graph different types of inequalities by graphing the corresponding equation and then testing points in the regions of the graph to find solutions for the inequality. When graphing systems of inequalities involving conics, the procedure is the same.

Example 3

Graph the solutions for the system of inequalities below.

$$x^2 + y^2 < 36$$
$$\frac{y^2}{9} - \frac{x^2}{4} \leq 0$$

First, graph $x^2 + y^2 = 36$. The circle should be dashed. Test a point either inside or outside the circle to see if its coordinates satisfy the inequality.

Test $(3, 2)$: $3^2 + 2^2 \overset{?}{<} 36$
$\qquad\qquad\quad 9 + 4 \, < 36$

Since $(3, 2)$ satisfies the inequality, shade the interior of the circle. Then, graph $\frac{y^2}{9} - \frac{x^2}{4} = 0$. The hyperbola should be a solid curve. Test a point inside the branches of the hyperbola, outside its branches, or on its branches. *Since the hyperbola is symmetric, you need not test points within both branches.*

Test $(0, 0)$: $\qquad 4x^2 - 9y^2 \overset{?}{\leq} 36$
$\qquad\qquad 4(0^2) - 9(0^2) \leq 36$
$\qquad\qquad\qquad\qquad 0 \leq 36$

Since $(0, 0)$ satisfies the inequality, the region outside the branches should be shaded. The intersection of the graphs of the two inequalities represents the solution set of the system.

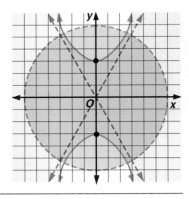

CHECKING FOR UNDERSTANDING

Communicating Mathematics

Read and study the lesson to answer each question.

1. **Draw** figures to show each of the possible numbers of solutions to a system that involves the equations of a circle and a line.

2. **Draw** figures to show each of the possible numbers of solutions to a system that involves the equations of a parabola and a hyperbola.

3. **Explain** why the negative values of w^2 and w were not used in finding the solution to the problem in Example 2.

4. **Describe** how you solve a system of inequalities by graphing.

Guided Practice

Use the graphs to determine the number of solutions for each system of equations.

5.

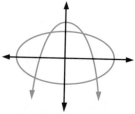

6.

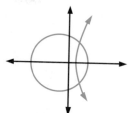

7.

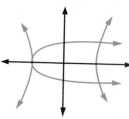

8.

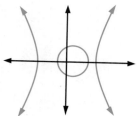

9.

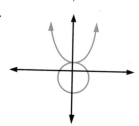

10.

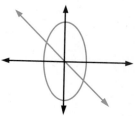

Graph each system of equations. Then solve. Round the coordinates of the solutions to the nearest tenth.

11. $x^2 + y^2 = 16$
    $x^2 + y^2 = 9$

12. $x^2 + y^2 = 49$
    $x + 1 = 0$

13. $2y - x + 3 = 0$
    $x^2 = 16 - y^2$

Graph each system of inequalities.

14. $y + x^2 < 0$
    $x^2 + y^2 < 9$

15. $x^2 - 16y^2 \geq 16$
    $x^2 + y^2 \geq 49$

EXERCISES

Practice

Graph each system of equations. Then solve. Round the coordinates of the solutions to the nearest tenth.

16. $y^2 = 100 - x^2$
    $x - y = 2$

17. $y = 7 - x$
    $y^2 + x^2 = 9$

18. $y^2 - x^2 = 4$
    $y - 5 = 0$

19. $x - y = 0$
    $\dfrac{(x-1)^2}{20} + \dfrac{(y-1)^2}{5} = 1$

20. $xy = 2$
    $x^2 = 3 + y^2$

21. $x^2 + 2y^2 = 10$
    $3x^2 - y^2 = 9$

22. $\dfrac{(x-2)^2}{36} - \dfrac{(y+3)^2}{4} = 1$
    $x - y = 0$

23. $5x^2 + y^2 = 30$
    $9x^2 - y^2 + 16 = 0$

24. $x^2 + y^2 = 64$
    $x^2 + 64y^2 = 64$

25. $9x^2 + 4y^2 = 36$
    $9x^2 - 4y^2 = 36$

26. $(y-1)^2 = x + 4$
    $y + 1 = -x$

27. $xy = -4$
    $x^2 + 9y^2 = 25$

Graph each system of inequalities.

28. $10 \geq (x-5)^2 + 2y$
$y \geq -2x + 9$

29. $(x+1)^2 + (y+1)^2 > 16$
$9x^2 + y^2 < 81$

30. $x^2 + y^2 \geq 25$
$x^2 + y^2 \leq 100$

31. $(x+3)^2 + (y+2)^2 \leq 36$
$y + 4 = 0$

32. $16x^2 - 25y^2 \geq 400$
$xy \geq 2$

33. $x^2 - 16 < 4y^2$
$x - (y-1)^2 > 0$

34. $x^2 + y^2 \geq 16$
$x + y = 2$

35. $9x^2 + 4y^2 \leq 36$
$4x^2 + 9y^2 \geq 36$

Critical Thinking

36. Solve the system of equations below.
$x^2 + y^2 = 1$
$y = 3x + 1$
$x^2 + (y+1)^2 = 4$

Applications and Problem Solving

37. Agriculture Mr. Wise owns a dairy farm in Plains, Georgia. He wishes to section off a small portion of a field to contain newly weaned calves. He has 56 feet of fencing and a 4-foot gate. He would like to enclose an area of 216 square feet. Use a system of equations to find the dimensions of the fenced area.

38. Space Science A satellite is placed in a geosynchronous orbit at a height of 35,800 kilometers above the equator. The radius of Earth is about 6400 kilometers. Another satellite is placed in an elliptical orbit such that its closest point to Earth is 10,000 kilometers.

   a. Draw a model of this situation.

   b. Determine the possible number of times that these two satellites might collide.

   c. How do you think these types of collisions are prevented?

Mixed Review

39. Surveying A surveying crew is studying a housing project for possible relocation for the airport expansion. They are located on the ground, level with the houses. If the distance to one of the houses is 253 meters and the distance to the other is 319 meters, what is the distance between the houses if the angle subtended by them at the point of observation is $42°12'$? **(Lesson 5-7)**

40. Write the equation for the inverse of $y = \cos x$. **(Lesson 6-6)**

41. Given points $A(3, 3, -1)$ and $B(5, 3, 2)$, find an ordered triple that represents $\overrightarrow{AB}$. **(Lesson 8-3)**

42. Find $6\left(\cos\dfrac{5\pi}{8} + i\sin\dfrac{5\pi}{8}\right) \div 12\left(\cos\dfrac{\pi}{2} + i\sin\dfrac{\pi}{2}\right)$. Then express the quotient in rectangular form. **(Lesson 9-7)**

43. Find the coordinates of the vertex of a parabola whose equation is $y - 3 = (x+4)^2$ with respect to $T_{(-3,5)}$. **(Lesson 10-6)**

44. College Entrance Exam The average of 8 numbers is 6. The average of 6 other numbers is 8. What is the average of all 14 numbers?

10-8 Tangents and Normals to the Conic Sections

Application

On April 25, 1990, the Hubble Space Telescope was deployed from the cargo bay of the space shuttle *Discovery*. When the first pictures were sent back to Earth, the stars had diffused halos of light around them. Such fuzzy images are caused when light from the inner and outer edges of the primary mirror are not brought to the same focal plane. Scientists have now concluded that the center of the primary mirror has the correct curvature, but the overall shape is more hyperbolic than it should be. At the edge of the mirror, the deviation from a parabolic curvature to a circular is serious enough to cause the halo effect. One solution to this problem is the installation of new optical instruments that compensate for the error in curvature.

Some of the rays of light that skim the surfaces of the mirrors of the Hubble Telescope can be modeled by lines that are tangent to the curves of the surface. Remember that a tangent to a curve intersects the curve in one point. The slope of the curve at a given point is also the slope of the tangent at that point.

Perpendicular lines have slopes that are negative reciprocals.

A line tangent to a circle is perpendicular to the radius at the point of tangency. Since the slope of the radius to $P(x_1, y_1)$ on a circle with center at the origin is $\dfrac{y_1}{x_1}$, the slope of the tangent is $-\dfrac{x_1}{y_1}$. You can write the point-slope form of the equation of the tangent using the coordinate of the point of tangency $P(x_1, y_1)$.

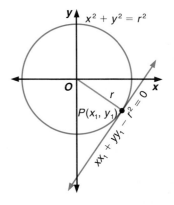

$$m = \frac{y - y_1}{x - x_1}$$

$$-\frac{x_1}{y_1} = \frac{y - y_1}{x - x_1} \qquad m = -\frac{x_1}{y_1}$$

$$xx_1 - x_1^2 = -yy_1 + y_1^2$$

$$xx_1 + yy_1 = x_1^2 + y_1^2 \qquad \textit{Since } (x_1, y_1) \textit{ is on the}$$

$$xx_1 + yy_1 = r^2 \qquad\qquad \textit{circle, } (x_1)^2 + (y_1)^2 = r^2.$$

Equation of the Tangent to a Circle	**The equation of the line tangent to the circle $x^2 + y^2 = r^2$ at (x_1, y_1) is $xx_1 + yy_1 - r^2 = 0$.**

Example 1

At the base of volcanic Mount Shasta is the small town of Mount Shasta, California. The town is 10 miles south and 7 miles west of the center of the volcano. A road that is tangent to the base of the volcano runs through the center of town. If the center of the volcano is at $(0, 0)$, determine the equation of the line representing the road.

The center of the circle is at $(0, 0)$ so the equation of the circle is $x^2 + y^2 = r^2$. The point on the circle representing the town of Mount Shasta is at $(-7, -10)$. The radius of the circle is $\sqrt{(-10)^2 + (-7)^2}$ or $\sqrt{149}$.

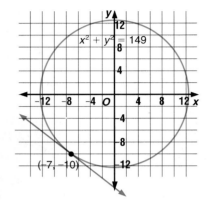

Substitute the values for $x_1, y_1,$ and r into the equation for the line tangent to a circle.

$$xx_1 + yy_1 - r^2 = 0$$
$$x(-7) + y(-10) - (\sqrt{149})^2 = 0$$
$$-7x - 10y - 149 = 0$$
$$7x + 10y + 149 = 0$$

The equation of the line representing the road is $7x + 10y = -149$.

Equations for lines tangent to other conic sections, including circles whose centers are not the origin, can be found. The chart below lists formulas for the slopes of lines tangent to the graphs of given equations.

You can review another method for finding the equation of the line tangent to a parabola in Lesson 3-6.

Conic Section	Standard Form of Equation	Slope of Tangent at (x, y)
circle	$(x - h)^2 + (y - k)^2 = r^2$	$-\dfrac{(x - h)}{(y - k)}$
parabola	$(y - k)^2 = 4p(x - h)$	$\dfrac{2p}{y - k}$
	$(x - h)^2 = 4p(y - k)$	$\dfrac{x - h}{2p}$
ellipse	$\dfrac{(x - h)^2}{a^2} + \dfrac{(y - k)^2}{b^2} = 1$	$-\dfrac{b^2(x - h)}{a^2(y - k)}$
	$\dfrac{(y - k)^2}{a^2} + \dfrac{(x - h)^2}{b^2} = 1$	$-\dfrac{a^2(x - h)}{b^2(y - k)}$
hyperbola	$\dfrac{(x - h)^2}{a^2} - \dfrac{(y - k)^2}{b^2} = 1$	$\dfrac{b^2(x - h)}{a^2(y - k)}$
	$\dfrac{(y - k)^2}{a^2} - \dfrac{(x - h)^2}{b^2} = 1$	$\dfrac{a^2(x - h)}{b^2(y - k)}$

When you have found the slope, you can then use the slope-intercept form to write the equation of the slope of the tangent.

Example 2

Find the equation of the line tangent to the graph of $\dfrac{x^2}{16} - \dfrac{y^2}{4} = 1$ at $(5, 1.5)$.

The equation represents a hyperbola with a horizontal transverse axis. Thus, use $\dfrac{b^2(x - h)}{a^2(y - k)}$ for the slope of the tangent. In this case, $a^2 = 16$, $b^2 = 4$, $h = 0$, $k = 0$, $x_1 = 5$, and $y_1 = 1.5$.

Find the slope.

$$m = \dfrac{b^2(x - h)}{a^2(y - k)}$$

$$= \dfrac{4(5 - 0)}{16(1.5 - 0)} \text{ or } \dfrac{5}{6}$$

Use the point-slope form.

$$y - y_1 = m(x - x_1)$$

$$y - \dfrac{3}{2} = \dfrac{5}{6}(x - 5)$$

The equation of the line tangent to the hyperbola at $(5, 1.5)$ is $y - \dfrac{3}{2} = \dfrac{5}{6}(x - 5)$ or $5x - 6y - 16 = 0$.

You can also determine which members of a family of lines are tangent to a given conic.

Example 3

A family of lines with slope 4 can be represented by the equation $y = 4x + k$, where k is a constant. Find which member(s) of this family of lines is tangent to the circle with equation $x^2 + y^2 = 153$.

The slope of the line is 4. The slope of the tangent to a circle is $-\dfrac{x - h}{y - k}$. Find the relationship of x and y at the point of tangency.

$$m = -\dfrac{x - h}{y - k}$$

$$4 = -\dfrac{x - 0}{y - 0} \qquad m = 4, (h, k) = (0, 0)$$

$$-4y = x$$

A point of tangency for a line in this family is a point on the circle for which $x = -4y$. Substitute $-4y$ for x in the equation of the circle.

$$x^2 + y^2 = 153$$
$$(-4y)^2 + y^2 = 153$$
$$17y^2 = 153$$
$$y = \pm 3$$

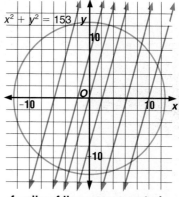

family of lines represented by $y = 4x + k$

If you substitute these values for y into the equation for the circle, you find there are four resulting ordered pairs: $(12, 3)$, $(-12, 3)$, $(12, -3)$, and $(-12, -3)$.

(continued on the next page)

However, since $-\dfrac{x}{y} = 4$, you can eliminate $(12, 3)$ and $(-12, -3)$. *Why?*

$(-12, 3)$ and $(12, -3)$ are the points of tangency.

You can use these points and the given slope to find the equations of the tangent lines.

$y - 3 = 4(x - (-12))$ or $y = 4x + 51$
$y + 3 = 4(x - 12)$ or $y = 4x - 51$

If $P(x_1, y_1)$ is a point in the plane of the circle with equation $(x - h)^2 + (y - k)^2 = r^2$, then P lies outside the circle if and only if $(x_1 - h)^2 + (y_1 - k)^2 > r^2$. This indicates that there are two lines from point P tangent to the circle. Sometimes it is necessary to find the length of the segment from that point to the point of tangency.

The length of the tangent segment PT can be found by using the Pythagorean theorem. Let $C(h, k)$ represent the center of the circle, $P(x_1, y_1)$ represent a point outside the circle, and $T(x_2, y_2)$ represent one of the points of tangency. Let $TP = t$, $CT = r$, and $CP = d$. Thus, $t = \sqrt{d^2 - r^2}$.

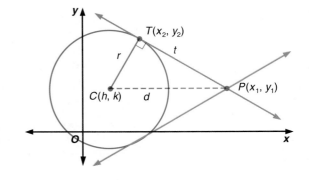

The distance formula can be used to find that $d^2 = (x_1 - h)^2 + (y_1 - k)^2$. You can substitute this value of d^2 into $t = \sqrt{d^2 - r^2}$ to obtain a formula for the length of a segment tangent to a circle.

Length of a Segment Tangent to a Circle	**The length t of a tangent segment from (x_1, y_1) to the circle with equation $(x - h)^2 + (y - k)^2 = r^2$ can be found by using the formula** $$t = \sqrt{(x_1 - h)^2 + (y_1 - k)^2 - r^2}.$$

Example 4

Find the length of the tangent segment from $(-4, 7)$ to the circle with equation $x^2 + y^2 = 10$.

$t = \sqrt{(x_1 - h)^2 + (y_1 - k)^2 - r^2}$
$ = \sqrt{(-4 - 0)^2 + (7 - 0)^2 - 10}$
$ = \sqrt{55}$

The length of the tangent segment is $\sqrt{55}$ or about 7.42 units.

The **normal** to a curve at any point on the curve is the line perpendicular to the tangent at that point. The slope of the normal is the negative reciprocal of the slope of the tangent.

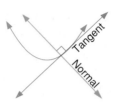

Example 5

Find the equations of the tangent and the normal to the graph of $\frac{y^2}{64} + \frac{x^2}{49} = 1$ at $\left(\frac{7\sqrt{3}}{2}, 4\right)$.

The graph is an ellipse with a horizontal major axis. Use $-\frac{b^2(x - h)}{a^2(y - k)}$ for the slope of the tangent.

The slope of the tangent is $\dfrac{-64\left(\frac{7\sqrt{3}}{2}\right)}{49(4)}$ or $-\dfrac{8\sqrt{3}}{7}$. The equation of the tangent is $y - 4 = -\dfrac{8\sqrt{3}}{7}\left(x - \dfrac{7\sqrt{3}}{2}\right)$ or $8\sqrt{3}x + 7y - 112 = 0$.

The slope of the normal is $\dfrac{7}{8\sqrt{3}}$. The equation of the normal is $y - 4 = \dfrac{7}{8\sqrt{3}}\left(x - \dfrac{7\sqrt{3}}{2}\right)$ or $14x - 16\sqrt{3}y + 15\sqrt{3} = 0$.

Portfolio

Select an item that shows something new you learned in this chapter and place it in your portfolio.

CHECKING FOR UNDERSTANDING

Communicating Mathematics

Read and study the lesson to answer each question.

1. **Illustrate** how there can be two lines tangent to a hyperbola from a given point outside the hyperbola.

2. **Explain** how you can determine which line in a family of lines is tangent to a circle.

Guided Practice

Find the slopes of the tangent and the normal to the graph of each equation at the given point.

3. $(x + 4)^2 - 8(y + 3) = 0,\ (0, -1)$

4. $\dfrac{(x - 2)^2}{10} + \dfrac{(y - 3)^2}{10} = 1,\ (-1, 2)$

5. $25(x - 3)^2 + 4(y - 1)^2 = 100,\ (3, -4)$

6. $(x + 2)^2 + 4(y - 3)^2 = 36,\ (-2, 6)$

7. $\dfrac{(y + 2)^2}{8} = x - 3,\ (5, 2)$

8. $9x^2 = 8y^2 + 72,\ (-4, -3)$

Find the equations of the tangent and the normal to the graphs of each equation at the given point. Write the equations in standard form.

9. $x^2 + y^2 = 145, (9, -8)$

10. $x^2 + 4x - y + 1 = 0, (0, 1)$

11. $\dfrac{x^2}{25} + \dfrac{y^2}{9} = 1, (4, 1.8)$

12. $\dfrac{x^2}{4} - y^2 = 1, (2, 0)$

EXERCISES

Practice

Find the equations of the tangent and the normal to the graphs of each equation at the point with the given coordinates. Write the equations in standard form.

13. $x^2 + y^2 = 25, (2, \sqrt{21})$

14. $\dfrac{x^2}{49} + \dfrac{y^2}{49} = 1, (\sqrt{13}, 6)$

15. $x^2 - 6 = 110 - y^2, (-10, -4)$

16. $3x^2 + 3y^2 = 5, \left(1, \dfrac{\sqrt{6}}{3}\right)$

17. $y(y + 1) - x = 5, (1, -3)$

18. $(y - 3)^2 + 9x^2 = 36, (\sqrt{3}, 6)$

19. $y^2 + (x + 2)^2 = 9, (-4, \sqrt{5})$

20. $2x^2 - 7x + 5y = 11, \left(-1, \dfrac{2}{5}\right)$

21. $(x - 4)^2 + 3y^2 = 12, (7, 1)$

22. $(x - 4)^2 + (y - 3)^2 = 16, (8, 3)$

23. $9(x - 1)^2 - 25(y + 3)^2 = 225, (-4, -3)$

24. $32(x - 2)^2 - 9(y - 1)^2 = 576, (-4, -7)$

Find the length of the tangent segment from each point to the graph of the given circle.

25. $(6, 2), x^2 + y^2 = 37$

26. $(4, -1), (x + 3)^2 + y^2 = 4$

27. $(10, 1), x(x - 6) + y(y - 8) = 0$

28. $(-7, 2), 4(x + 3)^2 + 4(y - 2)^2 = 16$

29. Find the equations of the horizontal lines tangent to the graph of circle $x^2 + y^2 = 25$.

30. Find the equations of the vertical lines tangent to the graph of circle $x^2 + y^2 = 49$.

31. Find the equations of the tangents if the lines tangent to the graph of $\dfrac{x^2}{25} + \dfrac{y^2}{64} = 1$ are horizontal.

32. Find the equations of the normals if the lines tangent to the graph of $x^2 + 4y^2 = 16$ are vertical.

33. The graph of $x^2 + y^2 = 64$ has two tangents that belong to the family of lines $y = -\dfrac{3}{4}x + k$. Find the equations of the tangents.

34. A line passes through the origin and $(2, 16)$. It is also tangent to the graph of $y = (x + 2)^2$.
   a. Name the family of lines to which this line belongs.
   b. Write the equation of the tangent.

35. If the line $6x - 2y = 5$ is parallel to the line tangent to the graph of $4y = x^2 + 4x - 16$ at a, find the value of a.

36. The line $y = 2x$ is tangent to the graph of $y = x^2 + ax + b$ at $(2, 4)$. Find the values of a and b.

37. Find the point on the parabola $x^2 - 6x - 2y + 1 = 0$ at which the tangent line makes a 45° angle with the positive x-axis.

38. Determine the positive value of k such that the line $x + y = 4$ is tangent to an equilateral hyperbola $xy = k$.

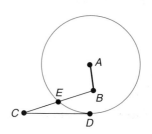

39. $\overline{CD}$ is tangent to circle A at D and $\overline{BC}$ intersects circle A at E. If $AB = 4, CE = EB = 6$, and $CD = 2\sqrt{30}$, find the radius of circle A.

40. Design For the cover design of a mathematics textbook, artist Cathy Watterson decides to make one of the shapes an ellipse with a horizontal major axis 8 centimeters long and a minor axis 6 centimeters long. She would like to place tangents at four different points around the ellipse so that the ellipse is enclosed in a parallelogram. The distance from each point of tangency to the major axis is 2 centimeters.

a. Make a sketch of this design.

b. Find the slope of each tangent.

c. Write the equations of the four tangents.

41. Communications A microwave signal from a television tower is tangent to the parabolic curve of a reflector at a point 2 meters right and 3 meters up from the vertex of the parabola. If a model of this reflector and signal were drawn on a coordinate plane, the parabola would have its vertex at $(0, 0)$, open to the right, and the distance from the vertex to the focus would be $\dfrac{9}{8}$ units.

a. Make a sketch of the model.

b. Determine the equation of the tangent at the given point.

42. The area of a triangle is 24 cm². One side is 6 cm long and another side is 10 cm long. Find the measure of the included angle of those two sides. **(Lesson 5-8)**

43. Graph $y = \cos\left(x - \dfrac{\pi}{6}\right)$. **(Lesson 6-1)**

44. Find the equation of the line that bisects the obtuse angle formed by the graphs of $2x - 3y + 9 = 0$ and $x + 4y + 4 = 0$. **(Lesson 7-7)**

45. Simplify $(3i + 5)^2$. **(Lesson 9-5)**

46. Express $5\left(\cos\dfrac{\pi}{6} + i\sin\dfrac{\pi}{6}\right)$ in rectangular form. **(Lesson 9-6)**

47. Graph the system of inequalities below. **(Lesson 10-7)**

$xy \geq 2$
$x - 3y = 2$

48. College Entrance Exam Choose the best answer.
The area of a square is $49x^2$ in². What is the length of the diagonal of the square?

(A) $7x$ in.   **(B)** $7x\sqrt{2}$ in.   **(C)** $14x$ in.   **(D)** $7x^2$ in.   **(E)** $\dfrac{7x}{\sqrt{2}}$ in.

VOCABULARY

Upon completing this chapter, you should be
familiar with the following terms:

asymptote	**551**	**534**	focus
axis	**534**	**551**	hyperbola
axis of symmetry	**534**	**524**	locus
center	**524**	**543**	major axis
circle	**524**	**543**	minor axis
concentric circles	**524**	**589**	normal
conic sections	**561**	**534**	parabola
conjugate axis	**551**	**524**	radius
degenerate case	**561**	**543**	semi-major axis
directrix	**534**	**543**	semi-minor axis
eccentricity	**563**	**570**	translation matrix
ellipse	**542**	**551**	transverse axis
equilateral hyperbola	**555**	**534**	vertex

SKILLS AND CONCEPTS

OBJECTIVES AND EXAMPLES

Upon completing this chapter, you should
be able to:

- graph circles   **(Lesson 10-1)**

  Graph $(x-2)^2+$
  $(y+1)^2 = 8$.

  center: $(2,-1)$

  radius: $\sqrt{8}$ or
  $2\sqrt{2}$ units

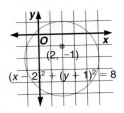

- graph parabolas   **(Lesson 10-2)**

  Graph
  $(x+1)^2 = 2(y-3)$.

  vertex: $(-1, 3)$

  focus: $\left(-1, \dfrac{7}{2}\right)$

  directrix: $y = \dfrac{5}{2}$

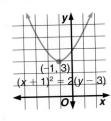

REVIEW EXERCISES

Use these exercises to review and prepare for
the chapter test.

**Write the standard form of each equation.
Then graph the equation.**

1. $3x^2 = 81 - 3y^2$

2. $x^2 = 6y - y^2$

3. $x^2 + y + y^2 = 12 - 3x$

4. $x^2 + 14x + y^2 + 6y = 23$

**Find the coordinates of the focus and vertex
and the equation of the directrix of each
parabola whose equation is given. Then
graph the equation.**

5. $(x-7)^2 = 8(y-3)$

6. $(y+4)^2 = -16(x-1)$

7. $y^2 + 6y - 4x + 25 = 0$

8. $x^2 + 4x - y + 8 = 0$

- graph ellipses **(Lesson 10-3)**

Graph $\dfrac{(x+1)^2}{16} + \dfrac{(y-3)^2}{4} = 1$.

center: $(-1, 3)$

foci:
$(-1 - 2\sqrt{3}, 3)$,
$(-1 + 2\sqrt{3}, 3)$

vertices:
$(3, 3), (-5, 3)$,
$(-1, 5), (-1, 1)$

For each equation, find the coordinates of the center, foci, and vertices of the ellipse. Then graph the equation.

9. $\dfrac{(x-3)^2}{25} + \dfrac{(y+1)^2}{4} = 1$

10. $\dfrac{(x-5)^2}{16} + \dfrac{(y-2)^2}{36} = 1$

11. $(x-4)^2 + 4(y-6)^2 = 36$

12. $3(x+3)^2 + 2(y-4)^2 = 12$

- graph hyperbolas **(Lesson 10-4)**

Graph $\dfrac{(y-2)^2}{4} - (x-5)^2 = 1$.

center: $(5, 2)$

foci:
$(5, 2 + \sqrt{5})$
$(5, 2 - \sqrt{5})$

vertices:
$(5, 4)$,
$(5, 0)$

asymptotes: $y - 2 = \pm\dfrac{1}{2}(x-5)$

Find the coordinates of the center, foci, and vertices and the equations of the asymptotes of the graph of each equation. Then graph the equation.

13. $x^2 - \dfrac{y^2}{2} = 4$

14. $2(x+5)^2 - (y-1)^2 = 8$

15. $y^2 - 5x^2 + 20x = 50$

16. $9x^2 - 16y^2 - 36x + 96y + 36 = 0$

- recognize conic sections by their equations **(Lesson 10-5)**

Identify the conic section represented by $3x^2 - 5x - y + 1 = 0$.

There is only one squared term. The conic is a parabola.

Identify the conic section represented by each equation. Write the equation in standard form and graph the equation.

17. $4x^2 - 8x + y^2 = 12$

18. $xy = 0$

19. $2(x-4)^2 = -2(y-1)^2 + 8$

20. $x^2 - 6y - 8x + 16 = 0$

- find the equations of conic sections that have been translated or rotated **(Lesson 10-6)**

Write the equation of the parabola $y = 2(x-1)^2$ for $T_{(-2,4)}$.

The vertex of $y = 2(x-1)^2$ is at $(1, 0)$.

$$\begin{bmatrix} 1 \\ 0 \end{bmatrix} + \begin{bmatrix} -2 \\ 4 \end{bmatrix} = \begin{bmatrix} -1 \\ 4 \end{bmatrix}$$

The vertex of the translated parabola is at $(-1, 4)$. The equation of the translated parabola is $(y - 4) = 2(x + 1)^2$.

Write the equation of each translated or rotated graph below. Then graph the equation.

21. $x^2 + y^2 = 3$ for $T_{(-3,2)}$

22. $4x^2 - 16(y-1)^2 = 64$ for $T_{(1,-2)}$

23. $4x^2 + 9y^2 = 36, \theta = \dfrac{\pi}{6}$

24. $y^2 - 4x = 0, \theta = 45°$

■ graph and solve systems of second-degree equations and inequalities **(Lesson 10-7)**

Graph the system of equations. Then solve.

$y = x^2 + 1$
$3y^2 = x^2 + 11$

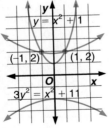

$$x^2 \quad\quad -y +1 = 0$$
$$-x^2 + 3y^2 \quad\quad -11 = 0$$
$$\overline{\quad 3y^2 - y - 10 = 0 \quad}$$

So, $y = -\dfrac{5}{3}$ or $y = 2$.

Substituting the values for y in one of the original equations, the solutions are $(1, 2)$ and $(-1, 2)$.

Graph each system of equations. Then solve.

25. $(x - 1)^2 + 4(y - 1)^2 = 20$
$\quad x = y$

26. $x^2 - 4x - 4y - 4 = 0$
$\quad (x - 2)^2 = -4y$

Graph each system of inequalities.

27. $y \geq x^2 + 4$
$\quad x^2 + y^2 < 49$

28. $x^2 + (y - 2)^2 \geq 0$
$\quad x^2 + 9(y - 2)^2 \leq 9$

■ write equations of tangents and normals to conic sections **(Lesson 10-8)**

Find the equation of the line tangent to the graph of $4x^2 + y^2 = 4$ at $(0.5, 2)$.

$$m = \frac{-b^2(x - h)}{a^2(y - k)}$$

$$= \frac{-4(0.5 - 0)}{1(2 - 0)} \text{ or } -1$$

The equation of the tangent line is

$y - 2 = -1\left(x - \dfrac{1}{2}\right)$ or $2x + 2y - 5 = 0$.

Find the equations of the tangent and the normal to the graphs of each equation at the given point. Write the equations in standard form.

29. $x^2 - 4y^2 + 2x + 8y - 7 = 0, (1, 1)$

30. $4x^2 - 9y^2 = 36, (6, 2\sqrt{3})$

31. $y = -4(x + 1)^2, (-2, -4)$

32. $9x^2 + 25y^2 = 225, \left(4, \dfrac{9}{5}\right)$

APPLICATIONS AND PROBLEM SOLVING

33. Construction Ms. Lopez, a construction worker, is building a bay window in the shape of a semi-ellipse for a new home. The width of the window is to be 5 feet and its depth is to be 2 feet. To sketch the arch of the bay window, Ms. Lopez uses a 5-foot string attached to two thumbtacks. Where should the thumbtacks be placed? *Hint: The thumbtacks are at the foci.* **(Lesson 10-3)**

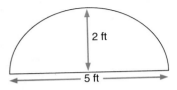

2 ft

5 ft

34. Astronomy A satellite orbiting Earth follows an elliptical path with Earth at its center. The eccentricity of the orbit is 0.2, and the major axis is 12,000 miles long. Assuming that the center of the ellipse is the origin and the foci lie on the x-axis, write the equation of the orbit of the satellite. **(Lesson 10-5)**

Identify the conic section represented by each equation. Write the equation in standard form and graph the equation.

1. $(x+3)^2 = 8(y+2)$

2. $4x^2 - y^2 = 0$

3. $2x^2 - 13y^2 + 5 = 0$

4. $16(x-3)^2 + 81(y+4)^2 = 1296$

5. $y^2 - 2x + 10y + 27 = 0$

6. $(x-4)^2 - 9(y-5)^2 = 36$

7. $x^2 + 2y^2 + 2x - 12y + 11 = 0$

8. $x^2 + 4x = -(y^2 - 6)$

9. Write the equation of the circle with center at $(-8, 3)$ that passes through $(-6, -4)$.

10. Write the equation of the parabola that has a focus at $(3, -5)$ and whose equation of directrix is $y = -2$.

11. Write the equation of an ellipse centered at the origin that has a horizontal major axis, $e = \dfrac{1}{2}$, and $2c = 1$.

12. Write the equation of the hyperbola that has eccentricity $\dfrac{3}{2}$ and foci at $(-5, -2)$ and $(-5, 4)$.

Write the equation of each translated or rotated graph below. Then graph the equation.

13. $4(x+1)^2 + (y-3)^2 = 36$ for $T_{(3, -5)}$

14. $2x^2 - y^2 = 8, \theta = 60°$

15. Graph the system of equations. Then solve.
$x^2 + 4y^2 = 4$
$(x-1)^2 + y^2 = 1$

16. Graph the system of inequalities.
$x^2 + y^2 - 2x - 4y + 1 \geq 0$
$x^2 - 4y - 2x + 5 \geq 0$

Find the equations of the tangent and the normal to the graphs of each equation at the given point. Write the equations in standard form.

17. $x^2 + y^2 + 6x - 10y = 0, (2, 1)$

18. $16x^2 - y^2 - 4y - 20 = 0, (\sqrt{2}, 2)$

19. Geometry The diameter of a circle has endpoints at $(-2, 3)$ and $(4, 5)$. Find the equation for the circle.

20. Technology Mr. Wilson bought a motion detector light and installed it in the center of his backyard. It can detect motion within a circle defined by the equation $x^2 + y^2 = 90$. If a person walks northeast through the backyard along a line defined by the equation $y = 2x - 3$, at what point will the motion detector light turn on?

Bonus

If the graph of the circle $x^2 + y^2 = k$ is tangent to the graph of the line $x + y = 2k$ where k is positive, find the value of k.

CHAPTER 11

EXPONENTIAL AND LOGARITHMIC FUNCTIONS

HISTORICAL SNAPSHOT

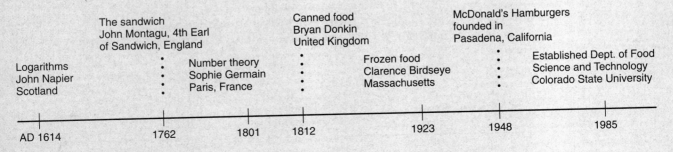

The sandwich
John Montagu, 4th Earl
of Sandwich, England

Canned food
Bryan Donkin
United Kingdom

McDonald's Hamburgers
founded in
Pasadena, California

Logarithms
John Napier
Scotland

Number theory
Sophie Germain
Paris, France

Frozen food
Clarence Birdseye
Massachusetts

Established Dept. of Food
Science and Technology
Colorado State University

AD 1614   1762   1801   1812   1923   1948   1985

CHAPTER OBJECTIVES

In this chapter, you will:

- Use the properties of exponents and logarithms.
- Solve equations by using the properties of exponents and logarithms.
- Graph exponential and logarithmic functions.

CAREER GOAL: Food Management

Sometimes inspiration can come from unexpected places. Hernan Rodriguez-Palacios's brother is a chef who has studied in Paris. As a result, he has always had an interest in food. His career goal, however, is much different from that of his brother. Hernan would like to own and operate a brewery some day.

Hernan's interest may seem narrow, but his training is preparing him for a number of different careers. "Food science and technology includes a broad spectrum of career opportunities," Hernan says. "These include food production, handling, processing, distribution, marketing, preservation, and consumption." Hernan's main interest is fermentation. He says, "The process of fermentation is involved in a wide spectrum of foods as well as in the production of antibiotics and alcohol fuels."

Food scientists use logarithms to determine the *pH balance* of foods. The pH balance is a measure of the basicity or acidity of a solution. A pH value of 7 is considered to be neutral, while a pH value below 7 indicates an acid and a pH value above 7 indicates a base. Each pH unit represents a factor of 10. So lemon juice, which has a pH balance of 2, is 10^{4-2}, or 100, times more acidic than tomato juice, which has a pH balance of 4.

While food has been around a long time, Hernan says, "Food science and technology is a very young field of study. A person has the opportunity to select from so many unanswered questions and investigate one or more that are of special interest. There are still many unsolved questions to explore."

For up-to-date information on food technology, visit:
www.glencoe.com/sec/math/amc/mathnet

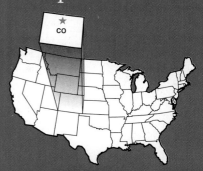

11-1 Rational Exponents

After studying this lesson, you should be able to:
- use the properties of exponents, and
- evaluate and simplify expressions containing rational exponents.

In your studies of biology, chemistry, and physics, you sometimes have to use very large numbers like Avogadro's number, 602,205,000,000,000,000,000,000, or the mass of Earth, 5,980,000,000,000,000,000,000,000 kg. Other times you must use very small numbers like Newton's universal gravitational constant, 0.0000000000667, or the diameter of a red blood cell, 0.000000775 m. Working with these numbers is a problem if your calculator displays less than ten digits. Expressing these numbers in **scientific notation** allows you to work with them more easily. A number is in scientific notation when it is in the form $a \times 10^n$, where $1 \leq a < 10$ and n is an integer. Written in scientific notation, Avogadro's number is 6.02205×10^{23} and Newton's constant is 6.67×10^{-11}.

Working with numbers in scientific notation requires a knowledge of the definitions and properties of integral exponents. For any real number x and a positive integer n, the following definitions hold.

Definition	**Example**
If $n = 1$, $x^n = x$.	$5^1 = 5$
If $n > 1$, $x^n = \overbrace{x \cdot x \cdot x \cdots x}^{n \text{ factors}}$.	$3^3 = 3 \cdot 3 \cdot 3$ or 27
If $x \neq 0$, $x^0 = 1$.	$6^0 = 1$
If $x \neq 0$, $x^{-n} = \dfrac{1}{x^n}$.	$2^{-4} = \dfrac{1}{2^4}$ or $\dfrac{1}{16}$

These definitions can be used to verify the properties of exponents for positive integers m and n and real numbers a and b.

Suppose m and n are positive integers, and a and b are real numbers. Then the following properties hold.

Product property:	$a^m a^n = a^{m+n}$
Power of a power property:	$(a^m)^n = a^{mn}$
Power of a quotient property:	$\left(\dfrac{a}{b}\right)^m = \dfrac{a^m}{b^m}$, where $b \neq 0$
Power of a product property:	$(ab)^m = a^m b^m$
Quotient property:	$\dfrac{a^m}{a^n} = a^{m-n}$, where $a \neq 0$

Example 1

Red blood cells are circular-shaped cells that carry oxygen through your bloodstream. The diameter of a red blood cell is about 7.75×10^{-7} m. Find the area of one of these cells. Express your answer in scientific notation.

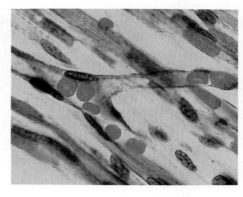

The radius of a circle is one-half its diameter. Thus, $r = \dfrac{7.75 \times 10^{-7}}{2}$ or 3.875×10^{-7} m.

$$
\begin{aligned}
A &= \pi r^2 & & \text{Formula for the area of a circle} \\
&= \pi(3.875 \times 10^{-7})^2 & & r = 3.875 \times 10^{-7} \\
&= \pi(3.875)^2(10^{-7})^2 & & \text{Properties of exponents} \\
&= \pi(3.875)^2(10^{-14})
\end{aligned}
$$

Use a calculator to evaluate. The calculator will give the solution in scientific notation if the number has more than eight digits.

$\boxed{\text{2nd}}\ \boxed{\pi}\ \boxed{\times}\ 3.875\ \boxed{x^2}\ \boxed{\times}\ 10\ \boxed{y^x}\ 14\ \boxed{+/-}\ \boxed{=}\ \textsf{4.71729719 – 13}$

The area of the blood cell is about 4.72×10^{-13} m^2.

FYI...

Red blood cells make up 44% of your blood. They are produced at the rate of 2.3 million per second by the red bone marrow of your long bones.

Expressions with rational exponents can be defined so that the properties of integral exponents are still valid. Consider the expressions $2^{\frac{1}{2}}$ and $9^{\frac{1}{3}}$. Extending the laws for integral exponents gives us the following equations.

$$
\begin{aligned}
2^{\frac{1}{2}} \cdot 2^{\frac{1}{2}} &= 2^{\frac{1}{2} + \frac{1}{2}} \\
&= 2^1 \text{ or } 2
\end{aligned}
\qquad
\begin{aligned}
9^{\frac{1}{3}} \cdot 9^{\frac{1}{3}} \cdot 9^{\frac{1}{3}} &= 9^{\frac{1}{3} + \frac{1}{3} + \frac{1}{3}} \\
&= 9^1 \text{ or } 9
\end{aligned}
$$

We know that, by definition, $\sqrt{2} \cdot \sqrt{2} = 2$. Therefore, it makes sense to define $2^{\frac{1}{2}}$ as $\sqrt{2}$.

But $\sqrt[3]{9} \cdot \sqrt[3]{9} \cdot \sqrt[3]{9} = 9$. So, it makes sense to define $9^{\frac{1}{3}}$ as $\sqrt[3]{9}$.

In general, let $y = b^{\frac{1}{n}}$ for a real number b and a positive integer n. Then, $y^n = (b^{\frac{1}{n}})^n = b^{\frac{n}{n}}$ or b. But $y^n = b$ if and only if $y = \sqrt[n]{b}$. Therefore, we can define $b^{\frac{1}{n}}$ as follows.

Definition of $b^{\frac{1}{n}}$

For any real number $b \geq 0$ and any integer $n > 1$,
$$b^{\frac{1}{n}} = \sqrt[n]{b}.$$
This also holds when $b < 0$ and n is odd.

Why is this definition not valid if $b < 0$ and n is even?

The properties of integral exponents given on page 598 can be extended to rational exponents.

Example 2

Evaluate.

a. $256^{\frac{1}{4}}$

a. $256^{\frac{1}{4}} = \left(4^4\right)^{\frac{1}{4}}$ *Rewrite 256 as 4^d.*

        $= 4^{\frac{4}{4}}$ $(a^m)^n = a^{mn}$

        $= 4^1$ or 4

b. $5^{\frac{1}{2}} \cdot 15^{\frac{1}{2}}$

b. $5^{\frac{1}{2}} \cdot 15^{\frac{1}{2}} = 5^{\frac{1}{2}} \cdot 5^{\frac{1}{2}} \cdot 3^{\frac{1}{2}}$ $(ab)^m = a^m b^m$

           $= 5^{\frac{1}{2} + \frac{1}{2}} \cdot 3^{\frac{1}{2}}$ $a^m a^n = a^{m+n}$

           $= 5^1 \cdot 3^{\frac{1}{2}}$ $b^{\frac{1}{n}} = \sqrt[n]{b}$

           $= 5\sqrt{3}$

How can you evaluate an expression with a rational exponent in which the numerator is not 1? Study the two methods for evaluating $7^{\frac{2}{3}}$ shown below.

$$\begin{array}{cc}
\textit{Method 1} & \textit{Method 2} \\
7^{\frac{2}{3}} = (7^{\frac{1}{3}})^2 & 7^{\frac{2}{3}} = \left(7^2\right)^{\frac{1}{3}} \\
= (\sqrt[3]{7})^2 & = \sqrt[3]{7^2}
\end{array}$$

Therefore, $\left(\sqrt[3]{7}\right)^2$ and $\sqrt[3]{7^2}$ both equal $7^{\frac{2}{3}}$. In general, we define $b^{\frac{m}{n}}$ as $(b^{\frac{1}{n}})^m$ or $(b^m)^{\frac{1}{n}}$. Now apply the definition of $b^{\frac{1}{n}}$ to $(b^{\frac{1}{n}})^m$ and $(b^m)^{\frac{1}{n}}$.

$$(b^{\frac{1}{n}})^m = (\sqrt[n]{b})^m \qquad\qquad (b^m)^{\frac{1}{n}} = \sqrt[n]{b^m}$$

Thus, $b^{\frac{m}{n}}$ is defined as $(\sqrt[n]{b})^m$ or $\sqrt[n]{b^m}$.

Definition of Rational Exponents	**For any nonzero number b, and any integers m and n with $n > 1$,** $$b^{\frac{m}{n}} = \sqrt[n]{b^m} = (\sqrt[n]{b})^m$$ **except when $\sqrt[n]{b}$ is not a real number.**

Example 3

Evaluate $64^{\frac{5}{6}}$.

$$64^{\frac{5}{6}} = \left(2^6\right)^{\frac{5}{6}} \qquad \textit{Rewrite 64 as } 2^6.$$
$$= 2^5 \qquad \quad (a^m)^n = a^{mn}$$
$$= 32$$

You can rewrite an expression containing radicals with rational exponents or vice versa.

Example 4

Express $\sqrt[4]{8x^2y^8}$ using rational exponents.

$$\sqrt[4]{8x^2y^8} = (8x^2y^8)^{\frac{1}{4}} \qquad b^{\frac{1}{n}} = \sqrt[n]{b}$$
$$= 8^{\frac{1}{4}}x^{\frac{2}{4}}y^{\frac{8}{4}} \qquad (ab)^m = a^m b^m$$
$$= 8^{\frac{1}{4}}x^{\frac{1}{2}}y^2$$

Example 5

Express $(7a)^{\frac{5}{8}}b^{\frac{3}{8}}$ using radicals.

$$(7a)^{\frac{5}{8}}b^{\frac{3}{8}} = \left((7a)^5b^3\right)^{\frac{1}{8}} \qquad a^{mn} = (a^m)^n$$
$$= \left(7^5a^5b^3\right)^{\frac{1}{8}} \qquad (ab)^m = a^m b^m$$
$$= \sqrt[8]{7^5a^5b^3} \qquad b^{\frac{1}{n}} = \sqrt[n]{b}$$

To simplify a radical, use the product property to factor out the nth roots and use the smallest index possible for the radical.

Example 6

Simplify $\sqrt{x^3y^9}$.

$$\sqrt{x^3y^9} = (x^3y^9)^{\frac{1}{2}} \qquad b^{\frac{1}{n}} = \sqrt[n]{b}$$
$$= x^{\frac{3}{2}}y^{\frac{9}{2}} \qquad (ab)^m = a^m b^m$$
$$= x^{\frac{2}{2}}x^{\frac{1}{2}}y^{\frac{8}{2}}y^{\frac{1}{2}}$$
$$= |x|y^4\sqrt{xy} \qquad \textit{Use } |x| \textit{ to denote the principal root for } x^{\frac{2}{2}}$$
$$\textit{since x may be negative.}$$

CHECKING FOR UNDERSTANDING

Communicating Mathematics

Read and study the lesson to answer each question.

1. **Explain** why rational exponents are not defined when the denominator of the exponent is even and the base is negative.

2. **Write** an expression that is equivalent to $\sqrt[3]{46}$.

3. Is $6.03 \times 10^{0.4}$ written in scientific notation? Explain.

4. When you are multiplying two expressions with like bases raised to different powers, you maintain the bases and __?__ the exponents. When you are dividing two expressions with like bases and different exponents, you maintain the bases and __?__ the exponents.

Guided Practice

Evaluate.

5. $81^{\frac{1}{2}}$

6. $27^{-\frac{2}{3}}$

7. $7^{\frac{1}{4}} \cdot 7^{\frac{7}{4}}$

Express using rational exponents.

8. $\sqrt[4]{a}$

9. $\sqrt{xy^3}$

10. $\sqrt[3]{8x^3y^6}$

Express using radicals.

11. $15^{\frac{1}{5}}$

12. $25^{\frac{1}{3}}$

13. $a^{\frac{3}{4}}y^{\frac{1}{4}}$

EXERCISES

Practice

Evaluate.

14. $\sqrt[3]{125}$

15. $3^{-4} \cdot 3^8$

16. $\sqrt[4]{16^2}$

17. $(5^{\frac{3}{4}})^4$

18. $(169^{\frac{1}{2}})^0$

19. $(8^{-\frac{1}{2}})^{-\frac{2}{3}}$

20. $\left(\sqrt[3]{216}\right)^2$

21. $(3^{-1} + 3^{-2})^{-1}$

22. $81^{\frac{1}{2}} - 81^{-\frac{1}{2}}$

23. $\dfrac{16^{\frac{3}{4}}}{16^{\frac{1}{4}}}$

24. $\left(\sqrt[3]{343}\right)^{-2}$

25. $\dfrac{27}{27^{\frac{2}{3}}}$

Express using rational exponents.

26. $\sqrt{a^6b^3}$

27. $\sqrt[6]{b^3}$

28. $\sqrt{25a^4b^{10}}$

29. $\sqrt[3]{125a^2b^3}$

30. $\sqrt[3]{64s^9t^{15}}$

31. $\sqrt[4]{24a^{12}b^{16}}$

32. $\sqrt{169x^5}$

33. $\sqrt[5]{32x^5y^8}$

34. $\sqrt[5]{15x^3y^{15}}$

Express using radicals.

35. $64^{\frac{1}{6}}$

36. $x^{\frac{2}{3}}$

37. $4^{\frac{1}{3}}a^{\frac{2}{3}}y^{\frac{4}{3}}$

38. $x^{\frac{4}{7}}y^{\frac{3}{7}}$

39. $(rt^2)^{\frac{1}{5}}v^{\frac{3}{5}}$

40. $a^{\frac{1}{6}}b^{\frac{4}{6}}c^{\frac{3}{6}}$

41. $\dfrac{x^{\frac{2}{3}}}{x^{\frac{1}{3}}}$

42. $15x^{\frac{1}{3}}y^{\frac{1}{5}}$

43. $(x^{10}y^2)^{\frac{1}{5}}a^{\frac{2}{5}}$

Simplify.

44. $4x^2(4x)^{-2}$

45. $x^6 \cdot x^{-3} \cdot x^2$

46. $(y^{-2})^4 \cdot y^8$

47. $(5x^{\frac{1}{3}})^3$,

48. $((2x)^4)^{-2}$

49. $(4y^4)^{\frac{3}{2}}$

50. $\sqrt{a^3 b^2} \cdot \sqrt{a^4 b^5}$

51. $(a^{\frac{1}{2}} b^{-2} c^{\frac{5}{4}})^{-4}$

52. $(5ac)^{\frac{1}{3}} (a^2 c^3)^{\frac{1}{3}}$

53. Find the simplest expression for $\dfrac{3^{60}}{9^{30}}$.

54. Architecture Mathematicians have shown that a soap bubble will enclose a maximum space with a minimum amount of material. Architects have used this property to create buildings that enclose a great amount of space while using a small amount of building material. If a soap bubble has a surface area of A, then its volume, V, is given by the equation $V = 0.094\sqrt{A^3}$. Find the surface area of a bubble with a volume of 7.5 cm³.

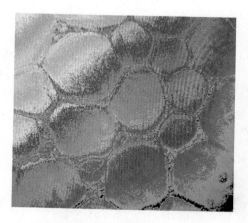

55. Aerospace The typical period of the orbit of a space shuttle around Earth is approximately 90 minutes. The radius of Earth is approximately 6400 km. Use the formula $r = \sqrt[3]{\dfrac{GM_e t^2}{4\pi^2}}$, where r represents the distance in meters from the center of Earth to the satellite, $G = 6.67 \times 10^{-11}$, t represents the time in seconds, and M_e represents the mass of Earth, which is 5.98×10^{24} kg, to determine how far the space shuttle is above Earth.

56. Chemistry All matter is composed of atoms. The nucleus of an atom is the center portion of the atom that contains most of the mass of the atom. A theoretical formula for the radius, r, of the nucleus of an atom is $r = (1.3 \times 10^{-15})A^{\frac{1}{3}}$ meters, where A is the mass number of the nucleus. Find the radius of the nucleus, if the mass number of an isotope of carbon is 12.

57. Graph $y > |x| - 2$. **(Lesson 3-5)**

58. Solve the equation $27x^2 + 15x - 8 = 0$. **(Lesson 4-1)**

59. Find all rational roots of the equation $2x^3 + x^2 - 8x - 4 = 0$. **(Lesson 4-4)**

60. Change $\dfrac{\pi}{24}$ to degree measure. **(Lesson 5-1)**

61. Aviation An airplane flying at an altitude of 9000 meters passes directly overhead. Fifteen seconds later, the angle of elevation to the plane is 60°. How fast is the airplane flying? **(Lesson 5-5)**

62. Graph the polar equation $r = -3$. **(Lesson 9-1)**

63. Find $(-3 + i)^4$. Express the result in rectangular form. **(Lesson 9-8)**

64. Name the coordinates of the center, foci, and vertices, and the equations of the asymptotes of the hyperbola that has the equation $x^2 - 4y^2 - 12x - 16y = -16$. **(Lesson 10-4)**

65. Find the length of the tangent segment from $(-1, -1)$ to the circle with equation $(x - 5)^2 + (y + 1)^2 = 4$. **(Lesson 10-8)**

66. College Entrance Exam Compare quantities A and B below.
Write A if quantity A is greater.
Write B if quantity B is greater.
Write C if the quantities are equal.
Write D if there is not enough information to determine the relationship.

(A) the sum of the measures of the angles in a scalene triangle

(B) the sum of the measures of the angles in an equilateral triangle

DECISION MAKING

Homelessness

America has always had homeless people, but in recent years, with the country experiencing a sluggish economy and rising unemployment, the number of homeless people has risen alarmingly. Exact figures are impossible to determine, but most authorities estimate that there are at least a million homeless Americans. Is there anything you can do to help alleviate the problem?

1. Research the extent of homelessness in your community. Are there shelters for the homeless? Are city or state funds allocated to provide for the needs of people without permanent shelter?

2. What are the causes of homelessness? What are the responsibilities of Americans who have homes toward those who do not? Talk to people who are involved in trying to solve the problem to learn their views on these questions.

3. What, if anything, can you do to help alleviate the problems of the homeless? Make a list of three viable options.

4. **Project** Work with your group to complete one of the projects listed at the right based on the information you gathered for Exercises 1–3.

PROJECTS

- *Conduct an interview.*
- *Write a book report.*
- *Write a proposal.*
- *Write an article for the school paper.*
- *Write a report.*
- *Make a display.*
- *Make a graph or chart.*
- *Plan an activity.*
- *Design a checklist.*

11-2A Graphing Calculators: Graphing Exponential Functions

In our society today, there are many applications that involve exponential functions. Exponential functions are functions of the form $y = a^x$, where $a > 0$ and $a \neq 1$. You can use a graphing calculator to draw all types of exponential functions, even those that would be extremely difficult to draw by hand.

Example 1

Graph $y = 3^x$ and $y = \left(\dfrac{1}{3}\right)^x$ on the same set of axes. Use the viewing window $[-5, 5]$ by $[-1, 9]$.

Enter the equations and graph.

Enter: [Y=] 3 [^] [X,T,θ] [ENTER] [(]
1 [÷] 3 [)] [^] [X,T,θ] [GRAPH]

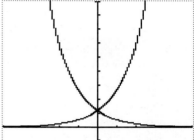

The intersection point is $(0, 1)$.

Example 2

Graph $y = 9^{2+x}$. Use the viewing window $[-7, 1]$ by $[-1, 9]$.

Enter the equations and graph.

Enter: [Y=] 9 [^] [(] 2 [+] [X,T,θ] [)]
[GRAPH]

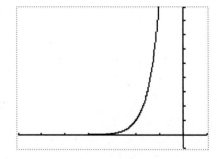

When entering this equation into your calculator, remember to use parentheses around the exponent. If you do not use parentheses, the calculator will graph the equation $y = 9^2 + x$.

EXERCISES

Graph each equation so that a complete graph is shown. Then sketch the graph on a piece of paper.

1. $y = 0.01^x$
2. $y = 5^x$
3. $y = -5^x$
4. $y = 5^{-x}$
5. $y = -5^{-x}$
6. $y = 0.5^{x-3}$
7. $y = -4^{x+6}$
8. $y = -\left(\dfrac{1}{2}\right)^{3-2x}$
9. $y = 7^{1-x}$

11-2 Exponential Functions

Objectives

After studying this lesson, you should be able to:
- evaluate expressions with irrational exponents,
- graph exponential functions, and
- graph exponential inequalities.

Application

In the United States, automobile accidents cause about 1.6 million injuries each year. It is estimated that the number of injuries could be cut in half if everyone wore seat belts whenever they are in a moving vehicle. However, less than 20% of drivers and passengers wear seat belts.

If the percent of people injured in auto accidents with and without seat belts is graphed against the speed at which the vehicle was traveling at the time of the accident, each graph resembles an **exponential curve**.

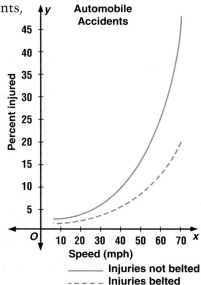

An exponential curve is the graph of a function of the form $y = a^x$. In the previous lesson, the expression a^x was defined for integral and rational exponents. The next step is to define a^x when x is an irrational number so that the properties of exponents remain valid.

Consider the graph of $y = 2^x$, where x is an integer. This is a function since there is a unique y-value for each x-value.

x	−4	−3	−2	−1	0	1	2	3	4	5
2^x	$\frac{1}{16}$	$\frac{1}{8}$	$\frac{1}{4}$	$\frac{1}{2}$	1	2	4	8	16	32

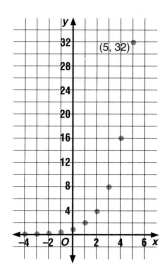

The graph shows that the function is increasing. That is, for any values x_1 and x_2, if $x_1 < x_2$, then $2^{x_1} < 2^{x_2}$.

Suppose the domain of $y = 2^x$ is expanded to include all rational numbers. The additional points graphed seem to "fill in" the graph of $y = 2^x$. That is, if k is between x_1 and x_2, then 2^k is between 2^{x_1} and 2^{x_2}. The graph of $y = 2^x$, when x is a rational number, is indicated by the broken line on the graph at the top of the next page.

Notice that the vertical scale is condensed.

x	2^x
-3.5	0.09
-2.5	0.18
-1.5	0.35
-0.5	0.71
0.5	1.41
1.5	2.83
2.5	5.66
3.5	11.31
4.5	22.63

Values given in the table are approximate.

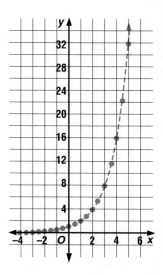

Since a^x has not been defined for irrational numbers, there are still "holes" in the graph of $y = 2^x$. How could we expand the domain of $y = 2^x$ to include both rational and irrational numbers? Consider a possible meaning for an expression such as $2^{\sqrt{3}}$. Since $1.7 < \sqrt{3} < 1.8$, it follows that $2^{1.7} < 2^{\sqrt{3}} < 2^{1.8}$. Closer and closer approximations for $\sqrt{3}$ allow us to find closer approximations for $2^{\sqrt{3}}$.

Therefore, we can now define a value for a^x when x is an irrational number.

Definition of Irrational Exponents	**If x is an irrational number and $a > 0$, then a^x is the real number between a^{x_1} and a^{x_2} for all possible choices of rational numbers x_1 and x_2, such that $x_1 < x < x_2$.**

You can use a calculator to find an approximate value of a^x for real values of x.

Example 1

Use a calculator to evaluate each expression to the nearest ten thousandth.

a. $3^{\sqrt{3}}$

a. 3 [y^x] 3 [2nd] [√x̄] [=] *6.7049919*

$3^{\sqrt{3}} \approx 6.7050$

b. 5^{π}

b. 5 [y^x] [2nd] [π] [=] *156.99255*

$5^{\pi} \approx 156.9926$

An **exponential function** has the form $y = a^x$, where a is a positive real number. The figure at the right shows the graphs of several exponential functions. Notice that the point with coordinates $(0, 1)$ is common to each function. Compare the graph of $y = 2^x$ to the graph of $y = \left(\frac{1}{2}\right)^x$. What do you notice? When $a > 1$ and x is increasing, is the graph of $y = a^x$ increasing or decreasing? When $a < 1$ and x is increasing, is the graph increasing or decreasing?

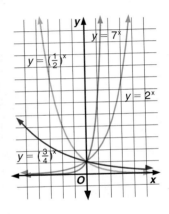

Example 2

Discuss the behavior of the graphs of $y = 2^x$ and $y = 3^x$. Compare the values of 2^x and 3^x on the intervals $-10 \le x < 0$ and $0 < x \le 10$.

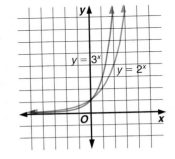

Both graphs have y-intercepts at $(0, 1)$. In the interval $-10 \le x < 0$, $2^x > 3^x$ and both graphs approach the x-axis as x approaches -10.

In the interval $0 < x \le 10$, $3^x > 2^x$.

The graphs of exponential functions can be transformed in the same way as the graphs of other functions.

Example 3

Graph the functions $y = 3^x$, $y = 3^x + 2$, $y = 3^x - 1$, and $y = 8(3^x)$ on the same set of axes. Then describe the transformations of the parent graph $y = 3^x$ that have taken place to form each of the other graphs.

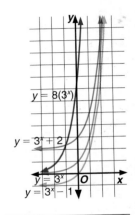

The horizontal asymptote of the graph of $y = 3^x + 2$ is the line $y = 2$ instead of $y = 0$, as for the graph of $y = 3^x$. The graph of $y = 3^x$ has been translated 2 units up.

The horizontal asymptote of the graph of $y = 3^x - 1$ is the line $y = -1$. The y-intercept is 0. The graph of $y = 3^x$ has been translated 1 unit down.

The horizontal asymptote of the graph of $y = 8(3^x)$ is the line $y = 0$ and the y-intercept is 8. The graph is much steeper than the parent graph.

Bankers call a series of payments made at equal intervals of time an **annuity**. The **present value of an annuity**, P_n, is the sum of the present values of all the periodic payments P. So an annuity of P dollars per period for n periods is worth P_n dollars now. In other words, a lump-sum investment of P_n dollars now will provide payments of P dollars per period for the next n periods. An example of this type of annuity is a retirement plan in which an employee saves money in an account that will provide monthly income after he or she retires.

The formula below can be used to find the present value of an annuity, P_n, if a series of n payments of P dollars each are made.

$$P_n = P\left[\frac{1 - (1 + i)^{-n}}{i}\right]$$

The **future value of an annuity**, F_n, is the sum of all of the annuity payments plus any accumulated interest. The formula below can be used to find the future value of an annuity, F_n, if a series of n payments of P dollars each are made.

$$F_n = P\left[\frac{(1 + i)^n - 1}{i}\right]$$

In both formulas, the variable i represents the interest rate per payment interval. Banks usually state interest rates as **annual percentage rates** or **APRs**. The interest rate per payment interval is equal to the APR divided by the number of payment intervals per year.

Example 4

APPLICATION

Finance

A monthly mortgage payment consists of an amount paid toward the principal and the interest on the loan. It may also contain an amount for the property taxes that the mortgage holder will pay from an escrow and an amount for insurance that protects the mortgage holder in case of default on the loan. The Wimberlys have taken a 30-year mortgage for $100,000, with an interest rate of 9.0%, on their new home.

a. What will the monthly payment for the principal and interest be?

b. How much will the Wimberlys pay in interest over the life of the loan?

a. Use the formula for the present value of an annuity to find the monthly payment. The APR is 9.0%, so the interest rate is $\frac{9.0\%}{12}$ or 0.75% per month. In the formula, $i = 0.0075$. *Why?*

The Wimberlys will make 12×30 or 360 payments on the loan. So the value of n in the formula is 360.

$$P_n = P\left[\frac{1 - (1 + i)^{-n}}{i}\right]$$

$$100,000 = P\left[\frac{1 - (1 + 0.0075)^{-360}}{0.0075}\right] \quad \textit{$P_n = 100,000$, $i = 0.0075$,}$$
and $n = 360$

$$\frac{750}{1 - (1.0075)^{-360}} = P$$

$$804.62262 = P \qquad \textit{Use a calculator to evaluate.}$$

The monthly payment will be $804.62. *(continued on the next page)*

b. The Wimberlys will pay 360 × 804.62 or $289,663.20 for the principal and interest on the loan. So, they will pay 289,663.20 − 100,000 or $189,663.20 in interest over the 30-year period.

As you can see in Example 4, a surprisingly large amount of interest is usually paid in a mortgage. Likewise, in an IRA, a large amount of interest can be earned.

Example 5

Finance

When Jim Baldini started his first job after he finished college, he opened an individual retirement account (IRA). He plans to contribute $2500 per year for 38 years until he reaches age 62. He hopes to earn an average APR of 8% over the 38-year period.

a. **If Mr. Baldini contributes to his IRA at the rate that he plans, how much will his account be worth when he is 62 years old?**

b. **How much interest will be earned on the account?**

a. Use the formula for the future value of an annuity.

$$F_n = P \left[\frac{(1 + i)^n - 1}{i} \right]$$

$$= 2500 \left[\frac{(1 + 0.08)^{38} - 1}{0.08} \right] \qquad i = 0.08, n = 38, P = 2500$$

$$= 2500 \left[\frac{(1.08)^{38} - 1}{0.08} \right]$$

$$= 550,789.86 \qquad \text{Use a calculator to evaluate.}$$

The value of Mr. Baldini's account will be $550,789.86.

b. Mr. Baldini will pay 2500 × 38 or $95,000 into his IRA. So, the account will earn 550, 789.86 − 95, 000 or $455, 789.86 in interest.

EXPLORATION: Graphing Calculator

The approximate world population, y, can be expressed as a function of the year, x, in the equation $y = 10^{0.00389x + 2}$.

1. Use the settings below to set up a viewing window.

XMIN = 1800   YMIN = 0
XMAX = 2180   YMAX = 7000000000
XSCL = 50   YSCL = 1000000000

2. Press [Y =] and enter the equation.

3. Use the TRACE function to determine the population growth from 1800 to the year 2000. According to the graph, what can we expect the population to be in the year 2000?

Graphing exponential inequalities is similar to graphing other inequalities.

Example 6

Graph $y < 2^x - 4$.

First, graph $y = 2^x - 4$. Since the points on this curve are not in the solution of the inequality, the graph of $y = 2^x - 4$ is shown as a dashed curve.

Then, use $(0, 0)$ as a test point to determine which area to shade.

$0 < 2^0 - 4$

$0 < 1 - 4$

$0 \nless -3$

Since $(0, 0)$ does not satisfy the inequality, the region that does not contain the origin should be shaded.

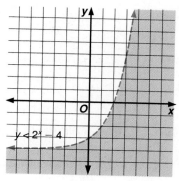

MATH JOURNAL

Describe how the graph of an exponential function differs from the other curved graphs you have studied.

CHECKING FOR UNDERSTANDING

Communicating Mathematics

Read and study the lesson to answer each question.

1. **Compare and contrast** the graphs of $y = 4^x$ and $y = 4^x - 3$.

2. **Compare and contrast** the graphs of $y = a^x$ and $y = \left(\frac{1}{a}\right)^x$, where a is an integer greater than 1.

3. **Describe** an annuity. Give examples of different types of annuities.

4. **Choose** the correct term.
   a. If $0 < a < 1$, a^x (decreases, increases) as x increases.
   b. If $a > 1$, a^x (decreases, increases) as x increases.

Guided Practice

Use a calculator to evaluate each expression to the nearest ten thousandth.

5. $2^{\sqrt{6}}$

6. $5^{\sqrt{2}}$

7. $\left(\frac{1}{3}\right)^{\pi}$

Graph each equation.

8. $y = 2^x$

9. $y = -2^x$

10. $y = 2^{-x}$

11. Compare the graphs for Exercises 8, 9, and 10. What do you notice?

12. Graph $y < 2^x + 1$.

13. **Finance** What is the present value of an annuity that pays $1000 every six months for nine years if the APR is 10%?

EXERCISES

Practice

Use a calculator to evaluate each expression to the nearest ten thousandth.

14. $7^{\sqrt{5}}$

15. $8^{\sqrt{3}}$

16. $5^{\sqrt{10}}$

Graph each equation.

17. $y = 3^x$

18. $y = \left(\frac{1}{3}\right)^x$

19. $y = 5^{-x}$

20. $y = 3^{-x}$

21. $y = -3^x$

22. $y = \left(\frac{1}{5}\right)^x$

23. $y = 2^{x+3}$

24. $y = -2^{x+3}$

25. $y = -2^{x-3}$

26. Compare and contrast the graphs for Exercises 17 and 18.

27. Compare and contrast the graphs for Exercises 19 and 20.

28. Compare and contrast the graphs for Exercises 23, 24, and 25.

Graph each inequality.

29. $y < 5^x$

30. $y > -4^x$

31. $y \le \left(\frac{1}{2}\right)^x$

32. $y < 2^{x-4}$

33. $y \ge 3^x - 4$

34. $y \le -4^x$

35. Without graphing, describe how the graphs of $y = -3^x$ and $y = 3^x$ are related.

36. Without graphing, describe how the graphs of $y = 5^x$ and $y = 5^{-x}$ are related.

Programming

37. The program below will calculate loan payments. You must input the amount of the loan, the annual percentage rate, the number of payments per year, and the number of years in the life of the loan.

```
Prgm 11:LOANPYMT
:ClrHome                             Clears the text screen.
:Disp "LOAN AMOUNT = "               Displays the quoted message.
:Input A                             Accepts a value for A.
:Disp "APR (AS A DECIMAL) = "
:Input R
:Disp "NO. OF PAYMENTS"
:Disp "PER YEAR = "
:Input N
:Disp "NO. OF YEARS = "
:Input Y
:A(R/N)/(1-(1+R/N)^(-NY))→ P         Calculates the payment, stores it in P.
:Disp "PAYMENT = "
:Disp P                              Displays the value stored in P.
```

Use the program above to find the monthly payment for each loan described below.

Exercise	Amt. of Loan (A)	Annual Percentage Rate (R)	No. of Payments per Year (N)	No. of Years (Y)
a.	$1500	18%	12	3
b.	$20,000	11%	12	5
c.	$125,000	9%	12	30

Critical Thinking

38. Graph the function $y = 2^x - x$.

39. Banking What is the future value of an annuity if $1000 is deposited into an account paying 6% every six months for 12 years?

40. Entertainment The state of Ohio offers its lottery winners a choice of prizes, with the jackpot amount paid in equal annual payments over 26 years or the present value of the annuity in one lump sum. If Mr. Arthur won a $4.5 million jackpot, how much would his lump-sum payment be? The interest rate used to find the present value is the yearly rate of inflation; assume this rate to be 5%.

41. Banking Ms. Bailey is considering two different investment structures for her IRA. One structure has her pay $500 each month into an account with an APR of 7.2%. The second structure has her pay $1500 every quarter into an account with an APR of 7.3%. Which structure will give Ms. Bailey a better return on her investment in thirty years? in five years?

42. Financial Planning The Sungs are saving for their daughter's college education. If they want to add $20,000 to her college fund at the end of five years, how much should they deposit each month into an account with an APR of 6.12%?

43. Construction A sinking fund is a fund to which regular payments are made in order to pay off a debt when it is due. The Gallway Construction Company foresees the need to buy a new cement truck in four years. At that time, the truck will probably cost $120,000. The firm sets up a sinking fund in order to accumulate the money. How much should the firm's semiannual payments to the fund be if the fund earns an APR of 7%?

44. State the domain and range of the relation $\{(0, 2), (4, -2), (9, -3), (-7, 11), (-2, 0)\}$. Is the relation a function? **(Lesson 1-1)**

45. Find the value of $\begin{vmatrix} 7 & -3 & 5 \\ 4 & 0 & -1 \\ 8 & 2 & 0 \end{vmatrix}$. **(Lesson 2-3)**

46. Find the critical points of the function $y = x^3 - 7x^2 + 8x - 2$. Then, determine whether each point represents a maximum, a minimum, or a point of inflection. **(Lesson 3-7)**

47. If $\sin \theta = \dfrac{7}{8}$ and the terminal side of θ is in the first quadrant, find $\cos 2\theta$. **(Lesson 7-4)**

48. Find an ordered pair that represents $\overrightarrow{AB}$ for $A(8, 3)$ and $B(0, -2)$. Then, find the magnitude of $\overrightarrow{AB}$. **(Lesson 8-2)**

49. Express $-6 + 6i$ in polar form. **(Lesson 9-6)**

50. Express $\sqrt{20a^4b^{12}}$ using rational exponents. **(Lesson 11-1)**

51. College Entrance Exam Choose the best answer.
A carpenter divides a board that is 7 feet 9 inches long into three equal parts. What is the length of each part?

(A) $2 \text{ ft } 6\dfrac{1}{3} \text{ in.}$ **(B)** $2 \text{ ft } 8\dfrac{1}{3} \text{ in.}$ **(C)** $2 \text{ ft } 7 \text{ in.}$ **(D)** $2 \text{ ft } 8 \text{ in.}$ **(E)** $2 \text{ ft } 9 \text{ in.}$

11-3 The Number *e*

Objective

After studying this lesson, you should be able to:
- use the exponential function $y = e^x$.

Application

Byron received a total of $700 in cash gifts for his high school graduation. He is going to invest his money in a savings account so that he can buy a car after he graduates from college. The Brentwood Savings and Loan offers a statement savings account that pays 4.5% APR compounded quarterly. At this rate, what would the balance of Byron's savings account be after four years?

Advertisements for savings accounts offer **compound interest**. When interest is compounded, it is added to the beginning principal at a specified time, thus forming the new principal. Suppose the beginning principal is P, the annual interest rate is r, interest is compounded n times per year, and A is the account balance after t years. Then the compound interest formula is $A = P \left(1 + \dfrac{r}{n}\right)^{nt}$.

Byron can find the amount that will be in his account after four years by applying the compound interest formula. Since Byron's account is compounded quarterly, $n = 4$.

$$A = P \left(1 + \frac{r}{n}\right)^{nt}$$
$$= 700 \left(1 + \frac{0.045}{4}\right)^{(4 \cdot 4)} \qquad P = 700, r = 0.045, n = 4, \text{ and } t = 4$$
$$= 700(1 + 0.01125)^{16}$$
$$= \$837.21$$

If Byron invests his money in the account that offers 4.5% interest compounded quarterly, his account will have a balance of $837.21 after four years.

Another bank in Byron's community offers an account with 4.4% APR with the interest compounded continuously. If interest is compounded continuously, the formula $A = Pe^{rt}$ can be used to find the amount in the account. The variables P, r, and t represent the principal, the annual interest rate, and the time in years, respectively. The number *e* is a special irrational number. This number is the sum of the infinite series shown below.

$$e = 1 + \frac{1}{1} + \frac{1}{1 \cdot 2} + \frac{1}{1 \cdot 2 \cdot 3} + \frac{1}{1 \cdot 2 \cdot 3 \cdot 4} + \cdots + \frac{1}{1 \cdot 2 \cdot 3 \cdot \cdots \cdot n} + \cdots$$

The following computation for e is correct to three decimal places.

$$e = 1 + 1 + \frac{1}{1 \cdot 2} + \frac{1}{1 \cdot 2 \cdot 3} + \frac{1}{1 \cdot 2 \cdot 3 \cdot 4} + \frac{1}{1 \cdot 2 \cdot 3 \cdot 4 \cdot 5} +$$
$$\frac{1}{1 \cdot 2 \cdot 3 \cdot 4 \cdot 5 \cdot 6} + \frac{1}{1 \cdot 2 \cdot 3 \cdot 4 \cdot 5 \cdot 6 \cdot 7}$$
$$= 1 + 1 + \frac{1}{2} + \frac{1}{6} + \frac{1}{24} + \frac{1}{120} + \frac{1}{720} + \frac{1}{5040}$$
$$= 1 + 1 + 0.5 + 0.16667 + 0.04167 + 0.00833 + 0.00139 + 0.000198$$
$$= 2.718$$

One of the most important exponential functions is $y = e^x$. A graph of $y = e^x$ is shown at the right.

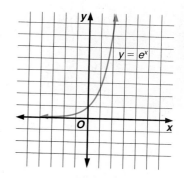

You can use a calculator to find values of $y = e^x$.

Example 1

Find the value of $e^{4.243}$ to the nearest ten thousandth.

4.243 [2nd] [LN] 69.616388   *The meaning of the term ln x will be discussed in the next lesson.*

Therefore, $e^{4.243} \approx 69.6164$.

You can check the solution by evaluating $2.718^{4.243}$ on your calculator.

Now we can use the formula for continuously compounded interest to find the amount in Byron's bank account after four years.

$$A = Pe^{rt}$$
$$= 700e^{(0.044 \cdot 4)} \qquad P = 700, r = 0.044, \text{ and } t = 4$$
$$= 700e^{0.176}$$
$$= 834.70664 \qquad \text{Use a calculator to evaluate.}$$

If Byron chooses the account with 4.4% APR with interest compounded continuously, the balance in his account will be $834.71 after four years.

There are many applications for the number e. Newton's Law of Cooling expresses the relationship between the temperature (in °F)of a cooling object, y, and the time elapsed since cooling began, t in minutes. This relationship is given by the following equation.

$$y = ae^{-kt} + c$$

In this equation, c represents the temperature of the medium surrounding the cooling object, and a and k represent constants related to the cooling object itself.

Example 2

APPLICATION

Physics

Janet heated vegetable soup to 210°F in the microwave at school. If the room temperature is 70°, what will the temperature of the soup be after 10 minutes? Assume that $a = 140$ and $k = 0.01$.

$y = ae^{-kt} + c$

$\quad = 140e^{-0.01(10)} + 70 \qquad$ *a = 140, k = 0.01, t = 10, and c = 70*

$\quad = 140e^{-0.1} + 70$

$\quad \approx 196.67724 \qquad\qquad$ *Use a calculator to evaluate.*

The soup will be about 197°F after 10 minutes.

Another application involving e has to do with the amount of knowledge that a person retains after a certain amount of time has passed.

Example 3

APPLICATION

Psychology

The *Ebbinghaus Model* for human memory gives the percent, p, of acquired knowledge that a person retains after an amount of time. The formula is $p = (100 - a)e^{-bt} + a$, where t is the time in weeks, and a and b vary from one person to another. If $a = 18$ and $b = 0.6$ for a certain student, how much information will the student retain two weeks after learning a new topic?

$p = (100 - a)e^{-bt} + a$

$\quad = (100 - 18)e^{-(0.6 \cdot 2)} + 18 \qquad$ *a = 18, b = 0.6, and t = 2*

$\quad = 82e^{-1.2} + 18$

$\quad \approx 42.7 \qquad\qquad\qquad$ *Use a calculator to evaluate.*

The student will retain about 43% of the information after two weeks.

CHECKING FOR UNDERSTANDING

Communicating Mathematics

Read and study the lesson to answer each question.

1. **Explain** how interest compounded monthly and interest compounded continuously are different.

2. The function $y = e^x$ is an exponential function. How do you expect that the graphs of $y = e^x$ and $y = e^{-x}$ would be related?

3. If the interest rates are the same, would you choose a savings account that compounded interest monthly or continuously? Why?

Use a calculator to evaluate each expression to the nearest tenth.

4. $e^{2.1}$

5. $e^{3.3}$

6. $e^{0.4}$

7. $\sqrt{e}$

8. $e^{-5.1} - 1$

9. $(2e)^{2.8}$

10. Use a graphing calculator to graph the equation $y = e^{2x}$. Over which intervals is the function increasing or decreasing?

11. Banking If interest is compounded semiannually, find the future value of $1200 invested for $4\frac{1}{2}$ years at 7.5% per year.

EXERCISES

Use a calculator to evaluate each expression to the nearest ten thousandth.

12. $e^{1.6}$

13. $e^{4.3}$

14. $\sqrt[3]{e}$

15. $2\sqrt[4]{e^3}$

16. $4\sqrt[3]{e^2}$

17. e^0

Use a graphing calculator to graph each equation. Then determine the interval in which each function is increasing or decreasing.

18. $y = xe^x$

19. $y = 2e^x$

20. $y = \left(\frac{1}{4}\right)e^x$

21. $y = 5e^{x-3}$

22. $y = -0.1e^x$

23. $y = e^{x^2}$

24. Statistics The graph of $y = e^{-x^2}$ approximates a *normal curve*. The normal curve is a bell-shaped curve that is often used to represent the distribution of a particular trait in a normal population. Graph the given function on a graphing calculator and sketch the curve on paper.

Given the original principal, the annual interest rate, the amount of time for each investment, and the type of compounded interest, find the amount at the end of the investment.

25. $P = \$1000, r = 10\%, t = 4$ years, monthly

26. $P = \$3200, r = 6\%, t = 5$ years 6 months, quarterly

27. $P = \$750, r = 5.5\%, t = 3$ years 2 months, continuously

28. $P = \$45,000, r = 7.2\%,$ $t = 30$ years, daily

29. Engineering A flexible cable suspended between two points, such as a suspension bridge cable, forms a catenary curve. Graph the catenary $y = \dfrac{e^x + e^{-x}}{2}$ on a graphing calculator and sketch the curve on paper.

30. Chemistry A beaker of water has been heated to 200°F in a room that is 72°F. If $k = 0.01$ and $a = 128$, use Newton's Law of Cooling to find the temperature of the water after 3 hours.

31. **Finance** The interest rate offered on a passbook savings account is 3.78% APR compounded continuously. Another account compounds interest monthly. What should the interest rate of the second account be so that the interest earned over one year is the same as the interest earned for the first account?

32. **Forestry** The yield, y, in millions of cubic feet of trees per acre for a forest stand that is t years old is given by $y = 6.7e^{\frac{-48.1}{t}}$.
    a. Find the yield after 15 years.
    b. Find the yield after 50 years.
    c. Graph the yield per year on a graphing calculator for 150 years. Does the yield ever decrease? If so, when?

33. **Political Science** One model used by political scientists to predict the number of legislators in the U.S. House of Representatives that will serve continuously is given by the exponential function $y = 434e^{-\frac{2t}{25}}$, where y is the number of legislators and t is the number of years since 1965.
    a. According to this model, how many of the legislators from 1965 were still in office in 1980?
    b. Predict the number of legislators from 1965 still in office in 1995.
    c. Graph the number of legislators from 1965 in office in the years 1965 to 2060 on a graphing calculator. When does the model predict that all of the members of the 1965 House will be out of office? Does this seem realistic? *Hint: Take the year that the predicted number of legislators becomes less that 0.5 to be the year that the last member of the 1965 House leaves office.*

34. **Medicine** Physicians measure the cardiac output, the amount of blood that a heart can pump in an amount of time, in order to determine how healthy the heart is. In the dye-dilution procedure for determining cardiac output, the amount of dye, D, in milligrams in the heart t minutes after injection is given by $D = de^{\frac{-rt}{V}}$, where d is the number of milligrams of dye injected, r is a constant representing the outflow of blood in liters per minute, and V is the volume of the heart in liters. Find the amount of dye in the heart after 30 seconds if $V = 0.4$ L, $r = 1.3$ liters per minute, and $d = 2.2$ mg.

35. **Biology** It has been observed that the rate of growth of a population of organisms will increase until the population is half of its maximum and then the rate will decrease. If M is the maximum population and b and c are constants determined by the type of organism, the population, n, after t years is given by $n = \dfrac{M}{1 + be^{-ct}}$. A certain organism yields the values of $M = 200$, $b = 20$, and $c = 0.35$. What is the shape of this curve?

36. **Banking** If your bank account earns interest that is compounded more than one time per year, the effective annual yield of the account, E, is the interest rate that would give the same amount of interest earnings if the interest were compounded once per year. If P dollars are invested, the value of the investment is $A = P(1 + E)$.

  a. Find a formula for the effective annual yield for an account with interest compounded n times per year.

  b. Find a formula for the effective annual yield for an account with interest compounded continuously.

  c. What is the effective annual yield of an account bearing 5.6% interest compounded monthly? What is the effective annual yield of an account with the same rate of interest compounded continuously?

Critical Thinking

37. The expression $\left(1 + \dfrac{1}{n}\right)^n$ can be used to approximate e. Greater values of n give closer approximations for e. Find the value of the expression for $n = 5, 100, 500$, and 10,000.

Mixed Review

38. Write the standard form of the equation of the line that is parallel to $y = 4x - 8$ and passes through $(-2, 1)$. **(Lesson 1-6)**

39. Transportation A boat trailer has wheels with a diameter of 14 inches. If the trailer is being pulled by a car going 45 miles per hour, what is the angular velocity of the wheels in revolutions per second? **(Lesson 5-2)**

40. Find the distance from the line $y = 9x - 3$ to the point at $(-3, 2)$. **(Lesson 7-7)**

41. Write the rectangular equation $y = 15$ in polar form. **(Lesson 9-4)**

42. Write the standard form of the equation $x^2 - 4x + y^2 - 6y - 3 = 0$. Then, graph the equation. **(Lesson 10-1)**

43. Finance What is the monthly principal and interest payment on a home mortgage of $90,000 for 30 years at 11.5%? **(Lesson 11-2)**

44. College Entrance Exam If $c + d = 12$ and $c^2 - d^2 = 48$, then what is the value of $(c - d)$?

CASE STUDY FOLLOW-UP

Refer to Case Study 4: *The U.S. Economy*, on pages 966–969.

Suppose the friendly and very wealthy foreign nation of Biparmala agreed to help the United States pay off its national debt. Also suppose that the terms of the agreement were as follows: first, the U.S. has to cut its annual budget deficit to zero immediately, preventing any further rise in the debt. Then, individual Americans would be asked to invest their shares of the national debt in Biparmala. In return, Biparmala would pay 20% interest on these investments, compounded continuously, for ten years.

1. About how much would each investor's investment be worth in ten years?

2. About how many Americans would need to become investors in order for the plan to succeed?

3. Research Read about the growth of the national debt and the many plans to reduce it. Choose the plan that seems most feasible to you and write about it in a 1-page paper.

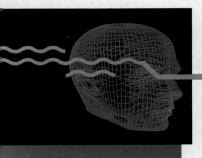

Technology

Compound Interest

You can use a spreadsheet to calculate how funds increase due to compound interest. Use the formula $A = P\left(1 + \frac{r}{n}\right)^{nt}$, where A is the amount in the account after t years, with an initial investment of P dollars at an annual interest rate r, if the interest is compounded n times per year. The spreadsheet below is set up to find the value of an investment after 1 to 20 years. The principal is entered in cell B1, the number of times that the interest is compounded per year is entered in cell B2, and the rate is entered in B3. Cells B5 to B24 contain a variation of the compound interest formula that will compute the balance in the account after successive years.

COMPOUND INTEREST		
	A	B
1	PRINCIPAL	
2	TIMES COMPOUNDED	
3	RATE	
4	YEAR	BALANCE
5	1	B1*(1 + B3/ B2) ∧ (B2 * A5)
6	2	B1*(1 + B3/ B2) ∧ (B2 * A6)
24	20	B1*(1 + B3/B2) ∧ (B2 * A24)

```
= = = = = = = = = = = = = = = = =
        CGMPOUND INTEREST
= = = = = = = = = = = = = = = = =
1    PRINCIPAL           9500
2    TIMES COMPOUNDED    12
3    RATE                0.055
4    YEAR                BALANCE
5    1                   10035.87
6    2                   10601.98
```

At the left is a partial printout from the spreadsheet. It shows the amount accumulated in an account that bears 5.5% interest compounded monthly if the principal is $9500.

EXERCISES

1. Describe how you could modify the spreadsheet program to print 50 years of account balances.

2. Describe how you could modify the spreadsheet program to print comparisons of the amount accumulated in accounts with two different interest rates.

11-4A Graphing Calculators: Graphing Logarithmic Functions

You can use a graphing calculator to graph logarithmic functions. A logarithmic function is a function of the form $x = a^y$, where $a > 0$ and $a \neq 1$. This function can also be written as $y = \log_a x$, and it is the inverse of an exponential function $y = a^x$.

Example 1

Graph $y = \log_4 x$. State the domain and range of the function.

To graph logarithmic functions with bases other than 10, you must first change the equation by using the change of base formula, $\log_a x = \dfrac{\log x}{\log a}$. In this example, $\log_4 x = \dfrac{\log x}{\log 4}$. Use the viewing window $[-1, 5]$ by $[-3, 3]$.

Enter: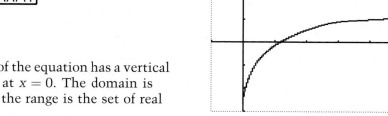

The graph of the equation has a vertical asymptote at $x = 0$. The domain is $x > 0$, and the range is the set of real numbers.

Example 2

Graph $y = \log (x + 6)$. State the domain and range of the function.

Use the viewing window $[-10, 10]$ by $[-2, 2]$.

Enter: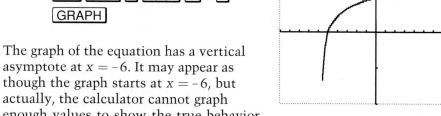

The graph of the equation has a vertical asymptote at $x = -6$. It may appear as though the graph starts at $x = -6$, but actually, the calculator cannot graph enough values to show the true behavior of the graph. In this case, it is important to know how to interpret the graph that the calculator draws. The domain is $x > -6$, and the range is the set of real numbers.

EXERCISES

Graph each equation so that a complete graph is shown. Then sketch the graph on a piece of paper and state the domain and range of each function.

1. $y = \log_6 x$

2. $y = \log(x + 2)$

3. $y = \log_5(-x)$

4. $y = \log_x 5$

5. $y = \log_3(x + 4)$

6. $y = \log_6(3 - x)$

7. $y = \log_{10}(x^2)$

8. $y = \log_{-4} x$

11-4 Logarithmic Functions

Objectives

After studying this lesson, you should be able to:
- evaluate expressions involving logarithms,
- solve equations involving logarithms, and
- graph logarithmic functions and inequalities.

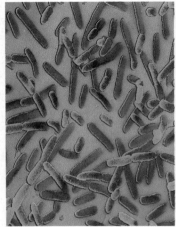

Application

When bacteria cells are placed in a medium that provides all of the nutrients needed for growth, the population increases according to a pattern called the bacterial growth curve. The growth occurs in four stages called the lag phase, the log phase, the stationary phase, and the decline phase. During the log phase, bacteria multiply almost exponentially.

Bacteria multiply by mitosis; that is, one cell divides into two new cells. This division process can take minutes or hours depending on the type of bacteria and the conditions of the medium. *Escherichia coli*, or *E. coli*, is one of the fastest growing bacteria. It can reproduce itself by mitosis in 15 minutes.

Suppose a petri dish contains 100 *E. coli*. The formula $b = 100(2^t)$ can be used to find the number of bacteria, b, in the dish after t fifteen-minute periods. If you wanted to know when there would be 12,800 bacteria in the dish, you would have to use an inverse function to find the value of t when $b = 12,800$. *This problem will be solved in Example 3.*

The exponential function $y = a^x$ increases for $a > 1$ and decreases for $0 < a < 1$. Since the function is one-to-one, its inverse is also a function. The inverse of $y = a^x$ is $x = a^y$. In the function $x = a^y$, y is called the **logarithm** of x. It is usually written as $y = \log_a x$ and is read *y equals the log, base a, of x*. The function $y = \log_a x$ is called a **logarithmic function**. Notice that since $y = a^x$ and $x = a^y$ are inverses, their graphs are reflections of each other over the line $y = x$.

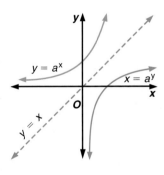

Definition of *Logarithmic* *Function*	The logarithmic function $y = \log_a x$, where $a > 0$ and $a \neq 1$, is the inverse of the exponential function $y = a^x$. Therefore, $y = \log_a x$ if and only if $x = a^y$.

Each logarithmic equation corresponds to an exponential equation. For example, the equation $y = \log_2 16$ can be written as $2^y = 16$. Since $2^4 = 16$, $\log_2 16 = 4$.

Example 1

> **Write $\log_{10} 0.01 = -2$ in exponential form.**
>
> The base is 10 and the exponent is -2.
> $0.01 = 10^{-2}$

Example 2

> **Evaluate the expression $\log_8 64$.**
>
> Let $x = \log_8 64$.
>
> $x = \log_8 64$
> $8^x = 64$ *Definition of logarithm*
> $8^x = 8^2$
> $x = 2$ *Property of equality for exponential functions*

Logarithms can help us solve the application problem from the beginning of the lesson.

Example 3

APPLICATION

Biology

> **If the population of 100 bacteria doubles every fifteen minutes, how long will it take for the population to reach 12,800?**
>
> $b = 100(2^t)$
> $12,800 = 100(2^t)$ *$b = 12,800$*
> $128 = 2^t$ *Divide each side by 100.*
> $\log_2 128 = t$ *Write the problem in logarithmic form.*
> $\log_2 2^7 = t$ *$128 = 2^7$*
> $7 = t$ *Definition of logarithm*
>
> The population will be 12,800 in 7 time periods, that is, in 105 minutes or 1 hour 45 minutes.

Since the logarithmic function and the exponential function are inverses of each other, both of their compositions yield the identity function.

$y = a^x$ $y = \log_a x$
$ = a^{\log_a x}$ *Let $x = \log_a x$.* $ = \log_a a^x$ *Let $x = a^x$.*
$ = x$ $ = x$

Since logarithms are exponents, the properties of logarithms can be derived from the properties of exponents.

Properties of Logarithms

Suppose m and n are positive numbers, b is a positive number other than 1, and p is any real number. Then the following properties hold.

Product property: $\log_b mn = \log_b m + \log_b n$

Quotient property: $\log_b \dfrac{m}{n} = \log_b m - \log_b n$

Power property: $\log_b m^p = p \cdot \log_b m$

Property of equality: If $\log_b m = \log_b n$, then $m = n$.

Example 4

Use the properties of exponents to show that the quotient property of logarithms is valid.

Let $\log_b m = x$ and let $\log_b n = y$.

Then $b^x = m$ and $b^y = n$.   *Write each equation in exponential form.*

$$\frac{m}{n} = \frac{b^x}{b^y}$$ *Find the quotient of m and n.*

$$\frac{m}{n} = b^{x-y}$$ $\dfrac{a^m}{a^n} = a^{m-n}$

$$\log_b \frac{m}{n} = \log_b b^{x-y}$$ *Take the logarithm of each side.*

$$\log_b \frac{m}{n} = x - y$$ $\log_a a^x = x$

$$\log_b \frac{m}{n} = \log_b m - \log_b n$$ *Substitute for x and y.*

Therefore, $\log_b \dfrac{m}{n} = \log_b m - \log_b n$.

Equations can be written involving logarithms. Use the properties of logarithms and the definition of a logarithm to solve these equations.

Example 5

Solve each equation.

a. $\log_b \sqrt{3} = \dfrac{1}{4}$

a. $\log_b \sqrt{3} = \dfrac{1}{4}$

$$b^{\frac{1}{4}} = \sqrt{3}$$ *Definition of logarithm*

$$\sqrt[4]{b} = \sqrt{3}$$ *Definition of $b^{\frac{1}{n}}$*

$$b = (\sqrt{3})^4$$ *Raise each side to the fourth power.*

$$b = 9$$

b. $\log_8 (5x + 3) = \log_8 (45 - 2x)$

b. $\log_8 (5x + 3) = \log_8 (45 - 2x)$

$$5x + 3 = 45 - 2x$$ *Property of Equality*

$$7x = 42$$

$$x = 6$$

c. $\log_2 (4x + 10) - \log_2 (x + 1) = 3$

c. $\log_2 (4x + 10) - \log_2 (x + 1) = 3$

$$\log_2 \frac{4x + 10}{x + 1} = 3$$ *Quotient property of logarithms*

$$\frac{4x + 10}{x + 1} = 2^3$$ *Definition of logarithm*

$$4x + 10 = 8(x + 1)$$

$$4x = 2$$

$$x = \frac{1}{2}$$

You can graph a logarithmic function by rewriting the logarithmic function as an exponential function and constructing a table of values.

Example 6

Graph $y = \log_2 (x - 2)$.

The equation $y = \log_2(x - 2)$ can be written as $2^y = x - 2$.

x	x − 2	y
2.125	0.125	−3
2.25	0.25	−2
2.5	0.5	−1
3	1	0
4	2	1
6	4	2
10	8	3

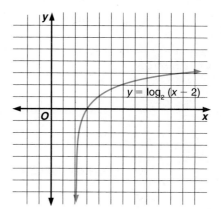

$y = \log_2 (x - 2)$

EXPLORATION: Graphing Calculator

In the Graphing Calculator Exploration in Lesson 11-2, an equation representing population growth was given as $y = 10^{0.00389x+2}$. Since the logarithmic function is the inverse of the exponential function, population growth can also be modeled with the logarithmic function $x = \dfrac{\log y - 2}{0.00389}$.

1. Use the settings below to set up a viewing window.

   XMIN=0   YMIN=1800
   XMAX=12000000000   YMAX=2180
   XSCL=1000000000   YSCL=25

2. Since graphing calculators graph y as a function of x, the equation will need to be rewritten as follows: $y = \dfrac{\log x - 2}{0.00389}$. Press $\boxed{Y =}$ and enter the equation.

3. Use the TRACE function to determine the population growth from 1800 to the year 2000. According to the graph, when can we expect the world population to be double the population in 1960?

You can graph logarithmic inequalities using the same graphing techniques, as shown in Example 6. Choose a test point to determine which region to shade.

CHECKING FOR UNDERSTANDING

Communicating Mathematics

Read and study the lesson to answer each question.

1. **Explain** how logarithms are related to exponents.

2. Why do you think that the second phase of growth for a colony of bacteria is called the log phase?

3. **Show** that the power property of logarithms is valid.

Guided Practice

Write each equation in logarithmic form.

4. $4^3 = 64$

5. $6^{-2} = \dfrac{1}{36}$

6. $49^{\frac{1}{2}} = 7$

Write each equation in exponential form.

7. $\log_{27} 3 = \dfrac{1}{3}$

8. $\log_{16} 4 = \dfrac{1}{2}$

9. $\log_9 27 = \dfrac{3}{2}$

Evaluate each expression.

10. $\log_2 32$

11. $\log_{10} 1000$

12. $\log_7 \dfrac{1}{343}$

Solve each equation.

13. $\log_6 x + \log_6 9 = \log_6 54$

14. $\log_7 n = \dfrac{2}{3} \log_7 8$

15. $4 \log_2 x + \log_2 5 = \log_2 405$

16. $\log_8 48 - \log_8 w = \log_8 4$

17. Graph $y = \log_3 (x + 1)$.

18. Graph $y < \log_2 x$.

EXERCISES

Practice

Write each equation in logarithmic form.

19. $2^4 = 16$

20. $5^{-2} = \dfrac{1}{25}$

21. $3^{-3} = \dfrac{1}{27}$

22. $10^6 = 1{,}000{,}000$

23. $8^{-\frac{2}{3}} = \dfrac{1}{4}$

24. $4^0 = 1$

Write each equation in exponential form.

25. $\log_2 8 = 3$

26. $\log_5 125 = 3$

27. $\log_{10} 10{,}000 = 4$

28. $\log_7 \dfrac{1}{2401} = -4$

29. $\log_8 2 = \dfrac{1}{3}$

30. $\log_{\sqrt{6}} 36 = 4$

Evaluate each expression.

31. $\log_9 9^6$

32. $\log_{10} 0.01$

33. $12^{\log_{12} 5}$

34. $\log_2 \dfrac{1}{16}$

35. $\log_6 6^5$

36. $\log_8 16$

37. $\log_a a^{10}$

38. $\log_{11} 11$

39. $10^{4 \log_{10} 2}$

Solve each equation.

40. $\log_x 49 = 2$

41. $\log_5 0.04 = x$

42. $\log_6 (4x + 4) = \log_6 64$

43. $\log_3 (3x) = \log_3 36$

44. $\log_x 16 = -4$

45. $\log_6 216 = x$

46. $\log_{10} \sqrt[3]{10} = x$

47. $\log_2 4 + \log_2 6 = \log_2 x$

48. $2 \log_6 4 - \dfrac{1}{4} \log_6 16 = \log_6 x$

49. $\log_3 12 - \log_3 x = \log_3 3$

50. $3 \log_7 4 + 4 \log_7 3 = \log_7 x$

51. $\log_4 (x - 3) + \log_4 (x + 3) = 2$

52. $\log_6 x = \dfrac{1}{2} \log_6 9 + \dfrac{1}{3} \log_6 27$

53. $\log_9 5x = \log_9 6 + \log_9 (x - 2)$

54. $\log_{10} x + \log_{10} x + \log_{10} x = \log_{10} 8$

Graph each equation or inequality.

55. $y = \log_4 x$

56. $y = \log_{10} x$

57. $y = \log_{\frac{1}{2}} x$

58. $y = \log_2 (x - 1)$

59. $y \le \log_6 x$

60. $y > \log_{10} (x + 1)$

Critical Thinking

61. Explain why 1 is excluded from the possible values of the base of the logarithmic function.

Applications and Problem Solving

62. **Biology** The generation time for bacteria is the time that it takes for the population to double. The generation time, G, can be found using experimental data and the formula $G = \dfrac{t}{3.3 \log_b f}$, where t is the time period, b is the number of bacteria at the beginning of the experiment, and f is the number of bacteria at the end of the experiment. The generation time for mycobacterium tuberculosis is 16 hours. How long will it take for two of these bacteria to multiply into 256 bacteria?

63. **Biology** Atlantic salmon swim up to 2000 miles upstream to spawn each year. Scientists who study salmon have found that the oxygen consumption of a yearling salmon, O, is given by the function $O = 100(3^{\frac{3s}{5}})$, where s is the speed that the fish is traveling in feet per second.

a. Find the oxygen consumption of a fish that is not moving.

b. How fast is a fish swimming when its oxygen consumption is 2700 units?

Mixed Review

64. Write the slope-intercept form of the equation of the line through the points with coordinates $(8, 5)$ and $(-6, 0)$. **(Lesson 1-5)**

65. **Business** Pristine Pipes Inc. produces plastic pipe for use in newly-built homes. Two of the basic types of pipe have different diameters, wall thicknesses, and strengths. The strength of a pipe is increased by mixing a special additive into the plastic before it is molded. The table below shows the resources needed to produce 100 feet of each type of pipe and the amount of the resource available each week.

Resource	Pipe A	Pipe B	Resource Availability
Extrusion Dept.	4 hours	6 hours	48 hours
Packaging Dept.	2 hours	2 hours	18 hours
Strengthening additive	2 pounds	1 pound	16 pounds

If the profit on 100 feet of type A pipe is \$34 and on type B pipe is \$40, how much of each should be produced? **(Lesson 2-6)**

66. Find the inverse of $y = 7 - 8x^2$. **(Lesson 3-3)**

67. Find the equation of the line tangent to the graph of $y = x^2 - 6x + 4$ at $(-1, 11)$. Write the equation in slope-intercept form. **(Lesson 3-6)**

68. Solve $\dfrac{-4}{x-1} = \dfrac{7}{2-x} + \dfrac{3}{x+1}$. **(Lesson 4-6)**

69. Find $\cos 90°$ without using a calculator. **(Lesson 5-4)**

70. Solve $\cos 2x + \sin x = 1$ for principal values of x. **(Lesson 7-5)**

71. Find the inner product of $\vec{u}$ and $\vec{v}$ if $\vec{u} = (9, 5, 3)$ and $\vec{v} = (-3, 2, 5)$. Are the vectors perpendicular? **(Lesson 8-4)**

72. Find $(8 - 2i)(2 + 6i)$. **(Lesson 9-5)**

73. Write the standard form of the equation $x^2 - 4x - 4y^2 = 0$. Then identify the conic section. **(Lesson 10-5)**

74. **Banking** Find the amount accumulated if $600 is invested at 6% for 15 years and interest is compounded continuously. **(Lesson 11-3)**

75. **College Entrance Exam** Choose the best answer.
    Divide $6\sqrt{45}$ by $3\sqrt{5}$.
    (A) 9 **(B)** 4 **(C)** 15 **(D)** 6 **(E)** 54

MID-CHAPTER REVIEW

Evaluate. (Lesson 11-1)

1. $81^{\frac{3}{4}}$

2. $\dfrac{25^{\frac{3}{4}}}{25^{\frac{1}{4}}}$

3. $\sqrt[3]{\sqrt{27}}$

4. Graph the equation $y = 0.1^x$. **(Lesson 11-2)**

5. **Financial Planning** Teresa Slane plans to contribute $2500 annually to a Keogh retirement account with an APR of 9%. Assuming the APR remains the same, what will the value of the account be in 25 years? **(Lesson 11-2)**

Evaluate each expression to the nearest ten thousandth. (Lesson 11-3)

6. $e^{3.8}$

7. $5\sqrt{3e}$

8. $e^{\frac{2}{5}}$

9. **Meteorology** The atmospheric pressure, p, varies with the altitude, a, above Earth. For altitudes up to about 10 km, the pressure in millimeters of mercury is given by $p = 760e^{-0.125a}$ when a is measured in kilometers. Find the atmospheric pressure 4.3 km above Earth. **(Lesson 11-3)**

Solve each equation. (Lesson 11-4)

10. $\log_x 512 = 3$

11. $\log_{\sqrt{5}} 5 = x$

12. $\log_5 (x^2 - 30) = \log_5 6$

11-5 Common Logarithms

Objectives

After studying this lesson, you should be able to:
- find common logarithms and antilogarithms of numbers, and
- use common logarithms to compute powers and roots.

Application

As people gain experience in performing a task, the time required for them to perform the task is reduced until they reach a standard minimum time. For businesses, this means that as a worker gains experience at his or her job, the amount of money spent on labor per item is reduced. This effect can be represented by a learning curve.

FYI...

The learning curve allows manufacturers to position themselves well in a competitive market. Companies with more experienced workers spend less on labor, so they can keep their prices low.

The learning curve was first developed in the aircraft industry before World War II. A survey of major airplane manufacturers showed that while the time required to build different types of airframes was very different, the rates at which workers increased their productivity was strikingly similar. They found that labor time for the eighth unit was 80% of the labor time for the fourth unit, and time for the twelfth unit was 80% of the

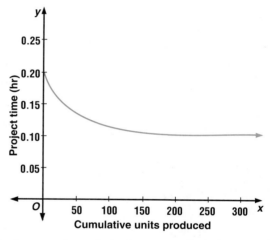

Cumulative units produced

labor for the sixth unit. Each time the number of airplanes produced was doubled, the time required was reduced by 20%. So the rate of learning for aircraft construction was 80%. While the rate of learning may be different for another product, the pattern of learning is the same.

The formula for a learning curve is $u_n = kn^b$, where u_n is the number of hours of labor for the nth product, k is the number of hours of labor for the first product, n is the number of products made, and $b = \dfrac{\log r}{\log 2}$, where r is the learning rate. Notice that no base is indicated for the logarithms in the expression for b. When no base is indicated, the base is assumed to be 10. So, $\log r$ means $\log_{10} r$. Logarithms with base 10 are called **common logarithms**.

You can find the common logarithms of integral powers of 10 easily.

$\log 1000 = 3$	since	$1000 = 10^3$
$\log 100 = 2$	since	$100 = 10^2$
$\log 10 = 1$	since	$10 = 10^1$
$\log 1 = 0$	since	$1 = 10^0$
$\log 0.1 = -1$	since	$0.1 = 10^{-1}$
$\log 0.01 = -2$	since	$0.01 = 10^{-2}$

The common logarithms of numbers that differ by integral powers of ten are closely related.

Example 1

Given that log 4 ≈ 0.6021, evaluate each logarithm.
a. log 40,000 **b. log 0.004**

a. $\log 40{,}000 = \log (10{,}000 \cdot 4)$

$= \log 10{,}000 + \log 4$

$\approx 4 + 0.6021$

≈ 4.6021

b. $\log 0.004 = \log \left(\dfrac{4}{1000} \right)$

$= \log 4 - \log 1000$

$\approx 0.6021 - 3$

≈ -2.3979

Look closely at the results of Example 1. Each of the common logarithms is made up of two parts, the **characteristic** and the **mantissa**. The mantissa is the logarithm of a number between 1 and 10. Thus, mantissas are greater than 0 and less than 1. In Example 1, the mantissa is log 4 or 0.6021. The characteristic is the exponent of ten that is used to write the number in scientific notation. In Example 1, the characteristic of 10,000 is 4, and the characteristic of $\dfrac{1}{1000}$ is –3. Traditionally, a logarithm is expressed as the indicated sum of the mantissa and the characteristic.

You can use a scientific calculator to solve equations containing common logarithms.

Example 2

APPLICATION

Psychology

Refer to the application presented at the beginning of the lesson for the equation of a learning curve.

a. The Carter Truck Company produces diesel engines for trucks. Find the number of hours required to build the forty-fifth engine, assuming that the learning rate is 75% and it took 48,000 hours to build the first one.
b. The company pays its workers an average of $11.50 per hour. How much less will the labor on the forty-fifth engine cost than the labor for the first engine?

a. $u_n = kn^b$

$u_{45} = 48{,}000(45)^{\frac{\log 0.75}{\log 2}}$ *n = 45, k = 45,000, and* $b = \dfrac{\log 0.75}{\log 2}$

Evaluate with a calculator.

48000 $\boxed{\times}$ 45 $\boxed{y^x}$ $\boxed{(}$ 0.75 $\boxed{\text{LOG}}$ $\boxed{\div}$ 2 $\boxed{\text{LOG}}$ $\boxed{)}$ $\boxed{=}$ *9887.7175*

The forty-fifth engine will take about 9888 hours to complete.

b. The first engine took $48{,}000 - 9888$ or 38,112 hours longer to make. This would save the company an additional $11.50 \times 38{,}112$ or $438,288.

Sometimes the logarithm of x is known to have a value of a, but x is not known. Then x is called the **antilogarithm** of a, written *antilog a*. If $\log x = a$, then $x =$ antilog a. *Remember that the inverse of a logarithmic function is an exponential function.*

Example 3

APPLICATION

Chemistry

The pH of a solution is a measure of its acidity. A low pH indicates an acidic solution, and a high pH indicates a basic solution. Neutral water has a pH of 7. The pH of a solution is related to the concentration of hydrogen ions it contains by the formula $pH = \log \dfrac{1}{H^+}$, where H^+ is the number of gram atoms of hydrogen ions per liter. If the pH of tomato juice is 4.1, what is the concentration of hydrogen ions?

$$pH = \log \frac{1}{H^+}$$

$$4.1 = \log \frac{1}{H^+} \qquad pH = 4.1$$

antilog $4.1 = \dfrac{1}{H^+}$ *Take the antilogarithm of each side.*

$$H^+ = \frac{1}{\text{antilog } 4.1}$$

Evaluate with a calculator.

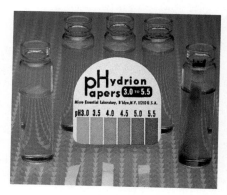

1 $\boxed{\div}$ 4.1 $\boxed{\text{2nd}}$ $\boxed{\text{LOG}}$ $\boxed{=}$ $\mathit{0.0000794}$

The concentration of hydrogen ions is about 7.94×10^{-5} gram atoms per liter.

Logarithms can be used to solve problems involving powers and roots. Before scientific calculators were invented, this was the most efficient method available.

Example 4

Use logarithms to evaluate $\dfrac{24.8\sqrt{451}}{(39.6)^3}$.

Let $A = \dfrac{24.8\sqrt{451}}{(39.6)^3}$. Then $\log A = \log\left[\dfrac{24.8\sqrt{451}}{(39.6)^3}\right]$.

$\log A = \log\left[\dfrac{24.8\sqrt{451}}{(39.6)^3}\right]$

$\qquad = \log 24.8 + \dfrac{1}{2}\log 451 - 3\log 39.6$ *What properties are used?*

$\qquad \approx 1.3945 + \dfrac{1}{2}(2.6542) - 3(1.5977)$

$\qquad \approx -2.0715$

$A \approx$ antilog -2.0715

$\qquad \approx 0.00848$

The value of the expression is approximately 0.00848.
Check the result with your calculator.

CHECKING FOR UNDERSTANDING

Read and study the lesson to answer each question.

1. **Explain** the relationship between the characteristic of a common logarithm and the exponent of 10 for a number expressed in scientific notation.

2. What kind of function are you using when you take the antilogarithm of a number?

3. If the characteristic for $\log x$ is 3, then __?__ $\leq x <$ __?__ .

Guided Practice

Use a calculator to find the common logarithm of each number to the nearest ten thousandth.

4. 98.2

5. 424

6. 2.43

Use a calculator to find the antilogarithm of each number to the nearest hundredth.

7. 2.5499

8. 0.4398

9. –1.8989

Between which two integral powers of ten is each number?

10. antilog 2.3456

11. antilog 4.8740

12. antilog –2.5232

Evaluate each expression by using logarithms. Round to the nearest tenth. Check your work with a calculator.

13. $784 \times 47.9 \times 0.0748$

14. $\dfrac{6.39 \times 1.54}{3.78}$

15. $(1.69)^4 \times 221$

16. $7.32 \div \sqrt[4]{0.0743}$

EXERCISES

Practice

Use a calculator to find the common logarithm of each number to the nearest ten thousandth.

17. 894.3

18. 0.7849

19. 0.0054

20. 0.871

21. 22,892

22. 4.4×10^7

Use a calculator to find the antilogarithm of each number to the nearest hundredth.

23. 0.2586

24. –0.2586

25. –1.9725

26. $2.2675 - 3$

27. 4.9243

28. 0.01

Evaluate each expression by using logarithms. Round to the nearest tenth. Check your work with a calculator.

29. $754 \times 24.5 \times 0.0128$

30. $\dfrac{5.43 \times 7.12}{2.28}$

31. $642 \times (2.01)^3$

32. $\sqrt{5.81 \times 71.1}$

33. $\sqrt[3]{(3.05)(730)}$

34. $8.83 \div \sqrt[5]{0.4218}$

35. $\left(\dfrac{1}{0.7891}\right)^4$

36. $\sqrt[6]{82.9}$

37. $\sqrt{\dfrac{8.4}{0.31}}$

38. $\dfrac{37.9\sqrt{488}}{(1.28)^3}$

Critical
Thinking

Applications
and Problem
Solving

Mixed
Review

39. Find the value of x in $(3x)^{\log 3} - (5x)^{\log 5} = 0$ if $x > 0$.

40. Astronomy The parallax of a star is the difference in direction of the star as seen from two widely separated points. Astronomers use the parallax of a star to determine its distance from Earth. The apparent magnitude of a star is its brightness as observed from Earth. Astronomers use parsecs to measure interstellar space. One parsec is about 3.26 light years or 19.2 trillion miles. The absolute magnitude is the magnitude a star would have if it were 10 parsecs from Earth. The greater the magnitude of a star the fainter the star appears. For stars more than 30 parsecs, or 576 trillion miles from Earth, the formula relating the parallax, p, the absolute magnitude, M, and the apparent magnitude, m, is $M = m + 5 + 5 \log p$.

   a. The star M35 in the constellation Gemini has an apparent magnitude of 5.3 and a parallax of about 0.018. Find the absolute magnitude of this star.

   b. Stars with apparent magnitudes greater than 5 can be seen only with a telescope. If a star has an apparent magnitude of 8.6 and an absolute magnitude of 5.3, find its parallax to four decimal places.

41. Geology The intensity of an earthquake is described by a number on the Richter scale. The Richter scale number, R, of an earthquake is given by the formula $R = \log\left(\dfrac{a}{T}\right) + B$, where a is the amplitude of the vertical ground motion in microns, T is the period of the seismic wave in seconds, and B is a factor that accounts for the weakening of seismic waves.

   a. Find the intensity of an earthquake to the nearest tenth if a recording station measured the magnitude as 200 microns and the period as 1.6 seconds, and $B = 4.2$.

   b. How much more intense is an earthquake of magnitude 5 on the Richter scale than an earthquake that measures 4 on the Richter scale?

42. Find the distance between the points with coordinates $(-9, 6)$ and $(11, -4)$. Then, find the slope of the line passing through the points. **(Lesson 1-4)**

43. Find $A + B$ if $A = \begin{bmatrix} 8 & 5 & -1 \\ 0 & -6 & -3 \end{bmatrix}$ and $B = \begin{bmatrix} 0 & 3 & 0 \\ 2 & -2 & 0 \end{bmatrix}$. **(Lesson 2-2)**

44. Geometry The sides of a parallelogram are 55 cm and 71 cm long. Find the length of each diagonal if the larger angle measures $106°$. **(Lesson 5-7)**

45. Verify that $\dfrac{\sin^2 x}{\cos^4 x + \cos^2 x \sin^2 x} = \tan^2 x$ is an identity. **(Lesson 7-2)**

46. If $A = (12, -5, 18)$ and $B = (0, -11, 21)$, find an ordered triple that represents $\overrightarrow{AB}$. **(Lesson 8-3)**

47. Write the equation of the ellipse with foci at $(3, 6)$ and $(1, 6)$ that passes through $(0, 6)$. **(Lesson 10-3)**

48. Solve $\log_5(7x) = \log_5(5x + 16)$. **(Lesson 11-4)**

49. College Entrance Exam Choose the best answer.
Fisher's Farm Market sells onions by the pound. The price of 3 pounds of onions is d dollars. At this rate, how many pounds of onions could you buy for 80 cents?

   (A) $2.4d$ **(B)** $\dfrac{2.4}{d}$ **(C)** $\dfrac{3d}{8}$ **(D)** $\dfrac{8d}{3}$ **(E)** $0.8d$

11-6A   Graphing Calculators: Exponential and Logarithmic Equations and Inequalities

You can use a graphing calculator to solve equations involving logarithms and exponents, just as you have used it to solve other types of equations. Once again, there are two methods that you can use to find the solution to an equation. The first method is to graph both equations and find the x-coordinate of the point of intersection. The second method is to transform the equation so that one side equals zero and find the x-intercept of the graph. For exponential and logarithmic equations, the second method is often easier because it is easier to see where the graph crosses the x-axis than to see the point of intersection. Often the point of intersection is near an asymptote and the gradual decrease or increase can make the point of intersection difficult to find.

Example 1

Solve $\left(\dfrac{1}{4}\right)^{3x} = 6^{x-2}$. Use the viewing window $[-2, 4]$ by $[-3, 3]$.

Let's use the second method to solve the equation. Transforming the original equation to get zero on one side yields $\left(\dfrac{1}{4}\right)^{3x} - 6^{x-2} = 0$. Transforming the equation so that either side is zero will produce the same solution.

Enter: [Y=] .25 [^] [(] 3 [X,T,θ] [)] [−] 6
[^] [(] [X,T,θ] [−] 2 [)] [GRAPH]

Using the TRACE and ZOOM functions on the calculator shows that the solution is about 0.6022.

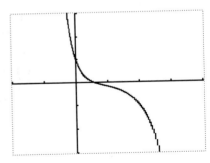

To solve logarithmic equations, use the same steps you used to solve exponential equations.

Example 2

Solve $\log(x + 3) = \log(7 - 4x)$. Use the viewing window $[-5, 5]$ by $[-3, 3]$.

Transform the equation so that $\log(x + 3) - \log(7 - 4x) = 0$. Graph this equation and look for the x-intercepts.

Enter: [Y=] [LOG] [(] [X,T,θ] [+] 3 [)] [−]
[LOG] [(] 7 [−] 4 [X,T,θ] [)] [GRAPH]

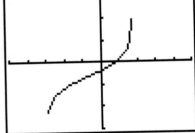

Using the TRACE and ZOOM functions on the calculator shows that the solution is 0.8.

You can also use a TI-82 graphing calculator to solve inequalities involving exponents and logarithms. The "Shade(" command can be used to graph inequalities and indicate areas that are solutions. The "Shade(" command can be found under the DRAW menu of the calculator.

Example 3

Solve the inequality $7^{x+4} \leq 0.1^{2x+8}$. Use the viewing window $[-10, 1]$ by $[-1, 3]$.

Decide which side of the inequality must be the upper bound and which side must be the lower bound. In this case, 7^{x+4} is the lower bound because it is less than or equal to 0.1^{2x+8}. Thus, 0.1^{2x+8} is the upper bound.

Next, clear all equations from the Y= list and clear the graphics screen by pressing [2nd] [DRAW] 1 [ENTER]. This must be done before graphing any other inequalities using the "Shade(" command.

Now, graph the inequality.

Enter: [2nd] [DRAW] 7

7 [^] [(] [(] [X,T,θ] [+] 4 [)]

[,] .1 [^] [(] [(] 2 [X,T,θ] [+] 8 [)] [)] [)]

[ENTER]

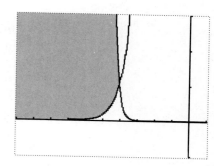

You can use the arrow keys and the ZOOM function of the calculator to find the solution. The graph shows this inequality is true for $x \leq -4.00$.

EXERCISES

Solve each equation to the nearest hundredth by using a graphing calculator.

1. $2^x = 95$
2. $4^x = x^4$
3. $5^x = 4^{x+3}$
4. $4 \log (x + 3) = 9$
5. $\log (2x + 3) = -\log(3 - x)$
6. $0.16^{4+3x} = 0.3^{8-x}$
7. $3 \log (x^3) = \log(x - 1)$
8. $12^{0.5x} = 8^{0.1x-4}$

Solve each inequality by using a TI-82 graphing calculator.

9. $4^{2x-5} \leq -3^{x-3}$
10. $0.5^{2x} - 4 \leq 0.1^{5-x}$
11. $\log (5x) > \log(x + 2)$
12. $\log_4 (x - 3) > \log_5 (2x - 7)$
13. $\log (3x - 1) \leq 2^{x-7}$
14. $5^x > \log_5 x$

11-6 Exponential and Logarithmic Equations

Objectives

After studying this lesson, you should be able to:
- solve exponential and logarithmic equations, and
- solve exponential and logarithmic inequalities.

Application

The ages of fossils were first estimated by their relative positions in layers of rock. While excavating for a canal, surveyor William Smith (1769–1839) observed that fossils occurred in layers. He reasoned that because rock forms in layers with younger rock forming on older rock, the fossils in lower layers must be older than fossils in upper layers.

Modern methods of estimating the age of a fossil or rock use the presence of naturally-occurring radioactive atoms. Radioactivity was discovered by Henri Becquerel in the late 1890s. Radioactive substances decay into more stable substances over time. The time it takes for one-half of a quantity of a radioactive element to decay is called the half-life of the element.

The element used most often in determining age is carbon-14, which is present in all living organisms. Radioactive carbon-14 decays, at a predictable rate, into the more stable carbon-12. In 1947, William Libby became the first person to use carbon-14 to date plant and animal remains. Since carbon dating is thought to be accurate only for fossils less than 50,000 years old, potassium-40 or rubidium-87 is used for fossils thought to be older than 50,000 years.

The formula $y = y_o \cdot c^{\frac{t}{T}}$ can be used to describe growth and decay in nature. The final amount is y, the initial amount is y_o, c is the constant of proportionality, t is time, and T is the time per cycle of c. When you use the growth and decay formula to solve a problem involving half-life, you will use $c = \frac{1}{2}$ and T is the half-life.

When finding the age of a fossil, you must find the value of t in the exponent. Logarithms can be used to solve equations in which variables appear as exponents. Such equations are called **exponential equations**. Solutions to equations involving exponential expressions are based on the fact that if $a^{x_1} = a^{x_2}$ for some base a and real numbers x_1 and x_2, then $x_1 = x_2$.

Example 1

Solve the equation $6.7^{x-2} = 42$.

$$6.7^{x-2} = 42$$
$$\log 6.7^{x-2} = \log 42 \qquad \textit{Take the logarithm of each side.}$$
$$(x-2)\log 6.7 = \log 42 \qquad \textit{Power property}$$
$$x = \frac{\log 42}{\log 6.7} + 2$$
$$x \approx 3.965$$

The solution is approximately 3.965.

Example 2

Stonehenge is an ancient megalithic site in southern England. Some believe it was designed to make astronomical observations, but archaeologists cannot agree on the reason that Stonehenge was constructed. Charcoal samples taken from a series of holes at Stonehenge have about 0.63 pounds of carbon-14 in a one-pound sample. Estimate the age of the charcoal pits at Stonehenge. Assume that the half-life of carbon-14 is 5730 years.

$$y = y_0 \cdot c^{\frac{t}{T}}$$

$$0.63 = 1 \cdot (0.5)^{\frac{t}{5730}} \qquad y = 0.63, y_0 = 1, c = 0.5, T = 5730$$

$$\log 0.63 = \log(0.5)^{\frac{t}{5730}} \qquad \text{Take the logarithm of each side.}$$

$$\log 0.63 = \frac{t}{5730} \log 0.5 \qquad \text{Power property}$$

$$t = \frac{5730 \log 0.63}{\log 0.5} \qquad \text{Solve for } t.$$

$$t \approx 3819.4820 \qquad \text{Evaluate with a calculator.}$$

The charcoal pits at Stonehenge are estimated to be 3819 years old.

The formula for growth and decay can also be used to describe the growth of a population. Choose the constant of proportionality as the ratio between the ending and beginning populations. For example, if you wish to know the time that it takes for a population to double, use $c = 2$ in the equation $y = y_0 \cdot c^{\frac{t}{T}}$.

Example 3

Under ideal conditions, the population of single-celled organisms in a pond will double in 5 days. How much time will it take for the original population to increase seven times?

At the ending time, there would be 7 organisms for every 1 organism at the starting time. So we can use $y = 7$ and $y_0 = 1$.

$$y = y_0 \cdot c^{\frac{t}{T}}$$

$$7 = 1 \cdot (2)^{\frac{t}{5}} \qquad c = 2 \text{ and } T = 5$$

$$\log 7 = \log(2)^{\frac{t}{5}} \qquad \text{Take the logarithm of each side.}$$

$$\log 7 = \frac{t}{5} \log 2 \qquad \text{Power property}$$

$$t = \frac{5 \log 7}{\log 2}$$

$$t \approx 14.04$$

The organisms will reach 7 times their original population in about 14 days.

Graphing is an alternate way of solving exponential or logarithmic equations. To do this, graph each side of the equation as a function and find the coordinates of the intersection of the graphs.

Example 4

Solve $5^x = 12^{3-x}$ by graphing.

Graph $y = 5^x$ and $y = 12^{3-x}$ on the same set of axes.

The graphs appear to intersect at about $(1.8, 18.1)$.

Therefore, $5^x = 12^{3-x}$ when $x \approx 1.8$.

Check the solution by graphing with a graphing calculator and using the TRACE function to find the intersection point.

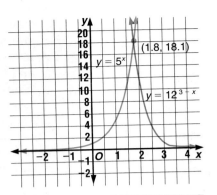

You can use the techniques for solving logarithmic or exponential equations to solve exponential and logarithmic inequalities.

Example 5

Solve $5^{x+3} > 10^{x-6}$.

$$5^{x+3} > 10^{x-6}$$
$$\log 5^{x+3} > \log 10^{x-6}$$
$$(x+3)\,\log 5 > (x-6)\,\log 10 \qquad \textit{Power property}$$
$$x \log 5 + 3 \log 5 > x - 6 \qquad\qquad \textit{log 10 = 1}$$
$$6 + 3\,\log 5 > x - x\,\log 5$$
$$6 + 3\,\log 5 > x(1 - \log 5)$$
$$\frac{6 + 3\,\log 5}{1 - \log 5} > x \qquad\qquad \textit{Solve for x.}$$
$$26.90 > x \qquad\qquad\qquad \textit{Evaluate with a calculator.}$$

Sometimes you want to find the logarithm of a number to a base other than 10. These problems can be solved with common logarithms and a calculator.

Example 6

Express $\log_8 92$ in terms of common logarithms. Then find its value.

Let $x = \log_8 92$.
$$8^x = 92 \qquad \textit{Write the equation in exponential form.}$$
$$\log 8^x = \log 92 \qquad \textit{Take the logarithm of each side.}$$
$$x \log 8 = \log 92 \qquad \textit{Power property}$$
$$x = \frac{\log 92}{\log 8} \qquad \textit{Solve for x.}$$

The value of the logarithm is $\dfrac{\log 92}{\log 8}$ or about 2.1745. Thus, $8^{2.1745} \approx 92$.

Any logarithm can be expressed in terms of a different base. However, the base is usually changed to base 10 or base e so that they may be evaluated with a calculator. *We will discuss logarithms to base e in the next lesson.*

Change of Base Formula	Suppose a, b, and n are positive numbers, and neither a nor b is 1. Then, the following equation is true. $\qquad \log_a n = \dfrac{\log_b n}{\log_b a}$

Example 7 | Find the value of $\log_3 14$ using the change of base formula.

$$\log_a n = \frac{\log_b n}{\log_b a}$$

$$\log_3 14 = \frac{\log 14}{\log 3} \qquad a = 3, b = 10, n = 14$$

$$\approx \frac{1.1461}{0.4771}$$

$$\approx 2.4022 \qquad \text{The value of } \log_3 14 \text{ is about } 2.4022.$$

CHECKING FOR UNDERSTANDING

Communicating Mathematics

Read and study the lesson to answer each question.

1. **Explain** why we can use logarithms to solve equations involving variables that are exponents.

2. When will the constant of proportionality in the growth and decay formula be positive and when will it be negative?

3. **Define** exponential equations. Is $x^6 = 248$ an exponential equation?

Guided Practice

State x in terms of common logarithms. Then find the value of x to the nearest hundredth.

4. $3^x = 72$
5. $7^x = 98$
6. $6^{2x} = 63$

7. $x = \log_5 121$
8. $2^{-x} = 10$
9. $x = \log_3 16$

10. $5^{3x} = 128$
11. $3^{-x} = 18$
12. $3^x = 3\sqrt{2}$

Solve each inequality by using logarithms.

13. $2^x > 14$
14. $5^x \leq 7\sqrt{6}$
15. $10^{x-3} \geq 52$

EXERCISES

Practice

Solve each equation or inequality by using logarithms. Round solutions to the nearest hundredth.

16. $6^x = 72$
17. $2^x = 27$
18. $4.3^x < 76.2$

19. $2.2^{x-5} = 9.32$
20. $9^{x-4} = 7.13$
21. $6^{3x} = 81$

22. $x < \log_3 52.7$
23. $x = \log_4 19.5$
24. $x^{\frac{2}{3}} \geq 27.6$

25. $x^{\frac{2}{5}} = 17.3$
26. $5^{x-1} = 2^x$
27. $3^{2x} = 7^{x-1}$

28. $6^{x-2} = 4^x$
29. $12^{x-4} = 3^{x-2}$
30. $\log_2 x = -3$

31. $\log_x 6 > 1$
32. $\log_{27} \frac{1}{3} = x$
33. $2^x > \sqrt{3^{x-2}}$

34. $6^{x^2-2} < 48$
35. $\log_3 \sqrt[4]{5} = x$
36. $\sqrt[3]{4^{x-1}} = 6^{x-2}$

Graphing Calculator

Use a graphing calculator to solve each equation or inequality by graphing. Round solutions to the nearest hundredth.

37. $6^{x+2} = 14^{x-3}$
38. $x^{\frac{2}{5}} = 2$
39. $12^{x-4} > 4^x$

40. Graph $y = \log_2 x$ on a graphing calculator. How does the graph differ from $y = \log_{10} x$?

41. Solve the equation $2^{2a+3} - 2^{a+3} - 2^a + 1 = 0$.

42. Accounting Assets such as houses, cars, business equipment, and art appreciate or depreciate with time. The formula we used to compute compound interest can be used to find the future value of these assets. Depreciation or appreciation is given in terms of one year, so $n = 1$ in the formula $A = P\left(1 + \dfrac{r}{n}\right)^{nt}$. This makes the formula for the future value of an object $A = P(1 + r)^t$, where P is the initial value, r is the rate of appreciation or depreciation, and t is the time in years. If the object is appreciating, then $r > 0$. If the object is depreciating, then $r < 0$.

a. You are considering buying a new car that has a list price of $12,000. A book that gives comparisons of different cars says that in past years this model has depreciated at a rate of 15% per year. If you are planning on keeping the car for five years, how much would you expect the car to be worth when you are ready to sell it?

b. Homes in the Sunbury area appreciate at a rate of about 4% per year. If a home sold for $89,000 this year, after how many years would you expect the home to be worth $124,000?

c. Inflation causes prices to rise over time. The rate of inflation in each country fluctuates in reaction to various economic conditions. Suppose the rate of inflation in the United States averages 3.5%. If a school lunch costs $1.85 today and a friend says he paid $1.35 for the same type of lunch some time ago, how long ago do you think he bought his lunch?

43. Chemistry After 13 years, 2.1 pounds of radioactive material remain from a 7-pound sample. What is the half-life of this material?

44. Sales After t years, the annual sales in hundreds of thousands of units of a product, q, is given by $q = \left(\dfrac{1}{2}\right)^{0.8^t}$. This equation is a Gompertz equation that describes growth in many areas of study.

a. Find the annual sales of the product after 7 years.

b. After how many years will the annual sales be about 64,171 units?

45. Graph $x + 4y < 9$. **(Lesson 1-3)**

46. Solve $2x^2 + 7x - 4 = 0$ by completing the square. **(Lesson 4-2)**

47. Write an equation of the sine function that has an amplitude of 17, a period of 45, and a phase shift of $-60°$. **(Lesson 6-2)**

48. Simplify $\dfrac{\tan^2 t}{1 - \sec^2 t}$. **(Lesson 7-1)**

49. Find the rectangular coordinates of the point with polar coordinates $(0.25, \pi)$. **(Lesson 9-3)**

50. Write the equation of the parabola with vertex at $(0, 0)$ and focus at $(0, -3)$. Then, draw the graph. **(Lesson 10-2)**

51. Use logarithms to evaluate $\left(\dfrac{1}{0.381}\right)^2$. **(Lesson 11-5)**

52. College Entrance Exam Choose the best answer.
Which is equivalent to 6.02×10^3?
(A) 602   **(B)** 6020   **(C)** 0.00602   **(D)** 0.0602   **(E)** none of these

11-7 Natural Logarithms

Objectives

After studying this lesson, you should be able to:
- find natural logarithms of numbers, and
- solve equations using natural logarithms.

Application

The city planners of Bakersfield, California, are studying the current and future needs of the city. The sewage system they have in place now can support 250,000 residents. The current population of the city is 175,000, up from 106,000 ten years ago. Assuming that the population will continue to grow at this rate, when will the sewage system need to be updated? How many people will the new system have to accommodate for the system to last 15 years before needing to be updated again? *This problem will be solved in Example 1.*

The growth or decay of a population can be modeled by using an exponential equation. A general formula for growth and decay is $y = ne^{kt}$, where y is the final amount, n is the initial amount, k is a constant, and t is time. If a population is growing, $k > 0$; if a population is decaying, $k < 0$.

Often when you solve problems involving the number e, it is best to use logarithms to the base e. These are called **natural logarithms** and are usually denoted **ln x**. The key marked LN on your calculator will display the natural logarithm of a number. Since e is a positive number between 2 and 3, all of the properties of logarithms also hold for natural logarithms. Note that if $\log_e e = x$ and $e^x = e$, then $x = 1$. Thus, $\ln e = 1$.

Example 1

APPLICATION

Demographics

Refer to the application above.

a. When will the sewage system need to be updated?

b. How many people will the new system have to accommodate for the upgraded system to last 15 years before needing to be updated again?

a. First, we must find the value of the constant k in the exponential growth formula. To do this, use the two known population values.

$$y = ne^{kt}$$
$$175,000 = 106,000e^{10k} \qquad \text{\textit{y = 175,000, n = 106,000, and t = 10}}$$
$$1.6509 \approx e^{10k}$$
$$\ln 1.6509 \approx \ln e^{10k} \qquad \text{\textit{Take the natural logarithm of each side.}}$$
$$\ln 1.6509 \approx 10k \ln e \qquad \text{\textit{Power property}}$$
$$0.5013 \approx 10k \qquad \text{\textit{ln e = 1}}$$
$$0.05013 \approx k$$

The value of k is about 0.05013.

(continued on the next page)

Then, we can find the time when the population will be 250,000.

$$y = ne^{kt}$$
$$250,000 \approx 175,000e^{0.05013t}$$
$$1.4286 \approx e^{0.05013t}$$
$$\ln 1.4286 \approx 0.05013t \ \ln e$$
$$t \approx 7.11$$

The current system must be updated in about 7 years.

b. We need to find the number of people that will live in Bakersfield 15 years after the system is updated. The population will be 250,000 when it is updated, so $n = 250,000$.

$$y = ne^{kt}$$
$$= 250,000e^{(0.05013)(15)}$$
$$= 530,283$$

To last 15 years, the updated system should be designed to accommodate about 530,300 people.

📁 **Portfolio**

Place your favorite word problem from this chapter in your portfolio with a note explaining why it is your favorite.

Sometimes you know the natural logarithm of a number x and must find x. The antilogarithm of a natural logarithm is written **antiln x**. If $\ln x = a$, then $x =$ antiln a. Use the INV or 2nd and LN keys of your calculator to evaluate the antilogarithm of a natural logarithm.

Example 2

APPLICATION

Linguistics

If two languages have evolved separately from a common ancestral language, the number of years since the split, $n(r)$, is given by the formula $n(r) = -5000 \ln r$, where r is the percent of the words from the ancestral language that are common to both languages now. If two languages split off from a common ancestral language about 2000 years ago, what portion of the words from the ancestral language would you expect to find in each language today?

$$n(r) = -5000 \ln r$$
$$2000 = -5000 \ln r$$
$$-0.4 = \ln r$$
$$\text{antiln} -0.4 = r$$
$$0.67 \approx r \qquad \textit{Use a calculator to evaluate.}$$

➡️ *FYI...*

The romance languages are all derived from Latin. The major romance languages that are in use today are French, Italian, Portuguese, Romanian, and Spanish.

One would expect about 67% of the words from the ancestral language to be common to both languages today.

CHECKING FOR UNDERSTANDING

Communicating Mathematics

Read and study the lesson to answer each question.

1. If the population of a city decreases over an amount of time, would you expect the value of k in $y = ne^{kt}$ to be positive or negative? Explain.

2. **Compare** the formula for growth and decay, $y = ne^{kt}$, to the formula for the balance of an account with interest compounded continually, $A = Pe^{rt}$. How are the formulas related?

Guided Practice

Use a calculator to find each value to the nearest ten thousandth.

3. $\ln 9.32$

4. $\ln 4.01$

5. $\ln 0.21$

6. antiln 2.84

7. antiln 0.7831

8. antiln –3.874

Solve each equation.

9. $9 = e^x$

10. $18 = e^{3x}$

11. $65 = e^{6x}$

12. **Biology** Kim is studying bacteria in her advanced biology class. She has calculated that $k = 0.658$ for a certain type of bacteria when t is measured in hours. How long would it take for 15 of these bacteria to multiply into 250 bacteria?

EXERCISES

Practice

Use a calculator to find each value to the nearest ten thousandth.

13. $\ln 56.8$

14. $\ln 0.0198$

15. $\ln 980$

16. $\ln 0.0089$

17. $\ln 1$

18. $\ln \left(\dfrac{1}{0.32} \right)$

19. antiln 4.987

20. antiln 2.94

21. antiln 0.62

22. antiln –0.053

23. antiln 0

24. antiln –2.81

Solve each equation. Round solutions to the nearest hundredth.

25. $1600 = 4e^{0.045t}$

26. $10 = 5e^{5k}$

27. $\ln 4.5 = \ln e^{0.031t}$

28. $25 = e^{0.075y}$

29. $\ln 40.5 = \ln e^{0.21t}$

30. $\ln 60.3 = \ln e^{0.21t}$

Applications and Problem Solving

31. **Chemistry** Radium 226, which used to be used for cancer treatment and as an ingredient in fluorescent paint, decomposes radioactively. Its half-life is 1800 years. Find the constant k you would use in the decay formula for radium. Use 1 gram as the original amount.

32. **Banking** Mr. Cuthbert invested a sum of money in a certificate of deposit that pays 8% interest compounded continuously. Recall that the formula for the amount in an account earning interest compounded continuously is $A = Pe^{rt}$. If Mr. Cuthbert made the investment on January 1, 1986 and the account is worth $10,000 on January 1, 2005, what was the original amount in the account?

33. **Ecology** DDT is an insecticide that has been used by farmers. It decays slowly and is sometimes absorbed by plants that animals and humans eat. DDT absorbed in the mud at the bottom of a lake is degraded into harmless products by bacterial action. Experimental data shows that 10% of the initial amount is eliminated in 5 years.

  a. Find the value of k in the decay formula.

  b. How much of the original amount of DDT is left after 10 years?

  c. The U.S. Environmental Protection Agency banned almost all use of DDT in the U.S. in 1972. If none has been used near the lake since then, in what year will the concentration of DDT fall below 25%?

34. **Sales** Sales of a product under relatively stable market conditions tend to decline at a constant annual rate in the absence of promotional activities. This sales decline can be expressed by the exponential function of the form $s = s_o e^{-at}$, where s is the sales at time t, t is time in years, s_o is the sales at time $t = 0$, and a is the sales decay constant. Suppose sales of On-Time Watches were 45,000 the first year and 37,000 the second year.

  a. Find the value of a in the equation for this sales decline.

  b. Find the projected sales for three years from now.

  c. If the trend continues, when would you expect sales to be 15,000 units?

35. **Finance** Mike Kallenberg deposited some money in a bank account that earns 5.6% interest compounded continuously.

  a. How long would it take to double the amount of money in Mr. Kallenberg's account?

  b. The Rule of 72 says that if you divide 72 by the interest rate of an account that compounds interest continuously, the result is the approximate number of years that it will take for the money in the account to double. Do you think that the rule of 72 is accurate? Explain.

36. **Meteorology** The atmospheric pressure varies with the altitude above the surface of Earth. Meteorologists have determined that for altitudes up to 10 km, the pressure p in millimeters of mercury is given by $p = 760e^{-0.125a}$, where a is the altitude in kilometers.

  a. What is the atmospheric pressure at an altitude of 3.3 km?

  b. At what altitude will the atmospheric pressure be 450 mm of mercury?

37. **Business** Have you ever wondered how long it will probably take for you to receive service when you are standing in a line? The probability that a customer will receive service after waiting T time periods can be described by an exponential function. If μ is the average number of customers serviced in a time period, then the probability p that a customer will be serviced in T time periods is given by $p = 1 - e^{-\mu T}$.

  a. The admissions clerk at Grady Memorial Hospital can service an average of 3 patients per hour. What is the probability that a customer will be serviced within 25 minutes?

  b. After how long is the probability that a patient in the admissions office will be serviced greater than 90%?

38. **Sociology** Rumors seem to spread like wild fire. Sociologists study the behavior of organized groups of people. A sociologist has shown that the fraction of people in a group, p, who have heard a rumor after t days can be approximated by
$p = \dfrac{p_o e^{kt}}{1 - p_o(1 - e^{kt})}$, where p_o is the fraction of people who had heard the rumor at time $t = 0$, and k is a constant.

  a. Suppose $k = 0.1$ and $p_o = \dfrac{1}{15}$. Find the fraction of the people who have heard the rumor after two weeks.

  b. At time $t = 0$, 15% of the people in a group had heard a certain rumor. If after 2 days 20% of the people had heard the rumor, what is the value of k in the equation?

39. **Medicine** Patients are often given glucose intravenously during surgery. When glucose is injected into the bloodstream at a constant rate of c grams per minute, the body converts the glucose and removes it from the bloodstream at a rate proportional to the amount present. The amount of glucose in the bloodstream at time t in minutes is given by $g(t) = \dfrac{c}{a} + \left(g_o - \dfrac{c}{a}\right) e^{-at}$, where a is a constant.

  a. Suppose at time $t = 0$, there are 0.07 grams of glucose in the bloodstream and glucose is being administered at a rate of 0.1 gram per minute. If $a = 1.25$, what is the amount of glucose in the bloodstream after 15 minutes?

  b. For a certain patient, there are 0.08 grams of glucose in the bloodstream at $t = 0$. After 2 minutes, the amount of glucose in the bloodstream is 0.046 grams. If the value of a in the equation is 1.28, how many grams of glucose are being administered each minute?

Critical Thinking

Solve for x. Round solutions to the nearest hundredth.

40. $e^{2x} - 2e^x + 1 = 0$

41. $e^{-2x} - 4e^{-x} + 3 = 0$

Mixed Review

42. Find $[f \circ g](x)$ and $[g \circ f](x)$ for $f(x) = 8x$ and $g(x) = 2 - x^2$. **(Lesson 1-2)**

43. Find the coordinates of P' if $P(4, 9)$ and P' are symmetric with respect to $M(-2, 9)$. **(Lesson 3-1)**

44. **Geometry** The length of the sides of a rhombus is 10 centimeters, and one diagonal is 12 centimeters long. Find the area of the rhombus. **(Lesson 5-8)**

45. Write an equation with phase shift 0 to represent simple harmonic motion if the initial position is 0, the amplitude is 12, and the period is 2. **(Lesson 6-7)**

46. Write an equation of the line with parametric equations $x = 11 - t$ and $y = 8 - 6t$ in slope-intercept form. **(Lesson 8-6)**

47. Graph $r = 2 + 2 \cos \Theta$. **(Lesson 9-2)**

48. Solve $3.6^x = 58.9$ by using logarithms. **(Lesson 11-6)**

49. **College Entrance Exam** Choose the best answer. Simplify $\dfrac{\frac{1}{x} + \frac{1}{y}}{3}$.

  (A) $\dfrac{3x + 3y}{xy}$ (B) $\dfrac{3xy}{x + y}$ (C) $\dfrac{xy}{3}$ (D) $\dfrac{x + y}{3xy}$ (E) none of these

VOCABULARY

Upon completing this chapter, you should be
familiar with the following terms:

annual percentage rates, APRs **609**	**608** exponential function
annuity **609**	**609** future value of an annuity
antilogarithm **631**	**622** logarithm
characteristic **630**	**622** logarithmic function
common logarithms **629**	**630** mantissa
compound interest **614**	**641** natural logarithms, ln x
exponential curve **606**	**609** present value of an annuity
exponential equations **636**	**598** scientific notation

SKILLS AND CONCEPTS

OBJECTIVES AND EXAMPLES	REVIEW EXERCISES

Upon completing this chapter, you should
be able to:

- evaluate and simplify expressions
  containing rational exponents
  (Lesson 11-1)

Simplify $(6c^4d)^2(3c)^{-2}$.

$$(6c^4d)^2(3c)^{-2} = \frac{36c^8d^2}{9c^2}$$

$$= 4c^6d^2$$

Use these exercises to review and prepare for
the chapter test.

Evaluate.

1. $\left(\dfrac{1}{64}\right)^{\frac{1}{6}}$ **2.** $27^{\frac{4}{3}}$

3. $\left(9^{\frac{3}{4}}\right)^{\frac{2}{3}}$ **4.** $\left(\dfrac{216}{729}\right)^{\frac{2}{3}}$

Simplify.

5. $3x^2(3x)^{-2}$ **6.** $\left(6a^{\frac{1}{3}}\right)^3$

7. $\left(\dfrac{1}{2}x^4\right)^3$ **8.** $(2a)^{\frac{1}{3}}\left(a^2b\right)^{\frac{1}{3}}$

- graph exponential functions
  (Lesson 11-2)

Graph $y = \left(\dfrac{1}{2}\right)^{x-1}$.

Graph each equation.

9. $y = 3^{-x}$ **10.** $y = \left(\dfrac{1}{2}\right)^x$

11. $y = 2^{x-1}$ **12.** $y = -4^{-x}$

OBJECTIVES AND EXAMPLES	REVIEW EXERCISES

■ use the exponential function $y = e^x$
(**Lesson 11-3**)

Evaluate $\sqrt{e^3} : \sqrt{e^3} = (e^3)^{\frac{1}{2}}$ or $e^{1.5}$

1.5 [2nd] [LN] 4.4816891

Use a calculator to evaluate each expression to the nearest ten thousandth.

13. $e^{2.34}$ **14.** $\sqrt[5]{e}$

15. $3\sqrt[4]{e^3}$ **16.** $\frac{1}{4}\sqrt{e^5}$

■ solve equations and evaluate expressions involving logarithms (**Lesson 11-4**)

Solve $\log_5 4 + \log_5 x = \log_5 36$.

$$\log_5 4 + \log_5 x = \log_5 36$$
$$\log_5 4x = \log_5 36$$
$$4x = 36$$
$$x = 9$$

Solve each equation.

17. $\log_x 81 = 4$ **18.** $\log_{\frac{1}{2}} x = -4$

19. $\log_3 3 + \log_3 x = \log_3 45$

20. $2\log_6 4 - \frac{1}{3}\log_6 8 = \log_6 x$

21. $\log_2 x = \frac{1}{3}\log_2 27$

■ find common logarithms and antilogarithms of numbers (**Lesson 11-5**)

Find the antilog of –1.3719.

1.3719 [+/-] [2nd] [LOG] 0.0424717

Use a calculator to find the common logarithm of each number to the nearest ten thousandth.

22. 0.0459   **23.** 42.8

Use a calculator to find the antilogarithm of each number to the nearest hundredth.

24. 1.6314   **25.** –0.1555

■ use common logarithms to compute powers and roots (**Lesson 11-5**)

Use logarithms to evaluate $\sqrt{\dfrac{5.1^3}{4.3}}$.

Let $A = \sqrt{\dfrac{5.1^3}{4.3}}$.

$\log A = \log \sqrt{\dfrac{5.1^3}{4.3}}$

$\log A = \dfrac{1}{2}(3\log 5.1 - \log 4.3)$

$\log A \approx 0.7446$

$A \approx 5.5542$

Evaluate each expression using logarithms. Round answers to the nearest hundredth.

26. $(4.22)^3 \times 0.629$

27. $\sqrt{9.12^2 \times 5.51}$

28. $\left(\dfrac{6.32}{8.67}\right)^3$

29. $\dfrac{43.9\sqrt{54.8}}{(1.29)^4}$

■ solve exponential and logarithmic equations (**Lesson 11-6**)

Solve $7^{x+1} = 18.6$.

$$7^{x+1} = 18.6$$
$$\log 7^{x+1} = \log 18.6$$
$$(x+1)\log 7 = \log 18.6$$
$$x = \dfrac{\log 18.6}{\log 7} - 1 \text{ or } 0.5022$$

Solve each equation using logarithms. Round answers to the nearest hundredth.

30. $2.5^x = 65.7$ **31.** $x = \log_3 8.9$

32. $4^{y+3} = 28.4$ **33.** $7^{x-2} = 5^{3-x}$

34. $2.3^{x^2} = 66.6$ **35.** $\sqrt{3^x} = 2^{x+1}$

■ find natural logarithms of numbers
(Lesson 11-7)

Find ln 9.65.

9.65 $\boxed{\text{LN}}$ 2.2669579

Use a calculator to find each value to the nearest ten thousandth.

36. ln 8.63

37. ln 403

38. antiln 3.7015

39. antiln 7.1121

■ use natural logarithms to solve problems
(Lesson 11-7)

Solve $400 = 2e^{0.032t}$.

$400 = 2e^{0.032t}$
$200 = e^{0.032t}$
$\ln 200 = \ln e^{0.032t}$
$\ln 200 = 0.032t \ln e$
$5.2983 = 0.032t$
$165.6 = t$

Solve each equation. Round solutions to the nearest hundredth.

40. $4500 = 3e^{0.061t}$

41. $16 = 5e^{0.4k}$

42. $\ln 19.8 = \ln e^{0.083t}$

43. $\ln 6.2 = \ln e^{0.55t}$

44. $6.6 = 1.5\,e^{210k}$

APPLICATIONS AND PROBLEM SOLVING

45. **Keyboarding** The keyboarding teacher, Mr. Eckert, found that the average number of words per minute, N, input by his students after t weeks was $N = 65 - 30e^{-0.20t}$. **(Lessons 11-3, 11-7)**
   a. What was the average number of words per minute after 2 weeks?
   b. What was the average number of words per minute at the end of the semester, 15 weeks?
   c. When will the average student in the class input 50 words per minute?

46. **Sound** The decibel, db, is the unit of measure for the intensity of sound. The equation to determine the decibel rating of a sound is $\beta = 10 \cdot \log \dfrac{I}{I_0}$, where $I_0 = 10^{-12}$ W/m^2, the reference intensity of the faintest audible sound. **(Lesson 11-5)**
   a. Find the decibel rating, in db, of a whisper that has an intensity of 1.15×10^{-10} W/m^2.
   b. Find the decibel rating of a teacher's voice in front of a classroom that has an intensity of 9×10^{-9} W/m^2.
   c. Find the decibel rating of a rock concert that has an intensity of 8.95×10^{-3} W/m^2.

47. **Archaeology** Carbon-14 tests are often performed to determine the age of an organism that died a long time ago. Carbon-14 has a half-life of 5730 years. If a turtle shell is found and tested to have 65% of its original carbon-14, how old is it? **(Lesson 11-6)**

48. **Population** A certain city has a population of $P = 142,000e^{0.014t}$ where t is the time in years and $t = 0$ is the year 1970. In what year will the city have a population of 200,000? **(Lesson 11-7)**

Evaluate.

1. $343^{\frac{2}{3}}$

2. $(0.064)^{-\frac{1}{3}}$

3. $49^{\frac{3}{2}} + 49^{-\frac{1}{2}}$

Simplify.

4. $((2a)^3)^{-2}$

5. $(x^{\frac{3}{2}} y^2 a^{\frac{5}{4}})^4$

6. $\sqrt{a^2 b} \cdot \sqrt{a^3 b^5}$

Graph each equation.

7. $y = \left(\dfrac{1}{3}\right)^{x-2}$

8. $y = 5^{x+1}$

Solve each equation.

9. $\log_x \sqrt[3]{8} = \dfrac{1}{3}$

10. $\log_5 (2x) = \log_5 (3x - 4)$

11. $\dfrac{1}{2} \log_3 64 - \log_3 x = \log_3 4$

12. $3.6^x = 72.4$

13. $4^{x+3} = 25.8$

14. $6^{x-1} = 8^{2-x}$

Find each value to the nearest ten thousandth.

15. $\dfrac{e^2}{5}$

16. $\dfrac{1}{2} \sqrt[5]{e^2}$

17. $\log 542$

18. $\ln 0.248$

19. antiln 1.1217

20. antilog -1.9101

Find the present value of each investment.

21. $3000 for 6 years at 8% compounded quarterly

22. $6000 for 2 years at 12% compounded continuously

23. **Traffic Engineering** The Ohio Department of Engineering is studying the traffic flow on State Route 23. They have found that you can predict the daily number of gaps in traffic greater than some time t by using the equation $n = 208(1.9)^{-0.11t}$, where t is measured in seconds. Find the number of gaps between vehicles that are greater than 3 seconds on any given day.

24. **Biology** A certain bacteria will triple in 6 hours. If the final count is 8 times the original, how much time has passed?

25. **Optics** The zoo has a large aquarium with several different types of fish and plants. One of the plants requires light at an intensity of 4.5 units in order to grow. The equation for the intensity of light passing through water is $I = 10e^{-0.3x}$, where x is the depth in meters. Use this equation to find the maximum depth of water at which the plant can grow.

Bonus

Find the value of x for $a > 1$ such that $\log_2 (\log_3 (\log_a x)) = 1$.

3 UNIT REVIEW

Graph the point that has the given polar coordinates. Then, name three other pairs of polar coordinates for each point. (Lesson 9-1)

1. $(2, 60°)$

2. $(-4, 45°)$

3. $(1.5, \frac{\pi}{6})$

4. $(-2, \frac{-2\pi}{3})$

Graph each polar equation. (Lessons 9-1 and 9-2)

5. $r = \sqrt{5}$

6. $\theta = 60°$

7. $r = 3\cos\theta$

8. $r = 2 + 2\sin\theta$

Find the polar coordinates of each point with the given rectangular coordinates. (Lesson 9-3)

9. $(-2, -2)$

10. $(2, 2)$

11. $(2, -3)$

12. $(-3, 1)$

Write each polar equation in rectangular form. (Lesson 9-4)

13. $2 = r\cos\left(\theta - \frac{\pi}{2}\right)$

14. $4 = r\cos\left(\theta + \frac{\pi}{3}\right)$

Simplify. (Lesson 9-5)

15. i^{45}

16. $(3 + 2i) + (3 - 3i)$

17. $i^4(3 + 3i)$

18. $(-i - 5)(i - 5)$

19. $\frac{2 + i}{2 - 3i}$

20. $\frac{3}{\sqrt{2} - 2i}$

Express each complex number in polar form. (Lesson 9-6)

21. $-3i$

22. $3 + 3i$

23. $-1 + 2i$

24. $4 - 5i$

Find each product. Then write the result in rectangular form. (Lesson 9-7)

25. $2\left(\cos\frac{\pi}{2} + i\sin\frac{\pi}{2}\right) \cdot 4\left(\cos\frac{\pi}{2} + i\sin\frac{\pi}{2}\right)$

26. $1.5(\cos 3.1 + i\sin 3.1) \cdot 2(\cos 0.5 + i\sin 0.5)$

Solve. (Lesson 9-8)

27. Find $(1 + i)^7$ by DeMoivre's Theorem. Express the result in rectangular form.

28. Solve the equation $x^5 - 1 = 0$ for all roots.

Find the coordinates of the center and the length of the radius of each circle whose equation is given. (Lesson 10-1)

29. $4x^2 + 4y^2 = 49$

30. $x^2 + 10x + y^2 + 8y = 20$

31. $x^2 + y^2 + 9x - 8y + 4 = 0$

Find the coordinates of the focus and vertex and the equation of the directrix and the axis of symmetry of each parabola whose equation is given. Then graph the equation. (Lesson 10-2)

32. $(x - 2)^2 = 2(y - 4)$

33. $y^2 + 2y - 5x + 18 = 0$

Find the coordinates of the center, foci, and vertices of each ellipse whose equation is given. Then graph the equation. (Lesson 10-3)

34. $(x - 1)^2 + 2(y - 3)^2 = 25$

35. $4(x + 2)^2 + 25(y - 2)^2 = 100$

Find the coordinates of the center, foci, and vertices and the equations of the asymptotes of each hyperbola whose equation is given. Then graph the equation.   (Lesson 10-4)

36. $4x^2 - y^2 = 27$

37. $\dfrac{(x+1)^2}{4} - \dfrac{(y+3)^2}{9} = 1$

Identify the conic section represented by each equation. Write the equation in standard form and graph the equation. (Lesson 10-5)

38. $12y - 3x + 2x^2 + 1 = 0$

39. $4(x-1)^2 = 25(y+3)^2 + 100$

40. $x^2 + 4x + y^2 - 12y + 4 = 0$

Write the equation of each translated or rotated graph below. Then draw the graph. (Lesson 10-6)

41. $x = 3(y+2)^2 + 1$ for $T_{(1, -5)}$

42. $x^2 - \dfrac{y^2}{16} = 1, \theta = \dfrac{\pi}{3}$

Graph each system of equations. Then solve.   (Lesson 10-7)

43. $x^2 - 2x - 2y - 2 = 0$
   $(x-4)^2 = -8y$

44. $x^2 - (y-1)^2 \geq 0$
   $x^2 + 2(y-3)^2 \leq 9$

Find the equations of the tangent and the normal to the graphs of each equation at the given point. Write the equations in standard form.   (Lesson 10-8)

45. $(x-1)^2 + (y+2)^2 = 2, (2, -1)$

46. $(y-4)^2 = 2(x+3), (5, 0)$

47. $\dfrac{(y-3)^2}{25} - \dfrac{(x-2)^2}{16} = 1, (2, 8)$

Simplify.   (Lesson 11-1)

48. $\sqrt{16x^2y^7}$

49. $\sqrt[3]{54a^4b^3c^8}$

50. $(3^2c^3d^5)^{\frac{1}{5}}$

51. $(3x)^2(3x^2)^{-2}$

Graph each equation.   (Lesson 11-2)

52. $y = 2^{-x}$

53. $y = 2^{x+2}$

Use a calculator to evaluate each expression to the nearest ten thousandth. (Lesson 11-3)

54. $e^{2.3}$

55. $\sqrt[3]{8e}$

56. $\dfrac{e^3}{4}$

57. $e^{-4.7}$

Solve each equation.   (Lesson 11-4)

58. $\log_x 36 = 2$

59. $\log_2(2x) = \log_2 27$

60. $\log_5 x = \dfrac{1}{3}\log_5 64 + 2\log_5 3$

Evaluate each expression using logarithms. (Lesson 11-5)

61. $(4.32)^2 \times (8.13)^3$

62. $\sqrt{(6.1)^3(10.3)}$

63. $\dfrac{5.4\sqrt{7.8}}{(1.4)^3}$

Solve each equation using logarithms. (Lesson 11-6)

64. $2.3^x = 23.4$

65. $x = \log_4 16$

66. $5^{x-2} = 2^x$

Solve each equation.   (Lesson 11-7)

67. $46 = e^x$

68. $18 = e^{4k}$

69. $519 = 3e^{0.035t}$

DISCRETE MATHEMATICS

Discrete mathematics is a branch of mathematics that deals with finite or discontinuous quantities. The distinction between continuous and discrete quantities is one that you have encountered throughout your life. For example, think of a staircase. You can slide your hand up the banister, but you have to climb the stairs one by one. The banister represents a continuous quantity, like a line graph. The stairs represent discrete quantities, like the points on a scatter plot. You will learn to use the skills you will acquire in this unit to solve problems that involve both discrete and continuous quantities.

Usually discrete mathematics is defined in terms of its key topics. These include graphs, certain functions, logic, combinatorics, sequences, iteration, algorithms, recursion, matrices, and induction. Some of these topics have already been introduced in this book. For example, linear functions are continuous, while step functions, which you examined in Chapter 3, are discrete.

In this unit, you will construct models, discover and use algorithms, and examine exciting new concepts as you solve real-world problems. It is our hope that you will greatly increase your understanding of this branch of mathematics and its limitations and advantages as a way of looking at the world.

CHAPTER 12

SEQUENCES AND SERIES

HISTORICAL SNAPSHOT

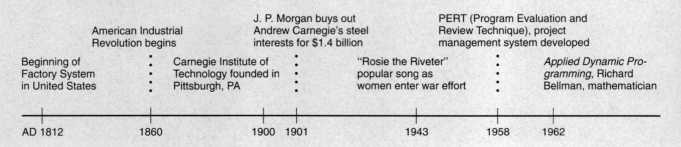

AD 1812	1860	1900	1901	1943	1958	1962
Beginning of Factory System in United States	American Industrial Revolution begins / Carnegie Institute of Technology founded in Pittsburgh, PA	J. P. Morgan buys out Andrew Carnegie's steel interests for $1.4 billion	"Rosie the Riveter" popular song as women enter war effort	PERT (Program Evaluation and Review Technique), project management system developed	*Applied Dynamic Programming*, Richard Bellman, mathematician	

CHAPTER OBJECTIVES

In this chapter, you will:
- Identify arithmetic, geometric, and infinite sequences.
- Find sums of arithmetic, geometric, and infinite series.
- Expand binomials.
- Use mathematical induction to prove the validity of formulas.

CAREER GOAL: Operations Research

What do a dairy farm, an aluminum foundry, and an international bank have in common? These are all places where Michael Gallmeyer has worked. Michael grew up on a dairy farm. He says, "In farming you are exposed to many things. For instance, when something breaks, you fix it. You get to see more in the way of technology, mechanics, and animal science." Michael originally considered a career in electrical engineering, but decided that it was too restrictive for him. He wanted a career that would allow him to enter several different areas. Industrial management has offered him that opportunity.

In industrial management, companies often have to make decisions about their investment in energy conservation. Suppose ABC, Inc. has $10,000 to invest. One option is to put it in a savings account at 6% per year for ten years. The *future value* of that investment is $17,910. Another option is to use that $10,000 to install solar energy panels. This option promises a $1000 per year savings in energy costs. Let's say they invest this $1000 savings in a savings account at 6%. The future value of that investment is $13,181. So far, it appears that it is more profitable to take the first option. However, without the conversion to solar energy, ABC would be paying $1000 more for energy each year. When you deduct the *present value* of this $1000 annual expenditure, their $17,910 return becomes only $17,910 − $7360 or $10,550. Therefore, ABC will have saved money by converting to solar energy.

Industrial management has provided Michael with some very marketable skills. In the summer of 1992, he interned with the Advanced Technology Group of J.P. Morgan, an international corporate bank. About this experience, he says, "I'm beginning to realize just how quantitative business has become and how important mathematics is to acquiring the toolkit of skills that the business world needs."

interNET CONNECTION

For up-to-date information on industrial management, visit:
www.glencoe.com/sec/math/amc/mathnet

12-1 Arithmetic Sequences and Series

Objectives

After studying this lesson, you should be able to:
- find the nth term and arithmetic means of an arithmetic sequence, and
- find the sum of n terms of an arithmetic series.

Application

Jason's younger brother, Jeffrey, is very interested in collecting basketball cards. Jason has quite a few basketball cards that he has collected over the years. As a birthday gift to Jeffrey, Jason has decided to give Jeffrey 3 basketball cards today, 6 basketball cards tomorrow, 9 basketball cards the next day, and so on each day for 14 days. How many cards will Jason give Jeffrey on the 14th day?

The chart below shows the number of Jason's cards that Jeffrey will receive each day.

Day	1	2	3	4	5	6	7	8	9	10	11	12	13	14
Number of Cards	3	6	9	12	15	18	21	24	27	30	33	36	39	42

Thus, Jeffrey will receive 42 of Jason's cards on the 14th day.

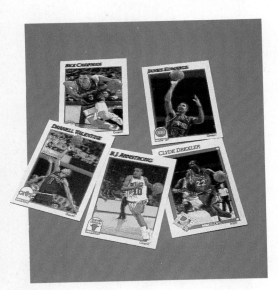

The set of numbers representing the number of Jason's cards that Jeffrey would receive each day is an example of a **sequence**. A sequence is a set of numbers in a specific order. The **terms** of a sequence are the numbers in it. The first term of a sequence is denoted a_1, or a, the second term is a_2, and so on up to the nth term, a_n.

Symbol	a_1	a_2	a_3	a_4	a_5	a_6
Term	13	21.5	30	38.5	47	55.5

In this sequence, how might a_7 be determined?

The sequences given in the tables above are both examples of **arithmetic sequences**. The difference between successive terms of an arithmetic sequence is a constant called the **common difference**, denoted d.

Arithmetic Sequence

An arithmetic sequence is a sequence in which each term after the first, a_1, is equal to the sum of the preceding term and the common difference, d. The terms of the sequence can be represented as follows.

$$a_1, a_1 + d, a_1 + 2d, \cdots$$

To find the next term in an arithmetic sequence, first find the common difference by subtracting any term from its succeeding term. Then add the common difference to the last term to find the next term in the sequence.

Example 1 | Find the next three terms in each sequence.

a. **-5, 7, 19, · · ·**

a. First, find the common difference.

$7 - (-5) = 12$ $19 - 7 = 12$

The common difference is 12.

Then, add 12 to the third term to get the fourth term, and so on.

$19 + 12 = 31$ $31 + 12 = 43$ $43 + 12 = 55$

The next three terms are 31, 43, and 55.

b. **$r + 15, r + 8, r + 1, · · ·$**

b. First, find the common difference.

$(r + 8) - (r + 15) = -7$ $(r + 1) - (r + 8) = -7$

The common difference is -7.

Then, add -7 to the third term to get the fourth term, and so on.

$r + 1 + (-7) = r - 6$ $r - 6 + (-7) = r - 13$ $r - 13 + (-7) = r - 20$

The next three terms are $r - 6$, $r - 13$, and $r - 20$.

The nth term of an arithmetic sequence can be found by using a formula if the first term and the common difference are known. This type of formula is called a **recursive** formula. This means that each succeeding term is formulated from one or more previous terms. Consider an arithmetic sequence in which $a = -11$ and $d = 4$. Notice the pattern in the way the terms are formed.

first term	a_1	a	-11
second term	a_2	$a + d$	$-11 + 1(4) = -7$
third term	a_3	$a + 2d$	$-11 + 2(4) = -3$
fourth term	a_4	$a + 3d$	$-11 + 3(4) = 1$
fifth term	a_5	$a + 4d$	$-11 + 4(4) = 5$
$\vdots$	$\vdots$	$\vdots$	$\vdots$
nth term	a_n	$a + (n - 1)d$	$-11 + (n - 1)4$

The nth Term of an Arithmetic Sequence

The **nth term of an arithmetic sequence with first term a_1 and common difference d is given by the following formula.**

$$a_n = a_1 + (n - 1)d$$

Notice that the preceding formula has four variables: $a_n, a_1, n,$ and d. If any three of these are known, the fourth can be found. By definition, the nth term is also equal to $a_{n-1} + d$, where a_{n-1} is the $(n - 1)$th term. That is, $a_n = a_{n-1} + d$.

Example 2

Find the 68th term in the sequence 16, 7, -2, · · · .

First, find the common difference.

$7 - 16 = -9$ $\qquad$ $-2 - 7 = -9$

The common difference is -9.

Then, use the formula for the nth term of an arithmetic sequence.

$a_n = a_1 + (n - 1)d$

$a_{68} = 16 + (68 - 1)(-9)$ $\qquad$ *$n = 68, a_1 = 16,$ and $d = -9$*

$\quad\;\; = -587$

Example 3

Find the first term in the sequence for which $a_{31} = 197$ and $d = 10$.

$a_n = a_1 + (n - 1)d$

$a_{31} = a_1 + (31 - 1)d$ $\qquad$ *$n = 31$ and $d = 10$*

$197 = a_1 + 30(10)$ $\qquad$ *$a_{31} = 197$*

$a_1 = -103$

Sometimes you may know two terms of an arithmetic sequence that are not in consecutive order. The terms between any two nonconsecutive terms of an arithmetic sequence are called **arithmetic means**. In the sequence below, 76 and 85 are the arithmetic means between 67 and 94.

$$49, 58, 67, 76, 85, 94$$

Example 4

Form an arithmetic sequence that has five arithmetic means between -11 and 19.

This sequence will have the form $-11, \underline{\;?\;}, \underline{\;?\;}, \underline{\;?\;}, \underline{\;?\;}, \underline{\;?\;}, 19$.

First, find the common difference.

$a_n = a_1 + (n - 1)d$ $\qquad$ *$n = 7$*

$19 = -11 + (7 - 1)d$ $\qquad$ *$a_7 = 19$ and $a_1 = -11$*

$19 = -11 + 6d$

$d = 5$

Then, determine the arithmetic means.

$-11 + 5 = -6$ $\qquad$ $-6 + 5 = -1$ $\qquad$ $-1 + 5 = 4$ $\qquad$ $4 + 5 = 9$ $\qquad$ $9 + 5 = 14$

The sequence is $-11, -6, -1, 4, 9, 14, 19$.

An indicated sum is $1 + 2 + 3 + 4$. The sum of $1 + 2 + 3 + 4$ is 10.

An **arithmetic series** is the indicated sum of the terms of an arithmetic sequence. The lists below show some examples of arithmetic sequences and their corresponding arithmetic series.

Arithmetic Sequence	**Arithmetic Series**
$3, 8, 13, 18, 23$	$3 + 8 + 13 + 18 + 23$
$\dfrac{1}{2}, \dfrac{1}{4}, 0, -\dfrac{1}{4}, -\dfrac{1}{2}$	$\dfrac{1}{2} + \dfrac{1}{4} + 0 + \left(-\dfrac{1}{4}\right) + \left(-\dfrac{1}{2}\right)$
$a_1, a_2, a_3, a_4, \cdots, a_n$	$a_1 + a_2 + a_3 + a_4 + \cdots + a_n$

The symbol S_n is used to represent the sum of the first n terms of a series. To develop a formula for S_n for an arithmetic series, a series can be written in two ways and added term by term, as shown below. The second equation for S_n given below is obtained by reversing the order of the terms in the series.

$$
\begin{array}{rl}
S_n = & a_1 + (a_1 + d) + (a_1 + 2d) + \cdots + (a_n - 2d) + (a_n - d) + a_n \\
+\ S_n = & a_n + (a_n - d) + (a_n - 2d) + \cdots + (a_1 + 2d) + (a_1 + d) + a_1 \\
\hline
2S_n = & (a_1 + a_n) + (a_1 + a_n) + (a_1 + a_n) + \cdots + (a_1 + a_n) + (a_1 + a_n) + (a_1 + a_n)
\end{array}
$$

$2S_n = n(a_1 + a_n)$ *There are n terms in the series, all of which are $(a_1 + a_n)$.*

Therefore, $S_n = \dfrac{n(a_1 + a_n)}{2}$.

Sum of an Arithmetic Series

The sum of the first n terms of an arithmetic series is given by the following formula.

$$S_n = \frac{n}{2}(a_1 + a_n)$$

Example 5

Find the sum of the 27 terms in the series $-14 - 8 - 2 - \cdots + 142$.

$S_n = \dfrac{n}{2}(a_1 + a_n)$

$S_{27} = \dfrac{27}{2}(-14 + 142)$ $n = 27, a_1 = -14,$ and $a_{27} = 142$

$= 1728$

When the value of the last term, a_n, is not known, you can still determine the sum of the series. Using the formula for an arithmetic sequence, you can derive another formula for the sum of an arithmetic series.

$S_n = \dfrac{n}{2}(a_1 + a_n)$

$S_n = \dfrac{n}{2}[a_1 + (a_1 + (n-1)d)]$ $a_n = a_1 + (n-1)d$

$S_n = \dfrac{n}{2}[2a_1 + (n-1)d]$

Example 6

APPLICATION

Savings

Celeste Bay will be a freshman at The University of Kentucky in the fall. Starting one month from today, she plans to withdraw $10 from her savings account, and then increase her withdrawal by $10 each week so that she can purchase items for college. If her account has a balance of $1200 today, how long will it take her to empty her savings account?

Let $S_n =$ the amount of money that is currently in the savings account, $1200.
Let $a_1 =$ the first withdrawal of $10. In this example, $d = 10$.
We want to find n, the number of weeks that it will take to empty the savings account.

(continued on the next page)

$$S_n = \frac{n}{2}[2a_1 + (n-1)d]$$

$1200 = \frac{n}{2}[2(10) + (n-1)(10)]$ $S_n = 1200, a_1 = 10,$ and $d = 10$

$2400 = n[20 + (n-1)(10)]$ *Multiply each side by 2.*

$2400 = n(20 + 10n - 10)$ *Simplify.*

$2400 = n(10 + 10n)$

$0 = 10n^2 + 10n - 2400$

$0 = n^2 + n - 240$ *Divide each side by 10.*

$0 = (n-15)(n+16)$ *Factor.*

$n - 15 = 0$ or $n + 16 = 0$

$n = 15$ $n = -16$ *-16 is not a possible answer.*

Thus, it would take Celeste 15 weeks to empty her account.

CHECKING FOR UNDERSTANDING

Communicating Mathematics

Read and study the lesson to answer each question.

1. Refer to the application at the beginning of the lesson.
   a. What formula would you use to determine how many of Jason's cards Jeffrey will have at the end of 14 days?
   b. How many of Jason's cards will Jeffrey have at the end of 14 days?
   c. Graph the numbers in the first table. Let days be the x-coordinate and let number of cards be the y-coordinate, and connect the points. Describe the graph.

2. Choose the correct term: If the numbers in a sequence are decreasing, then the common difference is (positive, negative).

3. In Example 4, why does n equal 7?

4. Refer to Example 6.
   a. If Celeste begins to withdraw the money from her savings account on March 1, will the account be empty by the end of August?
   b. Why is -16 *not* a possible answer?

Guided Practice

Find the next five terms in each arithmetic sequence.

5. $5, 9, 13, \cdots$

6. $-9, -2, 5, \cdots$

7. $5, -1, -7, \cdots$

8. $0, 7, 14, \cdots$

9. $1.5, 3, 4.5, \cdots$

10. $a, a+3, a+6, \cdots$

11. $-n, 0, n, \cdots$

12. $x, 2x, 3x, \cdots$

13. $b, -b, -3b, \cdots$

14. $5k, -k, \cdots$

15. Find the 79th term in the sequence $-7, -4, -1, \cdots$.

16. Find the first term in the sequence for which $a_{15} = 38$ and $d = -3$.

17. Form an arithmetic sequence that has one arithmetic mean between 12 and 21.

18. Find the sum of the first 63 terms in the series $-19 - 13 - 7 - \cdots$.

EXERCISES

Practice

Solve. Assume that each sequence is an arithmetic sequence.

19. Find the 19th term in the sequence for which $a_1 = 11$ and $d = -2$.

20. Find the 16th term in the sequence for which $a_1 = 1.5$ and $d = 0.5$.

21. Find n for the sequence for which $a_n = 37$, $a_1 = -13$, and $d = 5$.

22. Find n for the sequence for which $a_n = 633$, $a_1 = 9$, and $d = 24$.

23. Find the first term in the sequence for which $d = -2$ and $a_7 = 3$.

24. Find the first term in the sequence for which $d = \frac{2}{3}$ and $a_8 = 15$.

25. Find d for the sequence for which $a_1 = 4$ and $a_{11} = 64$.

26. Find d for the sequence for which $a_1 = -6$ and $a_{29} = 20$.

27. Find the sixth term in the sequence $-2 + \sqrt{3},\ -1,\ -\sqrt{3},\ \cdots$.

28. Find the seventh term in the sequence $1 + i,\ 2 - i,\ 3 - 3i,\ \cdots$.

29. Find the 43rd term in the sequence $-19,\ -15,\ -11,\ \cdots$.

30. Find the 58th term in the sequence $10,\ 4,\ -2,\ \cdots$.

31. Form a sequence that has one arithmetic mean between 36 and 48.

32. Form a sequence that has two arithmetic means between -4 and 5.

33. Form a sequence that has two arithmetic means between $\sqrt{2}$ and 10.

34. Form a sequence that has three arithmetic means between 1 and 4.

35. Find the sum of the first 11 terms in the series $-3 - 1 + 1 + 3 + \cdots$.

36. Find the sum of the first 32 terms in the series $0.5 + 0.75 + 1 + \cdots$.

37. Find n for a series for which $a_1 = -7$, $d = 1.5$, and $S_n = -14$.

38. Find n for a series for which $a_1 = 5$, $d = 3$, and $S_n = 440$.

Critical
Thinking

39. Geometry The measures of the angles of a convex polygon form an arithmetic sequence. The least measurement in the sequence is $129°$. The greatest measurement is $159°$. Find the number of sides in this polygon.

40. Find the sum of the first 102 terms in the sequence $5, 7, 2, \cdots$, where $a_n = a_{n-1} - a_{n-2}$ for all $n > 2$.

Applications
and Problem
Solving

41. Employment Terri works after school at the Fine Foods Supermarket. One day, Terri had to stack cans of soup in a grocery display in the form of a triangle. On the top row, there was only one can. Each row below it contained one more can than the one above it. On the bottom row, there were 21 cans. If all of the cans were the same size, how many cans were in the display?

42. Number Theory Find the sum of the first 100 positive even integers and the first 100 positive odd integers. Are the sums the same? If not, by how much do they differ? Explain your findings.

43. Food Michael is a chocoholic. On New Year's Day, he ate one piece of chocolate. On the next day, he ate 2 pieces. On each subsequent day, he ate one additional piece of candy.
 a. How many pieces of candy did he eat on the last day of January?
 b. How many pieces did he eat during the month of January?

44. Banking Juan García has $650 in a checking account, and he is closing out the account by writing one check against it each week. The first check is for $20, the second is for $25, and so on. Each check exceeds the previous one by $5. In how many weeks will the balance in Juan's account be $0 if there is no service charge?

Mixed Review

45. Find $[f \circ g](4)$ and $[g \circ f](4)$ if $f(x) = x^2 - 4x + 5$ and $g(x) = x - 2$. **(Lesson 1-2)**

46. Business A pharmaceutical company manufactures two drugs. Each case of drug A requires 3 hours of processing time and 1 hour of curing time per week. Each case of drug B requires 5 hours of processing time and 5 hours of curing time per week. The schedule allows 55 hours of processing time and 45 hours of curing time weekly. The company must produce no more than 10 cases of drug A and no more than 9 cases of drug B. If the company makes a profit of $320 on each case of drug A and $500 on each case of drug B, how many cases of each drug should be produced in order to maximize profit? **(Lesson 2-6)**

47. Geometry Find the area of a regular pentagon that is inscribed in a circle with a diameter of 7.3 cm. **(Lesson 5-5)**

48. Simplify $\dfrac{\tan^2 \theta - \sin^2 \theta}{\tan^2 \theta \, \sin^2 \theta}$. **(Lesson 7-1)**

49. Simplify $(2 + i)(3 - 4i)(1 + 2i)$. **(Lesson 9-5)**

50. Solve $45.9 = e^{0.075t}$. **(Lesson 11-7)**

51. College Entrance Exam Choose the best answer.

If $\dfrac{x}{y}$ is a fraction greater than 1, which of the following must be less than 1?

(A) $\dfrac{2y}{x}$ **(B)** $\dfrac{x}{2y}$ **(C)** $\sqrt{\dfrac{x}{y}}$ **(D)** $\dfrac{y}{x}$ **(E)** $\left(\dfrac{x}{y}\right)^2$

CASE STUDY FOLLOW-UP

Refer to Case Study 1, *Buying a Home,* on pages 956–958.

Rebecca and Tom Payton decide to buy a $200,000 home. They put 20% down and apply for a 15-year, 9% FRM to finance the balance. The Paytons have a combined gross annual income of $70,000.

1. Do the Paytons meet the 28% debt requirement? Explain.

2. Use an arithmetic series to find the total amount the Paytons will repay the mortgage lender during the life of the loan. Identify the first term, the common difference, and the number of terms.

3. Research current mortgage rates and the types of mortgages currently available in your city. Then choose a mortgage lender for a $150,000 mortgage. Write about your findings in a 1-page paper. Explain why you chose that particular lender.

12-2 Geometric Sequences and Series

Objectives
After studying this lesson, you should be able to:
- find the nth term and geometric means of a geometric sequence, and
- find the sum of n terms of a geometric series.

Application

For his science experiment, Adam exposed the plants he was growing to a grow light for 20 hours each day. During the course of the school year, he found that the plants grew 7% each month. When he began his experiment, the plants were 3 inches tall. Predict how tall they will be at the beginning of the ninth month. *This problem will be solved in Example 3.*

The following sequence is an example of a **geometric sequence**.

$$1, \quad 0.5, \quad 0.25, \quad 0.125, \cdots$$ *Can you find the next term?*

The ratio of successive terms in a geometric sequence is a constant called the **common ratio**, denoted r.

Geometric Sequence	A geometric sequence is a sequence in which each term after the first, a_1, is the product of the preceding term and the common ratio, r. The terms of the sequence can be represented as follows, where a_1 is nonzero and r is not equal to 1 or 0. $$a_1, a_1r, a_1r^2, \cdots.$$

The common ratio of a geometric sequence can be found by dividing any term by the preceding term. Then multiply the last term by the common ratio to find the next term in the sequence.

Example 1

Find the next three terms in the geometric sequence 27, 135, 675, $\cdots$.

First, find the common ratio.
$$135 \div 27 = 5 \qquad\qquad 675 \div 135 = 5$$
The common ratio is 5.

Then, multiply the third term by 5 to get the fourth term, and so on.
$$675 \cdot 5 = 3375 \qquad 3375 \cdot 5 = 16{,}875 \qquad 16{,}875 \cdot 5 = 84{,}375$$
The next three terms are 3375, 16,875, and 84,375.

As with arithmetic sequences, geometric sequences are also recursive. Successive terms of a geometric sequence can be expressed as the product of the common ratio and the previous term. Thus, it follows that each term can be expressed as the product of a_1 and a power of r. The terms of a geometric sequence for which $a_1 = 3$ and $r = 4$ can be represented as follows.

first term	a_1	a_1	3
second term	a_2	$a_1 r$	$3 \cdot 4^1 = 12$
third term	a_3	$a_1 r^2$	$3 \cdot 4^2 = 48$
fourth term	a_4	$a_1 r^3$	$3 \cdot 4^3 = 192$
fifth term	a_5	$a_1 r^4$	$3 \cdot 4^4 = 768$
$\vdots$	$\vdots$	$\vdots$	$\vdots$
nth term	a_n	$a_1 r^{n-1}$	$3 \cdot 4^{n-1}$

The nth Term of a Geometric Sequence	The *n*th term of a geometric sequence with first term a_1 and common ratio r is given by the following formula. $$a_n = a_1 r^{n-1}$$

By definition, the nth term is also equal to $a_{n-1} r$, where a_{n-1} is the $(n-1)$th term. That is, $a_n = a_{n-1} r$.

Example 2

Find the 14th term in the sequence $1, \dfrac{1}{3}, \dfrac{1}{9}, \cdots$.

First, find the common ratio.

$$\frac{1}{3} \div 1 = \frac{1}{3} \qquad\qquad \frac{1}{9} \div \frac{1}{3} = \frac{1}{3}$$

The common ratio is $\dfrac{1}{3}$.

Then, use the formula for the nth term of a geometric sequence.

$$a_n = a_1 r^{n-1}$$

$$a_{14} = 1 \left(\frac{1}{3}\right)^{14-1} \qquad n = 14, a_1 = 1, \text{ and } r = \frac{1}{3}$$

$$= \left(\frac{1}{3}\right)^{13}$$

$$= \frac{1}{3^{13}} \text{ or } \frac{1}{1,594,323} \qquad \textit{Use a calculator.}$$

Geometric sequences can represent growth and decay. For a common ratio greater than 1, a sequence denotes growth. Examples are compound interest, appreciation of personal property, and population. Sequences that represent decay have a common ratio that is less than 1. Examples include depreciation and radioactive behavior.

Example 3

APPLICATION

Biology

Refer to the application at the beginning of the lesson. Predict how tall the plants will be at the beginning of the ninth month.

At the beginning of the first month, the plants were 3 inches tall. At the beginning of the second month, the plants will be $3 + 3(0.07)$ or 3.21 inches tall. At the beginning of the third month, the plants will be $3.21 + 3.21(0.07)$ or 3.4347 inches tall. The sequence of heights at the beginning of the month may be generated by using the formula for the nth term of a geometric series.

$$a_n = a_1 r^{n-1}$$
$$a_9 = 3(1.07)^{9-1} \qquad n = 9, a_1 = 3, \text{ and } r = 1.07$$
$$= 5.15$$

Thus, the plants will be 5.15 inches tall at the beginning of the ninth month.

The terms between any two nonconsecutive terms of a geometric sequence are called **geometric means**.

Example 4

Form a sequence that has two geometric means between 136 and 459.

This sequence will have the form 136, __?__, __?__, 459.

First, find the common ratio.

$$a_n = a_1 r^{n-1}$$
$$a_4 = a_1 r^3 \qquad \text{Since there will be 4 terms in the sequence, } n = 4.$$
$$459 = 136 r^3 \qquad a_4 = 459 \text{ and } a_1 = 136$$
$$r^3 = \frac{459}{136} \text{ or } \frac{27}{8}$$
$$r = \frac{3}{2} \qquad \text{Take the cube root of each side.}$$

Then, determine the geometric means.

$$136 \cdot \frac{3}{2} = 204 \qquad\qquad 204 \cdot \frac{3}{2} = 306$$

The sequence is 136, 204, 306, 459.

A **geometric series** is the indicated sum of the terms of a geometric sequence. The lists below show some examples of geometric sequences and their corresponding geometric series.

Geometric Sequence
$1, 4, 16, 64, 256$
$2, 1, \frac{1}{2}, \frac{1}{4}, \frac{1}{8}$
$a_1, a_2, a_3, a_4, \cdots, a_n$

Geometric Series
$1 + 4 + 16 + 64 + 256$
$2 + 1 + \frac{1}{2} + \frac{1}{4} + \frac{1}{8}$
$a_1 + a_2 + a_3 + a_4 + \cdots + a_n$

To develop a formula for the sum of a series, S_n, write an expression for S_n and for rS_n, as shown below. Then subtract rS_n from S_n and solve for S_n.

$$S_n = a_1 + a_1r + a_1r^2 + \cdots + a_1r^{n-2} + a_1r^{n-1}$$
$$\underline{-(rS_n = \qquad a_1r + a_1r^2 + \cdots + a_1r^{n-2} + a_1r^{n-1} + a_1r^n)}$$
$$S_n - rS_n = a_1 \qquad\qquad\qquad\qquad\qquad\qquad - a_1r^n$$

$S_n(1 - r) = a_1 - a_1r^n$ *Factor.*

$S_n = \dfrac{a_1 - a_1r^n}{1 - r}$ *Divide each side by $1 - r, r \neq 1$.*

Sum of a Geometric Series

The sum of the first n terms of a geometric series is given by the following formula.

$$S_n = \frac{a_1 - a_1r^n}{1 - r}, r \neq 1$$

Example 5

Find the sum of the first eight terms of the series $3 - 6 + 12 + \cdots$.

First, find the common ratio.

$-6 \div 3 = -2$ $12 \div (-6) = -2$

The common ratio is -2.

Then, find the sum.

$S_n = \dfrac{a_1 - a_1r^n}{1 - r}$

$S_8 = \dfrac{3 - 3(-2)^8}{1 - (-2)}$ $n = 8, a_1 = 3,$ *and* $r = -2$

$ = \dfrac{3 - 3(256)}{3}$

$ = -255$

Thus, the sum of the first eight terms is -255.

In Chapter 11, you studied annuities and present and future value. Recall that the future value of an annuity is the sum of all payments plus the interest earned on each payment. Suppose an annuity pays an APR of 6% and you make 4 semiannual payments of $1000 into the account. The equation below gives the future value, F_n, of the account after the fourth payment.

The semiannual percentage rate is 3% since the APR is 6%.

$$F_n = 1000 + 1000(1.03)^1 + 1000(1.03)^2 + 1000(1.03)^3 = \$4183.63$$

As you learned in Chapter 11, this type of annuity can be represented by the formula $F_n = P\dfrac{(1 + i)^n - 1}{i}$. However, this expression is also a geometric series with $a_1 = \$1000$, $r = 1.03$, and $n = 4$.

Example 6

APPLICATION

Finance

Phillip Morrow is investing \$500 quarterly (January 1, April 1, July 1, and October 1) in a retirement account that pays an APR of 12%, compounded quarterly. Interest for each quarter is posted on the first day of the following quarter. Determine the value of the investment at the end of the year.

Since the interest is compounded quarterly, the interest rate per quarter is $12\% \div 4$ or 3%. The value of the common ratio, r, for each term in the series is $1 + 0.03$ or 1.03.

$$S_n = \frac{a_1 - a_1 r^n}{1 - r}$$

$$S_4 = \frac{500 - 500(1.03)^4}{1 - 1.03} \qquad a_1 = 500, r = 1.03, \text{ and } n = 4$$

$$= \frac{500 - 500(1.1255088)}{-0.03} \qquad \textit{Use a calculator to evaluate.}$$

$$= 2091.81$$

The value of the retirement account at the end of the year will be \$2091.81.

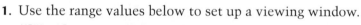

EXPLORATION: Graphing Calculator

You can use a graphing calculator to examine the growth of Mr. Morrow's investment over a 2-year period.

MATH JOURNAL

Write how you would explain the difference between a series and a sequence to another student and write the examples you would use.

1. Use the range values below to set up a viewing window.

   $XMIN = 0$ $YMIN = 500$

   $XMAX = 9.5$ $YMAX = 5000$

   $XSCL = 1$ $YSCL = 500$

2. Let $Y_1 = \dfrac{500 - 500(1.03)^x}{1 - 1.03}$, where x represents the quarters that have passed.

3. Graph the equation.

4. What would be the value of Mr. Morrow's investment after 2 years? Use the TRACE function.

CHECKING FOR UNDERSTANDING

Communicating Mathematics

Read and study the lesson to answer each question.

1. **Explain** why the common ratio is considered to be unequal to 1 or 0.

2. **Explain** why the first term in a geometric sequence must be nonzero.

3. **Explain** why arithmetic and geometric sequences are recursive.

4. Refer to Example 3.

    a. Make a table to represent the situation. In the first row, put the number of months, and in the second row, put the height of the plants.

    b. Graph the numbers in the table. Let months be the x-coordinate and let height of the plants be the y-coordinate, and connect the points. Describe the graph.

5. Show how the formula for the future value of an annuity can be derived from the formula for the sum of a geometric series.

Guided Practice

Find the next four terms of each geometric sequence.

6. $2, 3, \cdots$ **7.** $7, 3.5, \cdots$ **8.** $1.2, 3.6, \cdots$

Determine whether the given terms form a geometric sequence. Write *yes* or *no*.

9. $\sqrt{2}, 2, \sqrt{8}, \cdots$ **10.** $\sqrt[3]{3^2}, 3, 3\sqrt[3]{3}, \cdots$ **11.** $t^{-2}, t^{-1}, 1, \cdots$

12. The first term of a geometric sequence is $\frac{1}{2}$, and the common ratio is $\frac{2}{3}$. Find the ninth term of the sequence.

13. If $r = 2$ and $a_5 = 24$, find the first term of the geometric sequence.

14. Form a sequence that has two geometric means between –2 and 54.

15. Find the sum of the first five terms of the series $\frac{5}{3} + 5 + 15 + \cdots$.

EXERCISES

Practice

Solve.

16. The first term of a geometric sequence is –3, and the common ratio is $\frac{2}{3}$. Find the next four terms.

17. The first term of a geometric sequence is 8, and the common ratio is $\frac{3}{2}$. Find the next three terms.

18. Find the ninth term of the geometric sequence $\sqrt{2}, 2, 2\sqrt{2}, \cdots$.

19. Find the sixth term of the geometric sequence $10, 0.1, 0.001, \cdots$.

20. Find the first four terms of the geometric sequence for which $a_5 = 32\sqrt{2}$ and $r = -\sqrt{2}$.

21. Find the first three terms of the geometric sequence for which $a_4 = 2.5$ and $r = 2$.

22. Form a sequence that has one geometric mean between $\frac{1}{4}$ and 4.

23. Form a sequence that has two geometric means between 1 and 27.

24. Find the sum of the first seven terms of the series $\frac{1}{2} + \frac{1}{4} + \frac{1}{8} + \cdots$.

25. Find the sum of the first six terms of the series $2 + 3 + 4.5 + \cdots$.

26. Find the sum of the first nine terms of the series $0.5 + 1 + 2 + \cdots$.

27. Find the sum of the first ten terms of the series $1 + \sqrt{2} + 2 + 2\sqrt{2} + \cdots$.

Critical Thinking

28. The sum of the first six terms in a geometric sequence of real numbers is 252. Find the sum of the first four terms when the sum of the first two terms is 12.

29. Investment   The Landbury Museum has been investing in paintings for many years. Twenty years ago, the museum purchased a painting by one of the French impressionist painters for $180,000. The value of the painting has appreciated at a rate of 14% per year. Find the value of the painting after 10, 20, 30, 40, and 50 years, assuming that the rate of appreciation remains constant.

30. Finance   Eric Scheidt is purchasing a new car that costs $13,500. He estimates that the value of the car will depreciate at a rate of 18% per year. Determine the value of the car when he plans to trade it in 5 years.

31. Biology   The population of a certain bacteria doubles every 30 minutes. The initial population is 150. Determine the number of organisms that would exist after 12 hours and after 24 hours.

32. Finance   Mr. Leshnock invests $5000 annually in an insurance annuity for twenty years. What will the annuity be worth if it has an APR of 9.5%?

33. Number Theory   If you tear a piece of paper that is 0.005 centimeters thick in half, and place the two pieces on top of each other, the height of the pile of paper is 0.010 centimeters. Let's call this the second pile. If you tear these two pieces of paper in half, the third pile will have four pieces of paper in it.

   a. How high is the third pile? the fourth pile?

   b. Write a formula to determine how high the nth pile is.

   c. Use the formula to determine in theory how high the 10th pile would be and how high the 100th pile would be.

34. Geometry   The vertices of a parallelogram are at $(2, 4)$, $(5, 9)$, $(14, 9)$, and $(11, 4)$. Find the lengths of the diagonals.   **(Lesson 1-4)**

35. Graph $y = \dfrac{2}{3} \cos \theta$.   **(Lesson 6-2)**

36. Write the standard form of the equation of the circle that passes through the points with coordinates $(5, 0)$, $(1, -2)$, and $(4, -3)$.   **(Lesson 10-1)**

37. Physical Science   The adiabatic process is the cooling of air as it moves upward into the atmosphere. Dry air expands as it moves upward into the atmosphere. For each 1000 feet that it moves upward, the air cools 5°F. Suppose the temperature at ground level is 80°F.   **(Lesson 12-1)**

   a. Write a sequence representing the temperature decrease per 1000 feet.

   b. If n is the height of the air in thousands of feet, write a formula for the temperature T in terms of n.

   c. What is the ground level temperature if the air at 40,000 feet is $-125°F$?

38. College Entrance Exam   Choose the best answer.

$$\frac{1}{2} \cdot \frac{2}{3} \cdot \frac{3}{4} \cdot \frac{4}{5} \cdot \frac{5}{6} \cdot \frac{6}{7} =$$

   (A) $\dfrac{1}{7}$ **(B)** $\dfrac{3}{7}$ **(C)** $\dfrac{21}{27}$ **(D)** $\dfrac{6}{7}$ **(E)** $\dfrac{7}{8}$

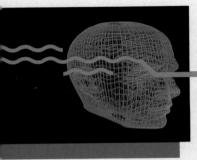

Technology

Amortization

Most consumer loans, like mortgages for homes, are paid in a series of equal payments made over a period of time. This is called **amortization**. The money from each payment includes an interest payment and a principal payment. The interest due for the period since the last payment was made is paid first, and then the remaining money from the payment goes toward reducing the principal. The spreadsheet program below constructs an amortization schedule for a two-year loan.

TECHNOLOGY *Tip*

You can use the amortization functions of the TI-83 graphing calculator to find the balance, sum of principal, and sum of interest for an amortization schedule.

AMORTIZATION SCHEDULE			
	A	B	C	D
1		LOAN AMOUNT	PAYMENT	INTEREST RATE
2				
3	PAYMENT	INTEREST PAID	PRINCIPAL PAID	BALANCE DUE
4	0	0	0	B2
5	1	D2/12*D4	C2-B5	D4-C5
6	2	D2/12*D5	C2-B6	D5-C6
28	24	D2/12*D27	D27	0

Lee is taking out a two-year loan to buy a computer for college. The loan amount is $1600, the monthly payment is $75.32, and the interest rate is 12.0%. A portion of the amortization table for Lee's loan is shown below.

```
= = = = = = = = = = = = = = = = = = = = = = = = = = = = = = = = = = = =
                    AMORTIZATION SCHEDULE
= = = = = = = = = = = = = = = = = = = = = = = = = = = = = = = = = = = =
1                   LOAN AMOUNT      PAYMENT          INTEREST RATE
2                   1600             75.32            0.120
3       PAYMENT     INTEREST PAID    PRINCIPAL PAID   BALANCE DUE
4       0           0                0                1600.00
5       1           16.00            59.32            1540.68
6       2           15.41            59.91            1480.77
```

EXERCISES

1. Explain why the amount of interest paid per month grows smaller.

2. Suppose Lee sends in $10 extra with his payment each month. Use the spreadsheet program to determine how much faster his loan will be paid.

12-3 Infinite Sequences and Series

Objectives

After studying this lesson, you should be able to:
- find the limit of the terms of an infinite sequence, and
- find the sum of an infinite geometric series.

Application

A vacuum pump removes 25% of the air from a sealed jar on each stroke of its piston. The jar contains 1 liter of air before the pump starts. Write a sequence to represent the amount of air remaining inside the jar after each stroke of the piston. Will there ever be no air inside the jar?

If 25% of the air is removed from the jar after each stroke, then 75% of the air remains in the jar. We can write a geometric sequence to represent the situation, with $a = 1$ and $r = 0.75$.

Stroke	Air Remaining	Terms of Sequence
1	$1(0.75)^1$	0.75
2	$1(0.75)^2$	0.5625
3	$1(0.75)^3$	0.4219
4	$1(0.75)^4$	0.3164
5	$1(0.75)^5$	0.2373
⋮	⋮	⋮
10	$1(0.75)^{10}$	0.0563
⋮	⋮	⋮
100	$1(0.75)^{100}$	3.2×10^{-13}
⋮	⋮	⋮
500	$1(0.75)^{500}$	3.4×10^{-63}
⋮	⋮	⋮
n	$1(0.75)^n$	ar^n

In theory, the sequence above can have infinitely many terms. Thus, it is called an **infinite sequence**. As n increases, the terms of the sequence decrease and get closer and closer to zero. The amount of air in the sealed jar will never actually be zero; however, the amount of air approaches zero as n increases without bound.

FYI...

The Thermos™ is actually the trade name for a vacuum flask in which the air between two bottles, one inside the other, has been drawn out by a vacuum pump.

Consider the infinite sequence $1, \frac{1}{2}, \frac{1}{3}, \frac{1}{4}, \frac{1}{5}, \cdots$, whose nth term is $\frac{1}{n}$. Several terms of this sequence are graphed below.

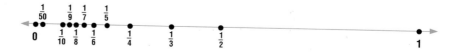

Notice that the terms approach 0 as n increases. Zero is called the *limit* of the terms in this sequence. This can be expressed as follows.

$$\lim_{n \to \infty} \frac{1}{n} = 0 \qquad \text{∞ is the symbol for infinity.}$$

This is read "the limit of 1 over n, as n approaches infinity, equals zero."

If a general expression for the nth term of a sequence is known, the limit can usually be found by substituting large values for n. Consider the following examples.

Example 1

Find the limit of each infinite sequence.

a. $2, \dfrac{2}{5}, \dfrac{2}{25}, \dfrac{2}{125}, \cdots, \dfrac{2}{5^{n-1}}, \cdots$

a. The 50th term is $\dfrac{2}{5^{49}}$, or about 1.13×10^{-34}. The 100th term is $\dfrac{2}{5^{99}}$, or about 1.27×10^{-69}. Notice that the values appear to approach 0. Therefore, $\lim\limits_{n \to \infty} \dfrac{2}{5^{n-1}} = 0$.

b. $\dfrac{7}{3}, \dfrac{20}{24}, \dfrac{45}{81}, \cdots, \dfrac{n^3 + 6n}{3n^3}, \cdots$

b. The 50th term is $\dfrac{(50)^3 + 6(50)}{3(50)^3}$ or 0.3341. The 100th term is $\dfrac{(100)^3 + 6(100)}{3(100)^3}$ or 0.3335. The 500th term is $\dfrac{(500)^3 + 6(500)}{3(500)^3}$ or 0.3333. Notice that the values appear to approach $\dfrac{1}{3}$. Therefore, $\lim\limits_{n \to \infty} \dfrac{n^3 + 6n}{3n^3} = \dfrac{1}{3}$.

The form of the expression for the nth term of a sequence can often be altered to make the limit easier to find.

Example 2

Evaluate each expression.

a. $\lim\limits_{n \to \infty} \dfrac{4n + 2}{n}$

a. $\lim\limits_{n \to \infty} \dfrac{4n + 2}{n} = \lim\limits_{n \to \infty} \left(4 + \dfrac{2}{n} \right)$ *Simplify.*

$\qquad\qquad\qquad = \lim\limits_{n \to \infty} 4 + \lim\limits_{n \to \infty} \dfrac{2}{n}$ *The limit of a sum equals the sum of the limits.*

But $\lim\limits_{n \to \infty} \dfrac{2}{n} = 0$. Therefore, $\lim\limits_{n \to \infty} 4 + \lim\limits_{n \to \infty} \dfrac{2}{n} = 4 + 0$ or 4.

Thus, the limit is 4.

b. $\lim\limits_{n\to\infty} \dfrac{n^2 + 2n - 5}{n^2 - 1}$

b. Divide the numerator and denominator by the highest power of n that occurs in either the numerator or the denominator, in this case, n^2.

$$\lim_{n\to\infty} \frac{n^2 + 2n - 5}{n^2 - 1} = \lim_{n\to\infty} \frac{\dfrac{n^2}{n^2} + \dfrac{2n}{n^2} - \dfrac{5}{n^2}}{\dfrac{n^2}{n^2} - \dfrac{1}{n^2}}$$

$$= \lim_{n\to\infty} \frac{1 + \dfrac{2}{n} - \dfrac{5}{n^2}}{1 - \dfrac{1}{n^2}} \qquad \textit{Simplify.}$$

$$= \frac{1 + 0 - 0}{1 - 0} \text{ or } 1 \qquad \lim_{n\to\infty} \frac{2}{n} = 0,\ \lim_{n\to\infty} \frac{5}{n^2} = 0,\ \lim_{n\to\infty} \frac{1}{n^2} = 0$$

Thus, the limit is 1.

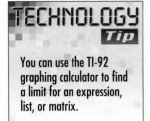
Limits do not exist for all sequences. If the absolute value of the terms of a sequence becomes arbitrarily great or if the terms do not approach a value, the sequence has no limit. Example 3 illustrates both of these cases.

Example 3

Evaluate each expression.

a. $\lim\limits_{n\to\infty} \dfrac{3n^2 + 4}{2n}$

a. $\lim\limits_{n\to\infty} \dfrac{3n^2 + 4}{2n} = \lim\limits_{n\to\infty} \left(\dfrac{3}{2}n + \dfrac{2}{n} \right)$

$\qquad\qquad = \lim\limits_{n\to\infty} \dfrac{3}{2}n + \lim\limits_{n\to\infty} \dfrac{2}{n}$

$\qquad\qquad = \lim\limits_{n\to\infty} \dfrac{3}{2}n$

Since $\dfrac{3}{2}n$ becomes increasingly large as n approaches infinity, the sequence has no limit.

b. $\lim\limits_{n\to\infty} \dfrac{(-1)^n n^3}{3n^3 + 4n}$

b. $\lim\limits_{n\to\infty} \dfrac{(-1)^n n^3}{3n^3 + 4n} = \lim\limits_{n\to\infty} (-1)^n \left(\dfrac{1}{3 + \dfrac{4}{n^2}} \right)$ *Divide the numerator and denominator by n^3.*

$\qquad\qquad\qquad = \lim\limits_{n\to\infty} (-1)^n \cdot \dfrac{1}{3}$ $\lim\limits_{n\to\infty}\left(\dfrac{1}{3 + \dfrac{4}{n^2}} \right) = \dfrac{1}{3}$

When n is even, $(-1)^n = 1$. When n is odd, $(-1)^n = -1$. Thus, the odd-numbered terms approach $-\dfrac{1}{3}$, and the even-numbered terms approach $\dfrac{1}{3}$. Therefore, the sequence has no limit.

An **infinite series** is the indicated sum of the terms of an infinite sequence. Consider the series $\frac{1}{2} + \frac{1}{4} + \frac{1}{8} + \frac{1}{16} + \cdots$. Since this is a geometric series, you can find the sum of the first 100 terms by using the formula $S_n = \frac{a_1 - a_1 r^n}{1 - r}$, where $r = \frac{1}{2}$.

$$S_{100} = \frac{\frac{1}{2} - \frac{1}{2}\left(\frac{1}{2}\right)^{100}}{1 - \frac{1}{2}} = \frac{\frac{1}{2} - \left(\frac{1}{2}\right)^{101}}{\frac{1}{2}} = 1 - \left(\frac{1}{2}\right)^{100}$$

Since $\left(\frac{1}{2}\right)^{100}$ is very close to 0, S_{100} is nearly equal to 1. No matter how many terms are added, the sum of the infinite series will never exceed 1. Thus, 1 is the sum of the infinite series.

Sum of an Infinite Series	**If S_n is the sum of n terms of a series, and S is a number such that $S > S_n$ for all n, and $S - S_n$ approaches zero as n increases without limit, then the sum of the infinite series is S.** $$\lim_{n \to \infty} S_n = S$$

If an infinite series has a limit, and thus, has a sum, the nth term of the series, a_n, must approach 0 as $n \to \infty$. Thus, if $\lim_{n \to \infty} a_n \neq 0$, the series has no sum. If $\lim_{n \to \infty} a_n = 0$, the series may or may not have a sum.

The formula for the sum of the first n terms of a geometric series can be written as follows.

$$S_n = \frac{a_1(1 - r^n)}{1 - r}, r \neq 1$$

Suppose $n \to \infty$; that is, the number of terms increases without limit. If $|r| > 1$, r^n increases without limit as $n \to \infty$. However, when $|r| < 1$, r^n approaches 0 as $n \to \infty$. Then, S_n approaches the value $\frac{a_1}{1 - r}$.

Sum of an Infinite Geometric Series	**The sum, S, of an infinite geometric series for which $|r| < 1$ is given by the following formula.** $$S_n = \frac{a_1}{1 - r}$$

Example 4

Find the sum of the series $\dfrac{1}{25} + \dfrac{1}{250} + \dfrac{1}{2500} + \cdots$.

In this series, $a_1 = \dfrac{1}{25}$ and $r = \dfrac{1}{10}$. Since $|r| < 1$, $S_n = \dfrac{a_1}{1 - r}$.

$$S_n = \frac{a_1}{1-r}$$

$$= \frac{\dfrac{1}{25}}{1 - \dfrac{1}{10}}$$

$$= \frac{2}{45}$$

The sum of the series is $\dfrac{2}{45}$.

Example 5

A tennis ball dropped from a height of 30 feet bounces 40% of the height from which it fell on each bounce. What is the vertical distance it travels before coming to rest?

The distance traveled is the actual path of the ball.

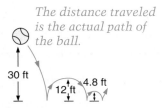

30 ft   12 ft   4.8 ft

The distance is given by the following series.

$30 + 12 + 12 + 4.8 + 4.8 + 1.92 + 1.92 + \cdots$

Notice that in this series, each term after the first term is repeated. This series can be rewritten as the sum of two infinite geometric series as follows.

$$\begin{aligned}
&30 + 12 + 4.8 + 1.92 + \cdots \\
+\ \ &\phantom{30 + {}} 12 + 4.8 + 1.92 + \cdots
\end{aligned}$$

For both series, $r = 0.4$, but $a_1 = 30$ in the first series and $a_1 = 12$ in the second series.

First Series

$S_n = \dfrac{30}{1 - 0.4}$ or 50

Second Series

$S_n = \dfrac{12}{1 - 0.4}$ or 20

Therefore, the ball travels $50 + 20$ or 70 feet before coming to rest.

You can use what you now know about infinite series to write repeating decimals as fractions. The first step is to write the repeating decimal as an infinite geometric series.

Example 6 | **Write 0.123123 · · · as a fraction.**

$$0.\overline{123} = \frac{123}{1000} + \frac{123}{1,000,000} + \frac{123}{1,000,000,000} + \cdots$$

$$S_n = \frac{a_1}{1 - r}$$

$$= \frac{\dfrac{123}{1000}}{1 - \dfrac{1}{1000}} \qquad a_1 = \frac{123}{1000} \text{ and } r = \frac{1}{1000}$$

$$= \frac{123}{999} \text{ or } \frac{41}{333}$$

Thus, $0.123123 \cdots = \dfrac{41}{333}$.

CHECKING FOR UNDERSTANDING

Communicating
Mathematics

Read and study the lesson to answer each question.

1. **Explain** why 0 is the limit of any geometric sequence for which $|r| < 1$.

2. A sequence has a limit if $|r|$ is
   a. greater than 1.   **b.** less than 1.

3. Choose the correct terms to make two true sentences. When determining the limit of a rational expression, if the highest power of the variable that appears in the numerator is (greater than, less than) the highest power of the variable that appears in the denominator, the limit (does, does not) exist. Use examples to explain your answer.

4. Do infinite arithmetic series have limits? Explain your answer.

Guided
Practice

Write an expression for the _n_th term of each sequence.

5. $1, 2, 4, 8, \cdots$

6. $5, 7, 9, 11, \cdots$

7. $3, 2, \dfrac{4}{3}, \dfrac{8}{9}, \cdots$

8. $3, \dfrac{5}{2}, \dfrac{7}{3}, \dfrac{9}{4}, \cdots$

Evaluate each limit, or state that the limit does not exist.

9. $\displaystyle\lim_{n \to \infty} \frac{1 - 2n}{5n}$

10. $\displaystyle\lim_{n \to \infty} \frac{n + 1}{n}$

11. $\displaystyle\lim_{n \to \infty} \frac{3n - 5}{n}$

Find the sum of each infinite series, or state that the sum does not exist.

12. $2\sqrt{2} + 8 + 16\sqrt{2} + \cdots$

13. $\dfrac{1}{20} + \dfrac{1}{40} + \dfrac{1}{80} + \cdots$

Write each repeating decimal as a fraction.

14. $0.454545 \cdots$

15. $0.888 \cdots$

16. $7.259259 \cdots$

EXERCISES

Practice

Evaluate each limit, or state that the limit does not exist.

17. $\lim\limits_{n\to\infty} \dfrac{1}{3^n}$

18. $\lim\limits_{n\to\infty} \dfrac{2n^2 - 6n}{5n^2}$

19. $\lim\limits_{n\to\infty} \dfrac{n^2 - 4}{2n}$

20. $\lim\limits_{n\to\infty} \dfrac{(n+2)(2n-1)}{n^2}$

21. $\lim\limits_{n\to\infty} \dfrac{n^2 + n - 3}{n^2}$

22. $\lim\limits_{n\to\infty} \dfrac{3n+1}{n-3}$

23. $\lim\limits_{n\to\infty} \dfrac{4n^2 + 5}{3n^2 + 2n}$

24. $\lim\limits_{n\to\infty} \dfrac{2n^3}{n^2 + 4n}$

25. $\lim\limits_{n\to\infty} \dfrac{2n + (-1)^n}{n^2}$

Write each repeating decimal as a fraction.

26. $0.555\cdots$

27. $0.2727\cdots$

28. $0.370370\cdots$

29. $0.3181818\cdots$

30. $3.242424\cdots$

31. $2.205205\cdots$

Find the sum of each infinite series, or state that the sum does not exist.

32. $\dfrac{2}{3} + \dfrac{1}{3} + \dfrac{1}{6} + \dfrac{1}{12} + \cdots$

33. $\sqrt{3} + 1 + \dfrac{\sqrt{3}}{3} + \cdots$

34. $\dfrac{2}{7} + \dfrac{4}{7} + \dfrac{8}{7} + \cdots$

35. $0.2 + 0.02 + 0.002 + \cdots$

36. $10 + 5 + 2.5 + \cdots$

37. $2 + 3 + \dfrac{9}{2} + \cdots$

Critical Thinking

38. Evaluate $\lim\limits_{n\to\infty} \dfrac{\sqrt{n}}{\sqrt{n}+1}$.

Applications and Problem Solving

39. Chemistry A vacuum pump removes 20% of the air in a sealed jar on each stroke of its piston. The jar contains 21 liters of air before the pump starts. After how many strokes will only 42% of the air remain?

40. Physics A tennis ball dropped from a height of 12 feet bounces 70% of the height from which it fell on each bounce. What is the vertical distance it travels before coming to rest?

41. Coordinate Geometry Two bugs start at the origin, one traveling along the positive half of the x-axis and the other traveling along the positive half of the y-axis. Each bug moves 1 unit on the first move, $\dfrac{1}{2}$ unit on the second move, and so on, where on any subsequent move the bug moves only half as far as the preceding move. The bugs always move at exactly the same time. How far apart will the bugs be after infinitely many moves?

Mixed Review

42. Find the product $\dfrac{1}{2}\begin{bmatrix} 9 & -3 \\ -6 & 6 \end{bmatrix}$. **(Lesson 2-2)**

43. Solve the equation $3k^2 + 3k + 2 = 0$ by using the quadratic formula. **(Lesson 4-2)**

44. **Aviation** An airplane flies at 150 km/h and heads 30° south of east. A 50-km/h wind blows in the direction 25° west of south. Find the ground speed and direction of the plane. **(Lesson 8-5)**

45. Find S_{14} for the arithmetic series for which $a_1 = 3.2$ and $d = 1.5$. **(Lesson 12-1)**

46. Form a sequence that has three geometric means between 2 and $\frac{1}{8}$. **(Lesson 12-2)**

47. **College Entrance Exam** Compare quantities A and B below.
Write A if quantity A is greater.
Write B if quantity B is greater.
Write C if the two quantities are equal.
Write D if there is not enough information to determine the relationship.

(A) the number of positive odd integer factors of 26

(B) the number of positive even integer factors of 26

DECISION MAKING

Finding Money for College

How much will it cost you to go to college? If you can get into the college of your choice, will you be able to afford to attend? The cost of going to college has skyrocketed in recent years, with expenses at a few schools topping $20,000 per year. Nevertheless, through smart planning and diligent searching for financial aid, you may be able to afford the school of your choice.

1. Use the list of colleges you compiled for the Decision Making feature in Chapter 2. Find the estimated annual cost of attending each school.

2. Research the financial aid that is available to students attending the schools on your list. What scholarships does the college offer? Will the school help you find a job? Can you get a loan from the college, the federal government, or some other agency? How much money can you earn on a part-time job?

3. Draw up a financial plan for meeting the expenses of the colleges on your list.

4. **Project** Work with your group to complete one of the projects listed at the right based on the information you gathered in Exercises 1–3.

PROJECTS

- *Conduct an interview.*
- *Write a book report.*
- *Write a proposal.*
- *Write an article for the school paper.*
- *Write a report.*
- *Make a display.*
- *Make a graph or chart.*
- *Plan an activity.*
- *Design a checklist.*

12-4 Convergent and Divergent Series

Objective

After studying this lesson, you should be able to:
- determine whether a series is convergent or divergent.

Application

To plan for his retirement, Al Early has been placing $1200 each year in an annuity that offers a 12% annual percentage rate. The table below shows the value of Mr. Early's annuity at the end of selected years.

Number of Years	Value of First $1200	Value of Annuity
1	$1200.00	$1200.00
2	$1200(1.12)^1 = \$1344.00$	$2544.00
3	$1200(1.12)^2 = \$1505.28$	$4049.28
4	$1200(1.12)^3 = \$1685.91$	$5735.19
5	$1200(1.12)^4 = \$1888.22$	$7623.42
⋮	⋮	⋮
10	$1200(1.12)^9 = \$3327.69$	$21,058.48
⋮	⋮	⋮
15	$1200(1.12)^{14} = \$5864.53$	$44,735.66
⋮	⋮	⋮
n	$1200(1.12)^{n-1}$	$1200\dfrac{(1.12)^n - 1}{0.12}$

The value of the annuity at the end of each year is a partial sum of the series that represents the value of the annuity. These partial sums form a sequence. As the number of terms used for the partial sums increases without bound, the value of the partial sums increases without bound. This result is to be expected, since the common ratio, r, of this sequence is 1.12 and $|r| > 1$. Since the sequence of partial sums does not approach a limit, the infinite series is said to **diverge**. If a sequence has a limit, then the related infinite series is said to **converge**.

Convergent and Divergent Series

> **If an infinite series has a sum, or limit, the series is convergent. If a series is not convergent, it is divergent.**

Example 1

Determine whether each series is convergent or divergent.

a. $\dfrac{2}{3} + \dfrac{1}{3} + \dfrac{1}{6} + \cdots$

a. This is a geometric series, with $r = \dfrac{1}{2}$. Since $|r| < 1$, the series has a limit. Therefore, the series is convergent.

(continued on the next page)

b. $-4 - 2 - 0 + 2 + \cdots$

b. This is an arithmetic series with $d = 2$. Since arithmetic series have no limits, the series is divergent.

c. $1 + 3 + 9 + 27 + \cdots$

c. This is a geometric series with $r = 3$. Since $|r| > 1$, the series has no limit. Therefore, the series is divergent.

When a series is neither arithmetic nor geometric, it is more difficult to determine whether the series is convergent or divergent. Several different techniques can be used. One test for convergence is the **ratio test**. This test can *only* be used when all terms of a series are positive. The test depends upon the ratio of consecutive terms of a series, which must be expressed in general form.

Ratio Test for Convergence of a Series

Let a_n and a_{n+1} represent two consecutive terms of a series of positive terms. Suppose $\lim\limits_{n \to \infty} \dfrac{a_{n+1}}{a_n}$ exists and that $r = \lim\limits_{n \to \infty} \dfrac{a_{n+1}}{a_n}$. The series is convergent if $r < 1$ and divergent if $r > 1$. If $r = 1$, the test provides no information.

Example 2

Use the ratio test to determine whether the following series are convergent or divergent.

a. $1 + \dfrac{1}{1 \cdot 2} + \dfrac{1}{1 \cdot 2 \cdot 3} + \dfrac{1}{1 \cdot 2 \cdot 3 \cdot 4} + \cdots$

a. First, find the nth term. Then, use the ratio test.

$$a_n = \frac{1}{1 \cdot 2 \cdots n}, a_{n+1} = \frac{1}{1 \cdot 2 \cdots (n+1)}$$

$$r = \lim_{n \to \infty} \frac{a_{n+1}}{a_n}$$

$$= \lim_{n \to \infty} \frac{\dfrac{1}{1 \cdot 2 \cdots (n+1)}}{\dfrac{1}{1 \cdot 2 \cdots n}}$$

$$= \lim_{n \to \infty} \frac{1 \cdot 2 \cdots n}{1 \cdot 2 \cdots (n+1)} \qquad \textit{Note that } 1 \cdot 2 \cdots (n+1) = 1 \cdot 2 \cdots n \cdot (n+1).$$

$$= \lim_{n \to \infty} \frac{1}{n+1} \text{ or } 0$$

Since $r < 1$, the series is convergent.

b. $\dfrac{3}{4} + \dfrac{7}{8} + \dfrac{11}{12} + \dfrac{15}{16} + \cdots$

b. $a_n = \dfrac{4n - 1}{4n}, \; a_{n+1} = \dfrac{4(n + 1) - 1}{4(n + 1)}$ or $\dfrac{4n + 3}{4n + 4}$

$$r = \lim_{n \to \infty} \dfrac{\dfrac{4n + 3}{4n + 4}}{\dfrac{4n - 1}{4n}}$$

$$= \lim_{n \to \infty} \dfrac{4n(4n + 3)}{(4n + 4)(4n - 1)}$$

$$= \lim_{n \to \infty} \dfrac{16n^2 + 12n}{16n^2 + 12n - 4}$$

$$= \lim_{n \to \infty} \dfrac{16 + \dfrac{12}{n}}{16 + \dfrac{12}{n} - \dfrac{4}{n^2}} \qquad \textit{Divide by the highest power of n.}$$

$$= \dfrac{16 + 0}{16 + 0 - 0} \text{ or } 1 \qquad \text{Since } r = 1, \text{ the test fails.}$$

When the ratio test does not determine if a series is convergent or divergent, other methods must be used.

Example 3

The series in this example is known as the harmonic series.

Determine whether the following series is convergent or divergent.

$$1 + \dfrac{1}{2} + \dfrac{1}{3} + \dfrac{1}{4} + \cdots$$

Suppose the terms are grouped as follows. Beginning after the second term, the number of terms in each successive group is doubled.

$$(1) + \left(\dfrac{1}{2}\right) + \left(\dfrac{1}{3} + \dfrac{1}{4}\right) + \left(\dfrac{1}{5} + \dfrac{1}{6} + \dfrac{1}{7} + \dfrac{1}{8}\right) + \left(\dfrac{1}{9} + \cdots + \dfrac{1}{16}\right) + \cdots$$

Notice that the first enclosed expression is greater than $\dfrac{1}{2}$, and the second is equal to $\dfrac{1}{2}$. Beginning with the third expression, each sum of enclosed terms is greater than $\dfrac{1}{2}$. Since there are an unlimited number of such expressions, the sum of the series is unlimited. Thus, the series is divergent.

A series can be compared to other series that are known to be convergent or divergent. The following list of series can be used for reference.

Summary of Series for Reference

1. **Convergent:** $a_1 + a_1 r + a_1 r^2 + \cdots + a_1 r^{n-1} + \cdots, |r| < 1$
2. **Divergent:** $a_1 + a_1 r + a_1 r^2 + \cdots + a_1 r^{n-1} + \cdots, |r| > 1$
3. **Divergent:** $a_1 + (a_1 + d) + (a_1 + 2d) + (a_1 + 3d) + \cdots$
4. **Divergent:** $1 + \dfrac{1}{2} + \dfrac{1}{3} + \dfrac{1}{4} + \dfrac{1}{5} + \cdots + \dfrac{1}{n} + \cdots$
5. **Convergent:** $1 + \dfrac{1}{2^p} + \dfrac{1}{3^p} + \cdots + \dfrac{1}{n^p} + \cdots, p > 1$

If a series has all positive terms, the **comparison test** can be used to determine whether the series is convergent or divergent.

Comparison Test	A series of positive terms is convergent if each term of the series is equal to or less than the value of the corresponding term of some convergent series of positive terms. The series is divergent if each term is equal to or greater than the corresponding term of some divergent series of positive terms.

Example 4

Use the comparison test to determine whether the following series are convergent or divergent.

a. $\dfrac{3}{4} + \dfrac{3}{5} + \dfrac{3}{6} + \dfrac{3}{7} + \cdots$

a. The general term of this series is $\dfrac{3}{n+3}$. The general term of the divergent series $1 + \dfrac{1}{2} + \dfrac{1}{3} + \dfrac{1}{4} + \dfrac{1}{5} + \cdots$ is $\dfrac{1}{n}$. Since $\dfrac{3}{n+3} > \dfrac{1}{n}$ for all $n > 1$, the series $\dfrac{3}{4} + \dfrac{3}{5} + \dfrac{3}{6} + \dfrac{3}{7} + \cdots$ is also divergent.

b. $\dfrac{1}{2 + 1^2} + \dfrac{1}{2 + 2^2} + \dfrac{1}{2 + 3^2} + \cdots$

b. The general term of this series is $\dfrac{1}{2 + n^2}$. The general term of the convergent series $1 + \dfrac{1}{2^2} + \dfrac{1}{3^2} + \dfrac{1}{4^2} + \cdots$ is $\dfrac{1}{n^2}$. Since $\dfrac{1}{2 + n^2} < \dfrac{1}{n^2}$ for all n, the series $\dfrac{1}{2 + 1^2} + \dfrac{1}{2 + 2^2} + \dfrac{1}{2 + 3^2} + \cdots$ is also convergent.

CHECKING FOR UNDERSTANDING

Communicating Mathematics

Read and study the lesson to answer each question.

1. a. **Make up** your own infinite arithmetic series.
   b. **State** the 50th and 100th terms of your series.
   c. **State** the sum of the first 50 terms and the sum of the first 100 terms of your series.
   d. Why do infinite arithmetic series not converge?

2. a. **Make up** an infinite geometric series in which $|r| > 1$.
   b. **State** the 50th and 100th terms of your series.
   c. **State** the sum of the first 50 terms and the sum of the first 100 terms of your series.
   d. Why does this type of infinite geometric series not converge?

3. Which test might you use if the ratio test finds the limit of the ratio of the $(n+1)$th term to the nth term is 1?

State whether each series is *arithmetic*, *geometric*, or *neither*. Then determine whether the series is *convergent* or *divergent*.

4. $1 + 3 + 5 + \cdots$

5. $\dfrac{1}{4} + \dfrac{3}{8} + \dfrac{5}{16} + \dfrac{7}{32} + \cdots$

6. $\dfrac{1}{2} + \dfrac{1}{8} + \dfrac{1}{32} + \cdots$

7. $\dfrac{1}{4} + \dfrac{5}{16} + \dfrac{3}{8} + \dfrac{7}{16} + \cdots$

8. $\dfrac{8}{3} + \dfrac{32}{9} + \dfrac{128}{27} + \cdots$

9. $64 + 32 + 16 + \cdots$

Use the ratio test to determine whether each series is *convergent* or *divergent*.

10. $1 + \dfrac{1}{2^2} + \dfrac{1}{3^3} + \dfrac{1}{4^4} + \cdots$

11. $1 + \dfrac{1}{2!} + \dfrac{1}{3!} + \cdots$

12. Use the comparison test to determine whether the series $\dfrac{2}{1} + \dfrac{3}{2} + \dfrac{4}{3} + \cdots$ is *convergent* or *divergent*.

EXERCISES

Use the ratio test to determine whether each series is *convergent* or *divergent*.

13. $\dfrac{1}{2} + \dfrac{2}{2^2} + \dfrac{3}{2^3} + \cdots$

14. $\dfrac{1}{1 \cdot 2} + \dfrac{1}{2 \cdot 2^2} + \dfrac{1}{3 \cdot 2^3} + \cdots$

15. $\dfrac{1}{1 \cdot 2} + \dfrac{1}{3 \cdot 4} + \dfrac{1}{5 \cdot 6} + \cdots$

16. $10 + \dfrac{10^2}{1 \cdot 2} + \dfrac{10^3}{1 \cdot 2 \cdot 3} + \cdots$

17. $1 + \dfrac{2}{1 \cdot 2 \cdot 3} + \dfrac{3}{1 \cdot 2 \cdot 3 \cdot 4 \cdot 5} + \cdots$

18. $\dfrac{1}{1 \cdot 2} + \dfrac{1}{1 \cdot 2 \cdot 3 \cdot 4} + \dfrac{1}{1 \cdot 2 \cdot 3 \cdot 4 \cdot 5 \cdot 6} + \cdots$

Use the comparison test to determine whether each series is *convergent* or *divergent*.

19. $\dfrac{1}{1^2} + \dfrac{1}{3^2} + \dfrac{1}{5^2} + \cdots$

20. $\dfrac{1}{2 \cdot 1} + \dfrac{1}{2 \cdot 2} + \dfrac{1}{2 \cdot 3} + \cdots$

21. $\dfrac{1}{2} + \dfrac{2}{3} + \dfrac{3}{4} + \cdots$

22. $\dfrac{2}{1} + \dfrac{3}{2} + \dfrac{4}{3} + \cdots$

Determine whether each series is *convergent* or *divergent*. Use the appropriate test.

23. $4 + 3 + \dfrac{9}{4} + \cdots$

24. $\dfrac{6}{5} + \dfrac{2}{5} + \dfrac{2}{15} + \cdots$

25. $\dfrac{1}{3 + 1^2} + \dfrac{1}{3 + 2^2} + \dfrac{1}{3 + 3^2} + \cdots$

26. $1 + \dfrac{1}{1 \cdot 2 \cdot 3} + \dfrac{1}{1 \cdot 2 \cdot 3 \cdot 4 \cdot 5} + \cdots$

27. Find the least positive number n such that the product of the first n terms exceeds $390{,}625$ when given the sequence $5^{\frac{1}{7}}, 5^{\frac{2}{7}}, 5^{\frac{3}{7}}, \cdots, 5^{\frac{n}{7}}$.

Applications and Problem Solving

28. **Employment** Matthew Stern and Beth Adams begin work and receive $2000 each the first month. Mr. Stern will receive a raise of $120 each month thereafter. Ms. Adams will receive a 5% raise each month thereafter.
    a. At the end of 12 months, how much are Mr. Stern and Ms. Adams making per month? Who has the higher monthly income?
    b. How much did Mr. Stern and Ms. Adams make during the 12-month period? Who has the higher yearly income?

29. **Physics** Kenneth drops a Super★Bouncer ball from his balcony. The balcony is 20 feet above the ground. If the ball rebounds 90% of the height from which it fell on each bounce, find the vertical distance that the ball travels before coming to rest.

Mixed Review

30. Determine whether the graph of $x^2 + y^2 = 16$ is symmetric with respect to the origin, the x-axis, the y-axis, the line $y = x$, or the line $y = -x$. **(Lesson 3-1)**

31. **Surveying** A building 60 feet tall is on top of a hill. A surveyor is at a point on the hill and observes that the angle of elevation to the top of the building measures 42° and the angle of elevation to the bottom measures 18°. How far is the surveyor from the bottom of the building? **(Lesson 5-6)**

32. Complete the table below for the equation $r = \dfrac{2}{\cos(\theta + 30°)}$. **(Lesson 9-4)**

θ	0°	15°	30°	70°
r				

33. Evaluate $\lim\limits_{n \to \infty} \dfrac{n-1}{n}$, or state that the limit does not exist. **(Lesson 12-3)**

34. **College Entrance Exam** Simplify the expression $\dfrac{0.25 + 0.25 + 0.25 + 0.25}{4}$.

MID-CHAPTER REVIEW

1. Form an arithmetic sequence that has seven arithmetic means between 5 and 17. **(Lesson 12-1)**

2. Find S_{23} for the arithmetic series for which $a_1 = -3$ and $d = 6$. **(Lesson 12-1)**

3. Form a geometric sequence that has three geometric means between 2 and $\dfrac{1}{8}$. **(Lesson 12-2)**

4. Find the sum of the first nine terms of the series $\dfrac{2}{3} + \dfrac{1}{3} + \dfrac{1}{6} + \cdots$. **(Lesson 12-2)**

5. Evaluate $\lim\limits_{n \to \infty} \dfrac{3n^2 + 4}{2n}$, or state that the limit does not exist. **(Lesson 12-3)**

6. Find the sum of the series $1 + \dfrac{2}{5} + \dfrac{4}{25} + \cdots$. **(Lesson 12-4)**

7. Write $0.3636\cdots$ as a fraction. **(Lesson 12-3)**

8. Determine if the series $1 + \dfrac{1}{5} + \dfrac{1}{25}\cdots$ is *convergent* or *divergent*. **(Lesson 12-4)**

9. **Finance** Carmella Jackson invests $2000 annually in an insurance annuity for 42 years. What will the annuity be worth if it has an APR of 8%? **(Lesson 12-2)**

12-5 Sigma Notation and the *n*th Term

Objective

After studying this lesson, you should be able to:

■ use sigma notation.

Application

The first electrically lighted theater was the Gaiety Theatre in London. It was lighted in 1878.

Stephanie Seim is responsible for purchasing the cushioned seats for the Theater in the Round. In the first row, there will be 10 seats. In each additional row, two more seats will be added. How many seats will there be in the 15th row? How many seats will there be in the *n*th row? How many seats does Ms. Seim need to order for the 15 rows that the theater has?

Since this sequence is arithmetic, we can use the formula for the *n*th term of an arithmetic sequence to find the number of seats in the 15th row.

$$a_n = a + (n-1)d$$
$$a_{15} = 10 + (15-1)2 \qquad \text{\textit{a = 10, n = 15, and d = 2}}$$
$$= 38 \qquad\qquad \text{There are 38 seats in the 15th row.}$$

To determine the number of seats in the *n*th row, we can use the same formula.

$$a_n = a + (n-1)d$$
$$= 10 + (n-1)2 \qquad \text{\textit{a = 10 and d = 2}}$$
$$= 10 + 2n - 2$$
$$= 2n + 8 \qquad\qquad \text{There are } 2n + 8 \text{ seats in the } n\text{th row.}$$

To determine the total number of seats in the theater, we can use the formula for the sum of an arithmetic series.

$$S_n = \frac{n}{2}(a + a_n)$$
$$= \frac{15}{2}(10 + 38) \qquad \text{\textit{n = 15, a = 10, and } } a_n = 38$$
$$= 360 \qquad\qquad \text{Ms. Seim should order 360 seats for the theater.}$$

The series $10 + 12 + 14 + \cdots + 38$ represents the number of seats in the 15 rows of the theater. In mathematics, the Greek letter sigma, Σ, is often used to indicate a sum or series. The series above can be written as $\sum_{n=1}^{15}(2n + 8)$. This represents a series of terms that are obtained by multiplying 2 by *n* and adding 8, first for $n = 1$, then for $n = 2$, and so on to $n = 15$.

$$\sum_{n=1}^{15}(2n + 8) = [2(1) + 8] + [2(2) + 8] + [2(3) + 8] + \cdots + [2(15) + 8]$$
$$= 10 + 12 + 14 + \cdots + 38 \text{ or } 360$$

$\sum_{n=1}^{15}(2n + 8)$ is read "the summation from n = 1 to 15 of 2 times n plus 8."

The variable used with the summation symbol is called the **index of summation.** In the previous example, the index of summation is n. The number above the sigma is the maximum value of the index of summation.

Example 1

Write each of the following in expanded form and find the sum.

a. $\displaystyle\sum_{b=3}^{5} (2^b - b)$

a. $\displaystyle\sum_{b=3}^{5} (2^b - b) = (2^3 - 3) + (2^4 - 4) + (2^5 - 5)$

$$= \quad 5 \quad + \quad 12 \quad + \quad 27$$

$$= \quad 44$$

b. $\displaystyle\sum_{n=1}^{\infty} 3\left(\frac{1}{2}\right)^{n+1}$

b. $\displaystyle\sum_{n=1}^{\infty} 3\left(\frac{1}{2}\right)^{n+1} = 3\left(\frac{1}{2}\right)^{1+1} + 3\left(\frac{1}{2}\right)^{2+1} + 3\left(\frac{1}{2}\right)^{3+1} + 3\left(\frac{1}{2}\right)^{4+1} + \cdots$

$$= \quad \frac{3}{4} \quad + \quad \frac{3}{8} \quad + \quad \frac{3}{16} \quad + \quad \frac{3}{32} \quad + \cdots$$

This is an infinite series. Use the formula $S_n = \dfrac{a}{1 - r}$.

$S_n = \dfrac{a}{1 - r}$

$\quad = \dfrac{\frac{3}{4}}{1 - \frac{1}{2}} \qquad a = \dfrac{3}{4} \text{ and } r = \dfrac{1}{2}$

$\quad = \dfrac{3}{2}$

Therefore, $\displaystyle\sum_{n=1}^{\infty} 3\left(\frac{1}{2}\right)^{n+1} = \frac{3}{2}$.

A series in expanded form can be written using sigma notation if a general formula can be written for the nth term of the series.

Example 2

Express the series $10 + 17 + 26 + 37 + \cdots + 122$ using sigma notation.

Notice that each term is 1 more than a perfect square. Thus, the nth term of the series is $n^2 + 1$. Since $10 = 3^2 + 1$ and $122 = 11^2 + 1$, the index of summation goes from $n = 3$ to $n = 11$.

Therefore, $10 + 17 + 26 + 37 + \cdots + 122 = \displaystyle\sum_{n=3}^{11} (n^2 + 1)$.

Not all sequences are arithmetic or geometric. Some important sequences are generated by products of consecutive integers. The product $n(n-1)(n-2)\cdots 3 \cdot 2 \cdot 1$ is called **n factorial** and is symbolized **n!**.

Definition of n Factorial	**The expression n! (n factorial) is defined as follows for n, an integer greater than zero.** $$n! = n(n-1)(n-2)\cdots 1$$

By definition, $0! = 1! = 1$.

Example 3

Evaluate each expression.

a. **6!**

b. $\dfrac{9!}{5!3!}$

a. $6! = 6 \cdot 5 \cdot 4 \cdot 3 \cdot 2 \cdot 1$

$= 720$

b. $\dfrac{9!}{5!3!} = \dfrac{9 \cdot 8 \cdot 7 \cdot 6 \cdot 5!}{5!3!}$

$= 504$

Example 4

Express the series $-\dfrac{3}{1} + \dfrac{9}{2} - \dfrac{27}{6} + \dfrac{81}{24}$ **using sigma notation.**

The sequence representing the numerators is 3, 9, 27, 81. This is a geometric sequence with a common ratio of 3. Thus, the nth term can be represented by 3^n.

Because the series has alternating signs, one factor for the general term of the series is $(-1)^n$. Thus, when n is odd, the terms are negative, and when n is even, the terms are positive.

The sequence representing the denominators is 1, 2, 6, 24. This sequence is generated by factorials.

$1! = 1$
$2! = 1 \cdot 2$ or 2
$3! = 1 \cdot 2 \cdot 3$ or 6
$4! = 1 \cdot 2 \cdot 3 \cdot 4$ or 24

Therefore, $-\dfrac{3}{1} + \dfrac{9}{2} - \dfrac{27}{6} + \dfrac{81}{24} = \displaystyle\sum_{n=1}^{4} \dfrac{(-1)^n 3^n}{n!}$.

You can check this answer by substituting values of n into the general term.

CHECKING FOR UNDERSTANDING

Communicating Mathematics

Read and study the lesson to answer each question.

1. In the application at the beginning of the lesson, we wrote the series $10 + 12 + 14 + \cdots + 38$ as $\displaystyle\sum_{n=1}^{15}(2n + 8)$. Could $\displaystyle\sum_{n=0}^{14}(2n + 10)$ represent the same series? Why or why not?

2. State the number of terms in each series.

a. $\displaystyle\sum_{n=0}^{5} 5n - 12$

b. $\displaystyle\sum_{k=25}^{125} 425(2)^{k-1}$

c. $\displaystyle\sum_{j=-4}^{4} \left(\frac{3}{4}\right)^{j}$

3. If the index of summation begins with 1, what does the number above the sigma represent?

4. Given the series $\displaystyle\sum_{k=1}^{6} (4k + 3)$, identify two other series expressed in sigma notation that would generate the same terms.

Guided Practice

Write each expression in expanded form and find the sum.

5. $\displaystyle\sum_{j=1}^{4} (j + 2)$

6. $\displaystyle\sum_{a=4}^{7} 2a$

7. $\displaystyle\sum_{p=5}^{7} (3p + 2)$

8. $\displaystyle\sum_{a=0}^{4} (0.5 + 2^{a})$

9. $\displaystyle\sum_{b=0}^{\infty} 6\left(\frac{2}{3}\right)^{b}$

Express each series using sigma notation.

10. $3 + 6 + 9 + 12$

11. $16 + 8 + 4 + 2 + 1$

12. $19 + 18 + 16 + 12 + 4$

13. $\dfrac{2}{5} + \dfrac{3}{5} + \dfrac{4}{5} + \cdots$

Evaluate each expression.

14. $7!$

15. $3(6!)$

16. $\dfrac{12!}{10!}$

EXERCISES

Practice

Evaluate each expression.

17. $9!$

18. $2(5!)$

19. $3!4!$

20. $5!3!$

21. $\dfrac{10!}{8!}$

22. $\dfrac{6!}{4!3!}$

Write each expression in expanded form and find the sum.

23. $\displaystyle\sum_{r=1}^{3} (r - 3)$

24. $\displaystyle\sum_{k=5}^{8} 3k$

25. $\displaystyle\sum_{b=4}^{8} (4 - 2b)$

26. $\displaystyle\sum_{z=1}^{9} (10 - z)$

27. $\displaystyle\sum_{b=2}^{5} (b^{2} + b)$

28. $\displaystyle\sum_{k=2}^{7} (5 - 2k)$

29. $\displaystyle\sum_{n=3}^{6} (3^{n} + 1)$

30. $\displaystyle\sum_{m=1}^{4} 4^{m}$

31. $\displaystyle\sum_{p=1}^{4} \left(3^{p-1} + \frac{1}{2}\right)$

32. $\displaystyle\sum_{r=0}^{\infty} 5(0.2)^{r}$

33. $\displaystyle\sum_{k=1}^{\infty} 4\left(\frac{1}{2}\right)^{k}$

34. $\displaystyle\sum_{j=0}^{\infty} 5\left(\frac{3}{4}\right)^{j}$

Express each series using sigma notation.

35. $10 + 20 + 30 + 40 + 50$

36. $2 + 4 + 8 + \cdots + 64$

37. $3 + 6 + 12 + \cdots + 48$

38. $3 + 30 + 300 + 3000$

39. $\dfrac{1}{2} + \dfrac{1}{3} + \dfrac{1}{4} + \cdots + \dfrac{1}{10}$

40. $\dfrac{1}{2} + \dfrac{1}{4} + \dfrac{1}{6} + \cdots + \dfrac{1}{14}$

41. $11 + 9 + 7 + 5$

42. $-8 + 4 - 2 + 1$

43. $2 + 4 + 6 + \cdots$

44. $4 + 9 + 14 + \cdots$

45. $\dfrac{2}{3} + \dfrac{2}{4} + \dfrac{2}{5} + \cdots$

46. $5 + 25 + 125 + \cdots$

47. $1 + \dfrac{1}{2} + \dfrac{2}{3} + \dfrac{6}{4} + \dfrac{24}{5} + \cdots$

48. $\dfrac{2 \cdot 4}{3} + \dfrac{6 \cdot 8}{8} + \dfrac{24 \cdot 16}{15} + \cdots$

Simplify as far as possible. Assume that x and y are positive integers, $x > y$, and $x > 2$.

49. $\dfrac{x!}{(x-2)!}$

50. $\dfrac{(x+1)!}{(x-1)!}$

51. $\dfrac{(x-y)!}{(x-y-1)!}$

Programming

52. The program below will evaluate a sum given in summation notation. The formula for the nth term must be stored in Y_1. You must input values for the lower and upper bounds of summation.

```
Prgm12:SIGMASUM
:ClrHome
:Disp "NTH TERM MUST"
:Disp "BE IN Y₁"
:Disp "LOWER BOUND ="
:Input L
:Disp "UPPER BOUND ="
:Input U
:0 → S
:L → X
:Lbl 1
:S + Y₁ → S
:X + 1 → X
:If X = U + 1
:Goto 2
:Goto 1
:Lbl 2
:Disp "SUM ="
:Disp S
```

Clears the text screen.
Displays the quoted message.

Accepts a value for L.

Sets the initial value of the sum to 0.
Stores the lower bound in X.
Sets a marker at position 1.
Adds the value of Y_1 to previous sum value.
Increases X by 1.
Checks if upper bound has been reached.
Program execution goes to line "Lbl 2."

Sets a marker at position 2.

Displays the value stored in S.

Use the program to find the sum of each sequence.

a. $\displaystyle\sum_{x=3}^{6} (x + 2)$

b. $\displaystyle\sum_{x=3}^{7} (2x - 1)$

c. $\displaystyle\sum_{x=3}^{5} (2^x - x)$

Determine whether each equation is *true* or *false*. Explain your answer.

53. $\displaystyle\sum_{k=0}^{5} a^k + \sum_{n=6}^{10} a^n = \sum_{b=0}^{10} a^b$

54. $\displaystyle\sum_{r=3}^{7} 3^r + \sum_{a=7}^{9} 3^a = \sum_{j=3}^{9} 3^j$

55. $\displaystyle\sum_{n=1}^{10} (5+n) = \sum_{m=0}^{9} (4+m)$

56. $\displaystyle\sum_{r=2}^{8} (2r-3) = \sum_{s=3}^{9} (2s-5)$

57. $\displaystyle 2\sum_{k=3}^{7} k^2 = \sum_{k=3}^{7} 2k^2$

58. $\displaystyle 3\sum_{n=1}^{5} (n+3) = \sum_{n=1}^{15} (n+3)$

59. Finance Suppose Dr. Cynthia Morgan invests $1500 in an annuity every six months for 30 years. What will the annuity be worth if it has an APR of 8% and is compounded annually?

60. Ballooning After one minute, a hot air balloon rose 90 feet. After that time, each succeeding minute the balloon rose 70% as far as it did the previous minute.
  a. How far above Earth was the balloon after 8 minutes?
  b. What was the maximum height of the balloon?

61. Agriculture If Ms. Huffman harvests her apple crop now, the yield will average 120 pounds per tree. Also, she will be able to sell the apples for $0.48 per pound. However, she knows that if she waits, her yield will increase by about 10 pounds per week, while the selling price will decrease by $0.03 per pound per week. **(Lesson 3-7)**
  a. How many weeks should Ms. Huffman wait in order to maximize profit?
  b. What is the maximum profit?

62. Find the value of k in $(x^2 + 8x + k) \div (x - 2)$ so that the remainder is zero. **(Lesson 4-3)**

63. Find the value of $\sin(\text{Tan}^{-1} 1 - \text{Sin}^{-1} 1)$. **(Lesson 6-5)**

64. Find the inner product $(1, 3) \cdot (3, -2)$ and state whether the vectors are perpendicular. Write *yes* or *no*. **(Lesson 8-4)**

65. Write the equation of the parabola with vertex at $(6, -1)$ and focus at $(3, -1)$. **(Lesson 10-2)**

66. Express $\sqrt[6]{64a^6 b^{-2}}$ using rational exponents. **(Lesson 11-1)**

67. Write $0.222\cdots$ as a fraction. **(Lesson 12-3)**

68. Determine whether the series $\dfrac{1}{9} + \dfrac{1}{27} + \dfrac{1}{81} + \cdots$ is *convergent* or *divergent*. **(Lesson 12-5)**

69. College Entrance Exam Choose the best answer. Which of the following equations are equivalent?
  I. $2x + 4y = 8$
  II. $3x + 6y = 12$
  III. $4x + 8y = 8$
  IV. $6x + 12 = 16$
  (A) I and II only
  (B) I and IV only
  (C) II and IV only
  (D) III and IV only
  (E) I, II, and III

12-6 The Binomial Theorem

Objective

After studying this lesson, you should be able to:
- use the binomial theorem to expand binomials.

Application

Blaise Pascal (1623–1662) was a French mathematician who was considered by many to be a true genius. At the age of 12, he discovered many of the theorems of elementary geometry without any assistance. By age 14, he was participating in weekly gatherings with a group of noted French mathematicians. At 18, he invented the first computing machine. Throughout his life, he made contributions to probability, calculus, and geometry.

In this lesson, you will learn about a sequence of numbers called **Pascal's triangle**. Pascal first used this triangle to answer a question posed to him by a gambler about how to divide the stakes when a dice game is interrupted. However, this triangle had appeared in China as early as A.D. 1303, when it was published in Chu Shih-chieh's *Precious Mirror of the Four Elements*. It is also believed that Omar Khayyam, a Persian mathematician who lived about A.D. 1100, had some knowledge of the coefficients of the triangle. In addition, Niccolo Tartaglia of Italy claimed the triangle as his own invention in 1556.

As you examine the triangle, note that if two consecutive numbers in any row are added, the sum is a number in the following row. The three numbers form a triangle, as shown below.

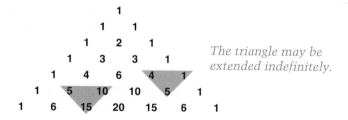

The triangle may be extended indefinitely.

One of the applications of Pascal's triangle is the expansion of binomial expressions of the form $(x + y)^n$. In the first row, $n = 0$. In any row, the second number indicates the power to which the binomial is raised. In the expansion process, the coefficients of the terms, written in decreasing order of x or y, are the same as the numbers in the triangle.

1	$(x + y)^0 =$	1
1 1	$(x + y)^1 =$	$1x + 1y$
1 2 1	$(x + y)^2 =$	$1x^2 + 2xy + 1y^2$
1 3 3 1	$(x + y)^3 =$	$1x^3 + 3x^2y + 3xy^2 + 1y^3$
1 4 6 4 1	$(x + y)^4 =$	$1x^4 + 4x^3y + 6x^2y^2 + 4xy^3 + 1y^4$
1 5 10 10 5 1	$(x + y)^5 =$	$1x^5 + 5x^4y + 10x^3y^2 + 10x^2y^3 + 5xy^4 + 1y^5$

The following patterns are seen in the expansions.

1. The expansion of $(x + y)^n$ has $n + 1$ terms.
2. The first term is x^n and the last term is y^n.
3. In successive terms, the exponent of x decreases by 1, and the exponent of y increases by 1.
4. The degree of each term is n.
5. In any term, if the coefficient is multiplied by the exponent of x and the product is divided by the number of that term, the result is the coefficient of the following term.
6. The coefficients are symmetric. That is, the first term and the last term have the same coefficient. The second term and the second from the last term have the same coefficient, and so on.

Example 1

Use Pascal's triangle to expand each binomial.
a. $(x + y)^6$

a. First, write the series without the coefficients. Recall that the expansion should have $6 + 1$ or 7 terms, with the first term being x^6 and the last term being y^6. Also note that the exponents of x should decrease from 6 to 0 while the exponents of y should increase from 0 to 6, while the degree of each term is 6.

$$x^6 + x^5y + x^4y^2 + x^3y^3 + x^2y^4 + xy^5 + y^6 \qquad y^0 = 1, x^0 = 1$$

Then, use the numbers in the seventh row of Pascal's triangle as the coefficients of the terms.

$$\begin{array}{ccccccc} 1 & 6 & 15 & 20 & 15 & 6 & 1 \end{array}$$
$$(x + y)^6 = x^6 + 6x^5y + 15x^4y^2 + 20x^3y^3 + 15x^2y^4 + 6xy^5 + y^6$$

b. $(2x + y)^7$

b. Extend Pascal's triangle to the eighth row.

$$\begin{array}{cccccccc} 1 & 7 & 21 & 35 & 35 & 21 & 7 & 1 \end{array}$$

Then, write the expansion and simplify each term. Replace each x with $2x$.

$$(2x + y)^7 = (2x)^7 + 7(2x)^6y + 21(2x)^5y^2 + 35(2x)^4y^3 + 35(2x)^3y^4 +$$
$$21(2x)^2y^5 + 7(2x)y^6 + y^7$$
$$= 128x^7 + 448x^6y + 672x^5y^2 + 560x^4y^3 + 280x^3y^4 +$$
$$84x^2y^5 + 14xy^6 + y^7$$

You can use Pascal's triangle to solve real-world problems in which there are only two outcomes for each event. For example, to determine the distribution of answers on true-false tests, to determine the combinations of heads and tails when tossing a coin, or to determine the possible sequences of boys and girls in a family.

Example 2

APPLICATION

Family

There are seven children in the Jones family. Of these seven children, there are at least four girls. How many of the possible groups of boys and girls have at least four girls?

To find the number of possible groups, expand $(g + b)^7$. Use the eighth row of Pascal's triangle for the expansion.

$$g^7 + 7g^6b + 21g^5b^2 + 35g^4b^3 + 35g^3b^4 + 21g^2b^5 + 7gb^6 + b^7$$

To have at least four girls means that there could be 4, 5, 6, or 7 girls in the family. The total number of ways to have at least four girls is the same as the sum of the coefficients of g^7, g^6b, g^5b^2, and g^4b^3. This sum is $1 + 7 + 21 + 35$ or 64.

Thus, there are 64 possible groups of boys and girls in which there are at least four girls.

The general form of the expansion of $(x + y)^n$ can also be determined by the **binomial theorem**.

The Binomial Theorem

If n is a positive integer, then the following is true.

$$(x + y)^n = x^n + nx^{n-1}y + \frac{n(n-1)}{1 \cdot 2}x^{n-2}y^2 +$$
$$\frac{n(n-1)(n-2)}{1 \cdot 2 \cdot 3}x^{n-3}y^3 + \cdots + y^n$$

Example 3

Use the binomial theorem to expand $(x - 2y)^4$.

$$(x - 2y)^4 = x^4 + 4x^3(-2y) + \frac{4 \cdot 3x^2(-2y)^2}{1 \cdot 2} + \frac{4 \cdot 3 \cdot 2x(-2y)^3}{1 \cdot 2 \cdot 3} + \frac{4 \cdot 3 \cdot 2(-2y)^4}{1 \cdot 2 \cdot 3 \cdot 4}$$
$$= x^4 - 8x^3y + 24x^2y^2 - 32xy^3 + 16y^4$$

An equivalent form of the binomial theorem uses both sigma and factorial notation. It is written as follows, where n is a positive integer and r is a positive integer or zero.

$$(x + y)^n = \sum_{r=0}^{n} \frac{n!}{r!(n-r)!}x^{n-r}y^r$$

You can use this form of the binomial theorem to find individual terms of an expansion.

Example 4

Find the third term of $(x - 3y)^5$.

$$(x - 3y)^5 = \sum_{r=0}^{5} \frac{5!}{r!(5-r)!} x^{5-r} (-3y)^r$$ *Since r increases from 0 to n, r is one*
less than the number of the term.

To find the third term, evaluate the general term for $r = 2$.

$$\frac{5!}{2!(5-2)!} x^{5-2} (-3y)^2 = \frac{5 \cdot 4 \cdot 3!}{2!3!} x^3 (-3y)^2$$

$$= 90x^3 y^2$$

The third term of $(x - 3y)^5$ is $90x^3 y^2$.

CHECKING FOR UNDERSTANDING

Communicating
Mathematics

Read and study the lesson to answer each question.

1. **Research** Pascal. Then write a one-page paper about his contributions to the field of mathematics.

2. **Write** the terms of the ninth and tenth rows of Pascal's triangle.

3. **Write** an expression to represent the sum of the terms in each row of Pascal's triangle.

4. The number of terms in the expansion of $(a + b)^n$ is __?__, and the degree of each term is __?__.

Guided
Practice

5. Use Pascal's triangle to expand $(a + b)^7$.

Use the binomial theorem to expand each binomial.

6. $(n + 2)^7$ 7. $(4 - b)^4$ 8. $(2x - 3y)^3$

Find the designated term of each binomial expansion.

9. 4th term of $(a + b)^7$ 10. 4th term of $(a - \sqrt{2})^8$

EXERCISES

Practice

Use Pascal's triangle to expand each binomial.

11. $(x + y)^5$ 12. $(r - s)^6$

13. $(x - 2)^7$ 14. $(2a + b)^7$

Use the binomial theorem to expand each binomial.

15. $(x + 3)^6$ 16. $(2 + d)^4$ 17. $(2x + y)^6$

18. $(3x - y)^5$ 19. $(2x + \sqrt{3})^4$ 20. $(3a^2 - 2b)^4$

Find the designated term of each binomial expansion.

21. 5th term of $(2x - y)^9$ 22. 5th term of $(2a - 3b)^8$

23. 7th term of $\left(x - \frac{1}{2}y\right)^{10}$

24. 6th term of $(3x - 2y)^{11}$

25. Find the sum of the coefficients for the variables in the expansion of $(4x - 1)^{16}$.

Critical Thinking **26.** Assume that the binomial theorem is true for all rational values of n. Find an approximate value of $\sqrt{6}$ by applying the binomial theorem to the equivalent form of $\sqrt{6}$ derived below.

$$\sqrt{6} = \sqrt{4 + 2} = 2\sqrt{1 + \frac{2}{4}} = 2\left(1 + \frac{1}{2}\right)^{\frac{1}{2}}$$

Applications and Problem Solving **27. Family** If six children are born into the same family, how many of the possible groups of boys and girls have at least two boys?

28. Coins A coin is flipped four times. How many of the possible sets of heads and tails have:

a. 0 tails

b. 1 tail

c. 2 tails

d. 3 tails

29. Chemistry If the half-life of Polonium-210 is 138 days, what percent of the element remains after 138 days, 276 days, 828 days, and d days?

Mixed Review **30.** Given $f(x) = 4 + 6x - x^3$, find $f(9)$. **(Lesson 1-1)**

31. Find the value of $\csc 180°$ without using a calculator. **(Lesson 5-4)**

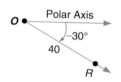

32. Name four different pairs of polar coordinates that represent point R. **(Lesson 9-1)**

33. Write $\sum_{n=1}^{\infty} 3\left(\frac{1}{2}\right)^{n+1}$ in expanded form and find the sum. **(Lesson 12-5)**

34. College Entrance Exam Choose the best answer.
For the triangles shown below, the perimeter of $\triangle ABC$ equals the perimeter of $\triangle XYZ$. If $\triangle ABC$ is equilateral, what is the length of side AB?

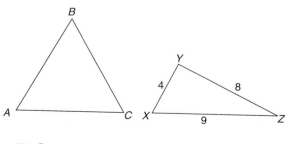

(A) 4   **(B)** 5   **(C)** 7   **(D)** 9   **(E)** 15

12-7 Special Sequences and Series

Objectives

After studying this lesson, you should be able to:
- evaluate e^x by using the exponential series,
- use Euler's formula to write the exponential form of a complex number, and
- evaluate the logarithms of negative numbers.

Application

The **Fibonacci sequence** describes patterns of numbers found in nature. This sequence was presented by an Italian mathematician named Leonardo de Pisa, also known as Fibonacci, in 1201. The first two numbers in the sequence are 1; that is, $a_1 = 1$ and $a_2 = 1$. Each additional term in the sequence is generated by adding the two previous terms. The first seven terms of the sequence are 1, 1, 2, 3, 5, 8, and 13.

FYI...

It is ironic that the sequence for which Fibonacci is famous is from one obscure problem in his book, *Liber Abaci*. This book is more well known for its role in introducing Hindu-Arabic numerals to Europeans.

In nature, plants with spiral leaf-growth patterns have spiral ratios that are adjacent terms in the Fibonacci sequence. In a daisy, for example, the ratio is 21 to 34, in a pine cone, the ratio is 5 to 8, and in a pineapple, the ratio is 8 to 13. Numbers in this sequence also describe other patterns in nature, such as spiral galaxies and the population growth of rabbits.

Another important series is the series that is used to define the irrational number e. In 1748, Leonhard Euler (pronounced OY ler), a Swiss mathematician, published a work in which he developed this irrational number. In his honor, the number is called e, the Euler number. The number is the sum of the following infinite series.

$$e = 1 + \frac{1}{1!} + \frac{1}{2!} + \frac{1}{3!} + \frac{1}{4!} + \frac{1}{5!} + \cdots + \frac{1}{n!} + \cdots$$

The binomial theorem can be used to derive the series for e, as follows. Let v be any number and apply the binomial theorem to $(1 + v)^n$.

$$(1 + v)^n = 1 + nv + \frac{n(n-1)}{2!}v^2 + \frac{n(n-1)(n-2)}{3!}v^3 + \cdots$$

Now let k be a variable such that $v = \frac{1}{k}$ and let x be a variable such that $kx = n$. Then, substitute these values for v and n.

$$\left(1 + \frac{1}{k}\right)^{kx} = 1 + kx\left(\frac{1}{k}\right) + \frac{kx(kx-1)}{2!}\left(\frac{1}{k}\right)^2 + \frac{kx(kx-1)(kx-2)}{3!}\left(\frac{1}{k}\right)^3 + \cdots$$

$$= 1 + x + \frac{x\left(x - \frac{1}{k}\right)}{2!} + \frac{x\left(x - \frac{1}{k}\right)\left(x - \frac{2}{k}\right)}{3!} + \cdots$$

Then, find the limit of $\left(1 + \dfrac{1}{k}\right)^{kx}$ as k increases without bound.

Recall that $\lim\limits_{k \to \infty} \dfrac{1}{k} = 0$.

$$\lim_{k \to \infty}\left(1 + \frac{1}{k}\right)^{kx} = 1 + x + \frac{x^2}{2!} + \frac{x^3}{3!} + \frac{x^4}{4!} + \frac{x^5}{5!} + \cdots$$

Let $x = 1$.

$$\lim_{k \to \infty}\left(1 + \frac{1}{k}\right)^{k} = 1 + 1 + \frac{1}{2!} + \frac{1}{3!} + \frac{1}{4!} + \cdots$$

Thus, e can be defined as follows.

| Definition of e | $$e = \lim_{k \to \infty}\left(1 + \frac{1}{k}\right)^{k} = 1 + 1 + \frac{1}{2!} + \frac{1}{3!} + \frac{1}{4!} + \cdots$$ |

e^x can be approximated by using the following series. This series is often called the **exponential series.**

$$e^x = 1 + x + \frac{x^2}{2!} + \frac{x^3}{3!} + \frac{x^4}{4!} + \frac{x^5}{5!} + \cdots$$

Euler's name is associated with a number of important mathematical relationships. Among these is the relationship between the exponential series and a series called the **trigonometric series.** The trigonometric series for $\cos x$ and $\sin x$ are given below.

$$\cos x = 1 - \frac{x^2}{2!} + \frac{x^4}{4!} - \frac{x^6}{6!} + \frac{x^8}{8!} - \cdots$$

$$\sin x = x - \frac{x^3}{3!} + \frac{x^5}{5!} - \frac{x^7}{7!} + \frac{x^9}{9!} - \cdots$$

These two series are convergent for all values of x. By replacing x with any angle measure expressed in radians and carrying out the computations, approximate values of the trigonometric functions can be found to any desired degree of accuracy.

Example 1

Use the first five terms of the trigonometric series to find the value of $\sin \dfrac{\pi}{4}$ to four decimal places.

$\dfrac{\pi}{4} \approx 0.7854$

$$\sin x \approx x - \frac{x^3}{3!} + \frac{x^5}{5!} - \frac{x^7}{7!} + \frac{x^9}{9!}$$

$$\sin (0.7854) \approx 0.7854 - \frac{(0.7854)^3}{3!} + \frac{(0.7854)^5}{5!} - \frac{(0.7854)^7}{7!} + \frac{(0.7854)^9}{9!}$$

$$\approx 0.7854 - \frac{0.4845}{6} + \frac{0.2989}{120} - \frac{0.1843}{5040} + \frac{0.1137}{362,880}$$

$$\approx 0.7071 \qquad \textit{Compare this result to the actual value.}$$

Another very important formula is derived by replacing x by $i\alpha$ in the exponential series, where i is the imaginary unit and α is the measure of an angle in radians.

$$e^{i\alpha} = 1 + i\alpha + \frac{(i\alpha)^2}{2!} + \frac{(i\alpha)^3}{3!} + \frac{(i\alpha)^4}{4!} + \cdots$$

$$= 1 + i\alpha - \frac{\alpha^2}{2!} - i\frac{\alpha^3}{3!} + \frac{\alpha^4}{4!} + \cdots$$

Group the terms according to whether or not they contain the factor i.

$$e^{i\alpha} = \left(1 - \frac{\alpha^2}{2!} + \frac{\alpha^4}{4!} - \frac{\alpha^6}{6!} + \cdots\right) + i\left(\alpha - \frac{\alpha^3}{3!} + \frac{\alpha^5}{5!} - \frac{\alpha^7}{7!} + \cdots\right)$$

Notice that the real part is exactly $\cos\ \alpha$ and the coefficient of i in the imaginary part is exactly $\sin\ \alpha$. This relationship is called **Euler's formula.**

Euler's Formula	$e^{i\alpha} = \cos\ \alpha + i\ \sin\ \alpha$

If $-i\alpha$ had been substituted for x rather than $i\alpha$, the result would have been $e^{-i\alpha} = \cos\alpha - i\sin\ \alpha$.

Euler's formula can be used to write the exponential form of a complex number, $x + yi$, where θ is expressed in radians.

$$x + yi = r(\cos\theta + i\sin\theta)$$
$$= re^{i\theta}$$

Example 2

You may wish to review the polar form of complex numbers in Lesson 9-6.

Write $\sqrt{3} + i$ in exponential form.

Write the polar form of $\sqrt{3} + i$. Recall that $x + yi = r(\cos\ \theta + i\sin\ \theta)$, where $r = \sqrt{x^2 + y^2}$ and $\theta = \text{Arctan}\ \frac{y}{x}$ when $x > 0$.

$r = \sqrt{(\sqrt{3})^2 + 1^2}$ or 2 and $\theta = \text{Arctan}\ \frac{1}{\sqrt{3}}$ or $\frac{\pi}{6}$ $x = \sqrt{3}$ and $y = 1$

$\sqrt{3} + i = 2\left(\cos\frac{\pi}{6} + i\sin\frac{\pi}{6}\right)$

$\qquad = 2e^{i\frac{\pi}{6}}$

Thus, the exponential form of $\sqrt{3} + i$ is $2e^{i\frac{\pi}{6}}$.

The equations for $e^{i\alpha}$ and $e^{-i\alpha}$ can be used to derive the exponential values of $\sin\ \alpha$ and $\cos\ \alpha$.

$$e^{i\alpha} - e^{-i\alpha} = (\cos\ \alpha + i\sin\alpha) - (\cos\alpha - i\sin\alpha)$$
$$e^{i\alpha} - e^{-i\alpha} = 2i\sin\ \alpha$$
$$\sin\ \alpha = \frac{e^{i\alpha} - e^{-i\alpha}}{2i}$$

$$e^{i\alpha} + e^{-i\alpha} = (\cos\,\alpha + i\sin\,\alpha) + (\cos\,\alpha - i\sin\,\alpha)$$
$$e^{i\alpha} + e^{-i\alpha} = 2\cos\,\alpha$$
$$\cos\,\alpha = \frac{e^{i\alpha} + e^{-i\alpha}}{2}$$

From your study of logarithms, you know that there is no real number that is the logarithm of a negative number. However, you can use a special case of Euler's formula to find a complex number that is the natural logarithm of a negative number.

$$e^{i\alpha} = \cos\alpha + i\,\sin\,\alpha$$
$$e^{i\pi} = \cos\,\pi + i\,\sin\,\pi \qquad \textit{Let } \alpha = \pi.$$
$$e^{i\pi} = -1 + i(0)$$
$$e^{i\pi} = -1 \qquad\qquad\qquad \textit{So, } e^{i\pi} + 1 = 0.$$

This equation has been called the most beautiful relationship in mathematics because it relates three of the most important numbers, e, π, and i.

If you take the natural logarithm of both sides of $e^{i\pi} = -1$, you will obtain a value for $\ln(-1)$.

$$\ln e^{i\pi} = \ln(-1)$$
$$i\pi = \ln(-1)$$

Thus, the natural logarithm of a negative number $-k$ exists since $\ln(-k) = \ln(-1)k = \ln(-1) + \ln k$, a complex number.

Example 3

Evaluate ln (-540).

$$\ln(-540) = \ln(-1) + \ln 540$$
$$\approx i\pi + 6.2916$$
Thus, $\ln(-540) \approx i\pi + 6.2916$.   *The logarithm is a complex number.*

CHECKING FOR UNDERSTANDING

Communicating Mathematics

Read and study the lesson to answer each question.

1. **Research** other examples, besides the ones given at the beginning of the lesson, of the Fibonacci sequence in nature.

2. **Write** a recursive formula for the terms of the Fibonacci sequence.

3. Euler's formula represents a relationship between what two series?

4. **State** the equation that relates e, π, and i.

5. **Research** the meaning of the term *transcendental number*.

Guided Practice

Answer each question about the Fibonacci sequence.

6. List the first 20 terms of the Fibonacci sequence.

7. Find the following quotients to the nearest thousandth.

$$\frac{\text{2nd term}}{\text{1st term}}, \frac{\text{3rd term}}{\text{2nd term}}, \frac{\text{4th term}}{\text{3rd term}}, \cdots, \frac{\text{10th term}}{\text{9th term}}$$

What do you notice?

8. Although the Fibonacci sequence is not a geometric sequence, the farther it is extended, the more closely it approximates a geometric series. What is the common ratio of this series?

Use the first five terms of the exponential series $e^x = 1 + x + \dfrac{x^2}{2!} + \dfrac{x^3}{3!} + \dfrac{x^4}{4!} + \cdots$
and a calculator to approximate each value to the nearest hundredth.

9. $e^{1.1}$ **10.** $e^{-0.5}$ **11.** $e^{0.95}$

Write each expression or complex number in exponential form.

12. $2\left(\cos \dfrac{\pi}{3} + i \sin \dfrac{\pi}{3}\right)$ **13.** $5\left(\cos \dfrac{5\pi}{3} + i \sin \dfrac{5\pi}{3}\right)$

14. $\sqrt{2}\left(\cos \dfrac{5\pi}{4} + i \sin \dfrac{5\pi}{4}\right)$ **15.** $12\left(\cos \dfrac{\pi}{6} + i \sin \dfrac{\pi}{6}\right)$

16. i **17.** $6i$

Find each value.

18. $\ln(-4)$ **19.** $\ln(-0.0082)$

EXERCISES

Practice

Use the first five terms of the exponential series $e^x = 1 + x + \dfrac{x^2}{2!} + \dfrac{x^3}{3!} + \dfrac{x^4}{4!} + \cdots$
and a calculator to approximate each value to the nearest hundredth.

20. $e^{1.5}$ **21.** $e^{0.45}$ **22.** $e^{4.6}$

23. $e^{2.3}$ **24.** $e^{0.8}$ **25.** $e^{1.9}$

Use the first five terms of the appropriate trigonometric series to approximate the value of each function to four decimal places. Then, compare the approximation to the actual value.

26. $\sin \pi$ **27.** $\cos \pi$ **28.** $\cos \dfrac{\pi}{4}$

Write each expression or complex number in exponential form.

29. $1\left(\cos \dfrac{\pi}{4} + i \sin \dfrac{\pi}{4}\right)$ **30.** $3\left(\cos \dfrac{3\pi}{4} + i \sin \dfrac{3\pi}{4}\right)$

31. $4 - 4i$ **32.** $-1 + i\sqrt{3}$

33. $3 + 3i\sqrt{3}$ **34.** $-2\sqrt{3} - 2i$

Find each value.

35. $\ln(-7)$ **36.** $\ln(-6.2)$ **37.** $\ln(-5.23)$

38. $\ln(-48.2)$ **39.** $\ln(-0.036)$ **40.** $\ln(-4320)$

41. Examine Pascal's triangle. What relationship can you find between Pascal's triangle and the Fibonacci sequence?

42. Biology A famous problem posed by Fibonacci concerned the breeding patterns of rabbits. He made the following assumptions: (1) newborn rabbits become adults in one month; (2) each pair of adult rabbits produces one pair of newborn rabbits each month; (3) no rabbits die; (4) there are an equal number of male and female rabbits.

a. Starting with a single pair of newborn rabbits, how many pairs of rabbits are there at the end of 6 months? 9 months? 12 months?

b. How does this relate to the Fibonacci sequence? Use a table to show your answer.

43. Number Theory The ancient Greeks were very interested in number patterns. Triangular numbers are numbers that can be represented by a triangular array of dots, with n dots on each side. The first five triangular numbers are 1, 3, 6, 10, and 15.

a. Is this sequence an arithmetic or geometric sequence, or neither?

b. What are the next five terms in this sequence?

c. Write the general formula for the nth term.

d. What is the 50th term in the sequence?

44. Create a function of the form $y = f(x)$ whose graph has holes at $x = -2$ and 0, and resembles the graph of $y = x^3$. **(Lesson 3-4)**

45. Solve $\sin^2 x = \cos x - 1$ for principal values of x. **(Lesson 7-5)**

46. Boating A boat heads due west across a lake at 8 m/s. If a current of 5 m/s moves due south, what is the boat's resultant velocity? **(Lesson 8-1)**

47. Finance Glen Scott invests $100 at 7% compounded continuously, how much will he have at the end of 15 years? **(Lesson 11-3)**

48. Find the third term of $(x - 3y)^5$. **(Lesson 12-6)**

49. College Entrance Exam Choose the best answer.
$\sqrt{2}\%$ of $3\sqrt{2}$ is

(A) 0.06   **(B)** 0.3   **(C)** $\frac{1}{3}$ **(D)** 6   **(E)** $33\frac{1}{3}$

12-8 Mathematical Induction

Objective

After studying this lesson, you should be able to:
- use mathematical induction to prove the validity of formulas.

Application

Consider the series $1 + 3 + 5 + 7 + \cdots$. What is the first term? What is the sum of the first two terms? the first three terms? the first four terms? Look for a pattern by finding the first five partial sums.

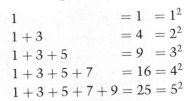

$$
\begin{aligned}
1 &= 1 &&= 1^2 \\
1 + 3 &= 4 &&= 2^2 \\
1 + 3 + 5 &= 9 &&= 3^2 \\
1 + 3 + 5 + 7 &= 16 &&= 4^2 \\
1 + 3 + 5 + 7 + 9 &= 25 &&= 5^2
\end{aligned}
$$

Notice that each partial sum is the square of the number of terms being added, or n^2. Even though we know that this rule is true for the first five partial sums, we do not know if this rule will be true for a partial sum having 1,000,000 odd numbers or n odd numbers.

A method of proof called **mathematical induction** can be used to prove certain formulas or conjectures. Mathematical induction depends on a recursive process that is much like climbing a ladder. First, you must climb onto the first step. Then, you must be able to climb from one step to the next. Thus, if you can get onto the first step, you can certainly climb to the second. If you are on the second step, you can climb to the third. If you are on the third step, you can climb to the fourth, and so on indefinitely, for all steps.

In general, the following steps are used to prove a result by mathematical induction.

Proof by Mathematical Induction	**1.** First, verify that S_n is valid for the first possible case, usually $n = 1$.
	2. Then, assume that S_n is valid for $n = k$ and prove that it is also valid for $n = k + 1$.

Thus, since S_n is valid for $n = 1$ (or any other first case), it is valid for $n = 2$. Since it is valid for $n = 2$, it is valid for $n = 3$, and so on, indefinitely.

Let's use mathematical induction to prove the conjecture that was made at the beginning of the lesson. Notice that this sequence of numbers is arithmetic with a common difference of 2. The general formula for the nth term is $2n - 1$.

S_n: $1 + 3 + 5 + 7 + \cdots + (2n - 1) = n^2$

1. First, verify that S_n is valid for the first possible case, $n = 1$.

If $n = 1$, then S_1 is 1. The sum is 1 and $1^2 = 1$, so the formula, S_n, is valid for the first case.

2. Then, assume that S_n is valid for $n = k$ and prove that it is also valid for $n = k + 1$.

It must be proven that if S_k: $1 + 3 + 5 + 7 + \cdots = k^2$, then S_{k+1}: $1 + 3 + 5 + 7 + \cdots = (k + 1)^2$ is also valid.

The kth term of the series S_k is $2k - 1$. Substitute $k + 1$ for k. Then $2(k + 1) - 1 = 2k + 1$. You can write a formula for S_{k+1} by adding the next term, $2k + 1$, to each side.

The $(k + 1)$th term is $2k + 1$.

$$S_k: \quad 1 + 3 + 5 + 7 + \cdots + (2k - 1) = k^2$$
$$S_{k+1}: 1 + 3 + 5 + 7 + \cdots + (2k - 1) + (2k + 1) = k^2 + (2k + 1)$$
$$= (k + 1)^2$$

Since the formula for the sum for $n = k + 1$ produces the same result as the direct computation of the sum of the series, the formula is valid for $n = k + 1$ if it is valid for $n = k$. Thus, it can be concluded that since the formula is valid for $n = 1$, it is also valid for $n = 2$. Since it is valid for $n = 2$, it is valid for $n = 3$, and so on, indefinitely. Therefore, the following formula is valid for any positive integer n.

$$S_n: 1 + 3 + 5 + 7 + \cdots + (2n - 1) = n^2$$

Example

Prove S_n: $1 + 4 + 7 + \cdots + (3n - 2) = \dfrac{n(3n - 1)}{2}$.

1. First, verify that S_n is valid for $n = 1$.

Since $S_1 = 1$ and $\dfrac{1[3(1) - 1]}{2} = 1$, the formula is valid for $n = 1$.

2. Then, assume that S_n is valid for $n = k$ and prove that it is also valid for $n = k + 1$.

$$S_k: 1 + 4 + 7 + \cdots + (3k - 2) = \frac{k(3k - 1)}{2} \qquad 3(k + 1) - 2 = 3k + 1$$

$$S_{k+1}: 1 + 4 + 7 + \cdots + (3k - 2) + (3k + 1) = \frac{k(3k - 1)}{2} + (3k + 1)$$

$$= \frac{k(3k - 1)}{2} + \frac{2(3k + 1)}{2}$$

$$= \frac{3k^2 - k + 6k + 2}{2}$$

$$= \frac{3k^2 + 5k + 2}{2}$$

$$= \frac{(k + 1)(3k + 2)}{2}$$

If $k + 1$ is substituted into the original formula, the same result is obtained.

$$S_{k+1} = \frac{(k + 1)[3(k + 1) - 1]}{2} \text{ or } \frac{(k + 1)(3k + 2)}{2}$$

Thus, if the formula is valid for $n = k$, it is also valid for $n = k + 1$. Since S_n is valid for $n = 1$, it is also valid for $n = 2$, $n = 3$, and so on.

Portfolio

Select some of your work from this chapter that shows how you used a calculator or computer. Place it in your portfolio.

CHECKING FOR UNDERSTANDING

Read and study the lesson to answer each question.

1. **Determine** the general formula for the sum of n terms of the series $2 + 4 + 6 + \cdots + 2n$.

2. **Describe** the initial step that you must perform when verifying a formula by mathematical induction.

3. In the second step of the mathematical induction process, you must assume that S_n is true for __?__ and prove that it is also valid for __?__.

4. **State** the formula for the sum of the first n positive integers.

Guided Practice

Find the sum of each series.

5. $1 + 4 + 7 + \cdots + 148$

6. the even integers between 29 and 51

7. the first nine terms of the series $1^2 + 3^2 + 5^2 + \cdots$

8. the first six terms of the series $2^5 + 2^6 + 2^7 + \cdots$

Use mathematical induction to prove that each formula is valid for all positive integral values of *n*.

9. $2 + 4 + 6 + \cdots + 2n = n(n + 1)$

10. $1 + 3 + 6 + \cdots + \dfrac{n(n + 1)}{2} = \dfrac{n(n + 1)(n + 2)}{6}$

11. $\dfrac{1}{2} + \dfrac{1}{2^2} + \dfrac{1}{2^3} + \cdots + \dfrac{1}{2^n} = 1 - \dfrac{1}{2^n}$

EXERCISES

Practice

Use mathematical induction to prove that each formula is valid for all positive integral values of *n*.

12. $1^2 + 2^2 + 3^2 + \cdots + n^2 = \dfrac{n(n + 1)(2n + 1)}{6}$

13. $1^3 + 2^3 + 3^3 + \cdots + n^3 = \dfrac{n^2(n + 1)^2}{4}$

14. $1^2 + 3^2 + 5^2 + \cdots + (2n - 1)^2 = \dfrac{n(2n - 1)(2n + 1)}{3}$

15. $1 + 2 + 4 + \cdots + 2^{n-1} = 2^n - 1$

16. $-\dfrac{1}{2} - \dfrac{1}{4} - \dfrac{1}{8} - \cdots - \dfrac{1}{2^n} = \dfrac{1}{2^n} - 1$

17. $1^4 + 2^4 + 3^4 + \cdots + n^4 = \dfrac{6n^5 + 15n^4 + 10n^3 - n}{30}$

18. $a + (a + d) + (a + 2d) + \cdots + [a + (n - 1)d] = \dfrac{n}{2}[2a + (n - 1)d]$

19. $\dfrac{1}{1 \cdot 2} + \dfrac{1}{2 \cdot 3} + \dfrac{1}{3 \cdot 4} + \cdots + \dfrac{1}{n(n + 1)} = \dfrac{n}{n + 1}$

Critical Thinking

20. Use mathematical induction to prove that the binomial theorem is valid for all positive integral values of n.

21. Time The Johnson family's grandfather clock strikes only on the hour.

  a. Can this situation be modeled by an arithmetic, geometric, or infinite sequence?

  b. If it strikes once at 1:00, twice at 2:00, and so on, how many times will it strike in a 24-hour period?

22. Economics Each time you buy something, it affects the nation's economy. The store from which the purchase is made now has money to spend. This trend continues and forms a sequence that is described in the multiplier doctrine of economics.

  a. Can this situation be modeled by an arithmetic, geometric, or infinite sequence?

  b. Suppose you receive a $600 refund on your income taxes and that you spend 70% of this refund. Also suppose that the people who receive your money spend 70% of what they receive, and so on indefinitely. How much money will eventually be spent?

23. Geneology The number of direct ancestors that you have had is given by the series $2 + 2^2 + 2^3 + 2^4 + \cdots + 2^n$, where n is the number of generations. Assume that you were not related to any of your ancestors in more than one way.

  a. Can this situation be modeled by an arithmetic, geometric, or infinite sequence?

  b. How many direct ancestors have you had since the year A.D. 1000? (Assume that a generation is 30 years.)

24. Find the value of the determinant $\begin{vmatrix} -1 & -2 \\ 3 & -6 \end{vmatrix}$. **(Lesson 2-3)**

25. Solve $\dfrac{2}{x+2} = \dfrac{x}{2-x} + \dfrac{x^2+4}{x^2-4}$. **(Lesson 4-6)**

26. Graph $y = -3\sin(\theta - 45°)$. **(Lesson 6-3)**

27. Name the coordinates of the center, foci, and vertices of the ellipse with equation $\dfrac{(y-3)^2}{16} + \dfrac{x^2}{9} = 1$. **(Lesson 10-3)**

28. Write $-1 + i$ in exponential form. **(Lesson 12-7)**

29. College Entrance Exam Compare quantities A and B below.
Write A if quantity A is greater.
Write B if quantity B is greater.
Write C if the two quantities are equal.
Write D if there is not enough information to determine the relationship.

Year	Cars Sold by Bob's Buicks
1992	🚗 🚗 🚗 🚗 🚗
1993	🚗 🚗 🚗

In 1992, Bob's Buicks sold 270 more cars than in 1993.

(A) the number of cars each represents   **(B)** 100

VOCABULARY

Upon completing this chapter, you should be
familiar with the following terms:

658 arithmetic mean	**698** Euler's formula	**702** mathematical induction
656 arithmetic sequence	**697** exponential series	**687** n factorial, $n!$
658 arithmetic series	**696** Fibonacci sequence	**691** Pascal's triangle
693 binomial theorem	**665** geometric mean	**680** ratio test
656 common difference	**663** geometric sequence	**657** recursive
663 common ratio	**665** geometric series	**656** sequence
682 comparison test	**686** index of summation	**656** term
679 converge	**671** infinite sequence	**697** trigonometric series
679 diverge	**674** infinite series	

SKILLS AND CONCEPTS

OBJECTIVES AND EXAMPLES

Upon completing this chapter, you should
be able to:

■ find the nth term and arithmetic means of
an arithmetic sequence **(Lesson 12-1)**

Find the 35th term in the sequence
$-5, -1, 3, \cdots$.

$d = -1 - (-5)$ or 4

$a_n = a + (n-1)d$

$a_{35} = -5 + (35-1)(4)$

$\quad = 131$

■ find the nth term and geometric means of
a geometric sequence **(Lesson 12-2)**

Find the 9th term of the sequence
$-8, 4, -2, 1, \cdots$.

$r = \dfrac{4}{-8}$ or $-\dfrac{1}{2}$

$a_n = ar^{n-1}$

$a_9 = -8\left(-\dfrac{1}{2}\right)^{9-1}$

$\quad = -\dfrac{1}{64}$

REVIEW EXERCISES

Use these exercises to review and prepare for
the chapter test.

1. Find the next five terms of the sequence
   $3, 4.3, 5.6, \cdots$.

2. Find the 20th term of the arithmetic
   sequence for which $a_1 = 7$ and $d = -4$.

3. Form a sequence that has three arithmetic
   means between 6 and -4.

4. Find the next four terms of the sequence
   $343, 49, 7, \cdots$.

5. Find the 7th term of the geometric
   sequence for which $a_1 = 2.2$ and $r = 2$.

6. Form a sequence that has three geometric
   means between 0.2 and 125.

■ find the limit of the terms of an infinite sequence **(Lesson 12-3)**

Evaluate $\lim\limits_{n\to\infty} \dfrac{n^2+3}{2n}$.

$$\lim_{n\to\infty} \frac{n^2+3}{2n} = \lim_{n\to\infty}\left(\frac{n}{2}+\frac{3}{2n}\right)$$
$$= \lim_{n\to\infty}\frac{n}{2} + \lim_{n\to\infty}\frac{3}{2n}$$
$$= \lim_{n\to\infty}\frac{n}{2} + 0$$

Since $\dfrac{n}{2}$ becomes infinitely large as n approaches infinity, the sequence has no limit.

Evaluate each limit, or state that the limit does not exist.

7. $\lim\limits_{n\to\infty} \dfrac{2n}{5n+1}$

8. $\lim\limits_{n\to\infty} \dfrac{4n+1}{n}$

9. $\lim\limits_{n\to\infty} \dfrac{(-1)^n n^2}{5n^2}$

10. $\lim\limits_{n\to\infty} \dfrac{4n^3-3n}{n^4-4n^3}$

■ determine whether a series is convergent or divergent **(Lesson 12-4)**

Determine whether $\dfrac{1}{2^2}+\dfrac{1}{4^2}+\dfrac{1}{6^2}+\cdots$ is convergent or divergent.

The general term of this series is $\dfrac{1}{(2n)^2}$.

Compare with $\dfrac{1}{n^2}$, the general term of a convergent series. Since $\dfrac{1}{(2n)^2} < \dfrac{1}{n^2}$, the series is convergent.

Use the comparison test to determine whether each series is *convergent* or *divergent*.

11. $\dfrac{6}{1}+\dfrac{7}{2}+\dfrac{8}{3}+\cdots$

12. $\dfrac{1}{1^3}+\dfrac{1}{2^3}+\dfrac{1}{3^3}+\cdots$

13. $2+1+\dfrac{2}{3}+\dfrac{1}{2}+\dfrac{2}{5}+\dfrac{1}{3}+\dfrac{2}{7}+\cdots$

■ use sigma notation **(Lesson 12-5)**

Write $\sum\limits_{n=1}^{3}(n^3-2)$ in expanded form and find the sum.

$$\sum_{n=1}^{3}(n^3-2) = (1^3-2)+(2^3-2)+(3^3-2)$$
$$= -1+6+25$$
$$= 30$$

Write each expression in expanded form and find the sum.

14. $\sum\limits_{a=5}^{11}(2a-4)$

15. $\sum\limits_{k=1}^{\infty}(0.4)^k$

Express each series using sigma notation.

16. $-1+1+3+5+\cdots$

17. $2+5+10+17+\cdots+82$

■ use the binomial theorem to expand binomials **(Lesson 12-6)**

Find the 4th term of $(2x+y)^7$.

$$(2x+y)^7 = \sum_{r=0}^{7}\frac{7!}{r!(7-r)!}(2x)^{7-r}y^r$$

The fourth term, $\dfrac{7!}{3!(7-3)!}(2x)^{7-3}y^3$, is

$\dfrac{7\cdot6\cdot5\cdot4!}{3!4!}(2x)^4 y^3$, or $560x^4 y^3$.

Use the binomial theorem to expand each binomial.

18. $(a-x)^6$

19. $(2r+3s)^4$

Find the designated term of each binomial expansion.

20. 5th term of $(x-1)^{15}$

21. 8th term of $(x+3y)^{10}$

OBJECTIVES AND EXAMPLES	REVIEW EXERCISES

■ use Euler's formula to write the exponential form of a complex number **(Lesson 12-7)**

Write $\sqrt{3} - i$ in exponential form.

$r = \sqrt{(\sqrt{3})^2 + (-1)^2}$ or 2

$\theta = \text{Arctan } \dfrac{-1}{\sqrt{3}}$ or $\dfrac{5\pi}{6}$

$\sqrt{3} - i = 2\left(\cos\dfrac{5\pi}{6} + i\sin\dfrac{5\pi}{6}\right) = 2e^{i\frac{5\pi}{6}}$

Write each complex number in exponential form.

22. $4i$

23. $2 - 2i$

24. $-5 - 5i\sqrt{3}$

25. $3\sqrt{3} + 3i$

■ use mathematical induction to prove the validity of formulas **(Lesson 12-8)**

Proof by mathematical induction:

1. First, verify that S_n is valid for the first possible case, usually $n = 1$.

2. Then, assume that S_n is valid for $n = k$ and prove that it is also valid for $n = k + 1$.

Use mathematical induction to prove that each formula is valid for all positive integral values of *n*.

26. $1 + 2 + 3 + \cdots + n = \dfrac{n(n + 1)}{2}$

27. $3 + 8 + 15 + \cdots + n(n + 2)$
$\qquad = \dfrac{n(n + 1)(2n + 7)}{6}$

APPLICATIONS AND PROBLEM SOLVING

28. Budgets The United States Department of Defense plans to cut the budget on one of its projects by 12 percent each year. If the current budget is $150 million, what will the budget be in 6 years? **(Lesson 12-2)**

29. Ballooning One minute after it is released, a gas-filled balloon rises 100 feet. In each succeeding minute, the balloon rises only 50% as far as it rose in the previous minute. How far will the balloon rise in 5 minutes? **(Lesson 12-5)**

1. Find the next five terms of the sequence $3, 4.5, 6, \cdots$.

2. Find the next four terms of the sequence $\dfrac{1}{4}, \dfrac{1}{10}, \dfrac{1}{25}, \dfrac{2}{125}, \cdots$.

3. Form a sequence that has three arithmetic means between -4 and 8.

4. Form a sequence that has three geometric means between 16 and 1.

5. Find the 24th term of the sequence $-6, -1, 4, \cdots$.

6. Find the eighth term of the sequence $\dfrac{1}{2}, \dfrac{3}{4}, \dfrac{9}{8}, \dfrac{27}{16}, \cdots$.

7. Find n for an arithmetic sequence for which $S_n = 345$, $a = 12$, and $d = 5$.

8. Find the sum of the first 10 terms of the geometric series $\dfrac{5}{2} + 5 + 10 + \cdots$.

Evaluate each limit, or state that the limit does not exist.

9. $\lim\limits_{n \to \infty} \dfrac{n^3 + 3}{3n^2 + 1}$

10. $\lim\limits_{n \to \infty} \dfrac{n^3 + 4}{2n^3 + 3n}$

Determine whether each series is *convergent* or *divergent*.

11. $\dfrac{1}{3 \cdot 1^2} + \dfrac{1}{3 \cdot 2^2} + \dfrac{1}{3 \cdot 3^2} + \cdots$

12. $\dfrac{1}{6} + \dfrac{1}{3} + \dfrac{1}{2} + \dfrac{2}{3} + \cdots$

Write each series using sigma notation.

13. $5 + 10 + 15 + \cdots + 95$

14. $7 + 9 + 11 + 13 + \cdots$

Find the designated term of each binomial expansion.

15. sixth term of $(a + 2)^{10}$

16. fifth term of $(3x - y)^8$

17. Write $-2 + 2i$ in exponential form.

18. Use mathematical induction to prove that the following formula is valid for all positive integral values of n.
$$2 \cdot 3 + 4 \cdot 5 + 6 \cdot 7 + \cdots + 2n(2n + 1) = \dfrac{n(n + 1)(4n + 5)}{3}$$

19. **Finance** Ms. Jackson deposits $200 into an annuity every 3 months. The annuity pays an APR of 8% and interest is compounded quarterly. What will be the total value of the annuity at the end of 10 years?

20. **Geneology** Using sigma notation, write a formula to find the number of direct ancestors you have had in the past 6 generations. Then, find this number.

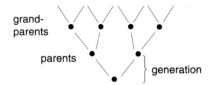

grand-parents

parents

generation

Bonus
Geometry   If the area of $\triangle ABC$ is 80 square units and the geometric mean between the lengths of sides $\overline{AB}$ and $\overline{AC}$ is 16 units, find $\sin A$.

CHAPTER 13

ITERATION AND FRACTALS

HISTORICAL SNAPSHOT

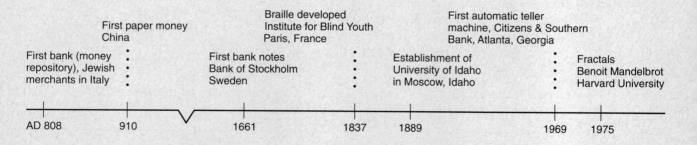

First bank (money repository), Jewish merchants in Italy

First paper money China

First bank notes Bank of Stockholm Sweden

Braille developed Institute for Blind Youth Paris, France

Establishment of University of Idaho in Moscow, Idaho

First automatic teller machine, Citizens & Southern Bank, Atlanta, Georgia

Fractals Benoit Mandelbrot Harvard University

AD 808   910   1661   1837   1889   1969   1975

CHAPTER OBJECTIVES

In this chapter, you will:
- Iterate functions numerically.
- Perform graphical iteration on both linear and quadratic functions.
- Develop an understanding of chaotic long-term behavior for some functions.
- Explore and examine fractals.

CAREER GOAL: Government Finance

In a recent evening news broadcast, the anchorman said, "The Dow Jones Industrial Average is up and stock prices are down in light trading." Did you ever wonder what this means for the nation's economy? J. Richard Rock became interested in finance from following the news. "Finance opens a lot of doors; it is one area where you can quickly see the fruits of your labor."

Richard would like to work in the public sector. "I would especially like to work in the treasurer's office for the state of Idaho, or in the state budget office." These offices serve the state in different ways. "All of the state's funds are invested by the treasurer, so they grow continually at minimum risk," Richard says. "The budget office, on the other hand, determines what the demand for the state's products and services will be in the future, and then structures the state budget accordingly."

"Mathematics plays a significant role in investment banking," Richard says. "The mathematical model used to determine the appropriate return on an investment is the *capital asset pricing model* (CAPM), $R_f + \beta (R_m - R_f)$. In this model, R_f represents the risk-free rate, β is the expression of risk, and R_m is the average market rate. A β of 0 means that the investment carries little or no risk, while a β greater than 1 represents a riskier investment. Suppose the risk-free rate on investments is currently 8%, the average market return is 12%, and a stock has been assigned a β of 1.5. Using the CAPM, the appropriate return that the company should offer its investors is 14%."

Richard has a genetic disease called retinitus pigmentosas and is legally blind. "As a blind person," he says, "I've had a lot of opportunities. I want to give back more than I've been given."

13-1 Iterating Functions with Real Numbers

Objective

After studying this lesson, you should be able to:
- iterate functions using real numbers.

Application

The population of a species in a defined area changes over time. Changes in the availability of food, good or bad weather conditions, the amount of hunting allowed, disease, and the presence or absence of predators can all affect the population of a species. We can use a mathematical equation to model the changes in a population.

FYI...

Pierre François Verhulst was a Belgian mathematician. He is credited with developing this dynamics model in 1845.

The Verhulst population model is one tool used to represent the changes in the population of a species. The model uses the recursive formula $p_{n+1} = p_n + rp_n(1 - p_n)$, where n represents the number of time periods that have passed, p_n represents the percentage of the maximum sustainable population that exists at time n, and r represents the growth rate.

The table and graph on the next page show the changes that the Verhulst model generates for the population of elk on the Bridger Range in the Rocky Mountains of western Montana. We assumed that the population in 1984 was 10% of the maximum population sustainable and that the growth rate was 2.5. A growth rate of 2.5 means that if there were no restrictions on the growth of the population, it would grow 2.5% each year. The method used to find the first few iterates is shown.

You may wish to review the composition of functions in Lesson 1-2.

$$p_{n+1} = p_n + rp_n(1 - p_n)$$
$$p_{1985} = 0.1000 + (2.5)(0.1000)(1 - 0.1000) \qquad p_{1984} = 10\% \text{ or } 0.1000, r = 2.5$$
$$= 0.3250$$

$$p_{1986} = 0.3250 + (2.5)(0.3250)(1 - 0.3250)$$
$$\approx 0.8734$$

$$p_{1987} = 0.8734 + (2.5)(0.8734)(1 - 0.8734)$$
$$\approx 1.1498$$

Year	P_n	Year	P_n
1984	0.1000	1990	0.5383
1985	0.3250	1991	1.1596
1986	0.8734	1992	0.6969
1987	1.1498	1993	1.2250
1988	0.7192	1994	0.5359
1989	1.2241	1995	1.1577

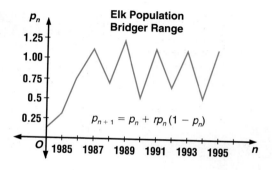

Notice that to determine the percentage of the maximum sustainable population for a year, you must use the percentage from the previous year. This process of composing a function with itself repeatedly is called **iteration**. Each output is called an **iterate**. To iterate a function $f(x)$, find the function value, $f(x_0)$, of the initial value x_0. The second iterate is the value of the function performed on the output, that is $f(f(x_0))$. A shorthand notation for iterates is $x_0, f(x_0), f^2(x_0), f^3(x_0), \cdots, f^n(x_0)$. Be sure not to confuse this type of notation with exponents.

Example 1

Find the first three iterates, $x_1, x_2,$ and x_3, of the function $f(x) = 2x + 1$ for an initial value of $x_0 = 1$.

To obtain the first iterate, find the value of the function for $x_0 = 1$.

$f(x_0) = f(1)$
$\qquad = 2(1) + 1$ or 3 $\qquad$ So, $x_1 = 3$.

To obtain the second iterate, x_2, substitute the function value for the first iterate, x_1, for x.

$f(x_1) = f(3)$
$\qquad = 2(3) + 1$ or 7 $\qquad$ So, $x_2 = 7$.

Now find the third iterate, x_3, by substituting x_2 for x.

$f(x_2) = f(7)$
$\qquad = 2(7) + 1$ or 15 $\qquad$ So, $x_3 = 15$.

Therefore, the first three iterates of the function $f(x) = 2x + 1$ for an initial value of $x_0 = 1$ are 3, 7, and 15.

Iteration sequences may represent a wide variety of real-world situations. One use of iteration is amortization. Amortization is the process of repaying a loan by paying a series of equal payments over time. Home mortgages and some consumer loans are amortized. In an amortized loan, the money from each payment consists of an interest payment and a principal payment. After the necessary interest is paid, the rest of the payment goes toward reducing the principal.

Example 2

APPLICATION

Finance

Charlotte is taking out a two-year, simple-interest loan for a used car. The loan amount is $2500 with a monthly payment of $115.94. The interest rate on the loan is 10.5%.

a. **Derive an equation for the balance of the loan after _n_ payments.**

b. **Find the balance of Charlotte's loan after each of the first five payments.**

a. Let b_n represent the balance after the nth payment. So $b_0 = 2500$.
   The simple interest for a payment is found by multiplying the balance of the loan after the $(n-1)$st payment by the monthly interest rate, which is $\dfrac{10.5\%}{12}$ or 0.00875. Thus, if i_n represents the interest payment for the nth month, then $i_n = 0.00875(b_{n-1})$.

   The principal portion of each month's payment is the monthly payment minus the interest portion of the payment. So, if p_n is the principal portion of the payment for the nth month, $p_n = 115.94 - i_n$.

   The balance of the loan after a monthly payment is the previous month's balance minus the principal portion of the payment. Thus, $b_n = b_{n-1} - p_n$.

   Now we can write the equation.

   $b_n = b_{n-1} - p_n$ _balance = previous balance − principal payment_
   $b_n = b_{n-1} - (115.94 - i_n)$ _principal payment = payment − interest_
   $b_n = b_{n-1} - (115.94 - 0.00875(b_{n-1}))$ _interest = rate · previous balance_
   $b_n = b_{n-1} - 115.94 + 0.00875(b_{n-1})$ _Simplify._
   $b_n = 1.00875(b_{n-1}) - 115.94$

b. Use the formula to find the balance after each of the first five payments.

   $b_1 = 1.00875(2500.00) - 115.94$ or 2405.93
   $b_2 = 1.00875(2405.93) - 115.94$ or 2311.04
   $b_3 = 1.00875(2311.04) - 115.94$ or 2215.32
   $b_4 = 1.00875(2215.32) - 115.94$ or 2118.76
   $b_5 = 1.00875(2118.76) - 115.94$ or 2021.36

MATH JOURNAL

Copy and complete this sentence.
"Mathematics is important in the study of banking because _____."

Interest amounts were rounded to the nearest cent.

CHECKING FOR UNDERSTANDING

Communicating Mathematics

Read and study the lesson to answer each question.

1. **Define** and give an example of iteration.

2. **Explain** how iteration and composition of functions are related.

3. Refer to the application at the beginning of the lesson.
   a. The years when the percentage of the maximum sustainable population is greater than 1 are years in which the maximum sustainable population is exceeded. Explain what this means for the population and how the next year's population number will be affected.

b. Do you think that the percentage of the maximum population in 1996 will be greater than 1 or less than 1? Explain.

4. Does the principal portion of an amortized loan payment increase or decrease as more payments are made? Explain.

Guided Practice **Find the first three iterates of each function using the given initial value.**

5. $f(x) = 2x + 1; x_0 = 2$

6. $f(x) = 3x - 7; x_0 = 4$

7. $f(x) = x^2 - 1; x_0 = -1$

8. $f(x) = (x - 5)^2; x_0 = 6$

9. $f(x) = -2x^3 + 5x - 4; x_0 = 1$

Find the first ten iterates for $f(x) = 3x - 5$ for each initial value.

10. $x_0 = 2.2$

11. $x_0 = -1$

12. $x_0 = 4.9$

EXERCISES

Practice **Find the first three iterates of each function using the given initial value. If necessary, round your answers to the nearest hundredth.**

13. $f(x) = x^2; x_0 = -1$

14. $f(x) = x^2; x_0 = -2$

15. $f(x) = x^3 + 1; x_0 = 1.1$

16. $f(x) = x(3 - x); x_0 = 2.1$

Find the first ten iterates for $f(x) = 2.8x(1 - x)$ for each initial value. If necessary, round your answers to the nearest hundredth.

17. $x_0 = 0.2$

18. $x_0 = 0.5$

19. $x_0 = 0.9$

20. $x_0 = 0.01$

Find the first ten iterates for $f(x) = 3.1x(1 - x)$ for each initial value. If necessary, round your answers to the nearest hundredth.

21. $x_0 = 0.2$

22. $x_0 = 0.45$

23. $x_0 = 0.9$

24. $x_0 = 1$

25. Suppose $t_{n+1} = \dfrac{2}{t_n}$. Find the first ten iterates for each initial value.

    a. $t_0 = 1$ **b.** $t_0 = 4$ **c.** $t_0 = a$

    d. What do you observe about the iterates of this function?

    e. Is there any initial value you could use so that all of the iterates of this function are the same?

26. If the iterates of a function become the same value after many iterations, that value is called an *attractor*. A *period-2 attractor* emerges when the iterates begin to repeat a sequence of two values after some iterations. If no pattern emerges after thousands of iterations, the results are *chaotic*.

    a. What is the attractor for the function $f(x) = 2.8x(1 - x)$, if x_0 is any value between 0 and 1? *Hint: Find at least thirty iterates.*

    b. Does the function $f(x) = 3.1x(1 - x)$ have an attractor or a period-2 attractor, or is it chaotic when it is iterated for an initial value between 0 and 1?

    c. Find the first thirty iterates of the function $f(x) = 4x(1 - x)$ for an initial value between 0 and 1. Do the iterates seem to have an attractor, a period-2 attractor, or do they appear to be chaotic?

27. Iterate the function $f(x) = 0.1x + 0.6$ several times with the initial value of $x_0 = 0.6$. Describe the pattern that emerges.

28. Banking The balance of a savings account using simple interest compounded at the end of a period of time can be found by iterating the function $p_{n+1} = p_n + rp_n$, where p_n is the principal after n periods of time and r is the interest rate for a period of time.

   a. John Thomas has a savings account that has an annual yield of 6.3%. Find the balance of the account after each of the first three years if his initial balance is $4210.

   b. Find the balance in Mr. Thomas' account for the first three years using the formula $A = P(1 + r)^t$, where A is the final account balance, P is the principal, r is the annual interest rate, and t is the time in years. Are the results the same as the results of the iterated function?

29. Biology The population of grizzly bears on the high Rocky Mountain Front near Choteau, Montana, has a growth rate of 1.75. The maximum population of bears that can be sustained in the area is 500 bears, and the current population is 240 bears.

   a. Write the Verhulst model equation for the population.

   b. Find the population of bears at the end of each of the first fifteen years. *Hint: Remember that p_n represents the percentage of the maximum sustainable population at time n.*

   c. Graph the bear population for the first fifteen years.

   d. Describe the changes in the population of bears. Explain why these changes occur.

30. Find the area of triangle ABC if $a = 9.2$, $b = 12.5$, and $C = 98°$. Round your answer to the nearest tenth. **(Lesson 5-8)**

31. Find $(2 - i)^4$. Express the result in rectangular form. **(Lesson 9-8)**

32. Write the standard form of the equation of the circle that passes through $(0, 9)$, $(-7, 2)$, and $(0, -5)$. **(Lesson 10-1)**

33. Find the next five terms of the sequence $3, 2.4, 1.8, 1.2, \cdots$. **(Lesson 12-1)**

34. Prove that the sum of the cubes of three consecutive positive integers is divisible by 9 using mathematical induction. **(Lesson 12-8)**

35. College Entrance Exam Choose the best answer. Simplify $\dfrac{1 + \dfrac{1}{x}}{\dfrac{y}{x}}$.

   (A) $\dfrac{x+1}{y}$ 　　 **(B)** $\dfrac{x^2+1}{xy}$ 　　 **(C)** $\dfrac{x+1}{xy}$ 　　 **(D)** $\dfrac{x+1}{x}$ 　　 **(E)** none of these

13-2 Graphical Iteration of Linear Functions

Objectives

After studying this lesson, you should be able to:
- perform graphical iteration on a linear function, and
- identify a fixed point as an attractor or a repeller.

Application

Astronomers believe that when a large star collapses from its own weight, it becomes a black hole. Black holes are invisible stars that have such a strong gravitational pull that nothing, not even light, can escape their influence. Objects that pass near a black hole are pulled in closer and closer to the hole. An **attractor** in an iteration behaves in much the same way.

Graphing the iterations of a function can help us understand the process of iteration better. To do this, graph a function $g(x)$ and the line $f(x) = x$ on the same set of axes. Choose an initial value, x_0, and locate the point $(x_0, 0)$. Draw a vertical line from $(x_0, 0)$ to the graph of $g(x)$. This will be the segment from the point $(x_0, 0)$ to $(x_0, g(x_0))$. Now draw a horizontal segment from this point to the graph of the line $f(x) = x$. This will be the segment from $(x_0, g(x_0))$ to $(g(x_0), g(x_0))$. Repeat the process for many iterations. This process is called **graphical iteration**.

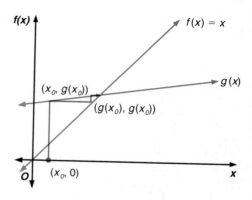

You can think of the line $f(x) = x$ as a mirror that reflects each function value to become the input for the next iteration of the function. The points where the graph of the function $g(x)$ intersects the graph of the line $f(x) = x$ are called **fixed points**. There is one fixed point for a linear function like the one illustrated above. If you try to iterate the initial value that corresponds to the x-coordinate of a fixed point, the iterates will all be the same.

Graphical iteration can generate many different paths for the points being iterated. Four basic paths are possible when a linear function is iterated. Two of the paths look like staircases and two look like spirals.

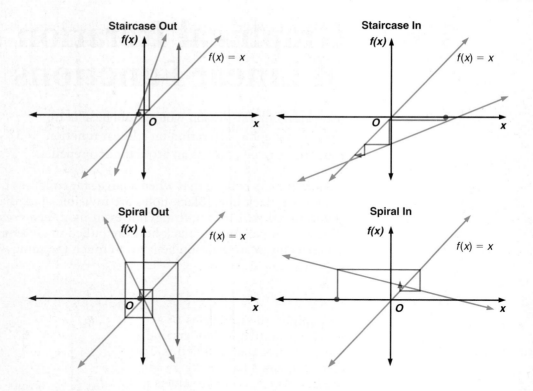

If a path staircases or spirals out from a fixed point, the x-coordinate of the fixed point is called a **repeller**. You can think of a repeller as a point that pushes everything away from itself. If a path staircases or spirals in to a fixed point, the x-coordinate of the point is an **attractor**. Attractors act like the black holes, pulling everything near them in. The graphs on the left above illustrate repellers, while the graphs on the right illustrate attractors.

Example 1

Perform graphical iteration on the function $g(x) = 3x$ for the first three iterates if the initial value is $x_0 = 0.2$. Which of the four types of paths does the iteration take?

To do the graphical iteration, first graph the functions $f(x) = x$ and $g(x) = 3x$.

Start at the point $(0.2, 0)$ and draw a vertical line to the graph of $g(x) = 3x$. From that point, draw a horizontal line to the graph of $f(x) = x$.

Repeat the process from the point on $f(x) = x$. Then repeat again.

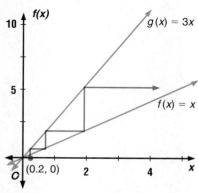

Notice that the scales on the axes are different.

The path of the iterations is a staircase out. So, the x-coordinate of the fixed point, 0, is a repeller.

Sometimes when you are analyzing the long-term behavior of the iterations of a function, it is helpful to know the fixed points. Since the fixed points occur where the graphs of the function and $f(x) = x$ intersect, you can find the fixed points by solving a system of equations.

Example 2

Find the fixed point for the function $g(x) = -0.5x + 6$. Then iterate the function graphically for $x_0 = -4$. Is the x-coordinate of the fixed point a repeller or an attractor?

We can find the coordinates of the fixed point by solving the following system of equations.

$$g(x) = -0.5x + 6$$
$$f(x) = x$$

Solve by substitution.
$$x = -0.5x + 6$$
$$1.5x = 6$$
$$x = 4$$

The coordinates of the fixed point are $(4, 4)$.

Now iterate the function graphically. Begin at the initial value of $x_0 = -4$. Draw a vertical line to the graph of $g(x) = -0.5x + 6$. From that point, draw a horizontal line to the graph of $f(x) = x$. Repeat the process from this point until the pattern becomes clear.

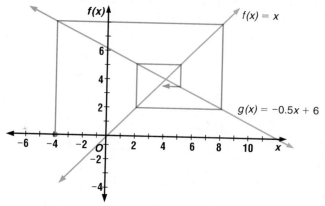

The graph is a spiral in toward $(4, 4)$, so the x-coordinate of the fixed point, 4, is an attractor. This means that the long-term behavior of iterating $g(x) = -0.5x + 6$ for any initial point is predictable. The iterates will become 4 as the number of iterations increases without bound.

CHECKING FOR UNDERSTANDING

Communicating
Mathematics

Read and study the lesson to answer each question.

1. **Define** a fixed point. How can you identify a fixed point?

2. Explain what the presence of a spiral in or staircase in path tells us about the fixed point.

Guided
Practice

Copy the graphs shown and perform graphical iteration for the first three iterates of the initial point shown.

3.

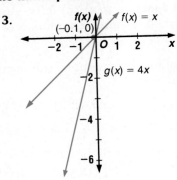

4.

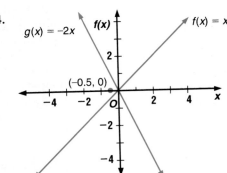

Find the fixed point for each function. Then perform graphical iteration for $x_0 = 2$. Is the x-coordinate of the fixed point a repeller or an attractor?

5. $f(x) = 6x + 2$

6. $f(x) = 0.2x - 1$

EXERCISES

Practice

Copy the graphs shown and perform graphical iteration for the first four iterates of the initial point shown.

7.

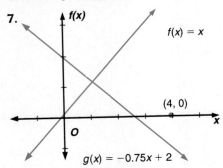

8.

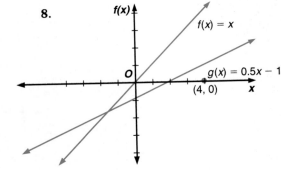

Graph each function and the function $f(x) = x$ on the same set of axes. Then perform graphical iteration for $x_0 = 1$. State the slope of the linear function and the type of path that the graphical iteration forms.

9. $g(x) = 4x + 2$

10. $g(x) = 10x - 6$

11. $g(x) = -2x$

12. $g(x) = 0.5x + 3$

13. $g(x) = -0.75x + 4$

14. $g(x) = -0.6x$

15. $g(x) = -6x - 1$

16. $g(x) = \frac{2}{3}x + 2$

17. $g(x) = 10x - 5$

18. $g(x) = -\frac{1}{4}x$

Study your answers for Exercises 9—18. Describe the slopes of functions that have the following iteration paths.

19. spiral in

20. spiral out

21. staircase in

22. staircase out

Find the coordinates of the fixed point for each function. Is the *x*-coordinate of the fixed point a repeller or an attractor?

23. $f(x) = 2.5x + 4$

24. $f(x) = -3x + 2$

25. $f(x) = 0.75x - 5$

26. $f(x) = -0.8x - 18$

27. Which functions in Exercises 23–26 have paths with a spiraling behavior?

28. Find a linear function that has an attractor of $\frac{7}{9}$ under iteration.

29. The value 0.4 is a repeller for iterations of a certain linear function. Find a possible equation for this function.

30. Find a linear function that has –2 as a repeller under iteration.

Critical Thinking

31. What linear function has all real numbers as attractors?

Applications and Problem Solving

32. Finance The function $p_{n+1} = p_n + rp_n$ gives the balance of a bank account after n time periods.

   a. Find the coordinates of the fixed point for the iteration of the function. Assume that $r > 0$.

   b. Is the *x*-coordinate of the fixed point a repeller or an attractor?

33. Banking The function $b_n = \left(1 + \frac{r}{12}\right) b_{n-1} - p$ gives the balance due on a student loan after n payments if r is the annual interest rate and p is the payment amount. *Hint: Solve a system of equations.*

   a. Find the coordinates of the fixed point for the iteration of the function. Assume that $r > 0$ and $p > 0$. Explain what is happening to the relationship between the interest and the payment at this point.

   b. Is the *x*-coordinate of the fixed point a repeller or an attractor?

Mixed Review

34. Find the product $-2 \begin{bmatrix} 6 & -2 \\ 0 & -1 \end{bmatrix}$. **(Lesson 2-2)**

35. Find all rational roots of the equation $12x^4 - 11x^3 - 54x^2 - 18x + 8 = 0$. **(Lesson 4-4)**

36. Entertainment Ryan observes that the angle of elevation to the top of the Sky Needle, a ride at Magic Planet Amusement Park, is 81° when he stands 50 feet from its base. If Ryan is 6 feet tall, about how tall is the Sky Needle? **(Lesson 5-5)**

37. Find the sum of the first six terms of the series $8 + 6 + 4.5 + \cdots$. **(Lesson 12-2)**

38. Find the first four iterates for $f(x) = 6.5x(1 - x)$ for $x_0 = 0.7$. If necessary, round your answers to the nearest hundredth. **(Lesson 13-1)**

39. College Entrance Exam In Captain Harper's battalion, 30% of the enlistees are from Texas, and 10% of these are from Dallas. What percent of the enlistees in the battalion are from Dallas?

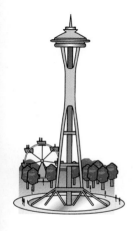

13-3 Graphical Iteration of the Logistic Function

Objectives

After studying this lesson, you should be able to:
- perform graphical iteration on the logistic function, and
- determine if the behavior of a function is predictable or unpredictable.

Application

The state of the environment has become a hotly debated topic in the 1990s. Scientists and citizens' groups are all pressing for more action from governments concerning issues such as global warming and the deterioration of the ozone layer. Changes in industries and political policies are aimed at cleaning up the environment and restoring habitat for wildlife. However, the suggested changes may not have the effect that is anticipated.

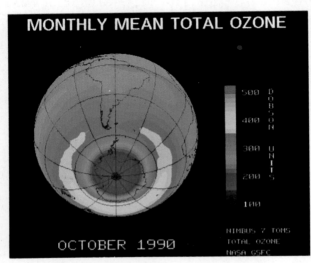

MONTHLY MEAN TOTAL OZONE

OCTOBER 1990

NIMBUS 7 TOMS
TOTAL OZONE
NASA GSFC

Up until this point, we have studied the behavior of linear functions under iteration. The work of researchers such as Edward N. Lorenz and Mitchel Feigenbaum has shown that most natural phenomena, like global weather, river currents, and population growth, are described most effectively by nonlinear functions. When these functions are iterated, their behavior may or may not be predictable. Very slight changes in the parameter values of a function can cause the iterated functions to change from stable and predictable behavior to unpredictable behavior which is known as **chaos**. This means that seemingly slight changes in the world's environment may have unpredictable results.

FYI...

Overproduction of the greenhouse gases causes the heating of Earth's surface that is called global warming. Greenhouse gases are produced by termites, wetlands, coal mining, the burning of fossil fuels, and cud-chewing animals.

One of the nonlinear functions used to describe natural phenomena is the **logistic function**, $f(x) = ax(1 - x)$. The graph of the logistic function is a parabola with x-intercepts at 0 and 1, as shown at the right. These will be the x-intercepts regardless of the value of a. The vertex of the graph of the logistic function is always the point with coordinates $\left(\dfrac{1}{2}, \dfrac{a}{4}\right)$.

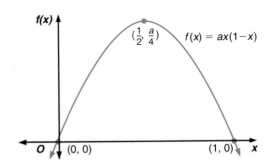

$\left(\dfrac{1}{2}, \dfrac{a}{4}\right)$

$f(x) = ax(1-x)$

$(0, 0)$ $(1, 0)$

Example 1
> **Without graphing, find the vertex of the graph of $y = 6x(1 - x)$.**
>
> In this function, $a = 6$ and the vertex of every logistic function is $\left(\frac{1}{2}, \frac{a}{4}\right)$.
> Therefore, the vertex is at $\left(\frac{1}{2}, \frac{6}{4}\right)$ or $(0.5, 1.5)$.

The behavior of the logistic function under iteration varies greatly as the parameter value of a changes. Let's look closely at the behavior of this function if $0 < a \leq 4$ and the initial value is between 0 and 1.

The graphical iterations for an initial value of 0.1 and three different values of a are shown below. For $a = 1.5$, the graphical iteration is a staircase that converges on a fixed point attractor. For $a = 2.6$, the iteration pattern ends in a spiral that converges on a fixed point attractor. The iteration pattern for $a = 3.3$ moves away from a fixed point repeller and settles into a cyclical pattern. A cyclical pattern is one where the iterates begin to repeat in a predictable pattern over time. Even though the three patterns are very different, all three have predictable long-term behavior.

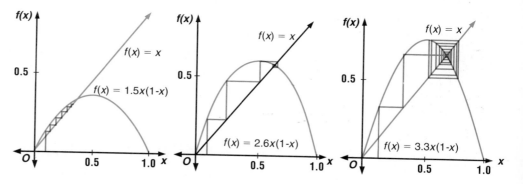

If $a = 4$ in the logistic function, the iteration pattern is quite different. The graphs below show the graphical iteration for the initial value $x_0 = 0.24$ after 10, 40, and 100 iterations.

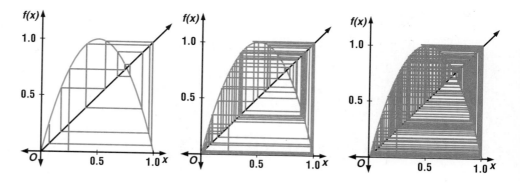

The behavior of the iteration pattern never seems to converge on one point or pattern. The third graph shows the iteration pattern filling the entire interval from 0 to 1 on the x-axis. This visual filling outcome is called **mixing**. Mixing is a good indicator that a system is unstable.

LESSON 13-3   GRAPHICAL ITERATION OF THE LOGISTIC FUNCTION   723

Example 2

Find the coordinates of the fixed points for the function $f(x) = 1.5x(1-x)$. From the graph on the previous page, which point appears to be an attractor?

Like linear functions, the coordinates of the fixed points of a nonlinear iteration can be found by solving a system of equations.

$$f(x) = x$$
$$f(x) = 1.5x(1-x)$$

Substitute.

$$x = 1.5x(1-x)$$
$$x = 1.5x - 1.5x^2$$
$$0 = 0.5x - 1.5x^2$$
$$0 = 0.5x(1 - 3x)$$
$$0 = 0.5x \quad \text{or} \quad 0 = 1 - 3x$$
$$x = 0 \quad \quad \text{or} \quad x = \frac{1}{3}$$

The fixed points are at $(0,0)$ and $\left(\frac{1}{3}, \frac{1}{3}\right)$. By observing the graph, it appears that $\frac{1}{3}$ is an attractor.

Observing the long-term behavior of iteration requires performing many steps of graphical iteration. Performing these iterations by hand is very time consuming and tedious. Technology can help us perform these steps to observe their behavior.

EXPLORATION: Graphing Calculator

The program below will perform graphical iteration of a logistic function. To enter the program into the program memory, select the EDIT option in the program menu. Refer to the manual for menus to locate unfamiliar keystrokes. You must give the program a name to begin. Names can be up to 8 characters long and must begin with a letter.

```
Line 1   :ClrDraw
Line 2   :Fix 3
Line 3   :0→Xmin
Line 4   :1→Xmax
Line 5   :0.1→Xscl
Line 6   :0→Ymin
Line 7   :1→Ymax
Line 8   :0.1→Yscl
Line 9   :Disp "A ="
Line 10  :Input A
Line 11  :Disp "I ="
Line 12  :Input I
Line 13  :DrawF AX(1 − X)
```

```
Line 14  :DrawF X
Line 15  :0→M
Line 16  :0→C
Line 17  :If C = 0
Line 18  :Goto 2
Line 19  :Lbl 1
Line 20  :AI − AII→I
Line 21  :M + 1→M
Line 22  :If M < C
Line 23  :Goto 1
Line 24  :Goto 3
Line 25  :Lbl 2
Line 26  :Line(I,0,I,I)
```

Line 27 :Lbl 3
Line 28 :AI − AII → J
Line 29 :Line (I,I,I,J)
Line 30 :Line (I,J,J,J)
Line 31 :Pause
Line 32 :Disp J

Line 33 :Pause
Line 34 :J→I
Line 35 :M + 1→M
Line 36 :If M < C + 15
Line 37 :Goto 3

Use the program to perform the first five steps of iteration of the function $f(x) = 2.5x(1 − x)$ for an initial value of 0.250.

1. Enter $a = 2.5$ and $i = 0.250$.

2. The first three iterates are as follows, to the nearest thousandth. *Press* ENTER *to display the iterates.*

$x_0 = i = 0.250$

$x_1 = 0.469$

$x_2 = 0.623$

3. Find the next three iterates.

$x_3 = 0.587$

$x_4 = 0.606$

$x_5 = 0.597$

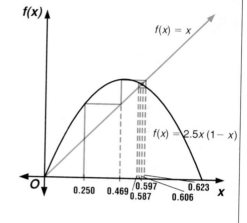

The program given in the Exploration can be used to examine the long-term iterative behavior of a function. To do this, replace Line 16 with "100 → C". This modification will make the calculator begin graphing and displaying values at the 100th iterate. If the graph is approaching an attractor, the values of the iterates will all be getting closer to it. Some forms of stable long-term behavior cause the iterates to repeat in a pattern. If the iterates repeat every n steps, the function is said to have a **period-n attractor** and the fixed point is a repeller. For example, if the iterates eventually settle into the pattern $0.12, 0.45, 1.23, 0.12, 0.45, 1.23, \cdots$, this function has a period-3 attractor.

Example 3

Modify the program in the Graphing Calculator Exploration so that the iterates will begin printing at the 100th iterate. Use the modified graphical iteration program to determine the long-term iterative behavior of the function $f(x) = 3.1x(1 − x)$ for any initial value in the interval 0 to 1.

The iterates form the pattern $0.765, 0.558, 0.765, 0.558, \ldots$. So the function has a period-2 attractor, 0.558 and 0.765.

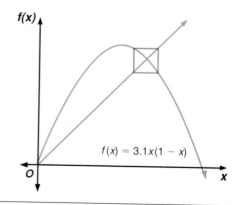

CHECKING FOR UNDERSTANDING

Read and study the lesson to answer each question.

1. Can the long-term iterative behavior of the logistic function be predicted if the parameter value a is changed? Why or why not?

2. **Explain** how you can find the coordinates of the fixed points of a logistic function.

3. **Describe** the long-term iterative behavior of a function with a period-4 attractor.

Find the coordinates of the vertex of the graph of each logistic function.

4. $f(x) = 2x(1 - x)$ 5. $f(x) = 3x(1 - x)$ 6. $f(x) = 1.2x(1 - x)$

Find the coordinates of the fixed points for each function. Use the graphical iteration program to graph the function and determine if the x-coordinate of each nonzero fixed point is a repeller or an attractor.

7. $f(x) = 2x(1 - x)$ 8. $f(x) = 3x(1 - x)$

9. $f(x) = 1.2x(1 - x)$ 10. $f(x) = x(1 - x)$

Use the modified graphical iteration program to determine the long-term iterative behavior of the function $f(x) = ax(1 - x)$ for each value of a. Write *period-n attractor*, *fixed point attractor*, or *chaos*.

11. $a = 2.95$ 12. $a = 3.05$

EXERCISES

Find the coordinates of the vertex of the graph of each logistic function.

13. $f(x) = 2.4x(1 - x)$ 14. $f(x) = 0.4x(1 - x)$

15. $f(x) = 4x(1 - x)$ 16. $f(x) = 0.64x(1 - x)$

Find the coordinates of the fixed points for each of the following functions. Use the graphical iteration program to graph the function and determine if the x-coordinate of each nonzero fixed point is a repeller or an attractor.

17. $f(x) = 1.6x(1 - x)$ 18. $f(x) = 2.9x(1 - x)$

19. $f(x) = 3.2x(1 - x)$ 20. $f(x) = 3.4x(1 - x)$

Use the modified graphical iteration program to determine the long-term iterative behavior of the function $f(x) = ax(1 - x)$ for each value of a. Write *period-n attractor*, *fixed point attractor*, or *chaos*.

21. $a = 3.24$ 22. $a = 3.50$ 23. $a = 3.68$

24. $a = 2.2$ 25. $a = 3.74$ 26. $a = 3.8$

27. The long-term iterative behavior of the logistic function forms patterns according to the value of *a*. That is, if *a* is in a certain range, the long-term behavior is a fixed point attractor. The next range shows that the long-term behavior is a period-2 attractor. The values of *a* that are the transition points between one range and the next are called *bifurcation points*, and the phenomenon is called *period doubling*.

   a. Find the value of *a* for which the long-term iterative behavior of $f(x) = ax(1 - x)$ makes a transition from a fixed point attractor to a period-2 attractor.

   b. Find the value of *a* for which the long-term iterative behavior of $f(x) = ax(1 - x)$ makes a transition from a period-2 attractor to a period-4 attractor.

   c. What is the value of *a* for which the long-term iterative behavior of $f(x) = ax(1 - x)$ makes a transition from a period-4 attractor to a period-8 attractor?

   d. Describe the long-term iterative behavior of the logistic function as the value of *a* increases just past 3.569946.

Critical Thinking

28. Dynamic systems like those found in nature are very sensitive to initial conditions. MIT meteorologist Edward N. Lorenz discussed this dependence in his paper "The Butterfly Effect," which asked the question, "Can the flap of a butterfly's wings in Brazil cause a tornado in Texas?"

   a. Find the first ten iterates of the function $f(x) = 4x(1 - x)$ for both of the initial values $x_0 = 0.200$ and 0.201. Compare the tenth iterates. Did the change in initial value change the iterates greatly?

   b. Explain why you think this is called the Butterfly Effect.

Applications and Problem Solving

29. **Biology** The Verhulst population model describing the population of antelope in an area is $p_{n+1} = p_n + 1.75p_n(1 - p_n)$. The maximum population sustainable in the area is 40 and the current population is 24.

   a. Find the population of antelope after each of the first ten years.

   b. Determine the long-term behavior of the population.

   c. Suppose the food supply has changed so that the growth rate for antelope has increased to 2.6 instead of 1.75. What will the long-term behavior of the population be?

30. **Research** Find articles on the Butterfly Effect in periodicals like *Discover*, *Omni*, or *Scientific American*. Write a one-page paper on the research that lead to the discovery of the Butterfly Effect. In what fields have scientists found applications for these discoveries?

31. Solve $\dfrac{x}{x-5} + \dfrac{17}{25-x^2} = \dfrac{1}{x+5}$. **(Lesson 4-6)**

32. Geometry The sides of a parallelogram are 20 cm and 32 cm long. If the longer diagonal measures 40 cm, find the measures of the angles of the parallelogram. Round measures to the nearest minute. **(Lesson 5-7)**

33. Write the standard form of the equation $25x^2 + 4y^2 - 100x - 40y + 100 = 0$. Then identify the conic section. **(Lesson 10-5)**

34. Find the sum of the infinite series $\dfrac{1}{16} + \dfrac{1}{8} + \dfrac{1}{4} + \cdots$, if it exists. **(Lesson 12-3)**

35. Find the coordinates of the fixed point for the function $f(x) = 1.36x + 2$. Is the x-coordinate of the fixed point a repeller or an attractor? **(Lesson 13-2)**

36. College Entrance Exam Compare quantities A and B below.
Write A if quantity A is greater.
Write B if quantity B is greater.
Write C if the two quantities are equal.
Write D if there is not enough information to determine the relationship.

Given: $x = 1, \quad 1 > y > 0$

(A) $\dfrac{1}{y}$ **(B)** x

MID-CHAPTER REVIEW

Find the first three iterates of each function using the given initial value. If necessary, round your answers to the nearest hundredth. (Lesson 13-1)

1. $f(x) = 5x - 3; x_0 = 8$

2. $f(x) = x^3; x_0 = 0.9$

3. $f(x) = x^2 - 4x + 2; x_0 = 2$

4. $f(x) = x^2(1 - 2x); x_0 = 0.6$

Graph each function and the function $f(x) = x$ on the same set of axes. Then perform graphical iteration for $x_0 = 2$. State the type of path that the iteration forms. (Lesson 13-2)

5. $f(x) = 2x - 4$

6. $f(x) = 6 - \dfrac{x}{2}$

7. $f(x) = 8 - 3x$

8. Is the fixed point for the function $f(x) = -5x + 9$ an attractor or a repeller? **(Lesson 13-2)**

9. Use the modified graphical iteration program to determine the long-term iterative behavior of the function $f(x) = 0.5x(1 - x)$ for each value of a. Write *period-n attractor*, *fixed point attractor*, or *chaos*. **(Lesson 13-3)**

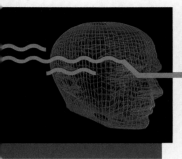

Technology

Graphical Iteration

IBM's *Mathematics Exploration Toolkit (MET)* allows you to define a sequence of commands and expressions to be stored and used as often as you like. These sequences are called *defer sequences*. Use the commands listed below to define a defer sequence to perform graphical iteration on the function $f(x) = 3.3x(1 - x)$.

Enter	Result
lims 0 1 0 1	*Sets the range of viewing window.*
color 3	*Sets the color to white.*
$y = 3.3x(1 - x)$	*Creates the function to be iterated.*
store q	*Stores the function as q.*
graph	*Graphs the function.*
$y = x$	
graph	*Graphs $y = x$.*
0.1	*Sets the starting point to 0.1.*
defer iterate	*Creates a defer sequence called "iterate."*
store x	*Stores the current value, now 0.1, as x.*
q	*Retrieves the function stored as q.*
replace	*Replaces x in the function with 0.1.*
getright	*Retrieves the right side of the equation.*
value	*Finds the value of the expression.*
store y	*Stores the values as y.*
color 2	*Changes the color to red.*
segment x x x y	*Draws a segment from (x, x) to (x, y).*
segment x y y y	*Draws a segment from (x, y) to (y, y).*
y	*Retrieves the value of y.*
replace	*Makes the value of y active.*
stop	*Ends the defer sequence.*

The graphics screen shows one iteration of the function. Type **repeat iterate 30 [Enter]** to perform 30 more iterates. Notice the pattern of iterates. The function $f(x) = 3.3x(1 - x)$ has a period-2 attractor.

EXERCISES

Use a defer sequence to perform graphical iteration on each function. Then describe the long-term iterative behavior of the function. Write *period-n attractor, fixed point attractor, fixed point repeller,* or *chaos.*

1. $f(x) = 3x + 2$
2. $f(x) = -0.3x + 1$
3. $f(x) = 0.4x - 5$
4. $f(x) = 3.4x(1 - x)$
5. $f(x) = 3.9x(1 - x)$
6. $f(x) = 3.5x(1 - x)$
7. $f(x) = 2.75x(1 - x)$
8. $f(x) = 4.0x(1 - x)$

13-4 Complex Numbers and Iteration

Objectives

After studying this lesson, you should be able to:
- iterate complex numbers, and
- plot the orbit of a complex number under iteration in the complex plane.

Application

One of the most exciting areas of modern mathematics is fractal geometry. This area of study was virtually unknown until Benoit B. Mandelbrot of the Physics Department at IBM's T.J. Watson Research Center made a mathematical breakthrough in 1980. Mandelbrot discovered the complex geometric structure that is called the Mandelbrot set. The Mandelbrot set and other fractal pictures are the result of iterating complex numbers through a quadratic equation.

As you know from your previous studies of complex numbers, every complex number has a real part and an imaginary part. At the right, the complex number $a + bi$ has been graphed on the complex plane. The horizontal axis of the complex plane represents the real part of the number, and the vertical axis represents the imaginary part.

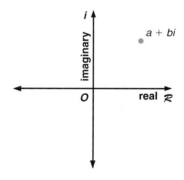

Functions can be iterated over the complex numbers.

Example 1

f(z) is used to denote a function on the complex number plane.

Find the first four iterates of the function $f(z) = 0.5z + 1$, if the initial value is $18 + 16i$.

$z_0 = 18 + 16i$
$z_1 = 0.5(18 + 16i) + 1$ or $10 + 8i$
$z_2 = 0.5(10 + 8i) + 1$ or $6 + 4i$
$z_3 = 0.5(6 + 4i) + 1$ or $4 + 2i$
$z_4 = 0.5(4 + 2i) + 1$ or $3 + i$

Just as we can represent the iteration of a real number through a function by graphing, we can graph the iteration sequence of complex numbers on the complex plane. The graph at the right shows the **orbit**, or sequence of successive iterates, of the initial value of $z_0 = 18 + 16i$ from Example 1.

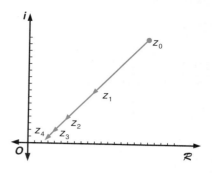

The function $f(z) = z^2 + c$, where c and z are complex numbers is central to the study of fractal geometry.

Example 2

Consider the function $f(z) = z^2 + c$.
a. Find the first six iterates of the function for $z_0 = 1 + i$ and $c = -i$.
b. Plot the orbit of the initial point at $1 + i$ under iteration of the function $f(z) = z^2 - i$ for six iterations.
c. Describe the long-term behavior of the function under iteration.

a. $z_1 = (1 + i)^2 - i$ or i
$z_2 = (i)^2 - i$ or $-1 - i$
$z_3 = (-1 - i)^2 - i$ or i
$z_4 = (i)^2 - i$ or $-1 - i$
$z_5 = (-1 - i)^2 - i$ or i
$z_6 = (i)^2 - i$ or $-1 - i$

b.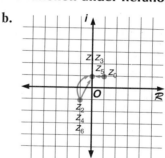

c. The iterates repeat after some time. This is a period-2 attractor.

CHECKING FOR UNDERSTANDING

Communicating Mathematics

Read and study the lesson to answer each question.

1. **Define** the orbit of a complex number under iteration.
2. Is the set of complex numbers closed to squaring? That is, will the square of a complex number always be a complex number?
3. **Compare** the method of graphing the orbit of a function $f(z) = z^2 + c$ for a complex number with the method of graphing orbits of the logistic function for some real number.

Guided Practice

Find the first three iterates of the function $f(z) = 0.6z + 2i$ for each initial value.

4. $z_0 = 25 + 40i$ 5. $z_0 = 5 - 10i$ 6. $z_0 = 6i$

Find the first three iterates of the function $f(z) = z^2 - 3i$ for each initial value.

7. $z_0 = 2i$ 8. $z_0 = 1 - i$ 9. $z_0 = 1$

10. Plot the orbit of the initial value $z_0 = 0 - i$ under iteration for the function $f(z) = z^2 - 1$ for four iterations.

EXERCISES

Practice

Find the first three iterates of the function $f(z) = 2z + i$ for each initial value.

11. $z_0 = 1 + 2i$

12. $z_0 = 5i$

13. $z_0 = 3$

14. $z_0 = 4 - i$

15. $z_0 = \dfrac{1}{2} - \dfrac{1}{4}i$

16. $z_0 = 0.1i$

Find the first three iterates of the function $f(z) = 3z + (2 - 3i)$ for each initial value.

17. $z_0 = 1 + 2i$

18. $z_0 = -2 + 3i$

19. $z_0 = 4 + i$

20. $z_0 = -1 + 2i$

21. $z_0 = 0.5 - i$

22. $z_0 = \dfrac{1}{3} + \dfrac{2}{3}i$

Find the first four iterates of the function $f(z) = z^2 + c$ for each given value of c and each initial value.

23. $c = 1 + 2i; z_0 = 0$

24. $c = 0; z_0 = \dfrac{\sqrt{2}}{2} - \dfrac{\sqrt{2}}{2}i$

25. $c = 2 - 3i; z_0 = 1 + 2i$

26. $c = 5 + 3i; z_0 = i$

Plot the orbit of each initial value for four iterations of the function $f(z) = z^2 + (1 - i)$.

27. $z_0 = 1 + i$

28. $z_0 = 0$

29. $z_0 = i$

30. $z_0 = -2$

Critical Thinking

31. If $f(z) = z^2 + c$ is iterated with an initial value of $z_0 = 2 + 3i$ and $z_1 = -1 + 15i$, find c.

Applications and Problem Solving

32. Research   Use a book or periodicals from a local library to research the work of Benoit Mandelbrot. Are there any new discoveries being made in the field of fractal geometry? Write a one-page paper about your findings.

Mixed Review

33. Find the value of $\begin{vmatrix} 5 & 3 & -2 \\ 0 & 2 & -2 \\ -3 & 0 & -1 \end{vmatrix}$.   **(Lesson 2-3)**

34. Security   A motion-detector light in the Gallagher's yard turns on when it senses motion within 25 feet of the light. The side boundaries of the monitored area form an angle of 135°. What is the area of the portion of the yard that is within the range of the motion sensor? Round to the nearest square foot.   **(Lesson 5-2)**

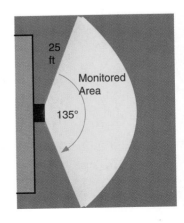

35. Express $\sqrt[4]{81x^4yz}$ using rational exponents.   **(Lesson 11-1)**

36. Write the series $1 + 5 + 25 + 125 + \cdots$ using sigma notation.   **(Lesson 12-5)**

37. Use the modified graphical iteration program to determine the long-term iterative behavior of the function $f(x) = 1.6x(1 - x)$. Write *period-n attractor, fixed point attractor,* or *chaos.*   **(Lesson 13-3)**

38. **College Entrance Exam**   Choose the best answer.
Layne answered all of the questions on a 25-question test. The test is scored by adding 4 points for each correct answer and subtracting 1 point for each incorrect answer. If Layne scored 70 on the test, how many questions did she answer correctly?

(A) 17   **(B)** 18   **(C)** 19   **(D)** 20   **(E)** 22

DECISION MAKING

Crime

The number of crimes committed in the United States increased by an average of more than 100,000 annually during the 1980s. The nation's prison population increased 134% during that decade. How should a citizen respond to the huge increase in crime we are witnessing? Does the answer lie in tougher laws and stricter punishment? Or should we concentrate on eliminating poverty, providing more job opportunities, increasing funding for education, and attacking other social problems that may lead to future crime?

1. Research the extent of the crime problem in your community. How big is the problem? What is being done to reduce the amount of crime?

2. Describe what you think should be done to attack the problem of crime in your area. Contact law enforcement officers, attorneys, people in social service agencies, and others who deal with the problem locally to learn their views and to find out how citizens can become involved. Investigate any neighborhood crime watch program in your area. Are there experts who could come to your school to discuss crime prevention?

3. Draw up a list of three ways you can help to lower the crime rate in your community.

4. **Project**   Work with your group to complete one of the projects listed at the right based on the information you gathered in Exercises 1–3.

PROJECTS

- *Conduct an interview.*
- *Write a book report.*
- *Write a proposal.*
- *Write an article for the school paper.*
- *Write a report.*
- *Make a display.*
- *Make a graph or chart.*
- *Plan an activity.*
- *Design a checklist.*

13-5 Escape Points, Prisoner Points, and Julia Sets

Objectives

After studying this lesson, you should be able to:
- determine if a point escapes or is a prisoner under iteration, and
- determine if a Julia set is connected or is a dust of points.

Application

For many centuries, people have used science and mathematics to better understand the world. However, Euclidean geometry, the geometry of points, lines, and planes that you have probably already studied, is not adequate to describe the natural world. Things like coastlines, clouds, and mountain ranges require a new type of geometry. This type of geometry is called **fractal geometry**.

Richard Voss, a physicist at the IBM T.J. Watson Research Center, created the image of the mountain range shown at the right using fractal geometry. Dr. Voss has found that he can use fractal geometry to simulate some of the beauty of natural forms. This relationship to nature has enabled scientists to use fractal geometry to study natural phenomena that had seemed out of reach. Currently, Dr. Voss is studying how fractal geometry can be used to date Chinese art and to help doctors diagnose breast cancer from mammograms.

→ *FYI...*

Gaston Julia was severely wounded while serving in the French Army in World War I. He lost his nose and required several surgeries. He conducted much of his mathematical research in a hospital between these operations.

At the heart of fractal geometry are the Julia sets. Named for mathematician Gaston Julia, Julia sets involve graphing the behavior of a function that is iterated in the complex plane. As a function is iterated, the iterates either escape or are held prisoner. If the iterates of a function approach infinity for some initial value, the initial point is called an escaping point. If the iterates do not approach infinity for some initial value, that point is called a prisoner point. Suppose the function $f(z) = z^2 + c$ is iterated for two different initial values. If the first one is an escaping point and the second one is a prisoner point, the graphs of the orbits may look like those shown on the next page.

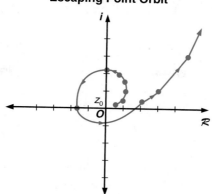

Escaping Point Orbit

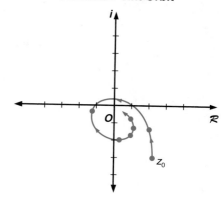

Prisoner Point Orbit

Some initial points will neither escape nor be held prisoner. These points will orbit about in a cyclical pattern.

Example 1

Recall from Chapter 9 that the modulus of a complex number is the distance that the graph of the complex number is from the origin. So $|a + bi| = \sqrt{a^2 + b^2}$.

Find the first four iterates of the function $f(z) = z^2 + c$ where $c = 0 + 0i$ for initial values whose moduli are in the regions $|z_0| < 1$, $|z_0| = 1$, and $|z_0| > 1$. Graph the orbits and describe the results.

Choose an initial value in each of the intervals. For $|z_0| < 1$, we will use $z_0 = 0.5 + 0.5i$, for $|z_0| = 1$, $0.5 + 0.5\sqrt{3}i$, and for $|z_0| > 1$, $0.75 + 0.75i$.

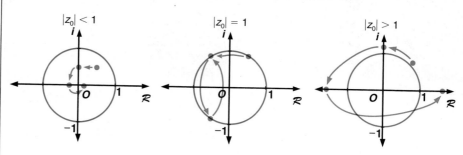

$z_0 = 0.5 + 0.5i$

$z_1 = (0.5 + 0.5i)^2$
  or $0.5i$

$z_2 = (0.5i)^2$ or -0.25

$z_3 = (-0.25)^2$ or 0.0625

$z_4 = 0.0625^2$ or 0.0039

$z_0 = 0.5 + 0.5\sqrt{3}i$

$z_1 = -0.5 + 0.5\sqrt{3}i$

$z_2 = -0.5 - 0.5\sqrt{3}i$

$z_3 = -0.5 + 0.5\sqrt{3}i$

$z_4 = -0.5 - 0.5\sqrt{3}i$

$z_0 = 0.75 + 0.75i$

$z_1 = 1.125i$

$z_2 = -1.265$

$z_3 = 1.602$

$z_4 = 2.565$

In the first case, where $|z_0| < 1$, the iterates will approach 0, since each successive square will be smaller than the one before. Thus, the orbit in this case approaches 0, a fixed point. In the second case, where $|z_0| = 1$, the iterates are orbiting around on the unit circle. When $|z_0| > 1$, the iterates approach infinity, since each square will be greater than the one before.

The orbits in Example 1 demonstrate that the iterative behavior of various initial values in different regions behave in different ways. An initial value whose graph is inside the unit circle is a prisoner point. A point chosen on the unit circle stays on the unit circle, and a point outside of the unit circle escapes. Other functions of the form $f(z) = z^2 + c$ when iterated over the complex plane also have regions in which the points behave this way; however, they are usually not circles.

All of the initial points for a function on the complex plane are split into three sets, those that escape, called the **escape set E**, and those that do not escape, called the **prisoner set P**. The boundary between the escape set and the prisoner set is called the **Julia set**. The escape set, the prisoner set, and the Julia set for the function in Example 1 are graphed at the right. The Julia set in Example 1 is the unit circle.

The Julia set for $f(z) = z^2$ is unusual since it is not a **fractal**. One of the characteristics of a fractal is that it exhibits **self-similarity**. Self-similar objects are those in which we can find replicas of the entire shape or object imbedded over and over again inside the object in different sizes. For example, if you look at the branch of a tree, it resembles the entire tree, only smaller. The broccoli shown at the left is one plant that exhibits fractal self-similarity.

Julia sets can be divided into two groups: those that are connected, like the unit circle, and those that are disconnected. Examples of the graphs of connected and disconnected Julia sets are shown below.

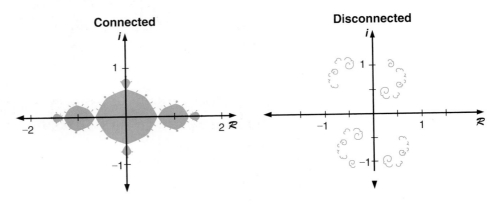

Julia sets like these can be generated by using powerful computers. You can determine if a Julia set for a function is connected or disconnected by applying the following test.

If the initial point $0 + 0i$ escapes to infinity under iteration in some function $f(z) = z^2 + c$, where c is a complex number, then the prisoner set of that function is disconnected. If $0 + 0i$ does not escape to infinity, the prisoner set is connected.

$0 + 0i$ is known as the critical point.

Example 2

Determine if the prisoner set for the function $f(z) = z^2 + 1 - i$ is connected or disconnected.

Test the initial value $0 + 0i$. Find the first five iterates and the distance from each iterate to the origin to determine if the value is escaping.

$z_0 = 0 + 0i$ $\qquad\qquad$ $|z_0| = \sqrt{0^2 + 0^2}$ or 0

$z_1 = 1 - i$ $\qquad\qquad$ $|z_1| = \sqrt{(1)^2 + (-1)^2} \approx 1.414$

$z_2 = 1 - 3i$ $\qquad\qquad$ $|z_2| = \sqrt{(1)^2 + (-3)^2} \approx 3.162$

$z_3 = -7 - 7i$ $\qquad\qquad$ $|z_3| = \sqrt{(-7)^2 + (-7)^2} \approx 9.899$

$z_4 = 1 + 97i$ $\qquad\qquad$ $|z_4| = \sqrt{(1)^2 + (97)^2} \approx 97.005$

$z_5 = -9407 + 193i$ $\qquad$ $|z_5| = \sqrt{(-9407)^2 + (193)^2} \approx 9408.980$

Since the iterates are getting farther away from the origin, the critical point will escape to infinity. Therefore, the Julia set for the function $f(z) = z^2 + 1 - i$ is disconnected.

Since the Julia set of a function is the set of points that lie on the boundary between the prisoner set and the escape set, the Julia set for a function with a connected prisoner set is the boundary of the prisoner set. If the prisoner set for a function is disconnected, the Julia set is disconnected and is the prisoner set itself. Disconnected Julia sets are called **dust** since they look like a dusting of points scattered in a fractal pattern on the plane.

The graphing calculator program listed below can be used to help you test to see if an initial point is in the escape set or the prisoner set of some function. It will display the sequence of iterates for a given initial point. You can also use this program to determine if a Julia set is connected or disconnected by using the critical point as your initial point.

EXPLORATION: Graphing Calculator

1. Enter the program below into the program memory of a TI-82 graphing calculator.

Line 1 : ClrHome

Line 2 : Disp "A IN C = A + BI"

Line 3 : Input A

Line 4 : Disp "B IN C = A + BI"

(continued on the next page)

```
Line 5   :Input B                    Line 14  :X→R
Line 6   :Disp " "                    Line 15  :Y→S
Line 7   :Disp "X IN THE INITIAL      Line 16  :R² − S² + A→X
          POINT"                      Line 17  :2RS + B→Y
Line 8   :Input X                     Line 18  :Disp X
Line 9   :Disp "Y IN THE INITIAL      Line 19  :Disp Y
          POINT"                      Line 20  :Pause
Line 10  :Input Y                     Line 21  :M + 1→M
Line 11  :0→M                         Line 22  :If M < 15
Line 12  :Lbl 1                       Line 23  :Goto 1
Line 13  :Disp " "
```

2. Use the program to iterate the function $f(z) = z^2 + (-1 + 0i)$ for the initial value $1 + 0.5i$.

3. Observe the iterates. $(1 + 0.5i)$ is in the escape set for the function since its iterates escape to infinity.

CHECKING FOR UNDERSTANDING

Communicating Mathematics

Read and study the lesson to answer each question.

1. **Define** the prisoner set of a function.

2. **Describe** how the Julia set of a function is related to the prisoner set of the function.

3. **Explain** how you can determine if the Julia set of a function is connected or disconnected.

Guided Practice

Determine whether the graph of each value is in the prisoner set, the escape set, or the Julia set of the function $f(z) = z^2$.

4. $0.5 - 0.5i$ 5. $1 + 2i$ 6. $-0.25 - 0.2i$

7. $0.5 - 0.5\sqrt{3}i$ 8. $-1.5 + 0i$ 9. $\dfrac{2\sqrt{2}}{3} + \dfrac{1}{3}i$

10. Is the Julia set for the function $f(z) = z^2 + 2 + 2i$ connected or disconnected?

EXERCISES

Practice

Determine whether the graph of each value is in the prisoner set or the escape set for the function $f(z) = z^2 + (-5 + 0i)$.

11. $1 + 2i$ 12. $4 - 2i$ 13. $\dfrac{1 + \sqrt{21}}{2} + 0i$

Determine whether the graph of each value is in the prisoner set or the escape set for the function $f(z) = z^2 + (-1 + 0i)$.

14. $1 + 1i$ 15. $1 - 2i$ 16. $0.5 + 0i$

Determine whether the Julia set for each function is connected or disconnected.

17. $f(z) = z^2 + (-1 + 0i)$

18. $f(z) = z^2 + (-0.3 + 0.7i)$

19. $f(z) = z^2 + (-1.25 + 0i)$

20. $f(z) = z^2 + (-0.3 - 0.3i)$

21. $f(z) = z^2 + (-1 + 0.5i)$

22. $f(z) = z^2 + (0.11 - 0.7i)$

23. $f(z) = z^2 + (-1.2 + 0i)$

24. $f(z) = z^2 + (0.4 + 0.5i)$

25. $f(z) = z^2 + (4 - 0.5i)$

26. $f(z) = z^2 + (0.31 + 0.04i)$

Find one value whose graph is in the prisoner set and one value whose graph is in the escape set of each function.

27. $f(z) = z^2 + (-0.2 + 0i)$

28. $f(z) = z^2 + (-0.75 + 0.25i)$

29. $f(z) = z^2 + (0.02 + 0.25i)$

Critical Thinking

30. As you know, a fixed point is mapped to itself under a function. Since fixed points do not escape to infinity, they must be in the prisoner set of the function. Find a fixed point for the function $f(z) = z^2 + (-3 + 0i)$. *Hint: $f(z) = z$ for a fixed point.*

Applications and Problem Solving

31. Language Arts The book *Jurassic Park* by Michael Crichton involves a mathematician who uses the principles of fractal geometry to describe why a park of dinosaurs would not work. Find a copy of the book or a book review and describe the role that fractal geometry plays in the book.

32. Medicine Investigate the use of fractal geometry to diagnose breast cancer that Dr. Voss is studying. Is the technique in use today? How accurate are the results?

Mixed Review

33. Graph the rational function $y = \dfrac{2}{x - 4}$. **(Lesson 3-4)**

34. Find values of x in the interval $0° \leq x \leq 360°$ such that $x = \cos^{-1} -\dfrac{1}{\sqrt{2}}$. **(Lesson 6-4)**

35. If $\sin r = \dfrac{3}{5}$ and r is in the first quadrant, find $\cos 2r$. **(Lesson 7-4)**

36. Use the binomial theorem to expand $(n + 2)^4$. **(Lesson 12-6)**

37. Find the first four iterates of the function $f(z) = z^2 + (1 - 3i)$ for $z_0 = 0$. **(Lesson 13-4)**

38. College Entrance Exam Choose the best answer.

If $\dfrac{a}{b} = \dfrac{4}{5}$, what is the value of $2a + b$?

(A) 3

(B) 13

(C) 14

(D) 26

(E) cannot be determined from given information

13-6 The Mandelbrot Set

After studying this lesson, you should be able to:
- determine if a complex number is inside the Mandelbrot set, and
- determine the color of a point outside the Mandelbrot set.

Application

    Gaston Julia was a world-famous mathematician in the 1920s, but his work was all but forgotten until Benoit Mandelbrot brought it back to light in the late 1970s. Mandelbrot was first introduced to Julia's work by his uncle Szolem Mandelbrojt, who was a mathematics professor at the prestigious Collège de France in Paris. But Mandelbrot didn't like Julia's work and chose another area of study, which eventually led him back to Julia's work. Mandelbrot's study of Julia's work led him to discover the **Mandelbrot set** and to invent fractal geometry. The Mandelbrot set is a complex mathematical picture. The set of pictures below shows successive zoom-ins on the Mandelbrot set. The rectangular window in each figure shows the region in the next zoom-in.

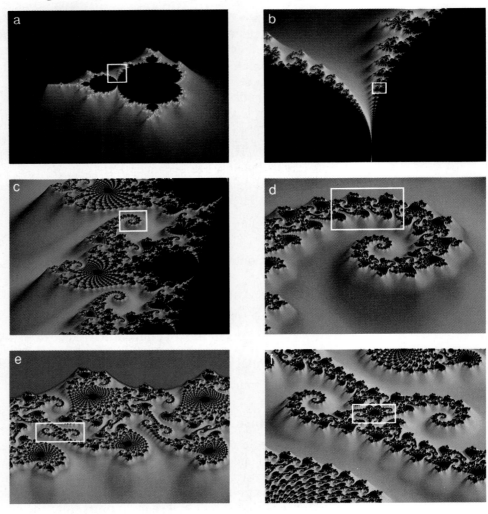

Both the Julia sets and the Mandelbrot set are created by iterating the function $f(z) = z^2 + c$ with z and c representing complex numbers. To define a Julia set, the parameter c is fixed and the initial value, z_0, is allowed to vary. The Mandelbrot set is defined by fixing $z_0 = 0 + 0i$ and allowing c to vary. If the iterations of the function $f(z) = z^2 + c$ do not approach infinity for some value of c, then the point c is in the Mandelbrot set. That is, if the Julia set for the function $f(z) = z^2 + c_1$ is connected, then c_1 is in the Mandelbrot set.

Test for Points in the Mandelbrot Set	**A point c_1 lies in the Mandelbrot set if the Julia set associated with that point is connected. That is, if the iterates of $f(z) = z^2 + c_1$ do not escape to infinity, then the point c_1 is in the Mandelbrot set.**

The figure below shows the Mandelbrot set and the Julia sets that are related to some of the points inside and outside of the Mandelbrot set. Notice that the Julia sets related to points inside of the Mandelbrot set are connected and the sets related to points outside of the Mandelbrot set are disconnected.

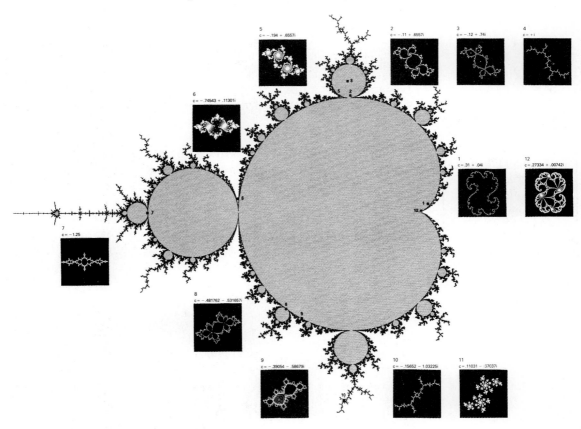

Performing the iterations to determine if a point is in the Mandelbrot set can be tedious. The program shown on page 742 can be used with a TI-81 graphing calculator to perform these iterations. The program displays each iterate and checks its distance from the origin. If the iterates from the critical point do not travel at least 80 units from the origin in 20 iterations, the critical point is assumed to be a prisoner point. If the point does escape, the number of iterations that it took to travel 80 units from the origin is displayed.

EXPLORATION: Graphing Calculator

We assume that if a point does not move at least 80 units from the origin in less than 20 iterations the point is a prisoner point.

Use the program below to determine if the point $-0.5 + i$ is in the Mandelbrot set.

```
Line 1    :ClrHome
Line 2    :Disp "A IN
            C = A + BI"
Line 3    :Input A
Line 4    :Disp "B IN
            C = A + BI"
Line 5    :Input B
Line 6    :Disp" "
Line 7    :A→R
Line 8    :B→S
Line 9    :1→M
Line 10   :Lbl 1
Line 11   :R² - S² + A → X
Line 12   :2RS + B→S
Line 13   :X→R
Line 14   :Disp R
```

```
Line 15   :Disp S
Line 16   :Pause
Line 17   :Disp " "
Line 18   :M + 1→M
Line 19   :If M > 20
Line 20   :Goto 2
Line 21   :If (R² + S²) < 6400
Line 22   :Goto 1
Line 23   :Disp " "
Line 24   :Disp "THE ESCAPING
            ITERATION COUNT IS "
Line 25   :Disp M
Line 26   :Lbl 2
Line 27   :Disp " "
Line 28   :Disp "C IS A
            PRISONER POINT"
```

The result of running the program for $-0.5 + i$ is that the point escapes after 6 iterations. Therefore, the point is outside of the Mandelbrot set.

Portfolio

Select an item from your work in this chapter that shows your creativity and place it in your portfolio.

Mapmakers use colors to help distinguish between areas with different characteristics. For example, some maps use blue to denote water, white to denote land, and yellow for metropolitan areas. Mathematicians use colors to describe different areas of fractal pictures as well. The different colors are used to represent the different speeds at which a point iterates to infinity. In the picture of a portion of the Mandelbrot set below, the points colored black are in the Mandelbrot set and the points that are not black iterate to infinity.

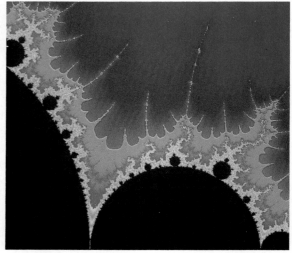

The following coloring scheme was used to color the Mandelbrot set picture.

Color a point:	If:
Black	Its iterates fail to escape to infinity.
Dark blue	Iterates are 80 units from origin in 1 to 3 iterations.
Orange	Iterates are 80 units from origin in 4 to 6 iterations.
Light blue	Iterates are 80 units from origin in 7 to 9 iterations.
Red	Iterates are 80 units from origin in 10 to 12 iterations.
Yellow	Iterates are 80 units from origin in 13 to 15 iterations.
Purple	Iterates are 80 units from origin in 16 or more iterations.

Example

For each value of c given, complete the following.
a. Determine whether the Julia set associated with each value of c is connected or disconnected.
b. Determine if the point is inside or outside of the Mandelbrot set.
c. Use the color scheme given above to assign a color to the point.

1. $c = 0.8 + 0.1i$ 2. $c = -1.5 + 0.01i$ 3. $c = 0.1 + 0.05i$

1. The point escapes after 5 iterations. Since it is an escaping point, the associated Julia set is disconnected. The point is outside of the Mandelbrot set. According to the color scheme, it should be orange.

2. The point escapes after 16 iterations, so the associated Julia set is disconnected. The point is outside of the Mandelbrot set and should be colored purple.

3. The point $0.1 + 0.05i$ is a prisoner point. Its associated Julia set is connected. Thus, the point is inside of the Mandelbrot set and should be colored black.

CHECKING FOR UNDERSTANDING

Communicating Mathematics

Read and study the lesson to answer each question.

1. **Describe** how you determine if a point is in the Mandelbrot set.
2. **Explain** how the Julia sets and the Mandelbrot set are related.
3. What is the significance of the colors in the Mandelbrot set?
4. **Compare** the use of color on the Mandelbrot set to the color used on a map.

Guided Practice

For each value of c, complete the following.
a. Determine whether the Julia set associated with each value of c is connected or disconnected.
b. Determine if the point is inside or outside of the Mandelbrot set.
c. Use the color scheme given in the lesson to assign a color to the point.

5. $c = 1 + i$
6. $c = -1 + 0i$
7. $c = -1 - 0.65i$
8. $c = -1 + i$
9. $c = 0.57 + 0.25i$
10. $c = -1 + 0.25i$

EXERCISES

Practice **For each value of c given, complete the following.**
   a. Determine whether the Julia set associated with each value of c is connected or disconnected.
   b. Determine if the point is inside or outside of the Mandelbrot set.
   c. Use the color scheme given in the lesson to assign a color to the point.

11. $c = -1.25 + 0.5i$ **12.** $c = -0.6 - 0.05i$

13. $c = -1 + 0i$ **14.** $c = 2 + 0i$

15. $c = -0.5 + 0.5i$ **16.** $c = -2.01 + 0i$

17. $c = -0.75 + 0.5i$ **18.** $c = -1 + 0.75i$

19. $c = -1.3 + 0.1i$ **20.** $c = -1.3 + 0.3i$

21. $c = -0.8 + 0.5i$ **22.** $c = 0.6 - 1.8i$

23. $c = 0.5 + 0.5i$ **24.** $c = 0.5 + 0.6i$

25. A function of the form $f(z) = z^2 + c$ has a disconnected prisoner set. Where is c located in relation to the Mandelbrot set?

26. One of the most interesting characteristics of fractals is that they are *self-similar*. That is, small replicas of the image are repeated again and again in a fractal. Describe some of the self-similarity you see in the Mandelbrot set.

Programming

27. Use the following program to draw a rough picture of the Mandelbrot set on a TI-82 graphing calculator. *Note: This program will take approximately 20 minutes to run.*

```
Line 1   : FnOff              Line 20  : 1 → K
Line 2   : ClrDraw            Line 21  : Lbl 1
Line 3   : -2.6 → Xmin        Line 22  : X² → U
Line 4   : 1.0 → Xmax         Line 23  : Y² → V
Line 5   : 0.5 → Xscl         Line 24  : 2XY + B → Y
Line 6   : -1.2 → Ymin        Line 25  : U - V + A → X
Line 7   : 1.2 → Ymax         Line 26  : K + 1 → K
Line 8   : 0.5 → Yscl         Line 27  : If U + V > 100
Line 9   : DispGraph          Line 28  : Goto 2
Line 10  : 0.0379 → C         Line 29  : If K < 15
Line 11  : 0.0381 → D         Line 30  : Goto 1
Line 12  : 0 → J              Line 31  : PT-On(A,B)
Line 13  : Lbl J              Line 32  : PT-On(A,-B)
Line 14  : -1.2 + JD → B      Line 33  : Lbl 2
Line 15  : 0 → I              Line 34  : IS>(I,95)
Line 16  : Lbl I              Line 35  : Goto I
Line 17  : -2.6 + IC → A      Line 36  : IS>(J,31)
Line 18  : A → X              Line 37  : Goto J
Line 19  : B → Y              Line 38
```

Critical Thinking **28.** Based on the graphing calculator graph of the Mandelbrot set, what are the dimensions of the smallest rectangle that would enclose the Mandelbrot set?

29. **Geology**   The Landsat-1 satellite took the photograph of the foothills of the Himalayas shown at the right on December 14, 1972. Describe the fractal self-similarity of the foothills.

30. **Botany**   Botanists have found that the angle between the main branches of a tree and its trunk remains constant in each species. How is this finding related to fractal geometry?

Mixed Review

31. Find an ordered pair that represents $\overrightarrow{AB}$ for $A(8, -3)$ and $B(5, -1)$.   **(Lesson 8-2)**

32. **Physics**   Don and Joyce are standing on a cliff that is 150 feet high. At the same time, Joyce drops a stone and Don throws a stone horizontally at a velocity of 35 ft/s. About how far apart will the stones be when they land?   **(Lesson 8-7)**

33. Solve the equation $\log_{\frac{1}{3}} x = -3$.   **(Lesson 11-4)**

34. Use the ratio test to determine if the series $\frac{1}{3} + \frac{2}{3^2} + \frac{3}{3^3} + \cdots + \frac{n}{3^n} + \cdots$ is convergent or divergent.   **(Lesson 12-4)**

35. Determine whether the point at $2 + 4i$ is in the prisoner set or the escape set for the function $f(z) = z^2 + (0 - i)$.   **(Lesson 13-5)**

36. **College Entrance Exam**   Choose the best answer.
The sum of three consecutive odd integers is always divisible by

    (A) 2   **(B)** 3   **(C)** 5   **(D)** 6   **(E)** 2 and 3

CASE STUDY FOLLOW-UP

Refer to Case Study 2: *Trashing the Planet* on pages 959–961.

Suppose you have three maps of Saudi Arabia that have the following scales.
    A: 1 inch = 200 miles   B: 1 inch = 50 miles   C: 1 inch = 15 miles
The length of the oil spill created during the Gulf War along the Persian Gulf coastline of Saudi Arabia is measured on each map. The measurements on maps A, B, and C were 1.8 inches, 7.3 inches and 26.7 inches, respectively.

1. Use the measurement from each map to find the length of the oil spill on the coastline. Are the lengths you found all the same?

2. Mathematician Benoit B. Mandelbrot discovered that the smaller the scale of a map, the greater the measurement of the length of a feature on the map. Use the measures of the length of the oil spill found in Exercise 1 to estimate the scale of the map of Saudi Arabia in the case study.

3. **Research**   Read about Dr. Mandelbrot's study of the coastline of Britain. Summarize your findings in a 1–page paper.

VOCABULARY

Upon completing this chapter, you should be familiar with the following terms:

attractor	**718**	**736**	Julia set
dust	**737**	**722**	logistic function
escape set	**736**	**740**	Mandelbrot set
fixed points	**717**	**723**	mixing
fractal	**736**	**731**	orbit
fractal geometry	**734**	**725**	period-n attractor
graphical iteration	**717**	**736**	prisoner set
iterate	**713**	**718**	repeller
iteration	**713**	**736**	self-similarity

SKILLS AND CONCEPTS

OBJECTIVES AND EXAMPLES	REVIEW EXERCISES

Upon completing this chapter, you should be able to:

- iterate functions using real numbers **(Lesson 13-1)**

  Find the first three iterates of $f(x) = (x + 1)^2$ for an initial value of $x_0 = -1$.

  $f(x_0) = f(-1) = (-1 + 1)^2 = 0$
  $f(x_1) = f(0) = (0 + 1)^2 = 1$
  $f(x_2) = f(1) = (1 + 1)^2 = 4$

Use these exercises to review and prepare for the chapter test.

Find the first three iterates of each function using the given initial value. If necessary, round your answers to the nearest hundredth.

1. $f(x) = 5 - 2x, x_0 = 3$
2. $f(x) = x^2 + 3, x_0 = 0$
3. $f(x) = (x + 2)^3, x_0 = -1.3$
4. $f(x) = x^2 + x - 6, x_0 = 2.5$

- identify a fixed point as an attractor or a repeller   **(Lesson 13-2)**

  Find the x-coordinate of the fixed point for $f(x) = 3 - 2x$. Is it a repeller or an attractor?

  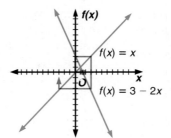

  1 is a repeller.

Find the x-coordinate of the fixed point for each function. Is it a repeller or an attractor?

5. $f(x) = 3x + 1$
6. $f(x) = 0.5x - 3$
7. $f(x) = -6x + 5$
8. $f(x) = -2x + 0.3$

OBJECTIVES AND EXAMPLES	REVIEW EXERCISES

OBJECTIVES AND EXAMPLES

- perform graphical iteration on the logistic function **(Lesson 13-3)**

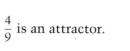

$\frac{4}{9}$ is an attractor.

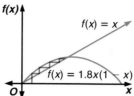

- determine if the behavior of a function is predictable or unpredictable **(Lesson 13-3)**

Describe the long-term iterative behavior of the function $f(x) = 3.49x(1 - x)$.

The iterates form the repeating pattern 0.829, 0.494, 0.872, 0.389. So the function has a period-4 attractor.

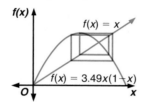

- iterate complex numbers **(Lesson 13-4)**

Find the first three iterates of the function $f(z) = 2z - (4 + i)$ if the initial value is $3 - i$.

$z_1 = 2(3 - i) - (4 + i)$ or $2 - 3i$
$z_2 = 2(2 - 3i) - (4 + i)$ or $-7i$
$z_3 = 2(-7i) - (4 + i)$ or $-4 - 15i$

- determine if a point escapes or is a prisoner under iteration **(Lesson 13-5)**

Is the point $1 - 2i$ in the prisoner set or the escape set for the function $f(z) = z^2 + (3 + 0i)$?

$z_0 = 1 - 2i$ $z_3 = 172$
$z_1 = -4i$ $z_4 = 29,587$
$z_2 = -13$

The sequence of iterates approaches infinity, so $1 - 2i$ is in the escape set.

REVIEW EXERCISES

Find the coordinates of the fixed points for each of the following functions. Use the graphical iteration program on pages 724–725 to graph the function and determine if the x-coordinate of a non-zero fixed point is a repeller or an attractor.

9. $f(x) = 2.7x(1 - x)$

10. $f(x) = 1.9x(1 - x)$

11. $f(x) = 3.19x(1 - x)$

12. $f(x) = 3.35x(1 - x)$

Describe the long-term iterative behavior of the function $f(x) = ax(1 - x)$ for each value of a.

13. $a = 3.33$

14. $a = 2.0$

15. $a = 3.52$

16. $a = 3.78$

Find the first three iterates of the function $f(z) = 0.5z + (4 - 2i)$ for each initial value.

17. $z_0 = 4i$

18. $z_0 = -8$

19. $z_0 = -4 + 6i$

20. $z_0 = 12 - 8i$

Is each point in the prisoner set or the escape set for the function $f(z) = z^2 + (-1 + 0i)$?

21. $0 + 2i$

22. $0 + 0.7i$

23. $1.628 + 0i$

24. $0.5 - 0.5i$

OBJECTIVES AND EXAMPLES	REVIEW EXERCISES

■ determine if a Julia set is connected or is a dust of points   (**Lesson 13-5**)

Determine if the Julia set for the function $f(z) = z^2 + (-1 + 0.2i)$ is connected or disconnected. Use the initial value $0 + 0i$.

$$z_0 = 0 + 0i \qquad\qquad |z_0| = 0$$
$$z_1 = -0.04 - 0.2i \qquad |z_1| \approx 0.204$$
$$z_2 = -1.038 - 0.216i \qquad |z_2| \approx 1.060$$
$$z_3 = 0.032 - 0.249i \qquad |z_3| \approx 0.251$$

Since the iterates do not escape to infinity, the Julia set is connected.

Determine whether the Julia set for each function is connected or disconnected.

25. $f(z) = z^2 + (-1 - 1i)$

26. $f(z) = z^2 + (0.2 + 0.2i)$

27. $f(z) = z^2 + (0.5 + 0.1i)$

28. $f(z) = z^2 + (-0.3 - 0.2)$

■ determine the color of a point outside the Mandelbrot set   (**Lesson 13-6**)

Determine the color that should be assigned to the point $0.2 - 1.4i$.

The point escapes after 5 iterations, so it should be colored orange.

Use the program on page 742 and the color scheme on page 743 to assign a color to each point.

29. $c = 0.5 + 0.8i$

30. $c = 0.6 - 0.3i$

31. $c = -0.7 + 0.5i$

32. $c = 0.12 - 0.3i$

APPLICATIONS AND PROBLEM SOLVING

33. **Banking**   Marcos and Carla Rodriguez have a 30-year mortgage of $85,000 on their home. The interest rate is 9.5% and the monthly principal and interest payment is $714.73.   (**Lesson 13-1**)

   a. Derive an equation for the balance of the mortgage after n payments.

   b. Find the balance of the mortgage after each of the first 4 payments.

34. **Biology**   The Verhulst population model equation that describes the population of otters in a certain area is $p_{n+1} = p_n + 1.9p_n (1 - p_n)$. The maximum population sustainable in the area is 72, and the current population is 54. (**Lesson 13-3**)

   a. Find the population of otters at the end of each of the first ten years.

   b. Determine the long-term iterative behavior.

Find the first three iterates of each function using the given initial value. If necessary, round your answers to the nearest hundredth.

1. $f(x) = 4x - 5, x_0 = 2$ **2.** $f(x) = (x-2)^2, x_0 = 1.1$ **3.** $f(x) = x^3 + 7, x_0 = -2$

Find the *x*-coordinate of the fixed point for each function. Is it a repeller or an attractor?

4. $f(x) = 2x - 5$ **5.** $f(x) = 0.2x - 2$ **6.** $f(x) = -2x + 0.4$

7. Describe the slope of the graph of a function whose iterations spiral inward.

Find the coordinates of the vertex and the fixed points for the graph of each logistic function. Then, use a graphing calculator to describe the long-term iterative behavior of the function.

8. $f(x) = 1.6x(1-x)$ **9.** $f(x) = 3.48x(1-x)$ **10.** $f(x) = 3.88x(1-x)$

Find the first three iterates of the function $f(z) = 2z + (3 - i)$ for each initial value.

11. $z_0 = 2i$ **12.** $z_0 = -1 + 2i$ **13.** $z_0 = 0.5 + i$

Is each point in the prisoner set or the escape set for the function $f(z) = z^2 + (-2 + 0i)$?

14. $1 + i$ **15.** $0.8 + 0i$ **16.** $0.5 - 0.5i$

Determine whether the Julia set for each function is connected or disconnected.

17. $f(z) = z^2 + (-1 + 1i)$ **18.** $f(z) = z^2 + (0.2 - 0.2i)$ **19.** $f(z) = z^2 + (-0.6 + 0.4i)$

Use a graphing calculator to determine if each point is inside or outside of the Mandelbrot set. If the point is outside, state the number of iterations it takes for the iterations to exceed 80.

20. $c = -0.6 + 0.5i$ **21.** $c = 0.4 + 0.1i$ **22.** $c = -0.7 + 0.8$ **23.** $c = 0.5 + 0.7i$

24. If a function of the form $f(z) = z^2 + c$ has a connected prisoner set, where is c located in relation to the Mandelbrot set?

25. Biology   The Verhulst population model equation that describes the population of bald eagles in an area is $p_{n+1} = p_n + 2.1p_n(1 - p_n)$. The maximum population sustainable in the area is 120 and the current population is 90. Find the population of bald eagles after each of the first 5 years.

Bonus   If $f(z) = z^2 + (2 + 2i)$ is iterated for some z_0 and $z_1 = 7 - 10i$, find z_0.

CHAPTER 14

COMBINATORICS AND PROBABILITY

HISTORICAL SNAPSHOT

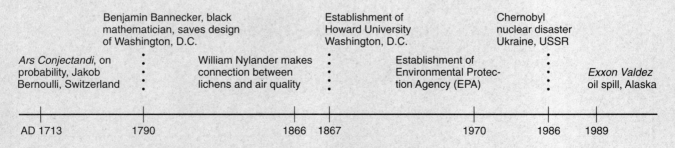

	Benjamin Bannecker, black mathematician, saves design of Washington, D.C.		Establishment of Howard University Washington, D.C.		Chernobyl nuclear disaster Ukraine, USSR	
Ars Conjectandi, on probability, Jakob Bernoulli, Switzerland		William Nylander makes connection between lichens and air quality		Establishment of Environmental Protection Agency (EPA)		*Exxon Valdez* oil spill, Alaska
AD 1713	1790		1866 1867		1970	1986 1989

CHAPTER OBJECTIVES

In this chapter, you will:

- Solve problems involving combinations and permutations.
- Distinguish between dependent or independent and mutually exclusive or inclusive events.
- Find probabilities.
- Find odds for the success or failure of an event.

CAREER GOAL: Special Education with Math and Science

With majors in environmental science and special education, and minors in civil engineering and allied science, Heather Katz stays pretty busy. Though her interests are varied, her goal remains clear. "I want to teach mathematics and science to students with disabilities." Eventually, Heather wants to earn a PhD in special education and help develop and implement programs for students with disabilities.

Heather was recently diagnosed with a mild reading comprehension disability. "If I'd been diagnosed in elementary school," Heather says, "I probably would have received the help that I needed. But when I got to high school, a teacher told me that I would never get through college math courses." Heather remained firm. "If I had listened to that teacher, I wouldn't be where I am today."

Today, Heather uses mathematics in many ways. "In environmental science, we use probability to analyze underground storage tanks. We want to examine the danger that gasoline or other chemicals will leak into the groundwater. This type of analysis begins by designating a *plume*. This is an area of land within which several underground storage tanks are located. To determine the probability of chemicals getting into the groundwater, wells are drilled within the plume, samples are taken of the groundwater, and then the levels of contamination in the samples are measured. The levels of contamination of the sample wells will help you to then calculate the probability that your plume is contaminated."

Heather says, "Special education can be very frustrating. But I want to teach disabled children that they can succeed in spite of their disabilities, and that they shouldn't let their disabilities stop them from achieving their goals."

For up-to-date information on environmental science, visit:
www.glencoe.com/sec/math/amc/mathnet

14-1 Permutations

Objectives

After studying this lesson, you should be able to:
- solve problems related to the basic counting principle, and
- distinguish between dependent and independent events.

Application

There are two tunnels under the Hudson River by which you can enter the borough of Manhattan in New York City from the west, and three bridges and one tunnel by which you can leave Manhattan to go south to Brooklyn. After entering Manhattan by the first tunnel, you have the choice of going south to Brooklyn by four exits. So there are four exits by way of the first tunnel. Likewise, there are four exits by way of the second tunnel. Altogether, there are 2×4 or 8 possible routes through Manhattan from west to south.

The choice of ways to enter Manhattan does *not* affect the choice of ways to exit to Brooklyn. Thus, these two choices are called **independent events**. Events that do affect each other are called **dependent events**.

The investigation of the different possibilities for the arrangement of objects is called **combinatorics**. This example of choosing possible routes illustrates a rule of combinatorics known as the **basic counting principle**.

Basic Counting Principle	Suppose one event can be chosen in p different ways, and another independent event can be chosen in q different ways. Then the two events can be chosen successively in $p \cdot q$ ways.

This principle can be extended to any number of independent events.

Example 1

How many seven-digit phone numbers can begin with the prefix 827?

Since the phone numbers all begin with 827, there is only one choice for each of the first three digits. There are 10 choices for each of the last four digits.

digit in phone number:	1st	2nd	3rd	4th	5th	6th	7th
ways to choose:	1	1	1	10	10	10	10

Thus, there are $1 \cdot 1 \cdot 1 \cdot 10 \cdot 10 \cdot 10 \cdot 10$ or 10,000 numbers.

A **tree diagram** is often used to show all of the choices.

Example 2

FYI...

Chevrolet Caprice police cars are assembled at the General Motors assembly plant in Arlington, Texas. For police cars alone, there are 125 available color combinations.

Mrs. Wilson has decided to buy a Chevrolet Caprice or a Buick Roadmaster. She has narrowed down the choices of exterior colors to silver metallic, white, or dark cherry metallic, and interior colors to grey or red. How many different selections of these models and colors are possible?

There are $2 \cdot 3 \cdot 2$ or 12 different selections possible. The possible selections are shown in the tree diagram below.

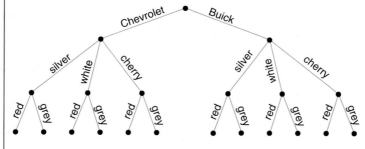

Notice that there are 12 choices, each of which differs from the others in some way.

Suppose Maria, Chiyo, and Wayne have been selected to debate another team at the annual debate contest. In how many different orders can they debate? If Maria goes first, then either Chiyo or Wayne can go second, and the other will go third. The possible arrangements of debates are listed below.

1st	Maria	Maria	Chiyo	Chiyo	Wayne	Wayne
2nd	Chiyo	Wayne	Maria	Wayne	Maria	Chiyo
3rd	Wayne	Chiyo	Wayne	Maria	Chiyo	Maria

There are six distinct orders in which the debates can be scheduled.

The arrangement of objects in a certain order is called a **permutation.** In a permutation, the order of the objects is very important.

Suppose there is time for only two debates at the contest. How many ways can the debates be scheduled if two of the three team members are asked to debate? There are three persons who can be scheduled to debate first, and then either of two persons can be scheduled to debate second. Therefore, there are six ways to schedule the three team members taken two at a time, as can be verified by the basic counting principle. The possible arrangements, or permutations, are listed below.

1st	Maria	Maria	Chiyo	Chiyo	Wayne	Wayne
2nd	Chiyo	Wayne	Maria	Wayne	Maria	Chiyo

The symbol $P(n, n)$ denotes the number of permutations of n objects taken all at once. The symbol $P(n, r)$ denotes the number of permutations of n objects taken r at a time.

Definition of $P(n, n)$ *and* $P(n, r)$	**The number of permutations of *n* objects, taken *n* at a time, is defined as** $$P(n, n) = n!.$$ **The number of permutations of *n* objects, taken *r* at a time, is defined as** $$P(n, r) = \frac{n!}{(n - r)!}.$$

Recall that $n! = n(n - 1)(n - 2)\ldots(1)$.

Example 3

How many ways can eight different cans of soup be displayed in a row on a shelf?

Since order is important, this situation is a permutation.

$P(8, 8) = 8!$
$\qquad = 8 \cdot 7 \cdot 6 \cdot 5 \cdot 4 \cdot 3 \cdot 2 \cdot 1$ or 40,320

There are 40,320 ways that the cans can be displayed.

Example 4

At the 1992 United States Olympic Track and Field Trials, Mark Witherspoon, Leroy Burrell, Dennis Mitchell, Mike Marsh, James Jett, and Carl Lewis qualified to run in the 4 × 100 meter relay. Only four of them were allowed to run in the race. How many different line-ups were possible from these qualifiers?

$P(6, 4) = \dfrac{6!}{(6 - 4)!}$

$\qquad = \dfrac{6 \cdot 5 \cdot 4 \cdot 3 \cdot 2 \cdot 1}{2 \cdot 1}$ or 360

There were 360 ways that the coach could have arranged the runners.

MATH JOURNAL

Suppose you forgot how the counting principle works. How could you figure out a problem without it?

CHECKING FOR UNDERSTANDING

Read and study the lesson to answer each question.

1. **Describe** the conditions under which two given events are considered to be independent of each other.

2. **Explain** how the basic counting principle works. Use an example in your answer.

3. What is a permutation?

4. **Write** an expression for the number of ways 5 books can be stacked from a group of 11 different books.

Guided Practice

State whether the events are *independent* or *dependent*.

5. tossing 3 coins one at a time

6. choosing 5 numbers in a bingo game

7. choosing color and size when ordering an item of clothing

8. choosing a president, secretary, and treasurer for a club

State whether each statement is *true* or *false*.

9. $6! - 3! = 3!$

10. $5 \cdot 4! = 5!$

11. $\dfrac{8!}{4!} = 2!$

12. $(5 - 3)! = 5! - 3!$

Find each value.

13. $P(4, 2)$

14. $P(9, 1)$

15. $P(6, 3)$

EXERCISES

Practice

Find each value.

16. $P(5, 3)$

17. $P(5, 5)$

18. $P(7, 4)$

19. $P(11, 10)$

20. $\dfrac{P(6, 4)}{P(5, 3)}$

21. $\dfrac{P(6, 3) \cdot P(4, 2)}{P(5, 2)}$

22. How many ways can 7 different books be stacked on a shelf?

23. A penny, a nickel, and a dime are tossed simultaneously. How many different ways can the coins land?

24. There are four roads from Erie to Mead, three from Mead to Titus, and four from Titus to Corry. How many different routes are there from Erie to Corry?

25. Regular license plates in Ohio have three letters followed by three digits. How many possible plates are there?

26. There are 10 students in a class that meets in a room that has 12 chairs arranged in a row. How many ways is it possible for the students to be seated?

Cup A contains two yellow counters, cup B contains one yellow and one red counter, and cup C contains two yellow counters and one red counter. A die is thrown. If a 1 or 2 shows, a counter is drawn from cup A. If a 3 or 4 shows, a counter is drawn from cup B. If a 5 or 6 shows, a counter is drawn from cup C. An outcome is a combination of a number on a die and a color of a counter, where counters of the same color are distinguishable.

27. How many possible outcomes are there?

28. Make a tree diagram to show all possible outcomes.

29. In how many of the possible outcomes is a red counter drawn?

30. In how many of the possible outcomes is a yellow counter drawn?

Find the number of different ways the letters of the word *pairs* can be arranged given the following.

31. The first letter must be p.

32. The first letter must be a vowel.

33. The first letter cannot be a vowel.

34. The letter r must be in the middle place.

35. Using the letters from the word *equation*, how many five-letter patterns can be formed in which q is followed immediately by u?

36. How many five-digit whole numbers between and including 56,000 and 59,999 can be formed if no digit is repeated?

Truck license plate numbers in a certain state consist of five digits followed by two letters. Find the number of possible license plates for each situation.

37. The letters O and I cannot be used.

38. The letters must be different.

39. The letter O cannot be used.

40. The five digits cannot be 00000.

Find the value of *n* in each equation.

41. $n[P(5,3)] = P(7,5)$

42. $P(n,4) = 3[P(n,3)]$

43. $7[P(n,5)] = P(n,3) \cdot P(9,3)$

44. $P(n,4) = 40[P(n-1,2)]$

Critical Thinking

45. The digits 2, 5, and 8 can be arranged in six ways:

    258   285   528   582   825   852

The average of these six numbers is $\dfrac{3330}{6} = 555 = 37(2+5+8)$.

If the digits are 1, 4, and 5, then the average of the six arrangements is

$\dfrac{2220}{6} = 370 = 37(1+4+5)$.

a. Use this pattern to find the average of the six arrangements of 7, 3, and 6.

b. Will this pattern hold for all triples of digits? If so, prove it.

Applications and Problem Solving

46. Sports The eight finalists in the men's 100-meter breaststroke at the 1992 Summer Olympics were Nick Gillingham of Great Britain, Vassili Iranov of the Unified Team, Phillip Rogers of Australia, Nelson Diebel of the United States, Norbert Rozsa of Hungary, Akira Hayashi of Japan, Dmitri Volkov of the Unified Team, and Adrian Moorehouse of Great Britain. Medals were given to the first three swimmers to finish the race.

a. How many ways could the medals be awarded in such a race?

b. If all eight swimmers finished the race, how many different orders of finish were possible?

47. Finance The state of Ohio has a Super Lotto drawing twice a week in which 6 numbers (1 through 46) are drawn at random. The proceeds from the lottery help to finance education in the state. How many ways can 6 numbers be drawn?

Mixed
Review

48. Line ℓ_1 has a slope of $\frac{1}{4}$, and line ℓ_2 has a slope of 4. Are the lines parallel, perpendicular, or neither? **(Lesson 1-6)**

49. Find the slope of the line tangent to the graph of $y = 2x^2$ at $(-1, 2)$. **(Lesson 3-6)**

50. Evaluate $\tan(\cos^{-1}\frac{3}{5})$. **(Lesson 6-5)**

51. Simplify $(4 - 3i)(-4 + 3i)$. **(Lesson 9-5)**

52. Finance Jill wants to invest a sum of money at 9% interest compounded continuously. How much must she invest now to have a total of $25,000 in 5 years? **(Lesson 11-7)**

53. Determine whether the Julia set associated with $C = 1 - 0.5i$ is connected or disconnected. Is the point in the Mandelbrot set? **(Lesson 13-6)**

54. College Entrance Exam If $x^2 + y^2 = 16$ and $xy = 8$, what is $(x + y)^2$?

DECISION MAKING

Choosing Car Insurance

More than 46,000 people were killed in car accidents in 1990. Nearly 150,000 suffered permanent disabilities. Medical expenses resulting from car accidents amounted to more than $6 billion. What kinds of insurance should you purchase to protect yourself against theft, medical costs, and damage your car might sustain in an accident? How can you choose the right insurance company? How can you be sure you are buying the right level of protection without overinsuring yourself?

1. Describe the insurance protection you think you should have as a driver. If you are unsure of terminology or the kinds of insurance that are available, talk to an insurance agent or read a consumer article on car insurance.

2. Find out how you can choose the right insurance company. Talk to car owners about their experiences, good and bad, with insurers. Learn about how insurers are rated by consumer groups and state insurance commissions.

3. Make a list of three companies and get a quote from each for the type of car insurance you think you should have.

4. Project Work with your group to complete one of the projects listed at the right based on the information you gathered in Exercises 1–3.

PROJECTS

- *Conduct an interview.*
- *Write a book report.*
- *Write a proposal.*
- *Write an article for the school paper.*
- *Write a report.*
- *Make a display.*
- *Make a graph or chart.*
- *Plan an activity.*
- *Design a checklist.*

14-2 Permutations with Repetitions and Circular Permutations

Objectives

After studying this lesson, you should be able to:
- solve problems involving permutations with repetitions, and
- solve problems involving circular permutations.

Application

Spell-checking software is designed to check the spelling in a document as well as to look for double words, words with numbers and some types of capitalization errors. Other software options include looking up words phonetically, counting words, and displaying all the words in the dictionary that match a pattern.

Suppose you meant to type the word *tooth* in a health report but accidentally entered *tootj*. How many different patterns could spell-checking software check using the five letters of *tootj*? *This problem will be solved in Example 1.*

In a simpler case, how many permutations of the letters of the word *too* are there? By tagging the o's as o_1 and o_2, the following arrangements are possible.

to_1o_2	o_1to_2	o_1o_2t
to_2o_1	o_2to_1	o_2o_1t

There are six arrangements when subscripts are taken into account, but only three arrangements without subscripts, too, oto, and oot.

When some objects are alike, use the following rule to find the number of permutations of those objects.

Permutations with Repetitions	The number of permutations of n objects of which p are alike and q are alike is $$\frac{n!}{p!q!}.$$

Example 1

How many five-letter patterns can be formed from the letters of the word *tootj*?

The five letters can be arranged in $P(5, 5)$ or $5!$ ways. However, several of these 120 arrangements have the same appearance since there are two o's and two t's. The arrangements are exactly the same unless the o's and t's are "tagged."

Find the number of permutations of 5 objects of which 2 are t's and 2 are o's.

$$\frac{5!}{2!2!} = \frac{5 \cdot 4 \cdot 3 \cdot 2 \cdot 1}{2 \cdot 1 \cdot 2 \cdot 1}$$

$$= 30$$

The spell-checking software would have to check 30 five-letter patterns.

Example 2

How many nine-letter patterns can be formed from the letters of the word Tennessee?

$$\frac{9!}{4!2!2!} = \frac{9 \cdot 8 \cdot 7 \cdot 6 \cdot 5 \cdot 4 \cdot 3 \cdot 2 \cdot 1}{4 \cdot 3 \cdot 2 \cdot 1 \cdot 2 \cdot 1 \cdot 2 \cdot 1}$$ *There are 4 e's, 2 n's, and 2 s's in Tennessee.*

$$= 3780$$

There are 3780 nine-letter patterns.

So far, you have been studying arrangements of objects that are in a line. Consider the problem of making distinct arrangements of five people eating at a round table. How many seating arrangements are possible?

Let the letters A, B, C, D, and E represent the five people. Three possible arrangements are shown below.

Sometimes it is helpful to treat a circular permutation as though it were a linear permutation. Then, think of the linear permutation as being wrapped around a circle like a string around a tree trunk.

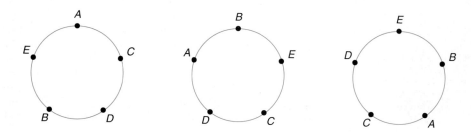

How does the first arrangement change if each person moves two places to the left? How do the arrangements differ?

When objects are arranged in a circle, some of the arrangements are alike. In the seating situation above, these arrangements fall into groups of five, each of which can be found by rotating the circle $\frac{1}{5}$ of a revolution. Thus, the number of distinct arrangements around the table is $\frac{1}{5}$ of the total number of arrangements in a line.

$$\frac{1}{5} \cdot 5! = \frac{5 \cdot 4 \cdot 3 \cdot 2 \cdot 1}{5}$$

$$= 4 \cdot 3 \cdot 2 \cdot 1$$

$$= 4! \text{ or } (5-1)!$$

Thus, there are $(5-1)!$ arrangements of 5 objects around a circle.

If *n* objects are arranged in a circle, then there are $\dfrac{n!}{n}$ or $(n-1)!$ permutations of the *n* objects around the circle.

Example 3

An ice cream parlor has 8 different toppings from which to choose. This week, you can create your own sundae. The eight toppings are placed on a revolving tray. How many ways can the toppings be arranged?

$(8 - 1)! = 7!$
$= 7 \cdot 6 \cdot 5 \cdot 4 \cdot 3 \cdot 2 \cdot 1$ or 5040

There are 5040 ways in which the toppings can be arranged on the tray.

Suppose five people are to be seated at a round table where one person is seated next to the door. Let each circle below represent a table and the labeled points represent the people at that table. Let the arrow represent the seat next to the door.

These arrangements are different. In each one, a different person sits closest to the door. Thus, there are $P(5, 5)$ or $5!$ arrangements relative to the door.

Circular arrangements with fixed points of reference are treated as linear permutations.

If *n* objects on a circle are arranged in relation to a fixed point, then there are *n*! permutations.

Suppose three charms are placed on a bracelet that has no clasp. It appears that there are at most $(3 - 1)!$ or 2 different arrangements of charms on the bracelet.

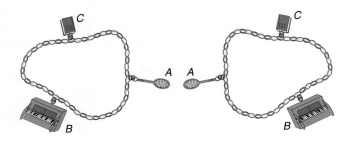

What happens if the bracelet with the first arrangement is turned over? The second arrangement appears. Thus, there is really only one arrangement of the three charms. These two arrangements are reflections of each other. There are only half as many arrangements when reflections are possible.

$$\frac{(3-1)!}{2} = \frac{2}{2} \text{ or } 1$$

Example 4

Six charms are to be placed on a bracelet.
a. How many ways can they be placed if the bracelet has no clasp?

a. This is a circular permutation. Because the bracelet can be turned over, the arrangement is also reflective.

$$\frac{(6-1)!}{2} = \frac{5 \cdot 4 \cdot 3 \cdot 2 \cdot 1}{2} \text{ or } 60$$

There are 60 ways to arrange the charms.

b. How many ways can they be placed if the bracelet has a clasp?

b. This is no longer a circular permutation since objects are arranged with respect to a fixed point, the clasp. However, it is still reflective.

$$\frac{6!}{2} = \frac{6 \cdot 5 \cdot 4 \cdot 3 \cdot 2 \cdot 1}{2} \text{ or } 360$$

There are 360 ways to arrange the charms.

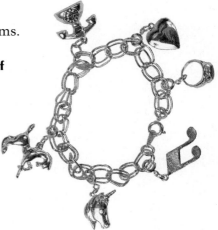

CHECKING FOR UNDERSTANDING

Communicating Mathematics

Read and study the lesson to answer each question.

1. **Explain** what a permutation with repetitions is.

2. **Describe** the difference between a linear permutation and a circular permutation.

3. **Explain** how to determine whether circular arrangements are alike.

4. **Explain** how to determine if a reflection is possible for an arrangement.

5. Under what conditions is a circular arrangement treated as a linear permutation?

Guided Practice

Determine whether each arrangement of objects is *linear* or *circular*. Then state whether it is *reflective* and find the number of arrangements.

6. a football huddle of 11 players

7. 6 chairs arranged in a circle

8. 8 beads on a necklace with no clasp

9. 10 chairs arranged in a row

EXERCISES

Practice **Determine whether each arrangement of objects is *linear* or *circular*. Then state whether it is *reflective*.**

10. people seated around a square table relative to each other

11. people seated around a square table relative to one chair

12. a list of students in a given class

13. placing coins in a circle on a desk

How many different ways can the letters of each word be arranged?

14. FLOWER

15. ALGEBRA

16. PARALLEL

17. CANDIDATE

18. ARREARS

19. MONOPOLY

20. QUADRATIC

21. BASKETBALL

22. CLOCKMAKER

23. How many ways can 7 keys be arranged on a key ring?

24. How many ways can 7 people be seated around a campfire?

25. How many ways can 8 charms be arranged on a bracelet that has no clasp?

26. How many ways can 4 congressmen and 4 congresswomen be seated alternately at a round table?

27. How many different arrangements can be made with ten pieces of silverware laid in a row if three are identical spoons, four are identical forks, and three are identical knives?

28. How many different six-digit license plates of the same state can have the digits 3, 5, 5, 6, 2, and 6?

29. Five algebra books and four geometry books are to be placed on a shelf. How many ways can they be arranged if all of the algebra books must be together?

30. How many ways can 6 points be labeled U through Z on the circle at the right, relative to the *y*-intercept?

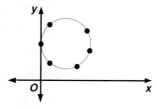

31. How many ways can 5 people be seated around a circular table if 2 of the people must be seated next to each other?

32. Twenty beads are strung on a necklace with no clasp. Fourteen are natural wood and 6 are red. How many ways can the beads be strung on the necklace?

Three men and three women are to be seated in a row containing six chairs. Find the number of seating arrangements for each situation.

33. The men and the women are to sit in alternate chairs.

34. The men are to sit in three adjoining chairs.

35. The men are to sit in three adjoining chairs and the women are to sit in three adjoining chairs.

36. To win a math contest, Joyce must determine how many marbles are in a box. She is told that there are 3 identical red marbles and some number of identical white marbles in the box. She is also told that there are 35 linear permutations of the marbles. What number should Joyce choose?

37. Communication Morse code is a system of dots, dashes, and spaces that telegraphers in the United States and Canada once used to send messages by wire. How many different arrangements are there of 4 dots and 2 dashes?

38. Entertainment A Sony® CD changer holds 5 CDs on a circular platter. How many different ways can 5 CDs be arranged on the changer?

39. Food Processing A meat packer makes a kind of sausage using beef, pork, cereal, fat, water, and spices. The minimum cereal content is 12%, the minimum fat content is 15%, the minimum water content is 6.5%, and the spices are 0.5%. The remaining ingredients are beef and pork. There must be at least 30% beef for flavor and at least 20% pork for texture. The beef content must equal or exceed the pork content. The cost of all of the ingredients except beef and pork is $32 per 100 pounds. Beef can be purchased for $140 per 100 pounds and pork for $90 per 100 pounds. Find the combination of beef and pork for the minimum cost. What is the minimum cost per 100 pounds? **(Lesson 2-6)**

40. Carpentry A cross brace is installed as a gate, as shown in the diagram at the right.
  a. Find the measure of the angle that the cross brace makes with the rail.
  b. How many inches long is the brace?
  (Lesson 5-5)

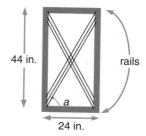

44 in.   rails

a

24 in.

41. Solve $\triangle ABC$ if $C = 105°18'$, $a = 6.11$, and $b = 5.84$. **(Lesson 5-7)**

42. Solve $\cos 2x + \sin x = 1$ for principal values of x. **(Lesson 7-5)**

43. Write the equation of the hyperbola with center at $(3, -1)$, focus at $(3, -4)$, and eccentricity $\frac{3}{2}$. **(Lesson 10-4)**

44. Akiko took 6 dresses, 5 pairs of shoes, and 2 coats on a trip. How many different outfits consisting of these items of clothing are possible? **(Lesson 14-1)**

45. College Entrance Exam Compare quantities A and B below.
Write A if quantity A is greater.
Write B if quantity B is greater.
Write C if the two quantities are equal.
Write D if there is not enough information to determine the relationship.
  (A) $\sqrt{0.4}$ **(B)** $(0.4)2$

14-3 Combinations

Objective

After studying this lesson, you should be able to:
■ solve problems involving combinations.

Application

Three students are chosen from the 5 juniors on the Student Council to study recycling opportunities in the high school. The first student can be selected in five ways, the second student in four ways, and the third student in three ways, for a total of $5 \cdot 4 \cdot 3$ or 60 ways. But consider the following six possible selections: (Andy, Bev, Cita), (Andy, Cita, Bev), (Bev, Cita, Andy), (Bev, Andy, Cita), (Cita, Andy, Bev), and (Cita, Bev, Andy). All these are considered the same. For each group of three people, there are 3! or 6 ways that they can be arranged in order. Thus, if order is disregarded, there are $\frac{60}{3!}$ or 10 different groups of three that can be selected from the five students. In this situation, the order in which the people are chosen is *not* a consideration.

The selection above is called a **combination** of five things taken three at a time. It is denoted by $C(5, 3)$. Evaluate $C(5, 3)$ as follows.

$$C(5, 3) = \frac{5!}{2!3!}$$

$$= \frac{5 \cdot 4 \cdot 3 \cdot 2 \cdot 1}{2 \cdot 1 \cdot 3 \cdot 2 \cdot 1} \text{ or } 10$$

Definition of $C(n, r)$

The number of combinations of n objects, taken r at a time, is written $C(n, r)$.

$$C(n, r) = \frac{n!}{(n - r)!r!}$$

The main difference between a permutation and a combination is whether order is considered (as in permutation) or not (as in combination).

Example 1

From a list of 10 different books, how many groups of 5 books can be selected?

Since order is not important, this selection is a combination of 10 things taken 5 at a time.

$$C(10, 5) = \frac{10!}{(10 - 5)!5!}$$

$$= \frac{10!}{5!5!}$$

$$= \frac{10 \cdot 9 \cdot 8 \cdot 7 \cdot 6 \cdot \cancel{5!}}{\cancel{5!}5!}$$

$$= \frac{10 \cdot 9 \cdot 8 \cdot 7 \cdot 6}{5 \cdot 4 \cdot 3 \cdot 2 \cdot 1} \text{ or } 252$$

There are 252 groups.

Example 2

APPLICATION

Business

A pizza delivery shop has 12 different toppings from which to choose. This week, if you buy a 2-topping pizza, you get 2 more toppings free. How many different ways can the special 4-topping pizza be made?

$$C(12, 4) = \frac{12!}{(12 - 4)!4!}$$
$$= \frac{12!}{8!4!}$$
$$= \frac{12 \cdot 11 \cdot 10 \cdot 9}{4 \cdot 3 \cdot 2 \cdot 1} \text{ or } 495$$

There are 495 ways the special pizza can be made.

Example 3

From a group of 4 men and 5 women, how many committees of 3 men and 2 women can be formed?

Order is not important. There are two questions to consider.
• How many ways can 3 men be chosen from 4?
• How many ways can 2 women be chosen from 5?

The answer is the product of two combinations, $C(4, 3)$ and $C(5, 2)$.

$$C(4, 3) \cdot C(5, 2) = \frac{4!}{(4 - 3)!3!} \cdot \frac{5!}{(5 - 2)!2!}$$
$$= \frac{4!}{1!3!} \cdot \frac{5!}{3!2!}$$
$$= 4 \cdot 10 \text{ or } 40$$

The combinations are multiplied to use the basic counting principle.

There are 40 possible committees.

Example 4

A bag contains 3 red, 5 white, and 8 blue marbles. How many ways can 2 red, 1 white, and 2 blue marbles be chosen?

$C(3, 2)$ — Two of 3 red marbles will be chosen.
$C(5, 1)$ — One of 5 white marbles will be chosen.
$C(8, 2)$ — Two of 8 blue marbles will be chosen.

$$C(3, 2) \cdot C(5, 1) \cdot C(8, 2) = \frac{3}{1} \cdot \frac{5}{1} \cdot \frac{8 \cdot 7}{2 \cdot 1} \text{ or } 420$$

There are 420 different ways.

CHECKING FOR UNDERSTANDING

Communicating Mathematics

Read and study the lesson to answer each question.

1. **Describe** the difference between a permutation and a combination.

2. **Write** an expression to represent the number of different groups of 5 basketball players that could be formed from a team of 9 players.

3. **Explain** how you would find how many bouquets of 2 red carnations and 3 white daisies can be formed from a group of 8 red carnations and 9 white daisies.

Guided Practice

State whether each arrangement represents a *permutation* or a *combination*.

4. 10 books on a shelf

5. a subset of 12 elements contained in a set of 26

6. a hand of 7 cards from a deck of 52 cards

7. 8 people seated around a circular table

Evaluate each expression.

8. $C(4, 2)$

9. $C(12, 7)$

10. $C(6, 6)$

11. $C(3, 2) \cdot C(8, 3)$

EXERCISES

Practice

Evaluate each expression.

12. $C(20, 15)$

13. $C(8, 5) \cdot C(7, 3)$

14. $C(8, 2) \cdot C(5, 1) \cdot C(4, 2)$

15. $P(4, 2) \cdot C(13, 3) \cdot C(13, 2)$

16. From a list of 10 books, how many groups of 4 books can be selected?

17. There are 85 telephones in the editorial department of Glencoe Publishing Company. How many 2-way connections can be made among the office phones?

18. How many baseball teams of 9 members can be formed from 14 players?

19. The cast of a school play requires 4 girls and 3 boys. They will be selected from 7 eligible girls and 9 eligible boys. How many ways can the cast be selected?

20. Suppose there are 8 points in a plane, no 3 of which are collinear. How many distinct triangles could be formed with these points as vertices?

21. Consider a deck of 52 cards.
    a. How many different 5-card hands can have 5 cards of the same suit?
    b. How many different 4-card hands can have each card from a different suit?

Find the value of *n* in each equation.

22. $C(n, 12) = C(30, 18)$

23. $C(14, 3) = C(n, 11)$

24. $C(11, 8) = C(11, n)$

25. $C(n, 5) = C(n, 7)$

A bag contains 4 red, 6 white, and 9 blue marbles. How many ways can 5 marbles be selected to meet each condition?

26. all white

27. all blue

28. exactly 2 are blue

29. 2 one color, 3 another color

From a group of 8 juniors and 10 seniors, a committee of 5 is to be formed to discuss plans for the prom. How many committees can be formed given each condition?

30. all juniors

31. 3 juniors, 2 seniors

32. 1 junior, 4 seniors

33. all seniors

Critical Thinking

34. Prove $C(n, r) = \dfrac{P(n, r)}{r!}$.

35. Describe the pattern of outcome values produced when you find each number of combinations of n objects taken r at a time, $C(n, r)$, as r goes from 0 to n.

Applications and Problem Solving

36. Geometry   A decagon is a polygon with 10 vertices and 10 sides. A diagonal is a segment that connects any two nonconsecutive vertices of a polygon. Find the total number of diagonals that can be drawn in a decagon.

37. Sports   How many different baseball teams of nine players can Defiance High School put on the field if they have 2 players that can only play catcher, 6 players that can only be pitchers, and 17 that can play any of the other seven positions?

Mixed Review

38. Solve $18x^3 - 34x^2 + 16x = 0$.   **(Lesson 4-1)**

39. Find an ordered triple that represents $\overrightarrow{AB}$, if $\vec{A} = (-2, 5, 8)$ and $\vec{B} = (3, 9, -3)$. **(Lesson 8-3)**

40. Write an expression for the nth term of the sequence $\dfrac{4}{2}, \dfrac{9}{4}, \dfrac{14}{6}, \dfrac{19}{8}, \cdots$. **(Lesson 12-3)**

41. A food vending machine has 6 different items on a revolving tray. How many ways can the items be arranged on the tray?   **(Lesson 14-2)**

42. College Entrance Exam   Choose the best answer.

If k is any odd integer and $x = 6k$, then $\dfrac{x}{2}$ will always be

(A) odd   **(B)** even   **(C)** positive   **(D)** negative   **(E)** zero

MID-CHAPTER REVIEW

1. At the Burger Hut, you can order your hamburger with or without cheese, with or without onions or pickles, and either rare, medium, or well-done. How many different hamburgers are possible?   **(Lesson 14-1)**

2. How many 7-letter patterns can be formed from the letters of *benzene*? **(Lesson 14-2)**

3. How many ways can 5 people be seated at a round table relative to each other? **(Lesson 14-2)**

4. How many ways can a club of 13 members choose 4 different officers? **(Lesson 14-3)**

5. How many ways can a club of 13 members choose a 4-person committee? **(Lesson 14-3)**

14-4 Probability and Odds

Objectives

After studying this lesson, you should be able to:
- find the probability of an event, and
- find the odds for the success and failure of an event.

Application

The cover story in the October 19, 1992 issue of *USA Today* reported that if you were a registered voter, there was a 1 in 1140 chance that a national political pollster would call you before Election Day. After reading this, few registered voters would linger by the phone from 6 to 9 every night waiting to be part of the next day's political headlines. When we are uncertain about the occurrence of an event, we can measure the chances of its happening with **probability**.

When you roll a die, there are six possible outcomes. The die will show either 1, 2, 3, 4, 5, or 6. A desired outcome is called a **success**. Any other outcome is called a **failure**.

Probability of Success and of Failure

If an event can succeed in *s* ways and fail in *f* ways, then the probability of success $P(s)$ and the probability of failure $P(f)$ are as follows.

$$P(s) = \frac{s}{s+f} \qquad P(f) = \frac{f}{s+f}$$

FYI...

An event that cannot fail has a probability of 1. An event that cannot succeed has a probability of 0. Thus, the probability of success, $P(s)$, is always between 0 and 1 inclusive.

A description of probability was first presented by the French mathematician Pierre-Simon Laplace in 1795.

The sum of the probability of success and the probability of failure for any event is always equal to 1.

$$P(s) + P(f) = \frac{s}{s+f} + \frac{f}{s+f}$$
$$= \frac{s+f}{s+f} \text{ or } 1$$

This property is often used in finding the probability of events. For example, if $P(s) = \frac{1}{4}$, then the $P(f) = 1 - \frac{1}{4}$ or $\frac{3}{4}$. Because their sum is 1, $P(s)$ and $P(f)$ are called **complements**.

Example 1

The term at random means that an outcome is chosen without any preference.

A box contains 3 baseballs, 7 softballs, and 11 tennis balls. What is the probability that a ball selected at random will be a tennis ball?

The probability of selecting a tennis ball is written P(tennis ball).

There are 11 ways to select a tennis ball from the box and $3 + 7$ or 10 ways not to select a tennis ball from the box. So, $s = 11$ and $f = 10$.

$$P(\text{tennis ball}) = \frac{s}{s+f}$$

$$= \frac{11}{11 + (3+7)} \text{ or } \frac{11}{21}$$

The probability of selecting a tennis ball is $\frac{11}{21}$.

The counting methods you used for permutations and combinations are often used in determining probability.

Example 2

Two cards are drawn at random from a standard deck of 52 cards. What is the probability that both are hearts?

$$P(\text{two hearts}) = \frac{C(13, 2)}{C(52, 2)} \qquad \textit{There are } C(13, 2) \textit{ ways to select 2 of 13 hearts.}$$
$$\textit{There are } C(52, 2) \textit{ ways to select 2 of 52 cards.}$$

$$= \frac{\frac{13!}{11!2!}}{\frac{52!}{50!2!}}$$

$$= \frac{78}{1326} \text{ or } \frac{1}{17}$$

The probability of selecting two hearts is $\frac{1}{17}$.

> *FYI...*
>
> Playing cards are often used in probability and combinatorics. There are 52 playing cards in a standard deck, 13 in each of the four suits: hearts, diamonds, spades, and clubs. There are 3 face cards in each suit: the jack, the queen, and the king.

Some probability applications are more easily solved by using complements.

Example 3

APPLICATION

Quality
Control

A collection of 15 transistors contains 3 that are defective. If 2 transistors are selected at random, what is the probability that at least one of them is good?

The complement of selecting at least 1 good transistor is selecting 2 defective transistors.

$$P(\text{2 defective transistors}) = \frac{C(3, 2)}{C(15, 2)}$$

$$= \frac{3}{105} \text{ or } \frac{1}{35}$$

Thus, the probability of selecting at least one good transistor is $1 - \frac{1}{35}$ or $\frac{34}{35}$.

Another way to measure the chance of an event occurring is with **odds**. The probability of the success of an event and its complement are used when computing the odds of an event.

Definition of Odds	The odds of the successful outcome of an event is the ratio of the probability of its success to the probability of its failure. $$\text{Odds} = \frac{P(s)}{P(f)}$$

Example 4

For the latest weather statistics, visit:
www.glencoe.com/sec/math/amc/mathnet

The Channel 11 weather forecaster announces that the probability of rain tomorrow is 60%. Find the odds that it will *not* rain tomorrow.

The probability of rain tomorrow is 60% or $\frac{3}{5}$.

$P(f) = \frac{3}{5}$ $\qquad$ $P(s) = 1 - \frac{3}{5}$ or $\frac{2}{5}$

$\text{Odds} = \frac{P(s)}{P(f)}$

$= \frac{\frac{2}{5}}{\frac{3}{5}}$ or $\frac{2}{3}$

The odds that it will not rain is $\frac{2}{3}$. *This is read "two to three."*

Sometimes when computing odds, you must find the total number of possible outcomes. This can involve finding permutations or combinations.

Example 5

Suppose Judi draws 5 cards from a standard deck of 52 cards. What are the odds that 4 cards will be of one suit and the other card will be of another suit?

First, find how many 5-card hands meet these conditions.

$P(4, 2)$ $\qquad$ Select 2 suits among 4. Order is important since different numbers of cards are to come from each suit.

$C(13, 4)$ $\qquad$ Select 4 cards from a suit containing 13 cards.

$C(13, 1)$ $\qquad$ Select 1 card from the other suit.

Then, use the basic counting principle.

$P(4, 2) \cdot C(13, 4) \cdot C(13, 1) = \frac{4!}{2!} \cdot \frac{13!}{9!4!} \cdot \frac{13!}{12!1!}$ or 111,540

The number of successes is 111,540.

Find the total number of 5-card hands.

$C(52, 5) = \frac{52!}{47!5!}$ or 2,598,960

The number of 5-card hands that do not meet the conditions is
2,598,960 − 111,540 or 2,487,420.

Finally, find the odds.

$$P(s) = \frac{111{,}540}{2{,}598{,}960} \qquad P(f) = \frac{2{,}487{,}420}{2{,}598{,}960}$$

$$\text{Odds} = \frac{\dfrac{111{,}540}{2{,}598{,}960}}{\dfrac{2{,}487{,}420}{2{,}598{,}960}} \text{ or } \frac{143}{3189}$$

Thus, the odds are $\dfrac{143}{3189}$ or about $\dfrac{1}{22}$.

CHECKING FOR UNDERSTANDING

Communicating Mathematics

Read and study the lesson to answer each question.

1. **Explain** what is meant by probability.
2. **Describe** the relationship between the probability of success and the probability of failure for any event.
3. **Write** an example of an outcome with a probability of 0.
4. **Describe** the difference between the probability of the successful outcome of an event and the odds of the successful outcome of an event.

Guided Practice

State the odds of an event occurring given the probability of the event.

5. $\dfrac{1}{2}$

6. $\dfrac{3}{4}$

7. $\dfrac{7}{15}$

8. $\dfrac{3}{20}$

State the probability of an event occurring given the odds of the event.

9. $\dfrac{3}{4}$

10. $\dfrac{6}{5}$

11. $\dfrac{4}{9}$

12. $\dfrac{1}{1}$

Suppose you select 2 letters at random from the word *algebra*. Find each probability.

13. $P2$ (consonants)

14. $P(2 \text{ vowels})$

15. $P(1 \text{ vowel and 1 consonant})$

EXERCISES

16. The odds are 7 to 5 that the Boston Celtics win in the conference championship game on Saturday night. What is the probability that they will win?

17. The probability of Jason getting accepted at the University of Michigan is $\dfrac{3}{4}$. What are the odds that he will *not* get accepted?

John has the following variety of CDs, in no particular order, in a carrying case: 5 rap CDs, 9 rock CDs, 4 country CDs, and 2 top 40 CDs. Two are selected at random. Find each probability.

18. $P(2 \text{ rap})$

19. $P(2 \text{ rock})$

20. $P(2 \text{ country})$

21. $P(1 \text{ rap and 1 rock})$

A die is thrown two times. What are the odds of each event occurring?

22. no fives

23. at least one five

24. both fives

Of seventeen students in a class, five have blue eyes. Two students are chosen at random. Find the probability of each selection. Then find the odds of that selection.

25. P(both have blue eyes)

26. P(neither has blue eyes)

27. P(at least one has blue eyes)

From a deck of 52 playing cards, 7 cards are dealt. What are the odds of each event occurring?

28. all kings

29. all face cards

30. all from one suit

31. 4 from one suit and 3 from another

Programming

32. The following program will generate random numbers between two values. You input a seed number, which determines the way in which the numbers are generated. Note that the program will not eliminate duplicate numbers, and the factory set seed number is 0.

```
Prgm 14: RANDNUM
:ClrHome
:Disp "LEAST INTEGER"
:Input S
:Disp "GREATEST INTEGER"
:Input L
:Disp "SEED NUMBER"
:Input N
:Disp "NUMBER OF VALUES TO GENERATE"
:Input A
:0 → B
:N → rand      Sets a starting point in the random number table.
:Lbl 1
:B+1 → B
:int ((L−S+1)rand + S) → R     Calculates value of desired random
:Disp R                        number, and stores in R.
:Pause                         Waits until user presses ENTER.
:If A ≠ B
:Goto 1
```

Run each program.

a. Let $S = 1$ and $L = 6$ to simulate rolls of a die.

b. Let $S = 1$ and $L = 2$ to simulate a coin toss.

c. Let $S = 1$ and $L = 54$ to simulate a lottery with 54 numbers.

Critical Thinking

33. A magician cuts a rope in two pieces at a point selected at random. What is the probability that the length of the longer rope is at least 8 times the length of the shorter rope?

Applications and Problem Solving

34. Finance The state of Pennsylvania has a Wild Card Lotto drawing twice a week in which 6 numbers out of 48 are drawn at random. The proceeds from the lottery help to finance programs for senior citizens in the state. What is the probability of winning one of the weekly Lotto drawings if you buy one ticket?

35. Education The table on the next page gives the status of employment of 500 students at Hawkins College.

Employment Status	Class				
	Freshman	Sophomore	Junior	Senior	Totals
Unemployed	75	56	35	18	184
Part-time	55	46	45	45	191
Full-time	30	28	30	37	125
Totals	160	130	110	100	500

a. If a student is selected at random, find the probability that the student is unemployed.

b. If a student is selected at random, find the probability that the student is a freshman employed part-time.

c. Find the probability that an unemployed student selected at random is a senior.

Mixed Review

36. Write an equation of the sine function with an amplitude 2, period 180°, and phase shift 45°. **(Lesson 6-2)**

37. If α and β are measures of two first quadrant angles, find $\cos(\alpha + \beta)$ if $\tan \alpha = \frac{4}{3}$ and $\cot \beta = \frac{5}{12}$. **(Lesson 7-3)**

38. Suppose the function $f(x) = x^2 - 1$ is to be iterated to produce a fractal. Find the first four iterates of the function if the initial value of x is $(1 + i)$. **(Lesson 13-4)**

39. Suppose there are 9 points on a circle. How many 4-sided closed figures can be formed by joining any 4 of these points? **(Lesson 14-3)**

40. **College Entrance Exam** Choose the best answer.
If $a + b = 8$ and $3b - 4 = -13a$, then what is the value of a?

(A) $\frac{-13}{3}$ **(B)** -3 **(C)** -2 **(D)** 10 **(E)** 2

CASE STUDY FOLLOW-UP

Refer to Case Study 3: *The Legal System*, on pages 962–965.

While walking one evening, Detective Odds heard a scream and looked up to see a thief running by. The thief was carrying something, but it was too dark for Odds to tell if the object was a car stereo or a purse. "It's got to be a stereo," said Odds. Three suspects were rounded up: Pilfer, whose 29 previous arrests included 8 for stereo theft; Filch, who had 15 stereo thefts among his 55 arrests; and Plunder, who had been nabbed 81 times, 22 on a stereo rap. "It's Pilfer," said Odds.

1. Explain Detective Odds' reasoning.

2. **Research** Talk to a police detective or district attorney to find out how probability and statistics are used in crime detection. Write about your findings in a 1-page paper.

14-5 Probability of Independent and Dependent Events

Objective

After studying this lesson, you should be able to:
- find the probability of independent and dependent events.

Application

As you made your course selections for this current school year, you could select certain courses only if certain other courses had been taken previously; that is, if you satisfied the prerequisite requirements. However, you could select other courses regardless of the courses previously taken. Thus, the selection of a specific course may or may not be dependent upon the selection of another. The idea that some events are independent and others are dependent on other events is an important one in the study of probability.

Suppose a die is tossed twice. The probability that the first toss shows a 5 is $\frac{1}{6}$. The probability that the second toss shows a 5 is $\frac{1}{6}$. By using the basic counting principle, the probability that both tosses show 5s is $\frac{1}{6} \cdot \frac{1}{6}$ or $\frac{1}{36}$. Since the outcome of the first toss does not affect the outcome of the second toss, the events are independent.

Probability of Two Independent Events	**If two events, A and B, are independent, then the probability of both events occurring is the product of each individual probability.** $$P(A \text{ and } B) = P(A) \cdot P(B)$$

Example 1

Find the probability of getting a sum of 7 on the first toss of two dice and a sum of 4 on the second toss.

Let A be a sum of 7 on the first toss. Let B be a sum of 4 on the second toss.

$$P(A) = \frac{6}{36} \qquad P(B) = \frac{3}{36} \qquad \textit{Why?}$$

The two tosses are independent.

$$P(A \text{ and } B) = P(A) \cdot P(B)$$
$$= \frac{6}{36} \cdot \frac{3}{36} \text{ or } \frac{1}{72}$$

The probability of a 7 on the first toss and a 4 on the second toss is $\frac{1}{72}$.

Example 2

APPLICATION

School

Students in Mr. Meyer's physics lab often refer to the *CRC Handbook of Chemistry and Physics.* Even though they contain the same information, 4 of the handbooks have brown covers, and 3 have red. One student selects a handbook at random from the shelf, looks up the heat of combustion for an organic compound, and places the book back on the shelf. A second student does the same thing. What is the probability that both students looked at a brown handbook?

The events are independent since the first handbook is placed back on the shelf. The outcome of the second selection is not affected by the results of the first selection.

$$P(\text{both brown}) = P(\text{brown}) \cdot P(\text{brown})$$
$$= \frac{4}{7} \cdot \frac{4}{7} \text{ or } \frac{16}{49}$$

The probability that both students selected a brown handbook is $\frac{16}{49}$ or about $\frac{1}{3}$.

In the example above, what is the probability of both students selecting a brown handbook if the first selection is not put back on the shelf? These events are dependent because the outcome of the first event affects the second selection. Suppose the first selection is brown.

first selection   *second selection*

$P(\text{brown}) = \frac{4}{7}$ $P(\text{brown}) = \frac{3}{6}$ *Notice that when the brown book is removed, there is not only one less brown book but also one less book on the shelf.*

$$P(\text{both brown}) = P(\text{brown}) \cdot P(\text{brown})$$
$$= \frac{4}{7} \cdot \frac{3}{6} \text{ or } \frac{2}{7}$$

The probability that both students selected a brown handbook is $\frac{2}{7}$.

Probability of Two Dependent Events	If two events, *A* and *B*, are dependent, then the probability of both events occurring is found as follows. $$P(A \text{ and } B) = P(A) \cdot P(B \text{ following } A)$$

Example 3

Kwag has 4 navy socks and 6 black socks in a drawer. One dark morning he randomly pulls out 2 socks. What is the probability that he will select a pair of navy socks?

(continued on the next page)

This situation is equivalent to pulling out 1 sock, not replacing it, then pulling out another.

$P(\text{navy, navy}) = P(\text{navy}) \cdot P(\text{navy following navy})$

$$= \frac{4}{10} \cdot \frac{3}{9} \text{ or } \frac{2}{15}$$

The probability that Kwag will select a pair of navy socks is $\frac{2}{15}$.

CHECKING FOR UNDERSTANDING

Communicating Mathematics

Read and study the lesson to answer each question.

1. **Write** an example of two independent events.

2. **Describe** what is meant by dependent events.

3. **Explain** how to find the probability of two dependent events.

Guided Practice

Determine if each event is *independent* or *dependent*. Then find the probability.

4. There are 2 glasses of root beer and 4 glasses of cola on the counter. Dave drinks two of them at random. What is the probability that he drank 2 glasses of cola?

5. A bowl contains 5 oranges and 4 tangerines. Noelle randomly selects one, puts it back, and then selects another. What is the probability that both selections were oranges?

6. A green die and a red die are tossed. What is the probability that a 4 shows on the green die and a 5 shows on the red die?

7. A bag contains 4 red, 4 green, and 7 blue marbles. Three are selected in sequence without replacement. What is the probability of selecting a red, a green, and a blue marble in that order?

EXERCISES

Practice

Determine if each event is *independent* or *dependent*. Then find the probability.

8. When Luis plays José on his video game, the odds are 4 to 3 that Luis will win. What is the probability that he will win the next four games?

9. There are two traffic lights along the route that Laura rides from home to school. One traffic light is red 50% of the time. The next traffic light is red 60% of the time. The lights operate on separate timers. Find the probability that these lights will both be red on Laura's way home from school.

There are 5 pennies, 7 nickels, and 9 dimes in an antique coin collection. Suppose two coins are to be selected at random from the collection. Find each probability.

10. $P(\text{selecting 2 pennies})$, if no replacement occurs

11. $P(\text{selecting 2 pennies})$, if replacement occurs

12. $P(\text{selecting the same coin twice})$, if no replacement occurs

Michael's family is preparing to move to a new home. Michael is helping his mother do some packing. There are 5 clocks, 5 candles, and 6 picture frames randomly placed on a table waiting to be boxed. Michael accidentally knocks two items off the table and breaks them. Find each probability.

13. P(breaking 2 picture frames)

14. P(breaking 2 clocks)

15. P(breaking a clock, then a candle)

16. P(breaking a clock and a candle)

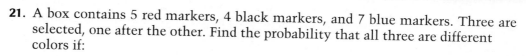

Two dice are tossed. Find each probability.

17. P(no 2s)

18. P(two numbers alike)

19. P(two different numbers)

20. P(2 and any other number)

21. A box contains 5 red markers, 4 black markers, and 7 blue markers. Three are selected, one after the other. Find the probability that all three are different colors if:

  a. no replacement occurs.   **b.** replacement occurs each time.

22. Brian's wallet contains four $1 bills, three $5 bills, and two $10 bills. If 3 bills are selected in succession, find the probability of selecting one of each if:

  a. each bill is replaced.   **b.** no bills are replaced.

For a bingo game, wooden balls numbered consecutively from 1 to 75 are placed in a box. Five balls are drawn at random. Find each probability.

23. P(selecting 5 even numbers), if replacement occurs

24. P(selecting 5 even numbers), if no replacement occurs

25. P(selecting 5 consecutive numbers), if no replacement occurs

A standard deck of 52 cards contains 4 suits of 13 cards each. Find each probability if 13 cards are drawn and no replacement occurs.

26. P(all diamonds)   **27.** P(all one suit)

28. P(all red cards)   **29.** P(all face cards)

30. For two events, X and Y, $P(X) = \frac{3}{10}$ and $P(Y) = \frac{1}{2}$. Are X and Y independent or dependent events, if $P(X \text{ and } Y) = \frac{1}{5}$? Explain.

31. A group of people was asked to choose a letter from the alphabet. How many people would have to respond in order for the probability of two persons choosing the same letter to be greater than $\frac{1}{2}$?

32. A room full of people are surveyed to determine their birthday (day and month only). How many people would have to respond for the probability of any two of them having the same birthday to be greater than $\frac{1}{2}$? Assume 366 different birthdays are possible.

33. Business A furniture buyer has ordered 100 grandfather clocks from an overseas manufacturer. Four clocks are damaged in shipment, but the packaging shows no indication of such damage. If a dealer buys 6 clocks without examining the contents, what is the probability that she does not have a damaged clock?

34. Political Science A 1992 poll based on returns from 2,035 registered Ohioans who intended to vote found that 37% were Republican, 40% were Democrat, and 21% were independent. Mike DeWine, a Republican, and John Glenn, a Democrat, were senatorial candidates. In the poll, 77% of the Republicans and 83% of the Democrats said they would vote for their party's candidate, while 41% of the independents said they would vote Democratic and 33% said they would vote Republican. If these people are representative of the entire state, which candidate had the better chance of winning Ohio's senate race, and by what odds?

35. Determine whether the graph of $f(x) = -\frac{1}{2}x^3$ is symmetric with respect to the x-axis, the y-axis, the line $y = x$, the line $y = -x$, or the origin. **(Lesson 3-1)**

36. Write the polar equation $r\sin\theta = 2$ in rectangular form. **(Lesson 9-3)**

37. Write the series $3 + 9 + 27 + \cdots$ using sigma notation. **(Lesson 12-5)**

38. Two dice are tossed and their sum is 6. Find the probability that each die shows a 3. **(Lesson 14-4)**

39. College Entrance Exam Choose the best answer.
The length of a rectangle is 1 unit and the width is w units. If the width is increased by 3 units, by how many units will the perimeter be increased?
(A) 4   **(B)** 6   **(C)** 12   **(D)** $2w + 3$ **(E)** $2w + 6$

14-6 Probability of Mutually Exclusive or Inclusive Events

Objectives

After studying this lesson, you should be able to:
- identify mutually exclusive events, and
- find the probability of mutually exclusive and inclusive events.

Application

In August, 1992, Hurricane Andrew caused billions of dollars of damage to south Florida and Louisiana. After a relief effort, 200 people in one community were surveyed to determine whether they donated food or money. Of the sample, 65 people said they donated food and 50 people said they donated money. Of these people, 30 people said they donated both. If a member of the community were selected at random, what is the probability that he or she donated food *or* money? *This problem will be solved in Example 2.*

FYI...

A hurricane originates as a rotation of strong winds over tropical ocean water. When the wind speed exceeds 74 miles per hour, the storm is classified as a hurricane.

Mutually exclusive events cannot occur simultaneously.

When tossing a die, what is the probability of tossing a 3 or a 4? In this case, both events cannot happen at the same time. That is, the events are **mutually exclusive**. The probability of tossing a 3 or a 4 is $\frac{1}{6} + \frac{1}{6}$ or $\frac{2}{6}$.

Probability of Mutually Exclusive Events	If two events, *A* and *B*, are mutually exclusive, then the probability that either *A* or *B* occurs is the sum of their probabilities. $$P(A \text{ or } B) = P(A) + P(B)$$

Inclusive events can occur simultaneously.

What is the probability of drawing a king or a black card from a deck of cards? It is possible to draw a card that is both a king and a black card. Therefore, these events are *not* mutually exclusive. They are called **inclusive** events. In this case, you must adjust the formula for mutually exclusive events much like you did to account for duplication in some permutations.

$P(\text{king})$

$\frac{4}{52}$

1 king in each suit

$P(\text{black card})$

$\frac{26}{52}$

clubs and spades

$P(\text{black king})$

$\frac{2}{52}$

king of clubs and king of spades

The probability of drawing a black king is counted twice, once for a king and once for a black card. To find the correct probability you must subtract P(black king) from the sum of P(king) and P(black card).

$$P(\text{king or black card}) = \frac{4}{52} + \frac{26}{52} - \frac{2}{52} \text{ or } \frac{7}{13}$$

The probability of drawing a king or a black card is $\frac{7}{13}$.

Probability of Inclusive Events	If two events, A and B, are inclusive, then the probability that either A or B occurs is the sum of their probabilities decreased by the probability of both occurring. $$P(A \text{ or } B) = P(A) + P(B) - P(A \text{ and } B)$$

Example 1

Leroy has 4 pennies, 3 nickels, and 6 dimes in his pocket. He takes one coin from his pocket at random. What is the probability that it is a penny or a dime?

These are mutually exclusive events since a coin cannot be a penny *and* a dime. Find the sum of the individual probabilities.

$$P(\text{penny}) = \frac{4}{13} \qquad P(\text{dime}) = \frac{6}{13}$$

$$P(\text{penny or dime}) = \frac{4}{13} + \frac{6}{13} \text{ or } \frac{10}{13}$$

The probability of selecting a penny or a dime is $\frac{10}{13}$.

Example 2

APPLICATION

Fund-Raising

In the application at the beginning of the lesson, determine the probability that a member of the community chosen at random donated food or money for the victims of Hurricane Andrew.

Since it is possible to donate both food and money, these events are inclusive.

$$P(\text{donate food}) = \frac{65}{200}$$

$$P(\text{donate money}) = \frac{50}{200}$$

$$P(\text{donate both}) = \frac{30}{200}$$

$$P(\text{donate food or money}) = \frac{65}{200} + \frac{50}{200} - \frac{30}{200} \text{ or } \frac{85}{200}$$

The probability that someone donated food or money is $\frac{85}{200}$ or $\frac{17}{40}$.

Example 3

There are 5 men and 4 women on the school board for the Titusville Area School District. A committee of 5 members is being selected at random to study the feasibility of building a new elementary school. What is the probability that the committee will have at least 3 men?

At least 3 men means that the committee may have 3, 4, or 5 men. It is not possible to select a group of 3 men, a group of 4 men, and a group of 5 men all in the same 5-member committee. Thus, the events are mutually exclusive.

$P(\text{at least 3 men}) = P(\text{3 men}) + P(\text{4 men}) + P(\text{5 men})$

$$= \frac{C(5,3) \cdot C(4,2)}{C(9,5)} + \frac{C(5,4) \cdot C(4,1)}{C(9,5)} + \frac{C(5,5) \cdot C(4,0)}{C(9,5)}$$

$$= \frac{60}{126} + \frac{20}{126} + \frac{1}{126} \text{ or } \frac{9}{14}$$

The probability of at least 3 men on the committee is $\frac{9}{14}$.

CHECKING FOR UNDERSTANDING

Communicating Mathematics

Read and study the lesson to answer each question.

1. **Describe** the difference between mutually exclusive and inclusive events.

2. **Write** an example of two mutually exclusive events in your own life.

3. **Write** an example of two events in your life that are not mutually exclusive.

4. **Draw** a Venn diagram to illustrate the events in the application at the beginning of the lesson.

Guided Practice

Determine if each event is *mutually exclusive* or *inclusive.* Then find the probability.

5. Two dice are tossed.
   a. $P(\text{sum is 6 or sum is 9})$
   b. $P(\text{either die shows a 2})$

6. A card is drawn from a standard deck of cards.
   a. $P(\text{drawing an ace or a red card})$
   b. $P(\text{drawing a king or a queen})$

7. **Business** An auto club's emergency service team knows from experience that when a member calls to report that their car won't start, the probability that the engine is flooded is $\frac{1}{2}$, the probability that the battery is dead is $\frac{2}{5}$, and the probability that both the engine is flooded and the battery is dead is $\frac{1}{10}$. What is the probability that the next member to report that their car won't start has a flooded engine or a dead battery?

EXERCISES

Determine if each event is *mutually exclusive* or *inclusive*. Then find the probability.

8. Five coins are dropped onto the floor. What is the probability that at least three of them land tails-up?

9. In homeroom, 5 of the 12 girls have blonde hair and 6 of the 15 boys have blonde hair. What is the probability of randomly selecting a boy or a blonde-haired person as homeroom representative to the student council?

10. A card is drawn from a deck of cards. What is the probability that it is a red card or a face card?

Two faces of a die are red, two are blue, and two are white. The die is tossed. Find each probability.

11. P(shows either red or blue)
12. P(does not show red)

Kevin has 11 CDs in the glove compartment of his car. Six are rock and 5 are rap. He selects three at random to take to a party. Find each probability.

13. P(all 3 rock or all 3 rap)
14. P(exactly 2 rap)

15. P(at least 2 rap)
16. P(at least 2 rock)

Two cards are drawn from a standard deck of cards. Find each probability.

17. P(both aces or both face cards)
18. P(both aces or both black)

19. P(both black or both face cards)
20. P(both either red or an ace)

The athletic director is counting the ticket money after the basketball game on Tuesday. Six coins fall from her desk to the floor. Find each probability.

21. P(landing 3 heads or 2 tails)
22. P(landing at least 4 tails)

23. P(landing 4 heads or 1 tail)
24. P(landing all heads or all tails)

There are 7 men and 7 women on the city council. A committee of 6 members is to be selected at random to attend a state convention. Find the probability of each group being selected.

25. P(all men or all women)
26. P(5 men or 5 women)

27. P(3 men and 3 women)
28. P(at least 4 women)

The numbers 1 through 25 are written on Ping-Pong® balls and placed in a wire cage. The numbers 20 through 40 are written on Ping-Pong® balls and placed in a different wire cage. One ball is chosen at random from each spinning cage. Find each probability.

29. $P(\text{each is a 22})$

30. $P(\text{neither is a 25})$

31. $P(\text{at least one is a 23})$

32. $P(\text{each is greater than 10})$

Critical Thinking

33. A bag contains 15 billiard balls, numbered consecutively from 1 to 15. Six balls are selected at random. What is the probability that the sum of the numbers on the billiard balls will be odd?

Applications and Problem Solving

34. Sports The Texas Ranger pitching staff has 5 left-handers and 8 right-handers. If 2 pitchers are selected at random to warm up, what is the probability that at least one of them is a right-hander?

35. Medicine Mrs. Sopher visits two doctors, Dr. Briggs and Dr. Villa, for a diagnosis. The probability that Dr. Briggs correctly diagnoses a disease is $\frac{93}{100}$, and the probability that Dr. Villa correctly diagnoses a disease is $\frac{97}{100}$. What is the probability that at least one of the two doctors makes the correct diagnosis?

Mixed Review

36. Construction A highway curve, in the shape of an arc of a circle, is 0.25 miles. The direction of the highway changes 45° from one end of the curve to the other. Find the radius of the circle in feet that the curve follows. **(Lesson 5-2)**

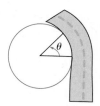

37. Write the equation of the ellipse with center at $(3, 1)$, semi-major axis of length 6 units, and eccentricity $\frac{1}{3}$. The major axis is parallel to the y-axis. **(Lesson 10-3)**

38. Finance Mrs. Williams invests $2000 annually in an insurance annuity for 42 years. What will the annuity be worth if it has an APR of 8%? **(Lesson 12-2)**

39. Fifty tickets, numbered consecutively 1 to 50, are placed in a box. Four tickets are drawn without replacement. What is the probability that 4 odd numbers are drawn? **(Lesson 14-5)**

40. College Entrance Exam Choose the best answer.
The distance between $P(4, 0)$ and Q is 8. The coordinates of point Q could be any of the following except

(A) $(-4, 0)$ **(B)** $(0, 4\sqrt{3})$ **(C)** $(4, 8)$ **(D)** $(8, 0)$ **(E)** $(4, -8)$

14-7 Conditional Probability

Objective

After studying this lesson, you should be able to:
■ find the probability of an event given the occurrence of another event.

Application

Suppose Jhan draws a card from a standard deck of 52 cards. What is the probability that he picks a queen? By forming the ratio of the number of successful outcomes to the number of all possible outcomes, we know that $P(\text{queen}) = \frac{4}{52}$. But now, suppose Jhan draws the card and tells you that he has drawn a red card. With this new information, what is the probability that he has drawn a queen? When you need to find the probability of an event under the condition that some preceding event has occurred, you are finding the **conditional probability**. The conditional probability of a queen given that it is known to be a red card can be symbolized as $P(\text{queen/red card})$.

Under the conditions of the problem, successful outcomes must be red *and* they must be queens. Therefore, the successful outcomes are red queens. There are two of these. So, $P(\text{queen and red card}) = \frac{2}{52}$. Also, the probability of choosing a red card is $\frac{26}{52}$.

$$P(\text{queen/red}) = \frac{P(\text{queen and red})}{P(\text{red})}$$

$$= \frac{\frac{2}{52}}{\frac{26}{52}} \text{ or } \frac{1}{13}$$

The probability of drawing a queen, given that the card is red, is $\frac{1}{13}$.

Conditional Probability

The conditional probability of event A, given event B, is defined as

$$P(A/B) = \frac{P(A \text{ and } B)}{P(B)} \text{ where } P(B) \neq 0.$$

$P(A/B)$ is read "the probability of A given B."

Example 1

In the game of Yahtzee®, players toss 5 dice and receive points for certain combinations. To score 25 points in the "Full House" box, three dice must show the same number and two dice must show another number. Suzan tosses the dice. Three of the five dice each show a three. Neither of the other two show a three, and they do not match. Suzan tosses the two dice that do not match. Find the probability that the numbers of the dice match, given that their sum is greater than 7.

Let event A be that the numbers match.

Let event B be that their sum is greater than 7.

$$P(B) = \frac{15}{36} \qquad P(A \text{ and } B) = \frac{3}{36}$$

$$P(A/B) = \frac{P(A \text{ and } B)}{P(B)}$$

$$= \frac{\frac{3}{36}}{\frac{15}{36}} \text{ or } \frac{1}{5}$$

The probability of tossing matching numbers given that their sum is greater than 7 is $\frac{1}{5}$.

The set of all possible outcomes for an event is called the **sample space** for the event. A **reduced sample space** is the subset of a sample space that contains only those outcomes that satisfy a given condition. For example, if a pair of dice is thrown, the sample space would contain 36 outcomes of the form $(1, 1), (1, 2), (1, 3), \cdots, (6, 6)$. Under the condition that the dice must match, the reduced sample space would contain only 6 outcomes: $(1, 1), (2, 2), (3, 3), (4, 4), (5, 5), (6, 6)$.

Example 2 illustrates how to find conditional probabilities using sample spaces and reduced sample spaces.

Example 2

A pair of dice is thrown. Find the probability that the dice match, given that their sum is greater than five.

Let event A be that the numbers match.

Let event B be that the sum of the dice is greater than 5.

The sample space includes all of the possible outcomes of throwing a pair of dice. These are shown in the box at the right. Using the basic counting principle, there are 36 outcomes.

(1, 1)	(1, 2)	(1, 3)	(1, 4)	(1, 5)	(1, 6)
(2, 1)	(2, 2)	(2, 3)	(2, 4)	(2, 5)	(2, 6)
(3, 1)	(3, 2)	(3, 3)	(3, 4)	(3, 5)	(3, 6)
(4, 1)	(4, 2)	(4, 3)	(4, 4)	(4, 5)	(4, 6)
(5, 1)	(5, 2)	(5, 3)	(5, 4)	(5, 5)	(5, 6)
(6, 1)	(6, 2)	(6, 3)	(6, 4)	(6, 5)	(6, 6)

(continued on the next page)

The condition that their sum is greater than five causes the reduced sample space to include only pairs in the shaded region. The reduced sample space contains 26 outcomes.

(1, 1)	(1, 2)	(1, 3)	(1, 4)	(1, 5)	(1, 6)
(2, 1)	(2, 2)	(2, 3)	(2, 4)	(2, 5)	(2, 6)
(3, 1)	(3, 2)	(3, 3)	(3, 4)	(3, 5)	(3, 6)
(4, 1)	(4, 2)	(4, 3)	(4, 4)	(4, 5)	(4, 6)
(5, 1)	(5, 2)	(5, 3)	(5, 4)	(5, 5)	(5, 6)
(6, 1)	(6, 2)	(6, 3)	(6, 4)	(6, 5)	(6, 6)

To find the probability of a match given a sum greater than five, circle the number of favorable outcomes in the reduced sample space. There are four such outcomes. Thus, the probability of A given B is $\frac{4}{26}$ or $\frac{2}{13}$.

(1, 1)	(1, 2)	(1, 3)	(1, 4)	(1, 5)	(1, 6)
(2, 1)	(2, 2)	(2, 3)	(2, 4)	(2, 5)	(2, 6)
(3, 1)	(3, 2)	(3, 3)	(3, 4)	(3, 5)	(3, 6)
(4, 1)	(4, 2)	(4, 3)	(4, 4)	(4, 5)	(4, 6)
(5, 1)	(5, 2)	(5, 3)	(5, 4)	(5, 5)	(5, 6)
(6, 1)	(6, 2)	(6, 3)	(6, 4)	(6, 5)	(6, 6)

In some situations, event A is a subset of event B. When this occurs, the probability that both event A and event B occur is the same as the probability of event A occurring. Thus, in these situations, $P(A/B) = \dfrac{P(A)}{P(B)}$. Consider the following example.

Example 3

One card is drawn from a standard deck of 52 cards. What is the probability that it is a queen if it is known to be a face card?

Let event A be that a queen is drawn.   *There are 4 queens.*

Let event B be that a face card is drawn.   *There are 12 face cards.*

Since a queen is a face card, A is a subset of B.

$$P(A \text{ and } B) = P(A) = \frac{4}{52} \text{ or } \frac{1}{13} \qquad\qquad P(B) = \frac{12}{52} \text{ or } \frac{3}{13}$$

$$P(A/B) = \frac{P(A)}{P(B)}$$

$$= \frac{\frac{1}{13}}{\frac{3}{13}} \text{ or } \frac{1}{3}$$

The probability that a queen is drawn, given that a face card is drawn, is $\frac{1}{3}$.

CHECKING FOR UNDERSTANDING

Communicating Mathematics

Read and study the lesson to answer each question.

1. **Explain** what is meant by conditional probability.

2. Are $P(A/B)$ and $P(B/A)$ the same? Illustrate your answer.

3. **Describe** an example of a reduced sample space that is different from Example 2.

4. Under what conditions does $P(A/B) = \dfrac{P(A)}{P(B)}$?

Identify events A and B in each problem. Then find the probability.

5. Two coins are tossed. What is the probability that one coin shows tails if it is known that at least one coin shows heads?

6. A city council consists of six Democrats, two of whom are women, and six Republicans, four of whom are men. A member is chosen at random. If the member chosen is a man, what is the probability that he is a Democrat?

A card is chosen at random from a standard deck of 52. Find each probability given that the card is red.

7. $P(\text{heart})$

8. $P(\text{ace})$

9. $P(\text{face card})$

10. $P(\text{six of spades})$

11. $P(\text{six of hearts})$

12. $P(\text{red six})$

Identify events A and B in each problem. Then find the probability.

13. Two boys and two girls are lined up at random. What is the probability that the girls are separated if a girl is at an end?

14. A bag contains 4 red marbles and 4 blue marbles. Another bag contains 2 red marbles and 6 blue marbles. A marble is drawn from one of the bags at random and found to be blue. What is the probability that the marble is from the first bag?

EXERCISES

A pair of dice is thrown. Find each probability given that their sum is greater than or equal to 9.

15. $P(\text{sum is 8})$

16. $P(\text{numbers match})$

17. $P(\text{sum is 12})$

18. $P(\text{sum is even})$

19. $P(\text{sum is 9 or 10})$

20. $P(\text{numbers match or sum is even})$

Three coins are dropped. Find the probability that they all land tails up for each known condition.

21. The first coin shows a tail.

22. One of the coins is a head.

23. At least one coin shows a tail.

24. At least two coins show tails.

A committee of 3 student council members is randomly selected from Andy, Becky, Cho, Dan, Emily, and Fernando. Draw a sample space to represent all of the possible outcomes. Then find each conditional probability.

25. $P(\text{Fernando was selected, given that Becky was selected})$

26. $P(\text{Emily was not selected, given that Andy and Cho were selected})$

27. $P(\text{Becky and Cho were selected, given that Dan was not selected})$

28. $P(\text{Cho and Andy were selected, given that neither Becky nor Emily were selected})$

29. A five-digit number is formed from the digits 1, 2, 3, 4, and 5. What is the probability that the number ends in the digits 52, given that it is even?

30. Two numbers are selected at random from the numbers 1 through 9 without replacement. If their sum is even, what is the probability that both numbers are odd?

In an advanced biology class, 60% of the students have brown hair, 30% have brown eyes, and 10% have both brown hair and eyes. A student is excused to go to the state track meet.

31. If the student has brown hair, what is the probability that the student also has brown eyes?

32. If the student has brown eyes, what is the probability that the student does not have brown hair?

In a card game played with a deck of 52 cards, each face card has a value of 10 points, each ace has a value of 1 point, and each number card has a value equal to the number. Two cards are drawn at random.

33. One card is the queen of hearts. What is the probability that the sum of the cards is greater than 18?

34. At least one card is an ace. What is the probability that the sum of the cards is 7 or less?

Critical Thinking

35. When would using the sample space/reduced sample space approach for solving a conditional probability problem not be the most advantageous one?

36. The probability of an event A is equal to the probability of the same event, given that event B has already occurred. Prove that A and B are independent events.

Applications and Problem Solving

37. Business The manager of a paint supply store wants to know whether people who come in and ask questions are more likely to make a purchase than the average customer. A survey of 200 people exiting the store found that 100 people bought something, 48 asked questions and bought something, and 12 people asked questions but did not buy anything. Based on the survey, determine whether a person who asks questions is more likely to buy something than the average person entering the store.

38. Medicine To test the effectiveness of a new vaccine, the developers gave 100 volunteers the conventional treatment and gave the new vaccine to another 100 volunteers. The results are shown in the table at the top of the next page.

Treatment	Disease Prevented	Disease Not Prevented
New Vaccine	68	32
Conventional Treatment	62	38

a. What is the probability that the disease is prevented in a volunteer chosen at random?

b. What is the probability that the disease is prevented in a volunteer who was given the new vaccine?

c. What is the probability that the disease is prevented in a volunteer who was not given the new vaccine?

d. What conclusion can the developers of the new vaccine make?

39. Money A dollar-bill changer in a soft drink machine was tested with 100 $1-bills. Twenty-five of the bills were counterfeit. The results of the test are shown in the chart at the right.

Bill	Accepted	Rejected
Legal	69	6
Counterfeit	1	24

a. What is the probability that a bill accepted by the changer is legal?

b. What is the probability that a bill is rejected given that it is legal?

c. What is the probability that a counterfeit bill is not rejected?

Mixed Review

40. Find $f(-2)$ if $f(x) = 2x^2 - 2x + 8$. **(Lesson 1-1)**

41. Use synthetic division to divide $x^5 - 3x^2 - 20$ by $x - 2$. **(Lesson 4-3)**

42. Finance What does the state of Illinois need to deposit into an account with an APR of 12% in order for the state to pay $1,000,000 to a single lottery winner in equal annual payments over the next twenty years? **(Lesson 11-2)**

43. Find the first three iterates of the function $f(x) = x^2 + 1$ for an initial value of $x_0 = -1$. **(Lesson 13-1)**

44. A letter is picked at random from the alphabet. Find the probability that the letter is contained in the word *house* or in the word *phone*. **(Lesson 14-6)**

45. College Entrance Exam Compare quantities A and B below.
Write A if quantity A is greater.
Write B if quantity B is greater.
Write C if the two quantities are equal.
Write D if there is not enough information to determine the relationship.

(A) 15% of 1400   **(B)** 14% of 1500

14-8 The Binomial Theorem and Probability

Objective

After studying this lesson, you should be able to:
- find the probability of an event by using the binomial theorem.

Application

Suppose you didn't study for your 20th-century U.S. history quiz, so you decide to guess on all 10 true/false questions. What is the probability that exactly half of your answers are correct? *This problem will be solved in Example 2.*

You may wish to review the binomial theorem in Lesson 12-6.

In a simpler case, suppose there are only 5 questions on the quiz. What is the probability that 3 of your answers are correct? The number of ways that this can happen is $C(5, 2)$ or 10. Let p_c represent the probability of guessing the answer correctly, and let p_i represent the probability of guessing the answer incorrectly. The terms of the binomial expansion of $(p_c + p_i)^5$ can be used to find the probabilities of each combination of correct and incorrect answers.

$$(p_c + p_i)^5 = 1p_c^5 + 5p_c^4p_i + 10p_c^3p_i^2 + 10p_c^2p_i^3 + 5p_cp_i^4 + 1p_i^5$$

coefficient	term	meaning
$C(5, 0) = 1$	$1p_c^5$	1 way to have all 5 correct
$C(5, 1) = 5$	$5p_c^4p_i$	5 ways to have 4 correct and 1 incorrect
$C(5, 2) = 10$	$10p_c^3p_i^2$	10 ways to have 3 correct and 2 incorrect
$C(5, 3) = 10$	$10p_c^2p_i^3$	10 ways to have 2 correct and 3 incorrect
$C(5, 4) = 5$	$5p_cp_i^4$	5 ways to have 1 correct and 4 incorrect
$C(5, 5) = 1$	$1p_i^5$	1 way to have all 5 incorrect

On a true/false test, the probability that you guess the correct answer is $\frac{1}{2}$. So, the probability that your guess is incorrect is also $\frac{1}{2}$. To find the probability of guessing exactly 3 out of 5 answers correctly, substitute $\frac{1}{2}$ for p_c and $\frac{1}{2}$ for p_i in the term $10p_c^3p_i^2$. Thus, $10p_c^3p_i^2 = 10\left(\frac{1}{2}\right)^3 \left(\frac{1}{2}\right)^2$ or $\frac{5}{16}$. Therefore, the probability that you guess exactly 3 answers correctly is $\frac{5}{16}$.

Other probabilities can be determined from the expansion on the previous page. For example, what is the probability of guessing at least 3 questions correctly? The first, second, and third terms represent the condition that three or more guesses are correct. Therefore, the probability of this happening is the sum of the probabilities of these terms.

$$P(\text{at least 3 correct}) = 1p_c^5 + 5p_c^4p_i + 10p_c^3p_i^2$$

$$= \frac{1}{32} + \frac{5}{32} + \frac{5}{16}$$

$$= \frac{16}{32} \text{ or } \frac{1}{2}$$

The probability that at least 3 guesses are correct is $\frac{1}{2}$.

Problems that can be solved using binomial expansion are called **binomial experiments**.

Conditions of a Binomial Experiment	A binomial experiment exists *if and only if* these conditions occur. • **The experiment consists of n identical trials.** • **Each trial results in *one* of two possible outcomes.** • **The trials are independent.**

Example 1

APPLICATION

Sports

While pitching for the Toronto Blue Jays, 4 of every 7 pitches Juan Guzman threw in the first 5 innings were strikes. What is the probability that 3 of the next 4 pitches will be strikes?

There are 4 pitches involved. Let's assume that there are only two possible outcomes, strikes (S) or balls (B), and that the 4 pitches are independent events. This is a binomial experiment.

When $(S + B)^4$ is expanded, the term S^3B represents 3 strikes and 1 ball. The coefficient of S^3B is $C(4, 3)$ or 4.

TECHNOLOGY Tip

You can use the **binompdf** (binomial probability function) of the TI-83 graphing calculator to determine the probabilities of a binomial experiment.

$P\,(3 \text{ strikes, 1 ball}) = 4S^3B$

$$= 4 \left(\frac{4}{7}\right)^3 \left(\frac{3}{7}\right) \text{ or } \frac{768}{2401} \qquad S = \frac{4}{7} \text{ and } B = \frac{3}{7}$$

The probability of 3 strikes and 1 ball is $\frac{768}{2401}$ or about 0.32.

The binomial theorem can be used to find probability when the number of trials makes working with the binomial expansion unrealistic.

Example 2

APPLICATION

School

In the application at the beginning of the lesson, determine the probability that exactly half of the answers are correct.

Let p_c be the probability that the answer is correct, and let p_i be the probability that the answer is incorrect. Since there are 10 questions, we can use the binomial theorem to find any term in the expansion of $(p_c + p_i)^{10}$.

$$(p_c + p_i)^{10} = \sum_{r=0}^{10} \frac{10!}{r!(10-r)!} p_c^{10-r} p_i^{r}$$

Getting half of the answers correct means that you would get exactly 5 correct and 5 incorrect. So, the probability can be found using the term where $r = 5$, the sixth term.

$$\frac{10!}{5!(10-5)!} p_c^5 p_i^5 = 252 p_c^5 p_i^5$$

$$= 252 \left(\frac{1}{2}\right)^5 \left(\frac{1}{2}\right)^5 \text{ or } \frac{63}{256}$$

Portfolio

Select one of the assignments from this chapter that you found especially challenging and place it in your portfolio.

The probability that you guess exactly 5 answers correctly is $\frac{63}{256}$ or about $\frac{1}{4}$.

CHECKING FOR UNDERSTANDING

Communicating Mathematics

Read and study the lesson to answer each question.

1. **Explain** how to find the probability of exactly 4 heads showing when 5 coins are tossed.

2. **List** the conditions that must be satisfied for a problem to be classified as a binomial experiment.

Guided Practice

In Exercises 3—5, determine whether each situation represents a binomial experiment. Solve those that represent a binomial experiment.

3. What is the probability of 2 heads and 2 tails if Sergio tosses a coin 4 times?

4. What is the probability of Bianca drawing 4 kings from a deck of cards for each condition?

   a. She replaces the card each time. **b.** She does not replace the card.

5. There are 8 pennies, 4 nickels, and 6 dimes in an antique coin collection. Two coins are selected with replacement after the first selection. Find each probability.

   a. $P(\text{both pennies})$ **b.** $P(\text{both nickels})$ **c.** $P(\text{both dimes})$

   d. $P(\text{1 penny, 1 dime})$ **e.** $P(\text{1 penny, 1 nickel})$ **f.** $P(\text{1 nickel, 1 dime})$

EXERCISES

Find each probability if a coin is tossed three times.

   6. P(all heads)   **7.** P(exactly 2 tails)   **8.** P(at least 2 heads)

Find each probability if a die is tossed five times.

   9. P(only one 4)   **10.** P(at least three 4s)

  11. P(no more than two 4s)   **12.** P(exactly five 4s)

Angie carries tubes of lipstick in a bag in her purse. The probability of pulling out the color she wants is $\frac{2}{3}$. Suppose she uses her lipstick 4 times in a day. Find each probability.

  13. P(never the correct color)

  14. P(correct at least 3 times)

  15. P(no more than 3 times correct)

  16. P(correct exactly 2 times)

Maroa guesses at all 10 true/false questions on her psychology test. Find each probability.

  17. P(7 correct)   **18.** P(all incorrect)

  19. P(at least 6 correct)   **20.** P(at least half correct)

Jojo MacMahon plays for the Worthington Wolves softball team. She is now batting 0.200 (meaning 200 hits in 1000 at bats). Find each probability for the next 5 times at bat.

  21. P(exactly 1 hit)   **22.** P(exactly 3 hits)   **23.** P(at least 4 hits)

Find each probability if three coins are tossed.

  24. P(3 heads)   **25.** P(3 tails)

  26. P(at least 2 heads)   **27.** P(exactly 2 tails)

If a thumbtack is dropped, the probability of its landing point up is $\frac{2}{5}$. Miss Hostetler drops 10 tacks while putting up the lunch menus for the week on the bulletin board. Find each probability.

  28. P(all point up)   **29.** P(exactly 3 point up)

  30. P(exactly 5 point up)   **31.** P(at least 6 point up)

32. In baseball, the World Series matches the champion from the American League against the champion from the National League in a best of 7 series of games (meaning the first team to win 4 games is the World Series champion).

    a. If the two teams are evenly matched, what is the probability that the World Series will end in exactly 5 games?

    b. If the probability that the team from the National League will win is $\frac{3}{5}$ for each game, what is the probability that the National League will win the World Series in exactly 5 games?

33. Cooking In cooking class, 1 out of 5 souffles that Sabrina makes will collapse. She is preparing 6 souffles to serve at a party for her parents. What is the probability that exactly 4 of them do not collapse?

34. Skeet Shooting Skeet shooting involves shooting at discs, called clay pigeons, propelled into the air by a machine. Heather usually hits 9 out of 10 clay pigeons. If she shoots 12 times, find each probability.

    a. $P(\text{all misses})$

    b. $P(\text{exactly 7 hits})$

    c. $P(\text{all hits})$

    d. $P(\text{at least 10 hits})$

35. Finance A stock broker is researching 13 independent stocks. An investment in each stock will either make money or lose money. The probability that each stock will make money is $\frac{5}{8}$. What is the probability that exactly 10 of the stocks make money?

36. Find the product of A and B if $A = \begin{bmatrix} 4 & 3 \\ 7 & 2 \end{bmatrix}$ and $B = \begin{bmatrix} 8 & 5 \\ 9 & 6 \end{bmatrix}$. **(Lesson 2-2)**

37. Find the inner product of $\vec{u}$ and $\vec{v}$ if $\vec{u} = (2, -1, 3)$ and $\vec{v} = (5, 3, 0)$. **(Lesson 8-4)**

38. Simplify $\dfrac{x!(x-3)!}{(x-2)!(x-1)!}$. Assume x and y are positive integers, $x > y$, $x > 2$. **(Lesson 12-6)**

39. A pair of dice is tossed. Find the probability that their sum is greater than 7 given that the numbers match. **(Lesson 14-7)**

40. College Entrance Exam Choose the best answer.

    If $a > b$ and $c < 0$, which of the following are true?

    I. $ac < bc$

    II. $a + c > b + c$

    III. $a - c < b - c$

    (A) I only   **(B)** II only   **(C)** I and II only   **(D)** I, II, and III

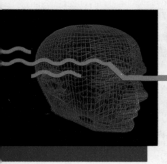

Technology

Dice Roll Simulation

```
  5 DIM A(12)
 10 INPUT "ENTER NUMBER OF
    ROLLS DESIRED: " ;R
 20 FOR N = 2 TO 12
 30 LET A(N) = 0
 40 NEXT N
 45 PRINT: PRINT
 50 FOR I = 1 TO R
 60 LET D1 = INT(6*RND(1) + 1)
 70 LET D2 = INT(6*RND(1) + 1)
 80 LET T = D1 + D2
 90 PRINT T:" ";
100 LET C = 2
110 IF T = C THEN GOTO 140
120 LET C = C + 1
130 GOTO 110
140 LET A(C) = A(C) + 1
150 NEXT I
155 PRINT: PRINT
160 FOR M = 2 TO 12
170 LET P = INT(((A(M)/R)*1000) + .5)/10
180 LET E = INT((250 * (6 - ABS(7 - M))/9)
    + .5)/10
190 PRINT "#";M:"TIMES ROLLED:";A(M):
    "EXPERIMENTAL PROBABILITY";P;"%"
200 PRINT "THEORETICAL PROBABILITY: ";E;
    "%":PRINT
210 NEXT M
220 END
```

Probability is used to predict the outcome in games of chance involving rolling dice, tossing coins, selecting cards, or winning sweepstakes or lotteries. A computer program can be used to simulate rolling dice, tossing coins, and so on.

The BASIC program at the left simulates rolling two dice, prints the number of times a sum was rolled, experimental probability, and the theoretical probability. Each time the program is run, the experimental probability may be different due to the random selection of values in lines 60 and 70.

Each random number selected by the computer lies between 0 and 1. Then it is multiplied by 6 to simulate the six faces of a die. Finally, the INT (integer) function eliminates any decimal places, and a result of 0, 1, 2, 3, 4, 5, or 6 is chosen randomly.

Enter the program and run it several times. Notice the changes in the experimental probability. When you use the program for the exercises below, it may take several minutes to complete the large samples.

EXERCISES

1. Run the program once. Enter 20 for the input.

2. Run the program four more times to simulate 40, 60, 80, and 100 rolls.

3. Run the program to simulate 1000 rolls.

4. Make a conjecture about the number of rolls in relation to the theoretical probability.

VOCABULARY

Upon completing this chapter, you should be
familiar with the following terms:

basic counting principle	**752**	**779**	mutually exclusive
binomial experiment	**791**	**779**	inclusive
combination	**764**	**769**	odds
combinatorics	**752**	**753**	permutation
complement	**768**	**768**	probability
conditional probability	**784**	**785**	reduced sample space
dependent event	**752**	**785**	sample space
failure	**768**	**768**	success
independent event	**752**	**753**	tree diagram

SKILLS AND CONCEPTS

OBJECTIVES AND EXAMPLES

Upon completing this chapter, you should
be able to:

- solve problems related to the basic
counting principle   **(Lesson 14-1)**

Using the letters from the word *compute,*
how many four-letter patterns can be
formed?

$$P(7,4) = \frac{7!}{(7-4)!}$$
$$= \frac{7 \cdot 6 \cdot 5 \cdot 4 \cdot 3 \cdot 2 \cdot 1}{3 \cdot 2 \cdot 1} \text{ or 840 patterns}$$

- solve problems involving permutations
with repetitions   **(Lesson 14-2)**

How many ways can the letters of
Mississippi be arranged?

$$\frac{11!}{4!4!2!} = 34,650 \text{ ways}$$

REVIEW EXERCISES

Use these exercises to review and prepare for
the chapter test.

Find each value.

1. $P(6,3)$ **2.** $P(8,6)$

3. $\dfrac{P(4,\ 2) - P(6,\ 3)}{P(5,\ 3)}$

4. How many ways can six different books
be placed on a shelf if the only dictionary
must be on an end?

**How many different ways can the letters of
each word be arranged?**

5. LEVEL   **6.** CINCINNATI

7. GRADUATE   **8.** BANANA

OBJECTIVES AND EXAMPLES	REVIEW EXERCISES

■ **solve problems involving combinations (Lesson 14-3)**

From a standard deck of cards, how many different 4-card hands can have 4 cards of the same suit?

$$4[C(13, 4)] = 4\left(\frac{13!}{(13-4)!4!}\right) \text{ or } 2860 \text{ hands}$$

Find each value.

9. $C(5, 3)$

10. $C(11, 8)$

11. $C(5, 5) \cdot C(3, 2)$

12. From a group of 3 men and 7 women, how many committees of 2 men and 2 women can be formed?

■ **find the probability and the odds for the success and failure of an event (Lesson 14-4)**

Find the probability of randomly selecting 3 consonants from the word *algebra* without replacement.

$$P(3 \text{ consonants}) = \frac{C(4, 3)}{C(7, 3)}$$
$$= \frac{\frac{4!}{1!3!}}{\frac{7!}{4!3!}} \text{ or } \frac{4}{35}$$

State the odds of an event occurring given the probability of the event.

13. $\frac{4}{9}$

14. $\frac{1}{12}$

A bag contains 7 pennies, 4 nickels, and 5 dimes. Three coins are selected at random with replacement. What is the probability of each of the following selections?

15. all 3 pennies

16. 2 pennies, 1 nickel

■ **find the probability of independent and dependent events (Lesson 14-5)**

Three yellow and 5 black marbles are placed in a bag. What is the probability of drawing a black marble, not replacing it, then drawing a yellow marble?

$P(\text{black, yellow})$
$\quad = P(\text{black}) \cdot P(\text{yellow following black})$
$\quad = \frac{5}{8} \cdot \frac{3}{7} \text{ or } \frac{15}{56}$

A green die and a red die are thrown. What is the probability that each of the following occurs?

17. The red die shows a 1 and the green die shows any other number.

18. Neither die shows a 1.

■ **find the probability of mutually exclusive and inclusive events (Lesson 14-6)**

On a school board, 2 of the 4 female members are over 40 years old and 5 of the 6 male members are over 40. If one person didn't attend the meeting, what is the probability that it was a male or a member over 40?

$P(\text{male or over 40}) = P(\text{male}) + P(\text{over 40}) -$
$\qquad\qquad\qquad\qquad\quad P(\text{male and over 40})$
$$= \frac{6}{10} + \frac{7}{10} - \frac{5}{10} \text{ or } \frac{4}{5}$$

A box contains slips of paper numbered from 1 to 14. A slip of paper is drawn at random. What is the probability that each of the following occurs?

19. The number is prime or is a multiple of 4.

20. The number is a multiple of 2 or a multiple of 3.

■ find the probability of an event given the occurrence of another event
(Lesson 14-7)

A coin is tossed three times. What is the probability that at the most 2 heads show given that at least 1 head shows?

Let event A be that at most 2 heads show.

Let event B be that at least 1 head shows.

$$P(A/B) = \frac{P(A \text{ and } B)}{P(B)}$$

$$= \frac{\frac{6}{8}}{\frac{7}{8}} \text{ or } \frac{6}{7}$$

A green die and a red die are thrown.

21. What is the probability that the sum of the dice is less than 5 if the green die shows a 1?

22. What is the probability that the sum of the dice is 9 if the two numbers showing are different?

■ find the probability of an event by using the binomial theorem   **(Lesson 14-8)**

If you guess on all 8 answers of a true/false quiz, what is the probability that exactly 5 of your answers will be correct?

$$(p + q)^8 = \sum_{r=0}^{8} \frac{8!}{r!(8 - r)!} p^{8-r} q^r$$

$$\frac{8!}{5!(8 - 5)!} \left(\frac{1}{2}\right)^5 \left(\frac{1}{2}\right)^3 = \frac{56}{256} \text{ or } \frac{7}{32}$$

A coin is tossed 4 times. What is the probability that each of the following occurs?

23. exactly 1 head

24. no heads

25. 2 heads and 2 tails

26. at least 3 tails

APPLICATIONS AND PROBLEM SOLVING

27. **Employment**   Ms. Sommers, the new school custodian, is given 7 keys for the school building. How many ways can she place these keys on a key ring? **(Lesson 14-2)**

28. **Tourism**   A taxi in The Netherlands can hold 4 passengers safely. A party of 6 people wish to travel by taxi to The Hague. How many different groups can occupy the first taxi if it will be full?   **(Lesson 14-3)**

29. **Gifts**   The Blackburn family is drawing names from a bag for their holiday gift exchange. There are 7 males and 8 females in the family. If family members draw their own name, then they must draw again before replacing their name. **(Lessons 14-4 and 14-5)**

    a. What is the probability that Marie Blackburn will draw a female's name that is not her own?

    b. What is the probability that Marie will draw her own name, not replace it, then draw a male's name?

Find each value.

1. $P(6, 2)$ 2. $P(7, 5)$ 3. $C(8, 3)$ 4. $C(5, 4)$

5. The letters r, s, t, u, and v are to be used to form five-letter patterns. How many patterns can be formed if repetitions are not allowed?

6. **Employment** There are 5 persons who are applicants for 3 different positions in a store, each person being qualified for each position. In how many ways is it possible to fill the positions?

7. How many ways can 7 people be seated at a round table relative to each other?

8. **Sports** How many baseball teams can be formed from 15 players if only 3 pitch and the others play the remaining 8 positions?

9. A bag contains 4 red and 6 white marbles. How many ways can 5 marbles be selected if exactly 2 must be red?

10. **Geometry** Find the number of possible diagonals in a polygon that has 20 sides.

11. **Sports** The probability that the Pirates will win a game against the Hornets is $\frac{4}{7}$. What are the odds that the Pirates will beat the Hornets?

12. Five cards are drawn from a standard deck of cards. What is the probability that they are all from one suit?

13. Find the probability of getting a sum of 8 on the first throw of two dice and a sum of 4 on the second throw.

14. **Telephones** A new phone is being installed at the Collier residence. What is the probability that the final 3 digits in the telephone number will be odd?

15. A bag contains 3 red, 4 white, and 5 blue marbles. If 3 marbles are selected at random, what is the probability that all are red or all are blue?

16. A card is drawn from a standard deck of cards. What is the probability of having drawn an ace or a black card?

17. Four-digit numbers are formed from the numbers 7, 3, 3, 2, and 2. If a number is odd, what is the probability that the 3s are together?

18. Two numbers are selected at random from the numbers 1 through 9. If their product is even, what is the probability that both numbers are even?

19. Five bent coins are tossed. The probability of heads is $\frac{2}{3}$ for each of them. What is the probability that no more than 2 will show heads?

20. **Archery** While shooting arrows, Akira can hit the center of the target 4 out of 5 times. What is the probability that he will hit it exactly 4 out of the next 7 times?

Bonus
A positive integer $n \leq 120$ is chosen in such a way that if $n \leq 80$, then the probability of choosing n is p, and if $n > 80$, then the probability of choosing n is $2p$. What is the probability that a perfect square is chosen?

CHAPTER 15

STATISTICS AND DATA ANALYSIS

HISTORICAL SNAPSHOT

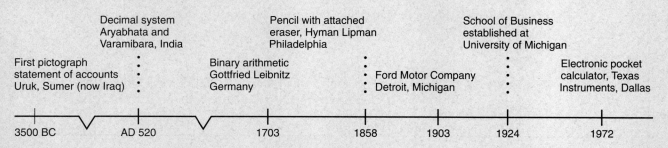

	Decimal system Aryabhata and Varamibara, India		Pencil with attached eraser, Hyman Lipman Philadelphia		School of Business established at University of Michigan	
First pictograph statement of accounts Uruk, Sumer (now Iraq)		Binary arithmetic Gottfried Leibnitz Germany		Ford Motor Company Detroit, Michigan		Electronic pocket calculator, Texas Instruments, Dallas
3500 BC	AD 520	1703	1858	1903	1924	1972

CHAPTER OBJECTIVES

In this chapter, you will:
- Make and use graphs and plots to analyze data.
- Find measures of central tendency and variability.
- Organize data into frequency distributions.
- Use the normal distribution curve.
- Fit curves to curvilinear data.

CAREER GOAL: Public Accounting

When most people think of accounting, they think of keeping track of account balances in a ledger. But much more is involved, especially for the public accountant. Darlene Helzerman would like to become a certified public accountant because "in public accounting I would gain experience with the operations of several corporations of different sizes and complexity. If I were to work as a corporate accountant, I would only experience the operation of a single corporation."

Darlene was inspired to pursue a career in accounting by her professors. "At the University of Michigan, our professors come from the real world of business. As a result, they are able to relate real-world experiences and opportunities. I would say that my professors have been the greatest inspiration for me."

Darlene emphasizes the role that mathematics plays in accounting. "Mathematics is so analytical that a knowledge of mathematics helps you to analyze and adapt to any new situation," she says. She also uses mathematics in her work. "For manufacturing corporations, we find variances in order to reduce errors. For example, consider the process of manufacturing cereal boxes. If variances related to the size of a cereal box are within a given range, it would be acceptable to continue manufacturing them. If the variances fall outside the given range, however, the process of manufacturing would have to be changed in order to avoid costly errors."

Darlene looks forward to being a positive role model for other young women. "It is important for women to be seen in successful professional roles. I find it exciting to think that I might help change society's view of women."

For up-to-date information on accounting, visit:
www.glencoe.com/sec/math/amc/mathnet

15-1 The Frequency Distribution

Objectives

After studying this lesson, you should be able to:
- draw, analyze, and use bar graphs and histograms, and
- organize data into a frequency distribution table.

Application

Compact discs (CDs) first appeared in music stores in 1983. A compact disc is $4\frac{2}{3}$ inches in diameter, and the music on a CD is recorded by sampling sounds from a master tape at a rate of 44,100 times per second. The samples are translated into a binary code that is stored on the disc. A laser in the CD player translates the codes back into musical sounds. How many compact discs do you think are made each year?

The table below shows different types of recording media and how many millions of units were shipped in 1973 compared to 1990. The **back-to-back bar graph** shows the comparisons for each recording medium. A back-to-back bar graph is plotted on a two-quadrant coordinate system with the horizontal scale repeated in each direction from the central axis.

Manufacturers' Shipments of Recording Media (in millions of units)		
Medium	1973	1990
albums	280.0	11.7
45s	228.0	27.6
cassette tapes	15.0	442.2
compact discs	0.0	286.5

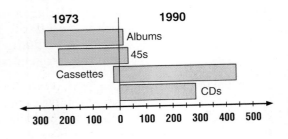

You can see from the graph that the popularity of cassettes and compact discs has increased dramatically from 1973 to 1990 while the popularity of albums and 45s has plummeted.

Example 1

APPLICATION

Civics

Make a back-to-back bar graph of the following data which compares the percent of the U. S. population in each age group over the century.

Age (yr)	0—9	10—19	20—29	30—39	40—49	50—59	60—69	70+
1890	25%	22%	18%	14%	9%	6%	4%	2%
1990	16%	14%	16%	17%	12%	9%	8%	8%

Let the age groups be the central axis. Draw a horizontal axis that is scaled from 0 to 20 in each direction. Let the left side of the graph represent the population groups from 1890 and the right side of the graph be those from 1990. Draw bars the appropriate lengths for the data.

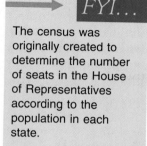

FYI...

The census was originally created to determine the number of seats in the House of Representatives according to the population in each state.

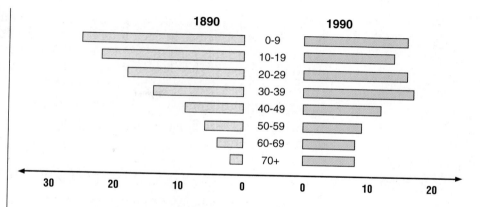

You can see from the graph that a greater percent of the population in 1890 was very young, and very few people lived past the age of 50. In 1990, that trend was very different. 25% of our population was 50 or older.

Wheat, Rice, and Corn Production, 1990
(in millions of metric tons)
Source: U.N. Food and Agriculture Organization

Sometimes it is desirable to show three or more aspects of a set of data at the same time. To present data in this way, a **three-dimensional bar graph** is often used. The graph at the left represents the production of wheat, rice, and corn in several countries. The grid defines the crop and the country. The height of each bar is drawn proportional to the millions of metric tons of the crop produced in each country.

Sometimes the amount of data you wish to represent in a bar graph is too great for each item of data to be considered individually. In this case, a **frequency distribution** is a convenient system for organizing the data. A number of classes are determined, and all values in a class are tallied and grouped together. The intervals are often named by a range of values. In the table, the interval described by 2.0–4.0 means all tree diameters d such that $2.0 \leq d < 4.0$. The **class interval** is the range of each class. The class intervals in a frequency distribution should all be equal. In the table at the right, the class interval for each range is 2.0.

The range of a set of data is the difference between the greatest and least values in the set.

Sample Tree Diameters from the Cumberland National Forest, 1992

Tree Diameters (in inches)	Frequency
2.0–4.0	6
4.0–6.0	30
6.0–8.0	38
8.0–10.0	33
10.0–12.0	9
12.0–14.0	4

The **class limits** of a set of data organized in a frequency distribution are the upper and lower values in each interval. The class limits in the tree sample data on the previous page are 2.0, 4.0, 6.0, 8.0, 10.0, 12.0, and 14.0. If an item of data falls on one of the class limits, it is tallied in the higher class. The **class marks** are the midpoints of the classes; that is, the average of the upper and lower limit in each interval. The class mark for 2.0–4.0 is $\frac{2.0 + 4.0}{2}$ or 3.0. Notice that the difference in the consecutive class marks is the same number as the class interval.

The most common way of displaying frequency distributions is by using a histogram. A **histogram** is a type of bar graph in which the width of each bar represents a class interval and the height of the bar represents the frequency in that interval.

Example 2

A complete frequency distribution includes the class limits, the class marks, the tallies, and the frequency of each class.

The weights of 200 boys trying out for football at Indian Hills High School have been tallied in the chart below. The actual range of weights was from 99 to 253.

Class Limits	Tally	Frequency f
95–110	|||	3
110–125	卌 ||	7
125–140	卌 卌 卌	15
140–155	卌 卌 卌 卌 卌 卌 |||	34
155–170	卌 卌 卌 卌 卌 卌 卌 卌 ||	42
170–185	卌 卌 卌 卌 卌 卌 卌 |||	38
185–200	卌 卌 卌 卌 ||||	24
200–215	卌 卌 卌	15
215–230	卌 卌		11
230–245	卌 ||	7
245–260	||||	4
	Total	$\overline{200}$

a. **Determine if the class intervals of 15 are appropriate for this data.**

b. **Find the class marks.**

c. **Draw a histogram of the data.**

a. The actual range of the data is $253 - 99$ or 154. Class intervals of 15 pounds, beginning with 95 and ending with 260, seem appropriate.

b. The class marks are the averages of the class limits of each interval. The class marks are 102.5, 117.5, 132.5, 147.5, 162.5, 177.5, 192.5, 207.5, 222.5, 237.5, and 252.5.

c. Label the horizontal axis with the class limits. A jagged line is often used to show the values omitted from 0 to the first limit. The vertical axis should be labeled from 0 to a value that will allow for the greatest frequency. Draw the bars side by side so that the height of each bar corresponds to its interval's frequency.

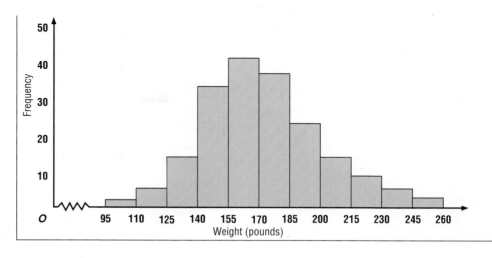

Notice that a single item of data loses its identity in a frequency distribution. For example, a weight of 138 is grouped with all the others in the 125–140 class. In working with such a distribution, it is assumed that the data in any class are uniformly distributed over the class.

Usually, only a slight error may be introduced by this assumption.

EXPLORATION: Graphing Calculator

Use a TI-82 graphing calculator to create a histogram of the data in Example 2.

1. Clear List 1 and List 2.
   - Press STAT 4 2ND L1 , 2ND L2 ENTER.
     The home screen will display "Done".
2. Enter the data.
   - Press STAT 1.
     The screen will show columns for L_1, L_2, and L_3.
   - List 1 is the class mark and List 2 is the frequency for the class. Enter each class mark and then press ENTER. Press the right arrow key and enter each number for the frequency and then press ENTER.
     102.5 ENTER 117.5 ENTER 132.5 ENTER ... 252.5 ENTER
     3 ENTER 7 ENTER 15 ENTER ... 4 ENTER
     132.5 ENTER ... 252.
   - Press 2ND QUIT. *You will return to the home screen.*

(continued on the next page)

3. Set the range for your graph.

- Press WINDOW . Set the range as follows:

  XMIN: 95   XMAX: 260   XSCL: 15
  YMIN: 0   YMAX: 50   YSCL: 5

- Press 2ND QUIT to return to the home screen.

4. Graph the histogram.

- Press 2ND STAT PLOT 1.
- Select "On", histogram, "L1" as the Xlist, and "L2" as the frequency.
- Press GRAPH .

5. Write a sentence to compare this graph to the one in Example 2.
Note: You will want to turn the STAT PLOT off when finished.

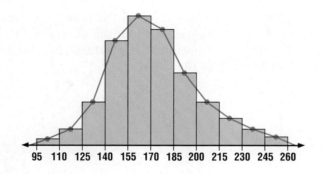

Another type of graph can be created from a histogram. A broken line graph, often called a **frequency polygon**, can be drawn by connecting the class marks on the graph. The class marks are graphed as the midpoints of the top edge of each bar. The frequency polygon for the histogram in Example 2 is shown at the left.

CHECKING FOR UNDERSTANDING

Communicating Mathematics

Read and study the lesson to answer each question.

1. **Explain** why the back-to-back bar graph on page 803 might be considered a variation of a histogram. Why might it not?

2. **Explain** when you are most likely to use a histogram.

3. **Demonstrate** how you would draw a frequency polygon for the histogram shown at the right.

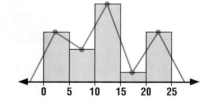

Guided Practice

Use the bar graph to answer each question.

4. Which cable channel has the greatest number of subscribers? the least number of subscribers?

5. What is the trend over time of the number of subscribers?

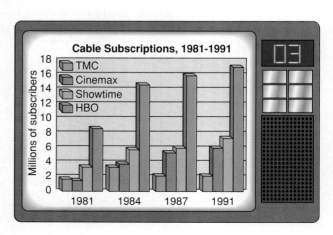

The table at the right gives the annual earnings of a random sample of 100 U.S. high school teachers.

Annual Salary	Frequency
Less than $26,000	9
$26,000–$32,000	21
$32,000–$38,000	31
$38,000–$44,000	26
$44,000–$50,000	13

6. State the class interval of the frequency distribution.

7. State the class limits of the frequency distribution.

8. State the class marks of the frequency distribution.

9. What percent of the teachers polled earn $38,000 to $44,000?

10. What percent of the teachers polled earn at least $38,000?

11. Draw a histogram of this data.

EXERCISES

Practice   Find the class interval and the class limits for each set of class marks of a frequency distribution.

12. {10, 20, 30, 40, 50}

13. {1.1, 1.2, 1.3, 1.4, 1.5, 1.6, 1.7}

14. {2.5, 5, 7.5, 10, 12.5}

15. {25, 26, 27, 28, 29, 30, 31, 32}

The table at the right lists the salaries of males and females and the corresponding educational backgrounds.

Average Monthly Earnings, 1992		
Education	Males	Females
Professional Degree	$4,480	$2,311
Master's Degree	$2,901	$1,733
Bachelor's Degree	$2,471	$1,136
Associate Degree	$1,977	$1,022
Vocational School	$1,699	$773
Some College-no degree	$1,483	$710
High School Degree	$1,350	$583
High School Non-graduate	$709	$207

16. Make a back-to-back bar graph of this information.

17. Write a statement that describes the comparative salaries.

The IQ's of a random sample of 100 seniors are given below.

```
 84 120  78  89 107 116  73  88 106 117
144  92 100 124  84 100 115  76  93 112
 89 109 110 128 101 109 135 100 112  81
 99 119  88 117 110  81 103 127  97 120
 93 115  92 116  68  97  66 102  84 108
 95  72 104  95  80  85 106  99  87 115
104 100  95  85 121 112  97 106 113  82
104 102 111 103 125  95 102  88 102  97
 99 100 103  78  99  99  91  97  99 103
102 110  88 100  90 100 103  92 103 100
```

18. Make a frequency distribution of the data using a class interval of 10. Then make a histogram of the data.

19. Make a frequency distribution of the data using a class interval of 12. Then make a histogram of the data.

20. Make a frequency distribution of the data using 16 classes. Then make a histogram of the data.

21. Make a frequency distribution of the data using 10 classes. Then make a histogram of the data.

22. What interval would you use if you wanted to create a histogram to show that the IQ of seniors is higher than average?

As an experiment in precision and gathering data, the 24 members of a pre-calculus class were given the assignment of measuring the length of the new driveway into the school's campus. Their measurements, rounded to the nearest tenth of a meter, are listed below.

2013.3	2012.6	2012.3	2012.8
2013.5	2011.4	2013.4	2012.4
2012.3	2012.2	2012.1	2011.6
2012.0	2012.6	2012.2	2012.2
2013.9	2012.7	2011.8	2013.7
2012.4	2011.7	2012.8	2012.7

23. Make a frequency distribution of the data using an appropriate class interval.

24. Make a histogram of the data.

25. Make a frequency polygon of the data.

26. How would you handle the data if one student reported a length of 2112.7 meters instead of 2012.7 meters?

27. Make a three-dimensional bar graph of the data on the population change of four regions of the United States according to the Census reports of 1970, 1980, and 1990 shown below.

Population Change (in millions)				
Year	Northeast	Midwest	South	West
1970	49	57	63	35
1980	49	59	75	43
1990	51	60	85	53

Graphing Calculator

Use a graphing calculator to make a histogram for the data in each exercise.

28. Exercise 18   **29.** Exercise 19   **30.** Exercise 20   **31.** Exercise 21

32. Read the graphing calculator manual to find how you could create a frequency polygon after you have graphed a histogram. Write a paragraph describing the process.

Critical Thinking

33. Find a graph in a newspaper or magazine.
  a. Does the graph accurately represent the data?
  b. Could the data have been presented effectively without a graph?
  c. Write a summary of what the graph says to you.

34. Sports The following table lists the yearly incomes in millions of dollars of some of the highest paid athletes in the world in 1991. The *Other Income* category includes endorsements, exhibition fees, and incentive bonuses. (Money made from businesses owned by the athletes is not included.)

Athlete	Sport	Salary	Other Income
Larry Bird	Basketball	$7.4	$0.5
Stefan Edberg	Tennis	1.4	6.0
Steffi Graf	Tennis	1.3	6.0
Evander Holyfield	Boxing	60.0	0.5
Michael Jordan	Basketball	2.8	13.2
Joe Montana	Football	3.5	4.0
Jack Nicklaus	Golf	0.5	8.0
Monica Selles	Tennis	1.6	6.0

a. Make a back-to-back bar graph that compares the salary of each athlete with their other income.

b. If you were an athlete, which appears to be the most lucrative profession if only salary is considered?

c. If you were an athlete, which appears to be the most lucrative profession if other income is considered?

d. Who has the greatest total salary? the least?

35. Demographics The number of participants to the 1992 Summer Olympics by state are listed below.

State	Participants	State	Participants	State	Participants
AL	4	LA	8	OH	15
AK	1	ME	2	OK	5
AZ	13	MD	16	OR	12
AR	4	MA	19	PA	18
CA	140	MI	15	RI	6
CO	15	MN	11	SC	4
CT	8	MS	2	SD	1
DE	0	MO	4	TN	4
FL	41	MT	1	TX	43
GA	6	NE	1	UT	3
HI	6	NV	5	VT	2
ID	1	NH	5	VA	13
IL	17	NJ	18	WA	5
IN	8	NM	7	WV	2
IA	6	NY	28	WI	14
KS	5	NC	7	WY	0
KY	3	ND	1		

a. Make a frequency distribution showing the number of states for each interval of participants, excluding California. Use a class interval of 10.

b. Make a histogram using the information in part a. Then draw the frequency polygon.

c. Why do you think California's data was excluded from this distribution? How does the inclusion of California change the distribution?

36. Find the measure of the reference angle for an angle of $-\frac{13\pi}{3}$ radians. **(Lesson 5-1)**

37. State whether $\overrightarrow{PQ}$ and $\overrightarrow{RS}$ are *equal, opposite, parallel,* or *none of these* for points $P(8,-7), Q(-2,5), R(8,-7),$ and $S(7,0)$. **(Lesson 8-1)**

38. Find the coordinates of the focus and the equation of the directrix of the parabola with equation $y^2 = 12x$. Then graph the equation. **(Lesson 10-2)**

39. Form a sequence that has two geometric means between 125 and 216. **(Lesson 12-2)**

40. Probability Tess is running a "Wheel of Fortune" booth to raise money for a muscular dystrophy charity. The wheel has the numbers 1 to 10 on it. What is the probability of 7 never coming up in five spins of the wheel? **(Lesson 14-8)**

41. College Entrance Exam Compare quantities A and B below. $(x \neq 0)$
Write A if quantity A is greater.
Write B if quantity B is greater.
Write C if the two quantities are equal.
Write D if there is not enough information to determine the relationship.

(A) $\left(\frac{3}{8} + \frac{1}{6} + \frac{5}{12}\right)x^2$ **(B)** $\left(\frac{4}{15} + \frac{2}{25} + \frac{1}{3}\right)x^2$

DECISION MAKING

Safety: Car Airbags

More than 40 states now have laws requiring car passengers to wear seat belts. But because they are not yet readily available, airbags remain a luxury for most Americans. Statistics show that an airbag provides a passenger with a better chance of surviving a serious accident than a seat belt does. Is the amount of added protection great enough to justify the added cost? If airbags become readily available, should people who do not want them in their cars be required to have them?

1. Research the safety record of cars with airbags. Has enough data been collected to draw any conclusions about their safety relative to the safety of seat belts?

2. What is your opinion of airbags? Would you like to have them installed in your car? Do you think that a date should be set when all new cars should be equipped with airbags?

3. Project Work with your group to complete one of the projects listed at the right based on the information you gathered in Exercises 1–2.

PROJECTS

- *Conduct an interview.*
- *Write a book report.*
- *Write a proposal.*
- *Write an article for the school paper.*
- *Write a report.*
- *Make a display.*
- *Make a graph or chart.*
- *Plan an activity.*
- *Design a checklist.*

15-2 Measures of Central Tendency

Objectives

After studying this lesson, you should be able to:
- find the mean, mode, and median of a set of data, and
- find the mean and median of data organized in a stem-and-leaf plot or a frequency distribution table.

Application

The tennis shoes sold at Hackett's Racquets are lightweight for playing tennis. Below is a table of various tennis shoes and the weight (in ounces) of men's and women's shoes. Which weight is most representative of the weights of men's tennis shoes?

Company/ Shoe name	Men's Weight	Women's Weight
Adidas Score Lite	10.7	9.1
Avia 758	13.7	11.3
Avia 748	13.5	10.9
Champion Unlimited	15.9	12.3
Converse Light 5000	12.8	11.1
Converse Light 3000	13.4	10.6
Head Radial	15.1	12.5
K-Swiss Si-18	16.5	13.5
K-Swiss Grancourt Gstaad	16.0	13.0
K-Swiss CT 500	14.0	10.8
Nike Air Tech Challenger	14.0	11.0
Nike Air Supreme Low	12.3	9.7
Nike Air Challenge Pro	12.8	11.0

Company/ Shoe name	Men's Weight	Women's Weight
Nike Air Courtlite	13.6	10.5
Nike Air Tour Challenge	12.8	10.4
Prince FST	15.8	10.0
Reebok Pump Court Victory II	16.0	14.0
Reebok Club Pump Low	14.5	11.5
Reebok Victory LWT Low	15.0	11.5
Reebok Club Low	13.5	10.5
Reebok Victory 7000	14.0	10.5
Wilson Perfecta	12.3	10.3
Wilson Plexus	15.4	11.7
Wilson Pro Staff	14.7	11.6
Wilson Staff	14.5	11.0

The average weight of men's shoes is a value that is the most representative of the entire set of men's shoe data. A **measure of central tendency** is a type of average. It is a number that represents a set of data. The most commonly used averages are the median, mode, and mean. These measures may be used to describe and summarize a set of data, or to compare one set of data with another.

The **arithmetic mean** is symbolized by $\overline{X}$ and is often referred to as the average or, simply, the **mean**. The mean is found by adding the values in the set of data and dividing the sum by the number of values in that set. Every number in a set of data affects the value of the mean. Consequently, the mean is generally a good representative measure of central tendency. However, the mean can be considerably influenced by extreme values.

Example 1

APPLICATION

Sportswear

Refer to the application at the beginning of the lesson. Find the mean of the weights of the men's tennis shoes.

$$\overline{X} = \frac{\text{sum of the weights of the men's shoes}}{\text{number of weights being added}}$$

$$= \frac{10.7 + 13.7 + 13.5 + 15.9 + \cdots + 15.4 + 14.7 + 14.5}{25}$$

$$= \frac{352.8}{25} \text{ or about } 14.1$$

The mean weight of the men's tennis shoes is about 14.1 ounces.

Notice that the mean is not necessarily a member of the set of data.

EXPLORATION: Graphing Calculator

Refer to Example 1. Use a TI-82 graphing calculator to find the mean of the weights of the men's tennis shoes.

1. Clear the statistical memory.
2. Enter the data. List 1 is the weight of the shoe, and List 2 is the frequency for the weight.
3. Press [STAT] [▶] 1.
4. Press [ENTER] to display the value of n and $\overline{x}$, which is the mean.
5. Write a sentence to compare this mean to the one in Example 1.

Refer to the Exploration in Lesson 15–1 for a review of data input.

A general formula for the mean of any set of data can be written using special notation. If X is a variable used to represent any value in a set of data containing n items, then the arithmetic mean $\overline{X}$ of the n values is given by the formula

$$\overline{X} = \frac{X_1 + X_2 + X_3 + \cdots + X_n}{n}.$$

The numerator of the fraction can be abbreviated using the summation symbol, $\sum$. *Recall that $\sum$ is the uppercase Greek letter sigma.*

$$\sum_{i=1}^{n} X_i = X_1 + X_2 + X_3 + \cdots + X_n$$

The symbol X_i represents successive values of the set of data as i assumes successive integral values from 1 to n. By substituting the sigma notation into the formula for the mean, the formula becomes

$$\overline{X} = \frac{\sum_{i=1}^{n} X_i}{n} \text{ or } \overline{X} = \frac{1}{n} \sum_{i=1}^{n} X_i.$$

Arithmetic Mean

If a set of data has n values given by X_i such that i is an integer and $1 \leq i \leq n$, then the arithmetic mean, $\overline{X}$, can be found as follows.

$$\overline{X} = \frac{1}{n} \sum_{i=1}^{n} X_i$$

Another measure of central tendency is the **median**, symbolized by M_d. It is the middle value of a set of data. Before the median can be found, the data must be organized in an **array**. An array is formed by arranging the data into an ordered sequence, usually from least to greatest. The median of an odd number of data is the middle value of the array. For example, in $\{2, 3, 4, 6, 7\}$, the median is 4. The median of an even number of data is the mean of the two middle values of the array. For example, in $\{2, 3, 4, 6, 8, 9\}$, $M_d = \dfrac{4+6}{2}$ or 5. *Notice that the median is not necessarily a member of the set of data.*

The median is preferable to the mean as a measure of central tendency when there are a few extreme values or when some of the values cannot be determined. Unlike the mean, the median is influenced very little by extreme values.

The **mode** of a set of data is the item of data that appears more frequently than any other in the set. Data with two modes are **bimodal**. Sets have no mode when each item of the set has equal frequency. The value of the mode is not affected by extreme values. *Notice that the mode, if it exists, is always a member of the set of data.*

Example 2

Find the mean, median, and mode of the salaries for the corporate employees listed below. State which measures of central tendency seem most representative of the set of data.

Employee	Salary	Employee	Salary
Abner	$54,000	Iron	$65,000
Adams	$75,000	Johnston	$59,000
Conners	$55,000	Jones	$61,000
Harrison	$62,000	Ortez	$162,000
Heiko	$226,000	Newman	$59,000

Since there are 10 salaries, $n = 10$.

$$\sum_{i=1}^{10} X_i = \frac{1}{10}(54{,}000 + 75{,}000 + 55{,}000 + 62{,}000 + 226{,}000 + 65{,}000 + 59{,}000 + 61{,}000 + 162{,}000 + 59{,}000) \text{ or } \$87{,}800$$

To find the median, place the data in an array. Since all of the salaries are multiples of $1000, you can order the set by thousands.

54   55   59   59   61   62   65   75   162   226

Since there are an even number of data, the median is the mean of the two middle numbers, 61,000 and 62,000. The median salary is $61,500.

The mode is $59,000.

Notice that the mean is affected by the extreme values $162,000 and $226,000 and does not accurately represent the data. The median and the mode are more representative measures of central tendency in this case.

When you have a large number of data, it is often helpful to use a **stem-and-leaf plot** to organize your data in an array. In a stem-and-leaf plot, each item of data is separated into two parts that are used to form a stem and a leaf. The parts are organized into two columns. The column on the left shows the **stems**. Stems usually consist of the digits in the greatest common place value of all the data. For example, if the set of data includes the numbers 980 and 1132, the greatest common place value is hundreds. Therefore, the stem of 980 is 9, and the stem of 1132 is 11. The column on the right contains the **leaves**. The leaves consist of the other part of each item of data. The leaves are always one-digit numbers. The leaf of 980 is 8, and the leaf of 1132 is 3. The stems and leaves are usually arranged from least to greatest.

Example 3

> **Refer to the application at the beginning of the lesson. Make a stem-and-leaf plot of the weights of women's tennis shoes. Then determine the mean, median, and mode.**
>
> Since the weights range from 9.1 to 14.0, we will use the units place for the stems. List the stems and draw a vertical line to the right of the stems. Then list the leaves, which in this case will be the tenths digit of each weight. As shown below it is often helpful to list the leaves as you come to them and then rewrite the plot with the leaves in order from least to greatest.
>
> | stem | leaf |
> |------|------|
> | 9 | 1 7 |
> | 10 | 9 6 8 5 4 0 5 5 3 |
> | 11 | 3 1 0 0 5 5 7 6 0 |
> | 12 | 3 5 |
> | 13 | 5 0 |
> | 14 | 0 |
>
> | stem | leaf |
> |------|------|
> | 9 | 1 7 |
> | 10 | 0 3 4 5 5 5 6 8 9 |
> | 11 | 0 0 0 1 3 5 5 6 7 |
> | 12 | 3 5 |
> | 13 | 0 5 |
> | 14 | 0 |
>
> 10|6 means 10.6 ounces
>
> Find the sum of the weights and divide by 25 to find the mean. The mean is about 11.2 ounces.
>
> Since the median is the middle value, it is the 13th leaf on the plot. The median is about 11.0 ounces.
>
> The stem-and-leaf plot shows the modes by repeated digits for a particular stem. There are three 5s with 10 and three 0s with 11. The set is bimodal with 10.5 ounces and 11.0 ounces as the modes.

An annotation usually accompanies a stem-and-leaf plot to give meaning to the representation.

In a frequency distribution containing large amounts of data, each individual value in the set of data loses its identity. The data in each class are assumed to be uniformly distributed over the class. Thus, the class mark is assumed to be the mean of the data tallied in its class. For example, the mean of the data in a class with limits 22.5–27.5 is assumed to be 25, the class mark.

In a frequency distribution, the sum of the values in a class is found by multiplying the class mark X, by the frequency, f, of that class. The sum of all the values in a given set of data is found by adding the sums of the values of each class in the frequency distribution. The sum of all values in the set can be represented by $\sum_{i=1}^{k} (f_i \cdot X_i)$, where k is the number of classes in the frequency distribution. Thus, the arithmetic mean of n values in a frequency distribution is found by dividing the sum of the values in the set by n or an expression equivalent to n, such as $\sum_{i=1}^{k} f_i$.

Remember that the measures of central tendency are only representations of the set of data.

Mean of the Data in a Frequency Distribution

If $X_1, X_2, \ldots, X_k$ are the class marks in a frequency distribution with k classes and $f_1, f_2, \ldots, f_k$ are the corresponding frequencies, then the arithmetic mean, $\overline{X}$, can be approximated as follows.

$$\overline{X} = \frac{\sum\limits_{i=1}^{k}(f_i \cdot X_i)}{\sum\limits_{i=1}^{k} f_i}$$

Example 4

Find the mean of the scores of 90 students on a geometry test given in the following frequency distribution.

Class Limits	Class Marks (X)	Frequency (f_i)	$f_i X_i$
93–101	97	12	1164
85–93	89	19	1691
77–85	81	38	3078
69–77	73	11	803
61–69	65	7	455
53–61	57	3	171

$$\sum_{i=1}^{6} f_i = 90 \qquad \sum_{i=1}^{6}(f_i \cdot X_i) = 7362$$

$\overline{X} \approx \dfrac{7362}{90}$ or 81.8   The mean is approximately 81.8.

Class Limits	Frequency f	Cumulative Frequency
93–101	12	90
85–93	19	78
77–85	38	59
69–77	11	21
61–69	7	10
53–61	3	3

It is often helpful to calculate the cumulative frequency from the last interval to the first.

The median, M_d, of the data in a frequency distribution is found from a cumulative frequency distribution. The chart at the left shows the cumulative frequency for the data in Example 4. The cumulative frequency of each class is the sum of the frequency of the class and the frequencies of the previous classes. So, for the class limit 77–85, the cumulative frequency equals $38 + 11 + 7 + 3$ or 59. This means that 59 geometry test scores fall below the upper class limit of 85.

Since the median is the value below which (50%) of the data lie, the class in which the median lies can be located. This class is called the **median class**. The median can be found by using an estimation technique called *interpolation*. *This method can also be used to find any percent other than 50%.*

Example 5

Estimate the median of the data in the frequency distribution in Example 4.

Since there are 90 scores in this frequency distribution, 45 scores are below the median and 45 are above. From the chart on the previous page, find the least cumulative frequency that is greater than or equal to 45. That cumulative frequency is 59. So, the median class is 77–85.

You can use a proportion to find the value of M_d by finding the ratios of the differences in the cumulative frequencies and the upper limits of the classes.

$$59 - 21 = 38 \begin{bmatrix} \text{59 test scores lie below 85} \\ \begin{bmatrix} 45 - 21 = 24 \begin{bmatrix} \text{45 test scores lie below } M_d \\ \text{21 test scores lie below 77} \end{bmatrix} \end{bmatrix} \end{bmatrix} \begin{matrix} 85 - 77 = 8 \\ M_d - 77 = x \end{matrix}$$

$$\frac{38}{24} = \frac{8}{x}$$
$$x \approx 5.1$$

$$M_d - 77 = x$$
$$M_d - 77 \approx 5.1 \qquad x \approx 5.1$$
$$M_d \approx 82.1 \qquad \text{The median of the data is approximately 82.1.}$$

CHECKING FOR UNDERSTANDING

Communicating Mathematics

Read and study the lesson to answer each question.

1. **Write** an argument you, as an employee, would use in Example 2 to convince management that you deserve a raise because your salary is so far below the average.

2. **Write** the sigma notation for the mean of a set of data containing 25 values.

3. **Explain** how to make a stem-and-leaf plot for a set of data whose greatest value is 135 and least value is 82.

4. Which measure of central tendency is most affected by extreme values?

Guided Practice

5. **Make** a stem-and-leaf plot of the following ages of people attending a Paula Abdul concert.

| 15 | 55 | 35 | 46 | 28 | 35 | 25 | 17 | 30 | 30 | 27 | 35 |
| 15 | 25 | 25 | 20 | 20 | 15 | 20 | 17 | 15 | 25 | 10 |

Find the mean, median, and mode(s) for each set of data.

6. $\{3, 3, 6, 12, 3\}$

7. $\{10, 45, 58, 10\}$

8. $\{140, 150, 160, 170\}$

9. $\{6, 9, 11, 11, 12, 7, 6, 11, 5, 8, 10, 6\}$

10. The store manager of a discount department store is studying the weekly wages of her employees and the effects of an increase in health insurance. The chart at the right profiles her employees.

Weekly Wages	Frequency
$130–$140	11
$140–$150	24
$150–$160	30
$160–$170	10
$170–$180	13
$180–$190	8
$190–$200	4

   a. Find the sum of the wages in each class.

   b. Find the sum of all of the wages in the frequency distribution.

   c. Find the number of employees in the frequency distribution.

   d. Find the mean weekly wage in the frequency distribution.

   e. Make a column showing the cumulative frequency of the weekly wage classes.

   f. Find the median class of the frequency distribution.

   g. Find the median weekly wage in the frequency distribution.

EXERCISES

Practice   **11.** The table shows the numbers of live births (in millions) in the United States between 1930 and 1990.

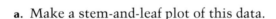

Year	1930	'35	'40	'45	'50	'55	'60	'65	'70	'75	'80	'85	'90
Live Births	2.6	2.4	2.6	2.9	3.6	4.1	4.3	3.8	3.7	3.1	3.6	3.8	4.2

   a. Make a stem-and-leaf plot of this data.

   b. Find the mean, median, and mode of the data.

   c. The term "baby boomers" is used to describe people born from 1946 to 1964. Explain how baby boomers are significant in the United States.

12. Emmanuel has scores of 84, 72, 91, 64, and 83. The class mean for the grading period was 80. How did Emmanuel compare?

13. Consider the heights of seven students chosen at random: 5'7", 4'8", 6'1", 5'4", 7'0", 6'7", and 5'4".

   a. Find the mean of this set of data.

   b. How representative of the data is the mean? Explain.

14. Crates of books are being stored for later use. The weights in pounds of the crates are 142, 160, 151, 139, 145, 117, 172, 155, and 124.

   a. Find the difference between the mean and the median of their weights.

   b. If each crate is reinforced with a 5-pound metal strip, how are the mean and median affected?

15. The Laketown Fitness Club is keeping a frequency table of how many times its members use the facilities in the club during a typical month. They organized their data into the following chart.

Visits	1–5	5–9	9–13	13–17	17–21	21–25	25–29	29–33
Members	2	8	15	6	38	31	13	7

    a. Find the mean of the data.
    b. Find the median class of the data.
    c. Find the median of the data.

16. Find the value of x so that the mean of $\{2, 4, 5, 8, x\}$ is 7.5.

17. Find the value of x so that the mean of $\{x, 2x - 1, 2x, 3x + 1\}$ is 6.

18. Find the value of x so that the median of $\{11, 2, 3, 3.2, 13, 14, 8, x\}$ is 8.

19. The mean height of five boys is 68 inches. If one boy is 5 feet tall and another is 6 feet tall, give an equation to describe the possible heights for the other three boys.

A 1-meter rod is suspended at its middle so that it balances. Suppose 1-gram weights are hung on the rod at the following distances from one end.

5 cm   20 cm   37 cm   44 cm   52 cm   68 cm   71 cm   85 cm

The rod does not balance at the 50-cm mark.

20. Where must one more 1-gram weight be hung so that the rod will balance at the 50-cm mark?

21. Where must a 2-gram weight be hung so that the rod will balance at the 50-cm mark?

Critical Thinking

22. The model below is a histogram created by letting coins represent the data. The letters A, B, and C represent the mean, median, and mode, but not necessarily in that order. Determine which is which.

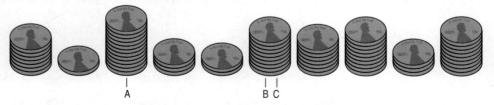

Applications and Problem Solving

23. **Testing**   The frequency distribution of the Verbal Scores of Females on the 1990 SAT test are shown below.

Scores	Number of Females (in thousands)	Scores	Number of Females (in thousands)
200–250	26.6	500–550	55.1
250–300	43.3	550–600	34.8
300–350	69.3	600–650	20.5
350–400	89.0	650–700	9.9
400–450	98.8	700–750	4.0
450–500	83.1	750–800	0.6

a. Find the mean of the data in the frequency distribution.

b. Make a column showing the cumulative frequency of the classes.

c. Find the median class of the frequency distribution.

d. Find the median of the data.

e. The published average verbal score for all females taking the SAT in 1990 was 422. Compare this score with your calculations and explain any differences that might exist.

24. **Sports**   The table lists the height and weight of the 11 professional basketball players who played on the Dream Team in the 1992 Summer Olympics in Barcelona.

Name	Height	Weight (lb)
Charles Barkley	6'6"	250
Larry Bird	6'9"	220
Clyde Drexler	6'7"	222
Patrick Ewing	7'0"	240
Earvin "Magic" Johnson	6'9"	220
Michael Jordan	6'6"	198
Karl Malone	6'9"	256
Chris Mullin	6'7"	215
Scottie Pippen	6'7"	210
David Robinson	7'1"	235
John Stockton	6'1"	175

a. Determine the mean, median, and mode of the heights of the Dream Team.

b. Determine the mean, median, and mode of the weights of the Dream Team.

c. Which measure(s) of central tendency is most representative of the data in parts a and b?

Mixed Review

25. Simplify $\dfrac{2 + i\sqrt{3}}{2 - i\sqrt{3}}$.   **(Lesson 9-5)**

26. Evaluate $\log_{10} 0.001$.   **(Lesson 11-4)**

27. Describe what is meant by iteration and give an example.   **(Lesson 13-1)**

28. **City Planning**   There are six women and seven men on the committee for city park enhancement. A subcommittee of five members is being selected at random to study the feasibility of redoing the landscaping in one of the parks. What is the probability that the committee will have at least three women?   **(Lesson 14-6)**

29. **Sports**   Refer to the application at the beginning of this lesson. Make a histogram of the weights of the women's tennis shoes. Use a class interval of 0.5 ounces. **(Lesson 15-1)**

30. **College Entrance Exam**   A rectangular sign is cut down by 10% of its height and 30% of its width. What percent of the original area remains?

15-3 Measures of Variability

Objectives After studying this lesson, you should be able to:
- find the mean deviation, semi-interquartile range, and standard deviation of a set of data, and
- organize and compare data using box-and-whisker plots.

Application Do you know what's in the food you eat? Today, people are more aware of the importance of eating a well-balanced diet than ever before. The chart below lists some common cereals and the number of calories and the grams of carbohydrates in a single serving of cereal (about $\frac{3}{4}$ cup). *You will use these data in Examples 1 through 4.*

Cereal Name	Calories	Carbohydrates (in grams)	Cereal Name	Calories	Carbohydrates (in grams)
Alpha-Bits	110	24	Alpen Natural Cereal with Raisins	440	80
Apple Jacks	110	26	Quaker 100% Natural Cereal	536	71
Cracklin' Oat Bran	330	57	Raisin Bran	200	44
Cookie Crisp	200	44	Rice Krispies	110	25
Cheerios	88	16	Shredded Wheat	165	35
Kellogg's Corn Flakes	110	24	Special K	88	17
Post Corn Flakes	88	19	Sugar Smacks	147	25
Froot Loops	110	25	Trix	110	25
Frosted Flakes	165	39	Wheat Chex	165	35
Hearty Life Fruit and Nut Natural 100	390	51			
Life	150	33			

Measures of central tendency, such as the mean, median, and mode of these data, are statistics that describe certain important characteristics of

data. However, they do not indicate anything about the variability of the data. For example, 40 is the mean of both {35, 40, 45} and {10, 40, 70}. The variability is much greater in the second set of data than in the first since 70 − 10 is much greater than 45 − 35.

One measure of variability is the **range**. The range is the difference of the greatest and least values in a set of data. In the table above, the range of calories is 536 − 88 or 448, and the range of carbohydrates is 80 − 16 or 64.

Suppose a set of data is rewritten in an array and the median is found, which separates the data into two groups. Then the median of each group is found, separating the data into four groups. Each of these groups is called a **quartile**. There are three quartile points, Q_1, Q_2, and Q_3, that denote the breaks in the data for each quartile. The median is the second quartile point Q_2. The medians of the two groups defined by the median are the first quartile point Q_1 and the third quartile point Q_3, respectively.

One fourth of the data is less than the first quartile point, Q_1, and three fourths of the data is less than the third quartile point, Q_3. The difference between the first quartile point and the third quartile point is called the **interquartile range**. When the interquartile range is divided by 2, the quotient is called the **semi-interquartile range**.

Semi-Interquartile Range	If a set of data has first quartile point, Q_1, and third quartile point, Q_3, the semi-interquartile range, Q_R, can be found as follows. $$Q_R = \frac{Q_3 - Q_1}{2}$$

Example 1

Refer to the application at the beginning of the lesson. Find the interquartile range and the semi-interquartile range of the calories in the breakfast cereals.

First, make an array.

Q_1 median Q_3
↓   ↓   ↓

88 88 88 110 110 110 110 110 110 147 150 165 165 165 200 200 330 390 440 536

$$Q_1 = \frac{110 + 110}{2} \text{ or } 110 \quad M_d = \frac{147 + 150}{2} \text{ or } 148.5 \quad Q_3 = \frac{200 + 200}{2} \text{ or } 200$$

The interquartile range is $200 - 110$ or 90. This means that half of the cereals are between 110 and 200 calories per serving and are within 90 calories per serving of each other. The semi-interquartile range is $\frac{90}{2}$ or 45. This means that one fourth of the cereals are within 45 calories per serving of each other.

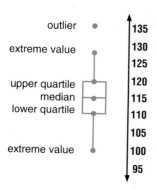

Box-and-whisker plots are used to summarize data and to illustrate the variability of the data. These plots graphically display the median, quartiles, interquartile range, and extreme values in a set of data. They can be drawn vertically, as shown at the left, or horizontally. A box-and-whisker plot consists of a rectangular box with the ends, or **hinges**, located at the first and third quartiles. The segments extending from the ends of the box are called **whiskers**. The whiskers stop at the extreme values of the set, unless the set contains **outliers**. Outliers are extreme values that are more than 1.5 of the interquartile range beyond the upper or lower quartiles. Outliers are represented by single points. If an outlier exists, each whisker is extended to the last value of the data that is not an outlier.

The dimensions of the box-and-whisker plot can help you characterize the data. Each whisker and each small box contains 25% of the data. If the whisker or box is short, the data are concentrated over a narrower range of values. The longer the whisker or box, the more diverse the data. Thus, the box-and-whisker is a pictorial representation of the variability of the data.

Example 2

Draw a box-and-whisker plot of the calorie data of the breakfast cereals.

In Example 1, you found that the median is 148.5, Q_1 is 110, and Q_3 is 200. The extreme values are the least value, 88, and the greatest value, 536.

Box-and-whisker plots are sometimes called box plots.

Draw a number line and plot the quartiles, the median, and the extreme values. Draw a box to show the interquartile range. Draw a segment through the median to divide the box into two smaller boxes.

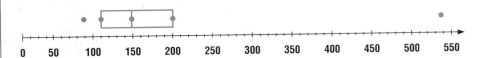

Before drawing the whiskers, determine if there are any outliers.

The interquartile range is $200 - 110$ or 90. An outlier is any value that lies more than $1.5(90)$ units below Q_1 or above Q_3.

$$Q_1 - 1.5(90) = 110 - 135 \qquad\qquad Q_3 + 1.5(90) = 200 + 135$$
$$= -25 \qquad\qquad\qquad\qquad\qquad\qquad = 335$$

The lower extreme, 88, is within the limits. However, 390, 440, and 536 are not within the limits. They are all outliers. Graph these points on the plot. Then draw the left whisker from 88 to 110 and the right whisker from 200 to the greatest value that is not an outlier, 330.

TECHNOLOGY *Tip*

You can use the TI-83 graphing calculator to create a box-and-whisker plot that will plot outliers. These are called **ModBoxplot** (modified boxplot) in the user's guide.

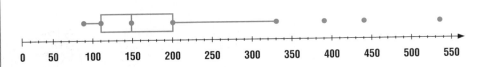

The box-and-whisker plot shows that the three lower quartiles of data are fairly close together in range. However, the upper quartile of data is more diverse.

Another measure of variability can be found by examining deviations from the mean, symbolized by $X_i - \overline{X}$. The sum of the deviations from the mean is zero. That is, $\sum_{i=1}^{n}(X_i - \overline{X}) = 0$. For example, the mean of the data set $\{13, 20, 17, 12, 38\}$ is 20. The sum of the deviations from the mean is shown in the chart at the right.

X_i	$\overline{X}$	$X_i - \overline{X}$
13	20	–7
20	20	0
17	20	–3
12	20	–8
38	20	18
$\sum(X_1 - \overline{X})$		0

To indicate how far individual items vary from the mean, we use the absolute values of the deviations. The arithmetic mean of the absolute values of the deviations from the mean of a set of data is called the **mean deviation**, symbolized by MD.

Mean Deviation	**If a set of data has n values given by X_i, such that $1 \leq i \leq n$, with arithmetic mean $\overline{X}$, then the mean deviation MD can be found as follows.** $$MD = \frac{1}{n} \sum_{i=1}^{n} |X_i - \overline{X}|$$

In sigma notation for statistical data, i is always an integer.

Example 3

FYI...

A body-builder should eat the same high-carbohydrate diet as a marathoner. Exercise builds muscle, but to have enough energy to exercise, you need carbohydrates.

Find the mean deviation of the grams of carbohydrates found in breakfast cereals.

There are 20 cereals listed and the mean is 35.75 grams of carbohydrates.

$$MD = \frac{1}{20} \sum_{i=1}^{20} |X_i - 35.75|$$

$$= \frac{1}{20}(|24 - 35.75| + |26 - 35.75| + \ldots + |25 - 35.75| + |35 - 35.75|)$$

$$= \frac{1}{20}(|\text{-}11.75| + |\text{-}13.75| + \ldots + |\text{-}10.75| + |\text{-}0.75|)$$

$$= 13.575$$

The mean deviation of the grams of carbohydrates is 13.575 grams. This means that the number of grams of carbohydrates per serving is an average of about 13.575 grams above or below the mean of 35.75.

A measure of variability that is often associated with the arithmetic mean is the **standard deviation**. Like the mean deviation, the standard deviation is a measure of the average amount by which individual items of data deviate from the arithmetic mean of all the data. Each individual deviation can be found by subtracting the arithmetic mean from each individual value, $X_i - \overline{X}$. Some of these differences will be negative, but since they are squared, the results are positive. The standard deviation is the square root of the mean of the squares of the deviations from the arithmetic mean. *The square of the standard deviation, called the variance, is often used as a measure of variability.*

Standard Deviation	**If a set of data has n values, given by X_i such that $1 \leq i \leq n$, with arithmetic mean $\overline{X}$, the standard deviation, σ, can be found as follows.** $$\sigma = \sqrt{\frac{1}{n} \sum_{i=1}^{n} (X_i - \overline{X})^2}$$

σ is the Greek lowercase letter sigma.

The standard deviation is the most important and widely used measure of variability.

Example 4

Refer to the application at the beginning of the lesson. Find the standard deviation of the grams of carbohydrates in the breakfast cereals.

The mean is 35.75 and there are 20 cereals.

$$\sigma = \sqrt{\frac{1}{20}[(24 - 35.75)^2 + (26 - 35.75)^2 + \cdots + (25 - 35.75)^2 + (35 - 35.75)^2]}$$

$$= \sqrt{294.7875}$$

$$\approx 17.169 \qquad \textit{Use your calculator.}$$

You can also use a TI-81 graphing calculator to find the standard deviation of this set of data. *Refer to the Exploration in Lesson 15-1 for a review of data input.*

- Clear the data memory.
- Input the new data. The data are not frequency based. The values for X are the data and the value for Y is always 1.
- Press 2nd STAT. Select 1:1 - Var. 1-Var will appear on the home screen. Press ENTER.
- The calculated mean, $\overline{X}$, appears on the first line.
- The calculated standard deviation, σX, appears on the next to last line.

The mean is 35.75 and the standard deviation is 17.16937681, which agree with the calculations using the formula.

When studying the standard deviation of a set of data, it is important to consider the mean. For example, compare a standard deviation of 50 with a mean of 100 to a standard deviation of 50 with a mean of 10,000. The latter indicates very little variation, while the former indicates a great deal of variation since 50 is 50% of 100 while 50 is only 0.5% of 10,000.

The standard deviation of a frequency distribution is the square root of the mean of the squares of the deviations of the class marks from the frequency.

Standard Deviation of the Data in a Frequency Distribution

If $X_1, X_2, \cdots, X_k$ are the class marks in a frequency distribution with k classes, and $f_1, f_2, \cdots, f_k$ are the corresponding frequencies, then the standard deviation σ of the data in the frequency distribution is found as follows.

$$\sigma = \sqrt{\frac{\sum\limits_{i=1}^{k} (X_i - \overline{X})^2 \cdot f_i}{\sum\limits_{i=1}^{k} f_i}}$$

The standard deviation of a frequency distribution is an approximate number.

Example 5

Use the frequency distribution data below to find the arithmetic mean and the standard deviation of the number of home runs scored by the National League home run leaders between 1901 and 1991.

To use the TI-81 graphing calculator to find the standard deviation of a frequency distribution, let X be the class mark and let Y be the frequency. Then follow the same procedures as in Example 4.

Class Limits	Class Marks	f	$f \cdot X$	$(X - \overline{X})$	$(X - \overline{X})^2$	$(X - \overline{X})^2 \cdot f$
0–10	5	6	30	–29.01	841.5801	5049.4806
10–20	15	12	180	–19.01	361.3801	4336.5612
20–30	25	10	250	–9.01	81.1801	811.801
30–40	35	26	910	0.99	0.9801	25.4826
40–50	45	31	1395	10.99	120.7801	3744.1831
50–60	55	6	330	20.99	440.5801	2643.4806
		91	3095			1,6610.9891

The mean, $\overline{X}$, is $\dfrac{3095}{91}$ or 34.01.

The standard deviation is $\sqrt{\dfrac{16,610.9891}{91}}$ or approximately 13.51.

Since the mean number of home runs is 34.01 and the standard deviation is 13.51, this indicates a great amount of variability in the data.

CHECKING FOR UNDERSTANDING

Communicating Mathematics

Read and study the lesson to answer each question.

1. **Determine** if there are any outliers for the carbohydrate data in the application at the beginning of the lesson. If they exist, state them.

2. **Explain** how calculating the standard deviation of data in a frequency table differs from calculating the standard deviation of data not in a frequency table.

3. **Demonstrate** how to make a box-and-whisker plot for a set containing the numbers 1 through 12.

4. Refer to Example 3. Suppose that the mean deviation of the grams of carbohydrates found in another sample of breakfast cereal was 3.5 with a mean of 37.75. How is this data different from that in Example 3?

Guided Practice

Use the box-and-whisker plot below for Exercises 5—11.

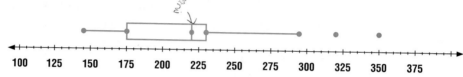

5. What is the range of the data?

6. What is the median of the data?

7. What percent of the data is greater than the third quartile point?

8. What values are outliers of the data?

9. What percent of the data is less than 220?

10. What percent of the data is greater than 175?

11. What percent of the data is greater than 175 and less than 230?

12. The table shows the resulting scores of the final exam in history for 90 students at Madrid High School.

Scores	53–61	61–69	69–77	77–85	85–93	93–101
Frequency	3	7	11	38	19	12

a. Find the mean of the test scores.

b. Find the median class of the distribution.

c. Find the median of the test scores.

d. Find the standard deviation of the test scores.

EXERCISES

Applications and Problem Solving

13. **Entertainment**   The best selling pre-recorded videos of all time as of 1991 are listed in the table. The number of units sold are in millions.

Video	Distributor	Units Sold
E.T. the Extra-Terrestrial	MCA	15.1
Batman	Warner	11.5
Bambi	Disney	10.5
The Little Mermaid	Disney	9.0
Teenage Mutant Ninja Turtles	LIVE	8.8
Who Framed Roger Rabbit?	Disney	8.5
Cinderella	Touchtone	8.5
Peter Pan (animated)	Disney	7.6
Pretty Woman	Paramount	6.2
Honey, I Shrunk the Kids	Disney	5.8

a. Find the mean of the units sold.

b. Find the deviation from the mean for each value in the set of data.

c. Find the mean deviation of the units sold.

d. Find the values of Q_1, Q_2, and Q_3 for the data.

e. Find the interquartile range of the data.

f. Find the semi-interquartile range for the data.

g. Find the sum of the squares of the individual deviations for the data.

h. Find the standard deviation of the data.

i. Make a box-and-whisker plot of the data.

j. State any outliers.

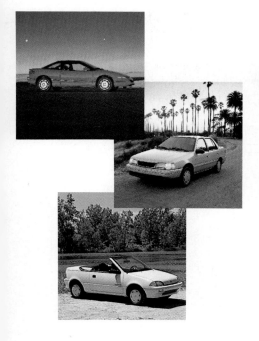

The miles per gallon for city travel and the fuel tank capacity in gallons of twelve 1991-model cars are given below.

Cars	MPG	Fuel Tank Capacity	Cars	MPG	Fuel Tank Capacity
Geo Metro	46	10.6	Eagle Summit	31	13.2
Honda Civic CX	42	11.9	Nissan Sentra E	29	13.0
Hyundai Excel GS	29	11.9	Ford Escort	30	11.9
Mazda 323	29	13.2	Ford Festiva GL	35	10.0
Plymouth Sundance	26	14.0	Subaru Justi GL	33	9.2
Saturn SL	28	12.8	Toyota Tercel	32	11.9

14. **Energy** Let Group 1 = {Geo Metro, Honda Civic CX, Hyundai Excel GS, Mazda 323, Plymouth Sundance, Saturn SL}. Let Group 2 = {Eagle Summit, Nissan Sentra E, Ford Escort, Ford Festiva GL, Subaru Justi GL, Toyota Tercel}.

    a. Find the mean, median, and the standard deviation of the miles per gallon of the cars in Group 1.

    b. Find the mean, median, and the standard deviation of the miles per gallon of the cars in Group 2.

  c. Make a box-and-whisker plot of each group of miles per gallon data.

    (1) How do the means of the two groups of data compare?

    (2) How do the standard deviations compare?

    (3) Which group of data has the smaller variability in miles per gallon?

15. **Energy** Let Group 3 = {Geo Metro, Mazda 323, Plymouth Sundance, Eagle Summit, Ford Festiva GL, Subaru Justi GL}. Let Group 4 = {Honda Civic CX, Hyundai Excel GS, Saturn SL, Nissan Sentra E, Ford Escort, Toyota Tercel}.

  a. Find the mean, median, and the standard deviation of the fuel tank capacities of the cars in Group 3.

  b. Find the mean, median, and the standard deviation of the fuel tank capacities of the cars in Group 4.

  c. Make a box-and-whisker plot of the two groups of data.

    (1) How do the means of the two groups compare?

    (2) How do the standard deviations compare?

    (3) Which group has the greater variability in fuel tank capacity?

16. **Sports** The chart at the right shows the frequency of runs batted in (RBI) by the American League batting leaders between 1907 and 1991.

  a. Find the mean of the data.

  b. Find the standard deviation.

  c. In 1991, the RBI champion in the American League was Cecil Fielder of the Detroit Tigers with 133 RBIs. Write a sentence to compare this number with the mean of the data.

RBI	Frequency
70–90	2
90–110	11
110–130	39
130–150	17
150–170	9
170–190	7

17. **Sports** The medals won by the top 20 countries at the 1992 Summer Olympics at Barcelona, Spain, are shown below.

Country	Gold	Silver	Bronze	Total
Unified Team	45	38	29	112
United States	37	34	37	108
Germany	33	21	28	82
China	16	22	16	54
Cuba	14	6	11	31
Hungary	11	12	7	30
South Korea	12	5	12	29
France	8	5	16	29
Australia	7	9	11	27
Spain	13	7	2	22
Japan	3	8	11	22
Britain	5	3	12	20
Italy	6	5	8	19
Poland	3	6	10	19
Canada	6	5	7	18
Romania	4	6	8	18
Bulgaria	3	7	6	16
Netherlands	2	6	7	15
Sweden	1	7	4	12
New Zealand	1	4	5	10

a. Find the range, mean, mean deviation, interquartile range, and the standard deviation of the number of gold medals.

b. Find the range, mean, mean deviation, interquartile range, and the standard deviation of the number of silver medals.

c. Which set of data demonstrated greater variability: the gold medals or the silver medals? Explain what this means.

18. Is it possible for the variance to be less than the standard deviation for a set of data? If so, explain when this would occur.

19. When would the variance for a set of data be equal to the standard deviation of the set?

20. Find all possible rational roots of $2x^3 + 3x^2 - 8x + 3 = 0$. **(Lesson 4-4)**

21. Aviation Two airplanes leave an airport at the same time. Each flies at a speed of 110 mph. One flies in the direction 60° east of north. The other flies in the direction 40° east of south. How far apart are the planes after 3 hours? **(Lesson 5-6)**

22. Find the value of x to the nearest tenth such that $x = e^{0.346}$. **(Lesson 11-3)**

23. Find the value of x so that the set $\{x, 2x - 1, 2x, 3x + 1\}$ has a mean of 6. **(Lesson 15-2)**

24. College Entrance Exam Choose the best answer.
In the figure at the right, the largest possible circle is cut out of a square piece of tin. The total area of the remaining pieces of tin is approximately

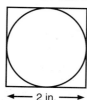

← 2 in. →

(A) 0.14 in²   **(B)** 0.75 in²   **(C)** 0.86 in²   **(D)** 1.0 in²   **(E)** 3.14 in²

CASE STUDY FOLLOW-UP

Refer to Case Study 3: *The Legal System*, on pages 962–965.

The Census Bureau reported a United States population of 248.7 million in 1990.

1. About how many of the crimes committed in 1990 occurred in the South?

2. Approximate the mean annual number of crimes per 100,000 population in the period from 1981 to 1990.

3. Find the standard deviation of the above set of data.

4. Research Compile a set of data relating to crime in your city. Use statistics to analyze the data and report your analysis in a 1-page paper.

15-4 The Normal Distribution

Objective

After studying this lesson, you should be able to:
- use the normal distribution curve.

Application

One of the many uses of census data is by special interest groups that target certain populations for their particular cause. Suppose a group wanted to know the proportion of the states in the United States (including the District of Columbia) that had large populations of people of Hispanic origin. The table below shows portions of a frequency distribution of data from the 1990 census. The graph beside it shows a histogram and frequency polygon of those states having less than 1,000,000 people of Hispanic origin.

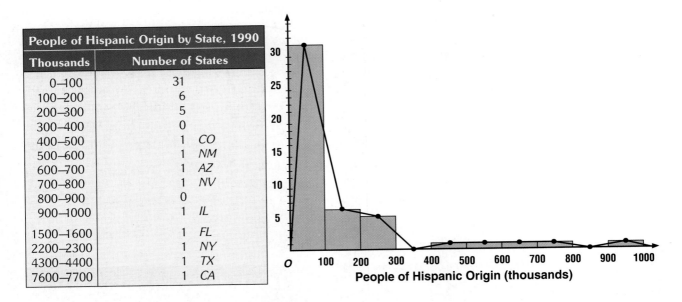

People of Hispanic Origin by State, 1990

Thousands	Number of States	
0–100	31	
100–200	6	
200–300	5	
300–400	0	
400–500	1	CO
500–600	1	NM
600–700	1	AZ
700–800	1	NV
800–900	0	
900–1000	1	IL
1500–1600	1	FL
2200–2300	1	NY
4300–4400	1	TX
7600–7700	1	CA

People of Hispanic Origin (thousands)

A frequency polygon displays a limited number of data and does not represent an entire population. To display the frequency of an entire population, a smooth curve is used rather than a polygon.

If the curve is symmetric, then information about the measures of central tendency can be gathered from the graph. Study the graphs below.

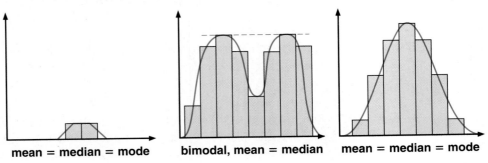

mean = median = mode   bimodal, mean = median   mean = median = mode

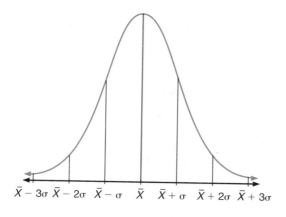

$\overline{X} - 3\sigma \ \ \overline{X} - 2\sigma \ \ \overline{X} - \sigma \ \ \ \ \overline{X} \ \ \ \ \overline{X} + \sigma \ \ \ \overline{X} + 2\sigma \ \ \overline{X} + 3\sigma$

A **normal distribution** is a frequency distribution that often occurs when there is a large number of values in a set of data. The graph of this distribution is a symmetric, bell-shaped curve, shown at the left. This is known as a **normal curve**. The shape of the curve indicates that the frequencies in a normal distribution are concentrated around the center portion of the distribution. A small portion of the population occurs at the extreme values.

In a normal distribution, small deviations are much more frequent than larger ones. Negative deviations and positive deviations occur with the same frequency. The points on the horizontal axis represent values that are a certain number of standard deviations from the mean, $\overline{X}$. In the curve shown above, each interval represents one standard deviation. So, the section from σ to $\overline{X} + \sigma$ represents those values between the mean and one standard deviation greater than the mean, the section from $\overline{X} + \sigma$ to $\overline{X} + 2\sigma$ represents the interval one standard deviation greater than the mean to two standard deviations greater than the mean, and so on. The total area under the normal curve and above the horizontal axis represents the total probability of the distribution, which is 1.

Example 1

APPLICATION

Quality Control

The lifetimes of 10,000 watch batteries are normally distributed. The mean lifetime is 500 days. The standard deviation is 60 days. Sketch a normal curve that represents the frequency of lifetimes of the batteries.

First, find the values defined by the standard deviation in a normal distribution.

$\overline{X} - 1\sigma = 500 - 1(60)$ or 440 $\overline{X} + 1\sigma = 500 + 1(60)$ or 560

$\overline{X} - 2\sigma = 500 - 2(60)$ or 380 $\overline{X} + 2\sigma = 500 + 2(60)$ or 620

$\overline{X} - 3\sigma = 500 - 3(60)$ or 320 $\overline{X} + 3\sigma = 500 + 3(60)$ or 680

Sketch the general shape of a normal curve. Then, replace the horizontal scale with the values you have calculated.

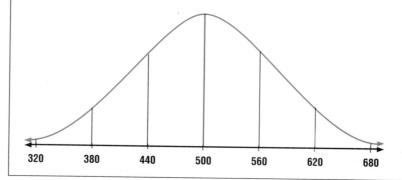

The tables below give the fractional parts of a normally distributed set of data for selected areas about the mean. The letter t represents the number of standard deviations from the mean (that is, $\overline{X} \pm t\sigma$). When $t = 1$, t means 1 standard deviation above and below the mean.

P represents the fractional part of the data that lies in the interval $\overline{X} \pm t\sigma$. The percent of the data within these limits is $100P$.

t	P	t	P	t	P	t	P
0.0	0.000	0.9	0.632	1.7	0.911	2.5	0.988
0.1	0.080	1.0	0.683	1.8	0.928	2.58	0.990
0.2	0.159	1.1	0.729	1.9	0.943	2.6	0.991
0.3	0.236	1.2	0.770	1.96	0.950	2.7	0.993
0.4	0.311	1.3	0.807	2.0	0.955	2.8	0.995
0.5	0.383	1.4	0.838	2.1	0.964	2.9	0.996
0.6	0.451	1.5	0.866	2.2	0.972	3.0	0.997
0.7	0.516	1.6	0.891	2.3	0.979	3.5	0.9995
0.8	0.576	1.65	0.900	2.4	0.984	4.0	0.9999

The P value also corresponds to the probability that a randomly selected member of the sample lies within t standard deviation units of the mean. For example, suppose the mean of a set of data is 60 and the standard deviation is 6.

Boundaries: $\overline{X} - t\sigma$ to $\overline{X} + t\sigma$

$$60 - t(6) \text{ to } 60 + t(6)$$
$$60 - 1(6) \text{ to } 60 + 1(6)$$
$$54 \quad \text{to} \quad 66$$

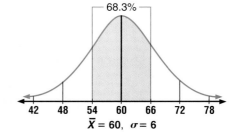

68.3% of the values in this set of data lie within 1 standard deviation of 60, that is, between 54 and 66.

If you randomly select one item from the sample, the probability that the one you pick will be between 54 and 66 is 0.683. If you repeat this process 1000 times, approximately 68.3% (about 683) of those selected will be between 54 and 66.

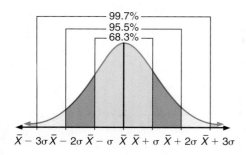

Thus, normal distributions have the following properties.

1. The maximum point of the curve is at the mean.

2. About 68.3% of the data are within one standard deviation from the mean.

3. About 95.5% of the data are within two standard deviations from the mean. *When $t = 2.0$, $P = 0.955$.*

4. About 99.7% of the data are within three standard deviations from the mean. *When $t = 3.0$, $P = 0.997$.*

Example 2

Refer to the information in Example 1. Estimate how many watch batteries will last for each of the following intervals.
a. 440—560 days

a. The interval 440–560 represents $\overline{X} \pm 1\sigma$, which represents a probability of 68.3%.

68.3%(10,000) = 6830

Approximately 6830 batteries in the sample will last for 440–560 days.

b. 380—620 days

b. The interval 380–620 represents $\overline{X} \pm 2\sigma$, which represents a probability of 95.5%.

95.5%(10,000) = 9550

Approximately 9550 batteries in the sample will last 380–620 days.

c. 320—680 days

c. The interval 320–680 represents $\overline{X} \pm 3\sigma$, which represents a probability of 99.7%.

99.7%(10,000) = 9970

Approximately 9970 batteries in the sample will last 320–680 days.

The TI-83 graphing calculator can draw normal curves given the mean and standard deviation. You can also shade the area between an interval of the curve and the percent of the area shaded will be displayed.

If you know the mean and the standard deviation, you can find a range of values for a given probability.

Example 3

Find the upper and lower limits of an interval about the mean within which 80% of the values of a set of normally distributed data can be found if $\overline{X} = 65$ and $\sigma = 6$.

Use the tables to find the value of t that most closely approximates $P = 0.80$. For $t = 1.3, P = 0.807$. Choose $t = 1.3$. Now find the limits.

$$\overline{X} \pm t\sigma = 65 \pm (1.3)(6) \qquad \overline{X} = 65, t = 1.3, \text{ and } \sigma = 6$$
$$= 72.8 \text{ and } 57.2$$

The interval in which 80% of the data lies is 57.2–72.8.

If you know the mean and standard deviation, you can also find the probability that a certain value lies within a given range of values.

Example 4

A person is selected at random from a population of retired people whose ages are normally distributed. The mean age of this population is 65, and the standard deviation is 5 years. What is the probability that the age of the person selected lies in the interval from 61 to 69 years of age?

Write each of the limits in terms of the mean.

$61 = 65 - 4$ and $69 = 65 + 4$

Therefore, $\overline{X} \pm t\sigma = 65 \pm 4$ and $t\sigma = 4$. Solve for t.

$$t\sigma = 4$$
$$t(5) = 4$$
$$t = 0.8$$

(continued on the next page)

If $t = 0.8$, then $P = 0.576$.   *Use the tables.*

The probability that the age of a retired person selected at random lies in the interval 61–69 is 57.6%.

Many colleges and universities require their entering students to take a college entrance exam, such as the SAT(Scholastic Aptitude Test), College Boards, and ACT(American College Test). Because of the large numbers of students tested each year, testing institutions assume that the test scores form a normal distribution.

Example 5

APPLICATION

Academics

The scores on a standardized college entrance examination form a normal distribution with a mean score of 480 and a standard deviation of 72. Nationally, what is the probability that a senior chosen at random scores between 516 and 600 on the mathematics portion of the test?

The graph shows that 516–600 does not define an interval that can be represented by $\overline{X} \pm t\sigma$. However, the interval can be defined as the difference between the intervals 480–600 and 480–516.

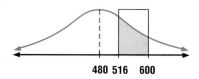

First, find the probability that the score is between the mean, 480, and the upper limit, 600.

$$\overline{X} + t\sigma = 600$$
$$480 + t(72) = 600$$
$$t \approx 1.7$$

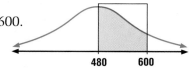

The value of P that corresponds to $t = 1.7$ is 0.911.

$P = 0.911$ describes the probability that a student's score falls $\pm 1.7(72)$ points about the mean, or between about 358 and 602, but we are only considering half that interval. So, the probability that a student's score is between 480 and 600 is $\frac{1}{2}(0.911)$ or about 0.456.

Next, find the probability that the score is between the mean and the lower limit, 516.

$$\overline{X} + t\sigma = 516$$
$$480 + t(72) = 516$$
$$t = 0.5$$

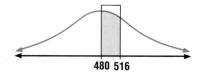

For $t = 0.5, P = 0.383$.

Likewise, we will only consider half of this probability or 0.192.

Now find the probability that a student's score falls in the interval 480–600.
$$P(516 \text{ to } 600) = P(480 \text{ to } 600) - P(480 \text{ to } 516)$$
$$P = 0.456 - 0.192$$
$$= 0.264 \text{ or } 26.4\%$$

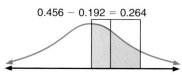

0.456 − 0.192 = 0.264

The probability that a student's score is between 516 and 600 is 26.4%.

CHECKING FOR UNDERSTANDING

Communicating Mathematics

Read and study the lesson to answer each question.

1. **Describe** the shape of the graph of a normal distribution.

2. **Explain** in your own words what a normal distribution is.

3. **Sketch** a normal distribution curve to describe the scores on a physics test that ranged from 56 to 96 and had an average score of 89. The mode is also the mean and 70% of the scores are in the interval around the mean.

Guided Practice

4. Sketch a normal curve with a mean of 75 and a standard deviation of 10. Label the points that would be 1, 2, and 3 standard deviations from the mean.

5. Sketch a normal curve with a mean of 75 and a standard deviation of 5. Label the points that would be 1, 2, and 3 standard deviations from the mean.

6. Compare the graphs in Exercises 4 and 5. Which curve displays less variability?

Suppose 200 values in a set of data are normally distributed.

7. How many values are within one standard deviation of the mean?

8. How many values are within two standard deviations of the mean?

9. How many values fall in the interval between the mean and one standard deviation above the mean?

10. How many values are between two and three standard deviations less than the mean?

A set of 500 values is normally distributed with a mean of 24 and a standard deviation of 2.

11. What percent of the data is in the interval 22–26?

12. What percent of the data is in the interval 20.5–27.5?

13. Find the range about the mean that includes 95% of the data.

14. Find the range about the mean that includes 50% of the data.

EXERCISES

Practice

A set of data is normally distributed with a mean of 140 and a standard deviation of 20. Assume values are selected at random.

15. What percent of the data is in the interval 120–160?

16. What percent of the data is between 130 to 150?

17. Find the probability that a value from the data lies in the interval 110–170.

18. Find the probability that a value from the data is less than 100.

19. Find the probability that a value from the data is greater than 160.

20. Find the probability that a value from the data lies in the interval 110–200.

21. Find the interval about the mean that includes 90% of the data.

22. Find the limit below which 90% of the data lies.

The volumes of the liquid content of soft drink cans are normally distributed. The mean volume is 355 mL with a standard deviation of 2 mL.

23. What percent of the cans contain between 353 mL and 357 mL?

24. What percent of the cans contain between 353 and 355 mL?

25. What percent of the cans contain more than 351mL?

26. Out of 1000 cans, how many will contain between 349 mL and 363 mL?

The lifetimes of a certain type of car tire are normally distributed. The mean lifetime is 40,000 miles, and the standard deviation is 5000 miles. Answer each of the following questions for a sample of 10,000 of these tires.

27. How many tires will last between 35,000 miles and 45,000 miles?

28. How many tires will last between 30,000 and 40,000 miles?

29. How many tires will last less than 40,000 miles?

30. How many tires will last more than 50,000 miles?

31. How many tires will last less than 25,000 miles?

32. The burning time of a synthetic fireplace log is normally distributed with a mean of 5 hours and a standard deviation of 15 minutes. A store has 1000 logs to sell. How many of these logs will burn longer than 4 hours 30 minutes?

Critical Thinking

33. The grading scale at a local university is 90–100 = A, 80 – 89 = B, 70 – 79 = C, 60 – 69 = D, and below 60 = F. The calculus class has just taken a test. Will the grades be normally distributed? Write a paragraph to defend your answer and draw a curve to represent the distribution.

Applications and Problem Solving

34. **Obstetrics** The longest baby born in the United States was a boy, born to Mrs. Anna Bates in Seville, Ohio, in 1879. The baby weighed 23.75 pounds and was about 30 inches long. However, the lengths of the 2400 babies born at Valley Hospital in the past year were normally distributed. The average length was 19.88 inches and the standard deviation was 1.42 inches.

   a. Find the probability that the length of a baby selected at random is less than 17.72 inches.

   b. What percent of the babies were between 18.9 and 19.88 inches long?

   c. What percent of the babies were between 20.47 and 21.65 inches long?

   d. How many standard deviations from the mean birth length at Valley Hospital was the length of Mrs. Bates' baby?

35. **Evaluation** Teachers sometimes change the grading scale for a given test because of student performance. At Jennings High School, the physics teacher decided to grade a test "on the curve." The teacher defined the grades in the following way: A(grades above $\overline{X} + 1.5\sigma$), B(grades from $\overline{X} + 0.5\sigma$ to $\overline{X} + 1.5\sigma$), C(grades from $\overline{X} + 0.5\sigma$ to $\overline{X} - 0.5\sigma$), D(grades from $\overline{X} - 0.5\sigma$ to $\overline{X} - 1.5\sigma$), and F(grades below $\overline{X} - 1.5\sigma$). What percent of the class will receive each letter grade if the grades are normally distributed?

Mixed Review

36. Find matrix X in $\begin{bmatrix} 3 & 7 \\ 8 & 9 \end{bmatrix} X - \begin{bmatrix} 2 & 7 \\ 6 & 9 \end{bmatrix} = \begin{bmatrix} 3 & 6 \\ 9 & 6 \end{bmatrix}$. **(Lesson 2-3)**

37. Verify that $\sin^2 \phi \cot^2 \phi = (1 - \sin \phi)(1 + \sin \phi)$ is an identity. **(Lesson 7-2)**

38. Write the equation of a circle that is tangent to the x-axis and has its center at $(2, -4)$. **(Lesson 10-1)**

39. A pair of dice is thrown. Find the probability that their sum is greater than 7 if both dice show the same number. **(Lesson 14-7)**

40. Nutrition The amounts of sodium, in milligrams, present in the top brands of peanut butter are given below.

195 210 180 225 255 225 195 225 203 225 195 195 188 191
210 233 225 248 225 210 240 180 225 240 180 225 240 240
195 189 178 255 225 225 225 194 210 225 195 188 205

a. Make a box-and-whisker plot of the data.
b. Write a paragraph describing the variability of the data. **(Lesson 15-3)**

41. College Entrance Exam Choose the best answer.
The arithmetic mean of a series of numbers is 20 and their sum is 160. How many numbers are in the series?

(A) 8   **(B)** 16   **(C)** 32   **(D)** 48   **(E)** 80

MID-CHAPTER REVIEW

The following list contains the all-time record high temperature (in degrees Fahrenheit) for each of the 50 states.

112 100 127 120 134 118 105 110 109 112 100 118 117
116 118 121 114 114 105 109 107 112 114 115 118 117
118 122 106 110 116 108 110 121 113 120 119 111 104
111 120 113 120 117 105 110 118 112 114 114

1. Make a frequency distribution of the data above. **(Lesson 15-1)**

2. Make a histogram of the frequency distribution. **(Lesson 15-1)**

3. Make a stem-and-leaf plot of the data. **(Lesson 15-2)**

4. Find the mean, median, and mode of the data. **(Lesson 15-2)**

5. Make an array of the data. **(Lesson 15-3)**
a. Find the upper and lower quartile marks.
b. Find the range of the data.
c. Make a box-and-whisker plot of the data.

6. Find the standard deviation of the data. **(Lesson 15-3)**

7. Assume that the data are normally distributed. Find the interval in which you would expect about 68% of the data to lie. **(Lesson 15-4)**

15-5 Sample Sets of Data

Objective

After studying this lesson, you should be able to:
■ find the standard error of the mean to predict the true mean of a population with a certain level of confidence.

Application

As part of his physiology project, Ward Quan must determine the mean height of all 1285 male high school seniors attending school in his school district. His resources and the amount of time that he has to collect the measurements of the seniors are limited. The 1285 senior males are the total population. The population in a statistical study is all of the items or individuals in the group being considered. Rarely will a researcher find 100% of a population accessible as a source of data. Therefore, a random sample of the population is selected so that it is representative of the entire population.

The characteristics of the population pertinent to the study should be found in the sample in about the same ratio as they exist in the total population. When Ward selects the seniors in the sample, there should be males that are tall, short, and of medium height from each of the senior high schools in his school district. Then, based upon a random sample of the population, certain inferences can be made about the population in general. The major purpose of **inferential statistics** is to use the information gathered in a sample to make predictions about a population.

Ward decides that the number of male seniors in his sample should be 100. So that the sample is a random sample of all of the male seniors in the district, he decides to measure the 100 male seniors that were selected for the All-District Band. All senior high schools are represented in proportion to their enrollment of boys. He assumes that musical talent has no relationship to height, and, therefore, that the sample is random with respect to height.

At a band rehearsal, the heights of the 100 males are measured to the nearest tenth of an inch. After tabulating the data, the mean and standard deviation are computed for that sample. Ward found that the mean height of the sample was 69.2 inches with a standard deviation of 2.7 inches.

Ward is not sure that the mean height of this one sample is a truly representative mean height of all male seniors. He knows that there is an Honor Society convention scheduled for the next weekend. There are also 100 senior males from his school district attending the conference in which all senior high schools are represented in proportion to their enrollment of boys. He again measures the 100 seniors, and this time he finds that the mean height of the sample is 68.4 inches with a standard deviation of 2.4.

μ is the lowercase Greek letter mu.

The discrepancies that Ward found in his two samples are common when taking random samples. Large companies and statistical organizations often take hundreds of samples to find the "average" they are seeking. For this reason, a sample mean is assumed to be near its true population mean, symbolized by μ. The standard deviation of the distribution of the sample means is known as the **standard error of the mean**.

Standard Error of the Mean	**If a sample set of data has N values and σ is the standard deviation, the standard error of the mean, $\sigma_{\overline{X}}$, is** $$\sigma_{\overline{X}} = \frac{\sigma}{\sqrt{N}}.$$

The symbol $\sigma_{\overline{X}}$ is read "sigma sub x bar."

The standard error of the mean is a measurement of how well a sample mean selected at random estimates the true mean. If the standard error of the mean is small, then the sample means are closer to (or approximately the same value as) the true mean. If the standard error of the mean is large, many of the sample means would be far from the true mean. The greater the number of items or subjects in the sample, the closer the sample mean reflects the true mean and the smaller the standard error of the mean.

Example 1

> **Refer to the application at the beginning of the lesson. The mean height of Ward's first sample containing 100 senior males was 69.2 inches and the standard deviation was 2.7 inches. Determine the standard error of the mean.**
>
> For the sample, $N = 100$. Find $\sigma_{\overline{X}}$.
>
> $$\sigma_{\overline{X}} = \frac{2.7}{\sqrt{100}} \text{ or } 0.27$$
>
> The standard error of the mean for Ward's first sample was 0.27.
>
> This standard error is very small and indicates that Ward's first mean is probably very close to the true mean.

The sample mean is only an estimate of the true mean of the population. Sample means of various random samples of the same population are normally distributed about the true mean with the standard error of the mean as a measure of their variability. Thus, the standard error of the mean behaves like the standard deviation. Using the standard error of the mean and the sample mean, we can state a range about the sample mean in which we think the true mean lies. Probabilities of the occurrence of sample means and true means may be determined by referring to the tables on page 832.

Example 2

The mean height of Ward's first sample containing 100 senior males was 69.2 inches, the standard deviation was 2.7 inches, and, from Example 1, the standard error of the mean is 0.27. Determine the range of heights such that the probability is 90% that the mean height of the entire population of senior males lies within that range.

When $P = 0.90$ or 90%, the value of $t = 1.65$. *Refer to the tables on page 832.*

To find the range, use a technique similar to finding the interval for a normal distribution.

$$\overline{X} \pm t\sigma_{\overline{X}} = 69.2 \pm (1.65)(0.27)$$
$$= 68.7545 \text{ and } 69.6455$$

The probability is 90% that the true mean, μ, is within the range of about 68.75 to 69.65 inches.

Portfolio

Select an item from this chapter that you feel shows your best work and place it in your portfolio. Explain why you selected it.

The probability of the true mean being within a certain range of a sample mean may be expressed as a **level of confidence**. The most commonly used levels of confidence are 1% and 5%. A 1% level of confidence means that there is less than a 1% chance that the true mean differs from the sample mean by a certain amount. That is, you are 99% confident that the true mean is within a certain range of the sample mean. A 5% level of confidence means that the probability of the true mean being within a certain range of the sample mean is 95%.

If a higher level of confidence is desired for the same number of values, accuracy must be sacrificed by providing a wider range. However, if the number of values in the sample is larger, the range for a given level of confidence is smaller.

Example 3

APPLICATION

Quality Control

The longevity of light bulbs is normally distributed. It is your job to determine the mean lifetime of the light bulbs being produced in your factory. You are to take a sample of light bulbs, test them, and record how many hours each light bulb lasts. Then you will calculate the mean lifetime. You wish to have a 1% level of confidence that the range containing the mean lifetime of the sample also contains the true mean.

a. One hundred light bulbs are randomly selected and illuminated. The time for each bulb to burn out is recorded. From this sample, the average life is 350 days with a standard deviation of 45 days. Determine the range of the sample mean that has a 1% level of confidence.

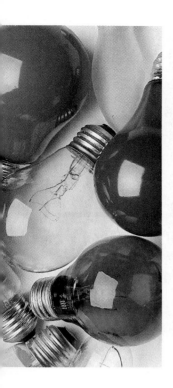

b. One thousand light bulbs are randomly selected and their lifetimes are recorded. From this sample, the average life is 350 days with a standard deviation of 45 days. Determine the range of the sample mean that has a 1% level of confidence.

c. Compare the results of parts a and b.

a. A 1% level of confidence is given when $P = 99\%$.

When $P = 0.99$, $t = 2.58$.

Find $\sigma_{\overline{X}}$. $\quad \sigma_{\overline{X}} = \dfrac{45}{\sqrt{100}}$ or 4.5

Find the range. $\quad \overline{X} = 350 \pm (2.58)(4.5)$
$$= 338.39 \text{ and } 361.61$$

Thus, the range of the sample means is 338.39 days to 361.61 days, for a 1% level of confidence.

b. Determine the range. As in part a, $t = 2.58$.

Find $\sigma_{\overline{X}}$. $\quad \sigma_{\overline{X}} = \dfrac{45}{\sqrt{1000}}$ or 1.42

Find the range. $\quad \overline{X} = 350 \pm (2.58)(1.42)$
$$= 346.34 \text{ and } 353.66$$

Thus, the range of the sample means is 346.34 days to 353.66 days, for a 1% level of confidence.

c. By increasing the number of items in the sample, the range has decreased substantially, from about 23 days to about 8 days.

CHECKING FOR UNDERSTANDING

Communicating Mathematics

Read and study the lesson to answer each question.

1. **Explain** the difference between a population and a sample.

2. **State** which group you might sample if you wished to collect data to determine the average number of hours of classes in which a college freshman enrolls.
   a. students in the advanced physics classes
   b. students who are also on the college football team
   c. students in the Tuesday and Thursday beginning composition classes
   d. students from the third floor of the freshman girls dormitory
   Explain your answer.

3. **Define** the standard error of the mean.

4. **Describe** what a 1% level of confidence means.

Guided Practice

Find the standard error of the mean for each sample. Then find the range of the sample mean that has a 1% level of confidence and the range for a 5% level of confidence.

5. $\sigma = 40, N = 64, \overline{X} = 200$

6. $\sigma = 5, N = 36, \overline{X} = 45$

7. In a random sample of 100 families in Waco, Texas, a television reporter found that children watched an average of 4.6 hours of television programming a day. The standard deviation was 1.4 hours.

   a. Find the standard error of the mean.

   b. Find the range about the sample mean that has a 1% level of confidence.

   c. Find the range about the sample mean that reflects a 50% chance that the true mean lies within that range.

   d. Find the range about the sample mean so that a probability of 0.90 exists that the true mean will lie within the range.

EXERCISES

Practice   **Find the standard error of the mean for each sample. Then find the range for a 1% level of confidence and the range for a 5% level of confidence.**

8. $\sigma = 2.4, N = 100, \overline{X} = 24$ 9. $\sigma = 12, N = 200, \overline{X} = 80$

10. Grapefruit were first planted in the United States in 1820. Grapefruit, unlike oranges, are native to the western hemisphere. Suppose 50 grapefruit were gathered at random from a grove of premium grapefruit trees. The mean diameter of the sample is 16.2 cm with a standard deviation of 1.4 cm.

    a. Find the standard error of the mean.

    b. Find the range about the sample mean that has a 5% level of confidence.

    c. Find the range about the sample mean that gives a 99% chance that the true mean lies within that range.

    d. Find the range about the sample mean such that the probability is 0.80 that the true mean lies within it.

11. The table shows a frequency distribution of the time in minutes required for a shopper to stand in line at a local supermarket. The distribution is a random sample from 2361 shoppers on a given weekend.

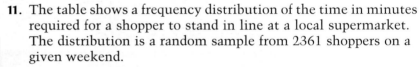

Number of Minutes	4	6	8	10	12	14	16	18	20
Number of Shoppers	1	3	5	12	17	13	7	4	2

    a. Find the standard deviation of the data in the frequency distribution.

    b. Find the standard error of the mean.

    c. Find the range about the sample mean such that the probability is 0.95 that the true mean lies within the range.

    d. Find the probability that the mean of the population will be less than one minute from the mean of the sample.

12. The standard deviation of the weights of 36 seven-year-olds in the United States is 8 pounds. What is the probability that the mean weight of the random sample will differ by more than one pound from the mean weight of all seven-year-olds?

13. Eighty-one beef cattle were fed a special diet for six weeks. The mean gain in weight was 40 pounds with a standard deviation of 5 pounds. With what level of confidence can it be said that the diet will cause an average gain of between 30 to 50 pounds in a similar group?

14. Describe an industrial situation in which decreasing the range for a given level of confidence by increasing the sample size would be prohibitive.

15. Manufacturing The tensile strength of a certain type of wire is normally distributed with a mean of 99.8 and a standard deviation of 5.48.
  a. Find the standard error of the mean for a sample of 100 wires.
  b. What is the probability that the mean of this sample will be between 98.8 and 100.9?

16. Business A steel manufacturing firm employs 1500 people. During a given year, the mean amount contributed to a charity drive for the homeless was $25.75 per employee with a standard deviation of $5.25. What is the probability that a random sample of 100 employees yields a mean between $25 and $27?

17. Employment An employment agency has found that the mean time required for an applicant to take an aptitude test is 24.5 minutes, with a standard deviation of 4.5 minutes. The time required to take the test is recorded on the applicant's file. If a random sample of 81 applicants' files are pulled, what is the probability that the mean time for test-taking in this sample is greater than 25 minutes?

18. Manufacturing Icon, Inc. manufactures two types of computer graphics cards, Model 28 and Model 74. There are three stations, A, B, and C, on the assembly line. The assembly of Model 28 graphics card requires 30 minutes at station A, 20 minutes at station B, and 12 minutes at station C. Model 74 requires 15 minutes at station A, 30 minutes at station B, and 10 minutes at station C. Station A can be operated for no more than 4 hours a day, station B can be operated for no more than 6 hours a day, and station C can be operated for no more than 8 hours. **(Lesson 2-6)**
  a. If the profit on Model 28 is $100 and on Model 74 is $60, how many of each model should be assembled each day to provide maximum profit?
  b. What is the maximum daily profit?

19. Transportation The change in the direction of a highway is given by the central angle θ, subtended by the arc of the curve. A curve on Interstate 64, modeled by an arc of a circle, is 500 feet long. The radius of the circle that the curve follows is 2500 feet. Find the change in direction of the highway in degrees formed by the curve. **(Lesson 5-2)**

20. Find the sum of the first ten terms of the series $1^3 + 2^3 + 3^3 + \cdots$. **(Lesson 12-8)**

21. Find the probability that a value selected at random from a set of data will be greater than 160 if the data are normally distributed, the mean is 140, and the standard deviation is 20. **(Lesson 15-4)**

22. College Entrance Exam Choose the best answer.
  Which of the following numbers is the least?

  (A) $\sqrt{3}$ **(B)** $\dfrac{1}{\sqrt{3}}$ **(C)** $\dfrac{\sqrt{3}}{3}$ **(D)** $\dfrac{1}{3}$ **(E)** $\dfrac{1}{3\sqrt{3}}$

15-6A Graphing Calculators: Scatter Plots and Regression Lines

Graphs are often used to organize and display data. We can use information obtained from graphs in many different ways, and we can display the data on many different graphs or plots as well. You have already studied several types of graphs and plots that are used to display data. Yet another type of plot is called a **scatter plot**. A scatter plot is used when data will *almost* form a straight line when the data points are graphed. In this case, a **best-fit line**, or **regression line**, can be drawn and used to approximate solutions to other problems.

Example 1

Main sequence stars are stars that are in the neighborhood of the sun. Draw a scatter plot and regression line for the data about the radius of eight main sequence stars that are given in the following table.

Stars	MU-1 Scorpii	Sirius A	Altair	Polycon A	Sun	61 Cygni A	Krueger 60	Barnard's Star
Surface Temp. in °C (in thousands)	20.0	10.2	7.3	6.8	5.9	2.6	2.8	2.7
Radius (in Suns)	5.2	1.9	1.6	2.6	1.0	0.7	0.35	0.15

First, set the window parameters. Use the viewing window [0, 25] by [0, 6] with scale factors of 5 for the *x*-axis and 1 for the *y*-axis.

Next, deselect any equations in the Y=list, clear List 1 and List 2, and clear the graphics screen.

Enter: [2nd] [Y-VARS] 5 2 [ENTER] [STAT] 4 [2nd] [L1] [,] [2nd] [L2]
[ENTER] [2nd] [DRAW] 1 [ENTER]

Now, input your data. List 1 will contain the surface temperatures, and List 2 will contain the radii.

Enter: [STAT] 1 20 [ENTER] 10.2 [ENTER] 7.3 [ENTER] ... 2.7 [ENTER]
[▶] 5.2 [ENTER] 1.9 [ENTER] 1.6 [ENTER] ... .15 [ENTER]

Now draw the scatter plot and regression line.

Enter: [2nd] [STAT PLOT] 1.   *Chooses the statplot menu.*

Highlight "On", the scatter plot, "L1", "L2", and "●".
[GRAPH]   *Draws the scatter plot.*
[STAT] [▶] 9 [ENTER]   *Calculates coefficients of regression line.*
[Y=] [VARS] 5 [▶] [▶] 7   *Writes the equation of the regression line.*
[GRAPH]   *Graphs the regression line.*

Notice that when you calculated the coefficients of the regression line by pressing $\boxed{\blacktriangleright}$ $\boxed{\text{STAT}}$ 5 $\boxed{\text{ENTER}}$, you were calculating the slope and y-intercept of the regression line. The value a is the y-intercept of the regression equation, and b is the slope of the regression equation. The value r is the correlation value or the value that tells you how well the regression line "fits" the data. In this case, with $r = 0.9534$, this is a good fit and shows a positive correlation.

Example 2

Use the TRACE function on the regression line in Example 1 to predict the radius of a main sequence star that has a surface temperature of 12,500 °C.

First activate the TRACE function.

Enter: $\boxed{\text{TRACE}}$

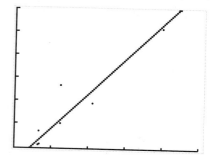

You can use the TRACE function to find that a main sequence star with a surface temperature of about 12,500 °C will have a radius of about 3.1 suns based on the regression line.

When finished, remember to turn off the statplot.

EXERCISES

Use a graphing calculator to draw a scatter plot and a regression line for the data in the following tables.

1.

x	1	2	3	4	5
y	0.4	1.1	1.9	3.2	4.4

2.

x	-4	-1	0	0.5	3	3.6
y	20	12.2	4	-0.3	-3	-6.8

3. The following table shows the average heights in feet of 30 men and 30 women at various ages.

Age (yr)	1	3	5	10	12	15	18	20	22
Men	2.4	3.2	3.8	4.5	4.8	5.3	5.7	5.9	6.0
Women	2.5	3.3	3.7	4.4	4.9	5.2	5.3	5.4	5.5

a. Use a graphing calculator to draw a scatter plot and regression line to show how age is related to the height of the 30 men.
b. Predict the height of one of the men at age 7.
c. Use a graphing calculator to draw a scatter plot and regression line to show how age is related to the height of the 30 women.
d. Predict the height of one of the women at age 17.
e. What do you think would happen if the data included ages beyond 22?

15-6 Scatter Plots

Objectives

After studying this lesson, you should be able to:
- draw and analyze scatter plots,
- draw regression lines, and
- compute correlation values to determine goodness of fit.

Application

Companies spend hundreds of thousands of dollars each year in focus groups to discover if consumers like their products. Do you ever wonder to what extent you like the same things as someone else? One device that is used to determine how groups of data are related is a **scatter plot**. Such a diagram visually shows the nature of a relationship. The table below shows how five students ranked a list of 10 of the most popular films to rent. Each person ranked the 10 films from 1 to 10, with their favorite film ranked 1.

Rating of Favorite Films					
Film	Lei	Maria	Brian	Lucia	Randy
Anastasia	10	4	10	1	4
As Good As It Gets	4	8	5	7	3
Batman & Robin	7	7	7	3	10
Blues Brothers 2000	8	1	9	4	9
Flubber	5	10	4	6	2
For Richer or Poorer	3	5	3	8	6
Home Alone 3	1	6	2	10	1
Men in Black	6	2	6	5	5
Spice World	9	3	8	2	8
Titanic	2	9	1	9	7

As you look at the rankings, it is difficult to make any quick comparisons. A statistician who is looking for relationships among these rankings may construct a scatter plot of the data.

Each scatter plot shown below illustrates the relationship between two people's rankings. The ordered pair for each point represents the rating of a single film by two people. For example, the ordered pair (1, 2) represents Lei and Brian's ranking of *Home Alone 3*.

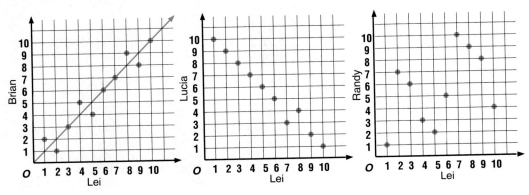

The left scatter plot on the previous page suggests a linear relationship. Notice that several of the points are on the line, with the rest being very close to it. These rankings are said to have a strong positive relationship, or correlation. In other words, Lei and Brian have similar tastes in films. Note that the slope of the linear pattern in this scatter plot is positive.

The middle scatter plot also has a linear pattern. However, the slope is negative. Thus, the correlation between the ranking is negative, or opposite. It could be said that Lei and Lucia have very different, or opposite, tastes in films. The films that Lei likes are those that Lucia dislikes, and vice versa.

The points in the right scatter plot are very scattered. The points have little or no correlation. Lei and Randy like some of the same movies and dislike some of the same movies. However, they also have differing opinions on other films.

In many cases, the analysis of data involves exactly two variables. When this occurs, the data are said to be **bivariate**. Bivariate data can be examined through the use of scatter plots. When data appear to be linear in a scatter plot, it is often helpful to fit a line to the data in order to define the apparent linear relationship between the two variables. Statisticians often study this relationship in the hope that it may be used to make predictions about one of the variables.

One informal technique used to fit a line is to choose several points on a line that is suggested by the pattern of points in the scatter plot. Then find the equation of that line. This equation is called a **prediction equation** for the relationship. Its graph is often referred to as the **best-fit line**, or the line of best fit. Techniques used to find lines of best fit are called **regression methods**.

Example 1

Find a prediction equation for the line of regression representing the ranking of films by Lei and Brian.

Select the points at $(6, 6)$ and $(10, 10)$.
Determine the slope.

$$m = \frac{y_2 - y_1}{x_2 - x_1} \qquad \textit{Defintion of slope}$$

$$= \frac{10 - 6}{10 - 6} \qquad \textit{(x}_1, y_1) = (6, 6), (x_2, y_2) = (10, 10)$$

$$= \frac{4}{4} \text{ or } 1$$

Now use one of the ordered pairs, such as $(10, 10)$, and the slope in the point-slope form of an equation.

$$y - y_1 = m(x - x_1) \qquad \textit{Point-slope form of an equation}$$
$$y - 10 = 1(x - 10) \qquad \textit{(x}_1, y_1) = (10, 10) \textit{ and } m = 1$$
$$y = x$$

Thus, a prediction equation is $y = x$. So you may predict that Lei's rating of films will generally be the same as Brian's.

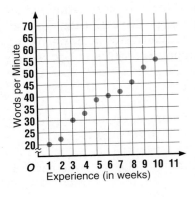

A more accurate method for determining a line of best fit is to find a **median-fit line**. Consider the data about an adult typing class and its graph shown below.

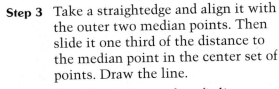

Experience (in weeks)	1	2	3	4	5	6	7	8	9	10
Typing Speed (in wpm)	20	22	30	33	38	40	42	46	52	55

Experience has shown that this trend does not continue much beyond 10 weeks. Some students would go on to excel by reaching speeds of 80 or more words per minute. However, these are exceptions to the results usually found in the typing class. By finding the median-fit line, we can approximate the linear relationship that might exist for this set of data.

Example 2

Find the median-fit line for the typing data given above. Then make a statement about how representative the line is of the data.

Step 1 Use two vertical lines to separate the data into three sets of equal size, if possible. The outer sets should have the same number of points in them.

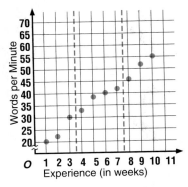

Step 2 Locate a point in each set whose *x*- and *y*-coordinates represent the medians of the *x*- and *y*-coordinates of all the points in that set. This can be done by examining the points in the set, both vertically and horizontally. Notice that in this example, the middle points in the three sets have been marked with a "+" to indicate the median point.

The points are not necessarily points in the set of data.

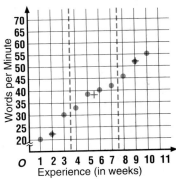

Step 3 Take a straightedge and align it with the outer two median points. Then slide it one third of the distance to the median point in the center set of points. Draw the line.

This line is the median-fit line.

These points cluster very closely to the median-fit line, so it is very representative of the data.

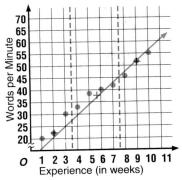

Data that are linear in nature will have varying degrees of **goodness of fit** to the lines of fit. Various formulas are often used to find a correlation value that describes the nature of the data. The more closely the data fit a line, the closer the correlation value, r, approaches 1 or –1. Positive correlation values are associated with linear data having positive slopes, and negative correlation values are associated with negative slopes. Thus, the more linear the data, the more closely the correlation value approaches 1 or –1.

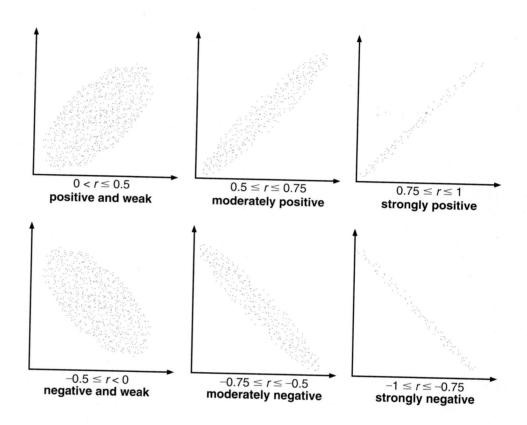

With the aid of technology, correlation values can easily be computed. The TI-81 graphing calculator uses the **Pearson product-moment correlation**. The Pearson product-moment correlation is labeled r and is the ratio of the degree of variation of X and Y together to the degree of the separate variation of X and Y.

The formula that is used to determine the Pearson product-moment correlation is $r = \dfrac{SP}{\sqrt{SS_x SS_y}}$, where $SP = \sum(X - \overline{X})(Y - \overline{Y}) = \sum XY - \dfrac{(\sum X)(\sum Y)}{n}$, $SS_X = \sum(X - \overline{X})^2$, $SS_Y = \sum(Y - \overline{Y})^2$, and n is the number of items being ranked. You can see how software or a graphing calculator that are programmed to compute this formula makes the use of correlation values much easier.

Example 3

Refer to the application at the beginning of the lesson. Use a graphing calculator to find the Pearson product-moment correlation for the film ratings of Lei and Maria.

Clear the statistical memory.

Enter: [STAT] 4 [2ND] [L1] [,] [2ND] [L2] [ENTER]

Set the window.

XMIN: 0   *XMAX: 11*   *XSCL: 1*   *YMIN: 0*   *YMAX: 11*   *YSCL: 1*

Input the data. List 1 will be the rankings of Lei, and List 2 will be the rankings of Maria.

Enter: [STAT] 1 10 [ENTER] 4 [ENTER] 7 [ENTER] ... 2 [ENTER]
[▶] 4 [ENTER] 8 [ENTER] 7 [ENTER] ... 9 [ENTER]

Now draw the scatter plot and regression line.

Enter: [2nd] [STAT PLOT] 1.

Highlight "On", the scatter plot, "L1", "L2", and "•".
[GRAPH]
[STAT] [▶] 9 [ENTER]
[Y=] [VARS] 5 [▶] [▶] 7
[GRAPH]

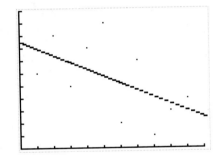

Find the value of r.

Enter: [STAT] [▶] 5 [ENTER]

On the screen you will see the display shown at the right.

```
LINREG
 Y= AX+B
 A= -0.5515151515
 B= 8.533333333
 R= -0.5515151515
```

MATH JOURNAL

Write a few sentences to tell how changing the scale on either axis might make the scatter plot appear differently.

The Pearson product-moment correlation value is about −0.55. The correlation between the rankings of Lei and Maria is moderately negative.

CHECKING FOR UNDERSTANDING

Communicating Mathematics

Read and study the lesson to answer each question.

1. **Explain** two methods for finding the line of best fit for a given set of paired data.
2. **Explain** what the Pearson product-moment correlation means in relation to the graph of the ordered pairs.

Guided Practice

Determine whether you would expect the scatter plot of each set of data to show a *positive*, *negative*, or *no correlation* between the variables.

3. heights of mothers (m) and heights of daughters (d)
4. age of a television set (a) and its current value (v)
5. weight of a person (w) and calories consumed (c)
6. height of a person (h) and the person's birth month (b)
7. money spent each week (m) and money saved each week (s)
8. driving skill (d) and experience (e)

9. The Wakeside Deli Corporation was informed by its marketing department that if they spend $5 million on product development, sales will reach $100 million, and if they spend $12 million on product development, sales will be $149 million. The marketing department said they based this information on a prediction equation in which p represented product development and s represented sales.

   a. Find the slope of a prediction equation.
   b. Find the s-intercept of a prediction equation.
   c. Find a prediction equation.
   d. What would you predict sales would be if $3 million is spent on product development?

Make a sketch of bivariate data that has each Pearson product-moment correlation value.

10. near 1.0

11. near –1.0

12. near 0

EXERCISES

13. The number of chirps per minute that one cricket makes as the temperature changed during a period of several hours is recorded below.

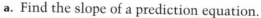

Chirps per minute	105	105	125	125	130	149	153	152	164	171	175
Temperature (°C)	18	19	20	21	21	23	24	24	24	26	26

   a. Draw a scatter plot and find a median-fit line for the data.
   b. Find the equation of the line of regression.

(continued on the next page)

c. What is the Pearson product-moment correlation value?

d. What does the correlation value in Exercise 13c mean?

At the regional music festival, ten of the top teenage pianists competed for a music scholarship. The chart below shows how the three judges ranked the ten competitors.

Contestant	A	B	C	D	E	F	G	H	I	J
Judge #1	1	3	5	4	2	9	10	6	8	7
Judge #2	2	4	6	3	1	10	9	5	7	8
Judge #3	9	8	6	7	10	1	2	5	3	4

14. Consider the scores awarded by judges 1 and 2.

a. Draw a scatter plot to show how these rankings are related.

b. Describe the relationship.

c. Determine the Pearson product-moment correlation value for the data.

15. Consider the scores awarded by judges 2 and 3.

a. Draw a scatter plot to show how these rankings are related.

b. Describe the relationship.

c. Determine the Pearson product-moment correlation value for the data.

d. Consider the correlation values in Exercises 14c and 15c. How would you analyze the consistency of the three judges?

16. A garment manufacturer wants to know the relationship between the age and annual maintenance costs of the sewing machines in their factory. A sample of 16 machines reveals the following information.

Age (years)	8	3	1	9	5	7	5	2
Maintenance Costs ($)	109	75	21	135	67	125	71	52

Age (years)	1	3	6	2	1	2	6	8
Maintenance Costs ($)	25	70	126	58	30	47	120	105

a. Draw a scatter plot to show how age is related to maintenance cost.

b. Describe the relationship.

c. Find an equation for the line of regression.

d. Determine the Pearson product-moment correlation value for the data.

e. What does the correlation value in part d mean?

Use a graphing calculator to draw the scatter plot and line of regression for each of the following exercises.

17. Exercise 13   **18.** Exercise 14   **19.** Exercise 15   **20.** Exercise 16

21. Refer to Example 3. Three values appear on the screen of the TI-81 graphing calculator when you access the linear regression menu. What do the values of *a* and *b* represent?

Critical
Thinking

22. Different correlation values are acceptable for different situations. For each situation give a specific example and explain your reasoning.

   a. When would a correlation value less than 0.99 be considered unsatisfactory?

   b. When would a correlation value of 0.60 be considered as good?

   c. When would a strong negative correlation value be desirable?

Applications
and Problem
Solving

23. Nutrition   The table below deals with the nutritional value of America's favorite snack foods, as reported by the USDA.

Snack Food	% Calories from Fat	% Calories from Carbohydrates
Apple (medium, Delicious)	9%	89%
Bagel (plain)	6%	76%
Banana (medium)	2%	93%
Bran Muffin (large)	40%	53%
Fig Newtons (4 cookies)	18%	75%
PowerBar (any flavor)	8%	76%
Snickers bar (regular size)	42%	51%
Ultra Slim-Fast Bar (one)	30%	63%

a. Draw a scatter plot to show the relationship of the percentage of the calories from fat and the percentage of calories from carbohydrates.

b. Find an equation for the line of regression.

c. Describe the relationship.

d. Find the Pearson product-moment correlation for the data.

e. In your own words, write what this correlation means.

24. Health The table below shows the ideal weight for each body frame for men and women in relationship to their height.

Ideal Weights for Women			
Height	Small Frame	Medium Frame	Large Frame
4'10"	95	100	111
4'11"	98	102	114
5'0"	101	107	117
5'1"	104	110	120
5'2"	106	113	124
5'3"	109	115	126
5'4"	112	119	129
5'5"	115	122	134
5'6"	118	127	137
5'7"	122	131	141
5'8"	126	135	145
5'9"	130	139	149
5'10"	135	143	154
5'11"	139	147	159
6'0"	143	151	163

Ideal Weights for Men			
Height	Small Frame	Medium Frame	Large Frame
5'2"	126	134	144
5'3"	129	137	147
5'4"	132	140	150
5'5"	135	143	154
5'6"	139	147	157
5'7"	143	152	162
5'8"	147	156	167
5'9"	151	159	171
5'10"	155	163	175
5'11"	159	168	180
6'0"	163	172	185
6'1"	167	177	190
6'2"	173	181	195
6'3"	176	186	199
6'4"	180	191	205

a. Draw scatter plots to compare the height to the ideal weight for each frame group for men and women.

b. Find an equation for the line of regression for each plot.

c. Find the Pearson product-moment correlation for each plot.

Mixed Review

25. Find $[f \circ g](4)$ and $[g \circ f](4)$ for $f(x) = 5x + 9$ and $g(x) = 0.5x - 1$. **(Lesson 1-2)**

26. Photography A photographer observes a 35-foot totem pole that stands vertically on a uniformly-sloped hillside and the shadow cast by it at different times of day. At a time when the angle of elevation of the sun is $37°12'$, the shadow of the pole extends directly down the slope. This is the effect that the photographer is seeking. If the hillside has an angle of inclination of $6°40'$, find the length of the shadow. **(Lesson 5-6)**

27. Find the first five iterates for $f(x) = 3x^2 - 10$ for $x_0 = 3.4$. **(Lesson 13-1)**

28. Botany A random sample of fifty acorns from an oak tree in the park reveals a mean diameter of 16.2 mm and a standard deviation of 1.4 mm. Find the range about the sample mean that gives a 99% chance that the true mean lies within it. **(Lesson 15-5)**

29. College Entrance Exam Choose the best answer.
Suppose every letter in the alphabet has a number value that is equal to its place in the alphabet: the letter A has a value of 1, B a value of 2, and so on. The number value of a word is obtained by adding up the values of the letters in the word and then multiplying the sum by the number of letters of the word. The "word" *DFGH* would have a number value of

(A) 22   **(B)** 44   **(C)** 66   **(D)** 100   **(E)** 108

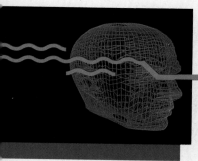

Technology

Median-Fit Lines

In Lesson 15-6, you learned to graph scatter plots and find the prediction equation that can describe a line of regression for the set of data. *Data Insights*, a software product from Sunburst, Inc., will graph the data you input as ordered pairs and then draw a line suggested by the points, if one exists. *Data Insights* will also give you the equation of the line, which it calls the **median-fit line**.

You can also customize your graph by choosing the scale increments that you want for your axes and the titles that you wish to appear.

The data below include the year and the millions of students in elementary and secondary schools in the United States. The graph done by *Data Insights* shows the graphed data and the median-fit line. Notice that *Data Insights* labels the axes for you.

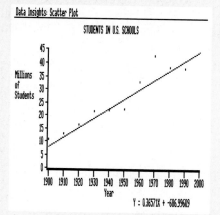

Year	Millions of Students
1900	10.6
1910	12.6
1920	16.2
1930	21.3
1940	22.0
1950	22.3
1960	32.5
1970	42.5
1980	38.2
1990	38.0

In the median-fit line equation, $y = 0.36571x + (-686.99609)$, x represents the year and y represents the millions of students.

EXERCISES

1. Use the median-fit line given above and your calculator to estimate the number of students in each year.
   a. 1922   b. 1955   c. 1978   d. 1991   e. 2005

2. Use *Data Insights* to enter the data below for the number of calories and grams of fat for certain bread products. What is the equation of the median-fit line comparing calories to grams of fat?

Item	Calories	Fat (g)	Item	Calories	Fat (g)
Corn tortilla	45	1	Plain bagel	150	1
Croissant	240	14	Pita bread	63	1
English muffin	120	1	Wheat bread	60	1
Flour tortilla	90	2	White bread	60	1
Oat-bran bagel	179	2	Honey wheat bread	70	1

15-6B   Graphing Calculators: Curve Fitting

The table below shows population data from the U. S. Census for the years 1790 through 1990. A scatter plot of the data is shown below the table.

Years since 1780	Population (in millions)	Years since 1780	Population (in millions)
10	3.9	120	76.2
20	5.3	130	92.2
30	7.2	140	106.1
40	9.6	150	123.2
50	12.8	160	132.2
60	17.0	170	151.3
70	23.2	180	179.3
80	31.4	190	203.3
90	38.6	200	226.5
100	50.2	210	248.7
110	63.0		

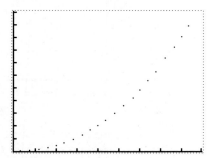

Rather than a linear scatter plot, a curved plot results. Much of the real-world data that are collected and analyzed are nonlinear in nature. A set of data that best fits a curve is called **curvilinear data**. Since most real-world data are positive, most scatter plots are graphs in the first quadrant. If the points were connected, the graph would resemble that of an exponential graph of the form $y = Ae^{Bx}$, or that of a polynomial graph of the form $y = Ax^B$.

The graphs of exponential functions and polynomial functions have similar characteristics in the first quadrant. As a result, it is difficult to distinguish between a quadratic, cubic, or an exponential function upon casual observation. Since it is too difficult to determine what type of function a scatter plot may resemble simply by looking at the data points, it is best to try both equations and see which one fits best.

The TI-82 graphing calculator will draw exponential and polynomial regression curves.

Example 1

APPLICATION

Population

Use the population data at the beginning of the lesson to determine the exponential equation that best fits the data and describe its goodness of fit.

First, set the window parameters. The values of data suggest that we use the viewing window [0, 225] by [0, 275] with scale factors of 25 for both axes.

Next, clear List 1 and List 2, and the graphics screen. Also clear the first equation from the Y= list.

Enter: STAT 4 2ND L1 , 2ND L2 ENTER 2ND DRAW 1 ENTER
Y= CLEAR ENTER

Then, input your data. List 1 will be the years since 1780, and List 2 will be the population in millions.

Enter: STAT 1 10 ENTER 20 ENTER 30 ENTER ... 210 ENTER
▶ 3.9 ENTER 5.3 ENTER 7.2 ENTER ... 248.7 ENTER

Now draw the scatter plot and regression curve or line.

Enter: 2nd STAT PLOT 1.

Highlight "On", the scatter plot, "L1", "L2", and "•".

GRAPH
STAT ▶ ALPHA A ENTER
Y= VARS 5 ▶ ▶ 7
GRAPH

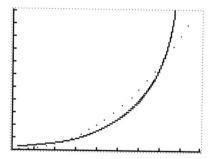

On the TI-82 graphing calculator, the equation for the regression curve is $y = 4.721 \times 1.0210^x$, where $a = A$, and $b = e^B$. Pressing LN VARS 5 ▶ ▶ 2 will give the value of B. The equation then can be written as $y = 4.721e^{0.0208x}$.

Looking at the third line of the viewing window shows the Pearson product-moment correlation value of +0.9836. From this, we can conclude that the regression curve is an extremely good fit to the data.

To determine if a polynomial equation would better fit the data in Example 1, you can press ▶ ALPHA STAT B ENTER and check the value *r* on the TI-82 graphing calculator. For Example 1, the exponential regression equation gives a better fit to the curve.

Example 2

The data in the table at the right were recorded in a physics class at West Chicago High School. The table shows the length of a pendulum (in meters) and the time (in seconds) it took to make the pendulum swing back and forth one time. Find the polynomial equation that best fits the data and describe its goodness of fit.

Time	Length	Time	Length
0.453	0.05	1.419	0.50
0.628	0.10	1.552	0.60
0.776	0.15	1.679	0.70
0.891	0.20	2.005	1.00
1.100	0.30	2.196	1.20
1.270	0.40	2.371	1.40

First, set the window parameters. Use the viewing window [0, 3] by [0, 2] with scale factors of .25 for both axes. Next, clear List 1 and List 2, and the graphics screen. Clear the first equation from the Y = list also.

Enter: STAT 4 2ND L1 , 2ND L2 ENTER 2ND DRAW 1 ENTER
Y= CLEAR ENTER

Then, input your data. List 1 will be the time, and List 2 will be the length.

Enter: STAT 1 .453 ENTER .628 ENTER .776 ENTER ... 2.371 ENTER
▶ .05 ENTER .1 ENTER .15 ENTER ... 1.4 ENTER

Now draw the scatter plot and regression curve or line.

Enter: 2nd STAT PLOT 1.

Highlight "On", the scatter plot, "L1", "L2", and "●".
GRAPH
STAT ▶ 6 ENTER
Y= VARS 5 ▶ ▶ 7
GRAPH

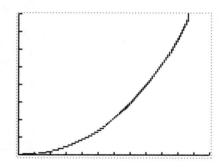

We see that using "QuadReg" or quadratic regression gives a graph that fits the data very well. The equation $y = ax^2 + bx + c$ with $a = 0.250$, $b = -0.004$, and $c = 0.002$ seems to go through most of the points from the scatter plot. A correlation value is not available for this regression model. However, we can check the graphs generated by "LinReg," "CubicReg," and "QuartReg" to see if any of these are a better fit visually. Observation will suggest that $y = 0.250x^2 - 0.004x + 0.002$ is a reasonable model for this data.

Remember that since it is difficult to determine whether a scatter plot is best fitted by a polynomial equation or an exponential equation, it is advisable to apply both processes to the data and observe the Pearson product-moment correlation values to determine which one fits the data best.

EXERCISES

Solve each problem using your graphing calculator.

1. A set of data has Pearson product-moment correlation values of $+0.964$ for $(x, \ln y)$ and $+0.962$ for $(\ln x, \ln y)$. Which type of regression equation fits the data best, exponential or polynomial?

2. Would you use a polynomial regression equation or an exponential regression equation for the data $(2, 121.40)$, $(3, 56.27)$, $(4, 32.11)$, $(5, 23.06)$, $(6, 18.27)$, $(7, 11.14)$? Why?

3. Find the exponential function that best fits the data $(2.2, 6.04)$, $(2.8, 10.65)$, $(3.6, 25.8)$, $(4.1, 43.57)$, $(4.4, 58.21)$, $(5.1, 129.98)$. What is the Pearson product-moment correlation value? Sketch the exponential regression equation and scatter plot, or the linearized regression equation and scatter plot.

4. Find the function of the form $y = Ax^B$ that best fits the data $(27, 2390)$, $(29, 3003)$, $(32, 3875)$, $(35, 4461)$, $(37, 5698)$. What is the Pearson product-moment correlation value? Sketch the polynomial regression equation and scatter plot, or the linearized regression equation and scatter plot.

VOCABULARY

Upon completing this chapter, you should be familiar with the following terms:

SKILLS AND CONCEPTS

OBJECTIVES AND EXAMPLES

Upon completing this chapter, you should be able to:

■ draw, analyze, and use bar graphs and histograms **(Lesson 15-1)**

Draw a histogram of the data below.

Score	Freq.
60—70	2
70—80	8
80—90	11
90—100	6

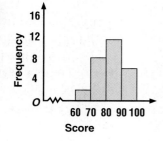

REVIEW EXERCISES

Use these exercises to review and prepare for the chapter test.

The table below gives the grade point averages of 500 college freshmen.

Point Average	Frequency
0.0-1.0	54
1.0-2.0	92
2.0-3.0	124
3.0-4.0	145
4.0-5.0	85

1. What are the class marks?

2. Find the class interval.

3. Draw a histogram of this data.

OBJECTIVES AND EXAMPLES	REVIEW EXERCISES

■ find the mean, mode, and median of a set of data **(Lesson 15-2)**

Find the mean, mode, and median of the set $\{46, 47, 48, 49, 50, 53, 54, 56, 58, 58, 64\}$.

$$\text{mean} = \frac{46 + 47 + \cdots + 58 + 64}{11} \text{ or } 53$$

mode $= 58$

median $= 53$ (the 6th value)

Find the mean, median, and mode(s) for each set of data.

4. $\{4, 8, 2, 4, 5, 5, 6, 7, 4\}$

5. $\{19, 11, 13, 15, 16\}$

6. $\{6.6, 6.3, 6.8, 6.6, 6.7, 5.9, 6.4, 6.3\}$

7. $\{130, 135, 131, 128, 122, 141, 133, 146\}$

■ find the mean deviation, semi-interquartile range, and standard deviation of a set of data **(Lesson 15-3)**

semi-interquartile range: $Q_R = \dfrac{Q_3 - Q_1}{2}$

mean deviation: $MD = \dfrac{1}{n} \sum\limits_{i=1}^{n} |X_i - \overline{X}|$

standard deviation: $\sigma = \sqrt{\dfrac{1}{n} \sum\limits_{i=1}^{n} (X_i - \overline{X})^2}$

A die was tossed 10 times with these results.

  5 1 5 4 2 3 6 2 5 1

8. Find the mean deviation of the data.

9. Find the standard deviation of the data.

10. Find the semi-interquartile range of the data.

■ use the normal distribution curve **(Lesson 15-4)**

A Normal Distribution

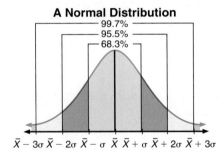

A set of data has a normal distribution with a mean of 88 and a standard deviation of 5.

11. What percent of the data is in the interval 78–98?

12. Find the probability that a value selected at random from the data lies in the interval 86–90.

13. Find the interval about the mean that includes 90% of the data.

■ find the standard error of the mean to predict the true mean of a population with a certain level of confidence **(Lesson 15-5)**

Find the standard error of the mean for $\sigma = 12$, $N = 100$, and $\overline{X} = 75$. Then find the range for a 1% level of confidence.

$$\sigma_{\overline{X}} = \frac{12}{\sqrt{100}} \text{ or } 1.2$$

1% level of confidence: $P = 0.99$, so $t = 2.58$.

$$\overline{X} = 75 \pm (2.58)(1.2)$$
$$= 71.90 \text{ and } 78.10$$

In a random sample of 200 adults, it was found that the average number of hours per week spent cleaning their home was 1.8, with a standard deviation of 0.5.

14. Find the standard error of the mean.

15. Find the range about the mean such that the probability is 0.90 that the true mean lies within the range.

16. Find the range about the sample mean that has a 5% level of confidence.

OBJECTIVES AND EXAMPLES

■ draw regression lines **(Lesson 15-6)**

Find the median fit line for the data.

Number of People at the Pool

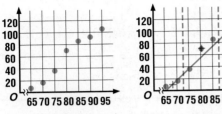

High Temperature (°F)   High Temperature (°F)

■ compute correlation values to determine goodness of fit **(Lesson 15-6)**

Pearson product-moment correlation:

$$r = \frac{SP}{\sqrt{SS_x SS_y}}, \text{ where } SP = \Sigma(X - \overline{X})(Y - \overline{Y}),$$

$SS_X = \Sigma(X - \overline{X})^2$, and $SS_Y = \Sigma(Y - \overline{Y})^2$,
and n is the number of items being ranked.

REVIEW EXERCISES

Use the chart to complete Exercises 17—18.

Student	A	B	C	D	E	F	G
Days Absent	0	3	6	1	2	2	4
Final Grade	95	88	69	89	90	86	77

17. Draw a scatter plot and find a median fit line for the data.

18. Find an equation of the line of regression.

Use the chart above to complete Exercises 19—20.

19. Determine the Pearson product-moment correlation value for the data.

20. Describe the relationship between the days absent and the final grade.

APPLICATIONS AND PROBLEM SOLVING

21. **Safety** The numbers of job-related injuries at a construction site for each month of 1992 are listed below. **(Lesson 15-2)**

    10   13   15   39   21   24   19   16   39   17   23   25

    a. Make a stem-and-leaf plot of the numbers of injuries.

    b. Find the median, mode, and mean of the numbers of injuries.

22. **Transportation** The length, in miles, of the eight longest underwater vehicular tunnels in North America are listed below. **(Lesson 15-3)**

    3.60   1.73   1.62   1.56   1.55   1.50   1.45   1.42

    a. Find the mean and median of the tunnel lengths.

    b. Find the mean deviation and the standard deviation of the tunnel lengths.

    c. Make a box-and-whisker plot of the data. Name any outliers.

23. **Recreation** Juan wants to know the relationship between the number of hours students spend watching TV each week and the number of hours students spend reading each week. A sample of 10 students reveals the following. **(Lesson 15-6)**

Student	A	B	C	D	E	F	G	H	I	J
Watching TV	20	32	42	12	5	28	33	18	30	25
Reading	8.5	3.0	1.0	4.0	14.0	4.5	7.0	12.0	3.0	3.0

   a. Draw a scatter plot to see how the data is related.

   b. Determine the Pearson product-moment correlation value for the data.

   c. Describe the relationship.

Attendance   The days missed for a random sample of 80 high school students at Dover High School in a certain school year are given.

6	16	12	7	7	9	13	12	7	7
19	4	9	6	4	11	13	10	16	20
10	17	11	12	6	9	10	14	3	8
8	12	13	8	8	11	12	1	11	5
11	16	13	5	10	1	10	8	15	10
13	9	20	5	9	15	11	18	12	14
10	16	8	10	2	11	19	10	12	17
14	6	9	12	10	14	8	9	7	9

1. Make a frequency distribution of the data using a class interval of 3.
2. Make a histogram of the data.
3. Find the mean of the data.
4. Find the median of the data.
5. Find the standard deviation of the data.

Use the data below for Exercises 6-12.

Physics   A small metal object is weighed on a laboratory balance by each of 15 pupils in a physics class. The weight of the object in grams is reported as 2.341, 2.347, 2.338, 2.350, 2.344, 2.342, 2.345, 2.348, 2.340, 2.345, 2.343, 2.344, 2.347, 2.341, and 2.344.

6. Find the mean of the data.
7. Find the median of the data.
8. Find the mode of the data.
9. Find the semi-interquartile range of the data.
10. Make a box-and-whisker plot of the data.
11. Find the mean deviation of the data.
12. Find the standard deviation of the data.

Assume the data given for Exercises 1–5 are normally distributed.

13. Find the lower and upper limits within which 90% of the data lie.
14. Find the probability that a value selected at random from the data will be between 8 and 12.
15. Find the standard error of the mean.
16. Find the range about the sample mean that has a 5% level of confidence.

Grades   The table below shows the statistics grades and the economics grades for a group of college students at the end of a semester.

Statistics Grades	95	51	49	27	42	52	67	48	46
Economics Grades	88	70	65	50	60	80	68	49	40

17. Draw a scatter plot and find a median fit line for the data.
18. What is an equation of the line of regression?
19. Find the Pearson product-moment correlation value.
20. Describe the relationship between the statistics grades and the economics grades.

Bonus   The *variance* of a set of data can be found using the formula $\sigma^2 = \frac{1}{n} \sum_{i=1}^{n} (X_i - \overline{X})^2$. How is this formula related to the formula for the standard deviation? Find the variance of the data for Exercises 6–12.

CHAPTER 16

GRAPH THEORY

HISTORICAL SNAPSHOT

First calculating machine Charles Babbage and Ada Byron Lovelace, London	UNIVAC 1 (Universal Automatic Computer) Philadelphia, Pennsylvania	Minicomputer (PDP-5) Digital Corporation Maynard, Massachusetts
Binary arithmetic Gottfried Leibnitz Germany	First Data-Processing Computer, Dr. Herman Hollerith, New York	FORTRAN (computer programming language) IBM, NYC	Department of Computer Science established at Univ. of Southern California

AD 1703	1843	1889	1952	1957	1962	1976

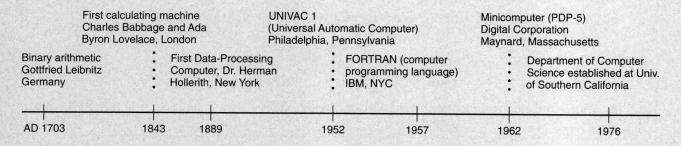

CHAPTER OBJECTIVES

In this chapter, you will:

- Define graphs, multigraphs, planar graphs, and digraphs.
- Find walks, paths, and circuits.
- Use matrices to draw graphs.
- Draw and use trees.

CAREER GOAL: Computer Consulting

Computers are used in every phase of every business you can think of. Mary Ann Terrazas says she chose computer science as a career because "I knew that it would be a career that would keep growing and one in which I would be needed."

Mary Ann has already spent time working in her field. "During the summer of 1989, I worked for Pacific Bell, programming with their data base. During the summer of 1990, I worked for the Central Intelligence Agency (CIA) in a computer science position. During the summer of 1992, I am working for Loral Electro-Optics Systems, programming for their Defense Department projects."

Mary Ann is glad that she was able to take algebra, geometry, trigonometry, mathematical analysis, and a full year of calculus in high school. Computer science is closely tied to mathematics. For example, "In order for a computer process to use a piece of hardware, like a monitor, a printer, or a modem, that hardware has to be free. Before a process will release the hardware it has been using, it will first seek out another piece of hardware that is in use. This may create a chain of processes that are waiting on each other for different resources. Such a chain is called *deadlock*. In computer science, we use a *dependency graph* to diagnose problems like deadlock. Once a dependency graph identifies a cycle that would create deadlock, you can eliminate one of the processes to avoid a system crash."

Mary Ann says that she dislikes the repetitive motion and the sitting for long hours that are involved when you work with computers. However, Mary Ann says, "I like the versatility of computer science. It will allow me to get a job almost anywhere, anytime."

16-1 Graphs

Objectives

After studying this lesson, you should be able to:
- distinguish between multigraphs and simple graphs, and
- determine whether a graph is complete.

Application

Airline magazines often include maps of the routes flown by the airline. On these maps, the cities are usually represented by points and the routes by line segments or arcs. The diagram at the right represents a portion of such a map.

The diagram above is a model of the relationship between cities and airline routes. Such a model is called a **graph.** This type of graph is quite different from the graphs of functions with which you are familiar. In this chapter, a graph is a collection of points in which a pair of points called **vertices,** or **nodes,** are connected by a set of segments or arcs, called **edges.**

Definition of Graph	**A graph G is a finite, nonempty set V of vertices, along with a set E of edges.**

The simplest graph has one vertex and no edges. There cannot be a graph of edges only, because edges must connect vertices. In a **simple graph,** only one edge is allowed between any two vertices.

Not all situations can be modeled by a simple graph. Consider the bus route maps illustrated below. The map on the left represents the routes of a single bus company. It is an example of a simple graph. The map on the right represents the routes of several bus companies. This map represents another type of graph called a **multigraph.**

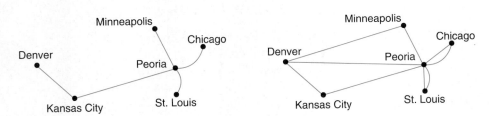

A multigraph is a graph with a finite set of vertices and a finite set of edges. Unlike a simple graph, a multigraph may have several edges connecting the same pair of vertices. These are called **parallel edges.** A multigraph may also contain edges that connect a vertex to itself. These are called **loops.** Since simple graphs are by definition multigraphs, all of the properties of multigraphs apply to simple graphs.

Example 1

In the multigraph at the right, name the parallel edges and loops.

The parallel edges are b and c.
The loops are edges f and h.

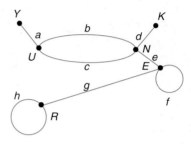

A graph can be described in three ways. One way to describe a graph is by the number of vertices and edges. We will use the notation $G(n, m)$, where n is the number of vertices and m is the number of edges of a graph G. The graph at the right can be described as $G(4, 5)$.

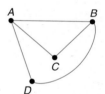

This way of describing a graph is not specific because in some cases, different graphs can be drawn with the same number of vertices and edges. For example, the graphs below also represent $G(4, 5)$.

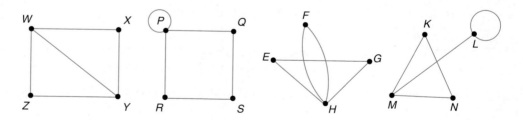

Edge $\{Y, Z\}$ is the same as edge $\{Z, Y\}$

A more specific way to describe a graph is by listing the vertices and edges. The graph on the left above can be described using the sets $V = \{W, X, Y, Z\}$ and $E = \{\{W, X\}, \{W, Y\}, \{Y, Z\}, \{Z, W\}, \{X, Y\}\}$. This is one way to ensure that the graph drawn more closely represents the correct relationships. Do not assume that when two edges cross, there will be a vertex at their crossing. In the last graph above, $\{M, L\}$ and $\{K, N\}$ cross, but there is no vertex. Unless a vertex is actually indicated and labeled, there is not a vertex where the two edges cross.

Example 2

Draw a diagram representing a graph with $V = \{R, S, T, M\}$ and $E = \{\{R, S\}, \{S, T\}, \{T, M\}\}$.

Two graphs that satisfy the given conditions are shown below.

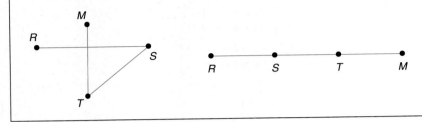

Vertices are **connected** or **adjacent** if there is an edge joining them. An edge that is connected to a vertex is **incident** on that vertex. Edges are considered adjacent if they share a vertex. In the graph at the right, vertices D and E are adjacent, but vertices A and C are not. Edge y is incident on vertices D and E, and edge w is adjacent to edge v since they share vertex A.

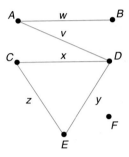

The **degree** of a vertex C, denoted deg (C), is the number of edges incident on C. In the graph above, deg $(C) = 2$ and deg $(B) = 1$. A vertex that has a degree of 0 is **isolated.** In the graph above, vertex F is isolated.

Degree Theorem

For any graph $G(n, m)$, where n is the number of vertices and m is the number of edges, the sum of the degrees of the vertices is twice the number of edges.

In a multigraph, a vertex can have any degree.

You can now describe a graph a third way by indicating the number of vertices and the degrees of each vertex.

Example 3

Draw a graph with five vertices, where deg $(A) = 1$, deg $(B) = 1$, deg $(C) = 2$, deg $(D) = 2$, and deg $(E) = 2$.

Two graphs that satisfy the given conditions are shown below.

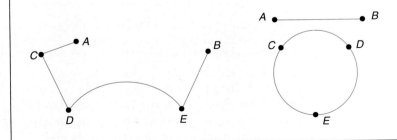

A **complete** graph with n vertices, denoted K_n, is a graph in which each pair of vertices is connected by exactly one edge. The graph at the right is an example of a complete graph. You can use a formula to find the number of edges in a complete graph.

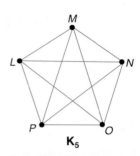

K_5

Complete Graph Theorem	For any complete graph K_n, 1. there are $\dfrac{n(n-1)}{2}$ edges; that is, $G(n, m) = G\left(n, \dfrac{n(n-1)}{2}\right)$, and 2. the maximum degree of every vertex is $n - 1$.

Example 4

Determine if each graph could be complete.
a. $G(7, 21)$ b. $G(6, 12)$

a. $\dfrac{n(n-1)}{2} = \dfrac{7(6)}{2}$
$= 21$

Since $21 = 21$, the graph could be complete.

b. $\dfrac{n(n-1)}{2} = \dfrac{6(5)}{2}$
$= 15$

Since $15 \neq 12$, the graph is not complete.

Properties of complete graphs are often applied in communication relationships.

Example 5

APPLICATION

Communications

The director of sales for Glencoe Publishing Company in Columbus, Ohio, is planning a teleconference with regional vice-presidents from New Jersey, Illinois, South Carolina, Texas, and California. How many telephone lines does the telephone company need to reserve for the teleconference?

Each person in the teleconference can be represented by a vertex of a graph. Since each person will have a direct line with everyone else, a model of this situation would be a complete graph.

$G(n, m) = G(6, m)$ *$n = 6$, the director plus 5 vice-presidents*

$\qquad = G\left(6, \dfrac{6(6-1)}{2}\right)$ *Definition of complete graph*

$\qquad = G(6, 15)$

The company needs to reserve 15 lines for the teleconference.

CHECKING FOR UNDERSTANDING

Communicating Mathematics

Read and study the lesson to answer each question.

1. **Describe** the difference between a simple graph and a multigraph.

2. *True* or *false:* The sum of the degrees of the vertices of a graph equals the number of edges.

3. **Draw** a graph that represents the relationships in Example 5.

4. **Explain** why the graph used for Exercises 5–12 is not complete.

Use the graph at the right to answer each question.

5. Which edges are incident on vertex C?

6. What is deg (E)?

7. Which vertices are adjacent to vertex D?

8. How many parallel edges are there?

9. How many loops are there? Name them.

10. Is this a simple graph or a multigraph? Explain.

11. Name the set of vertices in the graph.

12. Name the set of edges in the graph.

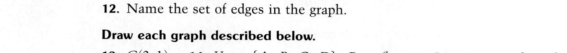

Draw each graph described below.

13. $G(2, 1)$ 14. $V = \{A, B, C, D\}$, $E = \emptyset$ 15. 5 vertices, each with degree 2

EXERCISES

Draw each graph described below.

16. $G(3, 4)$ 17. $G(4, 3)$ 18. $G(2, 0)$ 19. $G(3, 3)$

20. $V = \{M, N, O, P\}, E = \{\{M, O\}, \{O, P\}\}$

21. $V = \{L, M, N\}, E = \{\{L, M\}, \{M, N\}\}$

22. $V = \{R, S, T, U, V\}, E = \{\{R, T\}, \{S, U\}, \{U, V\}, \{S, V\}\}$

23. 3 vertices; deg $(A) = 0$, deg $(B) = 1$, deg $(C) = 1$

24. 4 vertices; deg $(A) = 1$, deg $(B) = 2$, deg $(C) = 2$, deg $(D) = 1$

25. 5 vertices; deg $(A) = 4$, deg $(B) = 1$, deg $(C) = 2$, deg $(D) = 1$, deg $(E) = 2$

26. 6 vertices, each with degree 1

Draw a graph to model each situation.

27. Abby, Brenda, Caitlin, Debbie, Evie, and Francine are good friends. Brenda, Debbie, and Evie can only walk to each others' houses because the others are too far. Caitlin can walk to everybody's house but Brenda's.

28. State Route 181 goes from San Antonio to Floresville and Corpus Christi. Route 97 goes through Floresville, Pleasanton, Jourdanton, and Charlotte. Interstate 37 connects San Antonio, Pleasanton, and Corpus Christi.

Draw each graph described below.

29. K_2 30. K_4 31. K_6

32. How many edges does the graph of K_5 have?

33. How many edges does the graph of K_7 have?

34. How many edges does the graph of K_{10} have?

35. How many edges does the graph of K_n have?

Determine whether each graph could be complete. Write *yes* or *no*.

36. $G(3, 4)$ **37.** $G(4, 5)$ **38.** $G(6, 15)$

39. $G(10, 45)$ **40.** $G(14, 91)$ **41.** $G(100, 4900)$

42. What is the maximum degree of any one vertex in $G(5,2)$?

43. Can you draw a graph with exactly 8 vertices, each with a degree of 1? Explain your answer.

44. Can you draw a graph with exactly 7 vertices, each with degree 3? Explain your answer.

45. How many vertices are there in a graph with 11 edges if each vertex has degree 2?

46. Draw a graph in which each vertex is adjacent to exactly two other vertices and each edge is adjacent to exactly two other edges.

Critical Thinking

47. Could $G\left(x + 3, \dfrac{x^2 + 6x + 9}{2}\right)$ be complete? Why or why not?

Applications and Problem Solving

Draw a graph to model each situation. Then write a paragraph that completely describes each graph.

48. BASIC programming

```
A 10 INPUT N
B 20 IF N = 1000 THEN 80
C 30 IF N > 0 THEN 60
D 40 PRINT N;" IS NEGATIVE"
E 50 GOTO 10
F 60 PRINT N:" IS POSITIVE"
G 70 GOTO 10
H 80 END
```

49. Airlines

A	New York	B	Los Angeles
A	New York	C	Chicago
B	Los Angeles	C	Chicago
B	Los Angeles	D	Dallas
E	Atlanta	D	Dallas
E	Atlanta	C	Chicago
E	Atlanta	A	New York
D	Dallas	C	Chicago

Mixed Review

50. Find an equation of a parabola that contains the points at $(1, 9), (4, 6),$ and $(6, 14)$. **(Lesson 2-4)**

51. Find all rational roots of $x^3 - x^2 - x - 2 = 0$. **(Lesson 4-4)**

52. Graph $y = 3\cos(2\theta + 180°)$. **(Lesson 6-3)**

53. Physics Two forces, one of 30 N and the other of 50 N, act on an object. If the angle between the forces is 40°, find the magnitude and the direction of the resultant force. *N is the symbol for the metric unit of force, the newton.* **(Lesson 8-5)**

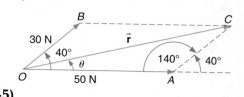

54. Find an equation of the circle that passes through the points at $(2, -1), (-3, 0),$ and $(1, 4)$. **(Lesson 10-1)**

55. Draw a scatter plot and find a prediction equation to show how study time and test scores are related. Use the data in the table below. **(Lesson 15-6)**

Study time (in minutes)	15	75	60	45	90	60	30	120	10	120
Test scores (percentages)	68	87	92	73	95	83	77	98	65	94

56. College Entrance Exam If $\dfrac{x}{8} = 8$, what would be the value of x?

16-2 Walks and Paths

Objectives

After studying this lesson, you should be able to:
- distinguish among walks, trails, paths, circuits, and cycles, and
- determine whether a graph is connected.

Application

A common problem in graph theory is finding a route that satisfies a given set of conditions.

For example, in some cities it is necessary to develop a spraying route for mosquito abatement. A graph can be used to model the streets of the city. In this case, the intersections would be modeled by vertices, and the streets would be modeled by edges. To minimize cost and resources, the route should include each street only once, if possible. In order to satisfy these conditions, a route is needed that includes each edge exactly once.

In a graph, a **walk** is the course taken from one vertex to another vertex along edges of a graph. For example, a walk from vertex U to vertex V (called a U–V walk) is an alternating sequence of vertices and edges beginning with U and ending with V. This sequence is of the form $V_1, e_1, V_2, e_2, V_3, e_3, \ldots, V_n, e_n, V_{n+1}$, where $V_1 = U$ and $V_{n+1} = V$. The length of this walk is n, the number of edges listed.

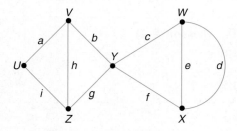

In the graph at the left, W, d, X, e, W, c, Y is a walk of length 3 from W to Y. This W–Y walk can also be written d, e, c. Notice that vertex W appears twice in this walk. Walks may repeat the same edges and vertices. A **trail** is a walk in which no edge is used more than once. In the graph at the left, Y, c, W, d, X, e, W is a Y–W trail. A **path** is a walk in which no vertex nor edge is repeated. Paths also do not contain loops or parallel edges. In the graph at the left, U, a, V, h, Z is a U–Z path.

Example 1

Find a *U–V* walk, a *U–V* trail, and a *U–V* path for the multigraph at the right.

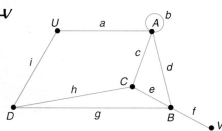

One U–V walk is a, b, c, h, g, f.

One U–V trail is a, d, g, h, e, f.

One U–V path is i, g, f.

In Example 1, another U–V path is a, c, h, g, f. It can be found in the U–V walk given if loop b is deleted. This observation leads to the following theorem.

Walks and Paths Theorem	**Every walk contains a path.**

A closed walk starts and ends at the same vertex. In a closed trail, or **circuit,** only vertices may be repeated. In a closed path, or **cycle,** no edges may be repeated and only the beginning and ending vertices may be the same.

Example 2

Suppose a chemical company hires an efficiency expert to create a weekly production schedule of seven chemicals. We'll refer to them as *A, B, C, D, E, F, G.* The schedule must start by producing chemical *A,* include each of the six other chemicals, and end the week in readiness to begin production of *A* the next week. The efficiency expert drew the graph below. The vertices represent the chemicals, and the edges represent the shortest times it takes to change the calibrations for production between each pair of chemicals. What schedule did the efficiency expert recommend?

To create the most efficient schedule, the efficiency expert must find a path where only the beginning and ending vertices are the same. In other words, the efficiency expert must find a cycle that begins and ends with A.

The path a, b, e, k, h, i, j satisfies the given condition.

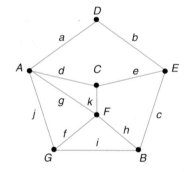

The production schedule would be $A, D, E, C, F, B, G,$ and A.

A graph is **connected** if there is a path between every two vertices. That is, a graph is connected if it is possible to go from any vertex to any other vertex by following the edges. Complete graphs are always connected. Of the graphs shown below, the graph on the left is connected. The graph on the right is **disconnected.**

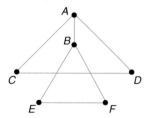

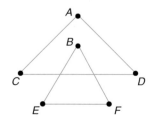

For $G(n, m)$ to be connected, there must be at least $n - 1$ edges. However, even if there are at least $n - 1$ edges, they may not be arranged in such a way for the graph to be connected.

Example 3

Determine if each graph is connected.
a. $G(12, 18)$
b. $G(12, 66)$

a. Since there are more than $12 - 1$ or 11 edges, it is possible that $G(12, 18)$ is connected. However, it is not possible to tell if the graph is connected without seeing the graph.

b. This could be a complete graph because $\dfrac{n(n-1)}{2} = \dfrac{12(11)}{2}$ or 66. If it is a complete graph, then it is connected.

In a connected graph, a **bridge** is an edge whose removal would cause the graph to no longer be connected. In the graphs shown at the bottom of the previous page, $\{A, B\}$ is a bridge.

CHECKING FOR UNDERSTANDING

Communicating Mathematics

Read and study the lesson to answer each question.

1. **Describe** the differences among walks, trails, and paths.

2. **Explain** why every walk contains a path.

3. **Draw** an example of a graph that is not connected.

4. **Explain** the relationship between a connected graph and a bridge.

Guided Practice

Use the graph at the right to answer each question.

5. Determine whether a, b, c, d, f, g, h is a walk, a trail, or a path. Then state its length.

6. What is the length of the shortest U–X path?

7. Is this a simple graph or a multigraph? Explain.

8. Is the graph connected?

9. Does the graph have a bridge? If so, name it.

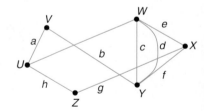

EXERCISES

Practice

Use the graph at the right to determine whether each walk is a circuit, cycle, path, trail, or walk. Use the most specific name.

10. a, b, c, k, ℓ, d, c

11. a, h, g

12. i, j, c, k, ℓ, e

13. k, d, e, f

14. i, k, d, ℓ, j, b

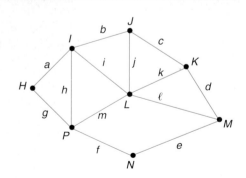

Use the graph at the right to answer each question.

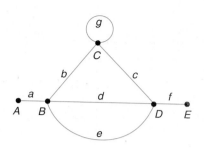

15. How many different A–E walks are there in the graph?

16. How many different A–E paths are there in the graph?

17. Name an A–E walk of length 5 in this graph.

18. Name an A–E path of length 4 in this graph.

For each multigraph in Exercises 19–21,
a. list three different walks from A to B and state the length of each,
b. list all of the paths from A to B and state the length of each, and
c. for each walk in part a, find a path from A to B that is contained in it.

19.

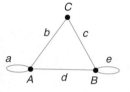

20.

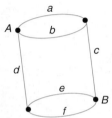

21.

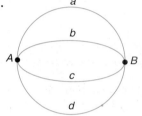

22. What is the greatest number of edges possible in a disconnected simple graph with 5 vertices?

23. What is the greatest number of edges possible in a disconnected simple graph with 8 vertices?

24. What is the least number of edges possible in a connected graph with 5 vertices?

25. What is the least number of edges possible in a connected graph with 8 vertices?

26. Is every complete graph a connected graph?

27. **Transportation** An agricultural chemical supplier needs to ship six chemicals from a refinery to a processing plant. We'll refer to the chemicals as U, V, W, X, Y, and Z. Environmental Protection Agency rail shipping regulations require that certain chemicals must be shipped in separate tank cars due to the possibility of explosion. In the graph at the right, chemicals linked by an edge may not be shipped together. What is the minimum number of tank cars needed to ship equal quantities of each chemical, and what chemicals will be in each car?

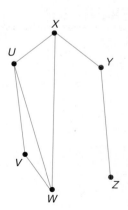

28. **Design** Jack Hanna, the director emeritus for the Columbus Zoo, wishes to provide enclosures for 10 new animals. We'll refer to them as $A, B, C, D, E, F, J, K, L$, and M. Because some animals are natural enemies, they must be kept in different enclosures. The table at the right lists the enemies of each animal.

Animal	Enemies
A	D, F, J, L
B	C, K
C	B, D
D	A, C, L
E	K, M
F	A, J, K
J	A, F
K	B, E, F, M
L	A, D
M	E, K

   a. Represent the information in the table with a graph. Let the animals represent the vertices and let the edges represent animals that may *not* be enclosed together. Try to position the vertices so that edges cross as few times as possible.

   b. How many enclosures does Mr. Hanna need and what animals will be in each enclosure?

Mixed
Review

29. Find the critical points of the graph of $f(x) = x^4 - 8x^2 + 16$. Then, determine whether each point represents a maximum, minimum, or a point of inflection. **(Lesson 3-7)**

30. Write $3x - 5y + 5 = 0$ in polar form. **(Lesson 9-4)**

31. **Food** Classic Pizza offers pepperoni, mushrooms, sausage, onions, peppers, and olives as toppings for their seven-inch pizza. How many different 3-topping pizzas can be made? **(Lesson 14-3)**

32. List the vertices adjacent to A and state the degree of A in the figure at the right. **(Lesson 16-1)**

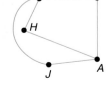

33. **College Entrance Exam** Choose the best answer.
    If $x^2 = 36$, then 2^{x-1} could equal

    (A) 4   (B) 6   (C) 8   (D) 16   (E) 32

16-3 Euler Paths and Circuits

Objective

After studying this lesson, you should be able to:
- find Euler paths and circuits.

Application

Before World War II, the city of Kaliningrad, in the former Soviet Union, was known as Königsberg. The Pregel River flowed through the city, and there were seven bridges that connected two islands and the shore. In the early 1700s, the residents of Königsberg tried to solve a problem that is now considered one of the oldest problems involving graph theory. The problem, known as the Königsberg bridge problem, asked whether it was possible to walk through the city while crossing each bridge exactly once. Consider the illustration of the city shown below.

FYI...

Kaliningrad has been almost entirely rebuilt after near total destruction in World War II.

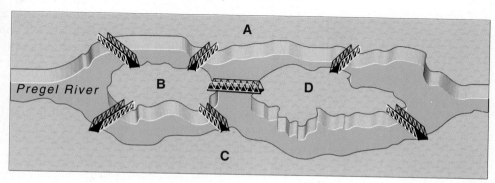

Tracing the graph is similar to walking the path.

In 1736, Leonhard Euler solved this problem by using the graph shown at the right. Each vertex represents a land mass and each edge represents a bridge. Euler showed that walking through Königsberg while crossing each bridge exactly once was equivalent to tracing the graph without retracing any edge.

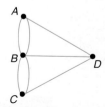

EXPLORATION: Modeling

Materials: different colored opaque markers

1. Copy the graph modeling the Königsberg bridge problem.
2. Choose a starting vertex and try to trace the graph without retracing any edge.
3. Using the same starting vertex and different color markers, repeat the process starting with each different edge.
4. Repeat the process, with each vertex as a starting point.
5. Use the results of your tracings to answer the questions below.
   a. In how many different ways did you start your tracings?
   b. Were you able to trace all of the edges of the graph without retracing any edges?
   c. What was Euler's solution to the Königsberg bridge problem?

In honor of Euler's work, two important characteristics of a connected multigraph bear his name. A path that includes each edge exactly once and has the same starting and ending vertex is called an **Euler circuit**. A path that includes each edge exactly once, but has different starting and ending vertices is called an **Euler path.**

Example 1

Determine whether each graph contains an Euler circuit or an Euler path. If so, name the circuit or path.

a. 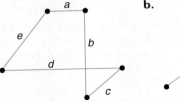 **b.** 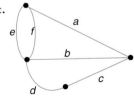 **c.**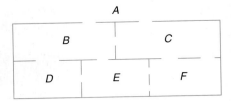

a. The path a, b, c, d, e is an Euler circuit since it contains all of the edges exactly once and returns to the starting vertex.
b. This graph contains neither an Euler circuit nor an Euler path.
c. The path c, d, b, a, e, f is an Euler path since it contains all the edges exactly once and ends at a vertex that is different from the starting vertex.

While working on the Königsberg bridge problem, Euler discovered that for a path to include each edge exactly once and return to the starting vertex, the degree of each vertex must be even.

Euler Circuit Test	**A connected multigraph contains an Euler circuit if and only if the degree of each vertex is even.**

Example 2

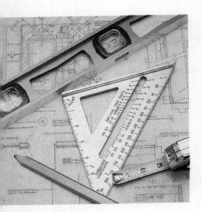

The figure at the right is a floor plan for a house. Is it possible for a person starting at A to take a tour of the house and walk through each doorway exactly once? If so, state the path.

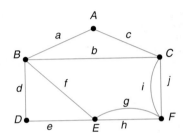

First, develop a multigraph to model the floor plan. Let each vertex represent a room and let each edge represent a doorway.

Since each vertex has even degree, the Euler circuit test is satisfied. Thus, it is possible for a person to tour the house and walk through each doorway exactly once.

One possible path is $a, d, e, h, g, f, b, i, j, c.$

Once it has been determined that a connected multigraph does contain an Euler circuit, the circuit can be found. Sometimes the circuit is easily found through observation, but often this is not possible in a more complicated graph. As is the case with any complex problem, a systematic process is helpful. An **algorithm** is a sequence of instructions that solves all cases of a certain type of problem. **Fleury's algorithm** gives a method for finding an Euler circuit in a connected multigraph that has a large number of vertices, all of even degree.

Fleury's Algorithm

If a connected multigraph contains an Euler circuit, the circuit can be located by the following process.

Step 1 Choose a starting vertex and walk the edges. Mark off the edges walked as you go.

Step 2 Do not choose a bridge, unless not choosing a bridge will disconnect the circuit.

Example 3

Find an Euler circuit in the connected multigraph at the right.

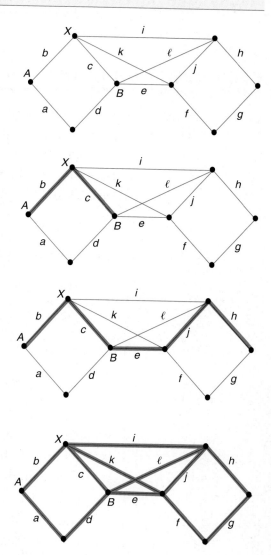

Using Fleury's algorithm, begin at vertex *A* and walk edge *b*. At vertex *X*, you can choose among edges *c*, *k*, and *i*. Since none of these edges are bridges, any one can be chosen. Let's choose *c*.

At vertex *B*, the choices include *l*, *e*, and *d*. Edges *b* and *c* have been eliminated, so *d* is now a bridge back to *A*. Since there are other choices, do *not* choose *d*. Choose *e*. Then choose *j* and *h*.

Continue to check for bridges at each vertex. To complete the circuit, choose *g*, *f*, *k*, *i*, *l*, *d*, and *a*, respectively.

An Euler circuit for this multigraph is *b*, *c*, *e*, *j*, *h*, *g*, *f*, *k*, *i*, *l*, *d*, *a*.

Since the degree of every vertex in the multigraph that models the city of Königsberg is odd, the entire city of Königsberg could not be explored without crossing one bridge at least twice. That is, no Euler circuit could be found. The following test will help you determine whether a graph has an Euler path.

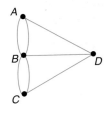

| *Euler Path Test* | **A connected multigraph has an Euler path if and only if each vertex has even degree, except for exactly two. Euler paths always start with one of the vertices of odd degree and end with the other.** |

To find an Euler path from one vertex of odd degree to the other, you can add a temporary edge connecting the two vertices. Then, find an Euler circuit using Fleury's Algorithm. When the temporary edge is removed from the circuit, the result is an Euler path.

CHECKING FOR UNDERSTANDING

Communicating Mathematics

Read and study the lesson to answer each question.

1. **Explain** why a graph must be connected in order to contain an Euler path or an Euler circuit.

2. **Describe** the difference between an Euler circuit and an Euler path.

3. **Describe** the conditions under which a connected graph will contain an Euler circuit.

Guided Practice

Use the figures below to answer Exercises 4—10.

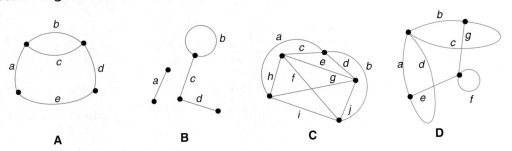

4. Which of the figures are connected multigraphs?

5. Which of the multigraphs contain Euler circuits?

6. Which of the multigraphs contain Euler paths?

7. Name an Euler circuit in *C*.

8. Is *C* a complete graph?

9. Name an Euler path in *D*.

10. What could you do to make *B* a connected multigraph?

EXERCISES

Determine whether each multigraph is a simple graph. Write *yes* or *no*.

11.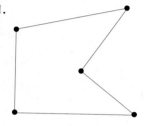

12.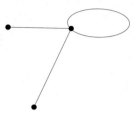

13.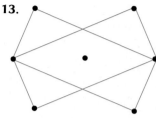

14.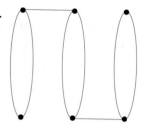

15.

16.

17. Which of the multigraphs in Exercises 11–16 are connected?

18. Which of the multigraphs in Exercises 11–16 have an Euler circuit?

19. Which of the multigraphs in Exercises 11–16 have an Euler path?

20. Does the graph below have an Euler circuit? Why or why not?

21. Does the graph below have an Euler circuit? Why or why not?

22. Draw a graph with 5 vertices that contains an Euler circuit.

23. Draw a graph with 6 vertices that contains an Euler circuit.

24. Draw a graph with 7 vertices that contains an Euler circuit.

25. Do all graphs of the form K_n, where $n \geq 2$, contain an Euler circuit? Explain.

26. Transportation Ed Bailey drives a snowplow for Harrison Township. Using the road map of the township shown at the right, his supervisor wants him to find a route that goes over each roadway once. Is such a route possible? If so, find a route that begins and ends at the township office, *T*.

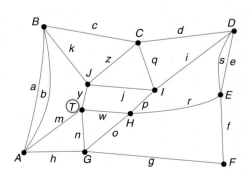

27. **Tourism** Staten Island, New York, is connected to New Jersey by three bridges and to Brooklyn, New York, by one bridge. Can a tour bus from New Jersey cross all four bridges exactly once and end the tour back in New Jersey? If not, what is the least number of additional bridges needed to complete the tour and where should they be located?

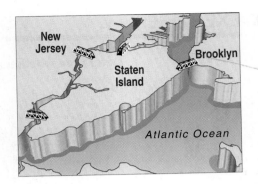

Mixed Review

28. Find the distance between the points with coordinates $(3, -5)$ and $(-1, 2)$. Then find the slope of the line containing these points. **(Lesson 1-4)**

29. Without using a calculator, find the value of $\sin 450°$. **(Lesson 5-4)**

30. Use the ratio test to determine whether the following series is *convergent* or *divergent:* $1 + \dfrac{1}{2^2} + \dfrac{1}{3^3} + \dfrac{1}{4^4} + \cdots + \dfrac{1}{n^n} + \cdots.$ **(Lesson 12-4)**

31. For the multigraph at the right, list three different walks from A to B and state the length of each. **(Lesson 16-2)**

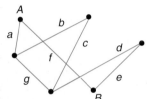

32. **College Entrance Exam** Compare quantities A and B below.
Write A if quantity A is greater.
Write B if quantity B is greater.
Write C if the two quantities are equal.
Write D if there is not enough information to determine the relationship.

　　(A) $\dfrac{8a + 12}{2}$ 　　　　(B) $4a + 6$

MID-CHAPTER REVIEW

Use the graph at the right to answer each question.

1. Is the figure a multigraph? **(Lesson 16-1)**
2. Is the figure a complete graph? **(Lesson 16-1)**
3. What is deg (E)? **(Lesson 16-1)**
4. Which edges are incident on D? **(Lesson 16-1)**
5. Which vertices are adjacent to D? **(Lesson 16-1)**
6. Is r, x, z, t, v, w a path, a circuit, or a cycle? **(Lesson 16-2)**
7. Is the figure a connected graph? **(Lesson 16-2)**
8. Does the graph contain an Euler circuit? If so, name it. **(Lesson 16-3)**

16-4 Shortest Paths and Minimal Distances

Objective

After studying this lesson, you should be able to:
- use the breadth-first search algorithm and Dijkstra's algorithm to find a shortest path.

Application

Ohio Electric has a fossil fuel powered generating plant in northeastern Ohio that serves several surrounding communities. Some of the electrical power is sent along older transmission lines that have a problem of power loss due to their deteriorating condition. Based on resistance tests, electrical engineers determined the percent of power lost along each line between these communities as shown in the graph below.

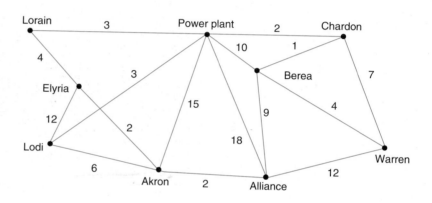

The engineers' final task is to determine the best electrical power route, that is, one with the least power loss, between any two cities. Can you find such a route and the minimum power lost between the power plant and Alliance? *This problem will be solved in Example 3.*

In this lesson, you will explore ways to find the shortest paths, or paths with the fewest edges, and the minimal paths, or paths with the least total weight, in multigraphs and weighted multigraphs. **Weighted multigraphs** are multigraphs in which a value is assigned to each edge.

To find the shortest path in a multigraph means to find the path that contains the fewest edges. One way to find the shortest path is to list all of the possible paths and then choose the shortest one. However, since some graphs are very complex, an algorithm called the **breadth-first search algorithm** can be used that will ensure finding the shortest path. In this algorithm, each vertex is assigned a number followed by a letter. An example is $3(F)$, where 3 is the sum of the fewest edges to this vertex and F represents the preceding vertex in this path.

**Breadth-First
Search Algorithm**

Step 1 Assign U, the starting vertex, the label 0(–). The (–) means that there is no preceding vertex in the path. Then label the vertices adjacent to U as 1(U).

Step 2 Label all of the vertices adjacent to the already labeled vertices. Assign these vertices a number, $n + 1$, based on the preceding vertex having been assigned the number n, followed by the name of the preceding vertex. If one of the unlabeled vertices is adjacent to more than one labeled vertex, then choose one of them at random.

Step 3 If you have reached V, then stop (if not, go back to Step 2). The label assigned to V describes the distance from U to V. Listing the sequence of preceding vertices produces one of the shortest paths from U to V.

Example 1

Find the shortest path and the distance from A to K in the multigraph at the right.

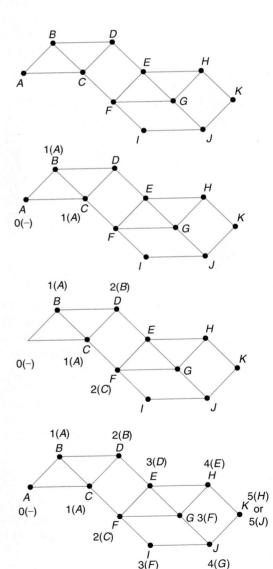

Step 1 Assign A the label 0(–). Assign vertices B and C the label 1(A) because they are adjacent to 0(–).

Step 2 Since vertex D is adjacent to labeled vertices B and C, vertex D is assigned either label 2(B) or 2(C) at random. Since vertex F is adjacent to labeled vertex C, assign F the label 2(C).

Step 3 Repeat this process until K is reached.

Two of the possible shortest paths are A, B, D, E, H, K and A, C, F, G, J, K. Both have a distance of 5.

In a weighted graph, the value assigned to each edge is called the **weight of the edge**. The **weight of a path** is the sum of the weights of the edges along a path. The path with the minimum weight is called the **minimal path**. The weight of the minimal path is called the **minimal distance**.

In a weighted multigraph, **Dijkstra's algorithm** can help to determine the minimal path and distance from any vertex to all others. In this algorithm, each vertex is assigned a label, sometimes temporarily. An example is $4(A)$, where 4 is the sum of the weights of all previous edges and A represents the preceding vertex in the path. When a label becomes permanent, a check ($\checkmark$) is placed next to it.

Dijkstra's Algorithm

Step 1 Assign U, the starting vertex, the label $0(-)$. Place a check beside it.

Step 2 Assign all vertices adjacent to U the temporary label $w(U)$, where w is the weight of the edge from U, the only permanently labeled vertex.

Step 3 Of all the vertices with temporary labels, choose the one with the least weight (if there are more than one, choose one at random). Make this label permanent by placing a check beside it. Then, assign (or revise, if necessary) temporary labels to each vertex adjacent to the latest permanently labeled vertex. Assign the vertex the label $w(X)$, where w is the sum of the weights of the preceding permanently labeled vertex and the weight of the new edge connected to X, the preceding vertex. Do not change the label of any temporarily labeled vertex if its current weight is less than what the new weight would be.

Step 4 Continue Step 3 until all of the vertices are permanently labeled with a check. The weight on a vertex is the minimal distance from vertex U. The list of preceding vertices will provide the minimal path from U.

Example 2

Find the minimal path and distance in the weighted multigraph at the right from A to D.

Step 1 Assign vertex A the permanent label $0(-)$.

Step 2 Assign vertices $B, G,$ and F temporary labels since they are adjacent to vertex A.

(continued on the next page)

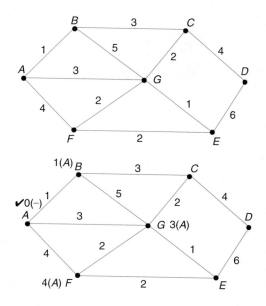

Step 3 Since vertex B has the least weight among B, G, and F, place a check beside it. Then, since vertices C and G are adjacent to B, assign label $4(B)$ to vertex C and leave label $3(A)$ on vertex G, since $3 < 1 + 5$ (the distance from A to B to G). Now G has the least weight of vertices C, G, and F, so mark vertex G with a check. Then repeat Step 3.

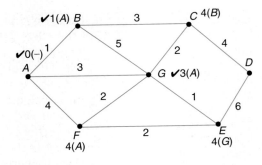

Step 4 Since vertices C, E, and F have the same weight, choose a vertex with which to continue. Let's choose F. Mark vertex F with a check, then repeat Step 3.

Investigate the labels next to C and E next. Repeat Step 3.

Finally, do vertex D.

The minimal distance from vertex A to vertex D is 8. The minimal path is A, B, C, D.

Example 3

APPLICATION

Energy

Find the route with the least power loss from the power plant to Alliance, using the weighted multigraph shown in the application at the beginning of the lesson. Also state the route.

Using Dijkstra's Algorithm resulted in the figure at the right.

There are two possible routes of least power loss. The route *power plant, Lorain, Elyria, Akron, Alliance* and the route *power plant, Lodi, Akron, Alliance* both have power losses of 11%.

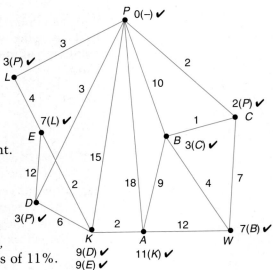

CHECKING FOR UNDERSTANDING

Communicating Mathematics

Read and study the lesson to answer each question.

1. **Explain** why a shortest path is not necessarily a minimal path.

2. **Describe** the difference between a multigraph and a weighted multigraph.

3. **Describe** the conditions under which a path of a weighted multigraph is the minimal path.

Guided Practice

Applying the breadth-first search algorithm to a multigraph resulted in the figure at the right. Use this figure to answer each question.

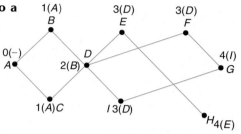

4. What is the distance from A to H?

5. What is the distance from A to F?

6. What is the shortest path from A to G?

7. What is the shortest path from C to G?

8. Could vertex D have been labeled $2(C)$?

Applying Dijkstra's algorithm to a weighted multigraph resulted in the figure at the right. Use this figure to answer each question.

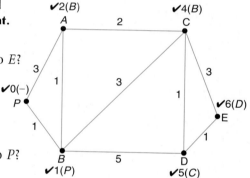

9. What is the minimal distance from P to E?

10. What is the minimal path from P to E?

11. Why does vertex A not have permanent label $3(P)$?

12. What is the minimal distance from E to P?

13. Which vertex, C or D, is closer to P?

EXERCISES

Practice

Determine the distance and a shortest path from A to Z in the graphs below by using the breadth-first search algorithm.

14.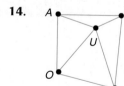

15.

16.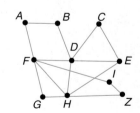

17.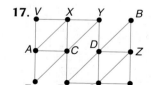

18.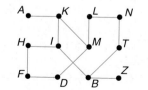

19.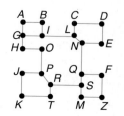

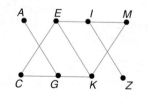
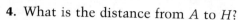

Determine the minimal distance from *A* to all of the other vertices. Then find a minimal path and the minimal distance from *A* to *Z* by using Dijkstra's algorithm.

20.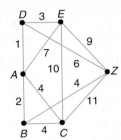

21.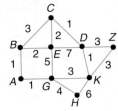

22.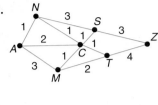

23.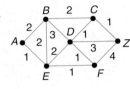

24.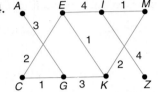

25.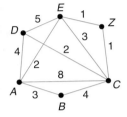

Critical Thinking

26. Eileen Snow needs to leave Salt Lake City, visit 3 cities exactly once, and return to Salt Lake City. It takes Ms. Snow 6 minutes to find the shortest route possible. When two more cities are added to the itinerary, it takes her 2 hours to find the shortest route from Salt Lake City through the other 5 cities and back to Salt Lake City. How long would it take Ms. Snow to find the shortest route from Salt Lake City through 9 cities and back to Salt Lake City?

Applications and Problem Solving

27. **Transportation** A truck driver makes deliveries to the cities listed in the mileage table below. She wants to develop a schedule where no city-to-city trip exceeds 450 miles.

    a. On the basis of the mileages given in the table, draw a graph with vertices representing cities and edges representing trips not exceeding 450 miles. If your initial graph has edge crossings, reposition the vertices and redraw the graph to eliminate edge crossings.

    b. Using your graph, find the shortest path from Buffalo to Atlanta.

    c. Using your graph, find the minimal path and distance from Mobile to Nashville.

	Atlanta	Birmingham	Buffalo	Charleston	Cleveland	Memphis	Mobile	Nashville	Norfolk
Atlanta	—	161	896	515	774	394	345	250	572
Birmingham	161	—	901	535	718	244	274	195	789
Buffalo	896	901	—	864	200	922	1246	721	657
Charleston	515	535	864	—	256	609	808	395	407
Cleveland	774	718	200	256	—	731	1030	530	571
Memphis	394	244	922	609	731	—	402	208	924
Mobile	345	274	1246	808	1030	402	—	467	893
Nashville	250	195	721	395	530	208	467	—	708
Norfolk	572	789	657	407	571	924	893	708	—

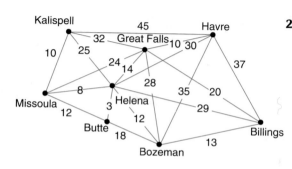

28. **Communication** A telephone company in a mountainous region of Montana is looking for ways to improve the efficiency of their communications system. Based on an analysis of the set-up costs to provide direct communications between several cities, they set the costs (in cents) for a three-minute call between the cities as shown on the graph at the left. Find the minimum cost of a three-minute call from Kalispell to Bozeman. Also state the switching path for routing this telephone call.

Mixed Review

29. Graph $y = \dfrac{4x}{x-1}$. **(Lesson 3-4)**

30. If $\sin\theta = \dfrac{3}{4}$, find $\sec\theta$ for values of θ between $0°$ and $360°$. **(Lesson 7-1)**

31. Use mathematical induction to prove that the following formula is valid for all positive integral values of n: $2 + 2^2 + 2^3 + \cdots + 2^n = 2^{n+1} - 2$. **(Lesson 12-8)**

32. Find an Euler path in the graph at the left. **(Lesson 16-3)**

33. **College Entrance Exam** Choose the best answer.

If $x = -7$ and $\dfrac{1}{2y} = -14$, what is the value of y in terms of x?

(A) $x - 7$ (B) $2x - 7$ (C) $\dfrac{1}{2x}$ (D) $\dfrac{1}{4x}$ (E) $\dfrac{2x}{7}$

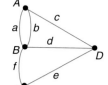

CASE STUDY FOLLOW-UP

Refer to Case Study 2: *Trashing the Planet*, on pages 959–961.

John Meadows is a planner for the Middleboro Waste Management System (MWMS). In addition to trash pickup, MWMS offers curbside recycling for newspapers, aluminum cans, plastics, and glass bottles. Mr. Meadows handles the recycling pickup for the new Montreal Park subdivision, whose streets are illustrated at the right.

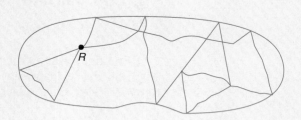

1. Middleboro has a 95% participation rate for curbside recycling. How does this compare with the national average of households that recycle at least one item?

2. Mr. Meadows wants to create a route that goes down each street exactly once.
   a. Is this possible? Explain.
   b. Find a route that begins and ends at the regional recycling center at R.

3. **Research** ways in which glass, newspapers, aluminum cans, or plastics are recycled. Write a 1-page paper that describes the process.

16-5 Trees

Objectives

After studying this lesson, you should be able to:
- find spanning trees, and
- find minimal spanning trees.

Application

The graph on the left below is a model of all of the possible communication lines linking seven cities in Colorado. Due to the enormous cost of building such a network, the communications company is seeking a more efficient way to link all seven cities together. The graph on the right accomplishes this task at a much lower cost. This graph is called a **tree**.

> **FYI...**
>
> Today's VCR Plus+™, a device that automatically programs a VCR, operates by using a coding tree invented in 1951 by a 25-year-old MIT student, David A. Huffman.

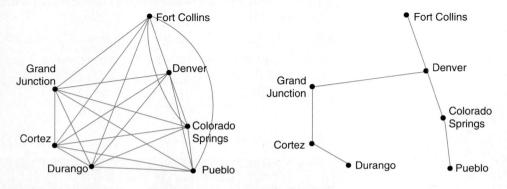

Properties of Trees

Trees are connected graphs that have the following properties.
1. **They contain no cycles.**
2. **There is exactly one path between any two vertices.**
3. **If a tree contains more than one vertex, then it will have at least two vertices of degree 1.**
4. **A tree with n vertices will always have $n - 1$ edges.**

A tree diagram, which you used in Chapter 14, is a type of tree.

Of the figures shown below, only the graph in the center is a tree. The graph on the left contains a cycle, so it is not a tree. The graph on the right has 7 vertices and 7 edges, so it is also not a tree.

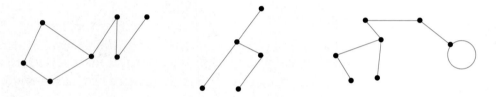

A **subgraph** is a graph that contains a subset of the edges and vertices of a larger graph. A tree that is a subgraph of a connected graph and includes all of the vertices of that graph is called a **spanning tree**. Graphs may contain more than one spanning tree.

Two different spanning trees are shown below for the same graph.

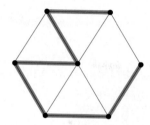

Since finding spanning trees in a large graph could be difficult, an algorithm exists to help you find them.

Spanning Tree Algorithm

To find a spanning tree in a graph, use the following process.
Step 1 Locate a cycle in the graph.
Step 2 Remove an edge from the cycle that will not disconnect the graph.
Step 3 If there are cycles left, then repeat Step 2. Otherwise, stop.

Example 1

Find a spanning tree in the graph at the right.

Remove *g* from cycle *a*, *b*, *g*.

Remove *a* from cycle *a*, *b*, *h*, *i*.

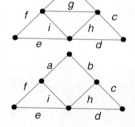

Remove *f* from cycle *f*, *i*, *e*.

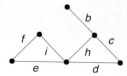

Remove *c* from cycle *c*, *d*, *h*.

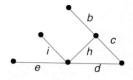

No cycles remain, so the result is a spanning tree.

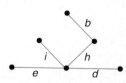

Since there are often several spanning trees for each graph, one of them must have a weight that is less than or equal to the weight of all the rest. This tree is called the **minimal spanning tree**. There may be more than one minimal spanning tree for a graph, but their weights will be the same.

You can use **Kruskal's algorithm** to find a minimal spanning tree.

Kruskal's Algorithm	**To find a minimal spanning tree in a weighted graph, use the following process.** **Step 1 Choose the edge with the least weight.** **Step 2 Choose the next edge with the least weight, as long as it does not create a cycle.** **Step 3 Continue as in Step 2 until all of the vertices are connected.**

Example 2

At a campsite in northern Michigan, there are seven main areas: 2 lodges, 2 cabin areas, a restaurant, the caretaker's home, and a recreation center. Tired of driving on dirt roads, the owner decides to put in paved roads. The owner must decide the most economical way to put the roads in so that people can get from one location to another. The firm that will do the paving has drawn the weighted graph at the right to show the roads that are possible. The cost to put in each road is indicated on the graph in thousands of dollars. Find the least expensive way to build the roads and still connect all of the main areas.

Since we can begin at any vertex, let's start at CH. Choose the edge of least weight incident on CH, the edge between CH and C_2. Next, choose the edge of least weight incident on C_2, the edge between C_2 and R.

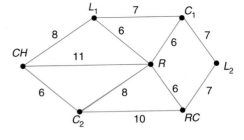

Since there are 3 edges incident on R with weight 6, choose one at random. Let's choose the edge between R and L_1. Since the edge between R and C_1 has less weight than the edge between L_1 and C_1, choose it.

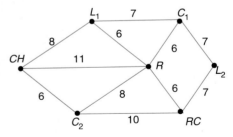

The next edge of least weight is the edge between R and RC, so choose it. Finally, choose the edge between L_2 and RC.

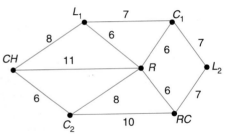

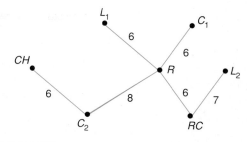

The roads to be constructed are shown at the right, at a minimal cost of $39,000.

MATH JOURNAL

Describe something new you learned in this lesson. Be sure to give examples.

CHECKING FOR UNDERSTANDING

Communicating Mathematics

Read and study the lesson to answer each question.

1. **Draw** a graph with a cycle.

2. **Draw** a graph without a cycle that is not a tree.

3. **Draw** a graph that is a tree.

4. **Draw** a graph that does not have a spanning tree.

5. **Design** a telephone tree for you and five of your closest friends. Is your tree a spanning tree?

Guided Practice

6. How many vertices are there in a tree with 21 edges?

7. How many edges are there in a tree with 18 vertices?

8. How many vertices of a graph does its spanning tree contain?

9. How many edges will a spanning tree have if the graph in which it is contained has 7 vertices?

10. How many vertices of degree 1 does a tree of 2 or more vertices have?

11. Find a spanning tree for the graph below.

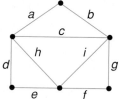

12. Find a minimal spanning tree for the weighted graph below.

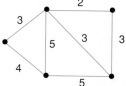

EXERCISES

Practice

Determine whether each graph is a tree. Write *yes* or *no*. If no, explain.

13.

14.

15.

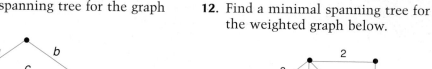

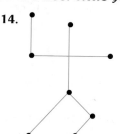

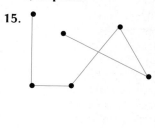

Find a spanning tree for each graph.

16.
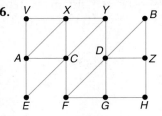

17.
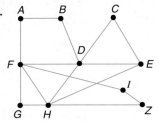

18.
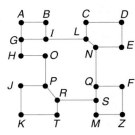

Find a minimal spanning tree for each weighted graph. State the weight of the tree.

19.
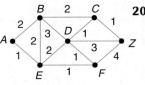

20.
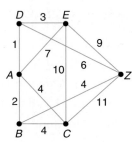

21.
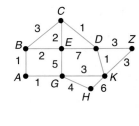

Critical Thinking

22. Draw a tree with at least 7 vertices that has exactly 2 vertices of degree 1.

23. Draw a weighted graph in which the minimal spanning tree must contain the edge of greatest weight.

Applications and Problem Solving

24. **Communications**
A new fiber optics communication system is to be installed that will serve the cities in the map at the right. The communications company has determined that the costs to connect each pair of cities are those that are shown on the map (weights are in millions of dollars).
Find the minimal cost to establish such a system and show its design.

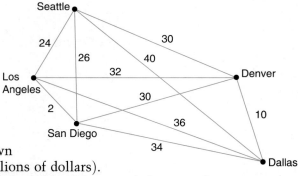

interNET CONNECTION

For the latest developments in fiber optic technology, visit:
www.glencoe.com/ sec/math/amc/ mathnet

25. **Computer Technology** The mathematics editorial department of Glencoe Publishing Company plans to network fifteen computers to the same laser printer. How might this be done using the fewest communications lines possible?

Mixed Review

26. Determine if $\text{Tan}^{-1}x = \dfrac{1}{\text{Tan}\,x}$ for all x. If false, give a counterexample. **(Lesson 6-6)**

27. Find the length of the tangent segment from the point with coordinates $(4, -1)$ to the graph of a circle with the equation $(x + 3)^2 + y^2 - 4 = 0$. **(Lesson 10-8)**

28. **Business**   A buyer for a department store will accept a carton containing 10 clocks if neither of two samples, chosen at random, is defective. What is the probability that she will accept a carton of 10 if it contains 4 defective clocks? **(Lesson 14-8)**

29. Find a minimal path and the minimal distance from A to D in the weighted multigraph at the right.   **(Lesson 16-4)**

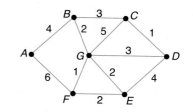

30. **College Entrance Exam**   Compare quantities A and B below.
Write A if quantity A is greater.
Write B if quantity B is greater.
Write C if the two quantities are equal.
Write D if there is not enough information to determine the relationship.
(A) $(-8)^{62}$ **(B)** $(-8)^{75}$

DECISION MAKING

Planning a Party

In the movie *Father of the Bride*, the title character, played by Steve Martin, was ridiculed for being reluctant to spend $50,000 on his daughter's wedding. Martin had good reasons for balking, but he was quickly talked out of them. How can you decide on the right amount of money to spend on a party and then, keep your expenses under the limit you have set? What arrangements can you take care of yourself and which ones should you leave to professionals?

1. Describe the ideal party. Where would you hold it? What food and entertainment would you provide? How many people would be on your guest list? List all of the ingredients you think are important.

2. Decide on the amount of money you can spend on a party. Then go back to the list in Exercise 1 and decide how you can hold the best possible party for the money. Which elements of the perfect party must you eliminate? Which ones can you provide at a level that is slightly less than "perfect?"

3. Plan your party, keeping within your budget. Include such factors as location, food, entertainment, number of people, and other essential details.

4. **Project**   Work with your group to complete one of the projects listed at the right based on the information you gathered in Exercises 1–3.

PROJECTS

- *Conduct an interview.*
- *Write a book report.*
- *Write a proposal.*
- *Write an article for the school paper.*
- *Write a report.*
- *Make a display.*
- *Make a graph or chart.*
- *Plan an activity.*
- *Design a checklist.*

16-6 Graphs and Matrices

Objectives

After studying this lesson, you should be able to:
- use digraphs to solve problems, and
- represent digraphs as matrices and vice versa.

Application

Biologists use graphs called food webs to model the predatory relationships among members of a particular ecosystem. In the food web below, arrows point from each member of the ecosystem to its predator.

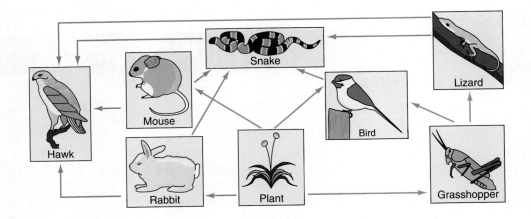

The graph at the right represents the food web relationships. Notice that arrowheads have been placed along the edges to represent the direction of each relationship. A graph with arrowheads on the edges is called a **directed graph,** or **digraph.**

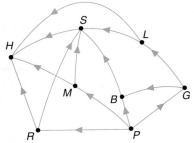

In a simple graph, you counted the number of incident edges to describe the degree of a vertex. Similarly, arrowheads are counted to describe a vertex in a digraph. The **indegree** of a vertex in a digraph is the number of edges pointing into the vertex. The **outdegree** of a vertex is the number of edges pointing away from the vertex. In the digraph above, the indegree of vertex G is 1 and the outdegree is 2.

Digraphs are often used to determine whether a location is **reachable.** In the graph above, *H* is reachable from *P* because there is a **directed path** from *P* to *H*. One such path is {*P,M*}, {*M,H*}. A directed path occurs when you can walk each edge in the direction of the arrows.

Example 1

The Beaumont Museum of Art is being renovated. In order to remain open during the renovation, one-way corridors will be used to control the flow of visitors. The digraph at the right represents the floor plan of the museum's corridors.

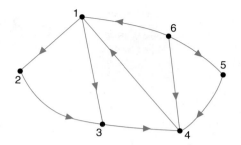

a. Are all of the exhibits reachable?
b. If not, modify the graph to make each exhibit reachable and simplify traffic flow.

a. Currently, Exhibit 6 is not reachable, because it has indegree 0.
b. To solve the problem, change the direction of the edge between Exhibit 6 and Exhibit 4.

Digraphs can also be used in scheduling a complex task. Suppose you are planning to cook dinner for friends, but you only have one hour in which to prepare. The tasks that you need to complete are shown in the chart at the right.

Task	Time Required
U: Set Table	5 minutes
V: Marinate Steaks	15 minutes
W: Grill Steaks	8 minutes
X: Bake Potatoes	40 minutes
Y: Make Salad	10 minutes
Z: Change Clothes	10 minutes

If you did one task at a time, it would take you 88 minutes to prepare. However, some of the tasks could be done simultaneously. For example, you could marinate the steaks, bake the potatoes, and set the table at the same time. However, you can't grill the steaks until you have marinated them. The digraph below is one way to model the situation.

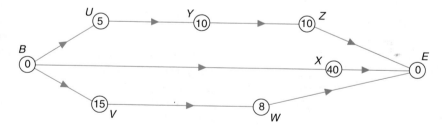

Notice that the times required are shown inside the vertices. Two vertices B and E have been added to signify when the tasks begin and end. You want to find the shortest time to complete all of the tasks. That time is shown by the longest path from B to E. This is called the **critical path.** There are three paths from B to E. The longest path from B to E is $B \rightarrow X \rightarrow E$, which requires 40 minutes. That means that the least amount of time needed to prepare the meal is 40 minutes.

Example 2

APPLICATION

Manufacturing

Suppose tasks *T* through *Z* are required to manufacture athletic shoes. The tasks, the time each task requires, and the tasks that must precede each task are given in the chart below.

Task	Time Required	Prerequisite Tasks
T	11 minutes	none
U	6 minutes	none
V	8 minutes	none
W	16 minutes	*T, U*
X	4 minutes	*U*
Y	3 minutes	*V, W, X*
Z	9 minutes	*W*

Draw a digraph to model this situation and find the critical path. Then find the minimum amount of time needed to complete all of the tasks.

One view of the digraph is shown below.

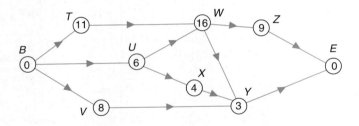

The critical path is $B \rightarrow T \rightarrow W \rightarrow Z \rightarrow E$. The minimum amount of time needed is 36 minutes.

In many situations, graphs can be very complex and difficult to analyze. Therefore, it is advantageous to describe a graph numerically so you could use a calculator or a computer. This can be done by using a matrix. Since we can count the number of edges joining the vertices of a graph, a matrix of a graph can be formed. This matrix is called an **adjacency matrix**.

*Definition of
Adjacency Matrix*

For any graph $G(n, m)$,

1. the adjacency matrix representing it, denoted $M(G)$, is an $n \times n$ matrix, and

2. the entry in the *i*th row and the *j*th column of the matrix represents the number of edges from V_i to V_j.

If the graph is a digraph, then the (i, j)th entry describes the number of directed edges from V_i to V_j.

Example 3 | **Find $M(G)$ for each graph.**

a.

b.

a. Since there are 4 vertices, the adjacency matrix will have 4 rows and 4 columns. There are no edges from R to R, so the entry in row 1, column 1 is 0. There are 2 edges from R to S, so the entry in row 1, column 2 is 2. Continuing the process gives the matrix below.

$$\begin{array}{c} \\ R \\ S \\ T \\ U \end{array}\begin{array}{c} \begin{array}{cccc} R & S & T & U \end{array} \\ \begin{bmatrix} 0 & 2 & 0 & 1 \\ 2 & 0 & 0 & 0 \\ 0 & 0 & 0 & 1 \\ 1 & 0 & 1 & 0 \end{bmatrix} \end{array}$$

b. Since there are 3 vertices, the adjacency matrix will have 3 rows and 3 columns. There is 1 directed edge from X to X, so the entry in row 1, column 1 is 1. There is 1 directed edge from X to Y, so the entry in row 1, column 2 is 1. Continuing the process gives the matrix below.

$$\begin{array}{c} \\ X \\ Y \\ Z \end{array}\begin{array}{c} \begin{array}{ccc} X & Y & Z \end{array} \\ \begin{bmatrix} 1 & 1 & 0 \\ 1 & 0 & 1 \\ 1 & 1 & 0 \end{bmatrix} \end{array}$$

You can use matrices and the theorems below to solve problems.

Matrix Theorem

If G is a digraph and $M(G)$ is its matrix, and if n is a positive integer, then the (i, j)th entry of $[M(G)]^n$ is the number of directed paths in G that go from V_i to V_j in exactly n steps.

If we look at $[M(G)]^2$, that will tell us how many 2-step paths there are between vertices. $[M(G)]^3$ will tell us how many 3-step paths there are, and so on. But how do we know the maximum number of steps that can be checked? The theorem below determines how many powers of $M(G)$ we need to check.

Path Theorem

In a digraph that has n vertices, if two distinct vertices can be joined by a path, then they can be joined by a path that is no longer than $n - 1$ edges.

Example 4

APPLICATION

Biology

The digraph at the right is a subgraph of the food web shown at the beginning of the lesson. Suppose this ecosystem is accidently contaminated by a poisonous substance, like a pesticide. When this occurs, one species becomes contaminated and then others that prey on that species also become contaminated, and so on. If the plant is contaminated, will the contamination spread to all of the others in two steps or less?

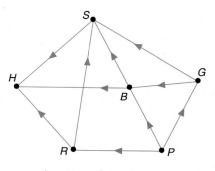

(continued on the next page)

Since there are 6 vertices, the adjacency matrix will have 6 rows and 6 columns. Label the rows and columns with the vertex names. To fill the entries, count the number of directed edges from a row to a column. For example, there are no edges from H to H, so the entry in row 1, column 1 is 0. The entire adjacency matrix is shown at the right.

$$\begin{array}{c} \\ H \\ S \\ B \\ G \\ P \\ R \end{array} \begin{array}{cccccc} H & S & B & G & P & R \\ \left[\begin{array}{cccccc} 0 & 0 & 0 & 0 & 0 & 0 \\ 1 & 0 & 0 & 0 & 0 & 0 \\ 1 & 1 & 0 & 0 & 0 & 0 \\ 0 & 1 & 1 & 0 & 0 & 0 \\ 0 & 0 & 1 & 1 & 0 & 1 \\ 1 & 1 & 0 & 0 & 0 & 0 \end{array}\right] \end{array}$$

Using a TI-82 graphing calculator, enter the above values in matrix [A]. You want to determine whether the contamination will spread to all the members in two steps or less. The sum $M(G) + M(G)^2$ represents the total number of paths of length less than or equal to 2.

To find $M(G) + M(G)^2$, press
$\boxed{\text{MATRX}}$ 1 $\boxed{+}$ $\boxed{\text{MATRX}}$ 1 $\boxed{x^2}$ $\boxed{\text{ENTER}}$.

$$\begin{array}{c} \\ H \\ S \\ B \\ G \\ P \\ R \end{array} \begin{array}{cccccc} H & S & B & G & P & R \\ \left[\begin{array}{cccccc} 0 & 0 & 0 & 0 & 0 & 0 \\ 1 & 0 & 0 & 0 & 0 & 0 \\ 2 & 1 & 0 & 0 & 0 & 0 \\ 2 & 2 & 1 & 0 & 0 & 0 \\ 2 & 3 & 2 & 1 & 0 & 1 \\ 2 & 1 & 0 & 0 & 0 & 0 \end{array}\right] \end{array}$$

$M(G) + M(G)^2$ is shown at the right. Notice that the row corresponding to the plant (P) shows paths to every member of the ecosystem except itself. Therefore, the contamination can spread from the plant to all the other members of the ecosystem in two steps or less.

CHECKING FOR UNDERSTANDING

Communicating Mathematics

Read and study the lesson to answer each question.

1. **Describe** a situation that was not discussed in this lesson that may be modeled by a digraph.

2. **Explain** how to find the critical path in a digraph.

3. **Construct** an adjacency matrix to describe the airline service digraph shown at the right.

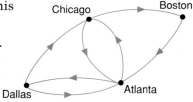

Guided Practice

4. **Catering** JoAnn's Catering is catering a retirement party. JoAnn's employees will have to complete the following tasks.

Task	Time Required
P: Decide on menu	30 minutes
Q: Purchase food	60 minutes
R: Set up kitchen on site	15 minutes
S: Cook food	50 minutes
T: Set tables	10 minutes
U: Serve food	40 minutes

Draw a digraph to model the situation. Then find its critical path and the minimum amount of time needed to complete all of the tasks. You may decide which activities should precede others.

Find $M(G)$ for each graph.

5.

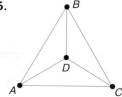

6.

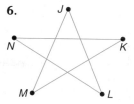

7.

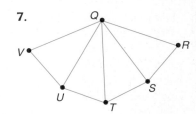

Draw the graph for which each matrix is $M(G)$.

8.

$$\begin{array}{c c c} & S & T \\ S & \begin{bmatrix} 0 & 1 \\ 1 & 0 \end{bmatrix} \\ T & \end{array}$$

9.

$$\begin{array}{c c c c} & P & Q & R \\ P & \begin{bmatrix} 0 & 1 & 1 \\ 1 & 0 & 0 \\ 1 & 0 & 0 \end{bmatrix} \\ Q & \\ R & \end{array}$$

10.

$$\begin{array}{c c c c c} & A & B & C & D \\ A & \begin{bmatrix} 0 & 1 & 2 & 1 \\ 1 & 0 & 1 & 0 \\ 2 & 1 & 0 & 1 \\ 1 & 0 & 1 & 0 \end{bmatrix} \\ B & \\ C & \\ D & \end{array}$$

EXERCISES

Practice

Draw a digraph to represent each situation.

11. In a volleyball tournament, the Lions defeated the Tigers and the Bulls, the Bulls defeated the Tigers and the Rams, the Rams defeated the Lions, and the Tigers defeated the Rams.

12. Kama knows the locker number of Ursula. Beth knows the locker numbers of Sam and Kama. Sam knows the locker numbers of Ursula and Tammy. Tammy knows the locker number of Ursula, who knows the locker numbers of Kama and Ray.

Find $M(G)$ for each graph.

13.

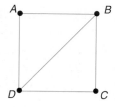

14.

15.

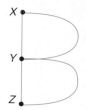

Find $M(G)$ for each digraph.

16.

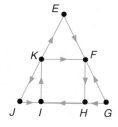

17.

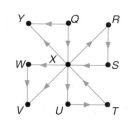

18.

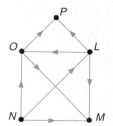

Draw the graph for which each matrix is $M(G)$.

19.
$$\begin{array}{c} \\ A \\ B \\ C \end{array} \begin{array}{ccc} A & B & C \\ \begin{bmatrix} 0 & 1 & 1 \\ 1 & 0 & 1 \\ 1 & 1 & 0 \end{bmatrix} \end{array}$$

20.
$$\begin{array}{c} \\ D \\ E \\ F \\ G \end{array} \begin{array}{cccc} D & E & F & G \\ \begin{bmatrix} 0 & 1 & 0 & 1 \\ 1 & 0 & 1 & 1 \\ 0 & 1 & 0 & 1 \\ 1 & 1 & 1 & 0 \end{bmatrix} \end{array}$$

21.
$$\begin{array}{c} \\ J \\ K \\ L \\ M \\ N \end{array} \begin{array}{ccccc} J & K & L & M & N \\ \begin{bmatrix} 0 & 2 & 1 & 0 & 1 \\ 2 & 0 & 1 & 0 & 2 \\ 1 & 1 & 0 & 2 & 1 \\ 0 & 0 & 2 & 0 & 1 \\ 1 & 2 & 1 & 1 & 0 \end{bmatrix} \end{array}$$

Draw the digraph for which each matrix is $M(G)$.

22.
$$\begin{array}{c} \\ X \\ Y \\ Z \end{array} \begin{array}{ccc} X & Y & Z \\ \begin{bmatrix} 0 & 1 & 1 \\ 1 & 0 & 1 \\ 1 & 1 & 0 \end{bmatrix} \end{array}$$

23.
$$\begin{array}{c} \\ A \\ B \\ C \\ D \end{array} \begin{array}{cccc} A & B & C & D \\ \begin{bmatrix} 0 & 1 & 1 & 1 \\ 0 & 0 & 1 & 1 \\ 1 & 0 & 0 & 0 \\ 0 & 0 & 1 & 0 \end{bmatrix} \end{array}$$

24.
$$\begin{array}{c} \\ R \\ S \\ T \\ U \end{array} \begin{array}{cccc} R & S & T & U \\ \begin{bmatrix} 0 & 1 & 0 & 1 \\ 1 & 0 & 1 & 0 \\ 1 & 1 & 0 & 1 \\ 0 & 1 & 0 & 0 \end{bmatrix} \end{array}$$

Draw a digraph to model each situation. Then find its critical path and the minimum time needed to complete all of the tasks. If precedence is not given, you may decide which tasks should precede others.

25. **Manufacturing**   A clothing manufacturer uses the schedule below to introduce a new line of sportswear.

Task	Time Required
R: Decide on the pieces in the new collection	1 week
S: Get a celebrity to endorse the line	2 weeks
T: Create advertising copy, plan commercials, and so on	4 weeks
U: Set manufacturer's suggested prices	1 week
V: Plan and hold fashion shows for buyers and the press	4 weeks
W: Choose department stores that will sell the line	1 week
X: Ship the line to retail outlets	3 weeks

26. **Construction**   Canarsie Builders builds homes using the schedule below.

Task	Time	Prerequisite Tasks
L: Site preparation	5 days	none
M: Laying the foundation	8 days	Site preparation
N: Drains and utilities	3 days	Site preparation
O: Framing	12 days	Laying the foundation
P: Roofing and gutters	3 days	Framing
Q: Windows and exterior doors	2 days	Roofing and gutters
R: Heating and cooling system	4 days	Windows and exterior doors
S: Plumbing	4 days	Windows and exterior doors
T: Electrical wiring	4 days	Heating, cooling, and plumbing
U: Insulation and drywall	12 days	Electrical wiring
V: Siding and painting exterior	8 days	Windows and exterior doors
W: Interior painting and finishing	24 days	Insulation and drywall
X: Landscaping	3 days	Siding and painting exterior

27. Suppose G is a connected graph. If you are given $M(G)$, how can you tell whether G contains an Euler circuit without drawing it?

28. Sociology Suppose the digraph at the right is a model for the way rumors spread among a group of friends in the Drama Club.

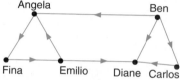

  a. Represent the data in the graph in matrix form.

  b. If you wanted to start a rumor, what one person in this group could you tell and expect the rumor to spread to all the others?

29. Transportation Suppose an airline's service routes are represented in the digraph at the right.

  a. Represent the data in the digraph in matrix form.

  b. In how many ways can a traveller get from Denver to St. Louis in three or fewer flights?

  c. Is it possible for a traveller to get from Chicago to Houston? If so, what is the minimum number of flights needed to accomplish this?

30. Find the distance from the line $3x - 7y - 1 = 0$ to the point with coordinates $(-1, 4)$. **(Lesson 7-7)**

31. Use logarithms to evaluate $\dfrac{\sqrt[4]{0.0063}}{6.73}$. **(Lesson 11-5)**

32. Find a set of numbers with a mean of 6, a median of 5.5, and a mode of 9. **(Lesson 15-2)**

33. Find a minimal spanning tree for the weighted graph at the right. **(Lesson 16-5)**

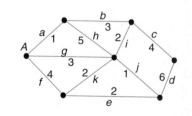

34. College Entrance Exam Choose the best answer.
  If $(a - b)^2 = 64$, and $ab = 3$, find $a^2 + b^2$.
  (A) 58 **(B)** 61 **(C)** 67 **(D)** 70 **(E)** 75

VOCABULARY

Upon completing this chapter, you should be
familiar with the following terms:

adjacency matrix **898**

adjacent **868**

algorithm **879**

breadth-first search algorithm **883**

bridge **874**

circuit **873**

complete graph **869**

connected **868, 873**

critical path **897**

cycle **873**

degree **868**

digraph **896**

Dijkstra's algorithm **885**

directed path **896**

disconnected **873**

edges **866**

Euler circuit **878**

Euler path **878**

Fleury's algorithm **879**

graph **866**

incident **868**

indegree **896**

868 isolated

892 Kruskal's algorithm

866 loops

885 minimal distance

885 minimal path

891 minimal spanning tree

866 multigraph

896 outdegree

866 parallel edges

872 path

896 reachable

866 simple graph

890 spanning tree

890 subgraph

872 trail

890 tree

866 vertex

872 walk

885 weight

885 weight of a path

883 weighted multigraph

SKILLS AND CONCEPTS

OBJECTIVES AND EXAMPLES

Upon completing this chapter, you should
be able to:

- determine whether a graph is complete
  (Lesson 16-1)

  Is a graph with 4 vertices and 6 edges a
  complete graph?

  $$\frac{n(n-1)}{2} = \frac{4(3)}{2}$$
  $$= 6$$

  Since $6 = 6$, the graph is complete.

REVIEW EXERCISES

Use these exercises to review and prepare
for the chapter test.

**Use the graph below to answer each
question.**

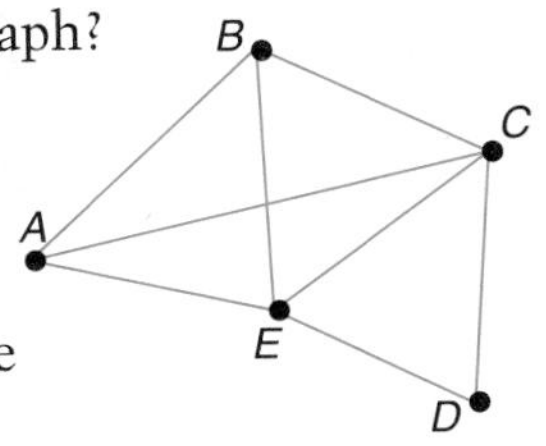

1. Is it a complete graph?

2. What is $\deg(E)$?

3. Which edges are
   incident on A?

4. Which vertices are
   adjacent to D?

5. Is the figure a multigraph?

- distinguish among walks, paths, trails, closed walks, circuits, and cycles
(**Lesson 16-2**)

Find an $A-E$ walk, an $A-E$ trail, and an $A-E$ path for the multigraph below.

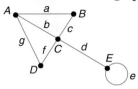

One $A-E$ walk is $a, c, b, g, f, b, a, c, d$.
One $A-E$ trail is g, f, b, a, c, d.
One $A-E$ path is b, d.

- find Euler paths and circuits
(**Lesson 16-3**)

Determine whether the graph below contains an Euler circuit or an Euler path. If so, name the circuit or path.

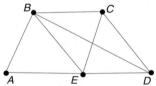

This graph does not contain an Euler circuit because the degree of each vertex is not even. The path $C, D, B, E, C, B, A, E, D$ is an Euler path since it contains all of the edges exactly once and ends at a vertex that is different from the starting vertex.

- use the breadth-first search algorithm and Dijkstra's algorithm to find a shortest path (**Lesson 16-4**)

Find a shortest path from A to E in the graph below. Then state its length.

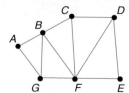

Paths A, G, F, E and A, B, F, E have length 3.

Use the graph below to determine whether each walk is a circuit, cycle, path, trail, or walk. Use the most specific name.

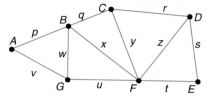

6. p, q, y, x, q, r, s

7. p, w, u, z, s, t

8. p, q, r, s, t, u, v

9. x, y, r, s, t, u, v, p

10. Find an Euler path in the graph below.

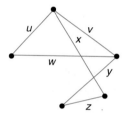

11. How could you alter the graph in Exercise 10 so it would contain an Euler circuit?

12. Find the minimal path and minimal distance from A to E in the weighted multigraph below.

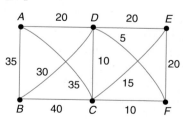

- find minimal spanning trees
(Lesson 16-5)

Find a minimal spanning tree and its weight for the graph below.

One minimal spanning tree is $\{S, P\}, \{P, Q\}, \{Q, T\}, \{T, V\}, \{V, W\}, \{W, U\}, \{T, R\}$ with a weight of $5 + 5 + 7 + 6 + 4 + 3 + 5$ or 35.

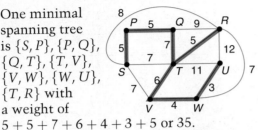

13. Find a spanning tree for the graph below.

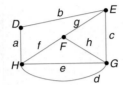

14. Find a minimal spanning tree and its weight for the graph below.

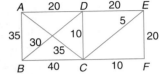

- represent digraphs as matrices and vice versa **(Lesson 16-6)**

Find $M(G)$ for the digraph at the right.

Since there are 4 vertices, the matrix will have 4 rows and 4 columns. There are no edges from P to P, so the entry in row 1, column 1 is 0. There is 1 edge from P to Q, so the entry in row 1, column 2 is 1. Continuing the process gives the matrix at the right.

$$\begin{array}{c} \\ P \\ Q \\ R \\ S \end{array} \begin{array}{cccc} P & Q & R & S \\ \left[\begin{array}{cccc} 0 & 1 & 0 & 0 \\ 0 & 0 & 1 & 0 \\ 2 & 0 & 0 & 0 \\ 1 & 0 & 1 & 0 \end{array}\right] \end{array}$$

Find $M(G)$ for each graph or digraph below.

15.

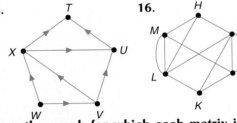

16.

Draw the graph for which each matrix is $M(G)$.

17.
$$\begin{array}{c} \\ A \\ B \\ C \\ D \\ E \end{array} \begin{array}{ccccc} A & B & C & D & E \\ \left[\begin{array}{ccccc} 0 & 0 & 0 & 1 & 0 \\ 1 & 0 & 1 & 0 & 1 \\ 0 & 0 & 0 & 0 & 0 \\ 0 & 0 & 1 & 0 & 2 \\ 2 & 1 & 2 & 0 & 0 \end{array}\right] \end{array}$$

18.
$$\begin{array}{c} \\ F \\ G \\ H \\ I \\ J \\ K \end{array} \begin{array}{cccccc} F & G & H & I & J & K \\ \left[\begin{array}{cccccc} 0 & 1 & 0 & 0 & 0 & 0 \\ 1 & 0 & 0 & 0 & 0 & 0 \\ 0 & 0 & 0 & 1 & 0 & 1 \\ 0 & 0 & 1 & 0 & 1 & 0 \\ 0 & 0 & 0 & 1 & 0 & 1 \\ 0 & 0 & 1 & 0 & 1 & 0 \end{array}\right] \end{array}$$

APPLICATIONS AND PROBLEM SOLVING

19. Transportation The Lopez family is moving to a new city. They need to do the things in the chart to make the move. Draw a digraph to model the situation and find a critical path. Then find the minimum amount of time needed to complete all of the tasks. **(Lesson 16-6)**

Task	Time Required
G: Get packing boxes	1 day
H: Secure a new home	30 days
I: Pack	7 days
J: Sell current home	30 days
K: Buy clothing appropriate for new climate	3 days
L: Send boxes to new address	5 days
M: Secure a moving company	3 days

Use the graph to complete Exercises 1—1.

1. Is this a simple graph or a multigraph? Explain.

2. What is deg (I)?

3. Which edges are incident on D?

4. Is this a complete graph?

5. Which vertices are adjacent to A?

6. Is this graph connected?

7. Determine whether l, m, n, s, t, u, v, y is a walk, trail, path, circuit, or cycle.

8. Determine whether the graph contains an Euler circuit or an Euler path. If so, name it.

9. Find a shortest path from A to K.

10. Find a spanning tree for the graph.

11. Find $M(G)$ for the graph.

12. What is the difference between a graph and a multigraph?

13. Draw K_6.

14. Draw a connected graph with four vertices, two of even degree and two of odd degree.

15. What is the minimum number of edges in a connected graph that has six vertices?

16. *True* or *false:* Every $U-V$ walk contains a $U-V$ path.

17. What is the difference between a path and an Euler path?

18. If a graph contains an Euler path, where does it begin and end?

19. **Recreation** The honors club is planning a trip to a nearby amusement park. Their faculty sponsor wants to make sure that only boys who are friends go in the same car. If the list below names the boys and their friends, what is the least number of cars needed?

Adam: Bob, Frank, Ho-Chin
Bob: Adam, Carlos, Dave, Ho-Chin
Carlos: Bob, Frank
Dave: Bob, Evan, Frank

Evan: Dave, Greg
Frank: Adam, Carlos, Dave, Greg, Ho-Chin
Greg: Evan, Frank
Ho-Chin: Adam, Bob, Frank

20. **Telecommunications** A telecommunications company wants to develop a network to connect six cities. The table below lists the cities and the cost (in thousands of dollars) to connect each pair. What is the minimum cost, and which cities should be connected?

City	Sarasota	St. Augustine	Lakeland	Jacksonville	Daytona Beach
St. Augustine	18				
Lakeland	9	35			
Jacksonville	12	14	11		
Daytona Beach	16	9	6	19	
Bunnell	23	4	18	26	14

Bonus Find the value of k when
$m\angle U + m\angle V + m\angle W + m\angle X + m\angle Y + m\angle Z = 45k.$

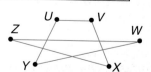

4 UNIT REVIEW

Solve. (Lessons 12-1 and 12-2)

1. Find the 20th term of the arithmetic sequence for which $a_1 = 3$ and $d = -2$.

2. Find the sum of the first nine terms of the geometric series $2 + 4 + 8 + \cdots$.

Evaluate each limit, or state that the limit does not exist. (Lesson 12-3)

3. $\lim\limits_{n\to\infty} \dfrac{4n + 1}{3n}$

4. $\lim\limits_{n\to\infty} \dfrac{n^2 - 1}{n}$

Determine whether each series is *convergent* or *divergent*. (Lesson 12-4)

5. $1 + 4 + 7 + 10 + \cdots$

6. $6 + 2 + \dfrac{2}{3} + \dfrac{2}{9} + \cdots$

7. $\dfrac{3}{1^1} + \dfrac{3}{2^2} + \dfrac{3}{3^3} + \cdots$

Write each expression in expanded form and find the sum. (Lesson 12-5)

8. $\displaystyle\sum_{a=2}^{8} (3a - 6)$

9. $\displaystyle\sum_{k=0}^{6} 7\left(\dfrac{1}{2}\right)^k$

Find the designated term of each binomial expansion. (Lesson 12-6)

10. 4th term of $(x + 1)^9$

11. 7th term of $(x - 2y)^{12}$

Find the first three iterates of each function using the given initial value. If necessary, round your answers to the nearest hundredth. (Lesson 13-1)

12. $f(x) = 3x + 1, x_0 = 1$

13. $f(x) = x^2 - 5, x_0 = -2$

Find the x-coordinate of the fixed point for each function. Is it a repeller or an attractor? (Lesson 13-2)

14. $f(x) = 1.5x + 2$

15. $f(x) = -0.5x + 3$

Find the first three iterates of the function $f(z) = 2z + 3i$ for each initial value. (Lesson 13-4)

16. $z_0 = -i$

17. $z_0 = 3 - i$

Determine whether the graph of each value is in the prisoner set or the escape set for the function $f(z) = z^2$. (Lesson 13-5)

18. $-0.5 + 0.5i$

19. $2 + i$

Solve. (Lessons 14-1, 14-2, 14-3)

20. The letters a, b, c, d, and e are to be used to form five letter patterns. How many patterns can be formed if repetitions are not allowed?

21. How many ways can the letters of the word *COLOR* be arranged?

22. From a group of 3 men and 5 women, how many committees of 2 men and 2 women can be formed?

State the odds of an event occurring given the probability of the event. (Lesson 14-4)

23. $\dfrac{2}{7}$

24. $\dfrac{1}{14}$

Solve. (Lessons 14-5, 14-6, 14-7, and 14-8)

25. Three cards are drawn at random from a standard deck of 52 cards. What is the probability that all are clubs?

26. Find the probability of getting a sum of 6 on the first throw of two dice and a sum of 2 on the second throw.

27. Find the probability of a sum of 7 or 9 on a single throw of two dice.

28. One card is drawn from a standard deck of 52 cards. What is the probability that it is a king if it is known to be a face card?

The test scores for Mrs. Humphrey's math class are listed.   (Lessons 15-2 and 15-3)

89	95	65	70	77	82	66
69	91	82	77	99	65	89
72	80	42	76	86	77

29. Make an array of the data.

30. Find the range of the data.

31. Find the mean of the data.

32. Find the median of the data.

33. Find the mode of the data.

34. Find the mean deviation of the data.

35. Find the semi-interquartile range of the data.

36. Find the standard deviation of the data.

Solve.   (Lessons 15-4 and 15-5)

37. A value is selected at random from a normally distributed set of data. Suppose the mean, $\bar{x}$, is 70 and the standard deviation, σ, is 6. What is the probability that the selected value lies within the limits 65 and 75?

38. The mean height of a random sample of 169 junior girls is 65.2 inches, and the standard deviation is 2.2 inches. Find the range of the sample mean that has a 5% level of confidence.

The table below shows the calculus grades and the physics grades for a group of college students at the end of a quarter.   (Lesson 15-6)

Student	A	B	C	D	E	F	G	H
Calculus Grade	90	78	56	32	45	69	75	79
Physics Grade	85	65	78	60	72	62	70	85

39. Draw a scatter plot relating calculus grades to physics grades and find a median-fit line for the data.

40. Use a graphing calculator to plot the data and draw a line of regression.

Use the graph at the right to answer each question.   (Lesson 16-1)

41. Is it a complete graph?

42. What is deg (D)?

43. Which edges are incident on A?

Use the graph below to determine whether each walk is a *circuit, cycle, path, trail,* or *walk.* Use the most specific name. (Lesson 16-2)

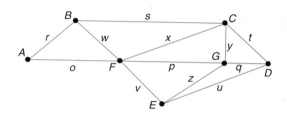

44. p, q, t, s, r

45. v, u, t, y, p

46. w, s, x, v, u

Solve.   (Lessons 16-3 and 16-4)

47. Does the multigraph below contain an Euler Circuit? Explain.

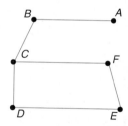

48. Use the breadth-first search algorithm to find a shortest path from A to E in the graph below. State its length.

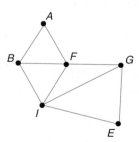

UNIT 5

AN INTRODUCTION TO CALCULUS

Chapter 17 Limits, Derivatives, and Integrals

For many of you, calculus will be largely uncharted territory. However, calculus is one of the most important areas of mathematics. There are two branches of calculus, *differential* calculus and *integral* calculus. Differential calculus deals mainly with variable, or changing, quantities. In this chapter, you will study problems of optimization, where you will find the maximum or minimum value, and problems of motion, where you will study acceleration and velocity.

Integral calculus deals mainly with finding sums of infinitesimally small quantities. This generally involves finding a limit. German astronomer Johannes Kepler used this method to calculate the area of the ellipses formed by the orbits of the planets. In this chapter, you will use calculus to find the areas and volumes of many objects.

The relationship between differential and integral calculus is not obvious, but it does exist. Differentiation and integration are actually inverse operations, which you studied in algebra. In order to succeed in your study of calculus, you will need to use your knowledge of algebra and geometry. Upon completion of your study of calculus, however, you will be ready to embark on studies of many different fields in mathematics. It is our hope that you will enjoy learning about this wondrous field of mathematics.

CHAPTER 17

LIMITS, DERIVATIVES, AND INTEGRALS

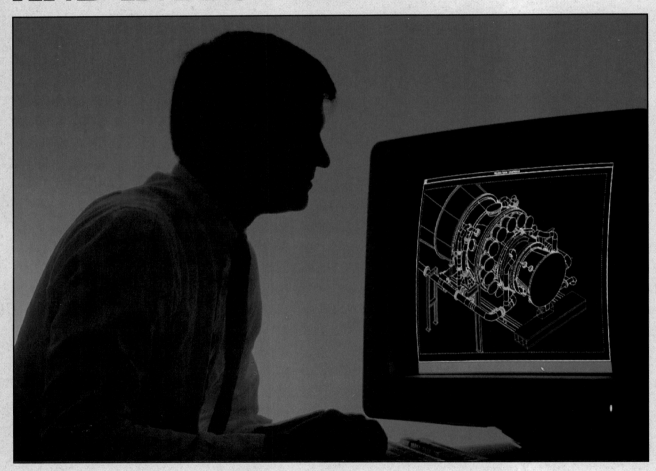

HISTORICAL SNAPSHOT

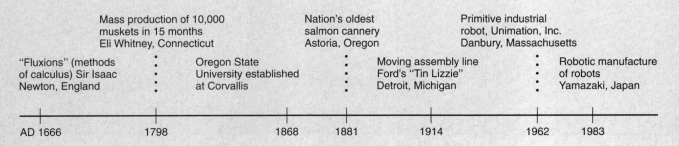

	Mass production of 10,000 muskets in 15 months Eli Whitney, Connecticut		Nation's oldest salmon cannery Astoria, Oregon		Primitive industrial robot, Unimation, Inc. Danbury, Massachusetts	
"Fluxions" (methods of calculus) Sir Isaac Newton, England		Oregon State University established at Corvallis		Moving assembly line Ford's "Tin Lizzie" Detroit, Michigan		Robotic manufacture of robots Yamazaki, Japan

| AD 1666 | 1798 | 1868 | 1881 | 1914 | 1962 | 1983 |

912

CHAPTER OBJECTIVES

In this chapter, you will:
- Define the limit of a function.
- Find the derivative of a function.
- Find the integral of a function by finding the area under the curve.
- Develop and use the fundamental theorem of calculus.

CAREER GOAL: Manufacturing Engineering

Robert Wickwire began his working life by selling heavy equipment for a mill supply company, but soon decided that he wanted to be an engineer. "When customers would call with problems or needing information, I would have to refer them to the engineering department of the company that manufactured the equipment. I wanted to be a problem-solver myself, so I decided to become a manufacturing engineer."

"Manufacturing engineering deals with production processes," Robert explains. "It's such a broad field that there are a lot of opportunities and different directions that I can pursue. During one internship, for example, I worked for a company that makes dental equipment, from small tools to dental chairs. I redesigned work centers where people make cabinetry. I also worked to reduce carpal tunnel syndrome, a wrist disorder that affects people who use their hands all day."

Robert took algebra, geometry, trigonometry, and pre-calculus in high school, but wishes he had also taken a full year of calculus. "Manufacturing engineering uses calculus a lot to control time-dependent operations and positioning. Whenever objects are in motion, like on a production line, their position, velocity, and acceleration are time-dependent, and require calculus for measuring each accurately."

Robert hopes to use his skills in manufacturing engineering to help revive the American economy. "Many people believe that the United States will become a totally service-oriented country. I don't agree with that; many areas of manufacturing need to improve to compete more effectively internationally. People in manufacturing engineering will be the people who give United States manufacturing back its competitive edge in the world market. That's something I want to be a part of."

inter NET CONNECTION

For up-to-date information on manufacturing engineering, visit:
www.glencoe.com/sec/math/amc/mathnet

17-1 Limits

After studying this lesson, you should be able to:

■ use limit theorems to evaluate the limit of a polynomial function.

Application

If a pendulum is moved from its equilibrium position and released, it will oscillate back and forth. Each swing will be shorter than the last as the forces of friction and gravity slow the pendulum. Assume that the length of the first swing was $\frac{1}{2}$ unit and each subsequent swing was half as long as the one before. Let ℓ_n represent the length of the nth swing. Find the lengths of the first five swings.

$\ell_1 = \dfrac{1}{2}$

$\ell_2 = \left(\dfrac{1}{2}\right)\dfrac{1}{2}$ or $\dfrac{1}{2^2}$

$\ell_3 = \dfrac{1}{2}\left(\dfrac{1}{2^2}\right)$ or $\dfrac{1}{2^3}$

$\ell_4 = \dfrac{1}{2}\left(\dfrac{1}{2^3}\right)$ or $\dfrac{1}{2^4}$

$\ell_5 = \dfrac{1}{2}\left(\dfrac{1}{2^4}\right)$ or $\dfrac{1}{2^5}$

The length of the nth swing, ℓ_n, will be $\dfrac{1}{2^n}$ unit.

Some of the terms of the sequence of swing lengths are graphed on the number line below.

As n increases without limit, the points representing the length of the swing become closer and closer to 0, but always remain greater than 0. Thus, the limit of $\dfrac{1}{2^n}$ as n increases without bound is 0.

In Chapter 12, we found the limits of sequences such as the one above and limits of series. The concept of a limit can be extended to a function. What is the limit of $5x - 3$ as x approaches 0? What is the limit as x approaches 6? The limit of a function $f(x)$ as x approaches a is represented by $\lim\limits_{x \to a} f(x)$. So $\lim\limits_{x \to 0} (5x - 3)$ is $5(0) - 3$ or -3, and $\lim\limits_{x \to 6} (5x - 3)$ is $5(6) - 3$ or 27.

Example 1

Show that $\lim\limits_{x \to 3} (x^2 + 1) = 10$ by selecting replacements for x near 3.

Replace x with values close to 3 and observe the patterns of $x^2 + 1$.

x	2.8	2.9	2.99	2.999
$x^2 + 1$	8.84	9.41	9.9401	9.994001

x	3.1	3.01	3.001	3.0001
$x^2 + 1$	10.61	10.0601	10.006001	10.00060001

The limit of $x^2 + 1$ as x approaches 3 is 10.

Consider any polynomial function with real coefficients of the form $P(x) = a_0 x^n + a_1 x^{n-1} + a_2 x^{n-2} + \cdots + a_{n-1}x + a_n$. Let the domain of P be the set of all real numbers. Then, for any real number r, there exists a real number $P(r)$. The limit of such a polynomial function as x approaches r is $P(r)$. For example, the limit of $P(x) = x^2 + 2x - 4$ as x approaches 2 is $P(2) = 2^2 + 2(2) - 4$ or 4.

Limit of a Polynomial Function

The limit of a polynomial function, $P(x)$, as x approaches r, is $P(r)$.
$$\lim_{x \to r} P(x) = P(r)$$

Example 2

The velocity of a molecule of liquid flowing through a pipe varies depending upon the distance that the molecule is from the center of the pipe. The velocity in inches per second of a certain molecule is given by the function $v(r) = k(R^2 - r^2)$, where r is the distance the molecule is from the center of the pipe in inches, R is the radius of the pipe in inches, and k is a constant. Suppose $k = 0.65$ and $R = 0.5$ for a certain situation.
a. Use a limit to find the velocity of a molecule at the center of the pipe.
b. What is the limiting velocity of a molecule as the molecule approaches the wall of the pipe?

FYI...

Civil engineers study hydraulics to help them design levees, irrigation systems, and water supply systems. Mechanical engineers use the principles of hydraulics to design hydraulic machines such as construction equipment and power steering for cars.

a. If the molecule is at the center of the pipe, then $r = 0$. Evaluate $\lim\limits_{r \to 0} v(r)$ to find the velocity of a molecule at the center.

$$v(r) = 0.65(0.5^2 - r^2) \qquad k = 0.65 \text{ and } R = 0.5$$
$$\lim_{r \to 0} v(r) = v(0)$$
$$= 0.65(0.5^2 - 0^2) \qquad r = 0$$
$$= 0.1625$$

The limit of the function as r approaches 0 is 0.1625. Thus, the velocity of a molecule at the center of the pipe is 0.1625 inches per second.

(continued on the next page)

b. If the molecule is approaching the wall of the pipe, then r is approaching 0.5. Find $\lim\limits_{r \to 0.5} v(r)$ to find the velocity of the molecule.

$$v(r) = 0.65(0.5^2 - r^2) \qquad k = 0.65 \text{ and } R = 0.5$$

$$\lim_{r \to 0.5} v(r) = v(0.5)$$

$$= 0.65(0.5^2 - 0.5^2) \qquad r = 0.5$$

$$= 0$$

The limit of the function as r approaches 0.5 is 0. Thus, the limiting velocity of a molecule that is approaching the wall of the pipe is 0 inches per second.

Now consider the rational function $f(x) = \dfrac{x^3}{x}$. Find the limit of this function as x approaches 2. Let $r = 2$.

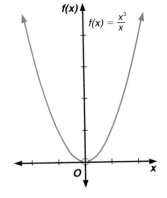

$$\lim_{x \to r} f(x) = f(r)$$

$$\lim_{x \to 2} \frac{x^3}{x} = f(2)$$

$$= \frac{8}{2} \text{ or } 4$$

What is $\lim\limits_{x \to 0} \dfrac{x^3}{x}$? Then $r = 0$. In this case, $f(r)$ is undefined. It appears from the graph above that $\lim\limits_{x \to 0} \dfrac{x^3}{x} = 0$. Replace x with values closer and closer to zero to observe the behavior of $\dfrac{x^3}{x}$.

x	1	0.1	0.01	0.001	0.0001
$\dfrac{x^3}{x}$	1	0.01	0.0001	0.000001	0.00000001

It appears that $\lim\limits_{x \to 0} \dfrac{x^3}{x} = 0$. Even though the function is undefined at zero, the limit of the function exists as x approaches zero. The limit is zero.

You can also show that $\lim\limits_{x \to 0} \dfrac{x^3}{x} = 0$ by transforming $\dfrac{x^3}{x}$ into x^2, since the two functions are equal for all values of x except for $x = 0$. Since x^2 is a polynomial, $\lim\limits_{x \to 0} x^2 = 0$. This method will work whenever a function can be simplified, even if the given function is *not* defined for the value that x approaches.

Change rational functions to polynomial functions whenever possible.

Example 3

Find $\lim\limits_{x \to 4} \dfrac{x^2 - 16}{x - 4}$.

$$\lim_{x \to 4} \frac{x^2 - 16}{x - 4} = \lim_{x \to 4} \frac{(x - 4)(x + 4)}{x - 4} \qquad \textit{Factor.}$$

$$= \lim_{x \to 4} (x + 4)$$

$$= 8$$

Note that $\dfrac{x^2 - 16}{x - 4} = x + 4$ *for all real values of x except 4. However, the fact that the function is not defined for x = 4 does not affect the value of the limit as x approaches 4.*

The limit of $\dfrac{x^2 - 16}{x - 4}$ as x approaches 4 is 8.

It is useful to have principles that define operations involving limits when attempting to find the limits of some functions. These operations include addition, subtraction, multiplication, and division, as well as using powers and roots. The principles are stated as theorems without proofs and are shown in the table below.

The following theorems are given for $f(x)$ and $g(x)$, where $\lim\limits_{x \to a} f(x) = F$, $\lim\limits_{x \to a} g(x) = G$, a and c are real numbers, and n is a positive integer.

Limit of a Constant Function	If $f(x) = c$, then $\lim\limits_{x \to a} f(x) = c$.	Theorem 1
Addition	$\lim\limits_{x \to a} [f(x) + g(x)] = \lim\limits_{x \to a} f(x) + \lim\limits_{x \to a} g(x) = F + G$	Theorem 2
Subtraction	$\lim\limits_{x \to a} [f(x) - g(x)] = \lim\limits_{x \to a} f(x) - \lim\limits_{x \to a} g(x) = F - G$	Theorem 3
Multiplication	$\lim\limits_{x \to a} [f(x) \cdot g(x)] = \left[\lim\limits_{x \to a} f(x)\right]\left[\lim\limits_{x \to a} g(x)\right] = F \cdot G$	Theorem 4
Division	$\lim\limits_{x \to a} \dfrac{f(x)}{g(x)} = \dfrac{\lim\limits_{x \to a} f(x)}{\lim\limits_{x \to a} g(x)} = \dfrac{F}{G} \; (G \neq 0)$	Theorem 5
Product of a Constant and a Limit	$\lim\limits_{x \to a} [c \cdot g(x)] = c \lim\limits_{x \to a} g(x) = cG$	Theorem 6
Powers	$\lim\limits_{x \to a} [f(x)^n] = \left[\lim\limits_{x \to a} f(x)\right]^n = F^n$	Theorem 7
Roots	$\lim\limits_{x \to a} \sqrt[n]{f(x)} = \sqrt[n]{\lim\limits_{x \to a} f(x)} = \sqrt[n]{F}$ ($F > 0$ if n is even.)	Theorem 8

The limit theorems make it easy to find the limit of almost any function.

Example 4

Use the limit theorems to evaluate $\displaystyle\lim_{x\to-1}\frac{\sqrt[3]{x^2+6x-22}}{\sqrt{10x^4-1}}.$

$$\lim_{x\to-1}\frac{\sqrt[3]{x^2+6x-22}}{\sqrt{10x^4-1}} = \frac{\displaystyle\lim_{x\to-1}\sqrt[3]{x^2+6x-22}}{\displaystyle\lim_{x\to-1}\sqrt{10x^4-1}} \qquad \textit{Theorem 5}$$

$$= \frac{\sqrt[3]{\displaystyle\lim_{x\to-1}(x^2+6x-22)}}{\sqrt{\displaystyle\lim_{x\to-1}(10x^4-1)}} \qquad \textit{Theorem 8}$$

$$= \frac{\sqrt[3]{-27}}{\sqrt{9}} \text{ or } -1$$

Recall that if $f(x)$ and $g(x)$ are two functions of x, the composite of f and g is defined to be $f(g(x))$. For example, if $f(x) = x^2$ and $g(x) = x + 2$, $f(g(x)) = f(x+2)$ or $(x+2)^2$. In general, composition of functions is *not* a commutative operation. In other words, the value of $f(g(x))$ is usually not the same as $g(f(x))$.

Theorem 9

Given functions $f(x)$ and $g(x)$, $\displaystyle\lim_{x\to a} g(x) = G$, f is continuous at $x = G$, and any real number a, the limit of the composite of functions $f(x)$ and $g(x)$ is as follows.

$$\lim_{x\to a} f[g(x)] = f\left(\lim_{x\to a} g(x)\right) = f(G)$$

Example 5

Evaluate $\displaystyle\lim_{x\to4} f[g(x)]$ **if** $f(x) = x^2$ **and** $g(x) = 6 - x.$

$$\lim_{x\to4} f[g(x)] = f\left[\lim_{x\to4} g(x)\right]$$

$$= f\left[\lim_{x\to4}(6-x)\right]$$

$$= f(2)$$

$$= 2^2 \text{ or } 4 \qquad f(x) = x^2$$

You can also evaluate limits of trigonometric functions.

Theorem 10

$\displaystyle\lim_{x\to0}\frac{\sin x}{x} = 1$, **where x is the radian measure of an angle.**

Theorem 10 can be demonstrated as follows. Suppose $0 < x < \frac{\pi}{2}$. Arc AB is part of a circle with center O and radius of 1 unit. As x approaches 0, the length of $\overline{AD}$ approaches the length of $\overparen{AB}$. Thus, $\dfrac{AD}{\text{measure of } \overparen{AB}}$ approaches 1 as x approaches 0. But, $AD = \sin x$ and the measure of $\overparen{AB} = x$, so $\lim\limits_{x \to 0} \dfrac{\sin x}{x} = 1$.

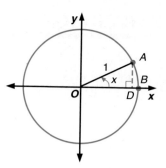

Example 6

Evaluate $\lim\limits_{x \to 0} \dfrac{\sin 2x}{x}$.

$\lim\limits_{x \to 0} \dfrac{\sin 2x}{x} = \lim\limits_{x \to 0} \dfrac{2 \cdot \sin 2x}{2x}$ *Multiply the numerator and denominator by 2.*

$\qquad\qquad = 2 \cdot \lim\limits_{x \to 0} \dfrac{\sin 2x}{2x}$ *Theorem 6*

Let $y = 2x$. Then $\lim\limits_{x \to 0} \dfrac{\sin 2x}{2x} = \lim\limits_{y \to 0} \dfrac{\sin y}{y}$ because as x approaches 0, y also approaches 0.

Since $\lim\limits_{y \to 0} \dfrac{\sin y}{y} = 1$, $\lim\limits_{x \to 0} \dfrac{\sin 2x}{2x} = 1$.

Then $2 \cdot \lim\limits_{x \to 0} \dfrac{\sin 2x}{2x} = 2 \cdot 1$ or 2.

EXPLORATION: Graphing Calculator

A graphing calculator can be helpful when you are evaluating the limit of a function. You can graph the function and use the TRACE function to estimate the value of the limit.

1. Use Theorem 10 to evaluate $\lim\limits_{x \to 0} \dfrac{\cos x - 1}{x}$.

2. Graph $y = \dfrac{\cos x - 1}{x}$ on a graphing calculator.

3. Use the TRACE function to estimate the value of $\lim\limits_{x \to 0} \dfrac{\cos x - 1}{x}$. Does the estimate support the conclusion you made using Theorem 10?

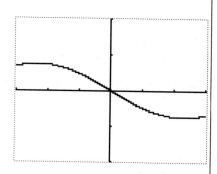

CHECKING FOR UNDERSTANDING

Communicating Mathematics

Read and study the lesson to answer each question.

1. **Define** the limit of a function.

2. **Sketch** the graph of $y = \dfrac{x^2 - 16}{x - 4}$. Does the graph confirm that $\lim\limits_{x \to 4} \dfrac{x^2 - 16}{x - 4} = 8$?

3. **Explain** the steps you follow to find $\lim\limits_{x \to s} f(x)$.

4. A student in your class says, "The limit of the sum of two functions is the sum of the limits of the functions." When is the statement *not* correct?

5. **Show** that $\lim\limits_{x \to 0} x^2 = 0$ by using the fact that $x \cdot x = x^2$.

6. **Express** $\lim\limits_{x \to 3} (x - 1)^3$ in two different ways.

Guided Practice

Evaluate each limit.

7. $\lim\limits_{x \to 2} 6x$

8. $\lim\limits_{x \to 3} (x^2 + 4x - 5)$

9. $\lim\limits_{x \to 1} \dfrac{x - 2}{x + 2}$

10. $\lim\limits_{x \to 5} \sqrt{25 - x^2}$

11. $\lim\limits_{x \to 3} \dfrac{x - 4}{x + 1}$

12. $\lim\limits_{x \to -2} \dfrac{x^2 - 4}{x^2 + 4}$

13. $\lim\limits_{x \to 3} \dfrac{x^2 - x - 6}{x - 3}$

14. $\lim\limits_{x \to 0} \dfrac{2x^3}{x}$

15. $\lim\limits_{x \to -1} \dfrac{x^2 + 3x + 2}{x^2 + 4x + 3}$

Evaluate the limit of $f(g(x))$ as x approaches 0 for each $f(x)$ and $g(x)$.

16. $f(x) = 4x - 4$
$g(x) = 8x + 1$

17. $f(x) = 5x + 2$
$g(x) = x^2 - 1$

Use Theorem 10 to evaluate each limit.

18. $\lim\limits_{x \to 0} \dfrac{\sin^2 x}{x}$

19. $\lim\limits_{x \to 0} \dfrac{\sin 3x}{5x}$

EXERCISES

Practice

Evaluate each limit.

20. $\lim\limits_{x \to 2} (x^2 - 4x + 1)$

21. $\lim\limits_{x \to 0} (4x + 1)$

22. $\lim\limits_{x \to 2} x^2$

23. $\lim\limits_{x \to 1} \dfrac{x + 1}{x + 2}$

24. $\lim\limits_{x \to 6} (7x - 22)$

25. $\lim\limits_{x \to 1} (x^2 + 4x + 3)$

26. $\lim\limits_{n \to 0} \left(5^n + \dfrac{1}{5^n}\right)$

27. $\lim\limits_{x \to 3} \dfrac{x^2 - 9}{x + 3}$

28. $\lim\limits_{n \to 3} \dfrac{n^2 - 9}{n - 3}$

29. $\lim\limits_{x \to 2} \dfrac{x^2 - 4}{x^3 - 8}$

30. $\lim\limits_{x \to 4} \dfrac{x - 4}{x^2 - 16}$

31. $\lim\limits_{x \to -2} (x^4 - x^2 + x - 2)$

32. $\lim\limits_{x \to -2} \dfrac{x^3 - 8}{x - 2}$

33. $\lim\limits_{x \to 2} \dfrac{x^2 - x - 2}{x^2 - 4}$

34. $\lim\limits_{x \to 0} \dfrac{(1 + x)^2 - 1}{x}$

35. $\lim\limits_{n \to 0} \dfrac{n^2}{n^4 + 1}$

36. $\lim\limits_{x \to -1} \sqrt{x^2 - 1}$

37. $\lim\limits_{x \to 1} \sqrt{\dfrac{2x + 1}{2x - 1}}$

Evaluate the limit of $f[g(x)]$ as x approaches 1 for each $f(x)$ and $g(x)$.

38. $f(x) = 2x + 1$
$g(x) = x - 3$

39. $f(x) = 7x - 2$
$g(x) = 9x + 2$

40. $f(x) = 3x - 4$
$g(x) = 2x + 5$

41. $f(x) = x^2 + 3$
$g(x) = 2x - 1$

Use Theorem 10 to evaluate each limit.

42. $\lim\limits_{x \to 0} \dfrac{\sin(-x)}{x}$

43. $\lim\limits_{x \to 0} \dfrac{1 - \cos x}{x^2}$

44. $\lim\limits_{x \to 0} \dfrac{\sin \sqrt[3]{x}}{\sqrt[3]{x}}$

Graphing Calculator

Evaluate each limit. Then use a graphing calculator to verify your answer.

45. $\lim\limits_{x \to \frac{1}{2}} \dfrac{6x - 3}{x(1 - 2x)}$

46. $\lim\limits_{x \to 0} \dfrac{\sin 6x}{x}$

47. $\lim\limits_{x \to 4} \dfrac{\sqrt{x} - 2}{4 - x}$

Critical Thinking

48. A function is **bounded** if there is a real number M such that $|f(x)| < M$ for all x in the domain of $f(x)$. Suppose a certain function $f(x)$ is defined for all real numbers and is bounded.

  a. Do you think that $f(x)$ has a limit as x approaches infinity? If so, what could the value of that limit be? Explain.

  b. If $f(x)$ is strictly increasing, do you think that $f(x)$ has a limit as x approaches infinity? If so, what could the value of that limit be? Explain.

Applications and Problem Solving

49. Physics Refer to the application at the beginning of the lesson. The limit of the sum of the lengths of the pendulum swings as n approaches infinity is the distance the pendulum will travel before it comes to rest. Find that limit.

50. Chemistry Acetylcholine is a chemical that is released at automatic nerve endings to transmit nerve impulses. The chemical is formed in body tissue and one of its many functions is to slow the heart rate. It has been found that the effect of acetylcholine on a frog's heart is given by the formula $f(x) = \dfrac{x}{a + bx}$ for $0 \leq x \leq 1$, where a and b are positive constants, x is the concentration of acetylcholine, and $f(x)$ is a measure of the degree of response. Find $\lim\limits_{x \to \frac{2}{3}} f(x)$ when $a = 8$ and $b = 6$.

51. Demographics The population of a certain small city is predicted to be given by $P(t) = 30,000 + \dfrac{15,000}{(t+3)^2}$, t years from now. Determine the population in the long run. That is, find $\lim\limits_{t \to \infty} P(t)$.

Mixed Review

52. Solve the matrix equation $\begin{bmatrix} 5 & -3 \\ -1 & 0 \end{bmatrix} \cdot \begin{bmatrix} x \\ y \end{bmatrix} = \begin{bmatrix} 9 \\ -2 \end{bmatrix}$. **(Lesson 2-3)**

53. Find the derivative of the function $f(x) = 4x^2 - 8x + 9$. **(Lesson 3-6)**

54. Find the values of the six trigonometric functions of an angle in standard position if the point at $(2, -4)$ is on its terminal side. **(Lesson 5-3)**

55. Find the distance between $P(8, -3)$ and the line $6x + 2y - 1 = 0$. **(Lesson 7-7)**

56. Evaluate $\left(6^{\frac{2}{3}}\right)^3$. **(Lesson 11-1)**

57. Use the first five terms of the exponential series $e^x = 1 + x + \dfrac{x^2}{2!} + \dfrac{x^3}{3!} + \dfrac{x^4}{4!} + \cdots$ and a calculator to approximate $e^{3.3}$ to the nearest hundredth. **(Lesson 12-7)**

58. Management The Foxtrail Condominium Association is electing board members. How many groups of four board members can be chosen from the ten candidates who are running? **(Lesson 14-3)**

59. Find the mean, median, and mode of the set $\{7.3, 8.6, 5.4, 8.6, 8.9, 10.1, 6.4\}$. **(Lesson 15-2)**

60. For any graph $G(n, m)$ the matrix representing it, $M(G)$, is an $n \times m$ matrix. Draw a graph for which the matrix $\begin{bmatrix} 0 & 1 & 1 \\ 1 & 0 & 0 \\ 1 & 0 & 0 \end{bmatrix}$ is $M(G)$. **(Lesson 16-6)**

61. College Entrance Exam Choose the best answer.
Triangle ABC has sides that are 6, 8, and 10 inches long. A rectangle that has area equal to that of the triangle has a width of 3 inches. Find the perimeter of the rectangle in inches.

(A) 30 **(B)** 24 **(C)** 22 **(D)** 16 **(E)** 11

17-2 Derivatives and Differentiation Techniques

Objective

After studying this lesson, you should be able to:
- find the derivative of a function.

Application

Howell Hosiery makes panty hose and stockings to be sold in department stores. The total cost function for Howell is $C(q) = -10,484.69 + 6.750q - 0.000328q^2$, where q is the number of dozens of pairs of stockings made and $C(q)$ is the cost in dollars. The marginal cost function indicates the rate at which the cost changes with respect to the number of units produced. Find the marginal cost function and evaluate it when $q = 4500$. *This problem will be solved in Example 4.*

Recall that in Chapter 3, a derivative was defined as the slope of the line tangent to a curve at a given point.

To solve this problem, we need to find the derivative of the function. The process of finding derivatives is called **differentiation**. A function is **differentiable** at a if the derivative of the function exists at a and is finite.

Recall from Chapter 3 that the definition of the derivative of a function $f(x)$ at $x = a$ is $f'(a) = \lim\limits_{h \to 0} \dfrac{f(a+h) - f(a)}{h}$. The symbol h in the definition is often replaced with Δx, read "delta x", to show the change in x. Then, the difference quotient is $\dfrac{f(x + \Delta x) - f(x)}{\Delta x}$. But, $f(x + \Delta x) - f(x)$ can be denoted as Δy since it is actually the change in y. Thus, the derivative of $f(x)$ may be written with delta notation as follows.

$f'(x) = \dfrac{dy}{dx} = \lim\limits_{\Delta x \to 0} \dfrac{\Delta y}{\Delta x}$

$$f'(x) = \lim_{\Delta x \to 0} \frac{f(x + \Delta x) - f(x)}{\Delta x} \text{ or } \lim_{\Delta x \to 0} \frac{\Delta y}{\Delta x}$$

$\dfrac{dy}{dx}$ *is the notation for a derivative. It is not a fraction.*

When the function is written in the form $y = f(x)$, the derivative is sometimes written as $\dfrac{dy}{dx}$. Thus, $\dfrac{dy}{dx} = \lim\limits_{\Delta x \to 0} \dfrac{f(x + \Delta x) - f(x)}{\Delta x}$ or $\lim\limits_{\Delta x \to 0} \dfrac{\Delta y}{\Delta x}$.

The notation $\dfrac{dy}{dx}$ is called **derivative notation** and is read "the derivative of y with respect to x."

Example 1

If $y = 5x^2 + 1$, find $\dfrac{dy}{dx}$ at $x = 3$.

First, find $\dfrac{\Delta y}{\Delta x}$.

$$\frac{\Delta y}{\Delta x} = \frac{f(x + \Delta x) - f(x)}{\Delta x}$$

$$= \frac{[5(x + \Delta x)^2 + 1] - (5x^2 + 1)}{\Delta x}$$

(continued on the next page)

$$= \frac{5(x^2 + 2x\Delta x + \Delta x^2) + 1 - 5x^2 - 1}{\Delta x}$$

$$= \frac{10x\Delta x + 5\Delta x^2}{\Delta x}$$

$$= 10x + 5\Delta x$$

Then, find $\dfrac{dy}{dx}$.

$$\frac{dy}{dx} = \lim_{\Delta x \to 0} \frac{\Delta y}{\Delta x}$$

$$= \lim_{\Delta x \to 0} (10x + 5\Delta x)$$

$$= 10x$$

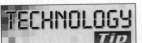
Thus, $\dfrac{dy}{dx}$ is $10(3)$ or 30 at $x = 3$.

The slope of the line tangent to the graph of $y = 5x^2 + 1$ at $x = 3$ is 30.

As you can see from Example 1, the process of finding the derivative of a function by setting up a difference quotient and then finding its limit can be rather involved. To shorten this process, formulas can be used to find derivatives of functions. You learned some of these formulas when you studied the tangents of a curve in Chapter 3.

Consider the function $f(x) = c$, where c is a constant.

$$f'(x) = \lim_{h \to 0} \frac{f(x + h) - f(x)}{h}$$

$$= \lim_{h \to 0} \frac{c - c}{h} \text{ or } 0$$

Thus, the derivative of a constant function, $f(x) = c$, is zero. For example, if $f(x) = 6$, then $f'(x) = 0$.

One of the most important differentiation formulas is the power formula.

Theorem 11 ***Power Formula***	**If $f(x) = cx^n$, where n is a real number and c is a constant, then $f'(x) = cnx^{n-1}$.**

Example 2

Find the derivative of each function.
a. $f(x) = x^4$ **b.** $f(x) = 4x^6 - 4$

a. $f(x) = x^4$
$f'(x) = 4x^{4-1}$ or $4x^3$

b. $f(x) = 4x^6 - 4$
$f'(x) = 4 \cdot 6x^{6-1} - 0$
$= 24x^5$

The notation $\dfrac{dy}{dx}$ is used in the statement of the following theorem. This notation can be used with other functions of x. For example, $\dfrac{du}{dx}$ is the derivative of u with respect to x.

Theorem 12

If $u = f(x)$ and $v = g(x)$ are differentiable functions of x, then
$$\frac{d(u+v)}{dx} = \frac{du}{dx} + \frac{dv}{dx}.$$

Example 3

Find the derivative of $f(x) = x^4 + (x^2 - 3)$.

Use $\dfrac{d(u+v)}{dx} = \dfrac{du}{dx} + \dfrac{dv}{dx}$.

$$\frac{d}{dx}[x^4 + (x^2 - 3)] = \frac{d}{dx}x^4 + \frac{d}{dx}(x^2 - 3) \qquad u = x^4 \text{ and } v = x^2 - 3$$
$$= (4x^3) + (2x - 0)$$
$$= 4x^3 + 2x$$

The derivative of a function can be thought of as the rate of change of the function when $x = a$. In business, the derivative is found in order to determine the rate of change of a company's cost or revenue. Suppose the equation $y = f(x)$ denotes the total cost or total revenue of a company when x units are produced or sold. Then the rate at which the cost or revenue changes with respect to the number of units produced or sold is $\dfrac{dy}{dx}$, called the **marginal cost function** or **marginal revenue function**.

Example 4

Refer to the application at the beginning of the lesson. Find the marginal cost function and evaluate it for $q = 4500$.

The total cost function is $C(q)$. To find the marginal cost function, we must find the derivative of the total cost function. Use Theorem 12.

$$\frac{d}{dq}C(q) = \frac{d}{dq}(-10{,}484.69 + 6.750q - 0.000328q^2)$$
$$= \frac{d}{dq}(-10{,}484.69) + \frac{d}{dq}(6.750q) - \frac{d}{dq}(0.000328q^2)$$
$$= 0 + 6.750 - 0.000656q$$
$$= 6.750 - 0.000656q$$

The marginal cost function is $\dfrac{dC(q)}{dq} = 6.750 - 0.000656q$. When $q = 4500$, the marginal cost is $6.750 - 0.000656(4500)$ or 3.798. This means that when production is at 4500 dozens of pairs, making an additional dozen will cost \$3.798.

FYI...

Modern panty hose and stockings are made of nylon, a man-made fiber invented by DuPont chemist Wallace Carothers in 1935. The chemicals used to make nylon come from petroleum, natural gas, and agricultural byproducts.

Theorem 13

If $u = f(x)$ is a differentiable function of x and c is a constant, then
$$\frac{d(cu)}{dx} = c\frac{du}{dx}.$$

Example 5

Find the derivative of $f(x) = 5(x^2 + 5x - 4)$.

Use $\dfrac{d(cu)}{dx} = c\dfrac{du}{dx}$.

$$\dfrac{d}{dx}[5(x^2 + 5x - 4)] = 5\dfrac{d}{dx}(x^2 + 5x - 4) \qquad c = 5 \text{ and } u = x^2 + 5x - 4$$
$$= 5(2x + 5)$$
$$= 10x + 25$$

The formulas for the derivatives of the sum of two functions and for a multiple of a function were relatively simple. The formula for the derivative of the product of two functions is more complicated.

Theorem 14
Product Rule

If $u = f(x)$ and $v = g(x)$ are differentiable functions of x, then
$$\dfrac{d(uv)}{dx} = u\dfrac{dv}{dx} + v\dfrac{du}{dx}.$$

Example 6

Find the derivative of $f(x) = (x^4 + 7x)(x^5)$.

Use $\dfrac{d(uv)}{dx} = u\dfrac{dv}{dx} + v\dfrac{du}{dx}$.   *You could also find the product before differentiating.*

$$\dfrac{d}{dx}[(x^4 + 7x)(x^5)] = (x^4 + 7x)\dfrac{d}{dx}x^5 + x^5\dfrac{d}{dx}(x^4 + 7x) \qquad u = x^4 + 7x \text{ and } v = x^5$$
$$= (x^4 + 7x)(5x^4) + (x^5)(4x^3 + 7)$$
$$= 5x^8 + 35x^5 + 4x^8 + 7x^5$$
$$= 9x^8 + 42x^5$$

Theorem 15

If $u = f(x)$ is a differentiable function of x and n is a nonzero rational number, then $\dfrac{d(u^n)}{dx} = nu^{n-1}\dfrac{du}{dx}$.

Example 7

Find the derivative of each function.
a. $f(x) = \sqrt{x^2 + 2x - 1}$

a. Use $\dfrac{d(u^n)}{dx} = nu^{n-1}\dfrac{du}{dx}$. Remember, $\sqrt{x^2 + 2x - 1} = (x^2 + 2x - 1)^{\frac{1}{2}}$.

$$\dfrac{d}{dx}(\sqrt{x^2 + 2x - 1}) = \dfrac{1}{2}(x^2 + 2x - 1)^{-\frac{1}{2}}(2x + 2) \qquad u = x^2 + 2x - 1 \text{ and } n = \dfrac{1}{2}$$
$$= \dfrac{2x + 2}{2\sqrt{x^2 + 2x - 1}}$$
$$= \dfrac{2(x + 1)}{2\sqrt{x^2 + 2x - 1}}$$
$$= \dfrac{x + 1}{\sqrt{x^2 + 2x - 1}}$$

b. $f(x) = x^3(x^2 - 3)^{-4}$

b. Use $\dfrac{d(uv)}{dx} = u\dfrac{dv}{dx} + v\dfrac{du}{dx}$ and $\dfrac{d(u^n)}{dx} = nu^{n-1}\dfrac{du}{dx}$.

$\dfrac{d}{dx}[x^3(x^2 - 3)^{-4}] = x^3\dfrac{d}{dx}[(x^2 - 3)^{-4}] + [(x^2 - 3)^{-4}]\dfrac{d}{dx}(x^3)$ $\quad\quad u = x^3,$
$\quad\quad\quad\quad\quad\quad\quad\quad\quad\quad\quad\quad\quad\quad\quad\quad\quad\quad\quad\quad v = (x^2 - 3)^{-4}$

$\quad\quad\quad\quad\quad\quad = x^3[-4(x^2 - 3)^{-5}(2x)] + [(x^2 - 3)^{-4}](3x^2)$ $\quad\quad \dfrac{du}{dx} = 3x^2,$

$\quad\quad\quad\quad\quad\quad = -8x^4(x^2 - 3)^{-5} + 3x^2(x^2 - 3)^{-4}$ $\quad\quad \dfrac{dv}{dx} = -4(x^2 - 3)^{-5}(2x)$

$\quad\quad\quad\quad\quad\quad = \dfrac{-8x^4}{(x^2 - 3)^5} + \dfrac{3x^2}{(x^2 - 3)^4}$

$\quad\quad\quad\quad\quad\quad = \dfrac{-8x^4 + 3x^2(x^2 - 3)}{(x^2 - 3)^5}$

$\quad\quad\quad\quad\quad\quad = \dfrac{-5x^4 - 9x^2}{(x^2 - 3)^5}$

Theorem 16
Quotient Rule

If $u = f(x)$ and $v = g(x)$ are differentiable functions of x at a point where $v \neq 0$, then $\dfrac{d\left(\dfrac{u}{v}\right)}{dx} = \dfrac{v\dfrac{du}{dx} - u\dfrac{dv}{dx}}{v^2}$.

Example 8

Find the derivative of each function.

a. $f(x) = \dfrac{4x^2 - 2x + 3}{2x - 1}$

a. Use $\dfrac{d\left(\dfrac{u}{v}\right)}{dx} = \dfrac{v\dfrac{du}{dx} - u\dfrac{dv}{dx}}{v^2}$. Let $u = 4x^2 - 2x + 3$ and $v = 2x - 1$. Then
$\dfrac{du}{dx} = 8x - 2$ and $\dfrac{dv}{dx} = 2$.

$\dfrac{d}{dx}\left(\dfrac{4x^2 - 2x + 3}{2x - 1}\right) = \dfrac{(2x - 1)(8x - 2) - (4x^2 - 2x + 3)(2)}{(2x - 1)^2}$

$\quad\quad\quad\quad\quad\quad\quad\quad\quad = \dfrac{(16x^2 - 12x + 2) - (8x^2 - 4x + 6)}{4x^2 - 4x + 1}$

$\quad\quad\quad\quad\quad\quad\quad\quad\quad = \dfrac{8x^2 - 8x - 4}{4x^2 - 4x + 1}$

b. $f(x) = \dfrac{(x + 1)^2}{(x - 1)^2}$

b. Use Theorems 15 and 16. Let $u = (x + 1)^2$ and $v = (x - 1)^2$. Then
$\dfrac{du}{dx} = 2(x + 1)(1)$ and $\dfrac{dv}{dx} = 2(x - 1)(1)$.

(continued on the next page)

$$\frac{d}{dx}\frac{(x+1)^2}{(x-1)^2} = \frac{(x-1)^2(2)(x+1) - (x+1)^2(2)(x-1)}{(x-1)^4}$$

$$= \frac{(x^2-2x+1)(2x+2) - (x^2+2x+1)(2x-2)}{(x-1)^4}$$

$$= \frac{-4(x^2-1)}{(x-1)^4} \text{ or } \frac{-4(x+1)}{(x-1)^3} \qquad x^2-1 = (x+1)(x-1)$$

CHECKING FOR UNDERSTANDING

Communicating Mathematics

Read and study the lesson to answer each question.

1. **Explain** how the formula $\dfrac{f(a+h)-f(a)}{h}$ is related to the tangent line of a function at a point $(a, f(a))$.

2. Write three different expressions for the derivative of $y = f(x)$.

3. **Explain** the product rule for differentiation in your own words.

4. Find the derivative of $x^4(3x^5 - 2x^3)$ using two different methods.

5. What does the symbol $\dfrac{du}{dx}$ represent?

Guided Practice

Find the derivative of each function.

6. $f(x) = 2x$
7. $f(x) = 7x - 3$
8. $f(x) = -x$
9. $f(x) = 2x^2$
10. $f(x) = x^3$
11. $f(x) = 6x^2$
12. $f(x) = (2x-3)(x+5)$
13. $f(x) = (3x+1)(2x^2-5x)$
14. $f(x) = (x - 2x^2)^2$
15. $f(x) = x^2(x^2+1)^{-3}$
16. $f(x) = \sqrt{4x^2 - 1}$
17. $f(x) = \dfrac{2x+3}{4x-1}$
18. $f(x) = \dfrac{x^3-1}{x^4+1}$
19. $f(x) = -3x - \dfrac{6}{x+2}$

EXERCISES

Practice

Find the derivative of each function.

20. $f(x) = x$
21. $f(x) = 6x - 4$
22. $f(x) = -4x - 2$
23. $f(x) = 5x^2 - x$
24. $f(x) = x^4 - 2x^2$
25. $f(x) = x^5 + 3$
26. $f(x) = x^{\frac{1}{2}}$
27. $f(x) = \sqrt[3]{x}$
28. $f(x) = x^2(x^2 - 3)$
29. $f(x) = (x^3 - 2x)(3x^2)$
30. $f(x) = 8x^4(1 - 9x^2)$
31. $f(x) = (x^2 + 4)^3$
32. $f(x) = (x^3 - 2x + 1)^4$
33. $f(x) = \sqrt{x^2 - 1}$
34. $f(x) = x^2(x+1)^{-1}$
35. $f(x) = (x^2 - 4)^{-\frac{1}{2}}$
36. $f(x) = \dfrac{x+1}{x^2-4}$
37. $f(x) = \left(\dfrac{x+1}{x-1}\right)^2$
38. $f(x) = x\sqrt{1 - x^3}$

39. $f(x) = 2x^3 + \dfrac{2}{x^3}$

40. $f(x) = \sqrt{2x} - \sqrt{2}x$

41. $f(x) = \left(1 + \dfrac{1}{x}\right)\left(2 - \dfrac{1}{x}\right)$

42. $f(x) = 3x - \dfrac{\dfrac{2}{x} - \dfrac{3}{x-1}}{x-2}$

43. Suppose $v = f(x)$ is a differentiable function such that $v \neq 0$. Derive a formula for $\dfrac{d\left(\dfrac{1}{v}\right)}{dx}$.

44. Business The total cost function for an electric company is estimated to be $C(t) = 32.07 - 0.79t + 0.02142t^2 - 0.0001t^3$, where t is the total output and $C(t)$ is total fuel cost in dollars.

   a. Find the marginal cost function.

   b. Evaluate the marginal cost function when $t = 70$.

   c. If possible, find the number of units produced when the marginal cost is 0.6.

45. Physics Suppose a ball has been shot upward with an initial velocity of 96 feet per second. Then the equation $h(t) = 256 + 96t - 16t^2$ gives the height of the ball in feet after t seconds.

   a. The derivative of the function for the height of the ball gives the rate of change of the height, or the velocity of the ball. Find the velocity function.

   b. Find the velocity of the ball after 2 seconds.

   c. Is the velocity increasing or decreasing?

   d. What is the velocity of the ball when it hits the ground?

46. Economics The consumption function expresses the relationship between the national income, I, and the national consumption, $C(I)$. The marginal propensity to consume, $\dfrac{dC}{dI}$, is the rate of change of consumption with respect to income. For a certain period of time, the consumption function for the United States was $C(I) = \dfrac{5(2\sqrt{I^3} + 3)}{I + 10}$, where C and I are measured in billions of dollars.

   a. Write a formula for the marginal propensity to consume for this time.

   b. What was the marginal propensity to consume at this time if the national income was 100 billion dollars?

   c. The marginal propensity to save, $\dfrac{dS}{dI}$, indicates how fast savings change with respect to income. The formula for the marginal propensity to save is $1 - \dfrac{dC}{dI}$. What was the marginal propensity to save for the given consumption function if the national income was 150 billion dollars?

47. **Research**   The notation for a derivative $\frac{dy}{dx}$ is often called *Leibniz notation* after German mathematician Gottfried Wilhelm Leibniz. Some claim that Leibniz was the inventor of calculus, while others credit Isaac Newton with the invention. Research the controversy and its effects on the development of mathematics. Write a one-page paper on your findings.

48. **Geometry**   Determine whether the figure with vertices at $(0, 3), (8, 4), (2, -5)$, and $(10, -4)$ is a parallelogram.   **(Lesson 1-4)**

49. List all possible rational roots of the function $f(x) = 8x^3 + 3x - 2$. **(Lesson 4-4)**

50. Use Hero's formula to find the area of $\triangle ABC$ if $a = 2.1, b = 3.2$, and $c = 4.4$. Round your answer to the nearest tenth.   **(Lesson 5-8)**

51. Write an equation with phase shift 0 to represent simple harmonic motion if the initial position is 0, the amplitude is 12, and the period is 4.   **(Lesson 6-7)**

52. Find the quotient of $16\left(\cos \frac{\pi}{8} + i \sin \frac{\pi}{8}\right)$ divided by $4\left(\cos \frac{\pi}{4} + i \sin \frac{\pi}{4}\right)$. **(Lesson 9-7)**

53. The first term of a geometric sequence is 9, and the common ratio is $-\frac{1}{3}$. Find the sixth term of the sequence.   **(Lesson 12-2)**

54. Shawn guessed on all of the five multiple choice questions on a quiz. If there were four choices for each question, what is the probability that he got all five questions correct?   **(Lesson 14-8)**

55. Evaluate $\lim\limits_{x \to -1} \sqrt{x^2 - 1}$.   **(Lesson 17-1)**

56. **College Entrance Exam**   If $2^n = 8$, what is the value of 3^{n+2}?

CASE STUDY FOLLOW-UP

Refer to Case Study 4, *The U.S. Economy*, on pages 966–969.

To approximate unemployment figures for the years 1980–1985, an economist used the function $f(x) = -0.42x^2 + 2.1x + 7.2$, where x is the number of years since 1980 and $f(x)$ is the percent of Americans unemployed.

1. How well does the function approximate the unemployment curve?
2. The derivative of a function is the rate of change of a function $f(x)$ when $x = a$. Find the derivative of the function used to describe the unemployment curve.
3. Find the rate of change of the unemployment rate for the year 1982. Explain what the rate of change means.
4. **Research**   Use an economics text to find one aspect of the economy that can be analyzed by using calculus. Write about your findings in a 1-page paper.

17-3 Area Under a Curve

Objective

After studying this lesson, you should be able to:
- find the area between a curve and the x-axis by using the limit of areas of rectangles.

Application

Traci is stopped at the traffic light at High Street and Wilson-Bridge Road. When the light turns green, Traci accelerates for the first 8 seconds. The velocity of the car in feet per second can be described by the equation $v(t) = 16t - t^2$, where t is time in seconds. How far will Traci's car travel in 8 seconds? *This problem will be solved in Example 3.*

When $f(x)$ is a function in terms of x, it is often useful to be able to find the area of a region bounded by the graph of $f(x)$. Consider the area between the graph of a function $f(x)$ and the x-axis for an interval from $x = a$ to $x = b$. We don't know a formula for the area of a figure with a curved edge, but we can approximate the area by drawing rectangles. Suppose the interval from a to b is separated into n subintervals of equal width, and vertical lines are drawn at each subinterval to form rectangles, as shown below. The sum of the areas of the rectangles drawn under the curve, A_u, must be less than the area of the region. The sum of the areas of the rectangles drawn above the curve, A_o, must be greater than the area of the region. A_u and A_o are called the **lower** and **upper sums**, respectively.

The area of a rectangle is length times width. The length of each of these rectangles is $f(x)$, and the width is $x_{i+1} - x_i$.

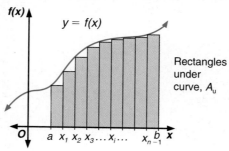

Rectangles under curve, A_u

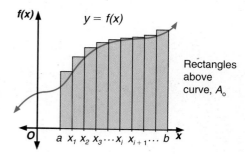

Rectangles above curve, A_o

area of 1st rectangle $= f(a)(x_1 - a)$
area of 2nd rectangle $= f(x_1)(x_2 - x_1)$
area of 3rd rectangle $= f(x_2)(x_3 - x_2)$
$\vdots$
area of $(i + 1)$th rectangle
$$= f(x_i)(x_{i+1} - x_i)$$
$\vdots$
area of nth rectangle $= f(x_{n-1})(b - x_{n-1})$

$A_u = $ *sum of areas of the rectangles*
$A_u = f(a)(x_1 - a) + f(x_1)(x_2 - x_1) +$
$\qquad f(x_2)(x_3 - x_2) + \cdots +$
$\qquad f(x_i)(x_{i+1} - x_i) + \cdots +$
$\qquad f(x_{n-1})(b - x_{n-1})$

area of 1st rectangle $= f(x_1)(x_1 - a)$
area of 2nd rectangle $= f(x_2)(x_2 - x_1)$
area of 3rd rectangle $= f(x_3)(x_3 - x_2)$
$\vdots$
area of $(i + 1)$th rectangle
$$= f(x_{i+1})(x_{i+1} - x_i)$$
$\vdots$
area of nth rectangle $= f(b)(b - x_{n-1})$

$A_o = $ *sum of areas of the rectangles*
$A_o = f(x_1)(x_1 - a) + f(x_2)(x_2 - x_1) +$
$\qquad f(x_3)(x_3 - x_2) + \cdots +$
$\qquad f(x_{i+1})(x_{i+1} - x_i) + \cdots +$
$\qquad f(b)(b - x_{n-1})$

Thus, the actual area, A, is between A_u and A_o. We can write this as $A_u \le A \le A_o$. As the number of subintervals, n, is increased, the areas A_u and A_o approach the actual area A. So, A is the limit of A_o as n increases without bound. The area A_o may be written using summation notation.

$$A_o = \sum_{i=1}^{n} f(x_i)(x_i - x_{i-1})$$

Area Under a Curve

The area A between the graph of $y = f(x)$ and the x-axis from $x = a$ to $x = b$ is as follows:

$$A = \lim_{n \to \infty} \sum_{i=1}^{n} f(x_i)(x_i - x_{i-1}),$$

where the width of each rectangle, $x_i - x_{i-1}$, approaches zero, $x_0 = a$, and $x_n = b$.

Example 1

Find the area of the region between the graph of $y = x^3$ and the x-axis from $x = 0$ to $x = 1$.

Form n equal intervals on the x-axis such that

$$0 < \frac{1}{n} < \frac{2}{n} < \frac{3}{n} < \cdots < \frac{i}{n} < \cdots < \frac{n-1}{n} < 1.$$

The area of each rectangle can be represented by

$$f\left(\frac{i}{n}\right)(x_i - x_{i-1}) \text{ or } \left(\frac{i}{n}\right)^3 \left(\frac{1}{n}\right). \quad \textit{Why is } (x_i - x_{i-1}) = \frac{1}{n}\text{?}$$

Use the definition of the area under a curve.

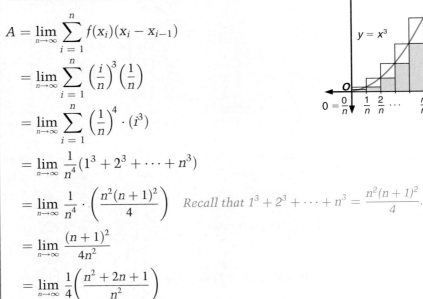

$$A = \lim_{n \to \infty} \sum_{i=1}^{n} f(x_i)(x_i - x_{i-1})$$

$$= \lim_{n \to \infty} \sum_{i=1}^{n} \left(\frac{i}{n}\right)^3 \left(\frac{1}{n}\right)$$

$$= \lim_{n \to \infty} \sum_{i=1}^{n} \left(\frac{1}{n}\right)^4 \cdot (i^3)$$

$$= \lim_{n \to \infty} \frac{1}{n^4}(1^3 + 2^3 + \cdots + n^3)$$

$$= \lim_{n \to \infty} \frac{1}{n^4} \cdot \left(\frac{n^2(n+1)^2}{4}\right) \quad \textit{Recall that } 1^3 + 2^3 + \cdots + n^3 = \frac{n^2(n+1)^2}{4}.$$

$$= \lim_{n \to \infty} \frac{(n+1)^2}{4n^2}$$

$$= \lim_{n \to \infty} \frac{1}{4}\left(\frac{n^2 + 2n + 1}{n^2}\right)$$

$$= \lim_{n \to \infty} \frac{1}{4}\left(1 + \frac{2}{n} + \frac{1}{n^2}\right) \text{ or } \frac{1}{4}$$

Thus, the area of the region under the curve is $\frac{1}{4}$ square unit.

When finding the area under a curve from $x = a$ to $x = b$, it may be simpler to find the area under the curve for $x = 0$ to $x = b$ and then subtract the area under the curve from $x = 0$ to $x = a$.

Example 2

Find the area of the region between the graph of $y = x^2$ and the x-axis from $x = 2$ to $x = 5$.

First, find the area under the curve from $x = 0$ to $x = 5$. Form n equal subintervals on the x-axis such that

$$0 < \frac{5 \cdot 1}{n} < \frac{5 \cdot 2}{n} < \frac{5 \cdot 3}{n} < \cdots < \frac{5 \cdot i}{n} < \cdots$$
$$< \frac{5(n-1)}{n} < \frac{5n}{n} \text{ or } 5.$$

The area of each rectangle can be represented as follows.

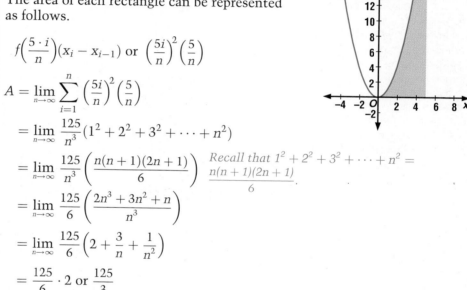

$$f\left(\frac{5 \cdot i}{n}\right)(x_i - x_{i-1}) \text{ or } \left(\frac{5i}{n}\right)^2\left(\frac{5}{n}\right)$$

$$A = \lim_{n \to \infty} \sum_{i=1}^{n} \left(\frac{5i}{n}\right)^2\left(\frac{5}{n}\right)$$

$$= \lim_{n \to \infty} \frac{125}{n^3}(1^2 + 2^2 + 3^2 + \cdots + n^2)$$

$$= \lim_{n \to \infty} \frac{125}{n^3}\left(\frac{n(n+1)(2n+1)}{6}\right) \quad \textit{Recall that } 1^2 + 2^2 + 3^2 + \cdots + n^2 = \frac{n(n+1)(2n+1)}{6}.$$

$$= \lim_{n \to \infty} \frac{125}{6}\left(\frac{2n^3 + 3n^2 + n}{n^3}\right)$$

$$= \lim_{n \to \infty} \frac{125}{6}\left(2 + \frac{3}{n} + \frac{1}{n^2}\right)$$

$$= \frac{125}{6} \cdot 2 \text{ or } \frac{125}{3}$$

Now, find the area under the curve from $x = 0$ to $x = 2$.

$$A = \lim_{n \to \infty} \sum_{i=1}^{n} \left(\frac{2i}{n}\right)^2\left(\frac{2}{n}\right)$$

$$= \lim_{n \to \infty} \frac{8}{n^3}(1^2 + 2^2 + 3^2 + \cdots + n^2)$$

$$= \lim_{n \to \infty} \frac{8}{n^3}\left(\frac{n(n+1)(2n+1)}{6}\right)$$

$$= \lim_{n \to \infty} \frac{8}{6}\left(2 + \frac{3}{n} + \frac{1}{n^2}\right)$$

$$= \frac{8}{6} \cdot 2 \text{ or } \frac{8}{3}$$

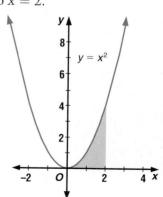

Thus, the area of the region between $y = x^2$ and the x-axis from $x = 2$ to $x = 5$ is $\frac{125}{3} - \frac{8}{3}$ or 39 square units.

In physics, when the velocity of an object is graphed with respect to time, the area under the curve represents the displacement of the object.

Example 3

APPLICATION

Physics

Refer to the application at the beginning of the lesson. Find the area between the graph of $v(t) = 16t - t^2$ and the x-axis to find the distance that the car traveled in the first 8 seconds after the light turned green.

In order to find the distance the car traveled, find the area under the curve from $x = 0$ to $x = 8$. Form n equal intervals on the x-axis such that

$$0 < \frac{8 \cdot 1}{n} < \frac{8 \cdot 2}{n} < \frac{8 \cdot 3}{n} < \cdots < \frac{8 \cdot i}{n} < \cdots < \frac{8(n-1)}{n} < \frac{8n}{n} \text{ or } 8. \text{ Then the}$$

area of each rectangle is $f\left(\dfrac{8 \cdot i}{n}\right)\left(\dfrac{8}{n}\right)$.

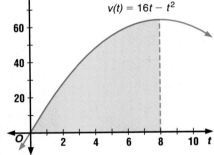

$$A = \lim_{n \to \infty} \sum_{i=1}^{n} \left(16\left(\frac{8i}{n}\right) - \left(\frac{8i}{n}\right)^2\right)\left(\frac{8}{n}\right)$$

$$= \lim_{n \to \infty} \sum_{i=1}^{n} \left(\frac{1024i}{n^2} - \frac{512i^2}{n^3}\right)$$

$$= \lim_{n \to \infty} \frac{1024}{n^2}\left(\frac{n(n+1)}{2}\right) - \frac{512}{n^3}\left(\frac{n(n+1)(2n+1)}{6}\right) \qquad 1+2+3+\cdots+n = \frac{n(n+1)}{2}$$

$$= \lim_{n \to \infty} \frac{1024n^2 + 1024n}{2n^2} - \frac{1024n^3 + 1536n^2 + 512n}{6n^3}$$

$$= \lim_{n \to \infty} 512 + \frac{512}{n} - \frac{512}{3} - \frac{256}{n} - \frac{256}{3n^2}$$

$$= \lim_{n \to \infty} \frac{1024}{3} + \frac{256}{n} - \frac{256}{3n^2}$$

$$= \frac{1024}{3} \text{ or } 341\frac{1}{3}$$

Traci's car traveled $341\frac{1}{3}$ feet in the first 8 seconds after the traffic light turned green.

FYI...

The electric traffic signal was invented by African-American inventor Garrett Morgan. Morgan also invented the gas mask that protected Allied soldiers from the deadly gases used in combat in World War I.

You may find the following formulas for the sums of series helpful when solving the problems in the exercises.

Mathematical induction can be used to prove these formulas.

$$1 + 2 + 3 + \cdots + n = \frac{n(n+1)}{2}$$

$$1^2 + 2^2 + 3^2 + \cdots + n^2 = \frac{n(n+1)(2n+1)}{6}$$

$$1^3 + 2^3 + 3^3 + \cdots + n^3 = \frac{n^2(n+1)^2}{4}$$

$$1^4 + 2^4 + 3^4 + \cdots + n^4 = \frac{6n^5 + 15n^4 + 10n^3 - n}{30}$$

$$1^5 + 2^5 + 3^5 + \cdots + n^5 = \frac{2n^6 + 6n^5 + 5n^4 - n^2}{12}$$

CHECKING FOR UNDERSTANDING

Communicating Mathematics

Read and study the lesson to answer each question.

1. **Describe** the upper and lower sums used to find the area under a curve. How are they similar and how are they different?

2. **Explain** why the width of the rectangles considered in Example 1 was $\frac{1}{n}$.

Guided Practice

Write a limit to find the area between each curve and the x-axis for the given interval. Then find the area.

3. $y = x^2$ from $x = 0$ to $x = 1$

4. $y = x^2$ from $x = 0$ to $x = 4$

5. $y = x$ from $x = 0$ to $x = 1$

6. $y = x^2$ from $x = 1$ to $x = 4$

7. $y = x$ from $x = 2$ to $x = 5$

8. $y = x^3$ from $x = 0$ to $x = 2$

EXERCISES

Practice

Write a limit to find the area between each curve and the x-axis for the given interval. Then find the area.

9. $y = x^2$ from $x = 0$ to $x = 2$

10. $y = x$ from $x = 0$ to $x = 10$

11. $y = x^5$ from $x = 0$ to $x = 1$

12. $y = x^2$ from $x = 0$ to $x = a, a > 0$

13. $y = x^4$ from $x = 4$ to $x = 7$

14. $y = x^3$ from $x = 0$ to $x = a, a > 0$

15. $y = x^2$ from $x = a$ to $x = b, 0 < a < b$

16. $y = x^2$ from $x = -3$ to $x = 2$

17. $y = |x|$ from $x = -2$ to $x = 4$

Find the area of each shaded region.

18.

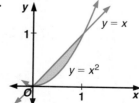

19.

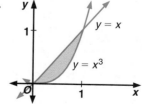

Write a formula for the area between each curve and the x-axis for the given interval. Assume that $0 < a < b$.

20. $y = x$ from $x = a$ to $x = b$

21. $y = x^3$ from $x = a$ to $x = b$

22. The following program will approximate the area between the graphs of two functions by dividing the region into rectangles. Store the two functions as Y_1 and Y_2 in the Y= list. The program requires you to input the lower and upper bounds and the number of rectangles to use for the calculations.

```
Prgm 17: AreaCrve
:ClrHome                          Clears the text screen.
:"abs (Y₁−Y₂)"→Y₃                 Finds the average height of a rectangle.
:Disp "LOWER BOUND ="             Displays the message.
:Input L                          Accepts a value for L.
:Disp "UPPER BOUND ="
:Input U
:Disp "NUMBER OF RECTANGLES"
:Input R
:(U−L)/R→W                        Finds the width of each rectangle.
:L→X                              Sets the first value of X equal to L.
:Y₃→C                             Stores height of rectangle at X = L as C.
:0→N                              Sets rectangle counter to 0.
:0→A                              Sets area to 0.
:Lbl 1                            Sets a marker at position 1.
:N + 1→N                          Increases the rectangle counter by 1.
:X + W→X                          Adds width to X to determine next X
:A + WY₃→A                        Adds area to current total area.
:If N = R                         Tests whether all rectangles were added.
:Goto 2                           Sends program to line Lbl 2".
:Goto 1
:Lbl 2
:U→X                              Sets X to value of upper bound.
:Y₃→D                             Stores height at X + U in D.
:A + WC−WD→B                      Calculates area using left endpoints.
:(A + B)/2→E                      Averages two areas.
:Disp "AREA = "
:Disp E
:End
```

Find the area between the graphs of each pair of functions for the given interval.

a. $y = x^2$

$y = 0$

$x = 0$ to $x = 4$

b. $y = x^4$

$y = x$

$x = 1$ to $x = 3$

c. $y = 4x^5$

$y = 12x^2$

$x = 0.5$ to $x = 1.5$

Critical Thinking

23. For what type of function would the upper and lower sums be equal?

Applications and Problem Solving

24. Business If $f(x)$ is the marginal cost function for a company, then the area under the curve from $x = a$ to $x = b$ is the amount it would cost to increase production from a units to b units. The Penguin Umbrella Company has a marginal cost function of $f(x) = 3 + 0.1x$ for $x < 30$ umbrellas. If they are now producing 20 umbrellas each day, how much would it cost to increase production to 30?

25. **Sports** A bicyclist accelerates from a resting position to a velocity of 18 feet per second in 6 seconds. His velocity over that time is given by $v(t) = 0.5t^2$. Find the distance that the bicyclist traveled in the first 6 seconds.

Mixed Review

26. Find the inner product of the vectors $(-3, 9)$ and $(2, 1)$. Are the vectors perpendicular? Explain. **(Lesson 8-4)**

27. Write an equation of the parabola that has a horizontal axis and passes through $(0, 0)$, $(2, -1)$, and $(4, -4)$. **(Lesson 10-2)**

28. Determine whether the point $0.3 + 0.4i$ is in the prisoner set or the escape set for the function $f(z) = z^2$. Explain. **(Lesson 13-5)**

29. Find the derivative of $f(x) = \left(1 + \frac{1}{x}\right)\left(2 - \frac{1}{x}\right)$. **(Lesson 17-2)**

30. **College Entrance Exam** Compare quantities A and B below.
    Write A if quantity A is greater.
    Write B if quantity B is greater.
    Write C if the quantities are equal.
    Write D if there is not enough information to determine the relationship.
    (A) The area of a square with sides s units long
    (B) The area of an equilateral triangle with sides $2s$ units long

MID-CHAPTER REVIEW

Evaluate each limit. **(Lesson 17-1)**

1. $\lim\limits_{x \to 5}(x^2 - 4x + 1)$

2. $\lim\limits_{n \to 3} \dfrac{n^2 - 3n + 2}{n - 1}$

3. $\lim\limits_{x \to -3} \dfrac{x^2 - 9}{x + 3}$

4. $\lim\limits_{n \to \infty} \dfrac{5n^2 - 1}{n^2}$

5. $\lim\limits_{x \to \infty} \dfrac{(x - 4)(x + 1)}{x^2}$

6. $\lim\limits_{x \to 0} \dfrac{(x + 2)^2 - 4}{x}$

7. $\lim\limits_{n \to \frac{1}{2}} \dfrac{2x^2 + 5x - 3}{x^2 - x}$

8. $\lim\limits_{n \to 3} \sqrt{\dfrac{2n + 3}{3n - 5}}$

9. $\lim\limits_{x \to 1} \dfrac{x^2}{\sqrt[3]{(x^2 - 2)^2}}$

Find the derivative of each function. **(Lesson 17-2)**

10. $f(x) = 4x^2$

11. $f(x) = (x - 1)^2$

12. $f(x) = \dfrac{5}{1 - 3x}$

13. $f(x) = -\dfrac{4}{x^9}$

14. $f(x) = \dfrac{x^2 - 5x + 4}{x^2 + x - 20}$

15. $f(x) = \dfrac{\sqrt{x^2 - 1}}{x}$

16. **Sports** A walker accelerates from a resting position to a velocity of 10.8 feet per second in 6 seconds. Her velocity is given by $v(t) = 0.3t^2$. Find the distance that the walker traveled in the first 6 seconds. **(Lesson 17-3)**

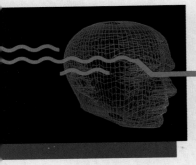

Technology

BASIC
Spreadsheets
▶ **Software**

Riemann Sums

*Riemann is
pronounced REE mahn.*

There are several methods used to approximate the area between the
x-axis and a curve on some interval $x = a$ to $x = b$. One method involves
partitioning the area into several rectangles and calculating the sum of
the areas of these rectangles. This sum is called a **Riemann sum** for the
nineteenth-century German mathematician Georg Bernhard Riemann. The
upper and lower sums that were described in Lesson 17-3 are two special
Riemann sums. IBM's *Mathematics Exploration Toolkit (MET)* has a defer
sequence that calculates Riemann sums. We will use this defer sequence
to examine the upper and lower sums for the function $f(x) = \frac{1}{5}x^2$ on the
interval $x = 0$ to $x = 5$.

Enter	Result
load numint	*Pulls up the defer sequence.*
run start	*Begins running the defer sequence.*
$1/5\, x \wedge 2$	*Enters the function*
go	$f(x) = \frac{1}{5}x^2.$
0	*Enters the limits of the*
5	*interval.*
go	
10	*Determines the number*
go	*of rectangles.*
run r	*Starts the calculations.*
lims 0 5 0 5	*Sets the range of the*
go	*viewing window.*

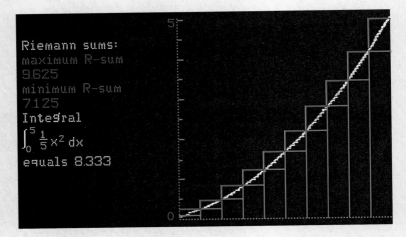

Riemann sums:
maximum R-sum
9.625
minimum R-sum
7.125
Integral
$\int_0^5 \frac{1}{5} x^2 \, dx$
equals 8.333

The program will draw rectangles and find the upper sum. Press ENTER.
After lower sum is displayed, the program will find the actual area of the
region, called the **definite integral.** Press ENTER to return to the menu.

EXERCISES

1. Find the upper and lower sums for the function $f(x) = \frac{1}{5}x^2$ on the interval $x = 0$
   to $x = 5$ using 20, 50 and 100 rectangles. As the number of rectangles is increased,
   what happens to the upper and lower sums?

2. Is the upper or the lower sum a better approximation of the area of a region for
   this function? Try different numbers of rectangles and different intervals. Is this
   sum always the better approximation?

3. The graph of $f(x) = \frac{1}{5}x^2$ is concave up. Find upper and lower sums for other
   functions that are concave up. Which sum is the better approximation of the
   area for these functions? Make a conjecture as to which sum gives the better
   approximation for the area under a curve that is concave down.

17-4 Integration

Objective

After studying this lesson, you should be able to:
- use integration formulas.

Application

"Seven out of ten American families own their own homes." "People in the District of Columbia purchase 1.3 newspapers each day." "On the average, each American spends 5¢ of each food dollar at a vending machine." Demographers study the habits and trends of populations to produce statistics like these. The statistics are used by businesses to identify the number and habits of their potential customers.

College Connections Inc. is a scholarship-search service that helps high school students find college scholarships. In developing their marketing plan for the next year, the marketing manager of College Connections needs to determine the number of people that will be in the college-age bracket in this year. They will use a life table function, a function ℓ such that $\ell(x)$ is the number of people in the population who reach the age of x years at any time during the year, to find this population. Life table functions involve **integrals**.

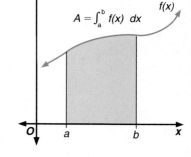

The integral of a function $f(x)$ is the area of the region bounded by the graph of $f(x)$, the x-axis and the lines $x = a$ and $x = b$. The area is symbolized as follows.

The expression dx in the integral notation tells you to integrate with respect to x.

$$A = \int_a^b f(x)\ dx$$

The notation is read the "integral of f, with respect to x, from a to b." By definition, $\int_a^b f(x)\ dx$ is equal to $\lim_{n \to \infty} \sum_{i=1}^{n} f(x_i)(x_i - x_{i-1})$.

To understand the concept of an integral, consider the function $f(x)$ in the following equation.

$$\frac{dy}{dx} = f(x), \text{ where } a < x < b$$

Is it possible to find a function $y = F(x)$ for which $f(x)$ is the derivative? This is the "anti" or inverse problem of finding the derivative. Thus, the function $F(x)$ is called an **antiderivative** of $f(x)$ if and only if $F'(x) = f(x)$.

Is it possible to find an antiderivative of $f(x)$, represented by $F(x)$, if $f(x) = 3x^2$? A few possibilities are $F(x) = x^3 + 3$, $F(x) = x^3 + \pi$, and $F(x) = x^3 - 5$. These are all valid answers and can be summarized as $F(x) = x^3 + C$, where C is a constant.

The following definition makes the connection between the antiderivative and the integral.

Definition of an Integral	**The function $F(x)$ is an integral of $f(x)$ with respect to x if and only if $F(x)$ is an antiderivative of $f(x)$. That is,** $$F(x) = \int f(x)\,dx \text{ if and only if } F'(x) = f(x).$$

Thus, in the previous example, $\int 3x^2\,dx = x^3 + C$ since $F'(x^3 + C) = 3x^2$.

Example 1 | **Find each integral.**

a. $\displaystyle\int 1\ dx$

b. $\displaystyle\int x^2\ dx$

a. Since $F'(x) = 1$ when $F(x) = x + C$, then
$$\int 1\ dx = x + C.$$

b. Since $F'(x) = x^2$ when $F(x) = \dfrac{x^3}{3} + C$, then
$$\int x^2\ dx = \dfrac{x^3}{3} + C.$$

Some of the formulas that are useful in finding integrals are listed below.

1. If k is a constant, $\displaystyle\int k\ dx = kx + C$.

2. If $n \neq -1$, $\displaystyle\int u^n\ du = \dfrac{u^{n+1}}{n+1} + C$ where u is a differentiable function.

3. By definition, $\displaystyle\int x^{-1}dx = \ln |x| + C$.

4. The integral of a sum of functions is the sum of the integrals of the functions.
$$\int (f(x) + g(x))\ dx = \int f(x)\ dx + \int g(x)\ dx$$

5. The integral of the product of a constant, a, and a function, $f(x)$, is the product of the constant and the integral of the function.
$$\int af(x)\ dx = a \int f(x)\ dx$$

Example 2 | **Find each integral.**

a. $\displaystyle\int (4x + 7)\ dx$

a. Use the formula that says the integral of a sum of functions is the sum of the integrals.

$$\int (4x + 7)\, dx = \int 4x\, dx + \int 7dx$$

$$= 4 \int x\, dx + \int 7dx \qquad \textit{Use the fifth formula.}$$

$$= 4 \left(\frac{x^2}{2} + C_1 \right) + 7x + C_2 \qquad \textit{Use the second and first formulas respectively.}$$

$$= 2x^2 + 4C_1 + 7x + C_2 \qquad \textit{Let } C = 4C_1 + C_2.$$

$$= 2x^2 + 7x + C$$

Thus, $\int (4x + 7)\, dx = 2x^2 + 7x + C$.

b. $\int (3x^2 + 6x + 1)\, dx$

b. $\int (3x^2 + 6x + 1)\, dx = \int 3x^2\, dx + \int 6x\, dx + \int 1\, dx$

$$= 3 \int x^2\, dx + 6 \int x\, dx + \int 1\, dx$$

$$= 3 \left(\frac{x^3}{3} + C_1 \right) + 6 \left(\frac{x^2}{2} + C_2 \right) + (x + C_3)$$

$$= x^3 + 3x^2 + x + C \qquad \textit{Let } C = 3C_1 + 6C_2 + C_3.$$

Thus, $\int (3x^2 + 6x + 1)\, dx = x^3 + 3x^2 + x + C$.

Sometimes it is possible to rewrite the integral by using u and $\frac{du}{dx}$. To do this, use the formula $\int f(x)\, dx = \int u \frac{du}{dx} \cdot dx = \int u\, du$.

Example 3

Find $\int (x^2 + 1)2x\, dx$.

Let $u = (x^2 + 1)$. Then $du = 2x\, dx$.

$$\int (x^2 + 1)2x\, dx = \int u\, du$$

$$= \frac{u^2}{2} + C \qquad \textit{Use the formula } \int u^n du = \frac{u^{n+1}}{n+1} + C.$$

$$= \frac{(x^2 + 1)^2}{2} + C \qquad \textit{Substitute.}$$

Therefore, $\int (x^2 + 1)2x\, dx = \frac{(x^2 + 1)^2}{2} + C$.

TECHNOLOGY Tip

You can use the TI-92 graphing calculator to find an integral of a function. The result will not display the constant, C.

Integration is used often in the sciences and in business.

Example 4

Refer to the application at the beginning of the lesson. The marketing manager for College Connections, Inc. has determined that the life table function for the region of the country that she is considering is $\ell(t) = 12{,}000\sqrt{100 - t}$. Under the appropriate conditions, the integral of the life table function gives a function that can be used to determine the number of people in the population that are t years old or younger. Find the integral of $\ell(t)$.

$$\int 12{,}000\sqrt{100 - t}\ dt = 12{,}000 \int \sqrt{100 - t}\ dt$$

Let $u = (100 - t)$, then $\dfrac{du}{dt} = -1$.

$$12{,}000 \int \sqrt{100 - t}\ dt = 12{,}000 \int \sqrt{u} \cdot -\dfrac{du}{dt} \cdot dt \qquad \sqrt{100 - t} = \sqrt{u},$$
$$\qquad\qquad\qquad\qquad\qquad\qquad\qquad\qquad 1 = -(-1) \text{ or } -\dfrac{du}{dt}$$
$$= 12{,}000 \int -\sqrt{u}\ du \qquad\qquad \text{Simplify.}$$
$$= 12{,}000 \int -u^{\frac{1}{2}}\ du \qquad\qquad \sqrt{u} = u^{\frac{1}{2}}$$
$$= 12{,}000 \left(-\dfrac{u^{\frac{1}{2}+1}}{\frac{1}{2}+1} \right) + C \qquad \text{Use the second formula.}$$
$$= 12{,}000 \left(-\dfrac{2}{3}u^{\frac{3}{2}} \right) + C$$
$$= -8000u^{\frac{3}{2}} + C$$
$$= -8000(100 - t)^{\frac{3}{2}} + C$$

MATH JOURNAL

Describe the most challenging topic you studied this year. Explain why it was a challenge.

The integral of the life table function is $-8000(100 - t)^{\frac{3}{2}} + C$.

CHECKING FOR UNDERSTANDING

Communicating Mathematics

Read and study the lesson to answer each question.

1. **Explain** what the notation $\int_a^b f(x)\ dx$ means in relation to $f(x)$.

2. **Describe** the relationship between integration and differentiation.

Guided Practice

Find each integral.

3. $\displaystyle\int 5\ dx$

4. $\displaystyle\int 2x\ dx$

5. $\displaystyle\int 3x^2\ dx$

6. $\displaystyle\int (2x - 3)\ dx$

7. $\displaystyle\int (5x^4 + 2x)\ dx$

8. $\displaystyle\int \sqrt{2x}\ dx$

EXERCISES

Practice   **Find each integral.**

9. $\int 10 \, dx$

10. $\int (2x - 12) \, dx$

11. $\int (3x^2 - 8x^3 + 5x^4) \, dx$

12. $\int (x^4 - 5) \, dx$

13. $\int 5x^3 \, dx$

14. $\int (\pi x + \sqrt{x}) \, dx$

15. $\int (x + 5)^{20} \, dx$

16. $\int \sqrt{1 + x} \, dx$

17. $\int (-2x + 3) \, dx$

18. $\int \frac{-2x}{\sqrt{1 - x^2}} \, dx$

19. $\int \frac{1}{x - 1} \, dx$

20. $\int \left(\frac{x + 1}{\sqrt[3]{x^2 + 2x + 2}} \right) dx$

Find the antiderivative of each function.

21. $f(x) = 8x^4$

22. $f(x) = 4\sqrt[3]{x}$

23. $f(x) = x^5 - \frac{1}{x^4}$

24. $f(x) = \frac{2}{x^3}$

25. $f(x) = \frac{2}{\sqrt{x}}$

26. $f(x) = \frac{2}{1 - 4x}$

Critical
Thinking   **27.** Find $\int \frac{1}{x^2} \left(\frac{x + 1}{x} \right)^{\frac{1}{3}} dx.$ $\left(\textit{Hint: Substitute } u = \frac{x + 1}{x}. \right)$

Applications
and Problem
Solving   **28. Physics**   If a function $v(t)$ describes the velocity of a moving object at time t, then the function $s(t)$ that describes the distance from a given point to the object at time t can be found by integrating $v(t)$.

a. A car is moving at a constant velocity of 45 miles per hour. Write an equation that describes the distance that the car has traveled after t hours.

b. If the car described in part a is 104 miles from the given point at time $t = 2$, write an equation that describes the position of the car at time t.

c. The velocity of a rocket is given by $v(t) = -32t + 100$. If the rocket is launched from a point 50 feet above the ground, write an equation that describes the distance the rocket has traveled after t seconds.

29. **Business** For Suds Inc., the marginal revenue from the sale of the xth pound of soap is found by $m(x) = 4 - 0.02x$ when $0 \le x \le 400$. If Suds sells no soap, they will lose \$300. Write an equation that describes the total revenue for Suds Inc.

30. **Business** The buyer for Food King Grocery has found that the marginal daily demand for their home made French bread can be described by the function $D'(x) = -3.2x + 20$ where x is the price charged for a loaf of bread and $D'(x)$ is the number of loaves.
    a. Find a daily demand function for this product.
    b. If the daily demand for French bread is 65 loaves when the price is 85¢ per loaf, what will the daily demand be if the price is increased to 95¢ per loaf?

Mixed Review

31. Michael Thomas, the manager of the paint department at Builder's Headquarters, is mixing paint for a spring sale. There are 32 units of yellow dye, 54 units of brown dye, and an unlimited supply of base paint available. Mr. Thomas plans to mix as many gallons as possible of Autumn Wheat and Harvest Brown paint. Each gallon of Autumn Wheat requires 4 units of yellow dye and 1 unit of brown dye. Each gallon of Harvest Brown paint requires 1 unit of yellow dye and 6 units of brown dye. Find the maximum number of gallons of paint that Mr. Thomas could mix. **(Lesson 2-6)**

32. Verify that $1 + \sin 2x = (\sin x + \cos x)^2$ is an identity. **(Lesson 7-4)**

33. Simplify $7^{\log_7 2x}$. **(Lesson 11-4)**

34. Find the first three iterates of the function $f(x) = 0.5x - 1$ using $x_o = 8$. **(Lesson 13-1)**

35. How many vertices are there in a graph with 15 edges if each vertex has degree 2? **(Lesson 16-1)**

36. Find the area between the graph of $y = x^2$ and the x-axis for the interval from $x = 2$ to $x = 7$. **(Lesson 17-3)**.

37. **College Entrance Exam** Choose the best answer.
    Sherry mixes a pounds of peanuts that cost b cents per pound with c pounds of rice crackers that cost d cents per pound to make Oriental Peanut Mix. What should the price in cents for a pound of Oriental Peanut Mix be if Sherry is to make a profit of 10¢ per pound?

    (A) $\dfrac{ab + cd}{a + c} + 10$ (B) $\dfrac{b + d}{a + c} + 10$ (C) $\dfrac{ab + cd}{a + c} + 0.10$

    (D) $\dfrac{b + d}{a + c} + 0.10$ (E) $\dfrac{b + d + 10}{a + c}$

17-5 The Fundamental Theorem of Calculus

Objective

After studying this lesson, you should be able to:
- use the fundamental theorem of calculus to evaluate definite integrals and to find area.

Application

Magic Memories Inc. (MMI) makes photo albums and scrapbooks. Their marginal cost function is $m(x) = \dfrac{1}{\sqrt{x}} + 2$, where x is thousands of books produced and $m(x)$ is measured in thousands of dollars. MMI has the opportunity to bid on a contract for producing senior memory books for a company that sells high school rings, key chains, and jackets. The contract would require MMI to produce between 1000 and 4000 scrapbooks. Producing the scrapbooks would require a fixed cost of $1000. Their sales manager would like to find the average cost for producing the scrapbooks in order to make the bid accurate. *This problem will be solved in Example 2.*

The derivative has been defined as the limit of the difference quotients.

$$f'(x) = \lim_{h \to 0} \frac{f(x+h) - f(x)}{h}$$

The integral has been defined as the limit of the sum of the areas.

$$A = \lim_{n \to \infty} \sum_{i=1}^{n} f(x_i)(x_i - x_{i-1}), \text{ where } (x_i - x_{i-1}) \text{ approaches zero}$$

The **fundamental theorem of calculus** formally states that these limiting processes are inverse operations.

Fundamental Theorem of Calculus	**If the function $f(x)$ is continuous and $F(x)$ is such that $F'(x) = f(x)$, then $\displaystyle\int_a^b f(x)\, dx = F(b) - F(a)$.**

a and b are called the lower and upper limits or bounds of the integration.

The fundamental theorem of calculus provides a way to evaluate the **definite integral** $\displaystyle\int_a^b f(x)\, dx$ if an antiderivative $F(x)$ can be found. A vertical line on the right side is used to abbreviate $F(b) - F(a)$. Thus, the principal statement of the theorem may be written as follows.

$$\int_a^b f(x)\, dx = F(x) \Big|_a^b = F(b) - F(a)$$

Example 1

Evaluate $\int_1^4 (5x+1)^2 \, dx$.

$$\int_1^4 (5x+1)^2 \, dx = \int_1^4 (25x^2 + 10x + 1) \, dx \qquad \textit{Expand } (5x+1)^2.$$

$$= \frac{25}{3}x^3 + 5x^2 + x \Big|_1^4 \qquad \textit{Integrate.}$$

$$= \left(\frac{25}{3}(4^3) + 5(4^2) + 4\right) - \left(\frac{25}{3}(1^3) + 5(1^2) + 1\right) \qquad \begin{array}{l} F(x)\big|_a^b = \\ F(b) - F(a) \end{array}$$

$$= 603$$

Since $F(x)$ is any antiderivative of $f(x)$, the constant C is omitted when using the integration formulas to find a definite integral. Therefore, the area between the x-axis and the graph of $f(x) = (5x+1)^2$ for the interval $x = 1$ to $x = 4$ is 603 units^2.

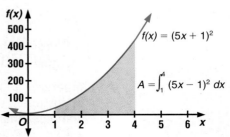

Definite integrals can be used to find the average value of a function over an interval. If $\bar{f}$ is the average value of the function $f(x)$, then

$$\bar{f} = \left(\frac{1}{b-a}\right) \int_a^b f(x) \, dx.$$

Example 2

Notice that C(x) is defined as the number of thousands of dollars to produce x thousands of scrapbooks.

Refer to the application at the beginning of the lesson. Find the average cost of producing each scrapbook if between 1000 and 4000 scrapbooks are produced.

The marginal cost function is $m(x) = \dfrac{1}{\sqrt{x}} + 2$, so the cost function is $\int m(x)dx$ or $2\sqrt{x} + 2x + C$. The fixed cost is \$1000, so the cost function for this contract would be $C(x) = 2\sqrt{x} + 2x + 1$.

Now use the formula $\bar{f} = \left(\dfrac{1}{b-a}\right) \int_a^b f(x) \, dx$ for $f(x) = C(x)$.

$$\overline{C(x)} = \left(\frac{1}{b-a}\right) \int_a^b C(x) \, dx$$

$$= \left(\frac{1}{4-1}\right) \int_1^4 (2\sqrt{x} + 2x + 1) \, dx \qquad a = 1, b = 4, C(x) = 2\sqrt{x} + 2x + 1$$

$$= \frac{1}{3}\left(\frac{4}{3}x^{\frac{3}{2}} + x^2 + x\right)\Big|_1^4 \qquad \textit{Integrate.}$$

$$= \frac{1}{3}\left(\frac{4}{3}(4)^{\frac{3}{2}} + (4)^2 + (4)\right) - \frac{1}{3}\left(\frac{4}{3}(1)^{\frac{3}{2}} + (1)^2 + (1)\right)$$

$$= \frac{92}{9} - \frac{10}{9} \text{ or } \frac{82}{9}$$

It will cost an average of $\dfrac{82}{9}$ or about 9.11 thousands of dollars to produce one thousand scrapbooks if between 1000 and 4000 scrapbooks are produced. Therefore, it will cost about \$9.11 to produce each scrapbook.

The definite integral will produce a negative value if $f(x) < 0$ in the interval from $x = a$ to $x = b$. It will produce a positive value if $f(x) > 0$ in the same interval. Therefore, if the integral is being used to find the area between a curve and the x-axis, the absolute value of the integral is used.

Example 3

Find the area between the x-axis and the graph of the function $f(x) = x^2 - 9$ from $x = -3$ to $x = 3$.

Use the absolute value of the definite integral since we are finding an area.

$$A = \left| \int_{-3}^{3} (x^2 - 9) \, dx \right|$$

$$= \left| \frac{x^3}{3} - 9x \right|_{-3}^{3}$$

$$= \left| \left[\frac{27}{3} - 9(3) \right] - \left[\frac{-27}{3} - 9(-3) \right] \right|$$

$$= |-36|$$

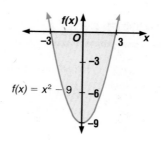

The area is 36 square units.

When a function is both positive and negative in the interval from $x = a$ to $x = b$, and the area is to be found, the limits of integration must be split at the zeros of the function.

Example 4

Find the area between the x-axis and the function $f(x) = x^3 - 3x^2 + 2x$ from $x = 0$ to $x = 2$.

The area is the sum of the areas of A_1 and A_2.

$$f(x) = x^3 - 3x^2 + 2x$$
$$= x(x - 1)(x - 2)$$

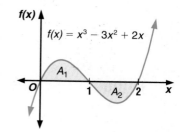

The zeros of the function are 0, 1, and 2. So evaluate the integral for the intervals $x = 0$ to $x = 1$ and $x = 1$ to $x = 2$.

$$A_1 = \left| \int_0^1 (x^3 - 3x^2 + 2x) \, dx \right|$$

$$= \left| \frac{1}{4}x^4 - x^3 + x^2 \right|_0^1$$

$$= \left| \left(\frac{1}{4}(1)^4 - (1)^3 + (1)^2 \right) - \left(\frac{1}{4}(0)^4 - (0)^3 + (0)^2 \right) \right| \text{ or } \left| \frac{1}{4} \right|$$

$$A_2 = \left| \int_1^2 (x^3 - 3x^2 + 2x) \, dx \right|$$

$$= \left| \frac{1}{4}x^4 - x^3 + x^2 \right|_1^2$$

$$= \left| \left(\frac{1}{4}(2)^4 - (2)^3 + (2)^2 \right) - \left(\frac{1}{4}(1)^4 - (1)^3 + (1)^2 \right) \right| \text{ or } \left| -\frac{1}{4} \right|$$

(continued on the next page)

Portfolio

..ew items in your portfolio. Make a table of contents of the items, noting why each item was chosen. Replace any items that are no longer appropriate.

$$A = A_1 + A_2$$

$$= \left|\frac{1}{4}\right| + \left|-\frac{1}{4}\right|$$

$$= \frac{1}{2}$$

The area is $\frac{1}{2}$ square unit.

CHECKING FOR UNDERSTANDING

Communicating Mathematics

Read and study the lesson to answer each question.

1. What relationship does the fundamental theorem of calculus establish for differentiation and integration?

2. When you find an indefinite integral of a function, a constant C is added to the function. Why is it not necessary to add the constant when you are finding a definite integral?

3. **Explain** why you must take the absolute value when finding the area between a curve and the x-axis.

Guided Practice

Use integration to find each area.

4.
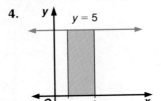

5.
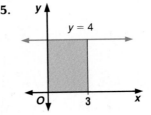

6.
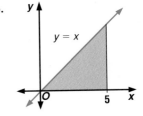

7.
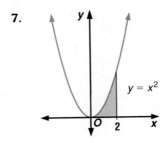

8.
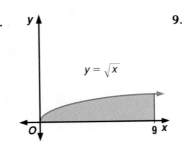

9.
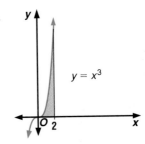

Evaluate each definite integral.

10. $\int_0^1 (2x + 3)\, dx$

11. $\int_0^1 (3x^2 + 6x + 1)\, dx$

12. $\int_0^3 \left(\frac{1}{2}x - 4\right) dx$

13. $\int_{-4}^{-1} (5x + 14)\, dx$

Graph each function. Then, find the area between the function and the x-axis for the given interval using integration.

14. $f(x) = 2x + 3$ for $x = 1$ to $x = 4$ **15.** $f(x) = x^2$ for $x = -2$ to $x = 2$

EXERCISES

Practice   Use integration to find the area of each shaded region.

16.

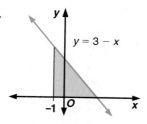

$y = 3 - x$

17.

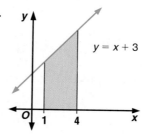

$y = x + 3$

18.

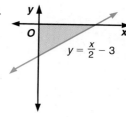

$y = \dfrac{x}{2} - 3$

19.

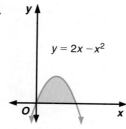

$y = 2x - x^2$

20.

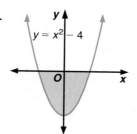

$y = x^2 - 4$

21.

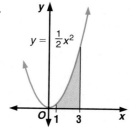

$y = \left| \dfrac{1}{2} x^2 \right|$

Evaluate each definite integral.

22. $\displaystyle\int_0^3 x \, dx$ **23.** $\displaystyle\int_{-1}^1 (x + 1)^2 \, dx$

24. $\displaystyle\int_{-1}^1 (4x^3 + 3x^2) \, dx$ **25.** $\displaystyle\int_1^4 \left(x^2 + \dfrac{2}{x^2} \right) dx$

26. $\displaystyle\int_{-1}^1 12x(x + 1)(x - 1) \, dx$ **27.** $\displaystyle\int_4^5 (x^2 + 6x - 7) \, dx$

28. $\displaystyle\int_0^2 (x - 4x^2) \, dx$ **29.** $\displaystyle\int_{-1}^0 (1 - x^2) \, dx$

30. $\displaystyle\int_1^4 (3x^2 - 6x) \, dx$ **31.** $\displaystyle\int_{-2}^3 (x + 2)(x - 3) \, dx$

Graph each function. Then, find the area between the graph of the function and the x-axis for the given interval by integrating.

32. $f(x) = -x$ for $x = 1$ to $x = 4$

33. $f(x) = x^3$ for $x = -1$ to $x = 2$

34. $f(x) = -x^2$ for $x = 0$ to $x = 5$

35. $f(x) = -x^3$ for $x = -4$ to $x = 0$

36. $f(x) = \dfrac{3x^2 - 18x + 15}{5}$ for $x = 0$ to $x = 6$

37. $f(x) = 9 - 3x^2$ for $x = 0$ to $x = 3$

38. a. Use integration and the fundamental theorem of calculus to prove that for any numbers a and b, $\int_a^b x^3\,dx = \frac{1}{4}(b^4 - a^4)$.

b. Find a formula for $\int_a^b x^n\,dx$ that would hold for any integer n.

39. Business The life table function of a population is a function ℓ such that $\ell(x)$ is the number of people in the population who reach the age of x years at any time during the year. Under the appropriate conditions, $\int_a^b \ell(x)\,dx$ gives the number of people in the populations between the exact ages of a and b years, inclusive. The marketing manager for College Connections Inc. has determined that the life table function for the region of the country she is considering is $\ell(x) = 12{,}000\sqrt{100 - x}$.

a. Find the number of people in the region that are between 18 and 21 years old.

b. A sister company, Career Connections Inc., finds jobs for clients and also finds people to fill jobs at local companies. Their largest market is recent college graduates. The marketing manager at Career Connections would like to know the number of people in this region that are between the ages of 21 and 23. Find this number.

40. Medicine Arteries carry blood away from the heart to the other parts of the body. Let r represent the radius of a cylindrical artery that is l unit long. A certain blood cell is x units from the center of a cross section of the artery. Then, the volume of blood that can flow through the artery in a unit of time is given by $V = \int_0^r \frac{k}{\ell}x(r^2 - x^2)\,dx$, where k is a constant depending on the difference in pressure at the two ends of the artery and on the viscosity of the blood. Find V.

41. Hydraulics A cylindrical water tank is 100 feet high and has a diameter of 100 feet. The work, W, in foot-pounds that is required to pump all the water out of the tank is given by $W = (156{,}250\pi)\int_0^{100}(100 - y)\,dy$. How much work is required?

42. Find the inverse of $y = 7 - x^2$. **(Lesson 3-3)**

43. Write the polynomial equation of least degree with roots of $-3, 0.5, 6$, and 2. **(Lesson 4-1)**

44. Write an equation of the line with parametric equations $x = -5t - 1$ and $y = 2t + 10$ in slope-intercept form. **(Lesson 8-6)**

45. Write the equation $2x^2 - y^2 - 16x + 4y + 24 = 0$ in standard form. Then find the coordinates of the center, the foci, the vertices, and the equations of the asymptotes of the graph and sketch the graph. **(Lesson 10-4)**

46. Use the binomial theorem to expand $(5x - 1)^3$. **(Lesson 12-6)**

47. Education The scores of a national achievement test are normally distributed with a mean of 500 and a standard deviation of 100. What percentage of those who took the test had a score more than 100 points above or below the mean? **(Lesson 15-4)**

48. Find $\int x^3(2 - 5x^4)^7 \, dx$. **(Lesson 17-4)**

49. College Entrance Exam Choose the best answer.
Triangle ABC is inscribed in circle O, and $\overleftrightarrow{CD}$ is tangent to $\odot O$ at point C. If $m\angle BCD = 40°$, find $m\angle A$.

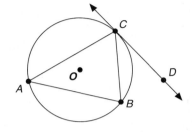

(A) $60°$ (B) $50°$ (C) $40°$
(D) $30°$ (E) $20°$

DECISION MAKING

Planning a Move

The average American moves 10 times in a lifetime. Every move brings with it new hopes and opportunities, but also some outlay of money, loss of friends and security, and temporary feelings of dislocation and confusion. The key to a successful move is careful planning to assure that the confusion is brief and the expense is minimal. What can you do to accomplish these goals?

1. Describe a place, other than where you live now, where you think you would like to live. What advantages does that place have over your present home? What disadvantages are there?

2. Plan a move to the place described in Exercise 1. What services will you have to have cut off in your present home and which ones will you have to arrange for your new home? How will you pack and transport your belongings? As you draw up your plan, look for ways you can cut expenses, minimize your feelings of dislocation, and pave the way for a smooth transition to life in your new home.

3. Project Work with your group to complete one of the projects listed at the right based on the information you gathered in Exercises 1–2.

PROJECTS

- *Conduct an interview.*
- *Write a book report.*
- *Write a proposal.*
- *Write an article for the school paper.*
- *Write a report.*
- *Make a display.*
- *Make a graph or chart.*
- *Plan an activity.*
- *Design a checklist.*

VOCABULARY

Upon completing this chapter, you should be
familiar with the following terms:

antiderivative	**939**	**939**	integral
definite integral	**945**	**931**	lower sums
derivative notation	**923**	**925**	marginal cost function
differentiable	**923**	**925**	marginal revenue function
differentiation	**923**	**938**	Riemann sum
fundamental theorem of calculus	**945**	**931**	upper sums

SKILLS AND CONCEPTS

OBJECTIVES AND EXAMPLES

Upon completing this chapter, you should
be able to:

- use limit theorems to evaluate the limit
  of a function.   **(Lesson 17-1)**

  Evaluate $\lim\limits_{n\to\infty} \dfrac{n^2 + n - 6}{n^2}$.

  $$\lim_{n\to\infty} \frac{n^2 + n - 6}{n^2} = \lim_{n\to\infty} \frac{n^2}{n^2} + \lim_{n\to\infty} \frac{n}{n^2} - \lim_{n\to\infty} \frac{6}{n^2}$$

  $$= \lim_{n\to\infty} 1 + \lim_{n\to\infty} \frac{1}{n} - \lim_{n\to\infty} \frac{6}{n^2}$$

  $$= 1$$

  Use Theorem 10 to evaluate $\lim\limits_{x\to 0} \dfrac{\sin x}{2x}$.

  $$\lim_{x\to 0} \frac{\sin x}{2x} = \lim_{x\to 0} \frac{1}{2} \cdot \lim_{x\to 0} \frac{\sin x}{x}$$

  $$= \frac{1}{2} \cdot 1 \text{ or } \frac{1}{2}$$

REVIEW EXERCISES

Use these exercises to review and prepare
for the chapter test.

Evaluate each limit.

1. $\lim\limits_{x\to 0} \left(4^x + \dfrac{1}{4^x} \right)$

2. $\lim\limits_{x\to 0} \dfrac{\sqrt{x+4} - 2}{x}$

3. $\lim\limits_{x\to 0} \dfrac{3x^3 - 2x}{2x^2 - 3x}$

4. $\lim\limits_{x\to 0} \dfrac{\sqrt{3x^2 + x + 1}}{\sqrt[3]{3x^3 - x + 8}}$

**Evaluate the limit of $f[g(x)]$ as x
approaches 1 for each $f(x)$ and $g(x)$.**

5. $f(x) = x + 2$
   $g(x) = -3x$

6. $f(x) = x^2 - 1$
   $g(x) = 2x + 1$

Use Theorem 10 to evaluate each limit.

7. $\lim\limits_{x\to 0} \dfrac{1 - \cos^2 x}{x^2}$

8. $\lim\limits_{x\to 0} \left(x - \dfrac{\sin 3x}{x} \right)$

■ find the derivative of a function
(Lesson 17-2)

Find the derivative of
$f(x) = (2x^3)(x^2 + 1)$.

$\frac{d}{dx}[2x^3(x^2 + 1)]$

$= 2x^3 \frac{d}{dx}(x^2 + 1) + (x^2 + 1)\frac{d}{dx}2x^3$

$= (2x^3)(2x) + (x^2 + 1)(6x^2)$

$= 4x^4 + 6x^4 + 6x^2$

$= 10x^4 + 6x^2$

Find the derivative of each function.

9. $f(x) = x^6$

10. $f(x) = 4x^3$

11. $f(x) = 3x + 4x^2$

12. $f(x) = (x^4 - 3x^2)(5x^3)$

13. $f(x) = \sqrt{2x^3 - 6x}$

14. $f(x) = 4x + \frac{(x-1)^2}{2x}$

■ find the area between a curve and the
x-axis by using the limit of areas of
rectangles **(Lesson 17-3)**

Find the area of the region between the
graph of $y = 3x^2$ and the x-axis from
$x = 0$ to $x = 1$.

$A = \lim_{n \to \infty} \sum_{i=1}^{n} f(x_i)(x_i - x_{i-1})$

$= \lim_{n \to \infty} \sum_{i=1}^{n} 3\left(\frac{i}{n}\right)^2 \left(\frac{1}{n}\right)$

$= \lim_{n \to \infty} \frac{3}{n^3}(1^2 + 2^2 + \ldots + n^2)$

$= \lim_{n \to \infty} \frac{3}{n^3}\left(\frac{n(n+1)(2n+1)}{6}\right)$

$= \lim_{n \to \infty} \frac{1}{2}(2 + \frac{3}{n} + \frac{1}{n^3})$

$= 1$

**Write a limit to find the area between each
curve and the x-axis for the given interval.
Then find the area.**

15. $y = 2x$ from $x = 0$ to $x = 2$

16. $y = x^3$ from $x = 0$ to $x = 1$

17. $y = x^2$ from $x = 3$ to $x = 4$

18. $y = 6x^2$ from $x = 1$ to $x = 2$

■ use integration formulas **(Lesson 17-4)**

Find $\int (6x - 4)\, dx$.

$\int (6x - 4)dx = \int 6x\, dx - \int 4\, dx$

$= 6 \int x\, dx - \int 4\, dx$

$= 6\left(\frac{x^2}{2}\right) + C_1 - (4x + C_2)$

$= 3x^2 - 4x + C$

Find each integral.

19. $\int \frac{4}{x^2} dx$

20. $\int 5x^3\, dx$

21. $\int (1 - x)\, dx$

22. $\int \frac{1}{\sqrt{x + 3}}\, dx$

Find the antiderivative of each function.

23. $f(x) = 5(x + 3)^9$

24. $f(x) = 1 - \frac{1}{x^2}$

OBJECTIVES AND EXAMPLES	REVIEW EXERCISES

■ use the fundamental theorem of calculus to evaluate definite integrals and to find area **(Lesson 17-5)**

Evaluate $\int_4^7 (x^2 - 3)\, dx$.

$$\int_4^7 (x^2 - 3)\, dx = \frac{x^3}{3} - 3x \Big|_4^7$$
$$= \left(\frac{7^3}{3} - 3(7)\right) - \left(\frac{4^3}{3} - 3(4)\right)$$
$$= 84$$

Evaluate each definite integral.

25. $\int_2^4 6x\, dx$

26. $\int_{-3}^2 3x^2\, dx$

27. $\int_{-2}^2 (3x^2 - x + 5)\, dx$

28. $\int_0^4 (x-2)(2x+3)\, dx$

APPLICATIONS AND PROBLEM SOLVING

29. **Physics** The kinetic energy of an object with mass m is given by the formula $k(t) = \frac{1}{2}m[v(t)]^2$, where $v(t)$ is the velocity of the object at time t. Suppose $v(t) = \dfrac{50}{1 + t^2}$ for all $t \geq 0$. What does the measure of the kinetic energy of an object approach as time approaches 100? **(Lesson 17-1)**

30. **Business** The controller for the AMC Electronics Company has used the production figures for the last few months to determine that the function $c(x) = -9x^5 + 135x^3 + 10{,}000$ approximates the cost of producing x thousands of one of their products. Find the marginal cost per unit production if they are now producing 2600 units. **(Lesson 17-2)**

31. **Automobile Industry** An advertisement for a new sports car claims that the car can accelerate from 0 to 50 miles per hour in 7 seconds. **(Lesson 17-4)**

    a. Find the acceleration of the sports car in feet per second squared.

    b. Write an equation for the velocity of the sports car at t seconds.

    c. Write an equation for the distance traveled in t seconds.

32. **Geometry** Find the area inside the parabola with equation $f(x) = 3x^2 - 2x + 1$ between $x = -1$ and $x = 2$. **(Lesson 17-5)**

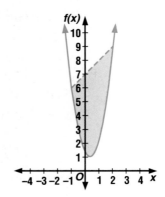

Evaluate each limit.

1. $\lim\limits_{x \to 3} \dfrac{x^2 - 9}{x^3 - 27}$

2. $\lim\limits_{x \to 1} \dfrac{x^2 - 2x + 3}{3x^2 - 5}$

3. $\lim\limits_{x \to \infty} \dfrac{(x - 2)(x + 1)}{x^2}$

4. $\lim\limits_{x \to 4} \sqrt{x}(x - 1)^2$

5. $\lim\limits_{x \to 0} \dfrac{\sin(-2x)}{3x}$

6. $\lim\limits_{x \to 1} \dfrac{x - 1}{x^3 - 1}$

7. Evaluate the limit of $f[g(x)]$ as x approaches 1 for $f(x) = x^3$ and $g(x) = 3x^2$.

Find the derivative of each function.

8. $f(x) = 4x^3 - 4$

9. $f(x) = (x + 3)^2$

10. $f(x) = 6(x^4 - 5)$

11. $f(x) = (2x^4)(x^3 + 3x^2)$

12. $f(x) = \dfrac{2x}{1 + x^2}$

13. $f(x) = \sqrt{4x^2 - 1}$

Write a limit to find the area between each curve and the x-axis for the given interval. Then find the area.

14. $y = x^3$ from $x = 0$ to $x = 2$

15. $y = 3x^2$ from $x = 1$ to $x = 3$

Find each integral.

16. $\displaystyle\int (1 - 2x)\, dx$

17. $\displaystyle\int (3x^2 + 4x + 7)\, dx$

18. $\displaystyle\int \dfrac{1}{x^2}\, dx$

Find the antiderivative of each function.

19. $f(x) = \sqrt[3]{x^2}$

20. $f(x) = \dfrac{1}{2x^3}$

Evaluate each definite integral.

21. $\displaystyle\int_0^1 (2x + 3)\, dx$

22. $\displaystyle\int_1^3 (-x^2 - x + 3)\, dx$

23. $\displaystyle\int_1^4 \left(x^2 + \dfrac{2}{x^2}\right) dx$

24. Physics The period of a simple pendulum is given by the formula $T(\ell) = 2\pi\sqrt{\dfrac{\ell}{g}}$, where ℓ is the length of the pendulum and g is the acceleration due to gravity. When ℓ is measured in feet, g is 32 ft/s^2. Find a formula for the rate of change of the period of a pendulum with respect to the length of the pendulum.

25. Geometry The volume of a sphere is given by $V = \pi \displaystyle\int_{-r}^{r} (r^2 - x^2)\, dx$, where r is the radius. Use this formula to find the volume of a sphere with a radius of 2 units.

Bonus Evaluate $\lim\limits_{x \to 0} \dfrac{\tan x}{x}$.

Buying a Home

Owning a home has always been part of the American dream. In recent years, however, as the cost of buying a home has risen markedly in comparison with personal income, the dream has begun to fade. During the 1980s, the median price of a new single-family house in the United States rose from $64,600 to $123,000, a 90.4% increase. New home prices in 1990 were highest in the Northeast, averaging $157,000, and lowest in the South, averaging $99,000. Average monthly mortgage payments rose from $599 in 1980 to $1054 in 1989.

Figure 1–1 Median sales price of single-family houses, 1970–1990

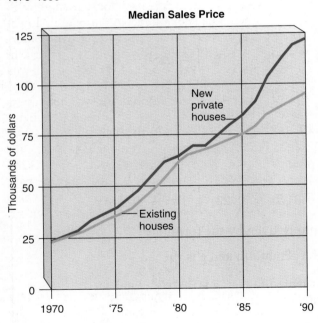

While housing costs were soaring, the after-tax income of low- and middle-income families remained almost unchanged. Only the top one fifth of income earners gained substantially during the 1980s. Yet even the average after-tax rise in income posted by that group—34%—paled in comparison with the overall increase in housing prices.

One reason for the huge rise in prices was a steadily increasing demand for existing houses. Sales of existing houses fell off between 1978 and 1982. But beginning in 1982, when 1,990,000 houses were sold, annual sales rose steadily, peaking at 3,594,000 in 1988. Prices have leveled off since then, although in selected areas they continue to skyrocket. In Honolulu, for example, the median price of a new home rose from $215,000 in 1988 to $345,000 in 1990, a 60% increase in just two years.

Figure 1–2 Single-family houses sold, 1970–1990

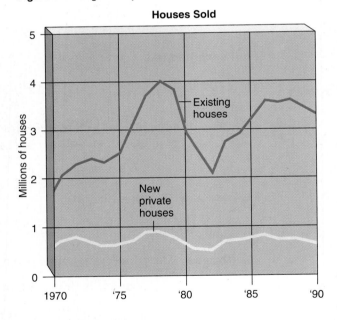

Homes and Homeowners

U.S. Census Bureau figures show that there were 105.8 million housing units in the United States in 1990. Of these, 60.0 million were occupied by their owners, 33.7 million were occupied by renters, and 12.2 million were vacant. Only 8% of all home owners are African-American, and 4% are Hispanic. This compares unfavorably to the general population that is 12% African-American and 9% Hispanic.

The great majority of American homes—72%, including mobile homes—are traditional single-unit structures. The typical home has 4 to 7 rooms (only 12% have 8 or more rooms; about the same number have 3 or fewer rooms). In 1989, the median age of all houses was 27 years. Only 10% of all houses are 70 or more years of age.

Along with price, several other features of new homes have changed greatly in recent years. Between 1970 and 1990, the average size of new single-family dwellings increased from 1500 square feet to 2080 square feet, a 38.7% increase. During the same period, the percent of homes with central air conditioning increased from 34% to 76%, and those with one or more fireplaces increased from 35% to 66%. By contrast, the portion with one or no bathrooms decreased from 32% to 8%.

Figure 1–3 Gas was the heating fuel of choice in single-family homes in 1990.

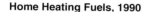

Home Heating Fuels, 1990

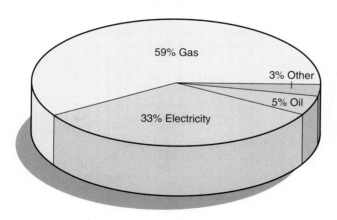

59% Gas

3% Other

5% Oil

33% Electricity

A House As an Investment

For those who can afford it, a house is one of the best investments they can make. Even in weak economic climates, home values tend to *appreciate* (increase) steadily. At a modest 5% rate of appreciation, a new home purchased for $120,000 will be worth $153,000 in only five years, a $33,000 profit should the owner decide to sell.

Since few people have the cash to purchase a home outright, most make only a down payment in cash, financing the balance with a loan, called a *mortgage*. Borrowers agree to pay back the loan, with interest, in monthly installments over a specified period of time, usually 15, 20, or 30 years. The down payment may be as low as 5% of the selling price, but is generally higher. In 1989, the average down payment for first-time buyers was 15.8% of the sales price; for repeat buyers it was 30.3%. The advantage of a high down payment is that it will lower the monthly mortgage payments. But it carries a risk, too: the higher the down payment, the less cash the buyer will have on hand to deal with emergencies like illness or job loss.

Two principal types of mortgages are available. A *fixed-rate mortgage* (FRM) is paid off at a fixed rate of interest. Monthly payments remain the same throughout the period of repayment regardless of fluctuations in interest rates.

In the beginning, an *adjustable-rate mortgage* (ARM) is paid off at an interest rate that is lower than that of a fixed-rate mortgage. The rate, however, may be raised or lowered several times during the life of the loan. An agreed-upon *cap* limits the total amount the rate may be raised and also limits the yearly rate increase. Home owners with ARMs have the advantage of initially lower monthly payments than those of FRM holders who borrowed the same amount to begin with. In later years, however, ARM monthly payments will increase steadily, eventually surpassing FRM payments. ARMs are tailored for young buyers with relatively low incomes who expect their incomes to increase steadily through the years. The risk is that if, due to unforeseen circumstances, incomes do not increase, borrowers carrying ARMs may be unable to make their higher monthly payments and could lose their homes. ARMs are also a good alternative if a buyer plans to move within a short period of time, for example, 5 years.

A Case History

George and Melinda Bauer have decided to buy their first home. They are each 30 years old, the average age of first-time buyers in the United States, and have a combined gross annual income of $42,000. The home they want to buy is on the market for $128,000, but through a series of offers and counter-offers, they are able to negotiate a price of $120,000 with the owner. They make a $20,000 down payment ($16\frac{2}{3}$%), leaving $100,000 to be financed by a mortgage lender.

the help of their realtor, the Bauers are ... to find a mortgage company offering a 30-year, 9% FRM. Before they can obtain the money, however, they must provide the company with tax records, banking records, proof of employment, credit histories, and a myriad of other records to prove that they are good credit risks.

Figure 1–4 Monthly payments on 9% fixed rate mortgages

	9% Annual Percent Rate				
	Monthly Payments (Payments and Interest)*				
Amount Financed	10 Years	15 Years	20 Years	25 Years	30 Years
$ 25,000	316.69	253.57	224.93	209.80	201.16
30,000	380.03	304.28	269.92	251.76	241.39
35,000	443.36	354.99	314.90	293.72	281.62
40,000	506.70	405.71	359.89	335.68	321.85
45,000	570.04	466.42	404.88	377.64	362.08
50,000	633.38	507.13	449.86	419.60	402.31
60,000	760.05	608.56	539.84	503.52	482.77
70,000	886.73	709.99	629.81	587.44	563.24
80,000	1013.41	811.41	719.78	671.36	643.70
90,000	1140.08	912.84	809.75	755.28	724.16
100,000	1266.76	1014.27	899.73	839.20	804.62
120,000	1520.10	1217.12	1079.68	1007.04	965.54
140,000	1773.46	1419.98	1259.62	1174.88	1126.48
160,000	2026.82	1622.82	1439.56	1342.72	1297.40
180,000	2280.16	1825.68	1619.50	1510.56	1448.32
200,000	2533.52	2028.54	1799.46	1678.40	1609.24

Mortgage companies require that borrowers earn enough so that monthly payments do not exceed 28% of their gross income. Twenty-eight percent of the Bauers' monthly income of $3500 is $980, well above the calculated $804.62 monthly mortgage payment. Therefore, they easily meet the 28% requirement. Furthermore, they are judged excellent risks by the company because of their credit history. Five weeks after applying for the loan, they learn that, like 85% of those who apply for mortgages, they have been approved for the loan. One-time costs relating to the purchase of the house, called *closing costs*, amount to $2645. Moving into their new home from their apartment costs $1580. Lastly, the mortgage company charges 1 *point* (1 percent of the loan), or $1000, for its services.

Cost to Purchase a Home

Down payment	$20,000
Closing costs	2,645
Moving costs	1,580
Mortgage fee	1,000
Total	$25,225

Is It Worth It?

Home owners face substantially greater operating costs than renters face. Furthermore, the cost of borrowing money is enormous. Over 30 years, the Bauers will pay a whopping $189,663.20 in interest on their $100,000 loan, repaying a total of $289,663.20 to the mortgage company.

On the other hand, most Americans move long before they pay off their mortgages. Through the sale of their homes they are able to repay mortgage lenders and earn a profit, which most then reinvest as down payments on more expensive homes. Renters never earn profits and can afford more expensive housing only if their financial situation improves markedly.

In addition, mortgage holders have an enormous tax advantage over renters. Mortgage interest payments are not subject to federal income taxes. The result is large tax savings to the mortgage holder, especially in the early years of repayment on a loan, when almost the entire monthly installment consists of interest.

Figure 1–5 The portion of a monthly mortgage payment that is interest remains high during the first years of payoff, then drops off steadily to zero.

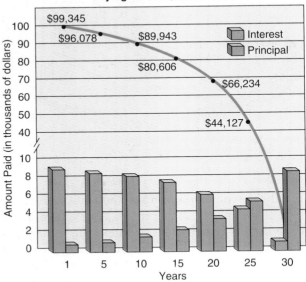

The major reasons that people want to own their own homes, however, are probably intangible. Home ownership conveys a sense of pride, security, and independence to the owner. Even if prices of new homes continue to skyrocket, owning one will remain an essential part of the American dream.

Case Study 2

Trashing the Planet

Can you imagine the archaeologists of the future digging into the ruins of the late 1900s to analyze what civilization was like then? Instead of buried cities, the memorials to American society would be its landfills. For example, between 1960 and 1986, the amount of plastic in trash in the United States increased by a factor of 26, effectively describing our lifestyle as that of a "throwaway" society.

Trash (technically, *municipal solid waste*) is becoming more problematic to the American public than air and water pollution. In reality, trash also affects air and water. Nearly one half of all landfills are within one mile of a drinking-water resource. Emissions escaping from decaying materials can contain 20 different toxic synthetic organic chemicals, some of which are carcinogens, such as vinyl chloride and carbon tetrachloride.

Americans dispose of 450,000 tons of residential and commercial trash per day. Landfills receive about 80% of the trash generated in the United States. We are running out of places to dump this trash. In 1986, there were approximately 6000 active trash landfills in the United States. In 1993, the number had been reduced by half, and by 2006, it is estimated that only about 20% of the 6000 will still be in operation.

Recycling to Reduce Waste

Experts estimate that ideally, 85–90% of the trash generated by households could be recycled. However, less than 65% of U.S. households participate in the recycling effort in some way. Only a few industries recycle materials to any great extent.

Recycling programs require a great deal of education, training, and up-front money to become effective. It's just too easy to throw everything away. Some basic separation of your trash can make recycling efficient and economical for your community. However, recycling does not solve all trash problems.

Not everything can be recycled effectively. Some items such as glass, aluminum, and steel can be recycled again and again without degeneration of the quality of the product. However, when paper is recycled, about 10% of the fibers are lost during processing and the end product is of lesser quality than the original material. The fibers in the highest quality of paper can usually last through 10 generations of recycling before the fibers are useless. Plastic products have the shortest recyclable life. The chemical properties of plastics change in the recycling process.

Figure 2–1 Experts estimate that 85–90% of household trash can be recycled. However, larger cities, like New York City, recycle only about 6% of their residential solid waste.

Average Composition of US. Trash

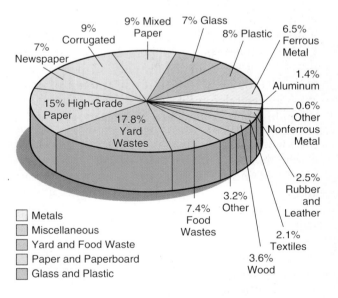

Another drawback to the recycling effort is the actual cost of recycling the materials. Unlike aluminum, the market for some types of materials,

The data below illustrate how wasteful Americans have become.

- Every two weeks, we throw away enough glass bottles and jars to fill the 1350-foot twin towers of the World Trade Center in New York.

- Americans use 2.5 million plastic bottles every hour, of which only a small portion are recycled.

- Households and industries throw away enough aluminum to rebuild the entire commercial airfleet every three months.

- Americans throw away enough iron and steel to continuously supply all of the nation's auto makers.

- Each year, we dispose of 24 million tons of leaves and grass clippings, which could be composted instead of taking up space in landfills.

- More than 500,000 trees are used each week to produce the newspapers that are never recycled.

like newspaper, is so glutted that the newspaper itself has little monetary value (less than $0.05 per ton). The storage, handling, and processing of these materials often cost the recycling company more than they yield. This is why many centers no longer accept some materials for recycling.

We as consumers can help the recycling effort by buying products made from recycled materials. In this way, our demands make recycled materials more attractive to manufacturers, causing a greater need for the supply of recycled materials. For example, McDonald's restaurants used to package all of their sandwiches in polystyrene boxes. Through awareness of the amount of space these boxes take up in landfills, they returned to wrapping their sandwiches in paper products.

International Environmental Concerns

While Americans may think they are making great strides toward saving the planet, they lag behind other countries in recycling efforts. The U.S. recycles about 10% of its glass and 27% of its paper. In Europe and Japan, an average of 31% of the glass and 40% of the paper is recycled. European countries are much more advanced in the effort to compost organic materials. For example, nearly one million tons of compost is produced in France each year and used in their vineyards.

While trash is not a great environmental concern in Europe and Asia, water and air pollution is. Until recently, there have been few regulations on industrial pollution and safety procedures. This has often meant that people near industrial plants and the workers in those plants have been exposed to toxic materials at levels not permitted in the United States. You may think, "Well, that doesn't affect me." Remember that the winds of Earth blow everywhere. Thus, what happens in one place does affect the world environmentally.

Scientists are closely watching the effects of the fires and oil spills in Kuwait as a result of the Gulf War in 1991. Many scientists feared the smoke from these fires would affect the weather worldwide. On November 6, 1991, the last of the 650 fires was extinguished. The task was accomplished much more quickly than experts had estimated. Even though the worldwide environmental effects of the smoke are not known, we do know of the trouble faced by the people of Kuwait. The acid rain resulting from water mixing with the toxic smoke has harmed much of the vegetation in the area. Because of the smoke cloud, the temperatures in Kuwait were often 20° below normal. The particles in the air have hampered breathing for many. Some experts estimate that as many as 1000 Kuwaitis died in 1992 from the polluted air.

Figure 2–3 About 350 miles of Saudi Arabia's coastline is oil-soaked and because the waters of the Gulf move so slowly, it may be a long time before the water is safe.

In addition to air pollution, the war brought about water pollution as well. The Persian Gulf is slowly recovering from a spill of 250 million gallons of oil—more than 20 times larger than the spill created near Alaska by the *Exxon Valdez* in 1989. The oil from the burning wells has also formed pools that are sinking into the sands of Kuwait, affecting the plant and animal life as well as the drinking water in the subterranean levels.

Only observations of future global weather patterns and studies of atmospheric conditions will determine if the Gulf War affected the far regions of Earth.

Protecting the Ozone Layer

Coal miners used to carry canaries with them into the mines to determine if there were toxic gases building up in the areas where they were working. If the canary died, they knew that the buildup of gases was dangerous. The hole in the ozone layer over the Antarctic has been the world's canary for the past two decades. The greater the hole grows, the more dangerous are the global effects of a thinner ozone layer.

Man-made chemicals, such as chlorofluorocarbons (CFCs), are attacking the ozone layer. While the effects are most dramatic over the Antarctic, the rest of Earth's atmosphere is also being affected. Because of wind patterns, pollutants, and atmospheric ice, the region over the Antarctic is most sensitive to the depletion of the ozone layer.

But why is the ozone so important? The ozone layer acts as a sunscreen for Earth. It protects the planet from the sun's harmful ultraviolet (UV) rays. With more UV rays being let through the atmosphere, there is a greater possibility for seasonal damage leading to reduced crop yields and increased skin cancer in humans. The Environmental Protection Agency (EPA) has estimated that the number of deaths due to skin cancer will increase from 9300 to 200,000 per year in the next 50 years.

Figure 2–4 The photos of the ozone layer above the Antarctic shown below show the increase in the hole over the last few years.

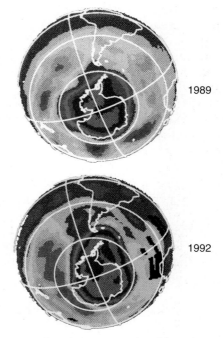

1989

1992

How can this be reversed? If we dramatically reduce the use of products containing CFCs, the damaging elements can be reduced in the atmosphere. Public awareness and demand have forced manufacturers to produce products without the use of CFCs.

Some proponents say emission controls are the best way to save the ozone layer. For example, the invention of the catalytic converter in cars was intended to reduce emissions of carbon monoxide by 90% and nitrogen oxide by 70% from 1975–1985. However, the carbon monoxide levels decreased only 19% in that period and nitrogen oxide increased by 4%. Thus, auto emissions are not the only contributor to the air pollution problem. Today, the EPA suggests that rather than concentrating on controlling the result of air pollution, we put our efforts toward avoiding the source of the pollution.

The Legal System

*L*aw is a set of rules by which a society governs itself. Through most of history, laws were set down by the king or queen or by the state. But beginning 200 years ago with the adoption of the United States Constitution and the Bill of Rights, the concept of law written and agreed to by the *citizens* of the state was established in this country. This is the foundation of a democratic society. Since, in general, it is not possible for all citizens to participate directly in the writing of law, the people of a democratic society elect *legislators*, or lawmakers, who create the law and are responsible to the people who elected them. Because a democracy operates on the rule of the majority of its citizens, the people may remove from office legislators whose performance is judged unsatisfactory. At the national level in the United States, this is rarely done, since more than 95% of legislators are routinely reelected by the voters.

Figure 3–1 The U.S. Capitol Building in Washington, D.C., is the symbol of the American legal system.

A Brief History

The law never stands still. Society changes constantly and each change brings with it the need for new laws. Congress considers more than 20,000 laws during every session. Legislative bodies at the state and local level consider thousands more. Only a fraction of these *bills*, or proposals, are enacted into law, but the large number that are proposed suggests that the law is a dynamic, constantly evolving system.

Some laws, like the law that makes murder illegal, are based on universally-accepted moral and ethical principles. They prescribe unvarying standards of behavior that are necessary for the smooth functioning of society. At the other extreme are laws that address minor problems temporarily affecting citizens' lives. Some of these remain in effect long after they have outlived their usefulness. One community in the United States still has a law stating that any family owning a bathtub must keep it in the yard rather than in the house. While most laws fall somewhere between these two extremes, all are derived from an understanding of how civilized people should behave that traces its roots back at least 4000 years.

The earliest law that has been preserved is the Code of Hammurabi, set down by Hammurabi, the ruler of the Sumerians, about 2000 B.C. It spelled out in great detail laws and penalties relating to business, real estate, wages, military service, loans, family rights, and many other areas. The code was based on the revolutionary premises that the individual has rights and that the strong shall not injure the weak.

Around 1200 B.C., the Hebrew leader Moses gave the world the Ten Commandments, which codified basic principles of human behavior and have been adopted by almost every succeeding civilized society. Eight hundred years later, the Greeks developed the concept of democracy and the idea that a nation should be governed by law rather than by rulers. The Greeks introduced the important notion that people can write their own laws and change them if the need arises.

At about the same time, the Romans published the Law of the Twelve Tables, which specified customary behavior in great detail. Modern American law is derived directly from Roman law, or *canon law*, which governed the Roman Catholic church during the Middle Ages, and from *common law*, a legal system used by the English and brought to America by the first colonists. Common law is based on customs and traditions relating to human behavior, rather than on written law.

The Need for Law

Laws would not be necessary if everyone agreed on proper behavior. No matter how strongly most people agree on what is acceptable practice, however, there are always some who disagree and who violate that practice. For that reason, laws need to be put into writing and enforced.

The clearest evidence of this is that violation of laws in the United States is widespread and, for most offenses, on the rise. Between 1981 and 1990, the overall crime rate rose 7.8%. The biggest increases during that period were in violent crimes, up 33.7%, and forcible rape, up 24.3%. (Greater willingness by women to report rape and by law enforcement agencies to prosecute the crime, compared with earlier years, partially explains the increase.) Burglary—breaking and entering to commit a felony—was the only crime to register a decrease during the same period.

The total number of crimes is huge and practically incomprehensible. In 1990, there were 14,475,630 crimes committed in the United States. These included 1,820,130 violent crimes and 12,655,500 property crimes. The violent crimes consisted of 23,440 murders, 102,560 forcible rapes, 639,270 robberies, and 1,054,860 aggravated assaults. The property crimes consisted of 3,073,900 burglaries, 7,945,700 larceny-thefts, and 1,635,900 motor vehicle thefts. Perhaps these numbers can best be grasped by considering their frequencies. In 1990, there were, on average:

- one violent crime every 17 seconds,
- one property crime every 2 seconds,
- one murder every 22 minutes,
- one forcible rape every 5 minutes,
- one robbery every 49 seconds,
- one aggravated assault every 30 seconds,
- one larceny (theft) every 4 seconds,
- one burglary every 10 seconds, and
- one motor vehicle theft every 19 seconds.

Figure 3–2 The Department of Justice collects statistics on the number and distribution of crimes in the United States.

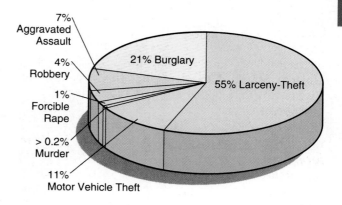

**Crime Index Offenses, 1990
Percent Distribution**

In 1990, an estimated 5820 crimes were committed for every 100,000 people in the United States. The figure was highest in the south, with 6334 crimes per 100,000 people, and lowest in the northeast, with 5194 per 100,000 people. With a population of 85,446,000, more crimes—37% of the total—were committed in the southern states than in any other region. The western states, with a population of 52,786,000, were second, with 23% of the total, followed by the midwestern states (59,669,000; 21%) and the northeastern states (50,809,000; 18%).

Figure 3–3 The number of crimes per 100,000 people reached its lowest point in recent history in 1984.

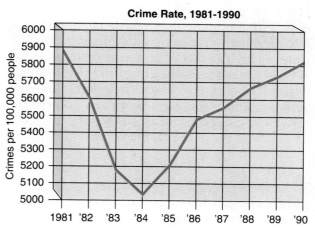

Crime Rate, 1981-1990

are victimized by crime more than age groups. In a given year, about 6% of teenagers, on average, experience violent crimes, compared with about 3% of persons age 20 or older. About half of violent crimes against teenagers occur in school buildings, on school property, or on the street.

Figure 3–4 There were nearly 8 million incidents of personal property larceny-theft in the United States in 1990.

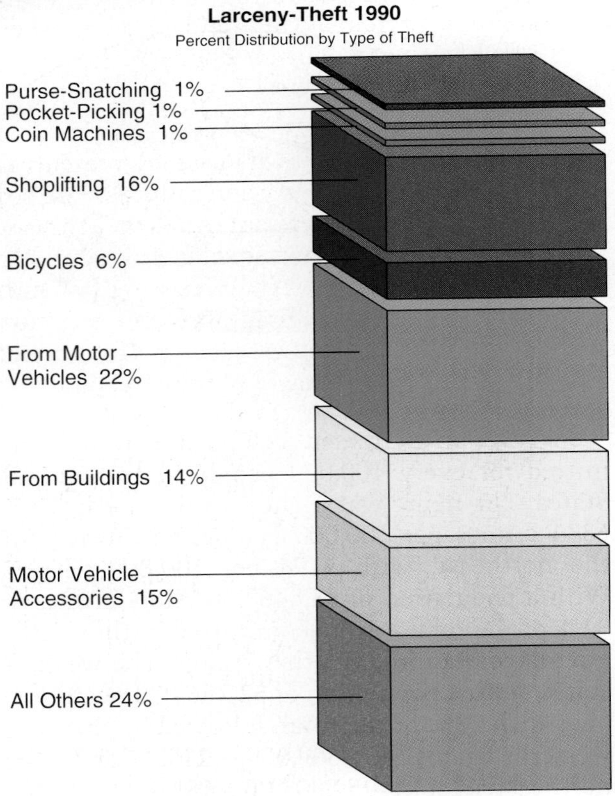

Larceny-Theft 1990
Percent Distribution by Type of Theft

Purse-Snatching 1%
Pocket-Picking 1%
Coin Machines 1%
Shoplifting 16%
Bicycles 6%
From Motor Vehicles 22%
From Buildings 14%
Motor Vehicle Accessories 15%
All Others 24%

Gangs are responsible for much of the teenage crime in America. A 1989 survey turned up evidence of some 1500 youth gangs nationwide, with a total of more than 120,000 members. The rate of violent crimes by gang members is three times as high as that of non-gang delinquents. In 1990 in Los Angeles, the city with the most highly publicized gang violence, there were 329 murders committed by gang members, 34% of the total homicides committed in the city that year. The average age of the murderers was nineteen.

The Justice System

There are two principal branches of the law. *Criminal law* addresses acts that are considered harmful to society as a whole. *Civil law* protects individual citizens in their dealings with other citizens.

Crimes can also be classified according to whether they violate federal or state laws. Federal laws apply to and generally affect the entire country. Alleged violators of federal law are tried in federal court. State laws apply only in the state in which they are passed. Generally they address crimes such as murder, robbery, drunken driving, assault, and other offenses that do not affect the entire nation or take place across state lines. Alleged violators of state law are tried in state court.

Like the number of crimes, the number of arrests and convictions in the United States and the number of inmates in the nation's prisons is on the rise. In 1990, there were about 14 million arrests in the United States. The largest numbers were for property crimes (2.2 million), larceny (1.6 million), drunken driving (1.8 million), and drug abuse violations (1.1 million). Of those arrested, 82% were males and 13% were under the age of 18.

The United States imprisons a greater proportion of its population than any other nation. Of every 100,000 Americans, 426 were in prison in 1990, a total of just over 1 million inmates. The state with the greatest imprisonment rate is South Carolina, followed by Nevada and Louisiana. The western states showed a 203% increase in the number of prisoners between 1980 and 1989, the biggest jump of any region. California's prison population increased by 11,000 in just one year, 1989. The annual cost of operating America's prisons is estimated at $16 billion.

Among foreign nations, South Africa has the second highest imprisonment rate behind the United States (333 per 100,000 people). European and Asian countries generally range from 20 to 140 prisoners per 100,000 people.

Efforts to use the prison system to rehabilitate criminals and prepare them to become contributing members of society upon their release have been only marginally successful. *Recidivism* is the tendency of criminals to return to crime after serving their prison sentences. The Bureau of Justice Statistics studied 108,000 men and women released from prison in 1983. Within three years,

31.9% of the burglars had been rearrested for burglary, 24.8% of the drug offenders had been rearrested for drug offenses, and 19.6% of the robbers had been rearrested for robbery.

Of all the former inmates, 62.5% had been rearrested for a felony or serious misdemeanor by 1986, 46.8% had been reconvicted, and 41.4% were back in prison. Released rapists were ten times more likely than nonrapists to be rearrested for rape. Released murderers were five times more likely than other offenders to be rearrested for homicide.

Trial by Jury

Television and movies have greatly exaggerated the drama of the courtroom. In reality, most trials are slow, meticulous, and sometimes boring, characterized more by tedious legal maneuvering than by startling developments or witnesses breaking down on the stand. While this may diminish the drama and excitement of a trial, it is necessary to assure that the accused receives the fair trial guaranteed by the Constitution.

Figure 3–5 The Constitution guarantees the right to a fair and speedy trial.

Before a trial begins, the prosecution and the defense choose a jury that faithfully represents the local community. The jury's duty is to listen to the arguments of both sides in the case and then to decide whether the defendant is innocent or guilty.

The trial begins with an opening statement by the prosecutor describing the evidence collected against the defendant and outlining the case the prosecution will make during the trial. This is followed by a statement by the defense attorney summarizing the evidence supporting the defendant's position.

The prosecution makes its case by presenting detailed evidence to the jury and calling witnesses to testify against the defendant. After the witnesses testify, the defense lawyer may *cross-examine* them to try to break down their testimony or turn it to the defendant's advantage. If the prosecution cannot establish a strong case, the judge can end the trial at this point in favor of the defendant.

If the case continues, the defense presents its evidence and witnesses, which the prosecutor may cross-examine. Then the prosecutor makes a *summation*, a closing statement summing up the case against the defendant. The defense attorney follows with the defense summation. The prosecution is entitled to make a final closing argument after the defense summation.

Then the judge presents the *charge* to the jury, an explanation of the relevant laws and a description of how the jury is to conduct its deliberations. The jury retires and discusses the case in seclusion. In a criminal case, the jury usually must agree unanimously on a verdict. If the members cannot agree, the result is a *hung jury*. A new trial, with a new jury, may be held.

If the jury comes to an agreement, they return to the courtroom and announce their verdict—guilty or not guilty. If the verdict is guilty, the judge will sentence the defendant to pay a fine, go to prison, or both. The judge also has the discretion to *suspend* the sentence, allowing the defendant to go free. Under a suspended sentence, a convicted criminal must obey all laws and conditions of the suspension, or else go directly to prison.

The U.S. Economy

The state of a nation's economy is a measure of the nation's health. If employment and personal income are high, if interest rates and inflation are low, if the nation is paying its bills at home and selling its goods abroad, if manufacturers are producing quality goods and selling them at prices that attract buyers, then the economy is healthy.

In practice, few national economies are ever so strong that they meet all of these goals at once or for an extended period of time. An economy is an enormously complex mixture of factors, each of which influences the others. Economists cannot always say with certainty how a change in one element of the economy—the amount of money in circulation, for example—will affect other key elements, such as unemployment or interest rates. Economics is not an exact science, but rather a dynamic system that changes constantly due to individual economic decisions of millions of people, all acting on their own self-interest. For this reason, most national economies follow cyclical patterns of growth, remaining relatively strong for a period of years, reaching a point of maximum strength, then falling into a period of relative weakness before beginning a new cycle.

Measuring the U.S. Economy

The United States has experienced its share of hard economic times, but overall, has remained one of the world's great economic powers throughout the twentieth century. There are a number of ways to measure the U.S. economy, both internally and in relation to that of other nations.

The Gross National Product   The Gross National Product (GNP) is the market value of all goods and services that have been bought for use during a year. It measures the value of purchases of everything from the services of baby-sitters and loaves of bread by 250 million individual Americans, to fighter planes and land for new parks by the government, to factories and machinery by manufacturers. Because the amount of money spent is an excellent indicator of the strength of an economy, the GNP is considered the most comprehensive measure of U.S. economic activity.

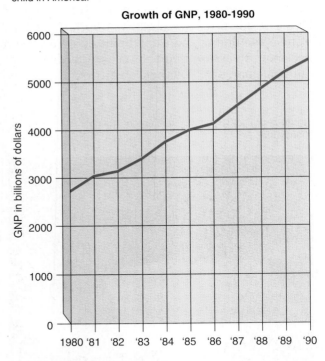

Figure 4–1 In 1990, the U.S. GNP amounted to about $22,000 in purchases of goods and services for every man, woman, and child in America.

Inflation   In 1920, it was possible to buy a new car for $400. Today the average new car costs almost forty times that much. In general, the price of goods does not remain constant. While the prices of some goods may decrease occasionally, most increase over time. An increase in the price of goods is called *inflation*.

In a strong economy, inflation is kept low. Consumers can count on stable price levels, allowing them to plan more easily for the future. This in turn gives them confidence in the economy, an intangible but extremely important component of a nation's economic strength. Suppose you found a CD player you liked that retailed for $200. Three months later, after saving $200, you discover that the CD now sells for $600. This is precisely the situation in some nations, where prices can multiply many times over in a single year. An extreme example of out-of-control prices occurred in 1989 in Argentina, when inflation reached 6000% that year. A quart of milk that cost $1.00 (in U.S. currency) in January cost $61.00 by year's end. How much confidence would you have in such an economy?

In the United States, the *Consumer Price Index* (CPI) measures the rate of inflation. The CPI is based on average prices of food, clothing, shelter, fuel, transportation, health care costs, and a variety of other goods and services. Index figures are given in relation to average prices for 1982–84, which are set at 100.0. Thus, the 1989 CPI for shoes, 114.4, means that, on average, a pair of shoes in 1989 cost $\frac{114.4}{100}$ as much as a pair of shoes in 1982–84. In other words, the price of shoes inflated 14.4% during that period. By 1990, the CPI for shoes had increased to 117.4. That means that shoe prices went up $\frac{117.4 - 114.4}{114.4}$ or about 2.6% from 1989 to 1990.

The CPI occasionally shows a drop in prices. For example, during 1986, the price of gasoline was 21.9% lower than in 1985 and the 1990 cost of televisions was 1.5% lower than in 1989. In general, however, prices have increased steadily since the 1982–84 base period began. The major categories showing the smallest changes during those years have been transportation, up 20.5% by 1990, wearing apparel, up 24.1%, and housing, up 28.5%. Two of the greatest changes from 1984 to 1990 were in the cost of personal and educational expenses, up 70.2% and in the cost of medical care, up 62.8%. Both of these expenses represent an annual inflation rate of more than 10%.

The CPI is calculated for geographical regions as well as for the entire country. The index shows that of the four principal regions, the northeast showed the greatest rate of inflation, followed by the west, the south, and the north central region. In 1990, prices in the northeast averaged about 9% higher than in the north central states.

Employment A strong economy will offer good jobs and decent wages to workers, and unemployment will be low. U.S. unemployment reached its highest point during the Depression, when for a period of four years (1929–1933), more than one out of five workers was unable to find a job. In 1933, unemployment peaked at 24.9%. More than 12 million Americans were unable to find work.

Bureau of Labor Statistics figures show that workers earned an average of $23,602 annually in the United States in 1990. The states whose workers received the highest pay were Alaska ($29,946) and Connecticut ($28,995). The states with the lowest rates of pay were South Dakota ($16,430) and North Dakota ($17,626). In general, men received significantly higher salaries than women, an average of 39% more according to the U.S. Census Bureau.

Congress specifies the minimum amount an employee can be paid. Beginning in 1991, the minimum wage was $4.25 per hour. An employee working 40 hours per week, 50 weeks a year, will earn $8,500 per year at minimum wage.

Figure 4–2 Energy prices have shown the most volatility in recent years.

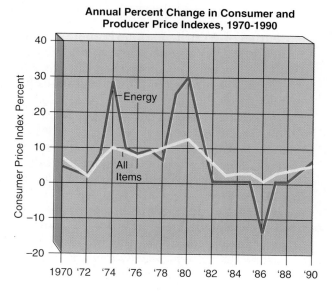

Annual Percent Change in Consumer and Producer Price Indexes, 1970-1990

In recent years, unemployment has fluctuated ...n 5% and 10% of the work force.

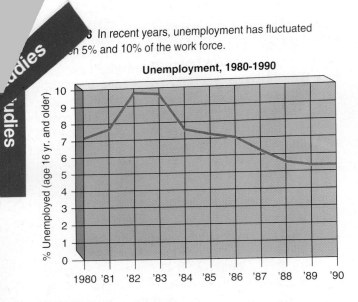

Unemployment, 1980-1990

Receipts versus Outlays   The graphs below show how the U.S. government makes its money and how it spends it. When the government makes more than it spends, it has a net *surplus* of funds. When outlays exceed receipts, there is a *deficit*. Since 1961, the government has registered a surplus only once, in 1969. This means that for more than thirty years the government has been falling further and further into debt. In 1990, the total national debt surpassed $3 trillion, which averages out to $13,000 for every person in the country.

Figure 4–4   Federal revenues and expenditures, 1988–89

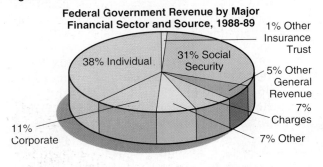

Federal Government Revenue by Major Financial Sector and Source, 1988-89

- 38% Individual
- 31% Social Security
- 1% Other Insurance Trust
- 5% Other General Revenue
- 7% Charges
- 7% Other
- 11% Corporate

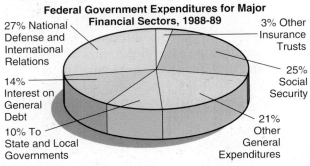

Federal Government Expenditures for Major Financial Sectors, 1988-89

- 27% National Defense and International Relations
- 14% Interest on General Debt
- 10% To State and Local Governments
- 3% Other Insurance Trusts
- 25% Social Security
- 21% Other General Expenditures

There are numerous problems associated with the federal deficit. The greater the deficit, the less likely the government will be able to provide needed services for its citizens. A huge amount of money—$265 billion in 1990—must be spent simply to pay the *interest* on the steadily increasing debt. The confidence of citizens, other nations, and foreign investors in the ability of the government to operate in a fiscally responsible manner erodes.

In 1982, the yearly federal deficit topped $100 billion for the first time. It has remained above that figure every succeeding year, several times topping $200 billion. Only through vastly increased revenues (taxes) and/or decreased expenditures will the government be able to bring the deficit under control. Both solutions require sacrifices that the American people have so far shown themselves unwilling to make.

A second problem of receipts versus outlays occurs when a nation sells less of its products abroad than it purchases from foreign nations. This relative dependency on goods produced by other countries is called a *trade deficit*. In 1990, U.S. companies sold $394 billion worth of goods abroad. That same year, American consumers purchased $495 billion in foreign-made products. This resulted in a trade deficit of $101 billion. The largest portion of the deficit—$41 billion—resulted from our trade imbalance with Japan. One of the world's most powerful economies and a leader in innovative technologies, Japan produces a wealth of goods desired by American consumers, especially cars and electronic equipment.

Distribution of Wealth   A strong economy should provide for the basic needs of everyone, the poor as well as the rich. During recent years, the gap between the two groups in the United States has grown wider by the year. In 1990, the top 20% of earners collected 45% of the income, while the bottom 20% collected only 5%. Figures from the Congressional Budget Office show that between 1977 and 1988, the yearly income of the wealthiest 1% of Americans rose, on average, from $203,000 to $451,000, a 122% increase. During the same period, the income of the poorest one-fifth of Americans declined by 10%. The amount of federal taxes paid by the top 1% dropped by 18%, while the amount paid by low- and middle-income families remained about the same.

In 1990, the *poverty line*—the minimum annual income needed to secure basic needs—was $13,359 for a family of four. The Census Bureau estimated that 33.6 million Americans—13.5% of the population—lived below the poverty line and that the number increased by 2.1 million persons during 1990. Children registered a poverty rate of 20.6%.

Figure 4–5 The overall poverty level was lowest in the early 1970s.

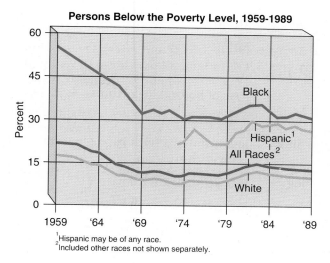

Persons Below the Poverty Level, 1959-1989

[1]Hispanic may be of any race.
[2]Included other races not shown separately.

The Economy in the 1990s

By most measures, the U.S. economy was extremely weak during the early years of the 1990s. On the positive side, inflation and interest rates remained low, although inflation showed signs of inching upward. The GNP continued to grow, but at a diminished rate.

Figure 4–6 The cost of borrowing money has dropped steadily in recent years.

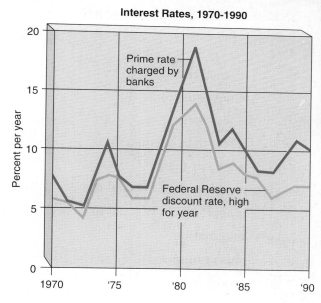

Interest Rates, 1970-1990

On the negative side, unemployment, the federal deficit, the trade deficit, and poverty remained high throughout the period. The Index of Leading Economic Indicators, a monthly figure published by the Commerce Department and designed to assess the overall economic health of the country, indicated that the nation had entered a *recession*, a period of general economic stagnation. The cyclical nature of economic health suggested that at some point in the future, the country would once again experience a robust economy. Exactly when that would occur neither economists nor anyone else could say with certainty.

Figure 4–7 The state of the economy was a major issue in the 1992 presidential campaign.

SYMBOLS

$=$	is equal to
$\neq$	is not equal to
$<$	is less than
$\leq$	is less than or equal to
$>$	is greater than
$\geq$	is greater than or equal to
$\approx$	is approximately equal to
$\{\ \}$	set notation
$\pm$	plus or minus
$\mp$	minus or plus
$f(x)$	f of x or the value of function f at x
$f'(x)$	f prime of x or the derivative of f at x
$f''(x)$	the second derivative of $f(x)$
$f \circ g$ or $f(g(x))$	composite of functions f and g
$\lim\limits_{x \to a}$	the limit as x approaches a
$\triangle ABC$	triangle ABC
$\overset{\frown}{RTS}$	arc RTS
$\angle ABC$	angle ABC
$m\angle ABC$	measure of angle ABC
AB	measure of line segment AB
$\overline{AB}$	line segment AB
$|n|$	the absolute value of n
x^n	the nth power of x
$\sqrt{x}$	the square root of x
$\sqrt[n]{x}$ or $x^{\frac{1}{n}}$	the nth root of x
$[x]$	greatest integer not greater than x
A^{-1}	inverse of A
a_{ij}	the element of the ith row and the jth column
$\begin{vmatrix} a_1 & b_1 \\ a_2 & b_2 \end{vmatrix}$	the determinant $a_1 b_2 - a_2 b_1$
$\deg(A)$	the degree of A
$\vec{v}$ or $\overrightarrow{AB}$	a vector or directed line segment

$|\vec{v}|$	magnitude of the vector $\mathbf{v}$
$\vec{a} \cdot \vec{b}$	inner product of dot product of vectors $\mathbf{a}$ and $\mathbf{b}$
$\vec{a} \times \vec{b}$	cross product of vectors $\mathbf{a}$ and $\mathbf{b}$
$\sin^{-1} x$	arcsin x
∞	infinity
i	$\sqrt{-1}$
e	base of natural logarithms; ≈ 2.718
$n!$	n factorial
$\ln x$	logarithm of x with base e; natural logarithm
$\log_a x$	logarithm of x with base a
$\log x$	logarithm of x with base 10
$P(n, r)$	permutation of n objects, taken r at a time
$C(n, r)$	combination of n objects, taken r at a time
$\overline{X}$	X bar or arithmetic mean
M_d	median
σ_X	standard error of the mean
MD	mean deviation
$\dfrac{dy}{dx}$	the derivative of y with respect to x
$\int$	integral
α	alpha
β	beta
Δ or δ	delta
ϵ	epsilon
θ	theta
λ	lambda
π	pi
σ	sigma; standard deviation
Σ	sigma; summation symbol
ϕ	phi

FORMULAS

Chapter 1
Distance Formula for Two Points (30)
$$d = \sqrt{(x_2 - x_1)^2 + (y_2 - y_1)^2}$$

Chapter 4
Quadratic Formula (188)
$$x = \frac{-b \pm \sqrt{b^2 - 4ac}}{2a}$$

Chapter 5
Degree/Radian Conversion Formulas (242)

1 radian $= \dfrac{180}{\pi}$ degrees

1 degree $= \dfrac{\pi}{180}$ radians

Length of an Arc (248)
$$s = r\theta$$

Linear and Angular Velocity (249)
$$v = r\frac{\theta}{t}$$

Area of a Circular Sector (250)
$$A = \frac{1}{2}r^2\theta$$

Area of a Triangle
$$K = \frac{1}{2}bc \sin A \quad (288)$$

$$K = \frac{1}{2}c^2 \frac{\sin A \sin B}{\sin C} \quad (289)$$

Hero's Formula (290)
$$K = \sqrt{s(s - a)(s - b)(s - c)}$$

Area of a Circular Segment (292)
$$S \doteq \frac{1}{2}r^2(\alpha - \sin \alpha)$$

Chapter 7
Distance from a Point to a Line (399)
$$d = \frac{Ax_1 + By_1 + C}{\pm\sqrt{A^2 + B^2}}$$

Chapter 9
Conversion from Polar Coordinates to Rectangular Coordinates (483)
$$x = r \cos \theta, \, y = r \sin \theta$$

Conversion from Rectangular Coordinates to Polar Coordinates (484)
$$r = \sqrt{x^2 + y^2}$$
$$\theta = \text{Arctan } \frac{y}{x} \text{ when } x > 0$$
$$\theta = \text{Arctan } \frac{y}{x} + \pi \text{ when } x < 0$$

Chapter 10
Length of a Segment Tangent to a Circle
$$t = \sqrt{(x_1 - h)^2 + (y_1 - k)^2 - r^2} \quad (588)$$

Chapter 11
Compound Interest Formulas
$$A = P\left(1 + \frac{r}{n}\right)^{nt} \quad (614)$$
$$A = Pe^{rt} \quad (615)$$

Change of Base Formula (638)
$$\log_a n = \frac{\log_b n}{\log_b a}$$

Chapter 12
nth Term of an Arithmetic Sequence (657)
$$a_n = a_1 + (n - 1)d$$

Sum of an Arithmetic Sequence (659)
$$S_n = \frac{n}{2}(a_1 + a_n)$$

nth Term of a Geometric Sequence (664)
$$a_n = a_1 r^{a-1}$$

Sum of a Geometric Sequence (666)
$$S_n = \frac{a_1 - a_1 r^n}{1 - r}$$

Sum of an Infinite Geometric Series (674)
$$S_n = \frac{a_1}{1 - r}$$

Euler's Formula (698)
$$e^{1\alpha} = \cos \alpha + i \sin \alpha$$

Chapter 13
Verhulst Population Model (712)
$$p_{n+1} = p_n + rp_n(1 - p_n)$$

Chapter 14
Definitions of Permutation (754)
$$P(n, n) = n!, \, P(n, r) = \frac{n!}{(n - r)!}$$

Definition of Combination (764)
$$C(n, r) = \frac{n!}{(n - r)!r!}$$

Chapter 15
Arithmetic Mean (812)
$$\overline{X} = \frac{1}{n}\sum_{i=1}^{n} X_i$$

Mean of the Data in a Frequency Distribution (815)
$$\overline{X} = \frac{\sum\limits_{i=1}^{k} f_i \cdot X_i}{\sum\limits_{i=1}^{k} f_i}$$

Semi-Interquartile Range (821)
$$Q_R = \frac{Q_3 - Q_1}{2}$$

Mean Deviation (823)
$$MD = \frac{1}{n}\sum_{i=1}^{n} |X_i - \overline{X}|$$

Standard Deviation (823)
$$\sigma = \sqrt{\frac{1}{n}\sum_{i=1}^{n} (X_i - \overline{X})^2}$$

Standard Deviation of the Data in a Frequency Distribution (824)
$$\sigma = \sqrt{\frac{\sum\limits_{i=1}^{k} (X_i - \overline{X})^2 \cdot f_i}{\sum\limits_{i=1}^{k} f_i}}$$

Standard Error of the Mean (839)
$$\sigma_{\overline{x}} = \frac{\sigma}{\sqrt{N}}$$

Pearson Product-Moment Correlation (849)
$$r = \frac{SP}{\sqrt{SS_x SS_y}}$$

Chapter 17
Limit of a Polynomial Function (915)
$$\lim_{x \to r} P(x) = P(r)$$

Limit of a Trigonometric Function (918)
$$\lim_{x \to 0} \frac{\sin x}{x} = 1$$

Power Formula (924)
$$f'(x) = cnx^{n-1}$$

Area Under a Curve (932)
$$A = \lim_{n \to \infty}\sum_{i=1}^{n} f(x_i)(x_i - x_{i-1})$$

Fundamental Theorem of Calculus (945)
$$\int_a^b f(x)dx = F(b) - F(a)$$

GLOSSARY

A

adjacency matrix   a square matrix whose entries are the number of edges from one vertex of a graph to another   (898)

adjacent vertices   vertices that are joined by an edge   (868)

algorithm   a sequence of instructions that solves all cases of a certain type of problem   (879)

alternate optimal solutions   When there are two or more optimal solutions for a linear programming problem, the problem is said to have alternate optimal solutions.   (93)

amplitude $|A|$ for functions in the form $y = A \sin k\alpha$ or $y = A \cos k\alpha$ (309)

amplitude of a complex number   the angle θ when a complex number is written in the form $r (\cos \theta + i \sin \theta)$ (501)

amplitude of a vector   the directed angle between the positive x-axis and the vector   (412)

analytic geometry   the study of coordinate geometry from an algebraic perspective   (31)

angle of depression   the angle between a horizontal line and the line of sight from the observer to an object at a lower level   (272)

angle of elevation   the angle between a horizontal line and the line of sight from an observer to an object at a higher level   (272)

angular velocity   the change in the central angle with respect to time as an object moves along a circular path   (249)

annual percentage rate   interest rate per year   (609)

annuity   a series of payments made at equal intervals of time   (609)

antiderivative $F(x)$ is an antiderivative of $f(x)$ if and only if $F'(x) = f(x)$.   (939)

antilogarithm   If $\log x = a$, then x is called the antilogarithm of a, abbreviated antilog a.   (631)

arcsine   the inverse of $y = \sin x$ (329)

arithmetic mean   1. the terms between any two nonconsecutive terms of an arithmetic sequence   (658)   2. a measure of central tendency found by dividing the sum of all values by the number of values   (811)

arithmetic sequence   a sequence in which the difference between successive terms is a constant   (656)

arithmetic series   the indicated sum of the terms of an arithmetic sequence   (658)

array   statistical data arranged in an ordered sequence   (813)

asymptote   lines that a curve approaches   (134)

asymptote of a hyperbola   the lines that the hyperbola approaches as it recedes from the center   (551)

attractor   the x-coordinate of the fixed point to which a path generated by iteration staircases or spirals   (718)

augmented matrix   an array of the coefficients and constants of a system of equations   (79)

axis   See axis of symmetry.

axis of symmetry   a line about which a figure is symmetric   (534)

B

back-to-back bar graph   a graph plotted on a two-quadrant coordinate system with the horizontal scale repeated in each direction from the central axis used to show comparisons   (802)

best-fit line   the graph of a prediction equation   (847)

bimodal data   data with two modes   (813)

binomial experiment   a problem that can be solved using binomial expansion   (791)

bivariate data   data that involve exactly two variables   (847)

boundary   a line or curve that separates the coordinate plane into two regions   (23)

box-and-whisker plot   a diagram that graphically displays the median, quartiles, extreme values, and outliers in a set of data   (821)

bridge   an edge of a connected graph whose removal would cause the graph to no longer be connected   (874)

C

center of a circle   See circle.

central angle   an angle whose vertex lies at the center of a circle   (247)

characteristic   the part of the logarithm of a number which is the exponent of 10 used to write the number in scientific notation   (630)

circle   the locus of all points in a plane at a given distance, called the radius, from a fixed point on the plane, called the center   (524)

circuit   a trail that starts and ends at the same vertex   (873)

circular functions   functions defined using a unit circle   (255)

class interval   the range of each class in a frequency distribution   (803)

class limits   the upper and lower values in each class in a frequency distribution   (804)

class marks   the means of the class limits in a frequency distribution   (804)

classic curves   special curves formed by graphing polar equations   (475)

column matrix   a matrix that has only one column   (64)

combination   an arrangement of objects where the order is not a consideration   (764)

combinatorics   the investigation of the different possibilities for the arrangement of objects   (752)

common difference   the difference between the successive terms of an arithmetic sequence   (656)

common logarithms   logarithms that use 10 as the base   (629)

common ratio   the ratio of successive terms of a geometric sequence   (663)

complements   Two events are complements if and only if the sum of their probabilities is 1.   (768)

complete graph a graph in which each pair of vertices is connected by exactly one edge (869)

complex number any number that can be written in the form $a + bi$, where a and b are real numbers and i is the imaginary unit (179)

components of a vector two or more vectors whose sum is a given vector (415)

composite Given functions f and g, the composite function $f \circ g$ can be described by $[f \circ g](x) = f(g(x))$. (13)

composition See composite. (13)

compound function a function consisting of sums or products of trigonometric functions (324)

compound interest interest computed on the sum of the original principal and any previously earned interest (614)

concentric circles circles with the same center (524)

conditional probability the probability of an event under the condition that some preceding event has occurred (784)

confidence interval an interval about the sample mean in which the population mean lies within a certain confidence level (840)

conic section a curve determined by the intersection of a plane with a double right cone (561)

conjugate axis the segment perpendicular to the transverse axis of a hyperbola through its center (551)

conjugate of a complex number The conjugate of the complex number $a + bi$ is $a - bi$. (190, 496)

connected graph a graph in which there is a path along an edge between each pair of two vertices (873)

connected vertices See adjacent vertices.

consistent system a system of equations that has at least one solution (57)

constant function a function of the form $f(x) = b$ (23)

constraints conditions given to variables, often expressed as linear inequalities (91)

continuous A function is said to be continuous at point (x_1, y_1) if it is defined at that point and passes through that point without a break. (164)

converge If a sequence has a limit, then the related infinite series is said to converge. (679)

coterminal angles two angles in standard position that have the same terminal side (242)

critical path the longest path between two vertices (897)

critical points points at which the nature of a graph changes (157)

cross product The cross product of $\vec{a}$ and $\vec{b}$ if $\vec{a} = (a_1, a_2, a_3)$ and $\vec{b} = (b_1, b_2, b_3)$ is defined as follows. (431)

$$\vec{a} \times \vec{b} = \begin{vmatrix} a_2 & a_3 \\ b_2 & c_2 \end{vmatrix} \vec{i} - \begin{vmatrix} a_1 & a_3 \\ b_1 & b_3 \end{vmatrix} \vec{j} + \begin{vmatrix} a_1 & a_2 \\ b_1 & b_2 \end{vmatrix} \vec{k}$$

cycle a path in which only the beginning and ending vertices are the same (873)

D

definite integral an integral that has lower and upper bounds (945)

degenerate case the intersection of a plane with a double right cone resulting in a point, a line, or two intersecting lines (561)

degree the measure of an angle that is $\frac{1}{360}$ of a complete rotation in the positive direction (241)

degree of a polynomial in one variable the greatest exponent of the variable of the polynomial (178)

degree of a vertex the number of edges incident on the vertex (868)

dependent events events that affect each other (752)

dependent system a system of equations that has infinitely many solutions (57)

depressed polynomial the quotient when a polynomial is divided by one of its binomial factors (197)

derivative notation $\frac{dy}{dx}$, read "the derivative of y with respect to x" (923)

derivative of $f(x)$ the function $f'(x)$, which is defined as follows. (150)

$$f'(x) = \lim_{h \to 0} \frac{f(x + h) - f(x)}{h}$$

determinant a square array of numbers having a numerical value; the numerical value of the square array of numbers (71)

differentiable A function is differentiable at a if the derivative of the function exists at a and is finite. (923)

differentiation the process of finding derivatives (923)

digraph a graph with arrowheads on the edges to represent the direction of each relationship; a directed graph (896)

dimensions of a matrix the number of rows, m, and the number of columns, n, of the matrix written as $m \times n$ (64)

directed path a path in a digraph that can be walked in the direction of the arrows (896)

direction vector a vector used to describe the slope of a line (442)

directrix See parabola.

disconnected graph a graph in which there is not a path along an edge between at least two vertices (873)

discontinuous A function is said to be discontinuous at point (x_1, y_1) if there is a break in the graph of the function at that point. (164)

discriminant in the quadratic formula, the expression under the radical sign, $b^2 - 4ac$ (189)

diverge If a sequence does not have a limit, then the related infinite series is said to diverge. (679)

domain the set of all abscissas of the ordered pairs of a relation (6)

dust a disconnected Julia set (737)

E

eccentricity the ratio of the distance between any point of a conic section and a fixed point to the distance between the same point of the conic section to a fixed line (563)

edge of a graph a line segment or arc that connects a pair of vertices of a graph (866)

element of a matrix any value in the array of values (64)

ellipse the locus of all points in a plane such that the sum of the distances from two given points in the plane, called foci, is constant (542)

end behavior the behavior of $f(x)$ as $|x|$ becomes very large (166)

equilateral hyperbola a hyperbola with perpendicular asymptotes (555)

escape set the set of initial values for which the iterates of a function approach infinity (736)

it   a path that includes each edge of a graph ... once and has the same starting and ending ... tex   (878)

...ler path   a path that includes each edge of a graph exactly once, but has different starting and ending vertices   (878)

even function   a function whose graph is symmetric with respect to the *y*-axis   (112)

exponential curve   the graph of a function of the form $y = a^x$ (606)

exponential equation   an equation in which variables appear as exponents   (636)

exponential function   an function in the form $y = a^x$, where a is a positive real number   (608)

exponential series   the series by which e^x may be approximated;
$$e^x = 1 + x + \frac{x^2}{2!} + \frac{x^3}{3!} + \frac{x^4}{4!} + \frac{x^5}{5!} + \cdots \quad (607)$$

F

failure   any outcome other than the desired outcome of an event   (768)

Fibonacci sequence   a sequence in which the first two terms are 1 and each of the additional terms is generated by adding the two previous terms   (696)

fixed points   the points where the graph of the function $g(x)$ intersects the graph of the line $f(x) = x$ (717)

foci   See ellipse, hyperbola.

focus   See parabola.

fractal   a complex geometric structure generated by the iteration of complex numbers through a quadratic equation having self-similarity   (736)

fractal geometry   the study of the properties of fractals and their application to natural phenomena   (734)

frequency   the number of cycles per unit of time   (347)

frequency distribution   a system of organizing data by determining classes and the frequency of values in each class   (803)

frequency polygon   a broken line graph drawn by connecting the class marks on a histogram   (806)

function   a relation in which each element of the domain is paired with exactly one element in the range   (7)

future value of an annuity   the sum of all of the annuity payments plus any accumulated interest   (609)

G

geometric mean   the terms between any two nonconsecutive terms of a geometric sequence   (665)

geometric sequence   a sequence in which the ratio between successive terms is a constant   (663)

geometric series   the indicated sum of the terms of a geometric sequence   (665)

geometric transformation   a transformation in which a nonlinear graph is stretched or shrunk   (120)

goodness of fit   the degree to which data fits a regression line   (849)

graph   in graph theory, a finite, nonempty set of points connected by a set of segments or arcs   (866)

graphical iteration   a process used to find the values of the iterates of a function and their path for a given initial value by graphing   (717)

greatest integer function   a step function, written as $f(x) = [x]$, where $f(x)$ is the greatest integer not greater than x (119)

H

hinges   in a box-and-whisker plot, the ends of the rectangular box which are located at the first and third quartiles   (821)

histogram   a type of bar graph in which the width of each bar represents a class interval and the height of the bar represents the frequency in that interval   (804)

hole   Whenever the denominator and numerator of a rational function contain a common factor, a hole appears in the graph of the function.   (138)

horizontal asymptote   The line $y = b$ is a horizontal asymptote for a function $f(x)$ if $f(x) \to b$ as $x \to \infty$ or as $x \to -\infty$.   (135)

hyperbola   the locus of all points in the plane such that the absolute value of the difference of the distances from two given points in the plane, called foci, is constant   (551)

I

identity   a statement of equality between two expressions that is true for all values of the variables for which the expressions are defined   (358)

imaginary number   a complex number of the form $a + b\boldsymbol{i}$ where $b \neq 0$ and $\boldsymbol{i}$ is the imaginary unit   (179)

incident   An edge that connects a vertex to another vertex is incident on each vertex.   (868)

inclusive events   two events whose outcomes may be the same   (779)

inconsistent system   a system of equations that has no solutions   (57)

indegree of a vertex   the number of edges pointing into a vertex in a digraph   (896)

independent events   events that do not affect each other   (752)

independent system   a system of equations that has exactly one solution   (57)

index of summation   the variable used with the summation symbol   (686)

infeasibility   When the constraints of a linear programming problem cannot be satisfied simultaneously, then infeasibility is said to occur.   (92)

inferential statistics   statistics based on information gathered in a sample to make predictions about a population   (838)

infinite sequence   a sequence which has infinitely many terms   (671)

infinite series   the indicated sum of the terms of an infinite sequence   (674)

inner product   If $\vec{a}$ and $\vec{b}$ are two vectors, (a_1, a_2, a_3) and (b_1, b_2, b_3), then the inner product of $\vec{a}$ and $\vec{b}$ is defined as follows.   (430)
$$\vec{a} \cdot \vec{b} = a_1 b_1 + a_2 b_2 + a_3 b_3$$

integral   the area between the curve of a function, the *x*-axis, and lines $x = a$ and $x = b$ (939)

interquartile range   the difference between the first quartile point and the third quartile point   (821)

Glossary

inverse function   Two functions are inverse functions if and only if $[f \circ g](x) = [g \circ f](x) = x$ for all values of x.   (15)

inverse relations   Two relations are inverse relations if and only if one relation contains the element (b, a), whenever the other relation contains the element (a, b).   (126)

isolated vertex   a vertex that has a degree of 0   (868)

iterate   the output of the composition of a function with itself   (713)

iteration   1. the composition of a function to itself   (15)
2. the repeated composition of a function with itself   (713)

J

Julia set   the boundary between the escape set and the prisoner set for the iteration of complex numbers through the quadratic equation $f(z) = z^2 + c$ where c is a complex number   (736)

L

line symmetry   Two distinct points P and P' are symmetric with respect to a line ℓ if and only if ℓ is the perpendicular bisector of $\overline{PP'}$. A point P is symmetric to itself with respect to a line ℓ if and only if P is on ℓ.   (108)

linear equation   an equation of the form $Ax + By = 0$ where A and B are not both 0   (22)

linear function   a function defined by $f(x) = mx + b$, where m and b are real numbers   (22)

linear inequality   a relation whose boundary is a straight line   (23)

linear programming   a procedure for finding the maximum or minimum value of a function in two variables subject to given constraints on the variables   (91)

linear transformation   a relocation of a graph on the coordinate plane that does not change the graph's size or shape   (118)

linear velocity   distance traveled per unit of time   (249)

locus   a set of points and only those points that satisfy a given set of conditions   (524)

logarithm   In the function $x = a^y$, y is called the logarithm, base a, of x.   (622)

logarithmic function $y = \log_a x$, $a > 0$ and $a \neq 1$ which is the inverse of the exponential function $y = a^x$ (622)

logistic function   the function $f(x) = ax(1 - x)$ (722)

loop   an edge that connects a vertex to itself   (866)

lower bound   the integer less than or equal to the least real zero of the polynomial $P(x)$ (212)

lower sum   a Riemann sum for rectangles which are always below the curve   (931)

M

$m \times n$ matrix   a matrix with m rows and n columns   (64)

magnitude of a vector   the length of the directed line segment   (412)

major axis   the axis of symmetry of an ellipse which contains the foci   (543)

Mandelbrot set   a complex mathematical picture generated by iterating the function $f(z) = z^2 + c$ where $z_0 = 0 + 0i$ and c is allowed to vary   (740)

mantissa   the common logarithm of a number between 1 and 10   (630)

marginal cost function   the rate at which the cost changes with respect to the number of units produced   (925)

marginal revenue function   the rate at which the revenue changes with respect to the number of units sold   (925)

mathematical induction   a method of proof that depends on a recursive process   (702)

matrix   any rectangular array of terms called elements   (64)

maximum   a critical point of a graph where the curve changes from an increasing curve to a decreasing curve   (157)

mean deviation   the arithmetic mean of the absolute value of the deviations from the mean of a set of data   (823)

median   the middle value of a set of data that has been arranged into an ordered sequence   (813)

median class   in a frequency distribution, the class in which the median of the data is located   (815)

median-fit line   a line drawn by using the medians of the x- and y-coordinates of groups of points in a scatter plot   (848)

minimal distance   the weight of the minimal path in a weighted graph   (885)

minimal path   the path with minimal weight in a weighted graph   (885)

minimal spanning tree   the spanning tree having a weight that is less than or equal to the weight of all other spanning trees in a graph   (891)

minimum   a critical point of a graph where the curve changes from a decreasing curve to an increasing curve   (157)

minor   The minor of an element of an nth-order matrix is the determinant of $(n - 1)$th order found by deleting the row and column containing the element.   (71)

minor axis   the axis of symmetry of an ellipse which does not contain the foci   (543)

minute   a unit of angle measure that is $\frac{1}{60}$ of a degree   (241)

mixing   the outcome of an iteration which fills the entire interval from 0 to 1   (723)

mode   the item of data that appears more frequently than any other in the set   (813)

modulus   the number r when a complex number is written in the form $r(\cos \theta + i \sin \theta)$ (501)

multigraph   a graph having loops or parallel edges   (866)

mutually exclusive events   two events whose outcomes can never be the same   (779)

N

n factorial   written $n!$, for n, an integer greater than zero, the product $n(n - 1)(n - 2) \cdots 1$ (687)

natural logarithm   logarithms that use e as the base, written $\ln x$ (641)

normal   1. a line that is perpendicular to another line, curve, or surface   (392)   2. the line that is perpendicular to the tangent of a curve at the point of tangency   (589)

normal curve   a symmetric bell-shaped graph of a normal distribution   (831)

normal distribution   a frequency distribution that often occurs when there is a large number of values in a set of data: about 68% of the values are within one standard of the mean, 95% of the values are within two ___d deviations of the mean, and 99% of the values are ___nin three standard deviations   (831)

___rmal form   the equation of a line that is written in terms of the length of the normal from the line to the origin   (392)

nth order matrix   a square matrix with n rows and n columns   (64)

O

odd function   a function whose graph is symmetric with respect to the origin   (112)

odds   the ratio of the probability of the success of an event to the probability of its complement   (769)

opposite vectors   two vectors that have the same magnitude and opposite directions   (414)

orbit   the graph of the sequence of successive iterates   (731)

outdegree of a vertex   the number of edges pointing away from a vertex in a digraph   (896)

outlier   a value of a set of data that is more than 1.5 interquartile ranges beyond the upper or lower quartiles   (821)

P

parabola   the locus of all points in a given plane that are the same distance from a given point, called the focus, and a given line, called the directrix   (534)

parallel edges   two edges that connect the same pair of vertices   (866)

parallel lines   nonvertical coplanar lines that have equal slopes; any two coplanar vertical lines   (42)

parallel vectors   two vectors that have the same or opposite directions   (414)

parameter   the independent variable t in the vector equation of a line   (443)

parametric equation of a line   the vector equation $(x - x_1, y - y_1) = t(a_1, a_2)$ written as the two equations $x = x_1 + ta_1$ and $y = y_1 + ta_2$ (443)

parent graph   an anchor graph from which other graphs in the family are derived   (117)

partial fraction   one of the fractions that were added or subtracted to result in a given rational expression   (217)

Pascal's triangle   a triangular array of numbers such that the $(n + 1)^{th}$ row is the coefficient of the terms of the expansion $(x + y)^n$ for $n = 0, 1, 2 \cdots$ (691)

path   a walk in which no vertex nor edge is repeated   (872)

Pearson product-moment correlation   a method of determining the goodness of fit   (849)

period of a trigonometric function   the least positive value of α for which $f(x) = f(x + \alpha)$ (311)

period-n attractor   If the iterates repeat every n steps, the function $f(x) = ax(1 - x)$ is said to have a period-n attractor.   (725)

permutation   the arrangement of objects in a certain order   (753)

perpendicular lines   any two nonvertical lines the product of whose slopes is -1; any vertical line and any horizontal line   (43)

phase shift   the least value of $|k\theta + c|$, for which the trigonometric function $f(k\theta + c) = 0$ (312)

point of inflection   a critical point of a graph where the graph changes its curvature from concave down to concave up or vice versa   (157)

point symmetry   Two distinct points P and P' are symmetric with respect to a point, M, if and only if M is the midpoint of $\overline{PP'}$. Point M is symmetric to itself.   (106)

point-slope form   the equation of the line that contains the point with coordinates (x_1, y_1) and having slope m written in the form $y - y_1 = m(x - x_1)$ (37)

polar axis   a ray whose initial point is the pole   (466)

polar coordinate system   a grid of concentric circles and their center, which is called the pole, whose radii are integral multiples of 1   (466)

polar equation   an equation that uses polar coordinates   (469)

polar form   the complex number $x + y\mathbf{i}$ written as $r(\cos \theta + \mathbf{i} \sin \theta)$ where $r = \sqrt{x^2 + y^2}$ and $\theta = \text{Arctan } \frac{y}{x}$ when $x > 0$ and $\theta = \text{Arctan } \frac{y}{x} + \pi$ when $x < 0$ (502)

polar graph   the representation of the solution set which is the set of points whose coordinates (r, θ) satisfy a given polar equation   (469)

pole   See polar coordinate system.

polygonal convex set   the solution of a system of linear inequalities   (87)

polynomial equation   a polynomial that is set equal to zero   (179)

polynomial function   a function $y = P(x)$ where $P(x)$ is a polynomial in one variable   (179)

polynomial in one variable   an expression of the form $a_0x^n + a_1x_{n-1} + \cdots + a_{n-1}x + a_n$ where the coefficients $a_0, a_1, \cdots, a_n$ represent complex numbers, a_0 is not zero, and n represents an nonnegative integer   (178)

prediction equation   an equation suggested by the points of a scatter plot used to predict other points   (847)

present value of an annuity   the sum of the present values of all the periodic payments   (609)

principal values   the unique solutions of a trigonometric equation if the values of the function is restricted to two adjacent quadrants   (334, 387)

prisoner set   the set of initial values for which the iterates of a function do not approach infinity   (736)

probability   the measure of the chance of a desired outcome happening   (768)

Q

quadrantal angle   an angle in standard position whose terminal side coincides with one of the axes   (240)

quadratic formula   the formula $x = \frac{-b \pm \sqrt{b^2 - 4ac}}{2a}$, that gives the roots of the quadratic equation of the form $ax^2 + bx + c = 0$ with $a \neq 0$ (188)

quadratic inequality   an inequality of the form $y > ax^2 + bx + c$, $y < ax^2 + bx + c$, $y \geq ax^2 + bx + c$, $y \leq ax^2 + bx + c$, where $a \neq 0$ (191)

quartile   one of four groupings of a set of data determined by the median of the set and the medians of the sets determined by the median   (821)

R

radian   the measure of a central angle whose sides intercept an arc that is the same length as the radius of the circle   (241)

radius   See circle.

range   the difference of the greatest and least values in a set of data   (820)

range of a relation   the set of all ordinates of the ordered pairs of a relation   (6)

rational equation   an equation that consists of one or more rational expressions   (216)

rational function   the quotient of two polynomials in the form $f(x) = \frac{g(x)}{h(x)}$, where $h(x) \neq 0$ (134)

reachable   Point A is reachable from point B if there is a directed path from B to A.   (896)

rectangular form   a complex number written as $x + yi$ where x is the real part and yi is the imaginary part   (501)

recursive   a formula for determining the next term of a sequence using one or more of the previous terms   (657)

reduced sample space   the subset of a sample space that contains only those outcomes that satisfy a given condition   (785)

reference angle   the acute angle formed by the terminal side of an angle in standard position and the x-axis   (243)

reflection   a linear transformation that flips a figure over a line called the line of symmetry   (118)

regression methods   techniques used to find lines of best fit   (847)

relation   a set of ordered pairs   (6)

relative maximum   a point that represents the maximum for a certain interval   (157)

relative minimum   a point that represents the minimum for a certain interval   (157)

repeller   the x-coordinate of the fixed point from which a path generated by iteration staircases or spirals   (718)

resultant of vectors   the sum of two or more vectors   (413)

Riemann sum   a sum of the areas of rectangles formed by partitioning the area between the x-axis and a curve on some interval $x = a$ to $x = b$ (938)

root   a solution of the equation $P(x) = 0$ (179)

row matrix   a matrix that has only one row   (64)

S

sample space   the set of all possible outcomes of an event   (785)

scalar   a real number   (66)

scientific notation   the expression of a number in the form $a \times 10^n$, where $1 \leq a < 10$ and n is an integer   (598)

secant line   a line that intersects a curve at two or more points   (149)

second   a unit of angle measure that is $\frac{1}{60}$ of a minute   (241)

sector of a circle   a region bounded by a central angle and the intercepted arc   (250)

segment of a circle   the region bounded by an arc and its chord   (291)

self-similarity   Self-similar objects contain replicas of the entire shape or object imbedded over and over again inside the object in different sizes.   (736)

semi-interquartile range   one-half the interquartile range of a set of data   (821)

semi-major axis   one of the two segments into which the center of an ellipse divides the major axis   (543)

semi-minor axis   one of the two segments into which the center of an ellipse divides the minor axis   (543)

sequence   a set of numbers in a specific order   (656)

simple graph   a graph having no loops or parallel edges   (866)

simple harmonic motion   the rhythmic motion of an object when friction and other factors affecting such motion are ignored   (345)

slant asymptote   The oblique line ℓ is a slant asymptote for a function $f(x)$ if the graph of $f(x)$ approaches ℓ as $x \to \infty$ or as $x \to -\infty$.   (137)

slope of a line   the value $m = \frac{y_2 - y_1}{x_2 - x_1}$, where (x_1, y_1) and (x_2, y_2), $x_2 \neq x_1$, are two points of the line   (30)

slope-intercept form   the equation of a line with slope, m, and y-intercept, b, written in the form $y = mx + b$ (36)

spanning tree   a tree that is a subgraph of a connected graph and includes all of the vertices of that graph   (890)

square matrix   a matrix with the same number of rows as columns   (64)

standard deviation   a measure of the average amount by which individual items of data deviate from the arithmetic mean of all the data   (823)

standard error of the mean   the standard deviation of the distribution of a sample mean   (839)

standard form   a linear equation written in the form $Ax + By + C = 0$, where A, B, and C are real numbers and A and B are not both zero   (42)

standard position   an angle with its vertex at the origin and its initial side along the positive x-axis   (240)

standard position of a vector   If a vector has its initial point at the origin, it is in standard position.   (412)

stem-and-leaf plot   a display of numerical data for which each value is separated into two numbers   (814)

subgraph   a graph that contains a subset of the edges and vertices of a larger graph   (890)

success   the desired outcome of an event   (768)

system of equations   a set of equations with the same variables   (56)

system of linear inequalities   a set of inequalities with the same variables   (86)

T

tangent   a line that intersects a curve at exactly one point   (149)

term of a sequence   a number in a sequence   (656)

trail   a walk in which no edge is used more than once   (872)

translation   a linear transformation that slides the graph vertically and/or horizontally on the coordinate plane, but does not change the shape   (118)

translation matrix   the matrix used to represent the translation of a set of points with respect to (h, k) which is equal to $\begin{bmatrix} h & h & h & h \\ k & k & k & k \end{bmatrix}$ (570)

transverse axis   the line segment that has as its endpoints the vertices of a hyperbola   (551)

nected graph in which there are no cycles, there is
y one path between any two vertices, there are at least
vertices of degree 1 if the graph contains more than one
vertex, and there is one less edge than vertices (890)

tree diagram   a diagram used to show the total number of
possible outcomes of an event (753)

trigonometric equation   an equation involving a
trigonometric function that is true for some, but not all,
values of the variable (387)

trigonometric form   See polar form.

trigonometric functions   For any angle with measure α, a
point $P(x, y)$ on its terminal side, and $r = \sqrt{x^2 + y^2}$, the
trigonometric functions of α are as follows. (258)

$$\sin \alpha = \frac{y}{r} \qquad \cos \alpha = \frac{x}{r} \qquad \tan \alpha = \frac{y}{x}$$

$$\csc \alpha = \frac{r}{y} \qquad \sec \alpha = \frac{r}{x} \qquad \cot \alpha = \frac{x}{y}$$

trigonometric identity   an equation involving a
trigonometric function that is true for all values of the
variable (358)

trigonometric series   infinite series that define the trigonometric functions sine and cosine (697)

U

unbounded   The solution of a linear programming problem
is unbounded if the region defined by the constraints is
infinitely large. (93)

unit circle   a circle of radius 1 unit whose center is at the
origin of a rectangular coordinate system (241)

unit vector   a vector of length 1 that is parallel to the x-, y-,
or z-axis (422)

upper bound   the integer greater than or equal to the greatest
zero of the polynomial $P(x)$ (212)

upper sum   a Riemann sum for rectangles which are always
above the curve (931)

V

vector   a quantity, or directed distance, that has both
magnitude and direction (412)

vertex of a conic section   a point at which a conic section
intersects its axis of symmetry (534, 543, 551)

vertex   in graph theory, a point of a graph; a node (866)

vertical asymptote   The line $x = a$ is a vertical asymptote for
a function $f(x)$ if $f(x) \to \infty$ or $f(x) \to -\infty$ as $x \to a$ from either
the left or the right. (135)

vertical displacement   the distance the graph of a function
has been translated vertically represented by h in
$y = A f(k\theta + c) + h$ when $y = f(\theta)$ is a trigonometric
function (322)

W

walk   the course taken from one vertex to another along an
edge (872)

weight of an edge   the value assigned to each edge in a
weighted graph (885)

weight of a path   the sum of the weights of the edges along a
path (885)

weighted multigraph   a multigraph in which a value is
assigned to each edge (883)

X

x-intercept   the x-coordinate of the point at which the graph
of an equation crosses the x-axis (22)

Y

y-intercept   the y-coordinate of the point at which the graph
of an equation crosses the y-axis (36)

Z

zero   a value of x for which $f(x) = 0$ (22, 179)

zero matrix   a matrix whose elements are all zero (65)

CHAPTER 1   LINEAR RELATIONS AND FUNCTIONS

Pages 10-11   Lesson 1-1

5. yes   **7.** yes   **9.** -9   **11.** -114   **13.** -6.69
15. $-29 - 12n - n^2$ **17.** $x \leq 4$ **19.** $\{-3, 0, 1, 2\}$, $\{-6, 0, 2, 4\}$, yes
21. $\{-2\}$, $\{7, 8, 9\}$, no   **23.** $\{1, 2, 3, 4\}$, $\{5, 6, 7, 8\}$, yes
25. $\{6, 7, 8\}$, $\{-3, 3\}$, yes   **27.** $\{(4, 10), (5, 12.5), (6, 15),$
$(7, 17.5), (8, 20)\}$, yes   **29.** $\{(11, 3), (11, -3)\}$, no   **31.** $\{(-2, 0.5),$
$(-1, -0.5), (0, -1.5), (1, -0.5), (2, 0.5), (3, 1.5)\}$, yes   **33.** 6
35. 5   **37.** 7   **39.** $[q] + 5$ **41.** 3   **43.** 9   **45.** $10\frac{3}{4}$
47. $|25m^2 - 13|$ **49.** -5   **51.** $\pm 4\frac{1}{2}$ **53.** $-\sqrt{7} \leq x \leq \sqrt{7}$
59. If the x-coordinate represents the pounds of aluminum produced and the y-coordinate represents the number of cans recycled, the relation is a function. If the x-coordinate represents the number of cans recycled and the y-coordinate represents the pounds of aluminum produced, the relation is not a function because the x-coordinate 75,000 would be paired with 2 y-coordinates, 3000 and 3100.

Pages 17-18   Lesson 1-2

5. $\frac{-x^3 - x^2 + 2x + 1}{x + 1}$, $x \neq -1$ **7.** $\frac{x}{x^3 + x^2 - x - 1}$, $x \neq 1$
9. $x^2 + 8x + 7$, $x^2 - 5$ **11.** no   **13.** $-\frac{x^3 - 2x^2 - 35x - 3}{x - 7}$, $x \neq 7$
15. $\frac{3}{x^3 - 2x^2 - 35x}$, $x \neq -5, 0, 7$ **17.** $3x^2 - 24x + 48$, $3x^2 - 4$
19. $5x^4 - 10x^2 + 5$, $25x^4 - 1$ **21.** $x^2 + 7x + 12$, $x^2 + 5x + 7$
23. yes   **25.** no   **27.** yes   **29.** $f^{-1}(x) = \sqrt[3]{x}$, function
37. \$65   **39.** $\{(-3, 14), (-2, 13), (-1, 12), (0, 11)\}$, yes
41. $\pm\sqrt{6}$

Pages 25-27   Lesson 1-3

3. 2   **5.** none   **7.** all points   **9.** $x > 2$ **11.** $y \leq x + 2$
13.
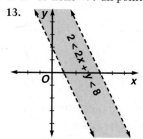
15. $1 \leq y \leq 4$ **17.** -12
19. $-\frac{5}{9}$ **21.** none

23.
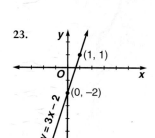

27.
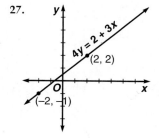

31.
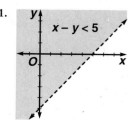

35.
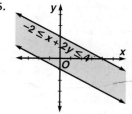

41a.
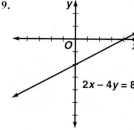

b. \$24 billion   **c.** If the nation had no disposable income, personal consumption expenditures would be \$24 billion.
43. $8 - 6a - 6a^2 - a^3$
45. $16x - 4$ **47.** A

Page 27   Mid-Chapter Review

1a. domain: $\{2, 3, 5, 6\}$, range: $\{-1, 0, 1, 2, 3\}$ **b.** No, the element 5 in the domain is paired with both 2 and -1 in the range.   **3.** -1   **5.** $4 + 18k - 27k^3$ **7.** $\frac{1}{x^2}$, $\frac{1}{x^2}$
9.

Pages 33-35   Lesson 1-4

5. 10, undefined   **7.** 5, $\frac{3}{4}$ **9.** $2\sqrt{2}$, 1   **11.** 12   **13.** 30
15. $(1, 1), (-2, -1), (2, -2)$ **17.** $6\sqrt{2}$, -1   **19.** $8\sqrt{5}$, -2   **21.** 3, undefined   **23.** $\sqrt{1 + 9n^2}$, -3n   **25.** no   **27.** yes   **29.** 8
31. -5   **33.** Prove that $DE = \frac{1}{2}AB = \frac{1}{2}a$.
$D = \left(\frac{b + 0}{2}, \frac{c + 0}{2}\right) = \left(\frac{b}{2}, \frac{c}{2}\right)$
$E = \left(\frac{a + b}{2}, \frac{0 + c}{2}\right) = \left(\frac{a + b}{2}, \frac{c}{2}\right)$
$DE = \sqrt{\left(\frac{b}{2} - \frac{a + b}{2}\right)^2 + \left(\frac{c}{2} - \frac{c}{2}\right)^2} = \sqrt{\left(-\frac{a}{2}\right)^2} = \frac{a}{2}$
37a. 0.2   **b.** \$361   **39.** -4   **41.** $x^5 - 3x^4 + 7x^3$, $\frac{x^3}{x^2 - 3x + 7}$

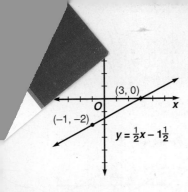

$y = \frac{1}{2}x - 1\frac{1}{2}$

(3, 0)

(−1, −2)

Pages 39-41   Lesson 1-5

7. $y = 4x + \frac{1}{2}, 4, \frac{1}{2}$ **9.** $y = 4x - 10$ **11.** $y = -6x - 22$
13. $y - 0 = -1(x + 2), y = -x - 2$ **15.** $y = 8x + 50$
17. $y = -3x + 14$ **19.** $y = -\frac{1}{4}x + \frac{5}{4}$ **21.** Point-slope form
is undefined, $x = -1$ **23.** $y - 2 = \frac{7}{2}(x - 5), y = \frac{7}{2}x - \frac{31}{2}$
25. $y - 1 = -\frac{3}{5}(x - 3), y = -\frac{3}{5}x + \frac{14}{5}$ **27.** The graphs are all
lines. When m is positive, the graph slopes up to the right.
When m is negative, the graph slopes down to the right. All
three graphs cross the y-axis at (0, -1).   **29.** The graphs are
all lines. When m is positive, the graph slopes up to the
right. When m is negative, the graph slopes down to the
right. All three graphs cross the y-axis at (0, 2).
33a. $C(x) = 0.5x + 1000$ **b.** $1000, $0.50
c.

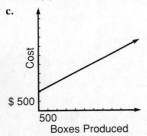

Cost

$ 500

500
Boxes Produced

35a. $y = \frac{7}{2}x + 83$
b. 21.286 cm
37. $4x^2 - 7, 4x^2 - 7$
39. 76,650 people per year

Pages 45-47   Lesson 1-6

5. 3, -3, neither   **7.** 1, -1, perpendicular   **9.** -4, -4, parallel
11. $2x - y - 6 = 0, x + 2y - 8 = 0$ **13.** $2x - y - 18 = 0,$
$x + 2y - 4 = 0$

15.

X(3, 6)

W(−1, 3)

Y(6, 2)

Z(2, −1)

$WZ = \sqrt{(2 + 1)^2 + (-1 - 3)^2} = 5$
$YZ = \sqrt{(2 - 6)^2 + (-1 - 2)^2} = 5$
$XY = \sqrt{(6 - 3)^2 + (2 - 6)^2} = 5$
$WX = \sqrt{(3 + 1)^2 + (6 - 3)^2} = 5$
Thus, $WXYZ$ is a rhombus. The
slope of $\overline{WX} = \frac{6 - 3}{3 + 1} = \frac{3}{4}$, and
the slope of $\overline{WZ} = \frac{3 + 1}{-1 - 2} = -\frac{4}{3}$.
Thus $\overline{WZ}$ is perpendicular to
$\overline{WX}$ and $WXYZ$ is a square.
17. $3x - y + 6 = 0$

19. $6x - y - 3 = 0$ **21.** $2x + 3y - 16 = 0$ **23.** $x - 2y - $
$6 = 0$ **25.** $5x + 4y + 43 = 0$ **27.** $2x + 3y - 40 = 0$
29. $4, -\frac{49}{4}$ **31.** slope of $\overline{AB} = \frac{2 - (-3)}{-1 - 4} = \frac{5}{-5} = -1$
slope of $\overline{BC} = \frac{-1 - (-3)}{-2 - 4} = \frac{2}{-6} = -\frac{1}{3}$
slope of $\overline{CA} = \frac{-1 - 2}{-2 - (-1)} = \frac{-3}{-1} = 3$
The slopes of $\overline{BC}$ and $\overline{CA}$ have a product of -1, so the lines
are perpendicular. Therefore, $\triangle ABC$ is a right triangle.

35. No, they are riding along parallel lines or the same line,
5 miles apart.
37.

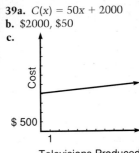

$-6 \leq 3x - y \leq 12$

39a. $C(x) = 50x + 2000$
b. $2000, $50
c.

Cost

$ 500

1
Televisions Produced

Pages 48-50   Chapter 1   Summary and Review

1. {(0, -7), (1, -2), (2, 3), (3, 8)}, yes   **3.** {(5, 1), (5, -1), (6, 2),
(6, -2)}, no   **5.** 0   **7.** -3   **9.** $3x + 1, 3x - 3$ **11.** $x^2 + 4x + 6,$
$x^2 + 2x + 4$ **13.** $f^{-1}(x) = \pm\sqrt{x + 12}$, no   **15.** $f^{-1}(x) = \frac{1}{2}\sqrt[3]{x}$,
yes   **17.** $\frac{8}{3}$ **19.** 20   **21.** 13   **23.** 5   **25.** $y = 2x - 5$
27. $y = \frac{3}{5}x$ **29.** $y = x + 4$ **31.** $y = \frac{7}{2}x - 10$
33. $4x - y + 11 = 0$ **35.** $x + 3y - 12 = 0$ **37a.** 10 m,
40 m, 90 m, 160 m, 250 m   **b.** Yes, each element of the
domain is paired with exactly one element of the range.
39. $y = 0.1x + 140$; $140, $0.10

Page 51   Chapter 1 Test

1. {(2, 5), (1, 2), (0, -1), (-1, -4), (-2, -7)}, yes   **2.** {(4, 13), (3, 11),
(2, 9), (1, 7), (0, 5), (4, -13), (3, -11), (2, -9), (1, -7), (0, -5)}, no
3. 0   **4.** -44   **5.** -114   **6.** -144.13   **7.** $\sqrt{2x^2 - 5}, 2x - 5$
8. $50x^2 + 120x + 72, 10x^2 + 6$ **9.** $3x - 7, 3x + 21$
10. $f^{-1}(x) = \pm\sqrt{3x - 5}$, no   **11.** $f^{-1}(x) = 4x - 4$, yes
12. $f^{-1}(x) = \sqrt[3]{-\frac{x}{2} + 1}$, yes

13.

$y = 3x - 6$

14.

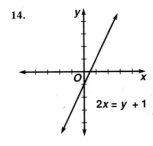

$2x = y + 1$

15.

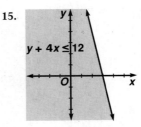

$y + 4x \leq 12$

16.

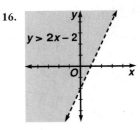

$y > 2x - 2$

17. $\sqrt{17}, -\frac{1}{4}$ **18.** $5\sqrt{2}, \frac{1}{7}$ **19.** $\sqrt{k^2 + 4}, \frac{2}{k}$
20. $y = \frac{5}{3}x + \frac{14}{3}$ **21.** $y = -\frac{3}{4}x + 4$ **22.** $3x + 2y - 17 = 0$
23. $5x + y = 0$

24.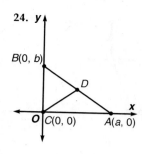

D is the midpoint of hypotenuse $\overline{AB}$. The coordinates of D are $\left(\frac{a}{2}, \frac{b}{2}\right)$. $BD = \sqrt{\left(0 - \frac{a}{2}\right)^2 + \left(b - \frac{b}{2}\right)^2} = \sqrt{\frac{a^2}{4} + \frac{b^2}{4}}$

$AD = \sqrt{\left(a - \frac{a}{2}\right)^2 + \left(0 - \frac{b}{2}\right)^2} = \sqrt{\frac{a^2}{4} + \frac{b^2}{4}}$

$CD = \sqrt{\left(\frac{a}{2} - 0\right)^2 + \left(\frac{b}{2} - 0\right)^2} = \sqrt{\frac{a^2}{4} + \frac{b^2}{4}}$

$BD = AD = CD$ **25a.** 9.5 1m/m² **b.** d cannot be negative or zero. **BONUS** -2

CHAPTER 2 SYSTEMS OF EQUATIONS AND INEQUALITIES

Pages 59-61 Lesson 2-1

5. yes **7.** no **9.** no **11.** (1, 2) **13.** I **15.** CI **17.** (0, 4)
19. (-2, 0) **21.** (4, 2) **23.** (-2, 3) **25.** (12, 2)
27. (5.25, 0.75) **31.** 70 soybeans, 30 corn **33.** {16}, {-4, 4}; no
35. $y = \frac{3}{4}x + 3\frac{1}{4}$ **37.** D

Pages 68-70 Lesson 2-2

5. $\begin{bmatrix} 8 & 1 \\ -4 & 9 \end{bmatrix}$ **7.** $\begin{bmatrix} 2 & 1 \\ -2 & -1 \end{bmatrix}$ **9.** $\begin{bmatrix} -15 & -3 \\ 9 & -12 \end{bmatrix}$ **11.** $\begin{bmatrix} 15 & 3 \\ -20 & 19 \end{bmatrix}$ **13.** $\begin{bmatrix} 22 & 9 \\ -27 & 13 \end{bmatrix}$

15. (7, 9) **17.** $\begin{bmatrix} 7 & 14 & 3 \\ -6 & 15 & -14 \\ 23 & 4 & -4 \end{bmatrix}$ **19.** $\begin{bmatrix} 4 & 20 & -6 \\ -2 & 12 & -14 \\ 31 & 2 & -1 \end{bmatrix}$ **21.** $\begin{bmatrix} 9 & 3 & 5 \\ -15 & 16 & -8 \\ 12 & 6 & -5 \end{bmatrix}$

23. $\begin{bmatrix} 5 & 4 & -11 \\ -16 & 11 & -2 \\ 17 & 4 & 0 \end{bmatrix}$ **25.** (7, 2) **27.** (4, 0) **29.** $\begin{bmatrix} 21 & 0 \\ 15 & 9 \end{bmatrix}$

31. $\begin{bmatrix} 6 & -6 & 12 \\ 10 & 8 & -4 \end{bmatrix}$ **33.** $\begin{bmatrix} 34 & 12 \\ 36 & -12 \\ 16 & 18 \end{bmatrix}$ **35.** $\begin{bmatrix} -30 & 60 \\ 46 & -8 \end{bmatrix}$ **37.** $\begin{bmatrix} 49 & 0 \\ 50 & 9 \end{bmatrix}$

39. $\begin{bmatrix} 162 & -54 & 180 \\ 48 & -156 & 240 \\ 138 & 24 & 60 \end{bmatrix}$ **41.** $\begin{bmatrix} 14 & -29 \\ 3 & 7 \end{bmatrix}$

43. Let $A = \begin{bmatrix} a_{11} & a_{12} \\ a_{21} & a_{22} \end{bmatrix}$, $B = \begin{bmatrix} b_{11} & b_{12} \\ b_{21} & b_{22} \end{bmatrix}$, and $C = \begin{bmatrix} c_{11} & c_{12} \\ c_{21} & c_{22} \end{bmatrix}$.

$(A + B) + C = \begin{bmatrix} a_{11}+b_{11} & a_{12}+b_{12} \\ a_{21}+b_{21} & a_{22}+b_{22} \end{bmatrix} + \begin{bmatrix} c_{11} & c_{12} \\ c_{21} & c_{22} \end{bmatrix}$

$= \begin{bmatrix} (a_{11}+b_{11})+c_{11} & (a_{12}+b_{12})+c_{12} \\ (a_{21}+b_{21})+c_{21} & (a_{22}+b_{22})+c_{22} \end{bmatrix}$

$= \begin{bmatrix} a_{11}+(b_{11}+c_{11}) & a_{12}+(b_{12}+c_{12}) \\ a_{21}+(b_{21}+c_{21}) & a_{22}+(b_{22}+c_{22}) \end{bmatrix}$

$= \begin{bmatrix} a_{11} & a_{12} \\ a_{21} & a_{22} \end{bmatrix} + \begin{bmatrix} b_{11}+c_{11} & b_{12}+c_{12} \\ b_{21}+c_{21} & b_{22}+c_{22} \end{bmatrix}$

$= A + (B + C)$

45. Let $A = \begin{bmatrix} a_{11} & a_{12} \\ a_{21} & a_{22} \end{bmatrix}$, $B = \begin{bmatrix} b_{11} & b_{12} \\ b_{21} & b_{22} \end{bmatrix}$, and $C = \begin{bmatrix} c_{11} & c_{12} \\ c_{21} & c_{22} \end{bmatrix}$.

$(AB)C = \begin{bmatrix} a_{11}b_{11}+a_{12}b_{21} & a_{11}b_{12}+a_{12}b_{22} \\ a_{21}b_{11}+a_{22}b_{21} & a_{21}b_{12}+a_{22}b_{22} \end{bmatrix} \begin{bmatrix} c_{11} & c_{12} \\ c_{21} & c_{22} \end{bmatrix}$

$= \begin{bmatrix} a_{11}b_{11}c_{11}+a_{12}b_{21}c_{11}+a_{11}b_{12}c_{21}+a_{12}b_{22}c_{21} \\ a_{21}b_{11}c_{11}+a_{21}b_{12}c_{21}+a_{22}b_{21}c_{11}+a_{22}b_{22}c_{21} \end{bmatrix}$

$A(BC) = \begin{bmatrix} a_{11} & a_{12} \\ a_{21} & a_{22} \end{bmatrix} \begin{bmatrix} b_{11}c_{11}+b_{12}c_{21} & b_{11}c_{12}+b_{12}c_{22} \\ b_{21}c_{11}+b_{22}c_{21} & b_{21}c_{12}+b_{22}c_{22} \end{bmatrix}$

$\begin{matrix} a_{11}b_{11}c_{12}+a_{12}b_{21}c_{12}+a_{11}b_{12}c_{22}+a_{12}b_{22}c_{22} \\ a_{21}b_{11}c_{12}+a_{22}b_{21}c_{12}+a_{21}b_{12}c_{22}+a_{22}b_{22}c_{22} \end{matrix}$

$= \begin{bmatrix} a_{11}b_{11}c_{11}+a_{11}b_{12}c_{21}+a_{12}b_{21}c_{11}+a_{12}b_{22}c_{21} \\ a_{21}b_{11}c_{11}+a_{21}b_{12}c_{21}+a_{22}b_{21}c_{11}+a_{22}b_{22}c_{21} \end{bmatrix}$

$\begin{matrix} a_{11}b_{11}c_{12}+a_{11}b_{12}c_{22}+a_{12}b_{21}c_{12}+a_{12}b_{22}c_{22} \\ a_{21}b_{11}c_{12}+a_{22}b_{21}c_{12}+a_{21}b_{12}c_{22}+a_{22}b_{22}c_{22} \end{matrix}$

Therefore, $(AB)C = A(BC)$.
47. TF, $8218; FG, $18,422; BR, $31,736
49. $[f \circ g] x = x^2 + x$, $[g \circ f] x = x^2 + 3x + 1$ **51.** $5|t|$, $\frac{4}{3}$
53. (45, 60)

Pages 75-77 Lesson 2-3

5. 62 **7.** -50 **9.** $\frac{1}{10}\begin{bmatrix} 1 & -3 \\ 4 & -2 \end{bmatrix}$ **11.** $\left(\frac{3}{4}, \frac{1}{2}\right)$ **13.** 8 **15.** 1

17. -93 **19.** $\begin{bmatrix} -1 & 2 \\ 0 & -1 \end{bmatrix}$ **21.** $-\frac{1}{5}\begin{bmatrix} -3 & -1 \\ -2 & 1 \end{bmatrix}$ **23.** $\frac{1}{7}\begin{bmatrix} 1 & -1 \\ 4 & 3 \end{bmatrix}$ **25.** (7, -9)
27. $\left(\frac{2}{3}, 1, -\frac{4}{3}\right)$ **29.** -109 **31.** 3411 **35.** 80 mL of the 60% solution, 120 mL of the 40% solution **37.** 24, 16 **39.** (3, 1)
41. B

Page 77 Mid-Chapter Review

1. (1, 2) **3.** (3, -1) **5.** $\begin{bmatrix} 8 & 1 \\ -4 & 9 \end{bmatrix}$ **7.** $\begin{bmatrix} 15 & 3 \\ -20 & 19 \end{bmatrix}$ **9.** (19, -11)

Pages 82-83 Lesson 2-4

3. Multiply row 2 by 5 and add it to row 1. Multiply row 1 by -2 and add it to row 2. **5.** Multiply row 1 by 2. Add row 2 to row 1. Multiply row 1 by -3 and row 2 by 5. Add row 1 to row 2. **7.** (-1, 5, -2) **9.** (3, 2) **11.** $\left(-\frac{97}{19}, \frac{155}{19}\right)$ **13.** (-1, 2, -3)
15. (7, 1, -2) **17.** (1, -1, 2, -2) **19.** $y = 2x^2 - 7x + 9$
23. blouse, $28; skirt, $32; jeans, $36 **25.** (7, 2) **27.** 0

Pages 89-90 Lesson 2-5

3. 14 **5.** 4.5 **7.** 32, 0 **9.** (-2, 0), (0, 3), (4, 2), (2, -4); maximum at (2, -4) = 8, minimum at (0, 3) = -1 **11.** (0, 0), (0, 4), (2, 0); maximum at (0, 4) = 4, minimum at (2, 0) = -2 **13.** (0, 1), (0, 4), (3, 1); maximum at (3, 1) = 21, minimum at (0, 1) = 9 **15.** (0, 0), (5, 0), (5, 1), (0, 2), (2, 4); maximum at (5, 0) = 15, minimum at (2, 4) = -14
17. $y \geq \frac{1}{5}x$, $y \leq -\frac{2}{5}x + \frac{32}{5}$, $y \geq 3x - 14$, $y \leq 6x$
21.

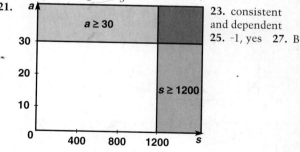

23. consistent and dependent
25. -1, yes **27.** B

Pages 94-97 Lesson 2-6

5. $x \geq 200$ **7.** $P(x, y) = 20x + 30y$ **9.** (200, 600), (500, 300), (200, 300) **11.** $22,000 **13.** infeasible **15.** alternate optimal **17.** $9\frac{2}{3}$ of Y and 0 of X **19.** unbounded

21. 1960 of #1, 1840 of #2 **23.** infeasible **25.** $\begin{bmatrix} 6 & -\frac{21}{4} \\ -3 & 0 \end{bmatrix}$
27. Multiply row 2 by -3 and add it to row 1. Multiply row 1 by 2 and row 2 by 3 and add row 1 to row 2. **29.** A

Pages 98-100   Chapter 2 Summary and Review

1. (2, -4)   **3.** $\left(-\dfrac{5}{11}, -\dfrac{2}{11}\right)$ **5.** $\begin{bmatrix} 4 & 3 \\ 2 & -6 \end{bmatrix}$ **7.** $\begin{bmatrix} -9 & -15 \\ 6 & -6 \end{bmatrix}$ **9.** $\begin{bmatrix} -5 & -51 \\ -8 & 8 \end{bmatrix}$

11. -1   **13.** 0   **15.** (13, -5)   **17.** (-7, -4)   **19.** $\left(-1, 3, \dfrac{1}{2}\right)$

21. (0, 9), (4, 7), (6, 5), (6, 4), (0, 4); maximum at (4, 7) = 25, minimum at (0, 4) = 8   **23.** infeasible   **25.** 39 inches, 31 inches, 13 inches

Page 101   Chapter 2 Test

1. (4, 0)   **2.** (2, 7)   **3.** (-1, 5)   **4.** $\left(\dfrac{3}{4}, \dfrac{1}{2}\right)$ **5.** $\begin{bmatrix} 5 & -10 \\ 0 & 20 \\ -15 & 20 \end{bmatrix}$

6. $\begin{bmatrix} 9 & 6 \\ 3 & 0 \end{bmatrix}$ **7.** $\begin{bmatrix} -7 & -8 \\ 11 & 10 \end{bmatrix}$ **8.** $\begin{bmatrix} -20 & 44 \\ 8 & -42 \end{bmatrix}$ **9.** $\begin{bmatrix} -5 & 50 \\ -1 & -48 \end{bmatrix}$ **10.** 7

11. -1   **12.** -4   **13.** -64   **14.** $\left(-2, -\dfrac{3}{4}\right)$ **15.** (-2, 1, 3)

16. (1, 6, -1)   **17.** (0, 0), (0, 3), $\left(\dfrac{5}{3}, \dfrac{14}{3}\right)$, (4, 0); $28\dfrac{1}{3}$; 0

17.
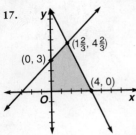

18.
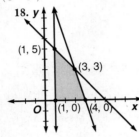

18. (1, 5), (3, 3), (1, 0), (4, 0); 28; 3
19. $\left(-\dfrac{5}{2}, 5\right)$, (3, 5), (3, 1), $\left(\dfrac{2}{3}, -\dfrac{4}{3}\right)$; 34, $-4\dfrac{2}{3}$
20. $460   **BONUS** pentagon

19.

CHAPTER 3   THE NATURE OF GRAPHS

Pages 113-116   Lesson 3-1

5. both   **7.** point   **9.** line   **11.** x = 2   **13.** x = 2, y = -2
15a. (-2, 5)   **b.** (2, -5)   **c.** (-5, -2)   **d.** (5, 2)   **17.** P'(4, 4)
19. P'(-3, -5)
21a.
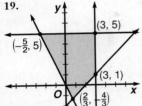
b.
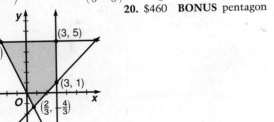

c.
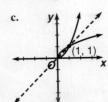
d.
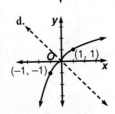

25a.
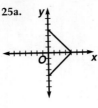
b.
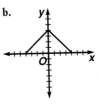

c.
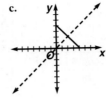
d.
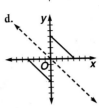

27. neither   **29.** even   **31.** odd   **33.** y-axis   **35.** no symmetry   **43.** x-intercept: (2, 0); other points: (-3, 4.5), (3, -4.5), (-3, -4.5)   **45.** 1, 3   **47.** Parallel; the lines have the same slope.   **49.** $\left(\dfrac{3}{4}, -\dfrac{2}{3}, \dfrac{1}{2}\right)$ **51.** 10

Pages 122-124   Lesson 3-2

5a. shrunk vertically   **b.** moved 2 units up   **c.** shrunk horizontally   **7.** f(x) + 2   **9.** -f(x - 4)   **11.** f(x + 5) - 2

13.
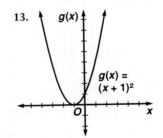
17. s(x) = -0.5(x - 2)²
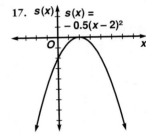

21a. narrower than the parent   **b.** narrower and reflected over the x-axis   **c.** narrower and moved 2 units right   **d.** wider and moved 1 unit left   **e.** narrower and moved 5 units down   **23a.** moved up 6 units   **b.** moved right 4 units   **c.** wider   **d.** narrower, moved 3 units left, and 8 units down   **e.** inverted, moved 3 units right and 6 units up

25.
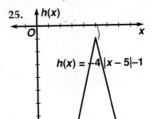
h(x) = -4|x - 5|-1
27. q(x) = -(x + 2)² + 5
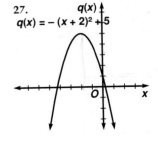

33a.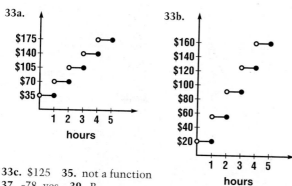

33b.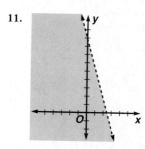

33c. $125   **35.** not a function
37. -78, yes   **39.** B

Pages 129-131   Lesson 3-3

5. $P'(5, -4)$ **7.** $P'(8, -2)$ **9.** yes   **11.** no   **13.** $y = \frac{x-3}{2}$
15. $y = \pm\frac{1}{3}\sqrt{3(4-x)}$ **17.** $y = \sqrt[3]{x} + 2$ **19.** $y = \pm\sqrt{x-2}$;
no   **21.** $y = \pm\sqrt[4]{x} + 4$; no   **23.** $y = \sqrt[3]{x-6} + 3$; yes
25a. reflection   **b.** reflection, moved 3 units right
c. reflection, moved 6 units up   **d.** wider version of the
reflection   **e.** narrower version of the reflection, moved 4
units up   **f.** reflection, moved 5 units right, and 3 units up
31. 93%
35. both   **37.** A

33.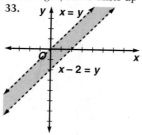

Pages 139-142   Lesson 3-4

5. $f(x) = \frac{1}{x-1}$ **7.** $f(x) = \frac{1}{x} + 5$ **9.** $f(x) = -\frac{1}{x}$ **11.** $x = -1$,
$y = 0$ **13.** $x = -5$ **15.** $x = 2, x = -3, y = 0$
17. $y = \frac{3}{2}x, x = 0$ **19a.** translated 1 unit down
b. translated 5 units left   **c.** closer to the axes   **d.** wider,
translated 8 units up   **e.** translated 3 units right, 7 units up

25.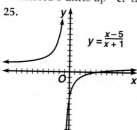

29.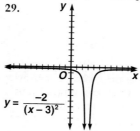

33.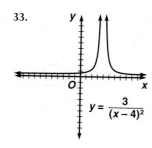

37.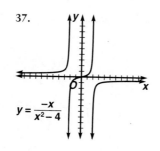

41a. x- and y-axes   **b.** It increases.   **c.** It is half.   **45.** $-\frac{4}{3}$
47. $y = \pm\sqrt{x + 9}$

Pages 147-148   Lesson 3-5
3. yes   **5.** no

7.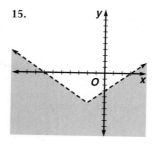

11.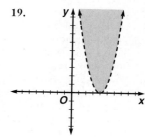

15.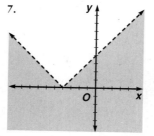

19.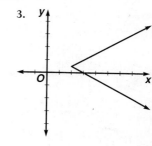

23. ϕ **25.** $\{x \mid x \geq 0\}$ **27.** all real numbers   **35.** $31.7\% < v$
$< 38.3\%$ **37.** $\frac{1}{2}$, -2   **39.** (-6, -3)   **41.** B

Page 148   Mid-Chapter Review
1. y-axis
5. $x = 1$

3.

Pages 153-154   Lesson 3-6

5. $f'(x) = 2x$ **7.** $f'(x) = 2x^3$ **9.** $f'(x) = 2.4x^3 - 0.8x$ **11.** 6
13. 0   **15.** $f'(x) = 3x^2 - 4x + 4$ **17.** $f'(x) = 0.9x^2 - 8x$
19. $f'(x) = -20x^4 + 21x^2 + 1$ **21.** -8   **23.** -1   **25.** $1\frac{1}{6}$
27. -11   **29.** $y = 6x - 12$ **31.** $y = -2x + 2.25$
33. $y = -3x - 9$ **35.** (-3, 1)   **37.** (2, -2), (-2, 2)   **45a.** 16π in³
b. For each inch of increase in the radius, the volume
increases 16π or about 50.3 cubic inches.   **47.** (-1, 2)
49. -42

Pages 162-163   Lesson 3-7

5. $\left(-\frac{3}{2}, \frac{19}{2}\right)$, max   **7.** (0, 0), inflection   **9.** x: 0, 1; y: 0
11. (-1, -16), min   **13.** $\left(\frac{2}{3}, -\frac{1}{3}\right)$, min   **15.** $\left(\frac{1}{3}, \frac{26}{27}\right)$, min;
(0, 1), max; $\left(\frac{1}{6}, \frac{53}{54}\right)$, inflection   **17.** (0, -28), inflection
19. x: $-\frac{3}{2}$, $-\frac{5}{2}$; y: 15   **21.** x: 0, 5, 7; y: 0   **23.** x: 2, -2, 3, -3; y: 36

25.

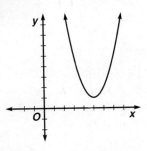

$D(t)$

33. 500 ft by 250 ft

35. P lies on $\overline{AB}$.
$$AP \stackrel{?}{=} PB$$
$$\sqrt{(9-4)^2 + (3-2)^2} \stackrel{?}{=} \sqrt{(4+1)^2 + (2-1)^2}$$
$$\sqrt{26} = \sqrt{26}$$
$$AP = PB$$

37. 0, no **39.** $y = -7x - 2$

Pages 169-171   Lesson 3-8

5. point discontinuity; As $x \to \infty$, $y \to -\infty$. As $x \to -\infty$, $y \to -\infty$.
7. infinite discontinuity; As $x \to \infty$, $y \to \pm\infty$. As $x \to -\infty$, $y \to \pm\infty$.
9. infinite discontinuity; As $x \to \infty$, $y \to 0$. As $x \to -\infty$, $y \to 0$.
11. all reals: increasing **13.** $x < 0$: decreasing, $x > 0$: increasing **15.** point discontinuity **17.** continuous
19. continuous **21.** infinite discontinuity **23.** infinite discontinuity **25.** $x \to \infty$, $f(x) \to \infty$; $x \to -\infty$, $f(x) \to \infty$ **27.** -3, 4 **29.** 64, 4 **35.** $r = 3.6$ cm, $h = 7.4$ cm, $S = 248.8$ cm²
37. 14 **39.** $f^{-1}(x) = \pm\sqrt{x} + 9$ **41.** max: (0, 3), min: (0.67, 2.85), inflection: (0.33, 2.93)

Pages 172-174   Chapter 3 Summary and Review

1. origin, $y = \pm x$, odd **3.** none, neither **5.** reflect across x-axis **7.** translate left 5 units **9.** $y = \frac{6-x}{2}$; yes
11. $y = \sqrt[3]{x + 2} - 1$; yes **13.** $x = -3, 4, y = 0$
15.

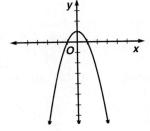

17.

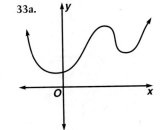

19. $y = 7x - 3$ **21.** $y = 10x - 5$ **23.** $\left(\frac{1}{2}, \frac{17}{4}\right)$, max
25. (0, -4), min; (-2, 0), max; (-1, -2), inflection **27.** infinite
29. continuous **31a.** 30.74 m/s **b.** 6 m/s²

Page 175   Chapter 3 Test

1. The graph is moved 5 units right. **2.** The graph is reflected over the x-axis and moved 2 units up.

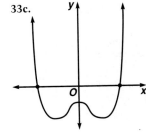

3.

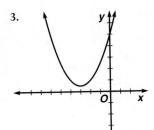

3. wider, shifted left 3 units
4. $y = \pm\sqrt{x} + 4$, no
5. $y = \frac{\sqrt[3]{x-1}}{2}$, yes
6. $y = \pm\sqrt[4]{x + 7} - 3$, no
7. $x = -1, 3, y = 0$; neither
8. $x = \pm2, y = 0$; odd
9. no asymptotes; neither

10.

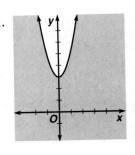

11.

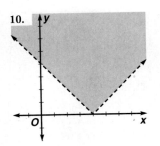

12. $y = -2x$ **13.** $y = 10x - 11$ **14.** (4, -12), min **15.** (0, 3), max; (-2, -1), min; (-1, 1), inflection **16.** point **17.** jump
18. x: -1, 0, $\frac{5}{3}$; y: 0 **19.** $x \to \infty$, $y \to \infty$; $x \to -\infty$, $y \to \infty$

18.

$y = 3x^3 - 2x^2 - 5x$

20a.

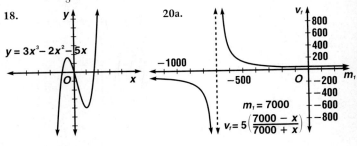

$m_1 = 7000$
$v_t = 5\left(\dfrac{7000 - x}{7000 + x}\right)$

b. 4.93 m/s **BONUS** $b = -3, c = 4$

CHAPTER 4   POLYNOMIAL AND RATIONAL FUNCTIONS

Pages 181-184   Lesson 4-1

5. yes, 3 **7.** yes, 7 **9.** yes **11.** no **13.** $x^4 - 5x^2 + 4 = 0$
15. 1; 2 **17.** 2; 7 **19.** 2; $\frac{5}{6}$, $-\frac{3}{2}$ **21.** 3; 0, 3, -2.5
23. $x^2 + x - 6 = 0$ **25.** $x^5 - 10x^4 + 25x^3 + 20x^2 - 80x - 64 = 0$ **27.** $x^4 - 2x^3 + x^2 + 2x - 2 = 0$ **29.** 4, $1 \pm i$
31. -1, 1, $\pm i\sqrt{2}$

33a.

33c.

33e.

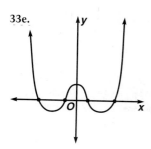

41a. $f(x) = 30{,}000x^3 + 55{,}000x^2 + 75{,}000x$; $198{,}266.41
b. $178{,}495.86; $19{,}770.55 less   **43.** $x^2 - 1$ **45.** $(-5, -2)$
47.

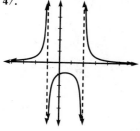

49. B

Pages 192-194   Lesson 4-2
5. $\frac{1}{4}$ **7.** $\frac{-5 \pm i\sqrt{11}}{2}$ **9.** -364, 2 imaginary roots

11. $5 + i\sqrt{2}$ **13.** $\frac{5 \pm i\sqrt{83}}{6}$ **15.** -8, 11   **17.** $\frac{3}{2}$, -7

19. 6, -4   **21.** $\frac{1}{4}$, -5   **23.** $\frac{-3 \pm i\sqrt{15}}{4}$ **25.** $\frac{3}{2}$, 5

27.   **31.**

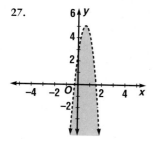

33. (4.1, -108.2), rel min; (-2.8, 60.4), rel max; (0.7, -24.3),
pt. of inf.   **35.** (4.2, -3.1), rel min; (1.8, 3.1), rel max; (3, 0),
pt. of inf.   **37a.** (-6, 0), (9, 0)

37b.

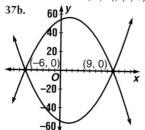

c. The two graphs have the
same shape and x-intercepts.
However, they are
reflections of each other
over the x-axis.   **43.** 18 ft
by 24 ft   **45.** 5 seconds
47. $-2x + y + 6 = 0$
49. y-axis
51. $x^3 + x^2 - 80x - 300 = 0$

Pages 199-200   Lesson 4-3
5. 0, yes   **7.** 88, no   **9.** $(x + 1)(x - 1)(x + 7)$ **11.** $x + 1$, R6
13. $x^3 + 3x^2 + 6x + 12$, R23   **15.** $2x^2 + 2x$, R-3   **17.** 0, yes
19. 0, yes   **21.** -2, no   **23.** 0, yes   **25.** 190, no   **27.** 4
29. twice   **33.** $r \approx 1$ inch, $h \approx 5$ inches   **35.** (0, -32),
inflection   **37.** 8 or -12

Pages 205-207   Lesson 4-4
5. ± 1, ± 2, ± 3, ± 6; -3, -1, 2   **7.** 1 pos; 2 or 0 neg; -3, -1, 2
9. 2 or 0 pos; 1 neg; -1.5, $\frac{1}{3}$, 3   **11.** ± 6, ± 3, ± 2, ± 1; -3, -1, 1, -2
13. ± 20, ± 10, ± 5, ± 4, ± 2, ± 1; 2, -2, 5   **15.** ± 2, ± 1; 1, 2
17. 1 pos and 1 neg; 0, 4, -2   **19.** 2 or 0 pos and 1 neg;
$\frac{1}{4}$, $-\frac{3}{2}$, 2   **21.** 1 pos and 2 or 0 neg; -0.5, -3, $\frac{1}{3}$ **23.** 2 or 0 pos
and 2 or 0 neg; -2, 2, -1, 1

31a.

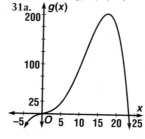

b. 0, 23.5   **33.** The three
types of discontinuity are
point, jump, and infinite.
Point discontinuity means
the function displays all the
characteristics of continuity
except for one point. Jump
discontinuity occurs when a
function is defined for all
values of x, but at one point
the graph jumps from (x, a)
to (x, b). Infinite discontinuity occurs for functions whose
graphs have asymptotes.   **35.** -95, 2 imaginary roots   **37.** B

Page 207   Mid-Chapter Review
1. 0, -10, 8   **3.** 113; 2 real roots; $\frac{13 \pm \sqrt{113}}{14}$ **5.** ± 12, ± 6,
± 4, ± 3, ± 2, ± 1, $\pm \frac{3}{2}$, $\pm \frac{1}{2}$; no rational zero

Pages 213-214   Lesson 4-5
5. 3 zeros; between -2 and -1, at 1, between 3 and 4; UB: 4
7. -0.4 and -2.6   **9.** 4 and 5, -1 and 0   **11.** 1 and 2   **13.** at 2,
at -1   **15.** -2, -1   **17.** -2.5   **19.** 1, -1   **21.** 2, -5   **23.** 1, 0
25. 3 complex zeros; 1 positive and 2 or 0 negative zeros;
zeros: -4.9, -1.8, 2.2; relative max: (-3.6, 26.5); relative min:
(0.6, -43.5)   **29.** 2 meters   **31.** $-\frac{8}{3}$

33.

35. 30

Pages 220-222   Lesson 4-6
5. $2m^2$; -34   **7.** $12(a + 1)(a - 1)$; 3 and $-\frac{1}{2}$ **9.** 31
11. $\frac{2}{x-1} - \frac{3}{x+1}$ **13.** -1, 5   **15.** -3   **17.** all reals except 1
19. $\frac{1 \pm \sqrt{145}}{4}$ **21.** $x < 0$, $x > 3$ **23.** $a < 1$, $a > 31$
25. $0 \le x \le 2$, $x \ge 9$ **27.** $\frac{2}{p+1} + \frac{1}{p-1}$
29. $\frac{1}{m+1} + \frac{1}{m}$ **31.** $x < -4$, $-1 < x < 4$, $x > 5$
35. $\frac{2}{5}$ or 10   **37.** 3.75 min   **39.** 24 multiple choice, 6 essay
41. ± 1, ± 2, $\pm \frac{1}{6}$, $\pm \frac{1}{2}$, $\pm \frac{1}{3}$, $\pm \frac{2}{3}$ **43.** E

Pages 228-229   Lesson 4-7
5. 17   **7.** $\frac{16}{3}$ **9.** $-\frac{3}{4}$ **11.** -20   **13.** no real solution
15. 15   **17.** -1   **19.** no real solution   **21.** $x \ge 16$
23. $1.8 \le x \le 5$ **25.** $\frac{3}{2}$ **27.** 0.64 meters   **29.** $7x + y = 2$
31. max: (-1, 7), min: (1, 3), inflection: (0, 5)   **33.** E

Pages 230-232 Chapter 4 Summary and Review

1. -4 **3.** $\frac{1}{2}, -\frac{2}{3}$ **5.** 81; 2 distinct real roots; 4, $-\frac{1}{2}$ **7.** -63;
no real roots; $\frac{-1 \pm 3i\sqrt{7}}{8}$ **9.** 0; yes **11.** 4; no **13.** 3; 1 pos;
2 or 0 neg; -2, -4, 7 **15.** 4; 2 or 0 pos, 2 or 0 neg; 2, -2, 3, -3
17. -1 and 0, 3 and 4 **19.** -2 and -1, 0 and 1, 1 and 2
21. -6, 1 **23.** $x \le -1, x \ge \frac{4}{3}$ **25.** 23 **27.** 1 **29.** Substitute
values for x and h(x); 30 ≠ -460.3; the ball hits the ground
about 337 feet from the plate.

Page 233 Chapter 4 Test

1. 1, 4 **2.** $-\frac{1}{3}, 0, \frac{3}{2}$ **3.** $\frac{5 \pm i\sqrt{7}}{4}$ **4.** $\frac{80}{7}$ **5.** $-3 \le y \le \frac{5}{3}$
6. $x < -17, -2 < x < 0$ **7.** 11 **8.** 7 **9.** $m < -7$
10.

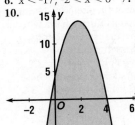

11. $2x^2 + x + 5$, R6
12. $x^3 - 6x^2 - 7x + 60$
13. 9; no **14.** 0; yes
15. 3; 1 pos, 2 or 0 neg; $\frac{1}{2}, -\frac{1}{3}, -2$
16. 4; 1 pos, 3 or 1 neg; -1, -1
17. 4; 1 pos, 2 or 0 neg; 0, 9
18. 3; 2 or 0 pos, 1 neg;
$-\frac{1}{2}, \frac{3}{2}, \frac{7}{2}$ **19.** $\frac{3}{z+2} - \frac{1}{2z-3}$
20. $\frac{6}{x+1} - \frac{3}{x+2} + \frac{4}{x-1}$

21. 3.8, -0.8 **22.** -1.3 **23.** -5 **24.** 5 cm by 12 cm by 2 cm
25. 40 km/h **BONUS** $3 < k < 20$

Pages 234-235 Unit 1 Review

1. {(-1, -2), (0, 1), (1, 4), (2, 7), (3, 10)}; yes **3.** $2x + 5, 2x + 2$
5. $4x^2 - 16x - 9, 2x^2 - 54$ **7.** 0
9.

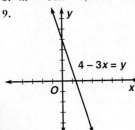

11.

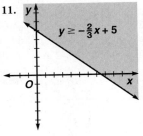

$y \ge -\frac{2}{3}x + 5$

$4 - 3x = y$

13. $\sqrt{58}, y = -\frac{7}{3}x - 5$ **15.** $3x - y + 7 = 0$ **17.** (1, -4)
19. (2, -2) **21.** $\begin{bmatrix} 16 & -2 \\ 1 & -13 \end{bmatrix}$ **23.** $\begin{bmatrix} 9 & 45 \\ -90 & 20 \end{bmatrix}$ **25.** 19 **27.** 1, $\frac{5}{12}, \frac{1}{4}$
29. (0, 0), $\left(0, \frac{1}{2}\right)$, (1, 0); -2; -5 **31.** odd **33.** neither
35.

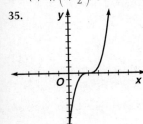

35. moved 2 units right
37. $y = \sqrt[3]{x - 2} + 1$; yes
39. hole: x = -3
41.

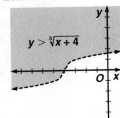

$y > \sqrt[3]{x + 4}$

43. $y = 9x + 3$ **45.** (1, 2), min; (-1, 6), max; (0, 4), inf
47. jump **49.** $\frac{1 \pm \sqrt{11}}{2}$ **51.** $x < \frac{1}{3}, x > 1$ **53.** $-8 \le x \le 1$
55. 11 **57.** 3 or 1; 1; none **59.** -1, -3.6, 0.6

CHAPTER 5   THE TRIGONOMETRIC FUNCTIONS

Pages 244-246 Lesson 5-1

5. IV **7.** II **9.** II **11.** II **13.** $\frac{\pi}{10}$ **15.** $\frac{\pi}{180}$ **17.** 180°
19. -210° **21.** yes **23.** yes **25.** III **27.** III **29.** IV **31.** IV
33. $-\frac{5\pi}{6}$ **35.** $\frac{7\pi}{12}$ **37.** $-\frac{5\pi}{2}$ **39.** $-\frac{125\pi}{18}$ **41.** -200°32'
43. -90° **45.** -105° **47.** 510° **53.** $\frac{2\pi}{5}$ **55.** 30° **57.** 60°
59. $\frac{\pi}{3}$ **61.** -1.935 **65a.** 3700 revolutions per minute
b. 7400π radians per minute **67.** $\frac{5}{8}$ **69.** (1, 3) **71.** 50
bicycles, 75 tricycles **73.** yes **75.** 23

Pages 251-254 Lesson 5-2

7. $\frac{5\pi}{2}$ **9.** 88 **11.** 2 **13.** 9.6 m² **15.** 530.1 cm² **17.** $\frac{20\pi}{3}$ cm
19. 4π cm **21.** $\frac{5\pi}{12}$ in. **23.** $\frac{191\pi}{40}$ in. **25.** 100.3° **27.** 89.5°
29. 17.0 in² **31.** 38.1 km² **33.** 20.5 yd² **35.** 157.5 m
37. 12.2 in., 2.4 in. **41.** 8 feet **43.** A, 0.39 ft²; B, 1.18 ft²;
C, 1.96 ft² **45.** 5.23 mi **47a.** B, clockwise; C,
counterclockwise **b.** 180 rpm, 75 rpm **49.** 0, no **51.** $-\frac{1}{2}, -\frac{2}{3}$
53. 3 **55.** A

Pages 260-262 Lesson 5-3

7. $\sin \theta = \frac{\sqrt{2}}{2}, \cos \theta = \frac{\sqrt{2}}{2}, \tan \theta = 1, \csc \theta = \sqrt{2}, \sec \theta = \sqrt{2},$
$\cot \theta = 1$ **9.** $-\frac{\sqrt{3}}{2}$ **11.** $-\frac{2\sqrt{3}}{3}$ **13.** II, IV **15.** III
23. $\sin \alpha = \frac{8}{17}, \cos \alpha = \frac{15}{17}, \tan \alpha = \frac{8}{15}, \csc \alpha = \frac{17}{8},$
$\sec \alpha = \frac{17}{15}, \cot \alpha = \frac{15}{8}$ **25.** $\sin \alpha = -\frac{8\sqrt{65}}{65}, \cos \alpha = \frac{\sqrt{65}}{65},$
$\tan \alpha = -8, \csc \alpha = -\frac{\sqrt{65}}{8}, \sec \alpha = \sqrt{65}, \cot \alpha = -\frac{1}{8}$
27. $\sin \alpha = \frac{\sqrt{2}}{2}, \cos \alpha = -\frac{\sqrt{2}}{2}, \tan \alpha = -1, \csc \alpha = \sqrt{2},$
$\sec \alpha = -\sqrt{2}, \cot \alpha = -1$ **29.** $\sin \alpha = 1, \cos \alpha = 0,$
$\tan \alpha = $ undefined, $\csc \alpha = 1, \sec \alpha = $ undefined, $\cot \alpha = 0$
31. $\sin \theta = \frac{\sqrt{3}}{2}, \tan \theta = -\sqrt{3}, \csc \theta = \frac{2\sqrt{3}}{3}, \sec \theta = -2,$
$\cot \theta = -\frac{\sqrt{3}}{3}$ **33.** $\sin \theta = -\frac{\sqrt{6}}{3}, \cos \theta = \frac{\sqrt{3}}{3}, \tan \theta = -\sqrt{2},$
$\csc \theta = -\frac{\sqrt{6}}{2}, \cot \theta = -\frac{\sqrt{2}}{2}$ **35.** + **37.** − **39.** undefined
41. − **49.** $\tan 35° = \frac{d}{250}$ **51.** $\pm\sqrt{5}$ **53.** 4, $-\frac{5}{3}$
55. $x^2 + 2x + 6$ R15 **57.** 3.14 m/s

Page 262 Mid-Chapter Review

1. -135° **5.** 6.1 cm **7.** 8700 ft²

Pages 267-268 Lesson 5-4

5. $\sqrt{3}$ **7.** $-\frac{\sqrt{2}}{2}$ **9.** $\frac{2\sqrt{3}}{3}$ **11.** 0.7431 **13.** 0.5774 **15.** 1
17. $\frac{\sqrt{3}}{2}$ **19.** 2 **21.** undefined **23.** $-\frac{1}{2}$ **25.** 0
27. -0.1763 **29.** -2.6695 **31.** -0.1584 **33.** -2.0809
35. 0.8288 **37.** 1.52 **39.** $y = -\frac{1}{2}x + 6$ **41.** $\frac{9}{2}$ **43.** 25π m/h
45. C

Pages 273-275 Lesson 5-5

5. $\tan 76° = \frac{a}{13}$ **7.** $\tan A = \frac{21.2}{9}$ **9.** $\tan 49°13' = \frac{10}{b}$
11. $\tan A = \frac{7}{12}$ **13.** 7.0 cm, 5.9 cm **15.** $B = 49°$,
$c = 9.9, a = 6.5$ **17.** $B = 67°38', c = 23.8, a = 9.1$
19. $B = 45°, a = 7, b = 7$ **21.** $B = 52°45', a = 8.4, c = 13.8$

23. $B = 34°5'$, $a = 13.3$, $b = 9.0$ **25.** 10.80 cm, 9.18 cm
27. 10.57 cm **29.** 4.47 m **31.** about 40° **33.** 29°32'
35. no **37.** I, III **39.** C

Pages 280-281 Lesson 5-6

5. $\dfrac{2.8}{\sin 61°} = \dfrac{a}{\sin 53°}$ **7.** $\dfrac{16}{\sin 42°} = \dfrac{12}{\sin C}$ **9.** one;
$A = 15°25'$, $B = 147°35'$, $b = 20.2$ **11.** none **13.** two;
$B = 50°19'$ or $129°41'$, $C_1 = 91°41'$, $c_1 = 13.0$, $C_2 = 12°19'$,
$c_2 = 2.8$ **15.** none **17.** one; $C = 80°$, $a = 13.1$, $b = 17.6$
19. none **21.** none **23.** none

25.
$$\frac{\sin A}{a} = \frac{\sin C}{c}$$
$$\frac{\sin A}{\sin C} = \frac{a}{c}$$
$$\frac{\sin A}{\sin C} - 1 = \frac{a}{c} - 1$$
$$\frac{\sin A - \sin C}{\sin C} = \frac{a - c}{c}$$

27.
$$\frac{\sin A}{a} = \frac{\sin B}{b}$$
$$\frac{a}{\sin A} = \frac{b}{\sin B}$$
$$\frac{a}{b} = \frac{\sin A}{\sin B}$$

31. 201.1 feet **33.** $45° + 360k°$, where k is an integer
35. 12.7 meters

Pages 286-287 Lesson 5-7

5. cosines; $A = 53°35'$, $B = 59°33'$, $C = 66°52'$ **7.** sines;
$C = 30°57'$, $B = 109°3'$, $b = 14.7$ **9.** cosines; $b = 18.5$,
$C = 79°3'$, $A = 40°57'$ **11.** sines; $A = 30°13'$, $C = 104°19'$,
$c = 23.1$ **13.** 247.3 miles **15.** $a = 7.8$, $B = 44°13'$,
$C = 84°47'$ **17.** $A = 44°25'$, $B = 57°7'$, $C = 78°28'$
19. $A = 21°47'$, $B = 120°$, $C = 38°13'$ **21.** $A = 51°50'$,
$B = 70°53'$, $C = 57°17'$ **23.** $b = 12.3$, $A = 39°19'$,
$C = 20°41'$ **25.** 101.1 cm, 76.9 cm **27.** 349.7 nautical
miles **29.** $\dfrac{5}{8}$ **31.** 30° **33.** 172.7 yd

Pages 293-295 Lesson 5-8

5. $K = \dfrac{1}{2} \cdot 20^2 \cdot \dfrac{\sin 45° \cdot \sin 30°}{\sin 105°}$; 73.2

7. $K = \dfrac{1}{2} \cdot 16 \cdot 12 \sin 43°$; 65.5

9. $K = \dfrac{1}{2} \cdot 12^2 \cdot \dfrac{\sin 30° \cdot \sin 15°}{\sin 135°}$; 13.2 **11.** 238.3 **13.** 6.4
15. 11,486.3 **17.** 21.1 **19.** 13,533.9 **21.** 474.9 **23.** 1.1
25. 13.6 **27.** 83.1 cm^2 **29.** 116.5 cm^2 **33a.** 302 in^2
b. 1208 ft^2 **35a.** 17 ft^2 **b.** 146 ft^3 **37.** y-axis
39. 139,000 cm/s **41.** A

Pages 296-298 Chapter 5 Summary and Review

1. 60° **3.** 240° **5.** $-\dfrac{7\pi}{4}$ **7.** $\dfrac{\pi}{4}$ **9.** $\dfrac{\pi}{3}$ **11.** $\dfrac{\pi}{6}$ **13.** $\dfrac{45\pi}{4}$
15. $\dfrac{25\pi}{4}$ **17.** $\dfrac{\sqrt{2}}{2}$ **19.** -4 **21.** $-\dfrac{\sqrt{2}}{2}$ **23.** $\dfrac{2\sqrt{3}}{3}$ **25.** 1
27. $B = 27°$, $c = 10.9$, $b = 4.9$ **29.** $A = 7°$, $c = 5.6$, $a = 0.7$
31. two; $C = 47°33'$, $B = 93°45'$, $b = 274.5$; $C = 132°27'$,
$B = 8°51'$, $b = 42.3$ **33.** two; $B = 37°18'$, $C = 113°42'$,
$c = 22.7$; $B = 142°42'$, $C = 8°18'$, $c = 3.6$ **35.** $a = 36.9$,
$B = 57°24'$, $C = 71°36'$ **37.** $A = 30°31'$, $B = 36°53'$,
$C = 112°36'$ **39.** 471.7 **41.** 78.1 **43.** 4.3 mph

Page 299 Chapter 5 Test

1. $\dfrac{3\pi}{4}$ **2.** -36° **3.** $\dfrac{8\pi}{3}$ **4.** 105° **5.** 16.5 cm **6.** 52.0 cm^2
7. 42.0 cm^2 **8.** $\dfrac{1}{2}$ **9.** $\sqrt{3}$ **10.** $-\sqrt{2}$ **11.** $\dfrac{\sqrt{3}}{2}$ **12.** 1
13. $B = 13°$, $c = 186.7$, $a = 181.9$ **14.** $A = 67°23'$,
$B = 22°37'$, $b = 5$ **15.** $A = 58°$, $b = 7.4$, $a = 11.9$ **16.** one;
$A = 44°46'$, $B = 37°14'$, $b = 55.0$ **17.** none **18.** one;
$A = 60°$, $B = 27°48'$, $C = 92°12'$ **19.** one; $b = 17.9$,
$C = 78°41'$, $A = 54°19'$ **20.** 172.7 **21.** 27.5 **22.** 656
hours or 27.3 days **23.** 61.1 meters **24.** 183.0 miles
25. 93.2 cm **BONUS** 30° or $\dfrac{\pi}{6}$

CHAPTER 6 GRAPHS AND INVERSES OF THE TRIGONOMETRIC FUNCTIONS

Pages 305-307 Lesson 6-1

5. 0 **7.** 0 **9.** 0 **11.** 0 **13.** -1 **15.** $45° + 180k°$, k is any
integer **17.** $90° + 180k°$, k is any integer **19.** 1 **21.** 1
23. -1 **25.** 0 **27.** undefined **29.** $180° + 360k°$, k is any
integer **31.** $135° + 180k°$, k is any integer **33.** $135° +
180k°$, k is any integer **35.** real numbers; $-1 \leq y \leq 1$
37. real numbers, except $\theta = k \cdot 180° + 90°$; $y \leq -1$, $y \geq 1$,
k is any integer **39.** real numbers, except $\theta = k \cdot 180°$;
$y \leq -1$, $y \geq 1$, k is any integer

41.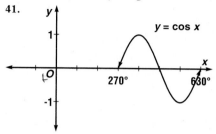

45.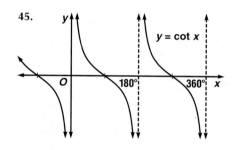

53a. $I_{eff} = 110\sqrt{2} \sin (120\pi t)$

53b.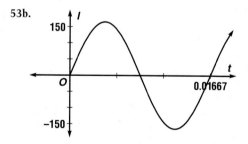

53c. $t = \dfrac{1}{120}k$, where k is any integer **55.** $y = -\dfrac{1}{3}x + \dfrac{11}{3}$
57. x-axis **59.** $x = -9$ **61.** No solution exists. **63.** 108
square units

Pages 315-318 Lesson 6-2

5. 2, 72°, 0° **7.** none, 90°, 90° **9.** 3, 360°, 90° **11.** 243,
24°, $2\dfrac{2}{3}$° **13.** $y = \pm 4 \cos\left(\dfrac{1}{2}\theta - \dfrac{\pi}{4}\right)$ **15.** 2, 360°, 0°
17. 110, 18°, 0° **19.** 7, 60°, 0° **21.** $\dfrac{1}{4}$, 720°, 0°
23. 6, -360°, 180° **25.** $y = \pm 5 \sin(x - 60°)$
27. $y = \pm 17 \sin(8x + 480°)$ **29.** $y = \pm 7 \sin\left(\dfrac{8}{5}x + 144°\right)$
31. $y = \pm 3 \cos(2x - 240°)$ **33.** $y = \pm\dfrac{7}{3} \cos\left(\dfrac{12}{5}x - 648°\right)$

35.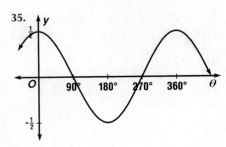

39.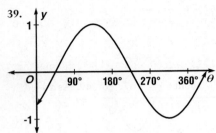

43a. $x = 135°$ **b.** $x = 30°$ **c.** $x = 90°$ **d.** $x = \left(\dfrac{90 + c}{k}\right)°$
47a. 323, 1388
47b.
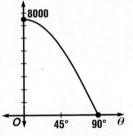

49a.
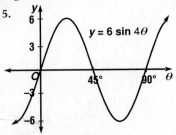

b. 7500 **51.** $\begin{bmatrix} 2 & -1 & 0 \\ 3 & -1 & 1 \\ 2 & 8 & -5 \end{bmatrix}$ **53.** 1.0 **55.** $\sin \alpha = -\dfrac{3\sqrt{58}}{58}$,

$\cos \alpha = \dfrac{7\sqrt{58}}{58}$, $\tan \alpha = -\dfrac{3}{7}$, $\csc \alpha = -\dfrac{\sqrt{58}}{3}$, $\sec \alpha = \dfrac{\sqrt{58}}{7}$,

$\cot \alpha = -\dfrac{7}{3}$ **57.** C

Pages 326-327   Lesson 6-3

5.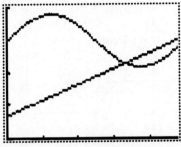

9.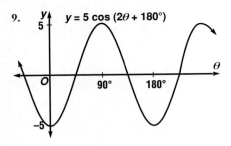

13.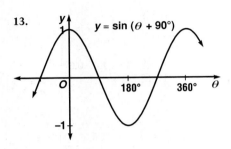

17.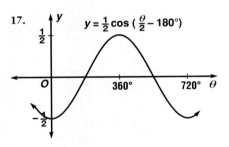

21.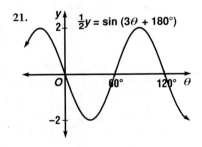

25.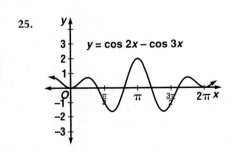

29.

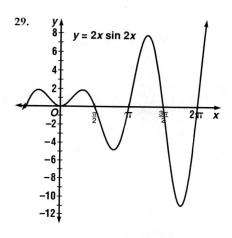
$y = 2x \sin 2x$

33. There is a 45° phase difference. **35.** (1.4, -0.04)
37a. 10π radians per second **b.** 15.7 ft/s
39. $y = \pm 4 \cos(2x - 40°)$

Pages 331-333 Lesson 6-4

3. $\theta = \arcsin x$ **5.** $y = \arctan -3$ **7.** $x = \arccos y$ **9.** 0°,
180°, 360° **11.** 60°, 120° **13.** 1 **15.** $\theta = \arcsin n$
17. $\delta = \arctan \frac{3}{2}$ **19.** $\theta = \arccos y$ **21.** 90°, 270° **23.** 30°,
210° **25.** 0°, 180° **27.** 25°, 205° **29.** 90°, 270° **31.** $\frac{5}{4}$
33. 2 **35.** $\sqrt{3}$ **37.** $\frac{\sqrt{2} + 1}{2}$ **39.** $\frac{\sqrt{3}}{3}$
41. $\arccos \frac{\sqrt{3}}{2} + \arcsin \frac{\sqrt{3}}{2} \stackrel{?}{=} \arctan 1 + \text{arccot } 1$

$$30° + 60° \stackrel{?}{=} 45° + 45°$$
$$90° = 90°$$

43. $\tan^{-1} 1 + \cos^{-1} \frac{\sqrt{3}}{2} \stackrel{?}{=} \sin^{-1} \frac{1}{2} + \sec^{-1} \sqrt{2}$

$$45° + 30° \stackrel{?}{=} 30° + 45°$$
$$75° = 75°$$

45. $\arcsin \frac{3}{5} + \arccos \frac{15}{17} \stackrel{?}{=} \arctan \frac{77}{36}$

$$36.9° + 28.1° \stackrel{?}{=} 65°$$
$$65° = 65°$$

47a. 82° **b.** 0 km; It is the north or south pole. **49.** about
11.5° **51.** -11 **53.** 2 or $\frac{1}{3}$ **55.** $a = 11.9$, $B = 70°$, $b = 32.9$
57.

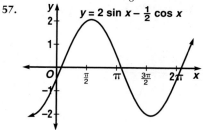
$y = 2 \sin x - \frac{1}{2} \cos x$

Page 333 Mid-Chapter Review

1. $n \cdot 180°$ **3.** $n \cdot 360° + 90°$ **5.** none, 180°, -60°
7.

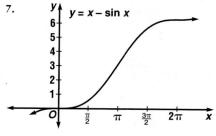
$y = x - \sin x$

9.

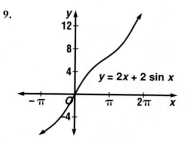
$y = 2x + 2 \sin x$

Pages 337-338 Lesson 6-5

5. -60° **7.** 0° **9.** 30° **11.** 90° **13.** $-\frac{4}{3}$ **15.** $-\frac{24}{25}$ **17.** 150°
19. 36°52' **21.** 60° **23.** 0° **25.** $\frac{1}{2}$ **27.** $\frac{\sqrt{3}}{2}$ **29.** $\frac{\sqrt{3}}{2}$
31. $-\frac{\sqrt{3}}{2}$ **33.** $\frac{\sqrt{3}}{2}$ **35.** 0.2669 **37.** $-\frac{1}{2}$ **39.** not possible
41. $-\frac{1}{2}$ **43.** $\frac{33}{56}$ **45.** about 85.4° **47.** no **49.** As $x \to \infty$,
$y \to \infty$, as $x \to -\infty$, $y \to -\infty$. **51.** no solution **53.** $\frac{\sqrt{3}}{2}$

Pages 342-344 Lesson 6-6

5. all real numbers except $90° + n \cdot 180°$; all real numbers
7. $-1 \le x \le 1$; all real numbers **9.** all real numbers;
$-90° < y < 90°$ **11.** true **13.** all real numbers; $-1 \le y \le 1$
15. $-90° < x < 90°$; all real numbers **17.** $-1 \le x \le 1$;
$-90° \le y \le 90°$ **19.** $y = \text{Tan } x$ **21.** $y = \text{Arcsin } x$
23. $y = \frac{1}{2} \text{Tan } x$ **25.** true **27.** false, $x = 1$
29. false, $x = \frac{\pi}{2}$ **31.**

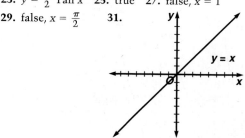
$y = x$

35a. about -5.43 **b.** about 63°, about -63°
37. $27x^3 - 1$; $3x^3 - 3$ **39.** $\frac{37}{8}$ **41.** $\frac{\sqrt{3}}{2}$

Pages 348-351 Lesson 6-7

5. 8, $\frac{\pi}{2}$, $\frac{2}{\pi}$, $\frac{\pi}{4}$ **7.** 4, $\frac{2\pi}{3}$, $\frac{3}{2\pi}$, $-\frac{\pi}{4}$ **9.** $y = 5 \sin \pi t$
11. $y = 7 \cos \frac{\pi t}{5}$ **13.** $y = -5 \cos \frac{\pi t}{6}$ **15.** 15, 6, $\frac{1}{6}$, 0
17. 150, $\frac{1}{40}$, 40, 0 **19.** 25, $\frac{\pi}{4}$, $\frac{4}{\pi}$, 0 **21.** $y = 0.5 \sin 2\pi t$
23. $y = 10 \cos 4\pi t$ **25a.** $\frac{30}{7}$ seconds **b.** $\left[3.5 \cos \left(\frac{7}{15} \pi t \right), 3.5 \sin \left(\frac{7}{15} \pi t \right) \right]$ **c.** (3.20, -1.42) **d.** (-3.50, 0.18)
e. $\left[3.5 \cos \left(\frac{7}{15} \pi t + \frac{\pi}{2} \right), 3.5 \sin \left(\frac{7}{15} \pi t + \frac{\pi}{2} \right) \right]$ **f.** ± 2.72
27a. pulled down **b.** 8 ft **c.** 2 ft **d.** $t = \frac{1}{2}$ second
e. 8 ft **f.** 2 seconds **g.** 3 ft **h.** $\frac{1}{2}$ cycle/second **i.** 5 ft

29.

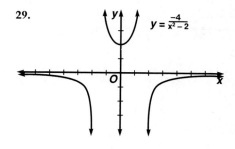
$y = \frac{-4}{x^2 - 2}$

31. yes   **33.** 21

Pages 352-354   Chapter 6 Summary and Review

1. 0   **3.** undefined

5.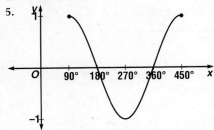

7.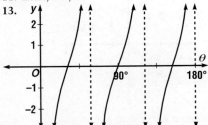

9. 15, 240°, -60°
11. none, 36°, 0°
13.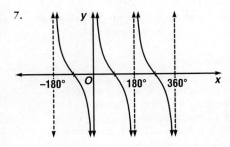

17. $\alpha = \arcsin y$
19. $\theta = \arccos n$
21. $\frac{1}{2}$ **23.** $\frac{\sqrt{2}}{2}$
25. $\frac{\sqrt{3}}{2}$ **27.** $\frac{\sqrt{3}}{2}$
29. 0
31. $y = \text{Arccsc } x$

31.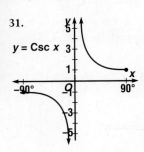

$y = \text{Csc } x$

33. $y = \sec x$

33.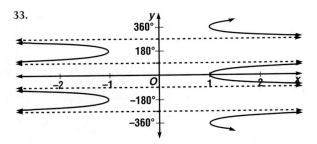

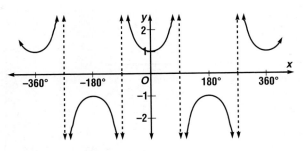

35. 4 cycles/second   **37a.** $y = -38 \cos \frac{\pi}{3} t$
b. $y = -38 \cos \frac{\pi}{3} t - 84$ **c.** 122 feet below the ledge

Page 355   Chapter 6 Test

1. $180k°$, k is any integer   **2.** $270° + 360k°$, k is any integer
3. $45° + 360k°$ and $315° + 360k°$, k is any integer

4.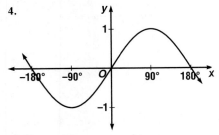

5.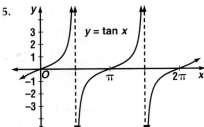

$y = \tan x$

6. 3, 90°, 0°
7. 110, 24°, $2\frac{2}{3}°$
8. 10, 360°, 180°

9.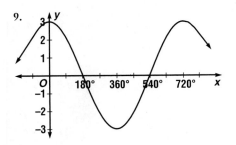

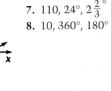

990

10.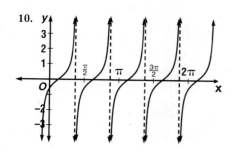

11.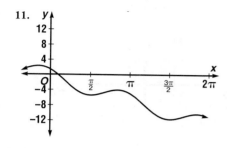

12. $\frac{1}{2}$ **13.** $\frac{12}{5}$ **14.** 0   **15.** $\frac{\sqrt{3}}{2}$ **16.** $\frac{3\sqrt{10}}{10}$ **17.** $\frac{2\sqrt{5}}{5}$
18. $y = \text{Csc } x$

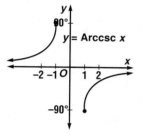 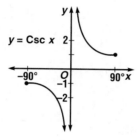

19. $y = \arctan x$

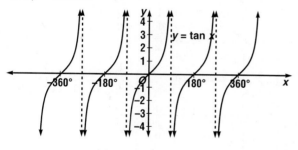

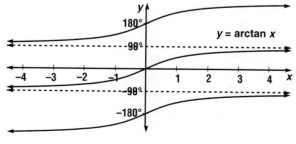

20. $x = 10 \cos 2t$ **21.** $y = \sin 2t$ **22.** 10   **23.** $\frac{1}{\pi}$
24. 5-foot mark   **25.** 13.3°   **BONUS** even

Pages 362-363   Lesson 7-1
5. $\frac{4}{3}$ **7.** $\frac{1}{3}$ **9.** $\frac{\sqrt{5}}{2}$ **11.** $\sin 40°$ **13.** $-\cos 40°$ **15.** $\sec^2 x$
17. 1   **19.** $\sec A$ **21.** undefined   **23.** 1   **25.** $\frac{\sqrt{7}}{2}$ **27.** $\frac{\sqrt{15}}{2}$
29. $\frac{3\sqrt{5}}{5}$ **31.** $\frac{\sqrt{53}}{2}$ **33.** $\frac{\sqrt{5}}{5}$ **35.** $-\cos 22°$ **37.** $-\sin 72°$
39. $-\sec 30°$ **41.** $1 + \sin\theta$ **43.** $\cos A$ **45.** 1   **47.** $1 - \sin A$
49. $\sin^2 A$ **51.** 1   **53.** $\pm\sqrt{1 - \sin^2\theta}$ **55.** $\pm\frac{\sqrt{1 - \cos^2\theta}}{\cos\theta}$
61. neither   **63.** 1; 3 or 1   **65.** 25.4 units and 54.4 units
67. 23.1 units^2 **69.** 15

Pages 366-368   Lesson 7-2
3. $\tan^2 x \cos^2 x \overset{?}{=} 1 - \cos^2 x$ **5.** $\tan\beta \csc\beta \overset{?}{=} \sec\beta$
$\frac{\sin^2 x}{\cos^2 x} \cdot \cos^2 x \overset{?}{=} 1 - \cos^2 x$ $\frac{\sin\beta}{\cos\beta} \cdot \frac{1}{\sin\beta} \overset{?}{=} \sec\beta$
$\sin^2 x \overset{?}{=} 1 - \cos^2 x$ $\frac{1}{\cos\beta} \overset{?}{=} \sec\beta$
$1 - \cos^2 x = 1 - \cos^2 x$ $\sec\beta = \sec\beta$

7. $\sin\theta \sec\theta \cot\theta \overset{?}{=} 1$ **9.** $\sec x = 1$ **11.** $\tan x = 2$
$\sin\theta \cdot \frac{1}{\cos\theta} \cdot \frac{\cos\theta}{\sin\theta} \overset{?}{=} 1$
$\frac{\sin\theta\cos\theta}{\sin\theta\cos\theta} \overset{?}{=} 1$
$1 = 1$

13. $\frac{1}{\sec^2\theta} + \frac{1}{\csc^2\theta} \overset{?}{=} 1$ **15.** $\frac{\sin A}{\csc A} + \frac{\cos A}{\sec A} \overset{?}{=} 1$
$\frac{1}{\frac{1}{\cos^2\theta}} + \frac{1}{\frac{1}{\sin^2\theta}} \overset{?}{=} 1$ $\frac{\sin A}{\frac{1}{\sin A}} + \frac{\cos A}{\frac{1}{\cos A}} \overset{?}{=} 1$
$\cos^2\theta + \sin^2\theta \overset{?}{=} 1$ $\sin^2 A + \cos^2 A \overset{?}{=} 1$
$1 = 1$ $1 = 1$

17. $\frac{1 + \tan\gamma}{1 + \cot\gamma} \overset{?}{=} \frac{\sin\gamma}{\cos\gamma}$
$\frac{1 + \frac{\sin\gamma}{\cos\gamma}}{1 + \frac{\cos\gamma}{\sin\gamma}} \overset{?}{=} \frac{\sin\gamma}{\cos\gamma}$
$\frac{\frac{\sin\gamma + \cos\gamma}{\cos\gamma}}{\frac{\sin\gamma + \cos\gamma}{\sin\gamma}} \overset{?}{=} \frac{\sin\gamma}{\cos\gamma}$
$\frac{\sin\gamma + \cos\gamma}{\cos\gamma} \cdot \frac{\sin\gamma}{\sin\gamma + \cos\gamma} \overset{?}{=} \frac{\sin\gamma}{\cos\gamma}$
$\frac{\sin\gamma}{\cos\gamma} = \frac{\sin\gamma}{\cos\gamma}$

19. $\cos^2 x + \tan^2 x \cos^2 x \overset{?}{=} 1$
$\cos^2 x + \frac{\sin^2 x}{\cos^2 x} \cdot \cos^2 x \overset{?}{=} 1$
$\cos^2 x + \sin^2 x \overset{?}{=} 1$
$1 = 1$

21. $1 - \cot^4 x \overset{?}{=} 2\csc^2 x - \csc^4 x$
$1 - \cot^4 x \overset{?}{=} \csc^2 x(2 - \csc^2 x)$
$1 - \cot^4 x \overset{?}{=} (1 + \cot^2 x)(2 - (1 + \cot^2 x))$
$1 - \cot^4 x \overset{?}{=} (1 + \cot^2 x)(1 - \cot^2 x)$
$1 - \cot^4 x = 1 - \cot^4 x$

23. $\frac{\sec x - 1}{\sec x + 1} + \frac{\cos x - 1}{\cos x + 1} \overset{?}{=} 0$
$\frac{\frac{1}{\cos x} - 1}{\frac{1}{\cos x} + 1} + \frac{\cos x - 1}{\cos x + 1} \overset{?}{=} 0$
$\frac{1 - \cos x}{1 + \cos x} + \frac{\cos x - 1}{\cos x + 1} \overset{?}{=} 0$
$\frac{1 - \cos x + \cos x - 1}{1 + \cos x} \overset{?}{=} 0$
$0 = 0$

25.
$$\frac{\cos x}{1 + \sin x} + \frac{\cos x}{1 - \sin x} \stackrel{?}{=} 2 \sec x$$
$$\frac{\cos x(1 - \sin x) + \cos x(1 + \sin x)}{1 - \sin^2 x} \stackrel{?}{=} 2 \sec x$$
$$\frac{\cos x(1 - \sin x) + \cos x(1 + \sin x)}{\cos^2 x} \stackrel{?}{=} 2 \sec x$$
$$\frac{(1 - \sin x) + (1 + \sin x)}{\cos x} \stackrel{?}{=} 2 \sec x$$
$$\frac{2}{\cos x} \stackrel{?}{=} 2 \sec x$$
$$2 \sec x = 2 \sec x$$

27. $1 + \sec^2 x \sin^2 x \stackrel{?}{=} \sec^2 x$
$$1 + \frac{1}{\cos^2 x} \cdot \sin^2 x \stackrel{?}{=} \sec^2 x$$
$$1 + \tan^2 x = \sec^2 x$$
$$\sec^2 x = \sec^2 x$$

39a.
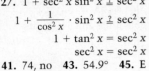
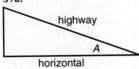
highway
A
horizontal

41. 74, no **43.** 54.9° **45.** E

Pages 375-376 Lesson 7-3

5. $\frac{1}{2}$ **7.** $-\frac{\sqrt{3}}{2}$ **9.** 1

11. $\sin \theta \stackrel{?}{=} \sin(180° - \theta)$
$$\sin \theta \stackrel{?}{=} \sin 180° \cos \theta - \cos 180° \sin \theta$$
$$\sin \theta \stackrel{?}{=} 0 \cdot \cos \theta - (-1) \sin \theta$$
$$\sin \theta = \sin \theta$$

13. $\cot x \stackrel{?}{=} \tan(270° - x)$
$$\cot x \stackrel{?}{=} \frac{\sin(270° - x)}{\cos(270° - x)}$$
$$\cot x \stackrel{?}{=} \frac{\sin 270° \cos x - \sin x \cos 270°}{\cos 270° \cos x + \sin 270° \sin x}$$
$$\cot x \stackrel{?}{=} \frac{-1 \cdot \cos x - \sin x \cdot 0}{0 \cdot \cos x + (-1) \sin x}$$
$$\cot x \stackrel{?}{=} \frac{-\cos x}{-\sin x}$$
$$\cot x = \cot x$$

15. $\frac{\sqrt{2} - \sqrt{6}}{4}$ **17.** $2 + \sqrt{3}$ **19.** $\sqrt{3} - 2$ **21.** $\sqrt{3} - 2$

23. $\frac{-\sqrt{6} - \sqrt{2}}{4}$ **25.** $-\frac{16}{65}$ **27.** $\frac{416}{425}$ **29.** $\frac{87}{425}$

31.
$$\cos(90° + \theta) \stackrel{?}{=} -\sin \theta$$
$$\cos 90° \cos \theta - \sin 90° \sin \theta \stackrel{?}{=} -\sin \theta$$
$$0 - 1 \cdot \sin \theta \stackrel{?}{=} -\sin \theta$$
$$-\sin \theta = -\sin \theta$$

33.
$$\sin\left(\frac{\pi}{2} + x\right) \stackrel{?}{=} \cos x$$
$$\sin \frac{\pi}{2} \cos x + \cos \frac{\pi}{2} \sin x \stackrel{?}{=} \cos x$$
$$1 \cdot \cos x + 0 \cdot \sin x \stackrel{?}{=} \cos x$$
$$\cos x = \cos x$$

35. $\tan(\pi - \theta) \stackrel{?}{=} -\tan \theta$
$$\frac{\tan \pi - \tan \theta}{1 + \tan \pi \tan \theta} \stackrel{?}{=} -\tan \theta$$
$$\frac{0 - \tan \theta}{1 + 0 \cdot \tan \theta} \stackrel{?}{=} -\tan \theta$$
$$-\tan \theta = -\tan \theta$$

37.
$$\frac{\sin(\beta - \alpha)}{\sin \alpha \sin \beta} \stackrel{?}{=} \cot \alpha - \cot \beta$$
$$\frac{\sin \beta \cos \alpha - \cos \beta \sin \alpha}{\sin \alpha \sin \beta} \stackrel{?}{=} \cot \alpha - \cot \beta$$
$$\frac{\cos \alpha}{\sin \alpha} - \frac{\cos \beta}{\sin \beta} \stackrel{?}{=} \cot \alpha - \cot \beta$$
$$\cot \alpha - \cot \beta = \cot \alpha - \cot \beta$$

39. $\cos(30° - x) + \cos(30° + x) \stackrel{?}{=} \sqrt{3} \cos x$
$$(\cos 30° \cos x + \sin 30° \sin x) +$$
$$(\cos 30° \cos x - \sin 30° \sin x) \stackrel{?}{=} \sqrt{3} \cos x$$
$$\frac{\sqrt{3}}{2} \cdot \cos x + \frac{\sqrt{3}}{2} \cdot \cos x \stackrel{?}{=} \sqrt{3} \cos x$$
$$\sqrt{3} \cos x = \sqrt{3} \cos x$$

41. $\cot(\alpha + \beta) = \frac{1}{\tan(\alpha + \beta)}$
$$= \frac{1 - \tan \alpha \tan \beta}{\tan \alpha + \tan \beta}$$

$$= \frac{1 - \frac{1}{\cot \alpha \cot \beta}}{\frac{1}{\cot \alpha} + \frac{1}{\cot \beta}}$$
$$= \frac{\cot \alpha \cot \beta - 1}{\cot \beta + \cot \alpha}$$

45. $\tan \theta = \frac{m_2 - m_1}{1 - m_1 m_2}$

47. $y = 2x - 7$ **49.** -1

51. $2 \sec^2 x \stackrel{?}{=} \frac{1}{1 + \sin x} + \frac{1}{1 - \sin x}$
$$2 \sec^2 x \stackrel{?}{=} \frac{(1 - \sin x) + (1 + \sin x)}{(1 + \sin x)(1 - \sin x)}$$
$$2 \sec^2 x \stackrel{?}{=} \frac{2}{1 - \sin^2 x}$$
$$2 \sec^2 x \stackrel{?}{=} \frac{2}{\cos^2 x}$$
$$2 \sec^2 x = 2 \sec^2 x$$

Pages 381-383 Lesson 7-4

5. $\frac{24}{7}$ **7.** $\frac{\sqrt{10}}{10}$ **9.** $\frac{1}{3}$ **11.** $\frac{\sqrt{2 + \sqrt{2}}}{2}$

13. $\frac{1}{2} \cdot \sin 2A \stackrel{?}{=} \frac{\tan A}{1 + \tan^2 A}$
$$\frac{1}{2} \cdot \sin 2A \stackrel{?}{=} \frac{\frac{\sin A}{\cos A}}{\sec^2 A}$$
$$\frac{1}{2} \cdot \sin 2A \stackrel{?}{=} \frac{\sin A}{\cos A} \cdot \cos^2 A$$
$$\frac{1}{2} \cdot \sin 2A \stackrel{?}{=} \frac{2 \sin A \cos A}{2}$$
$$\frac{1}{2} \cdot \sin 2A = \frac{1}{2} \cdot \sin 2A$$

15. $\sin 2x \stackrel{?}{=} 2 \cot x \sin^2 x$
$$\sin 2x \stackrel{?}{=} \frac{2 \cos x}{\sin x} \cdot \sin^2 x$$
$$\sin 2x \stackrel{?}{=} 2 \cos x \sin x$$
$$\sin 2x = \sin 2x$$

17. $\frac{120}{169}$ **19.** $\frac{5\sqrt{26}}{26}$ **21.** -5 **23.** $\frac{\sqrt{2 + \sqrt{3}}}{2}$ **25.** $2 - \sqrt{3}$

27. $\frac{\sqrt{2 - \sqrt{2}}}{2}$ **29.** $1 + \cos 2A \stackrel{?}{=} \frac{2}{1 + \tan^2 A}$
$$1 + \cos 2A \stackrel{?}{=} \frac{2}{\sec^2 A}$$
$$1 + \cos 2A \stackrel{?}{=} 2 \cos^2 A$$
$$1 + \cos 2A \stackrel{?}{=} 1 + 2 \cos^2 A - 1$$
$$1 + \cos 2A = 1 + \cos 2A$$

31. $\csc A \sec A \stackrel{?}{=} 2 \csc 2A$
$$\frac{1}{\sin A \cos A} \stackrel{?}{=} 2 \csc 2A$$
$$\frac{2}{2 \sin A \cos A} \stackrel{?}{=} 2 \csc 2A$$
$$\frac{2}{\sin 2A} \stackrel{?}{=} 2 \csc 2A$$
$$2 \csc 2A = 2 \csc 2A$$

33. $\cot X \stackrel{?}{=} \frac{\sin 2X}{1 - \cos 2X}$
$$\cot X \stackrel{?}{=} \frac{2 \sin X \cos X}{1 - (1 - 2 \sin^2 X)}$$
$$\cot X \stackrel{?}{=} \frac{2 \sin X \cos X}{2 \sin^2 X}$$
$$\cot X \stackrel{?}{=} \frac{\cos X}{\sin X}$$
$$\cot X = \cot X$$

35.
$$\sin 2B(\cot B + \tan B) \stackrel{?}{=} 2$$
$$2 \sin B \cos B\left(\frac{\cos B}{\sin B} + \frac{\sin B}{\cos B}\right) \stackrel{?}{=} 2$$
$$2 \sin B \cos B\left(\frac{\cos^2 B + \sin^2 B}{\sin B \cos B}\right) \stackrel{?}{=} 2$$
$$2 = 2$$

37.
$$\cot \frac{\alpha}{2} \stackrel{?}{=} \frac{\sin \alpha}{1 - \cos \alpha}$$
$$\frac{1}{\tan \frac{\alpha}{2}} \stackrel{?}{=} \frac{\sin \alpha}{1 - \cos \alpha}$$

$$\sqrt{\dfrac{1-\cos\alpha}{1+\cos\alpha}} \overset{?}{=} \dfrac{\sin\alpha}{1-\cos\alpha}$$

$$\sqrt{\dfrac{(1-\cos\alpha)^2}{1-\cos^2\alpha}} \overset{?}{=} \dfrac{\sin\alpha}{1-\cos\alpha}$$

$$\sqrt{\dfrac{(1-\cos\alpha)^2}{\sin^2\alpha}} \overset{?}{=} \dfrac{\sin\alpha}{1-\cos\alpha}$$

$$\dfrac{1-\cos\alpha}{\sin\alpha} \overset{?}{=} \dfrac{\sin\alpha}{1-\cos\alpha}$$

$$\dfrac{\sin\alpha}{1-\cos\alpha} = \dfrac{\sin\alpha}{1-\cos\alpha}$$

39.
$$\dfrac{\cos 2A}{1+\sin 2A} \overset{?}{=} \dfrac{\cot A - 1}{\cot A + 1}$$

$$\dfrac{\cos 2A}{1+\sin 2A} \overset{?}{=} \dfrac{\dfrac{\cos A}{\sin A} - 1}{\dfrac{\cos A}{\sin A} + 1}$$

$$\dfrac{\cos 2A}{1+\sin 2A} \overset{?}{=} \dfrac{\cos A - \sin A}{\cos A + \sin A}$$

$$\dfrac{\cos 2A}{1+\sin 2A} \overset{?}{=} \dfrac{\cos^2 A - \sin^2 A}{\cos^2 A + 2\sin A \cos A + \sin^2 A}$$

$$\dfrac{\cos 2A}{1+\sin 2A} \overset{?}{=} \dfrac{\cos 2A}{1 + 2\sin A \cos A}$$

$$\dfrac{\cos 2A}{1+\sin 2A} = \dfrac{\cos 2A}{1+\sin 2A}$$

41.
$$\sin 3\alpha = \sin(2\alpha + \alpha)$$
$$= \sin 2\alpha \cos\alpha + \sin\alpha \cos 2\alpha$$
$$= (2\sin\alpha\cos\alpha)\cos\alpha + \sin\alpha(1 - 2\sin^2\alpha)$$
$$= 2\sin\alpha\cos^2\alpha + \sin\alpha - 2\sin^3\alpha$$
$$= 2\sin\alpha(1 - \sin^2\alpha) + \sin\alpha - 2\sin^3\alpha$$
$$= 2\sin\alpha - 2\sin^3\alpha + \sin\alpha - 2\sin^3\alpha$$
$$= 3\sin\alpha - 4\sin^3\alpha$$

45.
$$n = \dfrac{\sin\left(\dfrac{\alpha+\beta}{2}\right)}{\sin\left(\dfrac{\alpha}{2}\right)}$$

$$= \dfrac{\sqrt{\dfrac{1-\cos(\alpha+\beta)}{2}}}{\sqrt{\dfrac{1-\cos\alpha}{2}}}$$

$$= \sqrt{\dfrac{\dfrac{1-(\cos\alpha\cos\beta - \sin\alpha\sin\beta)}{2}}{\dfrac{1-\cos\alpha}{2}}}$$

$$= \sqrt{\dfrac{\dfrac{1-\cos\alpha\cos\beta + \sin\alpha\sin\beta}{2}}{\dfrac{1-\cos\alpha}{2}}}$$

$$= \sqrt{\dfrac{1-\cos\alpha\cos\beta + \sin\alpha\sin\beta}{1-\cos\alpha}}$$

47. $3x - y - 2 = 0$ **49.** no triangle **51.** $-\dfrac{\sqrt{3}}{3}$

Page 383 Mid-Chapter Review

1. $\dfrac{3}{5}$ **3.** $\cot^2\alpha$ **5.** $\csc\theta \overset{?}{=} \sin\theta(1+\cot^2\theta)$
$$\csc\theta \overset{?}{=} \sin\theta(\csc^2\theta)$$
$$\csc\theta \overset{?}{=} \dfrac{1}{\csc\theta}(\csc^2\theta)$$
$$\csc\theta = \csc\theta$$

7. $-\cos\theta \overset{?}{=} \cos(180° - \theta)$
$$-\cos\theta \overset{?}{=} \cos 180° \cos\theta + \sin 180° \sin\theta$$
$$-\cos\theta \overset{?}{=} (-1)\cos\theta + (0)\sin\theta$$
$$-\cos\theta = -\cos\theta$$

9. $\sin^2\theta \overset{?}{=} \dfrac{1}{2}(1 - \cos 2\theta)$
$$\sin^2\theta \overset{?}{=} \dfrac{1}{2}[1 - (1 - 2\sin^2\theta)]$$
$$\sin^2\theta \overset{?}{=} \dfrac{1}{2}(2\sin^2\theta)$$
$$\sin^2\theta = \sin^2\theta$$

Pages 390-391 Lesson 7-5

3. $-30°$ **5.** $45°$ **7.** $30°$ **9.** $20°, 100°, 140°$ **11.** $0° + 120k°$
13. $0° + 180k°$ **15.** $90° + 180k°, 120° + 360k°, 240° + 360k°$
17. $60°, 120°$ **19.** $150°$ **21.** $30°, 90°, 150°$ **23.** $30°, 90°, 150°$
25. $135°$ **27.** $120°$ **29.** $180°$ **31.** $0°$ **33.** $60°$ **35.** $30°, 150°$
37. $45° + 90k°$ **39.** $270° + 360k°$ **41.** $0° + 180k°, 90° + 360k°$
43. $60° + 360k°, 300° + 360k°$ **47a.** $\approx 6.6°$ **b.** $\approx 7.5°$ **c.** $10°$
49. yes, yes, no **51.** $\left(\dfrac{2}{3}, 7\dfrac{5}{27}\right)$ max, $(2, 6)$ min, $\left(\dfrac{4}{3}, 6\dfrac{16}{27}\right)$

pt of inf **53.** $\dfrac{5}{2}$ **55.** $\dfrac{\sqrt{2 + \sqrt{2 + \sqrt{3}}}}{2}$

Pages 395-397 Lesson 7-6

3. $x + y - 11\sqrt{2} = 0$ **5.** $x - y = 0$ **7.** $\sqrt{3}x + y -$
$4 = 0$ **9.** $-y - 4 = 0$ **11.** $\dfrac{3x}{\sqrt{10}} - \dfrac{y}{\sqrt{10}} - \dfrac{4}{\sqrt{10}} = 0$;
$\dfrac{4}{\sqrt{10}} \approx 1.3$; $\approx 342°$ **13.** $x + y - 5\sqrt{2} = 0$
15. $x + y + 25\sqrt{2} = 0$ **17.** $x + \sqrt{3}y + 16 = 0$
19. $-\dfrac{x}{\sqrt{2}} + \dfrac{y}{\sqrt{2}} - \dfrac{6}{\sqrt{2}} = 0$; $\dfrac{6}{\sqrt{2}} \approx 4.24$; $135°$
21. $\dfrac{x}{\sqrt{2}} + \dfrac{y}{\sqrt{2}} - \dfrac{8}{\sqrt{2}} = 0$; $\dfrac{8}{\sqrt{2}} \approx 5.66$; $45°$
23. $\dfrac{x}{\sqrt{10}} - \dfrac{3y}{\sqrt{10}} - \dfrac{2}{\sqrt{10}} = 0$; $\dfrac{2}{\sqrt{10}} \approx 0.63$; $-71°34'$
25. $x - y + 8 = 0$ **27.** $x + \sqrt{3}y + 2 = 0$ and $x + \sqrt{3}y - 2 = 0$
31a.

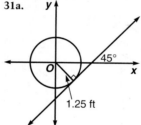

b. $x - y - 1.25\sqrt{2} = 0$
33. no **37.** D

Pages 402-405 Lesson 7-7

3. 0.9 **5.** $\dfrac{16\sqrt{58}}{29} \approx 4.2$ **11.** $27x - 99y + 161 = 0$
13. $x - (1 + \sqrt{2})y + 2 + 5\sqrt{2} = 0$ **15.** $\dfrac{7\sqrt{2}}{5} \approx 1.98$
17. 0; the point is on the line **19.** 2.4 **21.** $\dfrac{4\sqrt{13}}{13} \approx 1.11$
23. $(2\sqrt{2} + \sqrt{5})x + (\sqrt{5} - \sqrt{2})y - 4\sqrt{2} - 8\sqrt{5} = 0$
25. $(\sqrt{2} - \sqrt{17})x + (4\sqrt{2} + \sqrt{17})y - 4\sqrt{17} + 6\sqrt{2} = 0$
27. $(1 - \sqrt{2})x + y - 5 + 5\sqrt{2} = 0$
29. $(6\sqrt{10} - \sqrt{37})x + (\sqrt{10} - 3\sqrt{37})y - 3\sqrt{10} - \sqrt{37} = 0$
31. $x - 5y + 10 + 3\sqrt{26} = 0, x - 5y + 10 - 3\sqrt{26} = 0$
35a. $\dfrac{2\sqrt{10}}{5} \approx 1.26$ units; $\dfrac{\sqrt{10}}{5} \approx 0.63$ units; 0 units
b. $(16, 76)$ **c.** $(16, 72)$ **37.** $y = \dfrac{x - 2}{x(x + 2)(x - 2)}$ **39.** $\cot^2\alpha$
41. A

Pages 406-408 Chapter 7 Summary and Review

1. 2 **3.** $\dfrac{4}{5}$ **5.** $\cos^2 x + \tan^2 x \cos^2 x \overset{?}{=} 1$
$$\cos^2 x + \left(\dfrac{\sin^2 x}{\cos^2 x}\right)\cos^2 x \overset{?}{=} 1$$
$$\cos^2 x + \sin^2 x \overset{?}{=} 1$$
$$1 = 1$$

7. $\dfrac{\sec \theta + 1}{\tan \theta} \stackrel{?}{=} \dfrac{\tan \theta}{\sec \theta - 1}$ **9.** $-\dfrac{1}{2}$ **11.** $\dfrac{\sqrt{6} + \sqrt{2}}{4}$

$\dfrac{\sec \theta + 1}{\tan \theta} \stackrel{?}{=} \dfrac{\tan \theta (\sec \theta + 1)}{\sec^2 \theta - 1}$

$\dfrac{\sec \theta + 1}{\tan \theta} \stackrel{?}{=} \dfrac{\tan \theta (\sec \theta + 1)}{\tan^2 \theta}$

$\dfrac{\sec \theta + 1}{\tan \theta} = \dfrac{\sec \theta + 1}{\tan \theta}$

13. $\cos (90° - \theta) \stackrel{?}{=} \sin \theta$ **15.** $\dfrac{24}{25}$

$\cos 90° \cos \theta + \sin 90° \sin \theta \stackrel{?}{=} \sin \theta$

$(0) \cdot \cos \theta + (1) \cdot \sin \theta \stackrel{?}{=} \sin \theta$

$\sin \theta = \sin \theta$

17. $\dfrac{\sqrt{5}}{5}$ **19.** $-\dfrac{24}{7}$ **21.** $360k°$ **23.** $180k°$

25. $\dfrac{7x}{\sqrt{58}} + \dfrac{3y}{\sqrt{58}} - \dfrac{8}{\sqrt{58}} = 0$; $\dfrac{8}{\sqrt{58}} \approx 1.1$; $\approx 23°$

27. $\dfrac{9x}{\sqrt{106}} + \dfrac{5y}{\sqrt{106}} - \dfrac{3}{\sqrt{106}} = 0$; $\dfrac{3}{\sqrt{106}} \approx 0.3$; $\approx 29°$

29. 1.7 units **31.** 4.6 units **33.** $(3\sqrt{5} + \sqrt{10})x + (\sqrt{5} + 2\sqrt{10})y - 2\sqrt{5} - 3\sqrt{10} = 0$ **35.** The formulas are equivalent.

Page 409   Chapter 7 Test

1. $\dfrac{\sqrt{3}}{2}$ **2.** $\dfrac{4}{5}$ **3.** $2\sqrt{2}$ **4.** $\dfrac{5}{3}$

5. $\tan \theta (\cot \theta + \tan \theta) \stackrel{?}{=} \sec^2 \theta$

$\tan \theta \cot \theta + \tan^2 \theta \stackrel{?}{=} \sec^2 \theta$

$1 + \tan^2 \theta \stackrel{?}{=} \sec^2 \theta$

$\sec^2 \theta = \sec^2 \theta$

6. $\sin^2 A \cot^2 A \stackrel{?}{=} (1 - \sin A)(1 + \sin A)$

$\sin^2 A \cdot \dfrac{\cos^2 A}{\sin^2 A} \stackrel{?}{=} (1 - \sin A)(1 + \sin A)$

$\cos^2 A \stackrel{?}{=} (1 - \sin A)(1 + \sin A)$

$1 - \sin^2 A \stackrel{?}{=} (1 - \sin A)(1 + \sin A)$

$(1 - \sin A)(1 + \sin A) = (1 - \sin A)(1 + \sin A)$

7. $\dfrac{\sec x}{\sin x} - \dfrac{\sin x}{\cos x} \stackrel{?}{=} \cot x$

$\dfrac{\sec x \cos x - \sin^2 x}{\sin x \cos x} \stackrel{?}{=} \cot x$

$\dfrac{1 - \sin^2 x}{\sin x \cos x} \stackrel{?}{=} \cot x$

$\dfrac{\cos^2 x}{\sin x \cos x} \stackrel{?}{=} \cot x$

$\dfrac{\cos x}{\sin x} \stackrel{?}{=} \cot x$

$\cot x = \cot x$

8. $\dfrac{\cos x}{1 + \sin x} + \dfrac{\cos x}{1 - \sin x} \stackrel{?}{=} 2 \sec x$ **9.** $\dfrac{-\sqrt{6} - \sqrt{2}}{4}$

$\dfrac{\cos x - \cos x \sin x + \cos x + \sin x \cos x}{1 - \sin^2 x} \stackrel{?}{=} 2 \sec x$

$\dfrac{2 \cos x}{1 - \sin^2 x} \stackrel{?}{=} 2 \sec x$

$\dfrac{2 \cos x}{\cos^2 x} \stackrel{?}{=} 2 \sec x$

$\dfrac{2}{\cos x} \stackrel{?}{=} 2 \sec x$

$2 \sec x = 2 \sec x$

10. $\dfrac{-\sqrt{2} - \sqrt{6}}{4}$ **11.** $\dfrac{\sqrt{6} - \sqrt{2}}{4}$ **12.** $-\dfrac{3\sqrt{7}}{8}$ **13.** $-\dfrac{\sqrt{14}}{4}$

14. $-3\sqrt{7}$ **15.** $45°$ **16.** $30°, 90°, 150°$ **17.** $0°, 60°, 180°$

18. $30°, 150°$ **19.** $-\dfrac{x}{\sqrt{2}} + \dfrac{y}{\sqrt{2}} - \dfrac{3}{\sqrt{2}} = 0$; $\dfrac{3}{\sqrt{2}} \approx 2.1$; $135°$

20. $\dfrac{x}{\sqrt{5}} + \dfrac{2y}{\sqrt{5}} - \dfrac{7}{3\sqrt{5}} = 0$; $\dfrac{7}{3\sqrt{5}} \approx 1.0$; $\approx 63°$

21. $\dfrac{2x}{\sqrt{5}} - \dfrac{y}{\sqrt{5}} - \dfrac{1}{\sqrt{5}} = 0$; $\dfrac{1}{\sqrt{5}} \approx 0.4$; $\approx 333°$

22. 3.6 units **23.** 3.2 units **24.** $(25 + 3\sqrt{29})x + (10 + 4\sqrt{29})y - 35 - 4\sqrt{29} = 0$ **25.** 232.32 feet

BONUS $\dfrac{x^2 + y^2}{x^2 - y^2}$

994

Pages 417-418   Lesson 8-1

5.

9.

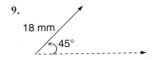

13.

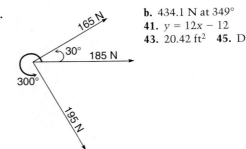

15. about 14.8 km **17.** 3.4 cm, 85° **19.** 5.0 cm, 3°
21. 8.1 cm, 60° **23.** 0.7 cm, 306° **25.** 2.7 cm, 14°
27. 2.9 cm, 286° **29.** 2.3 cm, 1.4 cm **31.** 1.8 cm, 1.8 cm
33. about 11.4 m **35.** the difference
37a.   **b.** about 100.12 m/s

39a.   **b.** 434.1 N at 349°
41. $y = 12x - 12$
43. 20.42 ft^2 **45.** D

Pages 422-424   Lesson 8-2

5. $5, 4\vec{i} + 3\vec{j}$ **7.** $\sqrt{13}, -2\vec{i} - 3\vec{j}$ **9.** $(-2, 6)$; $2\sqrt{10} \approx 6.325$
11. $(5, 9)$; $\sqrt{106} \approx 10.296$ **13.** $(6, 6)$ **15.** $(-9, 1)$ **17.** $(1, 1)$
19. $(5, 7)$ **21.** $(-9, -9)$; $9\sqrt{2} \approx 12.728$ **23.** $(-5, -5)$;
$5\sqrt{2} \approx 7.071$ **25.** $(1, 1)$; $\sqrt{2} \approx 1.414$ **27.** $(1, 0)$; 1
29. $(-2, -1)$ **31.** $(12, -9)$ **33.** $(-14, 8)$ **35.** $(-10, 0)$
37. $(-44, 18)$ **39.** $(0, 0)$ **41.** 81.7 N, 73.6 N **43.** The
magnitude of the horizontal component will increase and
the magnitude of the vertical component will decrease.
45. $\dfrac{4 \pm 2i\sqrt{11}}{5}$ **47.** $63°26'$ **49.** D

Pages 427-429   Lesson 8-3

5. $\sqrt{14} \approx 3.742$ **7.** $\sqrt{69} \approx 8.307$ **9.** $3\sqrt{3} \approx 5.196$
11. $(4, -8, -14)$; $4\vec{i} - 8\vec{j} - 14\vec{k}$ **13.** $(7, 3, -1)$; $7\vec{i} + 3\vec{j} - \vec{k}$
15. $(2, 0, 3)$; $\sqrt{13} \approx 3.606$ **17.** $(-1, 1, -5)$; $3\sqrt{3} \approx 5.196$
19. $(15, -24, 4)$; $\sqrt{817} \approx 28.583$ **21.** $3\vec{i} + \vec{k}$
23. $-5\vec{i} - 8\vec{j} + \vec{k}$ **25.** $8\vec{i} - 3\vec{j} + 11\vec{k}$ **27.** $(2, 12, 7)$
29. $(-5, -21, -6)$ **31.** $(1, 33, 38)$ **33.** If $\vec{a} = (a_1, a_2, a_3)$,
then $|\vec{a}| = \sqrt{(a_1)^2 + (a_2)^2 + (a_3)^2}$. If $-\vec{a} = (-a_1, -a_2, -a_3)$, then
$|\vec{a}| = \sqrt{(-a_1)^2 + (-a_2)^2 + (-a_3)^2}$. Since $a_1^2 = (-a_1)^2$, $a_2^2 = (-a_2)^2$,
and $a_3^2 = (-a_3)^2$, $|-\vec{a}| = |\vec{a}|$. **37a.** $(-132, -3454, 0)$
b. $(-9, 0, -9)$ **39.** no **41.** $\cos 20°$ **43.** $(1, 1)$; $\sqrt{2} \approx 1.414$

5. 32, no **7.** 0, yes **9.** -1, no **11.** 0, yes **13.** (9, 6, 0)
15. (-8, 19, -2) **17.** 0, yes **19.** 5, no **21.** 0, yes **23.** -10,
no **25.** 68, no **27.** (5, -4, -3), yes **29.** (2, 14, 4), yes
33. $\vec{a} \cdot \vec{b} = (a_1, a_2, a_3) \cdot (b_1, b_2, b_3)$
$= a_1 b_1 + a_2 b_2 + a_3 b_3$
$= b_1 a_1 + b_2 a_2 + b_3 a_3$
$= (b_1, b_2, b_3) \cdot (a_1, a_2, a_3)$
$= \vec{b} \cdot \vec{a}$

35.
$$\vec{a} \times (\vec{b} + \vec{c}) = \begin{vmatrix} \vec{i} & \vec{j} & \vec{j} \\ a_1 & a_2 & a_3 \\ (b_1 + c_1) & (b_2 + c_2) & (b_3 + c_3) \end{vmatrix}$$

$= \begin{vmatrix} a_2 & a_3 \\ (b_2 + c_2) & (b_3 + c_3) \end{vmatrix} \vec{i}$

$- \begin{vmatrix} a_1 & a_3 \\ (b_1 + c_1) & (b_3 + c_3) \end{vmatrix} \vec{j}$

$+ \begin{vmatrix} a_1 & a_2 \\ (b_1 + c_1) & (b_2 + c_2) \end{vmatrix} \vec{k}$

$= [a_2(b_3 + c_3) - a_3(b_2 + c_2)] \vec{i}$
$- [a_1(b_3 + c_3) - a_3(b_1 + c_1)] \vec{j}$
$+ [a_1(b_2 + c_2) - a_2(b_1 + c_1)] \vec{k}$

$= [(a_2 b_3 + a_2 c_3) - (a_3 b_2 + a_3 c_2)] \vec{i}$
$- [(a_1 b_3 + a_1 c_3) - (a_3 b_1 + a_3 c_1)] \vec{j}$
$+ [(a_1 b_2 + a_1 c_2) - (a_2 b_1 + a_2 c_1)] \vec{k}$

$= [(a_2 b_3 - a_3 b_2) + (a_2 c_3 - a_3 c_2)] \vec{i}$
$- [(a_1 b_3 - a_3 b_1) + (a_1 c_3 - a_3 c_1)] \vec{j}$
$+ [(a_1 b_2 - a_2 b_1) + (a_1 c_2 - a_2 c_1)] \vec{k}$

$= (a_2 b_3 - a_3 b_2) \vec{i} + (a_2 c_3 - a_3 c_2) \vec{i}$
$- (a_1 b_3 - a_3 b_1) \vec{j} - (a_1 c_3 - a_3 c_1) \vec{j}$
$+ (a_1 b_2 - a_2 b_1) \vec{k} + (a_1 c_2 - a_2 c_1) \vec{k}$

$= [(a_2 b_3 - a_3 b_2) \vec{i} - (a_1 b_3 - a_3 b_1) \vec{j} + (a_1 b_2 - a_2 b_1) \vec{k}]$
$+ [(a_2 c_3 - a_3 c_2) \vec{i} - (a_1 c_3 - a_3 c_1) \vec{j} + (a_1 c_2 - a_2 c_1) \vec{k}]$

$= \left[\begin{vmatrix} a_2 & a_3 \\ b_2 & b_3 \end{vmatrix} \vec{i} - \begin{vmatrix} a_1 & a_3 \\ b_1 & b_3 \end{vmatrix} \vec{j} + \begin{vmatrix} a_1 & a_2 \\ b_1 & b_2 \end{vmatrix} \vec{k} \right]$

$+ \left[\begin{vmatrix} a_2 & a_3 \\ c_2 & c_3 \end{vmatrix} \vec{i} - \begin{vmatrix} a_1 & a_3 \\ c_1 & c_3 \end{vmatrix} \vec{j} + \begin{vmatrix} a_1 & a_2 \\ c_1 & c_2 \end{vmatrix} \vec{k} \right]$

$= (\vec{a} \times \vec{b}) + (\vec{a} \times \vec{c})$

39a. (0, 71.5, 0) **b.** 71.5 lb/ft **c.** the y-axis
41. as $x \to +\infty$, $y \to -\infty$; as $x \to -\infty$, $y \to +\infty$ **43.** $\frac{4}{7}$ **45.** B

Page 435   Mid-Chapter Review

1.

1 cm

≈ 0.4 cm; ≈ 0.9 cm

3.

0.8 cm

≈ 0.4 cm; ≈ 0.7 cm

5. (-8, 1); $\sqrt{65} \approx 8.062$ **7.** (-10, 12); $2\sqrt{61} \approx 15.620$
9. (-8, -6, -1); $\sqrt{101} \approx 10.050$ **11.** (-32, 54) **13.** (-3, 0, 8)
15. (-4, -4, 20) **17.** 0, yes

Pages 438-441   Lesson 8-5

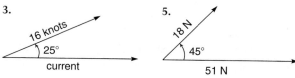

3.

16 knots
25°
current

5.

18 N
45°
51 N

7. 33.5 km/h, 333°26' **9.** 220.5 km/h, 16°42' **11.** 14.1 lb

13. 7.65 lb **15.** 0 lb **17.** 20.6 ft/s² **19.** 74 N, 253°
21. 112.4 lb, 278.2 lb **23.** 34 N, 223° **25.** 261.7 km,
21°3' SE **27.** 19°28' **29.** 171.5 N, 28°7' **31.** 132.4 lb
33. 66 **35.** 2, -1, $\frac{1}{3}$ **37.** -6, no

Pages 446-447   Lesson 8-6

5a. $(x + 1, y + 5) = t(3, 7)$ **b.** $x = -1 + 3t$, $y = -5 + 7t$
c. $y = \frac{7}{3}x - \frac{8}{3}$ **7a.** $(x - 5, y + 9) = t(3, 7)$ **b.** $x = 5 + 3t$,
$y = -9 + 7t$ **c.** $y = \frac{7}{3}x - \frac{62}{3}$ **9a.** $(x - 11, y + 4) = t(3, 7)$
b. $x = 3t + 11$, $y = 7t - 4$ **c.** $y = \frac{7}{3}x - \frac{89}{3}$ **11.** $x = t$;
$y = 3t + 11$ **13.** $(x + 4, y + 11) = t(-3, 8)$; $x = -3t - 4$,
$y = 8t - 11$ **15.** $(x + 1, y) = t(3, 2)$; $x = 3t - 1$, $y = 2t$
17. $x = t$, $y = -2t + 3$ **19.** $x = t$, $y = -\frac{3}{2}t + \frac{5}{2}$ **21.** $y = -x + 8$
23. $y = \frac{1}{4}x + \frac{23}{4}$ **25.** $x = 8$

27.

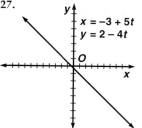

$x = -3 + 5t$
$y = 2 - 4t$

33a. 13.5 hours **b.** 675 miles
c. 9:30 A.M., 8:50 A.M.
d. at least 1.53 mph
35. $\begin{bmatrix} 28 & 20 \\ 10 & 14 \end{bmatrix}$
37. 8, 360°, 30° **39.** A

Pages 453-455   Lesson 8-7

7a. $x = 75t \cos 25°$, $y = 5 + 75t \sin 25° - 16t^2$ **c.** 145 ft
d. 20.7 ft **e.** no **9.** 46.07 ft/s **11a.** ≈ 107 ft, 1.1 s
b. ≈ 270 ft, 3.1 s **c.** ≈ 313 ft, 4.4 s **d.** ≈ 271 ft, 5.4 s
e. ≈ 107 ft, 6.2 s **f.** ≈ 0 ft, 6.25 s **13.** The range is the
same for projectiles fired at angles of $\alpha°$ and $(90 - \alpha)°$. But
the time in the air is greater when the angle is greater.
17a. $x = 155t \cos 22°$, $y = 155t \sin 22° - 16t^2 + 3$ **b.** about
36.12 feet, it will clear the fence **c.** 528.87 ft **19.** No, the
projectile will travel four times as far. **21a.** 127 ft/s
b. about 114.8 yards
23.

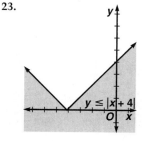

$y \le |x + 4|$

25. $5x - 4y - 33 = 0$

Pages 456-458   Chapter 8 Summary and Review

1. 4.1 cm, 25° **3.** 7.3 cm, 357° **5.** 1 cm, 0.8 cm **7.** (5, 12); 13
9. (-2, 12); $2\sqrt{37}$ **11.** (5, -6) **13.** (12, -17) **15.** (4, -1, -3),
$4\vec{i} - \vec{j} - 3\vec{k}$ **17.** (6, 2, 7), $6\vec{i} + 2\vec{j} + 7\vec{k}$ **19.** -16
21. (7, 22, 2) **23.** 43.95 lb, 35°52' **25.** $(x - 3, y + 5) = t(4, 2)$;
$x = 3 + 4t$, $y = -5 + 2t$ **27.** $(x - 4, y) = t(3, -6)$; $x = 4 + 3t$,
$y = -6t$ **29.** $x = 30t \cos 28°$, $y = 30t \sin 28° - 16t^2$
31. 25 lb/ft **33.** 10.2 feet

Page 459   Chapter 8 Test

1. ≈ 3.1 cm, $\approx 69°$ **2.** ≈ 9.5 cm, $\approx -8°$ **3.** ≈ 1.5 cm,
≈ 2.5 cm **4.** 1.4 cm, 1.4 cm **5.** (-4, 3) **6.** (5, 3)
7. (7, 1, 2) **8.** (-4, -2, 4) **9.** (-5, 0, 10) **10.** (14, 3, -26)
11. (11, 12, -14) **12.** $\sqrt{26}$

13. $\sqrt{61}$ 14. $-\vec{i} + 3\vec{j} + 4\vec{k}$ 15. $4\vec{i} + 3\vec{j} - 6\vec{k}$ 16. -19
17. (-30, 10, -15) 18. no 19. $x = 3 + 2t, y = 11 - 5t$
20. $x = -2 + t, y = 9t$ 21. $x = 12 - 4t, y = -8 - 7t$
22. $y = -\frac{9}{2}x + 19$ 23a. 13.7 knots b. 277.7 m
24. 82.5 lb/ft 25. 23 feet **BONUS** $90° < \theta < 270°$ or $\cos \theta < 0$

Pages 460-461 Unit 2 Review
1. 90° 3. 630° 5. 6.28 inches 7. $\frac{1}{5}$ 9. 1 11. $\frac{\sqrt{3}}{3}$
13. $-\frac{2\sqrt{3}}{3}$ 15. $A = 30°58', B = 59°2', c = 5.8$
17. no solution 19. One; $a = 40.1, c = 28.1, C = 42°$
21. 78.2 23. none, 36°, 0°

25.

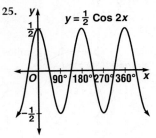

$y = \frac{1}{2} \text{Cos } 2x$

27.

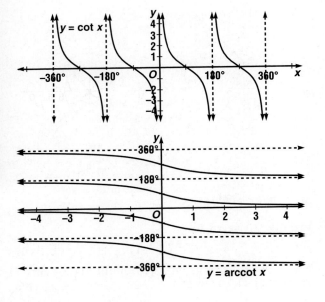

$y = x + 2 \sin 3x$

29. $\frac{2\sqrt{5}}{5}$ 31. $\frac{1}{2}$ 33. $\frac{\sqrt{3}}{2}$
35.

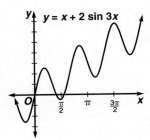

$y = \cot x$

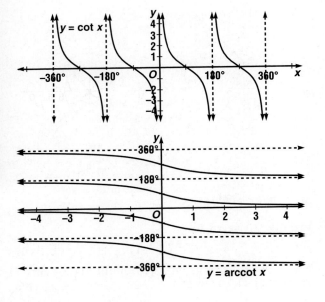

$y = \text{arccot } x$

37.

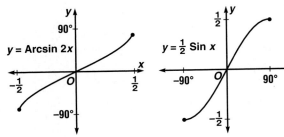

$y = \text{Arcsin } 2x$ $y = \frac{1}{2} \text{Sin } x$

39. $\frac{2\sqrt{2}}{3}$
41. Sample answer:
$$\tan x + \tan x \cot^2 x \stackrel{?}{=} \sec x \csc x$$
$$\tan x(1 + \cot^2 x) \stackrel{?}{=} \sec x \csc x$$
$$\tan x(\csc^2 x) \stackrel{?}{=} \sec x \csc x$$
$$\left(\frac{\sin x}{\cos x}\right)\left(\frac{1}{\sin^2 x}\right) \stackrel{?}{=} \sec x \csc x$$
$$\left(\frac{1}{\cos x}\right)\left(\frac{1}{\sin x}\right) \stackrel{?}{=} \sec x \csc x$$
$$\sec x \csc x = \sec x \csc x$$

43. $\frac{\sqrt{2} + \sqrt{6}}{4}$ 45. $\frac{\sqrt{2} + \sqrt{6}}{4}$ 47. $\frac{17}{25}$ 49. $\sqrt{\frac{5 - \sqrt{21}}{5 + \sqrt{21}}}$

51. 0°, 90°, 180° 53. 120° 55. $\frac{5x}{\sqrt{29}} + \frac{2y}{\sqrt{29}} - \frac{8}{\sqrt{29}} = 0$; $\frac{8}{\sqrt{29}} \approx 1.49, \approx 22°$ 57. 1.1 59. 1.5 61. 1 cm, 1.8 cm
63. (1, 1) 65. (-1, 7) 67. (13, -3, -7); $\vec{u} = 13\vec{i} - 3\vec{j} - 7\vec{k}$
69. (22, 0, -10); $\vec{u} = 22\vec{i} - 10\vec{k}$ 71. 22 73. $(x, y - 5) = t(-1, 5); x = -t, y = 5 + 5t$

CHAPTER 9 POLAR COORDINATES AND COMPLEX NUMBERS

Pages 470-472 Lesson 9-1
5.

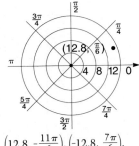

$\left(12.8, -\frac{11\pi}{6}\right)$, $\left(-12.8, \frac{7\pi}{6}\right)$, $\left(-12.8, -\frac{5\pi}{6}\right)$

9.

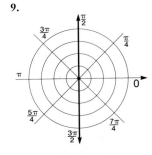

13.

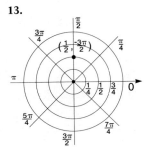

17.

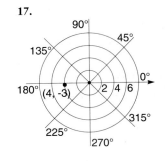

21.

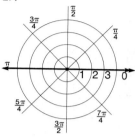

25.

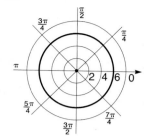

45.

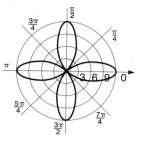

29.

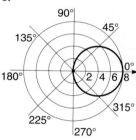

33.

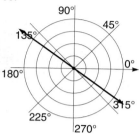

3. $\frac{2x}{\sqrt{13}} + \frac{3y}{\sqrt{13}} - \frac{5}{\sqrt{13}} = 0$ **5.** $\frac{2\sqrt{10}}{5} = r\cos(\theta + 72°)$

7.

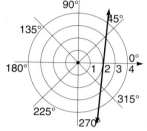

11.

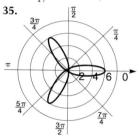

41. 1.25, 1 **43.** $\cos 20° + \sin 50°$ **45.** ≈ 15 feet horizontally, ≈ 22 feet vertically

15. $12 = r\cos\theta$ **17.** $\frac{2\sqrt{10}}{5} = r\cos(\theta - 162°)$ **19.** $y = 0$
21. $x = 3$ **23.** $\sqrt{2} = r\cos(\theta - 8°)$
25. $\frac{10\sqrt{34}}{17} = r\cos(\theta - 121°)$ **31.** Leonhard Euler **33.** 0.5
35.

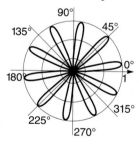

37. B

3.

5. cardioid **7.** spiral of Archimedes **9.** rose
11. limaçon **13.** lemniscate
15. cardioid **17.** rose
19. $(\sqrt{3}, 60°)$, $(-\sqrt{3}, 240°)$
21. $(2.1, 0.8)$ **23.** $(3.0, -0.9)$, $(1.8, 0.3)$, $(3.0, 0.9)$, $(2.2, -1.2)$

27. a rose with 10 petals

31. $\frac{-\sqrt{6} - \sqrt{2}}{4}$
33. $(10, 340°)$, $(-10, -200°)$, $(-10, 160°)$

1.

3.

$(6, 135°)$, $(6, -225°)$, $(-6, 315°)$ cardioid

5. $(2, 60°)$ or $\left(2, \frac{\pi}{3}\right)$

5. $7 + 7i$ **7.** $11 + 2i$ **9.** $\frac{5+i}{13}$ **11.** $-\frac{7}{5}$ or $-1\frac{2}{5}$ **13.** $7 + 2i$
15. $-7 + 9i\sqrt{7}$ **17.** $3 + 69i$ **19.** $17 + 17i$ **21.** $-7 - 24i$
23. 75 **25.** $-8 + 6i\sqrt{2}$ **27.** $\frac{7+i}{5}$ **29.** $-2 \pm 5i$ **31.** $x = 2\sqrt{3}$,

3. $\frac{\pi}{4} \approx 0.79$ **5.** 1.18 **7.** $(2\sqrt{2}, 3.93)$ **9.** $(0, 3)$
11. $(-1.04, 2.27)$ **13.** $x^2 + y^2 = 49$ **15.** $\left(3, \frac{\pi}{2}\right)$ **17.** $(2, 0)$
19. $\left(2, \frac{5\pi}{4}\right)$ **21.** $(2.99, 4.01)$ **23.** $(-0.45, -1.34)$ **25.** $(-1.72, 3.01)$
27. $r = 10\sec\theta$ **29.** $r = \frac{5}{2}\sin\theta$ **31.** $y = -x$ **33.** $x = -2$
35. $x^2 + y^2 = x + y$ **39.** about 219 meters **41.** $y = \frac{x^4(x+2)}{x(x+2)}$
43. equal

$y = -\frac{2}{7}$ **33.** $x = -3$, $y = -2$ **35.** $50 - 125i$ **37.** $\frac{27 + 10i\sqrt{3}}{147}$

39. $\frac{-1 + 2i}{2}$ **41a.** 3 volts   **b.** $0.75(1 - j\sqrt{3})$ ohms

43. $22.28 + 7.96j$ amps   **45.** $y = -\frac{2}{5}x + 2$, $m = -\frac{2}{5}$,

y-intercept $= 2$ **47a.** 18 SA, 2 essay for a score of 120
b. 12 SA, 8 essay for a score of 180 points

49.
$$\sin^4 A + \cos^2 A \stackrel{?}{=} \cos^4 A + \sin^2 A$$
$$(\sin^2 A)^2 + \cos^2 A \stackrel{?}{=} \cos^4 A + \sin^2 A$$
$$(1 - \cos^2 A)^2 + \cos^2 A \stackrel{?}{=} \cos^4 A + \sin^2 A$$
$$(1 - 2\cos^2 A + \cos^4 A) + \cos^2 A \stackrel{?}{=} \cos^4 A + \sin^2 A$$
$$\cos^4 A + 1 - \cos^2 A \stackrel{?}{=} \cos^4 A + \sin^2 A$$
$$\cos^4 A + \sin^2 A = \cos^4 A + \sin^2 A$$

51. $r = 5$ or $r = -5$ **53.** B

Pages 504-505   Lesson 9-6

5. $\sqrt{2}\left(\cos \frac{3\pi}{4} + i \sin \frac{3\pi}{4}\right)$ **7.** $\sqrt{70}\left(\cos \frac{\pi}{2} + i \sin \frac{\pi}{2}\right)$

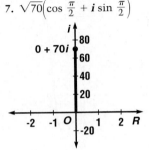

9. $2\sqrt{3}\left(\cos \frac{11\pi}{6} + i \sin \frac{11\pi}{6}\right)$

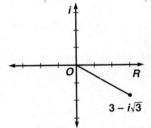

11. 4   **13.** $3i$
15. $\sqrt{2}\left(\cos \frac{5\pi}{4} + i \sin \frac{5\pi}{4}\right)$
17. $10\left(\cos \frac{\pi}{2} + i \sin \frac{\pi}{2}\right)$
19. $5(\cos 0.64 + i \sin 0.64)$
21. $2\left(\cos \frac{\pi}{3} + i \sin \frac{\pi}{3}\right)$
23. $5(\cos \pi + i \sin \pi)$
25. $12 - 12i\sqrt{3}$

27. $-1.25 + 2.73i$ **29.** $\sqrt{3} + i$ **31.** $\frac{-5\sqrt{3}}{2} + \frac{5}{2}i$

35. $I = 22.28 + 7.96j$; $E = 236.59(\cos 1.27 + i \sin 1.27)$,
$Z = 10(\cos 0.93 + i \sin 0.93)$, $I = 23.66(\cos 1.34 + i \sin 1.34)$
37. No, the height of the building is approximately 61.9 feet, making a total of 101.9 feet with the tower, which is over the city's limit.
39. $(-1, -1, -10)$, $-\vec{\mathbf{i}} - \vec{\mathbf{j}} - 10\vec{\mathbf{k}}$ **41.** $15i$ ohms

Pages 508-509   Lesson 9-7

3. $30(\cos 3\pi + i \sin 3\pi)$, -30 **5.** $3(\cos \pi + i \sin \pi)$, -3 **7.** $-2i$
9. $-\frac{3}{2}i$ **11.** $6(\cos 90° + i \sin 90°)$, $6i$
13. $2\sqrt{2}\left(\cos \frac{\pi}{12} + i \sin \frac{\pi}{12}\right)$, $\approx 2.7 + 0.7i$
15. $18\left(\cos \frac{4\pi}{3} + i \sin \frac{4\pi}{3}\right)$, $-9 - 9i\sqrt{3}$
17. $6\left(\cos \frac{3\pi}{4} + i \sin \frac{3\pi}{4}\right)$, $-3\sqrt{2} + 3i\sqrt{2}$
19. $2.98(\cos 1.1 + i \sin 1.1)$, $\approx 1.35 + 2.66i$
21. $\sqrt{3}\left(\cos \frac{5\pi}{8} + i \sin \frac{5\pi}{8}\right)$, $\approx -0.66 + 1.60i$
23. $10(\cos \pi + i \sin \pi)$, -10 **25.** Since $\cos \frac{\pi}{4} = \frac{\sqrt{2}}{2}$,

$\sin \frac{\pi}{4} = \frac{\sqrt{2}}{2}$, $\cos\left(-\frac{\pi}{4}\right) = \frac{\sqrt{2}}{2}$, $\sin\left(-\frac{\pi}{4}\right) = -\frac{\sqrt{2}}{2}$,

$s_1 = 4\sqrt{2}\left(\frac{\sqrt{2}}{2} + i\frac{\sqrt{2}}{2}\right)$ or $4 + 4i$ and $s_2 = 4\sqrt{2}\left(\frac{\sqrt{2}}{2} - i\frac{\sqrt{2}}{2}\right)$

or $4 - 4i$. $s_1 + s_2 = 4 + 4i + 4 - 4i$ or 8. $\frac{-b}{a} = \frac{8}{1} = 8$.

$s_1 s_2 = (4 + 4i)(4 - 4i) = 16 - 16i^2$ or 32. $\frac{c}{a} = \frac{32}{1} = 32$.
27. ≈ 1434 ft, $\approx 86{,}751$ ft^2 **29.** $44 - 58i$ **31.** A

Pages 516-517   Lesson 9-8

5. $-\frac{9}{2} + \frac{9\sqrt{3}}{2}i$ **7.** $-i$ **9.** $-\frac{1}{2} - \frac{\sqrt{3}}{2}i$ **11.** $-527 - 336i$
13. -1024 **15.** $-119 - 120i$ **17.** $0.72 + 1.08i$
19. $\frac{\sqrt{3}}{2} + \frac{1}{2}i$ or $0.87 + 0.5i$ **21.** $1.22 + 0.71i$
23. $1, 0.31 \pm 0.95i, -0.81 \pm 0.59i$ **25.** $\pm i$, $\frac{\sqrt{3}}{2} \pm \frac{1}{2}i$,
$-\frac{\sqrt{3}}{2} \pm \frac{1}{2}i$ **27.** $0.81 \pm 0.59i, -1, -0.31 \pm 0.95i$ **35.** -1
37a. 800.8 kg   **b.** 278.2 kg   **39.** $10\left(\cos \frac{17\pi}{12} + i \sin \frac{17\pi}{12}\right)$,
$-2.59 - 9.66i$

Pages 518-520   Chapter 9 Summary and Review

1. $(-3, -310°)$, $(3, 230°)$,   **5.**
$(3, -130°)$

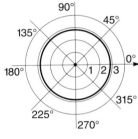

9. rose

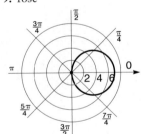

13. $(3\sqrt{2}, 3\sqrt{2})$
15. $(1.33, -1.49)$
17. $\left(2\sqrt{3}, \frac{4\pi}{3}\right)$
19. $(3.61, -0.59)$
21. $\frac{3\sqrt{5}}{5} = r\cos(\theta - 207°)$
23. $x + \sqrt{3}y - 6 = 0$
25. $6 - i$ **27.** $-i$ **29.** 50
31. $\frac{5\sqrt{2} + 20i}{18}$
33. $4\left(\cos \frac{2\pi}{3} + i \sin \frac{2\pi}{3}\right)$
35. $-2\sqrt{3} + 2i$

37. $-4 + 4i\sqrt{3}$ **39.** $-4i$ **41.** $0.31 + 0.39i$
43. $-64\sqrt{3} + 64i$ **45.** 64   **47.** $1.24 + 0.22i$
49. $-\frac{1}{2} \pm \frac{\sqrt{3}}{2}i, \frac{1}{2} \pm \frac{\sqrt{3}}{2}i, \pm 1$ **51.** $(22.44 + 16.95j)$ amps

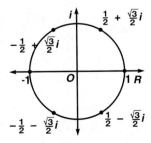

Page 521   Chapter 9 Test

1. $\left(-2, -\frac{3\pi}{4}\right), \left(2, \frac{\pi}{4}\right), \left(2, -\frac{7\pi}{4}\right)$ **2.** $\left(3, \frac{11\pi}{6}\right), \left(-3, -\frac{7\pi}{6}\right), \left(-3, \frac{5\pi}{6}\right)$

Selected Answers

1.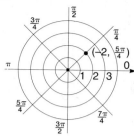

2.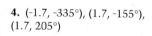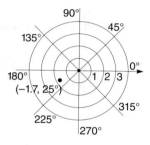

3. (2.5, -220°), (-2.5, -40°), (-2.5, 320°)

4. (-1.7, -335°), (1.7, -155°), (1.7, 205°)

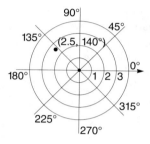

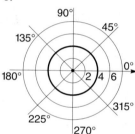

5.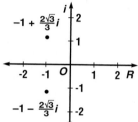

6.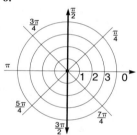

7.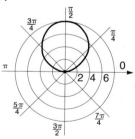

8.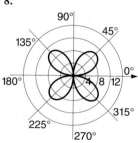

9.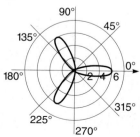

10.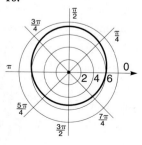

11. $\left(-\dfrac{3\sqrt{2}}{2}, \dfrac{3\sqrt{2}}{2}\right)$ **12.** (-0.68, -3.94) **13.** (2, 2)

14. $\dfrac{3\sqrt{61}}{61} = r\cos(\theta - 230°)$ **15.** $\dfrac{\sqrt{5}}{10} = r\cos(\theta + 63°)$

16. $\dfrac{3\sqrt{10}}{5} = r\cos(\theta - 72°)$ **17.** i **18.** -i **19.** -8 + 3i

20. 19 + 9i **21.** 48 + 14i **22.** 2 − 2i

23. $4\sqrt{2}\left(\cos\dfrac{3\pi}{4} + i\sin\dfrac{3\pi}{4}\right)$ **24.** 5(cos π + i sin π)

25. $12\left[\cos\left(-\dfrac{\pi}{3}\right) + i\sin\left(-\dfrac{\pi}{3}\right)\right]$ **26.** $12\left(\cos\dfrac{7\pi}{4} + i\sin\dfrac{7\pi}{4}\right)$;

$6\sqrt{2} - 6i\sqrt{2}$ **27.** $6 - 2i\sqrt{3}$ **28.** $2\left(\cos\dfrac{\pi}{2} + i\sin\dfrac{\pi}{2}\right)$; 2$i$

29. 16 **30.** $-\sqrt{3} - i$ or -1.73 − i **31.** 1, $\dfrac{\sqrt{2}}{2} + \dfrac{i\sqrt{2}}{2}$,

$i, -\dfrac{\sqrt{2}}{2} + \dfrac{i\sqrt{2}}{2}, -1, -\dfrac{\sqrt{2}}{2} - \dfrac{i\sqrt{2}}{2}, -i, \dfrac{\sqrt{2}}{2} - \dfrac{i\sqrt{2}}{2}$

32. 160(cos 62° + j sin 62°)

33. $-1 \pm \dfrac{2i\sqrt{3}}{3}$

BONUS

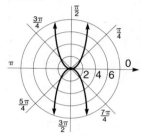

CHAPTER 10 CONICS

Pages 529-532 Lesson 10-1

5. $x^2 + y^2 = 81$ **7.** $(x - 2)^2 + (y + 7)^2 = 100$ **9.** 3, (4, 3)

11. $\dfrac{\sqrt{10}}{2}$, (3, 2) **13.** $(x - 1.4)^2 + (y - 3.75)^2 = 16.02$

15. $(x - 7)^2 + (y - 4)^2 = 5$ **17.** $\left(x + \dfrac{1}{2}\right)^2 + y^2 = 1$

19. $x^2 + y^2 = 9$ **21.** $(x + 0.25)^2 + (y - 1)^2 = 9$

23. $\left(x - \dfrac{3}{2}\right)^2 + \left(y + \dfrac{1}{4}\right)^2 = 7$ **25.** $(x + 7)^2 + (y + 12)^2 = 36$

27. $(x - 3)^2 + (y - 4)^2 = 5$, (3, 4), $\sqrt{5}$ **29.** $(x - 4)^2 +$

$(y - 2)^2 = 18$, (4, 2), $3\sqrt{2}$ **31.** $\left(x + \dfrac{1}{6}\right)^2 + \left(y - \dfrac{7}{6}\right)^2 =$

$\dfrac{169}{18}$, $\left(-\dfrac{1}{6}, \dfrac{7}{6}\right)$, $\dfrac{13\sqrt{2}}{6}$ **33.** $(x + 3)^2 + (y - 4)^2 = 25$

35. $(x + \dfrac{1}{2})^2 + (y - \dfrac{5}{2})^2 = \dfrac{34}{4}$ **37.** $(x - 5)^2 + (y - 12)^2 = \dfrac{1}{5}$

47. 1608.5 m/s **49.** sin 100° **51.** D(-8, 1) **53.** 1.31 + 0.14i

Pages 539-541 Lesson 10-2

5. (9, -12), 5, right **7.** (6, -8), 0.5, up **9.** vertex: (2, -1), focus: (2, 1), directrix: $y = -3$ **11.** $(y - 3)^2 = 4x$

13a. $(y - 2)^2 = x - 7$ **b.** vertex: (7, 2), focus: (7.25, 2), directrix: $x = 6.75$, axis: $y = 2$ **15a.** $(x + 4)^2 = -4(y - 2)$

b. vertex: (-4, 2), focus: (-4, 1), directrix: $y = 3$, axis: $x = -4$

17a. $y^2 = -2x$ **b.** vertex: (0, 0), focus: $\left(-\dfrac{1}{2}, 0\right)$, directrix: $x = \dfrac{1}{2}$,

axis: $y = 0$ **19a.** $(x - 3)^2 = 10(y + 1)$ **b.** vertex: (3, -1),

focus: $\left(3, \dfrac{3}{2}\right)$, directrix: $y = -\dfrac{7}{2}$, axis: $x = 3$

21a. $(y - 3)^2 = -3(x - 3)$ **b.** vertex: (3, 3), focus: $\left(\dfrac{9}{4}, 3\right)$,

directrix: $x = \dfrac{15}{4}$, axis: $y = 3$ **23.** $y^2 = -12x$

25. $(x - 4)^2 = -1(y - 3)$ **27.** $(x - 2)^2 = -8(y + 4)$

29. $(y - 2)^2 = 4(x + 1)$ **31a.** The opening becomes narrower. **b.** The opening becomes wider. **39a.** $7 **b.** $2450

41.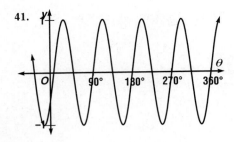

43. (-1, -3)   **45.** $(x - 7)^2 + (y - 4)^2 = 2$

Pages 548-550   Lesson 10-3

5. center: (0, 0), foci: (0, $\pm\sqrt{5}$), vertices: (0, $\pm$3), ($\pm$2, 0)
7. center: (-2, 0), foci: (-2 $\pm 4\sqrt{2}$, 0), vertices: (-11, 0), (7, 0),
(-2, $\pm$7)   **9.** center: (0, -9), foci: (0, -9 $\pm 4\sqrt{3}$), vertices: (0, -1),
(0, -17), ($\pm$4, -9)   **11.** $\frac{x^2}{9} + \frac{y^2}{36} = 1$ **13.** $\frac{x^2}{49} + \frac{(y + 5)^2}{36} = 1$
15. center: (3, 4), foci: (6, 4), (0, 4), vertices: (8, 4), (-2, 4),
(3, 0), (3, 8)   **17.** center: (0, 0), foci: ($\pm\sqrt{5}$, 0), vertices:
(-3, 0), (3, 0), (0, -2), (0, 2)   **19.** center: (3, 1), foci: (3, 1 $\pm\sqrt{5}$),
vertices: (3, 4), (3, -2), (1, 1), (5, 1)   **21.** center: (7, -6), foci:
(7 $\pm\sqrt{5}$, 6), vertices: (10, -6), (4, -6), (7, -4), (7, -8)
23. $\frac{y^2}{64} + \frac{x^2}{36} = 1$ **25.** $\frac{x^2}{36} + \frac{y^2}{16} = 1$ **27.** $\frac{x^2}{49} + \frac{y^2}{45} = 1$
29. ellipse; two squared terms, $A \neq C$ **31.** parabola; one
squared term   **33.** ellipse; two squared terms, $A \neq C$
35. $\frac{(y - 2)^2}{18} + \frac{(x - 1)^2}{9} = 1$
37. $49x^2 + 9y^2 + 294x - 126y + 441 = 0$ **43.** 116 million km

45.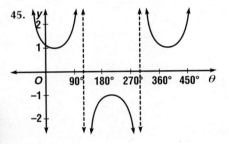

47. 334.3 mph, 9°28'SW   **49.** $(x + 5)^2 + (y - 3)^2 = 169$
51. 4

Pages 556-559   Lesson 10-4

5. center: (-3, 4), foci: (-3 $\pm\sqrt{74}$, 4), vertices: (2, 4), (-8, 4),
asymptotes: $y - 4 = \pm\frac{7}{5}(x + 3)$ **7.** center: (8, 5), foci:
(8, 5 $\pm\sqrt{104}$), vertices: (8, 7), (8, 3), asymptotes:
$y - 5 = \pm\frac{1}{5}(x - 8)$ **9.** center: (-11, -7), foci: (-11 $\pm 6\sqrt{5}$, -7),
vertices: (1, -7), (-23, -7), asymptotes: $y + 7 = \pm\frac{1}{2}(x + 11)$
11. $\frac{(y - 3)^2}{16} - \frac{(x - 4)^2}{9} = 1$ **13.** $\frac{x^2}{9} - \frac{y^2}{9} = 1$ **15.** center:
(0, 0), foci: ($\pm\sqrt{41}$, 0), vertices: (5, 0), (-5, 0), asymptotes:
$y = \pm\frac{4}{5}x$ **17.** center: (0, 0), foci: ($\pm\sqrt{117}$, 0), vertices:
(6, 0), (-6, 0), asymptotes: $y = \pm\frac{3}{2}x$ **19.** center: (3, -5), foci:
(3 $\pm\sqrt{13}$, -5), vertices: (5, -5), (1, -5), asymptotes:
$y + 5 = \pm\frac{3}{2}(x - 3)$ **21.** center: (5, -3), foci: (5 $\pm\sqrt{13}$, -3),
vertices: (7, -3), (3, -3), asymptotes: $y + 3 = \pm\frac{3}{2}(x - 5)$
23. ellipse   **25.** circle   **27.** ellipse

29.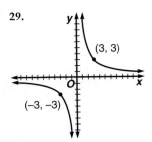

33.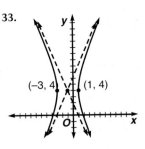

37. $\frac{y^2}{16} - \frac{x^2}{9} = 1$ **39.** $\frac{(x - 3)^2}{9} - \frac{(y - 3)^2}{16} = 1$
41. $\frac{2y^2}{25} - \frac{2x^2}{25} = 1$ **43.** $\frac{(y - 2)^2}{9} - \frac{(x - 4)^2}{16} = 1$
45. $\frac{5(y - 1)^2}{64} - \frac{5(x - 1)^2}{16} = 1$

53a.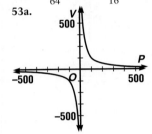

b. 116 cm³   **c.** 58 cm³
d. $V = 0.5$(original V)
55a. 9.77×10^4 hours or
4070 days   **b.** May, 1983
(actual: June 13, 1983)

57.
$$\frac{\sin 2X}{1 - \cos 2X} \stackrel{?}{=} \cot X$$
$$\frac{2 \sin X \cos X}{1 - \cos^2 X + \sin^2 X} \stackrel{?}{=} \cot X$$
$$\frac{2 \sin X \cos X}{2 \sin^2 X} \stackrel{?}{=} \cot X$$
$$\frac{\cos X}{\sin X} \stackrel{?}{=} \cot X$$
$$\cot X = \cot X$$

59. $0 = r \cos(\theta - 135°)$ **61.** $(y + 1)^2 = 4(x - 1)$ **63.** C

Page 559   Mid-Chapter Review

1. $(x - 2)^2 + (y + 7)^2 = 81$ **3.** vertex: (0, 3), focus: (1, 3),
directrix: $x = -1$ **5.** center: (3, 2), foci: (3 $\pm\sqrt{15}$, 2), 8 units,
2 units   **7.** $40,000x^2 + 90,000y^2 = 3,600,000,000$

9.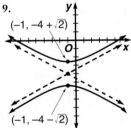

Pages 565-567   Lesson 10-5

5. parabola   **7.** hyperbola   **9.** circle, $(x - 4)^2 + y^2 = 5$
11. parabola, $\left(y + \frac{1}{2}\right)^2 = -2(x - 2)$ **13.** $\frac{y^2}{4} + \frac{x^2}{1.75} = 1$
15. $\frac{\sqrt{5}}{3}$ **17.** 0   **19.** $\sqrt{2}$ **21.** hyperbola, $xy = \frac{4}{9}$
23. parabola, $x^2 = 4(y + 7)$ **25.** hyperbola,
$\frac{(x - 2)^2}{5} - \frac{(y + 2)^2}{5} = 1$ **27.** parabola, $(x - 4)^2 = y$
29. point, (2, 3)   **31.** $\frac{(x - 2)^2}{16} + \frac{(y - 5)^2}{15} = 1$
33. $\frac{y^2}{36} - \frac{x^2}{28} = 1$ **35.** $\frac{y^2}{100} + \frac{(x - 3)^2}{51} = 1$

37. $\dfrac{(y-2)^2}{18} + \dfrac{(x-1)^2}{9} = 1$ **39.** 1.5 **41a.** 2095.2 miles, 424.8 miles **b.** $\dfrac{x^2}{27,248,400} + \dfrac{y^2}{26,550,840.96} = 1$

43. 3.05 cm **45.** 0° **47.** $\dfrac{(y+1)^2}{4} - \dfrac{(x-3)^2}{5} = 1$

Pages 575-576 Lesson 10-6

5. circle, $3x^2 + 3y^2 + 30x - 36y + 167 = 0$ **7.** parabola, $x^2 + 16x - 4y + 88 = 0$ **9.** $21x^2 + 2\sqrt{3}xy + 23y^2 = 120$
11. $x^2 + 16\sqrt{3}x - 2\sqrt{3}xy + 16y + 3y^2 = 0$ **13.** circle, $x^2 + y^2 - 4x - 6y + 7 = 0$ **15.** parabola, $2x^2 + 5x - y - 2 = 0$ **17.** $2xy = 9$ **19.** $x^2 - y^2 = 12$
21. hyperbola, 11° **23.** ellipse, -18° **25.** parabola, -30°
27. point **29.** line

39a.
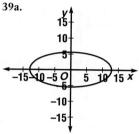

b. $\dfrac{(x+12)^2}{169} + \dfrac{y^2}{25} = 1$
c. $25y^2 - 600y + 169x^2 = 625$ **41.** no triangle
43. **45.** $(x-4)^2 + y^2 = 5$, circle

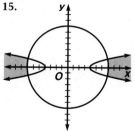

Pages 583-584 Lesson 10-7

5. 4 **7.** 3 **9.** 1 **11.** no solution **13.** (4.0, 0.5), (-2.8, -2.9)
15.
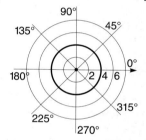

17. no solution **19.** (3, 3), (-1, -1) **21.** (±2, 1.7), (±2, -1.7) **23.** (±1, 5), (±1, -5) **25.** (±2, 0)
27. $\left(3, -\dfrac{4}{3}\right)$, (4, -1), $\left(-3, \dfrac{4}{3}\right)$, (-4, 1)

29. **33.**
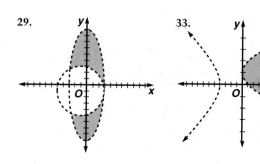

37. 12 ft by 18 ft **39.** 214.9 meters **41.** (2, 0, 3)
43. (-7, 8)

Pages 589-591 Lesson 10-8

3. 1, -1 **5.** 0, undefined **7.** 1, -1
9. $9x - 8y - 145 = 0$, $9y + 8x = 0$ **11.** $4x + 5y - 25 = 0$, $25x - 20y - 64 = 0$ **13.** $2x + \sqrt{21}y - 25 = 0$, $y - \dfrac{\sqrt{21}}{2}x = 0$ **15.** $10x + 4y + 116 = 0$, $5y - 2x = 0$
17. $x + 5y + 14 = 0$, $5x - y - 8 = 0$
19. $2x - \sqrt{5}y + 13 = 0$, $\sqrt{5}x + 2y + 2\sqrt{5} = 0$
21. $x + y - 8 = 0$, $x - y - 6 = 0$ **23.** $x = -4$, $y = -3$
25. $\sqrt{3}$ or 1.73 **27.** $\sqrt{33}$ or 5.74 **29.** $y = \pm5$ **31.** $y = \pm8$
33. $y = -\dfrac{3}{4}x + 10$, $y = -\dfrac{3}{4}x - 10$ **35.** $a = 4$

41a.
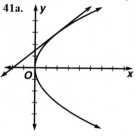
b. $4y - 3x - 6 = 0$

43.
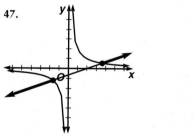

45. $16 + 30i$

47.
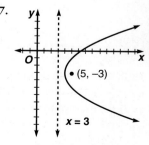

Pages 592-594 Chapter 10 Summary and Review

1. $x^2 + y^2 = 27$ **3.** $\left(x + \dfrac{3}{2}\right)^2 + \left(y + \dfrac{1}{2}\right)^2 = \dfrac{29}{2}$

5.
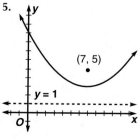
vertex: (7, 3), focus: (7, 5), directrix: $y = 1$

7.
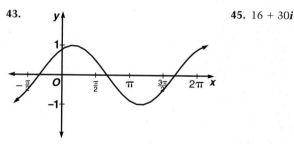
vertex: (4, -3), focus: (5, -3), directrix: $x = 3$

9. center: (3, -1), foci: $(3 \pm \sqrt{21}, -1)$, vertices: (-2, -1), (8, -1), (3, -3), (3, 1) **11.** center: (4, 6), foci: $(4 \pm 3\sqrt{3}, 6)$, vertices: (10, 6), (-2, 6), (4, 3), (4, 9) **13.** center: (0, 0), foci: $(\pm 2\sqrt{3}, 0)$, vertices: (2, 0), (-2, 0), asymptotes: $y = \pm\sqrt{2}x$ **15.** center: (2, 0), foci: (2, 6), (2, -6), vertices: $(2, \pm\sqrt{30})$, asymptotes: $y = \pm\sqrt{5}(x - 2)$ **17.** ellipse, $\frac{(x-1)^2}{4} + \frac{y^2}{16} = 1$ **19.** circle, $(x - 4)^2 + (y - 1)^2 = 4$ **21.** $(x + 3)^2 + (y - 2)^2 = 3$ **23.** $21x^2 - 10\sqrt{3}xy + 31y^2 = 144$ **25.** (3, 3), (-1, -1)

27.

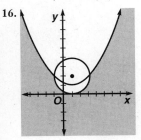

29. $x = 1, y = 1$
31. $8x - y + 12 = 0$, $x + 8y + 34 = 0$
33. 1.5 feet from the center

Page 595   Chapter 10 Test

1. parabola, $(x + 3)^2 = 8(y + 2)$ **2.** degenerate hyperbola, $4x^2 - y^2 = 0$ **3.** hyperbola, $\frac{y^2}{\frac{5}{13}} - \frac{x^2}{\frac{5}{2}} = 1$ **4.** ellipse, $\frac{(x-3)^2}{81} + \frac{(y+4)^2}{16} = 1$ **5.** parabola, $(y + 5)^2 = 2(x - 1)$ **6.** hyperbola, $\frac{(x-4)^2}{36} - \frac{(y-5)^2}{4} = 1$ **7.** ellipse, $\frac{(x+1)^2}{8} + \frac{(y-3)^2}{4} = 1$ **8.** circle, $(x + 2)^2 + y^2 = 10$ **9.** $(x + 8)^2 + (y - 3)^2 = 53$ **10.** $(x - 3)^2 = -6\left(y + \frac{7}{2}\right)$ **11.** $x^2 + \frac{4y^2}{3} = 1$ **12.** $\frac{(x-1)^2}{4} - \frac{(y+5)^2}{5} = 1$ **13.** $4(x - 2)^2 + (y + 2)^2 = 36$ **14.** $-x^2 + 6\sqrt{3}xy + 5y^2 = 32$ **15.** (2, 0), $\left(\frac{2}{3}, \frac{2\sqrt{2}}{3}\right)$, $\left(\frac{2}{3}, \frac{-2\sqrt{2}}{3}\right)$

16.

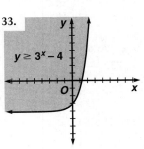

17. $x + 3y - 5 = 0, 3x - y - 5 = 0$
18. $x + 4\sqrt{2}y - 9\sqrt{2} = 0, 4\sqrt{2}x - y - 6 = 0$
19. $(x - 1)^2 + (y - 4)^2 = 10$ **20.** (-3, -9) **BONUS** $\frac{1}{2}$

CHAPTER 11   EXPONENTIAL AND LOGARITHMIC FUNCTIONS

Pages 602-604   Lesson 11-1

5. 9 **7.** 49 **9.** $x^{\frac{1}{2}}y^{\frac{3}{2}}$ **11.** $\sqrt[5]{15}$ **13.** $\sqrt[4]{a^3y}$ **15.** 81 **17.** 125 **19.** 2 **21.** $\frac{9}{4}$ **23.** 4 **25.** 3 **27.** $b^{\frac{3}{2}}$ **29.** $5a^{\frac{2}{3}}b$ **31.** $24^{\frac{1}{4}}a^3b^4$ **33.** $2xy^{\frac{8}{5}}$ **35.** $\sqrt[6]{64}$ or 2 **37.** $y\sqrt[3]{4a^2y}$ **39.** $\sqrt[5]{rt^2v^3}$ **41.** $\sqrt[3]{x}$ **43.** $x^2\sqrt[5]{y^2a^2}$ **45.** x^5 **47.** $125x$ **49.** $8y^6$ **51.** $a^{-2}b^8c^{-5}$ or $\frac{b^8}{a^2c^5}$ **55.** ≈ 254 km

57.

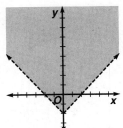

59. $-2, -\frac{1}{2}, 2$
61. about 346 m/s
63. $28 - 96i$
65. $4\sqrt{2}$ units

Pages 611-613   Lesson 11-2

5. 5.4622 **7.** 0.0317 **11.** The graph of Exercise 9 is the graph of Exercise 8 reflected over the x-axis. The graph of Exercise 10 is the graph of Exercise 8 reflected across the y-axis. **13.** $11,689.59 **15.** 36.6604

9.

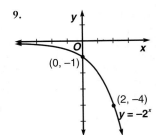

17.

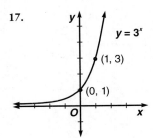

21.

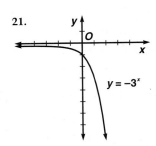

25.

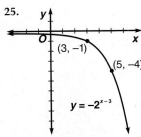

27. They have similar shape, but the graph of Exercise 19 is steeper.

29.

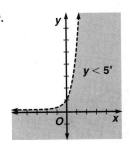

33.

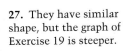

35. They are reflections over the x-axis. **39.** $34,426.47 **41.** over 30 years, #2 is better; over 5 years, #1 is better **43.** $13,257.20 **45.** 78 **47.** $-\frac{17}{32}$

49. $6\sqrt{2}\left(\cos\frac{3\pi}{4} + i\sin\frac{3\pi}{4}\right)$ **51.** C

Pages 617-619   Lesson 11-3

5. 27.1 **7.** 1.6 **9.** 114.5 **11.** $1671.38 **13.** 73.6998 **15.** 4.2340 **17.** 1.0000 **25.** $1489.35 **27.** $892.69

29.

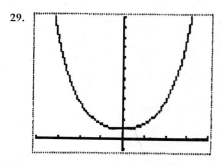

31. $\approx 3.786\%$ **33a.** 131   **b.** 39   **c.** 2050; No, this is 85 years and the legislator would have to be more than 100 years old then.   **35.** The graph is a logarithmic graph that approaches 0 as x approaches $-\infty$. It has a y-intercept of approximately 9.5 and curves sharply upward for the positive values of x.   **39.** 18 rps   **41.** $15 = r \sin \theta$ **43.** $891.26

Pages 626-628   Lesson 11-4

5. $\log_6 \left(\frac{1}{36}\right) = -2$ **7.** $27^{\frac{1}{3}} = 3$ **9.** $9^{\frac{3}{2}} = 27$ **11.** 3   **13.** 6

15. 3

19. $\log_2 16 = 4$

21. $\log_3 \left(\frac{1}{27}\right) = -3$

23. $\log_8 \frac{1}{4} = -\frac{2}{3}$

25. $2^3 = 8$ **27.** $10^4 = 10,000$

29. $8^{\frac{1}{3}} = 2$ **31.** 6   **33.** 5

35. 5   **37.** 10   **39.** 16   **41.** -2

43. 12   **45.** 3   **47.** 24

49. 4   **51.** 5   **53.** 12

17.

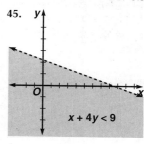

$y = \log_3(x + 1)$

55.

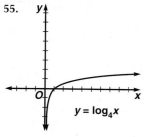

$y = \log_4 x$

59.

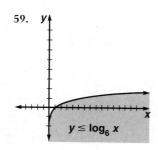

$y \leq \log_6 x$

63a. 100 units   **b.** 5 ft/s   **65.** 300 ft of A, 600 ft of B

67. $y = -8x + 3$ **69.** 0   **71.** -2, no   **73.** $\frac{(x-2)^2}{4} - \frac{y^2}{1} = 1$; hyperbola   **75.** D

Page 628   Mid-Chapter Review

1. 27   **3.** $\sqrt{3}$ **5.** $211,752.24   **7.** 14.2783   **9.** 444 mm of mercury   **11.** 2

Pages 632-633   Lesson 11-5

5. 2.6274   **7.** 354.73   **9.** 0.01   **11.** 10,000 and 100,000

13. 2809.0   **15.** 1802.8   **17.** 2.9515   **19.** -2.2676

21. 4.3597   **23.** 1.81   **25.** 0.01   **27.** 84,004.01   **29.** 236.5

31. 5213.4   **33.** 13.1   **35.** 2.6   **37.** 5.2   **41a.** 6.3

b. 10 times   **43.** $\begin{bmatrix} 8 & 8 & -1 \\ 2 & -8 & -3 \end{bmatrix}$

45. $\dfrac{\sin^2 x}{\cos^4 x + \cos^2 x \sin^2 x} \overset{?}{=} \tan^2 x$

$\dfrac{\sin^2 x}{\cos^2 x(\cos^2 x + \sin^2 x)} \overset{?}{=} \tan^2 x$

$\dfrac{\sin^2 x}{\cos^2 x} \overset{?}{=} \tan^2 x$

$\tan^2 x = \tan^2 x$

47. $\dfrac{(x-2)^2}{4} + \dfrac{(y-6)^2}{3} = 1$ **49.** B

Pages 639-640   Lesson 11-6

5. $x = \dfrac{\log 98}{\log 7}$; 2.36   **7.** $x = \dfrac{\log 121}{\log 5}$; 2.98   **9.** $x = \dfrac{\log 16}{\log 3}$; 2.52

11. $x = -\dfrac{\log 18}{\log 3}$; -2.63   **13.** $x > 3.807$ **15.** $x \geq 4.716$

17. 4.75   **19.** 7.83   **21.** 0.82   **23.** 2.14   **25.** 1244.84

27. -7.74   **29.** 5.58   **31.** $1 < x < 6$ **33.** $x > -7.64$ **35.** 0.37

43. 7.48 yr

47. $y = \pm 17 \sin (8x + 480°)$ **45.**

49. (-0.25, 0)   **51.** 6.89

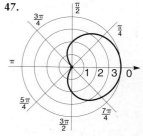

$x + 4y < 9$

Pages 643-645   Lesson 11-7

3. 2.2322   **5.** -1.5606   **7.** 2.1882   **9.** 2.1972   **11.** 0.6957

13. 4.0395   **15.** 6.8876   **17.** 0   **19.** 146.4963   **21.** 1.8589

23. 1   **25.** 133.14   **27.** 48.52   **29.** 17.63   **31.** -0.000385

33a. -0.0211   **b.** 81%   **c.** 2037   **35a.** about 12.4 years

37a. 0.713   **b.** about 46 minutes   **39a.** 0.08 g   **b.** 0.055 g

43. (-8, 9)

49. D

47.

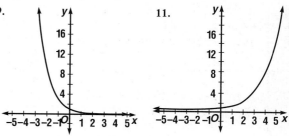

Pages 646-648   Chapter 11 Summary and Review

1. $\frac{1}{2}$ **3.** 3   **5.** $\frac{1}{3}$ **7.** $\frac{1}{8}x^{12}$

9.   **11.**

13. 10.3812   **15.** 6.3510   **17.** 3   **19.** 15   **21.** 3

23. 1.6314   **25.** 0.70   **27.** 21.41   **29.** 117.35   **31.** 1.99

33. 2.45   **35.** -4.82   **37.** 5.9989   **39.** 1226.7210   **41.** 2.91

43. 3.32   **45a.** 45 words per minute   **b.** 64 words per minute   **c.** at 3.5 weeks   **47.** 3561 years old

1. 49 **2.** 2.5 **3.** $343\frac{1}{7}$ **4.** $\frac{1}{64a^6}$ **5.** $x^6y^8a^5$ **6.** $a^{\frac{5}{2}}b^3$

7.

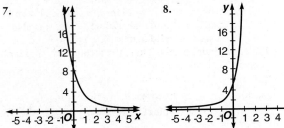

8.

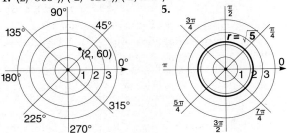

9. 8 **10.** 4 **11.** 2 **12.** 3.34 **13.** -0.66 **14.** 1.54
15. 1.4778 **16.** 0.7459 **17.** 2.7340 **18.** -1.3943
19. 3.0701 **20.** 0.0123 **21.** \$4825.31 **22.** \$7627.49
23. 168 gaps **24.** 11 hours, 21 minutes **25.** 2.66 meters
BONUS a^9

Pages 650-651 Unit 3 Review

1. (2, -300°), (-2, -120°), (-2, 240°)

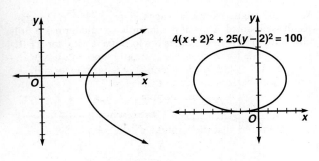

9. (2.83, 3.93) **11.** (3.61, -0.98) **13.** $y = 2$ **15.** i
17. $3 + 3i$ **19.** $\frac{1 + 8i}{13}$ **21.** $3\left(\cos\frac{3\pi}{2} + i\sin\frac{3\pi}{2}\right)$
23. $\sqrt{5}(\cos 2.03 + i\sin 2.03)$ **25.** -8 **27.** $8 - 8i$
29. (0, 0); $\frac{7}{2}$ **31.** $\left(-\frac{9}{2}, 4\right)$, $\frac{\sqrt{129}}{2}$
33. vertex: (3.4, -1), focus: **35.** center: (-2, 2),
(4.65, -1), directrix: $x = 2.15$, foci: $(-2 \pm \sqrt{21}, 2)$, vertices:
axis: $y = -1$ (-7, 2), (3, 2), (-2, 4), (-2, 0)

37. center: (-1, -3), foci: $(-1, -3 \pm \sqrt{13})$, vertices: (-1, 0),
(-1, -6), asymptotes: $y + 3 = \pm\frac{4}{9}(x + 1)$
(See art in next column.)

37.

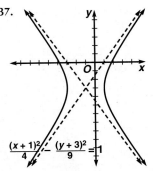

39. hyperbola;

$$\frac{(x - 1)^2}{25} - \frac{(y + 3)^2}{4} = 1$$

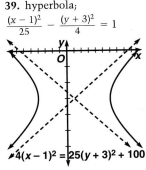

41. $\frac{x - 2}{3} = (y + 7)^2$

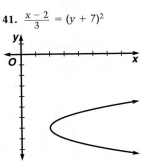

43.

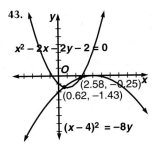

45. $x + y - 1 = 0$,
$x - y - 3 = 0$
47. $y - 8 = 0$, $x - 2 = 0$
49. $3abc^2\sqrt[3]{2ac^2}$ **51.** $\frac{1}{x^2}$
55. 2.7912 **57.** 0.0091
59. $\frac{27}{2}$ **61.** 10,028.573
63. 5.496 **65.** 2 **67.** 3.83
69. 147.24

53.

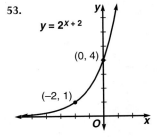

CHAPTER 12 SEQUENCES AND SERIES

Pages 660-662 Lesson 12-1

5. 17, 21, 25, 29, 33 **7.** -13, -19, -25, -31, -37 **9.** 6, 7.5, 9,
10.5, 12 **11.** 2n, 3n, 4n, 5n, 6n **13.** -5b, -7b, -9b, -11b, -13b
15. 227 **17.** 12, 16.5, 21 **19.** -25 **21.** 11 **23.** 15 **25.** 6
27. $3 - 4\sqrt{3}$ **29.** 149 **35.** 77 **37.** 8 **41.** 231 cans **43a.** 31
b. 496 **45.** 1, 3 **47.** 31.68 cm² **49.** $20 + 15i$ **51.** D

Pages 668-669 Lesson 12-2

7. 1.75, 0.875, 0.4375, 0.21875 **9.** yes **11.** yes **13.** $\frac{3}{2}$
15. $201\frac{2}{3}$ **17.** 12, 18, 27 **19.** 10^{-9} **21.** 0.3125, 0.625, 1.25
25. 41.5625 **27.** $31(1 + \sqrt{2})$ **29.** \$585,350.73,
\$2,170,024.72, \$8,044,761.88, \$29,823,712.71, \$110,563,103.40
31. 1.258×10^9, 2.111×10^{16} **33a.** 0.020 cm, 0.040 cm
b. $0.005(2)^{n-1}$ **c.** 2.56 cm, 3.169×10^{27} cm

35.

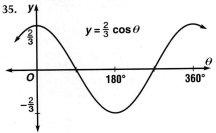

37a. 80, 75, 70, . . . **b.** $T = 80 - 5n$ **c.** 75°F

Pages 676-678 Lesson 12-3

5. 2^{n-1} **7.** $3 \cdot \left(\frac{2}{3}\right)^{n-1}$ **9.** $-\frac{2}{5}$ **11.** 3 **13.** $\frac{1}{10}$ **15.** $\frac{8}{9}$
17. 0 **19.** does not exist **21.** 1 **23.** $\frac{4}{3}$ **25.** 0 **27.** $\frac{3}{11}$
29. $\frac{7}{22}$ **31.** $2\frac{205}{999}$ **33.** $\frac{3}{2}(\sqrt{3} + 1)$ **35.** $\frac{2}{9}$ **37.** does not
exist **39.** 4 strokes **41.** $2\sqrt{2}$ units **43.** $k = \frac{-3 \pm i\sqrt{15}}{6}$
45. 181.3 **47.** C

Pages 683-684 Lesson 12-4

5. N, C **7.** A, D **9.** G, C **11.** C **13.** C **15.** cannot tell
17. C **19.** C **21.** D **23.** C **25.** C **29.** 380 feet
31. 109.63 ft **33.** 1

Page 684 Mid-Chapter Review

1. $5, 6\frac{1}{2}, 8, 9\frac{1}{2}, 11, 12\frac{1}{2}, 14, 15\frac{1}{2}, 17$ **3.** $2, \pm1, \frac{1}{2}, \pm\frac{1}{4}, \frac{1}{8}$
5. does not exist **7.** $\frac{4}{11}$ **9.** $608,487.05

Pages 688-690 Lesson 12-5

5. 18 **7.** 60 **9.** 18 **11.** $\sum\limits_{n=1}^{5} \frac{32}{2^n}$ **13.** $\sum\limits_{n=2}^{\infty} \frac{n}{5}$ **15.** 2160
17. 362,880 **19.** 144 **21.** 90 **23.** -3 **25.** -40 **27.** 68
29. 1084 **31.** 42 **33.** 4 **35.** $\sum\limits_{k=1}^{5} 10k$ **37.** $\sum\limits_{k=0}^{4} 3 \cdot 2^k$
39. $\sum\limits_{k=2}^{10} \frac{1}{k}$ **41.** $\sum\limits_{k=0}^{3} (11 - 2k)$ **43.** $\sum\limits_{k=1}^{\infty} 2k$ **45.** $\sum\limits_{k=3}^{\infty} \frac{2}{k}$
47. $\sum\limits_{k=0}^{\infty} \frac{k!}{k+1}$ **49.** $x(x - 1)$ **51.** $x - y$ **59.** $360,110.31
61a. 2 weeks **b.** $58.80 per tree **63.** $-\frac{\sqrt{2}}{2}$
65. $(y + 1)^2 = -12(x - 6)$ **67.** $\frac{2}{9}$ **69.** A

Pages 694-695 Lesson 12-6

5. $a^7 + 7a^6b + 21a^5b^2 + 35a^4b^3 + 35a^3b^4 + 21a^2b^5 + 7ab^6 + b^7$
7. $256 - 256b + 96b^2 - 16b^3 + b^4$ **9.** $35a^4b^3$
11. $x^5 + 5x^4y + 10x^3y^2 + 10x^2y^3 + 5xy^4 + y^5$
13. $x^7 - 14x^6 + 84x^5 - 280x^4 + 560x^3 - 672x^2 + 448x - 128$
15. $x^6 + 18x^5 + 135x^4 + 540x^3 + 1215x^2 + 1458x + 729$
17. $64x^6 + 192x^5y + 240x^4y^2 + 160x^3y^3 + 60x^2y^4 + 12xy^5 + y^6$
19. $16x^4 + 32\sqrt{3}x^3 + 72x^2 + 24\sqrt{3}x + 9$ **21.** $4032x^5y^4$
23. $\frac{105}{32}x^4y^6$ **25.** 43,046,721 **27.** 57 **29.** 50%, 25%,
1.56%, $100[1 - (0.5)^d]$% **31.** undefined **33.** 1.5

Pages 699-701 Lesson 12-7

7. $\frac{1}{1} = 1, \frac{2}{1} = 2, \frac{3}{2} = 1.5, \frac{5}{3} = 1.667, \frac{8}{5} = 1.6, \frac{13}{8} = 1.625,$
$\frac{21}{13} = 1.615, \frac{34}{21} = 1.619, \frac{55}{34} = 1.618$; They approach 1.618.
9. 2.99 **11.** 2.58 **13.** $5e^{i\frac{5\pi}{3}}$ **15.** $12e^{i\frac{\pi}{6}}$ **17.** $6e^{i\frac{\pi}{2}}$
19. $i\pi - 4.8036$ **21.** 1.57 **23.** 9.14 **25.** 6.39 **27.** -0.9760; -1
29. $e^{i\frac{\pi}{4}}$ **31.** $4\sqrt{2}e^{i\frac{7\pi}{4}}$ **33.** $6e^{i\frac{\pi}{3}}$ **35.** $i\pi + 1.9459$
37. $i\pi + 1.6544$ **39.** $i\pi - 3.3242$ **43a.** neither **b.** 21, 28, 36,
45, 55 **c.** $\frac{n(n + 1)}{2}$ **d.** 1275 **45.** 0° **47.** $285.77 **49.** A

Pages 704-705 Lesson 12-8

5. 3725 **7.** 969 **9.** Step 1: Verify that the formula is valid
for $n = 1$. Since $S_1 = 2$ and $1(1 + 1) = 2$, the formula is valid
for $n = 1$.
Step 2: Assume that the formula is valid for $n = k$ and derive
a formula for $n = k + 1$.
$2 + 4 + 6 + \cdots + 2k + 2(k + 1) = k(k + 1) + 2(k + 1)$
$2 + 4 + 6 + \cdots + 2k + 2(k + 1) = (k + 1)(k + 2)$

Apply the original formula for $n = k + 1$.
$S_{k+1}: (k + 1)[(k + 1) + 1] = (k + 1)(k + 2)$
The formula gives the same result as adding the $(k + 1)$ term
directly. Thus, if the formula is valid for $n = k$, it is also valid
for $n = k + 1$. Since the formula is valid for $n = 1$, it is also
valid for $n = 2$. Since it is valid for $n = 2$, it is also valid for
$n = 3$, and so on, indefinitely. Thus, the formula is valid for
all positive integral values of n.
11. Step 1: Verify that the formula is valid for $n = 1$. Since
$S_1 = \frac{1}{2}$ and $1 - \frac{1}{2^1} = \frac{1}{2}$, the formula is valid for $n = 1$.
Step 2: Assume that the formula is valid for $n = k$ and derive
a formula for $n = k + 1$.
$$\frac{1}{2} + \frac{1}{2^2} + \frac{1}{2^3} + \cdots + \frac{1}{2^k} = 1 - \frac{1}{2^k}$$
$$\frac{1}{2} + \frac{1}{2^2} + \frac{1}{2^3} + \cdots + \frac{1}{2^k} + \frac{1}{2^{k+1}} = 1 - \frac{1}{2^k} + \frac{1}{2^{k+1}}$$
$$= 1 - \frac{2}{2 \cdot 2^k} + \frac{1}{2^{k+1}}$$
$$= 1 - \frac{1}{2^{k+1}}$$
When the original formula is applied for $n = k + 1$, the same
result is obtained. Thus, if the formula is valid for $n = k$, it is
also valid for $n = k + 1$. Since the formula is valid for $n = 1$,
it is also valid for $n = 2$. Since it is valid for $n = 2$, it is also
valid for $n = 3$, and so on, indefinitely. Thus, the formula is
valid for all positive integral values of n.
13. Step 1: Verify that the formula is valid for $n = 1$. Since
$S_1 = 1$ and $\frac{1^2(1 + 1)^2}{4} = 1$, the formula is valid for $n = 1$.
Step 2: Assume that the formula is valid for $n = k$ and derive
a formula for $n = k + 1$.
$1^3 + 2^3 + 3^3 + \cdots + k^3 = \frac{k^2(k + 1)^2}{4}$
$1^3 + 2^3 + 3^3 + \cdots + k^3 + (k + 1)^3 = \frac{k^2(k + 1)^2}{4} + (k + 1)^3$
$= \frac{(k + 1)^2(k + 2)^2}{4}$
When the original formula is applied for $n = k + 1$, the same
result is obtained. Thus, if the formula is valid for $n = k$, it is
also valid for $n = k + 1$. Since the formula is valid for $n = 1$,
it is also valid for $n = 2$. Since it is valid for $n = 2$, it is also
valid for $n = 3$, and so on, indefinitely. Thus, the formula is
valid for all positive integral values of n.
15. Step 1: Verify that the formula is valid for $n = 1$. Since
$S_1 = 1$ and $2^1 - 1 = 1$, the formula is valid for $n = 1$.
Step 2: Assume that the formula is valid for $n = k$ and derive
a formula for $n = k + 1$.
$1 + 2 + 4 + \cdots + 2^{k-1} = 2^k - 1$
$1 + 2 + 4 + \cdots + 2^{k-1} + 2^k = 2^k - 1 + 2^k$
$= 2(2^k) - 1$ or $2^{k+1} - 1$
When the original formula is applied for $n = k + 1$, the same
result is obtained. Thus, if the formula is valid for $n = k$, it is
also valid for $n = k + 1$. Since the formula is valid for $n = 1$,
it is also valid for $n = 2$. Since it is valid for $n = 2$, it is also
valid for $n = 3$, and so on, indefinitely. Thus, the formula is
valid for all positive integral values of n.
17. Step 1: Verify that the formula is valid for $n = 1$. Since
$S_1 = 1$ and $\frac{6 + 15 + 10 - 1}{30} = 1$, the formula is valid for $n = 1$.
Step 2: Assume that the formula is valid for $n = k$ and derive
a formula for $n = k + 1$.
$1^4 + 2^4 + 3^4 + \cdots + k^4 = \frac{6k^5 + 15k^4 + 10k^3 - k}{30}$
$1^4 + 2^4 + 3^4 + \cdots + k^4 + (k + 1)^4$
$= \frac{6k^5 + 15k^4 + 10k^3 - k}{30} + (k + 1)^4$
$= \frac{6(k + 1)^5 + 15(k + 1)^4 + 10(k + 1)^3 - (k + 1)}{30}$

When the original formula is applied for $n = k + 1$, the same result is obtained. Thus, if the formula is valid for $n = k$, it is also valid for $n = k + 1$. Since the formula is valid for $n = 1$, it is also valid for $n = 2$. Since it is valid for $n = 2$, it is also valid for $n = 3$, and so on, indefinitely. Thus, the formula is valid for all positive integral values of n.

19. Step 1: Verify that the formula is valid for $n = 1$. Since $S_1 = \frac{1}{2}$ and $\frac{1}{1+1} = \frac{1}{2}$, the formula is valid for $n = 1$.
Step 2: Assume that the formula is valid for $n = k$ and derive a formula for $n = k + 1$.

$$\frac{1}{1 \cdot 2} + \frac{1}{2 \cdot 3} + \frac{1}{3 \cdot 4} + \cdots + \frac{1}{k(k+1)} = \frac{k}{k+1}$$

$$\frac{1}{1 \cdot 2} + \frac{1}{2 \cdot 3} + \frac{1}{3 \cdot 4} + \cdots + \frac{1}{k(k+1)} + \frac{1}{(k+1)(k+2)}$$

$$= \frac{k}{k+1} + \frac{1}{(k+1)(k+2)}$$

$$= \frac{k+1}{k+2}$$

When the original formula is applied for $n = k + 1$, the same result is obtained. Thus, if the formula is valid for $n = k$, it is also valid for $n = k + 1$. Since the formula is valid for $n = 1$, it is also valid for $n = 2$. Since it is valid for $n = 2$, it is also valid for $n = 3$, and so on, indefinitely. Thus, the formula is valid for all positive integral values of n. **21a.** arithmetic **b.** 156 times **23a.** geometric **b.** 1.718×10^{10} **25.** no solution **27.** $(0, 3)$, $(0, 3 + \sqrt{7})$, $(0, 3 - \sqrt{7})$; $(0, 7)$, $(0, -1)$, $(3, 3)$, $(-3, 3)$ **29.** A

Pages 706-708   Chapter 12 Summary and Review

1. 6.9, 8.2, 9.5, 10.8, 12.1 **5.** 140.8 **7.** $\frac{2}{5}$ **9.** does not exist **11.** divergent **13.** divergent **15.** $\frac{2}{3}$ **17.** $\sum_{k=1}^{9} (k^2 + 1)$ **19.** $16r^4 + 96r^3s + 216r^2s^2 + 216rs^3 + 81s^4$ **21.** $262{,}440x^3y^7$ **23.** $2\sqrt{2}e^{i\frac{7\pi}{4}}$ **25.** $6e^{i\frac{\pi}{6}}$ **27.** Step 1: Verify that the formula is valid for $n = 1$. Since $S_1 = 1(1 + 2) = 1(3) = 3$ and since $\frac{1(1+1)(2 \cdot 1 + 7)}{6} = \frac{1(2)(2 + 7)}{6} = \frac{1(2)(9)}{6} = \frac{18}{6} = 3$, the formula is valid for $n = 1$.
Step 2: Assume that the formula is valid for $n = k$ and derive a formula for $n = k + 1$.

$$3 + 8 + 15 + \cdots + k(k+2) = \frac{k(k+1)(2k+7)}{6}$$

$$3 + 8 + 15 + \cdots + k(k+2) + (k+1)(k+3)$$

$$= \frac{k(k+1)(2k+7)}{6} + (k+1)(k+3)$$

$$= \frac{(k+1)[(k+1)+1][2(k+1)+7]}{6}$$

When the original formula is applied for $n = k + 1$, the same result is obtained. Thus, if the formula is valid for $n = k$, it is also valid for $n = k + 1$. Since the formula is valid for $n = 1$, it is also valid for $n = 2$. Since it is valid for $n = 2$, it is also valid for $n = 3$, and so on, indefinitely. Thus, the formula is valid for all positive integral values of n.
29. 193.75 feet

Page 709   Chapter 12 Test

1. 7.5, 9, 10.5, 12, 13.5 **2.** $\frac{4}{625}$, $\frac{8}{3125}$, $\frac{16}{15{,}625}$, $\frac{32}{78{,}125}$ **3.** -4, -1, 2, 5, 8 **4.** Sample answer: 16, 8, 4, 2, 1 **5.** 109 **6.** $\frac{2187}{256}$ **7.** 10 **8.** $2557\frac{1}{2}$ **9.** does not exist **10.** $\frac{1}{2}$ **11.** convergent **12.** divergent **13.** $\sum_{k=1}^{19} 5k$ **14.** $\sum_{k=4}^{\infty} (2k - 1)$ **15.** $8064a^5$ **16.** $5670x^4y^4$ **17.** $2\sqrt{2}e^{i\frac{3\pi}{4}}$ **18.** Step 1: Verify that the formula is valid for $n = 1$. Since

$S_1 = 6$ and $\frac{1(1+1)(4+5)}{3} = 6$, the formula is valid for $n = 1$.
Step 2: Assume the formula is valid for $n = k$ and derive a formula for $n = k + 1$.

$$2 \cdot 3 + 4 \cdot 5 + 6 \cdot 7 + \cdots + 2k(2k+1) = \frac{k(k+1)(4k+5)}{3}$$

$$2 \cdot 3 + 4 \cdot 5 + \cdots + 2k(2k+1) + 2(k+1)(2k+3)$$

$$= \frac{k(k+1)(4k+5)}{3} + 2(k+1)(2k+3)$$

$$= \frac{(k+1)(k+2)(4k+9)}{3}$$

When the original formula is applied for $n = k + 1$, the same result is obtained. Thus, if the formula is valid for $n = k$, it is also valid for $n = k + 1$. Since the formula is valid for $n = 1$, it is also valid for $n = 2$. Since it is valid for $n = 2$, it is also valid for $n = 3$, and so on, indefinitely. Thus, the formula is valid for all positive integral values of n.

19. \$12,080.40 **20.** $\sum_{n=1}^{6} 2^n$; 126 ancestors **BONUS** 0.625

CHAPTER 13   ITERATION AND FRACTALS

Pages 715-716   Lesson 13-1

5. 5, 11, 23 **7.** 0, -1, 0 **9.** -1, -7, 647 **11.** -8, -29, -92, -281, -848, -2549, -7652, -22,961, -68,888, -206,669 **13.** 1, 1, 1 **15.** 2.33, 13.65, 2544.30 **17.** 0.45, 0.69, 0.60, 0.67, 0.62, 0.66, 0.63, 0.65, 0.64, 0.65 **19.** 0.25, 0.53, 0.70, 0.59, 0.68, 0.61, 0.67, 0.62, 0.66, 0.63 **21.** 0.50, 0.78, 0.53, 0.77, 0.55, 0.77, 0.55, 0.77, 0.55, 0.77 **23.** 0.28, 0.62, 0.73, 0.61, 0.74, 0.60, 0.74, 0.60, 0.74, 0.60 **25a.** 2, 1, 2, 1, 2, 1, 2, 1, 2, 1 **b.** $\frac{1}{2}$, 4, $\frac{1}{2}$, 4, $\frac{1}{2}$, 4, $\frac{1}{2}$, 4, $\frac{1}{2}$, 4 **c.** $\frac{2}{a}$, a, $\frac{2}{a}$, a, $\frac{2}{a}$, a, $\frac{2}{a}$, a, $\frac{2}{a}$, a **d.** They repeat in pairs. **e.** yes, $\pm\sqrt{2}$
29a. $p_{n+1} = p_n + 1.75p_n(1 - p_n)$ **b.** 458, 525, 479, 514, 489, 508, 494, 505, 496, 503, 498, 501, 499, 501, 499

29c.

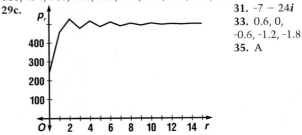

31. $-7 - 24i$ **33.** 0.6, 0, -0.6, -1.2, -1.8 **35.** A

Pages 720-721   Lesson 13-2

3.

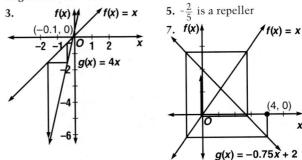

5. $-\frac{2}{5}$ is a repeller
7.

9. 4, staircase out **11.** -2, spiral out **13.** -0.75, spiral in **15.** -6, spiral out **19.** $-1 < m < 0$ **21.** $0 < m < 1$ **23.** $\left(-2\frac{2}{3}, -2\frac{2}{3}\right)$; repeller **25.** (-20, -20); attractor **27.** $f(x) = -3x + 2$ and $f(x) = -0.8x - 18$ **33a.** $\left(\frac{12p}{r}, \frac{12p}{r}\right)$;

the interest for a term is equal to the payment, so the principal is never reduced. **b.** repeller **35.** $\frac{1}{4}$, $-\frac{4}{3}$
37. $26\frac{39}{128}$ **39.** 3%

Pages 726-728 Lesson 13-3

5. (0.5, 0.75) **7.** (0, 0), (0.5, 0.5); attractor **9.** (0, 0), $\left(\frac{1}{6}, \frac{1}{6}\right)$; attractor **11.** fixed point attractor **13.** (0.5, 0.6) **15.** (0.5, 1)
17. (0, 0), $\left(\frac{3}{8}, \frac{3}{8}\right)$; attractor **19.** (0, 0), $\left(\frac{11}{16}, \frac{11}{16}\right)$; repeller
21. period-2 attractor **23.** chaos **25.** period-5 attractor
27a. $a \approx 3$ **b.** $a \approx 3.4495$ **c.** $a \approx 3.5441$ **d.** chaos
29a. 41, 39, 40, 40, 40, 40, 40, 40, 40, 40 **b.** fixed point attractor **c.** period-4 attractor **31.** -6, 2
33. $\frac{(x-2)^2}{4} + \frac{(y-5)^2}{25} = 1$; ellipse **35.** $\left(-\frac{50}{9}, -\frac{50}{9}\right)$; repeller

Page 728 Mid-Chapter Review

1. 37, 182, 907 **3.** -2, 14, 142 **5.** staircase out **7.** spiral out **9.** fixed point attractor

Pages 731-733 Lesson 13-4

5. $3 - 4i$, $1.8 - 0.4i$, $1.08 + 1.76i$ **7.** $-4 - 3i$, $7 + 21i$, $-392 + 291i$ **9.** $1 - 3i$, $-8 - 9i$, $-17 + 141i$ **11.** $2 + 5i$, $4 + 11i$, $8 + 23i$ **13.** $6 + i$, $12 + 3i$, $24 + 7i$ **15.** $1 + \frac{1}{2}i$, $2 + 2i$, $4 + 5i$ **17.** $5 + 3i$, $17 + 6i$, $53 + 15i$ **19.** 14, $44 - 3i$, $134 - 12i$ **21.** $3.5 - 6i$, $12.5 - 21i$, $39.5 - 66i$ **23.** $1 + 2i$, $-2 + 6i$, $-31 - 22i$, $478 + 1366i$ **25.** $-1 + i$, $2 - 5i$, $-19 - 23i$, $-166 + 871i$

27. 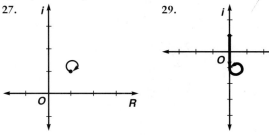 **29.**

33. -4 **35.** $3xy^{\frac{1}{4}}z^{\frac{1}{4}}$ **37.** fixed point attractor

Pages 738-739 Lesson 13-5

5. escape set **7.** Julia set **9.** Julia set **11.** escape set
13. prisoner set **15.** escape set **17.** connected
19. connected **21.** disconnected **23.** connected
25. disconnected **35.** $\frac{7}{25}$ **37.** $1 - 3i$, $-7 - 9i$, $-31 + 123i$, $-14,167 - 7629i$

Pages 743-745 Lesson 13-6

5a. disconnected **b.** outside **c.** orange **7a.** disconnected
b. outside **c.** light blue **9a.** disconnected **b.** outside
c. light blue **11a.** disconnected **b.** outside **c.** light blue
13a. connected **b.** inside **c.** black **15a.** connected
b. inside **c.** black **17a.** disconnected **b.** outside **c.** light blue **19a.** disconnected **b.** outside **c.** purple
21a. disconnected **b.** outside **c.** light blue
23a. disconnected **b.** outside **c.** light blue **25.** outside the set **31.** (-3, 2) **33.** 27 **35.** escape set

Pages 746-748 Chapter 13 Summary and Review

1. -1, 7, -9 **3.** 0.34, 12.81, 3248.37 **5.** $-\frac{1}{2}$ is a repeller
7. $\frac{5}{7}$ is a repeller **9.** (0, 0), (0.630, 0.630); attractor

11. (0, 0), (0.687, 0.687); repeller **13.** period-2 attractor
15. period-4 attractor **17.** 4, $6 - 2i$, $7 - 3i$ **19.** $2 + i$, $5 - 1.5i$, $6.5 - 2.75i$ **21.** escape **23.** escape
25. disconnected **27.** disconnected **29.** orange
31. red **33a.** $b_n = 1.00792b_{n-1} - 714.73$ **b.** $84,958.48, $84,916.63, $84,874.45, $84,831.94

Page 749 Chapter 13 Test

1. 3, 7, 23 **2.** 0.81, 1.42, 0.34 **3.** -1, 6, 223 **4.** (5, 5); repeller **5.** (-2.5, -2.5); attractor **6.** (0.133, 0.133); repeller
7. $-1 < m < 0$ **8.** (0.5, 0.4); (0, 0), (0.375, 0.375); fixed attractor **9.** (0.5, 0.87); (0, 0), (0.713, 0.713); period-4 attractor **10.** (0.5, 0.97); (0, 0), (0.742, 0.742); chaos
11. $3 + 3i$, $9 + 5i$, $21 + 9i$ **12.** $1 + 3i$, $5 + 5i$, $13 + 9i$
13. $4 + i$, $11 + i$, $25 + i$ **14.** escape **15.** prisoner
16. escape **17.** disconnected **18.** connected
19. connected **20.** outside; 15 **21.** outside; 11
22. outside; 7 **23.** outside; 6 **24.** inside the Mandelbrot set **25.** 137, 96, 136, 97, 136 **BONUS** $3 - 2i$

CHAPTER 14 COMBINATORICS AND PROBABILITY

Pages 755-757 Lesson 14-1

5. independent **7.** independent **9.** false **11.** false
13. 12 **15.** 120 **17.** 120 **19.** 39,916,800 **21.** 72 **23.** 8
25. 17,576,000 **27.** 14 **29.** 4 **31.** 24 **33.** 72 **35.** 480
37. 57,600,000 **39.** 62,500,000 **41.** 42 **43.** 12
47. 6,744,109,680 **49.** -4 **51.** $-7 + 24i$ **53.** disconnected; outside

Pages 761-763 Lesson 14-2

7. circular, not reflective, 120 **9.** linear, not reflective, 3,628,800 **11.** linear, not reflective **13.** circular, not reflective **15.** 2520 **17.** 90,720 **19.** 6720 **21.** 453,600
23. 360 **25.** 2520 **27.** 4200 **29.** 14,440 **31.** 12 **33.** 72
35. 72 **37.** 15 **39.** 30% beef, 20% pork; $76 **41.** $c = 9.5$, $A = 38°20'$, $B = 36°22'$ **43.** $\frac{(y+1)^2}{4} - \frac{(x-3)^2}{5} = 1$ **45.** B

Pages 766-767 Lesson 14-3

5. combination **7.** permutation **9.** 792 **11.** 168
13. 1960 **15.** 267,696 **17.** 3570 **19.** 2940 **21a.** 5148
b. 28,561 **23.** 14 **25.** 12 **27.** 126 **29.** 2808 **31.** 2520
33. 252 **37.** 233,376 **39.** (5, 4, -11) **41.** 120

Page 767 Mid-Chapter Review

1. 24 **3.** 24 **5.** 715

Pages 771-773 Lesson 14-4

5. $\frac{1}{1}$ **7.** $\frac{7}{8}$ **9.** $\frac{3}{7}$ **11.** $\frac{4}{13}$ **13.** $\frac{2}{7}$ **15.** $\frac{4}{7}$ **17.** $\frac{1}{3}$
19. $\frac{18}{95}$ **21.** $\frac{9}{38}$ **23.** $\frac{11}{25}$ **25.** $\frac{5}{68}$, $\frac{5}{63}$ **27.** $\frac{35}{68}$, $\frac{35}{33}$
29. $\frac{99}{16,722,971}$ **31.** $\frac{4719}{9,000,011}$ **35a.** $\frac{46}{125}$ **b.** $\frac{11}{100}$ **c.** $\frac{9}{92}$
37. $-\frac{33}{65}$ **39.** 126

Pages 776-778 Lesson 14-5

5. independent, $\frac{25}{81}$ **7.** dependent, $\frac{8}{195}$ **9.** independent, $\frac{3}{10}$
11. $\frac{25}{441}$ **13.** $\frac{1}{8}$ **15.** $\frac{5}{48}$ **17.** $\frac{25}{36}$ **19.** $\frac{5}{6}$ **21a.** $\frac{1}{24}$

b. $\frac{35}{1024}$ **23.** $\frac{69{,}343{,}957}{2{,}373{,}046{,}875} \approx 0.029$ **25.** $\frac{1}{29{,}170{,}800}$

27. $\frac{4}{635{,}013{,}559{,}600}$ **29.** 0 **33.** $\frac{435{,}643}{560{,}175}$ **35.** origin

37. $\sum_{k=1}^{\infty} 3^k$ **39.** B

Pages 781-783   Lesson 14-6

5a. exclusive, $\frac{1}{4}$ **b.** inclusive, $\frac{11}{36}$ **7.** inclusive, $\frac{4}{5}$

9. inclusive, $\frac{20}{27}$ **11.** $\frac{2}{3}$ **13.** $\frac{2}{11}$ **15.** $\frac{14}{33}$ **17.** $\frac{12}{221}$

19. $\frac{188}{663}$ **21.** $\frac{35}{64}$ **23.** $\frac{21}{64}$ **25.** $\frac{2}{429}$ **27.** $\frac{175}{429}$ **29.** $\frac{1}{525}$

31. $\frac{3}{35}$ **35.** $\frac{9979}{10{,}000}$ **37.** $\frac{(y-1)^2}{36} + \frac{(x-3)^2}{36} = 1$ **39.** $\frac{253}{4606}$

Pages 787-789   Lesson 14-7

5. Event A: one coin shows tails; Event B: at least one coin shows heads; $\frac{2}{3}$ **7.** $\frac{1}{2}$ **9.** $\frac{3}{13}$ **11.** $\frac{1}{26}$ **13.** Event A: the girls are separated; Event B: a girl is on an end; $\frac{3}{5}$

15. 0 **17.** $\frac{1}{10}$ **19.** $\frac{7}{10}$ **21.** $\frac{1}{4}$ **23.** $\frac{1}{7}$ **25.** $\frac{2}{5}$

27. $\frac{3}{10}$ **29.** $\frac{1}{8}$ **31.** $\frac{1}{6}$ **33.** $\frac{19}{51}$ **37.** Let event A be that a person buys something. Let event B be that a person asks questions.

$$P(A/B) = \frac{P(A \text{ and } B)}{P(B)}$$ Four out of five customers who ask questions will make a purchase. Therefore, they are more likely to buy something if they ask questions.
$$= \frac{\frac{48}{200}}{\frac{60}{200}} \text{ or } \frac{4}{5}$$

39a. $\frac{69}{70}$ **b.** $\frac{2}{25}$ **c.** $\frac{1}{25}$ **41.** $x^4 + 2x^3 + 4x^2 + 5x + 10$
43. 2, 5, 26 **45.** C

Pages 792-794   Lesson 14-8

3. binomial, $\frac{3}{8}$ **5a.** binomial, $\frac{16}{81}$ **b.** binomial, $\frac{4}{81}$

c. binomial, $\frac{1}{9}$ **d.** not binomial **e.** not binomial **f.** not binomial **7.** $\frac{3}{8}$ **9.** $\frac{3125}{7776}$ **11.** $\frac{625}{648}$ **13.** $\frac{1}{81}$ **15.** $\frac{65}{81}$

17. $\frac{15}{128}$ **19.** $\frac{193}{512}$ **21.** $\frac{256}{625}$ **23.** $\frac{21}{3125}$ **25.** $\frac{1}{8}$ **27.** $\frac{3}{8}$

29. $\frac{419{,}904}{1{,}953{,}125}$ **31.** $\frac{1{,}623{,}424}{9{,}765{,}625}$ **33.** $\frac{768}{3125}$ **35.** 0.1372

37. 7 **39.** $\frac{1}{2}$

Pages 796-798   Chapter 14 Summary and Review

1. 120 **3.** $-\frac{9}{5}$ **5.** 30 **7.** 20,160 **9.** 10 **11.** 3 **13.** $\frac{4}{5}$

15. $\frac{343}{4096}$ **17.** $\frac{5}{36}$ **19.** $\frac{9}{14}$ **21.** $\frac{1}{2}$ **23.** $\frac{1}{4}$ **25.** $\frac{3}{8}$

27. 360 ways **29a.** $\frac{7}{15}$ **b.** $\frac{1}{30}$

Page 799   Chapter 14 Test

1. 30 **2.** 2520 **3.** 56 **4.** 5 **5.** 120 patterns **6.** 60 ways
7. 720 ways **8.** 1485 teams **9.** 120 ways **10.** 170 diagonals

11. $\frac{4}{3}$ **12.** $\frac{33}{16{,}660}$ **13.** $\frac{5}{432}$ **14.** $\frac{1}{8}$ **15.** $\frac{1}{20}$ **16.** $\frac{7}{13}$

17. $\frac{5}{18}$ **18.** $\frac{3}{13}$ **19.** $\frac{17}{81}$ **20.** $\frac{1792}{15{,}625}$ **BONUS** 0.075

CHAPTER 15   STATISTICS AND DATA ANALYSIS

Pages 806-810   Lesson 15-1

5. HBO, Cinemax, and Showtime have shown an increase in the number of subscribers. TMC has shown a decrease in the number of subscribers. **7.** $0; $26,000; $32,000; $38,000; $44,000; $50,000 **9.** 26%

11.

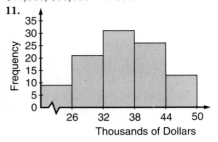

13. interval: 0.1, limits: 1.05, 1.15, 1.25, 1.35, 1.45, 1.55, 1.65, 1.75 **15.** interval: 1, limits: 24.5, 25.5, 26.5, 27.5, 28.5, 29.5, 30.5, 31.5, 32.5 **17.** Women make less money than men for the same education.

19.

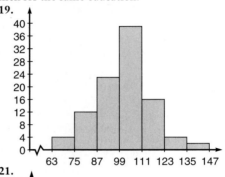

21.

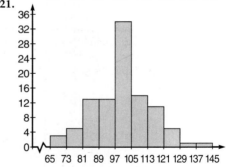

Population Change (in millions)

27.

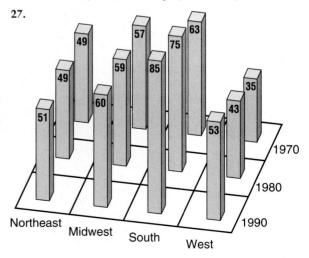

35a.

Class Limits	Class Marks	Tally	Frequency
0–10	5	JHT JHT JHT JHT JHT JHT III	33
10–20	15	JHT JHT III	13
20–30	25	I	1
30–40	35		0
40–50	45	II	2

35b.

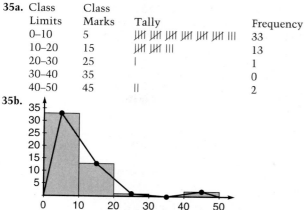

c. Because it differs greatly from the other data. If it was included, it would skew the intervals and the representation of the majority of the distribution.   **37.** none   **39.** 125, 150, 180, 216   **41.** A

Pages 816-819   Lesson 15-2

5.
```
1 | 0 5 5 5 5 7 7
2 | 0 0 0 5 5 5 5 7 8
3 | 0 0 5 5 5
4 | 6
5 | 5        1|0 means 10.
```

7. 30.75, 27.5, 10
9. 8.5, 8.5, 6 and 11

11a.
```
2 | 4 6 6 9
3 | 1 6 6 7 8 8
4 | 1 2 3
    2|4 means 2.4 million.
```

b. 3,438,500; 3,600,000; 2,600,000, 3,600,000, and 3,800,000

c. These are the years when the most babies were born. This large population is now adults in their 30s and 40s. They are an active group in commerce, politics, and other fields.

13a. about $5'9\frac{1}{2}''$ **13b.** Answers will vary. The mean may not be representative because the mean is greater than the median (5'7") and the mode (5'4").   **15a.** 19.3   **b.** 17–21
c. 20.1   **17.** 3   **21.** 34 cm from the end   **23a.** 424.4

b.	26.6	**c.** 400–450
	69.9	**d.** 419.9
	139.2	**e.** Lower; means differ because published
	228.2	average has access to each item of data
	327.0	whereas frequency distributions do not.
	410.1	
	465.2	**25.** $\dfrac{1 + 4i\sqrt{3}}{7}$
	500.0	
	520.5	
	530.4	
	534.4	
	535.0	

27. Iteration is the process of repeated evaluation of a function using the result of the previous evaluation. Example: Let $f(x) = x^2$. Let $x_1 = 2$. Results: 4, 16, 256, . . .

29.

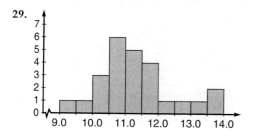

Pages 825-829   Lesson 15-3

5. 205   **7.** 25%   **9.** 50%   **11.** 50%   **13a.** 9.15   **b.** 5.95, 2.35, 1.35, -0.15, -0.35, -0.65, -0.65, -1.55, -2.95, -3.35   **c.** 1.93   **d.** 10.5, 8.65, 7.6   **e.** 2.9   **f.** 1.45   **g.** 66.065   **h.** about 2.57   **j.** 15.1   **15a.** mean: 11.7, median: 11.9, SD: 1.83   **b.** mean: 12.23, median: 11.9, SD: 0.47   **c.** (1) similar   (2) very different   (3) Group 3   **17a.** range: 44, mean: 11.5, MD: 8.95, IQ range: 10.5, SD: 12.2   **b.** range: 35, mean: 10.8, MD: 7.3, IQ range: 5.5, SD: 9.74   **c.** gold   **19.** when both have a value of 1   **21.** 424.24 miles   **23.** 3

Pages 835-837   Lesson 15-4

5.

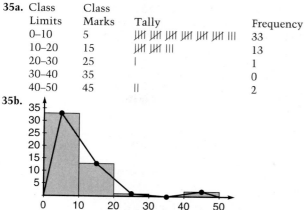

7. 137   **9.** 68   **11.** 68.3%
13. 20.1–27.9   **15.** 68.3%
17. 86.6%   **19.** 15.85%
21. 107–173   **23.** 68.3%
25. 97.8%   **27.** 6800
29. 5000   **31.** 15   **35.** A: 6.7%, B: 24.15%, C: 38.3%, D: 24.15%, F: 6.7%

37.
$$\sin^2 \Phi \cot^2 \Phi \stackrel{?}{=} (1 - \sin \Phi)(1 + \sin \Phi)$$
$$\sin^2 \Phi \left(\frac{\cos \Phi}{\sin \Phi} \right)^2 \stackrel{?}{=} (1 - \sin \Phi)(1 + \sin \Phi)$$
$$\cos^2 \Phi \stackrel{?}{=} (1 - \sin \Phi)(1 + \sin \Phi)$$
$$1 - \sin^2 \Phi \stackrel{?}{=} (1 - \sin \Phi)(1 + \sin \Phi)$$
$$(1 - \sin \Phi)(1 + \sin \Phi) = (1 - \sin \Phi)(1 + \sin \Phi)$$

39. $\frac{1}{2}$ **41.** A

Page 837   Mid-Chapter Review

1.

Class Limits	Class Marks	Tally	Frequency
100–105	102.5	III	3
105–110	107.5	JHT III	8
110–115	112.5	JHT JHT JHT II	17
115–120	117.5	JHT JHT III	13
120–125	122.5	JHT II	7
125–130	127.5	I	1
130–135	132.5	I	1

3.
```
10 | 0 0 4 5 5 5 6 7 8 9 9
11 | 0 0 0 0 1 1 2 2 2 2 3 3 4 4 4
 · | 4 4 5 6 6 7 7 7 8 8 8 8 8 8 9
12 | 0 0 0 0 1 1 2 7
13 | 4
10|9 means 109.
```

5a. 118, 110   **b.** 34

5c.

7. 107–120

Pages 841-843   Lesson 15-5

5. 5, 187.1–212.9, 190.2–209.8   **7a.** 0.14   **b.** 4.239–4.961
c. 4.506–4.694   **d.** 4.369–4.831   **9.** 0.85, 77.81–82.19, 78.33–81.67   **11a.** 3.4   **b.** 0.425   **c.** 11.547–13.213

1009

d. 0.982 **13.** 4.5% **15a.** 0.548 **b.** 0.9415 **17.** 0.1585
19. 11°28' **21.** 0.1585

Pages 851-854 Lesson 15-6

3. positive **5.** positive **7.** negative **9a.** 7 **b.** 65 million
c. $s = 7p + 65{,}000{,}000$ **d.** $86 million

11.
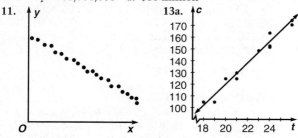

13a.
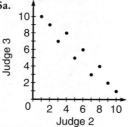

13b. $c = 9t - 59$ **c.** 0.9844291973 **d.** There is a strongly
positive correlation between the temperature and the chirps
per minute.
15b. strongly negative **c.** -0.963
d. Judges 1 and 2 seem very
close in the scoring while
judge 3 is not in synch with
their opinions.

15a.
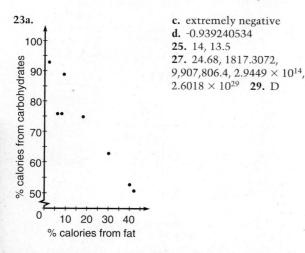

23a.
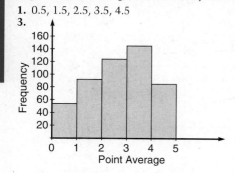

c. extremely negative
d. -0.939240534
25. 14, 13.5
27. 24.68, 1817.3072,
9,907,806.4, 2.9449×10^{14},
2.6018×10^{29} **29.** D

Pages 860-862 Chapter 15 Summary and Review
1. 0.5, 1.5, 2.5, 3.5, 4.5
3.

5. 14.8, 15, none **7.** 133.25, 132, none **9.** 1.74 **11.** 95.5%

13. 79.75–96.25 **15.** 1.742–1.858
17.
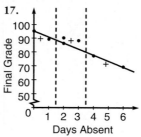

21a.
```
1 | 0 3 5 6 7 9
2 | 1 3 4 5
3 | 9 9
1|0 means 10.
```
b. 20; 39; 21.75

23a.
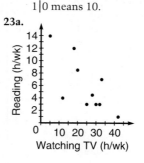

b. -0.72 **c.** moderately negative

Page 863 Chapter 15 Test

1.

Class Limits	Class Marks	Tally	Frequency
0–3	1.5	|||	3
3–6	4.5	JHT		6
6–9	7.5	JHT JHT JHT		16
9–12	10.5	JHT JHT JHT JHT JHT	25
12–15	13.5	JHT JHT JHT ||	17
15–18	16.5	JHT |||	8
18–21	19.5	JHT	5

2.
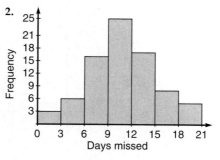

3. 10.438 **4.** 10 **5.** 4.210 **6.** 2.344 **7.** 2.344 **8.** 2.344
9. 0.003
10.
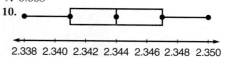

11. 0.0025 **12.** 0.0031 **13.** 3.47–17.33 **14.** 0.381

15. 0.4707   **16.** 9.515–11.361

17.

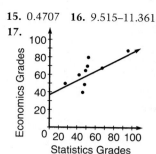

18. Let x = statistics grades and let
y = economics grades;
$y = 0.59x + 32.06$.

19. 0.724   **20.** moderately positive   **BONUS**   The formula is the radicand of the standard deviation formula, or it is the square of the standard deviation. variance = 0.00000961

CHAPTER 16   GRAPH THEORY

Pages 870-871   Lesson 16-1

5. $t, u,$ and v **7.** $B, C, E,$ and F **9.** 1: z
11. $V = \{A, B, C, D, E, F\}$ **13.**

17.

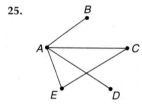

21.

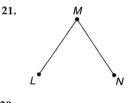

25.

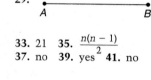

29.

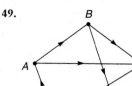

33. 21   **35.** $\dfrac{n(n-1)}{2}$
37. no   **39.** yes   **41.** no

43. Yes; vertices are connected by edges in pairs. Sample graph:

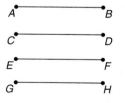

45. 11 vertices

49.

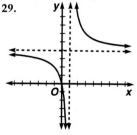

51. 2   **53.** 75.5 N, 14°50'

55.

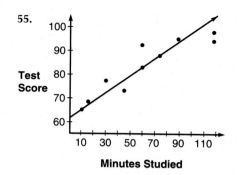

Minutes Studied

Pages 874-876   Lesson 16-2

5. trail, 7   **7.** Multigraph; it has parallel edges.   **9.** no
11. cycle   **13.** path   **15.** an infinite number   **19b.** d; 1; b, c; 2   **19c.** d; b, c; d **21b.** a; 1; b; 1; c; 1; d; 1   **21c.** a; b; c
23. 21   **25.** 7   **27.** 3 cars   **29.** (0, 16), maximum; (±2, 0), minima; $\left(\pm \dfrac{2\sqrt{3}}{3}, \dfrac{64}{9} \right)$, pts of inflection   **31.** 20   **33.** E

Pages 880-882   Lesson 16-3

5. C   **11.** yes   **13.** yes   **15.** yes   **17.** 11, 12, 14, 16
19. 12   **21.** Yes; each vertex has even degree.   **27.** No; one, between New Jersey and Brooklyn   **29.** 1

Page 882   Mid-Chapter Review

1. yes   **3.** 4   **5.** $B, C,$ and E **7.** no

Pages 887-889   Lesson 16-4

5. 3   **7.** C, D, F, G or C, D, I, G **9.** 6   **11.** because the path P, B, A is shorter   **13.** C **15.** 5; one path is A, G, K, M, I, Z.   **17.** 4; one path is A, C, Y, D, Z.   **19.** 7; one path is A, B, I, L, N, Q, F, Z **21.** 7; A, G, K, Z **23.** 5; A, B, C, Z
25. 3; A, E, Z
27a.

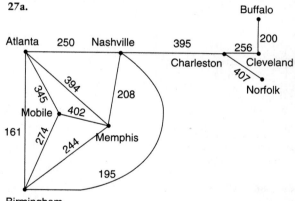

b. Buffalo, Cleveland, Charleston, Nashville, Atlanta
c. Mobile, Birmingham, Nashville; 469 miles
29.

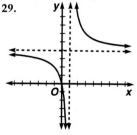

31. Since $S_1 = 2$ and $2^{1+1} - 2 = 2$, the formula is valid for $n = 1$. Assume the formula is valid for $n = k$ and derive a formula for $n = k + 1$.

$$2 + 2^2 + 2^3 + \cdots + 2^k = 2^{k+1} - 2$$
$$2 + 2^2 + 2^3 + \cdots + 2^k + 2^{k+1} = 2^{k+1} - 2 + 2^{k+1}$$
$$= 2(2^{k+1}) - 2$$
$$= 2^{k+2} - 2$$

33. D

Pages 893-895   Lesson 16-5

7. 17 edges   **9.** 6 edges   **13.** No; contains a cycle.   **15.** yes
17. $\{Z, I\}, \{I, F\}, \{F, A\}, \{A, B\}, \{B, D\}, \{D, H\}, \{D, C\}, \{D, E\}, \{G, H\}$
19. $\{B, A\}, \{A, E\}, \{E, F\}, \{F, D\}, \{D, C\}, \{C, Z\}; 7$ **21.** $\{H, G\},$
$\{G, A\}, \{A, B\}, \{B, E\}, \{E, C\}, \{C, D\}, \{D, Z\}, \{D, K\}; 15$
27. $\sqrt{46}; 6.78$ **29.** $A, B, C, D; 8$

Pages 901-903   Lesson 16-6

5.
$$\begin{array}{c|cccc} & A & B & C & D \\ \hline A & 0 & 1 & 1 & 1 \\ B & 1 & 0 & 1 & 1 \\ C & 1 & 1 & 0 & 1 \\ D & 1 & 1 & 1 & 0 \end{array}$$

7.
$$\begin{array}{c|cccccc} & Q & R & S & T & U & V \\ \hline Q & 0 & 1 & 1 & 1 & 1 & 1 \\ R & 1 & 0 & 1 & 0 & 0 & 0 \\ S & 1 & 1 & 0 & 1 & 0 & 0 \\ T & 1 & 0 & 1 & 0 & 1 & 0 \\ U & 1 & 0 & 0 & 1 & 0 & 1 \\ V & 1 & 0 & 0 & 0 & 1 & 0 \end{array}$$

9.

11.
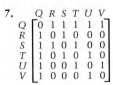

13.
$$\begin{array}{c|cccc} & A & B & C & D \\ \hline A & 0 & 1 & 0 & 1 \\ B & 1 & 0 & 1 & 1 \\ C & 0 & 1 & 0 & 1 \\ D & 1 & 1 & 1 & 0 \end{array}$$

15.
$$\begin{array}{c|ccc} & X & Y & Z \\ \hline X & 0 & 2 & 0 \\ Y & 2 & 0 & 2 \\ Z & 0 & 2 & 0 \end{array}$$

17.
$$\begin{array}{c|ccccccccc} & Q & R & S & T & U & V & W & X & Y \\ \hline Q & 0 & 0 & 0 & 0 & 0 & 0 & 0 & 1 & 1 \\ R & 0 & 0 & 1 & 0 & 0 & 0 & 0 & 0 & 0 \\ S & 0 & 0 & 0 & 0 & 0 & 0 & 1 & 0 & 0 \\ T & 0 & 0 & 0 & 0 & 0 & 0 & 1 & 0 & 0 \\ U & 0 & 0 & 0 & 1 & 0 & 0 & 0 & 0 & 0 \\ V & 0 & 0 & 0 & 0 & 0 & 0 & 0 & 0 & 0 \\ W & 0 & 0 & 0 & 0 & 1 & 0 & 0 & 0 & 0 \\ X & 0 & 1 & 0 & 0 & 1 & 1 & 1 & 0 & 1 \\ Y & 0 & 0 & 0 & 0 & 0 & 0 & 0 & 0 & 0 \end{array}$$

19.
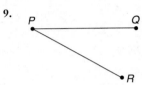

23.
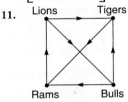

25. The paths $B \to R \to U \to W \to X \to E$ and $B \to S \to V \to E$ are both critical. The minimum amount of time needed is 6 weeks.

29a.
$$\begin{array}{c|cccccc} & A & C & D & H & M & S \\ \hline A & 0 & 1 & 0 & 0 & 0 & 1 \\ C & 0 & 0 & 1 & 0 & 1 & 0 \\ D & 0 & 0 & 0 & 1 & 1 & 0 \\ H & 1 & 0 & 0 & 0 & 0 & 1 \\ M & 0 & 0 & 0 & 0 & 0 & 1 \\ S & 0 & 0 & 1 & 0 & 1 & 0 \end{array}$$
b. 3   **c.** yes; 2 flights

31. 0.042

Pages 904-906   Chapter 16 Summary and Review

1. no   **3.** $\{A, B\}, \{A, C\}, \{A, E\}$ **5.** yes   **7.** trail   **9.** circuit
11. Add a temporary edge parallel to v.

13.
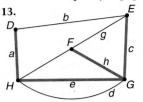

15.
$$\begin{array}{c|ccccc} & X & T & U & V & W \\ \hline X & 0 & 1 & 1 & 0 & 0 \\ T & 0 & 0 & 0 & 0 & 0 \\ U & 0 & 1 & 0 & 0 & 0 \\ V & 1 & 0 & 1 & 0 & 0 \\ W & 1 & 0 & 0 & 1 & 0 \end{array}$$

19. Critical path is $B \to H \to L \to E$; 35 days.

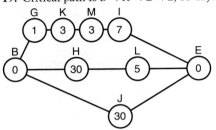

Page 907   Chapter 16 Test

1. simple graph; no loops or parallel edges   **2.** 5   **3.** $m, y,$
w, n, z, v **4.** no   **5.** B, C, D **6.** yes   **7.** circuit
8. Euler path; $o, p, q, r, s, t, k, n, z, u, v, w, x, y, m, l$
9. y, n, o, p **10.** $\{A, C\}, \{C, D\}, \{D, B\}, \{D, I\}, \{I, F\}, \{F, E\},$
$\{I, G\}, \{I, J\}, \{J, K\}, \{I, H\}$

11.
$$\begin{array}{c|ccccccccccc} & A & B & C & D & E & F & G & H & I & J & K \\ \hline A & 0 & 1 & 1 & 1 & 0 & 0 & 0 & 0 & 0 & 0 & 0 \\ B & 1 & 0 & 0 & 1 & 0 & 0 & 0 & 0 & 0 & 0 & 0 \\ C & 1 & 0 & 0 & 1 & 0 & 0 & 0 & 0 & 0 & 0 & 0 \\ D & 1 & 1 & 1 & 0 & 1 & 0 & 0 & 1 & 0 & 0 & 0 \\ E & 0 & 0 & 0 & 1 & 0 & 1 & 0 & 0 & 0 & 0 & 0 \\ F & 0 & 0 & 0 & 1 & 1 & 0 & 1 & 0 & 1 & 0 & 0 \\ G & 0 & 0 & 0 & 0 & 0 & 1 & 0 & 0 & 1 & 0 & 0 \\ H & 0 & 0 & 0 & 0 & 0 & 0 & 0 & 0 & 1 & 0 & 1 \\ I & 0 & 0 & 0 & 1 & 0 & 1 & 1 & 1 & 0 & 1 & 0 \\ J & 0 & 0 & 0 & 0 & 0 & 0 & 0 & 0 & 1 & 0 & 1 \\ K & 0 & 0 & 0 & 0 & 0 & 0 & 0 & 1 & 0 & 1 & 0 \end{array}$$

12. A multigraph may contain loops and/or parallel edges; a graph may not.

13.
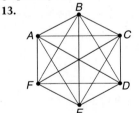

14. Sample answer:
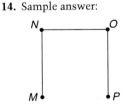

15. 5   **16.** true   **17.** A path may include any number of edges; an Euler path must include all of the edges.   **18.** at the vertices of odd degree   **19.** 3 cars   **20.** $39,000; Sarasota, Lakeland; Lakeland, Jacksonville; Lakeland, Daytona Beach; Daytona Beach, St. Augustine; St. Augustine, Bunnell
BONUS   8

Pages 908-909   Unit 4 Review

1. -35   **3.** $\frac{4}{3}$ **5.** divergent   **7.** convergent   **9.** $\frac{889}{64}$
11. $59{,}136x^6y^6$ **13.** -1, -4, 11   **15.** $(2, 2)$; attractor
17. $6 + i, 12 + 5i, 24 + 13i$ **19.** escape set   **21.** 60 ways

23. $\frac{2}{5}$ **25.** $\frac{11}{850}$ **27.** $\frac{5}{18}$ **29.** 42, 65, 65, 66, 69, 70, 72, 76, 77, 77, 77, 80, 82, 82, 86, 89, 89, 91, 95, 99 **31.** 77.45
33. 77 **35.** 9 **37.** about 0.595
39.

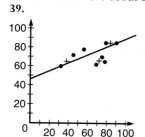

41. no **43.** $\{A, D\}$, $\{A, C\}$, $\{A, B\}$ **45.** cycle **47.** No; the degrees of vertices A and C are not even.

CHAPTER 17 LIMITS, DERIVATIVES, AND INTEGRALS

Pages 920-922 Lesson 17-1

7. 12 **9.** $-\frac{1}{3}$ **11.** $-\frac{1}{4}$ **13.** 5 **15.** $\frac{1}{2}$ **17.** -3 **19.** $\frac{3}{5}$
21. 1 **23.** $\frac{2}{3}$ **25.** 8 **27.** 0 **29.** $\frac{1}{3}$ **31.** 8 **33.** $\frac{3}{4}$ **35.** 0
37. $\sqrt{3}$ **39.** 75 **41.** 4 **43.** $\frac{1}{2}$ **45.** -6 **47.** $-\frac{1}{4}$ **49.** 1
51. 30,000 **53.** $f'(x) = 8x - 8$ **55.** $\frac{41\sqrt{10}}{20} \approx 6.48$
57. 20.68 **59.** 7.9, 8.6, 8.6 **61.** C

Pages 928-930 Lesson 17-2

7. 7 **9.** $4x$ **11.** $12x$ **13.** $18x^2 - 26x - 5$
15. $\frac{-2x(2x^2 - 1)}{(x^2 + 1)^4}$ **17.** $-\frac{14}{\sqrt{(4x - 1)^2}}$ **19.** $-3 + \frac{6}{(x + 2)^2}$
21. 6 **23.** $10X - 1$ **25.** $5x^4$ **27.** $\frac{1}{3\sqrt[3]{x^2}}$ **29.** $15x^4 - 18x^2$
31. $6x(x^2 + 4)^2$ **33.** $\frac{x}{\sqrt{x^2 - 1}}$ **35.** $\frac{-x}{\sqrt{(x^2 - 4)^3}}$ **37.** $\frac{-4(x + 1)}{(x - 1)^3}$
39. $6x^2 - \frac{6}{x^4}$ **41.** $-\frac{1}{x^2} + \frac{2}{x^3}$ **45a.** $h'(t) = 96 - 32t$ **b.** 32 ft/s
c. decreasing **d.** -160 ft/s **49.** $\pm 1, \pm 2, \pm\frac{1}{2}, \pm\frac{1}{4}, \pm\frac{1}{8}$ **53.** $-\frac{1}{27}$
55. 0

Pages 935-937 Lesson 17-3

3. $A = \lim\limits_{n\to\infty} \sum\limits_{i=1}^{n} \left(\frac{i}{n}\right)^2 \left(\frac{1}{n}\right)$; $\frac{1}{3}$ **5.** $A = \lim\limits_{n\to\infty} \sum\limits_{i=1}^{n} \left(\frac{i}{n}\right)\left(\frac{1}{n}\right)$; $\frac{1}{2}$
7. $A = \lim\limits_{n\to\infty} \sum\limits_{i=1}^{n} \left(\frac{5i}{n}\right)\left(\frac{5}{n}\right) - \lim\limits_{n\to\infty} \sum\limits_{i=1}^{n} \left(\frac{2i}{n}\right)\left(\frac{2}{n}\right)$; $10\frac{1}{2}$
9. $A = \lim\limits_{n\to\infty} \sum\limits_{i=1}^{n} \left(\frac{2i}{n}\right)^2 \left(\frac{2}{n}\right)$; $\frac{8}{3}$ **11.** $A = \lim\limits_{n\to\infty} \sum\limits_{i=1}^{n} \left(\frac{i}{n}\right)^5 \left(\frac{1}{n}\right)$; $\frac{1}{6}$
13. $A = \lim\limits_{n\to\infty} \sum\limits_{i=1}^{n} \left(\frac{7i}{n}\right)^4 \left(\frac{7}{n}\right) - \lim\limits_{n\to\infty} \sum\limits_{i=1}^{n} \left(\frac{4i}{n}\right)^4 \left(\frac{4}{n}\right)$; $\frac{15783}{5}$
15. $A = \lim\limits_{n\to\infty} \sum\limits_{i=1}^{n} \left(\frac{ai}{n}\right)^2 \left(\frac{a}{n}\right) - \lim\limits_{n\to\infty} \sum\limits_{i=1}^{n} \left(\frac{bi}{n}\right)^2 \left(\frac{b}{n}\right)$; $\frac{b^3 - a^3}{3}$
17. $A = \lim\limits_{n\to\infty} \sum\limits_{i=1}^{n} \left(\frac{2i}{n}\right)\left(\frac{2}{n}\right) + \lim\limits_{n\to\infty} \sum\limits_{i=1}^{n} \left(\frac{4i}{n}\right)\left(\frac{4}{n}\right)$; 10 **19.** $\frac{1}{4}$
21. $\frac{b^3 - a^3}{3}$ **25.** 36 feet **27.** $y^2 + 3x + 7y = 0$ **29.** $-\frac{1}{x^2} + \frac{2}{x^3}$

Page 937 Mid-Chapter Review

1. 6 **3.** -6 **5.** 1 **7.** 0 **9.** 1 **11.** $2x - 2$ **13.** $\frac{36}{x^{10}}$
15. $\frac{1}{x^2\sqrt{x^2 - 1}}$

Pages 942-944 Lesson 17-4

3. $5x + C$ **5.** $x^3 + C$ **7.** $x^5 + x^2 + C$ **9.** $10x + C$
11. $x^3 - 2x^4 + x^5 + C$ **13.** $\frac{5}{4}x^4 + C$ **15.** $\frac{(x + 5)^{21}}{21} + C$
17. $-x^2 + 3x + C$ **19.** $\ln |x - 1| + C$ **21.** $\frac{8}{5}x^5 + C$
23. $\frac{x^6}{6} + \frac{1}{3x^3} + C$ **25.** $4\sqrt{x} + C$
29. $r(x) = 4x - 0.01x^2 - 300$ **31.** 14 gallons **33.** $2x$
35. 15 **37.** A

Pages 948-951 Lesson 17-5

5. 12 sq units **7.** $2\frac{1}{3}$ sq units **9.** 4 sq units **11.** 5 **13.** $\frac{9}{2}$
15. $\frac{16}{3}$ sq units **17.** $16\frac{1}{2}$ sq units **19.** $1\frac{1}{3}$ sq units
21. $4\frac{1}{3}$ sq units **23.** $2\frac{2}{3}$ **25.** $22\frac{1}{2}$ **27.** $40\frac{1}{3}$ **29.** $\frac{2}{3}$
31. $-20\frac{5}{6}$ **33.** $4\frac{1}{4}$ sq units **35.** 64 sq units **37.** about 20.8
sq units **39a.** 322,994 **b.** 211,961 **41.** 781,250,000π or
2,454,369,261 foot-pounds
43. $2x^4 - 11x^3 - 19x^2 + 84x - 36 = 0$
45. equation: $\frac{(x - 4)^2}{2} - \frac{(y - 2)^2}{4} = 1$; center: (4, 2);
foci: $(4 \pm \sqrt{6}, 2)$; vertices: $(4 \pm \sqrt{2}, 2)$; asymptotes:
$y = 2 \pm \sqrt{2}(x - 4)$

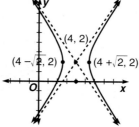

47. 31.7% **49.** C

Pages 952-954 Chapter 17 Summary and Review

1. 2 **3.** $\frac{2}{3}$ **5.** -1 **7.** 1 **9.** $6x^5$ **11.** $3 + 8x$
13. $\frac{3x^2 - 3}{\sqrt{2x^3 - 6x}}$ **15.** 4 unit^2 **17.** $\frac{37}{3}$ unit^2 **19.** $-\frac{4}{x} + C$
21. $x - \frac{x^2}{2} + C$ **23.** $\frac{(x + 3)^{10}}{2} + C$ **25.** 36 **27.** 36
29. 0.0000125 **31a.** 10.48 ft/s^2 **b.** $v(t) = 10.48t + C_1$
c. $d(t) = 5.24t^2 + C_1t + C_2$

Pages 955 Chapter 17 Test

1. $\frac{2}{9}$ **2.** -1 **3.** 1 **4.** 18 **5.** $-\frac{2}{3}$ **6.** $\frac{1}{3}$ **7.** 27
8. $12x^2$ **9.** $2(x + 3)$ **10.** $24x^3$ **11.** $14x^6 + 36x^5$
12. $\frac{2 - 2x^2}{1 + 2x^2 + x^4}$ **13.** $\frac{4x}{\sqrt{4x^2 - 1}}$ **14.** $\lim\limits_{n\to\infty} \sum\limits_{i=1}^{n} \left(\frac{2i}{n}\right)^3 \left(\frac{2}{n}\right)$; 4 unit^2
15. $\lim\limits_{n\to\infty} \sum\limits_{i=1}^{n} 3\left(\frac{3i}{n}\right)^2 \left(\frac{3}{n}\right) - \lim\limits_{n\to\infty} \sum\limits_{i=1}^{n} 3\left(\frac{i}{n}\right)^2 \left(\frac{1}{n}\right)$; 26 unit^2
16. $x - x^2 + C$ **17.** $x^3 + 2x^2 + 7x + C$ **18.** $-\frac{1}{x} + C$
19. $\frac{3}{5}\sqrt[3]{x^5} + C$ **20.** $-\frac{1}{4x^2} + C$ **21.** 4 **22.** $-\frac{20}{3}$ **23.** $22\frac{1}{2}$
24. $T'(\ell) = \frac{\pi}{\sqrt{32\ell}}$ **25.** $\frac{32\pi}{3}$ **BONUS** 1

INDEX

Index

Index

PHOTO CREDITS

Cover, H. Ross/H. Armstrong Roberts

vi, Andy Sacks/Tony Stone Worldwide; **vii,** (l) Mark J. Wilson, (r) Stephen Webster; **viii,** Elaine Shay; **ix,** (l) Stephen Webster, (r) NASA/Photo Researchers, Inc.; **x,** (l) Ed Degginger, (r) Barry Bomzer/Tony Stone Worldwide; **xi,** Elaine Shay; **xii,** (t) David L. Brown/The Stock Market, (c) Stephen Webster, (b) Aaron Haupt Photography; **xiii,** (l) Stephen Webster, (r) Brownie Harris/The Stock Market; **1,** Mark Burnett/Glencoe; **2–3,** Mak-I Photo Design; **4,** Andy Sacks/Tony Stone Worldwide; **5,** ©1992 Dede Hatch; **6,** NASA/Peter Arnold, Inc.; **11, 12, 14, 18,** Stephen Webster; **19,** Platinum Productions; **24,** Robert Landau/Westlight; **26,** (l) Daniel Wilson/Westlight, (r) Kenneth Garrett/Westlight; **29,** Tom Martin/The Stock Market; **35,** Rod Joslin/Kaplan Graphics; **38,** file photo; **39,** Platinum Productions; **40,** Howard Sochurek/Medichrome; **41,** Wes Thompson/The Stock Market; **42,** Ken Rogers/Westlight; **43,** Matt Meadows; **46,** Gail Meese; **52,** Kristian Hilson/Tony Stone Worldwide; **53,** Jim Lyle/Texas A&M University; **56, 58,** Matt Meadows; **61,** Aaron Haupt Photography; **64,** Larry Lefever from Grant Heilman; **67,** Elaine Shay; **74,** Chris Jones/The Stock Market; **76,** Tim Courlas; **79,** Lowell Georgia/Photo Researchers; **83,** Elaine Shay, courtesy Peasant on the Lane Restaurant; **86,** Platinum Productions; **90,** William Finch/Stock Boston; **91,** Brent Bear/Westlight; **93,** Stephen Webster; **94,** Superstock; **95,** Les Losego/Crown Studios; **96,** Stephen Webster; **102,** P. Petersen/Custom Medical Stock Photo; **103,** courtesy Les Todd, Duke University; **106,** (l) E. R. Degginger, (r) Superstock; **111,** Tracy I. Borland; **117,** Garrett H. Jones, Jr., C.P.C.; **124,** Stephen Webster; **126,** Paul Light/Lightwave; **131,** Mark Stephenson/Westlight; **134,** Mark Fortenberry/Transparencies; **137,** NASA; **141,** Platinum Productions; **142,** Earl Glass/Stock Boston; **144,** Manfred Kage/Peter Arnold, Inc.; **146,** file photo; **147,** Platinum Productions; **148,** Matthew McVay/Stock Boston; **149,** Superstock; **152,** courtesy Daniel L. Feicht, Cedar Point Amusement Park; **154,** Stephen Webster; **156,** AP/Wide World Photos; **160,** Elaine Shay; **162,** Randall L. Schieber; **171,** Chuck O'Rear/Westlight; **176, 177,** Mark J. Wilson; **178,** ©Cindy Lewis; **181,** Platinum Productions; **183,** Mark D. Phillips/Science Source/Photo Researchers; **187,** Aaron Haupt Photography; **190,** Stephen Webster; **193,** G & G Design/The Stock Market; **195,** Duomo; **199,** Superstock; **200,** NASA; **201,** Superstock; **205,** Matt Meadows; **207,** Superstock; **209, 210,** Platinum Productions; **214,** file photo; **216,** Superstock; **217,** file photo; **221,** Elaine Shay; **222,** Mak-I Photo Design; **222,** Stephen Webster; **225,** Comstock; **228,** Stock Montage, Inc.; **236–237,** Derek Bayes/The Stock Market; **238,** Superstock; **239,** courtesy Jerome Crump, Multimedia, University of California, Berkeley; **241,** file photo; **246,** Stephen Webster; **250,** Les Losego/Crown Studios; **251,** Gerard Vandystadt/Photo Researchers; **254, 255, 261,** Doug Martin; **262,** Les Losego/Crown Studios; **266, 268,** Superstock; **268,** Walter Hodges/Westlight; **269,** courtesy Carolyn S. Garcia, Dallas TX Fire Dept.; **274,** Superstock; **281,** Harvey Lloyd/The Stock Market; **286,** Doug Martin, courtesy Rickenbacher Air Force Base, Ohio Port Authority; **287,** E. R. Degginger; **289,** Culver Pictures; **292,** Craig Hammell/The Stock Market; **295,** Bill Gallery/Stock Boston; **300,** Stephen Webster; **301,** Gale Zucker; **302,** Mark J. Wilson; **306,** Platinum Productions; **307,** Comstock; **309,** NASA; **317,** Animals Animals/Mike Andrews; **317,** Bob Daemmrich/Stock Boston; **321,** Clyde H. Smith/Peter Arnold, Inc.; **324,** Les Losego/Crown Studios; **327,** Rod Joslin; **328,** Malcolm S. Kirk/Peter Arnold, Inc.; **330,** NASA; **333,** Royce Bair/The Stock Market; **334,** B. Ross/Westlight; **336, 338,** Les Losego/Crown Studios; **340,** file photo; **342,** Al Francekevich/The Stock Market; **343,** Stephen Webster; **345,** Animals Animals/Johnny Johnson; **347,** Culver Pictures; **348,** Les Losego/Crown Studios; **349,** Art Gingert/Comstock; **351,** Superstock; **356,** Stephen Webster; **357,** courtesy Brigham Young University; **358,** Superstock; **363,** Harvey Lloyd/The Stock Market; **364,** Les Losego/Crown Studios; **368,** George C. Anderson; **368,** Superstock; **376,** Les Losego/Crown Studios; **377,** AP/Wide World Photos; **380,** file photo; **382,** Comstock; **387,** file photo; **388,** Myron J. Dorf/The Stock Market; **391,** file photo; **392, 393,** Superstock; **398, 400,** file photo; **404,** Alvin E. Staffan; **405,** Elaine Shay; **410,** Mark M. Lawrence/The Stock Market; **411,** Doug Martin; **412, 414,** Mark E. Gibson; **416,** file photo; **418,** Platinum Productions; **419, 421,** Mark E. Gibson; **423,** David Madison/Duomo; **425,** Mark E. Gibson; **426,** Audrey Gibson; **428,** Bob Wahlgren/Pictorial History Research; **429,** Stephen Webster; **432,** Platinum Studios; **434,** Rod Joslin; **436,** David Madison/Duomo; **438,** file photo; **439, 440,** Mark E. Gibson; **442,** courtesy Daytona International Speedway; **445,** Richard Dole/Duomo; **447,** Superstock; **449,** file photo; **451,** Al Tielemans/Duomo; **452,** Aaron Haupt Photography; **452,** Dan Helms/Duomo; **454,** Al Tielemans/Duomo; **455,** Mark E. Gibson; **462–463,** Chris Bjornberg/Photo Researchers; **464,** NASA/Photo Researchers; **465,** Gale Zucker; **468,** Les Losego/Crown Studios; **471,** Tom Stack & Associates; **475,** Jeffry W. Myers/The Stock Market; **475,** Manfred Kage/Peter Arnold, Inc.; **481,** Les Losego/Crown Studios; **485,** Tony Stone Worldwide; **487,** courtesy Daniel L. Feicht, Cedar Point Amusement Park; **493,** Mark E. Gibson; **493,** Superstock; **494,** ©1992 R. F. Voss/IBM Research; **494,** Dietmar Saupe; **497,** Les Losego/Crown Studios; **501,** Ann Ronan Picture Library; **505,** Superstock; **506,** ©1991(2) R.F. Voss/IBM Research; **509,** D. Seawell/Westlight; **510,** courtesy the Royal Society; **517,** Superstock; **522,** Ed Degginger or NASA; **523,** John A. Miller; **524,** Mark E. Gibson; **531,** Frank P. Rossotto/The Stock Market; **534,** **537,** Mark E. Gibson; **539,** file photo; **541,** Mark E. Gibson; **542,** NASA; **550,** Chris Butler/Astrostock; **556,** file photo; **558,** NASA; **564,** John Sanford/Astrostock; **567,** Mark E. Gibson; **570,** courtesy GM Powertrain; **576,** Matt Meadows; **579,** Paul Trent/J. Allen Hynek Center for UFO Studies; **581,** Elaine Shay; **584,** Mark E. Gibson; **584,** Matt Meadows; **585,** NASA; **586, 591,** Mark E. Gibson; **596,** Barry Bomzer/Tony Stone Worldwide;